*Fritz Appel, Jonathan
David Heaton Paul,
and Michael Oehring*

**Gamma Titanium
Aluminide Alloys**

Related Titles

Kumar, C. S. S. R. (ed.)

Nanomaterials for the Life Sciences

10 Volume Set

approx. 4000 pages in 10 volumes with approx. 1800 figures
Hardcover
ISBN: 978-3-527-32261-9

Dubois, J.-M., Belin-Ferré, E. (eds.)

Complex Metallic Alloys

Fundamentals and Applications

434 pages with 232 figures and 20 tables
2011
Hardcover
ISBN: 978-3-527-32523-8

Krzyzanowski, M., Beynon, J. H., Farrugia, D. C. J.

Oxide Scale Behavior in High Temperature Metal Processing

386 pages with 327 figures and 17 tables
2010
Hardcover
ISBN: 978-3-527-32518-4

Pfeiler, W. (ed.)

Alloy Physics

A Comprehensive Reference

1003 pages with approx. 362 figures and approx. 50 tables
2007
Hardcover
ISBN: 978-3-527-31321-1

Leyens, C., Peters, M. (eds.)

Titanium and Titanium Alloys

Fundamentals and Applications

532 pages with 349 figures and 56 tables
2003
Hardcover
ISBN: 978-3-527-30534-6

Westbrook, J. H., Fleischer, R. L. (eds.)

Intermetallic Compounds

Principles and Applications. 2 Volume Set

1934 pages in 2 volumes
1994
Hardcover
ISBN: 978-0-471-93453-0

*Fritz Appel, Jonathan David Heaton Paul,
and Michael Oehring*

Gamma Titanium Aluminide Alloys

Science and Technology

WILEY-VCH Verlag GmbH & Co. KGaA

The Authors

Dr. habil. Fritz Appel
Helmholtz-Zentrum Geesthacht
Institute for Materials Research
Max-Planck-Str. 1
21502 Geesthacht

Dr. Jonathan David Heaton Paul
Helmholtz-Zentrum Geesthacht
Institute for Materials Research
Max-Planck-Str. 1
21502 Geesthacht

Dr. Michael Oehring
Helmholtz-Zentrum Geesthacht
Institute for Materials Research
Max-Planck-Str. 1
21502 Geesthacht

Cover
Foreground shows the General Electric GEnx-1B engine (photo courtesy of General Electric). The background is an artificially colored high resolution TEM image of a deformation twin/matrix interface in TiAl.

■ All books published by **Wiley-VCH** are carefully produced. Nevertheless, authors, editors, and publisher do not warrant the information contained in these books, including this book, to be free of errors. Readers are advised to keep in mind that statements, data, illustrations, procedural details or other items may inadvertently be inaccurate.

Library of Congress Card No.: applied for

British Library Cataloguing-in-Publication Data
A catalogue record for this book is available from the British Library.

Bibliographic information published by the Deutsche Nationalbibliothek
The Deutsche Nationalbibliothek lists this publication in the Deutsche Nationalbibliografie; detailed bibliographic data are available on the Internet at <http://dnb.d-nb.de>.

© 2011 Wiley-VCH Verlag & Co. KGaA, Boschstr. 12, 69469 Weinheim, Germany

All rights reserved (including those of translation into other languages). No part of this book may be reproduced in any form – by photoprinting, microfilm, or any other means – nor transmitted or translated into a machine language without written permission from the publishers. Registered names, trademarks, etc. used in this book, even when not specifically marked as such, are not to be considered unprotected by law.

Cover Design   Grafik-Design Schulz, Fußgönheim
Typesetting   Toppan Best-set Premedia Limited, Hong Kong
Printing and Binding   Fabulous Printers Pte Ltd

Printed in Singapore
Printed on acid-free paper

Print ISBN: 978-3-527-31525-3
ePDF ISBN: 978-3-527-63622-8
ePub ISBN: 978-3-527-63621-1
oBook ISBN: 978-3-527-63620-4

Contents

Preface *XIII*
Figures–Tables Acknowledgement List *XV*

1 Introduction *1*
References *3*

2 Constitution *5*
2.1   The Binary Ti–Al Phase Diagram *5*
2.2   Ternary and Multicomponent Alloy Systems *11*
References *20*

3 Thermophysical Constants *25*
3.1   Elastic and Thermal Properties *25*
3.2   Point Defects *27*
3.3   Diffusion *29*
References *30*

4 Phase Transformations and Microstructures *33*
4.1   Microstructure Formation on Solidification *33*
4.2   Solid-State Transformations *49*
4.2.1 $\beta \rightarrow \alpha$ Transformation *50*
4.2.2 Formation of $(\alpha_2 + \gamma)$ Lamellae Colonies *52*
4.2.3 Feathery Structures and Widmannstätten Colonies *60*
4.2.4 Massive Transformation *63*
References *64*

5 Deformation Behavior of Single-Phase Alloys *71*
5.1   Single-Phase γ(TiAl) Alloys *71*
5.1.1 Slip Systems and Deformation Kinematics *71*
5.1.2 Planar Faults *75*
5.1.3 Planar Dislocation Dissociations in γ(TiAl) *81*
5.1.4 Nonplanar Dissociations and Dislocation Locking *85*

5.1.5	Mechanical Twinning in γ(TiAl) 89
5.1.6	Effects of Orientation and Temperature on Deformation of γ Phase 95
5.2	Deformation Behavior of Single-Phase α_2(Ti$_3$Al) Alloys 106
5.2.1	Slip Systems and Deformation Kinematics 106
5.2.2	Effects of Orientation and Temperature on Deformation of α_2 Phase 112
5.3	β/B2 Phase Alloys 114
	References 118

6	**Deformation Behavior of Two-Phase α_2(Ti$_3$Al) + γ(TiAl) Alloys** 125
6.1	Lamellar Microstructures 125
6.1.1	Interface Structures in Lamellar TiAl Alloys 125
6.1.2	Energetic Aspects of Lamellar Interfaces 136
6.1.3	Coherent and Semicoherent Interfaces 139
6.1.4	Coherency Stresses 149
6.1.5	Plastic Anisotropy 156
6.1.6	Micromechanical Modeling 161
6.2	Deformation Mechanisms, Contrasting Single-Phase and Two-Phase Alloys 164
6.2.1	Methodical Aspects of TEM Characterization 164
6.2.2	Deformation of (α_2 + γ) Alloys at Room Temperature 165
6.2.3	Independent Slip Systems 169
6.2.4	High-Temperature Deformation of (α_2 + γ) Alloys 171
6.2.5	Slip Transfer through Lamellae 173
6.3	Generation of Dislocations and Mechanical Twins 178
6.3.1	Dislocation Source Operation in γ(TiAl) 178
6.3.2	Interface-Related Dislocation Generation 184
6.3.3	Twin Nucleation and Growth 186
6.3.4	Twin Intersections 197
6.3.5	Acoustic Emissions 204
6.3.6	Thermal Stability of Twin Structures 206
6.4	Glide Resistance and Dislocation Mobility 207
6.4.1	Thermally Activated Deformation 207
6.4.2	Glide Resistance at the Beginning of Deformation 217
6.4.3	Static and Dynamic Strain Aging of TiAl Alloys 222
6.4.4	Diffusion-Assisted Dislocation Climb, Recovery, and Recrystallization 232
6.5	Thermal and Athermal Stresses 234
	References 240

7	**Strengthening Mechanisms** 249
7.1	Grain Refinement 249
7.2	Work Hardening 254
7.2.1	Work-Hardening Phenomena 255

7.2.2	Athermal Contributions to Work Hardening	*257*
7.2.3	Jog Dragging and Debris Hardening	*259*
7.2.4	Thermal Stability of Deformation Structures	*262*
7.2.5	High-Temperature Flow Behavior	*268*
7.2.6	High Strain-Rate Deformation	*272*
7.3	Solution Hardening	*273*
7.3.1	Elemental Size Misfit of Solute Atoms with the TiAl Matrix	*273*
7.3.2	Survey of Observations	*276*
7.3.3	Effect of Solute Niobium	*277*
7.4	Precipitation Hardening	*282*
7.4.1	Carbide Precipitation in TiAl Alloys	*282*
7.4.2	Hardening by Carbides	*284*
7.4.3	Hardening by Borides, Nitrides, Oxides, and Silicides	*289*
7.5	Optimized Nb-Bearing Alloys	*292*
	References	*295*
8	**Deformation Behavior of Alloys with a Modulated Microstructure**	*301*
8.1	Modulated Microstructures	*301*
8.2	Misfitting Interfaces	*306*
8.3	Mechanical Properties	*310*
	References	*311*
9	**Creep**	*313*
9.1	Design Margins and Failure Mechanisms	*313*
9.2	General Creep Behavior	*314*
9.3	The Steady-State or Minimum Creep Rate	*316*
9.3.1	Single-Phase γ(TiAl) Alloys	*317*
9.3.2	Two-Phase α_2(Ti$_3$Al) + γ(TiAl) Alloys	*319*
9.3.3	Experimental Observation of Creep Structures	*320*
9.4	Effect of Microstructure	*322*
9.5	Primary Creep	*325*
9.6	Creep-Induced Degradation of Lamellar Structures	*329*
9.7	Precipitation Effects Associated with the $\alpha_2 \rightarrow \gamma$ Phase Transformation	*339*
9.8	Tertiary Creep	*340*
9.9	Optimized Alloys, Effect of Alloy Composition and Processing	*341*
9.10	Creep Properties of Alloys with a Modulated Microstructure	*346*
9.10.1	Effect of Stress and Temperature	*346*
9.10.2	Damage Mechanisms	*347*
	References	*352*
10	**Fracture Behavior**	*357*
10.1	Length Scales in the Fracture of TiAl Alloys	*357*
10.2	Cleavage Fracture	*360*

10.3	Crack-Tip Plasticity	362
10.3.1	Plastic Zone	362
10.3.2	Interaction of Cracks with Interfaces	365
10.3.3	Crack–Dislocation Interactions	367
10.3.4	Role of Twinning	369
10.4	Fracture Toughness, Strength, and Ductility	373
10.4.1	Methodical Aspects	373
10.4.2	Effects of Microstructure and Texture	376
10.4.3	Effect of Temperature and Loading Rate	383
10.4.4	Effect of Predeformation	387
10.5	Fracture Behavior of Modulated Alloys	388
10.6	Requirements for Ductility and Toughness	391
10.7	Assessment of Property Variability	393
10.7.1	Statistical Assessment	393
10.7.2	Variability in Strength and Ductility of TiAl	394
10.7.3	Fracture Toughness Variability of TiAl	396
	References	398
11	**Fatigue**	**403**
11.1	Definitions	403
11.2	The Stress–Life (S–N) Behavior	405
11.3	HCF	407
11.3.1	Fatigue Crack Growth	407
11.3.2	Crack-Closure Effects	409
11.3.3	Fatigue at the Threshold Stress Intensity	411
11.4	Effects of Temperature and Environment on the Cyclic Crack-Growth Resistance	413
11.5	LCF	418
11.5.1	General Considerations	418
11.5.2	Cyclic-Stress Response	419
11.5.3	Cyclic Plasticity	422
11.5.4	Stress-Induced Phase Transformation and Dynamic Recrystallization	426
11.6	Thermomechanical Fatigue and Creep Relaxation	428
	References	429
12	**Oxidation Behavior and Related Issues**	**433**
12.1	Kinetics and Thermodynamics	433
12.2	General Aspects Concerning Oxidation	437
12.2.1	Effect of Composition	437
12.2.2	Mechanical Aspects of Oxide Growth	439
12.2.3	Effect of Oxygen and Nitrogen	441
12.2.4	Effect of Other Environmental Factors	443
12.2.5	Subsurface Zone, the Z-Phase, and Silver Additions	446
12.2.6	Effect of Surface Finish	447

12.2.7	Ion Implantation *448*
12.2.8	Influence of Halogens on Oxidation *450*
12.2.9	Embrittlement after High-Temperature Exposure *450*
12.2.10	Coatings/Oxidation-Resistant Alloys *456*
12.3	Summary *458*
	References *459*

13 Alloy Design *465*
13.1 Effect of Aluminum Content *465*
13.2 Important Alloying Elements – General Remarks *467*
13.2.1 Cr, Mn, and V *468*
13.2.2 Nb, W, Mo, and Ta *469*
13.2.3 B, C, and Si *469*
13.3 Specific Alloy Systems *471*
13.3.1 Conventional Alloys *472*
13.3.2 High Niobium-Containing Alloys *472*
13.3.3 β-Solidifying Alloys *473*
13.3.4 Massively Transformed Alloys *474*
13.4 Summary *476*
References *477*

14 Ingot Production and Component Casting *479*
14.1 Ingot Production *479*
14.1.1 Vacuum Arc Melting (VAR) *480*
14.1.2 Plasma-Arc Melting (PAM) *483*
14.1.3 Induction Skull Melting (ISM) *489*
14.1.4 General Comments *492*
14.2 Casting *495*
14.2.1 Investment Casting *497*
14.2.2 Gravity Metal Mold Casting (GMM) *503*
14.2.3 Centrifugal Casting *506*
14.2.4 Countergravity Low-Pressure Casting *513*
14.2.5 Directional Casting *514*
14.3 Summary *515*
References *515*

15 Powder Metallurgy *521*
15.1 Prealloyed Powder Technology *522*
15.1.1 Gas Atomization *522*
15.1.1.1 Plasma Inert-Gas Atomization (PIGA) at GKSS *524*
15.1.1.2 Titanium Gas-Atomizer Process (TGA) *526*
15.1.1.3 Electrode Induction Melting Gas Atomization (EIGA) *527*
15.1.2 Rotating-Electrode Processes *529*
15.1.3 Rotating-Disc Atomization *531*
15.1.4 General Aspects of Atomization *532*

15.1.5	Postatomization Processing	542
15.1.5.1	Hot Isostatic Pressing (HIPing), Hot Working, and Properties	543
15.1.5.2	Laser-Based Rapid-Prototyping Techniques	547
15.1.5.3	Metal Injection Molding (MIM)	549
15.1.5.4	Spray Forming	551
15.1.5.5	Sheet/Foil Production through (i) HIP of Cast Tapes and (ii) Liquid-Phase Sintering	556
15.1.5.6	Spark Sintering	557
15.1.6	Summary	558
15.2	Elemental-Powder Technology	559
15.2.1	Reactive Sintering	559
15.2.1.1	Mechanical Properties of Reactive Sintered Material	563
15.2.1.2	Manufacture of Reactively Sintered Components/Parts	564
15.2.2	Summary	564
15.3	Mechanical Alloying	565
	References	566

16	**Wrought Processing**	**573**
16.1	Flow Behavior under Hot-Working Conditions	574
16.1.1	Flow Curves	574
16.1.2	Constitutive Analysis of the Flow Behavior	578
16.2	Conversion of Microstructure	585
16.2.1	Recrystallization of Single-Phase Alloys	585
16.2.2	Multiphase Alloys and Alloying Effects	587
16.2.3	Influence of Lamellar Interfaces	595
16.2.4	Microstructural Evolution during Hot Working above the Eutectoid Temperature	603
16.2.5	Technological Aspects	605
16.3	Workability and Primary Processing	607
16.3.1	Workability	607
16.3.2	Ingot Breakdown	617
16.4	Texture Evolution	642
16.5	Secondary Processing	658
16.5.1	Component Manufacture through Wrought Processing	658
16.5.2	Rolling – Sheet Production and Selected Mechanical Properties	662
16.5.2.1	Pack Rolling	663
16.5.2.2	Rolling Defects	666
16.5.2.3	Industrial Production of Sheet	667
16.5.2.4	Mechanical Properties of Sheet	668
16.5.2.5	Superplastic Behavior	669
16.5.3	Novel Techniques	671
16.5.3.1	Manufacture of Large "Defect-Free" Components	671
	References	673

17	**Joining** *683*	
17.1	Diffusion Bonding *683*	
17.1.1	Alloy Compositions and Microstructures *684*	
17.1.2	Microasperity Deformation *685*	
17.1.3	Diffusion Bonding; Experimental Setup *687*	
17.1.4	Metallographic Characterization of the Bonding Zone *687*	
17.1.5	Effect of Alloy Composition *696*	
17.1.6	Influence of Bonding Time and Stress *698*	
17.1.7	Mechanical Characterization of the Bonds *700*	
17.2	Brazing and Other Joining Technologies *702*	
17.2.1	Brazing and Transient Liquid-Phase Joining *702*	
17.2.2	Other Techniques *704*	
	References *704*	
18	**Surface Hardening** *707*	
18.1	Shot Peening and Roller Burnishing *707*	
18.2	Residual Stresses, Microhardness, and Surface Roughness *708*	
18.3	Surface Deformation Due to Shot Peening *712*	
18.4	Phase Transformation, Recrystallization, and Amorphization *716*	
18.5	Effect of Shot Peening on Fatigue Strength *721*	
18.6	Thermal Stability of the Surface Hardening *724*	
	References *726*	
19	**Applications, Component Assessment, and Outlook** *729*	
19.1	Aerospace *729*	
19.1.1	Aircraft-Engine Applications *729*	
19.1.2	Exotic Aerospace Applications *731*	
19.2	Automotive *732*	
19.3	Outlook *736*	
	References *737*	

Subject Index *739*

Preface

There is an ever-increasing demand for the development of energy-conversion systems towards improved thermodynamic efficiency and ecological compatibility. Advanced design concepts are based on higher service temperatures, lower weight, and higher operational speeds. For example, the operating efficiency of a gas-turbine engine will increase by over 1% for every 10 °C increase in the turbine-inlet temperature. Substantial fuel savings in aircraft and power generation can be achieved through the introduction of new materials that can provide higher temperatures or reduced component weight. The conventional metallic systems that are currently in use have been developed over the last 50 years to near the limits of their capability. If further advances are to be made, new classes of materials will be required.

Titanium aluminide alloys based on the intermetallic gamma phase are widely recognized as having the potential to meet the design requirements mentioned above. Undoubtedly, the development of such a material system has important implications for spin-offs to other high-temperature technologies, as well as for the general economy. For example, General Electric has recently made public that its most recent engine, the GEnx, includes the use of titanium aluminide as a blade material. This is a significant milestone for a relatively new, advanced engineering material.

Although there is a vast body of TiAl literature going back over 20 years, there have only been a few review articles published in the recent past, the latest nearly a decade ago. Since that time, considerable advances have been made, both in the basic understanding of the physical metallurgy and in processing technology. It is our intention that the publication of this book will, for the first time, give a wide-ranging interpretation and discussion of the voluminous amounts of data documented in the literature. For TiAl to be successfully employed as a structural material requires a comprehensive understanding of the complex microstructures, down to the nanometer scale, and knowledge concerning how the structure–property relationships are determined by, for example, the atomic details of interface-related phenomena.

The overview of all relevant research topics that are presented in this book is intended to form a link between scientific findings and alloy development, material properties, industrial processing technologies, and engineering applications.

The metallurgy of TiAl alloys undoubtedly has several features in common with other intermetallic system. Thus, in that we have chosen to emphasize the scientific principles, the book will provide a treatment of the subject for researchers and advanced students who need a more detailed coverage than is found in physical metallurgy textbooks. We expect that our compilation of the current state of titanium aluminide science and technology will not only serve as a guide through the huge body of literature to the TiAl community, but will also be of interest to materials scientists, engineers, and technical managers who are involved in areas where low-density, high-temperature resistant materials are required. The detailed description of interfaces and interface related phenomena will certainly be of interest to an extended scientific community.

It would not have been possible to write such a book without the help and support from numerous people and organizations. First, we would like to acknowledge the generous support and the excellent research conditions provided by the Helmholtz-Zentrum Geesthacht (formerly GKSS) under its Scientific Director Prof. Wolfgang Kaysser, Prof. Andreas Schreyer as the Director of the Institute for Materials Research, and Prof. Florian Pyczak as group leader.

We also thank the BMBF (German Ministry for Education and Research), DFG (German Science Foundation), Helmholtz Gemeinschaft (Helmholtz Association), Rolls-Royce Deutschland, and CBMM (Companhia Brasileira de Metalurgia e Mineração) for financial support through their funding of numerous research projects.

We would particularly like to thank Prof. Richard Wagner (now Director at the Institute Laue-Langevin, Grenoble, France) who initiated the work on TiAl in the late 1980s while he was director of our institute. Additionally, we would like to thank our colleagues and former students, Ulrich Brossmann, Stefan Eggert, Dirk Herrmann, Roland Hoppe, Ulrich Fröbel, Viola Küstner, Uwe Lorenz, the late Johann Müllauer, Thorsten Pfullmann, and Ulf Sparka for their interest, support, and for contributing to an excellent group atmosphere. The generous help from the HZG library personnel is also acknowledged.

A very special mention must be made to acknowledge Dr. Young-Won Kim (Universal Energy Systems, Dayton, USA) for his achievement in keeping the titanium aluminide community together for very many years and his friendship. Fritz Appel would like to thank his wife, Bärbel, for her support. Finally, the authors would like to expresss their gratitude to Wiley-VCH for the opportunity to write the book and in particular gratefully acknowledge the patient support by Waltraud Wüst and Ulrike Werner and careful copyediting of Bernadette Cabo.

Geesthacht, January 2011
Fritz Appel
Jonathan David Heaton Paul
Michael Oehring

Figures – Tables Acknowledgement List

In order to cover the wide range of literature, the authors have used copies or slightly modified copies of figures/micrographs from previously published work. Where this has been done the figure has been referenced so that the source paper, authors, and journal of the original work are credited. The table below is intended to indicate and thank the publishers, companies, or individuals who hold the copyright to these figures and acknowledges their generous permission for reuse.

Publishing source	Figures/tables
ACCESS e.V. Reprinted with permission, Copyright ACCESS.	Figure: 14.21.
American Physical Society (APS) Reprinted from *"Phys. Rev. B"* with permission, Copyright (1998) APS.	Figures: 3.1 and 3.2.
ASM International®. Reprinted from *"Castings, Metals Handbook"* with permission, Copyright ASM. www.asminternational.org	Figure: 14.25.
Cambridge University Press. Reprinted with permission, Copyright Cambridge University Press.	Figures: 5.7, 12.1, 15.15, and 16.47.
Deutsche Gesellschaft für Materialkunde (DGM). Reprinted with permission, Copyright DGM.	Figure: 15.12.
General Electric Company (Aviation). Reprinted with permission, Copyright GE.	Figure: 1.2 and cover.

(Continued)

Figures–Tables Acknowledgement List

Publishing source	Figures/tables
Elsevier Publishing. Reprinted with permission, Copyright Elsevier.	Figures: 1.1a, 2.5–2.10, 3.2, 4.1, 4.13–4.15, 6.6, 6.26, 6.30, 6.31, 6.60, 6.65, 6.69, 6.70, 6.72–6.74, 6.77, 6.78, 7.6, 7.12, 7.13, 7.20, 7.21, 7.23, 7.25, 8.1–8.3, 8.9, 8.10, 9.3, 9.27, 9.31–9.33, 10.14, 10.17, 10.25, 10.26, 11.5, 11.6, 11.8, 11.9, 11.11–11.13, 11.16, 11.17, 12.6, 12.8, 12.9, 12.11, 12.12, 12.16, 13.4, 13.5, 13.6, 14.12–14.15, 14.26, 15.7, 15.16, 16.57, 18.2–18.4, 18.5, 18.14, 18.15, 19.3, and 19.4. Tables: 6.3, 7.4, 11.1, and 11.2.
IOP Publishing. Reprinted with permission, Copyright IOP Publishing.	Figure: 7.2.
The Japanese Institute of Metals (JIM). Reprinted with permission, Copyright JIM.	Figure: 9.2.
Metal Powder Industries Federation (MPIF), 105 College Road East, Princeton, New Jersey, USA. Reprinted with permission, Copyright MPIF.	Figures: 7.1, 15.2, 15.6, 15.8, and 15.9.
Springer Publishing. Reprinted with permission, Copyright Springer.	Figures: 2.2–2.4, 2.11, 6.67, 7.15, 7.16, 10.16, 15.19–15.21, 16.3–16.5, 16.8–16.10, 16.12, 16.14–16.17, 16.29, 16.44–16.46, 16.48–16.50, 16.52, 16.58, and 17.4–17.16. Table: 7.6.
Taylor & Francis Publishing. Reprinted with permission, Copyright Taylor & Francis.	Figures: 5.11, 5.12, 5.16, 5.20–5.22, 6.11, 6.49, 6.50, 6.53, 6.59, 7.27, 10.8, 16.21, and 16.22.
The Minerals, Metals and Materials Society, Warrendale, PA, (TMS). Reprinted with permission, Copyright TMS.	Figures: 4.2–4.4, 4.6, 6.29, 6.49, 7.17, 9.1, 9.22, 9.24, 10.13, 10.15, 10.18, 10.21, 10.30, 10.31, 11.2–11.4, 11.7, 12.15, 13.2, 13.3, 14.2, 14.3, 14.5, 14.6, 14.8, 14.9, 14.16–14.20, 15.3, 15.11, 15.17, 15.18, 16.11, 16.18, 16.55, 19.1, and 19.2.
Wiley Publishing. Reprinted with permission, Copyright Wiley.	Figures: 6.40, 6.42, 10.2, 12.2–12.5, 12.7, 12.10, 12.14, 14.1, 15.10, 15.13, 15.14, 16.33, 16.35–16.37, 16.39 and 16.53. Table: 14.1.
G. Hug, PhD Thesis, Université de Paris-Sud, France, 1988.	Figure: 5.9.

1
Introduction

The reason why gamma TiAl has continued to attract so much attention from the research community including universities, publicly funded bodies, industrial manufactures, and end-product users is that it has a unique combination of mechanical properties when evaluated on a density-corrected basis. In particular, the elevated temperature properties of some alloys can be superior to those of superalloys.

Dimiduk [1] has assessed gamma TiAl with other aerospace structural materials and shown that new capabilities become available on account of its properties. The most important pay-offs involve

- high melting point;
- low density;
- high specific strengths and moduli;
- low diffusivity;
- good structural stability;
- good resistance against oxidation and corrosion;
- high ignition resistance (when compared with conventional titanium alloys).

Figure 1.1 shows how the specific modulus and specific strength of gamma TiAl alloys compare to other materials. As a result of these properties TiAl alloys could ultimately find use in a wide range of components in the automotive, aero-engine and power-plant turbine industries.

For a material to be ready for introduction, the whole production chain and supplier base, from material manufacture through processing and heat treatment must have achieved "readiness". This includes detailed knowledge of how component properties are related to alloy chemistry, microstructure, and processing technology. In addition, TiAl-specific component design and lifing methodologies need to be developed and give reliable predictions [2]. At the implementation stage no unforeseen technical problems concerning the processing route or component behavior, which may be very costly to remedy, should arise. In 1999, a time when fuel costs were relatively low compared to the current day, Austin [3] discussed how introduction of gamma would depend on economic viability. This was identified as the chief obstacle for the use of gamma, with marketplace factors dominating implementation decisions.

Gamma Titanium Aluminide Alloys: Science and Technology, First Edition. Fritz Appel, Jonathan David Heaton Paul, Michael Oehring.
© 2011 Wiley-VCH Verlag GmbH & Co. KGaA. Published 2011 by Wiley-VCH Verlag GmbH & Co. KGaA.

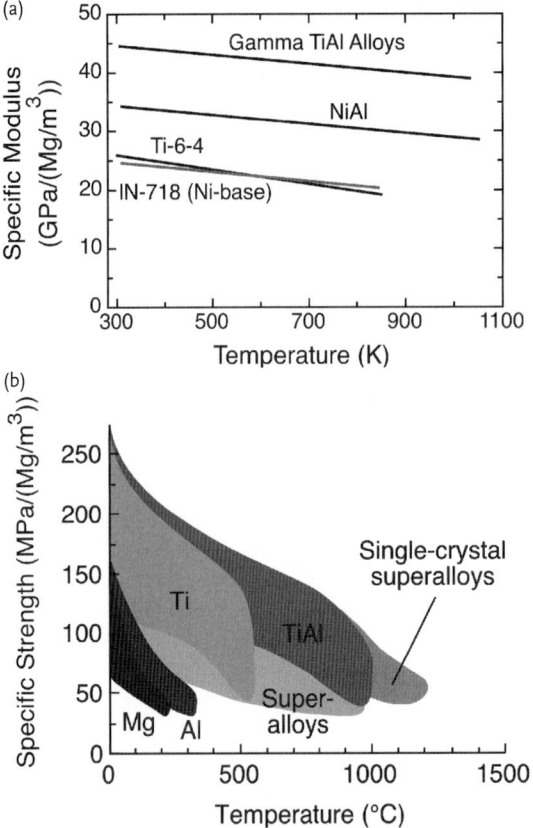

Figure 1.1 Graphs showing the (a) specific moduli and (b) specific strengths of TiAl and other structural materials, as a function of temperature [1]. The data indicates that TiAl compares favorably with the other materials. The data has been redrawn based on the original diagrams.

Due to its intermetallic nature, the complex constitution and microstructure, and the inherent brittleness, the physical metallurgy of TiAl alloys is very demanding. Nevertheless, we will attempt to discuss the broad literature that has been published over the last two decades concerning synthesis, processing and characterization. In our opinion, significant advances have been made, in particular General Electric has made public its intention [4, 5] to use gamma TiAl in its latest engine, the GEnx-1B (Figure 1.2), which best illustrates the present state that has been achieved in TiAl technology. Gamma TiAl has also been successfully introduced into at least one automotive series production, used in formula 1 racing engines, and a variety of components have been manufactured and successfully tested. In the following chapters we will present a comprehensive assessment of both the science and the related technology that has enabled TiAl to be used in the real world.

Figure 1.2 The General Electric GEnx-1B engine for the Boeing 787 Dreamliner. The blades in the last 2 stages of the low-pressure turbine in this engine are made from cast TiAl, making this engine the first to use TiAl in the real world. Photo courtesy of General Electric.

References

1 Dimiduk, D.M. (1999) *Mater. Sci. Eng.*, **A263**, 281.
2 Prihar, R.I. (2001) *Structural Intermetallics 2001* (eds K.J. Hemker, D.M. Dimiduk, H. Clemens, R. Darolia, H. Inui, J.M. Larsen, V.K. Sikka, M. Thomas, and J.D. Whittenberger), TMS, Warrendale, PA, p. 819.
3 Austin, C.M. (1999) *Curr. Opin. Solid State Mater. Sci.*, **4**, 239.
4 Weimer, M., and Kelly, T.J. Presented at the 3rd international workshop on γ-TiAl technologies, 29th to 31st May 2006, Bamberg, Germany.
5 Norris, G. (2006) Flight International Magazine.

2
Constitution

2.1
The Binary Ti–Al Phase Diagram

The binary Ti–Al phase diagram contains several intermetallic phases, which represent superlattices of the terminal solid solutions and have been recognized to be an attractive basis for lightweight high-temperature materials for many years [1]. However, over the past two decades of research, only alloys based on the α_2(Ti$_3$Al) phase with the hexagonal D0$_{19}$ structure or the γ(TiAl) phase with the tetragonal L1$_0$ structure (Figure 2.1) have emerged as structural materials. Among these, interest has been strongly focused on γ titanium aluminide alloys, which, for engineering applications always contain minor fractions of the α_2(Ti$_3$Al) phase. Further, the high-temperature β phase with the bcc A2 structure and its ordered B2 variant (Figure 2.1) play a significant role in some engineering alloys. Despite intensive research, the binary Ti–Al phase diagram still remains a matter of debate and thus has been the subject of experimental work and critical assessment in recent years [2–5]. The impact of such work cannot be overestimated as a full understanding of the microstructural evolution in an alloy is limited by knowledge of the relevant phase equilibria. The discrepancies between different versions of the phase diagram might predominantly have been caused by the high sensitivity of phase equilibria to nonmetallic impurities, in particular oxygen [6–8]; but experimental difficulties, for example, problems in the identification of superlattices and sluggish phase transformations [9] may also play a role. Historically, investigations date back to the 1920s, as reported by Mishurda and Perepezko [6] and Schuster and Palm [2]. The first phase diagrams that covered the whole concentration range were published in the 1950s [10, 11]. More information on the historical development can be found in the article by Mishurda and Perepezko [6] cited earlier. The first critical and thorough assessment of the binary phase diagram was published by Murray [12] and has been considered as a standard reference [2]. Although Murray's assessment does not reflect current knowledge on the phase equilibria in the Ti–Al system, it is a very useful compilation of phase diagram and physical data. Published experimental data was recently reassessed in a comprehensive study by Schuster and Palm [2]. This publication, together with the thermodynamic re-evaluations [3, 7–9, 13–15] constitute the current state of knowledge on

Gamma Titanium Aluminide Alloys: Science and Technology, First Edition. Fritz Appel, Jonathan David Heaton Paul, Michael Oehring.
© 2011 Wiley-VCH Verlag GmbH & Co. KGaA. Published 2011 by Wiley-VCH Verlag GmbH & Co. KGaA.

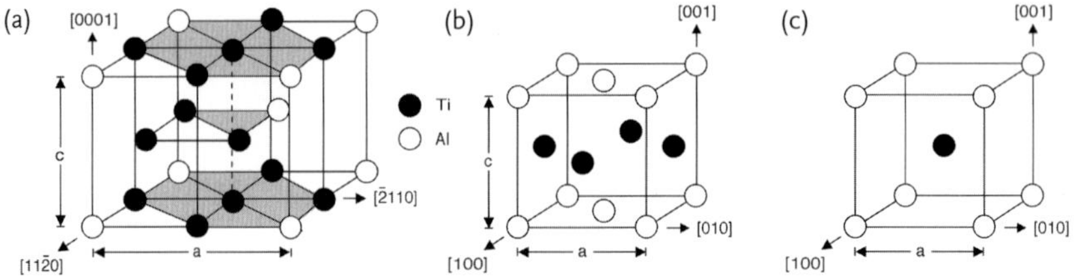

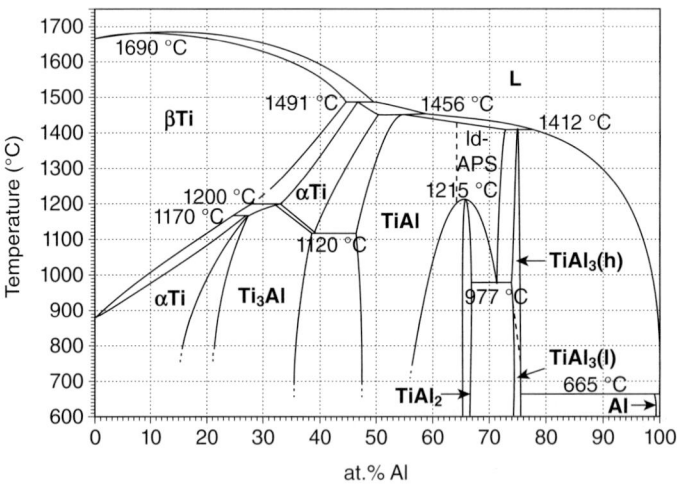

Figure 2.1 Crystal structures of binary Ti aluminide phases. (a) Hexagonal α_2(Ti$_3$Al) phase (Strukturbericht designation D0$_{19}$, prototype Ni$_3$Sn, Pearson symbol hP8, space group P6$_3$/mmc), (b) tetragonal γ(TiAl) phase (Strukturbericht designation L1$_0$, prototype AuCu, Pearson symbol tP4, space group P4/mmm), (c) cubic high temperature B2 phase (Strukturbericht designation B2, prototype CsCl, Pearson symbol cP2, space group Pm$\bar{3}$m). As explained in the text, and by comparing Figures 2.2 to 2.4 with Figure 2.5, the occurrence of the B2 phase in binary alloys is not yet fully clarified.

Figure 2.2 Binary Ti–Al phase diagram according to the assessment of Schuster and Palm [2].

this phase diagram. Figure 2.2 shows the phase diagram that was constructed by Schuster and Palm [2] after their critical assessment of all available experimental data. Figures 2.3 and 2.4 show sections of this diagram covering Al concentrations that are relevant for titanium aluminide alloys based on the γ(TiAl) phase. Experimental data has been plotted on these sections to give an impression on the reliability of phase boundaries. Figure 2.5 shows the most recent result for a thermodynamic evaluation of the binary Ti–Al phase diagram [3] obtained using the CALPHAD approach [16, 17]. Despite many similarities this diagram differs from that of Schuster and Palm [2], in particular for Al concentrations above 60 at.%. These discrepancies mainly arise due to the occurrence of the Ti$_3$Al$_5$ and Ti$_{2+x}$Al$_{5-x}$ phases in the diagram of Witusiewicz et al. [3] and will be briefly dis-

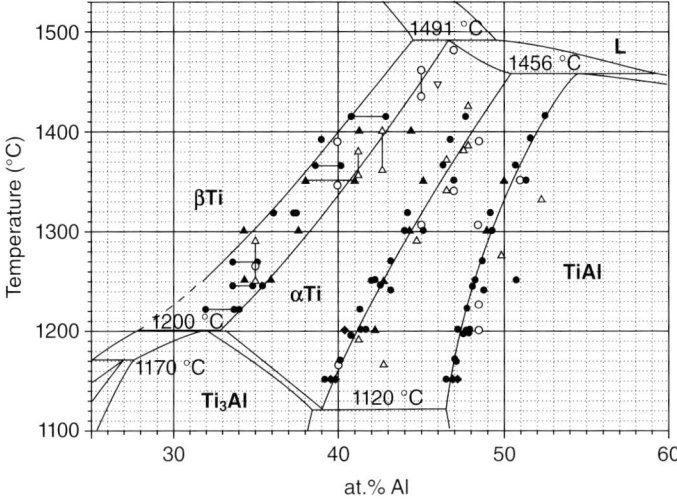

Figure 2.3 Section of the binary Ti–Al phase diagram according to the assessment of Schuster and Palm [2].

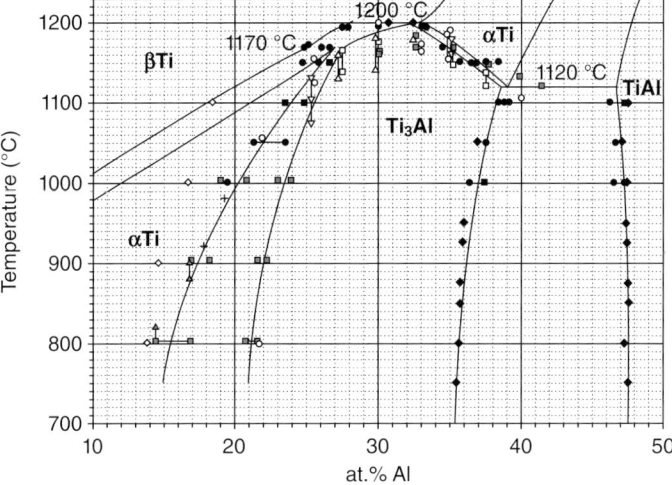

Figure 2.4 Section of the binary Ti–Al phase diagram according to the assessment of Schuster and Palm [2].

cussed below. The crystallographic data of the stable and metastable phases considered by Schuster and Palm [2] and Witusiewicz et al. [3] are given in Table 2.1. Thermodynamic data and data on phase equilibria can be found in these two recent publications as well as in the references cited therein.

As already mentioned, many details of the Ti–Al phase diagram have been debated for a long time. One prominent example is the peritectic reaction

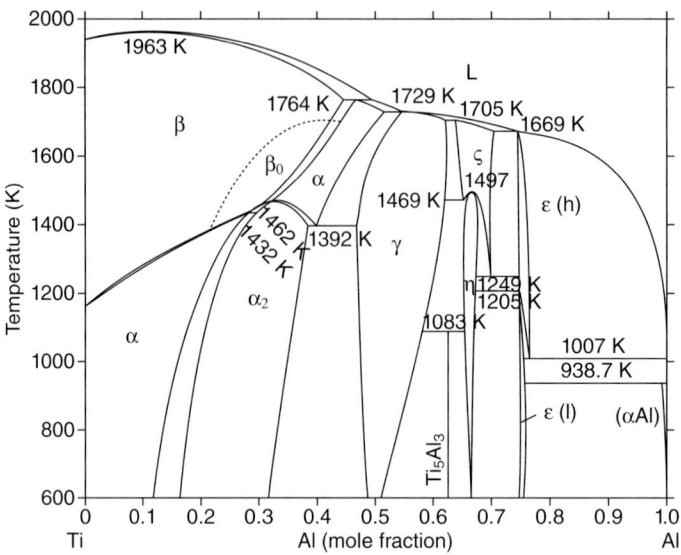

Figure 2.5 Calculated binary Ti–Al phase diagram according to the thermodynamic evaluation of Witusiewicz et al. [3].

L + β → α, which is particularly important for an understanding of the solidification of γ alloys but was not accepted in the assessment of Murray [12]. However, the work of McCullough et al. [32], Mishurda et al. [33], Mishurda and Perepezko [6], Kattner et al. [7] and Jung et al. [34] clearly confirmed this reaction. A further question that has attracted attention is whether the α_2 phase is formed congruently from α or in a peritectoid reaction from α + β as shown in Figures 2.2, 2.4 and 2.5. The work of Kainuma et al. [35] together with other studies, cited by Schuster and Palm [2], has shown that the peritectoid reaction and, as a consequence, a second peritectoid reaction β + α_2 → α occurs. However, other studies by Veeraghavan et al. [36] and Suzuki et al. [37] cast some doubts. Another issue that has been discussed in the literature concerns the question as to whether the β phase occurs as an ordered B2 variant in the binary phase diagram. The B2 ordering has been proposed by Kainuma et al. [35] and would result in a better agreement of the β phase boundary with the experimental results. The two possible versions of the phase diagram are seen in Figures 2.2 and 2.5. By combining DSC measurements with a theoretical extrapolation of the ordering temperature from ternary phase diagrams, Ohnuma et al. [8] confirmed that the β phase shows a second-order transition to the B2 phase. In contrast, Suzuki et al. [37] found no experimental evidence for B2 ordering. Schuster and Palm [2] concluded that B2 ordering in binary alloys is rather unlikely but that it could not be definitely ruled out. This subject could be the topic of future research work.

In Al-rich alloys containing 65 to 72 at.% Al that were quenched from temperatures around 1215 °C, the presence of so-called one-dimensional antiphase domain structures (1d APS) was observed and were designated as Ti_5Al_{11}, γ_2, Ti_2Al_5 or

2.1 The Binary Ti–Al Phase Diagram

Table 2.1 Crystallographic data of phases occurring in the binary Ti–Al system. The data has been taken from the publications of Schuster and Palm [2] and Witusiewicz et al. [3] and partially supplemented. 1d-APS: one-dimensional antiphase domain structures.

Designation	Pearson symbol	Space group	Strukturbericht designation	Prototype	Lattice parameters	Ref.
Phases within in the phase diagram published by Schuster and Palm [2] (Figs. 2.2 to 2.4)						
L (liquid)						
Al	cF4	$Fm\bar{3}m$	A1	Cu	$a = 4.0496$ nm	[18]
β, β-Ti	cI2	$Im\bar{3}m$	A2	W	$a = 0.33065$ nm	at 900 °C [19]
α, α-Ti	hP2	$P6_3/mmc$	A3	Mg	$a = 0.29504$ nm $c = 0.46833$ nm	[19]
α_2, Ti$_3$Al	hP8	$P6_3/mmc$	$D0_{19}$	Ni$_3$Sn	$a = 0.5765$ nm $c = 0.4625$ nm	at 25 at.% Al [20]
γ, TiAl, γ(TiAl)	tP4	$P4/mmm$	$L1_0$	AuCu	$a = 0.3997$ nm $c = 0.4062$ nm $a = 0.4000$ nm $c = 0.4075$ nm $a = 0.4016$ nm $c = 0.4068$ nm	at 50 at.% Al, [21] at 50 at.% Al [22] at 50 at.% Al [23]
η, TiAl$_2$	tI24	$I4_1/amd$		HfGa$_2$	$a = 0.3971$ nm $c/6 = 0.4052$ nm	[24]
ε(h), TiAl$_3$(h), TiAl$_3$(HT)	tI8	$I4/mmm$	$D0_{22}$	TiAl$_3$(h)	$a = 0.3849$ nm $c/2 = 0.4305$ nm	[25]
ε(l), TiAl$_3$(l), TiAl$_3$(LT)	tI32	$I4/mmm$		TiAl$_3$(l)	$a = 0.3877$ nm $c/2 = 0.4229$ nm	[25]

(Continued)

Table 2.1 (Continued)

Designation	Pearson symbol	Space group	Strukturbericht designation	Prototype	Lattice parameters	Ref.
Additional phases occurring in the phase diagram published by Witusiewicz et al. [3] (Fig. 2.5)						
β_o, β', B2	cP2	$Pm\bar{3}m$	B2	CsCl		
Ti_3Al_5	tP32	P4/mbm		Ti_3Ga_5	a = 1.1293 nm c = 0.4038 nm	[26]
ζ, $Ti_{2+x}Al_{5-x}$	tP28	P4/mmm		Ti_2Al_5		
Phases not included in the equilibrium diagrams of Schuster and Palm [2] or that of Witusiewicz et al. [3]						
1d-APS	Tetragonal ordered superstructures of $L1_0$					
Ti_5Al_{11}	tI16	I4/mmm	DO_{23}	$ZrAl_3$	a = 0.3923 nm c/4 = 0.41377 nm	[27]
Ti_2Al_5	tP28	P4/mmm		Ti_2Al_5	a = 0.39053 nm c/7 = 0.41703 nm	[27]
$Ti_{1-x}Al_{1+x}$	tP4 oP4	P4/mmm Pmmm	$L1_0$	AuCu $Ti_{1-x}Al_{1+x}$	a = 0.4030 nm c = 0.3955 nm a = 0.40262 nm b = 0.39617 c = 0.40262 nm	[28] [27]
$TiAl_2$, metastable	oC12	Cmmm		$ZrGa_2$	a/3 = 0.40315 nm b = 0.39591 c = 0.40315 nm	[27]
$TiAl_3$, metastable	cP4	$Pm\bar{3}m$	$L1_2$	$AuCu_3$	a = 0.3967 nm a = 0.3972 nm a = 0.4001 nm	mechanical alloying [29] splat cooling [30] mechanical alloying [31]

long-period superstructures [2]. However, according to the authors it is an open question as to whether the structures are the result of a second-order transition, or if narrow two-phase regions occur in the phase diagram, or if the 1d-APS are transient metastable structures. This open question, together with the not unambiguous existence of the Ti_3Al_5 phase [3] results in the discrepancies between the phase diagrams shown in Figures 2.2 and 2.5. Nevertheless, these phase diagrams reflect fairly well current knowledge concerning the constitution of binary Ti–Al alloys and the remaining uncertainties.

Finally, some specific features of the Ti–Al phase diagram should be noted. Engineering alloys based on the γ(TiAl) phase usually have Al concentrations of 44 to 48 at.% and thus solidify, according to the phase diagram, either through the β phase or peritectically. Depending on processing conditions and alloy composition, it is even possible that two peritectic reactions could occur. Indeed small differences in the Al concentration can result in very different solidification microstructures and textures. Further, the existence of one (α) or two (β and α) single-phase high-temperature regions is a characteristic feature of γ(TiAl) alloys. Similar to steels after heat treatment in the austenite region, a variety of different phase transformations can occur during cooling from high-temperatures or after subsequent heat treatments. In principle, this enables one to obtain a wide range of microstructures [38, 39]. The complexity of possible phase transformations is further increased in multicomponent alloys and has only just begun to be systematically studied. Thus, the mechanical properties may be tailored to some degree using conventional metallurgical processing and the limited damage tolerance of γ alloys may be controlled to some extent. It should be further mentioned that the eutectoid transformation $\alpha \rightarrow \alpha_2 + \gamma$ that takes place on cooling, occurs in all engineering γ(TiAl) alloys. The mechanism of this reaction seems to be identical to the reaction $\alpha \rightarrow \alpha + \gamma$ that proceeds via nucleation and growth of single γ lamellae. To maintain thermodynamic equilibrium in alloy compositions that deviate from the eutectoid composition, the volume fraction of the γ phase has to increase abruptly when the temperature falls below that of the eutectoid transformation. However, the cooling rates usually employed are often too fast to result in thermodynamic equilibrium, and therefore the microstructures obtained may be not stable at the intended service temperature of around 700 °C. Moreover, since the final microstructure in γ(TiAl) alloys is formed during cooling from high-temperature heat treatments, the variation of the volume fraction of γ and the other phases with temperature is a general problem that can result in nonequilibrium phase constitutions. For this reason, a microstructural stabilization treatment or appropriate cooling conditions should be considered.

2.2
Ternary and Multicomponent Alloy Systems

Over the past two decades a broad range of engineering alloys has emerged from a number of alloy development programs, each being developed with respect to different processing routes and applications. The alloys can be described by the general composition:

$$\text{Ti-}(42\text{--}49)\text{Al-}(0.1\text{--}10)\text{X (at.\%)} \qquad (2.1)$$

with X designating the elements Cr, Nb, V, Mn, Ta, Mo, Zr, W, Si, C, Y and B [40–43]. For the alloy compositions described by Equation 2.1, the α_2 phase usually exists in addition to the γ phase and sometimes other phases. Considering the effect of the various alloying elements on the constitution, two alloying strategies can be distinguished. Alloying elements that go into solid solution can be added to γ(TiAl) alloys. Such additions can influence the properties of the γ phase; like the energies of planar defects or the diffusion coefficient. In contrast, other alloying elements are aimed towards the formation of third (or even further) phases to obtain for example, precipitation hardening, grain refinement during casting, a stabilization of the microstructure against grain growth, or transient phases that decompose into fine structures. In Table 2.2 the crystallographic data of some stable and metastable phases occurring in ternary or multicomponent systems are given. When the effect of alloying additions on the intrinsic properties is considered, the solubility in the γ phase, partitioning of alloying elements between the α_2 and the γ phases, and the occupation of the alloying element on the two sublattices of the γ phase are of interest. Information on the site occupancy of different alloying elements in the γ phase can be found in studies by Mohandas and Beaven [58], Rossouw et al. [59] and Hao et al. [60]. In an interesting study, Hao et al. [61] have shown that the lattice occupancy of different alloying elements is related to the boundaries of the ($\alpha_2 + \gamma$) two-phase field in the respective (Ti–Al–X) systems. A more detailed discussion of site occupancy and its influence on mechanical properties can be found in Chapter 7. With respect to solubility in the γ phase, most of the metals mentioned above are only soluble up to a limited amount of around 2 to 3 at.%, as the respective ternary phase diagrams show [62–68]. For higher concentrations, the bcc β solid solution or its ordered variant with a B2 structure usually form as a third phase. A good overview on the constitution of ternary alloy systems in relevant composition ranges has been given by Kainuma et al. [68]. In a comprehensive study the authors investigated the phase equilibria between the α, β and γ phases for most of the Ti–Al–X systems according to the general composition described by Equation 2.1. Figure 2.6 shows the three-phase equilibria at 1200 °C obtained in this study. From this figure it is obvious that the alloying elements Zr, Nb and Ta are exceptional in respect that they are soluble in higher contents in the γ phase, in the case of Nb up to around 9 at.% at 1200 °C [68]. The study of Kainuma et al. [68] is also interesting because attention was paid to the partitioning coefficient between the different phases. With respect to partitioning between the α_2 and the γ phases, V, Cr, Mo, Ta and W were enriched in the α_2 phase, Nb and Mn were distributed in equal amounts between the α_2 and γ phases, whereas Zr was concentrated in the γ phase. The elements Fe, Cr, Mo and W strongly partitioned to the β phase, with respect to both the α_2 and the γ phases. In Figures 2.7 and 2.8 a number of isothermal sections from important ternary phase diagrams are displayed that were published by Kainuma et al. [68]. These diagrams give an impression regarding the constitution of the different

Table 2.2 Crystallographic data of some phases arising in ternary or multicomponent TiAl alloys of engineering relevance.

Phase	Pearson symbol	Space group	Strukturbericht designation	Prototype	Lattice parameters	Ref.
Ti_2AlNb, O-phase	oC16	Cmcm		NaHg, Cd_3Er	$a = 0.609$ nm $b = 0.957$ nm $c = 0.467$ nm	[44]
B19	oP4	Pmma	B19	AuCd	$a = 0.45$ nm $b = 0.28$ nm $c = 0.49$ nm	[45, 46]
Ti_4Al_3Nb (ω_o, τ)	hP6	$P6_3/mmc$	$B8_2$	Ni_2In	$a = 0.458$ nm $c = 0.552$ nm	[47]
ω-Ti	hP3	P6/mmm		ω-Ti	$a = 0.463$ nm $c = 0.281$ nm	[48]
ω''		$P\bar{3}m1$			$a = 0.456$ nm $c = 0.554$ nm	[47]
Ti_2AlC (H-phase)	hP8	$P6_3/mmc$		Cr_2AlC	$a = 0.306$ nm $c = 1.362$ nm	[49]
Ti_3AlC (perovskite)	cP5	$Pm\bar{3}m$	$E2_1$	$CaTiO_3$	$a = 0.416$ nm	[50]
Ti_2AlN	hP8	$P6_3/mmc$		Cr_2AlC		[51]
Ti_3AlN	cP5	$Pm\bar{3}m$	$E2_1$	$CaTiO_3$		[51]
TiO_2 (rutile)	tP6	$P4_2/mnm$	C4	TiO_2	$a = 0.459$ nm $c = 0.296$ nm	[52, 53]
β-Ti_2O_3	hR10	$R\bar{3}c$	$D5_1$	α-Al_2O_3	$a = 0.516$ nm $c = 1.361$ nm	[52]
β-TiO	cF180				$a = 1.254$ nm	[52]
α-Al_2O_3	hR10	$R\bar{3}c$	$D5_1$	α-Al_2O_3	$a = 0.475$ nm $c = 1.299$ nm	[52]
Ti_5Si_3	hP16	$P6_3/mcm$	$D8_8$	Mn_5Si_3	$a = 0.747$ nm $c = 0.516$ nm	[54]
TiB	oP8	Pnma	B27	FeB	$a = 0.611$ nm $b = 0.305$ nm $c = 0.456$ nm	[55–57]
TiB	oC8	Cmcm	B_f	CrB	$a = 0.323$ nm $b = 0.856$ nm $c = 0.305$ nm	[56, 57]
Ti_3B_4	oI14	Immm	$D7_b$	Ta_3B_4	$a = 0.326$ nm $b = 1.373$ nm $c = 0.304$ nm	[56, 57]
TiB_2	hP3	P6/mmm	C32	AlB_2	$a = 0.303$ nm $c = 0.323$ nm	[56, 57]

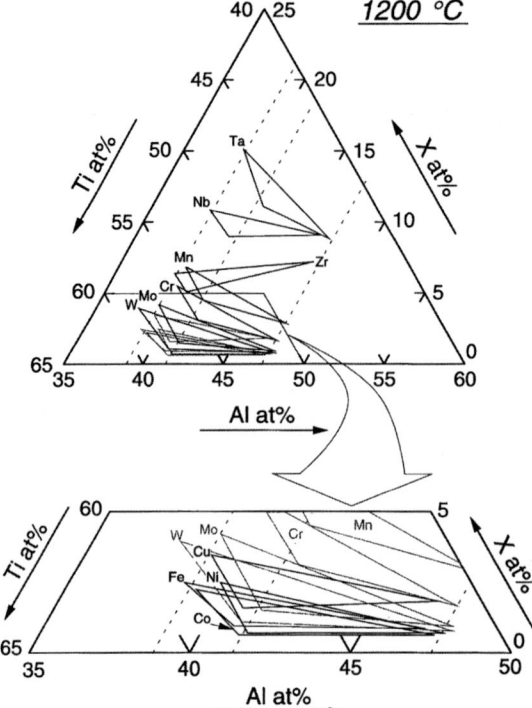

Figure 2.6 Position of the α + β + γ three-phase field at 1200 °C for different ternary Ti–Al–X systems [68].

ternary systems. In recent years, some alloy development programs have been directed to alloy systems in which higher amounts of β phase occur [69–76]. However, lack of knowledge regarding the constitution of such multicomponent systems is one of the major obstacles to alloy development and significant work has to be expended to obtain reliable information on phase relationships. This originates from the fact that the constitution of multicomponent systems can be intricate and rich in detail. Similar to the case for the binary Ti–Al system some ternary phase diagrams, such as the Ti–Al–Nb diagram, have remained a matter of debate for a long time [61, 66, 67, 77–88]. With regard to other ternary systems a vast body of valuable information can be found in the literature, for example, Ti–Al–Cr [62, 63, 67, 68, 89, 90], Ti–Al–Mo [64, 65, 68, 71, 91], Ti–Al–V [61, 68, 71, 92, 93], Ti–Al–Mn [61, 68, 94], Ti–Al–Ta [61, 68, 95], Ti–Al–W [68], Ti–Al–Fe [61, 68, 96, 97], Ti–Al–Si [98, 99], Ti–Al–O [67, 100, 101], Ti–Al–N [102], Ti–Al–C [103] and Ti–Al–B [104–106].

As mentioned above, Nb is soluble in comparatively large amounts in the γ as well as the α₂ phase and has been found to be particularly advantageous for γ(TiAl) engineering alloys [107–111]. The properties of such alloys with high Nb contents are addressed in Chapters 7 and 13. Here, it should be mentioned that

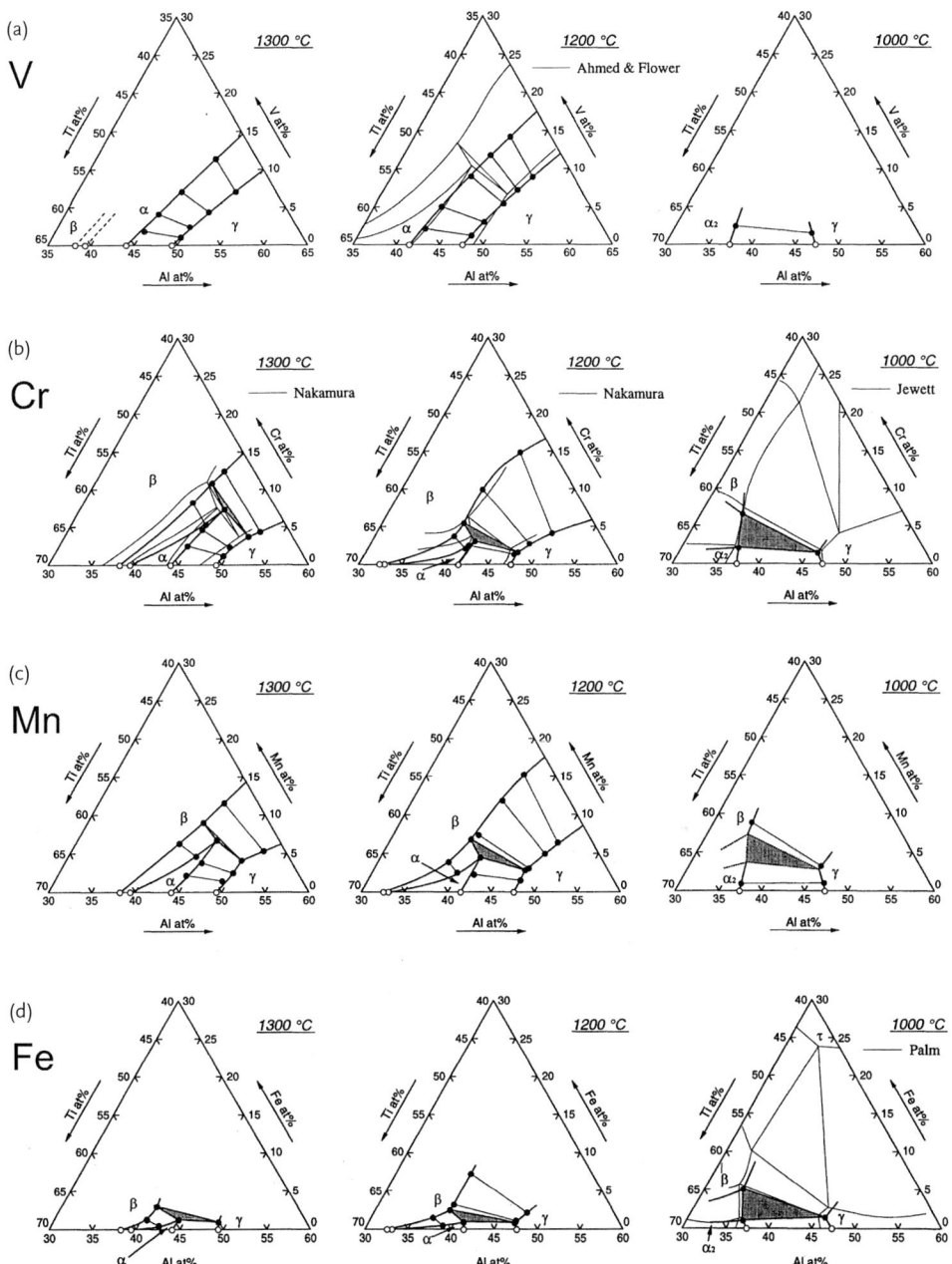

Figure 2.7 Isothermal sections of different ternary Ti–Al–X systems [68]. (a) Ti–Al–V, (b) Ti–Al–Cr, (c) Ti–Al–Mn, (d) Ti–Al–Fe.

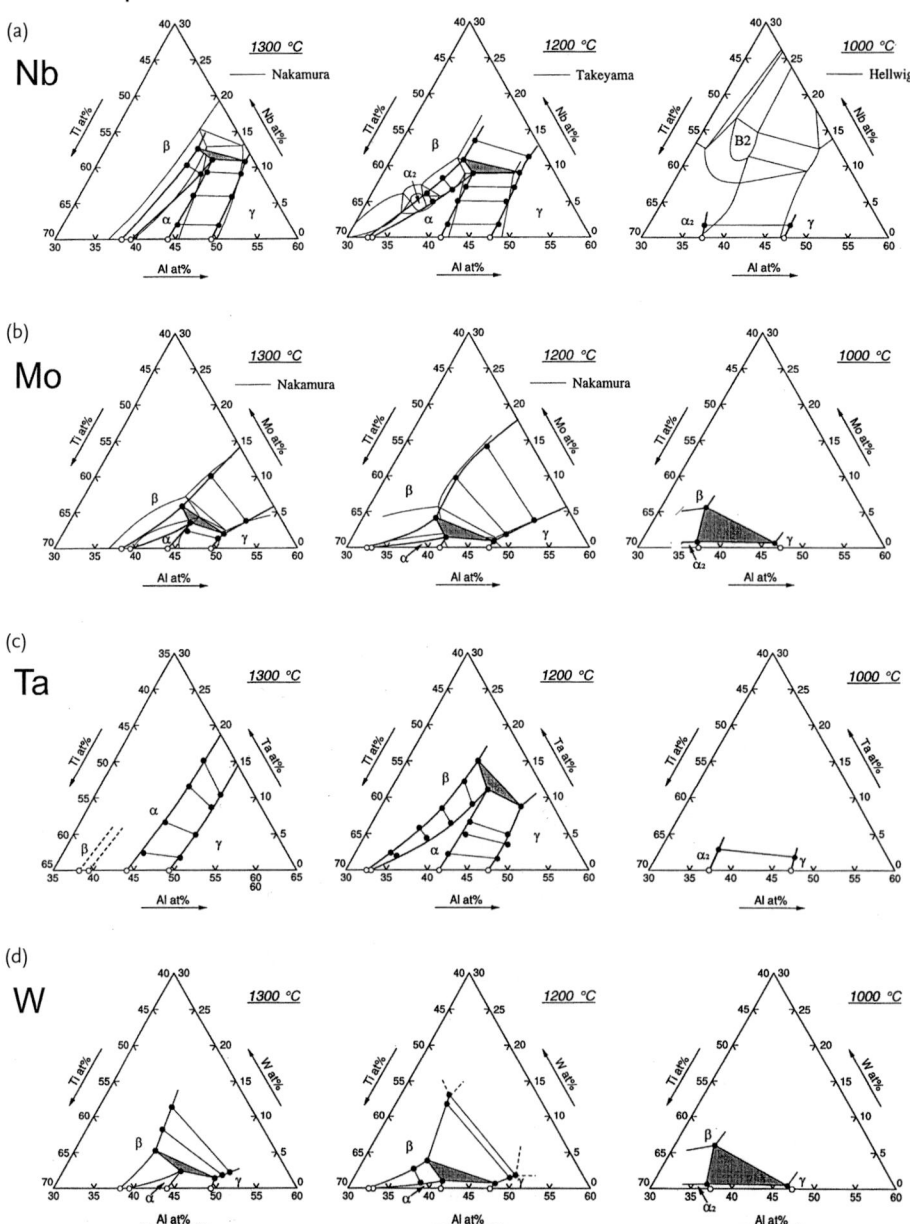

Figure 2.8 Isothermal sections of different ternary Ti–Al–X systems [68]. (a) Ti–Al–Nb, (b) Ti–Al–Mo, (c) Ti–Al–Ta, (d) Ti–Al–W.

2.2 Ternary and Multicomponent Alloy Systems

the Ti–Al–Nb phase diagram is complicated since several ternary compounds with compositions that are not far away from the $\alpha + \gamma + \beta(B2)$ and $\alpha_2 + \gamma + B2$ three-phase fields occur. The occurrence of these compounds is one reason for the discrepancies between the different constitutional studies mentioned above. It is also noteworthy to mention that multiphase assemblies including orthorhombic phases were observed in high Nb-containing alloys [73] and these were not predicted by the phase diagrams available at that time. See Chapter 8 for more details on such alloys. In Figures 2.9 and 2.10 the constitution of the ternary Ti–Al–Nb system is depicted for temperatures and compositions that are relevant for engineering applications. The isothermal and vertical sections that are shown were

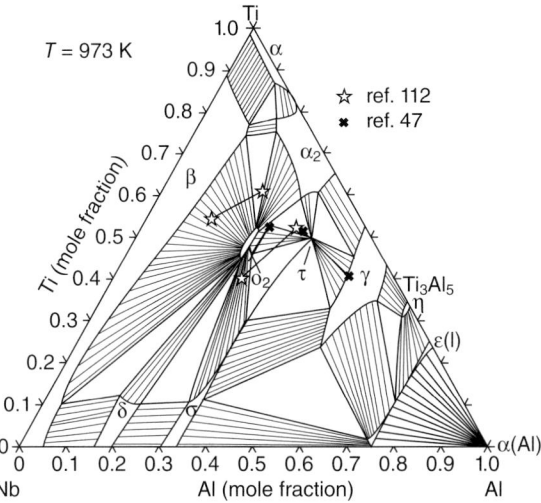

Figure 2.9 Calculated isothermal section of the Ti-Al-Nb system at 700 °C according to the most recent thermodynamic re-evaluation [86]. The table below indicates which phases are present in the phase diagram:

Designation	Pearson symbol	Space group	Strukturbericht designation
Al	cF4	Fm $\bar{3}$m	A1
α, α-Ti	hP2	P6$_3$/mmc	A3
β, β-Ti	cI2	Im $\bar{3}$m	A2
α$_2$ (Ti$_3$Al)	hP8	P6$_3$/mmc	D0$_{19}$
γ (TiAl)	tP4	P4/mmm	L1$_0$
O$_2$ (Ti$_2$AlNb)	oC16	Cmcm	
τ (Ti$_4$Al$_3$Nb, ω$_o$)	hP6	P6$_3$/mmc	B8$_2$
δ (Nb$_3$Al)	cP8	Pm $\bar{3}$n	A15
σ (Nb$_2$Al)	tP30	P4$_2$/mmm	D8$_b$
Ti$_3$Al$_5$	tP32	P4/mbm	
η (TiAl$_2$)	tI24	I4$_1$/amd	
ε(l) (TiAl$_3$ (l))	tI32	I4/mmm	

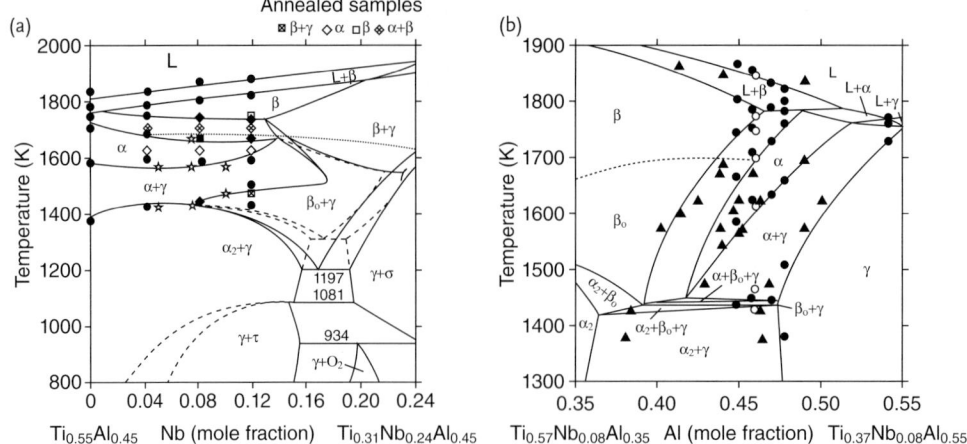

Figure 2.10 Calculated vertical sections close to the Ti-Al side of the Ti-Al-Nb system according to the most recent thermodynamic re-evaluation [86]. (a) Isopleth for 45 at. % Al, (b) isopleth for 8 at. % Nb. The table below indicates which phases are present in the phase diagrams:

Designation	Pearson symbol	Space group	Strukturbericht designation
α, α-Ti	hP2	P6$_3$/mmc	A3
β, β-Ti	cI2	Im $\bar{3}$m	A2
β$_o$	cP2	Pm $\bar{3}$m	B2
α$_2$ (Ti$_3$Al)	hP8	P6$_3$/mmc	DO$_{19}$
γ (TiAl)	tP4	P4/mmm	L1$_0$
O$_2$ (Ti$_2$AlNb)	oC16	Cmcm	
τ (Ti$_4$Al$_3$Nb, ω$_o$)	hP6	P6$_3$/mmc	B8$_2$
σ (Nb$_2$Al)	tP30	P4$_2$/mmm	D8$_b$

taken from the most recent thermodynamic re-evaluation of the Ti–Al–Nb and the constituent binary systems [86] and thus represent current knowledge on the constitution of Ti–Al–Nb alloys.

Besides Nb and Cr the most widely used alloying element in γ(TiAl) alloys is B. This is due to the grain-refining effect that B additions within the range of 0.1 to 2 at.% have on cast alloys [113, 114]. Additionally, B additions have also been found to be useful with respect to the hot-working behavior and the microstructure of wrought products (see Chapter 16). B is almost insoluble in all binary phases of the Ti–Al system [106, 115]. In an interesting study by Hyman et al. [104] it was shown that, depending on the Al concentration, either the β or α phase are the primary solidifying phase for B contents of 0.7 to 1.5 at.% (Figure 2.11). This finding is in agreement with very recent work by Witusiewicz et al. [106]. Taken together these results provide evidence that the simplest explanation for the mechanism of grain refinement in alloys with B concentrations less than 0.7 at.%, that is, that the borides serve as nuclei for the binary TiAl phases, is not correct. Thus,

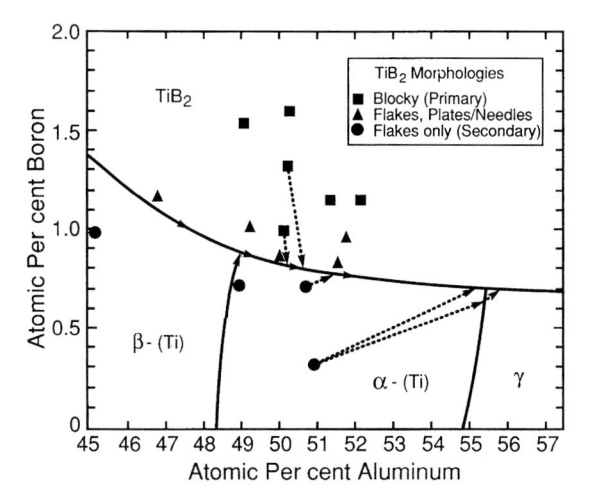

Figure 2.11 Suggested liquidus projection for the ternary Ti–Al–B system in the vicinity of the equiatomic TiAl composition [104]. Dashed lines denote schematic solidification of some alloys.

a full understanding of grain refinement in TiAl alloys, and higher multicomponent systems, requires detailed knowledge on the constitution. More on this subject can be found in Chapter 4.

To conclude this chapter, the influence of O on the constitution of TiAl alloys will be briefly considered. As the available ternary Ti–Al–O phase diagrams show [67, 100], the solubility of O is much higher in the α_2 phase as compared to the γ phase. This has been confirmed by atom probe (AP) analysis, a method that is capable of quantitatively determining the O content with a very high spatial resolution, in fine α_2 and γ lamellae for example. Using this technique, Nérac-Partaix et al. [116] showed that the solubility of O in the γ phase was about 300 at. ppm for Ti–Al two-phase alloys, with the α_2 phase containing between 8000 and 22 000 at. ppm O depending on the Al concentration [116, 117]. In alloys with an Al content of 52 at.% that did not contain any α_2 phase, the same solubility of O as in two-phase alloys was determined, even if the specimens contained 1 or 2 at.% O [116]. The authors proposed that the difference in the solubility of O in the α_2 and the γ phase might originate from the presence of octahedral sites, surrounded by six titanium atoms, in the structure of the α_2 phase [117]. The high solubility of O in the α_2 phase obviously prevents the formation of oxides, which occur in higher Al containing alloys when α_2 is not present, as early has been concluded by Vasudevan et al. [118]. The significantly different solubility of O in the two phases results in the phase boundaries between the α_2 and the γ phases depending on the O content of the material. Similar results were obtained for alloys that contained additions of Cr, Mn and Nb [117, 119]. Taken together, these findings confirm the high sensitivity of binary Ti–Al and ternary Ti–Al–X systems on the O content. This also has been concluded indirectly from other studies [6–8].

References

1. McAndrew, J.B. and Kessler, H.D. (1956) *Trans. AIME/J. Met.*, **8**, 1348.
2. Schuster, J.C. and Palm, M. (2006) *J. Phase Equil. Diff.*, **27**, 255.
3. Witusiewicz, V.T., Bondar, A.A., Hecht, U., Rex, S., and Velikanova, T.Ya. (2008) *J. Alloys Compd.*, **465**, 64.
4. Grytsiv, A., Rogl, P., Schmidt, H., and Giester, G. (2003) *J. Phase Equil.*, **24**, 511.
5. Raghavan, V. (2005) *J. Phase Equil. Diff.*, **26**, 171.
6. Mishurda, J.C. and Perepezko, J.H. (1991) *Microstructure/Property Relationships in Titanium Aluminide Alloys* (eds Y.-W. Kim and R.R. Boyer), TMS, Warrendale, PA, p. 3.
7. Kattner, U.R., Lin, J.-C., and Chang, Y.A. (1992) *Metall. Trans.*, **23A**, 2081.
8. Ohnuma, I., Fujita, Y., Mitsui, H., Ishikawa, K., Kainuma, R., and Ishida, K. (2000) *Acta Mater.*, **48**, 3113.
9. Jones, S.A. and Kaufman, M.J. (1993) *Acta Metall. Mater.*, **41**, 387.
10. Ogden, H.R., Maykuth, D.J., Finlay, W.L., and Jaffee, R.I. (1951) *Trans. AIME/J. Met.*, **3**, 1150.
11. Bumps, E.S., Kessler, H.D., and Hansen, M. (1952) *Trans. AIME/J. Met.*, **4**, 609.
12. Murray, J.L. (1987) *Phase Diagrams of Binary Titanium Alloys* (ed. J.L. Murray), ASM, Metals Park, OH, p. 12.
13. Okamoto, H. (1993) *J. Phase Equil.*, **14**, 120.
14. Zhang, F., Chen, S.L., Chang, Y.A., and Kattner, U.R. (1997) *Intermetallics*, **5**, 471.
15. Okamoto, H. (2000) *J. Phase Equil.*, **21**, 311.
16. Kaufman, L. (1969) *Prog. Mater. Sci.*, **14**, 57.
17. Kaufman, L. and Bernstein, H. (1970) *Computer Calculation of Phase Diagrams*, Academic Press, New York, NY.
18. Massalski, T.B. (1990) *Binary Alloy Phase Diagrams*, ASM International, Metals Park, OH, p. 2179 (data collected by H.W. King).
19. Murray, J.L. and Wriedt, H.A. (1987) *Phase Diagrams of Binary Titanium Alloys* (ed. J.L. Murray), ASM, Metals Park, OH, p. 1.
20. Blackburn, M.J. (1967) *Trans. Metall. Soc. AIME*, **239**, 1200.
21. Duwez, P. and Taylor, J.L. (1952) *Trans. AIME/J. Met.*, **4**, 70.
22. Braun, J., Ellner, M., and Predel, B. (in German), (1995) *Z. Metallkd.*, **86**, 870.
23. Pfullmann, T. and Beaven, P.A. (1993) *Scr. Metall. Mater.*, **28**, 275.
24. Mabuchi, H., Asai, T., and Nakayama, Y. (1989) *Scr. Metall.*, **23**, 685.
25. van Loo, F.J.J. and Rieck, G.D. (1973) *Acta Metall.*, **21**, 61.
26. Braun, J. and Ellner, M. (2001) *Metall. Trans.*, **32A**, 1037.
27. Schuster, J.C. and Ipser, H. (1990) *Z. Metallkd.*, **81**, 389.
28. Braun, J., Ellner, M., and Predel, B. (in German) (1994) *J. Alloys Compd.*, **203**, 189.
29. Srinivasan, S., Desh, P.B., and Schwarz, R.B. (1991) *Scr. Metall. Mater.*, **25**, 2513.
30. Braun, J., Ellner, M., and Predel, B. (in German) (1994) *Z. Metallkd.*, **85**, 855.
31. Klassen, T., Oehring, M., and Bormann, R. (1994) *J. Mater. Res.*, **9**, 47.
32. McCullough, C., Valencia, J.J., Levi, C.G., and Mehrabian, R. (1989) *Acta Metall.*, **37**, 1321.
33. Mishurda, J.C., Lin, J.C., Chang, Y.A., and Perepezko, J.H. (1989) *High-Temperature Ordered Intermetallic Alloys III*, vol. 133 (eds C.T. Liu, A.J. Taub, N.S. Stoloff, and C.C. Koch), *Mater. Res. Soc. Symp. Proc.*, Mater. Res. Soc., Pittsburgh, PA, p. 57.
34. Jung, I.-S., Kim, M.-C., Lee, J.-H., Oh, M.-H., and Wee, D.-M. (1999) *Intermetallics*, **7**, 1247.
35. Kainuma, R., Palm, M., and Inden, G. (1994) *Intermetallics*, **2**, 321.
36. Veeraghavan, D., Pilchowski, U., Natarajan, B., and Vasudevan, V.K. (1998) *Acta Mater.*, **46**, 405.
37. Suzuki, A., Takeyama, M., and Matsuo, T. (2002) *Intermetallics*, **10**, 915.
38. Kim, Y.-W. (1992) *Acta Metall. Mater.*, **40**, 1121.
39. Yamabe, Y., Takeyama, M., and Kikuchi, M. (1995) *Gamma Titanium*

Aluminides (eds Y.-W. Kim, R. Wagner, and M. Yamaguchi), TMS, Warrendale, PA, p. 111.

40 Kim, Y.-W. (1991) *JOM (J. Metals)*, **43** (August), 40.

41 Yamaguchi, M. and Inui, H. (1993) *Structural Intermetallics* (eds R. Darolia, J.J. Lewandowski, C.T. Liu, P.L. Martin, D.B. Miracle, and M.V. Nathal), TMS, Warrendale, PA, p. 127.

42 Kim, Y.-W. (1994) *JOM (J. Metals)*, **46** (July), 30.

43 Kim, Y.-W. and Dimiduk, D.M. (1997) *Structural Intermetallics 1997* (eds M.V. Nathal, R. Darolia, C.T. Liu, P.L. Martin, D.B. Miracle, R. Wagner, and M. Yamaguchi), TMS, Warrendale, PA, p. 531.

44 Mozer, B., Bendersky, L., Boettinger, W.J., and Grant Rowe, R. (1990) *Scr. Met. all.*, **24**, 2363.

45 Ohba, T., Emura, Y., Miyazaki, S., and Otsuka, K. (1990) *Mater. Trans. (Japanese Inst. Metals)*, **31**, 12.

46 Abe, E., Kumagai, T., and Nakamura, M. (1996) *Intermetallics*, **4**, 327.

47 Bendersky, L.A., Boettinger, W.J., Burton, B.P., Biancaniello, F.S., and Shoemaker, C.B. (1990) *Acta Metall. Mater.*, **38**, 931.

48 Banerjee, S. and Mukhopadhyay, P. (2007) *Phase Transformations – Examples from Titanium and Zirconium Alloys*, Pergamon Materials Series, vol. 12, Elsevier, Amsterdam.

49 Schuster, J.C. and Nowotny, H. (1980) *Z. Metallkd.*, **71**, 341.

50 Schuster, J.C., Nowotny, H., and Vaccaro, C. (1980) *J. Solid State Chem.*, **32**, 213.

51 Schuster, J.C. and Bauer, J. (1984) *J. Solid State Chem.*, **53**, 260.

52 Hoch, N. and Lin, R.Y. (1993) *Ternary Alloys: A Comprehensive Compendium of Evaluated Constitutional Data and Phase Diagrams*, vol. 8 (eds G. Petzow and G. Effenberg), VCH, Weinheim, Germany, p. 79.

53 Murray, J.L. and Wriedt, H.A. (1987) *Phase Diagrams of Binary Titanium Alloys* (ed. J.L. Murray), ASM, Metals Park, OH, p. 211.

54 Pietrowsky, P. and Duwez, P. (1951) *J. Metals*, **3**, 772.

55 Decker, B. (1954) *Acta Crystallogr.*, **7**, 77.

56 De Graef, M., Löfvander, J.P.A., McCullough, C., and Levi, C.G. (1992) *Acta Metall. Mater.*, **40**, 3395.

57 Kitkamthorn, U., Zhang, L.C., and Aindow, M. (2006) *Intermetallics*, **14**, 759.

58 Mohandas, E. and Beaven, P.A. (1991) *Scr. Metall. Mater.*, **25**, 2023.

59 Rossouw, C.T., Forwood, M.A., Gibson, M.A., and Miller, P.R. (1996) *Philos. Mag. A*, **74**, 77.

60 Hao, Y.L., Xu, D.S., Cui, Y.Y., Yang, R., and Li, D. (1999) *Acta Mater.*, **47**, 1129.

61 Hao, Y.L., Yang, R., Cui, Y.Y., and Li, D. (2000) *Acta Mater.*, **48**, 1313.

62 Jewett, T.J., Ahrens, B., and Dahms, M. (1996) *Intermetallics*, **4**, 543.

63 Palm, M. and Inden, G. (1997) *Structural Intermetallics 1997* (eds M.V. Nathal, R. Darolia, C.T. Liu, P.L. Martin, D.B. Miracle, R. Wagner, and M. Yamaguchi), TMS, Warrendale, PA, p. 73.

64 Singh, A.K. and Banerjee, D. (1997) *Metall. Mater. Trans.*, **28A**, 1735.

65 Singh, A.K. and Banerjee, D. (1997) *Metall. Mater. Trans.*, **28A**, 1745.

66 Hellwig, A., Palm, M., and Inden, G. (1998) *Intermetallics*, **6**, 79.

67 Saunders, N. (1999) *Gamma Titanium Aluminides 1999* (eds Y.-W. Kim, D.M. Dimiduk, and M.H. Loretto), TMS, Warrendale, PA, p. 183.

68 Kainuma, R., Fujita, Y., Mitsui, H., Ohnuma, I., and Ishida, K. (2000) *Intermetallics*, **8**, 855.

69 Naka, S., Thomas, M., Sanchez, C., and Khan, T. (1997) *Structural Intermetallics 1997* (eds M.V. Nathal, R. Darolia, C.T. Liu, P.L. Martin, D.B. Miracle, R. Wagner, and M. Yamaguchi), TMS, Warrendale, PA, p. 313.

70 Tetsui, T., Shindo, K., Kobayashi, S., and Takeyama, M. (2002) *Scr. Mater.*, **47**, 399.

71 Kobayashi, S., Takeyama, M., Motegi, T., Hirota, N., and Matsuo, T. (2003) *Gamma Titanium Aluminides 2003* (eds Y.-W. Kim, H. Clemens, and A.H. Rosenberger), TMS, Warrendale, PA, p. 165.

72 Takeyama, M. and Kobayashi, S. (2005) *Intermetallics*, **13**, 993.

73 Appel, F., Oehring, M., and Paul, J.D.H. (2006) *Adv. Eng. Mater.*, **8**, 371.

74 Imayev, R.M., Imayev, V.M., Oehring, M., and Appel, F. (2007) *Intermetallics*, **15**, 451.

75 Kremmer, S., Chladil, H.F., Clemens, H., Otto, A., and Güther, V. (2007) *Ti-2007, Science and Technology, Proc. of the 11th World Conference on Titanium* (eds M. Ninomi, S. Akiyama, M. Ikeda, M. Hagiwara, and K. Maruyama), The Japan Institute of Metals, Tokyo, Japan, p. 989.

76 Kim, Y.-W., Kim, S.-L., Dimiduk, D.M., and Woodward, C. (2008) *Structural Intermetallics for Elevated Temperatures* (eds Y.-W. Kim, D. Morris, R. Yang, and C. Leyens), TMS, Warrendale, PA, p. 215.

77 Perepezko, J.H., Chang, Y.A., Seitzman, L.E., Lin, J.C., Bonda, N.R., and Jewett, T.J. (1990) *High Temperature Aluminides and Intermetallics* (eds S.H. Whang, C.T. Liu, D.P. Pope, and J.O. Stiegler), TMS, Warrendale, PA, p. 19.

78 Kattner, U.R. and Boettinger, W.J. (1992) *Mater. Sci. Eng.*, **152**, 9.

79 Das, S., Jewett, T.J., and Perepezko, J.H. (1993) *Structural Intermetallics* (eds R. Darolia, J.J. Lewandowski, C.T. Liu, P.L. Martin, D.B. Miracle, and M.V. Nathal), TMS, Warrendale, PA, p. 35.

80 Nakamura, H., Takeyama, M., Yamabe, Y., and Kikuchi, M. (1993) *Scr. Metall. Mater.*, **28**, 997.

81 Zdziobek, A., Durand-Charre, M., Driole, J., and Durand, F. (1995) *Z. Metallkd.*, **86**, 334.

82 Chen, G.L., Wang, X.T., Ni, K.Q., Hao, S.M., Cao, J.X., Ding, J.J., and Zhang, X. (1996) *Intermetallics*, **6**, 13.

83 Chen, G.L., Zhang, W.J., Liu, Z.C., Li, S.J., and Kim, Y.-W. (1999) *Gamma Titanium Aluminides 1999* (eds Y.-W. Kim, D.M. Dimiduk, and M.H. Loretto), TMS, Warrendale, PA, p. 371.

84 Servant, I. and Ansara, I. (2001) *CALPHAD*, **25**, 509.

85 Chladil, H.F., Clemens, H., Zickler, G.A., Takeyama, M., Kozeschnik, E., Bartels, A., Buslaps, T., Gerling, R., Kremmer, S., Yeoh, L., and Liss, K.-D. (2007) *Int. J. Mater. Res. (formerly Z. Metallkd.)*, **98**, 1131.

86 Witusiewicz, V.T., Bondar, A.A., Hecht, U., and Velikanova, T.Ya. (2009) *J. Alloys Compd.*, **472**, 133.

87 Bystranowski, S., Bartels, A., Stark, A., Gerling, R., Schimansky, F.-P., and Clemens, H. (2010) *Intermetallics*, **18**, 1046.

88 Shuleshova, O., Holland-Moritz, D., Löser, W., Voss, A., Hartmann, H., Hecht, U., Witusiewicz, V.T., Herlach, D.M., and Büchner, B. (2010) *Acta Mater.*, **58**, 2408.

89 Nakamura, H., Takeyama, M., Yamabe, Y., and Kikuchi, M. (1993) *Proc. 3rd Japan International SAMPE Symposium and Exhibition* (eds M. Yamaguchi and H. Hukutomi), Society for the advancement of material and process engineering, Chiba, Japan, p. 1353.

90 Shao, G. and Tsakiropoulos, P. (1999) *Intermetallics*, **7**, 579.

91 Budberg, A. and Schmid-Fetzer, R. (1993) *Ternary Alloys: A Comprehensive Compendium of Evaluated Constitutional Data and Phase Diagrams*, vol. 7 (G. Petzow and G. Effenberg (eds) VCH, Weinheim, Germany, p. 229.

92 Ahmed, T. and Flower, H.M. (1992) *Mater. Sci. Eng.*, **A152**, 31.

93 Ahmed, T. and Flower, H.M. (1994) *Mater. Sci. Technol.*, **10**, 272.

94 Butler, C.J., McCartney, D.G., Small, C.J., Horrocks, F.J., and Saunders, N. (1997) *Acta Mater.*, **45**, 2931.

95 Weaver, M.L. and Kaufman, M.J. (1995) *Acta Metall. Mater.*, **43**, 2625.

96 Palm, M., Inden, G., and Thomas, N. (1995) *J. Phase Equil.*, **16**, 209.

97 Palm, M., Gorzel, A., Letzig, D., and Sauthoff, G. (1997) *Structural Intermetallics 1997* (eds M.V. Nathal, R. Darolia, C.T. Liu, P.L. Martin, D.B. Miracle, R. Wagner, and M. Yamaguchi), TMS, Warrendale, PA, p. 885.

98 Perrot, P. (1993) *Ternary Alloys: A Comprehensive Compendium of Evaluated Constitutional Data and Phase Diagrams*, vol. 8 (eds G. Petzow and G. Effenberg), VCH, Weinheim, Germany, p. 283.

99 Manesh, S.H. and Flower, H.M. (1994) *Mater. Sci. Technol.*, **10**, 674.

100 Li, X.L., Hillel, R., Teyssandier, F., Choi, S.K., and van Loo, F.J.J. (1992) *Acta Metall. Mater.*, **40**, 3149.

101 Lee, B.-J. and Saunders, N. (1997) *Z. Metallkd.*, **88**, 152.
102 Jehn, H.A. (1993) *Ternary Alloys: A Comprehensive Compendium of Evaluated Constitutional Data and Phase Diagrams*, vol. 7 (eds G. Petzow and G. Effenberg), VCH, Weinheim, Germany, p. 305.
103 Hayes, F.H. (1990) *Ternary Alloys: A Comprehensive Compendium of Evaluated Constitutional Data and Phase Diagrams*, vol. 3 (eds G. Petzow and G. Effenberg), VCH, Weinheim, Germany, p. 557.
104 Hyman, M.E., McCullough, C., Levi, C.G., and Mehrabian, R. (1991) *Metall. Trans.*, **22A**, 1647.
105 Gröbner, J., Mirkovic, D., and Schmid-Fetzer, R. (2005) *Mater. Sci. Eng.*, **395**, 10.
106 Witusiewicz, V.T., Bondar, A.A., Hecht, U., Zollinger, J., Artyukh, L.V., Rex, S., and Velikanova, T.Ya. (2009) *J. Alloys Compd.*, **474**, 86.
107 Huang, S.C. (1993) *Structural Intermetallics* (eds R. Darolia, J.J. Lewandowski, C.T. Liu, P.L. Martin, D.B. Miracle, and M.V. Nathal), TMS, Warrendale, PA, p. 299.
108 Chen, G.L., Zhang, W.J., Wang, Y.D.G., Sun, Z.Q., Wu, Y.Q., and Zhou, L. (1993) *Structural Intermetallics* (eds R. Darolia, J.J. Lewandowski, C.T. Liu, P.L. Martin, D.B. Miracle, and M.V. Nathal), TMS, Warrendale, PA, p. 319.
109 Paul, J.D.H., Appel, F., and Wagner, R. (1998) *Acta Mater.*, **46**, 1075.
110 Kim, Y.-W. (2003) *Niobium High Temperature Applications* (eds Y.-W. Kim and T. Carneiro), TMS, Warrendale, PA, p. 125.
111 Wu, X. (2006) *Intermetallics*, **14**, 1114.
112 Sadi, F.-A. (1997) *Etude du système Al-Nb-Ti au voisinage des compositions $AlNbTi_2$ et Al_3NbTi_4*. Ph.D. Thesis, Univ. de Paris-Sud.
113 Larsen, D.E., Kampe, S., and Christodoulou, L. (1990) *Intermetallic Matrix Composites*, vol. 194 (eds D.L. Anton, R. McMeeking, D. Miracle, and P. Martin), *Mater. Res. Soc. Symp. Proc.*. Mater. Res. Soc., Pittsburgh, PA, p. 285.
114 Cheng, T.T. (1999) *Gamma Titanium Aluminides 1999* (eds Y.-W. Kim, D.M. Dimiduk, and M.H. Loretto), TMS, Warrendale, PA, p. 389.
115 Kim, Y.-W. and Dimiduk, D.M. (2001) *Structural Intermetallics 2001* (eds K.J. Hemker, D.M. Dimiduk, H. Clemens, R. Darolia, H. Inui, J.M. Larsen, V.K. Sikka, M. Thomas, and J.D. Whittenberger), TMS, Warrendale, PA, p. 625.
116 Nérac-Partaix, A., Huguet, A., and Menand, A. (1995) *Gamma Titanium Aluminides* (eds Y.-W. Kim, R. Wagner, and M. Yamaguchi), TMS, Warrendale, PA, p. 197.
117 Menand, A., Huguet, A., and Nérac-Partaix, A. (1996) *Acta Mater.*, **44**, 4729.
118 Vasudevan, V.K., Stucke, M.A., Court, S.A., and Fraser, H.L. (1989) *Philos. Mag. Lett.*, **59**, 299–307.
119 Nérac-Partaix, A. and Menand, A. (1996) *Scr. Mater.*, **35**, 199.

3
Thermophysical Constants

TiAl alloys of technical significance have the general composition

$$\text{Ti-}(42\text{--}49)\text{Al-}(0.1\text{--}10)\text{X} \qquad (3.1)$$

with X designating alloying elements such as Cr, Nb, W, V, Ta, Si, B, and C. The constitution of these alloys may involve a variety of stable and metastable phases, depending on the detailed composition and processing conditions [1–6], as was described in Chapter 2. The crystallographic data of the most important phases are listed in Tables 2.1 and 2.2. It should be noted that, apart from γ(TiAl), significant alloy developments have been made around the α_2(Ti$_3$Al) and Ti$_2$AlNb phases, for a review see [7].

3.1
Elastic and Thermal Properties

Thermophysical properties that might be important for technical applications are listed in Table 3.1. The elastic modulus of γ(TiAl) alloys normalized by density is outstanding, when compared with other metallic systems [14, 15], see Figure 1.1. At the intended operating temperature of 750 °C, the elastic modulus of γ alloys approaches that of superalloys, while γ alloys have half the density. In aero engines this could be beneficial for the design of load-bearing elements, for which shape retention is required under loading. An appreciable advantage of the high elastic stiffness is an increase in the vibrational frequency of a component. This could be advantageous for engine design, since excitation of vibration seems to be easier to avoid at higher frequencies. Utilizing data determined by Schafrik [8], the temperature dependencies of the elastic moduli for γ(TiAl) and α_2(Ti$_3$Al) are given in Table 3.1. The six independent elastic constants C_{ij} were experimentally determined by Tanaka et al. [16] and He et al. [9] for Ti-56Al, utilizing ultrasonic resonance techniques. The shear anisotropy factors $A_{<110]}$ and $A_{<011]}$ with respect to the <110] and <011] directions are given by [16]

$$A_{<110]} = 2C_{66}/(C_{11} - C_{12}) = 1.44 \qquad (3.2)$$

Gamma Titanium Aluminide Alloys: Science and Technology, First Edition. Fritz Appel, Jonathan David Heaton Paul, Michael Oehring.
© 2011 Wiley-VCH Verlag GmbH & Co. KGaA. Published 2011 by Wiley-VCH Verlag GmbH & Co. KGaA.

Table 3.1 Thermophysical properties of Ti-Al alloys; E (Young's modulus), μ (shear modulus), ν (Poisson's ratio), ρ (density), α (linear thermal expansion coefficient), c_p (specific heat), λ (thermal conductivity), ρ_e (electrical resistivity), D_0 and Q see Equation (3.4).

Property		Ref.
Elastic moduli: γ(TiAl) (Ti-50Al) α_2(Ti$_3$Al) (Ti-26.7Al)	$E[\text{GPa}] = 173.59 - 0.0342\,T$, $\mu[\text{GPa}] = 70.39 - 0.0141\,T$, $\nu = 0.234 + 6.7 \times 10^{-6}\,T$, $T = 25\,°C$ to $847\,°C$ $E[\text{GPa}] = 147.05 - 0.0525\,T$, $\mu[\text{GPa}] = 57.09 - 0.0187\,T$, $\nu = 0.295 - 5.9 \times 10^{-5}\,T$, $T = 25\,°C$ to $954\,°C$	[8]
Density:	$\rho = 3.9\,\text{g cm}^{-3}$ to $4.3\,\text{g cm}^{-3}$	
Thermal expansion: (Ti-56Al)	in [100] direction: $\alpha[\text{K}^{-1}] = 9.77 \times 10^{-6} + 4.46 \times 10^{-9}\,T$ in [001] direction: $\alpha[\text{K}^{-1}] = 9.26 \times 10^{-6} + 3.36 \times 10^{-9}\,T$, $T = 293\,\text{K}$ to $750\,\text{K}$	[9]
Thermal expansion: (α_2 phase in Ti-46Al-1.9Cr-3Nb)	lattice parameter a: $\alpha[\text{K}^{-1}] = 3.2 \times 10^{-6}$; $T = 273\,\text{K}$ to $1873\,\text{K}$ lattice parameter c: $\alpha[\text{K}^{-1}] = 2.1 \times 10^{-6}$; $T = 273\,\text{K}$ to $1873\,\text{K}$	[10]
Thermal conductivity: (Ti-48Al-1V-0.2C)	$\lambda(\text{W m}^{-1}\text{K}^{-1}) = 21.7959 + 8.1633 \times 10^{-3}\,T$, $T = 25\,°C$ to $760\,°C$	[11]
Electrical resistivity: (Ti-46.5Al-8Nb)	$\rho[\text{n}\Omega\,\text{m}] = 665.88 + 0.79\,T - 8.16 \times 10^{-4}\,T^2 + 5.72 \times 10^{-3}\,T^3$, $T = 400\,°C$ to $1430\,°C$	[12]
Specific heat: (Ti-46.5Al-8Nb)	$c_p[\text{J K}^{-1}\text{g}^{-1}] =$ $= 0.6324 + 7.44 \times 10^{-5}\,T - 2.07 \times 10^{-7}\,T^2 + 2.97 \times 10^{-10}\,T^3$, $T = 400\,°C$ to $1430\,°C$	[12]
Diffusivity: TiAl (Ti-(53-56)Al) Ti$_3$Al (Ti-(25-35)Al)	Ti: $D_0 = 1.43 \times 10^{-6}\,\text{m}^2/\text{s}$, $Q = 2.59\,\text{eV}$ Al: $D_0 = 2.11 \times 10^{-2}\,\text{m}^2/\text{s}$, $Q = 3.71\,\text{eV}$ Ti: $D_0 = 2.24 \times 10^{-5}\,\text{m}^2/\text{s}$, $Q = 2.99\,\text{eV}$ Al: $D_0 = 2.32 \times 10^{-1}\,\text{m}^2/\text{s}$, $Q = 4.08\,\text{eV}$	[13]

and

$$A_{<011]} = 4C_{44}/(C_{11} + C_{12} - 2C_{13}) = 1.98 \tag{3.3}$$

The elastic stiffness constants C_{ij} of α_2(Ti$_3$Al) were determined by Tanaka et al. [17]. More information about this subject is provided in [18]. Based on this data, Yoo and Foo [19] have assessed the elastic incompatibility between α_2(Ti$_3$Al) and γ(TiAl) platelets of the lamellar morphology that occurs upon loading. The stress induced in the α_2 platelets was estimated to be (−9%) of a stress applied normal to the interfacial planes and (+4%) of a stress applied parallel to the interfaces.

The thermal expansion coefficients of γ(TiAl) [9] and α_2(Ti$_3$Al) [10] are listed in Table 3.1. The anisotropy of thermal expansion is remarkable because it leads to thermoelastic stresses during thermal treatments. The thermal expansion depends on alloy composition, as described in [10–12]. The thermal conductivity of TiAl alloys is comparable to that of superalloys but much higher than that of conventional Ti alloys. Thus, similar thermal gradients are expected in components when superalloys are replaced by TiAl alloys [11]. The electrical resistivity of TiAl alloys is lower than that of the superalloys and conventional Ti alloys. The specific heat of TiAl alloys is higher than that of superalloys, which may reduce thermal gradients and increase shock resistance [11].

3.2
Point Defects

Attempts have been made to rationalize the thermophysical properties in terms of point-defect characteristics and site occupancy of alloying elements [18]. There is good evidence that no structural vacancies are formed in TiAl alloys [20, 21]. Off-stoichiometric deviations are accommodated by substitutional antisite defects on both sublattices, that is, either Ti atoms on Al sites (Ti_{Al}), or Al atoms on Ti sites (Al_{Ti}). The concentration of the antisite defects symmetrically increases with deviation from stoichiometry [18, 22, 23]. Figure 3.1 demonstrates the density of these defects as predicted by first-principles calculations of Woodward et al. [22]. The density of vacancies is also affected by ternary additions, however, the effects that have been recognized for Si, Cr, Nb, Mo, W, and Ta are small compared to the change in vacancy density as a function of stoichiometry (Figure 3.1).

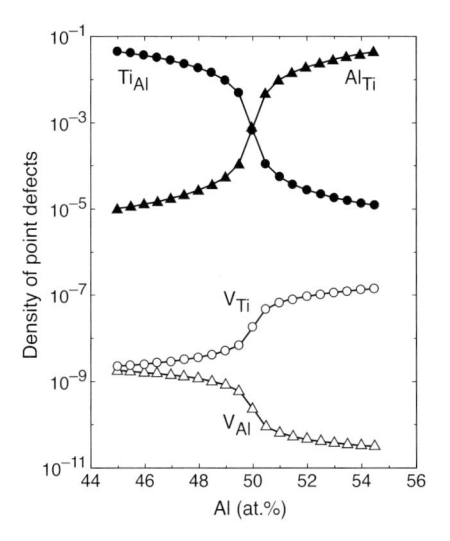

Figure 3.1 Predicted density of point defects in binary γ(TiAl) at 1073 K as a function of the Al content. Ti_{Al} (Ti atom situated on an Al site), Al_{Ti}, (Al atom situated on a Ti site), V_{Ti} (Ti vacancy), V_{Al} (Al vacancy). Data from Woodward et al. [22], redrawn.

Atom location by channeling-enhanced microanalysis (ALCHEMI) has shown that Nb, Hf, Zr, and Ta substitute exclusively or preferentially at Ti sites (Figure 3.2). Elements partitioning preferentially at the Al sites include Ga, Mn, W, Mo, and Cr [24]. These experimental findings agree well with first-principles calculations; however, these calculations have predicted that the site preference of Mo and W depends on the Al content of the host alloy [23].

The solubility limits of oxygen and carbon in γ(TiAl) were determined by Menand et al. [25] utilizing atom-probe field ion microscopy; this data is listed in Table 3.2. For the α_2 phase the solubility limits of these interstitial elements is much higher; for example, the solubility limit of oxygen is at least 2.1 at.%. The oxygen content in TiAl alloys is significantly higher than the solubility limit of the γ phase (Table 3.2). The amount of solute oxygen in the γ phase was always found to be at the saturation limit of 250 at. ppm. In single-phase γ alloys the excess oxygen is thought to precipitate as oxide or H phase, whereas in two-phase alloys the excess oxygen is taken up by the α_2 phase [25]. In broad terms, the different solubilities of the γ and α_2 phases can be rationalized by crystallographic arguments. While the sizes of the octahedral cavities in the two phases are identical, the chemical environments of the cavities are different. In TiAl, there are two types of octahedral sites: the octahedral cavity is either surrounded by two Al atoms and four Ti atoms, or by four Al atoms and two Ti atoms. In Ti$_3$Al there are also two types of octahedral sites; however, these are either surrounded by two Al atoms and four Ti atoms, or by six Ti atoms.

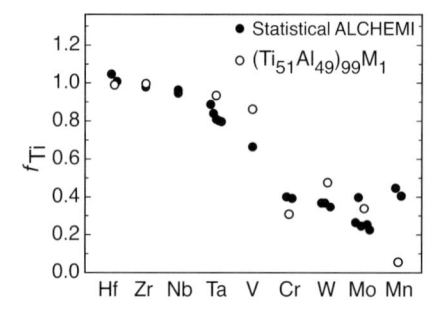

Figure 3.2 Partitioning of transition-metal solutes in γ(TiAl) with composition (Ti-49Al)-1M at 1473 K. M represents solid solutions at 1 at.%, and f_{Ti} indicates the fraction of solute that resides on the Ti sublattice. Results obtained from first-principles calculations [22] and statistical ALCHEMI [24]. Figure from Woodward et al. [22], redrawn.

Table 3.2 Variation of oxygen and carbon solubility limit in single-phase γ(TiAl) alloys that were subject to aging treatments at the temperatures indicated. Data from Menand et al. [25].

	800 °C	1000 °C	1300 °C
Oxygen (at. ppm)	120 ± 60	210 ± 80	250 ± 90
Carbon (at. ppm)	90 ± 50	220 ± 80	250 ± 90

3.3
Diffusion

As with other metals, the diffusion coefficients D of titanium aluminide alloys are described as [13, 26–29]

$$D = D_0 \exp(-Q/kT), \text{ with } Q = H_V^F + H_V^M \tag{3.4}$$

D_0 is the pre-exponential factor, k the Boltzmann constant and T the absolute temperature. The activation energy Q is the sum of the formation and migration energy H_V^F and H_V^M, respectively, for thermal vacancies. The formation energies for Ti vacancies are $H_V^F = 1.55 \pm 0.2$ eV in Ti_3Al and 1.41 ± 0.06 eV in TiAl [21, 30]. Molecular dynamic simulations of Wang et al. [31] have shown that the formation energies for all the stable interstitial configurations in TiAl are two to three times those of vacancies. The Ti self-diffusion data listed in Table 3.1 was determined by the radio-tracer method utilizing the isotope ^{44}Ti. The data for Al self-diffusion was calculated [32] from the Darken–Manning equation [33, 34] because no suitable radioisotope was available. In this estimation the diffusion data of Ga in TiAl was used. As mentioned in the previous section, Ga is known to exclusively occupy Al sites and is therefore utilized in diffusion studies as an Al-substituting element. Taken together, the data indicates that in TiAl and Ti_3Al the self-diffusion of Al is slower than that of Ti. Ikeda et al. [35] have recognized a remarkable anisotropy of the diffusion coefficient in TiAl single crystals. Diffusion in the direction perpendicular to the [001] axis was found to be about one order of magnitude faster that that parallel to the [001] direction. This result is not unexpected because in the first case diffusion can be accomplished by a sublattice self-diffusion mechanism, whereas in the second case intersublattice jumps are required.

There is supporting evidence that diffusion in TiAl and Ti_3Al could be enhanced by the presence of antisite defects [13]. Together with vacancies, the antisite defects form antistructural bridges. These particular atomic configurations allow sublattice and intersublattice jumps of the vacancies through the lattice, without disturbing the long-range order. Possible antistructural bridge (ASB) mechanisms that fit to the $L1_0$ and $D0_{19}$ structure were proposed by Herzig et al. [36] and Mishin and Herzig [13]. In the case of Ti-rich γ alloys, the so-called ASB-2 process could accomplish diffusion. The relevant elementary bridge involves a Ti vacancy V_{Ti} and a Ti_{Al} antisite atom, as illustrated in Figure 3.3 (stage (i)). The elementary bridge event is

$$V_{Ti} + Ti_{Al} \rightarrow V_{Al} \rightarrow Ti_{Al} + V_{Ti} \tag{3.5}$$

The initial stage (i) and final stage (iii) of the bridge event include the same defects and are crystallographically equivalent. Mishin and Herzig [13] have calculated the migration energy H_V^M (Equation 3.4) for different diffusion mechanisms in γ(TiAl) involving sublattice self-diffusion, sublattice antisite diffusion, three-jump cycles, six-jump cycles, and different ASB mechanisms. Among these mechanisms the ASB-2 process, as specified in Figure 3.3, requires by far the lowest migration

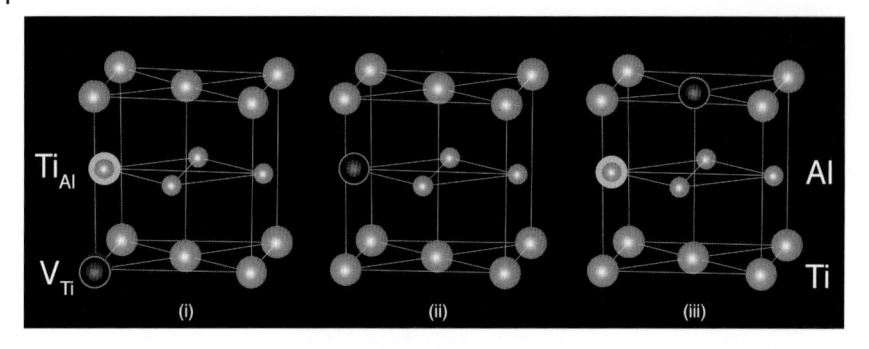

Figure 3.3 Diffusion in γ(TiAl) via antistructural bridges according to the so-called (ASB-2) mechanism [13]; see accompanying text.

energy $H_V^M = 0.72$ eV. It is interesting to note that the analogous ASB-2 mechanism for Al diffusion requires a significantly higher migration energy of $H_V^M = 1.323$ eV [13]. This finding indicates that diffusion in TiAl alloys depends on off-stoichiometric deviations. However, long-range diffusion via antistructural bridges is only possible if the concentration of antisite defects exceeds a certain critical value, the so-called percolation threshold. Below this critical value, the substructure of the ASB bridges is not continuous, and diffusion is confined to isolated clusters. Unfortunately, the percolation threshold concentrations of antisite defects are not known for TiAl. A theoretic investigation performed by Belova and Murch [37] for B2 ordered intermetallic compounds has shown that the threshold concentration required for an ASB mechanism can be relatively high, depending on the long-range order parameter. For high levels of order, no significant contribution of an ASB mechanism to long-range diffusion is expected. However, local diffusion by an ASB mechanism may still occur below the threshold concentration of antisite defects. A potential mechanism, for example, is diffusion-assisted reorientation of defect clusters in the stress field of dislocations.

The diffusion characteristics of solute elements in TiAl and Ti$_3$Al were investigated in several studies [13, 35, 38, 39]. Nb is a slow-diffusing element in both γ(TiAl) and α$_2$(Ti$_3$Al). Fe diffuses faster in γ(TiAl) than Ti, however, the ratio of the diffusion coefficients is only a factor of 2 to 3. Fe and Ni are fast diffusers in Ti$_3$Al; the diffusion coefficients are two to three orders of magnitude higher than that of Ti self-diffusion [13].

More information about thermophysical properties, involving bonding characteristics, defect structures and planar fault energies can be found in [18].

References

1 McCullough, C., Valencia, J.J., Levi, C.G., and Mehrabian, R. (1989) *Acta Metall.*, **37**, 1321.

2 Okamoto, H. (1993) *J. Phase Equil.*, **14**, 120.

3 Hellwig, A., Palm, M., and Inden, G. (1998) *Intermetallics*, **6**, 79.

4 Nguyen-Manh, D. and Pettifor, D.G. (1999) *Gamma Titanium Aluminides 1999*

(eds Y.-W. Kim, D.M. Dimiduk, and M.H. Loretto), TMS, Warrendale, PA, p. 175.
5. Ohnuma, I., Fujita, Y., Mitsui, H., Ishikawa, K., Kainuma, R., and Ishida, K. (2000) *Acta Mater.*, **48**, 3113.
6. Kattner, U.R., Lin, J.C., and Chang, Y.-A. (1992) *Metall. Trans. A*, **23A**, 2081.
7. Banerjee, S. and Mukhopadhyay, P. (2007) *Phase Transformations-Examples from Titanium and Zirconium Alloys*, Elsevier, Amsterdam.
8. Schafrik, R.E. (1977) *Metall. Trans. A*, **8A**, 1003.
9. He, Y., Schwarz, R.B., Darling, T., Hundley, M., Whang, S.H., and Wang, Z.M. (1997) *Mater. Sci. Eng. A*, **239–240**, 157.
10. Novoselova, T., Malinov, S., Sha, W., and Zhecheva, A. (2004) *Mater. Sci. Eng. A*, **371**, 103.
11. Zhang, W.J., Reddy, B.V., and Deevi, S.C. (2001) *Scr. Mater.*, **45**, 645.
12. Egry, I., Brooks, R., Holland-Moritz, D., Novakovic, R., Matsushita, T., Ricci, E., Seetharaman, S., Wunderlich, R., and Jarvis, D. (2007) *Int. J. Thermophys.*, **28**, 1026.
13. Mishin, Y. and Herzig, Chr. (2000) *Acta Mater.*, **48**, 589.
14. Huang, S.C. and Chesnutt, J.C. (1995) *Intermetallic Compounds. Vol. 2, Practice* (eds J.H. Westbrook and R.L. Fleischer), John Wiley & Sons, Ltd, Chichester, p. 73.
15. Austin, C.M., Kelly, T.J., McAllister, K.G., and Chesnutt, J.C. (1997) *Structural Intermetallics 1997* (eds M.V. Nathal, R. Darolia, C.T. Liu, P.L. Martin, D.B. Miracle, R. Wagner, and M. Yamaguchi), TMS, Warrendale, PA, p. 413.
16. Tanaka, K., Ichitsubo, T., Inui, H., Yamaguchi, M., and Koiwa, M. (1996) *Philos. Mag. Lett.*, **73**, 71.
17. Tanaka, K., Okamoto, K., Inui, H., Minonishi, Y., Yamaguchi, M., and Koiwa, M. (1996) *Philos. Mag. A*, **73**, 1475.
18. Yoo, M.H. and Fu, C.L. (1998) *Metall. Mater. Trans. A*, **29A**, 49.
19. Yoo, M.H. and Fu, C.L. (1995) *Mater. Sci. Eng. A*, **192–193**, 14.
20. Shirai, Y. and Yamaguchi, M. (1992) *Mater. Sci. Eng. A*, **152**, 173.
21. Brossmann, U., Würschum, R., Badura, K., and Schäfer, H.-E. (1994) *Phys. Rev. B*, **49**, 6457.
22. Woodward, C., Kajihara, S., and Lang, L.H. (1998) *Phys. Rev. B*, **57**, 13459.
23. Woodward, C., Kajihara, S.A., Rao, S.I., and Dimiduk, D.M. (1999) *Gamma Titanium Aluminides 1999* (eds Y.-W. Kim, D.M. Dimiduk, and M.H. Loretto), TMS, Warrendale, PA, p. 49.
24. Rossouw, C.J., Forwood, C.T., Gibson, M.A., and Miller, A.R. (1996) *Philos. Mag. A*, **74**, 77.
25. Menand, A., Huguet, A., and Nérac-Partaix, A. (1996) *Acta Mater.*, **44**, 4729.
26. Kroll, S., Mehrer, H., Stolwijk, N., Herzig, Chr., Rosenkranz, R., and Frommeyer, G. (1992) *Z. Metallkd.*, **83**, 591.
27. Nakajima, H., Sprengel, W., and Nonaka, K. (1996) *Intermetallics*, **4** (Suppl. 1), S17.
28. Sprengel, W., Nakajima, H., and Oikawa, H. (1996) *Mater. Sci. Eng. A*, **213**, 45.
29. Sprengel, W., Oikawa, N., and Nakajima, H. (1996) *Intermetallics*, **4**, 185.
30. Würschum, R., Kümmerle, E.A., Gergen, K.B., Seeger, A., Herzig, Chr., and Schaefer, H.-E. (1996) *J. Appl. Phys.*, **80**, 724.
31. Wang, B.-Y., Wang, Y.X., Gu, Q., and Wang, T.M. (1997) *Computational Mat. Sci.*, **8**, 267.
32. Herzig, Chr., Friesel, M., Derdau, D., and Divinski, S.V. (1999) *Intermetallics*, **7**, 1141.
33. Darken, L.S. (1948) *Trans. Am. Inst. Min. Metall. Engrs.*, **175**, 184.
34. Manning, J.R. (1968) *Diffusion Kinetics for Atoms in Crystals*, Van Nostrand, Princeton.
35. Ikeda, T., Kadowaki, H., and Nakajima, H. (2001) *Acta Mater.*, **49**, 3475.
36. Herzig, Chr., Przeorski, T., and Mishin, Y. (1999) *Intermetallics*, **7**, 389.
37. Belova, I.V. and Murch, G.E. (1998) *Intermetallics*, **6**, 115.
38. Breuer, J., Wilger, T., Friesel, M., and Herzig, Chr. (1999) *Intermetallics*, **3–4**, 381.
39. Herzig, Chr., Przeorski, T., Friesel, M., Hisher, F., and Divinski, S. (2001) *Intermetallics*, **9**, 461.

4
Phase Transformations and Microstructures

4.1
Microstructure Formation on Solidification

Almost all processing routes for metallic alloys involve melting and subsequent solidification of the material. Thus, solidification not only determines the microstructure, texture and distribution of alloying elements in cast materials but also plays a significant role in wrought or powder processing. The phase transformation processes during solidification often occur far from thermodynamic equilibrium and are governed both by the driving forces and by the complex heat transport and diffusion processes that depend on the solidification conditions. Over the Al concentration range relevant for binary two-phase γ alloys, two peritectic reactions occur within a rather narrow concentration range (Figure 4.1), and result in the solidification being highly sensitive to the alloy composition. It is further noteworthy that in two-phase TiAl alloys, the phases formed during solidification as well as their morphology are often difficult to recognize using microstructural analysis, since solid-state phase transformations alter or completely remove the initial microstructure. For this reason microstructural investigations often have to be supplemented by other characterization methods such as texture measurements, microchemical analyses and observation of dendrite morphology in shrinkage cavities. Metallographical investigation of specimens quenched during directional solidification also appears to be helpful with respect to gaining an understanding of microstructure formation during solidification [2].

In an early study, McCullough et al. [3] investigated the phase equilibria and solidification behavior of binary alloys in the concentration range 40–55 at.% Al. The authors identified the primary solidifying phase by direct observation of the dendrite symmetry and also employed *in situ* high-temperature X-ray diffraction for phase analysis. The study clearly showed that for alloys containing less than 45 at.% Al, solidification took place solely through the β phase. For such alloys microsegregation was significantly lower than predicted by the Gulliver–Scheil equation [4, 5]. This indicates backdiffusion in the β phase and solidification relatively close to equilibrium. For Al concentrations of 45–49 at.% Al, primary β dendrites form first followed by the peritectic formation of the α phase. However, above 49 at.% Al the hexagonal α phase is the first phase to form that later

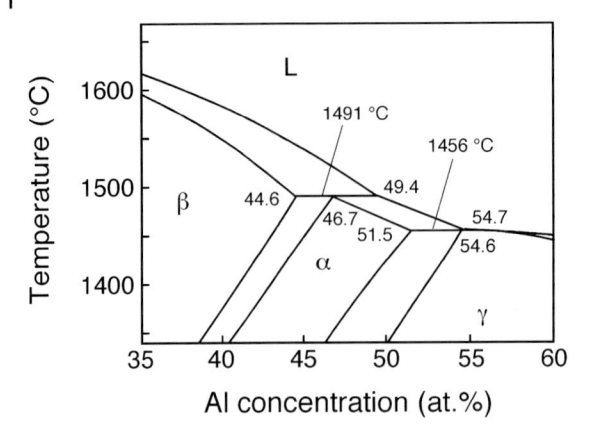

Figure 4.1 Section of the calculated binary Ti–Al phase diagram according to the thermodynamic evaluation of Witusiewicz et al. [1].

transforms to the γ phase through a peritectic reaction. The β dendrites grow with <100> alignment parallel to the heat-flow direction, whereas the α dendrites were found to have a [0001] growth direction with dendrite side arms along <10$\bar{1}$0>. Even though the phase equilibria involving the liquid phase are now well established for binary alloys, according to McCullough et al. [3] and other publications [6–8], the formation ranges of the solidifying phases are also determined by kinetic factors, and their analysis is essential in order to gain an understanding of the solidification microstructure. In this respect texture measurements have turned out to yield useful information.

In a study on the texture of cast alloys, Muraleedharan et al. [9] found that the alloy Ti-48Al-2Nb-2Cr exhibited a lamellar microstructure with <111> directions of the γ phase parallel to the growth direction of columnar grains, when the alloy was rapidly cooled during solidification. This can easily be explained by [0001] growth of α dendrites parallel to the direction of heat flow and the subsequent formation of γ lamellae according to the Blackburn-orientation relationship [10]

$$(0001)_{\alpha_2} \| \{111\}_\gamma \text{ and } <11\bar{2}0>_{\alpha_2} \| <1\bar{1}0>_\gamma \quad (4.1)$$

which describes the crystallographic alignment of γ lamellae with respect to α_2 lamellae in one prior α grain. Similar textures that correspond to the growth of α grains along the [0001] direction were observed by De Graef et al. [11] for Ti-46.5Al and Ti-48.5Al, by Küstner et al. [12] for the alloy Ti-48Al (Figure 4.2) and by Dey et al. [13] in Ti-46.8Al-1.7Cr-1.8Nb. This type of texture that results in lamellar interfaces lying perpendicular to the heat-flow direction has often been confirmed by microstructural analyses [14–18]. However, using texture measurements, De Graef et al. [11] observed that the growth of the α phase with a [0001] orientation only occurred for relatively high cooling rates, whereas for low cooling rates the α phase grew with a <11$\bar{2}$0> orientation that resulted in γ lamellae parallel to the heat-flow direction. Interestingly for all the alloys mentioned above, the β phase

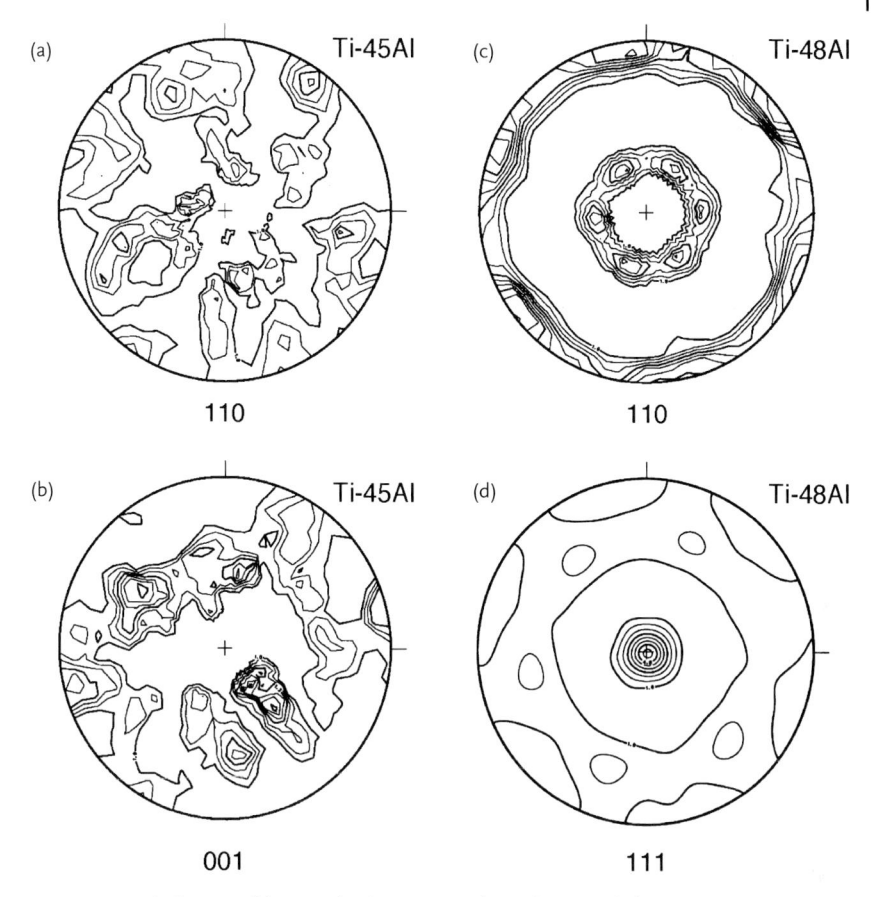

Figure 4.2 Pole figures of the γ(TiAl) phase determined on binary arc-melted buttons using neutron diffraction [12]. (a)–(c) Measured pole figures and (d) pole figure recalculated from the orientation distribution function (see Chapter 16.) (a) Ti-45Al, {110} pole figure, maximum pole density 1.66 random, (b) Ti-45Al, {001} pole figure, maximum pole density 1.44 random, (c) Ti-48Al, {110} pole figure, maximum pole density 2.67 random, (d) Ti-48Al, {111} pole figure, maximum pole density 4.67 random. The projection plane is perpendicular to the axis of symmetry through the buttons.

was unambiguously the primary phase but this was not reflected in the texture of the materials. Thus, it has to be concluded that the α phase is surprisingly not nucleated at the primary β phase. In this case one would expect that α would form with an orientation according to the Burgers orientation relationship [19]

$$\{110\}_\beta \parallel (0001)_\alpha \text{ and } <111>_\beta \parallel <11\bar{2}0>_\alpha \tag{4.2}$$

as is the case for the solid-state β/α transformation in Ti [20, 21] and TiAl alloys [3, 18, 22, 23]. In the literature, an exclusive <100> growth direction of the β phase has been reported and in this case the Burgers orientation relationship would result in 12 orientation variants with basal planes of the α phase at 0° and 45° to the heat-flow direction [3, 17, 18, 24–26]. Johnson et al. [27] and Küstner et al. [28] suggested that the peritectic α phase is nucleated in the melt and not at the primary β phase and thus it adopts orientations related to its preferential growth direction and not related to the β phase. Thus, it can be concluded that over a certain time period during the solidification of peritectic alloys, the α phase grows simultaneously with the β phase in the direction of heat flow. Here, it is worth mentioning that peritectic solidification is usually divided into three stages [29]: (i) peritectic reaction by direct interaction of the primary and the liquid phases, (ii) the peritectic transformation, during which the peritectic phase grows by solid-state diffusion and (iii) direct solidification of the peritectic phase. Accordingly, the texture investigations have shown the direct solidification of the peritectic α phase. After solidification has been completed, the α phase will grow into the primary β phase and suppress further nucleation of α from β according to the Burgers orientation relationship in the solid-state β/α transformation. In this way, the secondary α phase determines the final texture and microstructure of the material. Indeed, Küstner et al. [12] observed that a cast Ti-45Al alloy did not exhibit significant preferred orientations (Figure 4.2), which was attributed to complete solidification via the β phase and the multitude of orientation variants resulting from the solid-state β/α transformation.

The high sensitivity of the solidification path on the Al content for both binary as well as multicomponent alloys is also reflected in the micro- and macrostructure of cast materials. According to a study by Küstner et al. [12], binary alloys with Al concentrations ≤45 at.% exhibited structures consisting of relatively large equiaxed grains, whereas alloys with higher Al contents showed columnar grains that had grown in the opposite direction to that of heat extraction (Figure 4.3). Similar observations were also made for ternary and multicomponent alloys (Figure 4.3). Such differences between alloys solidifying via the β phase or peritectically via α have often been reported in the literature for cast specimens ranging from arc-melted buttons to large ingots [3, 12, 18, 30–33]. Fine-grained equiaxed microstructures were observed particularly for B-containing alloys with low Al concentrations (Figure 4.3).

At higher magnifications it can be seen that a morphology similar to that of lamellar near-α Ti alloys occurs in β-solidifying binary alloys, as shown in Figure 4.4 [12]. The plate-like constituent in the microstructure consists completely of α_2 and γ lamellae. This morphology can be explained by the phase transformation path L → L + β → β → β + α → α → α + γ. In the β → β + α transformation, crystallographically oriented α plates with orientations according to the Burgers relationship (Equation 4.2) are precipitated from the β phase. This leads to segregation of Ti to ribs of remaining β phase. Subsequently, γ lamellae are precipitated within the α plates. In particular, for alloys containing additions of heavy elements the ribs of retained β phase are clearly identifiable due to segregation (Figure 4.5). Such microstructures have often been reported to occur in cast alloys [2, 6, 12, 18, 27, 32–34] and exhibit a striking similarity to the basket-weave microstructure of conventional Ti alloys. In β-solidifying alloys the size of the lamellar colonies apparently primarily depends on the kinetics of the β/α solid-state transformation,

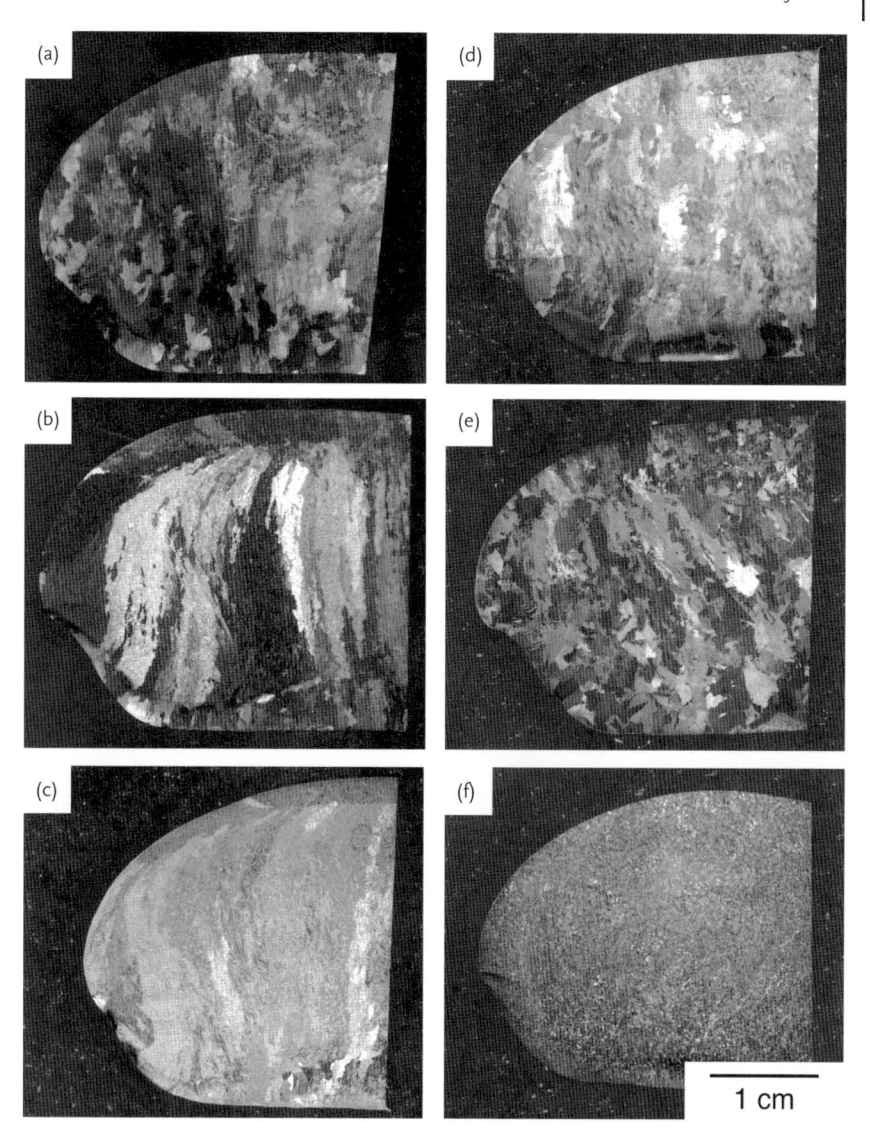

4.1 Microstructure Formation on Solidification | 37

Figure 4.3 Macrophotographs of etched arc-melted buttons: (a) Ti-45Al, (b) Ti-47Al, (c) Ti-48Al, (d) Ti-45Al-5Nb, (e) Ti-45Al-2Mo, (f) Ti-45Al-5Nb-1.0B [12]. The buttons were cut through the axis of symmetry. Solidification proceeded in the images from the lower edge to the upper edge.

thus offering the possibility of grain refining such alloys through heat treatments or alloying additions [2, 18, 32, 33, 35–38]. It should be mentioned that several other studies have highlighted both the role of the β phase in multicomponent alloys and the multitude of phase transformations involving this phase [23, 39–45]. To summarize, in β-solidifying alloys a single-phase β grain structure is formed

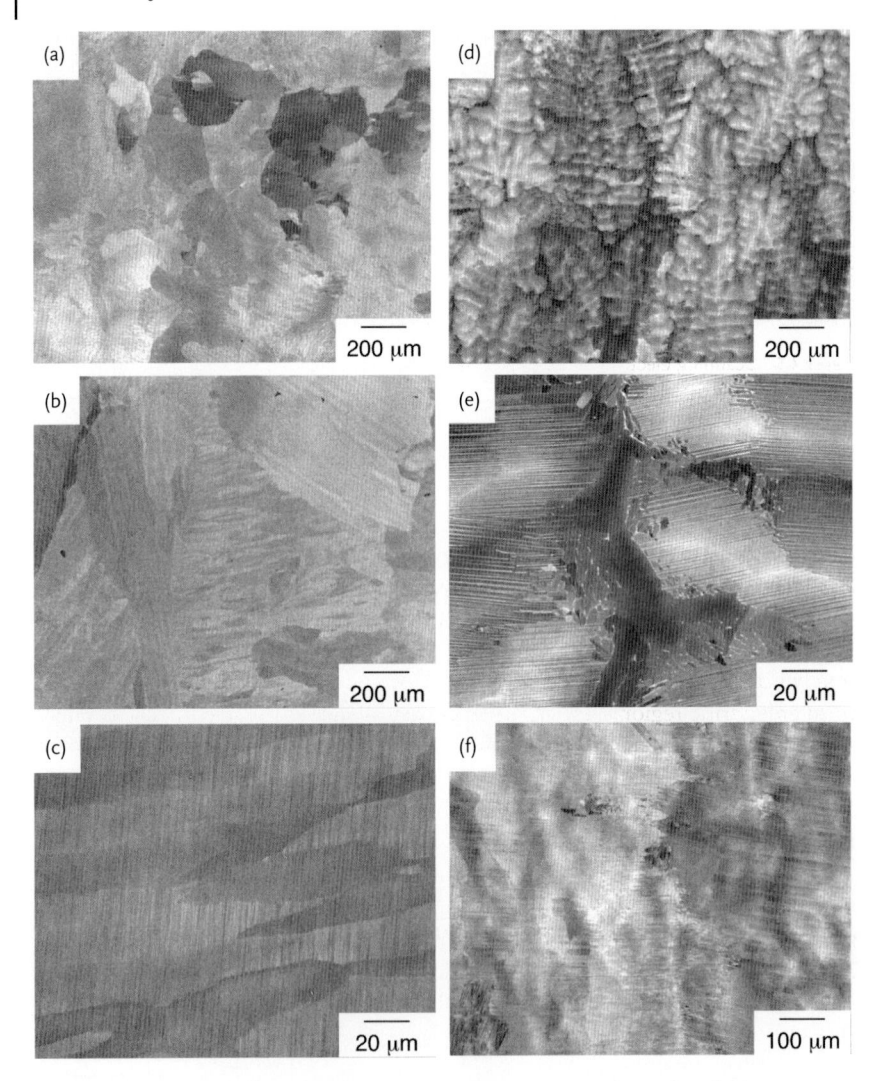

Figure 4.4 Scanning electron micrographs (using backscattered electrons) from arc-melted buttons: (a) Ti-45Al, upper part of the button, (b) Ti-45Al, lower part of the button, (c) region from (b) in higher magnification, (d) Ti-48Al, upper part of the button, (e) region from (d) in higher magnification, (f) Ti-48Al, lower part of the button [12]. Solidification proceeded in the images from the lower edge to the upper edge. Please note that some dendrites in (d) have orthogonal side arms.

directly after solidification, but the final microstructure and texture of cast materials is developed during the subsequent solid-state transformations.

As an example for peritectic alloys, the solidification microstructure of Ti-48Al will be described in more detail with reference to the study of Küstner et al. [12]. In the lower part of small arc-melted buttons the authors observed a nearly lamel-

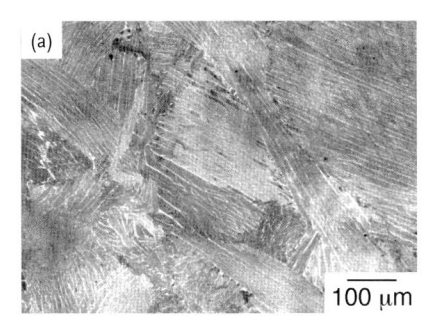

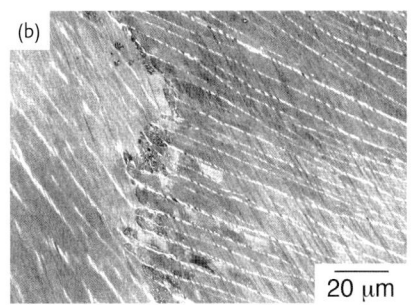

Figure 4.5 Scanning electron micrographs (using backscattered electrons) from an arc-melted button of the alloy Ti-45Al-7Nb-1Mo; (b) shows a section of (a) in higher magnification.

lar microstructure (Figure 4.4). The lamellar colonies were extended along the direction of heat flow and in most, but not all cases, the lamellae were aligned perpendicular to the temperature gradient. The dendrites in the upper part of the buttons were also lamellar (Figure 4.4) and exhibited dendrite cores that appeared light-colored using BSE imaging conditions, indicating the enrichment of Ti, while the interdendritic regions consisted of darker single-phase γ grains. A line profile taken across three dendrite arms in the upper part of a button showed a composition typical of the primary β phase in the dendrite cores. In adjacent dendrites and in interdendritic regions Al enrichments corresponding to the peritectic α phase and up to 54 at.% were observed, respectively (Figure 4.6). Interestingly, moving away from the core the Al concentration increases up to 49 at.%, which is expected for steady-state solidification of the β phase. The Al concentration then falls from this value to 47 at.% Al – the concentration of the α phase at the peritectic point – and then increases to 54 at.%. Thus, both phases grew simultaneously and each developed a segregation zone. This results in concentration variations of up to 9 at.% over relatively short distances because the alloy passed through both peritectic transformations in the Ti–Al system. Immediately after solidification is completed, the Ti-48Al alloy traverses the α single-phase field and growth of the pre-existing α grains can occur at low undercooling, which prevents further nucleation of the α phase from remaining β. The α grains grow from the interdendritic regions and meet each other at the center of the prior β dendrites, where Ti and other alloying elements that partition to the β phase are enriched. Examples of corresponding microstructures are shown in Figures 4.4 and 4.7.

In the alloy Ti-51Al, an Al concentration of 47–48 at.% was measured within dendrite cores and thus α was the primary solidifying phase, in agreement with the phase diagram (Figure 4.6). Similarly as for Ti-48Al, segregation of Al up to 55 at.% was found. As pointed out by Hecht *et al.* [46] the microsegregation behavior is very useful in helping to analyze the solidification path of an alloy and might be used to estimate the undercooling at which different phases grow during solidification. Accordingly, the authors performed EDX measurements in which the individual analysis spots were arranged in a quadratic grid on the specimen. The concentrations obtained were then sorted and the results plotted in curves of

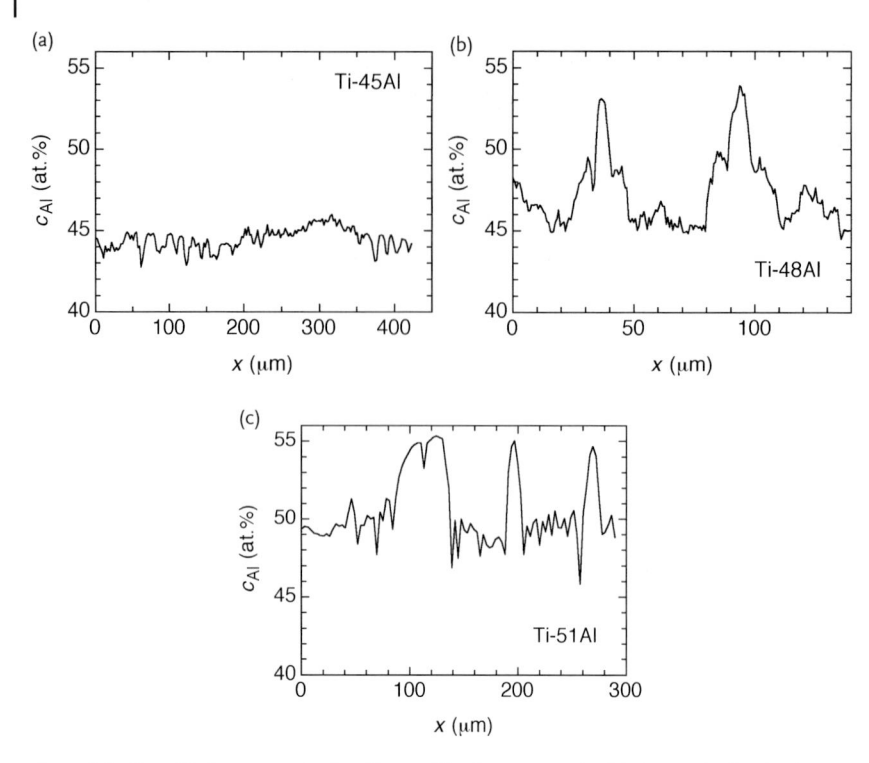

Figure 4.6 Quantitative concentration line profiles from arc-melted buttons obtained using EDX analysis in a scanning electron microscope. (a) Ti-45Al, (b) Ti-48Al, (c) Ti-51Al [12].

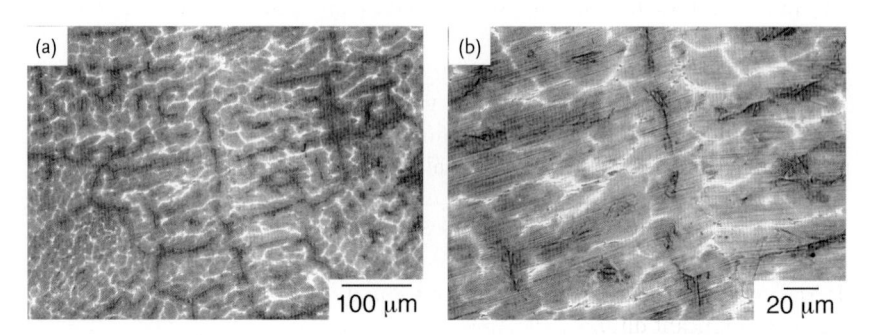

Figure 4.7 Scanning electron micrographs taken (using backscattered electrons) from an arc-melted button of the alloy Ti-46Al-5Nb-1Mo-0.2B-0.2C.

the concentration of an alloying element against the solid-phase fraction. Such curves can be compared with a Scheil analysis of the solidification. More on this technique can be found in the publication by Hecht *et al.* [46] and the references cited therein [47–49]. Figure 4.8 displays results from such grid measurements obtained on binary alloys. For the alloy Ti-51Al (Figure 4.8a) the measured data shows a very good fit to the Gulliver–Scheil equation [4, 5]

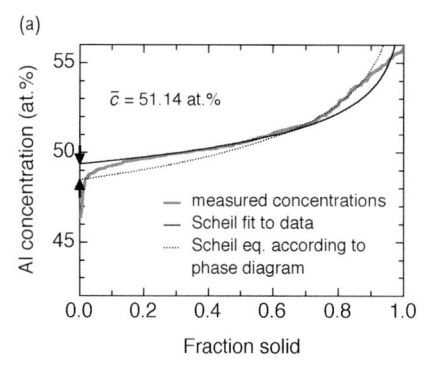

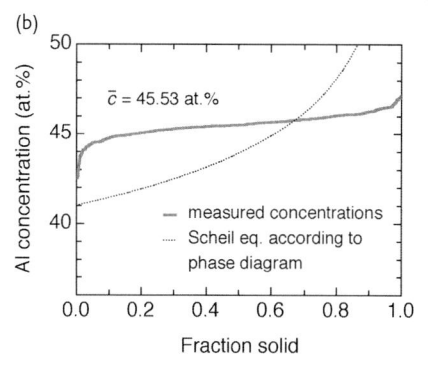

Figure 4.8 Al concentration versus the volume fraction of solid material as obtained from quantitative EDX grid measurements on binary alloy buttons in a scanning electron microscope. The compositions were determined at regularly spaced measurement positions within a rectangular area of around 10 mm², ranked by the Al concentration and assigned to the solid fraction. The technique is described in more detail by Hecht et al. [46] and the references cited therein [47–49]. In addition to the measured concentrations, the Al concentrations determined using the Scheil equation (Equation 4.3) according to the binary phase diagram, and the curve fitted to the measured concentrations on the basis of the Scheil equation are shown. The difference between these two curves can be used to estimate the undercooling at the beginning of solidification. (a) Ti-51Al, (b) Ti-45Al.

$$c_S = k\, c_L = k\, c_0 (1 - f_s)^{k-1} \qquad (4.3)$$

that describes the segregation curve when there is rapid mixing of the elements in the liquid phase and no diffusion in the solid phase. In this equation c_S is the solidus concentration, c_L the liquidus concentration, c_0 the alloy concentration, f_s the fraction solidified and $k = c_S/c_L$ the partitioning coefficient. It should be mentioned here that the steep rise in the segregation curve at the beginning occurs due to subsequent solid-state transformations and corresponds to the phase diagram (Figure 4.1). In addition to the (Scheil) curve fitted to the measured data, the Scheil curve according to the phase diagram is also shown in Figure 4.8a and it lies below the fitted curve. From the concentration difference at the beginning of solidification the undercooling might be estimated at which solidification occurred. In this case this would result in an undercooling of 4.6 K. In contrast to Ti-51Al, no Scheil behavior was found for the alloy Ti-45Al (Figure 4.8b), suggesting that significant diffusion occurred in the solid phase during solidification of this β-solidifying alloy, as observed by McCullough et al. [3]. This seems plausible when one considers that the diffusion coefficient of the β phase is more than two orders of magnitude higher than that of the α phase [50]. The results are in agreement with studies in the literature, in which strong backdiffusion in β-solidifying alloys [51], but only very limited backdiffusion, that is, Scheil behavior, in α-solidifying alloys was observed [51, 52]. Due to backdiffusion in the solid β phase nothing can be concluded from the segregation curve of Ti-45Al (Figure 4.8b) regarding the undercooling at solidification. Since solidification in this alloy occurred relatively close to the thermodynamic equilibrium, β-solidifying alloys appear preferable to peritectic alloys with respect to microsegregation. However, Hecht et al. [46], using grid measurements, observed that the solidification behavior of the β phase in Ti-45Al-8Nb corresponded to the Scheil equation, which was attributed to limited backdiffusion of Nb. From the segregation curves the authors estimated that the nucleation undercooling of the α phase exceeded 25 K. This high undercooling, that contradicts the value obtained above for the binary alloy, can explain why α grains are often large and elongated [46]. To summarize, only little is known about the real solidification path of TiAl alloys and, in particular, the mechanisms of microstructure formation during solidification and their dependence on the solidification conditions and alloy composition.

Generally, a wide spectrum of complex microstructures can be established in peritectic alloys that are determined by the nucleation kinetics and growth competition between the primary and the peritectic phases. At low values of G/V, with G being the thermal gradient and V the velocity of the solidification front, the primary phase will grow in a cellular or dendritic morphology if the constitutional undercooling criterion

$$G/V \leq m_L c_0 (k-1)/(k D_L) \tag{4.4}$$

is fulfilled [53, 54]. In this equation m_L is the slope of the liquidus curve and D_L the diffusion coefficient in the liquid. In the case of constitutional undercooling of the primary phase the peritectic phase is not necessarily constitutionally undercooled and can exhibit planar, cellular or dendritic growth. At higher values of G/V, i.e. above the limit of constitutional undercooling for both phases, complex microstructures can occur comprised of different forms of banding [55]. It has been shown that near the limit of constitutional undercooling the selection of the primary phase/morphology cannot be understood on the basis of growth at the highest interface temperature [55, 56]. This is because during the initial transient of plane-front primary phase growth, the second phase can be nucleated and in the initial transient of this phase nucleation of the primary phase can again take place [56]. In this way, a cycle is established that results in banded microstructures. Thus, in peritectic systems nucleation and the transient growth regimes have to be considered for the selection of phases and growth morphologies. Hunziker et al. [56] have presented an analytical model to predict phase and microstructure selection in such systems. The model considers constitutional undercooling and instantaneous nucleation of the phases at a constant undercooling ΔT_N with respect to the liquidus temperature. A phase transition occurs if another phase is nucleated ahead of the primary phase. If this is not possible, the growing phase and its growth morphology are assumed to be stable for the alloy composition and the given solidification conditions (growth velocity V and thermal gradient G). Hunziker et al. [56] have successfully applied this so-called nucleation and constitutional undercooling (NCU) model to the peritectic solidification of Fe–Ni alloys.

The NCU model by Hunziker et al. [56] was employed by Johnson et al. [57] and Oehring et al. [58] to describe the solidification of binary TiAl alloys. For the model, the slopes of the liquidus and solidus lines are determined from the phase diagram, that has to be linearized in the respective concentration range under the assumption that the liquidus and solidus lines cross at pure Ti. The diffusion coefficient of the melt $D_L = 2.8 \times 10^{-9}\,m^2 s^{-1}$ is given in a publication by Liu et al. [59]. Additionally, the nucleation undercoolings for the various phases, which to date are not reliably known, are required. From the results of containerless directional solidification experiments, Johnson et al. [57] concluded that the nucleation undercooling of the β phase is around 12 K. For solidification in a CaO crucible they assumed $\Delta T_{N,\beta}$ to be 6.3 K. In directional solidification experiments using alumina molds, Luo et al. [60] determined a value $\Delta T_{N,\beta} \leq 5.7$ K. As already discussed above, nucleation undercoolings ranging between 4.6 and over 25 K were estimated for the α phase. Figure 4.9 shows a microstructure formation map calculated according to the NCU model of Hunziker et al. [56], for which an arbitrary nucleation undercooling of 2 K has been assumed for both the α and β phases. The diagram also demonstrates how the lines separating different phase/morphology regimes would move for increasing nucleation undercoolings. Results from directional solidification experiments are also shown in the diagram and indicate that the map qualitatively describes the microstructural formation in binary TiAl alloys. Interestingly, it can be seen that a transition between a region where growth of β phase cells or dendrites to regions where the growth of β cells or dendrites occurs simultaneously with planar or cellular/dendritic growth of the α phase exists. The Al concentration at which this transition takes place depends on G/V and is 46.1–48.6 at.%. Higher nucleation undercoolings would reduce this transition to lower Al concentrations and widen the regions marked in Figure 4.9. In summary, it can be concluded that the observed microstructure selection can be qualitatively understood via the NCU model.

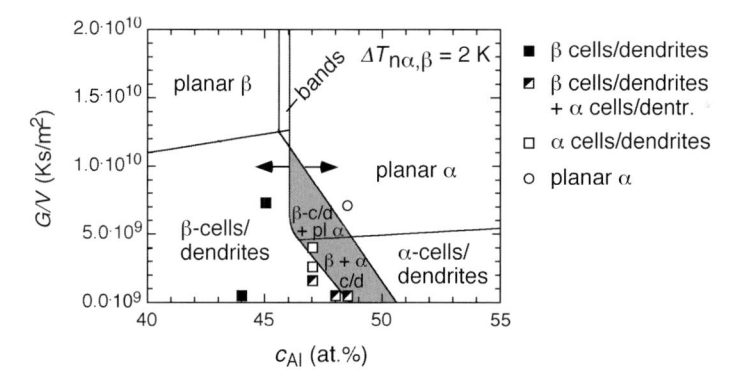

Figure 4.9 Microstructure formation map obtained for binary TiAl alloys obtained by applying the nucleation and constitutional undercooling (NCU) model developed by Hunziker et al. [56]. For the model, the calculated binary phase diagram by Ohnuma et al. [61] and the diffusion coefficient of the melt $D_L = 2.8 \times 10^{-9}\,m^2 s^{-1}$ as published by Liu et al. [59] were used. The nucleation undercooling ΔT_N of both the α and β phases has been assumed to be 2 K. The arrows in the diagram show how the lines separating different phase/morphology regimes would move for increasing nucleation undercoolings. The results of directional solidification experiments are also shown in the diagram [27, 60, 62, 63].

As noted above, it has been found that in peritectic alloys near the limit of constitutional undercooling, the phase/morphology selection criterion of the highest interface temperature does not adequately describe the solidification behavior [56]. Nevertheless, it is interesting to consider microstructural formation according to this criterion. Su et al. [63] modeled the phase/microstructure formation in binary TiAl alloys on the basis of this criterion with the aid of the so-called interface response function, that is, the temperature of the migrating interface of a phase as a function of the interface velocity. The resulting phase/microstructure formation map has some similarities to that shown in Figure 4.9, except that no regions occur in which both the β and α phases grow simultaneously. In this case, it could not be explained why β dendrites occurred in certain alloys but a texture indicating growth of the α phase was observed. Thus, the solidification behavior of TiAl alloys cannot be fully understood by the maximum growth temperature criterion; a quantitative understanding of solidification in this alloy system is only in its infancy.

Recently, Eiken et al. [64] performed phase-field simulations [65, 66] on the solidification of binary TiAl alloys and elucidated some interesting aspects. The authors varied the Al concentration from 43 to 47 at.% and assumed nucleation undercoolings ΔT_N for the secondary α phase between 2 and 10 K. For the highest ΔT_N nucleation of α occurred only during the early stages of solidification, even for an Al concentration of 47 at.%. This phase subsequently grew in a coupled peritectic reaction along the surface of the primary dendrites. During later stages of solidification the critical undercooling was never again exceeded. Due to the subsequent β/α transformation this results in large elongated α grains, which explains the frequently observed cast microstructure of TiAl alloys with near-peritectic compositions. When the authors decreased the nucleation undercooling to 2 K, frequent α phase nucleation at the primary β dendrites and a considerable grain refinement were obtained. The authors concluded that grain refinement in peritectic alloys can be achieved if either the growth temperature is decreased by applying a higher growth rate in order to exceed the nucleation undercooling, or the nucleation undercooling is reduced by adding a grain refining agent. The latter effect can be implemented by adding B to the alloys as has been known for a long time [30]. This grain-refining mechanism will be discussed in more detail later. The study by Eiken et al. [64] is also interesting since it indicates that the nucleation undercooling of the α phase might be in the range 5–10 K.

In the literature it has often been reported that cast alloys usually have coarse lamellar microstructures with colony sizes in the range of 100–1000 μm [30, 67–74] (see Figures 4.4, 4.5 and 4.7). Indeed even larger colony sizes of up to 6000 μm have been reported [30]. As should be evident from the discussion above, the constitution has a significant influence on the microstructural scale of cast alloys. Alloys that solidify through the β phase have the potential to exhibit fine solidification microstructures if they contain B as an alloying addition. The refining effect arises through the solid-state β/α transformation and will be discussed later in the context of grain refinement through B additions. However, the Al concentration seems to have only a modest effect on the dendrite size, which was found to decrease from 350 μm at 35 at.% Al to around 100 μm at 47 at.% in arc-melted buttons [6, 7]. For Al contents higher than 47 at.% the dendrite size increased [6, 7]. This is an interesting result since it indicates that peritectic alloys with Al concentrations in the range 46–48 at.% seem to be the optimum choice, disregarding solid-state transformations. Besides the Al concentration the solidification microstructure depends on the solidification conditions, that is, the thermal gradient G and the solidification front velocity V. Under normal casting conditions G and V are determined by the cooling rate, which is the only parameter that can be varied. Eylon *et al.* [70] compared the colony sizes of a Ti-47Al-2Nb-1.75Cr alloy cast into permanent steel molds and ceramic investment molds. The microstructure of the latter exhibited colony sizes of 100–1000 µm, significantly coarser than the 100–250 µm colony size range observed for the permanent-mold cast material. The authors also observed that preheating the molds results in slightly coarser microstructures. These observations demonstrate the influence of the cooling rate. In systematic studies by Rishel *et al.* [75] and Raban *et al.* [76] the cooling rates during investment casting of three alloys were varied by applying different mold preheat temperatures and varying the degree of mold wrapping. This resulted in solidification times between 10 and 235 s. Over the range of applied cooling rates, the authors observed columnar structures in the alloys Ti-48Al-2Cr-2Nb and Ti-47Al-2Cr-2Nb-0.5B, while equiaxed microstructures were found for Ti-45Al-2Cr-2Nb-0.9B. The coarsest microstructure occurred in slowly cooled material of the alloy Ti-48Al-2Cr-2Nb. All alloys exhibited an increase of the colony size with solidification time. For Ti-45Al-2Cr-2Nb-0.9B the colony size varied between 50 and 86 µm. These investigations clearly support the conclusions of Eiken *et al.* [64] and indicate the need for effective grain-refinement agents.

Grain-refining agents provide heterogeneous nucleation sites during solidification and are usually added to engineering materials such as Al and Mg alloys [77, 78]. For TiAl alloys both B [30] and N [31, 79–81] have been identified to act as grain-refining alloying elements. However, the addition of N results in a large loss of ductility [31, 79–81] and this approach has not been followed further. In contrast, B additions have been widely used for microstructural refinement; and a considerable fraction, if not the majority, of alloys investigated in the literature contain B as an alloying element. Although a number of studies were devoted to the grain-refining effect of B additions [2, 30, 32, 64, 67, 71, 82–88], the refinement mechanism still remains a matter of some debate [2, 32, 71, 84–86, 88]. This is because for B concentrations of 0.7–1.5 at.%, either the β or α phases are the primary solidifying phases, depending on the Al concentration (Figure 2.11) [89, 90], see Chapter 2; but grain refinement has also been observed for significantly lower B concentrations of 0.2–0.3 at.% [2, 32, 71, 84, 85]. Since B has often been added as a boride, for example, TiB_2 [30], it might be argued that the boride particles did not dissolve in the melt, even if they were not thermodynamically stable. However, the grain-refining effect also occurs if B is added in elemental form [81] or as AlB_{12} [83] and thus, nucleation on undissolved Ti-boride particles in the melt is impossible. Nevertheless, heterogeneous nucleation of the β phase has been observed to occur on TiB_2 particles in Ti-45Al-2B [87], that is, at a boron concentration for which the TiB_2 phase is the primary solidifying phase.

According to the most recent phase diagram calculation of the Ti–Al–B system [90], the primary phase on solidification changes above a certain B content from TiB to Ti$_3$B$_4$ at around 43 at.% Al and at only a slightly higher Al concentration to TiB$_2$. Extensive experimental work has shown that for alloys in which β is the primary phase to solidify, monoborides with the B27 and B$_f$ structures formed [91]. In α-solidifying alloys TiB$_2$ with C32 structure was found [83, 89]. Crystallographic data for the various boride phases is given Table 2.2. Gosslar et al. [87] have applied the plane-to-plane matching model and have shown that the misfit strains for nucleation of the α and β phases on TiB with the B27 structure, TiB with the B$_f$ structure, Ti$_3$B$_4$ (D7$_b$ structure) and TiB$_2$ (C32 structure) are far lower than those for α nucleation on β, according to the Burgers orientation relationship. Thus, such boride phases can serve as inoculants for both the α and β phases. However, the phenomenon is more complex since it is still not understood how B induces grain refinement when borides are not the primary solidifying phase at low B concentrations.

Cheng [84, 85] has suggested that under extreme constitutional undercooling conditions, nucleation can take place in the liquid ahead of the solidification front. The partitioning coefficient $k^{Al} = c_S^{Al}/c_L^{Al}$ is around 0.9 for the β and α phases in binary alloys, while $k^B = c_S^B/c_L^B$ is estimated to be below 0.1 due to the low solubility of B in the β and α phases (<0.2 at.%) at the melting temperature [90]. This leads to strong solute rejection at the solidification front and increases both the undercooling in the constitutionally undercooled zone and its width. Thus, nucleation of crystalline phases can take place at higher nucleation undercoolings. Cheng [84, 85] has argued that above a critical boron content, the β or α phases could be nucleated in the constitutionally undercooled zone, leading to even more B enrichment in the segregation zone. According to the work of Plaskett and Winegard [92] this would take place if the ratio $G/V^{1/2}$ is below a certain value, that is proportional to the boron concentration of the alloy. The mechanism proposed by Cheng [84, 85] explains why grain refinement occurs in alloys at B concentrations above a critical value but which is below the concentration at which the boride phase is the primary solidifying phase. It further accounts for the observation that B concentrations higher than a critical level do not intensify the effect and that grain refinement due to B can be deactivated by lowering the cooling rate, for example, when large ingots are produced. The mechanism suggested by Cheng [84, 85] has been treated in the literature [93–95] within the framework of the so-called growth-restriction factor Q, which is the inverse derivative of the fraction solid f_s with respect to solutal undercooling ΔT_S:

$$Q = \left(\frac{\partial(\Delta T_S)}{\partial f_s}\right)_{f_s \to 0} = m_L c_0 (k-1) \tag{4.5}$$

For binary alloys Q has been shown to be a measure for the inverse grain size [94, 95]. By comparing Equations 4.4 and 4.5 it is seen that Q is directly related to constitutional undercooling, that is, pronounced constitutional undercooling will result in a large growth-restriction factor Q and a small grain size, as suggested by Cheng [84, 85]. However, for ternary and multicomponent alloys the concept has to be extended. Quested et al. [93] have shown that under certain assumptions Q can be calculated in multicomponent alloys by summing the effect of each solute and thus, it is possible to apply the growth-restriction approach to B-containing TiAl alloys.

Despite the clear progress in understanding grain refinement due to B additions of ≤1 at.%, the mechanism suggested by Cheng [84, 85] cannot completely explain all experimental findings in low B-containing alloys. Hu [71] observed that the colony size in alloys with 1 at.% B decreased monotonically with decreasing Al concentration and explained this effect through a higher solubility of B in the α phase compared to the β phase. However, according to recent work on the Ti–Al–B phase diagram [90], B is more soluble in the β phase. Imayev et al. [32] studied the influence of alloying elements on microstructure formation in β-solidifying alloys and showed that grain refinement can be obtained by slowing down the transformation kinetics of the β/α solid-state transformation. It has been observed that boride particles in the alloy play a significant role with respect to grain refinement through the β/α solid-state transformation [32, 42]. This is remarkable since it implies an alternative grain-refining mechanism that does not originate from solidification. Indeed, Hecht et al. [2] have recently shown that fine α grains are formed with a random orientation from the solid β phase in boron-containing β-solidifying alloys. In contrast, the authors found coarse microstructures in a B-free alloy solidified under the same conditions. In the latter alloy, the α phase occurred with orientations according to the Burgers orientation relationship. This clearly indicates that the mechanism of refinement is based on heterogeneous nucleation of α on borides. In order to examine this hypothesis, Oehring et al. [88] examined a B-containing β-solidifying alloy (Ti-44.5Al-[5-7Nb]-[0.5-1.5Mo]-0.1B) with a fine cast microstructure. The material was heat treated in the α single-phase field and subsequently oil quenched. After this heat treatment the alloy showed a microstructure consisting of α grains with a size of 0.5–1 mm. This material was then subjected to a second heat treatment in the β phase field followed by either air cooling or furnace cooling with an initial cooling rate of 1 K/s. In the air-cooled specimen a relatively fine dispersion of β layers in lamellar colonies was observed, however, the colony size was in the range of 0.5–1 mm. In contrast, the furnace-cooled specimen exhibited a homogeneous microstructure of lamellar colonies with a size of around 50–100 µm. The orientation of the lamellae appeared to be randomly distributed. These results clearly show that grain refinement by B additions in β-solidifying alloys has to be attributed to the heterogeneous nucleation of the α phase during the β/α solid-state transformation, as indicated by Hecht et al. [2] and does not originate from solidification. Notably, refinement is only achieved if the cooling rate is not too fast. This means that the formation of Widmannstätten α colonies, nucleated at former β grain boundaries, seems to be faster than nucleation of α at borides inside β grains, although the former transformation mode apparently requires a higher undercooling. In peritectic alloys, the α phase existing prior to the β/α transformation impedes the establishment of sufficient undercooling for further nucleation of the α phase, as discussed earlier, and thus this mechanisms cannot operate in such alloys. Interestingly, the same effect has also been observed in Ti-6Al-4V (wt.%) [96]. In this conventional Ti alloy, equiaxed α grains appeared to nucleate at TiB particles during the β/α transformation on slow cooling from temperatures above the β-transus. On faster cooling, the α phase adopted a Widmannstätten lath-like morphology often referred to as the basket-weave microstructure with orientations according to the Burgers orientation relationship [96]. From a general point of view it seems worth mentioning that borides not only serve as nucleants for the α and β phases during solidification, and for the α phase during the β/α transformation, but also for the γ phase as found by Kim and Dimiduk [97]. Therefore, many aspects have to be taken into account regarding B additions. With respect to β-solidifying alloys, the β/α transformation kinetics are thought to be a key factor regarding microstructural refinement. Besides nucleation, retardation of growth through reduced diffusion kinetics or decreasing the transformation temperature have also been shown to induce microstructural refinement [32]. Figure 4.10 shows an example of the fine and particularly homogeneous cast microstructures that can be obtained in B-containing β-solidifying alloys. However, only limited room-temperature ductility (less than 0.45% plastic elongation to fracture) has been observed for such cast materials [38], which was attributed to the occurrence of adjacent lamellar colonies with nearly parallel lamellae [38]. Obviously, besides colony-size refinement, the colony-orientation distribution is important for ductility and other factors such as the presence of brittle phases could also play a role. This topic certainly deserves continuing research.

To summarize, β-solidifying alloys have the highest potential with regard to the formation of fine and homogeneous cast microstructures. However, peritectic solidification often results in a beneficial texture of cast components and thus, research on corresponding alloys also appears to be interesting. Alloying with B is the most effective means to obtain grain refinement in TiAl alloys, although the mechanism is complex and depends on the alloy constitution. Nucleation at borides in the melt during solidification and during the β/α solid-state trans-

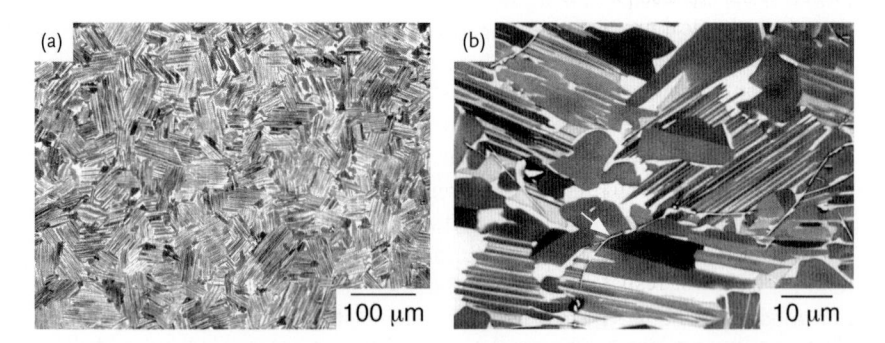

Figure 4.10 Scanning electron micrographs (using backscattered electrons) from cast Ti-45Al-7Nb-1Mo-0.2B in the HIPed condition; (b) shows a section of (a) in higher magnification. The arrow indicates an elongated boride particle in the material.

formation, as well as growth restriction, have been identified as grain-refining mechanisms. However, a full understanding of these mechanisms as well as the solidification behavior is still lacking.

4.2 Solid-State Transformations

As described in the previous section, after solidification engineering γ(TiAl) alloys pass through the single-phase α solid-solution region or, for lower Al contents, even through two single-phase regions, namely the β and subsequently the α region. A variety of different phase transformations can occur on either cooling from the high-temperature regions or on subsequent heat treatment. In Al-lean alloys the β phase transforms to the α phase, or can follow other decomposition paths, depending on the nature and amount of alloying additions. On moving out of the single-phase α field, different phase transformations are possible depending on the cooling rate. For the highest cooling rates the α phase cannot decompose but orders to the $α_2$ phase [98, 99]. With decreasing cooling rate a number of transformations are observed including i) the composition-invariant massive α → γ-transformation, ii) the lamellar reaction, in which crystallographically oriented plates of the γ phase are precipitated from α, and iii) at very low cooling rates the formation of γ grains [10, 13, 39, 42, 98, 100–106].

The lamellar reaction also shows two further variants at the highest cooling rates for which lamellae occur. These are the formation of Widmannstätten colonies in lamellar colonies [42, 98, 101, 103, 106–110] and formation of so-called "feathery structures" consisting of γ lamellae with a misorientation of 2°–15° compared to the Blackburn-orientation relationship (Equation 4.1) [42, 102, 104, 109, 111, 112]. Furthermore, after the formation of the lamellar microstructure a discontinuous coarsening reaction can occur, which will significantly alter the microstructure [113]. For higher Al concentrations of around 50 at.%, a secondary precipitation of $α_2$ Widmannstätten plates from the γ phase is possible on the four close-packed planes of the γ phase [98]. According to current knowledge a beneficial combination of mechanical properties is achieved in material with a lamellar microstructure that has both a fine lamellar spacing and small colony size. This is why alloy development is mainly focused on this type of microstructure. The potential multitude of other possible microstructures has only rarely been investigated and little is known with respect to mechanical properties. It should be mentioned, however, that some phase reactions might be very dependent on local conditions, for example if the cooling rate varies locally due to the different cross-sections of a component. Such challenges need to be addressed during alloy development and processing. In this respect it should be remembered that the kinetics of the diverse phase transformations can sensitively depend on alloy composition. For instance, alloying with even low B additions depresses the occurrence of a massive reaction even for the highest technically possible cooling rates, thus allowing the formation of fine lamellar microstructures [97, 114].

4.2.1
β → α Transformation

In binary β-solidifying γ(TiAl) alloys the high-temperature β phase transforms to the α phase when the alloy is cooled after solidification or after heat treatment in the β phase field (Figure 4.1). Similar to many Ti alloys, this transformation can occur in a diffusion-less martensitic mode or through a composition-invariant massive transformation involving short-range diffusion or also through the precipitation of the α phase from β [98]. Which type of transformation occurs depends on the alloy composition and the cooling rate. For very fast cooling rates, as was achieved during powder atomization, the martensitic transformation of the β phase was observed in Ti-48Al powder [115, 116]. However, this transformation could not be obtained in the solid state, even after ice-brine quenching [98]. This finding was attributed to the high T_0 temperature of the β/α transformation [98] that is above 1330 °C for the range of Al concentrations considered, which promotes the massive transformation [98]. Correspondingly, it was found that during water quenching of Ti-40Al, the β phase completely transformed to α in a massive transformation [98]. In this transformation, which was also observed in Ti alloys [21], nuclei of the product phase are formed with the same composition as the parent phase and grow by migration of the phase boundaries. The growth kinetics of a massive transformation are usually several orders of magnitude faster when compared to precipitation reactions. This is because only short-range diffusion at the migrating interfaces takes place and no partitioning between the phases is involved. Yamabe et al. [98] observed small antiphase domains in the $α_2$ phase of a Ti-40Al alloy indicating that the massive transformation preceded the ordering reaction. Besides the β to α massive transformation, transformation of the α phase to the γ phase can occur massively, as will be described later in more detail.

For relatively slow cooling rates Widmannstätten plates of the α phase are precipitated when the material is cooled from the β phase field, as has been found in several studies [2, 6, 12, 18, 27, 32–34, 117]. This precipitation reaction results in microstructures that resemble the so-called basket-weave microstructures seen in near α or (α + β) Ti alloys, which are apparently formed through a similar mechanism. An example of such a microstructure is shown in Figure 4.5. According to Banerjee and Mukhopadyay [21], in Ti alloys the reaction starts with the nucleation of α plates at β grain boundaries, so-called grain-boundary allotriomorphs, which have a Burgers orientation relationship (Equation 4.2) with one of the adjacent β grains. Groups of parallel Widmannstätten side plates with the same crystallographic orientation grow from the grain-boundary allotriomorphs or from the grain boundaries into the β grains. The individual Widmannstätten α plates are separated by retained β phase that is enriched in Ti and other alloying elements that partition to the β phase. The broad face of the α plates contains ledges of up to a few lattice planes in height and the long direction of the plates lies approximately parallel to $<335>_β$. The morphology and crystallographic orientation of such growing Widmannstätten plate colonies is sketched in Figure 4.11. The growth kinetics correspond to those of a cellular transformation, however, the β grain boundary does not propagate with the tips of the plates, as would occur in

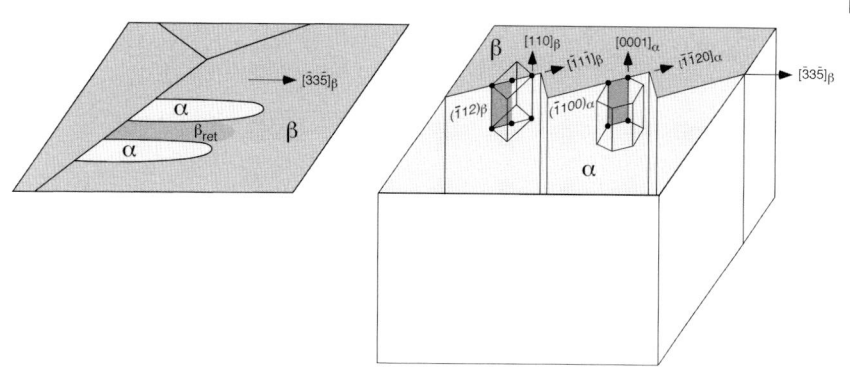

Figure 4.11 Morphology and crystallography of a colony of Widmannstätten α plates growing from a β grain boundary during the β → α transformation [21]. The β parent phase and the α plates have a Burgers orientation relationship {110}β ∥ (0001)α and <111>$_\beta$ ∥ <1120>$_\alpha$ (Equation 4.2). The interfaces contain {112}$_\beta$ and {1100}$_\alpha$ planes and ledges of some atomic distances in height. The broad faces of the α plates have an orientation of {11 11 13}$_\beta$ and the plates grow with a rough <335>orientation. β plates that are enriched in Ti and other alloying elements that segregate to the β phase grow in between and together with the α plates. For kinetic reasons, these β plates do not dissolve in Ti aluminide alloys on further cooling as the alloy passes through the single α-phase field. In an even later stage of cooling, each α plate transforms into an (α_2 + γ) lamellae colony. Since all α plates of a prior Widmannstätten α colony have the same orientation, only one orientation of (α_2 + γ) lamellae occurs in such a prior Widmannstätten colony. The microstructure resulting from this series of phase transformations is shown in Figure 4.5.

a cellular reaction. Besides the formation of side plates, colonies of intragranular Widmannstätten plates that nucleate inside the β grains and also obey the Burgers orientation relationship can be formed. Such intragranularly formed plates give rise to the typical "basket-weave" morphology. Some characteristics of Widmannstätten colony formation in Ti alloys, such as the Burgers orientation relationship with respect to the parent β phase [2, 3, 18, 22, 39], nucleation at grain boundaries, and the microstructural morphology, have also been observed in TiAl alloys. However, a detailed understanding of the nucleation and growth of Widmannstätten colonies in TiAl alloys is still lacking. As already described above, the kinetics of the β/α transformation determine the grain size of the α phase and thus, the size of the final (α_2 + γ) lamellar colonies. It should also be mentioned, as discussed in Section 4.1, that borides can serve as nucleants for the α phase and suppress the formation of Widmannstätten α plates during the β/α transformation. Correspondingly, Cheng and Loretto [39] have shown that the β/α transformation occurs after heat treatment in the (β + α) phase field by growth of existing α grains if the material is slowly cooled, whereas Widmannstätten colonies are formed for higher cooling rates. This observation can be explained if it is assumed that the Widmannstätten colony-formation kinetics are faster than those for the growth of isolated α grains, similar to the results obtained for B-containing β-solidifying alloys [88] as discussed in the foregoing section.

4.2.2
Formation of (α_2 + γ) Lamellae Colonies

Lamellar colonies consist of stacks of lamellae, or more strictly speaking, platelets of the α_2 and γ phases that form during the phase formation sequence $\alpha \rightarrow \alpha + \gamma \rightarrow \alpha_2 + \gamma$ at medium cooling rates [3, 10, 105, 118]. The formation of γ lamellae usually starts in the two-phase ($\alpha + \gamma$) region. At the eutectoid temperature the $\alpha \rightarrow \alpha_2$ ordering reaction begins and the equilibrium volume fraction of γ phase abruptly increases, resulting (in principle) in the precipitation of further γ lamellae. In Al-lean alloys the phase formation sequence $\alpha \rightarrow \alpha_2 \rightarrow \alpha_2 + \gamma$ takes place, but results in the same microstructure [98, 102, 119, 120]. The transition between the two phase formation sequences was determined on the basis of T_0 curves to be at 44 at.% Al in binary TiAl alloys [105, 119, 121]. Even for alloys with an Al concentration close to the eutectoid composition at 40 at.% Al [98, 102, 122], the eutectoid reaction $\alpha \rightarrow \alpha_2 + \gamma$ has never been observed to occur as a discontinuous transformation [123] such as pearlite formation in steels, where the two product phases grow together with a moving interface into the parent phase. On the contrary, in such eutectoid alloys the γ lamellae are precipitated from the α phase as described above [98]. The lamellar microstructure contains interfaces not only between the α_2 and γ phases, but also between differently oriented γ platelets. The interfaces are coherent or semicoherent and atomically flat over relatively large distances, but they contain ledges that are a few multiples of their {111} interplanar spacing in height [124–132]. The lamellar microstructure can be seen in Figure 4.12 and Figure 6.12 schematically illustrates this microstructure. TEM micrographs of this structure are shown in Figures 6.3 and 6.9. The lamellar microstructure is described in great detail in Chapter 6 with particular attention being paid to the lamellar interfaces. In the following, the focus will be directed to the phase-transformation processes and kinetics.

From a general point of view the formation of γ lamellae from the α phase can be subdivided into the hcp $\rightarrow$ fcc lattice transformation, a change of the composition by diffusion processes and the ordering of the fcc phase. As observed in the early work of Blackburn [10], the transformation begins with the formation of a stacking fault in the α phase, that is formed by the splitting of a 1/3 <11$\bar{2}$0> dislocation into two Shockley partials with $\mathbf{b}$ = 1/3 <10$\bar{1}$0> and 1/3 <01$\bar{1}$0>. The occurrence of stacking faults as a prenucleation stage of the transformation has been confirmed in TEM work by Xu et al. [133] and Denquin and Naka [134, 135]. In these studies the material was either quenched from the α phase and subsequently annealed or slowly cooled from the α phase and then quenched from 1250 °C in order to preserve the initial stage of the transformation. If stacking faults are formed on every second close-packed plane in the α phase, then the fcc structure is formed. There are two possibilities, glide of the partial dislocations on every odd-numbered basal plane (A|BA|BA|B), which transforms the hcp stacking sequence ABABAB into ACBACB, or glide on every even-numbered plane (AB|AB|AB|), resulting in the stacking sequence ABCABC [105, 134–136]. The occurrence of stacking faults in the prenucleation stage indicates that the phase

Figure 4.12 Scanning electron micrographs (using backscattered electrons) showing (a) a fully lamellar microstructure, (b) a nearly lamellar microstructure, (c) a duplex microstructure and (d) Widmannstätten γ plates. (a) Ti-47Al-1.5Nb-1Cr-1Mn-0.2Si-0.5B, extruded, heat-treated at the α-transus temperature followed by furnace cooling. (b) Ti-45Al-8Nb-0.2C, extruded, heat treated 20 K below the α-transus temperature followed by air cooling. (c) Ti-45Al-8Nb-0.2C, extruded, heat treated 40 K below the α-transus temperature followed by air cooling. (d) Ti-50Al, heat treated in the γ single-phase field (2h/1240 °C/furnace cooling) and subsequently in the ($\alpha + \gamma$) phase field (0.5h/1340 °C/air cooling, compare Figure 2.3).

transformation may indeed take place by this mechanism. Subsequent ordering of the fcc phase results in six orientation variants of the γ phase, since for each of the two fcc orientation variants there are three possible ways to align the c-axis of the L1$_0$ structure. The orientations are described by the Blackburn-orientation relationship $(0001)_{\alpha 2} \parallel \{111\}_\gamma$ and $<11\bar{2}0>_{\alpha 2} \parallel <1\bar{1}0>_\gamma$ (Equation 4.1), which corresponds to rotations of multiples of 60° around $<111>_\gamma$ or $[0001]_{\alpha,\alpha 2}$, as illustrated in Figures 6.2 and 6.6. Figure 6.7 shows the stacking sequences of these orientation variants. Following the work of Blackburn [10], the different orientation variants have been confirmed in numerous TEM studies [13, 118, 135–142]. As the α and α_2 phases only have one set of basal planes, only one colony of parallel ($\alpha_2 + \gamma$) lamellae can be formed from one prior α grain.

The orientation relationships lead to three different types of interface between adjacent γ lamellae. For one stacking sequence of (111) planes, either ABC or ACB, three orientations variants occur and are described by rotations of 0°, 120° and 240° around [111]. These orientation variants are separated by so-called 120°

rotational order fault interfaces [118]. The *c*-axes of these three variants are perpendicular to each other. If neighboring lamellae have different stacking sequences, then they are separated by either a true twin boundary (orientation 180°) or pseudotwin boundaries (orientations 60° and 300°) [118]. Here, the reader is again referred to Chapter 6 and Figures 6.6 and 6.7 for a more detailed description. In addition to the interfaces between adjacent γ lamellae, so-called domain boundaries divide single γ lamellae into domains of different orientations [136, 138, 139, 143]. The domain boundaries do not show any preference for a specific crystallographic habit plane [136, 138, 139, 144] and have been found to separate only 120° orientation variants, i.e. variants that have the same stacking sequence of {111} planes perpendicular to the lamellar interface [136, 138, 139, 144].

The stacking faults in the α phase described above as a prenucleation stage of the formation of γ lamellae, were reported to originate from α grain boundaries or γ grain-boundary allotriomorphs [133, 134, 145]. These stacking faults do not represent γ lamellae nuclei as they have neither the composition nor the chemical ordering of the γ phase. Denquin and Naka [134, 135] suggested that an embryonic platelet of a metastable fcc phase initially forms by the successive motion of Shockley partial dislocations in the α phase on every second plane, and then subsequently adopts the chemical ordering of the γ phase. Indeed, it was possible to obtain a metastable almost equiatomic fcc phase by melt spinning, thus demonstrating that at least for very high undercoolings a driving force exists for the precipitation of this phase [146]. The formation of fcc platelets would lower the nucleation barrier of the γ phase, which is certainly high given the high undercooling required for its formation. In the literature, values of at least 50 to 100 K have been reported for the undercooling, even if relatively low cooling rates were applied [97, 114, 141, 147]. The formation of the embryonic fcc platelets is then followed by ordering at different sites within the platelets. This leads to the formation of APBs and 120° rotational order fault domain boundaries and would explain why domain boundaries have always been observed to separate 120° rotational variants. The absence of APBs in γ lamellae was concluded to arise from the fact that they migrate at a significantly higher velocity than the 120° rotational boundaries. They thus form "mixed boundaries" consisting of an APB and a 120° order-domain boundary inside the lamellae as well as mixed lamellar interfaces, that is, APBs lying at the lamellar interface (Figure 4.13) [134, 135]. However, work by Zhang *et al.* [145] has cast some doubts on whether the formation of fcc platelets is a prenucleation stage of γ-phase formation. The authors quenched a TiAl alloy from the α-phase field and observed planar faults in the TEM that appeared to be stacking faults similar to those described above. However, HREM investigations showed that the faults were actually 6–8 atomic layer thick platelets of the γ phase with a $L1_0$ structure. Very thin embryonic γ lamellae were also observed in other HREM studies (Figure 6.17), clearly demonstrating that ordering of the fcc platelets occurs at a very early stage. Thus, the formation of order domains by ordering of embryonic fcc plates at different sites is not a fully established mechanism. Denquin and Naka [134, 135] have argued that it is rather improbable that γ lamellae encounter each other due to their extreme aspect ratio. For this reason it is difficult to understand how domain boundaries are formed in the lamellae if no precursor fcc phase

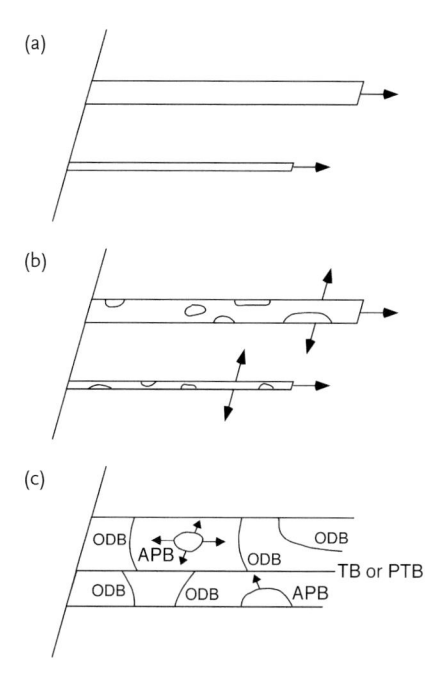

Figure 4.13 Schematic illustration of the formation of γ lamellae from the α phase following Denquin and Naka [134, 135]. (a) Propagation of Shockley partial dislocation generating a local fcc structure. (b) Nucleation of orientation variants of the L1$_0$ structure in the prenucleation fcc phase. (c) After growth of the orientation variants order-domain boundaries (ODB) and antiphase boundaries (APB) are formed inside lamellae. Growth of the lamellae results in the formation of lamellar γ/γ interfaces, which are twin (TB) or pseudotwin boundaries (PTB), if adjacent lamellae have a different stacking sequence of the close-packed planes. Due to rapid growth the APBs lie at the other boundaries resulting in so-called mixed boundaries that have been proven to exist in lamellar structures [135].

exists. In particular, this applies for lamellae that contain many domain boundaries. Despite this open question, there appears to be a consensus that γ lamellae are usually heterogeneously nucleated, in particularly at grain boundaries or grain-boundary allotriomorphs [133, 134, 141, 142, 145, 148]. However, borides have also been shown to act as nucleation sites for γ lamellae [97]. Zghal et al. [141] have shown that fine γ lamellae were presumably homogeneously nucleated inside α grains at temperatures below the α → α$_2$ ordering transition. Since an equal occurrence of both stacking sequences was observed for these fine lamellae, the authors concluded that the stress induced by the precipitation of a small γ lamella was partially relaxed by the nucleation of a lamella with the opposite stacking sequence in its vicinity. A population of very thin γ lamellae that did not contain domain boundaries, and that exhibited a twin-orientation relationship with the adjacent much thicker lamella, was also observed. This finding was explained by nucleation at the lamellar interface, a process that would be energetically beneficial for nuclei with twin orientation because the energy of lamellar interfaces is lowest for twin boundaries (see Table 6.1). This process where γ lamellae nucleate at a moving interface, also called sympathetic nucleation [13, 149, 150] has been observed in other work [151]. Dey et al. [13] concluded that γ(TiAl) alloys appear to be prone to sympathetic nucleation due to the anisotropic elastic behavior of the α phase and the availability of low-energy interfaces between the phases.

Once nucleated, the growth of γ lamellae requires both a lattice transformation and diffusion to equilibrate the composition between the α/α_2 and the γ phases. This process has been described as proceeding via the migration of partial dislocations accompanied by diffusion [126, 127, 133, 152]. Denquin et al. [134, 135] specified this process as a transition from the motion of partial dislocations to diffusion-controlled ledge migration at the interfaces, capable of providing both a change in the stacking sequence as well as in composition. This ledge mechanism was proposed by Aaronson [153, 154] to explain how a coherent or semicoherent boundary between phases of different crystal structure could migrate in a diffusional phase transformation, although the attachment of atoms at the interface is energetically very unfeasible. Shang et al. [155] and Pond et al. [132] investigated interfacial steps at lamellar interfaces using HREM and showed that the Burgers vectors at the interface steps were consistent with those expected for interfacial disconnections on the basis of Pond's theory [156, 157]. The authors concluded that the γ lamellae grow in a displacive-diffusive transformation by diffusion-controlled ledge migration along the interfaces rather than by the motion of partial dislocations accompanied by diffusion. In this respect it is interesting to note that the thickness of the γ lamellae observed by Shang et al. [155] was consistent with that estimated by the authors in a diffusion analysis. The concept of disconnections is described in more detail in Chapter 6 and the references cited therein. A HREM image of a γ lamella within α_2 phase with a number of interface steps is presented (Figure 6.17). In this context, it may be speculated that the tips of growing γ lamellae are preferential sites where composition exchange occurs, as long as the γ lamellae have not traversed the entire α/α_2 grain. It should also be noted that the motion of disconnections along interphase boundaries may not only transport material from one phase to the other, but also results in deformation. Sun [99] has actually shown that the formation of a lamellar microstructure leads to an undulated surface relief as earlier observed by Valencia et al. [158] and McCullough et al. [159]. This demonstrates the displacive nature of the phase transformation and resembles the displacive-diffusive bainitic transformation in steels [99].

In order to gain a deeper insight into the complex formation mechanism of the lamellar microstructure the phase-field simulation method [65, 66] has been found to be valuable. Wen et al. [160] have developed a three-dimensional phase-field model to simulate the evolution of the lamellar microstructure. The model included the bulk chemical free energy, the interfacial energies and the elastic energy and was able to predict the essential features of the lamellar structure. The authors concluded that the formation of lamellar structures is dominated by the accommodation of elastic coherency stresses. The higher observed frequency of adjacent twin-related compared to pseudotwin-related lamellae, could be explained by strain-induced correlated nucleation and elastic-energy minimization. Interestingly, this modeling study observed that some lamellae shrank in the growth process while others grew. The study did not include either the fcc prenucleation stage or heterogeneous nucleation of the γ phase. These were taken into account in a later two-dimensional phase-field simulation by Katzarov et al. [161] that described the heterogeneous formation of a fcc prenucleation phase as well as nucleation and growth of $L1_0$ orientation variants at different sites in the fcc precursor phase. The authors concluded that the formation mechanism of lamellar microstructures proposed by Denquin and Naka [134, 135] and described above, reflects the main features of the phase transformation. In agreement with Wen et al. [160], the critical role of the elastic interaction between lamellae was highlighted. To summarize, in view of these phase-field modeling studies the essential characteristics for the formation mechanisms of lamellar structures can be considered to be well established.

The lamellar microstructure can be characterized by the colony size, the α_2 volume fraction, the width of γ and α_2 lamellae, the spacing between α_2 lamellae and the distance between domain boundaries. These structural parameters depend on the alloy composition and the processing history, in particular on the heat-treatment temperature and the cooling rate. In lamellar alloys these parameters can have a wide range, for example, the width of lamellae can range from several nm up to several μm [3, 102, 119, 136, 141, 142, 147, 148, 162–165]. Statistical data on microstructural parameters are found in the work of Zghal et al. [136], Dimiduk et al. [162], Parthasaraty et al. [163], Maruyama et al. [164, 165], Zghal et al. [141, 142] and Charpentier et al. [147]. The kinetics of lamellar microstructure formation has been described by TTT curves (time–temperature–transformation). Such diagrams give valuable information about the range of cooling rate ranges required for specific transformations and the undercooling at which the transformations begin. TTT curves have been determined for several alloys using experimental techniques such as resistivity measurements and dilatometry [42, 103–105, 107, 109, 147, 166–169]. Figure 4.14 shows TTT diagrams for the alloys Ti-47.5Al and Ti-48Al-2Cr-2Nb. From the diagrams it is obvious that the formation of lamellae requires a significant undercooling, and that for rapid cooling the transformation start temperature is shifted to very low values. Charpentier et al. [147] not only determined the variation of lamellae volume fraction with the cooling rate, for example, the transformation start and finish temperatures, but also performed a quantitative analysis of the microstructural parameters in the alloy Ti-48Al-2Cr-2Nb after different heat treatment times and cooling rates. The results are displayed in Figure 4.15. From this figure it can be concluded that the equilibrium fraction of the α_2 phase is only obtained for particularly low cooling rates of below 10 K/min. Further, the diagrams in Figure 4.15 indicate that both the mean width and the mean spacing of α_2 lamellae decrease continuously with the cooling rate, and that the relationship depends linearly on the inverse square root of the cooling rate, as observed by Kim and Dimiduk [170], Perdrix et al. [171] and confirmed by Rostamian and Jacot [169]. In the latter work, phenomenological models were developed for binary alloys to describe the competing lamellar and massive transformations when the material is cooled from the α phase field. Coupling of these

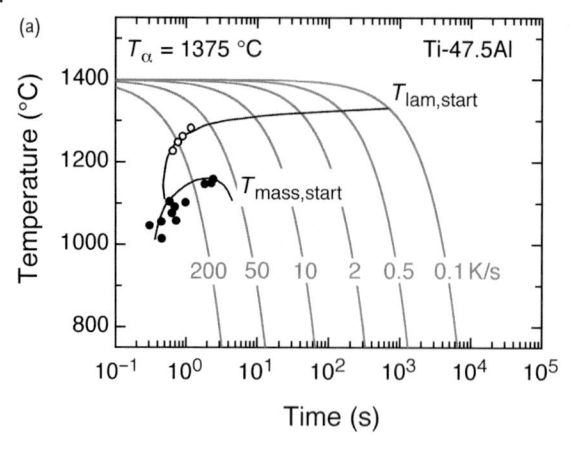

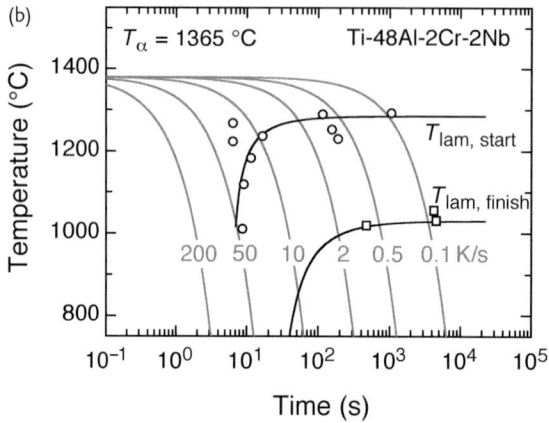

Figure 4.14 TTT curves for different phase transformations in two TiAl alloys. (a) Start temperatures for the formation of lamellar microstructures and the massive transformation in the alloy Ti-47.5Al calculated by Rostamian and Jacot [169] according to the model described in the text. In addition, experimental results determined by Veeraraghavan et al. [104] are shown. The material had been heat treated at 1400 °C, that is, 25 K above the α-transus temperature. (b) Experimental start and finish temperatures for the formation of lamellar microstructures in the alloy Ti-48Al-2Cr-2Nb after a heat treatment at 1380 °C [147].

models enabled the influence of alloy chemistry, cooling rate and α grain size on the microstructural features and the TTT diagrams to be calculated. With respect to the formation of lamellar structures the nucleation rate of γ lamellae was described by

$$J_{\gamma L} = J_{\gamma L}^0 s_\alpha \exp\left(-\frac{\Delta G_{\gamma L}^{nucl}(x,T)}{kT}\right)\exp\left(-\frac{Q_{at}}{kT}\right) \quad (4.6)$$

according to classical nucleation theories [172–177] with

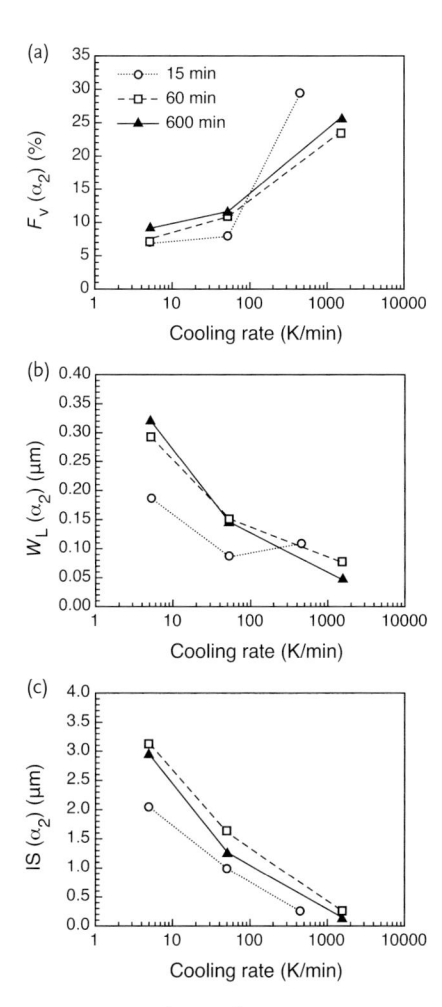

Figure 4.15 Evolution of microstructure parameters of the alloy Ti-48Al-2Nb-2Cr [147] as a function of the cooling rate after heat treating the material at 1380 °C (15 K above the α-transus temperature). As indicated in the diagrams, the duration of the heat treatment also had been varied. (a) Volume fraction of the α_2 phase, (b) mean width of α_2 lamellae and (c) mean interlamellar spacing of α_2 lamellae.

$$\Delta G_{\gamma L}^{\text{nucl}}(x,T) = f_\theta \frac{16\pi \sigma_{\alpha\gamma}^3}{3(\Delta G_V^{\alpha\gamma}(x,T))^2} \tag{4.7}$$

where $J_{\gamma L}^0$ is a prefactor corresponding to the density of nucleation sites, s_α is the grain boundary fraction still available for nucleation, Q_{at} the activation energy of atomic mobility, f_θ the wetting factor for heterogeneous nucleation, $\sigma_{\alpha\gamma}$ the interfacial energy, $\Delta G_V^{\alpha\gamma}(x,T)$ the driving force and x the Al concentration [169]. The model also included the longitudinal growth rate at lamellae edges [178, 179] which is given by

$$v_{\gamma L} = \frac{\tilde{D}_\alpha(x_0 - x_\alpha^*)}{b(x_\gamma^* - x_\alpha^*)} \cdot \frac{1}{r} \tag{4.8}$$

where r is the edge radius, b a geometrical factor, x_0 the nominal Al concentration, x^*_α and x^*_γ the equilibrium Al concentrations of the α and the γ phases, and $\tilde{D}_\alpha$ the interdiffusion coefficient in the α phase. The thickening of lamellae was treated as a one-dimensional diffusion problem using the Zener model [180], whereupon the ledge mechanism has been taken into account. The authors were able to describe the measured TTT curves for the lamellar and massive transformations (see Figure 4.14), and the dependence of lamellar spacing on cooling rate reasonably well using this model with only four adjustable nucleation parameters. The model can be considered as being the first analytical description for lamellar and massive transformations and the authors were able to draw a number of interesting conclusions. For example, the tendency of the alloy to transform in the massive mode increases with the Al concentration. The simulations also showed that the critical cooling rate for the massive transformation is almost independent of the Al content. Further, it was found that the lamellar spacing increases with the Al concentration and that a smaller prior α grain size accelerates both the lamellar and the massive transformation, but has almost no effect on the critical cooling rate separating the two transformation modes.

In addition to the phase transformations that occur during thermal treatments, the microstructure is also influenced by recrystallization during hot working. As described in Chapter 16, the lamellar microstructures present in ingot material can be fully converted to equiaxed microstructures with grain sizes in the range of some µm by wrought processing (see Chapter 16). Subsequent heat treatments at temperatures below the α-transus temperature result in microstructures consisting of different fractions of lamellar colonies and equiaxed γ grains [119]. The variety of microstructures that can be formed has been classified by Kim [119] into near-gamma, duplex, nearly lamellar and fully lamellar, as shown in Figure 4.12. The near-gamma microstructures are obtained by annealing close to the eutectoid temperature and consist of bands of relatively coarse γ grains. Duplex microstructures composed of γ grains and fine lamellae colonies arise through heat treatment at temperatures where approximately equal fractions of the α and γ phases occur. At higher temperatures, only a small fraction of γ grains is present, but prevents extensive grain growth of the α phase. During subsequent cooling nearly lamellar microstructures with a moderate colony size are formed. Fully lamellar microstructures result from heat treatments in the single α-phase field and can have quite large colony sizes of several 100 µm.

4.2.3
Feathery Structures and Widmannstätten Colonies

When γ alloys are cooled with intermediate rates from the α-phase field, characteristic alterations of the lamellar microstructure occur. These lead to the forma-

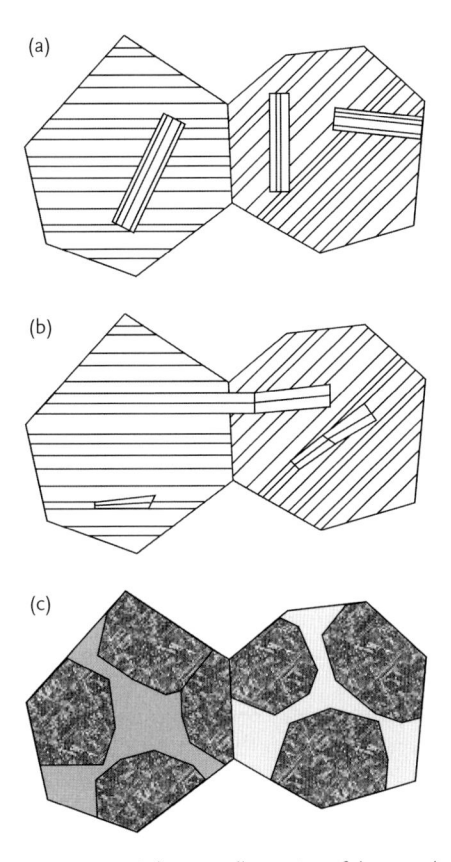

Figure 4.16 Schematic illustration of the morphology of (a) Widmannstätten colonies, (b) feathery structure and (c) the microstructure of massively transformed material.

tion of the so-called Widmanstätten lath colonies [42, 98, 103, 106–110, 181] and feathery structures [42, 102, 104, 109, 111, 112, 182]. Widmanstätten colonies are packets of parallel ($\alpha_2 + \gamma$) lamellae inside lamellar colonies and have a different interface orientation to the surrounding lamellae. The morphology of this microstructure is schematically shown in Figure 4.16. These Widmannstätten colonies should not be mixed up with Widmannstätten γ plates, which are formed in alloys with an Al concentration around 50 at.%. Such alloys can be heat treated in the γ single-phase field and Widmannstätten γ plates are precipitated with an orientation parallel to the 4 close-packed planes of the γ phase, when single-phase material is annealed in the ($\alpha + \gamma$) or ($\alpha_2 + \gamma$) phase field. An example is shown in Figure 4.12d. In contrast to Widmannstätten colonies feathery structures consist of groups of lamellae within a colony, which are characterized by a small misorientation with respect to the surrounding colony and typically are not exactly parallel to one another (Figure 4.16). Widmannstätten laths occur at somewhat lower cooling rates than feathery structures, the former developing at cooling rates between air and sand cooling, while the latter form after oil or water quenching [13]. In the following, a brief overview of the formation mechanisms for these variants of the lamellar microstructure will be given, for more detailed information the reader is referred to the articles of Dimiduk and Vasudevan [109], Zhang et al. [42], Hu et al. [106] and Dey et al. [13].

The lamellae inside Widmannstätten colonies show some similarities with ordinary lamellar colonies, in particular the α_2 and γ lamellae have crystallographic orientations according to the Blackburn-orientation relationship (Equation 4.1). Detailed TEM work has shown that the α_2 laths in Widmannstätten colonies have an orientation that corresponds to a rotation of close to 64° around $<1\bar{1}00>_{\alpha 2}$ [13, 106, 110] with respect to the α_2 lamellae in the surrounding colony. This orientation is consistent with twinning in the parent α phase by activation of the $\{11\bar{2}2\}<\bar{1}\bar{1}23>$ twinning system. This system has often been observed in Ti and Ti alloys. Thus, twinning of the α phase during cooling followed by transformation of the twin into a Widmannstätten lath appears to be the formation mechanism of Widmannstätten colonies [106] as proposed by Yamabe et al. [98]. A different mechanism has been suggested by Dey et al. [13] who noted that a rotation of α_2 lamellae by 64° around $<1\bar{1}00>_{\alpha 2}$ corresponds to a rotation of γ lamellae by 50° around $<110>_\gamma$ if the α_2 and γ lamellae are in Blackburn-orientation relationship. A $\Sigma 11$ CSL boundary exists between γ lamellae of this orientation. Thus, Widmannstätten laths could be nucleated directly on the face of existing γ lamellae generating a low-energy interface. Such sympathetic nucleation [149, 150] has also been observed to occur during the formation of common lamellar structures as discussed above, and appears plausible due to the anisotropic elastic behavior of the α phase and to the availability of low-energy interfaces [13]. To date, however, the formation mechanism of Widmannstätten lath colonies has not been definitely identified, both of the above described mechanisms possibly operating.

The feathery structures resemble lamellar structures and have the usual crystallographic-orientation relationship between the phases, but are misoriented by less than around 15° with respect to the surrounding colony [109]. However, the α_2 lamellae in feathery structures exhibit an irregular thickness and habit plane [109]. Dimiduk and Vasudevan [109] have proposed that this microstructural variant may either arise due to deformation of the α phase or result from massively formed γ followed by precipitation of α in γ. More recently, Zhang et al. [42] showed that the orientation of feathery lamellae packets continuously change their orientation and neither of these mechanisms was concluded to operate. Rather, the observed features were thought to indicate edge-to-edge and face-to-face sympathetic nucleation of γ lamellae during the growth of a previously formed lamellae packet. This conclusion was supported by Dey et al. [13] who suggested the misorientation to originate from elastic stresses during sympathetic nucleation. The authors further concluded that sympathetic nucleation could be a common feature in the formation of both feathery structures and Widmannstätten colonies and could explain their specific morphologies [13].

4.2.4
Massive Transformation

The massive transformation is a composition-invariant phase transformation that occurs discontinuously behind a migrating reaction front. This transformation only requires short-range diffusion since no composition change has to take place. Massive transformations have been observed in many systems, Cu- and Ti-based alloys [21], Fe–Ni and Ag–Al alloys [104] for example. They often involve a mobile incoherent interface between the product and the parent phase [104]. However, semicoherent interfaces that migrate by a ledge mechanism [104] can also occur. Concerning TiAl, several studies have observed the massive transformation of the α phase into the γ phase at sufficiently high cooling rates [13, 98, 101–105, 111, 181, 183–192]. According to Dey et al. [13] such cooling is similar to or even faster than that experienced during oil quenching. The critical cooling rate varies with alloy composition and alloys have been identified in which the α phase transforms massively to γ on air cooling [188] (see Chapter 13). The microstructure that results from the massive transformation consists of irregular, mottled γ grains that are usually easy to recognize (Figure 4.16).

The nucleation of massive γ grains was observed to take place heterogeneously not only at grain boundaries [184, 186, 189, 190] but also at grain-boundary triple points [191], intragranular twins in the α phase [190] and at already existing γ grains or lamellae [13, 184]. Using TEM, Zhang et al. [184] determined that massive γ grains were nucleated at α phase grain boundaries with the usual Blackburn-orientation relationship (Equation 4.1) with respect to the parent α grain. These nuclei then grew into the adjacent α grain. This explains why no specific orientation relationship was found between massive γ grains and the α grain into which they had grown. In other work, nucleation with a Blackburn-orientation relationship with respect to the parent α grain as well as growth into the adjacent α grain has clearly been confirmed [186, 189, 191]. Zhang et al. [184] further reported that the γ phase grew as a disordered fcc phase after initially growing with the $L1_0$ structure. This was deduced from the occurrence of APDB-like defects in the massive γ phase. Microtwinning of the γ phase and the formation of thin α_2 platelets during growth of the massive γ was also observed. Several more recent investigations have confirmed that successive twinning during growth of massive γ grains occurs [13, 186, 189, 191], and plays an important role regarding the phase transformation mechanism. Dey et al. [186] suggested that after nucleation in the Blackburn-orientation relationship, massive γ grains not only easily grow into the adjacent α grain, but also within the grain in which they were nucleated. Due to twinning and in particular after successive twinning, the massively formed γ phase has no simple orientation relationship with the parent α grain and forms a mobile incoherent phase boundary with the parent α grain. In conclusion, the massive transformation appears to be reasonably well understood although not all details of the transformation mechanism have been clarified. Since the massively formed γ phase is not thermodynamically stable, decomposing on subsequent aging into the α_2 and the equilibrium γ phases, it has attracted much attention. This is because of the possibility of using the precipitation reaction to obtain refined microstructure by applying a heat treatment, in cast materials for example. During such aging, Widmannstätten α_2 plates are formed parallel to the four {111} planes of the γ phase similarly as described for the microstructure shown in Figure 4.12, and results in the so-called convoluted microstructure. This topic, together with alloy design concepts, are described in Chapter 13.

References

1 Witusiewicz, V.T., Bondar, A.A., Hecht, U., Rex, S., and Velikanova, T.Ya. (2008) *J. Alloys Compd.*, **465**, 64.
2 Hecht, U., Witusiewicz, V., Drevermann, A., and Zollinger, J. (2008) *Intermetallics*, **16**, 969.
3 McCullough, C., Valencia, J.J., Levi, C.G., and Mehrabian, R. (1989) *Acta Metall.*, **37**, 1321.
4 Gulliver, G.H. (1922) *Metallic Alloys*, Griffin, London.
5 Scheil, E. (1942) *Z. Metallkd.*, **34**, 70.
6 Jung, J.Y., Park, J.K., and Chun, C.H. (1995) *Gamma Titanium Aluminides* (eds Y.-W. Kim, R. Wagner, and M. Yamaguchi), TMS, Warrendale, PA, p. 459.
7 Jung, J.Y., Park, J.K., and Chun, C.H. (1999) *Intermetallics*, **7**, 1033.
8 Jung, I.-S., Kim, M.-C., Lee, J.-H., Oh, M.-H., and Wee, D.-M. (1999) *Intermetallics*, **7**, 1247.
9 Muraleedharan, K., Rishel, L., De Graef, M., Cramb, A., Pollock, T., and Gray, T.I. (1997) *Structural Intermetallics 1997* (eds M.V. Nathal, R. Darolia, C.T. Liu, P.L. Martin, D.B. Miracle, R. Wagner, and M. Yamaguchi), TMS, Warrendale, PA, p. 215.
10 Blackburn, M.J. (1970) *The Science, Technology and Application of Titanium* (eds R.I. Jaffee and N.E. Promisel), Pergamon Press, Oxford, p. 633.
11 De Graef, M., Biery, N., Rishel, L., Pollock, T.M., and Cramb, A. (1999) *Gamma Titanium Aluminides 1999* (eds Y.-W. Kim, D.M. Dimiduk, and M.H. Loretto), TMS, Warrendale, PA, p. 247.
12 Küstner, V., Oehring, M., Chatterjee, A., Güther, V., Brokmeier, H.-G., Clemens, H., and Appel, F. (2003) *Gamma Titanium Aluminides 2003* (eds Y.-W. Kim, H. Clemens, and A.H. Rosenberger), TMS, Warrendale, PA, p. 89.
13 Dey, S.R., Hazotte, A., and Bouzy, E. (2009) *Intermetallics*, **17**, 1052.
14 Kishida, K., Johnson, D.R., Shimida, Y., Inui, H., Shirai, Y., and Yamaguchi, M. (1995) *Gamma Titanium Aluminides* (eds Y.-W. Kim, R. Wagner, and M. Yamaguchi), TMS, Warrendale, PA, p. 219.
15 Johnson, D.R., Inui, H., and Yamaguchi, M. (1996) *Acta Mater.*, **44**, 2523.
16 Johnson, D.R., Inui, H., and Yamaguchi, M. (1997) *Acta Mater.*, **45**, 2523.
17 Johnson, D.R., Masuda, Y., Shimada, Y., Inui, H., and Yamaguchi, M. (1997) *Structural Intermetallics 1997* (eds M.V. Nathal, R. Darolia, C.T. Liu, P.L. Martin, D.B. Miracle, R. Wagner, and M. Yamaguchi), TMS, Warrendale, PA, p. 287.
18 Naka, S., Thomas, M., Sanchez, C., and Khan, T. (1997) *Structural Intermetallics 1997* (eds M.V. Nathal, R. Darolia, C.T. Liu, P.L. Martin, D.B. Miracle, R. Wagner, and M. Yamaguchi), TMS, Warrendale, PA, p. 313.
19 Burgers, W.G. (1934) *Physica*, **1**, 561.
20 Lütjering, G. and Williams, J.C. (2003) *Titanium*, Springer-Verlag, Berlin.
21 Banerjee, S. and Mukhopadhyay, P. (2007) *Phase Transformations – Examples from Titanium and Zirconium Alloys*, Pergamon Materials Series, vol. 12, Elsevier, Amsterdam.

22 Das, S., Howe, J.M., and Perepezko, J.H. (1996) *Metall. Mater. Trans. A*, **27A**, 1623.
23 Cheng, T.T. and Loretto, M.H. (1997) *Structural Intermetallics 1997* (eds M.V. Nathal, R. Darolia, C.T. Liu, P.L. Martin, D.B. Miracle, R. Wagner, and M. Yamaguchi), TMS, Warrendale, PA, p. 253.
24 Johnson, D.R., Masuda, Y., Inui, H., and Yamaguchi, M. (1997) *Mater. Sci. Eng.*, **A239–240**, 577.
25 Johnson, D.R., Chihara, K., Inui, H., and Yamaguchi, M. (1998) *Acta Mater.*, **46**, 6529.
26 Jung, I.S., Yang, H.S., Oh, M.H., Lee, J.H., and Wee, D.M. (2002) *Mater. Sci. Eng.*, **A329–331**, 13.
27 Johnson, D.R., Inui, H., and Yamaguchi, M. (1998) *Intermetallics*, **6**, 647.
28 Küstner, V., Oehring, M., Chatterjee, A., Clemens, H., and Appel, F. (2004) *Solidification and Crystallization* (ed. D. Herlach), Wiley-VCH Verlag GmbH, Weinheim, Germany, p. 250.
29 Kerr, H.W. and Kurz, W. (1996) *Int. Mater. Rev.*, **41**, 129.
30 Larsen, D.E., Kampe, S., and Christodoulou, L. (1990) *Intermetallic Matrix Composites*, vol. 194 (eds D.L. Anton, R. McMeeking, D. Miracle, and P. Martin), Mater. Res. Soc. Symp. Proc., Materials Research Society, Pittsburgh, PA, p. 285.
31 Huang, S.C. (1993) *Structural Intermetallics* (eds R. Darolia, J.J. Lewandowski, C.T. Liu, P.L. Martin, D.B. Miracle, and M.V. Nathal), TMS, Warrendale, PA, p. 299.
32 Imayev, R.M., Imayev, V.M., Oehring, M., and Appel, F. (2007) *Intermetallics*, **15**, 451.
33 Thomas, M. (2008) *Structural Aluminides for Elevated Temperatures* (eds Y.-W. Kim, D. Morris, R. Yang, and C. Leyens), TMS, Warrendale, PA, p. 229.
34 Singh, A.K. and Banerjee, D. (1997) *Metall. Mater. Trans. A*, **28A**, 1735.
35 Jin, Y., Wang, J.N., Yang, J., and Wang, Y. (2004) *Scr. Mater.*, **51**, 113.
36 Wang, Y., Wang, J.N., Yang, J., and Zhang, B. (2005) *Mater. Sci. Eng. A*, **392**, 235.
37 Clemens, H., Chladil, H.F., Wallgram, W., Zickler, G.A., Gerling, R., Liss, K.-D., Kremmer, S., Güther, V., and Smarsly, W. (2008) *Intermetallics*, **16**, 827.
38 Hu, D., Jiang, H., and Wu, X. (2009) *Intermetallics*, **17**, 744.
39 Cheng, T.T. and Loretto, M.H. (1998) *Acta Mater.*, **46**, 4801.
40 Krishnan, M., Natarajan, B., Vasudevan, V.K., and Dimiduk, D.M. (1997) *Structural Intermetallics 1997* (eds M.V. Nathal, R. Darolia, C.T. Liu, P.L. Martin, D.B. Miracle, R. Wagner, and M. Yamaguchi), TMS, Warrendale, PA, p. 235.
41 Takeyama, M., Ohmura, Y., Kikuchi, M., and Matsuo, T. (1998) *Intermetallics*, **6**, 643.
42 Zhang, Z., Leonard, K.J., Dimiduk, D.M., and Vasudevan, V.K. (2001) *Structural Intermetallics 2001* (eds K.J. Hemker, D.M. Dimiduk, H. Clemens, R. Darolia, H. Inui, J.M. Larsen, V.K. Sikka, M. Thomas, and J.D. Whittenberger), TMS, Warrendale, PA, p. 515.
43 Kobayashi, S., Takeyama, M., Motegi, T., Hirota, N., and Matsuo, T. (2003) *Gamma Titanium Aluminides 2003* (eds Y.-W. Kim, H. Clemens, and A.H. Rosenberger), TMS, Warrendale, PA, p. 165.
44 Takeyama, M. and Kobayashi, S. (2005) *Intermetallics*, **13**, 993.
45 Appel, F., Oehring, M., and Paul, J.D.H. (2006) *Adv. Eng. Mater.*, **8**, 371.
46 Hecht, U., Daloz, D., Lapin, J., Drevermann, A., Witusiewicz, V.T., and Zollinger, J. (2009) *Advanced Intermetallic-Based Alloys for Extreme Environment and Energy Applications*, vol. 1128 (eds M. Palm, B.P. Bewlay, Y.-H. He, M. Takeyama, and J.M.K. Wiezorek), Mater. Res. Soc. Symp. Proc., MRS, Warrendale, PA, p. 79.
47 Gungor, M.N. (1989) *Metall. Trans. A*, **20A**, 2529.

48 Ganesan, M., Dye, D., and Lee, P. (2005) *Metall. Mater. Trans. A*, **36A**, 2191.
49 Hazotte, A., Lecomte, J., and Lacaze, J. (2005) *Mater. Sci. Eng.*, **A413–414**, 267.
50 Hirano, K. and Iijima, Y. (1984) *Diffusion in Solids: Recent Developments* (eds M.A. Dayananda and G.E. Murch), The Metallurgical Society of AIME, New York, NY, p. 141.
51 Zollinger, J., Lapin, J., Daloz, D., and Combeau, H. (2007) *Intermetallics*, **15**, 1343.
52 Charpentier, M., Daloz, D., Hazotte, A., Gauthier, E., Lesoult, G., and Grange, M. (2003) *Metall. Mater. Trans. A*, **34A**, 2139.
53 Tiller, W.A., Jackson, K.A., Rutter, J.W., and Chalmers, B. (1953) *Acta Metall.*, **1**, 42.
54 Chalmers, B. (1964) *Principles of Solidification*, John Wiley & Sons, Inc., New York, NY.
55 Boettinger, W.J., Coriell, S.R., Greer, A.L., Karma, A., Kurz, W., Rappaz, M., and Trivedi, R. (2000) *Acta Mater.*, **48**, 43.
56 Hunziker, O., Vandyoussefi, M., and Kurz, W. (1998) *Acta Mater.*, **46**, 6325.
57 Johnson, D.R., Inui, H., Muto, S., Omiya, Y., and Yamanaka, T. (2006) *Acta Mater.*, **54**, 1077.
58 Oehring, M., Küstner, V., Appel, F., and Lorenz, U. (2007) *Mater. Sci. Forum*, **539–543**, 1475.
59 Liu, Y., Yang, G., and Zhou, Y. (2002) *J. Cryst. Growth*, **240**, 603.
60 Luo, W., Shen, J., Min, Z., and Fu, H. (2008) *J. Cryst. Growth*, **310**, 5441.
61 Ohnuma, I., Fujita, Y., Mitsui, H., Ishikawa, K., Kainuma, R., and Ishida, K. (2000) *Acta Mater.*, **48**, 3113.
62 Kim, M.C., Oh, M.H., Lee, J.H., Inui, H., Yamaguchi, M., and Wee, D.M. (1997) *Mater. Sci. Eng.*, **A239–240**, 570.
63 Su, Y., Liu, C., Li, X., Guo, J., Li, B., Jia, J., and Fu, H. (2005) *Intermetallics*, **13**, 267.
64 Eiken, J., Apel, M., Witsuiewicz, V.T., Zollinger, J., and Hecht, U. (2009) *J. Phys. Condens. Matter*, **21**, 464104 (7pp).
65 Chen, L.Q. (2002) *Ann. Rev. Mater. Res.*, **32**, 113.
66 Boettinger, W.J., Warren, J.A., Beckermann, C., and Karma, A. (2002) *Ann. Rev. Mater. Res.*, **32**, 163.
67 Larsen, D.E., Christodoulou, L., Kampe, S.L., and Sadler, P. (1991) *Mater. Sci. Eng.*, **A144**, 45.
68 Kampe, L., Sadler, P., Christodoulou, L., and Larsen, D.E. (1994) *Metall. Mater. Trans. A*, **25A**, 2181.
69 Wagner, R., Appel, F., Dogan, B., Ennis, P.J., Lorenz, U., Müllauer, J., Nicolai, H.P., Quadakkers, W., Singheiser, L., Smarsly, W., Vaidya, W., and Wurzwallner, K. (1995) *Gamma Titanium Aluminides* (eds Y.-W. Kim, R. Wagner, and M. Yamaguchi), TMS, Warrendale, PA, p. 387.
70 Eylon, D., Keller, M.M., and Jones, P.E. (1998) *Intermetallics*, **6**, 703.
71 Hu, D. (2001) *Intermetallics*, **9**, 1037.
72 Hu, D. (2002) *Intermetallics*, **10**, 851.
73 Grange, M., Raviart, J.L., and Thomas, M. (2004) *Metall. Mater. Trans. A*, **35A**, 2087.
74 Wu, X., Hu, D., and Loretto, M.H. (2004) *Niobium High Temperature Applications* (eds Y.-W. Kim and T. Carneiro), TMS, Warrendale, PA, p. 183.
75 Rishel, L.L., Biery, N.E., Raban, R., Gandelsman, V.Z., Pollock, T.M., and Cramb, A.W. (1998) *Intermetallics*, **6**, 6129.
76 Raban, R., Rishel, L.L., and Pollock, T.M. (1999) *High-Temperature Ordered Intermetallic Alloys VIII*, vol. 552 (eds E.P. George, M.J. Mills, and M. Yamaguchi), Mater. Res. Soc. Symp. Proc., Materials Research Society, Pittsburgh, PA, p. K.K2.1.1.
77 Quested, T.E. and Greer, A.L. (2005) *Acta Mater.*, **53**, 4643.
78 StJohn, D., Qian, M., Easton, M.A., Cao, P., and Hildebrand, Z. (2005) *Metall. Mater. Trans. A*, **36A**, 1669.
79 Huang, S.C. and Hall, E.L. (1991) *High-Temperature Ordered Intermetallic Alloys IV*, vol. 213 (eds L.A. Johnson, D.P. Pope, and J.O. Stiegler), Mater. Res. Soc. Symp. Proc., Materials Research Society, Pittsburgh, PA, p. 827.

80 Yamaguchi, M. and Inui, H. (1993) *Structural Intermetallics* (eds R. Darolia, J.J. Lewandowski, C.T. Liu, P.L. Martin, D.B. Miracle, and M.V. Nathal), TMS, Warrendale, PA, p. 127.
81 Huang, S.C. and Chesnutt, J.C. (1994) *Intermetallics Compounds, Vol. 2, Practise* (eds J.H. Westbrook and L. Fleischer), Chapter 4 (John Wiley & Sons, Ltd, Chichester, UK, p. 617.
82 Inkson, B.J., Boothroyd, C.B., and Humphreys, C.J. (1995) *Acta Metall. Mater.*, **43**, 1429.
83 Godfrey, A.B. and Loretto, M.H. (1996) *Intermetallics*, **4**, 47–53.
84 Cheng, T.T. (1999) *Gamma Titanium Aluminides 1999* (eds Y.-W. Kim, D.M. Dimiduk, and M.H. Loretto), TMS, Warrendale, PA, p. 389.
85 Cheng, T.T. (2000) *Intermetallics*, **8**, 29.
86 Kitkamthorn, U., Zhang, L.C., and Aindow, M. (2006) *Intermetallics*, **14**, 759.
87 Gosslar, D., Hartig, C., Günther, R., Hecht, U., and Bormann, R. (2009) *J. Phys. Condens. Matter*, **21**, 464111 (7pp).
88 Oehring, M., Appel, F., Paul, J.D.H., Imayev, R.M., Imayev, V.M., and Lorenz, U. (2010) *Mater. Sci. Forum*, **638–642**, 1394.
89 Hyman, M.E., McCullough, C., Levi, C.G., and Mehrabian, R. (1991) *Metall. Trans. A*, **22A**, 1647.
90 Witusiewicz, V.T., Bondar, A.A., Hecht, U., Zollinger, J., Artyukh, L.V., Rex, S., and Velikanova, T.Ya. (2009) *J. Alloys Compd.*, **474**, 86.
91 De Graef, M., Löfvander, J.P.A., McCullough, C., and Levi, C.G. (1992) *Acta Metall. Mater.*, **40**, 3395.
92 Plaskett, T.S. and Winegard, W.C. (1959) *Trans. ASM*, **51**, 222.
93 Quested, T.E., Dinsdale, A.T., and Greer, A.L. (2005) *Acta Mater.*, **53**, 1323.
94 Maxwell, I. and Hellawell, A. (1975) *Acta Metall.*, **23**, 229.
95 Easton, M.A. and StJohn, D.H. (2001) *Acta Mater.*, **49**, 1867.
96 Hill, D., Banerjee, R., Huber, D., Tiley, J., and Fraser, H.L. (2005) *Scr. Mater.*, **52**, 387.
97 Kim, Y.-W. and Dimiduk, D.M. (2001) *Structural Intermetallics 2001* (eds K.J. Hemker, D.M. Dimiduk, H. Clemens, R. Darolia, H. Inui, J.M. Larsen, V.K. Sikka, M. Thomas, and J.D. Whittenberger), TMS, Warrendale, PA, p. 625.
98 Yamabe, Y., Takeyama, M., and Kikuchi, M. (1995) *Gamma Titanium Aluminides* (eds Y.-W. Kim, R. Wagner, and M. Yamaguchi), TMS, Warrendale, PA, p. 111.
99 Sun, Y.Q. (1998) *Philos. Mag. Lett.*, **78**, 305.
100 Sastry, S.M.L. and Lipsitt, H.A. (1977) *Metall. Trans. A*, **8A**, 299.
101 Wang, P., Viswanathan, G.B., and Vasudevan, V.K. (1992) *Metall. Mater. Trans. A*, **23A**, 690.
102 Jones, S.A. and Kaufman, M.J. (1993) *Acta Metall. Mater.*, **41**, 387.
103 Takeyama, M., Kuamagai, T., Nakamura, M., and Kikuchi, M. (1993) *Structural Intermetallics* (eds R. Darolia, J.J. Lewandowski, C.T. Liu, P.L. Martin, D.B. Miracle, and M.V. Nathal), TMS, Warrendale, PA, p. 167.
104 Veeraraghavan, D., Wang, P., and Vasudevan, V.K. (1999) *Acta Mater.*, **47**, 3313.
105 Ramanujan, R.V. (2000) *Intern. Mater. Rev.*, **45**, 217.
106 Hu, D., Huang, A.J., and Wu, X. (2005) *Intermetallics*, **13**, 211.
107 McQuay, P.A., Dimiduk, D.M., Lipsitt, H., and Semiatin, S.L. (1993) *Titanium '92, Science and Technology* (eds F.H. Froes and I.L. Caplan), TMS, Warrendale, PA, p. 1041.
108 Veeraraghavan, D. and Vasudevan, V.K. (1995) *Mater. Sci. Eng.*, **A192–193**, 950.
109 Dimiduk, D.M. and Vasudevan, V.K. (1999) *Gamma Titanium Aluminides 1999* (eds Y.-W. Kim, D.M. Dimiduk, and M.H. Loretto), TMS, Warrendale, PA, p. 239.
110 Dey, S.R., Hazotte, A., Bouzy, E., and Naka, S. (2005) *Acta Mater.*, **53**, 3783.
111 Veeraraghavan, D. and Vasudevan, V.K. (1995) *Gamma Titanium Aluminides* (eds Y.-W. Kim, R. Wagner, and M. Yamaguchi), TMS, Warrendale, PA, p. 157.
112 Hu, D. and Botten, R.R. (2002) *Intermetallics*, **10**, 701.

113 Mitao, S. and Bendersky, L.A. (1997) *Acta Mater.*, **45**, 4475.
114 Oehring, M., Lorenz, U., Appel, F., and Roth-Fagaraseanu, D. (2001) *Structural Intermetallics 2001* (eds K.J. Hemker, D.M. Dimiduk, H. Clemens, R. Darolia, H. Inui, J.M. Larsen, V.K. Sikka, M. Thomas, and J.D. Whittenberger), TMS, Warrendale, PA, p. 157.
115 McCullough, C., Valencia, J.J., Levi, C.G., and Mehrabian, R. (1990) *Mater. Sci. Eng.*, **A124**, 83.
116 Nishida, M., Tateyama, T., Tomoshige, R., Morita, K., and Chiba, A. (1992) *Scr. Metall. Mater.*, **27**, 335.
117 Shong, D.S. (1989) *Scr. Metall.*, **23**, 1181.
118 Yamaguchi, M. and Umakoshi, Y. (1990) *Prog. Mater. Sci.*, **34**, 1.
119 Kim, Y.-W. (1992) *Acta Metall. Mater.*, **40**, 1121.
120 Denquin, A., Naka, S., and Khan, T. (1993) *Titanium '92, Science and Technology* (eds F.H. Froes and I.L. Caplan), TMS, Warrendale, PA, p. 1017.
121 Yamabe, Y., Takeyama, M., and Kikuchi, M. (1994) *Scr. Metall. Mater.*, **30**, 533.
122 Takeyama, M. and Kikuchi, M. (1998) *Intermetallics*, **6**, 573.
123 Doherty, R.D. (1983) Diffusive Phase Transformations in the Solid State, in *Physical Metallurgy*, vol. 2 (eds R.W. Cahn and P. Haasen), North-Holland Physics Publishing, Amsterdam, The Netherlands, p. 933.
124 Zhao, L. and Tangri, K. (1991) *Acta Metall. Mater.*, **39**, 2209.
125 Zhao, L. and Tangri, K. (1991) *Philos. Mag. A*, **64**, 361.
126 Inui, H., Nakamura, A., Oh, M.H., and Yamaguchi, M. (1991) *Ultramicroscopy*, **39**, 268.
127 Singh, S.R. and Howe, J.M. (1992) *Philos. Mag. A*, **66**, 739.
128 Appel, F., Beaven, P.A., and Wagner, R. (1993) *Acta Metall. Mater.*, **41**, 1721.
129 Yang, Y.S., Wu, S.K., and Wang, J.Y. (1993) *Philos. Mag. A*, **67**, 463.
130 Rao, S., Woodward, C., and Hazzledine, P. (1994) *Defect Interface Interactions*, vol. 319 (eds E.P. Kvam, A.H. King, M.J. Mills, T.D. Sands, and V. Vitek), Mater. Res. Soc. Symp. Proc., Materials Research Society, Pittsburgh, PA, p. 285.
131 Zhang, L.C., Chen, G.L., Wang, J.G., and Ye, H.Q. (1997) *Intermetallics*, **5**, 289.
132 Pond, R.C., Shang, P., Cheng, T.T., and Aindow, M. (2000) *Acta Mater.*, **48**, 1047.
133 Xu, Q., Lei, C.H., and Zhang, Y.G. (1995) *Gamma Titanium Aluminides* (eds Y.-W. Kim, R. Wagner, and M. Yamaguchi), TMS, Warrendale, PA, p. 189.
134 Denquin, A. and Naka, S. (1995) *Gamma Titanium Aluminides* (eds Y.-W. Kim, R. Wagner, and M. Yamaguchi), TMS, Warrendale, PA, p. 141.
135 Denquin, A. and Naka, S. (1996) *Acta Mater.*, **44**, 343.
136 Zghal, S., Naka, S., and Couret, A. (1997) *Acta Mater.*, **45**, 3005.
137 Schwartz, D.S. and Sastry, S.M.L. (1989) *Scr. Metall.*, **23**, 1621.
138 Yang, Y.S. and Wu, S.K. (1991) *Scr. Metall. Mater.*, **25**, 255.
139 Inui, H., Oh, M.H., Nakamura, A., and Yamaguchi, M. (1992) *Philos. Mag. A*, **66**, 539.
140 Inui, H., Oh, M.H., Nakamura, A., and Yamaguchi, M. (1992) *Philos. Mag. A*, **66**, 557.
141 Zghal, S., Thomas, M., and Couret, A. (2005) *Intermetallics*, **13**, 1008.
142 Zghal, S., Thomas, M., Naka, S., Finel, A., and Couret, A. (2005) *Acta Mater.*, **53**, 2653.
143 Feng, C.R., Michel, D.J., and Crowe, C.R. (1989) *Scr. Metall.*, **23**, 1135.
144 Yamaguchi, M., Inui, H., and Ito, K. (2000) *Acta Mater.*, **48**, 307.
145 Zhang, L.C., Cheng, T.T., and Aindow, M. (2004) *Acta Mater.*, **52**, 191.
146 Shao, G., Grosdidier, T., and Tsakiropoulos, P. (1994) *Scr. Metall. Mater.*, **30**, 809.
147 Charpentier, M., Hazotte, A., and Daloz, D. (2008) *Mater. Sci. Eng. A*, **491**, 321.
148 Sun, Y.Q. (1998) *Metall. Mater. Trans. A*, **29A**, 2679.
149 Menon, E.S.K. and Aaronson, H.I. (1987) *Acta Metall.*, **35**, 549.

150 Aaronson, H.I., Spanos, G., Masamura, R.A., Vardiman, R.G., Moon, D.W., Menon, E.S.K., and Hall, M.G. (1995) *Mater. Sci. Eng. B*, **32**, 107.
151 Lefebvre, W., Loiseau, A., Thomas, M., and Menand, A. (2002) *Philos. Mag. A*, **82**, 2341.
152 Mahon, G.J. and Howe, J.M. (1990) *Metall. Trans. A*, **21A**, 1655.
153 Aaronson, H.I. (1962) *The Decomposition of Austenite by Diffusional Processes*, Interscience, New York, NY.
154 Aaronson, H.I. (1993) *Metall. Trans. A*, **24A**, 241.
155 Shang, P., Cheng, T.T., and Aindow, M. (1999) *Philos. Mag. A*, **79**, 2553.
156 Pond, R.C. (1989) *Dislocations in Solids*, vol. 8 (ed. F.R.N. Nabarro), North-Holland, Amsterdam, The Netherlands, p. 1.
157 Howe, J.M., Pond, R.C., and Hirth, J.P. (2009) *Prog. Mater. Sci.*, **54**, 792.
158 Valencia, J.J., McCullough, C., Levi, C.G., and Mehrabian, R. (1987) *Scr. Metall.*, **21**, 1341.
159 McCullough, C., Valencia, J.J., Mateos, H., Levi, C.G., Mehrabian, R., and Rhyne, K.A. (1988) *Scr. Metall.*, **22**, 1131.
160 Wen, Y.H., Chen, L.Q., Hazzledine, P.M., and Wang, Y. (2001) *Acta Mater.*, **49**, 2341.
161 Katzarov, I., Malinov, S., and Sha, W. (2006) *Acta Mater.*, **54**, 453.
162 Dimiduk, D.M., Hazzledine, P.M., Parthasararty, T.A., Seshagiri, S., and Mendiratta, M.G. (1998) *Metall. Mater. Trans. A*, **29A**, 37.
163 Parthasaraty, T.A., Mendiratta, M.G., and Dimiduk, D.M. (1998) *Acta Mater.*, **46**, 4005.
164 Maruyama, K., Suzuki, G., Kim, H.Y., Suzuki, M., and Sato, H. (2002) *Mater. Sci. Eng.*, **A329–331**, 190.
165 Maruyama, K., Yamaguchi, M., Suzuki, G., Zhu, H., Kim, H.Y., and Yoo, M.H. (2004) *Acta Mater.*, **52**, 5185.
166 Ramanath, G. and Vasudevan, V.K. (1993) *High-Temperature Ordered Intermetallic Alloys V*, vol. 288 (eds I. Baker, R. Darolia, J.D. Whittenberger, and M.H. Yoo), Mater. Res. Soc. Symp. Proc., Materials Research Society, Pittsburgh, PA, p. 223.
167 Pouly, P., Hua, M.J., Garcia, C., and Deardo, A.J. (1993) *Scr. Metall. Mater.*, **29**, 1529.
168 Dimiduk, D.M., Martin, P., and Kim, Y.-W. (1997) *Mater. Sci. Eng.*, **A226**, 127.
169 Rostamian, A. and Jacot, A. (2008) *Intermetallics*, **16**, 1227.
170 Kim, Y.-W. and Dimiduk, D.M. (1997) *Structural Intermetallics 1997* (eds M.V. Nathal, R. Darolia, C.T. Liu, P.L. Martin, D.B. Miracle, R. Wagner, and M. Yamaguchi), TMS, Warrendale, PA, p. 531.
171 Perdrix, F., Trichet, M.F., Bonnetien, J.L., Cornet, M., and Bigot, J. (1999) *Intermetallics*, **7**, 1323.
172 Farkas, L. (1927) *Z. Phys. Chem.*, **125**, 239.
173 Becker, R. and Döring, W. (1935) *Ann. Phys.*, **24**, 719.
174 Zeldovich, J.B. (1943) *Acta Physicochim. URSS*, **18**, 1.
175 Becker, R. (1938) *Ann. Phys.*, **32**, 128.
176 Turnbull, D. and Fisher, J.C. (1949) *J. Chem. Phys.*, **17**, 71.
177 Russell, K.C. (1970) *Phase Transformations* (ed. H.I. Aaronson), ASM, Metals Park, OH, p. 219.
178 Zener, C. (1946) *Trans. AIME*, **167**, 550.
179 Hillert, M. (1961) *Jernkontorets Ann.*, **141**, 757.
180 Zener, C. (1949) *J. Appl. Phys.*, **20**, 950.
181 Wang, P. and Vasudevan, V.K. (1992) *Scr. Metall. Mater.*, **27**, 89.
182 Dey, S.R., Bouzy, E., and Hazotte, A. (2008) *Acta Mater.*, **56**, 2051.
183 Ramanujan, R.V. (1995) *Acta Metall. Mater.*, **43**, 4439.
184 Zhang, X.D., Godfrey, S., Weaver, M., Strangwood, M., Threadgill, P., Kaufman, M.J., and Loretto, M.H. (1996) *Acta Mater.*, **44**, 3723.
185 Veeraraghavan, D., Wang, P., and Vasudevan, V.K. (2003) *Acta Mater.*, **51**, 1721.
186 Dey, S.R., Bouzy, E., and Hazotte, A. (2006) *Intermetallics*, **14**, 444.
187 Hu, D., Huang, A.J., and Wu, X. (2007) *Intermetallics*, **15**, 327.
188 Hu, D., Huang, A., Loretto, M.H., and Wu, X. (2007) *Ti-2007 Science and*

Technology, Proc. of the 11th World Conference on Titanium (eds M. Nimomi, S. Akiyama, M. Ikeda, M. Hagiwara, and K. Maruyama), The Japan Institute of Metals, Tokyo, Japan, p. 1317.
189 Huang, A., Hu, D., Wu, X., and Loretto, M.H. (2007) *Intermetallics*, **15**, 1147.
190 Dey, S.R., Bouzy, E., and Hazotte, A. (2007) *Scr. Mater.*, **57**, 365.
191 Sankaran, A., Bouzy, E., Humbert, M., and Hazotte, A. (2009) *Acta Mater.*, **57**, 1230.
192 Jiang, H., Zhang, K., Hao, X.J., Saage, H., Wain, N., Hu, D., Loretto, M.H., and Wu, X. (2010) *Intermetallics*, **18**, 938.

5
Deformation Behavior of Single-Phase Alloys

Titanium aluminide alloys are relatively brittle materials, exhibiting little plasticity at ambient temperatures. The deformation characteristics involve strong plastic anisotropy, anomalous increase of yield stress with temperature, breakdown of Schmid's law, and lack of independent slip systems. Such behavior is often governed by the fine structure of the dislocation cores and the kinematics of the dislocation propagation. In the present chapter these factors will be considered for the most important constituents of engineering alloys, namely γ(TiAl), $α_2$(Ti$_3$Al) and β/B2.

5.1
Single-Phase γ(TiAl) Alloys

5.1.1
Slip Systems and Deformation Kinematics

The glide elements of γ(TiAl) are determined by its tetragonal L1$_0$ structure [1, 2], which is related to the face centered cubic (f.c.c.) lattice. Since the c/a axial ratio of the unit cell of γ(TiAl) is close to unity, it is customary to describe dislocations in terms of a modified cubic notation, which was introduced by Hug *et al.* [3, 4]. In this notation Miller indices with mixed parentheses <*uvw*] and {*hkl*) are used in order to differentiate the first two (equivalent) indices from the third, meaning all permutations of the first two indices are allowed, whereas the third one is fixed.

According to classical arguments of crystal plasticity, the most favorable slip systems should involve close-packed lattice planes and Burgers vectors that correspond to the shortest translation vector in these planes. Dislocation glide in γ(TiAl) preferentially occurs on {111} planes along close- or relatively close-packed directions and is in this respect similar to that in face-centered cubic (f.c.c.) metals. However, the reduced symmetry of the L1$_0$ structure, together with the fact that the dislocation Burgers vector must be a lattice translation vector, imply several restrictions on dislocation mechanisms. Figure 5.1 shows the orientation of the possible Burgers vectors in a hard-sphere model of the L1$_0$ structure. The diameters of the spheres representing the atoms should merely express their nearest

Gamma Titanium Aluminide Alloys: Science and Technology, First Edition. Fritz Appel, Jonathan David Heaton Paul, Michael Oehring.
© 2011 Wiley-VCH Verlag GmbH & Co. KGaA. Published 2011 by Wiley-VCH Verlag GmbH & Co. KGaA.

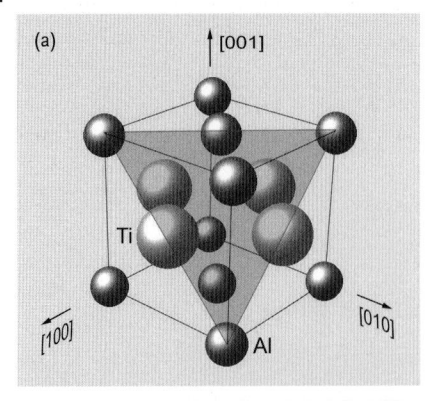

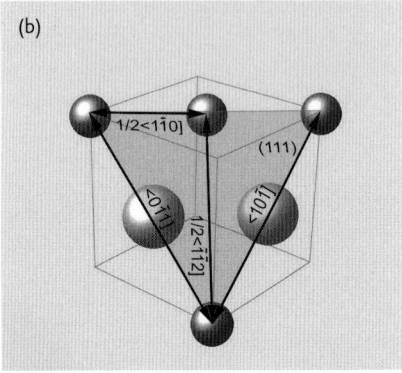

Figure 5.1 Potential slip systems in γ(TiAl). (a) Atomic arrangement of the L1₀ structure of γ(TiAl) with one of the four octahedral {111} planes. (b) Illustration of the Burgers vectors $\mathbf{b}_{<110]} = 1/2<\bar{1}10]$, $\mathbf{b}_{<112]} = 1/2<\bar{1}\bar{1}2]$, $\mathbf{b}_{<011]} = <0\bar{1}1]$, and $\mathbf{b}_{<101]} = <10\bar{1}]$ situated on the (111) plane. The drawing corresponds to the lattice constants $a = 0.3999$ nm and $c = 0.4077$ nm.

distance of approach, that is, an atom should be considered as a rigid sphere that has a radius determined by the condition that the sum of the radii of two bonded atoms corresponds to the length of the chemical bond between the same two atoms. It should be noted that in the models shown here the difference in the radii of Ti and Al atoms is exacerbated in order to visualize three-dimensional stacking sequences.

Because of the long-range order, the close-packed <110> directions are crystallographically not equivalent. The shortest lattice translation vector in these directions is $\mathbf{b}_{<110]} = 1/2<1\bar{1}0]$, which is situated on the low-index {111}, {110} and (001) planes. This translation corresponds to the Burgers vector of dislocations in disordered f.c.c. crystals. Thus, in γ(TiAl) dislocations with the Burgers vector $\mathbf{b}_{<110]} = 1/2<1\bar{1}0]$ are commonly referred to as ordinary or unit dislocations. The displacement $1/2<0\bar{1}1]$ involving a **c** component is not a lattice translation vector in the L1₀ structure. A moving dislocation having this Burgers vector leaves behind a surface of disorder, called an antiphase boundary (APB). The order may be restored by a second displacement of the same magnitude. Thus, the perfect dislocation involving shear components in the **c** direction of the $L1_0$ structure has the Burgers vector $\mathbf{b}_{<011]} = <0\bar{1}1]$ and consists of two identical $1/2<0\bar{1}1]$ dislocations joined by an APB. The constituent $1/2<0\bar{1}1]$ partial dislocations, which are perfect dislocations in the f.c.c. lattice, are commonly called superpartial dislocations, or just superpartials. The superdislocation is in a sense similar to two Shockley partials joined by a stacking fault in the f.c.c. lattice; the spacing in equilibrium is given by a balance between the elastic repulsive force between the two dislocations and the opposing force due to the APB energy. The energetics of the APB and other planar faults that exist in the $L1_0$ structure will be considered in the following section.

The other possible superdislocation in the $L1_0$ structure has the Burgers vector $\mathbf{b}_{<11\bar{2}]} = 1/2<11\bar{2}]$ and consists of the superpartials $1/2<10\bar{1}]$ and $1/2<01\bar{1}]$, joined by an APB. Thus, in γ(TiAl) glide on {111} planes can be accomplished by ordinary dislocations with the Burgers vector $\mathbf{b}_{<110]} = 1/2<1\bar{1}0]$ and superdislocations with the Burgers vectors $\mathbf{b}_{<011]} = <0\bar{1}1]$ and $\mathbf{b}_{<11\bar{2}]} = 1/2<11\bar{2}]$ [1–8]. The two other $1/2<1\bar{2}1]$ translations have nearly the same magnitude as that of $\mathbf{b}_{<11\bar{2}]} = 1/2<11\bar{2}]$, but lead to the formation of an APB.

Screw dislocations with their Burgers vector parallel to the dislocation line are in principle capable of gliding in any plane that passes through the dislocation line. This change of the slip plane is called cross-slip, a phenomenon that can occur when there are two or more slip planes with a common slip direction. Cross-slip plays an important role in the deformation, work hardening and recovery of most materials and does so in TiAl. The potential cross-slip planes, are the close-packed or relatively close-packed planes that intersect the primary (111) plane so that the line of intersection is parallel to the Burgers vector. Thus, the ordinary dislocation $1/2[1\bar{1}0]$ is capable of cross gliding onto $(\bar{1}\bar{1}1)$, (110) and (001) planes, which is illustrated in Figure 5.2a. Likewise, the $[0\bar{1}1]$ superdislocation may cross-slip from the primary (111) glide plane onto $(1\bar{1}\bar{1})$, (100) and (011) planes, respectively (Figure 5.2b). Contrary to the $1/2<1\bar{1}0]$ and $<0\bar{1}1]$ dislocations, there is only one octahedral plane available for glide of a $<11\bar{2}]$ dislocation; a possible low-index cross-slip plane of this dislocation is $\{1\bar{1}0)$.

The energetic driving force for cross-slip can be assessed by comparing the line tension of the screw dislocation on the respective planes. The line tension may be defined as the increase in energy per unit increase in dislocation length and is thus a measure of the resistance to dislocation bowing [9]. In an elastically anisotropic material, the line tension of a screw dislocation is expected to depend on the plane on which the dislocation chooses to glide. A high positive line tension on the cross-slip plane may counteract or even counterbalance any external forces that drive cross-slip. If the line tension on the cross-slip plane is negative, the dislocation would be elastically unstable on the cross-slip plane, that is, cross-slip would be spontaneous, requiring no external driving force. Sun [10] and Jiao *et al.* [11] have considered the line tension of $1/2<110]$ screw dislocations in γ(TiAl) and demonstrated the elastic stability of these dislocations on the {110}, {111} and (001) cross-slip planes. The line tension was found to be lowest for the {110) planes, intermediate for the {111} planes and highest for the (001) plane; hence,

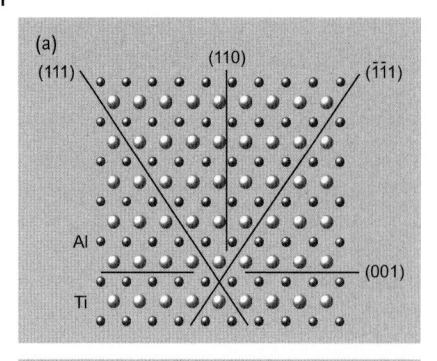

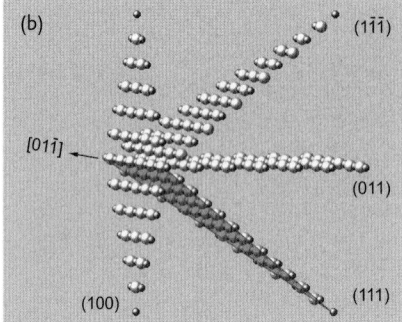

Figure 5.2 Potential cross-slip systems in γ(TiAl) for screw dislocations gliding on the primary (111) plane. (a) An ordinary screw dislocation with the Burgers vector $\mathbf{b}_{<110]} = 1/2<1\bar{1}0]$ may cross-slip onto the planes (110), ($1\bar{1}\bar{1}$) and (001), which are indicated by their traces. The γ(TiAl) lattice is shown in an $<1\bar{1}0]$ projection, that is, the Burgers vector $\mathbf{b}_{<110]} = 1/2<1\bar{1}0]$ is perpendicular to the plane of the drawing. (b) Possible cross-slip planes of the $<0\bar{1}1]$ superdislocation.

cross-slip should occur on {110} planes. This finding is in agreement with the TEM observations of Feng and Whang [12], who showed that ordinary dislocations preferentially cross-slip onto {110} planes and only rarely onto other octahedral planes or (001) planes. For $<0\bar{1}1]$ superdislocations, the choice of the cross-slip plane, depends on temperature [3, 13]. The cross-slip plane of these dislocations is {111} at intermediate temperatures and {010} at high temperatures. As will be discussed later, this behavior can be attributed to the anisotropy of the APB energy and to the elastic interaction forces that exist in an elastically anisotropic material between the leading and trailing superpartial [14].

In principle, shear may also occur along <100>, which is also a translation vector in the $L1_0$ structure, but not situated on the close-packed octahedral planes. A relatively high density of dislocations with <100] Burgers vectors was recognized by Whang and Hahn [15] in a polycrystalline Ti-55Al-5Nb alloy after compression at 1000 °C. Dislocations with [001] Burgers vector were found in Ti-54.5Al single crystals that were oriented for [001](110) glide and deformed at 950 °C [16]. Experimental evidence for slip of ordinary dislocations on (001) and {110} planes was also obtained on Ti-56Al single crystals deformed at elevated temperatures [17]. However, it should be noted that large-scale glide, even at high temperatures, only rarely occurs on slip planes other than {111}.

In summary, the perfect dislocations in γ(TiAl) that maintain the L1$_0$ structure have the Burgers vectors $\mathbf{b}_{<110]} = 1/2<110]$, $\mathbf{b}_{<011]} = <011]$ and $\mathbf{b}_{<112]} = 1/2<11\bar{2}]$. These dislocations preferentially glide on {111} planes and have the potential to cross-slip onto other low-index planes.

5.1.2
Planar Faults

There is good evidence that the mobility of the dislocations in γ(TiAl) is governed by their core structure. In intermetallic alloys with large translation vectors it is equally important that the Burgers vector **b** can dissociate into smaller components, which reduces the glide resistance of the lattice. This can be rationalized in terms of the classical Peierls concept [18]. Within this model the critical shear stress τ_p (Peierls stress) to propagate a dislocation is given by

$$\tau_p = \frac{2\mu}{1-\nu}\exp\left(-\frac{2\pi w}{b}\right) \qquad (5.1)$$

with $w = a(1-\nu)$ for edge dislocations and $w = a$ for screw dislocations. μ is the shear modulus, ν is Poisson's ratio, w the width of the dislocation core, and a the lattice constant. The two opposing factors affecting w are (i) the elastic energy of the crystal, which is reduced by spreading out the elastic strains, and, (ii) the misfit energy, which increases with the number of misaligned atomic bonds across the slip plane. Dislocations with a wide (and planar) core are expected to have a low friction stress τ_p. Moreover, the model predicts a low friction stress for dislocations gliding on close-packed, that is, widely spaced slip planes, which in terms of Equation (5.1) are characterized by a small b/a factor. Metals that exhibit these two attributes have highly mobile dislocations and are intrinsically soft. In γ(TiAl) the dissociated dislocations involve planar defects that are unique to the L1$_0$ structure and are not present in disordered f.c.c. metals. Thus, these defects will be considered in more detail.

In Figure 5.3 the L1$_0$ structure of γ(TiAl) is viewed as stacking of three close-packed (111) planes in the stacking sequence of **ABC**. It should be noted that the use of a hard-sphere crystal model to draw conclusions about the structure and motion of dislocations is subject to valid criticism. Strictly speaking, a hard-sphere model does not allow a dislocation because it does not allow elastic distortions. In the present considerations the hard-sphere model is used to investigate the changes in atom positions that must be accomplished by a gliding dislocation and also to look at the kinds of faulted structures that develop when dislocations dissociate. As is to be expected, details of the dislocation structure and the mechanisms of glide so derived are to some extent arbitrary. In the framework of this geometrical model dislocation glide may be considered as sliding these planes over each other along the Burgers vectors $\mathbf{b}_{<110]} = 1/2<1\bar{1}0]$, $\mathbf{b}_{<112]} = 1/2<11\bar{2}]$ and

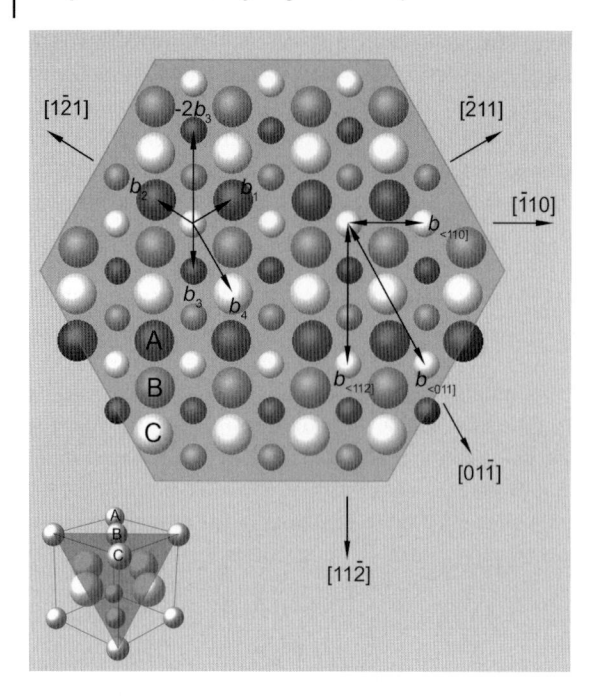

Figure 5.3 Dislocation glide on the (111) plane, illustrated by a hard-sphere model according to Figure 5.1. The figure shows the three-layer sequence **ABC** of atom stacking on the (111) plane. The slip plane between the layers **B** and **C** is imaged opaque so that the atoms below the slip plane appear darker. The shear vectors b_1, b_2, b_3, and b_4 represent partial dislocations, which in the L1$_0$ structure are associated with distinct planar faults; see the accompanying text.

$b_{<01\bar{1}]} = <0\bar{1}1]$, respectively. These shear displacements can be split up into the smaller translations b_1, b_2 and b_3, which leads to a zig-zag motion of the top plane **C** along the atom "valleys" in the lower plane **B** and probably makes slip easier. The shear vectors b_1, b_2 and b_3 represent partial dislocations, which in the L1$_0$ structure are associated with distinct planar faults [19]. Generally, these faults may be considered as interfaces where the regular stacking sequence is broken. Thus, the planar faults can be conservatively formed when one half of the crystal is sheared over the other half by a displacement vector in the slip plane. Since the atoms on either side of a planar fault are not at the positions they would normally occupy in a perfect lattice; a fault possesses a surface energy. This energy is defined as the energy difference between the ground-state energy of a system with a planar fault and that of a perfect system. A fault with a distinct displacement vector will only be stable if the energy of the faulted crystal is lower than that of the same crystal with a slightly smaller or larger displacement vector. Utilizing this concept, an energy-displacement surface can be established, commonly called a γ-surface. The local minima of this surface define the displacement vectors of stable faults. The concept of the γ-surface was introduced by Vitek [20] for the investigation of the dislocation core spreading in b.c.c. metals and subsequently applied to intermetallic alloys; for a review see [19]. It is understood that what is called a "stable fault" is, in a strict sense, a metastable fault, because a defect-free crystal has the lowest energy.

To a first approximation, the stability of planar faults can be assessed by symmetry consideration, utilizing the hard-sphere model of the structure. A complex stacking fault (CSF) is produced, when the top layer of the {111} planes and all planes above it are shifted along $\mathbf{b}_1 = 1/6<\bar{2}11]$ or $\mathbf{b}_2 = <1\bar{2}1]$, as shown in Figure 5.4a. This displacement results in a local hexagonal stacking sequence in that the Al atoms in the top layer C lie directly above the Ti atoms in the bottom layer A. The CSF can also be produced by the displacements $1/6<14\bar{5}]$ or $1/6<4\bar{5}1]$. The CSF destroys the chemical environment of the first neighbors for the atoms in the fault plane and is thus expected to have a relatively high energy.

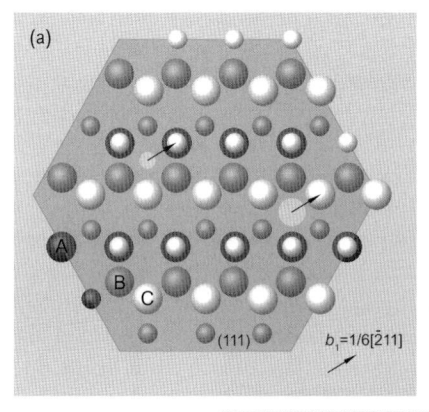

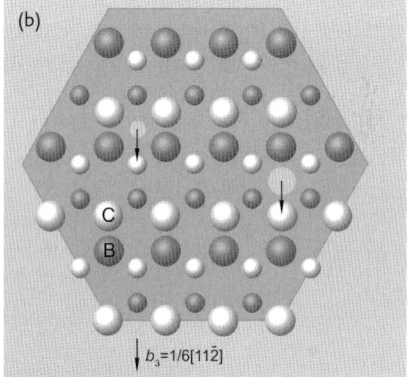

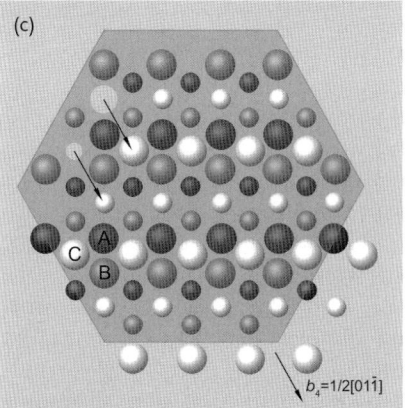

Figure 5.4 Planar faults in the L1$_0$ structure of γ(TiAl) associated with the shear vectors $\mathbf{b}_1$, $\mathbf{b}_2$, and $\mathbf{b}_4$ that were specified in Figure 5.3. The planar faults are generated by displacing the C layer along the vectors indicated. (a) Complex stacking fault (CSF), (b) superlattice intrinsic stacking fault (SISF) and (c) antiphase boundary (APB).

A displacement along $\mathbf{b}_3 = 1/6\langle11\bar{2}]$ generates a superlattice intrinsic stacking fault (SISF) with an **ABCBCA** stacking of {111} planes, which does not change the nearest-neighbor coordination (Figure 5.4b). A repetition of this shear on consecutive (111) planes produces a true twin [21–23], a mechanism that will be considered in the next section of this chapter. The SISF can also be produced by displacements along $-2\mathbf{b}_1$, $-2\mathbf{b}_2$, $-2\mathbf{b}_3$, and $1/6\langle1\bar{5}4]$, respectively. The displacement along $-\mathbf{b}_3$ generates an extrinsic stacking fault (SESF), which can formally be described by an **ABBCAB** stacking. As with the SISF, the SESF does not change the nearest-neighbor coordination. Electron microscope observations [24] and atomistic modeling [25] have shown that the SESF is formed by two displacements of adjacent {111} planes, which results in an **ABACAB** stacking and avoids the energetically unfavorable **BB** packing. Thus, the SESF may formally be described as a two-layer twin [26, 27]. This relaxation of a high-energy stacking fault was proposed by Hirth and Lothe [9] for f.c.c. crystals.

As shown in Figure 5.3, the displacement $\mathbf{b}_4 = 1/2\mathbf{b}_{\langle011]}$ generates an antiphase boundary (APB), the atomic details of which are illustrated in Figure 5.4c. The APB can also be produced by the displacements $3\mathbf{b}_1$ or $3\mathbf{b}_2$, that is, by shear along $1/2\langle1\bar{2}1]$. The APB destroys the nearest-neighbor coordination of the $L1_0$ superlattice and affects the directional bonding of the Ti and Al atoms, which implies a high formation energy. While the SISF and CSF are specific to the {111} planes, an APB can be introduced on any crystallographic plane [28]. In γ(TiAl) the most important APB planes are {111} and {010} [19].

The planar faults play an important role in the plasticity of γ(TiAl) because they control the glide and climb behavior of the dislocations. The faults themselves act as strong barriers to dislocation propagation. More importantly, the faults are involved in various dissociation reactions of dislocations, which often lead to dislocation locking. Given this importance, much effort has been spent in estimating of the planar fault energies Γ by simulation analysis [8, 25, 29–38] and weak-beam transmission electron microscope (TEM) examination of dissociated dislocations [3, 5, 7, 39–42]. For example, the APB energy has been experimentally determined by measuring the separation between APB-coupled superpartials. The advantage of this method is that the widths of the planar faults can be observed over a wide range of dislocation characters. However, analysis of the data is sophisticated because the difference between the image and the actual position of the partial dislocations has to be corrected. The image-shift correction proposed by Cockayne et al. [43] leads to a significant increase of the fault energies, which in the case of faults in Ni_3Al amounts up to 20%. The correction is most effectively achieved by comparing experimental and simulated WB-TEM images. Also, calculation of the fault energy has to be performed in conjunction with anisotropic elasticity theory. At the time of the earlier studies only three of the six elastic constants had been measured. This led Hug et al. [3] to determine the fault energies using elastic constants that represented a cubic structure. Later, the TEM observations were re-evaluated utilizing full anisotropic elastic crystal properties. If no direct experimental determination of the CSF energy is possible, the CSF energy Γ_{CSF} is often approximated as

Table 5.1 Planar fault energies of γ(TiAl) in mJ/m^2.

Composition (at.%)	Γ(APB) {010}	Γ(APB) {111}	Γ(CSF)	Γ(SISF)	Γ(SESF)	Method	Ref.
Ti-50Al	430	510	600	90	80	LDF/FLAPW	[29–31]
Ti-50Al	430	560	410	90	80	LDF/FLAPW[a]	[38]
Ti-50Al	438	667	363	172		LDF/FLAPW	[37]
Ti-50Al	350	670	280	110	110	LDF/LKKR	[32, 33]
Ti-50Al	330	672	294	123		LDF/LKKR	[36]
Ti-50Al	91	550	320	220		EAM	[34]
Ti-50Al	268	322	308	60		EAM	[35]
Ti-50Al	51	275	275	3		F-S	[25]
Ti-54Al		145		77		WB[b], 25 °C	[3, 39]
	100	120		60		WB[b], 600 °C	
Ti-54Al	210	253		143		WB[c], 25 °C	[3, 34, 39]
Ti-54Al	>250			140		WB[d], 25 °C	[3, 41]
Ti-56Al				116(185[d])		WB, 25 °C	[40]
	198(386[d])					WB, 600 °C	
Ti-52Al			470–620			HREM, 25 °C; EAM	[42]

a) Values obtained after atomic relaxations.
b) Analysis of WB-TEM data calculated utilizing cubic elastic constants.
c) Re-evaluation of WB-TEM data utilizing tetragonal elastic constants.
d) Image shift correction.

APB, antiphase boundary; CSF, complex stacking fault; SISF, superlattice intrinsic stacking fault; SESF, superlattice extrinsic stacking fault; LDF, local-density-functional theory; FLAPW, full-potential linearized augmented plane-wave; LKKR, layered Korringa–Kohn–Rostoker coherent potential; EAM, embedded atom method; F–S, Finnis–Sinclair method; WB, weak-beam transmission electron microscopy, performed on predeformed samples at the temperature indicated; HREM, high-resolution electron microscopy.

$$\Gamma(CSF) \approx \Gamma(APB_{(111)}) + \Gamma(SISF) \quad (5.2)$$

following classical geometrical and thermodynamic arguments [44, 45]. The results obtained by these methods are listed in Table 5.1.

The calculated fault energies are generally larger than those estimated using weak-beam WB-TEM. The detailed hierarchy of the fault energies seems not to be completely clear. Based on computational studies [30–33] the fault energies were ranked as

$$\Gamma(APB) > \Gamma(CSF) \gg \Gamma(SISF) \quad (5.3)$$

whereas the most recent analysis of experimental data suggests [41]

$$\Gamma(CSF) > \Gamma(APB) \gg \Gamma(SISF) \quad (5.4)$$

In spite of the differences, collectively the data indicates that the CSF and APB energies are substantially higher than the SISF energy. The atomistic calculations generally place the APB energy for the {010} plane lower than that of the {111} plane. Thus, cross-slip of the antiphase boundaries from {111} planes onto the {010} plane is thermodynamically justified. The ab-initio calculations of Ehmann and Fähnle [37] revealed a mechanical instability of the APB situated on {111} planes with respect to shear along <112> towards the CSF; in other words, there is no energy barrier between the APB and the CSF.

Theoretical studies using the layered Korringa–Kohn–Rostoker (LKKR) method have been performed in order to predict the effects of alloy composition on the fault energies [36]. Accordingly, in binary alloys containing 48 at.% to 51 at.% Al, the APB and SISF energies weakly increase with increasing Al content. For example, in Ti-51Al the APB energy is increased by about 1%, when compared with the value of the stoichiometric composition. At a given Al concentration, Nb additions reduce the APB and SISF energies. Cr additions appear to have little influence on the planar fault energies in Ti-rich alloys, while slightly reducing the energy for the APB on {111} planes in Al-rich alloys. The segregation of impurities to stacking faults in f.c.c. metals is well known as the Suzuki effect [46, 47]. Such segregation may also occur at all types of planar faults in γ(TiAl). Small concentrations of a solute can cause a significant decrease in the energy of an APB or an SISF if the solute atoms are attracted to the plane of the fault because the solute concentration near such planes can greatly exceed that in the bulk. Likewise, antisite atoms, which compensate for deviations from the stoichiometric composition of γ(TiAl), may segregate to the planar faults. As will be shown in Section 6.4.3, there is strong supporting evidence that Ti_{Al} antisite atoms form defect atmospheres around dislocations. The spurious effect of impurities and off-stoichiometric deviations may rationalize the differences that exist between the planar fault energies that has been published in the literature.

Antiphase boundaries can also be formed when a crystal transforms from a higher to a lower symmetry retaining a similar structure with the loss of some translation of symmetry element. The evolution of the microstructure proceeds by the nucleation and growth of ordered domains in a disordered crystal. These "grown-in" APBs are not generated by a shearing process and are thus not terminated by superpartials. Such interfaces may act as glide barriers to dislocations. A detailed account on the planar faults in intermetallic alloys is given in recent reviews by Paidar and Vitek [19] and Yoo and Fu [38].

To summarize, in γ(TiAl) different planar faults occur that are unique to the $L1_0$ structure. These are superlattice intrinsic and extrinsic stacking faults (SISF and SESF), complex stacking fault (CSF) and antiphase boundary (APB). Experimental and theoretical studies all conclude that the APB and CSF energies are substantially higher than the SISF energy. While the SISF, SESF and CSF are specific to {111} planes, an APB can be introduced on any crystallographic plane. The APB energy is higher on {111} planes than on {010} planes, which supports cross-slip of the APB from octahedral to cube planes.

5.1.3
Planar Dislocation Dissociations in γ(TiAl)

The basic reason for the dissociation or spreading of dislocation cores is that the energy of any dislocation is decreased if its core configuration is redistributed into dislocations with smaller Burgers vectors. The resulting partial dislocations repel each other by elastic forces. The formation of the fault between the partials produces an increase in energy, which leads to an attractive force between the partials. The separation width of the two partials thus represents an equilibrium between the repulsive energy of the dislocations and the surface energy of the fault. As with disordered alloys, the tendency that a dislocation will undergo a dissociation reaction can be assessed using Frank's rule, according to which the dislocation line energy is proportional to the square of the Burgers vector [9]. In γ(TiAl) the dislocation dissociation is more complex, depending on which faults are involved and which partial dislocation is leading.

Based on TEM studies and modeling of dislocation cores, various planar and nonplanar dissociation schemes for perfect dislocations have been proposed in the TiAl literature [3–7, 24, 38, 48–53]; for a brief review see [54]. A few possible planar dissociations are shown in Figure 5.5. The ordinary dislocations may reduce their energy by directly splitting into the related Shockley partials, according to the dissociation scheme

$$1/2[\bar{1}10] \rightarrow 1/6[\bar{2}11] + \text{CSF} + 1/6[\bar{1}2\bar{1}] \tag{5.5}$$

The hard-sphere model shown in Figure 5.6 illustrates the atomic configuration of a dissociated edge dislocation involving a CSF. The gray spheres **B** represent

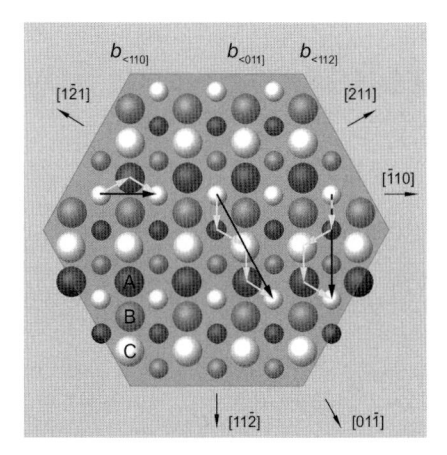

Figure 5.5 Planar dissociation of perfect dislocations in γ(TiAl) according to Equations 5.5, 5.6b, and 5.7b. The figure shows the three-layer sequence **ABC** of atom stacking on the (111) plane, as described in Figure 5.3.

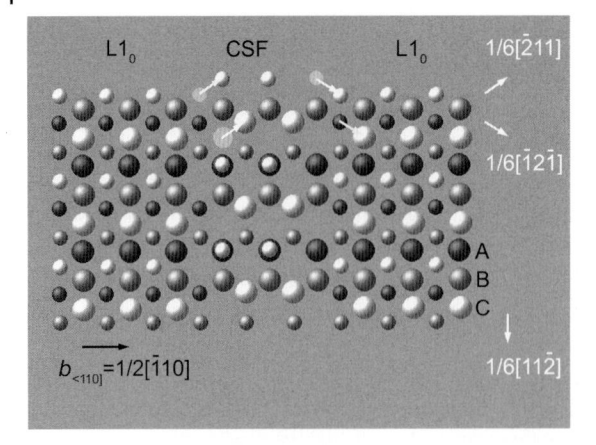

Figure 5.6 Atomic configuration of a dissociated ordinary edge dislocation in γ(TiAl) as viewed when looking down on the slip plane. Three-layer sequence **ABC** of atom stacking on the (111) plane, as described in Figure 5.3. The gray spheres **B** represent the close-packed (111) planes at which the extra plane of the dislocation ends, while the white spheres are the atoms in the next following (111) plane. The light colored **C** layer is displaced along the vectors indicated. The ordinary dislocation with the Burgers vector $b_{<110]} = 1/2<\bar{1}10]$ is dissociated according to Equation (5.5). The extension of the dissociation is exacerbated in order to make the CSF in between the partials visible.

the close-packed (111) planes at which the extra plane of the dislocation ends, while the white spheres are the atoms in the next following (111) plane. TEM weak-beam studies on a deformed Ti-54Al alloy revealed that the $1/2<\bar{1}10]$ ordinary dislocations have a compact core, that is, any possible separation of the $1/6[\bar{2}11]$ and $1/6[\bar{1}2\bar{1}]$ partial dislocation according to Equation (5.5) is below the resolution limit of this method [5]. This finding is in accordance with high-resolution TEM examinations of ordinary dislocations in ternary Ti-48.5Al-5Ga [55] and binary Ti-52Al [42]. In these two studies the experimental images of 60° mixed dislocations were matched against images that were simulated for differently spread core configurations. The evaluation suggests that the core of an ordinary dislocation is not dissociated. Figure 5.7 demonstrates this analysis of dislocation cores. The 1/2<110] dislocation becomes evident by the presence of one extra ($11\bar{1}$) plane. The projected Burgers vector of this dislocation is $b_{pr} = 1/4<12\bar{1}]$, which is consistent with a 60° ordinary dislocation. The ($\bar{1}11$) plane indicated in the micrograph is continuous and is therefore the slip plane of the dislocation. The compact core structure of the ordinary dislocations is probably a consequence of the high CSF energy, which, according to an estimation of Simmons *et al.* [42], is between 470 mJ/m² and 620 mJ/m². This compact core structure has important implications on the cross-glide and climb behavior of ordinary dislocations, which will be considered in a later section.

Possible planar dissociation schemes of the <01$\bar{1}$] superdislocations that appear to be important are listed below:

5.1 Single-Phase γ(TiAl) Alloys

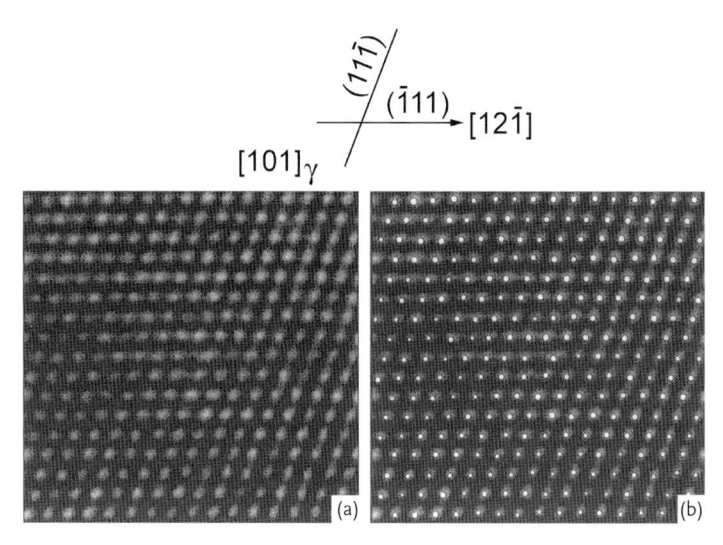

Figure 5.7 Core structure of ordinary dislocations in γ(TiAl). (a) High-resolution image of a 1/2 [110] dislocation taken along the [101] direction. (b) Experimental image with superimposed contrast dots that were simulated on the basis of embedded-atom calculations. Polycrystalline Ti-52 Al (at.%) deformed at room temperature. Micrographs and simulation by Simmons et al. [42].

$$[01\bar{1}] \rightarrow 1/2[01\bar{1}] + APB + 1/2[01\bar{1}] \quad (5.6a)$$

$$[01\bar{1}] \rightarrow 1/6[11\bar{2}] + SISF + 1/6[\bar{1}2\bar{1}] + APB + 1/6[11\bar{2}] + CSF + 1/6[\bar{1}2\bar{1}] \quad (5.6b)$$

$$[01\bar{1}] \rightarrow 1/6[11\bar{2}] + SISF + 1/6[\bar{1}2\bar{1}] + APB + 1/2[01\bar{1}] \quad (5.6c)$$

$$[01\bar{1}] \rightarrow 1/6[11\bar{2}] + SISF + 1/6[\bar{1}54] \quad (5.6d)$$

In a first step towards energy reduction, $[01\bar{1}]$ superdislocation may decompose into APB-coupled superpartials according to Equation (5.6a). The separation distance of the superpartials is given by a balance of the elastic repulsive force between the superpartials and the opposing force due to the APB energy. Evidence for this decomposition reaction is reported in [40, 56].

The fully dissociated state of the $[01\bar{1}]$ dislocation is described by Equation (5.6b) and illustrated in Figure 5.5. The atomic configuration of this dissociation involving the SISF, APB and CSF is shown in Figure 5.8. The equilibrium widths of the SISF, APB and CSF were depicted assuming the hierarchy of the planar fault energies Γ(SISF)<Γ(APB)<Γ(CSF) [41]. Because of the high energy of the APB and CSF and the resulting close spacing of the Shockley partials, it is often not possible to resolve the core spreading in full detail. Several authors [3, 5, 17, 57, 58] reported a three-fold dissociation that is consistent with Equation (5.6c). Figure 5.9 demonstrates this dissociation of a [011] dislocations in a weak-beam transmission electron micrograph [4–6]. In some cases the two $1/6[\bar{1}2\bar{1}]$ and $1/2[01\bar{1}]$ dislocations connected by the APB couldn't be separated by weak-beam TEM. Thus, following

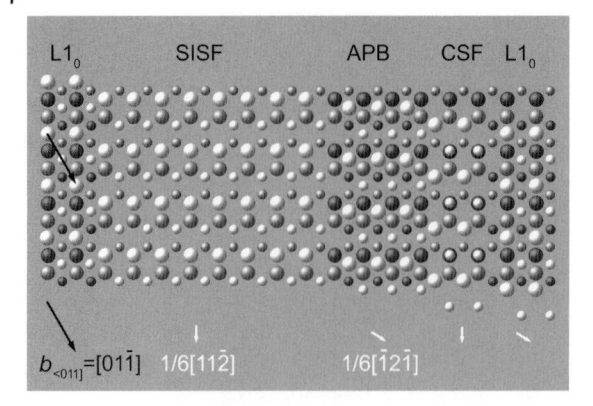

Figure 5.8 Atomic configuration of a superdislocation <0$\bar{1}$1] fully dissociated according to Equation 5.6b. Three-layer sequence **ABC** of atom stacking on the (111) plane, as described in Figure 5.3. The light-colored **C** layer is displaced along the vectors indicated. The widths of the planar faults correspond qualitatively to the hierarchy of fault energies $\Gamma(CSF) > \Gamma(APB) > \Gamma(SISF)$.

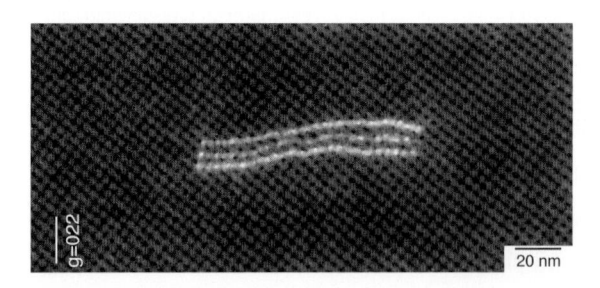

Figure 5.9 Weak-beam transmission electron micrograph of a near-edge [011] superdislocation in a Ti-54Al single-phase alloy that had been deformed at room temperature. The three-fold dissociation is consistent with Equation 5.6c. Micrograph from Hug [4].

a suggestion of Greenberg [49, 50], the core of the [01$\bar{1}$] dislocation may formally be described by Equation (5.6d) as one 1/6[11$\bar{2}$] Shockley partial dislocation connected with the rest of the dislocation 1/6[$\bar{1}$5$\bar{4}$] by an SISF. This two-fold dissociation was observed by Sriram et al. [59] in Ti-50Al deformed at room temperature. The asymmetric core spreading of the <01$\bar{1}$] dislocation (Figure 5.8) has led some investigators to believe that the glide resistance of the dislocation depends on the direction of motion, that is, on whether the SISF or CSF is leading [54, 55].

A few possible planar dissociation reactions of the 1/2[11$\bar{2}$] superdislocation are listed as follows:

$$1/2\,[11\bar{2}] \rightarrow 1/2[10\bar{1}] + APB + 1/2[01\bar{1}] \tag{5.7a}$$

$$1/2[11\bar{2}] \rightarrow 1/6[11\bar{2}] + SISF + 1/6[2\bar{1}\bar{1}] + APB + 1/6[11\bar{2}] + CSF + 1/6[\bar{1}2\bar{1}] \tag{5.7b}$$

$$1/2[11\bar{2}] \rightarrow 1/6[11\bar{2}] + \text{SISF} + 1/6[2\bar{1}\bar{1}] + \text{APB } 1/2[01\bar{1}] \tag{5.7c}$$

$$1/2[11\bar{2}] \rightarrow 1/6[11\bar{2}] + \text{SISF} + 1/3[11\bar{2}] \tag{5.7d}$$

$$1/2[11\bar{2}] \rightarrow 1/6[11\bar{2}] + \text{SISF} + 1/6[11\bar{2}] + \text{SESF} + 1/6[11\bar{2}] \tag{5.7e}$$

The fully dissociated state of the $1/2[11\bar{2}]$ dislocation according to Equation (5.7b) is illustrated in Figure 5.5. Because of the large differences in the planar fault energies, it might be expected that the dissociated core exhibits a similar asymmetry as described for the $<01\bar{1}]$ dislocation, that is, the Shockley partial dislocations separated by the SISF are widely spread in the {111} plane, whereas the rest of the dislocation is relatively compact. This view essentially coincides with weak-beam TEM analysis. Hug et al. [3, 5] observed a three-fold spreading in deformed Ti-54Al, which was discussed in terms of the reactions described by Equations (5.7c, e). More information on this subject is provided in the following section.

A striking feature in the deformation structure of single-phase γ(TiAl) alloys deformed at low and moderate (400°C to 500°C) temperatures are elongated faulted dipolar loops [1, 5]. The formation of the faulted dipoles has been ascribed to localized pinning of dissociated <001] or $1/2<112]$ superdislocations, bypassing of the pinned segments and the drawing out of a dipole. The energy of the dipole is subsequently reduced by the passage of the partial dislocations and the formation of an extrinsic stacking fault (SESF). There are two types of faulted dipoles that differ in the character of the dislocation bordering the SESF; the Burgers vector of this dislocation can be either $\mathbf{b} = 1/6<112]$ or $\mathbf{b} = 1/3<211]$ [5, 60–62]. For more mechanistic details concerning the formation of faulted dipoles see [63, 64].

In summary, dislocation dissociation in γ(TiAl) is very complex because different planar faults can be involved. The predicted core dissociations follow the plausible trend that the highest planar fault energies produce the smallest spreadings. The high APB and CSF energies imply that the dissociation involving these faults are often below the resolution limit of weak-beam TEM; in particular ordinary $1/2<110]$ dislocations have compact cores. The significant differences in the planar fault energies results in asymmetric core spreading of the superdislocations and makes the propagation of these dislocations sensitive to stress components other than the shear stress in the direction of the total Burgers vector.

5.1.4
Nonplanar Dissociations and Dislocation Locking

The planar dissociations described above result in core configurations, which in principle are glissile. The early investigations of Lipsitt et al. [2] and Hug et al. [3, 6] have shown, however, that the superdislocations in γ(TiAl) assume a locked configuration, which is manifested by long straight segments with distinct dislocation characters. This obvious similarity to the deformation behavior of γ'(Ni$_3$Al) suggests that the dislocation-locking mechanism in the L1$_0$ and L1$_2$ structures may have a common origin. Following this concept, Greenberg et al. [65] proposed several nonplanar sessile-core configurations for superdislocations in γ(TiAl). The common feature of all core modifications is that the screw component of the leading superpartial constricts and locally cross-slips from the original (111) slip plane onto a conjugated octahedral plane or a cube plane and then eventually redissociates on a parallel or oblique {111} plane. When the APB is completely on the cube plane, the dissociated configuration is known as a Kear–Wilsdorf lock [66]. As the partial dislocations involved have no common slip plane, the nonplanar cores are considered immobile with respect to glide. This pinning of screw dislocations leads to a strengthening of the material. Since cross-slip is thermally activated, the number of cross-slipped segments increases with increasing temperature. Such a mechanism may explain why the yield stress of γ'(Ni$_3$Al) exhibits an anomalous increase with temperature, which has also been observed for TiAl alloys and will be discussed later.

Woodward and MacLaren [36] have revisited the locking of <011] superdislocations in γ(TiAl) by considering the energetics of the anticipated mechanisms. Accordingly, the nonplanar dissociation of the <011] dislocations is thermodynamically driven because the APB energy for the cube plane is lower than that of the octahedral plane. There is also a torque force between the leading and trailing superpartials due to elastic anisotropy, which pushes the superpartials off the primary (111) glide plane and thus promotes cross-slip [14, 38]. The stresses associated with the torque are substantially larger than the shear stresses resulting from the externally applied load [67]. This finding implies that components of the stress tensor that exert no net resolved shear stress on the perfect dislocation can significantly affect the fine structure of the dislocation and thus the flow behavior [54]. Based on this cross-slip model several modifications for nonplanar <011] dislocation cores have been proposed, which differ in the core configuration that produces the pinning effect. For example, in a fashion similar to the Kear–Wilsdorf mechanism [66], the leading 1/2<01$\bar{1}$] superpartial dislocation formed by the decomposition reaction (Equation 5.6a) may cross-slip from the octahedral (111) plane onto the (100) cube plane, where the APB energy is lower [3, 49, 50]. The superpartial may then redissociate on the oblique ($\bar{1}\bar{1}\bar{1}$) plane, which leads to the Kear–Wilsdorf lock shown in Figure 5.10a. In a Ti-54Al alloy Hemker et al. [55] identified a nonplanar sessile-core configuration of the <01$\bar{1}$] dislocation that is similar to this Kear–Wilsdorf barrier. The structure is comprised of a SISF on the primary {111} glide plane, an APB on the (010) plane and an SESF on the intersecting octahedral plane. Kear–Wilsdorf locks of <01$\bar{1}$] dislocations were also observed by Jiao et al. [16] in Ti-54.5Al single crystals after deformation at 973 K.

Alternatively, the so-called "roof barriers" can be formed, where the APB is confined to the primary (111) glide plane (Figure 5.10b). TEM examination of Ti-56Al single crystals deformed at 573 K provided evidence for the existence of this type of lock [68]. The investigations of Woodward and MacLaren [36] suggest that the cross-slip driving forces generally favor the formation of roof barriers over Kear–Wilsdorf locks. Based on atomistic modeling Mahapatra et al. [69] proposed a nonplanar splitting of the <01$\bar{1}$] dislocation onto intersecting {111} planes forming SISF ribbons on these planes.

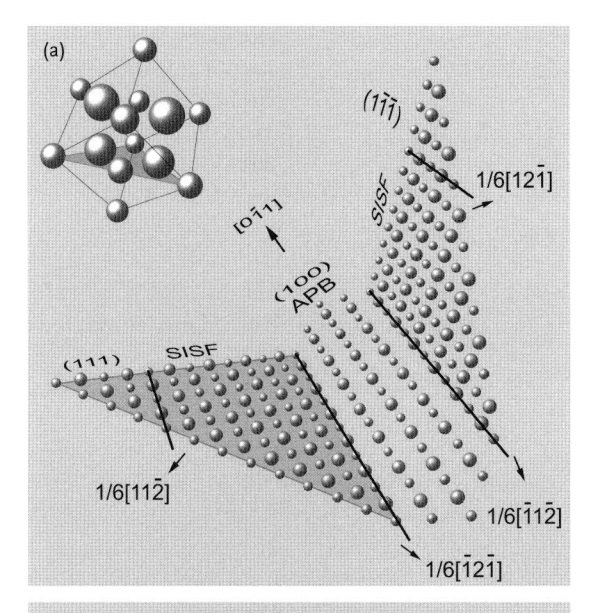

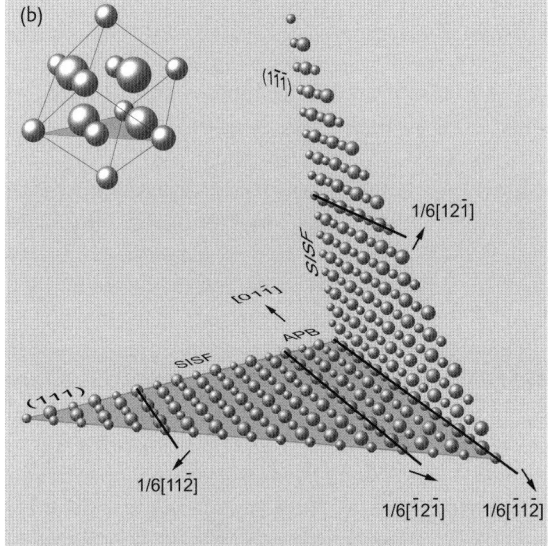

Figure 5.10 Nonplanar core structures of the <01$\bar{1}$] superdislocation formed by cross-slip of one of the 1/2<01$\bar{1}$] superpartials; perspective view along the screw [0$\bar{1}$1] orientation of the superpartials. (a) Formation of a Kear–Wilsdorf lock. Cross-slip of the leading 1/2[01$\bar{1}$] superpartial with the APB on the (010) plane and redissociation on the oblique (1$\bar{1}\bar{1}$) plane into the Shockley partial dislocations 1/6[$\bar{1}$1$\bar{2}$] and 1/6[12$\bar{1}$]. Dissociation of the trailing 1/2<011] superpartial on the primary (111) glide plane into the Shockley partial dislocations 1/6[11$\bar{2}$] and 1/6[$\bar{1}$2$\bar{1}$]. (b) Formation of a roof barrier. Dissociation of the trailing 1/2<0$\bar{1}$1] superpartial and formation of the APB on the primary (111) plane. Cross-slip of the leading 1/2<0$\bar{1}$1] superpartial onto the oblique (1$\bar{1}\bar{1}$) plane and redissociation into the Shockley partial dislocations 1/6[$\bar{1}$1$\bar{2}$] and 1/6[12$\bar{1}$].

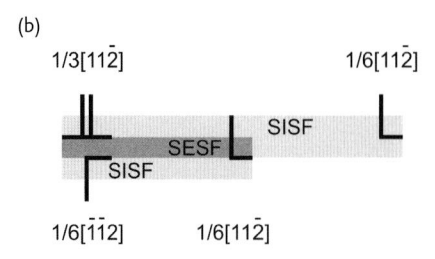

Figure 5.11 Nonplanar dissociation of 1/2[11$\bar{2}$] superdislocations; redrawn from illustrations in [72]. (a) Planar dissociation into 1/6[11$\bar{2}$] and 1/3[11$\bar{2}$] partial dislocations bounded by an SISF according to Equation 5.7d. (b) Core spreading on two adjacent (111) planes due to the formation of a dipole of two Shockley partial dislocations with the Burgers vectors $\mathbf{b} = \pm 1/6[1\bar{1}\bar{2}]$. The overlapping of the two SISFs generates an SESF. Final threefold dissociation into identical Shockley partial dislocations bounded by an SISF and SESF according to Equation 5.7e.

Delocalized nonplanar cores were also proposed for the 1/2<11$\bar{2}$> dislocations because these dislocations were found to be aligned along their edge direction, again suggesting a locked configuration [17, 52, 53]. Several authors [5, 24, 27, 50, 53, 59, 70–72] have considered the dissociation of the 1/2[11$\bar{2}$] dislocation into the partial dislocations 1/6[11$\bar{2}$] and 1/3[11$\bar{2}$], as described by Equation (5.7d) (Figure 5.11a). The transformation is thought to proceed by the nucleation of a dipole of partial dislocations ±1/6[11$\bar{2}$] on the adjacent (111) plane and the formation of a SESF-SISF pair (Figure 5.11b) [19, 41]. The region of overlap of the two SISFs represents the SESF [24, 52]. The net Burgers vector of the overlapping Shockley partial dislocations on the left-hand side of Figure 5.11b is 1/6[11$\bar{2}$]. The final dissociation can be described by Equation (5.7e) and consists of three identical 1/6[11$\bar{2}$] Shockley partials, which are delocalized on two {111} planes. The model was confirmed by high-resolution [24] and weak-beam [71] electron microscopy and rationalized by atomistic modeling [25]. It is further expected that the three Shockley partial dislocations involved may serve as a twin nucleus [25, 38].

Based on their electron microscope observations, Lang *et al.* [72] associated the locking of 1/2<11$\bar{2}$> dislocations with a dissociation into the stair-rod dislocations 1/3[00$\bar{1}$] and 1/3[11$\bar{1}$], as schematically illustrated in Figure 5.12. Climb and subsequent dissociation of the Frank partial dislocation 1/3[11$\bar{1}$] on a parallel (111) plane lead to a configuration that is locked by the stair-rod dislocation and an SISF. Because climb is involved, the mechanism is confined to edge dislocations, which explains why only edge dislocations are apparently locked.

In summary, superdislocations in γ(TiAl) are liable to adopt sessile nonplanar core configurations. The associated dislocation locking can give rise to an anoma-

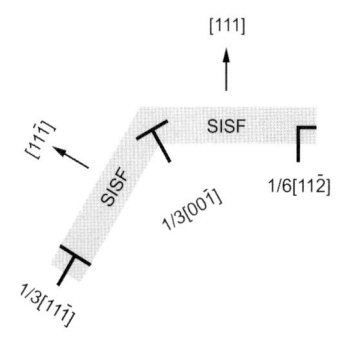

Figure 5.12 Locking of superdislocations due to nonplanar core spreading, redrawn from illustrations in [72]. Dissociation of an 1/2[11$\bar{2}$] dislocation by climb dissociation into a 1/3[11$\bar{1}$] Frank partial dislocation situated on the {11$\bar{1}$} plane, glissile Shockley partial dislocation 1/6 [11$\bar{2}$] on the (111) plane and a sessile stair-rod dislocation 1/3[00$\bar{1}$]. The three reactant dislocations are bounded by superlattice intrinsic stacking faults (SISF).

lous increase of the yield strength with temperature and strong work hardening and eventually makes the strength properties dependent on the deformation pathway. It should be noted that the dissociation of the superdislocations occurs on a scale smaller than that which is directly interpretable using weak-beam TEM. Adequate image simulation is difficult because of the close separation distance of the partial dislocations and the overlapping fault contrast. Thus, several details of sessile dislocation cores in γ(TiAl) remain to be determined. For more details on dislocation dissociations in γ(TiAl) the reader is referred to recent review articles of Veyssière and Douin [48], Vitek [54] and Paidar and Vitek [19].

5.1.5
Mechanical Twinning in γ(TiAl)

As recognized in the early studies of Shechtman et al. [1] and Lipsitt et al. [2] and more recent investigations [3, 7, 73–86], deformation of γ(TiAl) can also be accomplished by mechanical twinning. This deformation mode exhibits characteristic differences to dislocation glide with significant implications on the plasticity of polycrystalline material. The geometrical aspects of twinning are usually represented by the parameters K_1, K_2, S, η_1, and η_2 [82, 87, 88], which are illustrated in Figure 5.13. If all the four indices are rational, that is, whether or not they pass through sets of points of the Bravais lattice, the twin is called a compound twin. The plane containing η_1 and the normal to the plane K_1 is called the plane of shear S with the normal n_S. On twinning, all points of the lattice above the K_1 plane are displaced in the direction η_1 by the amount $e\eta_1$ that is proportional to their distance above K_1. The proportionality constant g is the strength of the simple shear and given by $g = 2 \cot \psi$. ψ is the acute angle between η_1 and η_2 or between the planes K_1 and K_2, that is, the vector η_2 in K_2 is rotated by $(180° - 2\psi)$ about n_S to become η'_2. Because the deformation is a pure shear, the plane K_1 is neither rotated nor distorted; K_1 is called the first undistorted plane, twin plane or composition plane.

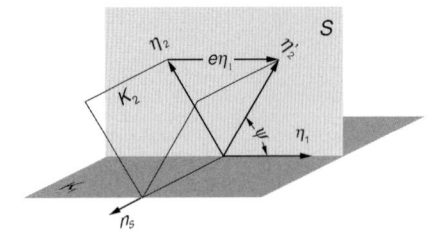

Figure 5.13 Crystallographic elements of mechanical twinning, see the accompanying text.

Table 5.2 Crystallographic elements of mechanical twinning in the L1$_0$ structure of γ(TiAl); $\lambda_T = c/a$; (after Yoo [22, 91] and Yoo et al. [90]).

Mode	K_1	K_2	η_1	η_2	g
Primary twinning	(111)	(11$\bar{1}$)	1/2[11$\bar{2}$]	1/2 [112]	$(2\lambda_T^2 - 1)/\lambda_T\sqrt{2}$
Complementary or antitwinning	(111)	(001)	1/2[$\bar{1}\bar{1}$2]	1/2 [110]	$\sqrt{2}/\lambda_T$

The plane K_2 is normal to the shear plane S and includes equal angles with K_1 before and after shear. Since all vectors in the K_2 plane are unchanged in length, K_2 is designated as the second undistorted plane. Thus, like slip, mechanical twinning leads to a permanent shape change of a crystal with negligible change in volume. However, unlike slip, mechanical twinning reproduces the crystal structure in a specific new orientation and involves a fixed amount of shear that is characteristic of the crystal structure.

Two twinning modes are said to be reciprocal or conjugate, if they have the same shear plane and g value but K_1 and K_2 and η_1 and η_2 are interchanged. Twinning systems in metals often have crystallographically equivalent K_1 and K_2 planes and shear directions η_1 and η_2. With regard to a primary twin characterized by the elements K_1 and η_1, the so-called complementary twinning system is defined by K_1 and $-\eta_1$, that is, by a shear in the reverse direction [87, 88]. It should be noted that in more complex twinning modes, which often occur in superlattice structures, a simple homogeneous shear translates only a fraction of the matrix atoms into positions appropriate to the twinned structure. The remaining atoms have to move into the correct twin position by shuffling. The total motion thus consists of shear plus shuffling [82, 87, 88].

The crystallography of deformation twinning in the L1$_0$ structure has been examined in the early study of Pashley et al. [89], Christian and Laughlin [21] and the group around Yoo [22, 23, 90, 91]. The twinning elements for the primary and complementary system are listed in Table 5.2 and illustrated in Figure 5.14. No atomic shuffles are required in these two twinning modes [22, 23, 92]. The primary twinning is often called order twinning or true twinning to refer to modes that restore the lattice and the order. As with the SISF and SESF, the twin boundary does not change the nearest-neighbor coordination [38]. The interfacial energy of

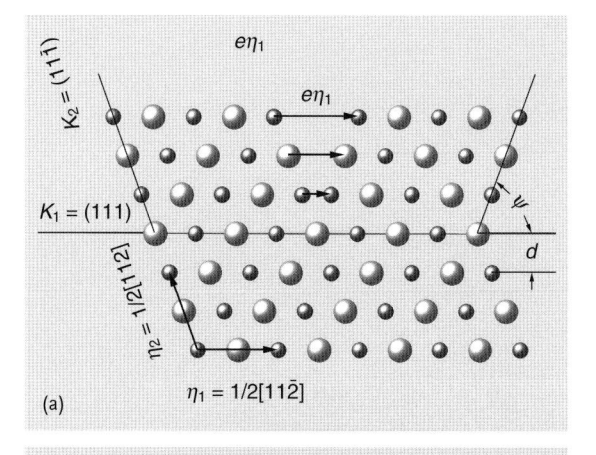

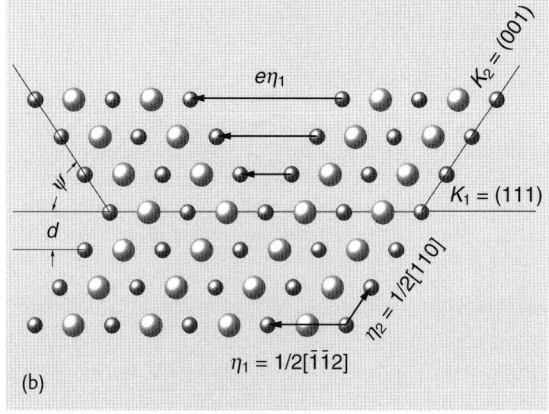

Figure 5.14 Structural model of mechanical twinning in γ(TiAl) according to Figure 5.13. (a) Primary twinning defined by the twinning elements $K_1 = (111)$, $K_2 = (11\bar{1})$, $\eta_1 = 1/2[11\bar{2}]$, $\eta_2 = 1/2\,[112]$, and shear $g = 0.707$. (b) Complementary (anti) twinning with $K_1 = (111)$, $K_2 = (001)$, $\eta_1 = 1/2[\bar{1}\bar{1}2]$, $\eta_2 = 1/2\,[110]$, and $g = 1.414$.

the twin arises from the change of the bond angle between second-nearest neighbors [29, 91]. Hence, the fault energies of the SISF, SESF and the twin are very similar. This twinning mode is associated with the shear $g = 0.707$, assuming $\lambda_T = c/a = 1$, which amounts to displacing each (111) layer in the twin by $1/6\,[11\bar{2}]$ over the layer underneath. In the notation of Figures 5.3 and 5.4b, this can be accomplished by passing a partial dislocation with Burgers vector $\mathbf{b}_3 = 1/6<11\bar{2}]$ over each (111) plane above the composition plane. These structural features can be directly observed by high-resolution transmission electron microscopy (HREM), as demonstrated in Figure 5.15 [86]. The micrograph shows the atomic structure of a narrow twin, which propagated at room temperature. The structure of the K_1 twinning plane is fully coherent apart from a one-plane ledge in the twin matrix interface. This ledge represents the Shockley partial dislocation, which accomplish

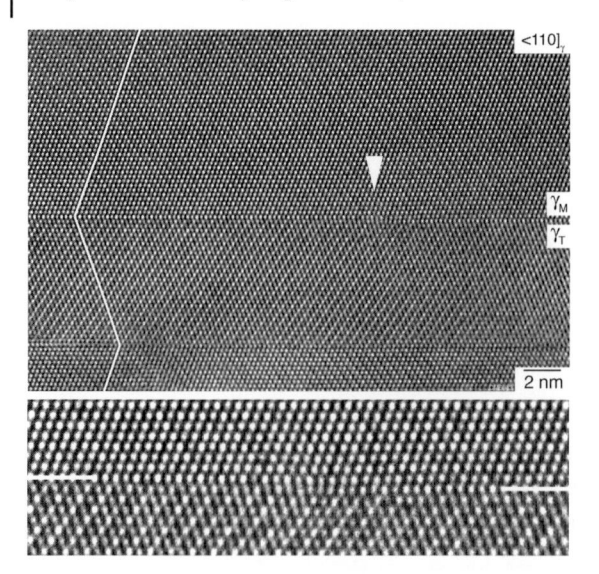

Figure 5.15 Atomic structure of a deformation twin observed in a Ti-48.5Al-0.37C (at.%) alloy after room-temperature compression to strain $\varepsilon = 3\%$. High resolution electron micrograph along the common $<\bar{1}10]$ direction. γ_M designates the matrix and γ_T the twin. White lines indicate the trace of the $(\bar{1}\bar{1}1)$ planes. Note the one-plane step in the (111) interface marked by the arrowhead, which is associated with a Shockley partial dislocation. The figure below shows the core of this dislocation in higher magnification; (Micrograph: Appel [86]).

the growth and broadening of the twin. The magnified figure below demonstrates that the core of the Shockley partial dislocation is relatively compact.

The conjugate twinning system is defined by $K_1 = (11\bar{1})$, $K_2 = (111)$, $\eta_1 = 1/2$ [112], $\eta_2 = 1/2[11\bar{2}]$, and the same amount of shear. Thus, a primary twin and its reciprocal are crystallographically equivalent but differently oriented with respect to the parent crystal. The crystallographic elements of the complementary twinning system are $K_1 = (111)$, $K_2 = (001)$, $\eta_1 = 1/2[\bar{1}\bar{1}2]$, $\eta_2 = 1/2$ [110], and $g = 1.414$, that is, the shear is twice as large as in the primary mode. In the TiAl literature complementary twinning is also referred to as antitwinning [22, 23]. The antitwinning operation corresponds to the displacement $-2\mathbf{b}_3 = 1/3[\bar{1}\bar{1}2]$ in Figure 5.3, which is probably very difficult because the high-energy **BB** stacking has to be overcome. Thus, twinning shear is unidirectional in the sense that the shear in one direction is not equivalent to the shear in the opposite direction; this is another difference to the bidirectional nature of dislocation glide. Yoo [22, 91] has pointed out that the conjugated deformation mode ($K_2 = (001)$, $\eta_2 = 1/2$ [110]) to the complementary twinning corresponds to the slip system 1/2<110](001), which apparently becomes active at elevated temperatures [17]; see Section 5.1.1. This slip/twin relationship might be important for the accommodation of constraint stresses, which often develop when a slip or twinning system is blocked.

To this point the description of mechanical twinning in γ(TiAl) is similar to that in f.c.c. metals. However, unlike f.c.c. metals, there is only one distinct shear direction $\mathbf{b}_3 = 1/6<11\bar{2}]$ per {111} plane that does not alter the ordered $L1_0$ structure of γ(TiAl), as described in Section 5.1.2. The two other displacements of the type $1/6<1\bar{2}1]$ destroy the chemical environment of first-order neighbors. When one of these displacements takes place at successively higher (111) layers, a so-called pseudotwin is generated [23]. Thus, there are only four true twinning systems in γ(TiAl), contrasting to twelve in f.c.c. metals. This crystallographic feature of the $L1_0$ structure, together with the fact that mechanical twinning is unidirectional, imposes significant restrictions upon possible twinning modes in γ(TiAl). The operation of twinning systems in samples under uniaxial load is expected to depend on the sense of the load, as well as on the axis orientation. Specifically, in a single crystal of given orientation some variants of a particular twin mode should operate only in compression, whereas others should operate only in tension.

Sun et al. [92] determined the Schmid factors for the $1/6<11\bar{2}]${111} twinning systems for uniaxial compression and tension. The stereographic projection shown in Figure 5.16 indicates the systems with the highest Schmid factor for load orientations in the [001]–[1$\bar{1}$0]–[010] extended triangle. Under compression, depending on the axis orientation, two primary twinning systems are expected to operate; a prediction that was essentially confirmed by an electron microscope analysis performed on a compressed Ti-50Al alloy [92]. In contrast, there is only one primary twinning system available for all tensile orientations. At some axis orientations the resolved shear stresses are in the antitwinning sense for all four twinning systems, and twinning is thus forbidden; these regions are shaded in Figure 5.16. However, there are secondary twinning systems with smaller Schmid factors that might be activated. This may occur during work hardening, when

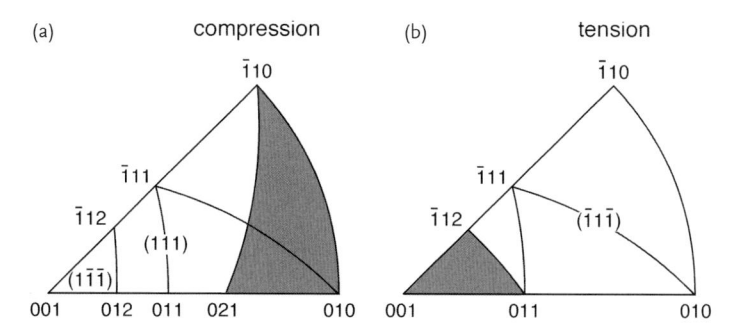

Figure 5.16 Twinning systems operating in γ(TiAl) under uniaxial compression and tension. The stereographic projection indicates the orientation of the loading axis for the primary twinning systems with the highest Schmid factor. The lines drawn within the triangle border regions with different operating systems. Only twin planes are indicated because the twinning direction is always of type $1/6<1\bar{1}\bar{2}]$. The shaded areas represent orientations for which twinning is forbidden. Figure redrawn from Sun et al. [92], slightly modified.

significant internal stresses are built-up. This multiple twinning leads to twin intersections, a problem that is described in Section 6.3.4.

Crystallographic theories [38, 82, 87, 88, 91] for the prediction of possible twinning modes are based on the hypothesis that twinning shear should be small; this ensures a small strain accommodation and a low lattice friction (Peierls stress) of the Shockley partial dislocations. This argument again implies that antitwinning operations in γ(TiAl) are extremely difficult and will not be activated unless appropriate internal stresses are present. The theory also suggests that any shuffle mechanism should be simple, and shuffles should be parallel to the twinning shear direction. While this condition is not significant for the mechanical twinning modes discussed above, it might be important for the so-called transformation twinning with $K_1 = (011)$, $\eta_1 = [01\bar{1}]$ and $g = 0.03$ (for $\lambda = 1.016$) because interchanges of Ti and Al atoms [90] are required to restore the original structure. However, owing to the very small shear value, the mechanism may operate at elevated temperatures, where shuffling can be supported by thermal activation.

In broad terms, twinning may start when the externally applied shear stress across the K_1 plane, resolved in the η_1 direction, reaches a critical value. Apart from twin nucleation, this stress is required to overcome the glide resistance of the twinning partial dislocations and to create the twin boundary. Thus, the stress required for twinning is often assumed to be higher than that for slip. The surface energy of a twin boundary in γ(TiAl) should be closely related to the SISF energy because the underlying elementary shear process $\mathbf{b}_3 = 1/6<11\bar{2}>$ produces an intrinsic stacking fault (Figure 5.4b). Fu and Yoo [29] have pointed out that the creation of a twin boundary only changes the bond angles between second-nearest neighbors, whereas the bonding of nearest neighbors is not affected. This reasoning may explain why the twin-boundary energies determined by ab-initio calculations for stoichiometric TiAl [29, 36] are about half the SISF energy (Table 5.1). This low energy, coupled with the relatively small shear may also explain why $1/6<11\bar{2}>\{111\}$ twinning can be a prominent deformation mechanism in γ(TiAl). The association of twinning with the SISF energy suggests that the conditions under which twinning occurs may sensitively depend on temperature and alloy composition. However, it is difficult to associate a certain twinning mode with a resolved shear stress because mechanical twinning is unidirectional and often masked by dislocation glide.

The four $1/6<11\bar{2}>\{111\}$ twinning systems together with the twelve $<110>\{111\}$ dislocation slip systems of γ(TiAl) formally provide five independent deformation modes, which, according to the von Mises criterion [93], are required for a general shape change in polycrystalline material. However, these deformation modes must operate at similar stresses; a condition that apparently is often not fulfilled in γ(TiAl). The relative ease and propensity of dislocation glide and mechanical twinning will be considered in Section 5.1.6.

In short, mechanical twinning along $1/6<11\bar{2}>\{111\}$ is a potential deformation mechanism in γ(TiAl) because the twinning shear is relatively small and no atomic shuffling is required. The mechanism provides auxiliary deformation modes with shear components in the **c** direction of the TiAl unit cell. Since twinning shear is unidirectional, the operating twinning systems vary with the sense of the load and the loading direction. There are crystal orientations for which twinning is forbidden.

5.1.6
Effects of Orientation and Temperature on Deformation of γ Phase

The ease and propensity of slip in a particular slip system is best expressed in terms of a shear stress τ, resolved on the slip plane in the slip direction. At a given normal stress $\sigma = F/A$ acting on the cross-sectional plane A of a sample, τ is defined as [94]

$$\tau = \sigma \cos\phi \cos\psi \tag{5.8}$$

F is the force applied along the sample axis, ψ is the angle between F and the slip direction, and ϕ is the angle between F and the normal to the slip plane. The term $S_G = \cos\phi \cos\psi$ is known as the Schmid factor. τ_c is called the critical resolved shear stress (CRSS) for the onset of slip. The Schmid law (5.8) implicitly states that τ_c is the only component of the stress tensor that activates glide in a particular direction on a particular plane. Hence, in crystals that deform on a single slip system, the critical resolved shear stress τ_c for the onset of plastic flow should be independent of the crystal orientation.

In conventional disordered materials τ_c decreases as the temperature rises, $d\tau_c/dT$ is negative. The negative temperature dependence of τ_c results basically from thermally activated processes, which support deformation in different ways and also make τ_c dependent on strain rate $\dot{\varepsilon}$. A higher temperature and lower strain rate generally results in a lower stress τ_c. Figure 5.17a schematically demonstrates the dependence of τ_c on T and $\dot{\varepsilon}$, which is typically observed in such materials. Contrary to this common experience, many intermetallic alloys exhibit the anomalous deformation behavior shown in Figure 5.17b, that is, $d\tau_c/dT$ is positive over a finite range of temperature. This anomalous yield behavior was first observed in hardness tests on Ni$_3$Al by Westbrook [95] and later confirmed by flow-stress measurements on this material [28, 96, 97]. Since that time, the anomalous deformation behavior has been observed in many other intermetallic alloys; for general information the reader is referred to review articles [48, 98–102].

The deformation kinematics of γ(TiAl) is difficult to assess because Al-rich alloys are often not single-phase. The unavoidable contamination with interstitial elements leads to the precipitation of carbides, nitrides and oxides. Carbon and nitrogen are either present as perovskite phase Ti$_3$AlX, (X = C, N) or as hexagonal H phase Ti$_2$AlX [103–106]. Wiezorek and Fraser have observed oxide precipitates at dislocations [107]. The importance of interstitial elements for the mechanical properties of γ(TiAl) was pointed out by Hug *et al.* [3], who also suggested a thermal treatment, in order to mitigate their deleterious effects on mechanical properties. The treatment consisted of homogenization at 1300 °C followed by quenching and aging at 1000 °C. The procedure aimed at the formation of large, widely separated precipitates, which have less influence on the mechanical properties. Several

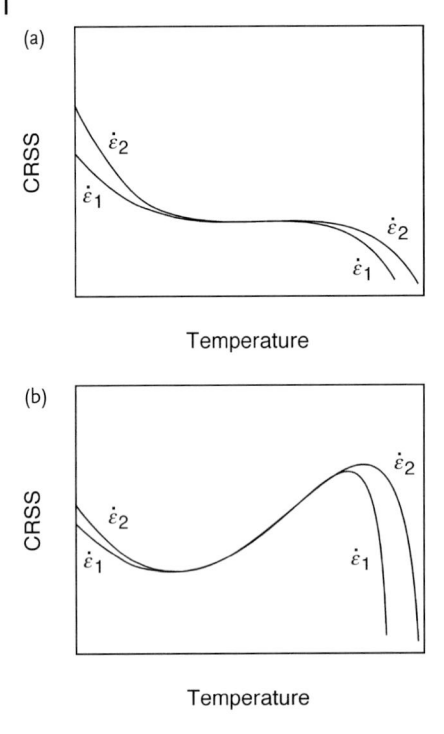

Figure 5.17 Schematic illustration of the dependence of the critical resolved shear stress on temperature and strain rate $\dot{\varepsilon}$, with $\dot{\varepsilon}_1 < \dot{\varepsilon}_2$. (a) Conventional disordered materials and (b) intermetallic alloys.

authors have since applied this thermal treatment. In Al-rich alloys the Ti_3Al_5 phase can also be formed and may significantly affect the mechanical properties [4, 108–111]. In a detailed review Grégori and Veyssière [112] have considered these so-called extrinsic factors on the yield behavior of TiAl and reconciled the significant differences, which obviously exist between the data of different groups.

The slip systems operating in γ(TiAl) have been identified by testing Al-rich TiAl single crystals in various orientations and at different temperatures [12, 16, 17, 40, 53, 57, 67, 68, 109, 112–123]. In most cases compression tests were performed at relatively low strain rates of $10^{-5}\,s^{-1}$ to $10^{-4}\,s^{-1}$. The effect of sample orientation has usually been classified with respect to the extended standard stereographic projection shown in Figure 5.18 [17, 113]. The domains I, II and III designate the range of orientations that might be favorable for the glide of <101] superdislocations and ordinary 1/2<110] dislocations on {111} planes. The diagram implies that the CRSS of these two deformation modes is the same and basically indicates which of the two systems has the larger Schmid factor in the orientation domains indicated. Glide of <101] superdislocations is favored for sample orientations situated in the domains I and III, whereas glide of 1/2<110] ordinary dislocations is expected in domain II. Table 5.3 lists the Schmid factors for the dislocation glide

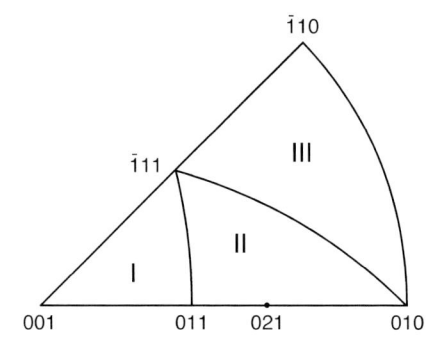

Figure 5.18 Axial orientations and selection of slip systems in TiAl single crystals. According to the relevant Schmid factors, glide of <101] superdislocations is expected for orientation domains I and III, whereas glide of 1/2<110] ordinary dislocations is favored in domain II.

Table 5.3 Schmid factors S_G for the slip systems in γ(TiAl) for the given axial orientations.

Slip system		Axial orientation			
		Schmid factor S_G			
		[001]	[021]	[010]	[$\bar{1}$10]
<110]{111}	[1$\bar{1}$0]{111}	0	0.490	0.408	0
	[110]{$\bar{1}$11}	0	0.490	0.408	0
	[110]{1$\bar{1}$1}	0	0.163	0.408	0
	[$\bar{1}$10]{11$\bar{1}$}	0	0.163	0.408	0
<101]{111}	[$\bar{1}$01]{111}	0.408	0.245	0	0
	[101]{$\bar{1}$11}	0.408	0.245	0	0.408
	[$\bar{1}$01]{1$\bar{1}$1}	0.408	0.082	0	0.408
	[101]{11$\bar{1}$}	0.408	0.082	0	0
<011]{111}	[0$\bar{1}$1]{111}	0.408	0.245	0.408	0
	[0$\bar{1}$1]{$\bar{1}$11}	0.408	0.245	0.408	0.408
	[011]{1$\bar{1}$1}	0.408	0.245	0.408	0.408
	[011]{11$\bar{1}$}	0.408	0.245	0.408	0
<112]{111}	[$\bar{1}\bar{1}$2]{111}	0.471	0	0.236	0
	[1$\bar{1}$2]{$\bar{1}$11}	0.471	0	0.236	0.471
	[$\bar{1}$12]{1$\bar{1}$1}	0.471	0.189	0.236	0.471
	[$\bar{1}\bar{1}$2]{11$\bar{1}$}	0.471	0.189	0.236	0

modes in γ(TiAl) for different sample orientations between the corners of the standard [001]–[010]–[$\bar{1}$10] triangle. Accordingly, several slip systems often receive equal components of shear stress; hence, under highly symmetrical orientations, such as [010], multiple slip is expected. In addition, mechanical twinning can occur and intervene with dislocation glide. The possible orientation ranges for twinning have been discussed in Section 5.1.5 (Figure 5.16). Under compression, which will be primarily discussed here, mechanical twinning is forbidden for the orientation domain between the orientations [021], [$\bar{1}$10] and [010] in Figure 5.18.

Inui *et al.* [17] have provided the most detailed information about the slip geometry by investigating dislocation structures and surface slip markings after compression of TiAl single crystals along seven orientations. For low and ambient temperatures, these observations essentially confirm the variation of the slip geometry with orientation sketched in Figure 5.18. At temperatures above 1100 K, deformation in most orientations is provided by ordinary dislocations gliding on {111}, {110} or {001} planes. In samples unfavorably oriented for ordinary slip, there is also evidence for mechanical twinning and glide of 1/2<112] superdislocations.

Figure 5.19 shows a selection of published critical resolved shear stresses τ_c [12, 17, 40, 67, 113, 114, 122] of Al-rich TiAl alloys, which were determined in compression. In most of these studies, prior to testing, the alloys had been subjected to the precipitation heat treatment proposed Hug et al. [3]. The CRSS decreases with

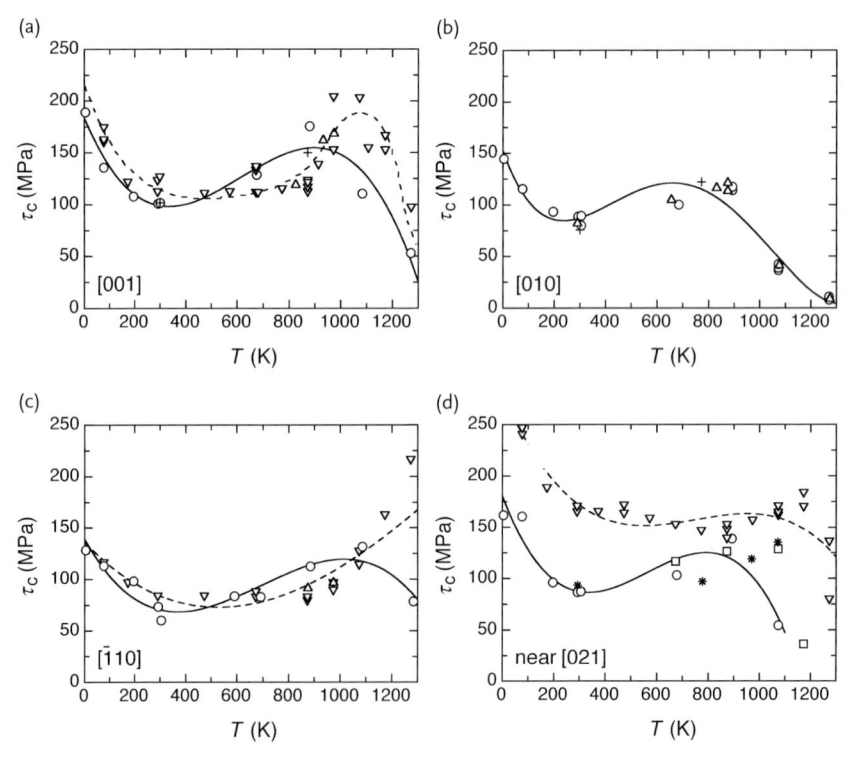

Figure 5.19 Dependence of the critical resolved shear stress τ_c on the absolute temperature T for the orientations (a) [001], (b) [010], (c) [$\bar{1}$10], and (d) near [021]. ○ Ti-56Al (at.%), [113, 114], ▽ Ti-56Al, [17], △ Ti-55.5Al, [67], + Ti-56Al, [40], □ Ti-56Al, [12], ✳ Ti-54.5 Al, [122].

temperature between 4.2 K and 300 K, varies little between 300 and 600 K, and then rises to a peak at 700 K to 1000 K, depending on orientation and alloy composition. The anomalous increase of the CRSS occurs for all the orientations studied; hence, it is manifested for both ordinary and superdislocations. This contrasts with the situation in Ni_3Al single crystals having the $L1_2$ structure, in which the yield-stress anomaly is associated with just one slip mode [48, 99, 123]. As seen in Figure 5.19a, above its peak τ_c falls rapidly to very low values. While the yield stress is rising with temperature, it is insensitive to changes of the strain rate. In the range of the anomalous increase of τ_c, the strain-rate sensitivity parameter, defined as $S_R = (d\ln \tau/d\ln \dot{\varepsilon})/T$, decreases with temperature and actually appears to be zero near the peak in τ_c. At higher temperatures S_R then again increases. It should be noted that this behavior has been recognized for both 1/2<110] and <101] slip [17, 113, 114, 122]. Hence, the dependence of the CRSS of γ(TiAl) on temperature corresponds relatively well to the profile observed in many other intermetallic alloys (Figure 5.17b). The τ_c values belonging to the orientations [001], [010] and [$\bar{1}$10], shown in Figures 5.19a–c, were calculated for the highest Schmid factors associated with <101]{111} glide, which, according to TEM observations, was the prevalent system for these orientations. These results essentially coincide with measurements of other authors [16, 40, 57, 68, 116, 118], who activated <101]{111} slip in other sample orientations.

The τ_c values determined for <101]{111} glide do not lie on a common curve; hence, at a given temperature, the CRSS still depends on orientation. The data of Inui et al. [17] (Figures 5.19a–c) shows, for example, that the CRSS for <101]{111} slip in the intermediate temperature range of 300 K to 800 K increases as the compression axis moves away from [$\bar{1}$10]. There is also a significant variation of the peak stress with orientation, which tends to be highest for [001]-oriented samples [17, 67, 113, 114]. Thus, the Schmid law for slip of <101] superdislocations is violated. As has been discussed in Section 5.1.4, this finding implies that the nonglide stress acting in the slip plane but normal to the total Burgers vector affects the yield stress and it does this through its effect on the dissociation mode of the partial dislocations.

In addition to the orientation dependence of the CRSS for <101] glide, there is an asymmetry between compression yield stress and tensile yield stress [67]. Crystals with deformation axes close to [010] or [$\bar{1}$10] are stronger in tension than in compression, whereas the reverse behavior is observed for the [001] orientation. An orientation-dependent tension/compression asymmetry for TiAl single crystals was predicted by Hazzledine and Sun [124]. The arguments essentially rely on the asymmetric core structure of the <101] superdislocations, which was described in Section 5.1.3 and Figure 5.8. The coupling of the Shockley partial dislocations by the fault sequence SISF–APB–CSF suggests that the slip and cross-slip behavior of the <101] superdislocation depends on its direction of motion. Hence, the CRSS should be prone to both the sense and orientation of the applied load. Given this model, Zupan and Hemker [67, 125] have explained the tension/compression asymmetries, which had been observed for the different sample orientations. The CRSS was found to be lowest, when the propagating superdislocation is led by the complex stacking fault (CSF) and highest when the CSF is trailed. Collectively these findings strongly support the operation of the dislocation-locking mechanisms described in Section 5.1.4, that is, the yield-stress anomaly for the <101] {111} slip system can be rationalized by the formation of Kear–Wilsdorf or roof-type locks. A detailed account of the dissociation and locking of <101] superdislocations in γ(TiAl) based on TEM investigations was given by Grégori and Veyssière in a recent collection of publications [62, 126, 127].

Jiao et al. [53] recognized a yield-stress anomaly for 1/2<112]{111} slip in Ti-54.5 Al single crystals with a sample orientation close to [$\bar{1}$10]. The authors ascribed the anomalous hardening to a climb dissociation of the edge dislocations on intersecting {111} planes so that sessile stair-rod dislocations are formed (Section 5.1.4, Figure 5.12). Grégori and Veyssière [126] have pointed out that the 1/2<112] dislocations in Ti-54.3 Al are often sessile byproducts of the decomposition of <101] dislocations into 1/2<1$\bar{1}$0] and 1/2<112] dislocations.

The CRSS associated with the 1/2<110]{111} slip systems also reveal a marked flow-stress anomaly, as demonstrated in Figure 5.19d. This observation was confirmed by the measurements of other authors, who activated ordinary slip by other sample orientations [109, 113, 114, 119, 121, 122]. Slip-line investigations and TEM inspection have shown that deformation in [021]-oriented samples is mainly carried by the two equally stressed 1/2<110]{111} slip systems [17], according to the Schmid factors listed in Table 5.3. It should be noted, however, that the exclusive activation of the 1/2<110]{111} slip systems is limited to a narrow range of sample orientations within a few degrees of the [021] orientation, where the Schmid factors of the other glide modes are very small (Table 5.3) [17, 121, 122]. Feng and Whang [12] recognized a marked change of the slip system in a Ti-56Al single crystal that had been oriented for glide of ordinary dislocations utilizing the axial orientations [$\bar{3}$ 12 7], [$\bar{1}$ 6 3] and [$\bar{1}$ 12 5], which are relatively close to [021]. Deformation at low temperatures (196 K to 463 K) was found to be carried by one of the <101] {111} superdislocation slip systems, whereas glide of ordinary dislocations predominated in the intermediate temperature interval of 673 K up to the peak temperature of 1073 K. The difficult activation of ordinary 1/2<110]{111} slip at low temperatures might be a consequence of the high friction stress of the ordinary dislocations in the Al-rich γ(TiAl) phase. The data determined by Inui et al. [17] on Ti-56Al single crystals indicates that the CRSS for the glide of ordinary dislocations at temperatures below 873 K is higher than that for glide of <101] superdislocations, as becomes evident by comparing Figures 5.19a–c with Figure 5.19d. This finding will be considered later in more detail.

A complete and consistent understanding of the yield-stress anomaly of the 1/2<110]{111} slip systems is not at hand. The most troubling aspect is that the stress–temperature profile is quite similar to that of the <101] {111} superdislocation slip system, although the core of the ordinary dislocations is characteristically different from that of the superdislocations. As has been discussed in Section 5.1.3, superdislocations are widely dissociated, whereas ordinary dislocations have a compact core and do not form antiphase boundaries. Thus, the yield-stress anomaly of the ordinary 1/2<110]{111} slip system cannot be explained by the formation of Kear–Wilsdorf or roof-type locks via thermodynamically driven cross-slip. The yield-stress anomaly also conflicts with the proposal of Greenberg et al. [128], according to which the mobility of the 1/2<110] dislocations is controlled by the lattice friction (Peierls stress) resulting from the directional bonding of TiAl. Due to thermal activation, with rising temperature the lattice friction would be expected to decrease, therefore allowing the dislocations to move at lower stresses. Thus, this mechanism is in itself not sufficient to explain the yield-stress anomaly.

However, the following factors might be related to the anomalous hardening. Below the peak temperature slip occurs on {111} planes with a distinct preference of screw dislocations [17, 80, 129–136]. The screw dislocations are bowed-out between pinning centers, which are roughly aligned along the Burgers vector direction. These observations imply that the glide resistance of the screw dislocations controls the yield stress. There are different and conflicting explanations about the origin of the pinning process. Several authors have proposed that localized extrinsic obstacles may act as pinning centers [129, 130], which appears plausible in view of the significant oxygen and nitrogen content of most alloys. However, there are many dipoles and small prismatic dislocations loops in the vicinity of the pinning centers, which are often connected with the dislocations [131–133]. These features are consistent with a classical jog-dragging mechanism, which was early proposed by Gilman and Johnston [137]. There is some structural evidence that the density of the pinning centers increases with temperature.

Jiao et al. [122] have also associated the yield-stress anomaly with double cross-slip of the screw components from the primary {111} plane to {110), a tendency that was observed to increase with temperature and with increasing Schmid factor for slip on {110} [12, 109, 120, 122]. The fact that the flow stress on the {111} plane is higher when there is a large component of stress on the {110) cross-slip plane was taken as evidence that it is this cross-slip that produces the anomalous flow behavior. The driving force for this process was attributed to the lower dislocation line tension on {110) than on {111} [10, 11], as described in Section 5.1.1. Above the peak temperature, unlocking of the dislocations occurs by diffusion-assisted climb of the trailed edge components on {110) planes.

Viguier et al. [138] and Louchet and Viguier [139] have proposed that an ordinary screw dislocation is subject to a Peierls friction and bulges kinks on the two {111} planes that intersect along the common <110] direction. The kinks are thought to move laterally along the dislocation and collide to form the pinning centers.

A common feature of these latter three models is that they relate the yield-stress anomaly with an intrinsic pinning process of the dislocations due to the generation of jogs or kinks. This appears plausible because the ordinary dislocations have a compact core, which makes cross-slip and kink-bulging easy. However, the proposed mechanisms do not adequately explain the variation of the strain-rate sensitivity with temperature. Thermally activated cross-slip or kink-bulging would imply a strain-rate sensitivity increasing with temperature in the range of the anomalous hardening, whereas experiments show the reverse [17, 113, 114, 122]. Furthermore, extensive jog dragging of ordinary dislocations has also been recognized after room-temperature deformation [131, 133] and thus is not specific to the temperature range of the anomalous hardening. Thus, the mechanism in itself is not sufficient to explain the yield-stress anomaly.

Nakano et al. [109] proposed an anomalous hardening mechanism for the 1/2<110]{111} slip system that is triggered by the presence of the Al_5Ti_3 phase. In alloys with Al concentrations higher than 54 at.% this phase can be formed by further ordering of the $L1_0$ structure of γ(TiAl) into a short- or long-range superstructure [4, 108–111]. The Al_5Ti_3 phase exists in the temperature range of the anomalous hardening and becomes unstable above the peak temperature [4, 109, 140]. A short-range ordered form of this phase was observed in Ti-54.7Al single crystals [109]. In this form Al atoms have a statistically greater number of Ti atoms as neighbors than would be expected for γ(TiAl) ordered in the $L1_0$ structure. The same authors identified in Ti-58Al single crystals the Al_5Ti_3 phase as 40-nm to 60-nm long precipitates. A single 1/2<110] dislocation intersecting such a precipitate on either the {111}, {110) or {001} plane produces an antiphase boundary within the precipitate. Subsequent intersection of the precipitate on the same slip plane by three other 1/2<110] dislocations completely restores the Al_5Ti_3 superstructure [109]. Supporting evidence for such a mechanism was provided by TEM observations of the dislocation structure in Al-rich TiAl crystals [4, 17, 141–143], which showed that the ordinary dislocations tend to propagate in groups of four dislocations. This coupled glide of ordinary dislocations probably makes the overcoming of the Al_5Ti_3 precipitates easier. Figure 5.20 demonstrates this dislocation behavior in a TEM micrograph.

Based on bonding and symmetry considerations, Nakano et al. [109] expected the APB energy on the {110) and {001} planes to be significantly lower than that on {111}, which implies that cross-slip of the 1/2<110] screw dislocations from

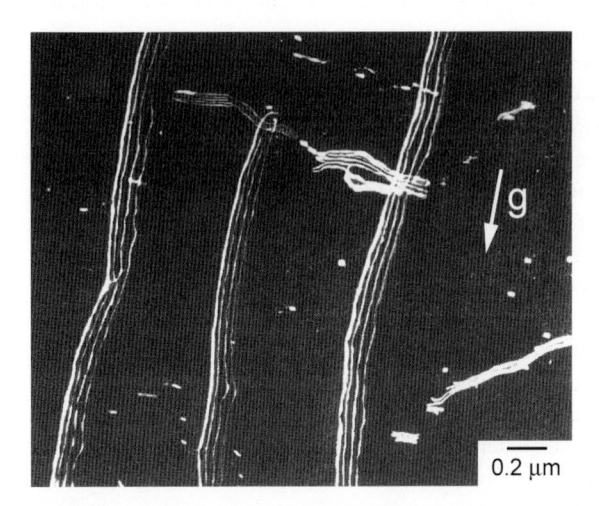

Figure 5.20 Weak-beam image of the dislocation structure in a Ti-58Al (at.%) single crystal. Deformation was performed at room temperature along the [201] direction in order to activate glide of 1/2<110] dislocations. Note the coupled glide of four ordinary dislocations; (Micrograph: Nakano et al. [143]).

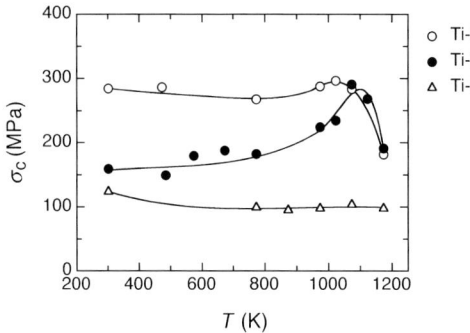

Figure 5.21 Dependence of the yield stress of TiAl crystals on the deformation temperature. The Ti-54.7Al and Ti-58Al crystals were deformed along the [201] direction so that glide of ordinary dislocation occurred. The PST crystal was oriented for shear along the lamellar boundaries. Data of Nakano et al. [109], redrawn.

the primary {111} plane to {110} or {001} planes is thermodynamically driven. With increasing temperature the cross-slip events progressively pin the screw components of the ordinary dislocation loops, which still primarily glide on the {111} planes. Likewise, the irregular structure of a short-range-ordered alloy may cause it to resist the glide of dislocations and to have a flow stress higher than that of $L1_0$-γ(TiAl). The pinning effect probably disappears at temperatures higher than 1100 K, where the Al_5Ti_3 precipitates begin to dissolve. Figure 5.21 demonstrates the variation of the yield stress with temperature observed on TiAl single crystals with different Al contents that were oriented for glide of ordinary dislocations [109]. The diagram contains the CRSS value of a TiAl-PST crystal with stoichiometric composition that had been oriented for shear along the lamellar boundaries. In agreement with the described mechanism, a yield-stress anomaly is only observed in Ti-54.7Al and Ti-58Al crystals, in which the Al_5Ti_3 phase exists. The authors ascribed the high stress level of the Ti-58Al crystal to the interaction of dislocations with the numerous APBs created at the precipitates, which probably masks the yield-stress anomaly. An anomalous behavior was not observed on the Ti-50.8Al polysynthetically twinned (PST) crystal, in which the Al_5Ti_3 phase should be absent because of the relatively low Al content of the γ(TiAl) phase. Another important result of the analysis by Nakano et al. [141] is that the energy of the APB created by ordinary dislocations in the Al_5Ti_3 superstructure is significantly higher than the energy of the APB trailed by the <101] superdislocations. Thus, when Al_5Ti_3 precipitates are present, the glide resistance of the <101] superdislocations is expected to be lower than that of ordinary dislocations. This concept could explain why in Al-rich alloys the CRSS for the glide of ordinary dislocations is significantly higher than that for glide of superdislocations over a wide temperature range (Figure 5.19). Moreover, it may be concluded that under these conditions the deformation of γ(TiAl) is mainly carried by <101] superdislocations. The variation of the CRSS for the glide of 1/2<110] and <101] dislocations with the Al concentration determined at 300 K by Nakano et al. [143] is shown in Figure 5.22.

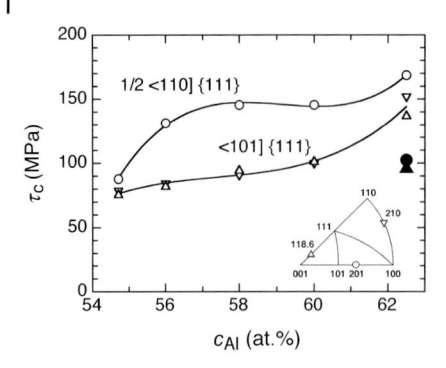

Figure 5.22 Dependence of the critical resolved shear stress for $1/2\langle110]\{111\}$ and $\langle101]\{111\}$ slip on the Al concentration of binary alloys measured in compression at room temperature. The compression axes are represented in a [001]–[100]–[110] unit triangle based on the $L1_0$ structure. The full symbols ● and ▲ represent the CRSS values of an Al_5Ti_3 single-phase single crystal that were produced by an aging heat treatment at 750 °C. Diagram redrawn from Nakano et al. [143].

The CRSS of the $1/2\langle110]\{111\}$ glide system is always higher than that of the $\langle101]\{111\}$ system. As expected from the considerations given above, the difference is manifested with the development of the Al_5Ti_3 structure, that is, for Al concentrations higher than 54 at.%. Figure 5.22 contains the CRSS value of an Al_5Ti_3 single-phase single crystal that was produced by long-term aging of the Ti-62.5Al alloy. In this long-term aged form the yield stress of the material is significantly reduced when compared with the starting material, and there is practically no difference between the $1/2\langle110]\{111\}$ and $\langle101]\{111\}$ systems. This observation indicates that the strong hardening of the Ti-62.5 Al alloy containing Al_5Ti_3 precipitates in an $L1_0$ matrix is mainly caused by antiphase boundaries. As discussed earlier, these APBs are formed within the ordered particles by intersecting dislocations and provide additional resistance to subsequent glide processes.

A flow-stress peak has also been reported for polycrystalline binary and ternary alloys [59, 61, 138, 144–148], which, at first glance, is not surprising because all the individual slip systems of γ(TiAl) single crystals behave anomalously. Vasudevan et al. [148] have early speculated that the stress–temperature profile may depend on the grain size, a hypothesis that has been confirmed in later studies [144–146, 149]. As demonstrated in Figure 5.23, a marked yield-stress anomaly is only seen in coarse-grained alloys, whereas fine-grained materials exhibit a weak peak or a yield-stress plateau in the corresponding temperature range. This observation can be rationalized by the following arguments. Grain refining is one of the most important hardening mechanisms for TiAl alloys, as will be described in Section 7.1. As the grain size decreases, the yield stress increases and then, because the general stress level at low and intermediate temperatures is high, any additional hardening by a locking mechanism is masked. The situation is similar to that in polycrystalline Ni_3Al, which exhibits a yield-stress anomaly for coarse-

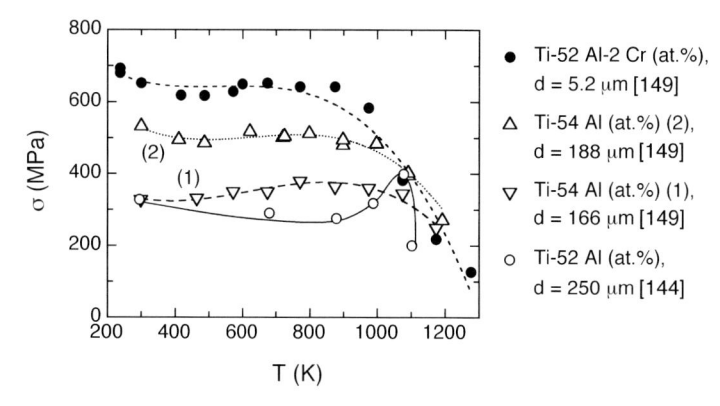

Figure 5.23 Dependence of the yield stress on temperature for different polycrystalline single-phase titanium aluminide alloys.

grained but not for fine-grained material [150]. The yield-stress anomaly also appears to be obscured by precipitation hardening, as becomes evident by comparing the stress–temperature profiles of the two Ti-54Al alloys, which have nearly the same grain size, but significantly different oxygen and nitrogen content. The impurity levels are 730 at. ppm oxygen and 30 at. ppm nitrogen for alloy 1 and 2600 at. ppm oxygen and 2600 at. ppm nitrogen for alloy 2. The high level of interstitial elements in alloy 2 certainly exceeds the solubility limit of γ(TiAl), which is about 250 at. ppm for oxygen [151]. It might thus be expected that oxides and nitrides are formed, which would give rise to significant precipitation hardening and enhances the stress level over a wide temperature range. Thus, although being coarse-grained, alloy 2 exhibits a weaker yield-stress anomaly than the relatively pure alloy 1.

The thermal stability of the locking mechanisms causing the yield-stress anomaly was investigated by reversibility test, which were proposed by Cottrell *et al.* [152]. In such experiments a sample is first deformed near the peak temperature, quenched and redeformed at room temperature. TiAl single crystals tested in this way mostly retained their high-temperature yield stress at lower temperature, that is, the resulting yield stress was higher than when tested at room temperature alone [40, 116, 153, 154]. This indicates that the dislocations in TiAl are permanently locked during the high-temperature deformation. The situation is unlike that in Ni_3Al, which exhibits a fully reversible yield-stress anomaly upon changes in temperature [96]. Loosely speaking, this means that the locked dislocation configurations in TiAl are less easily broken than in Ni_3Al. Stucke *et al.* [116] have rationalized this observation by noting that in TiAl interstitials may precipitate onto the dislocations at high temperatures thereby locking them at lower temperature. In some cases a reversible change of the yield stress with the temperature was observed [17, 155], however, this is not yet fully understood. In this context, Mahapatra *et al.* [154] have pointed out that the flow-stress anomaly on retesting at room temperature is irreversible when the high-temperature test is performed below the peak temperature, but reversible when it is performed above the peak temperature.

In summary, the slip systems operating in Al-rich γ(TiAl) exhibit a pronounced anomalous increase of the yield stress with temperature and often violate the Schmid law. In the case of the <101]{111} superdislocation glide systems the yield-stress anomaly can be attributed to the formation of thermally stable dislocation locks. The anomalous hardening of the 1/2<110]{111} ordinary glide system is probably caused by the formation of an ordered Al_5Ti_3 phase. Taken together, these results provide good supporting evidence that the yield-stress anomaly can be strongly affected by minor phases present in Al-rich alloys. These observations also cast some doubt on the transferability of dislocation properties determined on Al-rich single-phase alloys onto two-phase ($\alpha_2 + \gamma$) alloys. This is because the γ phase that coexists together with the α_2(Ti$_3$Al) phase certainly has a significantly lower Al concentration and a different content and state of interstitial impurities.

5.2
Deformation Behavior of Single-Phase α_2(Ti$_3$Al) Alloys

5.2.1
Slip Systems and Deformation Kinematics

Polycrystalline Ti$_3$Al alloys are known to be brittle; a property that is often associated with the slip geometry of the hexagonal ordered DO_{19} structure. Based on the most dense lattice planes and the shortest translation vectors, several possible slip systems may be deduced for α_2(Ti$_3$Al); however, only a few of these are really operative. These slip systems involve the prism, basal and pyramidal planes and are illustrated in Figure 5.24. Deformation is basically accomplished by two types of superdislocations; the so-called <a>-type dislocations with the Burgers vector $\mathbf{b} = 1/3<11\bar{2}0>$, and the <2c + a>-type dislocations with Burgers vector $\mathbf{b} = 1/3<\bar{1}\bar{1}26>$.

Under most conditions deformation is largely provided by $1/3<11\bar{2}0>$ superdislocations gliding on $\{10\bar{1}0\}$ prism planes, as shown in early studies [156, 157]. The $1/3<11\bar{2}0>\{10\bar{1}0\}$ slip system and the relevant Burgers vectors are illustrated in Figures 5.25 and 5.26. The $1/3<11\bar{2}0>$ dislocations propagate as coupled $1/6<11\bar{2}0>$ superpartials separated by an antiphase ribbon lying on the $\{10\bar{1}0\}$ prism plane [158–163], according to the decomposition reaction

$$1/3<11\bar{2}0> \rightarrow 1/6<11\bar{2}0> + APB + 1/6<11\bar{2}0> \tag{5.9}$$

TEM weak-beam images of the dissociated $1/3<11\bar{2}0>$ dislocation often exhibit a symmetrical four-fold peak in the intensity profile, suggesting a subsequent dissociation of the superpartials into Shockley partial dislocations [164, 165]. A detailed image analysis has revealed, however, that the contrast phenomena are

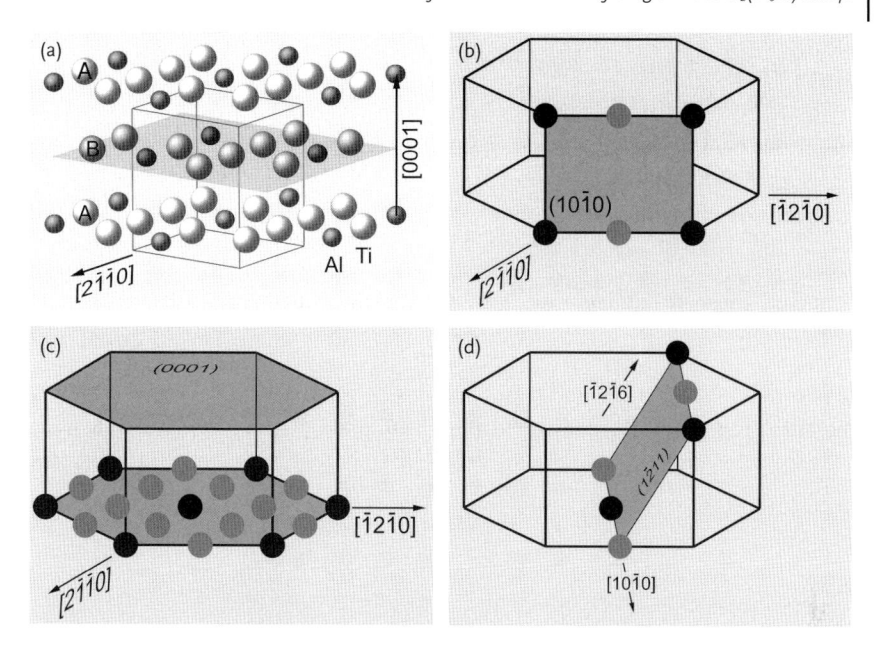

5.2 Deformation Behavior of Single-Phase α_2(Ti$_3$Al) Alloys

Figure 5.24 Slip geometry α_2(Ti$_3$Al). (a) DO$_{19}$ structure of Ti$_3$Al with the unit cell indicated. (b) Prismatic slip system [$\bar{1}2\bar{1}0$] ($10\bar{1}0$), (c) basal slip system [$\bar{1}2\bar{1}0$](0001), and (d) pyramidal slip system [$\bar{1}2\bar{1}6$] ($\bar{1}211$).

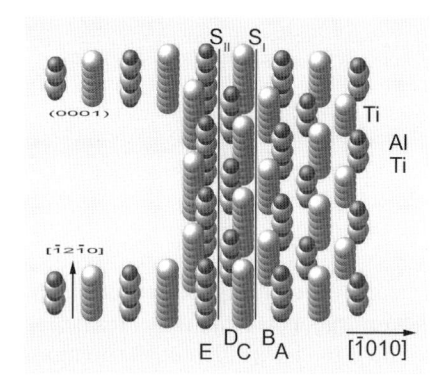

Figure 5.25 Prismatic <$\bar{1}2\bar{1}0$>{$10\bar{1}0$} glide of α_2(Ti$_3$Al), illustrated by a hard-sphere model. Perspective view of the ($10\bar{1}0$) planes along the [$\bar{1}21.\bar{6}$] direction. Note the different site occupancy of the widely spaced atomic layers adjacent to ($10\bar{1}0$) by either, only Ti atoms or a mixture of Ti and Al atoms, which gives rise to two distinct morphologies of the slip plane marked with S$_I$ and S$_{II}$.

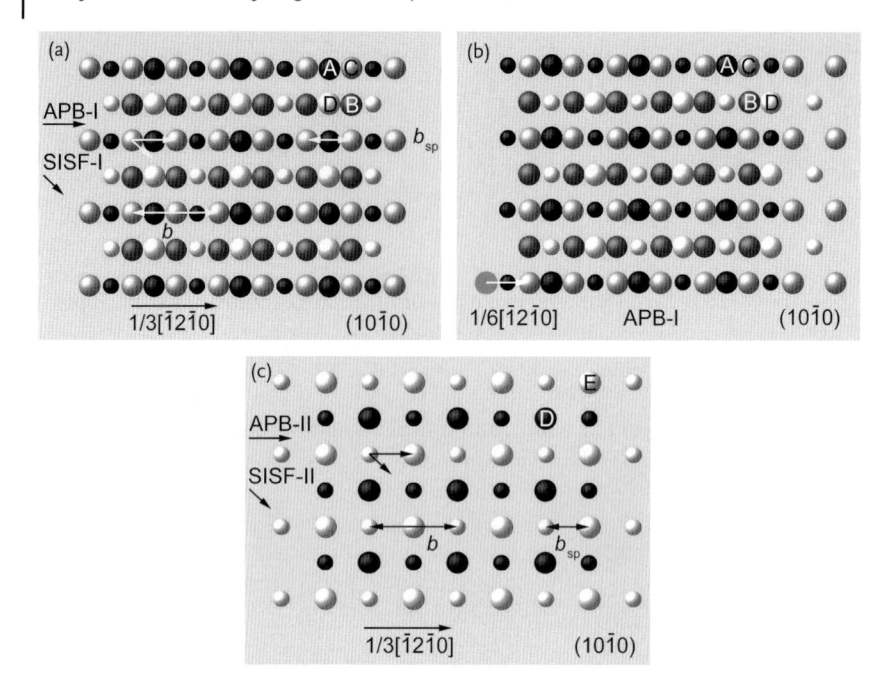

Figure 5.26 Burgers vectors and planar faults in $\alpha_2(Ti_3Al)$ associated with $1/3<\bar{1}2\bar{1}0>\{10\bar{1}0\}$ prismatic glide. (a) Atomic configuration of the $(10\bar{1}0)$ slip plane designated with S_I in Figure 5.25. Four adjacent $(10\bar{1}0)$ layers **ABCD** are imaged in different brightness so that the lowest **A** plane appears darkest. The slip and fault plane S_I is located between planes **B** and **C**. The shear displacements associated with the formation of the antiphase boundary APB-I and a superlattice intrinsic stacking fault SISF-I are indicated. $b = 1/3[\bar{1}2\bar{1}0]$ is the Burgers vector of the superdislocation and $b_{sp} = 1/6[\bar{1}2\bar{1}0]$ that of the related superpartials. (b) Structure of the APB-I formed by shifting the upper layers **C** and **D** along $1/6[\bar{1}2\bar{1}0]$. (c) Structure of the $(10\bar{1}0)$ slip plane S_{II} that is located between layers **D** and **E**, both composed of Ti and Al atoms and defined in Figure 5.25.

associated with a dipole configuration of two decomposed $1/3<11\bar{2}0>$ dislocations [161, 166].

For prismatic glide two different corrugated layers of atomic planes have to be considered, as early noted by Umakoshi and Yamaguchi [167]. This is because the widely spaced $\{10\bar{1}0\}$ layers adjacent to the slip plane have different site occupancies (Figure 5.25). In the first case the two neighboring planes contain only Ti atoms, whereas in the second case the planes consist of both Ti and Al atoms. This gives rise to two distinct morphologies of the slip plane that are marked with S_I and S_{II} in Figure 5.25. The two possible ways to choose the $\{10\bar{1}0\}$ slip plane are associated with two different APB and SISF types. The anticipated faults situated on the $\{10\bar{1}0\}$ plane and the associated shear displacements [19, 167, 168] are sketched in Figure 5.26. While the two APBs are identical and have the same displacement vector $1/6[\bar{1}2\bar{1}0]$, the atomic occupancy of the neighboring $\{10\bar{1}0\}$ planes is different. For the APB-I the separations and the local composition of the first- and second-nearest neighbors are the same as in the perfect lattice, which is not the case for the APB-II [19, 38, 167, 168]. Hence, the energy Γ(APB) of the APB_I is expected to be lower than that of the APB_{II}. Based on experimental data on Mg_3Cd [169], Umakoshi and Yamaguchi [167] have suggested that the energy ratio for the two APB types in Ti_3Al is

$$\Gamma(\text{APB-II})/\Gamma(\text{APB-I}) = 5 \tag{5.10}$$

a hypothesis that was supported by atomistic calculations [170, 171] and TEM observations [162, 163]. Likewise, two types of SISFs are expected for the two possible $\{10\bar{1}0\}$ glide planes, which have the same fault vectors. Atomistic calculations of Cserti et al. [170] have shown that the fault vectors are very close to the translations SISF-I and SISF-II depicted in Figures 5.26a and c. A general description of the fault vector is $a/6[\bar{1}2\bar{1}X]$, with X and α depending on the details of the atomic interactions. Thus, the authors have pointed out that the faults are not solely determined by the symmetry of the $D0_{19}$ structure and might be different in different $D0_{19}$ materials. The planar fault energies determined by TEM analysis of dissociated dislocations [162, 163] and atomic simulations [170–173] are listed in Table 5.4. Accordingly, the APB energies can be ranked as

$$\Gamma(\text{APB-I}) < \Gamma(\text{APB}_{(0001)}) < \Gamma(\text{APB-II}) \tag{5.11}$$

The different core structures of the $1/3{<}11\bar{2}0{>}\{10\bar{1}0\}$ dislocations apparently affect their glide behavior in different characteristic ways, according to which the dislocations can be classified. In situ straining experiments performed inside the electron microscope on a polycrystalline Ti-24Al alloy have indeed shown that two types of dislocations exist on this glide system [162, 163]. The $1/3{<}11\bar{2}0{>}\{10\bar{1}0\}$ superdislocations involving the low-energy APB-I are widely decomposed into superpartials along their screw orientation and preferentially propagate in planar groups. The dislocations glide by kink bulging in a jerky manner, which suggests that their

Table 5.4 Planar fault energies of $\alpha_2(Ti_3Al)$ in mJ/m^2.

Γ(CSF) (0001)	Γ(APB) (0001)	Γ(APB) $\{11\bar{2}1\}$	Γ(APB-I) $\{10\bar{1}0\}$	Γ(APB-II) $\{10\bar{1}0\}$	Γ(SISF) (0001) $\{10\bar{1}0\}$	Method	Ref.
	63		42	84	69	WB	[162, 163]
	63		11	101	44.7	F-S	[170]
320	300	293	133	506		LDF/FLAPW	[171]
	299		129.4	484.4	69	LDF	[172]
	270		108.4	446.6		LDF	[172]
			6	318		EAM	[173]

CSF, complex stacking fault; APB, antiphase boundary; SISF, superlattice intrinsic stacking fault; WB, weak-beam transmission electron microscopy; EAM, Embedded-atom method; F–S, Finnis–Sinclair method; LDF, local density-functional theory; FLAPW, full-potential linearized augmented plane-wave.

mobility is impeded by high friction forces. In contrast, the decomposition of the $1/3<11\bar{2}0>$ dislocation is significantly smaller, when the superpartials are coupled by the high-energy APB-II. Dislocations with this core configuration are smoothly bowed-out between weak obstacles and propagate steadily, which suggests a relatively low glide resistance. It might be expected that the $1/3<11\bar{2}0>$ superdislocations involving the APB-II may cross-slip into the energetically favorable configuration involving the APB-I, however, this has only rarely been observed [162, 163].

The $1/3<11\bar{2}0>$ superdislocations may also glide on the basal plane, which, however, occurs to a much lesser extent than glide on prism planes [159, 160, 162, 163, 174–176]. Significant basal slip can only be achieved when the related Schmid factor is at least twice that of prismatic slip; nevertheless basal slip is often accompanied by prismatic slip [163]. The atomic structure of the $1/3<11\bar{2}0>(0001)$ slip system and the relevant Burgers vectors of perfect and partial dislocations are illustrated in Figures 5.27a and b. The shear displacements associated with the APB, SISF and CSF that may exist on the basal plane are shown in Figure 5.27c [15]. The structure of the APB that is produced by a displacement of the upper B layer along $1/6[\bar{1}\bar{1}20]$ is illustrated in Figure 5.27d; the APB energies determined for the basal plane are given in Table 5.4. *In situ* straining experiments have shown that the $1/3<11\bar{2}0>$ superdislocations propagate jerkily as pairs of superpartials, which are coupled by an APB. At low and intermediate temperatures basal slip occurs very planar, that is, a large number of dislocations glide on the same slip plane. This strain localization apparently leads to the formation of shear cracks [159, 162].

The third potential slip system $1/3<\bar{1}2\bar{1}6>\{1\bar{2}11\}$ is illustrated in Figure 5.28. The $\{1\bar{2}11\}$ planes are also designated as type II pyramidal planes of the second order or as π_2 planes. The $<2c + a>$ dislocations, $1/3<\bar{1}2\bar{1}6>$, propagate as $1/6<\bar{1}2\bar{1}6>$ superpartials, which are coupled by a ribbon of antiphase boundary on the $\{1\bar{2}11\}$ slip plane [158–160, 177, 178]. Under certain conditions, the $1/3<\bar{1}2\bar{1}6>$ dislocations are also capable of gliding on $\{2\bar{2}01\}$ planes, as observed by Legros et al. [179, 180] during TEM *in situ* straining. The $\{2\bar{2}01\}$ planes are also called pyramidal planes of the first order or as π_1 planes. Legros et al. [179] ascribed the different choice of pyramidal slip planes to an intrinsic glide anisotropy of the dislocations with respect to the shear direction, that is, the glide planes are π_2 under compression and π_1 under tension. For completeness, it should be mentioned that in rare cases $<c>$ slip on the second-order prismatic planes $\{2\bar{1}\bar{1}0\}$ has been observed [181, 182]. Glide of these [0001] dislocations is probably impeded by high Peierls stresses because the dislocations were aligned along certain crystallographic orientations.

To date, no evidence of mechanical twinning in stoichiometric Ti$_3$Al single crystals has been reported, although deformation structures were examined over a wide temperature range and after shock loading [183]. This observation contrasts with the situation in α-Ti, which exhibits profuse twinning at low and elevated temperatures [82, 184]. The difficulty of twinning in stoichiometric Ti$_3$Al was attributed to a complex shuffling mechanism [185]. This involves the atomic shuffling required to twin the hexagonal structure and an additional site interchange

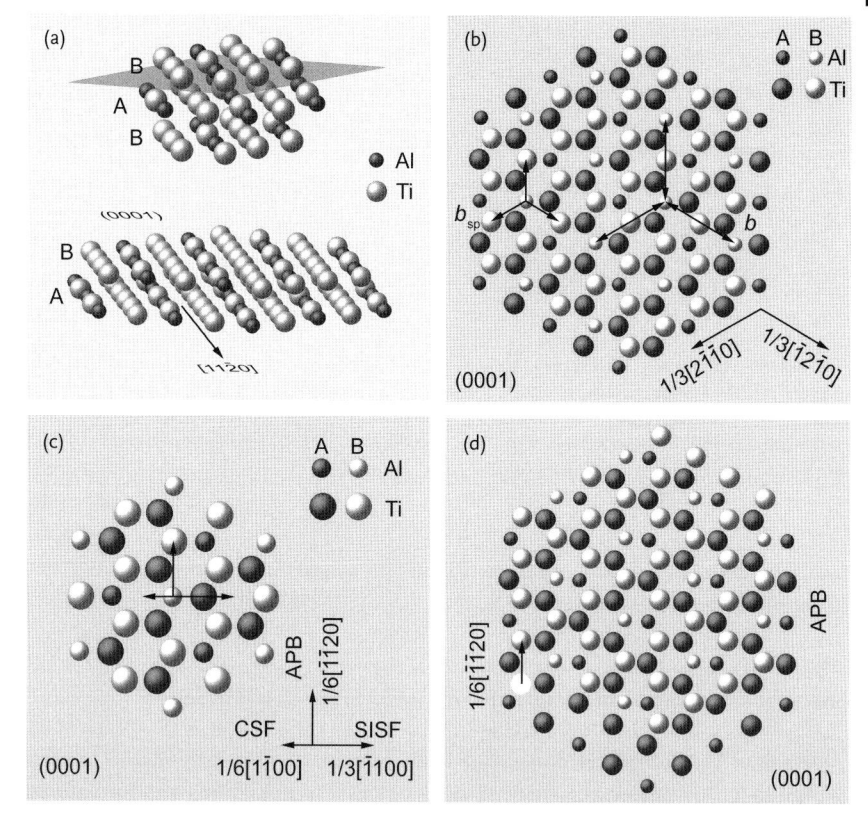

Figure 5.27 Basal glide $1/3<1\bar{1}20>(0001)$ of Ti$_3$Al. (a) Perspective view of two (0001) planes along the $[\bar{1}\bar{1}20]$ direction. (b) Atomic structure of the (0001) glide plane as seen from the [0001] direction. The figure shows two adjacent atomic layers **A** and **B** with the slip plane in between. $\mathbf{b} = 1/3[\bar{1}2\bar{1}0]$ is the Burgers vector of the superdislocation and $\mathbf{b}_{sp} = 1/6[\bar{1}2\bar{1}0]$ that of the related superpartials. (c) Planar faults in α_2(Ti$_3$Al) associated with $<1\bar{1}20>(0001)$ glide. Two adjacent (0001) planes, the lower plane **A** is imaged darker. The shear displacements associated with the formation of an antiphase boundary (APB), a superlattice intrinsic stacking fault (SISF) and a complex stacking fault (CSF) are indicated. (d) Structure of the APB formed by a displacement of the upper **B** layer along $1/6[\bar{1}\bar{1}20]$.

of Ti and Al atoms to preserve the order of the DO$_{19}$ structure. However, off-stoichiometric Ti$_3$Al can apparently be more easily forced to twin, as demonstrated by Lee et al. [186] on polycrystalline Ti-34Al. The authors pointed out that twin nucleation at the grain boundaries might be important for the activation of the mechanism.

In short, plastic deformation of α_2(Ti$_3$Al) may occur on prismatic, basal and pyramidal glide systems, a situation that is similar to the behavior of the closest related disordered metal α-Ti. However, deformation of α_2(Ti$_3$Al) is carried by superdislocations with complex core structures. Decomposition and dissociation

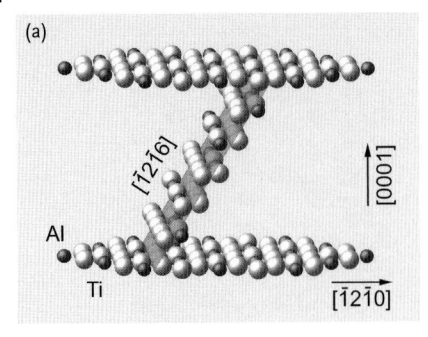

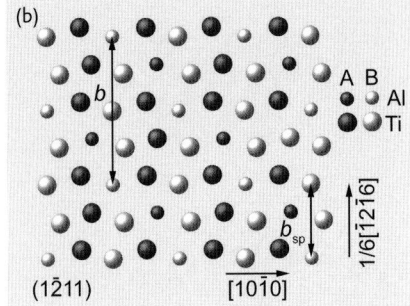

Figure 5.28 Pyramidal glide $<\bar{1}2\bar{1}6>\{\bar{1}2\bar{1}1\}$ of α_2(Ti$_3$Al). (a) Perspective view of the glide plane situated between two adjacent $(\bar{1}2\bar{1}1)$ planes seen from near the $[10\bar{1}0]$ direction. (b) Atomic structure of the $(\bar{1}2\bar{1}1)$ glide plane. The slip plane is situated between the two layers **A** and **B**. $\mathbf{b} = 1/3[\bar{1}2\bar{1}6]$ is the Burgers vector of the superdislocation and $\mathbf{b}_{sp} = 1/6[\bar{1}2\bar{1}6]$ that of the related superpartials.

reactions involving various types of planar faults apparently make the activation of certain slip systems difficult.

5.2.2
Effects of Orientation and Temperature on Deformation of α_2 Phase

The relative ease of the above-described slip systems was assessed by mechanical testing of Ti$_3$Al single crystals over a wide range of temperatures and for various sample orientations. Figure 5.29 demonstrates the critical resolved shear stresses τ_c (CRSS) for the onset of prism, basal and pyramidal slip as determined by different authors [160, 175, 177, 187, 188]. The data clearly indicates that the activation of prismatic slip is easiest, followed by basal and pyramidal slip; the ratio of the related τ_c values is approximately

$$\tau_{c\,prismatic} : \tau_{c\,basal} : \tau_{c\,pyramidal} = 1 : 3 : 9 \tag{5.12}$$

The prismatic and basal slip systems involving only <a>-type dislocations do not supply the five independent slip systems, which are required by the von Mises criterion [93] for general plasticity. For this criterion to be fulfilled, pyramidal slip is necessary to accommodate strains with components in the **c** direction of the hexagonal cell [189]. However, the large differences in the yield stresses suggest that pyramidal slip can only be activated in [0001] sample orientations, where the Schmid factors for prismatic and basal slips are zero. This plastic anisotropy is also manifested in the plastic deformability of Ti$_3$Al single crystals. When pure prismatic glide is activated in such crystals, extremely large tensile elongation can be achieved. For example, on Ti-rich Ti$_3$Al single crystals an elongation of more than 200% was measured in room-temperature tensile tests [159].

The CRSS values for prismatic slip are independent of sample orientation, which indicates that the Schmid law is obeyed. Prism and basal slip exhibit a

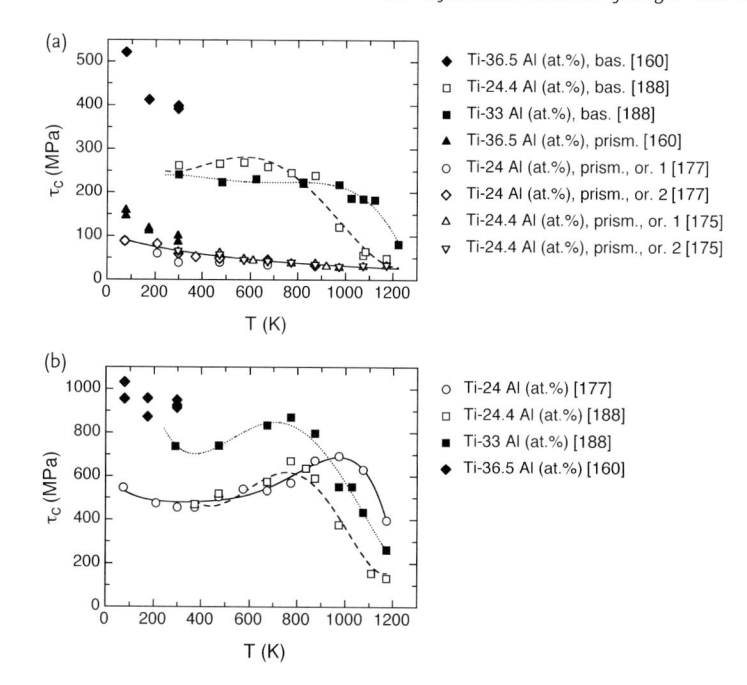

Figure 5.29 Deformation behavior of Ti$_3$Al single crystals; variation of the critical resolved shear stress with temperature. (a) Basal slip 1/3<11$\bar{2}$0>(0001) and prismatic slip 1/3<11$\bar{2}$0>{10$\bar{1}$0} (two sample orientations). (b) Pyramidal slip 1/3<$\bar{1}\bar{1}$26>{11$\bar{2}$1}.

normal flow-stress decrease with increasing temperature. This contrasts with the CRSS for pyramidal slip, which shows an anomalous dependence on temperature, with peak values at 500 °C to 600 °C (Figure 5.29b). This anomalous stress–temperature profile of the pyramidal slip system significantly enhances the plastic anisotropy of Ti$_3$Al alloys at intermediate temperatures. Thus, unlike common experience, the tendency for brittle fracture may increase with temperature. Off-stoichiometric crystals behave similarly; however, the yield stresses are generally higher [159, 160, 188]. The origin of this hardening is not yet clear, but might be related to the particular point-defect situation in Ti-rich Ti$_3$Al alloys [190, 191], as will be discussed in Section 6.4.3.

The slip geometry of Ti$_3$Al alloys is apparently not fully understood. The strong preference of prism glide in Ti$_3$Al is not in accordance with the classical argument of lattice friction, according to which the most widely separated planes should be the preferred slip planes [9]. In Ti$_3$Al the maximum separation distance between adjacent layers of (0001) planes is significantly larger than that between {10$\bar{1}$0} planes. Thus, contrary to the experimental observations, the 1/3<11$\bar{2}$0> superdislocations should be more mobile on the (0001) planes. Following the arguments of Greenberg et al. [128], Court et al. [158] ascribed the deformation behavior to the bonding character of Ti$_3$Al. Strong covalent bonding is expected between crystallographic planes containing only Ti atoms, which tend to immobilize specific dislocation characters. Such Ti layers exist, for example, in $\{10\bar{1}0\}$, $\{11\bar{2}0\}$ and $\{0\bar{1}11\}$ planes. Hence, the propagation of dislocation segments with orientations close to $<10\bar{1}0>$, $<11\bar{2}0>$ and $<0\bar{1}12>$ will be impeded by high Peierls forces. Legros et al. [162] have ruled out covalency as the only reason for the friction stress by noting that such a mechanism would provide the same friction force for both types of $1/3<11\bar{2}0>$ dislocations gliding on prism planes, which is clearly not the case.

Based on atomistic calculations Cserti et al. [170] have proposed that the mobility of the $1/3<11\bar{2}0>$ superdislocation is controlled by nonplanar core spreading. Accordingly, the $1/3<11\bar{2}0>$ superdislocations decompose more easily into superpartials on the prism plane than on the basal plane because the APB energy is lower on the prism plane. The calculations have shown that the $1/6<11\bar{2}0>$ superpartials can simultaneously spread into both the prism and basal plane, which results in a sessile nonplanar core configuration. Legros et al. [162] have noted that repeated locking and unlocking provides a natural explanation for the observed jerky motion of the dislocations. A similar mechanism has been proposed for the pyramidal slip system. Based on the observation that the $1/3<\bar{1}\bar{1}26>\{11\bar{2}1\}$ dislocations gliding on the $\{11\bar{2}1\}$ planes are significantly wider dissociated than the $1/3<11\bar{2}0>$ dislocations [192, 193], Loretto [194] has suggested that one of the $1/6<\bar{1}\bar{1}26>$ superpartials may further dissociate into $1/6<11\bar{2}0]$ and $[0001]$ dislocations, which leads to pinning.

In summary, Ti$_3$Al alloys have several potential slip systems, which in principle may provide sufficient shear components for the deformation of polycrystalline material. However, there is a strong predominance for prismatic slip, which makes plastic shear with **c** components of the hexagonal cell practically impossible. This plastic anisotropy appears to be even more enhanced at intermediate temperatures because the pyramidal slip systems exhibit an anomalous increase of the CRSS with temperature. Thus, the brittleness of polycrystalline Ti$_3$Al alloys can be attributed to the lack of independent slip systems that can operate at comparable stresses; hence, the von Mises criterion [93] is not satisfied.

5.3
β/B2 Phase Alloys

As has been discussed in Chapter 2, the body centered cubic (b.c.c.) β phase and its ordered counterpart B2 can be significant constituents in Al-lean TiAl alloys containing refractory metals such as Cr, W, Mo, Nb, Ta, and V. There is only little information available about the deformation behavior of these two phases, when they are constituents of multiphase assemblies. Thus, the kinematical aspects of the deformation of the b.c.c. and B2 structure will primarily be described.

The b.c.c. structure is shown in Figure 5.30 as hard spheres that touch along the <111> direction, that is, along the body-diagonal directions. The structure can be built up by the stacking of identical {112} layers in the sequence **ABCDEFAB-CDEF** with a repeat distance perpendicular to this plane of six layers. The structure can also be considered as the stacking of {110} layers in an **ABAB** sequence, that

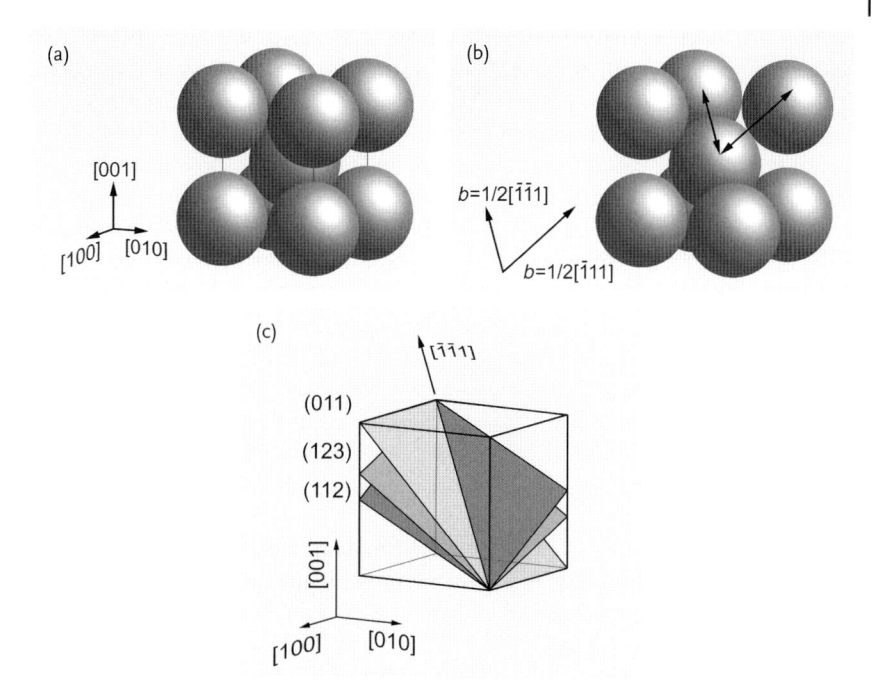

Figure 5.30 Slip geometry of the body centered cubic (b.c.c.) structure. (a) Unit cell represented as hard spheres touching along the <111>directions. (b) Illustration of the 1/2<111>Burgers vectors; one corner atom of the unit cell is removed. (c) Potential slip planes in the b.c.c. structure.

is, with a two-layer repeat distance. Shear deformation of b.c.c. metals in almost all cases occurs along close-packed <111> directions, which corresponds to the shortest lattice translation or Burgers vector **b** = 1/2<111> (Figure 5.30b). Unlike f.c.c. metals, there is no strong preference for a specific slip plane in b.c.c. metals because there is no fully closely packed plane. The commonly observed slip planes are the {110} planes, which are the most-densely packed planes. Slip has also been recognized on {112} and {123} planes. The three slip planes have a common <1$\bar{1}$1> zone axis, that is, three {110}, three {112} and six {123} planes intersect along the same <111> direction, which is parallel to the Burgers vector (Figure 5.30c). Thus, in view of the von Mises criterion [93] there are sufficient independent slip modes available for the deformation of polycrystalline b.c.c. metals [9]. If cross-slip is easy, the screw dislocations tend to propagate on different {110} planes or in combinations of {110}, {112} and {123} planes, often making the identification of the slip plane extremely difficult. It is often speculated that glide processes on {123} and perhaps {112} planes are actually composed of short "composite" slip steps on {110} planes. The apparent slip plane varies with composition, crystal orientation, temperature, and strain rate. At elevated temperatures noncrystallographic slip has been observed.

Atomistic calculations of Vitek [195] have shown that in the b.c.c. structure no metastable faults exist other than on {112} planes. The term "metastable" is meant in the sense that the fault has a sufficiently low energy to allow dislocation dissociation or fault growth. However, as noted by Hirsch [196], the 1/2<111> Burgers vector lies along an axis of three-fold symmetry of the b.c.c. lattice, which can give rise to a three-fold extension of the screw dislocation core into three {110} planes having the <111> zone. This nonplanar core extension was justified by atomistic calculations [197–200]. In particular the model explains the indeterminate nature of the slip plane, the high Peierls stress of b.c.c. metals, and that the dislocation structure after deformation tends to consist of straight screw dislocations. A very important consequence of the nonplanar cores is that the dislocation behavior is prone to nonglide components of the stress tensor [198]. The self-trapping of the screw dislocations and their release by thermal nucleation of double kinks is a commonly held concept to explain the rapid increase of the flow stress of b.c.c. metals at low temperatures [200]. For more information the reader is referred to several reviews [201–206].

Deformation twinning is a prominent deformation mode in all the b.c.c. transition metals, when deformed at low temperatures or high strain rates. The twinning elements of the most important system are $K_1 = \{112\}$, $K_2 = \{\bar{1}\bar{1}2\}$ $\eta_1 = <\bar{1}\bar{1}1>$, and $\eta_2 = <111>$, for reviews see [82, 91]. The successive twinning displacements can be accomplished by the movement of partial 1/6<111> dislocations. The shear displacement is asymmetric with respect to the twinning and antitwinning sense on the K_1 planes. There is an interesting analogy between slip and twinning, which might be related to the above sketched core structure of the 1/2<111> dislocations. Slip seems to be easier when the applied stress is such that a dislocation would move in the twinning sense on {112} planes rather than the antitwinning sense, even when the actual slip plane is not {112}.

Upon ordering, the symmetry of the b.c.c structure is reduced to the B2 structure, shown in Figure 5.31. The **A** and **B** atoms occupy two interpenetrating simple cubic lattices displaced from one another by (1/2, 1/2, 1/2). Examples of the B2 phase identified in ternary alloys are Ti$_2$AlCr [207] or Ti$_2$AlMo [208]. The shortest translation vectors in the order of increasing magnitude are <100>, <110> and

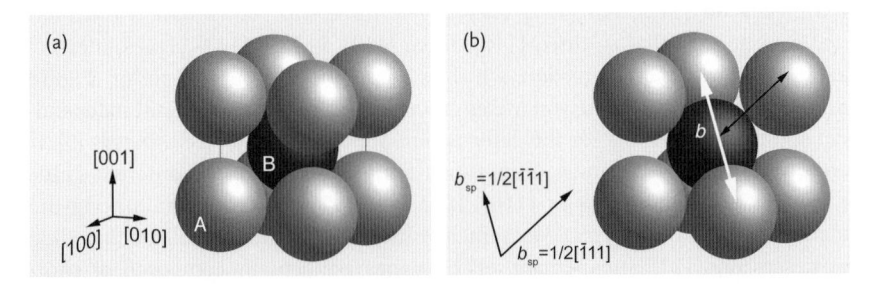

Figure 5.31 Slip geometry of the B2 structure. (a) Unit cell represented as hard spheres touching along the <111> directions. (b) Illustration of the Burgers vector of the superdislocations **b** = <111> and of the related superpartials **b** = 1/2<111>.

<111>. With rare exceptions, only dislocations with <100> and <111> Burgers vectors have been observed after low-temperature deformation. The 1/2<111> dislocation, being perfect in the b.c.c. structure, is an APB trailing superpartial in the B2 structure. Baker and Munroe [206] and Sauthoff [209] have given extensive surveys of the mechanical properties of B2 compounds. The commonly observed slip planes of the <100> dislocations are {011} and {001}, whereas the <111> dislocations normally slip on {110}, {112} and {123} planes. In the B2 structure no stacking fault is known to be stable [19, 210].

B2 alloys that solely deform by <100> slip have only three independent slip modes, which are insufficient for the deformation of polycrystalline material. B2 compounds gliding along <111> have five independent slip systems and are thus likely to be inherently ductile. The activation of <111> slip is therefore an important issue in the design of B2 compounds. For order–disorder B2 compounds the choice of the slip direction apparently depends on the degree of order. Alloys with a high ordering energy tend to glide along <100>; whereas <111> slip is favored in weakly ordered compounds [206, 211]. As pointed out by Baker [212], the underlying principles behind the choice of the Burgers vector, that is, <111> or <100>, have still not been explained adequately.

Fully ordered B2 compounds apparently cannot twin on the usual b.c.c. twinning system. Constraint stresses, which arise from the lack of independent slip modes, therefore, cannot relax by mechanical twinning. Ab-initio calculations of Nguyen-Man and Pettifor [213] have shown that the B2 phase is prone to decomposition into structurally related stable and metastable phases. In this context, pseudotwinning of B2 compounds has been discussed, which can formally be described as a martensitic transformation from the simple cubic structure $Pm\bar{3}m$ to an orthorhombic structure of space group Cmmm, that is, the product structure is not crystallographically identical to the matrix. Debate remains, whether or not such a mechanism occurs under usual stress levels. For more details the reader is referred to relevant review articles [21, 82, 91].

The little reported data about the deformation behavior of the β/B2 phase in the γ(TiAl)-based alloys might be summarized as follows. β-Ti deforms by 1/2<111> glide on {110}, {112} and {123} planes [214]. Banerjee et al. [215] investigated the deformation structure of a Ti_3Al–Nb alloy, in which the B2 phase was manifested as Ti_2AlNb. The B2 phase was found to slip along <111> on {110}, {112} and {123} planes, with a strong tendency for slip localization into slip bands. Morris and Li [216] have identified the B2 phase in a Ti-44Al-2Mo alloy that was strengthened by ω phase particles. TEM examination of samples compressed at room temperature showed that deformation mainly occurred in the γ phase. Deformation of the B2 phase was initiated by constraint stresses, which develop by the deformation and work hardening of adjacent γ grains. These glide processes of the B2 phase were found to be carried by <100> dislocations and dissociated <111> superdislocations on {110} planes. Thus, deformation of the B2 phase appears difficult, when compared with the γ phase.

Fu-Sheng Sun et al. [217] stabilized a significant amount of B2 phase in a two-phase ($α_2$ + γ) alloy by ternary and quaternary additions of Fe, Cr, V, and Nb. After room-temperature deformation, the authors recognized little dislocation activity in the B2 phase and cleavage fracture of B2 grains. This observation was taken as an indication that the B2 phase is relatively brittle at low temperatures.

Naka and Khan [218] compared the deformation behavior of single-phase β and B2 alloys. The disordered β phase alloy with the composition Ti-10Al-60Nb-10Mo exhibited <111>{110} slip after deformation at 25 °C and 800 °C. The 1/2<111> dislocations observed after room-temperature deformation were found to be aligned along their screw orientation, which was taken as an indication of a high Peierls stress. The deformation behavior of the ordered B2 alloy with composition Ti-17Al-32Nb-17Mo was found to depend on temperature. Deformation at room temperature was controlled by slip on the <111>{1$\bar{2}$1} and <001>{1$\bar{1}$0} systems, whereas the slip systems observed after deformation at 800 °C were <111>{1$\bar{1}$0} and <001>{1$\bar{1}$0}. The <111> dislocations were not dissociated, which suggests a relatively high APB energy. The observed activation of <111> glide might be advantageous for the deformability of polycrystalline material.

In summary, the b.c.c. structure of the disordered β phase formally exhibits sufficient independent glide and twinning systems to satisfy the von Mises criterion, which is a good precondition for the plasticity of polycrystalline material. In the ordered counterpart with the B2 structure this condition is only fulfilled if <111> shear can be activated. In both the β and B2 phase, the operation of a certain slip system strongly depends on the alloy composition, the degree of ordering, the point-defect situation, and the deformation conditions. Thus, the possible implications that these two phases may have on the deformability of titanium aluminide alloys remain somewhat vague.

References

1 Shechtman, D., Blackburn, M.J., and Lipsitt, H.A. (1974) *Metall. Trans. A*, **5A**, 1373.
2 Lipsitt, H.A., Shechtman, D., and Schafrik, R.E. (1975) *Metall. Trans. A*, **6A**, 1991.
3 Hug, G., Loiseau, A., and Veyssière, P. (1988) *Philos. Mag.*, **57**, 499.
4 Hug, G. (1988) *Etude par microscopie électronique en faisceau faible de la dissociation des dislocations dans TiAl: relation avec le comportement plastique.* PhD Thesis, Université de Paris-Sud, France.
5 Hug, G., Loiseau, A., and Lasalmonie, A. (1986) *Philos. Mag.*, **54**, 47.
6 Veyssière, P. (1988) *Rev. Phys. Appl.*, **23**, 431.
7 Court, S.A., Vasudevan, V.K., and Fraser, H.L. (1990) *Philos. Mag. A*, **1**, 141.
8 Yamaguchi, M. and Umakoshi, Y. (1990) *Prog. Mater. Sci.*, **34**, 1.
9 Hirth, J.P. and Lothe, J. (1992) *Theory of Dislocations*, Krieger, Melbourne.
10 Sun, Y.Q. (1999) *Philos. Mag. Lett.*, **79**, 539.
11 Jiao, Z., Whang, S.H., Yoo, M.H., and Feng, Q. (2002) *Mater. Sci. Eng. A*, **329–331**, 171.
12 Feng, Q. and Whang, S.H. (2000) *Acta Mater.*, **48**, 4307.
13 Jiao, Z., Whang, S.H., and Wang, Z. (2001) *Intermetallics*, **9**, 891.
14 Yoo, M.H. (1998) *Scr. Mater.*, **39**, 569.
15 Whang, S.H. and Hahn, Y.D. (1990) *Scr. Metall. Mater.*, **24**, 1679.

16 Jiao, S., Bird, N., Hirsch, P.B., and Taylor, G. (1998) *Philos. Mag. A*, **78**, 777.
17 Inui, H., Matsumoro, M., Wu, D.-H., and Yamaguchi, M. (1997) *Philos. Mag. A*, **75**, 395.
18 Peierls, R.E. (1940) *Proc. Phys. Soc.*, **52**, 34.
19 Paidar, V. and Vitek, V. (2002) *Intermetallic Compounds. Vol. 3, Principles and Practice* (eds J.H. Westbrook and R.L. Fleischer), John Wiley & Sons, Ltd, Chichester, p. 437.
20 Vitek, V. (1974) *Crystal Lattice Defects*, **5**, 1.
21 Christian, J.W. and Laughlin, D.E. (1988) *Acta Metall.*, **36**, 1617.
22 Yoo, M.H. (1989) *J. Mater. Res.*, **4**, 50.
23 Yoo, M.H., Fu, C.L., and Lee, J.K. (1994) *Twinning in Advanced Materials* (eds M.H. Yoo and M. Wuttig), TMS, Warrendale, PA, p. 97.
24 Inkson, B. and Humphreys, C.J. (1995) *Philos. Mag. Lett.*, **71**, 307.
25 Girshick, A. and Vitek, V. (1995) *High-Temperature Ordered Intermetallic Alloys VI, Materials Research Society Symposia Proceedings*, vol. 364 (eds J.A. Horton, I. Baker, S. Hanada, R.D. Noebe, and D.S. Schwartz), MRS, Pittsburgh, PA, p. 145.
26 Greenberg, B.A. (1970) *Phys. Status Solidi*, **42**, 459.
27 Hug, G. and Veyssière, P. (1989) *Electron Microscopy in Plasticity and Fracture Research of Materials* (eds U. Messerschmidt, F. Appel, and V. Schmidt), Akademie Verlag, Berlin, p. 451.
28 Flinn, P.A. (1960) *Trans. Metall. Soc. AIME*, **218**, 145.
29 Fu, C.L. and Yoo, M.H. (1990) *Philos. Mag. Lett.*, **62**, 159.
30 Fu, C.L. and Yoo, M.H. (1993) *Intermetallics*, **1**, 59.
31 Yoo, M.H. and Fu, C.L. (1993) *Structural Intermetallics* (eds R. Darolia, J.J. Lewandowski, C.T. Liu, P.L. Martin, D.B. Miracle, and M.V. Nathal), TMS, Warrendale, PA, p. 283.
32 Woodward, C., MacLaren, J.M., and Rao, S.I. (1991) *High-Temperature Ordered Intermetallic Alloys IV, Materials Research Society Symposia Proceedings*, vol. 213 (eds L.A. Johnson, D.P. Pope, and J.O. Stiegler), MRS, Pittsburgh, PA, p. 715.
33 Woodward, C., MacLaren, J.M., and Rao, S.I. (1992) *J. Mater. Res.*, **7**, 1735.
34 Simmons, J.P., Rao, S.I., and Dimiduk, D.M. (1993) *High-Temperature Ordered Intermetallic Alloys V, Materials Research Society Symposia Proceedings*, vol. 288 (eds I. Baker, R. Darolia, J.D. Whittenberger, and M.H. Yoo), MRS, Pittsburgh, PA, p. 335.
35 Panova, J. and Farkas, D. (1995) *Gamma Titanium Aluminides* (eds Y.-W. Kim, R. Wagner, and M. Yamaguchi), TMS, Warrendale, PA, p. 331.
36 Woodward, C. and MacLaren, J.M. (1996) *Philos. Mag. A*, **74**, 337.
37 Ehmann, J. and Fähnle, M. (1998) *Philos. Mag. A*, **77**, 701.
38 Yoo, M.H. and Fu, C.L. (1998) *Metall. Mater. Trans. A*, **29A**, 49.
39 Hug, G., Douin, J., and Veyssière, P. (1989) *High-Temperature Ordered Intermetallic Alloys III, Materials Research Society Symposia Proceedings*, vol. 133 (eds C.T. Liu, A.I. Taub, N.S. Stoloff, and C.C. Koch), MRS, Pittsburgh, PA, p. 125.
40 Stucke, M.A., Vasudevan, V.K., and Dimiduk, D.M. (1995) *Mater. Sci. Eng. A*, **192–193**, 111.
41 Wiezorek, J.M.K. and Humphreys, C.J. (1995) *Scr. Metall. Mater.*, **33**, 451.
42 Simmons, J.P., Mills, M.J., and Rao, S.I. (1995) *High-Temperature Ordered Intermetallic Alloys VI, Materials Research Society Symposia Proceedings*, vol. 364 (eds J.A. Horton, I. Baker, S. Hanada, R.D. Noebe, and D.S. Schwartz), MRS, Pittsburgh, PA, p. 137.
43 Cockayne, D.J.H., Ray, I.L.F., and Whelan, M.J. (1969) *Philos. Mag.*, **20**, 1265.
44 Marcinkowski, M.J. (1963) *Electron Microscopy and Strength of Crystals* (eds G. Thomas and J. Washburn), Interscience, New York, p. 431.
45 Suzuki, K., Ichihara, M., and Takeuchi, M. (1979) *Acta Metall.*, **27**, 193.
46 Suzuki, H. (1952) *Sci. Rep. Tohoku Univ.*, **A4**, 455.

47 Suzuki, H. (1962) *J. Phys. Soc. (Japan)*, **17**, 322.
48 Veyssière, P. and Douin, J. (1995) *Intermetallic Compounds. Vol. 1, Principles* (eds J.H. Westbrook and R.L. Fleischer), John Wiley & Sons, Ltd, Chichester, p. 519.
49 Greenberg, B.A. (1973) *Phys. Status Solidi*, **55**, 59.
50 Greenberg, B.A. and Gornostirev, Y.N. (1982) *Scr. Metall.*, **16**, 15.
51 Hahn, Y.D. and Whang, S.H. (1990) *Scr. Metall. Mater.*, **24**, 139.
52 Inkson, B. (1998) *Philos. Mag. A*, **77**, 715.
53 Jiao, S., Bird, N., Hirsch, P.B., and Taylor, G. (1999) *Philos. Mag. A*, **79**, 609.
54 Vitek, V. (1998) *Intermetallics*, **6**, 579.
55 Hemker, K.J., Viguier, B., and Mills, M.J. (1993) *Mater. Sci. Eng. A*, **164**, 391.
56 Farenc, S.V. and Couret, A. (1993) *High-Temperature Ordered Intermetallic Alloys V, Materials Research Society Symposia Proceedings*, vol. 288 (eds I. Baker, R. Darolia, J.D. Whittenberger, and M.H. Yoo), MRS, Pittsburgh, PA, p. 465.
57 Li, Z.X. and Whang, S.H. (1992) *Mater. Sci. Eng. A*, **152**, 182.
58 Morris, D.G., Günter, S., and Leboeuf, M. (1994) *Philos. Mag. A*, **69**, 527.
59 Sriram, S., Vasudevan, V.K., and Dimiduk, D.M. (1995) *Mater. Sci. Eng. A*, **192–193**, 217.
60 Zhang, Y.G., Xu, Q., Chen, C.Q., and Li, H.X. (1992) *Scr. Metall. Mater.*, **26**, 865.
61 Viguier, B. and Hemker, K. (1996) *Philos. Mag. A*, **73**, 575.
62 Grégori, F. and Veyssière, P. (2000) *Philos. Mag. A*, **80**, 2933.
63 Zhang, Y.G., Lei, C.H., Chen, C.Q., and Chaturvedi, M.C. (1998) *Philos. Mag. Lett.*, **77**, 33.
64 Chiu, Y.-L., Grégori, F., Nakano, T., Umakoshi, Y., and Veyssière, P. (2003) *Philos. Mag.*, **83**, 1347.
65 Greenberg, B.A., Antonova, O.V., Indenbaum, V.N., Karkina, L.E., Notkin, A.B., Pomarev, M.V., and Smirnov, L.V. (1991) *Acta Metall. Mater.*, **39**, 233.
66 Kear, B.H. and Wilsdorf, H.G. (1962) *Trans. Metall. Soc. AIME*, **224**, 382.
67 Zupan, M. and Hemker, K.J. (2003) *Acta Mater.*, **51**, 6277.
68 Wang, Z.M., Wei, C., Feng, Q., Whang, S.H., and Allard, L.F. (1998) *Intermetallics*, **6**, 131.
69 Mahapatra, R., Girschick, A., Pope, D.P., and Vitek, V. (1995) *Scr. Metall. Mater.*, **33**, 1921.
70 Rao, S., Woodward, C., Simmons, J., and Dimiduk, D.M. (1995) *High-Temperature Ordered Intermetallic Alloys VI, Materials Research Society Symposia Proceedings*, vol. 364 (eds J.A. Horton, I. Baker, S. Hanada, R.D. Noebe, and D.S. Schwartz), MRS, Pittsburgh, PA, p. 129.
71 Kumar, M., Sriram, S., Schwartz, A.J., and Vasudevan, V.K. (1999) *Philos. Mag. Lett.*, **79**, 315.
72 Lang, C., Hirsch, P.B., and Cockhayne, D.J.H. (2004) *Philos. Mag. Lett.*, **84**, 139.
73 Feng, C.R., Michel, D.J., and Crowe, C.R. (1988) *Scr. Metall.*, **22**, 1481.
74 Vasudevan, V.K., Stucke, M.A., Court, S.A., and Fraser, H.L. (1989) *Philos. Mag. Lett.*, **59**, 299.
75 Huang, S.C. and Hall, E.L. (1991) *Metall. Trans. A*, **22A**, 427.
76 Inui, H., Nakamura, A., Oh, M.H., and Yamaguchi, M. (1992) *Philos. Mag. A*, **66**, 557.
77 Farenc, S., Coujou, A., and Couret, A. (1993) *Philos. Mag. A*, **67**, 127.
78 Appel, F., Beaven, P.A., and Wagner, R. (1993) *Acta Metall. Mater.*, **6**, 1721.
79 Wardle, S., Phan, I., and Hug, G. (1993) *Philos. Mag. A*, **67**, 497.
80 Morris, M.A. (1993) *Philos. Mag. A*, **68**, 259.
81 Appel, F. and Wagner, R. (1994) *Twinning in Advanced Materials* (eds M.H. Yoo and M. Wuttig), TMS, Warrendale, PA, p. 317.
82 Christian, J.W. and Mahajan, S. (1995) *Prog. Mater. Sci.*, **39**, 1.
83 Jin, Z. and Bieler, T.R. (1995) *Philos. Mag. A*, **71**, 925.
84 Morris, M.A. and Leboeuf, M. (1997) *Intermetallics*, **5**, 339.
85 Cerreta, E. and Mahajan, S. (2001) *Acta Mater.*, **49**, 3803.
86 Appel, F. (2005) *Philos. Mag.*, **85**, 205.
87 Christian, J.W. (1975) *The Theory of Transformations in Metals and Alloys, Part I*, Pergamon Press, Oxford.

88 Bilby, B.A. and Crocker, A.G. (1965) *Proc. R. Soc. A*, **288**, 240.
89 Pashley, D.W., Robertson, T.L., and Stowell, M.J. (1969) *Philos. Mag.*, **19**, 83.
90 Yoo, M.H., Fu, C.L., and Lee, J.K. (1989) *High-Temperature Ordered Intermetallic Alloys III, Materials Research Society Symposia Proceedings*, vol. 133 (eds C.T. Liu, A.I. Taub, N.S. Stoloff, and C.C. Koch), MRS, Pittsburgh, PA, p. 189.
91 Yoo, M.H. (2002) *Intermetallic Compounds. Vol. 3, Principles and Practice* (eds J.H. Westbrook and R.L. Fleischer), John Wiley & Sons, Ltd, Chichester, p. 403.
92 Sun, Y.Q., Hazzledine, P.M., and Christian, J.W. (1993) *Philos. Mag. A*, **68**, 471.
93 von Mises, R. (1928) *Z. Angew. Meth. Mech.*, **8**, 161.
94 Schmid, E. and Boas, W. (1950) *Plasticity of Crystals*, Hughes, London.
95 Westbrook, J.H. (1957) *Trans. Metall. Soc. AIME*, **209**, 898.
96 Davies, R.G. and Stoloff, N.S. (1965) *Trans. Metall. Soc. AIME*, **233**, 714.
97 Copley, S.M. and Kear, B.H. (1967) *Trans. Metall. Soc. AIME*, **239**, 977.
98 Sauthoff, G. (1995) *Intermetallic Compounds. Vol. 1, Principles* (eds J.H. Westbrook and R.L. Fleischer), John Wiley & Sons, Ltd, Chichester, p. 911.
99 Liu, C.T. and Pope, D.P. (1995) *Intermetallic Compounds. Vol. 2, Practice* (eds J.H. Westbrook and R.L. Fleischer), John Wiley & Sons, Ltd, Chichester, p. 17.
100 Miracle, D.B. and Darolia, R. (1995) *Intermetallic Compounds. Vol. 2, Practice* (eds J.H. Westbrook and R.L. Fleischer), John Wiley & Sons, Ltd, Chichester, p. 53.
101 Huang, S.-C. and Chestnutt, J.C. (1995) *Intermetallic Compounds. Vol. 2, Practice* (eds J.H. Westbrook and R.L. Fleischer), John Wiley & Sons, Ltd, Chichester, p. 73.
102 Banerjee, D. (1995) *Intermetallic Compounds. Vol. 2, Practice* (eds J.H. Westbrook and R.L. Fleischer), John Wiley & Sons, Ltd, Chichester, p. 91.
103 Tian, W.H., Sano, T., and Nemoto, M. (1993) *Philos. Mag. A*, **68**, 965.
104 Tian, W.H. and Nemoto, M. (1995) *Gamma Titanium Aluminides* (eds Y.-W. Kim, R. Wagner, and M. Yamaguchi), TMS, Warrendale, PA, p. 689.
105 Kaufmann, M.J., Konitzer, D.G., Shull, R.D., and Fraser, H. (1986) *Scr. Metall.*, **20**, 103.
106 Hug, G. and Fries, E. (1999) *Gamma Titanium Aluminides 1999* (eds Y.-W. Kim, D.M. Dimiduk, and M.H. Loretto), TMS, Warrendale, PA, p. 125.
107 Wiezorek, J.M. and Fraser, H.L. (1998) *Philos. Mag. A*, **77**, 661.
108 Miida, R., Hashimoto, S., and Watanabe, D. (1982) *Jpn. J. Appl. Phys.*, **21**, L59.
109 Nakano, T., Hagihara, S.K., Seno, T., Sumida, N., Yamamoto, M., and Umakoshi, Y. (1998) *Philos. Mag. Lett.*, **78**, 385.
110 Inui, H., Chikugo, K., Nomura, K., and Yamaguchi, M. (2002) *Mater. Sci. Eng. A*, **329–331**, 377.
111 Doi, M., Koyama, T., Taniguchi, T., and Naito, S. (2002) *Mater. Sci. Eng. A*, **329–331**, 891.
112 Grégory, F. and Veyssière, P. (2001) *Philos. Mag. A*, **81**, 529.
113 Kawabata, T., Kanai, T., and Izumi, O. (1985) *Acta Metall.*, **33**, 1355.
114 Kawabata, T., Abumiya, T., Kanai, T., and Izumi, O. (1990) *Acta Metall. Mater.*, **38**, 1381.
115 Kawabata, T., Kanai, T., and Izumi, O. (1991) *Philos. Mag. A*, **63**, 1291.
116 Stucke, M.A., Dimiduk, D.M., and Hazzledine, P.M. (1993) *High-Temperature Ordered Intermetallic Alloys V, Materials Research Society Symposium Proceedings*, vol. 288 (eds I. Baker, R. Darolia, J.D. Whittenberger, and M.H. Yoo), Materials Research Society, Pittsburgh, PA, p. 471.
117 Kawabata, T., Kanai, T., and Izumi, O. (1994) *Philos. Mag. A*, **70**, 43.
118 Wang, Z.-M., Li, Z.-X., and Whang, S.H. (1995) *Mater. Sci. Eng. A*, **192–193**, 211.
119 Bird, N., Taylor, G., and Sun, Y.Q. (1995) *High-Temperature Ordered Intermetallic Alloys VI, Materials Research Society Symposium Proceedings*, vol. 364 (eds J.A. Horton, I. Baker, S. Hanada, R.D. Noebe, and D.S.

Schwartz), Materials Research Society, Pittsburgh, PA, p. 635.
120 Feng, Q. and Whang, S.H. (1999) *High-Temperature Ordered Intermetallic Alloys VIII, Materials Research Society Symposium Proceedings*, vol. 552 (eds E.P. George, M.J. Mills, and M. Yamaguchi), Materials Research Society, Warrendale, PA, p. KK1.10.1.
121 Grégory, F. and Veyssière, P. (1999) *Gamma Titanium Aluminides 1999* (eds Y.-W. Kim, D.M. Dimiduk, and M.H. Loretto), TMS, Warrendale, PA, p. 75.
122 Jiao, S., Bird, N., Hirsch, P.B., and Taylor, G. (2001) *Philos. Mag. A*, **81**, 213.
123 Pope, D. and Ezz, S.S. (1984) *Int. Met. Rev.*, **29**, 136.
124 Hazzledine, P.M. and Sun, Y.Q. (1991) *High-Temperature Ordered Intermetallic Alloys IV, Materials Research Society Symposium Proceedings*, vol. 213 (eds L.A. Johnson, D. Pope, and J.O. Stiegler), Materials Research Society, Pittsburgh, PA, p. 209.
125 Zupan, M. and Hemker, K.J. (1999) *Gamma Titanium Aluminides 1999* (eds Y.-W. Kim, D.M. Dimiduk, and M.H. Loretto), TMS, Warrendale, PA, p. 89.
126 Grégori, F. and Veyssière, P. (2000) *Philos. Mag. A*, **80**, 2913.
127 Grégori, F. and Veyssière, P. (2001) *Mater. Sci. Eng. A*, **309–310**, 87.
128 Greenberg, B.A., Anisimov, V.I., Gornostirev, Yu.N., and Taluts, G.G. (1988) *Scr. Metall.*, **22**, 859.
129 Kad, B.K. and Fraser, H.L. (1994) *Philos. Mag. A*, **69**, 689.
130 Messerschmidt, U., Bartsch, M., Häussler, D., Aindow, M., Hattenhauer, R., and Jones, I.P. (1995) *High-Temperature Ordered Intermetallic Alloys VI, Materials Research Society Symposia Proceedings*, vol. 364 (eds J.A. Horton, I. Baker, S. Hanada, R.D. Noebe, and D.S. Schwartz), MRS, Pittsburgh, PA, p. 47.
131 Appel, F. and Wagner, R. (1995) *High-Temperature Ordered Intermetallic Alloys VI, Materials Research Society Symposia Proceedings*, vol. 364 (eds J.A. Horton, I. Baker, S. Hanada, R.D. Noebe, and D.S. Schwartz), MRS, Pittsburgh, PA, p. 623.

132 Sriram, S., Dimiduk, D.M., Hazzledine, P.M., and Vasudevan, V.K. (1997) *Philos. Mag. A*, **76**, 965.
133 Appel, F. and Wagner, R. (1998) *Mater. Sci. Eng.*, **R22**, 187.
134 Veyssière, P. and Grégori, F. (2002) *Philos. Mag. A*, **82**, 553.
135 Veyssière, P. and Grégori, F. (2002) *Philos. Mag. A*, **82**, 567.
136 Veyssière, P. and Grégori, F. (2002) *Philos. Mag. A*, **82**, 579.
137 Gilman, J.J. and Johnston, W.G. (1962) *Solid State Phys.*, **13**, 147.
138 Viguier, B., Hemker, K.J., Boneville, J., Louchet, F., and Martin, J.L. (1995) *Philos. Mag. A*, **71**, 1295.
139 Louchet, F. and Viguier, B. (1995) *Philos. Mag. A*, **71**, 1313.
140 Grégori, F. and Veyssière, P. (1999) *Philos. Mag.*, **A79**, 403.
141 Nakano, T., Matsumoto, K., Seno, T., Oma, K., and Umakoshi, Y. (1996) *Philos. Mag. A*, **74**, 251.
142 Whang, S.H., Hahn, Y.D., and Li, Z.C. (1991) *Proceedings International Symposium on Intermetallics Compounds (JIMIS-6) – Structure and Mechanical Properties* (ed. O. Izumi), Japan Institute of Metals, Sendai, p. 763.
143 Nakano, T., Hayashi, K., Umakoshi, Y., Chiu, Y.-L., and Veyssière, P. (2005) *Philos. Mag.*, **85**, 2527.
144 Rao, P.P. and Tangri, K. (1991) *Mater. Sci. Eng. A*, **132**, 49.
145 Sriram, S., Vasudevan, V.K., and Dimiduk, D.M. (1993) *High-Temperature Ordered Intermetallic Alloys V, Materials Research Society Symposia Proceedings*, vol. 288 (eds I. Baker, R. Darolia, J.D. Whittenberger, and M.H. Yoo), MRS, Pittsburgh, PA, p. 737.
146 Sriram, S., Vasudevan, V.K., and Dimiduk, D.M. (1995) *High-Temperature Ordered Intermetallic Alloys VI, Materials Research Society Symposia Proceedings*, vol. 364 (eds J.A. Horton, I. Baker, S. Hanada, R.D. Noebe, and D.S. Schwartz), MRS, Pittsburgh, PA, p. 647.
147 Viguier, B., Boneville, J., and Martin, J.L. (1996) *Acta Mater.*, **44**, 4403.
148 Vasudevan, V.K., Court, S.A., Kurath, P., and Fraser, H.L. (1989) *Scr. Metall.*, **23**, 467.

149 Sparka, U. (1998) *Verformungs- und Verfestigungsverhalten in ein- und zweiphasigen Titanaluminid-Legierungen*. PhD Thesis, University Hamburg, Germany.
150 Weihs, T.P., Zinoview, V., Viens, D.V., and Schulson, E.M. (1987) *Acta Metall.*, **5**, 1109.
151 Menand, A., Huguet, A., and Nèrac-Partaix, A. (1996) *Acta Mater.*, **44**, 4729.
152 Cottrell, A.H., Stokes, F.R.S., and Stokes, J.R. (1955) *Proc. R. Soc. A*, **233**, 17.
153 Jiao, S., Bird, N., Hirsch, P.B., and Taylor, G. (1999) *High-Temperature Ordered Intermetallic Alloys VIII, Materials Research Society Symposium Proceedings*, vol. 552 (eds E.P. George, M.J. Mills, and M. Yamaguchi), MRS, Pittsburgh, PA, p. KK8.11.1.
154 Mahpatra, R., Chou, Y.T., Girschick, A., Pope, D., and Vitek, V. (1996) *Deformation and Fracture of Ordered Intermetallic Materials III* (eds W.O. Soboyejo, T.S. Srivatsan, and H.L. Fraser), TMS, Warrendale, PA, p. 623.
155 Lu, M. and Hemker, K.J. (1998) *Philos. Mag. A*, **77**, 325.
156 Williams, C.J. and Blackburn, M.J. (1970) *Ordered Alloys* (eds H. Kear, T. Sims, N.S. Stoloff, and J.H. Westbrook), Claitor's Publishing Division, Baton Rouge, Louisiana, p. 425.
157 Lipsitt, H.A., Shechtman, D., and Schafrik, R.E. (1980) *Metall. Trans. A*, **11A**, 1369.
158 Court, S.A., Löfvander, J.P.A., Loretto, M.H., and Fraser, H.L. (1990) *Philos. Mag. A*, **61**, 109.
159 Inui, H., Toda, Y., and Yamaguchi, M. (1993) *Philos. Mag. A*, **67**, 1315.
160 Inui, H., Toda, Y., Shirai, Y., and Yamaguchi, M. (1994) *Philos. Mag. A*, **69**, 1161.
161 Wiezorek, J.M.K., Court, S.A., and Humphreys, C.J. (1995) *Philos. Mag. Lett.*, **72**, 393.
162 Legros, M., Couret, A., and Caillard, D. (1996) *Philos. Mag. A*, **73**, 61.
163 Legros, M., Couret, A., and Caillard, D. (1996) *Philos. Mag. A*, **73**, 81.
164 Minonishi, Y. (1990) *Philos. Mag. Lett.*, **62**, 153.
165 Minonishi, Y. (1991) *Philos. Mag. A*, **63**, 1085.
166 Wiezorek, J.M.K., Court, S.A., and Humphreys, C.J. (1995) *High-Temperature Ordered Intermetallic Alloys VI, Materials Research Society Symposia Proceedings*, vol. 364 (eds J.A. Horton, I. Baker, S. Hanada, R.D. Noebe, and D.S. Schwartz), MRS, Pittsburgh, PA, p. 659.
167 Umakoshi, Y. and Yamaguchi, M. (1981) *Phys. Status Solidi A*, **68**, 45.
168 Yamaguchi, M., Pope, D.P., Vitek, V., and Umakoshi, Y. (1981) *Philos. Mag. A*, **43**, 1265.
169 Blackburn, M.J. (1967) *Trans. Metall. Soc. AIME*, **239**, 660.
170 Cserti, J., Khanta, M., Vitek, V., and Pope, D. (1992) *Mater. Sci. Eng. A*, **152**, 95.
171 Fu, C.L., Zou, J., and Yoo, M.H. (1995) *Scr. Metall. Mater.*, **33**, 885.
172 Koizumi, Y., Ogata, S., Minamino, Y., and Tsuji, N. (2006) *Philos. Mag.*, **86**, 1243.
173 Yakovenkova, L.I., Karkina, L.E., and Rabobovskaya, M.Y. (2003) *Tech. Phys.*, **48**, 56.
174 Löfvander, J.P., Court, S.A., Kurath, P., and Fraser, H.L. (1989) *Philos. Mag. Lett.*, **59**, 289.
175 Umakoshi, Y., Nakano, T., Takenaka, T., Sumimoto, K., and Yamane, T. (1993) *Acta Metall. Mater.*, **41**, 1149.
176 Nakano, T., Yanagisawa, E., and Umakoshi, Y. (1995) *ISIJ Int.*, **35**, 900.
177 Minonishi, Y. (1995) *Mater. Sci. Eng. A*, **192–193**, 830.
178 Wiezorek, J.M.K., Humphreys, C.J., and Fraser, H.L. (1997) *Philos. Mag. Lett.*, **75**, 281.
179 Legros, M., Minonishi, Y., and Caillard, D. (1997) *Philos. Mag. A*, **76**, 995.
180 Legros, M., Minonishi, Y., and Caillard, D. (1997) *Philos. Mag. A*, **76**, 1013.
181 Thomas, M., Vassel, A., and Veyssière, P. (1987) *Scr. Metall.*, **21**, 501.
182 Thomas, M., Vassel, A., and Veyssière, P. (1989) *Philos. Mag. A*, **59**, 1013.
183 Gray, G.T., III (1992) *Shock-Wave and High-Strain Rate Phenomena in Materials* (eds M.A. Meyers, L.E. Murr, and K.P. Staudhammer), Marcel Dekker, Inc., New York, p. 899.
184 Yoo, M.H. (1981) *Metall. Trans. A*, **12A**, 409.

185 Yoo, M.H., Fu, C.L., and Lee, J.K. (1991) *J. Phys.*, **III**, 1065.
186 Lee, J.W., Hanada, S., and Yoo, M.H. (1995) *Scr. Metall. Mater.*, **33**, 509.
187 Minonishi, Y. (1993) *Intermetallic Compounds for High-Temperature Structural Applications, SAMPE Symp. Proc.* (eds M. Yamaguchi and H. Fukutomi), SAMPE, Chiba, Japan, p. 1542.
188 Umakoshi, Y., Nakano, T., Sumimoto, K., and Maeda, Y. (1993) *High Temperature Ordered Intermetallic Alloys V, Materials Research Society Symposia Proceedings*, vol. 288 (eds I. Baker, R. Darolia, J.D. Whittenberger, and M.H. Yoo), MRS, Pittsburgh, PA, p. 441.
189 Minonishi, Y., Ishioka, S., Koiwa, M., Morozumi, S., and Yamaguchi, M. (1981) *Philos. Mag. A*, **4**, 1017.
190 Fröbel, U. and Appel, F. (2002) *Acta Mater.*, **50**, 3693.
191 Fröbel, U. and Appel, F. (2006) *Intermetallics*, **14**, 1187.
192 Court, S.A., Loretto, M.H., and Fraser, H.L. (1989) *Philos. Mag. A*, **61**, 379.
193 Minonishi, Y., Otsuka, M., and Tanaka, K. (1991) *Proceedings International Symposium on Intermetallics Compounds (JIMIS-6) – Structure and Mechanical Properties* (ed. O. Izumi), Japan Institute of Metals, Sendai, p. 534.
194 Loretto, M.H. (1992) *Philos. Mag. A*, **65**, 1095.
195 Vitek, V. (1992) *Prog. Mater. Sci.*, **36**, 1.
196 Hirsch, P.B. (1960) *Fifth Inter. Congress on Crystallography*, University Press, Cambridge, p. 139.
197 Vitek, V., Perrin, R.C., and Bowen, D.K. (1970) *Philos. Mag.*, **21**, 1049.
198 Vitek, V. (1985) *Dislocations and Properties of Real Crystals* (ed. M.H. Loretto), Institute of Metals, London, p. 30.
199 Bassinski, Z.S., Dusbery, M.S., and Taylor, R. (1970) *Philos. Mag.*, **21**, 1201.
200 Seeger, A. and Wüthrich, C. (1976) *Nuovo Cim.*, **33B**, 38.
201 Ohr, S.M. and Beshers, D.N. (1963) *Philos. Mag.*, **8**, 1343.
202 Hull, D. (1963) *Electron Microscopy and Strength of Crystals* (eds G. Thomas and J. Washburn), Interscience, New York, p. 291.
203 Keh, A.S. and Weissmann, S. (1963) *Electron Microscopy and Strength of Crystals* (eds G. Thomas and J. Washburn), Interscience, New York, p. 231.
204 Priestner, R. and Leslie, W.C. (1965) *Philos. Mag.*, **11**, 859.
205 Keh, A.S. (1965) *Philos. Mag.*, **12**, 9.
206 Baker, I. and Munroe, P.R. (1990) *High-Temperature Aluminides and Intermetallics* (eds S.H. Whang, C.T. Liu, D.P. Pope, and J.O. Stiegler), TMS Publication, New York, p. 425.
207 Huang, S.C. and Hall, E.L. (1991) *Metall. Trans. A*, **22A**, 2619.
208 Li, Y.G. and Loretto, M.H. (1994) *Acta Metall. Mater.*, **42**, 2913.
209 Sauthoff, G. (1995) *Intermetallics*, Wiley-VCH Verlag GmbH, Weinheim, p. 51.
210 Yamaguchi, M., Pope, D.P., Vitek, V., and Umakoshi, Y. (1981) *Philos. Mag. A*, **3**, 867.
211 Yoo, M.H., Takasuki, T., Hananda, S., and Izumi, O. (1990) *Mater. Trans. Jpn. Inst. Metals*, **31**, 435.
212 Baker, I. (1995) *Mater. Sci. Eng. A*, **192–193**, 1.
213 Nguyen-Man, D. and Pettifor, D.G. (1999) *Gamma Titanium Aluminides 1999* (eds Y.-W. Kim, D.M. Dimiduk, and M.H. Loretto), TMS, Warrendale, PA, p. 175.
214 Paton, N.E. and Williams, J.C. (1970) *Second International Conference on the Strength of Metals and Alloys*, ASM, Metals Park, OH, p. 108.
215 Banerjee, D., Gogia, A.K., and Nandy, T.K. (1990) *Metall. Trans. A*, **21A**, 627.
216 Morris, M.A. and Li, Y.G. (1995) *Mater. Sci. Eng. A*, **197**, 133.
217 Sun, F.-S., Cao, C.X., Yan, M.G., Lee, Y.T., and Kim, S.E. (1999) *Gamma Titanium Aluminides 1999* (eds Y.-W. Kim, D.M. Dimiduk, and M.H. Loretto), TMS, Warrendale, PA, p. 415.
218 Naka, S. and Khan, T. (2002) *Intermetallic Compounds. Vol. 3, Principles and Practice* (eds J.H. Westbrook and R.L. Fleischer), John Wiley & Sons, Ltd, Chichester, p. 842.

6
Deformation Behavior of Two-Phase α_2(Ti$_3$Al) + γ(TiAl) Alloys

Two-phase titanium aluminide alloys exhibit a much better mechanical performance than their monolithic constituents γ(TiAl) and α_2(Ti$_3$Al), provided that the phase distribution and grain size are suitably controlled. The synergistic effects of the two phases are undoubtedly associated with the many influences that the microstructure has on the kinematics and dynamics of dislocation propagation. Because of this engineering potential much effort has been expended on the characterization of the interfaces and the related deformation mechanisms in these alloys; these issues are the subjects of the present chapter.

6.1
Lamellar Microstructures

6.1.1
Interface Structures in Lamellar TiAl Alloys

The technologically most relevant α_2(Ti$_3$Al) + γ(TiAl) alloys contain a significant volume fraction of lamellar grains [1]. The morphology of these grains represents a multilayer system made of two phases. It is well documented in the literature [2–6] that such a system could exhibit extraordinary mechanical properties when the layer thickness is small enough. Equally important are the structure and chemical environment of the interfaces; thus these aspects of the lamellar α_2(Ti$_3$Al) + γ(TiAl) microstructure warrant special consideration.

The nature of the interface between two phases is determined by their structural relationships. There is a strong tendency for planes and directions with the highest atomic densities to align across the interface. This reduces the number of broken bonds at the interface and therefore lowers the interfacial energy. A coincident site lattice (CSL) is formed by those lattice points that are coincident, when the two misoriented lattices of neighboring lamellae are considered to interpenetrate. A particular CSL is specified by the parameter Σ, the inverse density of the lattice points that are common to both lattices. However, the occupation of coincident sites in an interface by atoms does not necessarily lead to an atomic configuration of the boundary with the lowest energy. As suggested by Chalmers and Gleiter [7],

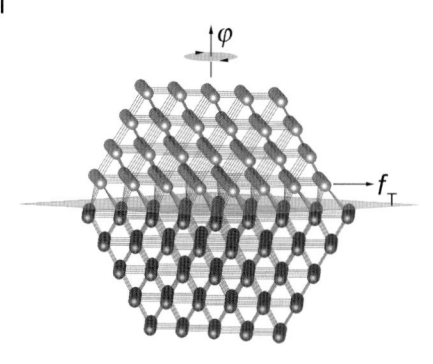

Figure 6.1 Schematic illustration of the construction of an interface between two crystals φ describes the rotation around the interface normal and f_T the rigid body translation of the two crystals.

a better atomic fit at a boundary could result if atoms were moved away from coincident sites by a rigid-body displacement of one grain relative to the other by a constant displacement vector. Thus, the degree of matching of close-packed planes can formally be described by a rigid-body rotation φ and a translation f_T of the two lattices relatively to each [6]. These operations are schematically sketched in Figure 6.1. The effect of rigid-body translations has been studied theoretically and experimentally in many interface systems. As will be shown later, such translations seem also to be a general feature of the lamellar interfaces in titanium aluminide alloys.

The lamellar morphology is the result of the $\alpha \rightarrow \gamma$ phase transformation and ordering reactions that occur upon solidification and cooling [8] (see Chapter 4). The γ lamellae can apparently be generated by several processes, which depend on the temperature and the degree of undercooling. Zghal et al. [9, 10] have investigated the implications of these processes on the structural details of lamellar Ti-47Al-2Cr-2Nb. At slow cooling rates, a primary high-temperature $\alpha \rightarrow \gamma$ transformation typically occurs at a temperature about 50 °C below the α-transus temperature. Thus, the undercooling and the chemical potential driving the transformation are relatively low, giving rise to the assumption that the γ lamellae are heterogeneously nucleated at grain boundaries. During this primary high-temperature transformation, isolated and relatively broad γ lamellae are formed. At faster cooling rates a secondary $\alpha_2 \rightarrow \gamma$ phase transformation occurs at lower temperatures, which is characterized by the generation of extremely fine γ lamellae. These lamellae are either homogeneously nucleated within the α_2 lamellae or heterogeneously nucleated at pre-existing α_2/γ interfaces.

As sketched in Figure 6.2, the γ and α_2 platelets are crystallographically aligned. The close-packed planes and directions match so that $\{111\}_\gamma \parallel (0001)_{\alpha 2}$ and $<1\bar{1}0>_\gamma \parallel <11\bar{2}0>_{\alpha 2}$. Figure 6.3 is a high-resolution electron micrograph of the lamellar microstructure. The image was obtained with the beam direction parallel to the $<10\bar{1}>_\gamma$ and $<11\bar{2}0>_{\alpha 2}$ directions, which lie in the interfacial plane. The micrograph reveals that the interfaces consist mainly of atomically flat terraces,

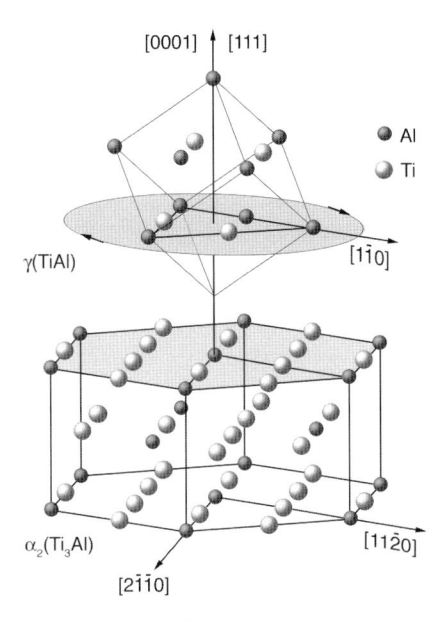

Figure 6.2 Crystallographic alignment of the α_2(Ti$_3$Al) and γ(TiAl) platelets within the lamellar microstructure. The matching close packed planes and directions are: $\{111\}_\gamma \| (0001)_{\alpha2}$ and $<1\bar{1}0>_\gamma \| <11\bar{2}0>_{\alpha2}$. Because of the symmetry of the γ phase, there are six variants of this orientation relationship, which can formally be described by rotations of the two phases relative to each other by a multiple of 60° around $[0001]_{\alpha2}$ and $<111>_\gamma$, respectively. The $(111)_\gamma$ and $(0001)_{\alpha2}$ planes are assigned to be parallel to the lamellar boundaries.

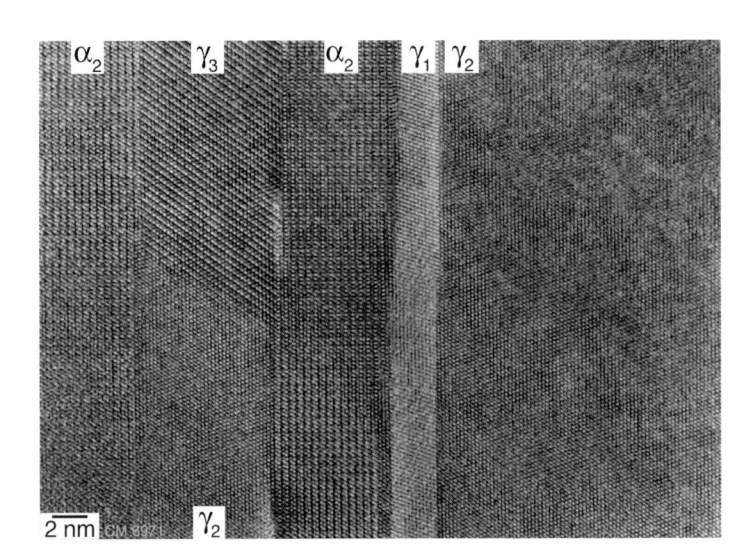

Figure 6.3 High-resolution micrograph of the lamellar microstructure obtained with the beam direction parallel to the $<10\bar{1}>_\gamma$ and $<11\bar{2}0>_{\alpha2}$ directions. Note the domain boundary present in one of the γ lamellae. Ti-45Al-10Nb, extruded at 1300°C, 2-h annealing at 1050°C, furnace cooling.

which are almost free of defects; hence, these terraces are locally coherent. Steps of different height, which will be described later, often delineate the terraces.

The atomic structure of the α_2/γ interface is geometrically detailed. At the interface the two crystals have lattice points in common, which may be occupied by different atomic species. Thus, as formally described in Figure 6.1, the lattice matching of two crystals can be modified by an additional translation f_T of the two lattices relatively to each other [6]. The possible translations that conserve a close-packed structure across the α_2/γ interface are the vectors f_{APB}, f_{SISF} and f_{CSF}. These translations correspond to the formation of an antiphase boundary (APB), a superlattice intrinsic stacking fault (SISF) and a complex stacking fault (CSF), respectively. Based on this concept, Lei Lu *et al.* [11] have investigated the structure and energy of different α_2/γ interfaces by atomistic modeling. Two of these configurations are sketched in Figure 6.4. The α_2 platelet is manifested by the upward **BABA**

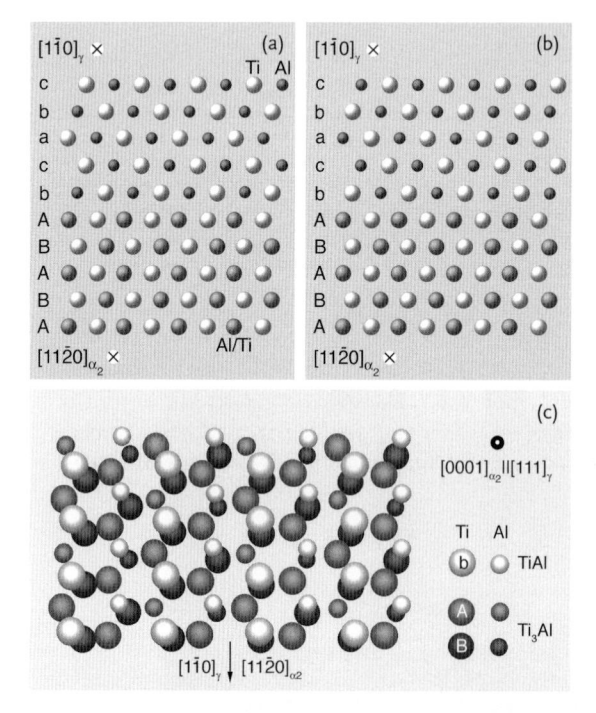

Figure 6.4 Two variants of geometrically constructed α_2/γ interfaces. Atomic configuration across the interface projected along the directions $<1\bar{1}0>_\gamma || <11\bar{2}0>_{\alpha_2}$. The stacking sequence of the α_2(Ti$_3$Al) is designated as **BABA** and that of γ(TiAl) as **bcabc**. The terminating **A** layer of the α_2 phase is bonded to a **b** layer of the γ phase. (a) Variant (1): the Ti rows of the interfacial **b** layer are above the mixed Ti/Al rows of the **B** layers. (b) Variant (2): the Ti atoms of the **b** layers are at the projected atomic positions of pure Ti rows in the **B** layers; the Al atoms of the **b** layers are above the Ti positions in the mixed Ti/Al rows of the **B** layers. (c) Atom positions of the $(0002)_{\alpha_2}$ and $\{111\}_\gamma$ planes forming the interface shown in (b). Slightly tilted perspective view along $[111]_\gamma || [0001]_{\alpha_2}$ onto the layers **B**, **A** and **b**.

stacking of the (0002) planes. The terminating **A** layer of the α_2 phase is bonded to the layer **b** of the γ phase with the subsequent (111) planes stacked in the sequence ***bcabca***. The Ti atoms of the interfacial **b** layer can be either situated above Al positions in the mixed Ti/Al rows of the **B** layers, as shown in Figure 6.4a. The formation of this interface involves a rigid-body shift of the adjacent lamellae against each other along an APB vector. Alternatively, the Ti atoms of the **b** layer can be above the atomic positions of pure Ti rows of the **B** layers (Figure 6.4b). Figure 6.4c is a perspective view on the matching $(0002)_{\alpha 2}$ and $\{111\}_\gamma$ planes forming this type of interface. Based on high-resolution electron microscopy, the interface model shown in Figure 6.4b, c was proposed by Mahon and Howe [12] and confirmed by Inui et al. [13]. As the α_2 phase has hexagonal symmetry, all the α_2/γ interfaces are thought to be structurally identical.

The α_2 lamellae occurring in a single lamellar colony all have the same orientation because they originate from the same α or α_2 grain. This is not the case for the γ lamellae present in the same colony. In the $L1_0$ structure, the $<1\bar{1}0]$ directions differ from the $<01\bar{1}]$ directions, whereas the three $<11\bar{2}0>$ directions in the α_2 phase are all equivalent. Thus, the γ phase can be nucleated in six orientation variants, as sketched in Figure 6.2. Three of these variants arise from the three cube axes of the $L1_0$ unit cell; each of them split into two subvariants because of the two possible stacking orders of the atomic planes parallel to the (111) interface [14–24]. If the first $\{111\}_\gamma$ layer of the $L1_0$ structure is designated ***a***, the second layer could occupy either all the ***b*** positions or all the ***c*** positions, that is, the stacking sequence could be either ***abca*** or ***acba***. In stacking the second layer, the choice of ***b*** or ***c*** is equivalent to merely rotating the structure through 180°. Thus, the respective γ variants are in twin orientation. Kad et al. [20] associated the formation of the two stacking sequences with the shear processes that occur during the $\alpha/\alpha_2 \rightarrow \gamma$ phase transformation. Stacking of the lattice in different directions is thought to induce strain fields of opposite signs, which eventually reduces the total strain energy. In confirmation of this model, Figure 6.5 demonstrates the nucleation of extremely thin twin-related γ lamellae within and adjacent to an α_2 lamella. As described above, these fine γ lamellae were probably nucleated in a later stage of the $\alpha \rightarrow \gamma$ phase transformation that occurs at relatively low temperatures and fast cooling rates.

The orientation of the individual γ domains can be described by rotations of the α_2 and γ platelets relatively to each other by a multiple of 60° around $[0001]_{\alpha 2}$ or $<111>_\gamma$, as illustrated in Figure 6.6 [23]. The stacking sequences of these orientation variants are shown in Figure 6.7. These orientation relationships lead to three distinct types of γ/γ interfaces with the following characteristics.

i) The pseudotwin boundary joining γ variants that are misoriented by a 60° rotation; the $<1\bar{1}0]$ direction of the matrix is antiparallel to the $<10\bar{1}]$ direction of the twin. At the pseudotwin interface the atomic stacking of the parent f.c.c. lattice is reversed and the order is rotated. With reference to Figure 6.6, a pseudotwin interface occurs between the variants (1)/(6), (2)/(1), (3)/(2), (4)/(3), (5)/(4), and (6)/(5).

ii) The interface joining 120° rotational variants, often designated as a 120° rotational fault; the [1$\bar{1}$0] direction of the matrix is antiparallel to the [$\bar{1}$10] direction of the adjacent γ variant. The interface retains the **ABC** stacking of the f.c.c. lattice (neglecting the tetragonality of the γ phase); the **c**-axes of the adjacent lamellae are perpendicular to each other. With reference to Figure 6.6, a 120° rotational fault occurs between the variants (1)/(5), (2)/(6), (3)/(1), (4)/(2), (5)/(3), and (6)/(4).

iii) The true twin boundary with a 180° rotation; the [1$\bar{1}$0] direction of the matrix is antiparallel to the [$\bar{1}$10] direction of the twin. At the interface only the stacking sequence of the parent L1$_0$ structure is changed. In Figure 6.6, the variants (1)/(4), (2)/(5) and (3)/(6) are joined by a true twin.

Figures 6.8a–c show geometrically constructed models of these lamellar γ/γ interfaces. Other variants of these interfaces can be generated by a rigid-body translation of the adjacent lamellae relative to each other, as has been described earlier for the α$_2$/γ interface. As an example, Figure 6.8d shows a 180° twin boundary that was modified by shifting the upper lamella along the APB vector $f_{APB} = 1/2<\bar{1}01]$. This type of interface was proposed by Denquin and Naka [25] and Ricolleau et al. [26] and observed by Siegl et al. [27]. It should be noted that the same APB displace-

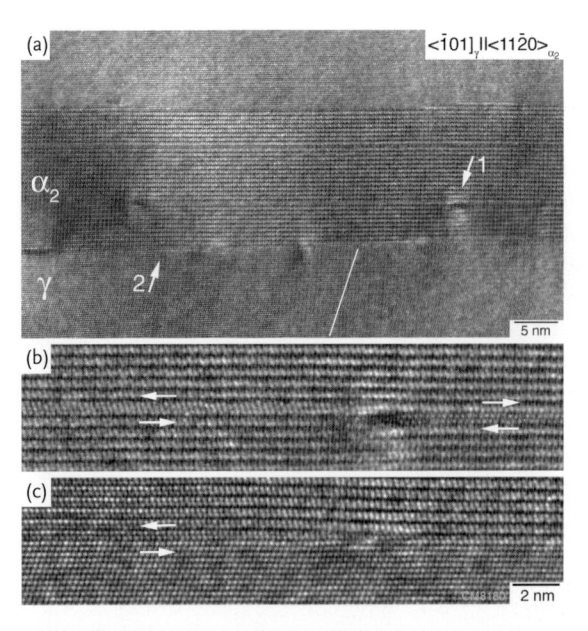

Figure 6.5 Phase transformation α$_2$→γ. Ti-46.5 Al-4(Cr, Ta, Mn, B), sheet material, thermal treatment after rolling at 1370°C, 20 min plus 1000°C, 60 min, air cooling. (a) Formation of twin-related γ variants within and adjacent to an α$_2$ lamella. The white line indicates the trace of the ($\bar{1}1\bar{1}$)$_γ$ plane. (b and c) Higher magnification of the areas marked with the arrows 1 and 2, respectively; arrow pairs indicate the shear processes involved in the phase transformation.

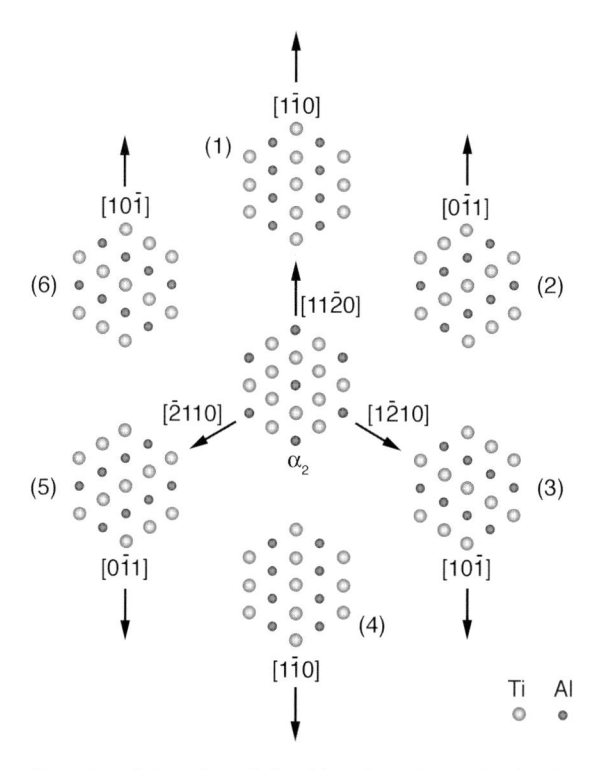

Figure 6.6 Orientation relationships of the six rotational variants of γ lamellae with respect to the parent α_2 phase (see Figure 6.2). Figure adapted from [23], modified.

ment does not change the structure of the pseudotwin or of the 120° boundary [6]. In addition to these lamellar γ/γ interfaces there are also order-domain boundaries present, which subdivide the individual γ lamellae into domains. In PST crystals the domain size is in the range of 10 μm to 100 μm [14, 18, 23]. The domain boundaries are often 120° rotational faults, but in contrast to the lamellar interfaces, the habit plane of the domain boundaries is not usually $\{111\}_\gamma$.

It should be noted that the ordering of the γ phase in high-resolution TEM images is only revealed when the beam direction is a zone axis of both, the interfacial $(111)_\gamma$ plane and the ordered $(002)_\gamma$ plane, that is, when the beam direction is parallel to $<1\bar{1}0]_\gamma$. Thus, with an $<10\bar{1}]_\gamma$ beam direction it is not possible to distinguish between a true twin and a pseudotwin relation of adjacent lamellae. To overcome this ambiguity and to sample a larger volume, the lamellar structure was mostly characterized by TEM diffraction analysis [10, 18, 23, 28]. Figure 6.9 is a TEM bright field image of the lamellar structure observed along the $<10\bar{1}>_\gamma \| <11\bar{2}0>_{\alpha 2}$ orientations, and Figure 6.10 is a simulated diffraction pattern that is characteristic for imaging the lamellar structure along these orientations. The pattern involves the $[11\bar{2}0]_{\alpha 2}$ reflections of the α_2 phase and of three different γ variants with the orientations $[1\bar{1}0]_{\gamma 1}$, $[\bar{1}01]_{\gamma 2}$, and $[0\bar{1}1]_{\gamma 3}$, respectively. The

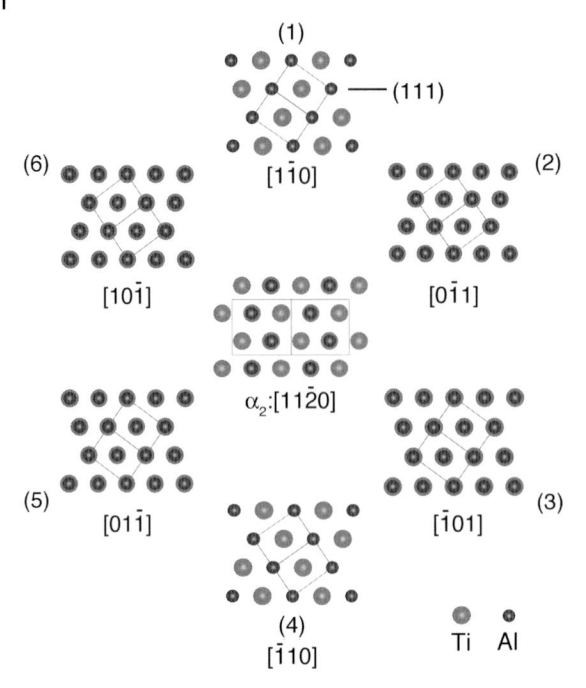

Figure 6.7 Stacking sequence of the six rotational γ variants and of the parent α_2 phase shown in Figure 6.6. Pseudotwin interfaces occur between the variants (1)/(6), (2)/(1), (3)/(2), (4)/(3), (5)/(4), and (6)/(5). 120° rotational faults occur between the variants (1)/(5), (2)/(6), (3)/(1), (4)/(2), (5)/(3), and (6)/(4). True twins occur between the variants (1)/(4), (2)/(5) and (3)/(6).

orientation of an individual γ lamella can unambiguously be determined by selected area diffraction along the low index directions of type <10$\bar{1}$> and <11$\bar{2}$> that are contained in the (111) interfacial plane. In order to distinguish between a twin and matrix variant, starting from the <112> beam direction, the sample could be tilted by ±30° around the [111] plane normal. The resulting sequence of <110> diffraction patterns is characteristic of the respective γ variant, as illustrated in Figure 6.11. More details are provided in [23, 28]. Based solely on the diffraction patterns, a lamellar 180° true twin boundary cannot be distinguished from the interface that a mechanical twin forms with the $L1_0$ matrix. However, mechanical twins are often related to internal deformation features such as dislocation glide bands or cracks. Also, mechanical twins exhibit a characteristic morphology; they are often lenticular, tapering to a point, and occur in clusters of variable thickness.

The characteristic features of the lamellar structure are sketched in Figure 6.12. The most important structural parameters of the lamellar morphology are the colony size, the lamellar spacing, the separation distance of the domain boundaries, and the separation distance of the α_2 lamellae. In this sense, the polysynthetically twinned (PST) crystal is a particular structural form of a lamellar alloy that

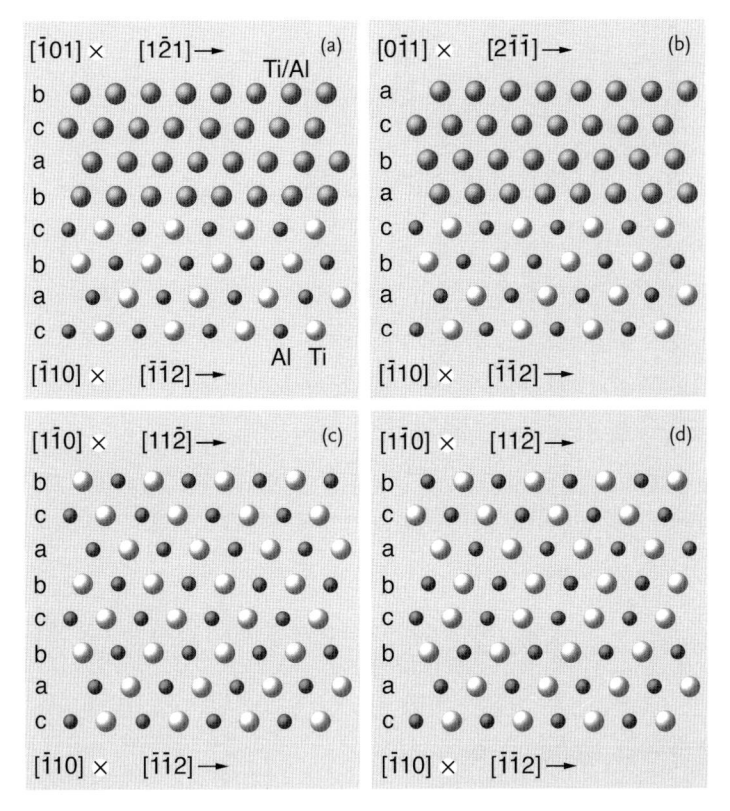

Figure 6.8 Geometrically constructed models of γ/γ interfaces in the lamellar microstructure. (a) Pseudotwin boundary joining γ variants that are misoriented by a 60° rotation, (b) interface joining 120° rotational variants, (c) true twin boundary with a 180° rotation of the adjacent γ variants, (d) 180° twin boundary containing an antiphase boundary that is produced by a rigid-body translation $f_T = f_{APB}$.

is comprised of a single set of lamellae; hence, a PST crystal does not contain colony boundaries.

The relative volume fraction and the width of α_2 and γ lamellae sensitively depend on the alloy composition and the cooling velocity after solidification. Thus, there is no hard and fast rule available for predicting these structural parameters. An analysis performed on polycrystalline Ti-46Al in the as-cast condition has shown that about 34% of the lamellae are α_2 phase, which corresponds to a volume fraction of 17% [23]. The average thickness of the α_2 lamellae was found to be 32 nm, while that of the γ lamellae was 73 nm with a strong preference of thin lamellae. The same analysis performed on a PST crystal of composition Ti-49.3 Al indicated a significantly smaller number of α_2 lamellae and relatively broad γ lamellae with an average thickness of about 2.5 μm. There is good consensus in the literature [23, 29–32] that the majority of lamellar γ/γ interfaces are true twin

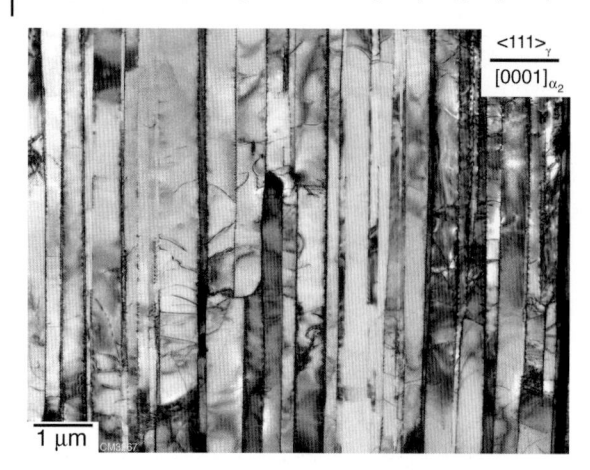

Figure 6.9 Multibeam TEM image of the lamellar microstructure imaged down the $<10\bar{1}>_\gamma$ and $<11\bar{2}0>_{\alpha 2}$ directions. Polycrystalline Ti-47Al-2Cr-0.2Si, as cast, 4-h annealing at 1220 °C, furnace cooling.

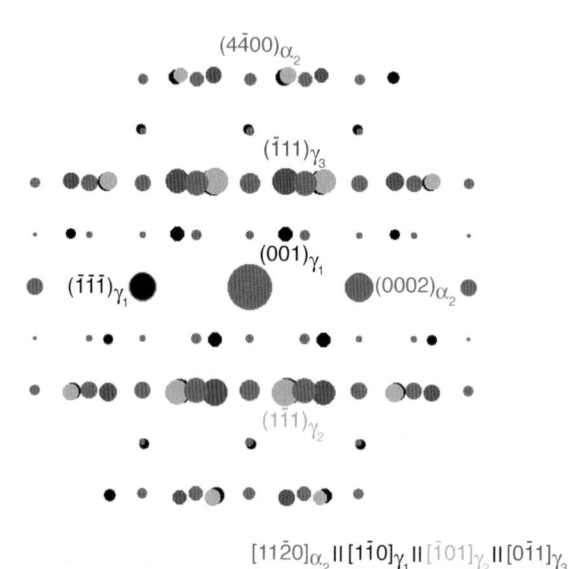

Figure 6.10 Diffraction pattern of the lamellar structure simulated for kinematical conditions for the zone axes $<10\bar{1}>_\gamma \parallel <11\bar{2}0>_{\alpha 2}$. The diagram involves the reflections of the α_2 phase and of three different γ variants with the orientations $[1\bar{1}0]_{\gamma 1}$, $[\bar{1}01]_{\gamma 2}$, and $[0\bar{1}1]_{\gamma 3}$, respectively.

boundaries, followed by 120° boundaries and the pseudotwin boundaries. At first sight, this is surprising because in the α→γ transformation no one γ variant is favored over any other; hence, the three γ variants should be present in equal volume fractions. There are at least two factors that may account for the preponderance of the true twin boundary. First, as described above, the γ phase is often

	Matrix			Twin		
1	[0̄11]	[1̄1̄2]	[1̄01]	[1̄01]	[1̄1̄2]	[0̄11]
2	[1̄10]	[1̄2̄1]	[01̄1]	[1̄10]	[21̄1̄]	[101̄]
3	[101̄]	[21̄1̄]	[1̄10]	[01̄1]	[1̄2̄1]	[1̄10]

Figure 6.11 Diffraction analysis for characterizing the rotational variants of γ lamellae. The beam directions of the simulated diffraction pattern are indicated. Black spots represent fundamental reflections and gray spots superlattice reflections. Figure adapted from Kishida et al. [28].

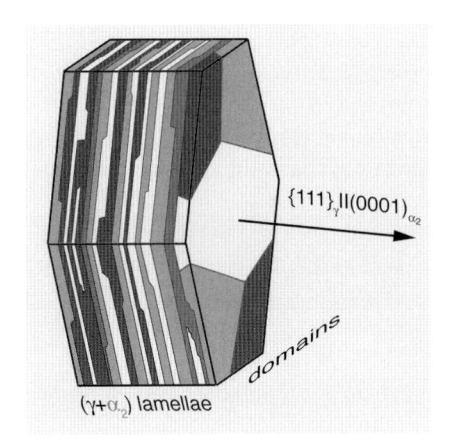

Figure 6.12 Schematic illustration of a lamellar colony. The structural parameters that are most important for the mechanical properties are the colony size, the lamellar spacing, the separation distance of the domain boundaries, and the separation distance of the α_2 lamellae.

nucleated in the form of two variants that have a true twin orientation; this naturally increases the relative frequency of lamellar true twin boundaries. Secondly, on the basis of the relative interface energies it is expected that the true twin boundary is easily formed, a tendency that acts in the same direction (see Section 6.1.2). As has been described above, several processes, depending on the composition and cooling rate, form the γ lamellae. This certainly gives rise to complex distributions of the widths of γ lamellae and of the relative frequency of the interface types. In a few cases a relatively high density of semicoherent γ/γ interfaces was recognized that was comparable with that of the true twin interfaces [33–36]. This disparity to the other observations is probably caused by different processing conditions. More information about the structural parameters of the lamellar morphology and the associated transformation pathways is provided in [9, 10, 20, 23, 30, 32]. Umakoshi and Nakano [37, 38] have given a detailed account on the influence of the Al content and growth conditions on the structural parameters of PST crystals.

To summarize, the lamellar grains consist of a set of γ lamellae; the lamellae are subdivided into domains and interspersed by α_2 lamellae. The volume fraction of the two phases is controlled by the composition on the basis of the phase diagram and the processing conditions of the alloy. There are four types of lamellar interfaces: the α_2/γ interface and three distinct γ/γ interfaces that are typified by rotations of 60°, 120° and 180° between adjacent lamellae.

6.1.2
Energetic Aspects of Lamellar Interfaces

At the lamellar boundaries the atomic arrangement across the interfaces is different from that in the bulk. This gives rise to a specific interface energy, which depends on the nature of the interface. Several attempts have been made to determine the energies of lamellar interfaces. Inui et al. [18] have estimated the energies of the γ/γ interfaces with the help of a bond-counting model. The interfacial energies were referred to the energies $\Gamma(SISF)$ and $\Gamma(APB_{111})$ of the superlattice intrinsic stacking fault and the antiphase boundary on the {111} plane in bulk material. In this approach the energies of the true twin boundary, the rotational fault and of the pseudotwin are

$$\Gamma(180°) = 1/2\Gamma(SISF) \tag{6.1}$$

$$\Gamma(120°) = 1/2\Gamma(APB_{111}) \tag{6.2}$$

$$\Gamma(60°) = \Gamma(180°) + \Gamma(120°) \tag{6.3}$$

Using the SISF and APB energies listed in Table 5.1, the authors have determined the ratios of the interfacial energies to be in the range

$$\Gamma(180°) : \Gamma(120°) : \Gamma(60°) = 1 : 2 : 3 \text{ to } 1 : 6 : 7 \tag{6.4}$$

In the framework of this model, the low energy of the 180° twin boundary is plausible because it only changes the stacking sequence of the atom planes and there are no missing bonds across the interface.

Rao et al. [39] investigated the pseudotwin and the rotational fault by embedded atom modeling (EAM) and recognized a substantial change of the γ-surface at these interfaces, when compared with that of the bulk. As described in Section 5.1.2, the γ-surface represents the total energy of the system as a function of a generalized displacement vector that characterizes the relative displacement of the two crystals above and below a planar interface. The group around Yoo [6, 40, 41] performed similar investigations for all types of lamellar interfaces, utilizing first-principles-local-density-functional (LDF) calculations. The resulting ratios of the energies are

$$\Gamma(180°) : \Gamma(120°) : \Gamma(60°) = 1 : 4.2 : 4.5 \tag{6.5}$$

All the calculated interfacial energies are listed in Table 6.1 under $f_T = 0$, together with the values determined for the bulk.

Lei Lu et al. [11] have experimentally determined the interface energies in a Ti-48Al PST crystal utilizing a thermal surface grooving technique. In this method, an alloy is annealed at high temperature in vacuum in order to equilibrate the phase boundary with the surface. At the intersection point with the surface the boundary forms a groove with a characteristic dihedral angle, which is determined by the balance between the interfacial energies and the surface energy at the triple point. Thus, the interfacial energy can be determined with reference to the surface energy. Lei Lu et al. [11] determined the dihedral angle by atomic force microscopy and coupled the experimental results with central force calculations of the interface energies. This analysis resulted in the ratios of the interfacial energies

$$\Gamma(180°) : \Gamma(120°) : \Gamma(60°) = 1 : 5.8 : 6.7 \tag{6.6}$$

Collectively, this data indicates that the lamellar true twin boundary has by far the lowest energy followed by the 120° and the 60° boundaries. This finding is a thermodynamic confirmation of the experimentally observed relative occurrence of the different interface types. Based on a bond-counting model Lei Lu et al. [11] described the energy of the α_2/γ interface as

$$\Gamma(\alpha_2/\gamma) = \Gamma(180°) + \Gamma(APB_{100})\sqrt{3}/4 \tag{6.7}$$

with $\Gamma(APB_{100})$ being the APB energy on {010} planes in bulk material. Atomic modeling performed by the same authors has revealed that $\Gamma(\alpha_2/\gamma)$ depends on the detailed atomic configuration of the interface. The structure shown in Figures 6.4b and c was found to have the lowest energy, when compared with the other configurations. As described in the previous section this finding corresponds well to the TEM observations. A detailed discussion of the planar fault energies determined by central force calculations led the authors to believe that the energy of the α_2/γ interface is higher than that of the γ/γ interfaces.

As has been mentioned in Sections 6.1.1 and 6.1.2, the energy of an interface could be minimized by a rigid-body translation of the two lattices relative to each other (Figure 6.1). In this context the effect of translations along the vectors $f_{APB} = 1/2<10\bar{1}]_\gamma$, $f_{SISF} = 1/6[11\bar{2}]_\gamma$ and $f_{CSF} = 1/6[\bar{2}11]_\gamma$ has been studied by atomistic modeling [6, 36–38]. These translations correspond to the formation of an APB, an SISF and a CSF, respectively, at the interface. Thus, the energies determined in this way are equivalent to the respective fault energies at the interface. The results of the calculations are listed in Table 6.1. Taken together, the atomic simulations indicate that the energies of the APB and CSF are markedly reduced, when these faults are located at the 60° γ/γ and 120° γ/γ interfaces or at an α_2/γ interface.

The reduction of the APB and CSF energies at the lamellar interfaces may significantly affect the dislocation glide behavior. Based on the classical Peierls model [42], Rao et al. [39] have suggested that this reduction of the planar fault energies

Table 6.1 Interfacial energies of lamellar boundaries in $\alpha_2(Ti_3Al) + \gamma(TiAl)$ alloys determined by atomistic modeling. f_T represents the rigid-body translations $f_{APB} = 1/2<10\bar{1}>_\gamma$, $f_{SISF} = 1/6[11\bar{2}]_\gamma$ and $f_{CSF} = 1/6[\bar{2}11]_\gamma$ of the adjacent γ lamellae relatively to each other, these correspond to the formation of an APB, an SISF or a CSF, respectively, at the $(111)_\gamma$ interfacial plane. $f_{APB} = 1/6[\bar{1}2\bar{1}0]_{\alpha_2}$, $f_{SISF} = 1/3[\bar{1}2\bar{1}0]_{\alpha_2}$ and $f_{CSF} = 1/6[01\bar{1}0]_{\alpha_2}$ are the translations of the α_2 lamella relative to the γ lamella; these correspond to the formation of an APB, an SISF and a CSF, respectively, at the $(0001)_{\alpha_2} \parallel (111)_\gamma$ interface.

Interface type	φ [deg.]	Γ [mJ/m²] $f_T = 0$	$f_T = f_{APB} = 1/2<10\bar{1}>$	$f_T = f_{SISF} = 1/6[11\bar{2}]$	$f_T = f_{CSF} = 1/6<2\bar{1}\bar{1}>$	Method	Ref.
Bulk γ(TiAl)	0		436	125	205	EAM	[39]
Pseudotwin	60	193	193	193	193		
Rotational fault	120	218	218	165	165		
Bulk γ(TiAl)	0		560	90	410	LDF/FLAPW	[6, 40, 41]
Pseudotwin	60	270	270	270	270		
Rotational fault	120	250	250	280	280		
True twin	180	60	550	60	550		

Interface type	φ [deg.]	$f_T = 0$	$f_T = f_{APB} = 1/6[\bar{1}2\bar{1}0]$	$f_T = f_{SISF} = 1/3[\bar{1}100]$	$f_T = f_{CSF} = 1/6[01\bar{1}0]$	Method	Ref.
Bulk α₂(Ti₃Al)	0		300		320	LDF/FLAPW	[40]
α₂/γ	n × 60	100	280	20	220		

EAM, embedded atom method; LDF, local-density-functional theory; FLAPW, full-potential linearized augmented plane-wave.

could increase the dislocation mobility. As outlined in Section 5.1.2 and described by Equation (5.1), the Peierls friction stress to propagate a dislocation is expected to decrease when its core width increases. Specifically, the reduction of the CSF energy could spread out the core of the $\mathbf{b} = 1/2<1\bar{1}0]$ ordinary dislocation, which in the bulk is not dissociated. In other words, a dissociation that does not occur in bulk TiAl may become favorable at the interface. Zhao and Tangri [43] have investigated the defect structure of α_2/γ interfaces in a lamellar Ti-42Al alloy and identified interfacial dislocations with the Burgers vectors 1/2<110] and 1/6<112], respectively. The 1/2<110] dislocations exhibited a significant planar dissociation on the interfaces of about 4.4 nm involving a complex stacking fault (CSF). Such dissociation has not been observed in bulk TiAl and may thus be attributed to the above-discussed change of the fault energies at the interfaces. This finding suggests that the mobility of ordinary dislocations is enhanced when their slip plane is confined to the 60° or 120° γ/γ interface or to an α_2/γ interface. Several authors [39, 40] have proposed that these highly glissile dislocations may support plastic shear along the lamellar interfaces. This could be an important factor in the deformation of suitably oriented PST crystals. However, this so-called supersoft glide mode, while perhaps present, seems to be less effective in the deformation of polycrystalline material than once expected. On the other hand, dislocations gliding on conjugate {111} glide planes can become trapped when intersecting the interfaces because the dislocation core energy is reduced [39]. For example, a 1/2<110] dislocation may cross-slip onto the interface rather than into the adjacent lamella. The behavior of the superdislocations is expected to be more complex because their dissociation modes involve all types of planar faults, not only the CSF. The climb dissociation of a superdislocation intersecting a 120° boundary was reported by Kad et al. [44].

Summary: The theoretical estimates of interfacial energies indicate that the lamellar 180° true twin boundary has by far the lowest energy followed by the 120° γ/γ, the 60° γ/γ and the α_2/γ boundaries. This finding is in good agreement with the experimentally observed relative occurrence of the different γ/γ interface types. The atomic bonding of some interfaces is such that fault translations could decrease the interfacial energy. Dislocations gliding in the close vicinity of these interfaces could therefore dissociate more widely, as compared to the bulk TiAl.

6.1.3
Coherent and Semicoherent Interfaces

Solid–solid interphase boundaries can be divided on the basis of their atomic structure into three classes: coherent, semicoherent and incoherent. Following Christian [45], a coherent interface between two crystals is one for which corresponding lattice planes and lines are continuous across the interface, that is, there will be a matching of lattice points at the interface. Any difference in the atomic spacings of adjacent crystals will develop elastic strains, called coherency strains. In this sense only the lamellar 180° true twin boundary is fully coherent because the adjacent lattices are symmetrically oriented. At all the other interfaces the matching is only imperfect; these interfaces are semicoherent. If there is no continuity of planes and lines across the interface, even in the sense of a local correspondence described above, the interface is then regarded as incoherent; a case that will not be considered here.

The mismatch at the α_2/γ interface arises from the differences in crystal structure and lattice parameters. For the close-packed $<11\bar{2}0>_{\alpha 2}$ and $<10\bar{1}>_\gamma$ directions considered here the mismatch between the α_2 and γ phases is $\varepsilon_m = 1\%$ to 2% depending on alloy composition, thermal expansion coefficients and temperature. Hazzledine [46] has analytically modeled the stress state of a constrained lamellar compound comprised of α_2 lamellae and equal volume fractions of the three different variants of γ lamellae. There is a biaxial tension in γ and a biaxial compression in α_2 because the atomic spacings in α_2 are larger than in γ. Furthermore, the γ phase does not posses a triad axis along <111>, while the α_2 phase has hexagonal symmetry along [0001]. Hence, the γ phase has to be sheared to three-fold symmetry to match the six-fold symmetry of the α_2 phase. This accommodation produces shear strains of opposite signs in the adjacent α_2 and γ lamellae. Because of the differences in the elastic constants the mismatch strains are unequally shared between the two phases.

The mismatches at the 60° and 120° γ/γ interfaces are solely caused by the tetragonality of the γ phase. According to an analysis by Kad and Hazzledine [47] the situation may be described as follows. Figure 6.13 shows an **AB** stack of two $(111)_\gamma$ planes that are rotated by 120° against each other, thus, representing the situation at the 120° lamellar boundary. The $[\bar{1}10]$ direction of the upper plane is extended because it is parallel to the $[10\bar{1}]$ direction of the lower plane. Likewise, the $[0\bar{1}1]$ direction of the upper plane will be compressed. These two strains are equivalent to two shears, one along $[0\bar{1}1]$ and the second along $[\bar{2}11]$, when referred to the lower plane γ_1. Thus, elastic strains of opposite sign occur in the upper and lower crystal plane. A systematic theoretical analysis performed by Saada and Couret [48] has basically confirmed this geometrical reasoning. Similar arguments hold for the 60° boundary. Thus, the mismatches occurring at the 60° and 120° γ/γ interfaces are pure shear strains. The mismatch strain is $\varepsilon_m = 1\%$ to 2%, as deduced from the tetragonality of the γ phase. It might be expected that ε_m depends on temperature because the thermal expansion coefficients of the γ phase in the **a** and **c** directions differ significantly.

In the literature, different modes of misfit accommodation have been discussed, which could be relevant for lamellar alloys. A thin plate sandwiched by two thick plates may be considered as a simple analog of misfitting lamellae. For such a system it is expected that the misfit strains and rotations tend to partition to the thinner plate. Similarly, in a two-phase compound the strains and rotations tend to partition to the softer elastic phase [49]. Up to a certain point, the strain could be solely taken up by elastic distortion, that is, the lamellae are uniformly strained to bring the atomic spacings into registry. This homogeneous strain accommodation leads to coherent interfaces but introduces lattice distortions that are known as coherency strains. Hazzledine [46] has shown that the elastic misfit accommodation in lamellar ($\alpha_2 + \gamma$) alloys is only possible if the lamellae are very thin. The predicted critical thicknesses are $d_c \leq 8$ nm for the mismatched γ/γ interfaces and

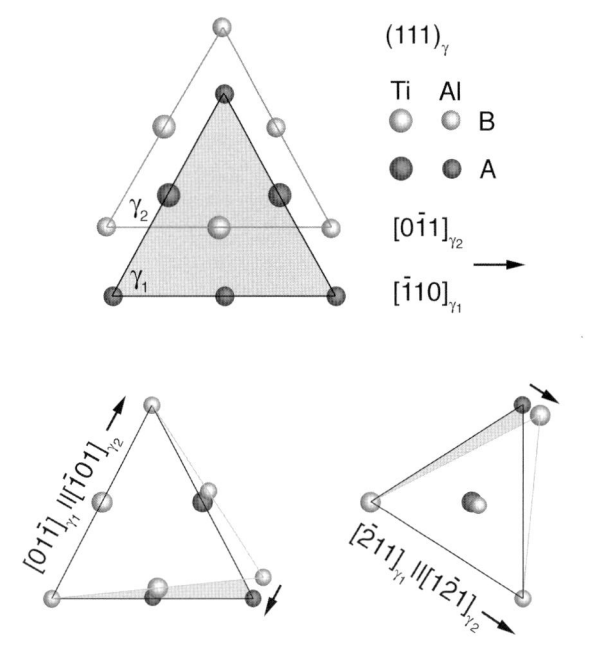

Figure 6.13 Misfit situation at a 120° γ/γ interface. A stack **AB** of two (111) planes of the variants γ₁ and γ₂ that are rotated by 120° against each other. The mismatch caused by the tetragonality of the γ phase is equivalent to shears along [0$\bar{1}$1] or [$\bar{2}$11] when referred to the lower plane γ₁. These shears could be taken up by a cross-grid of screw dislocations with Burgers vectors parallel to these directions.

$d_c \leq 0.8$ to 3.9 nm for the α_2/γ interfaces, depending on the volume content of α_2 phase. Based on TEM observations performed on a lamellar Ti-39.4Al alloy, Maruyama et al. [50] determined the critical thickness as $d_c \leq 50$ nm. Figure 6.14 shows a small Ti$_3$Al platelet embedded in γ phase with a thickness of 4.5 nm, which is just above the coherency limit predicted by Hazzledine [46]. While homogeneous strain accommodation is still recognizable, part of the misfit is already taken up by interfacial defects.

The coherency strains raise the total energy of the system. Thus, for a sufficiently large misfit, or lamellar spacing, it becomes energetically more favorable to replace the coherent interface by a semicoherent interface, a situation that is referred to as a loss of coherency. At semicoherent boundaries the misfit may be accommodated by different mechanisms, which have been reviewed by Aaronson [51]. In the early model of Frank and van der Merve [52] the structure of a semicoherent interface was described in terms of "misfit dislocations"; since then the term has come into common use in both diffusional and martensitic transformations [53]. The misfit dislocations partially take up the misfit, that is, the atoms at the interface adjust their positions to give regions of good and bad registry. In other words, the misfit is concentrated at the dislocations. The misfit parameter ε_m along a certain crystallographic direction can be defined as

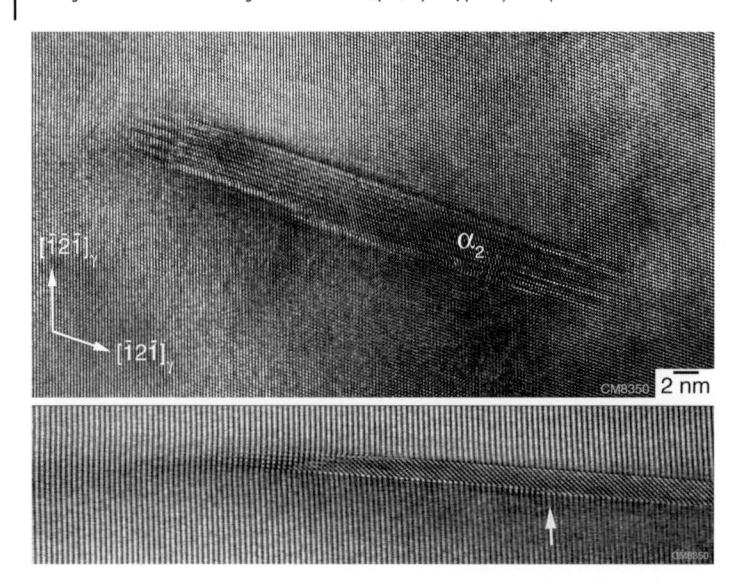

Figure 6.14 High-resolution electron micrograph showing a small Ti$_3$Al platelet embedded in γ phase. Interfacial steps, dislocations and homogeneous elastic straining accommodate the misfit between the particle and the matrix. In the compressed image below, the elastic straining at the tip of the particle is readily visible by the distortion of the $(\bar{1}1\bar{1})$ planes. A dislocation compensating the misfit between the $(\bar{1}1\bar{1})_\gamma$ and $\{2\bar{2}01\}_{\alpha 2}$ planes is arrowed. Ti-46.5Al-4(Cr, Ta, Mo, B), sheet material.

$$\varepsilon_m = \frac{a_2 - a_1}{a_1} \quad (6.8)$$

a_1 and a_2 are the lattice constants of the two crystals parallel to the interface. ε_m may be considered as the strain to make the adjacent crystals coherent. Unlike isolated dislocations in bulk material, which increase the energy, the misfit dislocations reduce the energy of the structure.

Figure 6.15 demonstrates a domain boundary that subdivides a γ lamella. The misfit between the $(020)_\gamma$ and $(002)_\gamma$ planes is accommodated by regularly spaced dislocations that are manifested by an additional $(020)_\gamma$ plane. The separation distance of these dislocations corresponds to the tetragonality of the γ phase according Equation (6.8).

Mismatch strains at interfaces can be also relieved by the formation of ledges or steps at the interface. Hall et al. [54] have introduced this concept in order to rationalize the structure of $\{111\}_{fcc} \parallel \{110\}_{bcc}$ interphase boundaries in Fe–C alloys. The authors were able to show that introduction of monoatomic steps significantly improve the atomic matching, thus preventing the disregistry from becoming large anywhere. Structural ledges may replace misfit dislocations as a way of retaining low-index terraces between the respective defects. This model, based on qualitative geometrical principles of good fit alone, was later justified on energetic grounds [55, 56]. The general view is that planar boundaries are favored for large

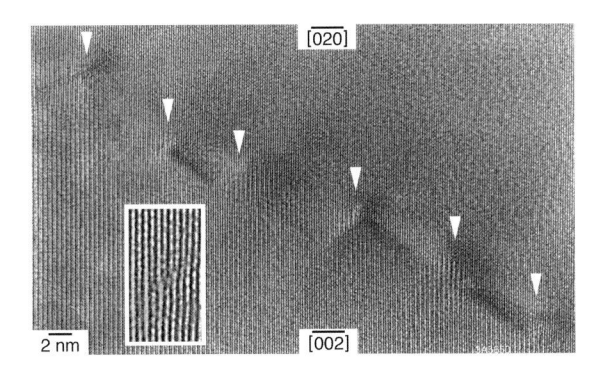

Figure 6.15 High-resolution electron micrograph showing a domain boundary subdividing a γ lamella. Note the regularly spaced interfacial dislocations that accommodate the misfit between the $(020)_\gamma$ and $(002)_\gamma$ planes. Cast Ti-48.5Al-1.5Mn, 24-h annealing at 850 °C.

misfits and small Burgers vectors of the misfit dislocations, whereas stepped boundaries are favored for small misfits and large values of the Burgers vector [57]. Steps may range in scale from atomic to multiatomic dimensions depending on energetic or kinetic factors.

However, it is very often the case that an interfacial defect exhibits both dislocation- and step-like character, thus, comprising a more general defect that has been defined as a disconnection [58–60]. The formation of a disconnection is schematically depicted in Figure 6.16. Two crystals λ and μ containing surface steps are rigidly brought into contact; **t**(λ) and **t**(μ) are vectors lying in the ledge risers. Figure 6.16a illustrates the general form of a disconnection that is produced by incompatible steps **t**(λ) and **t**(μ). The step height h of the disconnection is defined by the material overlap normal to the interface. Closing the gap on the left of the step produces the dislocation component of the disconnection. The Burgers vector of this dislocation component is defined by the difference in the two **t** vectors

$$\mathbf{b} = \mathbf{t}(\lambda) - \mathbf{Pt}(\mu) \tag{6.9}$$

P describes the relationships between the coordinate frames of the two crystals. As indicated in Figure 16a, $\mathbf{b}_n$ is the Burgers vector component perpendicular to the interface. Because of its step character, disconnection motion along an interface transports material from one phase to the other, the extent of which is essentially determined by the step height h. At the same time, the dislocation content of the disconnection leads to deformation. In this sense, disconnection motion couples interface migration with deformation. The different extents of symmetry breaking at the interfaces lead to a broad variety of disconnections [58]. This is reflected in different step heights and dislocation contents, which eventually determine the function of disconnections in phase transformations.

There are two limiting cases of disconnections, which could be important for the interfaces in lamellar $(\alpha_2 + \gamma)$ alloys. The disconnection sketched in Figure 6.16b is a pure step with $\mathbf{t}(\lambda) = \mathbf{Pt}(\mu)$, $h = h(\lambda) = h(\mu)$ and $\mathbf{b} = 0$. Propagation of this step along the interface accomplishes the transformation of the crystal λ into μ,

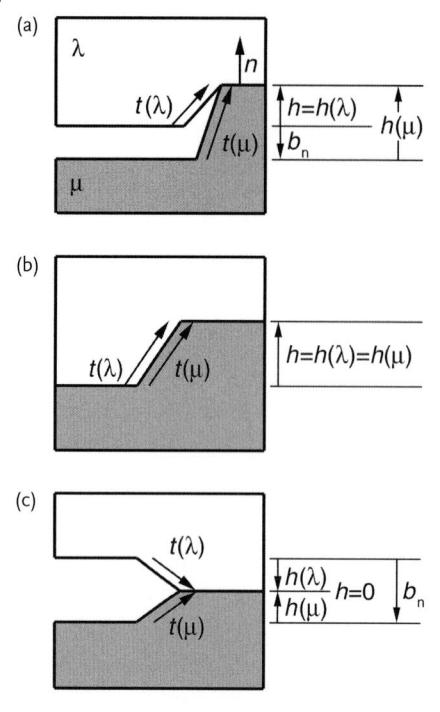

Figure 6.16 Schematic illustration of disconnections formed by surface steps on the crystals λ and μ. (a) A disconnection formed by incompatible surface steps **t**(λ) and **t**(μ). The step height $h = h(\lambda)$ defines the extent of material overlap normal to the interface; b_n is the Burgers vector component parallel to the interface normal **n**. (b) A pure step with **t**(λ) = **Pt**(μ), $h = h(\lambda) = h(\mu)$ and **b** = 0. (c) A pure interfacial dislocation with the Burgers vector **b** = **t**(λ)−**Pt**(μ). Figure adapted from [58].

or vice versa. The differential material flux associated with the step movement is proportional to h. The other limiting form of a disconnection (Figue 6.16c) is formed by two surface steps of opposite sign; hence, there is no material overlap and $h = 0$. Closing the gap produces a pure interfacial dislocation with the Burgers vector defined by Equation (6.9). Incorporation of lattice dislocations into the interface could also produce interfacial dislocations with the Burgers vectors **b** = **t**(λ) or **b** = **Pt**(μ), respectively. In the theory of Frank and van der Merwe [52], these dislocations are called misfit dislocations. The misfit is most effectively accommodated when **b** is parallel to the interface; no misfit is compensated if **b** is perpendicular to the interface. The motion of disconnections with $h = 0$ produces material flow only by the simultaneous growth (or dissolution) of the two phases, which is proportional to the Burgers vector component normal to the interface. However, no differential flux is provided from one phase to the other. Disconnection models have been developed for a variety of diffusional and diffusionless phase transformations in crystalline solids and an extensive body of literature has evolved. For more details the reader is referred to a review of Howe et al. [61]. In that article, the authors go well beyond of the limited treatment here, with a very readable account of the crystallographic aspects of disconnections and the role of the various interfacial defects in phase transformations.

In the framework of the above considerations, the situation in lamellar TiAl alloys can be described as follows [12, 13, 33, 46, 62]. The 180° true twin boundary was found to be fully coherent, whereas all the other interfaces, that is, the lamellar interfaces α_2/γ, 60° γ/γ, 120° γ/γ, and the domain boundaries, are semicoherent. The situation at the α_2/γ interface is complex because the misfit involves a dilatation component in addition to shear. Dislocations with edge character are needed in order to relieve the dilatational misfit [21]. This is in accordance with the early TEM work of Mahon and Howe [12], who observed Shockley partial dislocations with edge and mixed character at the α_2/γ interfaces in a Ti-50Al alloy. These dislocations were present as dense parallel arrays or as hexagonal and rectangular networks. Zhao and Tangri [43] observed two sets of closely spaced ($h_i < 50\,nm$) interfacial edge dislocations with the Burgers vectors 1/2<110] and 1/6[112] at the α_2/γ interfaces of lamellar Ti-42Al. As has been discussed in the previous section, the 1/2<110] dislocations were found to exhibit a significant core spreading, which does not occur in TiAl bulk material. Wunderlich et al. [63, 64] investigated lamellar α_2/γ boundaries in several two-phase alloys. Adjacent to the boundary, in both, the α_2 and the γ phase, independent dislocation ensembles were observed. Within the α_2 phase, networks occurred that involved three sets of dislocations with Burgers vectors parallel to <11$\bar{2}$0>. The mismatch structures observed in the γ phase were more complex and could involve one, two, or three sets of <110] and <101] dislocations, depending on the addition of ternary elements.

Several authors [9, 33, 39, 43, 65–69] observed interfacial steps at α_2/γ interfaces with heights that were always a multiple of $\{111\}_\gamma$ planes. Shang et al. [70–73] have analyzed the step character by high-resolution electron microscopy. Accordingly, the steps could be described as disconnections, as has been outlined at the beginning of this section. The topological parameters of the most commonly observed two-plane step are $\mathbf{b} = 1/6[11\bar{2}]$, $\mathbf{t}(\gamma) = 1/2[\bar{1}\bar{1}2]_\gamma$ and $\mathbf{t}(\alpha_2) = [000\bar{1}]_{\alpha 2}$ [61]. Since the step heights $h(\gamma)$ and $h(\alpha_2)$ are equal, their difference, $\mathbf{b_n}$, is zero, meaning this disconnection has no Burgers vector component perpendicular to the interface. The motion of the disconnection leads to growth of α_2 phase at the expense of γ phase, or vice versa, depending on the direction of motion. Equal and opposite fluxes of Ti and Al occur during these transformations [61]. In this sense the $\alpha_2 \to \gamma$ transformation could be described as diffusion-controlled step migration [70–73]. However, in a recent analysis Howe et al. [61] have pointed out that the change of the stacking sequence and interdiffusion can also sequentially occur, provided that no long-range diffusion is required. Figure 6.17 shows a γ(TiAl) lamella terminated within the α_2 phase of a two-phase alloy. The interface marked in the micrograph borders the crystal region in which the exact **ABC** stacking of the L1$_0$ structure is fulfilled. Outside of this exactly stacked region of γ phase, there is a two to three atomic plane thick layer in which neither the **ABC** stacking of the γ phase nor the **ABAB** stacking of the α_2 phase is correctly fulfilled. This becomes particularly evident at the tip of the γ lamellae and indicates a significant homogeneous straining of the lattice. This strain seems to locally relax by the formation of dislocations, as can be seen in the compressed form of the image. The other salient feature is

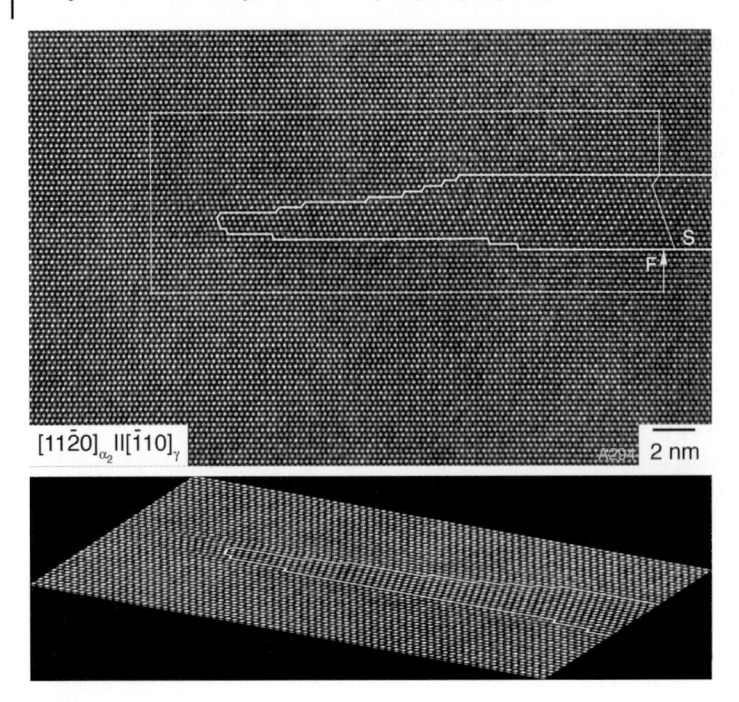

Figure 6.17 A high-resolution micrograph of a γ(TiAl) lamella terminated within α_2 phase imaged down the $<10\bar{1}>_\gamma$ and $<11\bar{2}0>_{\alpha_2}$ directions. The thick white line marks the position of the interface. The salient feature is the misfit accommodation by steps. S and F denote the start and finish, respectively, of the Burgers circuit constructed around the tip of the γ lamella. After elimination of all the canceling components in the initial circuit, and transforming the sequence of operations in the α_2 phase into the γ coordinate frame, the projected Burgers vector is $\mathbf{b}_p = 1/6[11\bar{2}]$. Note the dislocation in front of the γ tip. This can be recognized in the image below, which was compressed along the $(2\bar{2}01)_{\alpha 2}$ planes. TiAl–Nb, as cast.

the misfit accommodation by steps. A Burgers circuit constructed around the tip of the γ lamella results in a projected Burgers vector of $\mathbf{b}_p = 1/6[11\bar{2}]$. This indicates that the small misfit between the $(111)_\gamma$ and $(0002)_{\alpha 2}$ planes is elastically taken up. The observed misfit accommodation is pertinent to the issue of how differential material flux during the $\alpha_2 \rightarrow \gamma$ transformation is accomplished. There is ample evidence [74] of enhanced self-diffusion along dislocation cores. Likewise, interfacial ledges are envisaged as regions where deviation from the ideal structure is localized and that may provide paths of easy diffusion. This gives rise to the speculation that it is mainly the tip of a newly formed γ lamella where the atomic composition between the two phases is adjusted during transformation.

The processing routes for technical alloys often involve rapid quenching or intense hot working, which apparently results in more complex lamellar interfaces. One particular feature is that the interfaces often contain a significant tilt component that is accommodated by dislocations with a Burgers vector out of the

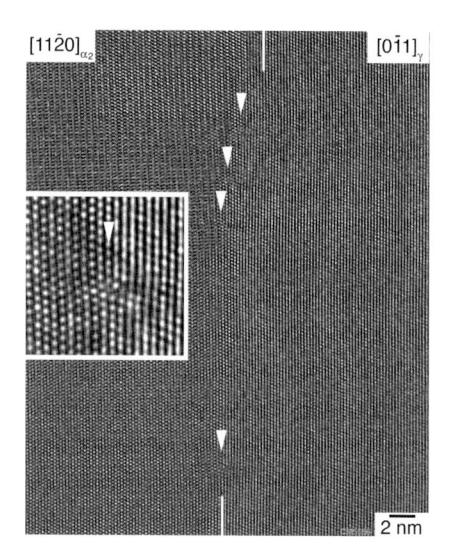

Figure 6.18 High-resolution TEM micrographs showing a misfitting α_2/γ interface in sheet material of a nearly lamellar TiAl alloy with composition Ti-47Al-3.7 (Nb, Cr, Mn, Si)-0.5B. Note the curved interface and the presence of interfacial dislocations that are manifested by an additional $\{111\}_\gamma$ plane parallel to the interface. The inset shows one of the interfacial dislocations in higher magnification.

interfacial plane [33, 68, 75, 76]. As demonstrated in Figure 6.18 for an α_2/γ interface, the misfit dislocations are manifested by an additional $(111)_\gamma$ plane parallel to the interface, which corresponds to a projected Burgers vector component $\mathbf{b}_p = 1/4[121]$. This is consistent, for example, with a mixed ordinary 1/2<110> dislocation and a 1/2<112] superdislocation. Due to the ambiguity of the high-resolution images the possible variants could not be distinguished. Thus, the Burgers vector $\mathbf{b} = 1/3[111]$ was attributed to these interfacial dislocations. The misfit that is accommodated in the interfacial plane by these dislocations is proportional to their edge component projected into the interface. Thus, the 1/3[111] dislocations are expected to be less efficient in accommodating the rotational misfit and may lead to high interfacial energies. These extraneous interfacial dislocations are probably formed as a result of phase transformations and migration of the lamellar boundaries, which occur during processing and probably change the misfit character of the interfaces. Furthermore, matrix dislocations can be incorporated into the interfaces and may react with pre-existing misfit dislocations to form stable configurations. For these reasons, the structure of the interfaces is expected to depend on the processing route and thermal history of the material, a view that is supported by TEM observations performed on hot-worked two-phase alloys [33, 68, 76]. As an example, Figure 6.19 shows a tilt boundary joining two misoriented α_2 lamellae. The tilt misfit of 10° is taken up by grain-boundary dislocations. It might be expected that this sub-boundary was formed during the rolling process by deformation and dynamic recovery of the α_2 phase.

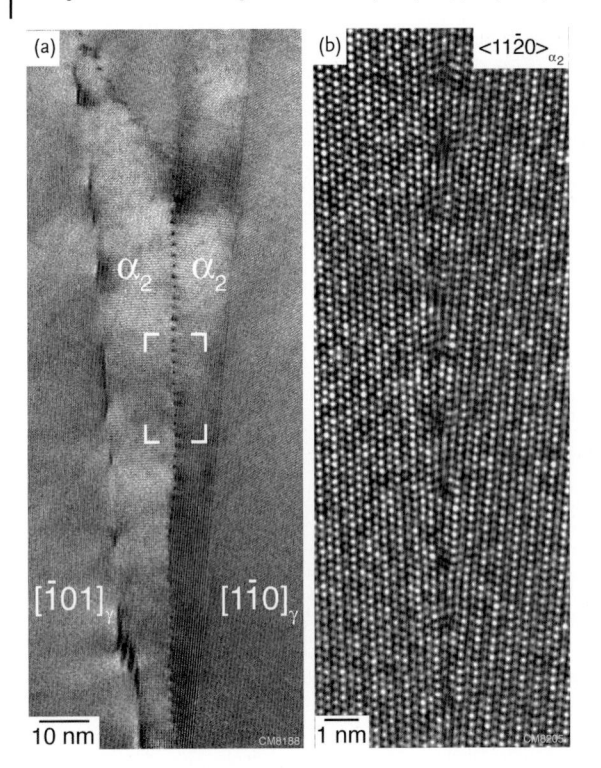

Figure 6.19 Structural features in engineering ($\alpha_2 + \gamma$) alloys. (a) A tilt boundary joining two α_2 lamellae. The tilt misfit of about 10° is accommodated by regularly spaced dislocations. Ti-46.5Al-(Cr, Nb, Ta, B) sheet material. Thermal treatment after rolling: annealing at 1400 °C, 1 h, followed by air cooling. (b) Higher magnification of the region marked in (a).

A misfitting lamellar γ/γ interface can be considered as a twist boundary because the adjacent lamellae are rotated relative to each other about the interface normal. The small angular deviation of these boundaries can in principle be accomplished by a single set of screw dislocations [77]. In such a dislocation array the strain fields of the individual dislocations are additive, which leads to a long-range stress field and to a relatively high boundary energy. Thus, the single array of screw dislocations is unstable, unless its stress field is compensated by a second set of screw dislocations 90° apart [78]. If h_i is the spacing and **b** the Burgers vector of the dislocations in the cross-grid then the twist angle Θ is approximately given by [78]

$$\Theta \approx \frac{b}{h_i} \tag{6.10}$$

The two sets of dislocations could react to form dislocations with a third Burgers vector. Along with this model, grids of two or three dislocation arrays have been observed on misfitting γ/γ interfaces. It is understood that the Burgers vectors of the dislocations lie entirely within the interface. All types of dislocations existing in the γ phase can probably relieve the misfit of the constrained lamellae [46, 67, 75, 79–82].

The details of the misfit accommodation have been thoroughly investigated at the 120° lamellar γ/γ interface. In a Ti-50Al alloy Kad and Hazzledine [47] identified single sets of parallel screw dislocations as the most common mismatch structure in this interface. The Burgers vector of these dislocations was found to be parallel to <211], presumably because 1/6<211] is the shortest Burgers vector in the {111} plane. Occasionally, cross-grids of screw dislocations with Burgers vectors **b** = 1/2<1$\bar{1}$0] and **b** = 1/2<11$\bar{2}$] have been detected. The separation distance of the misfit dislocations was found to be proportional to the Burgers vector, according to Equation (6.10). The authors have pointed out that the strain accommodation could also be achieved by a cross-grid of edge dislocations that is at 45° to the orientation of the grid of screw dislocations, as originally proposed by Matthews [77] for this type of boundary. This accommodation mechanism is unlikely, however, because the line energy of an edge dislocation is higher than that of a screw dislocation. Couret et al. [82] observed cross-grids of 1/2<1$\bar{1}$0] and 1/2<11$\bar{2}$] dislocations at the 120° boundaries in Ti-47Al-2Cr-2Nb.

Summary: Among the various interfaces occurring in the lamellar morphology only the 180° twin boundary is fully coherent because the adjacent lattices are symmetrically oriented. At all the other interfaces the matching is imperfect; these interfaces are semicoherent. The mismatch arises from the differences in crystal structure and lattice parameters and amounts to 1% to 2%, depending on alloy composition and processing conditions. Interfacial dislocations and steps accommodate the misfit. Only when the lamellae are very thin, can the mismatch be taken up by uniform elastic straining of the lamellae, which, however, would result in extremely high constraint stresses.

6.1.4
Coherency Stresses

Despite the misfit accommodation by dislocations and ledges it is often the case that significant coherency strains remain at the interfaces. Shoykhet et al. [83] have considered this situation in detail by determining the strain distribution in multilayer systems as a function of the misfit in lattice parameters and elastic constants between the lamellae. The residual coherency strain present in the individual lamellae was found to be inversely proportional to the lamellar spacing λ_L. However, when sampled over a sufficiently large volume, the average of the coherency strain was zero. Thus, the sign of the coherency strain alternates from lamella to lamella. This oscillating nature of the coherency strain is apparently reflected in the dislocation structure. Figure 6.20 shows a 1/3[111] dislocation that is anchored at two adjacent semicoherent interfaces. The configuration of the dislocation varies between concave and convex; hence, the sign of the dislocation curvature is inverted. In terms of the dislocation line tension model [74] this configuration could be rationalized by two shear stresses of opposite sign that act on the respective parts of the dislocation.

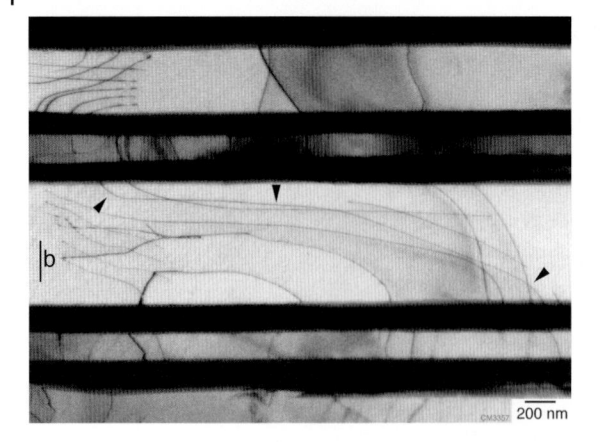

Figure 6.20 Dislocation configuration indicating the oscillating nature of coherency stresses present in the lamellar morphology. The dislocation with Burgers vector $b = 1/3[111]$ (arrowed) is anchored at two semicoherent interfaces and changes the sign of its curvature. This is indicative of the inversion of the coherency stress present between the adjacent interfaces. Investment cast Ti-48Al-2Cr, annealed at 1100 °C, followed by air cooling.

As with the coherency strain, the resulting coherency stress τ_i is proportional to $1/\lambda_L$. The misfit strain accommodated by the interfacial dislocations increases with λ_L, that is, the behavior is opposite to that expected for the residual coherency strains. In the design of lamellar alloys the thickness of the lamellae is often reduced in order to maximize the yield stress. The above consideration shows that at the same time the coherency stresses grow both in absolute magnitude and relative to the yield stress [46]. Thus, in high-strength alloys the coherency stresses can be very large and can affect deformation, phase transformations, recovery, and recrystallization in various ways. Given this importance the coherency stresses were determined by two methods. Hazzledine et al. [84] have investigated a lamellar Ti-50Al alloy by convergent beam electron diffraction (CBED). The method has the advantage that the lattice parameters can be precisely determined. Thus, by recording the CBED pattern in adjacent lamellae, the change of the lattice parameters could be monitored. From the differences in the lattice parameters the coherency stresses present at the interfaces were calculated by using Hooke's law. The coherency stresses were found to be in the region of 100 MPa [20, 84].

The coherency stresses are also manifested by dislocation loops that emerge from the interfaces [85, 86], as shown in Figure 6.21. It should be noted that these loops occurred at all types of semicoherent interfaces and were probably bowed out under the action of the coherency stresses. Trace analysis showed that the loops are mostly situated on $\{111\}_\gamma$ planes obliquely oriented to the interfaces. Invisibility criteria were used to determine the Burgers vector of these dislocations; part of this contrast analysis is demonstrated in Figure 6.21. The Burgers vector of the loops was identified as $b = 1/2<110>$, which often was not contained in the

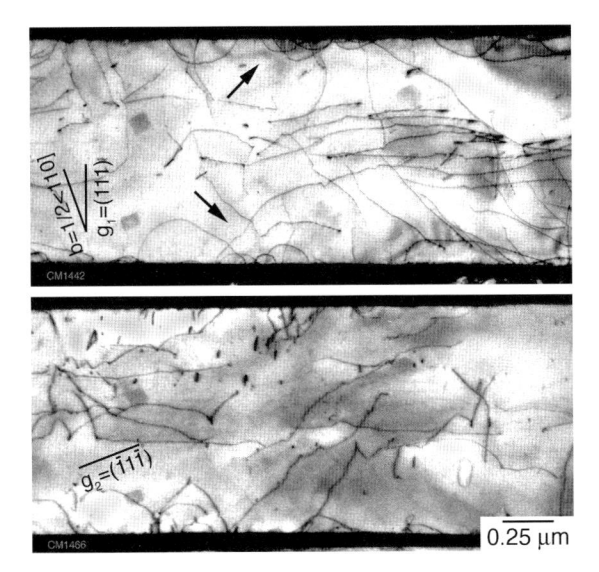

Figure 6.21 Dislocation structure adjacent to lamellar interfaces imaged with the diffraction conditions indicated. Note the dislocation loops anchored at the interfaces (arrowed), with Burgers vector $\mathbf{b} = 1/2<110]$ out of the interfacial plane. Foil orientation close to $<10\bar{1}]_\gamma$. Investment cast Ti-48Al-2Cr, hot isostatic pressing at 1200 °C, 1.7 kbar, 4 h.

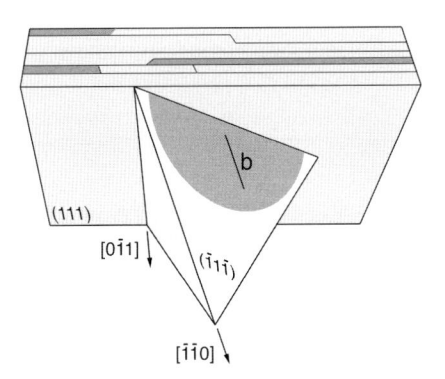

Figure 6.22 Schematic drawing of the geometrical configuration of dislocation loops emerging from a semicoherent interface. The loop shown in the drawing has the Burgers vector $\mathbf{b} = 1/2<110]$ out of the interfacial plane and is situated on a $(\bar{1}1\bar{1})$ plane obliquely oriented to the (111) interface plane.

interface. The geometrical arrangement of the loops at the interfaces is sketched in Figure 6.22. The mechanism by which the loops are formed is not completely clear. Figure 6.23 shows a network of interfacial dislocations with slightly bowed out segments, which gives the impression that the loops can be directly released from the network of ordinary dislocations that accommodates the misfit at the semicoherent interfaces. The loops may also be formed by a decomposition

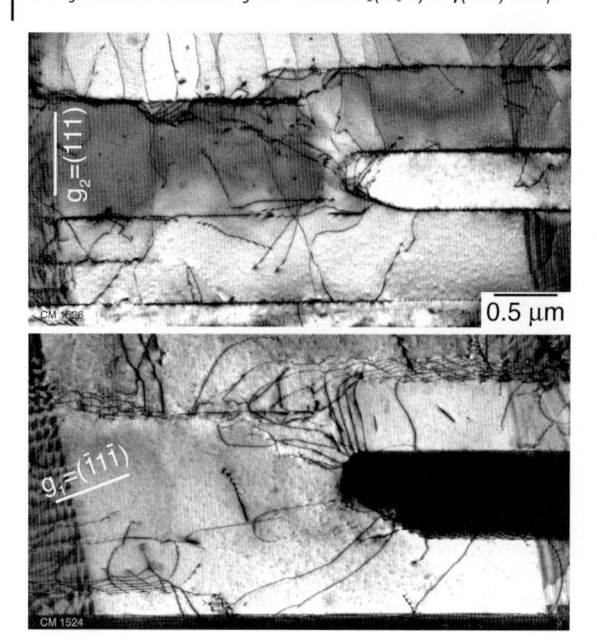

Figure 6.23 Generation of dislocation loops at lamellar interfaces under the action of coherency stresses. Investment cast Ti-48Al-2Cr, hot isostatic pressing at 1200 °C, 1.7 kbar, 4 h. (a) Lamellar structure imaged end on showing a terminated lamella and stepped interfaces. (b) The same area imaged with a different diffraction condition; with reference to Figure 6.23a the sample was tilted about $[1\bar{2}1]$ by 15°. Network of interfacial dislocations at the stepped interface with slightly bowed out segments, which can give rise to loop formation.

reaction of the misfit dislocations with Burgers vector $\mathbf{b} = 1/3[111]$. A possible mechanism could be described as

$$1/3[111] + 1/6[11\bar{2}] \rightarrow 1/2[110] \tag{6.11}$$

As judged from Frank's rule [74], reaction (6.11) is not energetically feasible, however, the coherency stresses may initiate the reaction and support the emission of the reactant dislocations. It is tempting to speculate that the formation of the loops at the interfaces can be supported by thermoelastic stresses. When the constituents of a two-phase material have different or anisotropic coefficients of thermal expansion, internal stresses are generated in the vicinity of the interfaces upon temperature change. It is well documented that, if these thermal stresses are high enough, dislocations are produced at the interfaces and punched into the matrix [87]. Mismatch structures and coherency stresses are therefore expected to be very sensitive to the processing conditions of the material.

The coherency stresses present at the interfaces were determined by analyzing the configuration of the dislocation loops [86]. There are various levels of approximation to an exact treatment analogous to the successive approximation of iso-

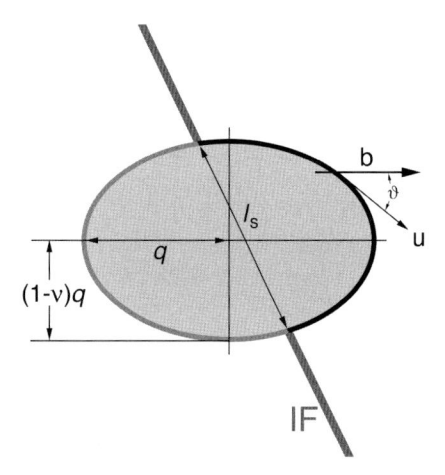

Figure 6.24 Geometrical parameters involved in line-tension calculations of stressed dislocation loops according to the DeWitt–Koehler model [89]. l_s length of the bowed out dislocation segment, q major semiaxis of the loop, ϑ angle between line element **u** and the Burgers vector **b** of the dislocation, IF trace of the interface.

tropic elasticity, anisotropic elasticity or self-stress calculations in deriving the elastic properties of dislocations [74]. These issues have been considered in a detailed study of stressed dislocation loops, which propagated in MgO single crystals during *in situ* deformation in a high-voltage electron microscope [88]. For the present analysis line-tension configurations of dislocations were described by the model of DeWitt and Koehler [89], which considers the dependence of the elastic energy on the orientation of the dislocation line. Elastic isotropy was assumed for the sake of simplicity. With the elastic constants calculated by Yoo et al. [90], the anisotropy factor of the [110] shear direction is

$$A_1 = \frac{2C_{66}}{C_{11} - C_{12}} = 1.18 \tag{6.12}$$

that indicates near isotropy. Thus, the approximation is relatively good for 1/2<110] shear loops in TiAl. Within the framework of this model the loops are apparently elliptical with the major semiaxis q (in µm) parallel to the Burgers vector. Loops of this configuration are related to the shear stress τ_i by

$$\tau_i(\text{MPa}) = (2.65/q)(\ln l_s + 8.64) \tag{6.13}$$

Equation (6.13) uses the shear modulus $\mu = 70.1\,\text{GPa}$ and Poisson's ration $v = 0.238$ determined by Schafrik [91]. The geometrical configuration of the loops and the quantities involved in Equation (6.13) are explained in Figure 6.24. l_s (in µm) is the length of the dislocation segments anchored at the interface. The loops were calculated for different stresses and projected into the foil plane. The shear stresses τ_i acting on the individual loops were determined by comparison with the line-tension configuration as demonstrated in Figure 6.25. The evaluation also

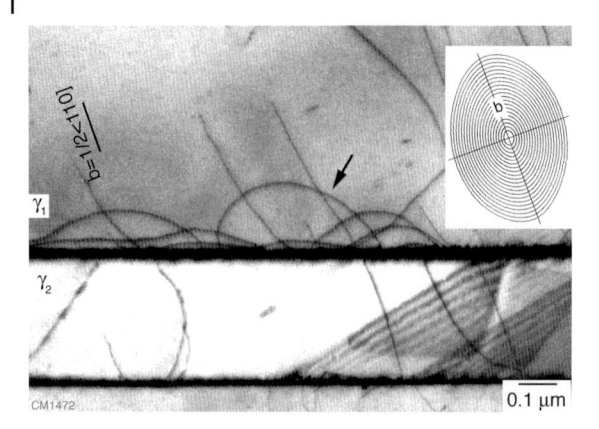

Figure 6.25 Dislocation loops with Burgers vector $\mathbf{b} = 1/2\langle110]$ bowed out from a semicoherent interface γ_1/γ_2 under the action of coherency stresses. The inset shows line-tension configurations calculated for different shear stresses and projected into the $\{10\bar{1}\}$ foil plane. By comparison, the shear stress $\tau_i = 30$ MPa was attributed to the loop marked with the arrow. Cast Ti-48Al-2Cr, hot isostatic pressing at 1200 °C, 1.7 kbar, 4 h; nearly lamellar microstructure.

gives the related values of q and l_s. Figure 6.26 demonstrates the frequency distribution of these three parameters. Due to the complexity of the diffraction analysis only 190 loops were used to plot the figure. Statistically, the validity of such a small population may be open to question, but it would appear that high coherency stresses are present at the interfaces. The average value of the coherency stresses is $\bar{\tau}_i = 130$ MPa. The room-temperature value of the flow stress of the material is $\sigma_a = 430$ MPa, from which a shear stress $\sigma_a/M_T = 143$ MPa can be deduced. M_T is a Taylor factor to convert normal stresses into average shear quantities. Thus, the coherency stresses are comparable with the shear stresses applied at the beginning of deformation. Because of their oscillating nature, the coherency stresses could either support or impede dislocation propagation. The presence of dislocation loops, which are subject to a high shear stress, gives rise to various interface-related deformation phenomena, which will be considered in a later section.

A frequent structural feature of the lamellar morphology is isolated loop structures (arrowed) that are present within γ lamellae (Figure 6.27). It might be expected that these loops remain from a thin α_2 lamellae that formerly interspersed the γ lamella and that was later removed by $\alpha_2 \rightarrow \gamma$ phase transformation. The position of the loops corresponds to the former α_2/γ interfaces. High internal stresses were probably left after the transformation, as indicated by the strong bowing of the loops.

Summary: In spite of the misfit accommodation by interfacial dislocations and ledges a significant elastic strain remains at the interfaces. The resulting coherency stresses are comparable with the yield stress of the material and may thus significantly affect the deformation behavior.

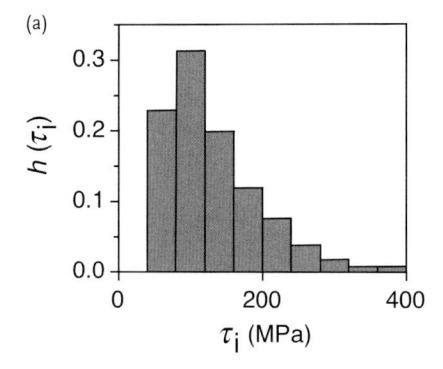

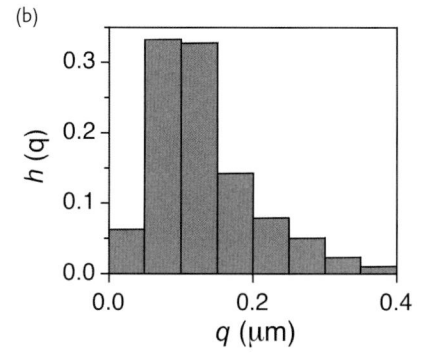

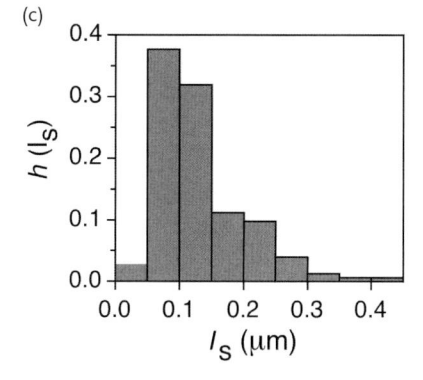

Figure 6.26 Line-tension analysis of dislocation loops adjacent to lamellar interfaces. Frequency distributions (a) $h(\tau_i)$ of the coherency stresses τ_i, (b) $h(q)$ of the major semiaxes q, (c) $h(l_s)$ of the segment lengths l_s. h is defined as $h = \Delta N/N$, with N being the number of segments found in a certain interval of τ_i, q and l_s, respectively. $n = 190$ is the total number of analyzed loops, and the average values are $\bar{\tau}_i = 130$ MPa, $\bar{q} = 0.128\,\mu m$, $\bar{l_s} = 0.132\,\mu m$ [86].

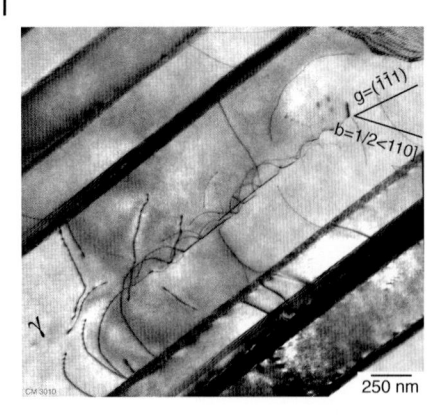

Figure 6.27 Dislocation loops in a γ lamella. The loops were probably left from an α_2/γ interface that was originally present at the position of the loops. The α_2 lamella was probably removed by phase transformation. Cast Ti-47Al-2Cr-0.2Si, annealed at 1220 °C, 4 h, furnace cooling; nearly lamellar microstructure.

6.1.5
Plastic Anisotropy

The lamellar microstructure exhibits a significant plastic anisotropy, which was early investigated by Fujiwara *et al.* [92] using a binary Ti-49.3Al PST crystal. Since then the effect has been recognized on various other (ternary) PST crystals for a wide range of lamellar spacings [38, 93–98]. The main characteristics of the deformation behavior are as follows: both the yield and fracture stresses are low when deformation occurs in the plane of the lamellae (soft mode) and are high when deformation occurs across the lamellae (hard mode). The tensile ductility is high when the tensile axis lies close to the lamellar plane and is low when the axis is nearly normal to the lamellar plane. There is no significant tension–compression asymmetry of the yield stress. Figure 6.28 demonstrates the variation of the yield stress measured in compression as a function of the orientation angle Φ of the lamellae with respect to the sample axis. The ratio of the yield stresses between the softest orientation close to $\Phi = 45°$ and the hardest orientation at $\Phi = 90°$ is 4 to 5. It should be noted that the variation of the yield stress is not symmetrical as Φ decreases or increases from 45°. For a given orientation angle Φ, the yield stress is almost insensitive to a rotation of the slip plane around its normal, meaning at fixed Φ nearly the same yield stress is measured, regardless of whether the *x*-face of the sample is parallel to $(11\bar{2})$ or $(1\bar{1}0)$. The obvious disparities between the measurements of different groups most probably arise from the differences in the Al content and the growth rate of the PST crystal, both of which are known to significantly affect the lamellar spacing.

In broad terms the anisotropy of the yield stress can be rationalized on the basis of factors that limit the slip path of the dislocations, that is, by classical Hall–Petch arguments. During soft-mode deformation ($\Phi \approx 45°$) dislocation glide is relatively easy because it is only impeded by the widely separated domain boundaries. When

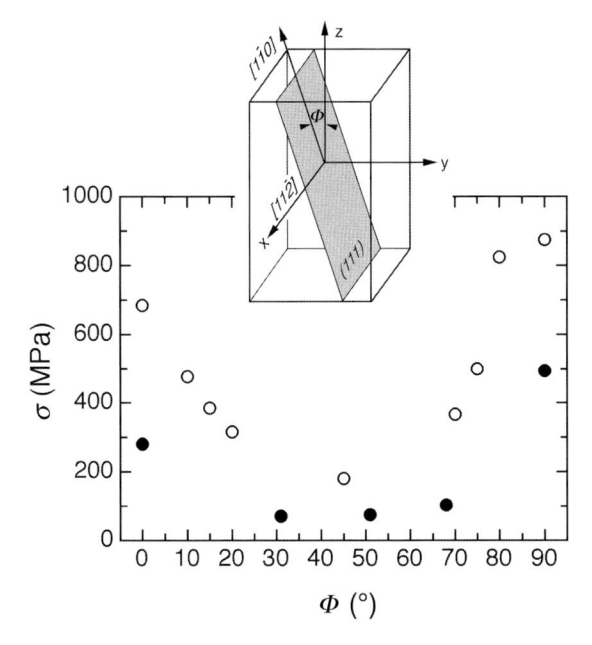

Figure 6.28 Plastic anisotropy of polysynthetically twinned (PST) crystals. Dependence of the yield stress measured in compression at room temperature as function of the angle Φ that describes the orientation of the lamellar (111) interfaces against the compression axis. All specimens were prepared with an [11$\bar{2}$] direction fixed as the x-axis of the coordinate system shown in the insert. (● Ti-49.3Al, Data of Fujiwara et al. [92]; ○ Ti-48 Al, Data of Nomura et al. [97]. The difference between the two sets of data is probably caused by the different Al contents.)

Φ is close to 0° or 90° deformation occurs along the maximum shear stress, that is, on the slip planes and slip directions oriented closest to the 45° direction from the deformation axis. Thus, shear must be traversed through the closely spaced lamellar boundaries. The barrier strength of the lamellar boundaries to dislocation glide could originate from different factors. The group around Hazzledine [39, 84, 99] has given a detailed consideration of the possible dislocation–interface interactions. Accordingly, for lamellar α_2(Ti$_3$Al) + γ(TiAl) alloys the following processes could be most important.

i) In many cases the orientation of slip is changed because the crystallographically available slip planes and directions are not continuous across the interface. This may significantly reduce the Schmid factor and thus impede slip transfer. At the γ/γ interfaces the orientation of the slip plane could change through a relatively large angle of about 39°. Reorientation of slip is always required at the α_2/γ interface; the smallest angle between the corresponding slip planes $\{111\}_\gamma$ and $\{10\bar{1}0\}_{\alpha2}$ is about 19° [44].

ii) The core of a dislocation intersecting an interface often needs to be transformed. For example, an ordinary 1/2<110] dislocation gliding in one γ

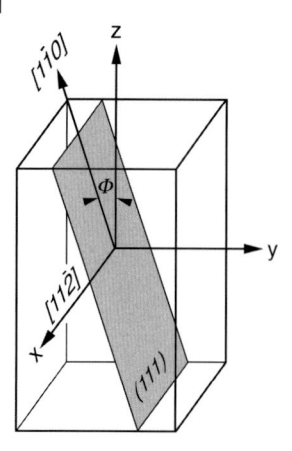

Figure 6.29 Deformation geometry of PST crystals to check the amount of strain that crosses the lamellar boundaries during compression. Figure adapted from Nomura et al. [97].

lamella has to be converted into a <101] superdislocation with the double Burgers vector gliding in an adjacent γ lamella. At the α_2/γ interface the dislocations existing in the DO_{19} structure have to be transformed into dislocations consistent with the $L1_0$ structure. These core transformations are associated with a change of the dislocation line energy because the lengths of the Burgers vectors and the shear moduli are different.

iii) Dislocations crossing semicoherent boundaries have to intersect the misfit dislocations, a process that involves elastic interaction, jog formation and the incorporation of gliding dislocations into the mismatch structure of the interface.

iv) When the slip is forced to cross α_2 lamellae, pyramidal slip of the α_2 phase is required, which needs an extremely high shear stress.

Although this picture provides a good outline for understanding the plastic anisotropy of PST crystals, the details could be subtler. This is suggested by studies of Kishida et al. [28], Nomura et al. [97] and Min-Chul Kim et al. [98], who observed the shape change of PST crystals occurring during room-temperature compression to plastic strains of 2% to 3%. As sketched in Figure 6.29, the authors recorded the strain components along the sample coordinates x, y and z for different orientations Φ of the lamellae with regard to the deformation axis. The results may be summarized as follows.

$\Phi = 0°$ to $10°$: the strain in the y direction was near to zero; hence, the net shear vector was parallel to the lamellar boundaries.

$\Phi = 15°$ to $75°$: the samples deformed by pure shear on $(111)_\gamma$ planes parallel to the lamellar boundaries because no straining along the x direction was observed.

$\Phi \geq 80°$: deformation was carried by strains along the x and y directions, an observation that is consistent with the onset of slip across the lamellar boundaries.

Nomura et al. [97] have suggested that the difficulty of this process is reflected in the asymmetric dependence of the yield stress on Φ, meaning that the yield stress increases more strongly for Φ approaching 90° compared to Φ approaching 0°.

While the shape changes observed for $\Phi = 15°$ to 75° and $\Phi \geq 80°$ corresponds largely to what is expected from the gross shear direction with respect to the lamellae orientation, the strain distribution observed for $\Phi = 0°$ to 10° warrants particular consideration. In this orientation range, deformation in all domains apparently occurred on planes that are parallel to the interfaces. This so-called channeling of plastic deformation between the lamellar boundaries is a remarkable phenomenon because it confines slip to planes with a near-zero Schmid factor, although other slip systems with relatively large Schmid factors are available. Kishida et al. [28], recognizing this problem, have carefully analyzed the deformation modes occurring in PST crystals that had been compressed along different loading axes. The deformation structure observed for $\Phi = 0°$ and the loading axis z parallel to <110> indicated that all six γ domains constituting the PST crystal were sheared on {111} planes that are inclined to the lamellar (111) planes. In two of these domains the loading axis is parallel to <1$\bar{1}$0], meaning ordinary slip and mechanical twinning cannot occur. These two domains deform by symmetrical double slip of superdislocations along [0$\bar{1}$1](1$\bar{1}\bar{1}$) and [$\bar{1}$01]($\bar{1}$1$\bar{1}$). As the superdislocations have the Burgers vector parallel to the lamellar planes, the respective domains are solely sheared along the lamellae, which is consistent with the macroscopic strain distribution.

The four remaining domains that have the <10$\bar{1}$] direction parallel to the loading axis deform on the (11$\bar{1}$) plane by glide of 1/2<1$\bar{1}$0] ordinary dislocations. The Burgers vector of these dislocations is parallel to the lamellar interfaces; hence, this glide mode is also consistent with the macroscopic strain distribution. However, this ordinary slip is always coupled with a combination of ordinary slip and mechanical twinning occurring on the other oblique {$\bar{1}$11) slip plane. The observed systems are either 1/2<110]($\bar{1}$11) glide plus 1/6<$\bar{1}$1$\bar{2}$]($\bar{1}$11) twinning or 1/2<110](1$\bar{1}$1) glide plus 1/6<1$\bar{1}\bar{2}$](1$\bar{1}$1) twinning. At first glance, this slip geometry of the four domains is not consistent with the macroscopic strain measurements described above because it would produce a shear component in the y direction perpendicular to the lamellar planes, which has not been observed. This apparent discrepancy can be solved by a detailed consideration of the shear directions involved [28, 97, 98]. The slip geometry is such that ordinary slip and twinning occur on the same slip plane, however, the related shear components perpendicular to the lamellar boundaries have opposite signs. At an appropriate combination of these mechanisms, these components normal to the interface can totally cancel each other, with the result that a net slip parallel to the lamellar boundaries is produced. For example, glide of three 1/6<$\bar{1}$1$\bar{2}$] twinning partial dislocations on successive ($\bar{1}$11) planes must be accompanied by glide of a 1/2<110] dislocation on every third ($\bar{1}$11) plane. This combination of slip and twinning is equivalent to double slip of the <101] superdislocations. The described glide geometry allows continuous shear translation through domain boundaries within the constraints set by the macroscopically observed shape change [28]. The close coupling of these processes was also suggested by atomistic simulations, which showed that ordinary dislocations and mechanical twins could propagate with nearly the same velocity [98].

For the sample orientation $\Phi = 0°$ and the loading axis parallel to $<110>_\gamma$, the α_2 platelets of a PST crystal are also expected to deform. Because of the orientation relationship between the α_2 and γ lamellae, the $<110>_\gamma$ loading axis corresponds to a $<11\bar{2}0>$ direction in α_2; hence, the α_2 platelets can deform by double prism slip. This deformation mode is considered to be relatively easy (see Section 5.2.2) and produces only a shear component parallel to the lamellae. Thus, no significant strain incompatibility between the α_2 and γ platelets is expected to occur. This particular feature may explain why PST crystals with a sample orientation of $\Phi = 0°$ exhibit a remarkable room-temperature tensile elongation.

It should be noted that in the above-described sample orientation some of the highly stressed slip systems present in the domains were not activated. For example, the described combination of ordinary slip and mechanical twinning could be replaced by glide of <101] superdislocations that have the Burgers vector parallel to the lamellar planes. The fact that this does not occur indicates that the critical resolved shear stress for superdislocation slip is significantly higher than those for ordinary slip and mechanical twinning [98]. Thus, the deformation modes operating in a certain domain cannot be solely predicted on the basis of Schmid factor considerations. In view of these results, Kishida et al. [28] have proposed that the maintenance of strain continuity at domain and lamellar boundaries is the most important factor controlling the activation of the slip and twinning systems in the individual domains.

There are also other factors that may favor the channeling of the deformation between the lamellae. As has been discussed in Section 6.1.2, the bonding of some lamellar interfaces is such that a fault translation reduces their interfacial energy. This means that the dislocations may dissociate at the interfaces into widely separated partials, which reduces the glide resistance during soft-mode deformation. The reduction of the CSF energy is such that the stress to move an ordinary dislocation at the interface could be about one order of magnitude smaller than in the bulk [39]. On the other hand, screw dislocations gliding in the hard mode on {111} planes inclined to the lamellar interfaces could reduce their energy by crossgliding onto the interface. These dislocations become immobilized and thus harden the hard deformation mode. Paidar et al. [100, 101] have pointed out that mechanical twins and superdislocations may not cross an interface, even if the adequate slip systems are available in the adjacent lamellae. This is because the propagation of superdislocations and mechanical twins is polarized, as has been described in Sections 5.1.4 and 5.1.5. These mechanisms are only activated if the sense of the shear superimposed by the external load coincides with their easy propagation direction.

As a final point of this discussion it is important to emphasize that the strong plastic anisotropy of PST crystals and the related deformation kinematics have been recognized after a relatively small plastic compression of 2% to 3%. At larger strains the situation is certainly more complex because high internal stress may develop upon strain hardening, which could give rise to the activation of additional deformation modes. In an analytical study, Paidar [102] has simulated the situation in which a colony is loaded parallel to the lamellae ($\Phi = 0°$), while deformation along the lamellae and perpendicular to the loading axis is blocked. Under these conditions significant constraint stresses develop, which redistribute the externally applied stress. The trends are such that the shear stresses on the slip systems with the highest Schmid factor are reduced, whereas those of the other systems are increased. Thus, in a constrained colony supplementary glide and twinning modes could be activated [102].

In summary, the data confirms to a high degree of confidence that slip transfer through lamellar boundaries is very difficult. Plastic strain resulting from a cooperative operation of several deformation modes can be localized between the lamellar boundaries. These facts, combined with the flat plate geometry of the lamellae, cause a marked plastic anisotropy of lamellar material. Both the yield and fracture stresses are low when deformation occurs in the plane of the lamellae (soft mode) and are high when deformation occurs across the lamellae (hard mode); the ductility is high when the tensile axis lies close to the lamellar plane and is low when the axis is nearly normal to the lamellar plane. In lamellar grains constrained within a polycrystal, the plastic anisotropy could develop internal stresses, which can easily exceed the fracture strength.

6.1.6
Micromechanical Modeling

With the advent of fast computational techniques, much effort has been made to analyze the flow characteristics of lamellar ($\alpha_2 + \gamma$) alloys by micromechanical modeling and finite-element codes. While it is now possible to construct accurate crystal plasticity models of the individual γ and α_2 phases, a direct extension of these models to simulate the yield behavior of the lamellar compound still remains difficult. This is mainly because the flow behavior of polycrystalline lamellar alloys is controlled by several parameters such as colony size, domain size, separation distance of α_2 lamellae, and thickness of lamellae. The length scales of these parameters range from mm down to nm, which, even in two dimensions, cannot be simulated simultaneously because of inadequate computer power. Furthermore, the relative importance of these structural parameters for the yield strength is not yet clearly established. Thus, in modeling the plasticity of lamellar alloys, simplifications are unavoidable. Naturally, the micromechanical models that have been proposed vary both in their approach and degree of sophistication.

Lebensohn *et al.* [103, 104] have developed a crystal plasticity model in which the lamellar morphology is reduced to a pair of twin-related lamellae. In order to simulate polycrystalline behavior, this aggregate is embedded in an effective medium, which has the average properties of the polycrystal. In spite of this simplification, the model represents the most important deformation modes and was extended to predict the yield surface of PST crystals.

Schlögl and Fischer [105–107] and Werwer and Cornec [108, 109] have described the anisotropic flow behavior of PST crystals by a three-dimensional model that involved all relevant slip systems within the framework of local continuum crystal-plasticity theory. The calculations accurately describe the experimentally observed orientation sensitivity of yielding in individual lamellar colonies and highlighted the importance of mechanical twinning and glide of superdislocations for achieving strain continuity.

Kad et al. [110, 111] and Dao et al. [112], utilizing two-dimensional finite-element computations, have simulated the deformation of polycrystalline lamellar and nearly lamellar material. The polycrystal was represented by a two-dimensional array of idealized hexagonal lamellar grains with small inset triangles or squares representing γ grains. Planar projections of the three-dimensional slip geometry were used in order to investigate the characteristics of the localized deformation-induced stress fields. The analysis suggests that polycrystalline lamellar material deforms very inhomogeneously with a strong tendency to develop shear bands, kink bands, lattice rotations, and internal buckling of lamellae. This led the authors to believe that constraints imposed by neighboring grains are a dominant factor in determining the yield stress. Even under compression, very high tensile hydrostatic stresses are generated at the triple points of colony boundaries. For obvious reasons, these factors could be detrimental for the deformation behavior of a brittle material like titanium aluminides.

More recently, Brockman and coworkers [113, 114] have performed three-dimensional finite-element simulations of polycrystalline nearly lamellar material. The investigations included an array of 512 randomly oriented grains. The volume fraction of the lamellar colonies was 0.96, with an assumed ratio of α_2 and γ lamellae of 1:5. The width of the colonies was about 145 μm. The remaining volume consisted of gamma grains filling the space between the colonies. Localized stress concentrations have been recognized, which developed within the polycrystal upon straining due to the variations of grain size, shape and orientation. The calculations reflect the smooth load-elongation response in the microyield region of lamellar material due to localized plastic flow, which has often been recognized in tensile or compression tests. At the microscopic scale this effect is associated with a remarkable heterogeneity in the deformed state, as demonstrated in Figures 6.30 and 6.31. For example, at the nominal (macroscopic) strain level of $\varepsilon = 0.2\%$, about five per cent of the polycrystal remains elastic, while in 80 per cent of the colonies the plastic strain ε_p is still smaller than the nominal strain ε. At the macroscopic strain $\varepsilon = 1\%$, nearly all of the colonies have yielded. However, the deformation is still inhomogeneous because 28 per cent of the grains exhibit a plastic strain of less than 1%. Colonies that contain undeformed material usually have their interfacial planes nearly perpendicular to the loading axes. The heterogeneity of the straining is accompanied by an adequate variation of the local stress. Dimiduk et al. [115] have reviewed the current status of modeling lamellar TiAl alloys.

In summary, the plasticity models that have been developed for lamellar alloys satisfactorily describe the plastic anisotropy by involving all the possible slip and

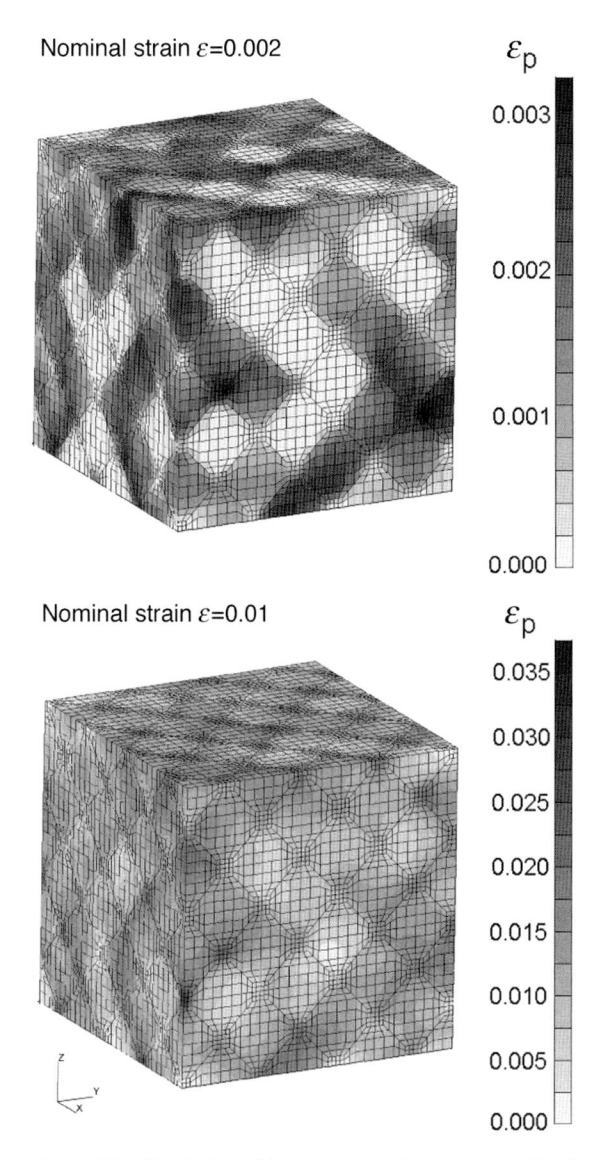

Figure 6.30 Simulation of the elastic/plastic response of a nearly lamellar TiAl polycrystalline material occurring upon deformation. The structural details of the material are defined in the text. Distribution of the effective plastic strains (normalized by the nominal strain) for the nominal strains indicated. Results provided by Brockman [113].

twinning systems. Most importantly, the models provide a quantitative description of the heterogeneity of the deformed state in polycrystalline lamellar material, which experimentally is difficult to assess. The major difficulties seen by the modeling groups are (i) scale-related constraints, which hinder a simultaneous description at the lamella and colony level, (ii) inadequate implementation of

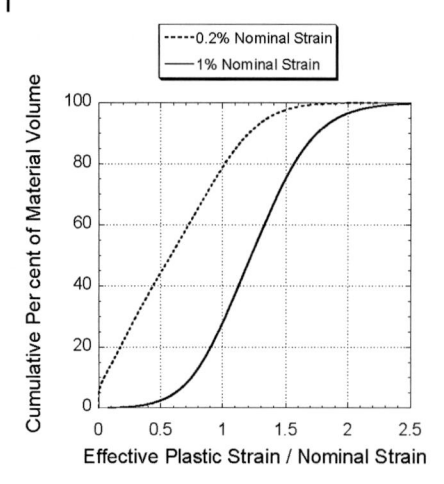

Figure 6.31 Quantitative representation of the simulations shown in Figure 6.30. At the nominal strain $\varepsilon = 0.002$, about 5 per cent of the volume remains elastic, while in 80 per cent of the grains the effective plastic strain ε_p is less than the nominal strain ε. At $\varepsilon = 0.01$, ε_p is higher than ε in about 72 per cent of the grains. Results provided by Brockman [113].

interface-related glide processes, and (iii) lack of information about the relative importance of lamellar spacing, colony size, domain size, lamellar spacing, and separation distance of α_2 lamellae.

6.2
Deformation Mechanisms, Contrasting Single-Phase and Two-Phase Alloys

Deformation phenomena in TiAl alloys have been widely studied in order to overcome the problems associated with the limited ductility and damage tolerance. The literature data covers a wide range of parameters such as alloy composition, microstructure and deformation temperature. Much of the work has been performed on single-phase γ alloys and PST crystals, which has been described in Sections 5.1 and 6.1.5. In order to avoid duplication with these sections, the following discussion concentrates on deformation phenomena that rely on the elastoplastic codeformation of the γ and α_2 phases and on the particular point defect situation occurring in two-phase alloys. Due to this effect $(\alpha_2 + \gamma)$ alloys exhibit some remarkable properties that are unlike those of either constituent.

6.2.1
Methodical Aspects of TEM Characterization

Deformation of TiAl alloys is governed by complex defect structures that occur at a very fine scale. Characterization has mostly been performed by TEM examination, which provides the required resolution, but involves inherent restrictions due to thin-foil effects, beam damage and the small volume that can be sampled. As these problems could significantly affect the analysis, a few methodical aspects will be discussed first.

An obvious problem arises from the dense process-related defect structures, which are usually present in engineering alloys. Processing of these alloys often comprises significant hot working due to the application of metallurgical techniques such as extrusion, forging or rolling. Although these thermomechanical treatments are accompanied by dynamic recovery and recrystallization, they often leave dense defect structures in the material; these could mask the defects generated by deformation. For example, the dislocation density in nominally undeformed material is often higher than $10^7\,\text{cm}^{-2}$. Thus, deformation-induced defect structures seem to be manifested only after significant straining of at least 2%.

Dislocation structures are often unstable and subject to internal stresses. Thus, during unloading and sample preparation rearrangement or loss of dislocations may occur. Generally, the demands on sample preparation become less strict as the beam voltage and the yield stress of the material increase [116, 117]. This is because thicker foils can be investigated and the dislocations can better sustain image forces. Because of the high yield stress of TiAl alloys, the rearrangement of dislocation structures is probably negligible when the samples are examined in medium-voltage microscopes with acceleration voltages between 200 kV and 400 kV. The advantage of higher acceleration voltages is offset by the danger of radiation damage.

In recent years, focused ion beam (FIB) instruments have been used for target preparation, in order to select areas of interest, such as slip bands or crack tips. Application of this technique often damages the sample due to ion impact, resulting in beam heating, formation of defect agglomerates, or even amorphization [118]. This could be a concern if the fine structure of dislocations is of interest, as dislocations easily recombine with the radiation damage.

Summary: Highly complementary to the surface view provided by scanning microscopy, transmission electron microscopy offers the ability to characterize deformation structures at the nanoscale level and in many cases even at the atomic scale.

6.2.2
Deformation of ($\alpha_2 + \gamma$) Alloys at Room Temperature

The main findings of TEM observations performed on ($\alpha_2 + \gamma$) alloys after room-temperature deformation can be summarized as follows. There is good consensus that the deformation is mainly confined to the majority γ(TiAl) phase, regardless of whether the alloy has a lamellar or equiaxed microstructure [32, 33, 75, 119–122]. As has been described in Section 5.1, γ(TiAl) deforms by octahedral glide of ordinary dislocations with the Burgers vector $\mathbf{b} = 1/2<110]$ and superdislocations with the Burgers vectors $\mathbf{b} = <101]$ and $\mathbf{b} = 1/2<11\bar{2}]$, respectively. The other potential deformation mode is mechanical twinning along $1/6<11\bar{2}]\{111\}$. The extent to

Figure 6.32 Deformation structure observed after room-temperature compression to strain $\varepsilon = 3.2\%$ showing glide of ordinary dislocations and $1/6<11\bar{2}]$ {111} order twinning. Ti-48Al-2Cr, as cast; fully lamellar microstructure with a preferred orientation of the lamellar interfaces parallel to the compression axis. Foil orientation axis close to <101].

which any of these deformation modes contributes to the total strain is the subject of ongoing discussion and seems to depend on alloy composition. Most investigations have shown that in the γ phase of two-phase alloys, glide of ordinary dislocations is the primary deformation mode, followed by mechanical twinning [32, 33, 75, 119–125]. This twinning activity is remarkably different from the deformation of single-phase γ alloys, which show a general reluctance to twin. As an example, Figure 6.32 shows the deformation structure that has been observed after room-temperature compression to a strain of $\varepsilon = 3\%$, involving ordinary dislocations and mechanical twins. Most, if not all, of the ordinary dislocations are lying near the screw orientation, whereas the superdislocations mostly have a mixed character. This observation indicates that the mobility of the ordinary dislocations depends on their character, a fact that will be discussed in a later section. Estimations of dislocation densities in such samples with the help of the line intersection method [126] typically yield dislocation densities of $10^8\,\text{cm}^{-2}$ to $10^9\,\text{cm}^{-2}$. Trace analysis and stereographic imaging have revealed that the ordinary dislocations mostly glide on oblique {111} planes [33, 75]; thus outside certain orientations, multiple slip of ordinary dislocations is the rule rather than the exception. As will be shown in a later section, multiple slip is probably an important mechanism contributing to work hardening. There are plenty of published micrographs showing evenly distributed ordinary screw dislocations, but as demonstrated in Figure 6.32, there are also specific instances of piled-up dislocations.

Couret et al. [82] reported an interesting case of self-organizing dislocation structures in lamellar Ti-47Al-2Cr-2Nb that had been deformed at room temperature. The structures were manifested as planar dislocation networks with their habit plane parallel to the lamellar boundaries. The networks occurred in γ lamellae that had an orientation for easy slip along the lamellar boundaries, that is, the primary slip plane was parallel to the habit plane of the network. The authors took

Figure 6.33 Glide of superdislocations in cast Ti-48Al-2Cr; fully lamellar microstructure with a preferred orientation of the lamellar interfaces parallel to the compression axis. Foil orientation and deformation axis close to <101>. Due to its <$\bar{1}$10] orientation, lamella γ_1 is unfavorably oriented for glide of 1/2<110] dislocations and mechanical twinning. The generation of the superdislocations (mostly of type 1/2<112]) in lamella γ_1 is probably supported by the high constraint stresses exerted by the deformation twins that propagated in the adjacent lamellae γ_2. Compression at room temperature to strain $\varepsilon = 2.7\%$.

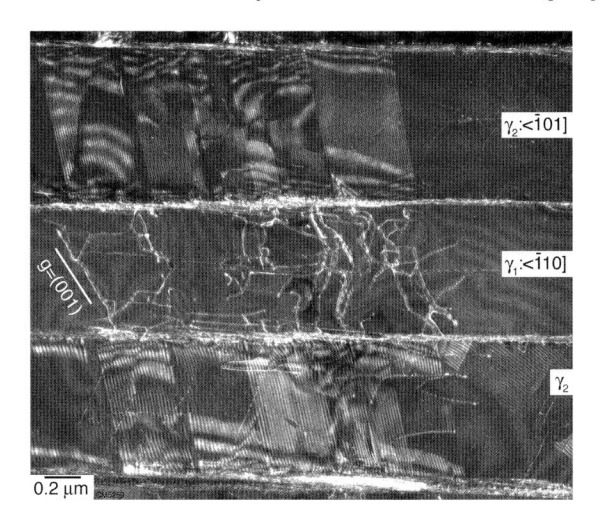

the view that the networks resulted from the intersection of <011] dislocations (propagating in the {$\bar{1}\bar{1}$1} network plane) and 1/2<110] dislocations (propagating on an oblique {111} slip plane). The resulting 1/2<112] junctions and the 1/2<110] dislocations form rectangular cross-grids, which were capable of accommodating the residual interfacial misfit. Thus, the network formation could be biased by mismatch strains.

Abundant glide of superdislocations has been only rarely observed after room-temperature deformation [127]. Thus, in the γ phase of two-phase alloys superlattice slip is apparently difficult to activate, a fact that had also been deduced from the deformation behavior of PST crystals (Section 6.1.5). However, in grains or lamellae that are unfavorably oriented for 1/2<110] glide or mechanical twinning, significant constraint stresses can develop due to the shape change of deformed adjacent grains. The constraint stresses can be high enough to activate glide of superdislocations. As an example, Figure 6.33 demonstrates localized glide of 1/2<112] superdislocations in a γ lamella that is unfavorably oriented for glide of ordinary dislocations and mechanical twinning. The difficulty of superdislocation slip imposes serious implications on the deformability of the γ phase, as the **c** component glide of the tetragonal unit cell is retarded. In this respect, the enhanced activity of twinning of the γ phase is certainly important, as it could provide such shear and reduces the requisite number of dislocation slip systems. More on this subject is provided in Section 6.2.3.

As mentioned above, of the two constituents of ($\alpha_2 + \gamma$) alloys, the α_2 phase is more difficult to deform. α_2 grains or lamellae that appear virtually free of defects are often sandwiched between heavily deformed γ phase [122]. This finding warrants particular consideration because both phases must deform, in order to ensure strain continuity. There are mainly two factors that could be responsible for the observed strain partitioning. First, TiAl alloys contain a significant amount of interstitial hardening elements, like oxygen, nitrogen and carbon, which partition preferentially to the α_2 phase [128]. Thus, in ($\alpha_2 + \gamma$) alloys, the γ phase may be softer while the α_2 phase may be harder than their isolated counterparts [119]. The group around Menand [129, 130], utilizing atom-probe field ion microscopy, has shown that this so-called scavenging effect is subtler. In these investigations, the same oxygen content of about 300 at. ppm was measured in the γ phase, regardless of whether the constitution of the alloy was single-phase γ or two-phase ($\alpha_2 + \gamma$). This oxygen level corresponds to the maximum oxygen solubility of the γ phase. In TiAl alloys the oxygen concentration is usually significantly higher than this solubility limit; a typical value is 600 at. ppm. The authors have proposed that the excess oxygen is taken up in different ways, depending on the constitution: oxide precipitates are formed in single-phase γ alloys, while in ($\alpha_2 + \gamma$) alloys the excess oxygen is absorbed by the α_2 phase. Thus, the γ phase of two-phase alloys is almost devoid of precipitates, which could be beneficial for its deformability. Similar arguments certainly hold for other interstitial impurity elements like nitrogen and carbon.

Another reason for the unequal strain partitioning between the α_2 and γ phase is certainly the strong plastic anisotropy of the α_2 phase (Section 5.2.2). TEM examinations performed on tensile-tested lamellar alloys have revealed that the limited plasticity of the α_2 phase is mainly carried by local slip of <a>-type dislocations with the Burgers vector $\mathbf{b} = 1/3<11\bar{2}0>$ on $\{1\bar{1}00\}$ prism planes [131–134], which is by far the easiest slip system in α_2 single crystals. To a lesser extent, glide of $1/3<11\bar{2}0>$ dislocations was observed on $\{2\bar{2}01\}$ pyramidal planes, that is, on π_1 pyramidal planes of the first order [135]. Another secondary glide system of the <a>-type dislocations is $1/3<11\bar{2}0>\{3\bar{3}01\}$, which has been observed under high levels of local strain [136]. Pyramidal slip seems to be triggered by stress concentrations exerted by deformation twins that propagated in adjacent γ lamellae and impinged against the γ/α_2 interface. It is important to note that no significant glide of <c>-type dislocations on pyramidal planes has been observed after tensile straining [133, 135]. Taken together, these findings again indicate that pyramidal glide requires extremely high yield stresses. Thus, there are certain grain orientations for which the α_2 phase is difficult to deform. For example, α_2 grains or lamellae that are loaded parallel to the [0001] direction cannot deform without glide of <c>-type dislocations on pyramidal planes. These α_2 grains are certainly constrained by the surrounding deformed grains. The resulting stresses could easily exceed the fracture stress; this could be a reason for premature failure.

At room temperature, glide of <c>-type dislocations probably can only be activated under compression, where crack propagation is largely suppressed by the superimposed hydrostatic stress. The strong work hardening occurring under these conditions probably provides high internal stresses, which enables pyramidal slip to be locally activated. The pyramidal slip systems that have been observed in heavily compressed (ε = 6 to 20%) samples are $1/3<\bar{1}2\bar{1}6>\{2\bar{2}01\}$ and $1/3<\bar{1}2\bar{1}6>\{1\bar{2}11\}$ [121, 134]. Again, this slip activity appeared to be initiated by stress concentrations produced by the incoming twins. As with monolithic Ti$_3$Al, the $1/3<\bar{1}2\bar{1}6>$ dislocations propagate as loosely coupled pairs of superpartials with parallel Burgers vectors $\mathbf{b} = 1/6<\bar{1}2\bar{1}6>$ (Section 5.2.1).

Summary: Room-temperature deformation of ($\gamma + \alpha_2$) alloys is mainly carried by the γ phase via ordinary slip and mechanical twinning. Localized glide of the α_2 phase is provided by prism glide of $1/3<11\bar{2}0>$ superdislocations. Essentially no glide of <c>-type dislocations has been observed within the α_2 phase. This glide geometry makes the individual phases plastically anisotropic. The problem is more accentuated in the α_2 phase.

6.2.3
Independent Slip Systems

During deformation, a grain in a polycrystal is constrained to undergo general plastic deformation, since it must remain in contact with and accommodate the shape changes of its neighbors. In theory, deformation is considered as a plastic strain tensor with six components. When the volume remains constant, as it does during plastic shear, a general shape change is fully specified by five independent strain components. Thus, for the plasticity of polycrystalline material, five independent slip systems must be able to operate, a result known as the von Mises condition [137]. Any slip or twin system corresponds to a vector in a given five-dimensional space. The five slip systems are not independent if any one can be expressed as a linear combination of the others. If the von Mises criterion is not satisfied, high constraint stresses could develop, which often lead to grain-boundary sliding, phase transformation, or fracture.

Application of the von Mises criterion to the deformation of polycrystalline $\gamma + \alpha_2$ alloys is more subtle because the γ phase is prone to twin. Furthermore, in lamellar alloys the two phases are aligned according to the Blackburn orientation relationship, which couples the deformation processes of the individual phases. This particular situation has been considered by Goo [138]. In terms of the von Mises criterion the most important difference between slip and twinning consists in the reversibility of the processes [139]. In the case of slip, the reversion of the slip direction provides an equivalent slip system, that is, each slip system produces two strain vectors of opposite sign. This is not the case for twinning; the reversal of primary twinning with the elements $K_1 = (111)_\gamma$, $K_2 = (11\bar{1})_\gamma$ and $\eta_1 = 1/2[11\bar{2}]$ leads to a complementary twin with $K_1 = (111)_\gamma$, $K_2 = (001)_\gamma$ and $\eta_1 = 1/2[\bar{1}\bar{1}2]$. As the twinning shear of the complementary twinning is about twice as large as that of primary twinning, it is considered not to occur (see Section 5.1.5).

The findings of Goo [138] could be summarized as follows. In the isolated γ phase, slip of ordinary dislocations provides four slip systems, only three of which are independent. Likewise, the four-order twinning systems $1/6<11\bar{2}]\{111\}$ by themselves do not fulfill the von Mises criterion. Also, the combination of ordinary slip and order twinning does not provide the five independent slip systems. Thus, in the isolated γ phase slip of superdislocations is also required in order to satisfy the von Mises criterion. Nevertheless, mechanical twinning certainly reduces the intensity of the requisite superslip. In the isolated α_2 phase dislocation slip on the prismatic, basal and pyramidal planes does not provide the five independent slip systems. The von Mises criterion could be satisfied, if this dislocation glide was accompanied by mechanical twinning along $<10\bar{1}2>\{10\bar{1}1\}$, which, however, has only been observed after high-temperature creep of hyperstoichiometric Ti-34Al [140]. Thus, together with the difficulty of basal and pyramidal glide, it is highly improbable that the von Mises criterion is satisfied in Ti_3Al.

In analyzing lamellar alloys, Goo [138, 139] has pointed out that in von Mises' theory no assumption is made about the local distribution of the slip systems, that is, they need not be available throughout the entire grain. Because slip, in general, is not homogeneously distributed, it is sufficient to consider volume elements with large dimensions compared to the slip-band spacings. In this sense, the lamellar morphology could provide a dispersion of shear processes that originate from the α_2 phase and the differently oriented γ variants. The evaluation of the combined strain tensor has indeed shown that the von Mises criterion in lamellar $(\alpha_2 + \gamma)$ alloys could be satisfied solely by ordinary slip of the γ phase, provided that all orientation variants of the γ phase are available. In this context, two points should be noted that could modify the conditions discussed above. First, the ordinary screw dislocations, which mainly contribute to the deformation of the γ phase, are capable of cross-gliding because of their compact core (Section 5.1.3). Cross-slipping of screw dislocations onto any number of planes can produce at most two additional independent slip systems [74]. Secondly, twinning can enhance the opportunity for slip by rotating the crystal structure into a more favorable orientation for slip. While these two processes are certainly limited to narrow regions, they could be important for the accommodation of stress concentrations. Taken together, these factors result in the much better mechanical performance of two-phase alloys, when compared with polycrystalline single-phase γ or α_2 alloys. As deformation is closely linked with fracture, two-phase lamellar alloys can be quite tough.

For polycrystalline material with randomly oriented grains Schmid's law may be generalized to

$$\sigma = M_T \tau \qquad (6.14)$$

M_T, known as Taylor factor [141], relates the yield stress σ of the polycrystal to the critical resolved shear stress τ of the constituent single crystals and can be determined by averaging the stress over the grains and considering the most favored slip systems. For f.c.c. metals with randomly oriented grains the Taylor factor is $M_T = 3.06$ [142, 143], however, a significant texture could increase the Taylor factor to $M_T = 3.67$ [144], since it further affects the degree of alignment from one grain to another and hence, the propagation of slip.

Mecking et al. [145] have calculated the Taylor factor for γ(TiAl) by considering ordinary glide, superdislocation glide and mechanical twinning. M_T depends on the relative strengths of these deformation modes. A large difference between the strengths of the different deformation modes tends to increase the Taylor factor. For example, if the yield stress ratio between ordinary and superslip is 0.6 and that between mechanical twinning and ordinary slip is 0.2, the Taylor factor is $M_T = 3.21$. An apparent Taylor factor can be deduced by comparing the work-hardening rate of polycrystalline material $\partial\sigma/\partial\varepsilon$ with that of single crystals $\partial\tau/\partial a_s$ [146]

$$M_T^2 = \frac{(\partial\sigma/\partial\varepsilon)}{(\partial\tau/\partial a_s)} \qquad (6.15)$$

The quantity a_s is the shear strain. Utilizing the data for polysynthetically twinned (PST) crystals and of polycrystalline lamellar Ti-48Al, Parthasarathy et al. [147] determined the values $M_T = 3.2$ to 3.8, which are consistent with the calculations of Mecking et al. [145]. Mecking et al. [148] have also calculated the morphology of the single-crystal yield surface for γ(TiAl) for various relative strengths of the deformation modes.

Summary: In lamellar ($\alpha_2 + \gamma$) alloys the von Mises criterion for general plastic deformation of polycrystalline material is satisfied by ordinary slip of the γ phase, provided that all orientation variants of the γ phase are available. This contrasts with single-phase γ(TiAl) or α_2(Ti$_3$Al) alloys, which exhibit a lack of independent slip systems that can operate at comparable stresses.

6.2.4
High-Temperature Deformation of ($\alpha_2 + \gamma$) Alloys

Deformation of two-phase alloys at elevated temperatures of 700–800 °C is mainly carried by the γ phase, similar to what has been observed for room-temperature deformation. However, there are qualitative and quantitative differences. While the glide modes of the γ phase are the same as those occurring at room temperature, additional strain accommodation is accomplished by climb of ordinary dislocations [149, 150]. The climb processes certainly relax the requirement for independent slip systems. The other significant difference compared to room-temperature deformation is the enhanced activation of mechanical twinning within the γ phase. This is a remarkable phenomenon since twinning is commonly a low-temperature mechanism [151]. As an example, Figure 6.34a demonstrates the deformation structure observed after tensile deformation of a Ti-48Al-2Cr alloy at 800 °C [152].

The twins are generally very narrow, as indicated by the frequency distribution of their widths (Figure 6.34b). Also, as has been discussed in Section 5.1.5, the twinning shear of γ(TiAl) is relatively small. Thus, despite their profusion in the microstructure, the contribution of mechanical twins to the overall deformation should not be overestimated. A quantitative evaluation of twin densities and widths in moderately deformed material ($\varepsilon = 3\%$) indicates that a twinning shear of up to 5% can be locally achieved [153]. It is interesting to note that the contribution of superdislocations to deformation of the γ phase is not significantly enhanced at

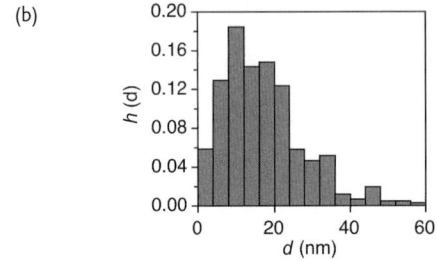

Figure 6.34 Mechanical twinning in a Ti-48Al-2Cr alloy observed after tensile deformation at 800 °C to strain $\varepsilon = 8.9\%$. (a) A TEM micrograph of the twin structure; the twin/matrix interfaces are imaged end on. (b) Frequency distribution $h(d) = \Delta N/N_0$ of the widths h of deformation twins observed under these experimental conditions. ΔN is the number of twins found in the width intervals, $N_0 = 884$ is the total number of twins investigated.

elevated temperatures. A possible explanation is that the superdislocations become locked due to the nonplanar dissociation reactions discussed in Section 5.1.4.

When compared with room-temperature deformation, the α_2 phase also seems to be more ductile. This is manifested by a homogeneous activation of prismatic glide and a relatively dense population of $1/3<11\bar{2}6>$ dislocations on pyramidal planes [131]. Climb of all these dislocations also contributes to deformation. These factors reduce the plastic incompatibility between the α_2 and γ phase, which is certainly beneficial for the materials plasticity. The observed changes of the deformation mechanisms coincide with the transition from brittle to ductile fracture, which in ($\gamma + \alpha_2$) alloys typically occurs between 700 and 800 °C. This observation leads some authors to believe that the plastic properties of the minority α_2 phase critically affects the deformation and fracture behavior of lamellar TiAl alloys [134].

Summary: At elevated temperatures the two constituents of ($\alpha_2 + \gamma$) alloys deform more easily. Deformation of the γ phase is supported by climb of ordinary dislocations and intensive mechanical twinning. The plasticity of the α_2 phase is enhanced by a significant glide activity in the prismatic and pyramidal glide systems coupled with nonconservative dislocation motion. Taken together, these factors make the codeformation of the two phases easier and ensure strain continuity.

6.2.5
Slip Transfer through Lamellae

Apart from the high intrinsic glide resistance and the plastic anisotropy, deformation in two-phase alloys is also impeded by the various interfaces occurring in the lamellar morphology. Overcoming these interfaces by perfect and twinning partial dislocations could involve all the processes that in broad terms have been described in Section 6.1.5. Thus, the barrier strength of an interface depends on its character, that is, whether it is a 60° γ/γ, 120° γ/γ, 180° γ/γ, or an α_2/γ interface. Of perhaps similar significance are the type and character of the incoming dislocations and the orientation of their slip plane relative to the interface. In general, the following scenario could occur during slip transfer through an interface:

i) overcoming of the long-range coherency stresses present at the interfaces;

ii) elastic interaction and intersection of the incoming dislocation with the interfacial misfit dislocations;

iii) reaction of the incoming dislocations with the interfacial dislocations and incorporation of the reactant dislocations into the network of interfacial dislocations;

iv) decomposition and dissociation reactions of the incoming dislocations onto new slip planes, in order to accommodate the gross shear direction;

v) activation of dislocation sources at the exit side of the boundary;

vi) elastic shear transfer through the lamella.

While most of these processes have been identified in deformed lamellar alloys, the relative importance of these factors is difficult to assess, especially since they interact with one another. However, it might be speculated that the extent to which these processes must occur at a particular boundary determines the barrier strength. For example, slip transfer through the 180° true twin boundary could be accomplished by the activation of a conjugate deformation system, such that the incoming and outgoing slip systems are in mirror symmetry. Ordinary screw dislocations with the Burgers vector parallel to the interface could easily overcome the twin boundary by cross-slip [154]. Thus, the barrier strength of the true twin boundary is expected to be relatively low. In most cases, however, the incoming and outgoing dislocations are linked by reactions. These are rich in detail, depending on the nature and character of the incoming dislocations. The formation of the resulting junctions could be an essential part of the transfer process and thus has been considered by several authors [136, 154–159]. The geometrically possible reactions are not always energetically favorable and sometimes violate Frank's law. There are also reactions at which the Burgers vector is not conserved; the Burgers vector of the incoming dislocations is different from the resultant Burgers vector of the outgoing dislocations. In these cases residual dislocations are left at the interface, which requires additional energy and makes the reactions unfavorable. The situation is further complicated as all types of misfit dislocations could be involved in the reactions. Furthermore, in the interface region where the constraints of the neighboring lamella are greatest, more dislocation reactions and slip systems could be activated than in the center of the lamella, which is less influenced by the boundary. Several authors have taken the view that the continuity of glide and twin planes determines the activated system, which implies that the resultant Burgers vector of the outgoing dislocations be parallel and equal in magnitude to that of the incoming dislocations [28, 160]. This view is supported by the findings of Singh et al. [32], who recognized that the slip mechanisms occurring in a lamellar colony are correlated. All the γ lamellae belonging to a certain orientation variant deform on the same slip system, regardless of the other lamellae interspersed in between. Depending on the Schmid factor, the deformation starts in a certain lamella on a "pilot system", which is either ordinary slip or mechanical twinning. The slip is symmetrically transferred into adjacent lamellae that have a true twin orientation. If ordinary slip is the "pilot system", then the "driven system" in the twin-related lamella deforms also by ordinary slip. Likewise, mechanical twinning is symmetrically transferred. The authors have noted that this symmetrical slip transfer also occurs if the Schmid factor in the new lamellae is unfavorable. Thus, the continuity of ordinary glide or twinning could be a predominant factor governing the slip transfer through lamellar true twin boundaries.

The complexity of the problem is illustrated in Figure 6.35, which demonstrates the early stage of shear translation through a 60° pseudotwin boundary. The deformation of the incoming twin T_1 is shared in γ_2 by a set of twinning partial dislocations and a set of $1/2<1\bar{1}0]$ ordinary dislocations propagating on the same $(1\bar{1}\bar{1})$ plane and another set of $1/2<110]$ ordinary dislocations propagating on the oblique $(\bar{1}\bar{1}1)$ plane. As some of the dislocations in lamella γ_2 are still anchored at the interface, the micrographs give the impression that the dislocations were emitted from the interface under the action of the localized stress exerted from the impinging twin. Thus, the activation of dislocation sources is undoubtedly an important factor for slip transfer through interfaces. It is reasonable to speculate that these sources are provided by the network of the interfacial dislocations, as has been described in Section 6.1.4.

There is good experimental evidence that the α_2/γ interface provides the highest barrier strength for incoming dislocations, when compared with the γ/γ interfaces [33, 75]. This can be rationalized by the fact that overcoming the α_2/γ interface is associated with all the energy-dissipating processes listed above. Thus, together with the high intrinsic glide resistance of the α_2 phase, it is often impossible to shear α_2 lamellae plastically. In such cases the strain could be elastically transferred, as has been proposed by Singh et al. [122]. The available TEM evidence suggests that this mechanism often occurs in front of incoming twins. The elastic stress field of the twin tip is thought to be elastically transferred into the α_2 lamella and could be high enough to activate dislocation sources at its exit side. As the

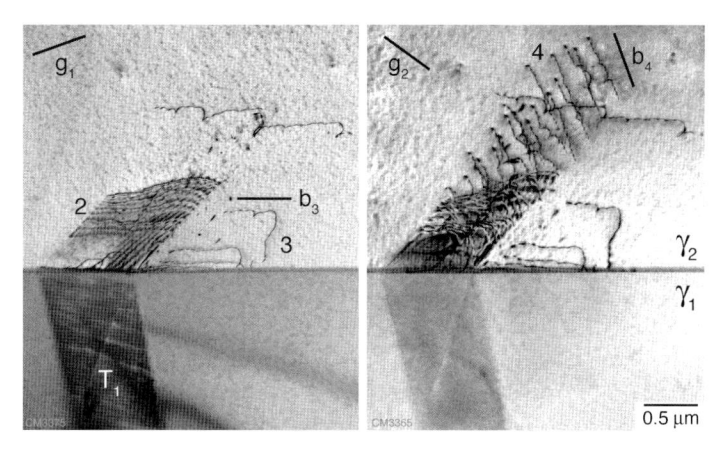

Figure 6.35 Slip/twin interactions at a 60° lamellar interface γ_1/γ_2. The twin T_1 propagated in lamella γ_1 was immobilized at the interface. Twinning partial dislocations (2) and two sets of 1/2<110> ordinary dislocations (3) and (4) situated on oblique {111} planes were generated in lamella γ_2. In this way, the high stress concentration ahead of the twin is shielded. Foil orientation close to <10$\bar{1}$]. Indicated are the Burgers vectors $\mathbf{b}_3 = 1/2<1\bar{1}0]$ and $\mathbf{b}_4 = 1/2<110]$; the diffraction vectors are $\mathbf{g}_1 = [\bar{1}1\bar{1}]$ and $\mathbf{g}_2 = [020]$. Investment cast Ti-48Al-2Cr, annealed at 1100 °C for 1 h, followed by air cooling. Tensile deformation at room temperature to strain $\varepsilon = 0.5\%$.

stress field decreases with increasing distance from the twin tip, this "elastically mediated" strain transfer is probably only possible through relatively thin α_2 lamella. The operation of such a mechanism is probably illustrated in Figure 6.36. Two twins (1) and (2) impinge an α_2 lamella that gradually decreases in width. Twin (1) hits a relatively thin part of the α_2 lamella; hence, the strain can be elastically transferred. This is manifested by the generation of dislocation loops in the adjacent lamella γ_2. Twin (2) impinging a thicker part of the α_2 lamella is blocked without evidence of glide in γ_2. Likewise, Forwood and Gibson [136] have proposed that the intersection of short α_2 lamellae could be avoided, in that the incoming twin bypasses the α_2 lamellae behind its termination.

Attempts to assess the relative strengths of the different interfaces by mechanical testing have been frustrated since the structural parameters of the lamellar morphology are interdependent. In polycrystalline material the colony size, domain size and lamellar thicknesses appear to be closely coupled. A detailed study of Dimiduk et al. [161] on fully lamellar Ti-45.3Al-2.1Cr-2Nb has shown that the lamellar spacings are proportional to the square root of the colony size. In this respect the investigations performed by Umakoshi and Nakano [37, 38] on PST crystals are most reliable, as the spurious effects of the colony size on the yield stress were eliminated. Different domain sizes, lamellar thickness and spacings of α_2 lamellae were established by different Al contents and growth rates of the PST crystals. By utilizing different sample orientations, plastic shear was forced to occur either parallel or perpendicular to the lamellae, as described in Section 6.1.5. The analysis performed in terms of a Hall–Petch relationship indicates that

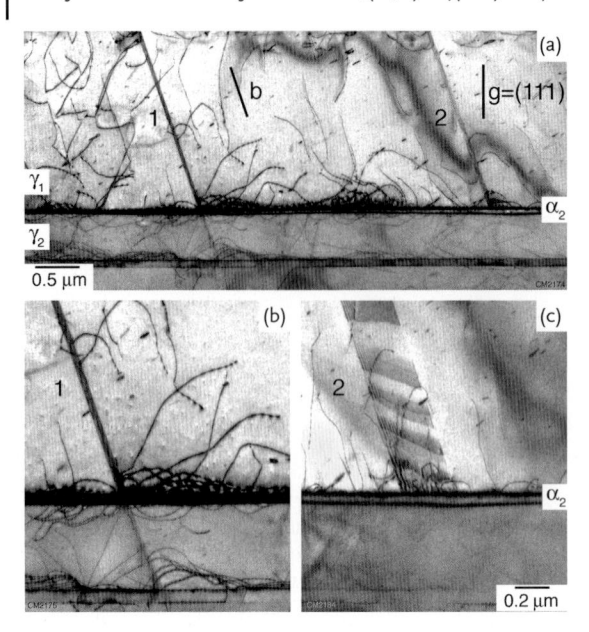

Figure 6.36 Elastically mediated strain transfer through an α_2 lamella that is sandwiched between two gamma lamellae and gradually decreases in width. Nearly lamellar Ti-48Al-2Cr, tensile deformation at room temperature beyond yield. (a and b) Twin (1) hits a relatively thin part of the α_2 lamella and elastically transfers its strain into the adjacent lamella γ_2. This is indicated by the nucleation of dislocations in γ_2. (c) Twin (2) impinges a thicker part of the α_2 lamella and is blocked without glide activity in γ_2. With respect to Figure 6.36a the foil was rotated around the interface normal by 30°.

the strengthening effect of the α_2/γ lamellar boundaries is significantly higher than that of the γ/γ interfaces and of the domain boundaries. This was inferred from the fact that the yield stress for lamellae orientations of $\Phi = 0°$ strongly increased with the volume fraction of α_2 lamellae.

Regardless of the details, the investigations show that the lamellar boundaries are strong barriers to the propagation of dislocations and mechanical twins. Slip transfer through lamellar boundaries entails mechanisms that cover a relatively large crystal volume. Typical examples are overcoming of the coherency stresses or forcing a boundary dislocation to act as a glide source under the stress concentration of a pile up. Such processes are not expected to be supported by thermal activation. Thus, the stress part associated with the overcoming of lamellar boundaries is almost independent of temperature and strain rate. Figure 6.37 supports this view. The micrographs are part of an *in situ* heating study performed on a nearly lamellar Ti-48Al-2Cr alloy. The material had been compressed at room temperature to a strain of $\varepsilon = 3\%$; this produced a moderate density of dislocations and a supersaturation of point defects, presumably vacancies. During *in situ* heating inside the microscope ordinary dislocations were emitted from an α_2/γ interface and piled up at the next interface. These moving dislocations trailed slip lines, which indicates that the dislocations must have screw character. The slip lines arise because the dislocations disturb the top and bottom of the foil as

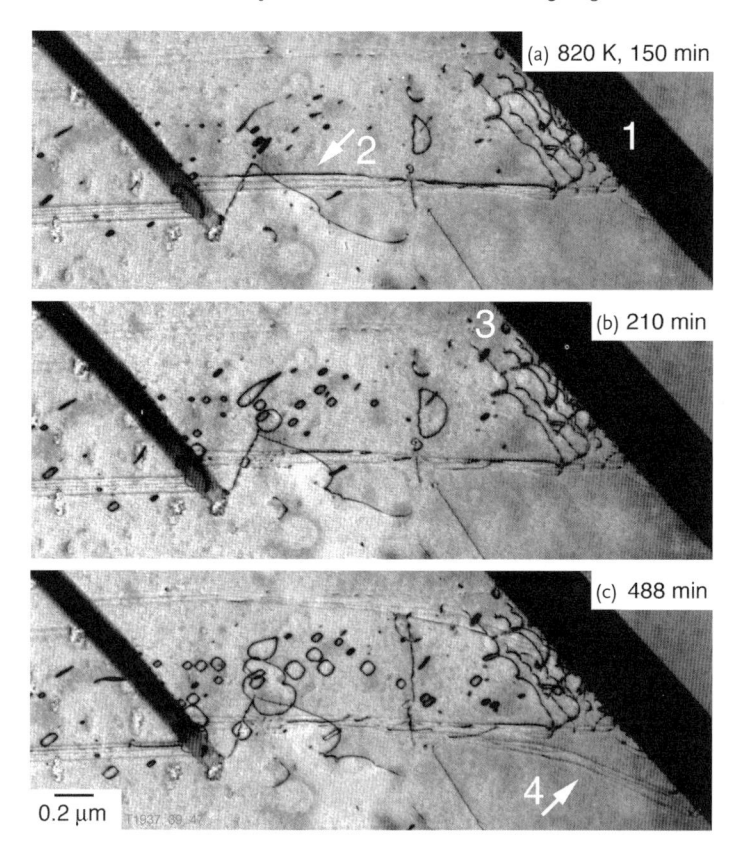

Figure 6.37 Interaction of gliding dislocations with interfaces. The micrographs, part of an *in situ* heating study, show a γ lamella that is interspersed by a terminated α_2 lamella; there is another unidentified lamella (1) on the right-hand side. During heating of the sample inside the TEM to $T = 820$ K, ordinary screw dislocations were released from the α_2/γ interface presumably under the action of the coherency stresses and moved from left to right, Figures (a) and (b). The screw dislocations trail slip lines, which are imaged as the nearly horizontal lines (arrow 2). The dislocations themselves are piled up at the next interface (3), indicating its high barrier strength. In Figure (c) another set of slip lines (arrow 4) is imaged (actually, there are two sets of slip lines that belong to the same dislocations but were trailed at the top and bottom sides of the foil). In the vicinity of lamella (1) the slip lines are strongly curved, which indicates that cross-slip out of the {111} slip plane occurred. The cross-slip starts a few tenths of a micrometer in front of the interface, indicating the long-range character of the interaction. Note also the growth of vacancy loops. Cast Ti-48Al-2Cr, predeformed in compression at room temperature to strain $\varepsilon = 3\%$.

they pass. The dislocations are bowed out in the direction of their motion with the ends lagging somewhat behind; this indicates that some energy is required for trailing the slip lines. In the vicinity of the next interface, the slip lines are curved; this indicates that the dislocations left their original {111} slip planes, presumably by cross-glide. It might be expected that the observed cross-slip and immobilization of the dislocations were caused by coherency stresses present at the barrier interface. There is a natural caution against TEM *in situ* experiments; the close proximity of the free surfaces in the thin foil makes it possible that the observed dislocation behavior is different from that in the bulk. Thus, during the *in situ* experiment surface-induced cross-slip could occur. However, against this is the fact that dislocations would have to cross-glide anywhere in the thin foil, not only in the vicinity of the interfaces. In Figure 6.37, there is also evidence of the growth of vacancy loops; this will be the subject of Section 7.2.4.

Summary: Overcoming lamellar boundaries by perfect and twinning partial dislocations involves several energy-consuming processes, which in detail depend on the interface type and character of the incoming dislocations. Lamellar true twin boundaries can probably be relatively easily overcome, while α_2/γ boundaries apparently provide the highest barrier strength. The barrier strength of the lamellae gives rise to an athermal stress part that is almost independent of temperature and strain rate. So, all in all, refinement of the lamellar spacing is an effective way to strengthen the material.

6.3
Generation of Dislocations and Mechanical Twins

There are well-established mechanisms in the literature explaining the generation of dislocations and mechanical twins in conventional metals. In TiAl alloys the mechanisms are more diverse because of the ordered $L1_0$ structure and the presence of dense interface structures; this demands particular consideration.

6.3.1
Dislocation Source Operation in γ(TiAl)

It is commonly accepted that dislocation multiplication in pure metals and disordered alloys mainly takes place through the operation of Frank–Read-type dislocation sources [162]. First, multiplication of ordinary dislocations with the Burgers vector **b** = 1/2<110] will be considered because these are the most frequently observed dislocations in γ(TiAl). Ordinary dislocations have a compact core, which makes cross-glide and climb relatively easy. Multiplication of these dislocations can therefore take place through the operation of dislocation sources incorporating stress-driven cross-slip or climb, as has been observed in conventional metals [74]. The different extent to which cross-slip and climb are involved in multiplication makes the process dependent on temperature.

At room temperature, the multiplication of ordinary dislocations in γ(TiAl) is closely related to jogs in screw dislocations [32, 75, 150]. Figure 6.38 shows ordinary screw dislocations that contain a high density of jogs. The jogs were probably formed by cross-slip; the details of this process have been discussed in Section 5.1. The possible implications of the jogs on dislocation multiplication depend on the jog height and are illustrated in Figure 6.39. Since the jogs are immobile in the direction of the motion of the screw dislocations, dipoles are trailed at the jogs

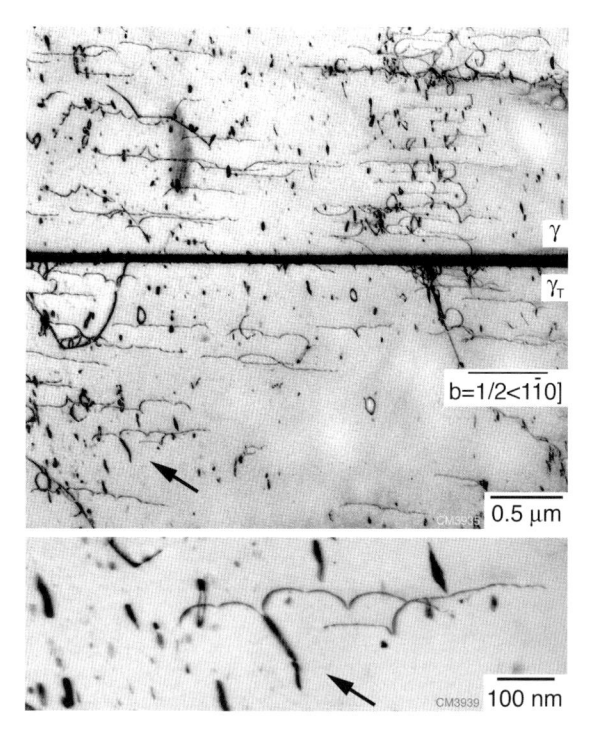

Figure 6.38 Structure of 1/2<110> dislocations generated during room-temperature compression to strain $\varepsilon = 3.2\%$. Ordinary screw dislocations with Burgers vector $\mathbf{b} = 1/2<1\bar{1}0]$ situated in adjacent γ lamellae. The screw dislocations bow between pinning points. A high resistance to glide is indicated by the dense obstacle ensemble and the observation that the dislocations remain bowed out in a smooth arc in the unloaded sample. The figure below shows the arrowed region in higher magnification. Note the dislocation dipoles and debris defects, which are trailed and terminated at jogs. This feature indicates the intrinsic character of the pinning centers. The dipole marked by the arrow is probably about to break up into smaller prismatic loops. Ti-48Al-2Cr, with a preferred orientation of the lamellae parallel to the compression axis.

(Figure 6.39, stage ii). The anchored segments bow out under the applied stress in a fashion similar to a Frank–Read source. The adjacent dipole arms can overcome their elastic interaction and pass each other, if the applied effective shear stress is larger than [163]

$$\tau_d = \frac{\mu b}{8\pi(1-\nu)h} \tag{6.16}$$

μ is the shear modulus, ν Poisson's ratio and h the height of the jog. If the bowing process continues, the expanding loops will annihilate over a portion of their lengths; this creates completely closed loops and restores the original configuration (Figure 6.39, stage iii). Figure 6.40 shows a dislocation dipole trailed at a high jog, apparently about to multiply.

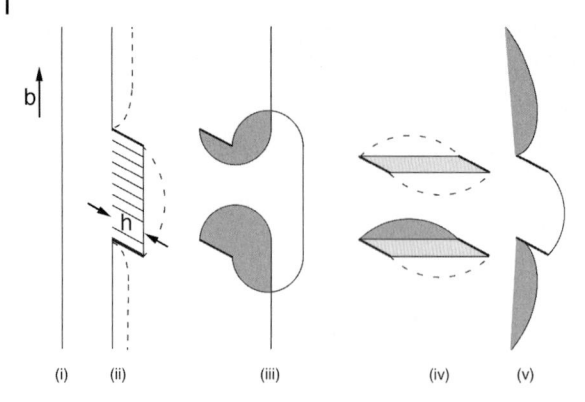

Figure 6.39 Behavior of a jogged screw dislocation [74, 163]; (i and ii) anchoring of the dislocation at jogs of height h, (iii) operation of the dipole arms as single- or double-ended dislocation source, (iv and v) trailing and termination of dipoles at small jogs. At higher stresses, the pinched off dipoles could multiply alike (dashed lines).

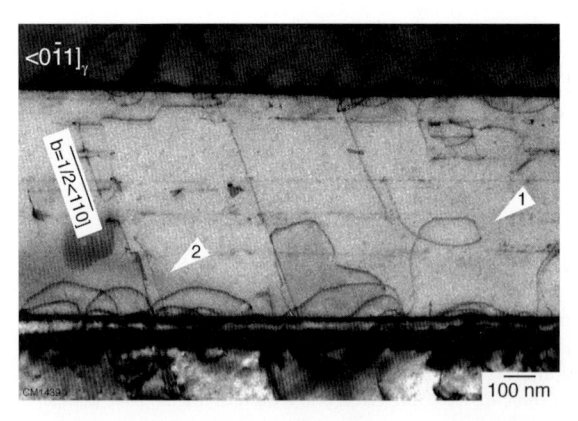

Figure 6.40 Initial stage of multiplication of an 1/2<110] dislocation corresponding to stage (iii) in Figure 6.39. The dipole arms trailed at the jog in the screw dislocation (arrow 1) could pass each other and apparently act as single-ended sources. Note the emission of the dislocation loops from the interface (arrow 2). High-order bright field image recorded from near the <0$\bar{1}$1] pole using **g** = (111)$_\gamma$ reflection. Nearly lamellar Ti-48Al-2Cr.

The effective stress τ_d involved in Equation (6.16) is the net stress or effective stress defined as the difference between the applied and the internal stress. As will be discussed in Section 6.5, τ_d is only a small part of the applied (normal) stress σ, at room temperature typically being $\tau_d = (0.1 \text{ to } 0.2)\sigma$. If $\tau_d = 30\,\text{MPa}$ is taken as a lower limit, a minimum dipole height $h = 130\mathbf{b}$ is calculated. This estimation suggests that very narrow dipoles can serve as dislocation sources. Dipoles of smaller height, which cannot multiply at a given stress, will be pinched off from the dislocation (Figure 6.39, stage iv). The process can be supported by

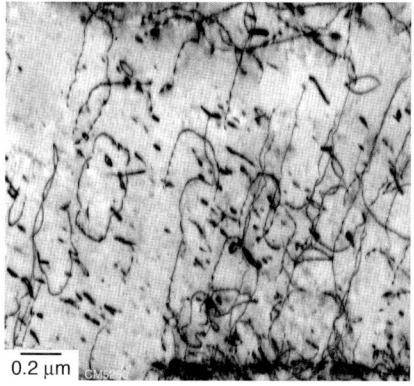

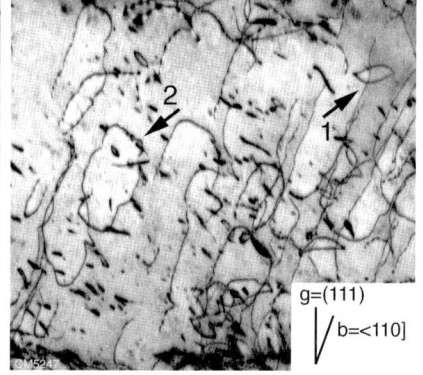

Figure 6.41 Stereo pair of micrographs showing dislocation multiplication starting from pinched off dipoles corresponding to stage (iv) in Figure 6.39. Details: (1) earlier stage of the mechanism; (2) final stage of the mechanism, two screw dislocations trailed by the dipole arms. Lamellar Ti-48Al-2Cr, deformation at room temperature to strain $\varepsilon = 3\%$.

glide of the jog along the dislocation due to unbalanced sideways components of line tension. The terminated dipoles could also operate as dislocation sources, when, in later stages of deformation the flow stress increases due to work hardening. Then the dipole arms can pass each other and operate as dislocation sources. Experimental evidence for this mechanism is suggested in Figure 6.41, which shows pinched off dipoles in different stages of reactivation. As can seen in the micrographs, the resulting loops are elongated along the screw components. This indicates a remarkable anisotropy in the velocity of ordinary dislocations with regard to their character. The loop shape indicates the ratio

$$r = v_{screw}/v_{edge} = 1:(3 \text{ to } 5) \tag{6.17}$$

v_{screw} is the velocity of screw dislocations and v_{edge} the velocity of edge dislocations. The factors governing the dislocation-glide resistance will be discussed in a later section. The observations are in accordance with the earlier models of dipole generation and dislocation multiplication [164–166]. When compared with the classical Frank–Read mechanism, the significant feature of the multiple cross-glide process is that the dislocation segments are intrinsically pinned. Any loop produced in the first cycle may, in turn, cross-slip and become a new source. Hence, the dislocations can expand and multiply in such a way that slip spreads from one slip plane to the next, thus, producing wide slip bands. This particular feature of the multiple cross-glide mechanism could explain why well-developed dislocation pile ups are seldom observed in TiAl alloys.

For multilayer systems it is often assumed that the dislocation source length is determined by the glide constraints set by internal boundaries. The increase of yield stress with decreasing layer thickness is then attributed to source hardening, meaning the activation of dislocation sources becomes increasingly difficult. This argument may not fully apply to lamellar ($\alpha_2 + \gamma$) alloys because the segments are intrinsically pinned by cross-slip and the evolution of the new loops occurs at the 30 nm scale. Thus, the figure of merit for multiplication in TiAl is the average dislocation slip path after which a sufficiently high jog is produced. In view of the small critical dipole height required for multiplication, it is concluded that under most circumstances sufficient sources for the generation of ordinary dislocations are available. It is sometimes argued that multiple cross-glide could be supported by thermal activation. Against this, is the fact that the multiplication rate increases with dislocation velocity [165]; the opposite behavior is expected, if it resulted from thermal activation.

Multiplication of superdislocations by this mechanism appears difficult for mainly two reasons. First, the superdislocations are widely dissociated. Hence, before cross-slip the dislocations would have to constrict [167, 168]; this requires additional energy but could be supported by thermal activation. Secondly, and perhaps more importantly, are geometrical constraints, which impede the Frank–Read-type mechanism. The argument bases on earlier models that have been developed for dissociated dislocations in metals and semiconductors [169, 170]. For the sake of argument, a pinned segment of a superdislocation is considered that is only decomposed into superpartials. Under the action of the resolved shear stress, each superpartial expands a loop and a faulted region between the superpartials is formed. The degree of expansion can be quite different, when the mobility of the leading superpartial is significantly larger than that of the trailing superpartial. Eventually, the leading superpartial loop reaches the critical configuration after which it rapidly wraps around the pinning points. At this stage, a complete faulted loop forms and surrounds the initial dissociated segment. However, in contrast to the initial stage, the arrangement of the two superpartials is reversed; the leading superpartial now comes behind the trailing superpartial. It might be expected that further operation of the source is extremely difficult. The problems associated with the source operation of superdislocations could be one of the reasons that these dislocations are seldom observed.

At elevated temperatures, dislocation multiplication can be supported by point defect supersaturations and diffusion. As has been described in Section 3.2, in γ(TiAl) vacancies predominate over interstitials because their formation energy is significantly lower than that of interstitials [171, 172]. Processing routes for TiAl alloys often involve thermal treatments followed by rapid cooling; this certainly leads to significant vacancy supersaturations. Nonequilibrium concentrations of point defects may also occur due to hot- or cold-working of the material. Off-stoichiometric deviations are accommodated by substitutional antisite defects on both sublattices. In the γ phase of ($\alpha_2 + \gamma$) alloys the relevant antisite defects are Ti atoms situated on Al sites, designated Ti_{Al}. The association of these antisite defects with vacancies leads to the formation of antistructural bridges that provide paths of easy diffusion. Depending on alloy composition and thermal treatment, the concentration of Ti_{Al} antisite defects could be up to 3 at.%. Dislocation motion in dense point defect concentrations seems therefore to be a common situation for TiAl alloys.

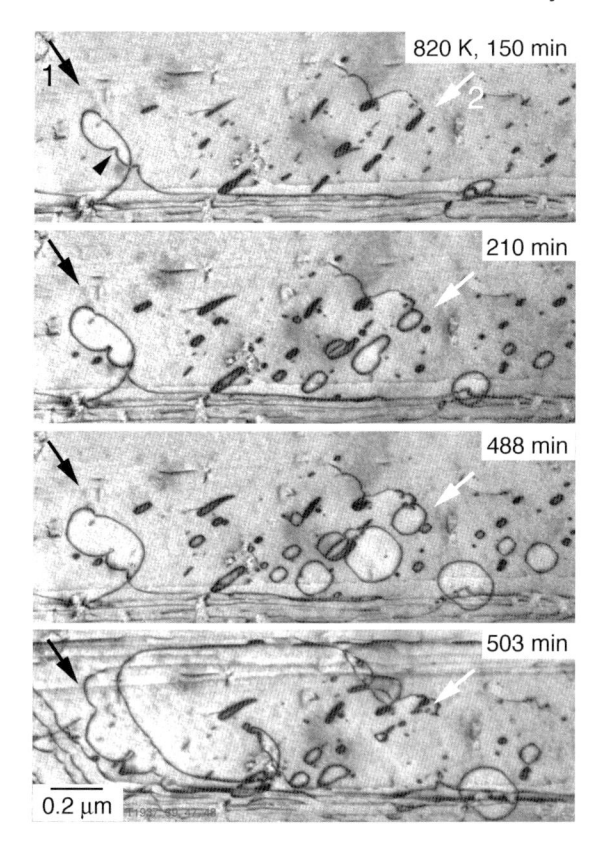

6.3 Generation of Dislocations and Mechanical Twins | 183

Figure 6.42 Generation of 1/2<110] ordinary dislocations at elevated temperature. Time sequence of an *in situ* heating experiment performed inside the TEM, showing the operation of Bardeen–Herring dislocation climb sources. Details: (1) expansion of a dislocation loop containing two jogs (one is marked by an arrowhead). After one cycle of the source a new dipole is generated so that the mechanism is regenerative. (2) Nucleation and growth of prismatic dislocation loops. Cast Ti-48Al-2Cr, predeformed at 300 K to strain $\varepsilon = 3\%$; acceleration voltage 120 kV.

The resulting climb processes can initiate multiplication, as proposed early by Bardeen and Herring [173]. The details of the mechanisms can hardly be deduced from post-mortem electron microscopy because of the complex climb kinematics. Direct evidence of climb-induced multiplication of ordinary dislocations was obtained from *in situ* heating experiments performed in the TEM. The samples used had been compressed at room temperature to a strain of $\varepsilon = 3\%$, as described in Section 6.2.5. During the *in situ* experiment, the dislocations moved under the combined action of thermomechanical stresses and osmotic climb forces due to the chemical potential of the excess vacancies. Figure 6.42 demonstrates the dynamical operation of a Bardeen–Herring climb source at 820 K over a period of about 350 min. The expanding loop designated with arrow 1 contains a jog so that climb on different atomic planes occurs. After one cycle of the source, a new dipole is generated, thus, the mechanism is regenerative. The climb processes often form spiral sources and interconnected multiple loops [75]. The dislocation loops marked by arrow 2 expand by the removal of one atomic plane, thus the mechanism is exhausted after only one cycle of the source. The critical vacancy supersaturation c^v/c_0^v required to operate a Bardeen–Herring source can be determined for the present case. The geometric situation occurring during the *in situ* experiments has been described in [174]. For a loop expanding from a source of length L, the critical value is

$$\ln\frac{c^v}{c_0^v} = \frac{\mu b \Omega}{L 2\pi(1-\nu)kT} \ln\frac{L\alpha}{1.8b} \qquad (6.18)$$

c^v is the nonequilibrium concentration of the vacancies, c_0^v is the equilibrium concentration of vacancies belonging to temperature T, Ω is the atomic volume, $\alpha = 4$, and ν is Poisson's ratio. For the present experimental conditions, $T = 820\,\text{K}$ and $L = 150\,b$ to $350\,b$, the values $c^v/c_0^v = 3$ to 1.7 were obtained, respectively [75]. These supersaturations are small in comparison to those produced initially after rapid cooling, which are easily on the order of 10^3 to 10^4 [74, 163]. Thus, Bardeen–Herring sources can probably operate throughout the entire period during annealing out excess vacancies. Such processes are expected to be particularly important for creep deformation, where only slow strain rates occur. The obvious conclusion is that rapid quenches from high temperatures are detrimental for the creep resistance of the material. It should be pointed out that under no circumstances were climb sources of superdislocations observed; this is certainly a consequence of their relatively wide dissociation, which impedes climb.

To summarize, the ordinary dislocations, which mainly carry the deformation in ($\alpha_2 + \gamma$) alloys, multiply by multiple cross-glide. Under usual deformation conditions, very narrow dislocation dipoles can serve as dislocations sources. The mechanism explains the occurrence of relatively broad slip bands. At elevated temperatures ordinary dislocations multiply through Bardeen–Herring climb sources. Multiplication of superdislocations by regenerative sources generally appears difficult because the source operation is kinematically constrained by the dissociation of the superdislocations.

6.3.2
Interface-Related Dislocation Generation

A mechanism common to both low and elevated temperatures is the emission of dislocations from the lamellar interfaces. The process was found to be closely related to the dislocation-loop structures and coherency stresses present at the semicoherent interfaces described in Section 6.1.4. At low temperatures the coherency stresses acting on the dislocation loops are certainly in equilibrium with the friction stresses impeding dislocation motion. However, a small superimposed applied stress could release the dislocation loops from the interfaces. Emission of dislocations from lamellar interfaces is a commonly made observation after room-temperature deformation [86, 175]; an example is shown in Figure 6.43.

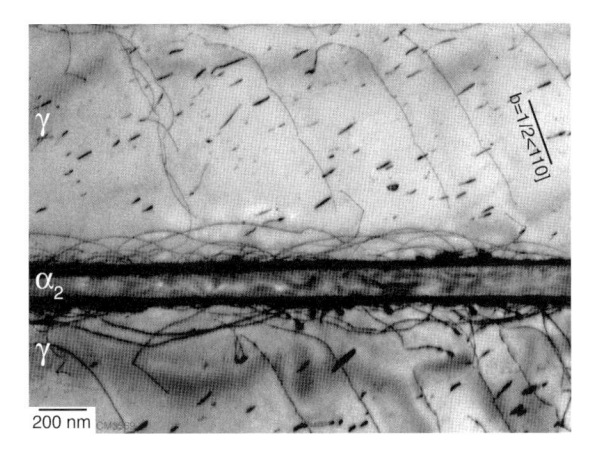

Figure 6.43 Initiation of glide processes of ordinary dislocations with Burgers vector $b = 1/2<110]$ at α_2/γ interfaces. Note the numerous dislocation loops emitted from the interfaces. Ti-47Al-2Cr-0.2Si, isothermal forging at 1185 °C, compression at room temperature to strain $\varepsilon = 3\%$.

Due to the geometrical and crystallographical constraints associated with the lamellar structure, a segment of the interfacial dislocation is obstructed from rotating about its pinning points. This is directly evident if an α_2 lamella is adjacent to the interface, but it also holds for neighboring γ lamellae. Thus, regenerative formation of complete loops will probably not occur. Nevertheless, activation of these dislocation nucleation centers could accommodate local stresses caused by impinging twins (Figures 6.35 and 6.36) or crack tips; this is certainly beneficial for the ductility and damage tolerance at low temperatures.

At elevated temperatures the glide resistance is reduced due to thermal activation. Under these conditions emission of dislocations may solely occur under the action of the coherency stresses. Direct evidence of the mechanism is demonstrated by the micrographs shown in Figure 6.44, part of an *in situ* heating experiment performed in the TEM. It might be expected that the dislocation emission could be enhanced if an external stress is superimposed, such as during creep. As will be seen in Section 9.5, emission of interfacial dislocations is detrimental for the high-temperature performance of lamellar alloys because it enhances primary creep.

The described dislocation emission is not restricted to ordinary dislocations. Large stress concentrations can help to overcome the high Peierls stresses, which are expected for the glide of superdislocations. Under these conditions superdislocations can be directly emitted from the interfacial dislocation network, a situation that is shown in Figure 6.45. In this case the generation of superdislocations (mostly of type $1/2<11\bar{2}]$) is probably supported by the high constraint stresses exerted by the impinging deformation twins. The inset shows the arrowed $1/2<11\bar{2}]$ dislocation loop emitted from the interface in higher magnification; note its dissociation.

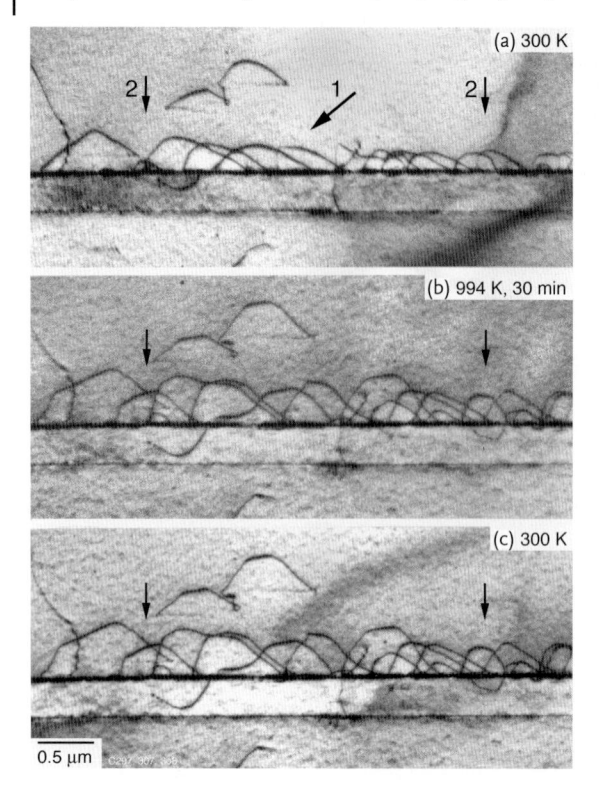

Figure 6.44 Generation of dislocation loops at a semicoherent γ/γ interface during *in situ* heating inside the TEM. The loops indicated by arrow (1) were probably generated in a first run of the experiment in which the foil was heated up to 994 K for 30 min and then cooled down to 300 K. Note the formation of new loops (arrows 2) in the second run of the experiment in which the sample was again heated up to 994 K. Cast Ti-48Al-2Cr, foil orientation close to <11$\bar{2}$], acceleration voltage 200 kV.

Summary: Mismatch structures and coherency stresses present at lamellar interfaces support dislocation nucleation. The mechanism is beneficial for the ductility and damage tolerance of the material at low and ambient temperatures, but detrimental for its high-temperature performance.

6.3.3
Twin Nucleation and Growth

Before describing the mechanistic details of twin nucleation, general information about the propensity of mechanical twinning in TiAl will be given. Following the pioneering work of the group of Yoo [90, 176–180], many researchers [151, 181–187] have studied the effects of alloy composition, constitution and microstructure on mechanical twinning. The experimental data and advances in mechanistic understanding were summarized in comprehensive reviews [151, 185]. In broad

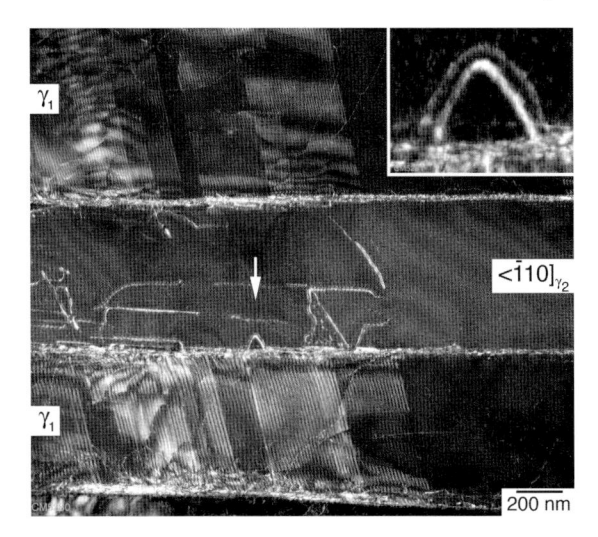

Figure 6.45 Emission of superdislocations from lamellar interfaces. Due to its <$\bar{1}10$] orientation parallel to the compression axis, lamella γ_2 is unfavorably oriented for glide of 1/2<110] dislocations and mechanical twinning. The generation of superdislocations (mostly of type 1/2<11$\bar{2}$] is probably supported by the high constraint stresses exerted by the impinging deformation twins. The inset shows the arrowed 1/2<11$\bar{2}$] dislocation loop emitted from the interface in higher magnification; note its dissociation. Ti-48A-2Cr, compression at room temperature to strain ε = 3%. Nearly lamellar Ti-48Al-2Cr, preferred lamellae orientation parallel to the compression axis.

terms, the propensity for twinning in γ(TiAl) is consistent with the relatively low superlattice intrinsic (SISF) stacking-fault energy (Section 5.1.2, Table 5.1), when compared with the APB and CSF energies. Several studies [188, 189] have shown that the twin shear fault energy is about half the SISF energy. Attempts have been made to correlate the twinning propensity with the effects of alloy composition, microstructure and deformation temperature on the SISF energy. The main findings may be summarized as follows.

The twinning propensity increases with decreasing Al content. Although Al-rich γ(TiAl) can be made to twin, it does so with reluctance, when compared with the γ phase in two-phase alloys. Investigations performed on Al-rich γ(TiAl) single crystals have revealed that dislocation glide is the primary deformation mode at any temperature [190–193]. Mahapatra *et al.* [194] have contrasted the different twinning behavior at room temperature of single-phase γ and (α_2 + γ) alloys. Whereas single-phase Ti-54Al deformed by slip, the two-phase alloy Ti-51Al deformed primarily by twinning. It seems that the effect of the Al concentration on the twinning propensity correlates with the composition dependence of the SISF energy Γ_{SISF}. First-principles electronic calculations [194–196] have predicted a significant increase of Γ_{SISF} with increasing Al content. Supporting experimental evidence of this effect was provided by the TEM analysis of Morris *et al.* [197].

There is a marked effect of the microstructure on the twinning propensity: twinning is easier in fully lamellar alloys than in their duplex or equiaxed counterparts [198]. Thus, the interfaces present in lamellar ($\alpha_2 + \gamma$) alloys seem to be beneficial for twin nucleation, a fact that will considered in more detail below.

In two-phase alloys the propensity for twinning increases with temperature [198, 199]. This observation is remarkable because twinning in disordered metals usually occurs at low temperatures or high strain rates, where the stress to produce slip is relatively high. As will be described below, the anomalous behavior of TiAl could be associated with climb-assisted twin nucleation.

In f.c.c. metals it is often the case that the stacking-fault energy decreases with increasing concentration of a substitutional solute so that twinning becomes more and more important the higher the solute concentration. Following these classical arguments, the effect of ternary alloying elements on the twinning propensity of γ(TiAl) has been investigated. The situation is not always clear because microalloying often leads to a change of the constitution and microstructure, and this, in turn, affects the propensity of twinning. Of the elements Cr, Nb and Mn considered in atomistic simulations [196], Mn has the strongest effect in reducing Γ_{SISF}. The beneficial effect of Mn in enhancing twinning has been reported by several authors [181, 200, 201], and the following factors have been considered in this context:

i) Mn segregation to interfacial boundaries [200];
ii) lowering of the SISF energy [181, 200];
iii) weakening of the Ti–Al covalency bonds by Mn replacing Al in the Al sites [201].

In Al-lean and high Nb-containing alloys an abundant activation of twinning has been recognized. The most convincing evidence of this effect was obtained from a diffusion-bonding study [202] (see Chapter 17), which reflects the characteristics of deformation and diffusion in a comprehensive manner. Figure 6.46 demonstrates the structural details in the bonding zone of diffusion couples of a binary Ti-45Al and a high Nb-containing alloy Ti-45Al-10Nb [202]. The diffusion zone of both couples consists of equiaxed γ grains interspersed with α_2 grains. In the Nb-bearing alloy the γ grains of the diffusion zone exhibit many annealing twins, which are almost absent in the binary alloy. Annealing twins and deformation twins are crystallographically identical; this suggests that mechanical twinning is an important deformation mode in Al-lean and Nb-bearing alloys. As will be detailed in Section 7.3.3, Nb solely occupies the Ti sublattice of TiAl with a very small misfit. Thus, a reasonable speculation is that Nb additions reduce the SISF energy. Figure 6.47 demonstrates dense twin structures in Ti-45Al-5Nb observed after low-temperature deformation. In Nb-bearing alloys the superdislocations were found to be widely dissociated, even after room-temperature deformation [203]. Such dislocations occur together with planar faults and twins, this gives rise to the speculation that the twins can originate from the superposition of planar faults stacked on alternate {111} planes. In an early study on a Ta-containing two-phase alloy Singh and Howe [204] have proposed this type of twin-nucleation mechanism.

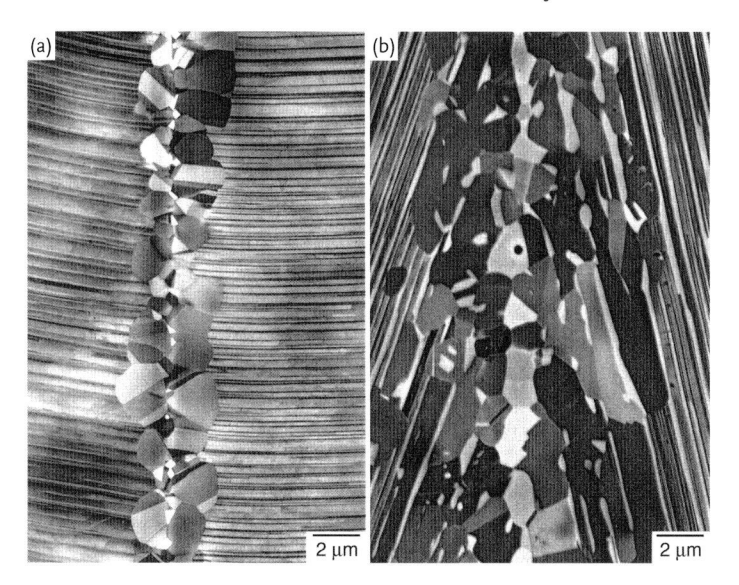

Figure 6.46 Recrystallization and diffusion characteristics of a binary and a Nb-bearing alloy observed after diffusion bonding for two hours at 950 °C with a normal stress of $\sigma = 60$ MPa. The bond layer of the diffusion couples is in the center of the micrographs. (a) Ti-45Al-10Nb and (b) Ti-45Al. Scanning electron micrographs obtained using backscattered electrons. Note that in the diffusion couple of the Nb-bearing alloy the diffusion zone is significantly smaller and that a high density of annealing twins occurs in the recrystallized γ grains [202].

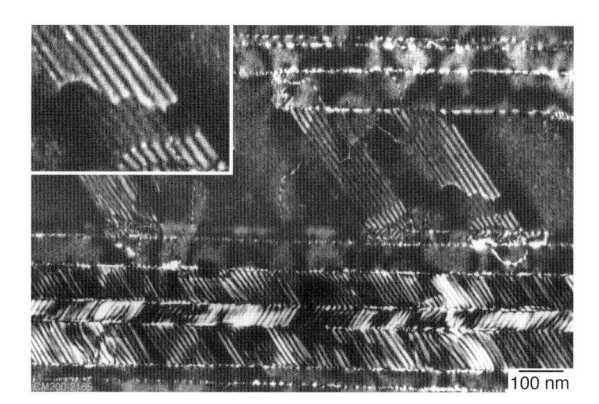

Figure 6.47 Deformation twins generated in Ti-45Al-5Nb alloy by compression at room temperature to strain $\varepsilon = 3\%$. Note the pinning of the Shockley partial dislocations (inset) [202].

The initial stage of twin nucleation is the passage of a first $1/6<11\bar{2}]$ partial dislocation loop on a $\{111\}_\gamma$ plane. The loop involves an intrinsic stacking fault, which increases the surface energy; hence, a large part of the work to form a twin goes into creating its boundaries [178, 185]. However, once the first loop has nucleated, succeeding loops are easier to form because these loops only propagate the coherent twin/matrix interface but do not create new interface area. For this reason, separate consideration is usually given to the nucleation of a small twinned volume, the embryonic twin, and to its subsequent growth into a large twin [185]. A three-layer fault is the smallest possible twin, which reproduces the $L1_0$ structure.

The homogeneous shearing of a lenticular-shaped twin volume would require unreasonably high stresses of five to ten per cent of the shear modulus [74]; hence, heterogeneous nucleation at a defect site is more likely. In accordance with this concept, twin nucleation in γ(TiAl) has been associated with energetically feasible dislocation dissociations and reactions [185, 187], which serve as nucleation sites and stress risers. The mechanisms that have been discussed involve:

i) dissociation of a 1/2<112> superdislocation into a two-layer twin [181, 187, 195, 205];

ii) expansion of a superlattice intrinsic stacking fault (SISF) loop [188];

iii) coincidental overlapping of wide stacking-fault ribbons [204];

iv) dissociation of misfit dislocations occurring at lamellar interfaces or precipitates [75, 152, 153];

v) recombination of Frank interstitial loops into embryonic twins [206];

vi) dissociation of a jog in an ordinary 1/2<110> dislocation into a Shockley and a Frank partial dislocation [180], a mechanism that is similar to Venables' model [169, 207] for twin nucleation in f.c.c. structures.

In the remainder of this section details will be provided that may typify these processes. A mechanism reported in [75, 153] relies on the dissociation of interfacial dislocations with a Burgers vector out of the interfacial plane, which has been discussed in Section 6.1.4. As demonstrated in Figure 6.48, these 1/3[111] interfacial dislocations often exhibit a significant core relaxation, which is manifested by a distortion of the fringe contrast that extends along $(\bar{1}1\bar{1})$ planes. The degree of relaxation varies considerably, and in some cases relatively wide stacking faults are formed (Figure 6.48b). When properly oriented with respect to the twinning shear direction, these stacking faults seem to provide nucleation sites for mechanical twins. An advanced stage of the process is demonstrated in Figure 6.49. The interface γ_1/γ_2 shown in the micrograph joins two γ lamellae with a true twin orientation. The tilt misfit between the γ lamellae is accomplished by a dense array of interfacial dislocations with the Burgers vector **b** = 1/3[111]. Two narrow twins T_1 and T_2, six and nine $(\bar{1}11)$ planes thick, respectively, were generated at the interfacial dislocations. Since there is only one true twinning system per octahedral plane, the twinning system of these two twins is defined as $1/6<\bar{1}\bar{1}2>(\bar{1}11)$ and the formation of the twinning partial dislocations can formally be described as

$$1/3[\bar{1}\bar{1}\bar{1}] \rightarrow 1/6<\bar{1}\bar{1}2> + 1/6[\bar{1}\bar{1}0] \tag{6.19}$$

The residual $1/6[\bar{1}\bar{1}0]$ dislocation may remain at the interface, but it seems to be more likely that several residuals recombine to form a perfect dislocation with Burgers vector 1/2[110]. A total reaction involving three interfacial dislocations

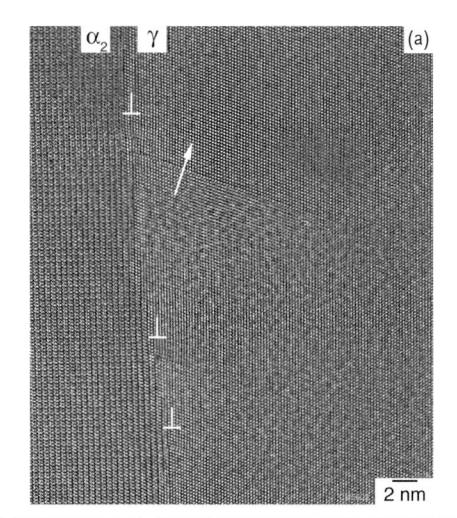

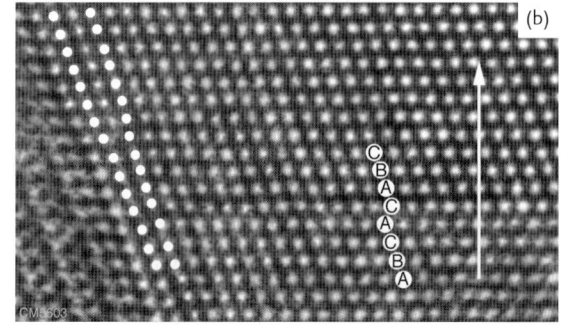

Figure 6.48 Nucleation of stacking-fault structures at lamellar interfaces. Sheet material of Ti-47Al-1Cr with additions of Nb, Mn, Si, and B. Tensile deformation at room temperature to failure at total fracture strain of $\varepsilon_f = 2.8\%$. (a) Atomic structure of a misfitting α_2/γ interface; high-resolution electron micrograph along coincident $<11\bar{2}0>_{\alpha_2}$ and $[\bar{1}01]_\gamma$. Note the stepped interface and interfacial dislocations (marked with dislocation symbols) with a Burgers vector out of the interfacial plane; these dislocations are manifested by an extra $(111)_\gamma$ plane parallel to the interface. An intrinsic stacking fault on the $(\bar{1}\bar{1}1)_\gamma$ plane is formed at one of these interfacial dislocations. (b) Stacking sequence across the fault along the direction of the arrow in (a). Note the close neighborhood of the interfacial dislocation marked by white circles.

would lead to three twinning partial dislocations and one ordinary dislocation. It might be expected that the recombined 1/2[110] dislocations propagate on the $(\bar{1}1\bar{1})$ planes, where the [110] shear direction is available. The extra half-plane of these dislocations can be recognized, if the micrograph is examined under a glancing angle along the $(\bar{1}\bar{1}1)$ planes (arrowed).

The dissociation described by Equation (6.19) is energetically feasible according to Frank's rule; nevertheless, it may be additionally supported by the high constraint stresses present at the interfaces. Thus, the formation of a small twinned region can be rationalized. Such a small nucleus can easily extend in all directions

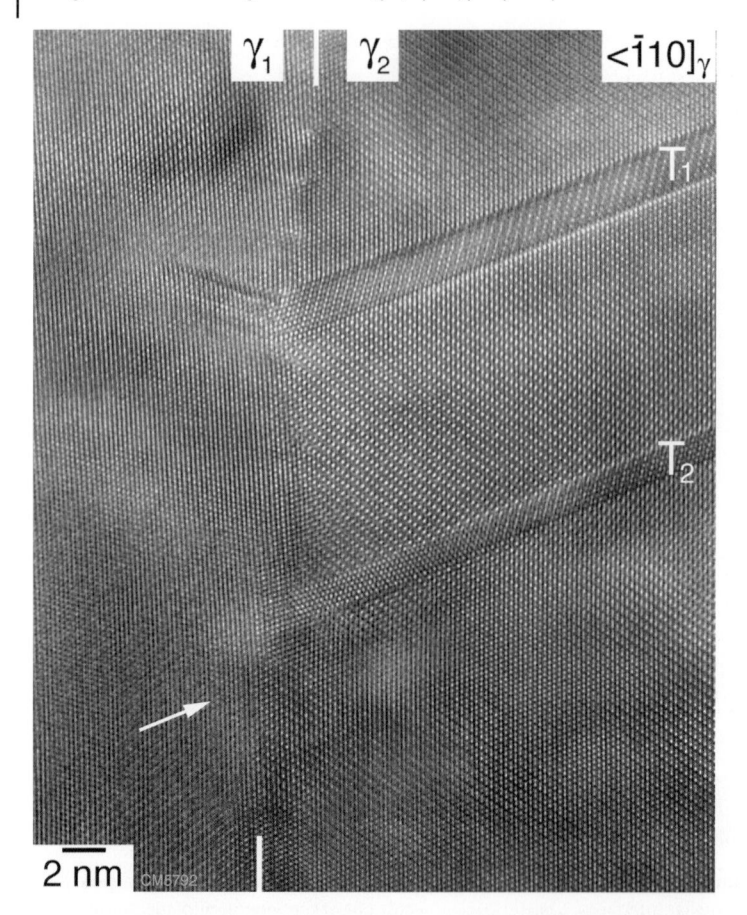

Figure 6.49 Nucleation of mechanical twins in lamellar TiAl alloys. High-resolution electron micrograph taken down the common $<\bar{1}10]$ direction of the two lamellae. Ti-48.5Al-0.37C, compressed at room temperature to strain $\varepsilon = 3\%$. Heterogeneous nucleation of embryonic twins at a γ_1/γ_2 interface between γ lamellae with true twin orientation. The tilt misfit between γ_1 and γ_2 is accommodated by an array of interfacial dislocations. The narrow twins T_1 and T_2 were nucleated at the interfacial dislocations. Two extra $(\bar{1}\bar{1}1)$ planes can be seen in lamella γ_2 if the micrograph is examined under a glancing angle along the arrow; these planes probably represent two 1/2[110] dislocations propagating on oblique $(\bar{1}1\bar{1})$ planes [153].

within the K_1 twinning plane by expansion of the Shockley partial dislocation loops, leaving flattened cylindrical plates of coherent K_1 interface. The thickening of twins through successive K_1 planes is usually described by a pole mechanism [151, 180, 181, 185, 187]. However, further thickening may also occur by a successive heterogeneous nucleation of K_1 interfaces at extended configurations of misfit dislocations. In loci with high tilt misfit, the separation distance of misfit dislocations is only 2 nm to 3 nm, and such regions often extend over several tens of nanometers (Figure 6.48). Twin thickening may therefore be considered as the cumulative result of many adjacent dissociation processes. In lamellar alloys the mean thickness of twins was estimated to be in the range of 15 to 20 nm (see Figure 6.34.), which coincides relatively well with the extension of the interface region that contains dense arrangements of interfacial dislocations. At elevated temperatures climb of the Frank partial dislocations along the interfacial boundary is possible. Thus, the interfacial dislocations can rearrange into a configuration that is favorable for the generation of a more perfect twin.

The group around Fischer [208–210] has justified the above-described twin nucleation at lamellar interfaces by micromechanical modeling, involving the strain energy of the system and twin surface energy. The calculation showed that the stored strain energy at the interface is reduced if a twin is generated at an interfacial dislocation. The model is able to explain why narrow twins with a distinct thickness are produced. Hsiung and Nieh [211] observed twin nucleation at lamellar interfaces due to the recombination of piled up misfit dislocations that propagated along α_2/γ interfaces.

As a second mechanism, twin nucleation at precipitates will be described. In advanced TiAl alloys precipitation reactions are often implemented in order to improve the high-temperature performance; the details of possible precipitation mechanisms will be described in Section 7.4.1. The twin nucleation has been observed in Ti-48.5Al-0.36C. In such alloys Ti_3AlC perovskite precipitates can be formed by homogenization and aging [75, 212, 213]. The anisotropic misfit of the precipitates with the γ matrix gives rise to a rod-like morphology of the precipitates and to high coherency stresses. The precipitation-hardened material exhibits an unusual propensity to profuse twinning. Faulted structures are frequently observed in these alloys, with fault planes parallel to the twin/matrix interface (Figure 6.50) [153]. The faults originate at the precipitates, as demonstrated in Figure 6.51a. The precipitates are associated with dislocations that are manifested by additional {111} planes (Figure 6.51b, arrows 1 and 3). The possibility that these dislocations are matrix dislocations cannot be ruled out because the projected Burgers vector is

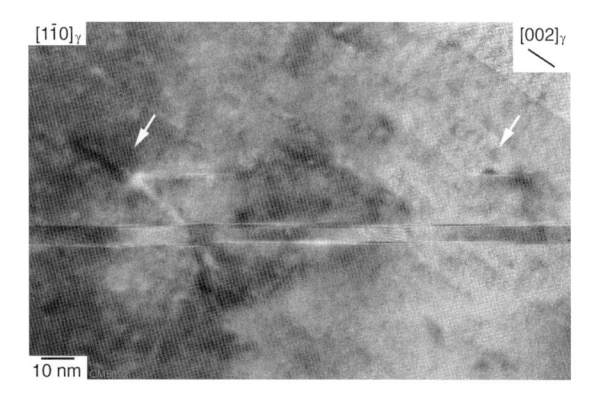

Figure 6.50 Nucleation of planar defects (arrowed) at rod-like perovskite precipitates with fault planes parallel to a deformation twin. Low-magnification high-resolution micrograph taken after room-temperature compression of a Ti-48.5Al-0.37C alloy to strain $\varepsilon = 3\%$ [153].

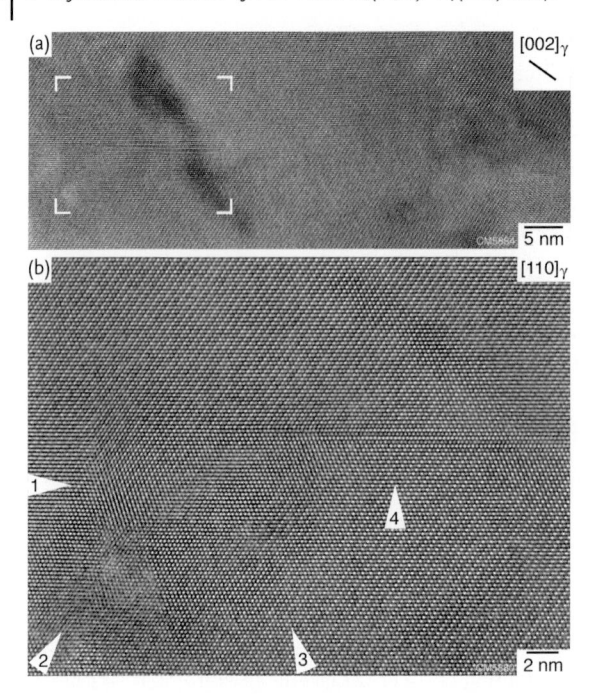

Figure 6.51 Heterogeneous nucleation of embryonic twins at perovskite precipitates, observed after room-temperature compression of a Ti-48.5Al-0.37C alloy to strain $\varepsilon = 3\%$. (a) Low-magnification high-resolution micrograph showing faulted defects adjacent to perovskite precipitates. The precipitates are manifested by washy strain contrast. (b) Magnified view of the boxed area in (a). Two dislocations that are represented by additional {111} planes are marked with arrows 1 and 3. Note the misfit dislocation at the precipitate with a resolved Burgers vector $\mathbf{b}_{res} = 1/2[001]$ that is manifested by an additional (002) plane and can be recognized by a shallow view along arrow 2.

consistent with 1/2⟨110] dislocations. However, there are also dislocations present at the precipitates that are manifested by an additional (002) plane (arrow 2). The resolved Burgers vector $\mathbf{b}_{res} = 1/2[001]$ is not consistent with the glide geometry of the L1$_0$ structure; this gives supporting evidence that these defects are misfit dislocations. This finding is contrary to a report in the literature [212] according to which perovskite precipitates in TiAl are fully coherent. The disparity may result from a larger size of the particles present in the material described here. The TiAl matrix surrounding the precipitates is subject to normal stresses acting in three mutually perpendicular directions, which corresponds to a triaxial stress state. The maximal shear stress occurs on elements oriented at angles of 45° to the **a**- and **c**-axis of the L1$_0$ structure. With growing precipitate size, the strain and surface energy of the particle can be reduced if dislocations are emitted; hence, the full coherency of the precipitate with the matrix is lost. According to this mechanism the dislocations can be located at a significant stand-off distance from the original interface, as has been observed. The stand-off distance of misfit dislocations is thought to be a general phenomenon whenever there is a substantial difference in the elastic constants of the materials at either side of the interface [214]. It is expected that the

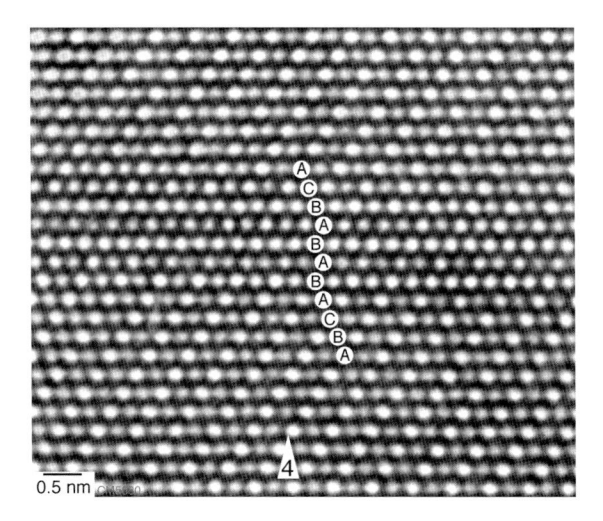

Figure 6.52 Stacking sequence across the fault marked by arrow 4 in (Figure 6.51b). Note the hexagonal stacking of the faulted structure [153].

misfit dislocations are situated in the elastically softer phase, which in the present case is γ(TiAl) [91, 215, 216]. Under the high constraint stresses, the interfacial dislocations may recombine so that perfect and twinning partial dislocations are released. Among the various mechanisms that may occur, the reaction

$$[00\bar{1}] \rightarrow 1/3[\bar{1}\bar{1}\bar{1}] + 2 \times 1/6[11\bar{2}] \qquad (6.20)$$

involving two 1/2[001] dislocations may account for the structure shown in Figure 6.51b. One of the 1/2[001] dislocations and the 1/3[111] dislocation can be seen by shallow views along arrows 2 and 1, respectively. The emission of the 1/6[11$\bar{2}$] partial dislocations forms faulted structures and accommodates the triaxial stress state.

Figure 6.52 shows the stacking sequence across the faulted structure marked by arrow 4 in Figure 6.51b. Locally, the fault has hexagonal stacking and may be described as superimposed intrinsic stacking faults. It is tempting to speculate that such defects can easily be rearranged into embryonic twins, which may coalesce to form incomplete thicker twins. Thus, in an early stage of this growing process the twins appear fragmented. The mechanism provides a natural explanation as to why the twins in the precipitation-hardened material are quite irregular and involve untwinned regions (see Section 7.4.1). As the rod-like precipitates obliquely intersect the {111} twin-habit plane, twin thickening by such a mechanism is basically restricted to some fraction of the precipitate length. Most of the observed twins indeed have a thickness below 10 nm, which is significantly smaller than the average length of the precipitates. This structural data has been used as input for modeling the process [217]. The calculation involves the elastic strain energy introduced by the precipitate, the twin surface energy and the elastic energy of the twin. The model is capable of predicting the aspect ratio of the twins formed. Interestingly, the model also suggests an element of instability, that is, a short twin could spontaneously extend to a long twin. The prediction coincides with the experimental observation that short and long twins coexist in the precipitation-hardened material.

Another twin-nucleation mechanism, closely related to that described earlier is associated with dislocation glide [153]. In the precipitation-hardened material the [011] superdislocations are widely dissociated at the precipitates, which might be a consequence of the high constraint stresses described above. Twin nuclei were often found together with these dislocations, giving rise to speculation that twins can also originate from overlapping faults trailed by the superdislocations.

Based on molecular dynamic simulation, Xu et al. [218] have proposed that twins in γ(TiAl) could be formed by five coordinated 1/6<112> shears on adjacent (111) planes, in the fashion of a synchroshear mechanism [219]. The five shear steps are equivalent to a composite dislocation that is comprised of a $1/6[\bar{5}14]$ superpartial at the twin/matrix interface and an ordinary $1/2[\bar{1}10]$ dislocation on the next (111) plane. It should be noted that 1/6<514] superpartials have been observed as one dissociation component of <101] superdislocations (see Section 5.1.3). The total Burgers vector of the synchroshear process is $\mathbf{b} = 2/3[\bar{2}11]$. The associated twinning shear $g = 2\sqrt{2}$ is four times that of conventional true twinning (Section 5.1.5). The advantage of the mechanism is that it does not require twinning dislocations to multiply by a pole mechanism. The authors have suggested that this type of twin formation could effectively accommodate stress concentrations or high strain rates. However, no experimental evidence for the mechanism has yet been observed.

As has been mentioned above, in two-phase alloys the twinning propensity increases with temperature. In this context, Yoo [180] has investigated the climb expansion of Frank partial dislocation segments as the nucleation stage of a pole mechanism. The analysis suggests that a high supersaturation of thermal vacancies, $c/c_0 \geq 13$, is required in order to drive the mechanism solely by chemical forces. This appears unrealistic in terms of the existing diffusion data of TiAl [41] (Sections 3.2 and 3.3). c_0 is the equilibrium concentration of vacancies, which at 1500 K is thought to be $c_0 = 10^{-7}$ to 10^{-6} [151]. The difficulty with this scenario can be overcome by considering the particular diffusion mechanism in two-phase alloys, as suggested in [153]. Diffusion in these alloys is probably supported by significant chemical disorder. There is good evidence that off-stoichiometric deviations in ($\alpha_2 + \gamma$) alloys are in part accommodated by the formation of Ti_{Al} antisite defects, that is, Ti atoms situated on the Al sublattice [41]. The essence of the argument is that the phase separation occurring upon cooling and described by the phase diagram often cannot be established within the constraints of processing routes because the $\alpha/\alpha_2 \rightarrow \gamma$ phase transformation is sluggish [153]. Microanalysis performed on two-phase alloys with the base-line composition Ti-47Al has shown that the concentration of Ti_{Al} antisite defects in the γ phase is in the order of 10^{-2} [153, 220]. At such a high concentration, the antisite defects form a percolating substructure of antistructural bridges along which diffusion can occur (see Section 3.3). The point to note is that the migration energy of Ti_{Al} antisite defects is expected to be significantly lower than the self-diffusion energy of Ti determined at high homologous temperatures [221]. Thus, via antistructural bridges high diffusion rates can be accomplished at moderately high temperatures, where the conventional vacancy mechanism is still ineffective. Under the same conditions, diffusion in Al-rich TiAl is probably very sluggish because the migration energy of Al_{Ti} antisite defects is significantly higher. Thus, climb-induced twin nucleation is expected to be less efficient in Al-rich TiAl. In PST crystals, no significant change of the deformation mode was observed up to 800 °C, that is, the relative contributions of dislocation glide and mechanical twinning remained constant [222]. This might be a consequence of the slow-growing process of PST crystals, which probably leads to a near-equilibrium phase constitution and a correspondingly low density of Ti_{Al} antisite defects. Thus, as with Al-rich TiAl, twin nucleation is not supported by dislocation climb. Taken together, these factors may account for the observed differences in the twinning behavior of Ti- and Al-rich TiAl alloys.

The lateral growth of an embryonic twin is considered as a coordinated movement of Shockley partial dislocations through successive {111} planes, as proposed for cubic and hexagonal metals [74, 151, 185, 207, 223, 224]. Most explanations of the origin of such a coordinated motion invoke the existence of a so-called pole dislocation, being partly or wholly of screw character. The screw component of the pole dislocation is perpendicular to the twin plane and is equal to the interplanar spacing of the {111} planes. A twinning partial dislocation winds around the pole dislocation in a fashion similar to a staircase. In doing so, the partial dislocation not only produces a stacking fault, but also climbs up the pole dislocation to the next layer. Repetition of this mechanism forms a thick twin. During TEM *in situ* deformation of Ti-54Al, Couret *et al.* [225] observed a pole mechanism that was probably anchored by a 1/2<110] screw dislocation. The twins observed in bulk material were generally very thin; this suggests that the operation of pole sources is limited to only a few turns. Thus, at least at low temperatures, the final thickness of the twins seems to be related to the geometrical extension of heterogeneous nucleation centers [217]. During the *in situ* observations of Farenc *et al.* [124] and Couret *et al.* [225] the twinning partial dislocations propagated at room temperature with a relatively constant velocity of about 10^{-2} μm/s. The authors suggested that the mobility of these dislocations is controlled by lattice friction. Other glide obstacles that have been identified are matrix dislocations and sessile Frank partial dislocations.

Summary: Mechanical twinning is a prominent deformation mechanism in the γ phase of two-phase titanium aluminide alloys. The factors supporting twinning are the relatively low Al content, the presence of substitutional alloying elements like Mn or Nb, the lamellar microstructure, and high deformation temperature. Twin nucleation is triggered by stress concentrations associated with structural heterogeneities; these are matrix dislocations, interfacial dislocations and precipitates. The lateral growth of the twins is apparently accomplished by a pole mechanism.

6.3.4
Twin Intersections

At the beginning of deformation the slip path of the twins is essentially identical with the grain size or lamellar spacing. As soon as multiple twinning with nonparallel shear vectors is activated, extensive intersections among twin bands occur. The intersection of a moving deformation twin with a barrier twin is generally expected to be difficult because the incorporation of an incident twinning system into the barrier twin may no longer constitute a crystallographically allowed twinning system. For example, a $1/6[\bar{1}\bar{1}2](\bar{1}\bar{1}1)$ twinning partial dislocation incorporated into a $[11\bar{2}](111)$ barrier twin will become $1/18[552](\bar{1}\bar{1}5)$. Similarly, a $1/2[110]\,(1\bar{1}\bar{1})$ ordinary dislocation is transformed into $1/6[114](\bar{5}11)$, when incorporated into the same barrier twin. As these slip systems are not normally observed in TiAl, the transformed dislocations are considered to be immobile. Thus, a mechanical twin provides an effective glide barrier to other dislocations or mechanical twins. Given this importance, several authors [33, 75, 153, 182–185, 226] have analyzed the mechanism and have proposed crystallographic relations by which an incident twin could intersect a barrier twin. In γ(TiAl), two different cases, designated as Type-I and Type-II intersections, can arise depending on the orientation of the intersection line, the common direction of the twin habit planes. A Type I-intersection occurs along $<\bar{1}10]$; both the shear of the incident twin and the shear of the barrier twin are normal to this intersection line. Type-II intersection occurs along $<0\bar{1}1]$; the shear of the incident twin and the shear of the barrier twin are at 30° to the intersection line. According to the Schmid factor analysis of the possible twinning systems in γ(TiAl) performed by Sun et al. [182] (Section 5.1.5), the sample orientations promoting a specific intersection under compression are as follows. Type-I intersections are favored if the $<\bar{1}10]$ intersection line is at 0° to 55° from the sample axis. Type-II intersections preferentially occur if the $<0\bar{1}\bar{1}]$ intersection line is at 45° to 90° from the compression axis.

Figure 6.53 illustrates two geometries of Type-I intersections, which have been reported in the TiAl literature. The Shockley partial dislocations of the incoming twin are in a pure edge orientation, hence, they do not carry shear components parallel to the $<\bar{1}10]$ intersection line. The geometry shown in Figure 6.53a is characterized by a deflection of both the barrier twin T_b and incident twin T_i. Several authors [184, 226] have proposed that in this case the strain of the incident twin could be accommodated by glide of perfect $1/2<110]$ dislocations on the $(001)_{Tb}$ basal plane of the barrier twin. A suitable reaction linking the incident twinning partials with the perfect dislocations is [184]

$$6\times 1/6[\bar{1}\bar{1}2]_{Ti}(\bar{1}\bar{1}1)_{Ti} \rightarrow 3\times 1/2[110]_{Tb}(001)_{Tb} + 1/2[110]_{Tb}(001)_M \qquad (6.21)$$

The $1/2[110]_{Tb}(001)_{Tb}$ shear deflects the barrier twin but does not change the crystal structure and orientation of the barrier twin in the intersection region. The $1/2[110]$ $(001)_M$ dislocation is emitted into the matrix. The height of the step in the barrier twin is equal to the half-width of the barrier twin. The mechanism appears plausible because the direction of the $1/2[110]_{Tb}$ Burgers vector is closest to the Burgers vector of the incoming partials $1/6[\bar{1}\bar{1}2]_{Ti}$. Furthermore, the (001) basal planes are the only slip planes, in both the barrier twin and the matrix. One problem with this model is the high Peierls stress, which is expected for the (001) basal slip of the $1/2<110]$ dislocations. This arises because Ti layers have to be displaced over Al layers without changing the bonding between nearest neighbors. Also, the

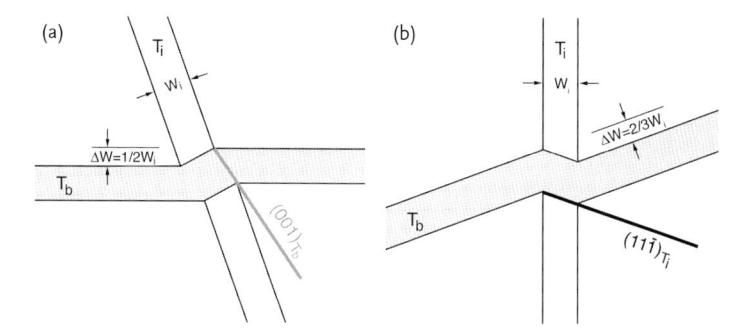

Figure 6.53 Geometry of Type-I intersection of mechanical twins in the $L1_0$ structure of γ(TiAl). The twins intersect along the common $<\bar{1}10>$ direction; both the shear of the incident twin and the shear of the barrier twin are normal to the intersection line. (a) Deflected intersection characterized by a deflection of both, the barrier twin T_b and an incident twin T_i. (b) Undeflected intersection; the incident twin T_i remains rectilinear after intersection and the deflection of the barrier twin can be geometrically described as following the $(11\bar{1})_{T_i}$ plane. Figure adapted from Sun et al. [182, 183].

$1/2[110]_{Tb}$ and $1/2[110]_M$ dislocations have pure edge character and could thus further dissociate to form sessile Lomer–Cottrell locks [184]. The emission of $1/2<110>$ dislocations on (001) basal planes at twin intersections has been observed by Morris [227] and can be considered as supporting evidence of the mechanism.

The mechanism shown in Figure 6.53b is the so-called undeflected Type-I intersection. The incident twin remains rectilinear after intersection and the deflection of the barrier twin can be geometrically described as following the $(11\bar{1})_{Ti}$ plane. The displacement of the deflected twin normal to its habit plane is about two thirds of the thickness of the incoming twin. Sun et al. [182, 183] have proposed that the $[\bar{1}\bar{1}2]_{Tb}(\bar{1}11)_{Tb}$ shear of an incoming Shockley partial is transformed into a $1/18[552]_{Tb}$ zonal displacement that occurs on every third $(\bar{1}\bar{1}5)_{Tb}$ plane. The problem with this scenario is that an unfavorable **AAA** stacking occurs within the intersection zone, which requires reshuffling. In view of this difficulty, the authors have proposed that the shear translation is accomplished by coordinated 1/2[552] glide steps that occur on every 27th $(\bar{1}\bar{1}5)_{Tb}$ plane. The mechanism has the advantage that the high-energy **AAA** stacking is avoided and that antiphase boundaries need not be formed. An incomplete operation of these mechanisms could give rise to a misorientation of the intersection zone of 7° to 9°, which in some cases has indeed been observed. The two geometries of Type-I intersections illustrated in Figure 6.53 are essentially consistent with the TEM observations reported in the literature [182–185, 226]. In their discussion the authors have pointed out that, while the above-described mechanisms may serve as a general guide, the processes occurring during twin intersections could actually be more complex. The strong stress concentration produced by the impinging twin and the close neighborhood of several interfaces could give rise to reaction schemes and activation of slip systems that usually do not occur. Furthermore, these processes occur at the atomic scale, which for observation require adequate resolution. The high-resolution electron micrograph of a Type-I intersection shown in Figure 6.54

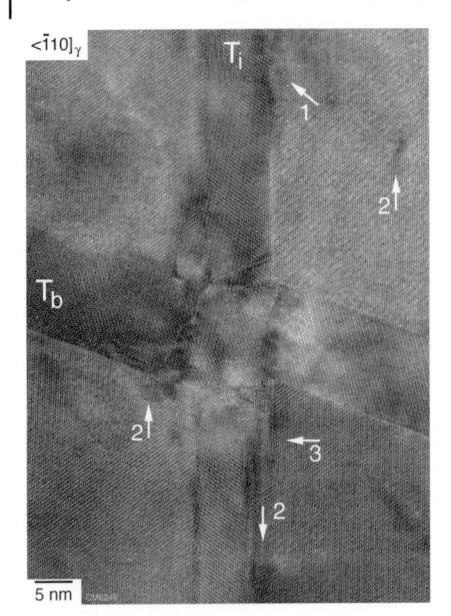

Figure 6.54 Type-I intersection of two deformation twins observed after room-temperature compression of a Ti-48.5Al-0.37 C alloy. Heat treatment after casting: annealing at 1250 °C followed by water quenching, and aging at 750 °C for 24 h. Deformation at room temperature to strain $\varepsilon = 3\%$. The structure is imaged along the common $<\bar{1}10]$ intersection line. The vertical twin T_i is considered to be the incident twin because its upper side is thicker than its lower side; T_b is the barrier twin. Figures 6.55 to 6.58 show the marked details in higher magnification.

illustrates these aspects. The intersection was observed after room-temperature compression of a cast Ti-48.5Al-0.37 C alloy. The structure is imaged along the common $<\bar{1}10]$ direction of the two twins; this can be recognized by the different contrast of the (002) planes, which are alternately occupied by Ti and Al atoms. On its upper side the vertical twin is thicker than on its lower side, thus this twin was considered as incident twin T_i; T_b is the barrier twin. The intersection leads to a significant deflection of the two twins, which, however, is more pronounced for the barrier twin. The difficulty in forming a twin intersection is manifested by structural details in the interface between the incident twin and the matrix that are marked with arrow 1. Figure 6.55 shows this region in higher magnification and reveals that the (002) planes are not continuous across the twin/matrix interface indicating a rigid-body translation relative to each other. The displacement corresponds to an APB vector $f_{APB} = 1/2<\bar{1}01]$. As has been discussed in Section 6.1, in a true twin boundary such a translation is crystallographically permissible, but, according to the authors knowledge, has not yet been observed before in the coherent interface of a mechanical twin. It might be expected that the related $1/2<\bar{1}01]$ superpartial was formed by a recombination of piled up $1/6[\bar{1}\bar{1}2]_{Ti}(\bar{1}11)_{Ti}$ twinning partials. The details of this mechanism are not clear; a speculation is that three identical partial dislocations recombine to a delocalized nonplanar core

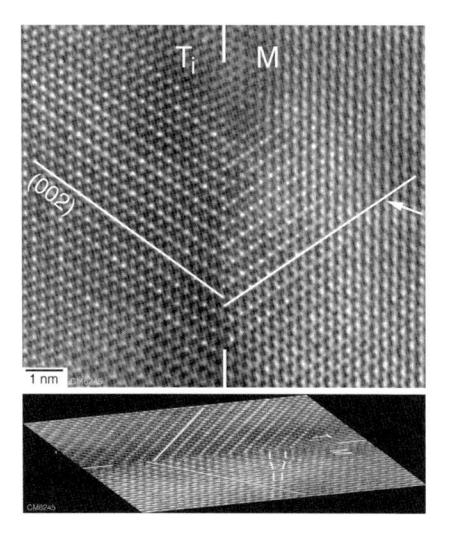

Figure 6.55 Structural detail marked by arrow 1 in Figure 6.54: rigid-body translation in the interface between the incident twin T_i and the matrix M indicated by the displacement of the (002) planes. The image below shows the micrograph compressed along the $(\bar{1}\bar{1}1)$ planes (arrowed). Note the dislocation adjacent to twin/matrix interface.

of a $1/2<11\bar{2}]$ dislocation, which further decomposes into $1/2<10\bar{1}]$ and $1/2<01\bar{1}]$ superpartials (Sections 5.1.3 and 5.1.4). In Figure 6.55 a dislocation in the γ matrix adjacent to the interface can be recognized that is manifested by an extra $(\bar{1}\bar{1}1)$ plane and could be consistent with a $1/2<10\bar{1}]$ superpartial. The lower image, compressed along the $(\bar{1}\bar{1}1)$ planes, shows these details more clearly. It might be expected that the reaction scheme is driven by the stress concentration from the piled up incident twinning partials.

Figure 6.56 shows the details of the intersection zone. The intersection zone remains in the $L1_0$ structure and seems to be relatively free of defects; however, highly defective regions border it. The $(002)_{Tb}$ planes of the central zone are not continuous with those of the barrier twin, but appear displaced along the $(\bar{1}\bar{1}1)_{Tb}$ planes of the barrier twin. This displacement is consistent with glide of $1/2<10\bar{1}]$ or $1/2<01\bar{1}]$ superpartials on $(\bar{1}\bar{1}1)_{Tb}$ planes. These dislocations become manifest by extra $(002)_{Tb}$ planes; a few of these dislocations and the orientation of their extra planes are indicated by dislocation symbols. It is tempting to speculate that these dislocations were generated under the high stress concentration acting at the corners of the intersection zone. It is worth adding that a shear accommodation by twinning along the $(\bar{1}\bar{1}1)_{Tb}$ planes is not possible because this would require antitwinning operations. The close distance of these dislocations explains why the lattice of the intersection zone is not congruent with that of the barrier twin, but rotated by about 10° against the barrier twin. The image below is the micrograph compressed along the (002) planes and shows these details more clearly. The observation largely reflects the strong rotation field that was generated by the incident twin. The various dislocation reactions that could be involved in the

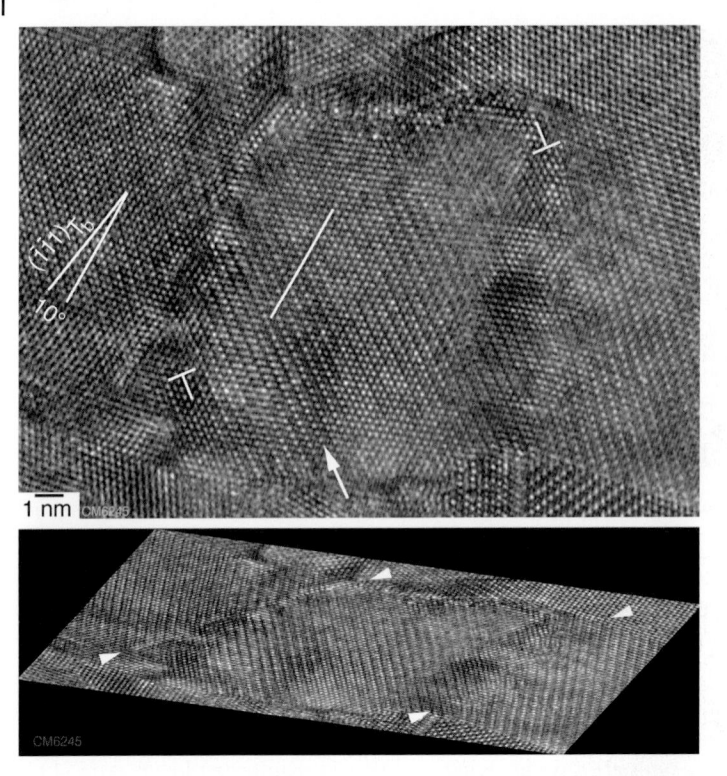

Figure 6.56 Structural details of the intersection zone: translation of the twinning shear along the $(\bar{1}\bar{1}1)_{Tb}$ planes by $1/2\langle 10\bar{1}]$ superpartials that become evident by extra $(002)_{Tb}$ planes. Two of the dislocations and their extra half-planes are indicated by dislocation symbols. Note the anticlockwise rotation of the intersection zone with respect to the barrier twin by 10°, which is indicated by the traces of the respective $(\bar{1}\bar{1}1)_{Tb}$ planes. The compressed image below shows these features in more detail. Arrowheads mark the dislocation walls on either side of the intersection zone.

intersection process probably give rise to dislocation emission (details 2). Figure 6.57 shows one of these areas at the exit side of the barrier twin in higher magnification. The strain accommodation seems not to be complete and significant internal stresses remain at the interface; this is indicated by the strong strain contrast of the image. Thus, new $1/6[\bar{1}\bar{1}2](\bar{1}11)$ twinning partial dislocations can probably be easily nucleated at the defect-rich exit side of the barrier twin, where many dislocations are present. Arrow 3 in Figure 6.54 indicates a region at the exit side of the barrier twin where a subsidiary twin adjacent to the main incident twin is formed; the details are shown in Figure 6.58.

Taken together, these observations suggest that the Type-I intersection could significantly differ from the mechanisms sketched in Figure 6.53. It might be concluded that the details of twin intersection, apart from the crystallographic constraints, depend on the width of the intersecting twins and the actual stress distribution around the intersection zone, that is, whether shear processes in adjacent grains or lamellae exert constraints.

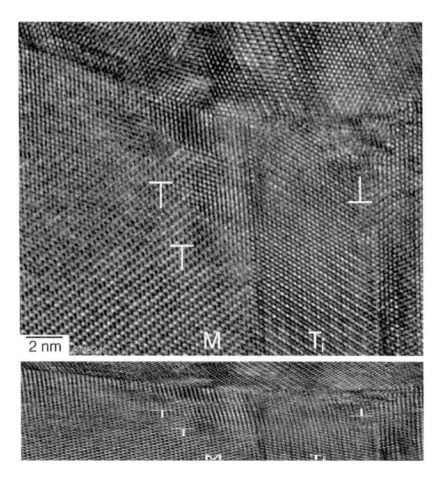

Figure 6.57 Detail 2 marked in Figure 6.54: dislocation structure at the exit side of the barrier twin T_b. The image below, compressed along the $(\bar{1}\bar{1}1)$ planes parallel to the twin matrix interface, shows this feature more clearly.

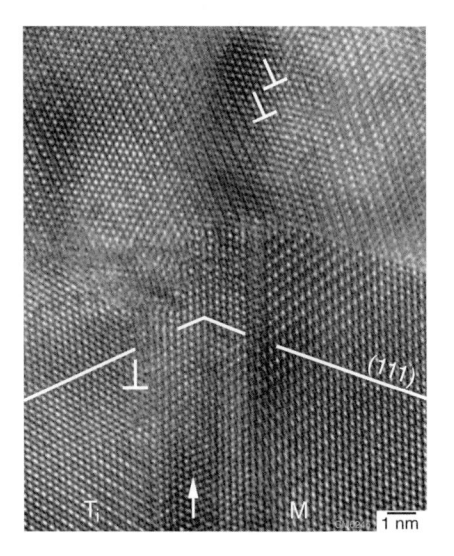

Figure 6.58 Detail 3 marked in Figure 6.54: formation of a subsidiary twin parallel to the incident twin (arrowed) at the exit side of the barrier twin T_b. Note the close distance between the neighboring dislocations.

The characteristics of Type-II intersections are schematically sketched in Figure 6.59. The geometry is characterized by a deflection of both, the barrier twin T_b and incident twin T_i. The deflection of the incident twin can formally be described as following the $(\bar{1}11)_{T_b}$ plane of the barrier twin. The characteristic features of the intersection zone are subsidiary twin lamellae, which alternate with layers of the barrier twin. A large number of mechanisms and dislocation reaction schemes have been discussed in the literature; for these details the reader is referred to the articles of Sun et al. [182, 183].

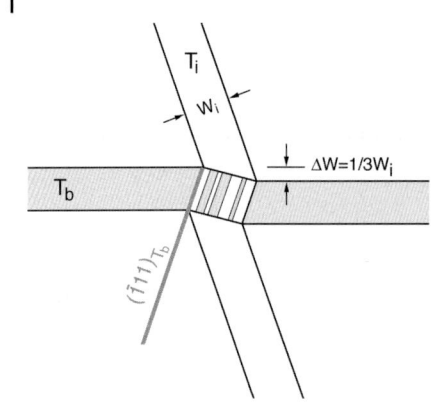

Figure 6.59 Schematic illustration of Type-II twin intersections in γ(TiAl). The deflection of the incident twin can formally be described as following the $(\bar{1}11)_{T_b}$ plane of the barrier twin. Subsidiary twin plates situated in the barrier twin T_b accommodate the shear of the incident twin T_i. The deflection of the barrier twin T_b is approximately 1/3 of the incident twin T_i. Figure adapted from Sun et al. [182, 183].

Regardless of the details, twin intersections undoubtedly leave significant internal stresses and dense defect arrangements bordering the intersection zone, which at low temperatures can give rise to crack formation [228]. Under hot-working conditions twin intersections provide heterogeneities in the deformed state, which can be the prevalent sites for recrystallization. At elevated temperatures rearrangement of the dislocation walls surrounding the misoriented zone may occur by climb so that the misorientation with respect to the surrounding matrix increases. The intersection zone is transformed into a new grain of low internal energy, growing into deformed material from which it is finally separated by a high-angle boundary. This process is certainly driven by the release of stored energy. Thus, it might be expected that the structural heterogeneities produced by a twin intersections act as precursor for recrystallization. Such processes are certainly beneficial for the conversion of the microstructure under hot-working conditions.

Summary: Mechanical twinning of two-phase titanium aluminide alloys occurs on multiple $1/6<11\bar{2}]\{111\}$ systems, which leads to intensive twin intersections. There is clear evidence that twin intersections are difficult to accomplish and probably contribute to work hardening.

6.3.5
Acoustic Emissions

To provide additional insight into the contribution of mechanical twinning to the deformation of TiAl alloys, acoustic emission has been studied [229–232]. Acoustic emissions are essentially elastic stress waves in the range of 30 kHz to 5 MHz, generated in a material by crack growth, plastic deformation, or phase transformation. Owing to its rapid energy release, twinning normally produces very discernible acoustic signals [233, 234]. Using electronic sensing and analyzing data from a multiplicity of sensors, the location and relative severity of deformation-induced defects can be determined. Careful interpretation of data is required to distinguish between mechanical twinning and contact-induced plasticity or cracking. For all the observations made on TiAl alloys, the overall trends in acoustic emission are similar: with increasing applied load, the number of signals increases gradually, reaching a peak at the onset of yielding; afterwards the acoustic emission rapidly declines to an ostensibly constant level. More acoustic events are apparently generated in lamellar alloys. Zhu *et al.* [229] recorded acoustic emissions during tensile tests performed at room temperature on four binary alloys Ti-(48-52)Al. The authors associated the acoustic signals with dislocation glide and the generation of microcracks. The authors noted a significant effect of the microstructure on the acoustic emission; the acoustic emission being strongest in the lamellar alloy. This finding was attributed to a different attenuation of the shear waves in the different microstructures. Kauffmann *et al.* [230] monitored acoustic emissions during room-temperature compression of a Ti-46.5 Al (Cr, Nb, Ta, B) alloy with a near γ microstructure. The coincidence of strong acoustic emission with the beginning of yielding led the authors to believe that nucleation of mechanical twins is the main source of the acoustic emissions. The decrease of the acoustic events at higher deformation was attributed to a change of the deformation mechanism; twin nucleation was thought to be substituted by twin growth and dislocation glide. It is generally assumed that these two latter mechanisms produce a continuous very low acoustic emission [235]. Kauffmann *et al.* [230] supported their interpretation by TEM examination of the deformation structure.

Botten *et al.* [231] investigated acoustic emission during room-temperature tensile deformation of fully lamellar Ti-44Al-8Nb-1B and Ti-44Al-4Nb-4Hf-0.2Si alloys. The generation of acoustic signals was attributed to interlamellar cracking. This conclusion was supported by metallographic inspection, which showed that the samples fractured at the location from which most of the acoustic signals were emitted. The point to note is that acoustic emission and cracking already occurred at about 70% of the 0.2% proof stress. The authors associated this early cracking with the plastic anisotropy of lamellar material. As has been described in Sections 6.1.5 and 6.1.6, the anisotropic flow behavior of lamellar alloys could give rise to significant constraint stresses and premature fracture. The implications of the early cracking events on the fatigue life are described in a companion paper [232] of the authors. Taken together, it seems that no clear position has emerged in the TiAl literature about the origin of acoustic emissions. As already mentioned, acoustic signals are generated by a variety of processes other than twinning. Compared to materials where plastic deformation is accommodated by slip only, the situation in TiAl alloys is even more complex because slip, twinning and microcrack propagation can occur concurrently. Since the investigations were performed on alloys with different microstructures, the question remains as to whether the disparity of the observations is caused by the influence that the microstructure has on slip, twinning and microcracking. In view of the results presented in the previous section, it may also be expected that twin intersections produce acoustic signals.

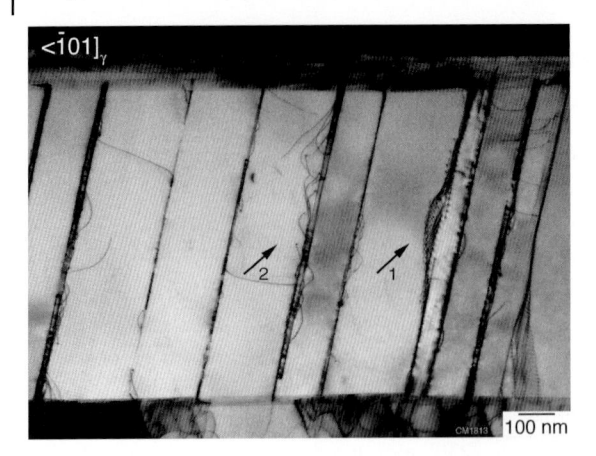

Figure 6.60 Thermal instability of mechanical twins. Twin structure observed after tensile deformation of lamellar Ti-48Al-2Cr at 800 °C to failure at a total fracture strain of $\varepsilon_f = 10.2\%$. The twin/matrix interfaces of the twins crossing the lamellae are imaged nearly end on. Note the formation of dislocation networks (arrow 1) and the emission of dislocation loops (arrow 2) at the twin/matrix interfaces [75].

Summary: Low-temperature deformation of TiAl alloys is accompanied by acoustic emissions, which have been attributed to slip, mechanical twinning or crack propagation.

6.3.6
Thermal Stability of Twin Structures

TEM observations performed after high-temperature deformation have revealed a remarkable instability of twin structures [75]. The twins are often fragmented and exhibit rough interfaces, as demonstrated in Figure 6.60. A plausible explanation is that reactions occur between the twinning partial and matrix dislocations that have been incorporated into the twin/matrix interface. In this way, the Shockley partial dislocations can be converted to perfect dislocations. Climb of these dislocations probably gives rise to the observed complex interfacial dislocation networks. As seen in the micrograph, the converted twin/matrix interfaces can act as dislocation sources.

The extent of climb of the interfacial dislocations certainly depends on temperature and strain rate; thus the balance between the generation and reconstruction of twins is obviously governed by the deformation conditions. In particular, long-term creep at low strain rates seems to be conducive for the reconstruction of twin structures. In the TiAl literature, some controversy has arisen over the point as to whether twinning may significantly contribute to creep. In view of the foregoing discussion, the disparity between the different observations could be rationalized by different degrees of twin reconstruction, depending on the actual creep conditions [236].

Summary: During high-temperature deformation, twin structures can easily recover due to the incorporation of matrix dislocations and subsequent climb of the reactant dislocations.

6.4
Glide Resistance and Dislocation Mobility

Dislocation motion has been established as a thermally activated process depending on time, temperature and strain rate. The factors controlling the dislocation mobility are of great interest because they not only govern the materials performance but also the design of various manufacturing processes. A wealth of experimental data has become available about the thermodynamics and kinetics of this rate-controlled deformation in TiAl alloys; this will be briefly reviewed in the present section.

6.4.1
Thermally Activated Deformation

Pioneered by Conrad and Wiedersich [237], thermally activated dislocation motion has been analyzed in many articles [238–242]. The treatment given here is not intended to be exhaustive; instead it will try to cover the subject in a selective but reasonably self-contained manner that is adapted to the particular situation in TiAl alloys. Overlap with the above-cited reviews is unavoidable because the necessary background must be laid.

The plastic shear produced by dislocation glide is usually expressed as

$$a = b \rho_m \bar{x} \tag{6.22}$$

b is the Burgers vector, ρ_m is the density of mobile dislocations and $\bar{x}$ is their average glide distance. Thus, the shear-strain rate $\dot{a}$ is

$$\dot{a} = b \rho_m v_d \tag{6.23}$$

v_d is the average dislocation velocity. An analogous relationship holds for climb-controlled strain rates. The dislocation velocity is determined by a variety of obstacles, which in broad terms may be classified by their interaction forces with the dislocations. There are obstacles with long-range forces f_μ, which vary only slowly with the distance of the obstacle from the dislocation; typical examples are grain boundaries or dislocations gliding on parallel slip planes. The other class of obstacles is characterized by short-range interaction forces f^*, which act over a few atomic distances only, such as solute atoms or jogs in screw dislocations. Overcoming the glide obstacles requires a certain amount of energy, which at 0 K must be supplied mechanically to the dislocations by the applied shear stress τ [243]

$$\tau = \frac{f_\mu + f^*}{l b} = \tau_\mu + \tau^* \tag{6.24}$$

l is the average length of the dislocation segments between the obstacles. At the atomic scale, dislocation movement is associated with the breaking and re-establishing of atomic bonds, which in principle can be assisted by thermal lattice vibrations. Overcoming glide obstacles by thermally activated shear requires a coordinated motion of several atoms that form the glide obstacle and the piece of dislocation being in contact with the obstacle. The probability for a simultaneous

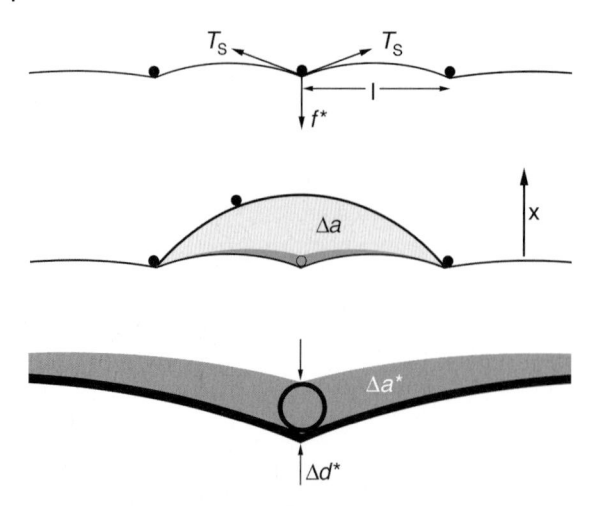

Figure 6.61 Point obstacles with separation distance l restraining a bowed-out dislocation. Each obstacle exerts a localized glide resistance force f^* balanced in equilibrium by dislocation line tension forces T_S. Δa^* is the area swept by the dislocation during the penetration through the central obstacle; Δd^* is the activation distance parallel to the reaction coordinate x along which the dislocation advances. When the dislocation unpin from the central obstacle it moves forward to encounter another glide obstacle; the area swept in so doing is Δa.

motion of these atoms through random fluctuations becomes very small as the number of atoms increases. Thus, for thermal activation to be significant the size of the activation complex is restricted to atomic dimensions. Also, the energy available from thermal fluctuations is very small and approximately given by kT; k is the Boltzmann constant. Hence, for thermal activation to be effective, the energy barrier formed by the glide obstacles should be sufficiently small (less than about 50 kT) [238]. In terms of the above classification, only short-range obstacles can be overcome with the assistance of thermal activation. The term "thermal obstacle" is synonymous to short-range obstacle. τ^* is the related thermal or effective stress. The stress part τ_μ to overcome long-range obstacles shields part of the total applied stress. τ_μ, often designated as athermal stress, is almost independent of temperature apart from the small variation of the shear modulus with temperature [244].

The energy that has to be provided to overcome the short-range obstacles determines the dependence of τ^* on temperature and strain rate. The geometrical principles involved in this consideration are illustrated in Figure 6.61 for the thermally activated penetration of a dislocation through localized obstacles. The obstacles exert a localized glide resistance force f^* that is balanced by line tension forces T_S. Gibbs [239] and Schoeck [238] have determined the thermal energy that has to be provided for dislocations to overcome the obstacles they have encountered. Figure 6.62a shows the variation of the Helmholtz free energy F^* and of

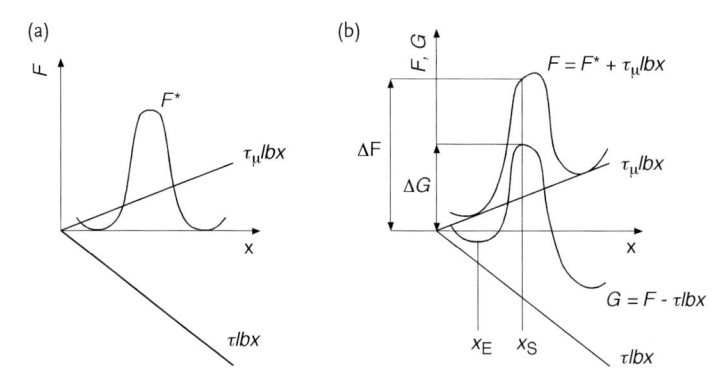

Figure 6.62 Thermally activated overcoming of localized glide obstacles. (a) Change of the Helmholtz free energy F and the deformation resistance $\tau l b x$ when a dislocation segment of length $2l$ glides through the central obstacle as illustrated in Figure 6.61. (b) Modification of the free energy by the work terms $\tau_\mu l b x$ and $\tau l b x$ done by the long-range internal stress τ_μ and the applied stress τ, respectively. ΔF is the total energy for overcoming the obstacle that opposes the dislocation movement. ΔG is the Gibbs free energy of activation that is supplied by thermal activation. Figure adapted from [239].

the external work $\tau l b x$ when the center of a pinned dislocation segment of length $2l$ glides through an obstacle along x. The free-energy profile is modified when a long-range stress τ_μ opposes the dislocation motion; $\tau_\mu l b x$ is the related work term (Figure 6.62b). The Gibbs free energy G is defined as

$$G = F - \tau l b x \tag{6.25}$$

At any stress level, the stress is in equilibrium with the deformation resistance at the points x_E and x_S, which correspond to the minimum and maximum in the $G(x)$ curve, respectively. x_E is the stable equilibrium position and x_S the metastable saddle point position. In order to overcome the obstacle, the dislocation must move from x_E to x_S. The total energy change associated with this process is ΔF. The activation distance is defined as

$$\Delta d^*(f^*) = x_S - x_E \tag{6.26}$$

and the activation area is

$$\Delta a^* = l \, \Delta d^* \tag{6.27}$$

The term

$$V = l \, b \, \Delta d^* \tag{6.28}$$

has the dimension of a volume. In keeping with what has become more or less accepted terminology, V is called activation volume, although a volume in a two-dimensional system is a geometrical absurdity. In the present context V is defined such that $V\tau^*$ is the mechanical work done by the thermal stress when the dislocation is propagated from x_E to x_S

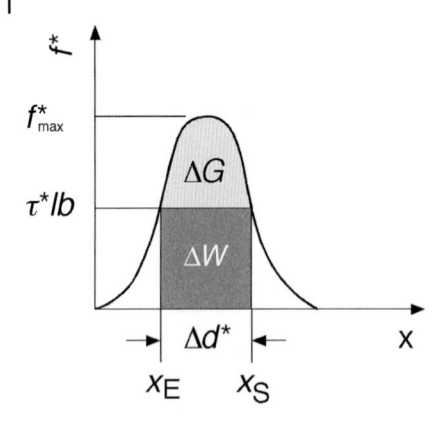

Figure 6.63 A force–distance profile describing the resistance force $f^* = \tau^* lb$ as a function of the reaction coordinate x. ΔG Gibbs free energy of activation, ΔW mechanical work done by τ^*, Δd^* activation distance, τ^* effective stress, l obstacle distance, x_e equilibrium position, and x_s saddle-point position of the dislocation at the obstacle.

$$\Delta W = \tau^* V = (\tau - \tau_\mu)V \tag{6.29}$$

The glide resistance resulting solely from the localized obstacle, the free energy of activation ΔF^*, is given by

$$\Delta F^* = \Delta G + V\tau^* \tag{6.30}$$

ΔG, the Gibbs free energy of activation, must be supplied by thermal fluctuations to permit the dislocation to move from the stable equilibrium position to the metastable saddle point position. In the analysis of thermally activated shear, the energies ΔF^*, ΔW and ΔG are often described in profiles of the resistance force $f^* = \partial F^*/\partial x = \tau^* lb$ versus the reaction coordinate x. If the obstacle to be overcome is repulsive, the plot has the form of Figure 6.63. In this description the obstacle can withstand the maximum force $f^*_{max} = \tau^*_{max} lb$. For any applied stress less than τ^*_{max} the dislocation waits at the equilibrium position x_E for a successful activation. If the effective stress increases, then the Gibbs free energy, represented by the light gray area, necessarily decreases; ΔF^*, ΔW and V change correspondingly.

It should be noted that the flexibility of the dislocations could give rise to a stress-dependent obstacle distance, a situation that was treated statistically by Friedel [163]. This variation of l is significant for dislocation propagation through an array of weak localized obstacles under a low stress. For other types of obstacles, for example, dislocation dragging by jogs, this so-called Friedel statistics is not applicable. Thus, this sophistication of the analysis will not be considered further.

The probability that a thermal fluctuation of ΔG can be supplied by the thermal reservoir is given by the Boltzmann factor $\exp(-\Delta G/kT)$. Thus, if a dislocation is effectively vibrating at a frequency ν, it successfully overcomes $\nu \exp(-\Delta G/kT)$ obstacles per second. The resulting dislocation velocity v_d is given by the Arrhenius-type equation

$$v_d = \kappa \, \nu \exp(-\Delta G/kT) \tag{6.31}$$

κ is the average glide distance of the dislocations after a successful fluctuation, ν is related to the Debye frequency by a factor of 10^{-2} to 10^{-1}, depending on the details of the obstacle ensemble [240]. It is worth adding that the equation only applies when there are no backfluctuations against the effective stress. From Equations (6.23) and (6.31), the macroscopic shear rate $\dot{a}$ is

$$\dot{a} = \rho_m \, b \, \kappa \, \nu \exp(-\Delta G/kT) = \dot{a}_0 \exp(-\Delta G/kT) \tag{6.32}$$

and τ can be expressed as

$$\tau = \tau_\mu + (1/V)(\Delta F^* + kT \ln \dot{a}/\dot{a}_0) \tag{6.33}$$

In this formulation a thermally activated process is characterized by ΔG, ΔF^* and V. The magnitude of these parameters and their dependence on stress, temperature and strain are characteristic for a certain rate-controlling mechanism and may help in identifying the glide obstacles controlling the dislocation velocity. For example, V can be considered as the number of atoms that have to be coherently thermally activated for overcoming the glide obstacles by the dislocations. It is therefore expected that V will undergo a significant change when there is a change of the mechanism that controls the dislocation glide resistance. Under the condition that τ_μ, $\dot{a}_0$, and l are independent of τ and T, the activation parameters can be related to the strain rate and temperature sensitivity of the flow stress by [241]

$$V = \frac{M_T kT}{(\Delta \sigma / \Delta \ln \dot{\varepsilon})_T} \tag{6.34}$$

$$\Delta G = \frac{Q_e + V\sigma(T/\mu M_T)(\partial \mu/\partial T)}{1 - (T/\mu)(\partial \mu/\partial T)} \tag{6.35}$$

$$Q_e \equiv \Delta H = -\frac{kT^2(\Delta \sigma/\Delta T)_{\dot{\varepsilon}}}{(\Delta \sigma/\Delta \ln \dot{\varepsilon})_T} \tag{6.36}$$

σ is the normal stress, $\dot{\varepsilon}$ the strain rate, μ the shear modulus. The Taylor factor $M_T = 3.06$ is used to convert σ and $\dot{\varepsilon}$ to average shear quantities. Q_e is the experimental activation energy, which, for a stress-independent obstacle distance, is identical to the activation enthalpy ΔH. The stress increments $(\Delta \sigma/\Delta T)_{\dot{\varepsilon}}$ and $(\Delta \sigma/\Delta \ln \dot{\varepsilon})_T$ can be determined by temperature and strain-rate cycling tests, as has been described in [126]. Figure 6.64 demonstrates the load–elongation traces of strain-rate cycling tests performed at different temperatures and illustrates the determination of the stress increments $\Delta \sigma$. The shape of the stress transients upon the strain-rate change was found to depend on the deformation temperature. Testing at room temperature resulted in step-like stress transients. On the contrary, in the temperature range 450 to 750 K, the stress responses exhibited yield-drop effects, a tendency that generally increased with strain. Strain-rate cycling tests performed at temperatures higher than 750 K exhibited smooth stress transients, which extend over strain intervals of about 0.5%. No yield-drop effects were superimposed under these conditions. The observed yield-drop effects and smooth

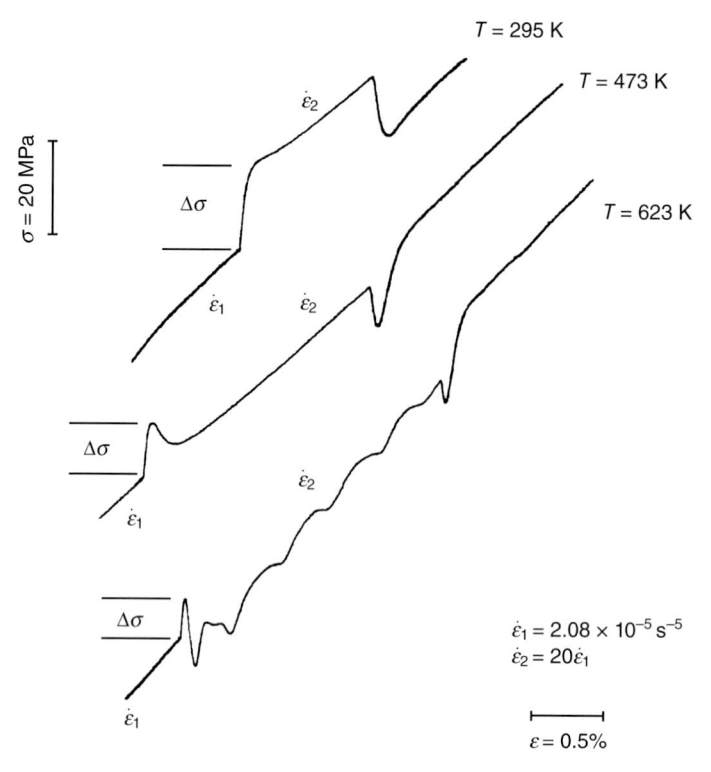

Figure 6.64 Determination of the strain rate sensitivity of the flow stress. Forged Ti-47Al-2Cr-0.2 Si, equiaxed microstructure. Load–elongation traces of strain-rate cycling tests performed at different temperatures demonstrating the estimation of the stress increments $\Delta\sigma$.

stress transients suggest that under the respective deformation conditions significant structural changes occurred during the incremental tests. The effects are expected to increase with increasing increment of strain rate or temperature. The implications of structural changes on the stress response have been assessed by measuring the stress increment $(\Delta\sigma/\ln(\dot{\varepsilon}_2/\dot{\varepsilon}_1))_T$ as function of the logarithm of the strain-rate ratio $n = \dot{\varepsilon}_2/\dot{\varepsilon}_1$. In the formulation used here, $(\Delta\sigma/\ln n)_T$ should be independent of n, if τ_μ and $\dot{a}_0$ remain constant during the strain-rate change. Figure 6.65 is a plot of the strain-rate sensitivities versus $\ln n$ with the two strains $\varepsilon_1 = 1.25\%$ and $\varepsilon_2 = 6\%$ as parameters. Whereas the strain-rate sensitivities estimated from the step-like transients at 295 K are almost constant, the values estimated at 973 K decrease strongly with increasing n. It is interesting to note that this tendency is almost independent of strain. The reasons for the variation of the strain-rate sensitivity with n are not altogether clear. The most probable variable in $\dot{a}_0$ that may alter with $\dot{\varepsilon}$ or T, is the mobile dislocation density ρ_m because the multiplication rate of the dislocations can be a sensitive function of the effective stress. A high deformation temperature may enhance these processes. Unlike

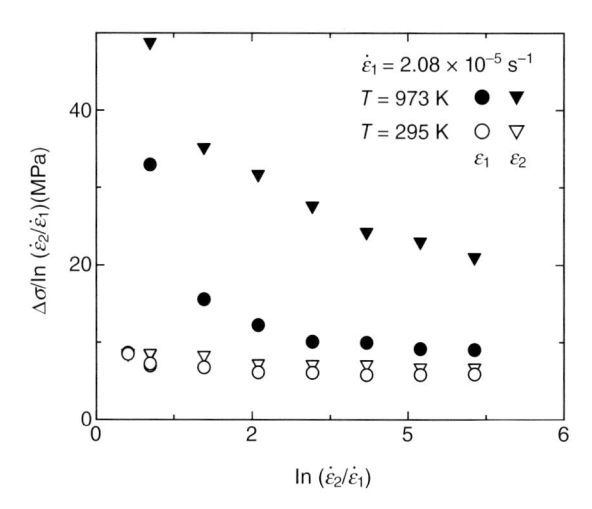

Figure 6.65 Dependence of the strain-rate sensitivity $(\Delta\sigma/\ln(\dot{\varepsilon}_2/\dot{\varepsilon}_1))_T$ on the logarithm of the strain-rate ratio $n = \dot{\varepsilon}_2/\dot{\varepsilon}_1$ determined for the temperatures $T_1 = 295$ K and $T_2 = 973$ K and strains $\varepsilon_1 = 1.25\%$ and $\varepsilon_2 = 6\%$. Forged Ti-47Al-2Cr-0.2 Si, equiaxed microstructure. Data from [126].

strain rate, temperature cannot be changed instantaneously, hence, obstacle structures and mobile dislocation densities can more easily adjust to the new deformation temperature. Structural changes became directly evident by the *in situ* heating study described in Sections 6.2.5 and 6.3.1, which followed a schedule similar to that of the temperature-cycling tests. Under these conditions dislocation loops were emitted from the interfaces, which suggests that ρ_m could significantly change upon a temperature change. It seems likely that the observed structural changes also have significant effects on athermal interactions between the gliding dislocations and the lamellar interfaces. Thus, τ_μ may change more strongly with temperature than expected from the temperature dependence of the shear modulus. In view of these problems, the analysis of temperature-cycling experiments is less satisfactory.

In order to minimize the implications of structural changes the following procedure has been proposed [126]. When the stress transients exhibited yield drops, the stress increments were determined from the upper yield point of the stress transients (Figure 6.64), which is thought to represent the initial response of the material to the strain-rate change and is probably less influenced by a variation of ρ_m. At higher temperatures ($T > 973$ K), strain-rate cycling tests were performed with a small strain-rate ratio of $n = 3$. Under these conditions structural changes during the strain rate jumps are believed to be relatively small, but still allow the determination of $\Delta\sigma$ with accuracy. A particular problem associated with strain-rate cycling tests is the reversibility of the stress response during upward and downward changes of the strain rate. Due to the limited elastic stiffness, a deformation machine expands or contracts in response to load changes; hence, a certain

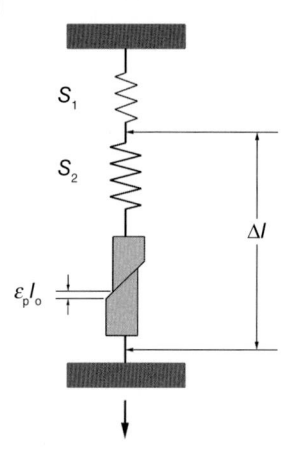

Figure 6.66 Schematic diagram showing the deformation components involved in a constant strain rate tensile test. Spring 1 represents the elastic deformation of the machine, including that of the load cell. Spring 2 represents the elastic deformation of the specimen. The displacement resulting from the plastic sample deformation is $\varepsilon_p l_0$. The deformation Δl used as feedback parameter in the strain-controlled closed-loop mode involve both the plastic elongation $\varepsilon_p l_0$ plus the elastic deformation of the sample.

amount of plastic deformation of the sample takes place before the new strain rate can be achieved. During downward changes plastic relaxation of the sample occurs, which makes the stress response difficult to interpret, in particular, when yield-drop effects are superimposed. In these cases only stress increments resulting from upward changes of $\dot{\varepsilon}$ should be evaluated. The problems can largely be overcome when the incremental tests are performed in closed-loop machine because the strain-controlled regime virtually enhances the elastic stiffness of the machine.

The strain-rate and temperature sensitivities can also be measured with the help of stress-relaxation experiments. Stress relaxation is the decay of stress over time when a constant strain rate test is stopped. The specimen continues to deform under the action of thermal activation, thereby expanding the load train. In other words, plastic deformation gradually substitutes the elastic deformation of the sample and the machine. Thus, apart from the thermodynamic factors controlling the sample plasticity, the kinetics of the stress relaxation is governed by the elastic response of both the deformation machine and the sample. The strain components involved in stress relaxation are illustrated in Figure 6.66. It is advantageous to perform relaxation tests in a closed-loop machine, utilizing the sample strain as feedback parameter. In the strain-controlled regime, a countermotion of the crosshead largely compensates for the elastic response of the machine to the plastic sample deformation. Thus, the total strain rate of the stress relaxation involves only the elastic and plastic deformation rates of the sample. Under these conditions, the kinetics of the stress rate $\dot{\sigma}$ can be described as [245, 246]

$$\dot{\sigma} = -M_E \dot{\varepsilon} = -M_E \dot{\varepsilon}_0 \exp-(\Delta F^* - V\sigma^*)/kT \qquad (6.37)$$

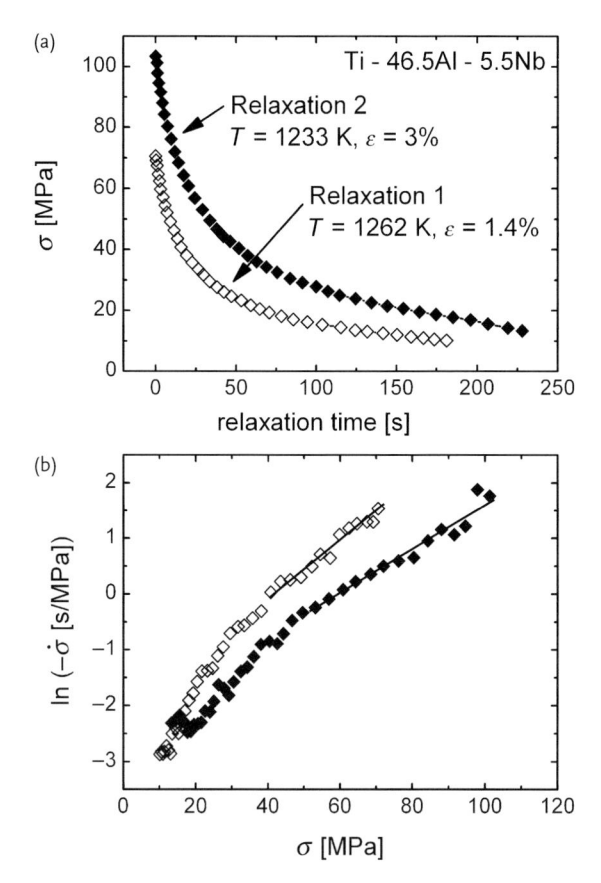

Figure 6.67 High-temperature stress-relaxation experiments performed during a constant strain-rate compression test at $T = 1262$ K. Extruded Ti-46.5Al-5.5Nb, nearly globular microstructure. (a) Decay of the stress after stopping the machine in two experiments that were performed at two different temperatures. (b) Evaluation of the kinetics according to Equation (6.37). Data from [245].

M_E is identical with the Young's modulus of the specimen. According to Equation (6.37), the strain-rate sensitivity can be determined as the inverse slope of the plot $\ln(-\dot{\sigma})$ vs. σ. Figure 6.67 [245] shows the kinetics of two relaxation experiments and the evaluation according to Equation (6.37). The experimental activation energy is defined as [245, 246]

$$Q_e = \frac{kT_1T_2}{T_1 - T_2}(\ln \dot{\sigma}_1 - \ln \dot{\sigma}_2) \qquad (6.38)$$

$\dot{\sigma}_1$ and $\dot{\sigma}_2$ are the stress rates measured at the temperatures T_1 and T_2, respectively; it is understood that these stress rates must be determined at the same stress. The main advantage of a stress relaxation test over strain rate or temperature jumps is that it may scan a broad range of stress rates while straining the sample by only

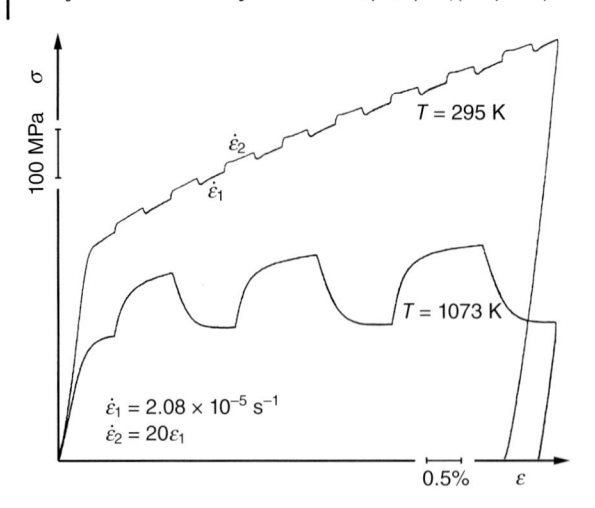

Figure 6.68 Load–elongation traces of strain-rate cycling tests performed at the temperatures indicated. Note the effect of temperature and strain on the magnitude of the stress increments occurring upon strain-rate change with the strain-rate ratio $\dot\varepsilon_2/\dot\varepsilon_1 = 20$. Forged Ti-47Al-2Cr-0.2 Si, equiaxed microstructure.

a small amount. For a brittle material like TiAl this is a major concern because it allows the activation parameters to be determined within a small strain interval. This feature of stress relaxation is particularly important for high-temperature testing because structural changes due to dynamic recovery and recrystallization can be minimized. It should be emphasized that for a successful analysis of stress relaxation the same suppositions have to be fulfilled that have been discussed earlier for strain-rate and temperature-cycling tests, namely the pre-exponential factor $\dot\varepsilon_0$ and the athermal stress component τ_μ have to remain constant during the relaxation. Spätig et al. [247] have recognized that substantial structural changes could occur during relaxation, which leads to an overestimation of the activation volume. To mitigate this problem, the authors proposed an analysis of repeated relaxations, which was successfully applied to TiAl alloys [248]. Unfortunately, no straightforward method has been established to directly measure the effective stress. However, for TiAl alloys it seems to be possible to distinguish τ^* and τ_μ by utilizing stress-reduction tests. Information about the relative value of these stresses can also be obtained from structural data. The problem will be discussed in greater detail in Section 6.5. Activation parameters can also be determined by creep tests; this will be described in Section 9.3.

Summary: The effects of temperature and strain rate on the flow stress can be coupled with an Arrhenius-type equation in order to characterize thermally activated glide processes. The most important parameters that enter this equation are the activation energy and the activation volume. These thermodynamic glide parameters can be determined by macroscopic deformation tests and could help in identifying the factors controlling the dislocation velocity.

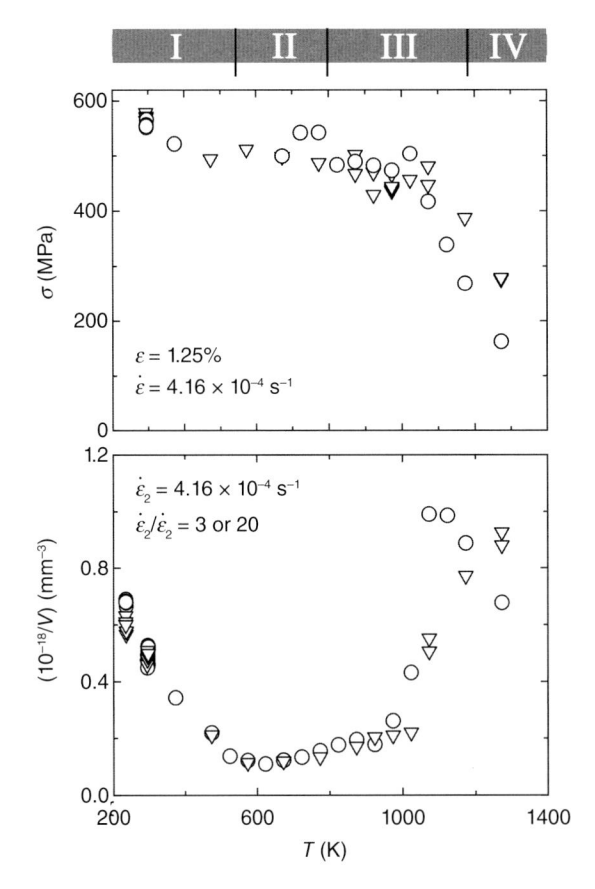

Figure 6.69 Dependence of the flow stress σ and the reciprocal activation $1/V$ at 1.25% plastic strain on the deformation temperature T. Ti-47Al-2Cr-0.2Si; ○ equiaxed microstructure, grain size 11 μm; ▽ nearly lamellar microstructure, colony size 326 μm, lamellar spacing 5 nm to 100 nm. Data from [126].

6.4.2
Glide Resistance at the Beginning of Deformation

The activation parameters have been determined over a wide temperature range for various TiAl alloys with different compositions and microstructures [126, 203, 245, 249, 250]. For a given alloy the strain rate and temperature sensitivity depend on strain; as an example Figure 6.68 demonstrates strain-rate cycling tests performed at different temperatures [126].

At first, the activation parameters determined at the beginning of deformation at $\varepsilon = 1.25\%$ will be discussed. At constant temperature and strain rate the flow stresses and reciprocal activation volumes are linearly related according to Equation 6.33. The characteristic temperature dependence of these two quantities is demonstrated in Figure 6.69. The data was determined on an equiaxed and a nearly lamellar form of a Ti-47Al-2Cr-0.2Si alloy [126]. The equiaxed microstructure with a grain size of about 11 μm was established by a two-step isothermal forging at 1220 °C; subsequent annealing at 1370 °C resulted in the nearly lamellar microstructure with a colony size of 326 μm and a lamellar spacing of 5 nm to 100 nm. The behavior is typical of $(\alpha_2 + \gamma)$ alloys [75, 126, 150, 203, 245, 249, 250]. The flow stress is almost independent of temperature up to about 1000 K and then decreases; in other words, there is no anomalous increase of the stress with temperature. This result is remarkable because both phases separately show an anomalous temperature dependence of the yield stress (Sections 5.1.6 and 5.2.2). The reciprocal activation volume passes through a broad minimum at $T = 600$ K, indicating that significant changes in the micromechanisms controlling the dislocation velocity occur. Thus, separate consideration will be given to the domains I, II, III, and IV designated in Figure 6.69.

The activation parameters determined for two-phase alloys in domain I ($T = 295$ K) are typically [75, 126, 150]

$$V = (70 \text{ to } 130)b^3, \Delta G = 0.7 \text{ to } 0.85 \text{ eV}, \Delta F^* = 1.3 \text{ eV} \qquad (6.39)$$

V was referred to the Burgers vector of ordinary dislocations, which in two-phase alloys mainly provide deformation of the γ phase. Whereas V and ΔG were directly measured with good accuracy, some uncertainty arises for ΔF^* because of the difficulties in determining τ^* and the related work term $V\tau^*$. For the equiaxed Ti-47Al-2Cr-0.2Si alloy a relatively large athermal stress $\tau_\mu = 430$ MPa was estimated from the measured grain size and dislocation density. This leads to an effective stress $\tau^* = 30$ MPa and a work term $V\tau^* = 0.5$ eV. The high athermal contribution to the total yield stress of TiAl alloys has been confirmed by stress-reduction tests (Section 6.5). The activation parameters determined for different two-phase alloys at the beginning of deformation are listed in Table 6.2.

The small value of V and the relatively high activation energies suggest that the glide resistance of the dislocations arise from a dense arrangement of relatively strong obstacles. This aspect of dislocation dynamics apparently becomes evident in the dislocation structure that is commonly observed after room-temperature deformation [75, 120, 126, 127, 150, 249–257]; see Figure 6.38. The salient feature is a preponderance of pinned ordinary screw dislocations. As has been described in Section 6.3.1, this finding indicates a remarkable anisotropy of the glide resistance for ordinary dislocations.

Some controversy has arisen in the TiAl literature over the nature of the glide obstacles. The ongoing discussion involves intrinsic and extrinsic pinning mechanisms. In deformed Ti-51Al-2Mn, Viguier et al. [254–256] observed a dislocation substructure that is consistent with the generation of kinks. The authors proposed that an ordinary screw dislocation is subject to a Peierls friction and bulges kinks on the two {111} planes that intersect along the common <110] direction. The kinks are thought to move laterally along the dislocation and collide to form the pinning centers. The characteristic feature of this mechanism is that the dislocation simultaneously bows out on the primary $(1\bar{1}1)$ plane and on the $(\bar{1}1\bar{1})$ cross-slip plane. The mean linear pinning point density was about $5 \mu m^{-1}$ at room

Table 6.2 Thermodynamic glide parameters for different microalloyed two-phase alloys determined from temperature and strain-rate cycling tests at the beginning of deformation $\varepsilon = 1.25\%$.

Alloy	σ (MPa)	V/b^3	ΔH (eV)	ΔG (eV)
$T = 295$ K				
Ti-48Al-2Cr, nearly lamellar	404	133	0.84	0.65
Ti-47Al-2Cr-0.2Si, equiaxed	538	91	0.98	0.81
Ti-47Al-2Cr-0.2Si, nearly lamellar	517	103	0.96	0.77
Ti-47Al-1.5Nb-1Mn-1Cr-0.2Si-0.5B, duplex	626	93	1.06	0.86
Ti-49Al-1V-0.3C, near gamma	570	116	1.09	0.86
$T = 910$ K				
Ti-48Al-2Cr, nearly lamellar	338	523		
Ti-47Al-2Cr-0.2Si, equiaxed	469	237	4.31	2.78
Ti-47Al-2Cr-0.2Si, nearly lamellar	460	320	4.38	2.56
Ti-47Al-1.5Nb-1Mn-1Cr-0.2Si-0.5B, duplex	549	245	3.60	3.0
Ti-49Al-1V-0.3C, near gamma	532	249	1.85	1.54
$T = 1100$ K				
Ti-48Al-2Cr, nearly lamellar	360	61	2.04	1.37
Ti-47Al-2Cr-0.2Si, equiaxed	368	47	3.16	2.40
Ti-47Al-2Cr-0.2Si, nearly lamellar	435	73	3.13	2.25
Ti-47Al-1.5Nb-1Mn-1Cr-0.2Si-0.5B, duplex	458	51	2.90	2.35
Ti-49Al-1V-0.3C, near gamma	476	70	3.07	2.46

σ, flow stress; ΔH, activation enthalpy; V/b^3, activation volume referred to the Burgers vector of ordinary dislocations; ΔG, Gibbs free energy of activation [126].

temperature, went through a maximum of about $13\,\mu m^{-1}$ at 500 °C and then decreased. Based on this data, Louchet and Viguier [256] have modeled the yield-stress anomaly that occurred for their material.

Sriram et al. [257] performed a detailed TEM study on Ti-50Al and Ti-52Al alloys that had been deformed at 573 K and 873 K. The alloys had a low oxygen level of 250 wt. ppm (about 575 at. ppm), which had been given a particular heat treatment in order to coarsen interstitial-containing precipitates. The authors have shown that the morphology of the pinned screw dislocations in these alloys corresponds to the classical cross-slip mechanism, meaning that the neighboring segments bow on parallel {111} planes. The double cross-slip mechanism is sketched in Figure 6.39. The driving forces for cross-slip may arise from dislocation intersections because deformation mostly occurs on multiple 1/2<110]{111} slip systems. Furthermore, the coherency stresses observed in the lamellar microstructure can locally provide shear components that cause dislocation segments to cross-slip. The drag of the jogs on the forward motion of the dislocations arises from the point defects and dipoles that are trailed at the jogs. The separation distance between the jogs observed after deformation at 573 K was close to 100 nm (350 b), and jog heights ranging from atomic dimensions up to about 40 nm were measured. The distribution of the jog heights and consequently that of the pinning forces were found to depend on temperature. These observations led the authors to believe that progressive pinning of the screw dislocations caused the yield-stress anomaly that occurred for their alloys. Unfortunately, no data about the pinning spectrum at room temperature was presented to support this hypothesis. Taken together, the investigations show that jog dragging occurs over a wide temperature range. It might be expected that the deformation structure partially recover at these temperatures in that small jogs coalesce or in that vacancy-producing jogs recombine with interstitial-producing jogs. The stability of deformation structures will be discussed in Section 7.2.4. The question arises whether the obvious disparity between the observations of Viguier et al. [253–255] and Sriram et al. [257] is caused by differences in the local orientation of the grains on which the observations were made. A symmetrical orientation of the load axis with respect to inclined $(\bar{1}1\bar{1})$ and $(1\bar{1}\bar{1})$ planes would favor simultaneous bow out of screw segments on these planes, as observed by Viguier et al. [253–255]. On the contrary, any deviation from this symmetrical orientation would support the double cross-slip on parallel {111} planes, which was recognized by Sriram et al. [257]. However, a general problem arises with the association of these two intrinsic mechanisms with the yield-stress anomaly because the same pinning features have been observed in the γ phase of two-phase alloys with the base-line composition Ti-48Al, which do not show a significant yield-stress anomaly (Figure 6.69). Thus, the mechanisms behind the yield-stress anomaly need further investigation.

Several authors have revealed that the glide resistance of ordinary dislocations originates from localized extrinsic obstacles. In TiAl alloys, oxygen, nitrogen and carbon are unavoidable impurity elements, which have a low solubility of about 300 at. ppm in the γ phase [129, 130]. Thus, these elements can form oxides, nitrides and carbides, and these precipitates may act as localized glide obstacles. Kad and Fraser [149] investigated the dislocation structure in binary Ti-52Al containing oxygen levels of either 750 wt. ppm (about 1730 at. ppm) or 1250 wt. ppm (about 2880 at. ppm). The alloys were examined after room-temperature deformation. Ordinary screw dislocations were found to be anchored at fine precipitates, presumably oxides. The authors proposed that this pinning process impedes glide and multiplication of the ordinary dislocations, with the result that the glide activity of these dislocations is strongly reduced in the higher oxygen-containing alloy.

Messerschmidt et al. [258] performed *in situ* straining experiments on Ti-52Al containing an oxygen level of 650 wt. ppm (about 1500 at. ppm) in a high-voltage (1000 kV) electron microscope. The kinematics of dislocation motion observed at room temperature was reminiscent of overcoming localized obstacles (presumably Al_2O_3), which were separated by about 100 nm. The idea of an external pinning process is also supported by the investigations of Zghal et al. [259], who performed TEM *in situ* straining experiments at room temperature in a 200-kV instrument and observed the same characteristics of dislocation dynamics. Localized obstacles with a separation distance of 140 nm were found to impede the screw dislocation motion in a single-phase Ti-54Al alloy and a Ti-49.3Al PST crystal. The authors also observed pinning of twinning partial dislocations, a process that should be attributed to external obstacles because an intrinsic pinning mechanism by cross-slip is difficult to envisage for partial dislocations.

However, there is a general problem with the hypothesis of localized pinning. Impurity-related defects with a spherical shape would only interact with the hydrostatic stress field of the edge dislocations. There is only a very weak interaction of spherical defects with the screw dislocations because their strain field is nearly pure shear. Thus, the question remains, why is it that screw dislocations and not edge dislocations are pinned? So, all in all, one is left with the impression that external pinning by impurity containing precipitates is only significant in single-phase γ alloys or, perhaps, in near-stoichiometric alloys with a low amount of $α_2$ phase. In Ti-rich two-phase alloys, the γ phase is devoid of precipitates because impurity elements exceeding the solubility limit of the γ phase are taken up by the $α_2$ phase. There is, however, one additional factor that should be taken into account. The antisite defects, which accommodate off-stoichiometric deviations in TiAl alloys, may interact in various ways with the dislocations; this mechanism will be considered in the next section. A salient feature of the above-described investigations is that the observed obstacle distances are often considerably larger than the value expected from the activation volume. This gives rise to the speculation that glide resistance in domain I may originate from lattice friction due to the directional bonding of TiAl, as has been proposed in [260]. Atomic calculations by Simmons *et al.* [261] revealed a high friction arising from a Peierls-type mechanism, particularly along the screw and 60°-mixed dislocation directions. In the TEM investigations the presence of lattice friction is suggested by the observation that the dislocations in the unloaded TEM samples are still bowed out in a smooth arc between the obstacles (Figure 6.38). This finding indicates that lattice friction forces occur on all dislocation characters and impede a complete relaxation of the bowed segments into the geometrically shortest configuration between the obstacles. The Peierls stress arises as a direct consequence of the lattice periodicity; the relevant structural unit, the kink, has atomic dimensions. Thus, the activation volume of the Peierls process is very small, typically $V = b^3$ to $20b^3$ [241]. The superposition of lattice friction with any of the above-discussed mechanisms could explain the relatively small activation volumes that have been experimentally determined. The glide resistance arising from intrinsic glide obstacles and lattice friction probably disappears at about 600 K, which is suggested by the small value of $1/V$. Nevertheless, the flow stress $σ$ is practically unchanged. This indicates that in domain II of the $σ(T)$ curve (Figure 6.69) a new friction mechanism appears, which substitutes for that occurring in domain I.

Summary: In γ(TiAl) the velocity of ordinary dislocations is controlled by several mechanisms, which superimpose in certain temperature ranges. However, there are three distinct temperature domains in which a certain mechanism is predominant. At room temperature, an intrinsic glide resistance due to jog dragging or the kink bulging on different octahedral planes determines dislocation glide. Other sources of the glide resistance are lattice friction and, perhaps, localized obstacles.

6.4.3
Static and Dynamic Strain Aging of TiAl Alloys

In domain II of the $\sigma(T)$ curve (550 to 800 K) deformation is characterized by discontinuous yielding and negative strain-rate sensitivity (Figure 6.64). These phenomena are usually associated with dislocation locking due to the formation of defect atmospheres around the dislocations [163]. In Figure 6.64 the serrated yielding is obvious in the load–elongation trace for 623 K, and the negative strain-rate sensitivity is readily seen when the steady-state portion of the load–elongation trace belonging to $\dot{\varepsilon}_1$ is backextrapolated to that of $\dot{\varepsilon}_2$. In disordered metals the formation of atmospheres has been associated with interstitial impurities that have a significant size misfit with the matrix. Solute atoms are drawn toward dislocations because the interaction of their strain fields results in a lowering of the strain energy of the system. The atmospheres exist only for a distinct temperature interval [74]. The condition for atmosphere formation is that the temperature is sufficiently high for defect migration to occur. However, further increasing the temperature tends to tear solute away from the dislocations, since this increases the entropy of the crystal. Thus, at significantly higher temperatures the atmospheres again disappear.

The mechanism is commonly characterized by strain aging experiments. Static strain aging refers to a situation in which a strained specimen is rested in either a stress-free or a stressed condition for a time that is sufficient to allow atmosphere formation and dislocation locking. At certain temperatures and strain rates, point-defect mobility may enable solutes to repeatedly lock dislocations during their motion. Then, yielding and aging occur successively and the stress–strain curve breaks up into serrations (Figure 6.64). This plastic instability is known as dynamic strain aging or the Portevin–LeChatelier effect [163]. It is expected that dynamic strain aging produces a friction drag on moving dislocations with a negative contribution to the total strain-rate sensitivity. According to theory, the diffusion rate of the defects forming the atmospheres must be seen relative to the dislocation velocity or strain rate. At a given temperature, one would expect to see serrations only over a certain range of strain rates, an aspect that is clearly recognizable in Figure 6.64. The load–elongation trace for $T = 623$ K shows serrations only at the higher strain rate $\dot{\varepsilon}_2$ but not at $\dot{\varepsilon}_1$. The higher strain rate apparently produces a dislocation velocity that is appropriate for a repeated locking and unlocking of the dislocations. At high enough temperature the diffusivity of solute atoms is high enough so that the solute atoms and the dislocations move together, thereby requiring no breakaway of the dislocations.

There were early reports about defect atmospheres and yield-point phenomena in TiAl alloys [150, 262–267]; however, only recently has detailed information about the relevant defects been obtained [220, 268–270]. In these studies, dislocation locking by defect atmospheres was investigated using the classical yield-point return technique [271, 272]. The samples were deformed to different levels of prestrain, aged *in situ* on the load frame for certain aging periods t_a, and then retested. This method ensures the specimen alignment and allows the stress level

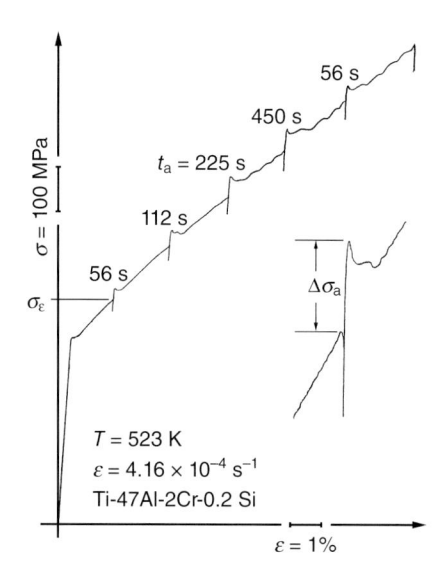

Figure 6.70 Sequence of a strain-aging experiment performed under a relaxing stress on a Ti-48.5Al-0.37C alloy. The stress increments $\Delta\sigma_a$ were measured as the difference in stress before aging and the upper yield point occurring on reloading. Different aging times t_a between unloading and subsequent reloading are indicated. Note the dependence of the stress increments on strain ε.

during aging to be kept constant. The aging of the samples was performed at different stresses σ_a. After yielding, the samples were unloaded to different stresses, which took a few seconds. However, in most cases the samples were aged under a relaxing stress starting from the stress σ_ε of the material at strain ε. On reloading distinct yield points $\Delta\sigma_a$ occurred, after which the original stress–strain curve was retraced, that is, there was no permanent hardening effect. This corresponds to a situation in which pinning of dislocations during the aging period occurred due to atmosphere formation. Upon reloading an extra stress $\Delta\sigma_a$ is required to move the dislocations away from the atmospheres; thus, this stress increment reflects the degree of dislocation locking. Figure 6.70 shows a load–elongation trace of a strain aging experiment and the definition of the aging parameters $\Delta\sigma_a$, t_a and σ_ε.

In Figure 6.71, the alloys investigated are depicted in the binary Ti–Al phase diagram [273] according to their Al content and the temperature of the final heat treatment; this may serve as a crude definition of the alloy constitution. Several alloys contained ternary or quaternary metallic elements. These alloys were chosen in order to study the influence of element partitioning on atmosphere formation. Likewise, alloys were involved that contained a relatively high concentration of oxygen, nitrogen, silicon, and carbon. The extent of strain aging was measured for all these alloys as a function of temperature, aging time, strain and stress.

Figure 6.72 demonstrates on two alloys the dependence of the stress increments $\Delta\sigma_a$ on aging time t_a for two alloys at different temperatures. In these plots only the values estimated at the beginning of deformation, at $\varepsilon = 1.25\%$, were involved

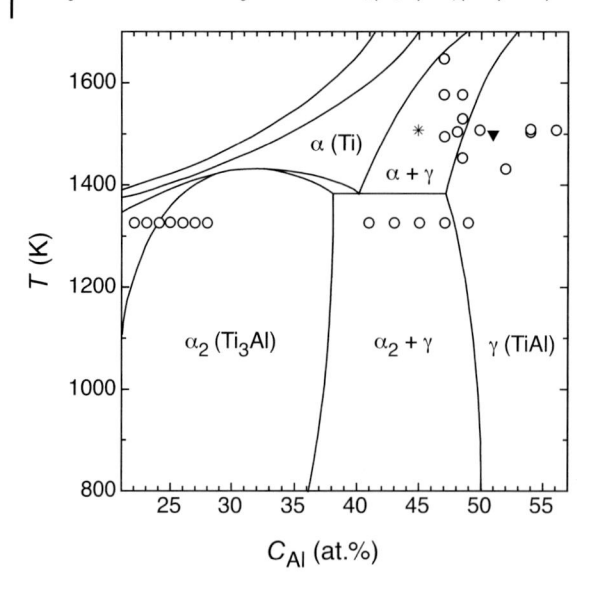

Figure 6.71 Strain aging of TiAl alloys. Constitution of the alloys investigated as defined by the binary Ti–Al phase diagram [273]. For discussion purposes, the two alloys ✶Ti-45Al-10Nb and ▼ Ti-47Al-4Ga are specifically identified.

in order to avoid any influence of the strain dependence of $\Delta\sigma_a$. As indicated by the double-logarithmic plots of the data, the value of the strain age yield point $\Delta\sigma_a$ becomes saturated over the time scale of the experiments leading to distinct saturation values $\Delta\sigma_s$. The kinetics of the process is obviously accelerated by a higher aging temperature; this manifests the fact that atmosphere formation is controlled by the rate at which the relevant defects can diffuse through the lattice. The stress response was found to depend on stress and strain. At a fixed ε, the highest values of $\Delta\sigma_a$ were measured when aging was performed under a relaxing stress. Unloading the sample to a lower aging stress σ_a resulted in a smaller $\Delta\sigma_a$, (Figure 6.73a). $\Delta\sigma_a$ increases almost linearly with strain ε, as shown in Figure 6.73b.

From the experimental data an activation energy can be determined if the time necessary to establish a certain degree of completeness of dislocation pinning and the related temperatures are combined to an Arrhenius plot. This evaluation of the data is shown in Figure 6.74 for the saturation times $t_s(T)$ and the shorter times $t_r(T)$ that correspond to the stress increments $\Delta\sigma_r = 0.8 \times \Delta\sigma_s$. The mean values of the activation energy determined from the slopes of the plots are listed in Table 6.3. For comparison, the self-diffusion energy for Ti in γ(TiAl) is $Q(Ti) = 2.59\,eV$ [221], Section 3.3. Unfortunately, there is no experimental data about the diffusion of Al available because of the lack of suitable radiotracers. The energy calculated from the Darken–Manning equation (using Ti self-diffusion and interdiffusion data) is $Q(Al) = 3.71\,eV$. This relatively high energy is in part compensated by a higher pre-exponential factor; nevertheless there is good evidence that Al is the slower-diffusing species in γ(TiAl). These energies are significantly higher than

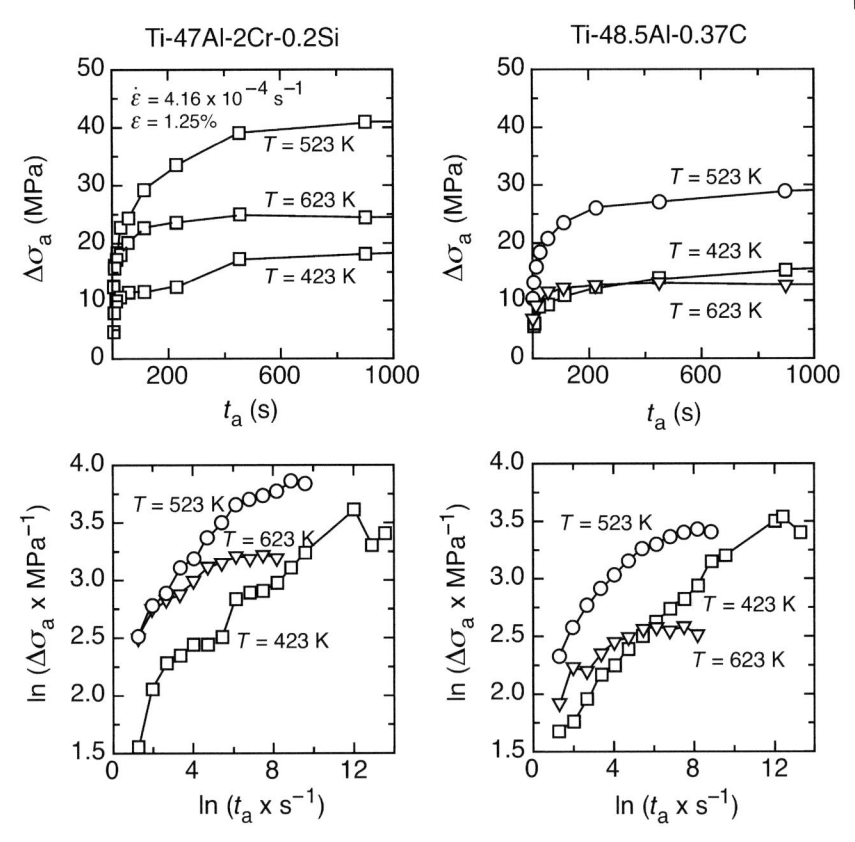

Figure 6.72 Kinetics of strain aging under relaxing stresses and the deformation parameters indicated. Dependence of the stress increments $\Delta\sigma_a$ on the aging time t_a for the alloys indicated. The double-logarithmic plots shown below indicate that the stress increments $\Delta\sigma_a$ become saturated; the saturation times t_s depend on temperature and alloy composition. Data from Fröbel and Appel [220].

Q_a, meaning that the defect transfer onto the mobile dislocations cannot be explained by a conventional vacancy-exchange mechanism. It might be expected that pipe diffusion along dislocation cores occurs. However, the energy for this process is about half that for bulk diffusion [74], that is, at least 1.3 eV. Thus, pipe diffusion is also not consistent with the activation energy measured for strain aging.

The effects of alloy composition on strain aging were assessed by comparing the saturation values $\Delta\sigma_s$ of the stress increments that were determined at $t_a = 7200$ s. $\Delta\sigma_s$ represents the maximum pinning strength of the dislocations that can be achieved under the respective aging conditions and may thus be characteristic of the underlying defect process. The data exhibits a complex variation that can be systematically ranked, however, if the data is plotted against the aluminum content (Figure 6.75). There is a well-expressed maximum of $\Delta\sigma_s$ at an aluminum

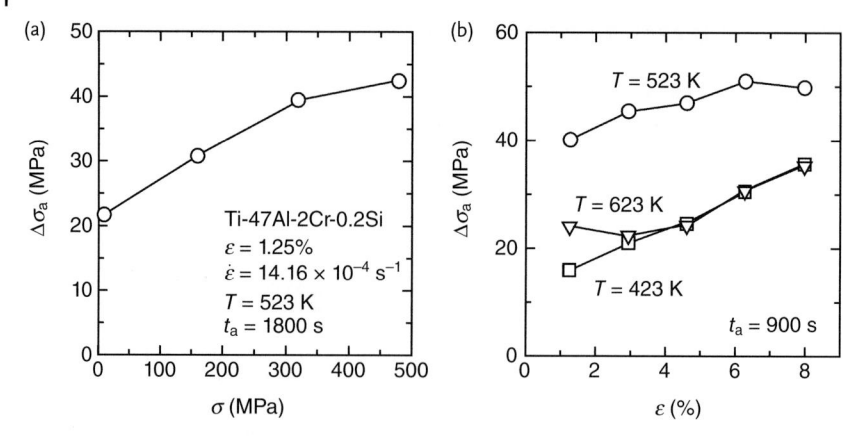

Figure 6.73 Strain-aging characteristics demonstrated on a two-phase alloy with the composition Ti-47Al-2Cr-0.2Si and an equiaxed microstructure. Dependence of the stress increments $\Delta\sigma_a$ (a) on stress σ_a maintained during aging and (b) on strain ε. Data from Fröbel and Appel [220].

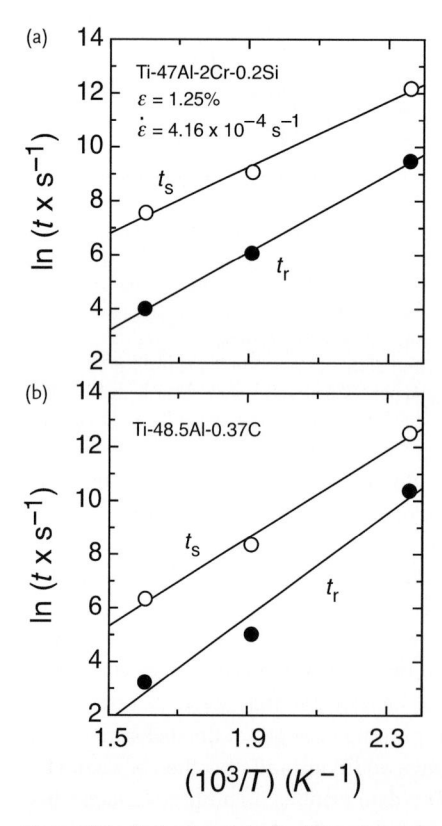

Figure 6.74 Arrhenius plots for determining the activation energy of strain-aging coupling the aging times t_s and t_r with the related temperatures. Data from Fröbel and Appel [220].

Table 6.3 Activation energies Q_a determined for static strain aging of TiAl alloys [220].

Alloy composition (at.%)	Q_a (eV)
Ti-47Al-2Cr-0.2Si	0.58
Ti-48.5Al-0.37C	0.77

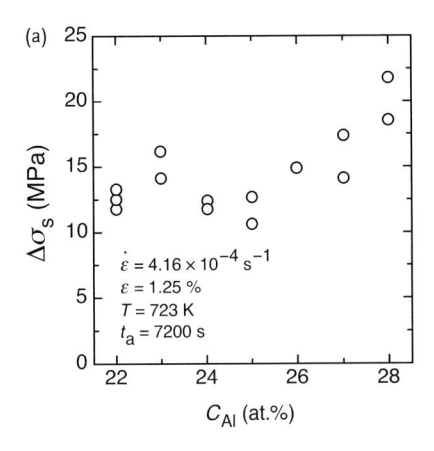

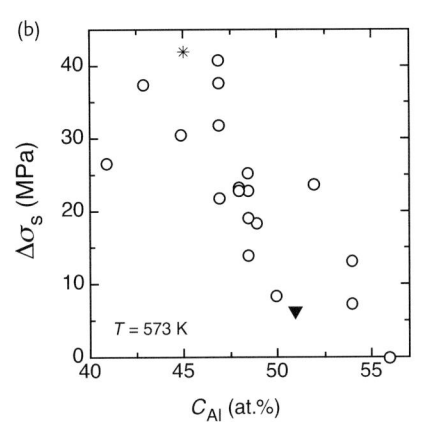

Figure 6.75 Dependence of the saturation values of the saturated stress increment $\Delta\sigma_s$ on the aluminum concentration. (a) α_2(Ti$_3$Al)-based alloys (individual values of $\Delta\sigma_s$). (b) γ(TiAl)-based alloys (mean values of $\Delta\sigma_s$). For discussion purposes, two alloys ∗ Ti-45Al-10Nb, ▼ Ti-47Al-4Ga are specifically identified. Data from Fröbel and Appel [220].

concentration of about 45 at.%, whereas the value is very small at the stoichiometric TiAl composition (Figure 6.75b). In case of the α_2 alloys, $\Delta\sigma_s$ increases on both sides of the stoichiometric composition. The results clearly indicate that off-stoichiometric deviations might be most important for the aging behavior of TiAl alloys. In this context also the effect of ternary or higher alloying elements on strain aging was investigated by comparing the microalloyed samples with binary alloys having the same Al content. No significant influence of the alloying elements Cr, Mn, Nb, Ga, C, and Si was recognized; the same holds for the interstitial impurities O_2 and N_2.

At first, the situation of Ti-rich γ alloys will be considered where the aging effect is strongest. The off-stoichiometric deviations of these alloys are largely accommodated by the formation of α_2 phase according to the phase diagram. However, in spite of the phase separation, a small portion of excess Ti is probably taken up by the γ phase. The essence of the arguments is that the phase separation occurring upon cooling and described by the phase diagram often cannot be established within the constraints of processing routes because the $\alpha/\alpha_2 \rightarrow \gamma$ transformation is sluggish [220]. It might be expected that the Ti concentration in the γ phase increases when the alloy becomes richer in Ti until the maximum solubility of Ti in the γ phase is reached. At the eutectoid point of the binary system the solubility of excess Ti is about 3 at.% [220]. There is good evidence that off-stoichiometric deviations in TiAl alloys are accommodated by the formation of Ti$_{Al}$ antisite defects; these are Ti atoms situated on the Al sublattice [41]. In view of the above considerations the concentration of the Ti$_{Al}$ antisite defects is expected to be in the order of 10^{-2}. First-principles calculations have shown [274] that antisite defects exhibit a lower symmetry than the γ(TiAl) matrix and are thus associated with noncentrosymmetric distortions. For a Ti$_{Al}$ antisite defect the predicted size misfits are [274]

$$\frac{1}{a}\frac{da}{dc_{AS}} = 2\times10^{-3} \text{ along the } \mathbf{a} \text{ direction} \tag{6.40}$$

and

$$\frac{1}{c}\frac{dc}{dc_{AS}} = 5\times10^{-3} \text{ along the } \mathbf{c} \text{ direction} \tag{6.41}$$

of the TiAl unit cell. c_{AS} is the concentration of the antisite defects. Because of this asymmetrical distortion the antisite defects are expected to interact with both hydrostatic and shear stress fields. This is important with regard to the dislocation glide resistance discussed in the previous section because the Ti$_{Al}$ antisite defects will interact with all dislocation characters, that is, also with screw dislocations. Association of the Ti$_{Al}$ antisite defects with vacancies leads to the formation of antistructural bridges that provide paths of easy diffusion, as previously proposed for B2 compounds [275, 276]. It is tempting to speculate that the defect agglomerate consisting of an antisite atom and a vacancy will produce an even stronger asymmetric distortion. There is a multiplicity of crystallographically equivalent vacancy positions around the antisite defect that gives rise to a variety of strain-field orientations of the composite defect. If such a defect interacts with the stress field of a dislocation, vacancy jumps may occur that reorient the vacancy–antisite complex. Herzig et al. [277] and Mishin and Herzig [221] have proposed antistructural bridge (ASB) mechanisms that are compatible with the L1$_0$ structure (see Section 3.3). The most favorable site of the vacancy is probably the one that lowers the dislocation strain energy over the defect volume, and thus causes a bonding between the dislocation and the defect. Figure 6.76 illustrates the reorientation of a vacancy–antisite defect adjacent to an ordinary dislocation via the so-called ASB-2 mechanism. It is important to note that only a few atomic jumps accomplish the reorientation of the defect complex and that, in particular, the process requires no long-range diffusion. Locking of the dislocations, therefore, occurs in a fashion that has been described for Snoek atmospheres [278].

The activation energy Q_{ASB} of the ASB diffusion mechanisms is determined by the formation energies E_f of the relevant point defects (that of the antisite atom and that of the vacancy) plus an effective migration energy E_m

$$Q_{ASB} = E_f + E_m \tag{6.42}$$

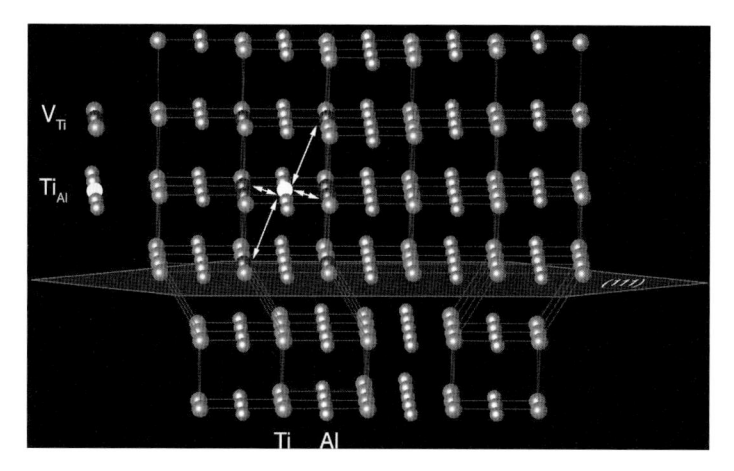

Figure 6.76 Possible orientations of the vacancy–antisite defect complex. The figure shows an ordinary edge dislocation in a projection near to <11$\bar{2}$>. In the dislocation stress field the complex may reorient via the antistructural bridge mechanism ASB-2 described in the text. Ti$_{Al}$ – Ti atom situated on an Al site, V$_{Ti}$ – vacancy on the Ti sublattice.

Specifically, the energy contributions for the Ti and Al diffusion via the ASB-2 mechanism are [221]

$$Q_{ASB2}(Ti) = E_f(Ti_{Al}) + E_f(V_{Ti}) + E_m \tag{6.43}$$

and

$$Q_{ASB2}(Al) = E_f(Al_{Ti}) + E_f(V_{Al}) + E_m \tag{6.44}$$

For the dislocation-locking mechanism considered here, E_m is the relevant energy. Using molecular dynamic calculations, Mishin and Herzig [221] have determined the migration energies of the different diffusion mechanisms in γ(TiAl) involving sublattice self-diffusion, sublattice antisite diffusion, three-jump cycles, six-jump cycles, and the antistructural bridge mechanisms considered here (see Section 3.3). Among all these mechanisms, the ASB-2 process (Equation 6.43) has by far the lowest migration energy for Ti diffusion

$$E_m = 0.712 \, eV \tag{6.45}$$

This migration energy is in reasonable agreement with the estimated activation energy for strain aging. It is important to note that Al diffusion by the analogous ASB-2 mechanism (Equation 6.44) requires significantly higher activation energy of

$$E_m = 1.323 \, eV \tag{6.46}$$

There is apparently no other mechanism in TiAl that provides a significantly easier diffusion path for Al [221]. The difference in the ASB migration energies of Ti and Al could rationalize that age hardening in Al-rich alloys is practically absent. The hypothesis of the atmosphere formation by Ti$_{Al}$ antisite defects is further supported by the data of the Ti-47Al-4Ga alloy (Figure 6.75b). Ga is known to exclusively occupy Al sites [279] and is therefore utilized in diffusion studies as an Al-substituting element. Given the site occupancy of Ga, Al_{Ti} (or Ga_{Ti}), antisite defects should be formed in Ti-47Al-4Ga. Thus, in spite of its low Al content, Ti-47Al-4Ga should exhibit similar defect characteristics as a binary Ti-51Al alloy. In confirmation of this hypothesis, as with Al-rich binary alloys, very small stress increments were determined on Ti-47Al-4Ga (Figure 6.75b). Likewise, the high Nb-containing alloy (Ti-45Al-10Nb) fits the trend shown in Figure 6.75b. Nb solely occupies the Ti sublattice [279] with small size misfit [274]. Thus, the alloy should behave like Ti-45Al, which has been indeed observed. Taken together, these observations give supporting evidence that the strain-aging phenomena in Ti-rich TiAl alloys are closely related to a diffusion mechanism involving antistructural bridges. The drag of the atmospheres on moving dislocations produces an additional friction stress, which has its maximum at about 500 K and then disappears. This additional stress part may substitute for the low-temperature glide resistance arising from intrinsic and extrinsic obstacles, which disappear at 600 K, as has been discussed in the previous section. Thus, the presence of Ti_{Al} antisite defects may explain why no yield-stress anomaly is observed in Ti-rich alloys.

The observation made on the Ti_3Al-based alloys, that is, the increase of the strain increments $\Delta\sigma_s$ at off-stoichiometric compositions (Figure 6.75a), seems to support the hypothesis of age hardening by antisite defects. Off-stoichiometric deviations in these alloys are also accommodated by antisite defects [41], and several diffusion paths via antistructural bridges have been proposed in addition to conventional vacancy-exchange mechanisms [221]. For the Al-rich side an ASB-2 mechanism analogous to Equation (6.44) has been proposed that requires an activation energy of $E_m = 0.761$ eV. For the Ti-rich side an ASB mechanism seems not to exist; however, other conventional diffusion mechanisms occur that have comparably low migration energies. This could explain why the stress increments $\Delta\sigma_s$ on the Al-rich side increase more strongly than on the Ti-rich side.

In the classical theories of defect accumulation at dislocations the resulting stress increments $\Delta\sigma_a$ are expected to increase proportionally with t_a^n, with the time exponent $n = 2/3$ for bulk diffusion and $n = 1/3$ for pipe diffusion [74]. The kinetics determined in the present study clearly deviates from these laws as can be seen from the double-logarithmic plots (Figure 6.72). In terms of the locking mechanism described earlier, the atmosphere formation is governed by vacancy jumps in the stress field of the dislocations, resulting in a reorientation of the antisite/vacancy complexes. In an attempt to describe the kinetics of the process it was assumed that the rate of defect reorientation dN/dt_a is proportional to the number N of defects that have not yet reoriented. Also, the assumption was made that the stress increments $\Delta\sigma_a(t_a)$ are proportional to the number of reoriented antisite/vacancy complexes. The evolution of $\Delta\sigma_a$ with t_a is then given by [280]

$$\Delta\sigma_a = \Delta\sigma_s \{1 - \exp(-t_a/t_0)\} \tag{6.47}$$

t_0 is the time corresponding to the stress increment $\Delta\sigma_a(t_0) = (1-e^{-1})\Delta\sigma_s$. There are mainly two shortcomings of this model. First, the reorientation rate of an antisite/vacancy complex certainly depends on its distance from the dislocation, and sec-

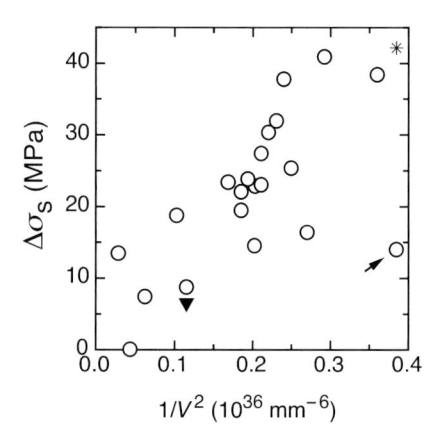

Figure 6.77 Dependence of the saturated aging stress $\Delta\sigma_s$ on the square of the reciprocal activation volume $1/V^2$ according to Equation (6.49). For symbols see Figure 6.71. The arrowed data point concerns a quenched Ti-48.5Al-0.37C alloy, which probably contained a high density of thermal vacancies, see accompanying text. Data from Fröbel and Appel [220].

ondly, the dislocation stress field is annulled by the reorientation of the complex. These two effects probably make the kinetics sluggish. Comparison with the experimental data suggests that the expression

$$\Delta\sigma_a = \Delta\sigma_s \{1 - \exp(-t_a/t_0)\}^{1/4} \quad (6.48)$$

could be a versatile description of the kinetics for the formation of ordered atmospheres.

As mentioned above, the antisite atoms generate a noncentrosymmetric distortion of the TiAl matrix, which gives rise to elastic interactions with all dislocation characters. At low temperatures, where atmosphere formation is not possible, the antisite atoms may act as localized obstacles. The structural parameters described by Equations (6.40) and (6.41) and the high concentration of the antisite defects suggest that a dense ensemble of relatively weak obstacles is formed. If this is so, then the aging phenomena should correlate with the low-temperature deformation characteristics. The aging stress increments are proportional to the concentration c_{AS} of the antisite defects. If, at lower temperatures, the same defects act as localized obstacles, then they will probably be overcome with the aid of thermal activation, which should be indicated in the activation volume. For this short-range interaction only those defects in the planes immediately adjacent to the slip plane have to be considered, thus, as with the relevant obstacle distance, V is proportional to $c_{AS}^{-1/2}$. This leads to the expression

$$\Delta\sigma_s = A/V^2 \quad (6.49)$$

A is a constant. As demonstrated in Figure 6.77, the present data is reasonably consistent with this interpretation. The deviation of the arrowed data point from the general trend concerns the alloy Ti-48.5Al-0.37C, which had been annealed at 1250 °C followed by water quenching. This thermal treatment resulted in a high concentration of carbon in solid solution leading to a small activation volume without affecting $\Delta\sigma_s$.

In the previous Section 6.4.2 the factors governing the glide resistance have been discussed; these involve lattice friction, jog dragging and kink collision, and perhaps extrinsic pinning by impurity related defects. The results presented in this section give supporting evidence to the belief that a significant part of the low-temperature glide resistance arises from antisite defects. This localized dislocation pinning is expected for both Ti_{Al} and Al_{Ti} antisite defects; thus, it is of concern in all off-stoichiometric alloys.

From the technical point of view, the formation of defect atmospheres is important because it provides fast and effective dislocation locking, which could be harmful for the materials performance. For example, the well-known blue brittleness of steels has been associated with such effects. In TiAl alloys it has been recognized that the tensile ductility is at a minimum when the samples are tested in the temperature range over which the defect atmospheres are formed [270]. The formation of defect atmospheres seems also to affect the fracture and fatigue of TiAl alloys, problems that will be addressed in Chapters 10 and 11.

Summary: Deformation of Ti-rich alloys in the intermediate temperature interval of 450 K to 750 K (domain II of the $\sigma(T)$ curve) (Fig. 6.69) is characterized by discontinuous yielding, negative strain-rate sensitivity and strain-aging phenomena. A fast and effective dislocation locking due to oriented defect atmospheres causes these attributes. The relevant defects are Ti_{Al} antisite atoms associated with Ti vacancies, which can easily reorient by vacancy jumps in the dislocation stress field.

6.4.4
Diffusion-Assisted Dislocation Climb, Recovery, and Recrystallization

Domain III of the $\sigma(T)$ curve (Figure 6.69) involves the transition from brittle to ductile material behavior (BDT). Decreasing stress and increasing reciprocal activation volume characterize deformation in this domain. The activation parameters determined at 910 and 1100 K are listed in Table 6.2. A great deal of scatter is present in the data determined at 910 K; this probably reflects the fact that the BDT transition temperature depends on the alloy composition and microstructure. The new thermally activated process seems to be established at 1100 K, as indicated by the relatively uniform activation parameters. For most alloys the activation enthalpy is close to $\Delta H = 3$ eV; for comparison the self-diffusion energy for Ti is $Q_{sd}(Ti) = 2.59$ eV, and that for Al is expected to be $Q_{sd}(Al) = 3.71$ eV [221]. This is indicative of a diffusion-assisted climb mechanism because climb of an edge dislocation depends on the diffusion of vacancies either towards the edge dislocation or away from it. Deformation by climb requires both elemental components in the alloy to be mobile; otherwise large gradients of the chemical potential are set up. Thus, the measured ΔH values probably represent an average of the Ti and Al self-diffusion energies. Kad and Fraser [149] observed climb of ordinary 1/2<110> dislocations in a single-phase Ti-52 Al alloy that had been deformed at elevated temperatures. Stereo-pair observations performed on a high strength Ti-45Al-5Nb

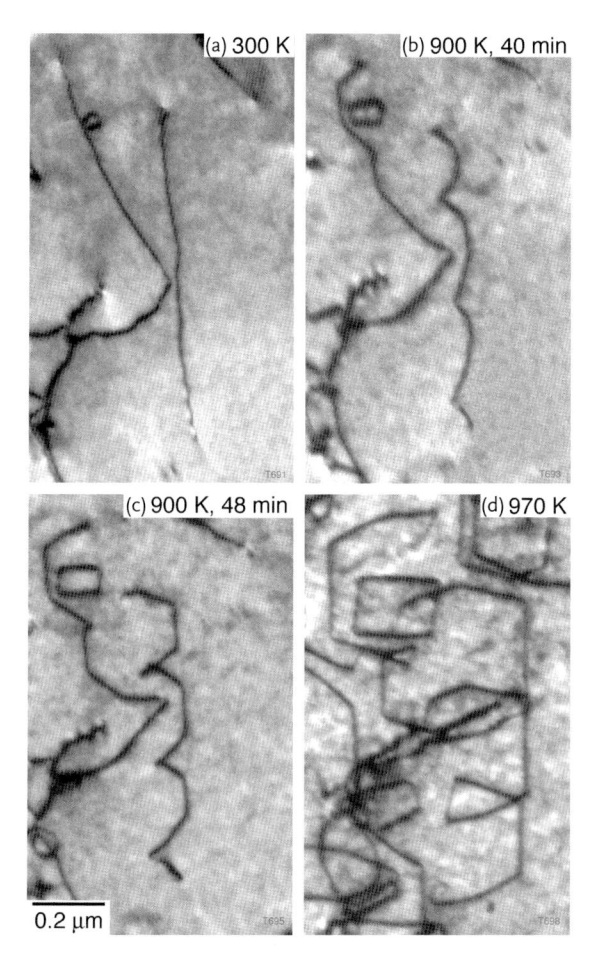

Figure 6.78 Deformation behavior in domain III of the $\sigma(T)$ curve (Figure 6.69). Climb of ordinary dislocations during *in situ* heating inside the TEM at an acceleration voltage of 120 kV. Note the formation of a helical dislocation configuration with segments of preferred <110]line orientations. Cast Ti-48Al-2Cr, predeformation at room temperature in compression to $\varepsilon = 3\%$. Micrographs from Appel and Wagner [75].

alloy deformed at 973 K have shown that ordinary dislocations are bowed out and not confined to a single slip plane, which was associated with climb [203]. Supporting evidence for climb has also been obtained from *in situ* heating experiments performed inside the TEM [126]. The samples used were predeformed at 300 K to a strain of $\varepsilon = 3\%$; this introduced sufficient dislocations for observation and certainly a small supersaturation of intrinsic point defects. Climb was exclusively observed on 1/2<110] dislocations, which is plausible because these dislocations have a compact core structure. As demonstrated in Figure 6.78, during the *in situ* experiment the dislocation form helical structures, which are characteristic of climb. The climb structure involves straight segments with <110] orientation; this is suggestive that at elevated temperatures a small nonplanar dissociation of these dislocation may occur, perhaps on intersecting {111} planes. Additional energy barriers are likely to impede climb of dissociated dislocations [74].

There is one major difficulty with the climb hypothesis, namely that the activation volumes of $V = 47b^3$ to $73b^3$ are too large for climb. From theory [238, 239, 241, 242] dislocation climb is associated with $V = 1b^3$ to $10b^3$. There are mainly three reasons that may account for this discrepancy. First, the activation volume was calculated according to Equation 6.34 assuming a Taylor factor of $M_T = 3.06$. However, at elevated temperatures, strain accommodation between differently oriented grains can in part be accomplished by climb of dislocations out of their {111} planes, which tend to reduce the Taylor factor and thus the values of V [281]. Secondly, the nonplanar dissociation mentioned above could give rise to a larger volume of the activation complex. Thirdly, it is also possible that deformation is primarily accomplished by slip, but the factor controlling the amount of slip is the climb of dislocations about strong obstacles. The relative contributions of the three processes certainly depend on the deformation conditions $\dot{\varepsilon}$ and T. At higher strain rates dislocation glide certainly significantly contributes to deformation. Thus, the total activation volume could reflect the contributions of both glide and climb. Concerning this complex situation, it is difficult to make generalizations about the ability of dislocations to climb in TiAl. Apart from the deformation conditions T, $\dot{\varepsilon}$ and ε, careful attention has to be paid in each case to other climb-controlling factors; these may involve alloy chemistry, constitution, microstructure, and impurity content. However, it is reasonable to assume that dislocation climb in the γ phase of ($\alpha_2 + \gamma$) alloys is fully established at $T = 1200$ K and $\dot{\varepsilon} \leq 10^{-4}$ s^{-1} Deformation at higher temperatures seems to be progressively governed by dynamic recovery and recrystallization (domain IV in Figure 6.69). These processes will be discussed in Chapter 16.

Summary: Starting at the transition from brittle to ductile material behavior, climb of ordinary dislocations becomes a prominent deformation mechanism in TiAl alloys.

6.5
Thermal and Athermal Stresses

As has been discussed in Section 6.4.1, the net stress τ^* to produce slip is the difference between the total shear stress τ and the long-range internal stress τ_μ. Knowledge about the relative contributions of these two stress parts is not only important for the identification of the factors governing the dislocation mechanisms, but also for the implementation of stable microstructures to tailor material properties. Several methods for the determination of internal stresses have been proposed, which have been reviewed by Kruml et al. [282]. Most of these methods are difficult to apply for TiAl alloys because of their complex stress–temperature profile. It is only recently that such information is available [246], and this will be reported in the present section.

The method used in this study is known as stress-reduction testing. The underlying concept is that the internal backstress τ_μ is generated during deformation

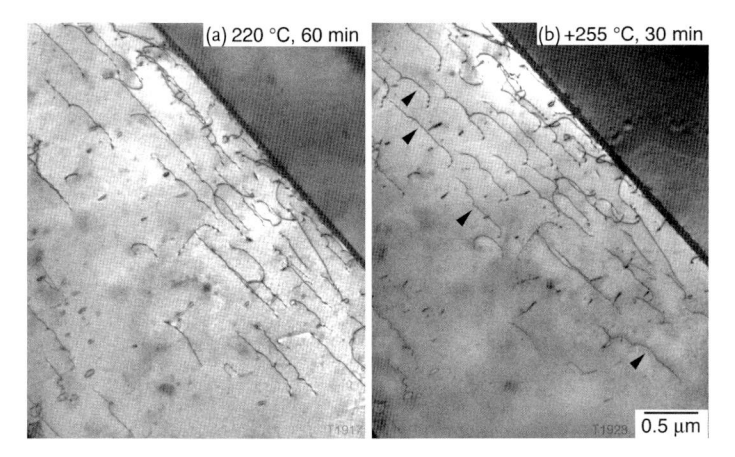

Figure 6.79 Ordinary dislocations piled up at a lamellar interface. A few of the new dislocations approaching the interface were marked with arrowheads in (b). The dislocations probably propagated under the action of thermal stresses that were released during *in situ* heating of the predeformed sample inside the TEM from 220 °C to 255 °C. Ti-48Al-2Cr, predeformed at room temperature to strain $\varepsilon = 3\%$.

because mobile dislocations are piled up at strong obstacles, like grain boundaries or precipitates. Under the action of the external shear stress in the slip direction, a first dislocation will propagate on its slip plane in the preferred slip direction. If this dislocation is stopped at a barrier, a second dislocation, following the same path, will encounter the pinned first dislocation and be repelled by elastic interaction forces. This gives rise to a backstress τ_μ, which increases as the number of dislocations in the pile up increases. In a sequence of two micrographs, part of an *in situ* study, Figure 6.79 shows ordinary dislocations that were successively piled up at a lamellar interface.

The backstress is opposite to the applied shear stress and must be overcome for slip to continue. On the other hand, since the net stress on a slip plane is the difference between the shear stress and the internal backstress, slip in the reverse direction can take place if the applied stress is reduced or removed. This anticipated dislocation behavior was used to determine the internal stress level. The investigations were performed in a strain-controlled closed-loop machine. During a constant strain-rate test the deformation was stopped and the sample was immediately unloaded to a certain stress level τ_x. The resulting flow behavior of the sample after unloading was monitored in the strain-controlled relaxation regime of the machine, meaning the total sample elongation was kept constant during the relaxation period. Application of the method requires a precise measurement of small stress rates, which in turn, requires good thermal stability and accurate strain control. In order to increase the signal-to-noise ratio, tensile tests were performed with a relatively large gage length of 20 mm. Figure 6.80 demonstrates the complex interaction of the sample and the deformation machine in a stress-reduction test. The sample was unloaded with a large decrement, which resulted in a backward motion of the dislocations and a corresponding inelastic

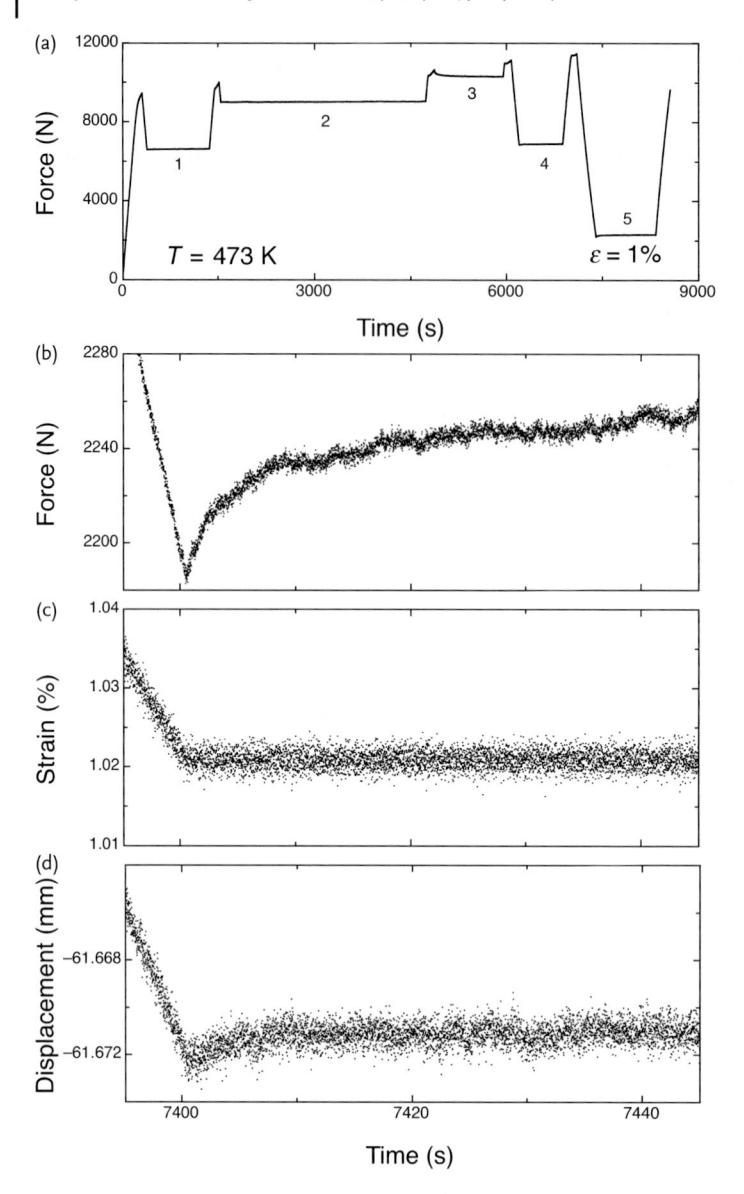

Figure 6.80 Deformation kinematics occurring during a strain-controlled stress-reduction test. Nearly lamellar Ti-45Al-8Nb-0.2C, deformation at 473 K. (a) Load-time schedule of the whole experiment. After plastic tensile deformation of $\varepsilon = 1\%$ the sample was unloaded with a large decrement to about 20% of the yield stress. The subsequent relaxation kinematics (marked with 5 in (a)) of the sample and the machine is shown in Figures. (b), (c) and (d). Unloading the sample resulted in an anomalous increase of the force as shown in (b) and a small inelastic compression of the sample. The adequate countermotion of the machine crosshead shown in (d) substitutes for this inelastic compression. In so doing, the total (elastic plus plastic) sample strain was kept constant (c). Data from Hoppe and Appel [246].

compression of the sample (Figure 6.80a). As shown in Figure 6.80c, an adequate countermotion of the crosshead compensates for this inelastic compression. It is understood that in the strain-controlled regime, the total strain measured on the sample gage has to remain constant (Figure 6.80b). Thus, as a measure of the experimental quality, the relaxation kinetics of the force acting on the sample was always registered together with the sample elongation and the displacement of the machine crosshead.

Figure 6.81 shows the complete schedule of such an experiment. The relaxation behavior occurring at different stress levels is demonstrated in Figure 6.82. At high

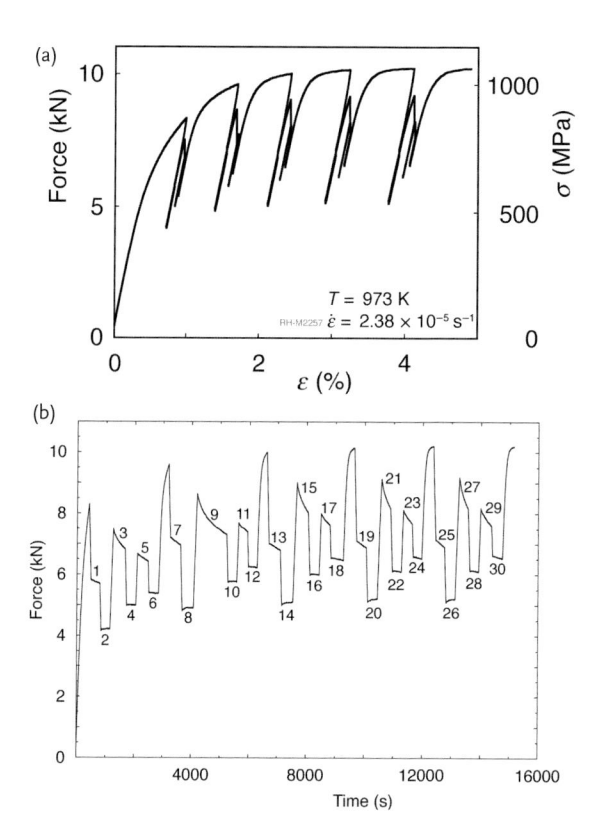

Figure 6.81 Stress-relaxation experiment for determining internal stresses; tensile test performed on nearly lamellar Ti-45Al-8Nb-0.2C alloy at 973 K. (a) Schedule of the experiment shown as load–elongation trace. (b) Load–time trace of the experiment; After yielding of the sample, deformation was stopped and the sample was immediately unloaded to different stresses. The subsequent stress relaxations (indicated by numbers) were recorded in the strain-controlled regime. Normal and anomalous relaxations occurred depending on the magnitude of the unloading stress decrement. In between, at a critical unloading decrement, zero stress rate occurs. Iteration of the loading and unloading procedures helped to identify the critical decrement. Figure 6.82 shows the kinetics of relaxations 3, 4, 6, and 8 in more detail. Data from Hoppe and Appel [246].

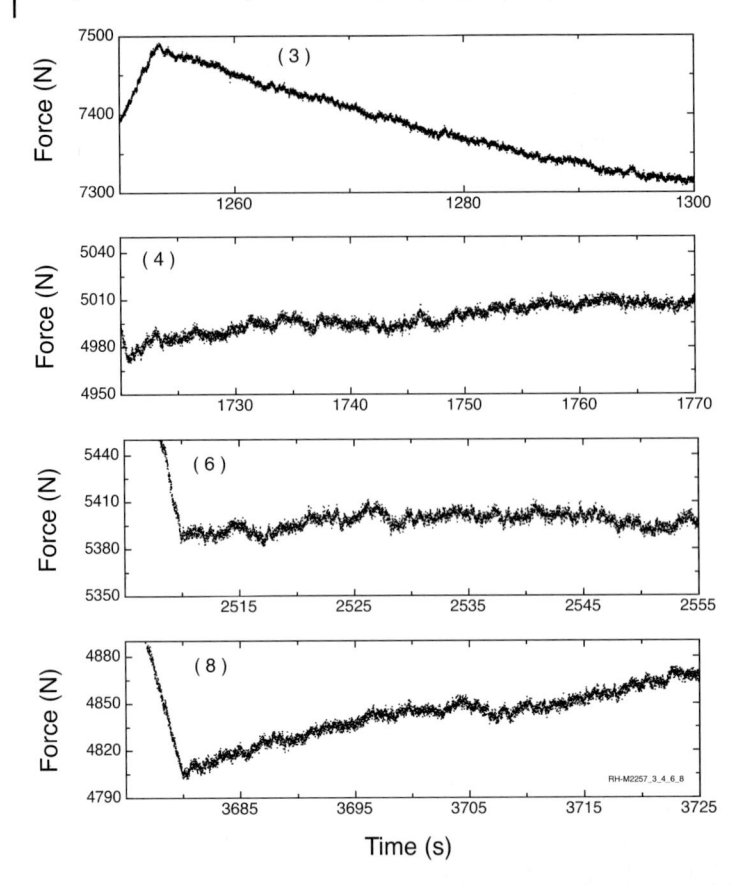

Figure 6.82 Load–time traces observed after different unloading decrements. The numbers refer to the relaxations designated in Figure 6.81. Note the "normal" stress decay after a small unloading decrement (relaxation 3) and anomalous increase of the stress at large load decrements (relaxations 4 and 8). In between, the stress rate is almost zero (relaxation 6); the related stress is considered to be the internal stress level. Nearly lamellar Ti-45Al-8Nb-0.2C. Data from Hoppe and Appel [246].

stress levels, normal stress relaxation occurred; the stress decayed with time. After unloading to a low stress level anomalous relaxation was observed that is manifested by an increase of the stress with time. In between, after a certain critical unloading decrement, zero stress rate occurs. It is then argued that in the material after the critical stress decrement the applied shear stress τ exactly balances the internal backstress τ_μ. Figure 6.83 shows the temperature profiles of the total (engineering) stress $\sigma_{0.2}$ and of the internal stress σ_μ that have been determined on an extruded Ti-45Al-8Nb-0.2C. The repeated relaxations, which are needed for the measurement of σ_μ, naturally consume strain, thus the σ_μ values given in

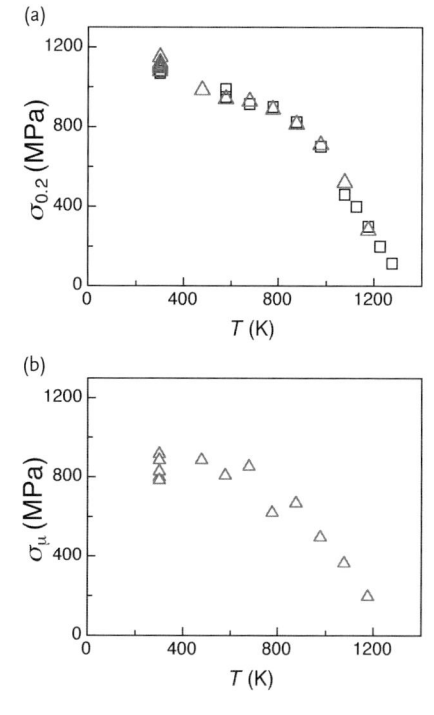

Figure 6.83 Temperature profiles of the yield stress $\sigma_{0.2}$ and of the internal (engineering) stress σ_μ determined by incremental unloading. Nearly lamellar Ti-45Al-8Nb-0.2C. Data from Hoppe and Appel [246].

Figure 6.83 correspond to about 1% strain. At room temperature the internal stress is found to be

$$\sigma_\mu(\varepsilon = 1\%) = 0.8\, \sigma_{0.2} \tag{6.50}$$

This corresponds to a relatively small effective shear stress of $\tau^* = (\sigma_{0.2} - \sigma_\mu)/M_T = 80\,\text{MPa}$. $M_T = 3.06$ is the Taylor factor. It is worth adding that the relation between σ_μ and $\sigma_{0.2}$ described by Equation (6.50) was also observed for other alloys. As can be recognized from Figure 6.83 the relative contribution of σ_μ decreases with temperature. More about internal stresses is provided in [246] and in Sections 7.1 and 7.2.

Summary: Estimations of the internal stresses performed on TiAl alloys by the incremental unloading technique suggest a relatively high athermal stress contribution, which at room temperature accounts for about 80% of the total yield stress. The temperature profiles of the yield stress and of the internal stress are similar, however the relative contribution of the internal stress decreases with temperature. Figure 6.84 is a graphical representation of the mechanisms determining the yield stress in two-phase γ alloys, which have been discussed in Sections 6.4 and 6.5. The intention of the diagram is to express that a substantial part of the yield stress

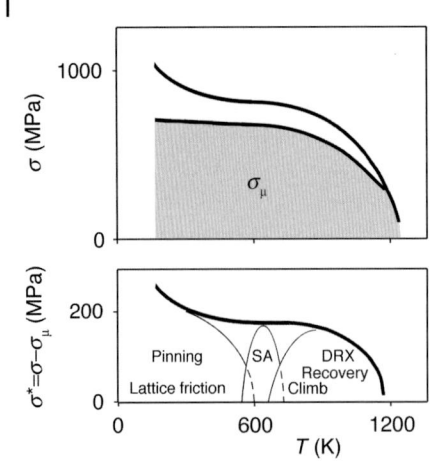

Figure 6.84 Possible mechanisms contributing to the yield stress of two-phase titanium aluminide alloys. The situation sketched in the figure corresponds to high Nb-containing alloys with the base-line composition Ti-45Al-(5-10)Nb. Note the high athermal contribution σ_μ to the total yield stress σ. σ_μ mainly results from dislocation and twin interactions with internal boundaries and basically reflects the fine microstructure of the alloys. $\sigma^* = \sigma - \sigma_\mu$ is the effective (normal) stress. SA (strain aging) refers to the stress contribution resulting from defect atmospheres consisting of Ti_{Al} antisite defects and vacancies. DRX designates dynamic recrystallization.

is generated by long-range interactions of the dislocations with the various internal boundaries constituting the microstructure. The mechanisms controlling the dislocation velocity relieve one another so that their total stress contribution of about 20% of the yield stress remains constant over a relatively wide temperature range.

References

1 Kim, Y.-W. and Dimiduk, D.M. (1991) *J. Met.*, **43**, 40.
2 Lehoczky, S.L. (1978) *J. Appl. Phys.*, **49**, 5479.
3 Bunshah, R.F., Nimmagadda, R., Doerr, H.J., Movchan, B.A., Grechanuk, N.I., and Dabizha, E.V. (1980) *Thin Solid Films*, **72**, 261.
4 Kelly, A. (1987) *Philos. Trans. R. Soc. A*, **322**, 409.
5 Was, G.S. and Foecke, T. (1996) *Thin Solid Films*, **286**, 1.
6 Fu, C.L. and Yoo, M.H. (1997) *Scr. Mater.*, **37**, 1453.
7 Chalmers, B. and Gleiter, H. (1971) *Philos. Mag.*, **23**, 1541.
8 Yamaguchi, M., Inui, H., and Ito, K. (2000) *Acta Mater.*, **47**, 307.
9 Zghal, S., Thomas, M., Naka, S., Finel, A., and Couret, A. (2005) *Acta Mater.*, **53**, 2653.
10 Zghal, S., Thomas, M., and Couret, A. (2005) *Intermetallics*, **13**, 1008.
11 Lu, L., Siegl, R., Girshick, A., Pope, D.P., and Vitek, V. (1996) *Scr. Mater.*, **34**, 971.
12 Mahon, G.J. and Howe, J.M. (1990) *Metall. Trans. A*, **21A**, 1655.
13 Inui, H., Nakamura, A., Oh, M.H., and Yamaguchi, M. (1991) *Ultramicroscopy*, **39**, 268.

14 Yamaguchi, M. and Umakoshi, Y. (1990) *Prog. Mater. Sci.*, **34**, 1.
15 Yang, Y.S. and Wu, S.K. (1990) *Scr. Metall. Mater.*, **24**, 1801.
16 Yang, Y.S. and Wu, S.K. (1991) *Scr. Metall. Mater.*, **25**, 255.
17 Yang, Y.S. and Wu, S.K. (1992) *Philos. Mag. A*, **65**, 15.
18 Inui, H., Oh, M.H., Nakamura, A., and Yamaguchi, M. (1992) *Philos. Mag. A*, **66**, 539.
19 Inui, H., Nakamura, A., Oh, M.H., and Yamaguchi, M. (1992) *Philos. Mag. A*, **66**, 557.
20 Kad, B.K., Hazzledine, P.M., and Fraser, H.L. (1994) *Defect Interface Interactions, Materials Research Society Symposium Proceedings*, vol. 319 (eds E.P. Kvam, A.H. King, M.J. Mills, T.D. Sands, and V. Vitek), MRS, Pittsburgh, PA, p. 311.
21 Hazzledine, P.M. and Kad, B.K. (1995) *Mater. Sci. Eng. A*, **192–193**, 340.
22 Denquin, A. and Naka, S. (1996) *Acta Mater.*, **44**, 343.
23 Zghal, S., Naka, S., and Couret, A. (1997) *Acta Mater.*, **45**, 3005.
24 Dey, S.R., Hazotte, A., and Bouzy, E. (2009) *Intermetallics*, **17**, 1052.
25 Denquin, A. and Naka, S. (1993) *Philos. Mag. Lett.*, **68**, 13.
26 Ricolleau, C., Denquin, A., and Naka, S. (1994) *Philos. Mag. Lett.*, **69**, 197.
27 Siegl, R., Vitek, V., Inui, H., Kishida, K., and Yamaguchi, M. (1997) *Philos. Mag. A*, **75**, 1447.
28 Kishida, K., Inui, H., and Yamaguchi, M. (1998) *Philos. Mag. A*, **78**, 1.
29 Dimiduk, D.M. and Hazzledine, P.M. (1995) *High-Temperature Ordered Intermetallic Alloys VI, Materials Research Society Symposia Proceedings*, vol. 364 (eds J.A. Horton, I. Baker, S. Hanada, R.D. Noebe, and D.S. Schwartz), MRS, Pittsburgh, PA, p. 145.
30 Zghal, S., Thomas, M., Naka, S., and Couret, A. (2001) *Philos. Mag. Lett.*, **81**, 537.
31 Chen, S.H., Schumacher, G., Mukherji, D., Frohberg, G., and Wahi, R.P. (2002) *Scr. Mater.*, **47**, 757.
32 Singh, J.B., Molénat, G., Sundararaman, M., Banerjee, S., Saada, G., Veyssière, P., and Couret, A. (2006) *Philos. Mag.*, **86**, 2429.
33 Appel, F., Beaven, P.A.B., and Wagner, R. (1993) *Acta Metall. Mater.*, **41**, 1721.
34 Jin, Z. and Gray, G.T., III (1997) *Mater. Sci. Eng. A*, **231**, 62.
35 Liang, W. (1999) *Scr. Mater.*, **40**, 1047.
36 Dey, S., Hazotte, A., and Bouzy, E. (2006) *Philos. Mag.*, **86**, 3089.
37 Umakoshi, Y. and Nakano, T. (1992) *ISIJ Int.*, **32**, 1339.
38 Umakoshi, Y. and Nakano, T. (1993) *Acta Metall. Mater.*, **41**, 1155.
39 Rao, S., Woodward, C., and Hazzledine, P. (1994) *Defect Interface Interactions, Materials Research Society Symposium Proceedings*, vol. 319 (eds E.P. Kvam, A.H. King, M.J. Mills, T.D. Sands, and V. Vitek), MRS, Pittsburgh, PA, p. 285.
40 Fu, C.L., Zou, J., and Yoo, M.H. (1995) *Scr. Metall. Mater.*, **33**, 885.
41 Yoo, M.H. and Fu, C.L. (1998) *Metall. Mater. Trans. A*, **29A**, 49.
42 Peierls, R.E. (1940) *Proc. Phys. Soc.*, **52**, 34.
43 Zhao, L. and Tangri, K. (1991) *Acta Metall. Mater.*, **39**, 2209.
44 Kad, B.K., Hazzledine, P.M., and Fraser, H.L. (1993) *High-Temperature Ordered Intermetallic Alloys V, Materials Research Society Symposia Proceedings*, vol. 288 (eds I. Baker, R. Darolia, J.D. Whittenberger, and M.H. Yoo), MRS, Pittsburgh, PA, p. 495.
45 Christian, W. (1975) *Transformations in Metals and Alloys-Part I, Equilibrium and General Kinetic Theory*, Pergamon Press, Oxford.
46 Hazzledine, P.M. (1998) *Intermetallics*, **6**, 673.
47 Kad, B.K. and Hazzledine, P.M. (1992) *Philos. Mag. Lett.*, **66**, 133.
48 Saada, G. and Couret, A. (2001) *Philos. Mag. A*, **81**, 2109.
49 Dregia, S.A. and Hirth, J.P. (1991) *J. Appl. Phys.*, **69**, 2169.
50 Maruyama, K., Yamaguchi, M., Suzuki, G., Zhu, H., Kim, H.Y., and Yoo, M.H. (2004) *Acta Mater.*, **52**, 5185.
51 Aaronson, H.I. (1993) *Metall. Trans. A*, **24A**, 241.
52 Frank, F.C. and van der Merwe, J.H. (1949) *Proc. R.. Soc. A*, **198**, 205.
53 Frank, F.C. (1953) *Acta Metall.*, **1**, 15.

54 Hall, M.G., Aaronson, H.I., and Kinsman, K.R. (1972) *Surf. Sci.*, **31**, 257.
55 van der Merwe, J.H. (1985) *S. Afr. J. Phys.*, **9**, 55.
56 Shiflet, G.J., Braun, M.W.H., and van der Merwe, J.H. (1988) *S. Afr. J. Sci.*, **84**, 653.
57 van der Merwe, J.H., Shiflet, G.J., and Stoop, P.M. (1991) *Metall. Trans. A*, **22A**, 1165.
58 Pond, R.C. (1989) *Dislocations in Solids*, vol. 8 (ed. F.R.N. Nabarro), North-Holland, Amsterdam, p. 1.
59 Hirth, J.P. (1994) *J. Phys. Chem. Solids*, **55**, 985.
60 Hirth, J.P. and Pond, R.C. (1996) *Acta Metall. Mater.*, **44**, 4749.
61 Howe, J.M., Pond, R.C., and Hirth, J.P. (2009) *Prog. Mater. Sci.*, **54**, 792.
62 He, L.L., Ye, H.Q., Ning, X.G., Cao, M.Z., and Han, D. (1993) *Philos. Mag. A*, **67**, 1161.
63 Wunderlich, W., Frommeyer, G., and Czarnowski, P. (1993) *Mater. Sci. Eng. A*, **164**, 421.
64 Wunderlich, W., Kremser, T., and Frommeyer, G. (1993) *Acta Metall. Mater.*, **41**, 1791.
65 Zhao, L. and Tangri, K. (1991) *Philos. Mag. A*, **64**, 361.
66 Singh, S.R. and Howe, J.M. (1992) *Philos. Mag. A*, **66**, 739.
67 Yang, Y.S., Wu, S.K., and Wang, J.Y. (1993) *Philos. Mag A*, **67**, 463.
68 Zhang, L.C., Chen, G.L., Wang, J.G., and Ye, H.Q. (1997) *Intermetallics*, **5**, 289.
69 Pond, R.C., Shang, P., Cheng, T.T., and Aindow, M. (2000) *Acta Mater.*, **48**, 1047.
70 Shang, P., Cheng, T.T., and Aindow, M. (1998) *Mater. Sci. Forum*, **294**, 239.
71 Shang, P., Cheng, T.T., and Aindow, M. (1999) *Philos. Mag. A*, **79**, 2553.
72 Shang, P., Cheng, T.T., and Aindow, M. (1999) *Gamma Titanium Aluminides 1999* (eds Y.-W. Kim, D.M. Dimiduk, and M.H. Loretto), TMS, Warrendale, PA, p. 59.
73 Shang, P., Cheng, T.T., and Aindow, M. (2000) *Philos. Mag. Lett.*, **80**, 1.
74 Hirth, J.P. and Lothe, J. (1992) *Theory of Dislocations*, Krieger, Melbourne.
75 Appel, F. and Wagner, R. (1998) *Mater. Sci. Eng. R*, **22**, 187.
76 Zhang, L.C., Chen, G.L., Wang, J.G., and Ye, H.Q. (1998) *Mater. Sci. Eng. A*, **247**, 1.
77 Matthews, J.W. (1974) *Philos. Mag.*, **29**, 797.
78 Read, W.T. (1953) *Dislocations in Crystals*, McGraw-Hill Book Company.
79 Paidar, V., Zghal, S., and Couret, A. (1999) *Mater. Sci. Forum*, **294**, 335.
80 Inkson, B.J. and Humphreys, C.J. (1996) *Philos. Mag. A*, **73**, 1333.
81 Paidar, V. (2002) *Interface Sci.*, **10**, 43.
82 Couret, A., Calderon, H.A., and Veyssière, P. (2003) *Philos. Mag.*, **83**, 1699.
83 Shoykhet, B., Grinfeld, M.A., and Hazzledine, P.M. (1998) *Acta Mater.*, **46**, 3761.
84 Hazzledine, P.M., Kad, B.K., Fraser, H.L., and Dimiduk, D.M. (1992) *Intermetallic Matrix Composites II*, Mater. Res. Soc. Symp. Proc., vol. 273 (eds D.B. Miracle, D.L. Anton, and J.A. Graves), MRS, Pittsburgh, PA, p. 81.
85 Appel, F. and Wagner, R. (1994) *Interface Control of Electrical, Chemical and Mechanical Properties Materials Research Society Symposia Proceedings*, vol. 318 (eds S.P. Murarka, K. Rose, T. Ohmi, and T. Seidel), , MRS, Pittsburgh, PA, p. 691.
86 Appel, F. and Christoph, U. (1999) *Intermetallics*, **7**, 1173.
87 Matthews, J.W. (1979) *Dislocations in Solids*, vol. 2 (ed. F.R.N. Nabarro), North-Holland, Amsterdam, p. 461.
88 Messerschmidt, U., Appel, F., and Schmid, H. (1985) *Philos. Mag. A*, **51**, 781.
89 DeWitt, G. and Koehler, J.S. (1959) *Phys. Rev.*, **116**, 1113.
90 Yoo, M.H., Fu, C.L., and Lee, J.K. (1990) *High-Temperature Ordered Intermetallic Alloys IV, Materials Research Society Symposia Proceedings*, vol. 213 (eds L.A. Johnson, D.P. Pope, and J.O. Stiegler), MRS, Pittsburgh, PA, p. 545.
91 Schafrik, R.E. (1977) *Metall. Trans. A*, **A8**, 1003.
92 Fujiwara, T., Nakamura, A., Hosomi, M., Nishitani, S.R., Shirai, Y., and

Yamaguchi, M. (1990) *Philos. Mag. A,* **61**, 591.

93 Inui, H., Oh, M.H., Nakamura, A., and Yamaguchi, M. (1992) *Acta Metall. Mater.*, **40**, 3095.

94 Inui, H., Kishida, K., Misaki, M., Kobayashi, M., Shirai, M., and Yamaguchi, M. (1995) *Philos. Mag. A,* **72**, 1609.

95 Umakoshi, Y., Nakano, T., and Yamane, T. (1992) *Mater. Sci. Eng. A,* **152**, 81.

96 Yao, K.-F., Inui, H., Kishida, K., and Yamaguchi, M. (1995) *Acta Metall. Mater.*, **43**, 1075.

97 Nomura, M., Kim, M.-C., Vitek, V., and Pope, D. (1999) *Gamma Titanium Aluminides 1999* (eds Y.-W. Kim, D.M. Dimiduk, and M.H. Loretto), TMS, Warrendale, PA, p. 67.

98 Kim, M.-C., Nomura, M., Vitek, V., and Pope, D.P. (1999) *High-Temperature Ordered Intermetallic Alloys VIII, Materials Research Society Symposia Proceedings,* vol. 552 (eds E.P. George, M.J. Mills, and M. Yamaguchi), MRS, Pittsburgh, PA, p. KK3.1.1.

99 Rao, S.I. and Hazzledine, P.M. (2000) *Philos. Mag. A,* **80**, 2011.

100 Paidar, V. (2004) *J. Alloys. Compd.*, **378**, 89.

101 Paidar, V., Imamura, D., Inui, H., and Yamaguchi, M. (2001) *Acta Mater.*, **49**, 1009.

102 Paidar, V. (2007) *Advanced Intermetallic-Based Alloys, Materials Research Society Symposia Proceedings,* vol. 980 (eds J. Wiezorek, C.L. Fu, M. Takeyama, D. Morris, and H. Clemens), MRS, Warrendale, PA, p. 95.

103 Lebensohn, R.A., Uhlenhut, H., Hartig, C., and Mecking, H. (1998) *Acta Mater.*, **46**, 4701.

104 Lebensohn, R.A., Turner, P.A., and Canova, G.R. (1997) *Comp. Mater. Sci.*, **9**, 229.

105 Schlögl, S.M. and Fischer, F.D. (1996) *Compos. Mater. Sci.*, **7**, 34.

106 Schlögl, S.M. and Fischer, F.D. (1997) *Philos. Mag. A,* **75**, 621.

107 Schlögl, S.M. and Fischer, F.D. (1997) *Mater. Sci. Eng. A,* **239**, 790.

108 Werwer, M. and Cornec, A. (2000) *Comput. Mater. Sci.*, **19**, 97.

109 Werwer, M. and Cornec, A. (2006) *Int. J. Plast.*, **22**, 1683.

110 Kad, B.K., Dao, M., and Asaro, R.J. (1995) *Philos. Mag. A,* **71**, 567.

111 Kad, B.K., Dao, M., and Asaro, R.J. (1995) *High-Temperature Ordered Intermetallic Alloys VI, Materials Research Society Symposia Proceedings,* vol. 364 (eds J.A. Horton, I. Baker, S. Hanada, R.D. Noebe, and D.S. Schwartz), MRS, Pittsburgh, PA, p. 169.

112 Dao, M., Kad, B.K., and Asaro, R.J. (1996) *Philos. Mag. A,* **74**, 569.

113 Brockman, R.A. (2003) *Int. J. Plast.*, **19**, 1749.

114 Frank, G.J., Olson, S.E., and Brockman, R.A. (2003) *Intermetallics*, **11**, 331.

115 Dimiduk, D.M., Parthasarathy, T.A., and Hazzledine, P.M. (2001) *Intermetallics*, **9**, 875.

116 Martin, J.L. and Kubin, L.P. (1978) *Ultramicroscopy*, **3**, 215.

117 Martin, J.L. and Kubin, L.P. (1979) *Phys. Status Solidi*, **56**, 487.

118 Mayer, J., Giannuzzi, L.A., Kamino, T., and Michael, J. (2007) *MRS Bull.*, **32**, 400.

119 Vasudevan, V.K., Stucke, M.A., Court, S.A., and Fraser, H.L. (1989) *Philos. Mag. Lett.*, **59**, 299.

120 Sriram, S., Dimiduk, D.M., and Hazzledine, P.M. (1997) *Structural Intermetallics 1997* (eds M.V. Nathal, R. Darolia, C.T. Liu, P.L. Martin, D.B. Miracle, R. Wagner, and M. Yamaguchi), TMS, Warrendale, PA, p. 157.

121 Wiezorek, J.M.K., Zhang, X.D., Godfrey, A., Hu, D., Loretto, M.H., and Fraser, H.L. (1998) *Scr. Mater.*, **38**, 811.

122 Singh, J.B., Molénat, G., Sundararaman, M., Banerjee, S., Saada, G., Veyssière, P., and Couret, A. (2006) *Philos. Mag. Lett.*, **86**, 47.

123 Hall, E.L. and Huang, S.C. (1989) *High-Temperature Ordered Intermetallic Alloys III, Materials Research Society Symposia Proceedings,* vol. 133 (eds C.T. Liu, A.I. Taub, N.S. Stoloff, and C.C. Koch), MRS, Pittsburgh, PA, p. 373.

124 Farenc, S., Coujou, A., and Couret, A. (1993) *Philos. Mag. A,* **67**, 127.

125 Li, Y.G. and Loretto, M.H. (1995) *Phys. Status Solidi*, **150**, 271.

126 Appel, F., Lorenz, U., Oehring, M., Sparka, U., and Wagner, R. (1997) *Mater. Sci. Eng. A*, **233**, 1.

127 Morris, M.A. (1994) *Philos. Mag. A*, **69**, 129.

128 Kaufman, M.J., Konitzer, D.G., Shull, R.D., and Fraser, H.L. (1986) *Scr. Metall.*, **20**, 103.

129 Menand, A., Huguet, A., and Nérac-Partaix, A. (1996) *Acta Mater.*, **44**, 4729.

130 Nérac-Partaix, A. and Menand, A. (1996) *Scr. Mater.*, **35**, 199.

131 Wiezorek, J.M.K., DeLuca, P.M., Mills, M.J., and Fraser, H.L. (1997) *Philos. Mag. Lett.*, **75**, 271.

132 Wiezorek, J.M.K., Mills, M.J., and Fraser, H.L. (1997) *Mater. Sci. Eng. A*, **234–236**, 1106.

133 Wiezorek, J.M.K., Zhang, X.D., Clark, W.A.T., and Fraser, H.L. (1998) *Philos. Mag. A*, **78**, 217.

134 Godfrey, A., Hu, D., and Loretto, M.H. (1998) *Philos. Mag. A*, **77**, 287.

135 Wiezorek, J.M.K., DeLuca, P.M., and Fraser, H.L. (2000) *Intermetallics*, **8**, 99.

136 Forwood, C.T. and Gibson, M.A. (2000) *Philos. Mag. A*, **80**, 2785.

137 von Mises, R. (1928) *Z. Angew. Math.*, **8**, 161.

138 Goo, E. (1998) *Scr. Mater.*, **38**, 1711.

139 Goo, E. and Park, K.T. (1989) *Scr. Metall.*, **23**, 1053.

140 Lee, J.W., Hanada, S., and Yoo, M.H. (1995) *Scr. Metall. Mater.*, **33**, 509.

141 Taylor, G.I. (1938) *Timoshenkow Anniversary Volume*, MacMillan, New York, p. 218.

142 Taylor, G.I. (1970) *J. Inst. Met.*, **62**, 1121.

143 Kocks, U.F. (1970) *Metall. Trans.*, **1**, 1121.

144 Kobrinski, M. and Thompson, C. (2000) *Acta Mater.*, **48**, 625.

145 Mecking, H., Kocks, U.F., and Hartig, Ch. (1996) *Scr. Mater.*, **35**, 465.

146 Dieter, G.E. (1986) *Mechanical Metallurgy*, 3rd edn, McGraw-Hill, New York.

147 Parthasarathy, T.A., Mendiratta, M.G., and Dimiduk, D.M. (1998) *Acta Mater.*, **46**, 4005.

148 Mecking, H., Hartig, Ch., and Kocks, U.F. (1996) *Acta Mater.*, **44**, 1309.

149 Kad, B.K. and Fraser, H.L. (1994) *Philos. Mag. A*, **69**, 689.

150 Appel, F. and Wagner, R. (1995) *Gamma Titanium Aluminides* (eds Y.-W. Kim, R. Wagner, and M. Yamaguchi), TMS, Warrendale, PA, p. 231.

151 Yoo, M.H. (2002) *Intermetallic Compounds. Vol. 3, Principles and Practice* (eds J.H. Westbrook and R.L. Fleischer), John Wiley & Sons, Ltd, Chichester, p. 403.

152 Appel, F. and Wagner, R. (1994) *Twinning in Advanced Materials* (eds M.H. Yoo and M. Wuttig), TMS, Warrendale PA, p. 317.

153 Appel, F. (2005) *Philos. Mag.*, **85**, 205.

154 Zghal, S. and Couret, A. (2001) *Philos. Mag. A*, **81**, 365.

155 Zghal, S. and Couret, A. (1997) *Mater. Sci. Eng. A*, **234–236**, 668.

156 Hu, D. and Loretto, M.H. (1999) *Intermetallics*, **7**, 1299.

157 Gibson, M.A. and Forwood, C.T. (2000) *Philos. Mag. A*, **80**, 2747.

158 Wiezorek, J.M.K., Zhang, X.D., Mills, M.J., and Fraser, H.L. (1999) *High-Temperature Ordered Intermetallic Alloys VIII, Materials Research Society Symposium Proceedings*, vol. 552 (eds E.P. George, M.J. Mills, and M. Yamaguchi), MRS, Warrendale, PA, p. KK3.5.1.

159 Zghal, S., Coujou, S., and Couret, A. (2001) *Philos. Mag. A*, **81**, 345.

160 Nakano, T., Biermann, H., Riemer, M., Mughrabi, H., Nakai, Y., and Umakoshi, Y. (2001) *Philos. Mag. A*, **81**, 1447.

161 Dimiduk, D.M., Hazzledine, P.M., Parthasarathy, T.A., Seshagiri, S., and Mendiratta, M.G. (1998) *Metall. Mater. Trans. A*, **29A**, 37.

162 Frank, F.C. and Read, W.T. (1950) *Phys. Rev.*, **79**, 722.

163 Friedel, J. (1964) *Dislocations*, Pergamon, Oxford.

164 Koehler, J.S. (1952) *Phys. Rev.*, **86**, 52.

165 Johnston, W.G. and Gilman, J.J. (1960) *J. Appl. Phys.*, **31**, 632.

166 Gilman, J.J. and Johnston, W.G. (1962) *Solid State Phys.*, **13**, 147.

167 Schoeck, G. and Seeger, A. (1955) *Rep. Conf. Defects in Crystalline Solids*, Phys. Soc., London, p. 340.

168 Escaig, B. (1968) *J. Phys. Paris*, **29**, 225.
169 Venables, J.A. (1961) *Philos. Mag.*, **6**, 379.
170 Pirouz, P. and Hazzledine, P.M. (1991) *Scr. Metall. Mater.*, **25**, 1167.
171 Shirai, Y. and Yamaguchi, M. (1992) *Mater. Sci. Eng. A*, **152**, 173.
172 Brossmann, U., Würschum, R., Badura, K., and Schaefer, H.E. (1994) *Phys. Rev. B*, **49**, 6457.
173 Bardeen, J. and Herring, C. (1952) *Imperfections in Nearly Perfect Crystals*, John Wiley & Sons, Inc., New York, p. 261.
174 Balluffi, R.W. and Granato, A.V. (1979) *Dislocations in Solids*, vol. 4 (ed. F.R.N. Nabarro), North-Holland, Amsterdam, p. 1.
175 Appel, F., Christoph, U., and Wagner, R. (1994) *Interface Control of Electrical, Chemical and Mechanical Properties Materials Research Society Symposium Proceedings*, vol. 318 (eds S.P. Murarka, K. Rose, T. Ohmi, and T. Seidel), , MRS, Pittsburgh, PA, p. 691.
176 Yoo, M.H. (1989) *J. Mater. Res.*, **4**, 50.
177 Yoo, M.H., Fu, C.L., and Lee, J.K. (1989) *High-Temperature Ordered Intermetallic Alloys III, Materials Research Society Symposia Proceedings*, vol. 133 (eds C.T. Liu, A.I. Taub, N.S. Stoloff, and C.C. Koch), MRS, Pittsburgh, PA, p. 189.
178 Lee, J.K. and Yoo, M.H. (1994) *Twinning in Advanced Materials* (eds M.H. Yoo and M. Wuttig), TMS, Warrendale, PA, p. 51.
179 Yoo, M.H., Fu, C.L., and Lee, J.K. (1994) *Twinning in Advanced Materials* (eds M.H. Yoo and M. Wuttig), TMS, Warrendale, PA, p. 97.
180 Yoo, M.H. (1997) *Philos. Mag. Lett.*, **76**, 259.
181 Hug, G. and Veyssière, P. (1989) *Electron Microscopy in Plasticity and Fracture* (eds U. Messerschmidt, F. Appel, J. Heidenreich, and V. Schmidt), Akademie Verlag, Berlin, p. 451.
182 Sun, Y.Q., Hazzledine, P.M., and Christian, J.W. (1993) *Philos. Mag. A*, **68**, 471.
183 Sun, Y.Q., Hazzledine, P.M., and Christian, J.W. (1993) *Philos. Mag. A*, **68**, 495.
184 Wardle, S., Phan, I., and Hug, G. (1993) *Philos. Mag. A*, **67**, 497.
185 Christian, J.W. and Mahajan, S. (1995) *Prog. Mater. Sci.*, **39**, 1.
186 Jin, Z. and Bieler, T.R. (1995) *Philos. Mag.*, **A71**, 925.
187 Cerreta, E. and Mahajan, S. (2001) *Acta Mater.*, **49**, 3803.
188 Fu, C.L. and Yoo, M.H. (1990) *Philos. Mag. Lett.*, **62**, 159.
189 Woodward, C., MacLaren, J., and Rao, S.I. (1992) *J. Mater. Res.*, **7**, 1735.
190 Kawabata, T., Kanai, T., and Izumi, O. (1985) *Acta Metall.*, **33**, 1355.
191 Kawabata, T., Abumya, T., Kanai, T., and Izumi, O. (1990) *Acta Metall.*, **38**, 1381.
192 Stucke, M.A., Vasudevan, V.K., and Dimiduk, D.M. (1995) *Mater. Sci. Eng. A*, **192–193**, 111.
193 Whang, S.H., Wang, Z.M., and Li, Z.X. (1995) *Gamma Titanium Aluminides* (eds Y.-W. Kim, R. Wagner, and M. Yamaguchi), TMS, Warrendale, PA, p. 245.
194 Mahapatra, R., Girshick, A., Pope, D.P., and Vitek, V. (1995) *Scr. Metall. Mater.*, **33**, 1921.
195 Girshick, A. and Vitek, V. (1995) *High-Temperature Ordered Intermetallic Alloys VI, Materials Research Society Symposia Proceedings*, vol. 364 (eds J.A. Horton, I. Baker, S. Hanada, R.D. Noebe, and D.S. Schwartz), MRS, Pittsburgh, PA, p. 145.
196 Woodward, C., MacLaren, J.M., and Dimiduk, D.M. (1993) *High-Temperature Ordered Intermetallic Alloys V, Materials Research Society Symposia Proceedings*, vol. 288 (eds I. Baker, R. Darolia, J.D. Whittenberger, and M.H. Yoo). MRS, Pittsburgh, PA, p. 171.
197 Morris, D.G., Günther, S., and Leboeuf, M. (1994) *Philos. Mag. A*, **69**, 527.
198 Sriram, S., Viswanathan, G.B., and Vasudevan, V.K. (1994) *Twinning in Advanced Materials* (eds M.H. Yoo and M. Wuttig), TMS, Warrendale, PA, p. 383.
199 Shechtman, D., Blackburn, M.J., and Lipsitt, H.A. (1975) *Metall. Trans. A*, **6A**, 1325.
200 Hanamura, T., Uemori, R., and Tanino, M. (1988) *J. Mater. Res.*, **3**, 656.

201 Tsujimoto, T. and Hashimoto, K. (1989) *High-Temperature Ordered Intermetallic Alloys III, Materials Research Society Symposia Proceedings*, vol. 133 (eds C.T. Liu, A.I. Taub, N.S. Stoloff, and C.C. Koch), MRS, Pittsburgh, PA, p. 391.

202 Appel, F., Paul, J.D.H., Oehring, M., and Buque, C. (2003) *Gamma Titanium Aluminides 2003* (eds Y.-W. Kim, H. Clemens, and A. Rosenberger), TMS, Warrendale, PA, p. 139.

203 Paul, J.D.H., Appel, F., and Wagner, R. (1998) *Acta Mater.*, **46**, 1075.

204 Singh, S.R. and Howe, J.M. (1991) *Scr. Metall. Mater.*, **25**, 485.

205 Greenberg, B.A. (1970) *Phys. Status Solidi*, **42**, 495.

206 Yoo, M.H. and Hishinuma, A. (1999) *Advances in Twinning* (eds S. Ankem and C.S. Pande), TMS, Warrendale, PA, p. 225.

207 Venables, J.A. (1963) *Acta Metall.*, **11**, 1368.

208 Fischer, F.D., Appel, F., and Clemens, H. (2003) *Acta Mater.*, **51**, 1249.

209 Fischer, F.D., Schaden, T., Appel, F., and Clemens, H. (2003) *Eur. J. Mech. A/Solids*, **22**, 709.

210 Petryk, H., Fischer, F.D., Marketz, W., Clemens, H., and Appel, F. (2003) *Metall. Mater. Trans. A*, **34A**, 2827.

211 Hsiung, L. and Nieh, T.G. (1997) *Mater. Sci. Eng. A*, **240**, 438.

212 Tian, W.H., Sano, T., and Nemoto, M. (1993) *Philos. Mag.*, **A68**, 965.

213 Christoph, U., Appel, F., and Wagner, R. (1997) *Mater. Sci. Eng. A*, **239–240**, 39.

214 Kamat, S.V., Hirth, J.P., and Carnahan, B. (1987) *Scr. Metall.*, **21**, 1587.

215 He, Y., Schwarz, R.B., Migliori, A., and Whang, S.H. (1995) *J. Mater. Res.*, **10**, 1187.

216 Wilhelmsson, O., Palmquist, J.P., Lewin, E., Emmerlich, J., Eklund, P., Persson, P.O.Å., Högberg, H., Li, S., Ahuja, R., Eriksson, O., Hultman, L., and Jansson, U. (2006) *J. Cryst. Growth*, **291**, 290.

217 Appel, F., Fischer, F.D., and Clemens, H. (2007) *Acta Mater.*, **55**, 4915.

218 Xu, D., Wang, H., Yang, R., and Veyssière, P. (2008) *Acta Mater.*, **56**, 1065.

219 Kronberg, M.L. (1957) *Acta Metall.*, **5**, 507.

220 Fröbel, U. and Appel, F. (2002) *Acta Mater.*, **50**, 3693.

221 Mishin, Y. and Herzig, Chr. (2000) *Acta Mater.*, **48**, 589.

222 Kishida, K., Inui, H., and Yamaguchi, M. (1999) *Intermetallics*, **7**, 1131.

223 Cottrell, A.H. and Bilby, B.A. (1951) *Philos. Mag.*, **42**, 573.

224 Sleeswyk, A.W. (1974) *Philos. Mag.*, **29**, 407.

225 Couret, A., Farenc, S., and Caillard, D. (1994) *Twinning in Advanced Materials* (eds M.H. Yoo and M. Wuttig), TMS, Warrendale, PA, p. 361.

226 Zhang, Y.G. and Chaturvedi, M.C. (1993) *Philos. Mag. A*, **68**, 915.

227 Morris, M.A. (1996) *Intermetallics*, **4**, 417.

228 Hull, D. (1961) *Acta Metall.*, **9**, 909.

229 Zhu, A., Yoshida, K., Tagaki, H., and Sakamoki, K. (1996) *Intermetallics*, **4**, 483.

230 Kauffmann, F., Bidlingmaier, T., Dehm, G., Wanner, A., and Clemens, H. (2000) *Intermetallics*, **8**, 823.

231 Botten, R., Wu, X., Hu, D., and Loretto, M. (2001) *Acta Mater.*, **49**, 1687.

232 Wu, X., Hu, D., Botten, R., and Loretto, M. (2001) *Acta Mater.*, **49**, 1693.

233 Heiple, C.R. and Carpenter, S.H. (1987) *J. Acoust. Emission*, **6**, 177.

234 Heiple, C.R. and Carpenter, S.H. (1987) *J. Acoust. Emission*, **6**, 215.

235 Lou, X.Y., Li, M., Boger, R.K., Agnew, S.R., and Wagoner, R.H. (2007) *Int. J. Plast.*, **23**, 44.

236 Appel, F., Paul, J.D.H., Oehring, M., Fröbel, U., and Lorenz, U. (2003) *Metall. Mater. Trans. A*, **34A**, 2149.

237 Conrad, H. and Wiedersich, H. (1960) *Acta Metall.*, **8**, 128.

238 Schoeck, G. (1965) *Phys. Status Solidi*, **8**, 499.

239 Gibbs, G.B. (1965) *Phys. Status Solidi*, **10**, 507.

240 Gibbs, G.B. (1967) *Philos. Mag.*, **16**, 97.

241 Evans, A.G. and Rawlings, R.D. (1969) *Phys. Status Solidi*, **34**, 9.

242 Kocks, U.F., Argon, A.S., and Ashby, M.F. (1975) *Prog. Mater. Sci.*, **19**, 1.

243 Seeger, A. (1957) *Dislocations and Mechanical Properties of Crystals* (eds J.C.

Fisher, W.G. Johnston, R. Thomson, and T. Vreeland, Jr.), John Wiley & Sons, Inc., New York, p. 243.
244 Seeger, A. (1958) *Handbuch Der Physik, Band VII/2*, Springer, Berlin.
245 Herrmann, D. and Appel, F. (2009) *Metall. Mater. Trans. A*, **40A**, 1881.
246 Hoppe, R. and Appel, F. (2011) Determination of internal streses in TiAl alloys by stress reduction tests. *Acta Mater.* To be published.
247 Spätig, P., Bonneville, J., and Martin, J.-L. (1993) *Mater. Sci. Eng. A*, **167**, 73.
248 Bonneville, J., Viguier, B., and Spätig, P. (1997) *Scr. Mater.*, **36**, 275.
249 Appel, F., Sparka, U., and Wagner, R. (1995) *High-Temperature Ordered Intermetallic Alloys VI, Materials Research Society Symposia Proceedings*, vol. 364 (eds J.A. Horton, I. Baker, S. Hanada, R.D. Noebe, and D.S. Schwartz), MRS, Pittsburgh, PA, p. 623.
250 Appel, F., Oehring, M., and Wagner, R. (2000) *Intermetallics*, **8**, 1283.
251 Morris, M.A. (1993) *Philos. Mag. A*, **68**, 259.
252 Morris, M.A. and Li, Y.G. (1995) *Gamma Titanium Aluminides* (eds Y.-W. Kim, R. Wagner, and M. Yamaguchi), TMS, Warrendale, PA, p. 353.
253 Viguier, B., Bonneville, J., Hemker, K.J., and Martin, J.L. (1995) *High-Temperature Ordered Intermetallic Alloys VI, Materials Research Society Symposium Proceedings*, vol. 364 (eds J.A. Horton, I. Baker, S. Hanada, R.D. Noebe, and D.S. Schwartz), MRS, Pittsburgh, PA, p. 629.
254 Viguier, B., Cieslar, M., Hemker, K.J., and Martin, J.L. (1995) *High-Temperature Ordered Intermetallic Alloys VI, Materials Research Society Symposium Proceedings*, vol. 364 (eds J.A. Horton, I. Baker, S. Hanada, R.D. Noebe, and D.S. Schwartz), MRS, Pittsburgh, PA, p. 653.
255 Viguier, B., Hemker, K.J., Bonneville, J., Louchet, F., and Martin, J.L. (1995) *Philos. Mag. A*, **71**, 1295.
256 Louchet, F. and Viguier, B. (1995) *Philos. Mag. A*, **71**, 1313.
257 Sriram, S., Dimiduk, D., Hazzledine, P.M., and Vasudevan, V.K. (1997) *Philos. Mag. A*, **76**, 965.
258 Messerschmidt, U., Bartsch, M., Häussler, D., Aindow, M., Hattenhauer, R., and Jones, I.P. (1995) *High-Temperature Ordered Intermetallic Alloys VI, Materials Research Society Symposium Proceedings*, vol. 364 (eds J.A. Horton, I. Baker, S. Hanada, R.D. Noebe, and D.S. Schwartz), MRS, Pittsburgh, PA, p. 47.
259 Zghal, S., Menand, A., and Couret, A. (1998) *Acta Metall.*, **46**, 5899.
260 Greenberg, B.A., Anisimov, V.J., Gornostirev, Yu.N., and Taluts, G.G. (1988) *Scr. Metall.*, **22**, 859.
261 Simmons, J.P., Rao, S.I., and Dimiduk, D.M. (1993) *High-Temperature Ordered Intermetallic Alloys V, Materials Research Society Symposia Proceedings*, vol. 288 (eds I. Baker, R. Darolia, J.D. Whittenberger, and M.H. Yoo), MRS, Pittsburgh, PA, p. 335.
262 Bartels, A., Koeppe, C., Zhang, T., and Mecking, H. (1995) *Gamma Titanium Aluminides* (eds Y.-W. Kim, R. Wagner, and M. Yamaguchi), TMS, Warrendale, PA, p. 655.
263 Morris, M.A., Lipe, T., and Morris, D.G. (1996) *Scr. Mater.*, **34**, 1337.
264 Christoph, U., Appel, F., and Wagner, R. (1997) *High-Temperature Ordered Intermetallic Alloys VII, Materials Research Society Symposium Proceedings*, vol. 460 (eds C.C. Koch, C.T. Liu, N.S. Stoloff, and A. Wanner), MRS, Warrendale, PA, p. 207.
265 Morris, D.G., Dadras, M.M., and Morris-Munoz, M.A. (1999) *Intermetallics*, **7**, 589.
266 Häussler, D., Bartsch, M., Aindow, M., Jones, I.P., and Messerschmidt, U. (1999) *Philos. Mag. A*, **79**, 1045.
267 Couret, A. (2001) *Intermetallics*, **9**, 899.
268 Christoph, U., Appel, F., and Wagner, R. (2001) *High-Temperature Ordered Intermetallic Alloys IX, Materials Research Society Symposium Proceedings*, vol. 646 (eds J.H. Schneibel, K.J. Hemker, R.D. Noebe, S. Hanada, and G. Sauthoff), MRS, Warrendale, PA, p. N7.1.1.
269 Fröbel, U. and Appel, F. (2003) *Gamma Titanium Aluminides 2003* (eds Y.-W. Kim, H. Clemens, and A.H. Rosenberger), TMS, Warendale, PA, p. 467.

270 Fröbel, U. and Appel, F. (2006) *Intermetallics*, **14**, 1187.
271 McCormick, P.G. (1972) *Acta Metall.*, **20**, 351.
272 Kalk, A. and Schwink, C. (1995) *Philos. Mag. A*, **72**, 315.
273 Kattner, U.R., Lin, J.C., and Chang, Y.A. (1992) *Metall. Trans. A*, **23A**, 2081.
274 Woodward, C., Kajihara, S.A., Rao, S.I., and Dimiduk, D.M. (1999) *Gamma Titanium Aluminides 1999* (eds Y.-W. Kim, D.M. Dimiduk, and M.H. Loretto), TMS, Warrendale, PA, p. 49.
275 Kao, C.R. and Chang, Y.A. (1993) *Intermetallics*, **1**, 237.
276 Belova, I.V. and Murch, G.E. (1998) *Intermetallics*, **6**, 115.
277 Herzig, C., Przeorski, T., and Mishin, Y. (1999) *Intermetallics*, **7**, 389.
278 Schoeck, G. and Seeger, A. (1959) *Acta Metall.*, **7**, 469.
279 Rossouw, C.J., Forwood, C.T., Gibson, M.A., and Miller, P.R. (1996) *Philos. Mag. A*, **74**, 77.
280 Appel, F. (1981) *Mater. Sci. Eng. A*, **50**, 199.
281 Appel, F. and Oehring, M. (2003) *Titanium and Titanium Alloys* (eds C. Leyens and M. Peters), Wiley-VCH Verlag GmbH, Weinheim, p. 89.
282 Kruml, T., Coddet, O., and Martin, J.L. (2008) *Acta Mater.*, **56**, 333.

7
Strengthening Mechanisms

Titanium aluminides are being considered to replace the more dense nickel-base superalloys currently in use, over certain ranges of temperature and stress. Thus, their mechanical properties have to be assessed against the high standard set by the superalloys. The strength and creep resistance of most titanium aluminides are inferior to those shown by the superalloys, even if density-corrected data are compared. The implementation of additional hardening mechanisms is therefore a major issue in the ongoing development of titanium aluminides. This chapter reviews the most important methods that have been applied to increase the yield and fracture strength. The accents are principally on grain refinement, work hardening, and strengthening by solute elements and precipitates.

7.1
Grain Refinement

Two-phase titanium aluminide alloys contain dense arrangements of internal boundaries due either to phase transformations or hot working. As has been described in Section 6.2.5, the boundaries are very effective barriers for all types of perfect and twinning partial dislocations. The strengthening effect resulting from internal boundaries is often described in terms of a Hall–Petch relationship [1, 2], and this has also been done for TiAl alloys. In the framework of this model, the yield stress of a polycrystal is given by

$$\sigma = \sigma_0 + \frac{k_y}{D^n} \tag{7.1}$$

One rationalization of Equation 7.1 is that a dislocation pileup at a grain boundary in one grain can generate a sufficiently large stress to operate dislocation sources in an adjacent grain. In Equation 7.1 this stress is described by the term $k_y D^{-n}$. D is the grain size or another structural parameter that determines the length of the dislocation slip path. The parameter k_y is a material constant that basically measures the difficulty with which slip penetrates from one grain to the next; in many

---

Gamma Titanium Aluminide Alloys: Science and Technology, First Edition. Fritz Appel, Jonathan David Heaton Paul, Michael Oehring.
© 2011 Wiley-VCH Verlag GmbH & Co. KGaA. Published 2011 by Wiley-VCH Verlag GmbH & Co. KGaA.

metals the exponent is approximately $n = 0.5$. It is understood that D is the only structural variable that determines the yield stress and that grain-boundary slip does not contribute to deformation. Overcoming of grain boundaries by dislocations certainly involves overcoming of long-range interaction forces, which cannot be supported by thermal activation. Thus, according to the classification made in Section 6.4.1, the term $k_y D^{-n}$ describes an athermal stress. σ_0 is a stress contribution that is independent of the grain size. A prime example of a mechanism that could be associated with σ_0 is the overcoming of localized glide obstacle with the aid of thermal activation (Sections 6.4.1 and 6.4.2). However, in multiphase materials, like TiAl alloys, constraint stresses may develop, which, irrespective of the term $k_y D^{-n}$, could also contribute to σ. The microstructure of two-phase titanium aluminide alloys is complex and rich in detail, encompassing more than one length parameter. In duplex alloys, the relevant structural parameters are the size of equiaxed α_2 and γ grains, the colony size, the lamellar spacing, the domain size, and the spacing of α_2 lamellae. The relative importance of these parameters is difficult to assess because they probably collectively affect the dislocation behavior. Also, it has been proven difficult to control one microstructural feature while others remain ostensibly constant. In view of these problems, the Hall–Petch model appears not to be appropriate for two-phase titanium aluminide alloys. Nevertheless, the strength data of TiAl alloys has been often presented in terms of a Hall–Petch relation, probably because it has the virtue of simplicity. References [3–19] provide a collection of data that has been assessed in this way. The outcome of the analysis is summarized in Table 7.1.

Figure 7.1 demonstrates the data of several TiAl alloys with equiaxed microstructures for which the Hall–Petch analysis is relatively straightforward. The flow-stress data beyond yield follows a similar form [10]. The diagram involves data of an alloy with the composition Ti-47Al-1.5Nb-1Cr-1Mn-0.2Si-0.5B, which was produced by powder metallurgical processing [11]. The use of powder metallurgy compacts is advantageous for analyzing the effects of grain refinement because fine-grained and texture-free materials with uniform microstructure and equiaxed grains can be obtained. By linear regression, the constants $\sigma_0 = 133\,\text{MPa}$ and $k_y = 0.91\,\text{MPa}\,\text{m}^{1/2}$ were determined for this alloy. It should be noted that the k_y value determined for the powder metallurgical alloy is at the lower limit of the values reported for equiaxed materials in the literature (Table 7.1).

Collectively, the data indicates a very small stress contribution σ_0. There is good reason to believe that a thermal stress part associated with the dislocation glide resistance contributes to σ_0 (see Sections 6.4 and 6.5). If this thermal stress part was the only cause of σ_0, then the effective stress is given by $\tau^* = \sigma_0/M_T$. From the data listed in Table 7.1 and using the Taylor factor $M_T = 3.06$, the range of the effective shear stresses for the equiaxed materials was calculated as

$$\tau^* = 44\,\text{MPa to } 67\,\text{MPa} \tag{7.2}$$

As noted above, other likely contributions to σ are athermal coherency stresses, which develop upon loading, but are difficult to quantify. Thus, the values defined by Equation 7.2 should be considered as the upper limit for the effective stress. Further work is required to elucidate the quantity σ_0.

7.1 Grain Refinement

Table 7.1 The effect of structural parameters on the yield stress σ_y of TiAl alloys.

Alloy composition (at.%)	Microstructure	σ_0 (MPa)	k_y (MPa m$^{1/2}$)	Symbol Ref.
Ti-52Al Ti-50Al-0.4Er	near-gamma, $\sigma_y = f(D)$	200	1.37	△ [5][a]
Ti-(50-54)Al	near-gamma, $\sigma_y = f(D)$	150	1.1	[6] ○
Ti-48Al-2Cr	equiaxed α_2 and γ grains, $\sigma_y = f(D)$	193	1.21	▽ [10]
Ti-47Al-1.5Nb-1Cr- -1Mn-0.2Si-0.5B	equiaxed α_2 and γ grains, $\sigma_y = f(D)$	133	0.91	● [11]
Ti-45Al-2.4Si Ti-48Al Ti-48.9Al Ti-46Al-5Si	equiaxed α_2, γ and Ti$_5$(Si,Al)$_3$ grains, $\sigma_y = f(D)$	125	1.00	[17]
Ti-(48.1-51.6)Al	PST, hard orientation; $\Phi = 0°$, $\sigma_y = f(\lambda_L)$ PST, hard orientation; $\Phi = 90°$, $\sigma_y = f(\lambda_L)$ PST, soft orientation; $\Phi = 45°$ $\sigma_y = f(\lambda_D)$		0.41 0.5 0.27	[9]
Ti-47Al-2Cr-1.8Nb-0.2W Ti-47Al-2Cr-2Nb Ti-46Al-2Cr-1.8Nb	lamellar, $\sigma_y = f(\lambda_L)$		0.22	[14, 15]
Ti-45.5Al-2Cr-2Nb Ti-47Al-2Cr-2Nb	fully lamellar, $\sigma_y = f(D_C)$	581	2.7	[16][b]
Ti-39.4Al	lamellar, $\sigma_y = f(\lambda_L)$	250	0.26	[18]

a) Evaluation includes data of [3–5].
b) Evaluation includes data of [14–16].
Analysis performed according to the Hall–Petch model describe by Equation 7.1. D, grain size; D_C, colony size; λ_L, lamellar spacing; λ_D, domain size.

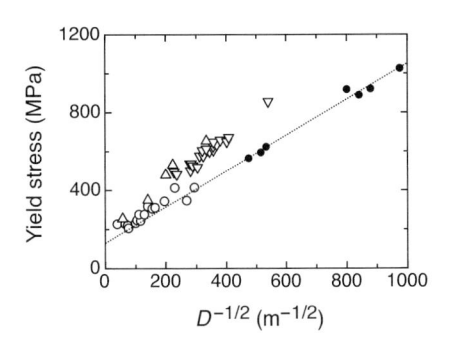

Figure 7.1 Hall–Petch plots of room-temperature yield stresses of equiaxed TiAl alloys as a function of $D^{-1/2}$. D (average grain diameter). The line drawn in the diagram represents a linear regression of the data presented by Gerling et al. [11]. For symbols and references see Table 7.1.

For obvious reasons a Hall–Petch analysis of the strength data of lamellar alloys is much more demanding. In an attempt to reduce the problems associated with the various structural parameters of the lamellar morphology, Umakoshi et al. [7–9] investigated the strengthening effects on differently oriented PST crystals. This has two advantages: first, the spurious influence of the colony size on the yield stress could be eliminated, and secondly, by deforming the PST crystal in specified orientations, the effects of the lamellae interfaces and domain boundaries could be separated. In the hard orientations, $\Phi = 0°$ and $\Phi = 90°$, the dislocation slip path is essentially determined by the lamellar spacing λ_L; for these orientations the dependence of the yield strength on λ_L was evaluated. In the soft 45° orientation the relevant structural parameter is the separation distance λ_D of the domain boundaries and this has been used for the evaluation. The results of the analysis are listed in Table 7.1. Unfortunately, in most of the published data concerning the yield stress as a function of the lamellar spacing, the range of lamellar spacing is not sufficient to permit an unambiguous determination of σ_0. Nevertheless, the investigations clearly reflect the plastic anisotropy of the lamellar morphology with respect to the lamellae orientation Φ against the load axis (Figure 6.28), which has been described in Section 6.1.5. Accordingly, overcoming the domain boundaries seems to be relatively easy, when compared with the lamellar interfaces. The analysis performed for the hard orientations is subtler because the α_2 lamellae affect the yield strength in different ways. In the 0° orientation, glide of the α_2 phase is accomplished by prismatic slip, which is relatively easy. On the contrary, in the 90° orientation pyramidal slip has to be activated in the α_2 phase, which requires an extremely high stress (Section 5.2). Thus, although the α_2 lamellae were widely separated in the PST crystals investigated, the difference in the k_y values observed for the two hard orientations probably reflects the different slip modes with which the α_2 lamellae are intersected. In supporting this hypothesis, Umakoshi et al. [9] have shown that the flow stress of PST crystals in the 90° orientation substantially increases with increasing volume fraction of α_2 lamellae. This interpretation is also consistent with the temperature dependence of the Hall–Petch parameters determined by Umakoshi and Nakano [20]. In the soft 45° orientation, the Hall–Petch slope k_y is almost independent of the test temperature up to 900 °C, as it is expected for athermal glide barriers. However, in the 90° orientation, k_y goes through a maximum of $k_y = 1.4\,\text{MPa}\,\text{m}^{1/2}$ at 500 °C and then decreases to a very low value of $k_y = 0.1\,\text{MPa}\,\text{m}^{1/2}$ at 900 °C. The maximum of k_y corresponds to the maximum in the temperature profile of the critical resolved shear stress of α_2 single crystals (Section 5.2.2).

Dimiduk et al. [16] have examined the yield strength of fully lamellar polycrystalline alloys with colony sizes ranging from $D_C = 55$ to 400 μm and lamellar spacings from $\lambda_L = 35$ to 150 nm. As noted in Table 7.1, an apparent Hall–Petch constant of $k_y = 2.7\,\text{MPa}\,\text{m}^{1/2}$ was determined when the yield stresses were ranked against the colony size (Table 7.1). The authors have recognized that the lamellar spacing is proportional to $D_C^{1/2}$ and have proposed that, irrespective of D_C, the yield stress is governed by λ_L. In order to make this idea quantitative, the authors proposed an analytical model for the yield stress of lamellar material that involves three

Hall–Petch constants, which account for the effects of the grain boundaries (k_y = 1 MPa m$^{1/2}$), the domain boundaries (k_y = 0.25 MPa m$^{1/2}$) and the lamellae interfaces (k_y = 0.45 MPa m$^{1/2}$). These findings agree with a pile-up model of Sun [21–23] according to which the lamellar spacing is the relevant structural parameter for describing the yield stress of fully lamellar alloys. An alternative evaluation of the Hall–Petch relation for dual-phase TiAl alloys was proposed by Jung et al. [12].

It is important to note that the plastic anisotropy of the lamellar morphology can also become evident in the strength properties of polycrystalline cast material. This was demonstrated on cast material that exhibited a columnar microstructure with a well-expressed casting texture. In such material, the lamellae in the columnar grains show a high degree of preferred orientation due to the radial dendrite growth of the α phase during solidification. Using this casting texture, three types of samples were prepared, in which the lamellae were oriented with respect to the deformation axis at Φ = 0°, 45° and 90° [24]. Compression tests were performed on these samples at room temperature and an initial strain rate $\dot{\varepsilon}_1 = 2.18 \times 10^{-5}$ s^{-1}. The strain rate and temperature sensitivity of the flow stress were measured so as to determine the activation volume and the activation energy, as described in Section 6.4.1. The results of this analysis are summarized in Table 7.2. The flow stresses and strain-rate sensitivities measured at the beginning of deformation of ε = 1.25% are shown in Figure 7.2. In spite of the relatively large scatter, the anisotropy of the yield stress corresponds to that measured on PST crystals (Figure 6.28). Although the strain-rate sensitivity $(\Delta\sigma/\Delta\ln\dot{\varepsilon})_T$ seems to correlate with $\sigma_{1.25}$, its variation with Φ is small. This indicates that the change of the yield stress between the soft and hard orientations is mainly athermal in nature [24]. The slightly higher increase of $(\Delta\sigma/\Delta\ln\dot{\varepsilon})_T$ at Φ = 90° might be attributed to the frequent intersection of the lamellar boundaries by dislocations, which is imposed for this orientation. This slip transfer probably involves thermally activated processes, such as dislocation intersection or jog dragging, which are reflected in higher values of $(\Delta\sigma/\Delta\ln\dot{\varepsilon})_T$. The interpretation is consistent with the observation that the activation energies are almost independent of Φ and fall in the range that has been determined for texture-free material (Table 6.2).

Summary: For TiAl alloys the resistance of internal boundaries to the spread of deformation can be expressed in a general way by the Hall–Petch equation. An

Table 7.2 Activation parameters determined for polycrystalline fully lamellar Ti-48Al-2Cr with a preferred orientation Φ of the lamellae against the compression axis.

Φ [deg.]	$\sigma_{1.25}$ [MPa]	V/b^3	Q_e [eV]	ΔG [eV]
0	404	136	0.84	0.65
45	284	169	0.71	0.56
90	518	93	0.91	0.73

Values determined at strain ε = 1.25%. V/b^3, activation volume referred to the Burgers vector **b** of ordinary dislocations; Q_e, experimental activation energy; ΔG, free energy of activation.

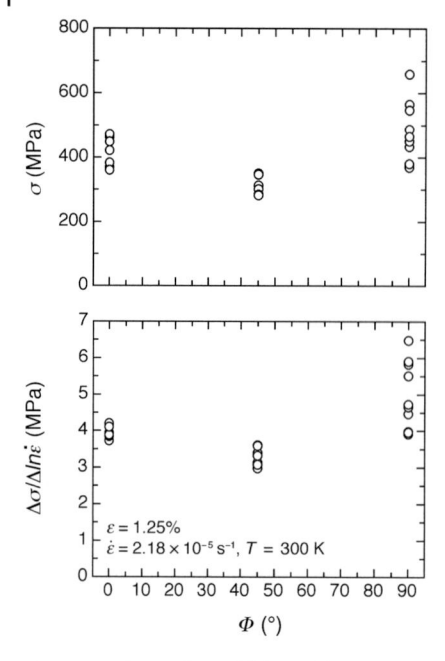

Figure 7.2 Dependence of the yield stress σ and of the strain-rate sensitivity $(\Delta\sigma/\Delta\ln\dot\varepsilon)_T$ on the average orientation Φ of the lamellae interfaces with respect to the compression axis. Fully lamellar Ti-48Al-2Cr with a preferred orientation of columnar lamellar grains. Compression at room temperature [24].

analysis performed on this basis clearly indicates that the yield strength of TiAl alloys is mainly determined by the microstructure. Thus, grain refinement is an important issue for improving the strength properties.

7.2
Work Hardening

Compared with the large amount of mechanical data, little information is available about the work hardening of TiAl alloys. This is certainly in part due to the brittleness of the material, which persists up to relatively high temperatures. However, in compression it is possible to produce work hardening over large plastic strains, which provides a significant potential to strengthen the material. The competitive processes to work hardening are recovery and recrystallization. Thus, the rate at which the material hardens depends on the imparted mechanical strain energy and its release by diffusion-assisted processes. Taken together, these factors lead to a complex dependence of the work hardening on temperature. From the technical point of view work hardening is important because it is involved in various metallurgical processes, such as forming, shaping, diffusion bonding, and surface hardening. For all these processes detailed knowledge about the deformation-induced defect structures and their thermal stability is required. The available information about this subject will be assessed in the following sections.

7.2.1
Work-Hardening Phenomena

Work-hardening mechanisms will be characterized in terms of glide obstacles controlling the velocity and the slip path of the dislocations, analogous to the procedure that has been described in Section 6.4. Within this approach the flow stress $\sigma(\varepsilon)$ beyond yielding may be described as [25]

$$\sigma(\varepsilon) = \sigma_0 + \sigma_\mu(\varepsilon) + \sigma^*(\varepsilon) = \sigma_0 + \sigma_\mu(\varepsilon) + \frac{1}{M_T V_D(\varepsilon)}(\Delta F_D^* + kT \ln \dot{\varepsilon}/\dot{\varepsilon}_0) \quad (7.3)$$

ΔG is the Gibbs free energy of activation, k the Boltzmann constant and M_T the Taylor factor. σ_0 represents a stress contribution from dislocation mechanisms operating at the onset of yielding and is considered to be independent of strain ε. Micromechanisms associated with σ_0 have been discussed in Section 6.4. $\sigma_\mu(\varepsilon)$ is an athermal stress contribution to work hardening representing long-range dislocation interactions. $\sigma^*(\varepsilon)$ is an effective or thermal stress component due to thermally assisted overcoming of deformation induced short-range glide obstacles. $V_D(\varepsilon)$ and ΔF_D^* are the activation volume and the free energy of activation of this thermally activated process. Thus, the variation of the reciprocal activation volume with ε will serve as a measure for the contribution of thermal glide obstacles to work hardening.

Figure 7.3 demonstrates the typical flow behavior that occurs under compression. Between room temperature and 873 K work hardening is insensitive to temperature, as indicated by the family of parallel stress–strain curves [25–29].

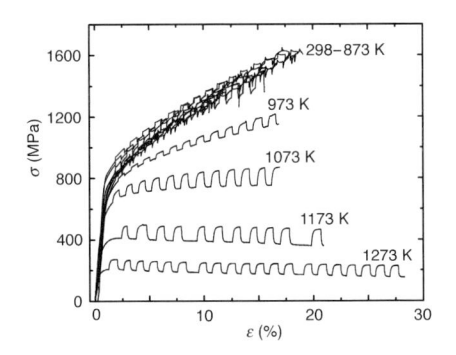

Figure 7.3 Load–elongation traces of compression tests demonstrating the work-hardening behavior of a nearly lamellar Ti-45Al-8Nb-0.2C alloy. Reversible strain-rate changes were performed between $\dot{\varepsilon}_1 = 2.3 \times 10^{-4}$ s^{-1} and $\dot{\varepsilon}_2 = 20\dot{\varepsilon}_1$; at $T = 1273$ K the strain-rate ratio was $\dot{\varepsilon}_2 = 3\dot{\varepsilon}_1$. Measurements performed by D. Herrmann [29].

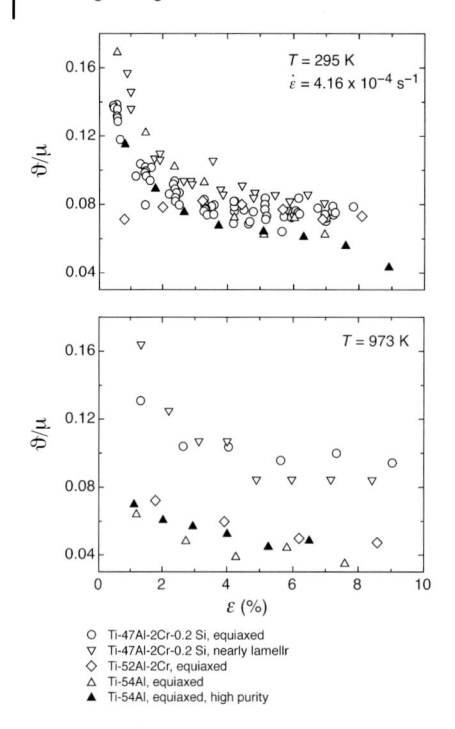

Figure 7.4 Dependence of the normalized work-hardening coefficient $\vartheta/\mu = (1/\mu)\mathrm{d}\sigma/\mathrm{d}\varepsilon$ on strain ε determined at $T = 295$ K and $T = 973$ K; μ is the shear modulus at the respective temperature.

Figure 7.4 shows the normalized work-hardening coefficient, $\vartheta/\mu = (1/\mu)\mathrm{d}\sigma/\mathrm{d}\varepsilon$, as a function of strain ε for two temperatures. The temperature dependent values of the shear modulus μ were taken from [30]. Coarse-grained alloys often exhibit a parabolic relation between σ and ε. In such material, localized flow probably starts in favorably oriented grains or colonies, and the small plastic deformation in the pre-macroyield region may be due entirely to such nonuniform deformation. The apparently high initial rate of work hardening is then only a result of the reduction of constraint stresses by localized flow and not a measure of conventional work hardening that is controlled by dislocation interactions. Under steady-state conditions, $\varepsilon = 6\%$ and $\dot\varepsilon = 10^{-4}$ s^{-1}, the normalized work-hardening coefficient is typically $\vartheta/\mu = 0.07$ to 0.08. It should be noted that this hardening rate is several times greater than that of a typical cubic metal. Kocks and Mecking [31] have provided a survey on the present position with regard to work-hardening models in conventional metals; Gray and Pollock [32] have assessed the work hardening of intermetallic alloys.

Summary: Under compression, TiAl alloys exhibit high rates of strain hardening, which are insensitive to temperature over a wide temperature range. Above the brittle to ductile transition temperature the strain-hardening rate decreases.

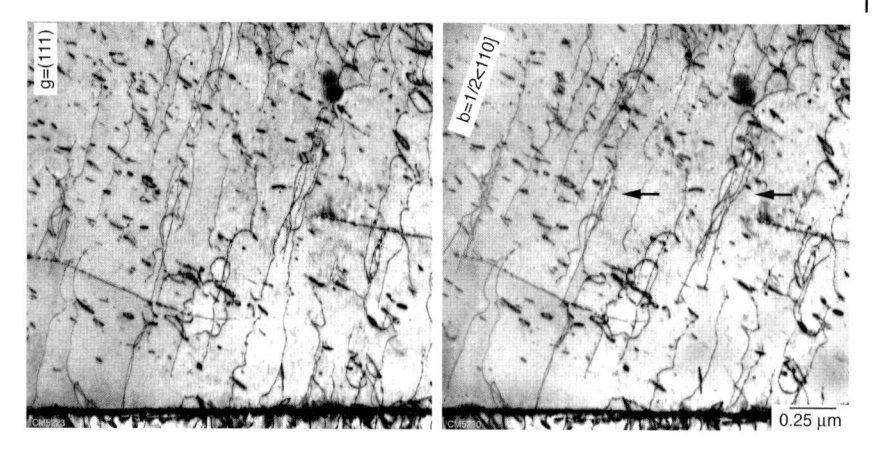

Figure 7.5 Stereo pair of micrographs showing dislocation multipoles (arrowed) of 1/2<110] screw dislocations situated on $(\bar{1}\bar{1}1)_\gamma$. High-order bright field images using $g = (111)_\gamma$ reflection from near the $[0\bar{1}1]$ pole. Ti-48Al-2Cr, compression at room temperature to strain $\varepsilon = 3\%$.

7.2.2
Athermal Contributions to Work Hardening

Athermal contributions to work hardening in terms of the stress part $\sigma_\mu(\varepsilon)$ in Equation 7.3 certainly arise due to the elastic interaction of dislocations propagating on parallel slip planes. This becomes evident by multipole structures, which have frequently been observed after room-temperature compression (Figure 7.5). The dislocations were probably trapped because of their elastic interaction and local recombination by cross-glide. It may be expected that the multipoles are relatively strong sessile obstacles, which effectively contribute to work hardening over a wide temperature range.

During deformation at room temperature, in most cases multiple slip occurs on obliquely oriented <110>{111} and <11$\bar{2}$]{111} systems. Dislocations gliding on these systems have to intersect each other. The mechanistic details of the intersection process depend on the mutual orientation and the type and character of the dislocations involved [33]. An analytical study has shown that the elastic stresses occurring between intersecting dislocations are as high as the yield stress of the material and extend over a relatively large distance [34]. At a separation distance between the dislocations of five Burgers vectors, the interaction stresses are of the order of a few per cent of the shear modulus. In most cases, there are stress components acting both within and perpendicular to the slip plane of the gliding dislocations. Thus, the elastic interaction not only tends to immobilize the dislocations, but also induces extended cross-slip [34]. Thus, the jogs that are formed in the intersecting dislocations could be significantly higher than those that are expected from the simple intersection geometry according to the mutual orientation of their Burgers vectors. Figure 7.6 demonstrates the intersection of two sets of 1/2<110] screw dislocations situated on oblique {111} planes. In the intersection

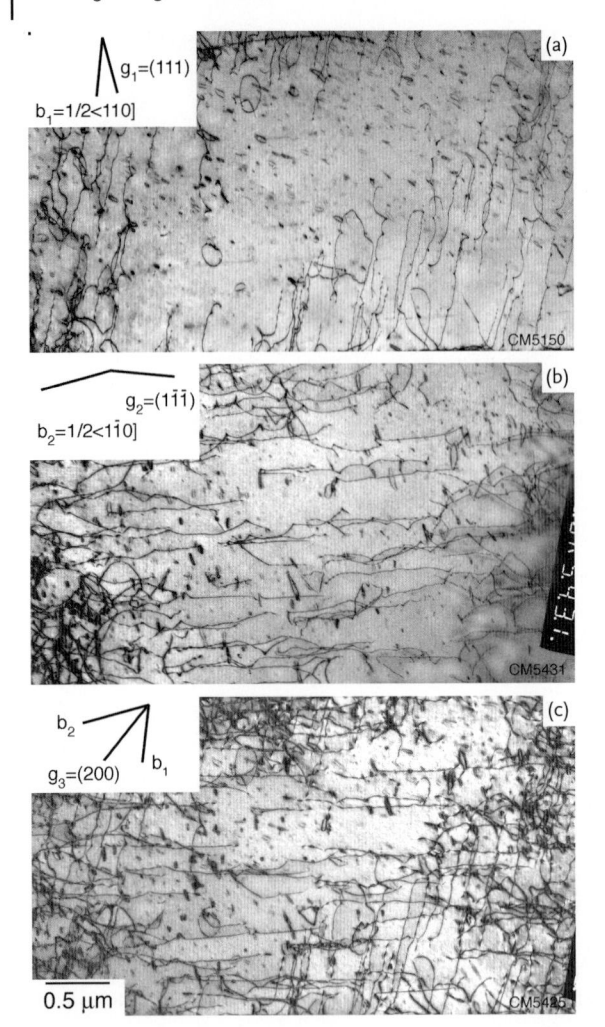

Figure 7.6 Intersection of two sets of 1/2<110] dislocations that propagated on oblique {111} planes. Ti-48Al-2Cr, compression at 300 K to strain $\varepsilon = 3\%$. High order bright-field image from near the $[0\bar{1}1]$ pole using the reflections g_i indicated. (a) set 1: screw dislocations with Burgers vector $\mathbf{b}_1 = 1/2$<110] situated on $(\bar{1}1\bar{1})$. (b) set 2: screw dislocations with Burgers vector $\mathbf{b}_1 = 1/2$ <1$\bar{1}$0] situated on $(\bar{1}\bar{1}1)$. (c) sets 1 and 2 of screw dislocations are simultaneously imaged. Note the high density of debris formed in the intersection zone.

zone numerous dipoles and debris defects are present. It is reasonable to speculate that these defect were trailed and terminated by the highly jogged dislocations; discussion of this mechanism can be found in Section 7.2.3.

The formation of junctions and dislocation intersections undoubtedly lead to the storage of dislocations. The common feature of the two mechanisms is that the related stress part σ_{DIS} varies inversely with the spacing of the dislocations, that is, σ_{DIS} is given by [35]

$$\sigma_{DIS} = \alpha\,\mu\,b\,\sqrt{\rho_{DIS}} \qquad (7.4)$$

ρ_{DIS} is the total density of the dislocations, **b** the Burgers vector and $\alpha = 0.5$ is a constant. At the beginning of deformation ($\varepsilon = 1.25\%$) the dislocation density is typically $\rho_{DIS} = 10^8\,\text{cm}^{-2}$; this leads to a stress contribution $\sigma_{DIS} = 30\,\text{MPa}$ [36]. Unfortunately, there is no published data about the evolution of the dislocation density with strain to check this hypothesis for TiAl alloys.

Another source of work hardening is probably mechanical twinning because the local crystal orientation is changed at the twin matrix interface and the interface itself may act as glide barrier. That is, profuse twin formation decreases the mean free path for slip, analogous to a Hall–Petch grain-size effect. Furthermore, twinning occurs on multiple $1/6 <11\bar{2}]\{111\}$ systems with nonparallel shear vectors, leading to extensive intersections among twin bands. As has been described in Section 6.3.4, twin intersections produce internal stresses and represent a potential locking mechanism for twinning dislocations. Finally, it should be noted that in γ(TiAl) sessile dislocation locks can be formed due to the nonplanar dissociation of the superdislocations (see Section 5.1.4). However, there is little evidence of such mechanisms in two-phase alloys. The salient feature of all these mechanisms is that the relevant defect structures have a relatively large size. Overcoming of such defects is expected to produce a stress part σ_μ that is independent of temperature and strain rate, since thermal fluctuation cannot supply sufficient energy to assist dislocation motion over such large distances.

Summary: Formation of dislocation multipoles, dislocation intersections and mechanical twinning are potential sources of work hardening. Overcoming these glide barriers is impeded by long-range internal stresses and is thus expected to provide an athermal contribution to work hardening.

7.2.3
Jog Dragging and Debris Hardening

Work hardening at room temperature is always accompanied by an increase of the reciprocal activation volume $1/V$ with strain ε, as shown in Figure 7.7. According to Equation 7.3, this suggests that new obstacles were generated during deformation. At the same time, the Gibbs free energy of activation ΔG is independent of strain (Figure 7.8), which implies that, in spite of the significant strain hardening, the average dislocation velocity does not strongly change with strain. The strain dependence of $1/V$ should most likely be ascribed to a jog-dragging mechanism. As has been described in Section 6.3.1, a jog on a screw dislocation has edge character and is unable to move conservatively together with the screw dislocation. At sufficiently high stress, movement of the jog may occur and will leave behind a trail of vacancies or interstitial atoms depending on the sign of the dislocation and the direction the dislocation is moving. The steady-state configuration of the pinning process is similar to that sketched in Figure 6.61, but with osmotic forces

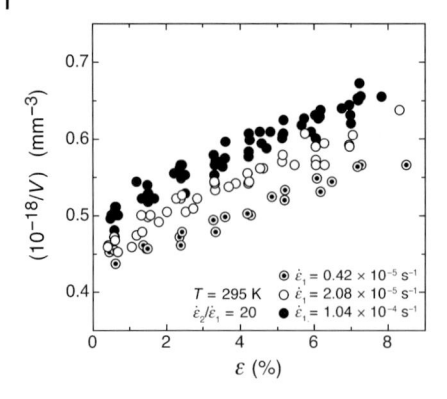

Figure 7.7 Dependence of the reciprocal activation volume $1/V$ on strain ε. Ti-47Al-2Cr-0.2Si, equiaxed microstructure [25].

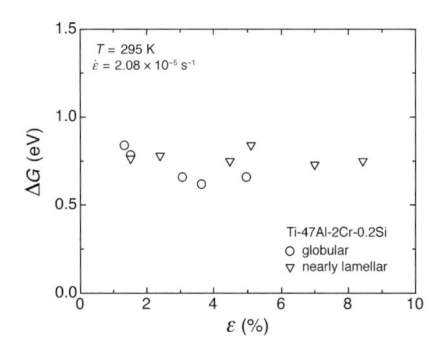

Figure 7.8 Dependence of the Gibbs free energy of activation ΔG on strain ε [25].

at the jogs balancing line tension forces from the bowed out adjacent dislocation segments. In principle, during deformation a screw dislocation can acquire both vacancy- and interstitial-producing jogs. However, the stress required for the forward motion of the jogged dislocation producing point defects is proportional to the formation energy E_F of the relevant point defect. Molecular dynamic simulations of Wang et al. [37] based on Finnis–Sinclair many-body potentials have shown that the formation energies for all the stable interstitial configurations in TiAl are two to three times those of vacancies. Thus, it is unlikely that jog dragging generates interstitial atoms. It is more likely that interstitial jogs conservatively glide along the screw dislocation and combine. Because of the relatively small formation energy for vacancies and the atomic dimensions of the activation complex, thermal activation is expected to support jog dragging. This gives rise to a thermal stress part $\sigma^*(\varepsilon)$ that depends on strain rate and temperature. Another consequence that might be expected from jog dragging is an increase of the vacancy concentration above that of thermal equilibrium, which in turn enhances the diffusivity. If a jog in a screw dislocation does not move, it is connected to the

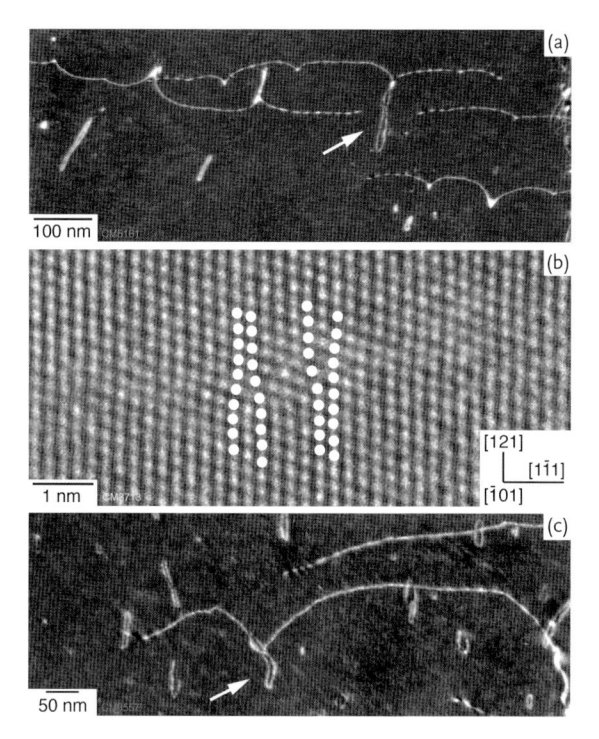

Figure 7.9 TEM micrographs illustrating debris hardening in two-phase titanium aluminide alloys. Compression at room temperature to strain $\varepsilon = 3\%$. (a) Ordinary screw dislocations trailing dipoles (arrowed). Note the terminated debris. (b) High-resolution image of a dislocation dipole of vacancy type. (c) Interaction of an ordinary screw dislocation with a debris defect (arrowed).

moving dislocation by two lengths of edge dislocations of opposite sign forming a dislocation dipole (Figure 6.61, stage iii). Because of the mutual attraction of the positive and negative edge dislocations forming the dipole, a dipole may break up in a row of prismatic loops. This process seems frequently to occur in TiAl alloys, as already demonstrated in Figure 6.38.

The debris and dipoles produced by jog dragging seem to be effective glide obstacles for other dislocations; Figure 7.9 shows the process in some detail. In part (a), ordinary screw dislocations are shown, which have just started to trail dipoles; terminated debris can also be seen. Figure 7.9b is a high-resolution image showing a small dipole across its two edge dislocation arms. The unlike edge dislocations are separated by a few atomic spacings indicating the vacancy character of the dipole. It should be noted that in the few cases that could be analyzed by high-resolution TEM, the dipoles were always of vacancy type. Figure 7.9c demonstrates the interaction of a screw dislocation with a debris defect. The dislocation is bowed out at the defect, which indicates its high obstacle strength. As can be seen in Figure 7.6 debris will also be formed on oblique {111} slip planes.

Intersection of this debris produces another contribution to work hardening. Gilman [38] and Chen et al. [39] have proposed a model of debris hardening, assuming that the debris production rate is proportional to the density of mobile dislocations. This leads to the prediction that the debris concentration increases linearly with strain. Early investigations of Kroupa [40–42], Bullough and Newman [43] and Bacon et al. [44] have shown that the strain field between a dislocation and a prismatic dislocation loop is very localized and that the interaction stresses are small. There is no long-range interaction between these defects. By virtue of these characteristics the debris can probably be overcome by the dislocations with the aid of thermal activation. Since the density of debris is expected to increase with strain, the mechanism should be manifested by an increase of the reciprocal activation volume with strain. As was demonstrated in Figure 7.7, this is indeed a salient feature of the room-temperature work hardening of TiAl alloys.

Summary: Jog dragging and debris defects provide a glide resistance to dislocation motion, which contributes with a thermal stress part to work hardening at room temperature. The mechanisms produce a supersaturation of point defects, most likely vacancies.

7.2.4
Thermal Stability of Deformation Structures

As described in the previous section, cold working of TiAl alloys significantly increases the density of dislocations and debris, which gives rise to a substantial strain energy stored in the material. Thus, the cold-worked condition is thermodynamically unstable relative to the undeformed state. The stored energy can be released if the dislocations rearrange themselves into configurations of lower energy and recombine with point defects. These processes require dislocation climb and will occur only when there is sufficient thermal activation to allow diffusion. There is good evidence that these recovery processes are determined by different energy barriers that are characteristic of the defects that are involved. Thus, recovery of cold-worked material can give additional information about the defect processes occurring during work hardening.

With this perspective in mind, static recovery of deformed samples was studied on several single-phase and two-phase alloys [25, 28] according to the schedule that is schematically sketched in Figure 7.10. The alloys investigated are listed in Table 7.3. The specimens were compressed at room temperature to a plastic strain of $\varepsilon_p = 6.5\%$ to 7.5%. The tests involved strain-rate changes for determining the activation volume. The deformed samples were then subject to either isochronal or isothermal annealing, in order to gradually recover the deformation-induced defect structure. Following each annealing cycle, the sample was retested at room temperature. After a total isothermal annealing time of $t_R^* = 225\,\text{min}$, the same sequence of measurements was made at the next-highest temperature. To measure the flow stress at the next temperature, the specimen had to undergo a small amount of plastic deformation that resulted in work hardening. This increase in stress was subtracted from the flow stress value for the annealing cycle. A load–

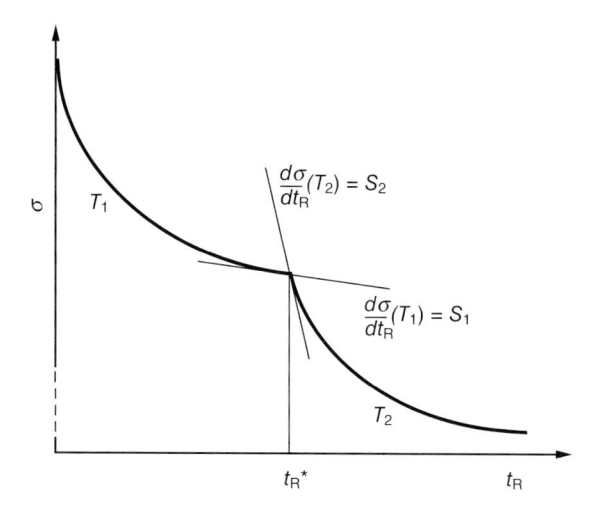

Figure 7.10 Schematic illustration of an annealing experiment performed on a deformed sample in order to characterize the recovery of deformation-induced defect structures. After a first period of isothermal annealing at T_1, the annealing temperature is increased to T_2.

Table 7.3 Composition, microstructure and mechanical properties of the alloy investigated in recovery experiments [25, 28].

Symbol	Composition (at.%)	Microstructure	σ_D (MPa)	$10^{-18}/V_D$ (mm^{-3})
	Ti-45Al-5Nb-0.2B-0.2C	nearly globular, $D = 1–8\,\mu m$	1350	1.03
	Ti-45Al-8Nb-0.2C	duplex, $D = 5–10\,\mu m$, $D_c = 10–30\,\mu m$,	1330	1.08
∗	Ti-48Al-2Cr	nearly lamellar, $D_c = 1\,mm$, $\lambda = 0.05–1\,\mu m$	1029	0.53
○	Ti-47Al-2Cr-0.2Si	equiaxed $\alpha_2 + \gamma$, $D = 11\,\mu m$	947	0.7
▽	Ti-47Al-2Cr-0.2Si	nearly lamellar, $D_c = 330\,\mu m$, $\lambda = 0.05–1\,\mu m$	1065	1.12
◇	Ti-52Al-2Cr	near γ, $D = 5.2\,\mu m$	1037	0.7
△	Ti-54Al	near γ, $D = 188\,\mu m$	901	0.75

σ_D, (flow stress) and $1/V_D$ (reciprocal activation volume) determined after compression to strain $\varepsilon_P = 7.5\%$ at $T = 295$ K and $\dot{\varepsilon} = 4.16 \times 10^{-4}$ s^{-1}. D, grain size; D_C, colony size; λ_L, lamellar spacing.

elongation trace of an isochronal annealing experiment is shown in Figure 7.11a. After 6.5% plastic strain the specimen was unloaded and then annealed at the temperature $T_R = 1023$ K for time $t_R = 50$ h, after which the specimen was allowed to cool before reloading. It can be seen that a significant amount of the work hardening developed during the initial deformation is lost on reloading after the annealing procedure; this is directly related to the annealing-out of glide obstacles. The related recovery behavior of the reciprocal activation volume is shown in Figure 7.11b. Similar experiments were performed on various alloys at annealing temperatures T_R between 673 K and 1133 K and with a fixed annealing time of

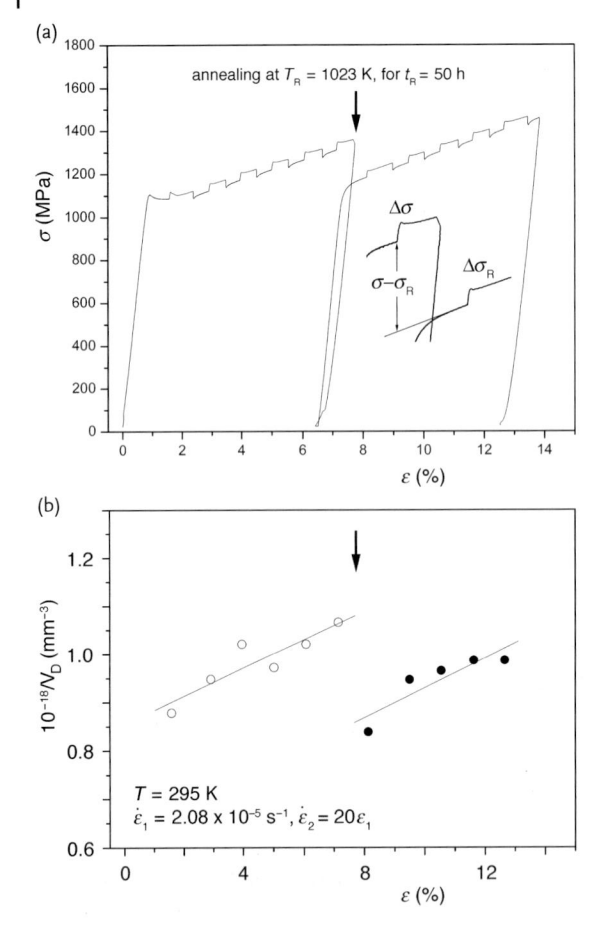

Figure 7.11 Recovery of work hardening observed on an extruded Ti-45Al-5Nb-0.2B-0.2C alloy (TNB-V5). σ flow stress and V_D activation volume determined during predeformation at 295 K. σ_R flow-stress measured after annealing. $\Delta\sigma$ and $\Delta\sigma_R$ are the stress increments of strain-rate cycling tests for determining the activation volume before and after annealing. (a) Load–elongation trace of a recovery experiment involving predeformation in compression at room temperature to strain $\varepsilon_P = 6.5\%$, followed by a 50-h annealing at 1023 K and subsequent retesting at room temperature. Reversible strain-rate changes were performed for determining the activation volume. (b) Reciprocal activation volume $1/V_D$ determined in this test before and after the annealing [29].

$t_R = 120$ min. The results of these experiments are shown in Figure 7.12. The upper graph shows that for annealing temperatures above 800 K, a significant fraction of the work hardening can be recovered. The flow-stress increase observed at $T_R < 800$ K probably results from dislocation locking due to the formation of defect atmospheres (Section 6.4.3), which occurs at these temperatures on the time scales used in the recovery experiments. In contrast, for all the annealing temperatures investigated, the reciprocal activation volume almost completely recovered

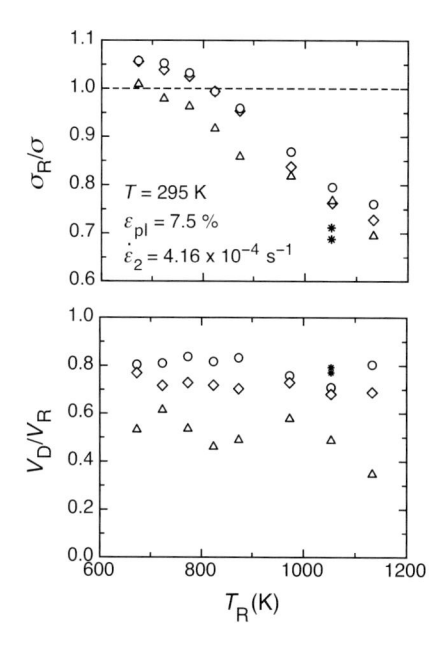

Figure 7.12 Influence of isochronal annealing on the flow stress σ and the reciprocal activation volume 1/V determined after room-temperature work hardening. Annealing performed at different temperatures T_R for a constant period of time $t_R = 120$ min. The flow stress σ and the activation volume V_D were determined at the end of the initial room-temperature deformation to $\varepsilon_P = 7.5\%$ plastic strain prior to annealing. The flow stress σ_R and the activation volume V_R were determined on reloading after annealing. For symbols see Table 7.3 [25].

to the values measured at the beginning of the predeformation, as seen in Figure 7.13. The result suggests that the deformation induced short-range glide obstacles can easily be annealed-out. The dipole and debris defects associated with the strain dependence of the activation volume are realized by a small amount of additional or missing material (Figure 7.9b). Thus, the rapid recovery of $1/V_D$ by annealing becomes plausible. The flow stress, however, is not completely recoverable; this indicates that long-range obstacles to dislocation motion, such as dislocation multipoles, are not removed at the relatively low annealing temperatures applied.

The activation energy Q_R of the recovery process was determined by analyzing the kinetics of isothermal annealing. The evaluation, performed according to the method proposed by Damask and Dienes [45], assumes that

i) the flow stress that can be recovered is proportional to number of defects n;

ii) annealing of defects occurs by a single thermally activated process with a constant activation energy;

iii) the change of the fractional defect concentration with annealing time is given by the equation $dn/dt_R = F(n) K_0 \exp(-Q_R/kT)$, where K_0 is constant and $F(n)$ is a continuous function of n.

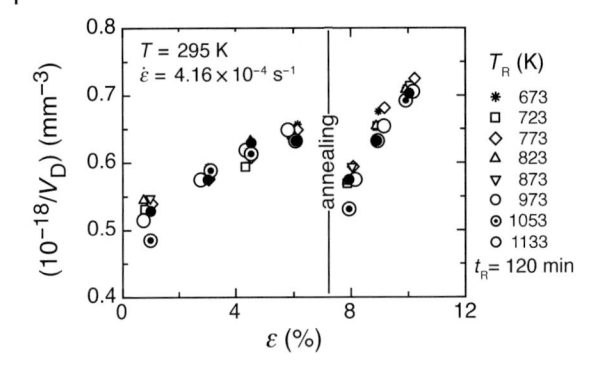

Figure 7.13 Strain dependence and recovery of the reciprocal activation volume. Predeformation at room temperature to $\varepsilon_P = 7.5\%$, followed by two-hour anneal at the annealing temperatures T_R indicated. Ti-47Al-2Cr-0.2Si, equiaxed microstructure [25].

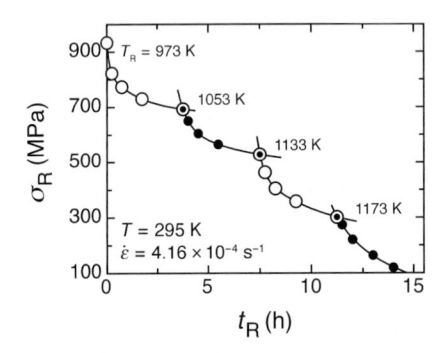

Figure 7.14 The kinetics of the flow stress recovery by isothermal annealing performed at different temperatures T_R after predeformation at room temperature to $\varepsilon_P = 7.5\%$. The flow stress σ_R was determined on reloading after annealing Ti-47Al-2Cr-0.2Si, equiaxed microstructure. For symbols see Table 7.3 [25].

Figure 7.14 shows the kinetics of flow stress recovery obtained from isothermal annealings performed at different temperatures. The activation energy was calculated from the equation [45]

$$Q_R = \frac{kT_1T_2}{T_2 - T_1} \ln \frac{S_2}{S_1} \tag{7.5}$$

The gradients $S = d\sigma/dt_R$ can be determined from isothermal annealings, as shown in Figure 7.10. The results of the evaluation are given in Table 7.4. For comparison the self-diffusion energies in γ(TiAl) are Q_{sd}(Ti) = 2.59 eV for Ti atoms and Q_{sd}(Al) = 3.71 eV for Al atoms [46]. The analysis performed at the lower temperatures T_R = 973 K/1053 K and T_R = 1053 K/1133 K gives Q_R values, which reasonably agree with the self-diffusion energy. This finding strongly supports the idea that the recovery of the work hardening is caused by the recovery of excess

Table 7.4 Compiled results for the activation energies Q_R for the recovery of the work hardening produced by predeformation at 300 K to strain $\varepsilon_P = 7.5\%$.

Alloy	Q_R (eV)		
	T_R: 973 K/1053 K	T_R: 1053 K/1133 K	T_R: 1133 K/1173 K
Ti-47Al-2Cr-0.2Si	3.6	3.3	5.1
Ti-52Al-2Cr	2.9	3.6	5.8
Ti-54Al	4.1	4.1	5.5

Evaluation of the recovery kinetics at the temperatures T_R indicated [25, 28].

vacancies and the debris produced during the predeformation. Direct evidence of the mechanism was provided by the *in situ* heating experiments that have been performed on predeformed samples (see Sections 6.2.5 and 6.3.1). The high Q_R values obtained at $T_R = 1133$ K/1173 K might be attributed to complicating factors that were not considered in the model of Damask and Dienes [45]; these involve

i) progressive change in the dislocation sink structure as a result of massive climb [47];
ii) participation of more than one type of defect in various stages of annealing,
iii) progressive changes of the microstructure at higher annealing temperatures.

It appears that the structural changes mentioned under (iii) become important at the highest annealing temperatures and lead to additional softening of the material. Possible processes are rearrangement of dislocations or formation of subboundaries. It is interesting to note that similar recovery characteristics of work hardening have also been observed on single-phase γ alloys [48]. This is surprising because the dislocation mechanisms in single-phase γ and two-phase alloys are known to be quite different (see Section 5.1 and Section 6.2). A prominent feature in the deformation structure of single-phase alloys is faulted dipoles, that is, extrinsic stacking faults bounded by Shockley partial dislocations. It might be speculated that these defects could also give rise to a deformation-induced thermally activated process, which, however, has yet to be confirmed. Viguier and Hemker [48] have recognized that the faulted dipoles are thermally unstable and attributed this instability to a localized reordering of the crystal structure by diffusive-like processes.

From the technical point of view, the susceptibility of the deformed state to recover could be a major concern in several applications. A significant amount of prior recovery lowers the driving force for static recrystallization, which might be important for the evolution of the microstructure upon annealing. Recovery, coupled with relaxation, is primarily responsible for loss of residual stress and loss of external load, which can occur in bolted fasteners at ambient temperatures. The early recovery is certainly detrimental for surface-hardening techniques, such as shot peening or roller burnishing, which are currently being explored for TiAl alloys to produce a crack-initiation-resistant outer layer. From the present investigations it might be expected that the surface hardening is only beneficial at low service temperatures, since the compressive stress will recover on high-temperature exposure.

Summary: Work hardening produced by room-temperature compression can be significantly recovered at moderately high temperatures. The low thermal stability of the deformed state can be attributed to annealing-out excess vacancies and dislocation debris.

7.2.5
High-Temperature Flow Behavior

The rate at which a material strain-hardens at elevated temperature is generally determined by the competition between mechanism that storage defects and the release of strain energy by dynamic recovery and recrystallization. There is good consensus that in TiAl alloys all these processes depend, apart from the deformation parameters, on alloy composition and microstructure [49–51]. The available information about this complex relationship will be discussed in the present section. The accent is on the physical mechanisms occurring during high-temperature flow. The discussion is complementary to Chapter 16, in which the structural evolution during hot working is considered.

The mechanisms governing flow behavior will be discussed on a series of single-phase and two-phase alloys that are briefly described in Table 7.5. Figure 7.15a shows the compression flow curves of these alloys measured at 1273 K [51]. Clearly, constitution and microstructure not only affect the yield strength, but the flow curve as a whole. The ingot material (alloy 2) has by far the highest strength with good temperature retention; this appears attributable to its coarse-grained lamellar microstructure. Another factor is probably the high Nb content, which reduces

Table 7.5 Composition and microstructure of the alloys investigated [51].

Alloy	Composition (at.%)	Microstructure
1	Ti-44.5Al	duplex, $D = 1$–$5\,\mu m$, $D_C = 20$–$30\,\mu m$
2	Ti-45Al-10Nb	fully lamellar, $D_C = 100\,\mu m$
3	Ti-46.5Al	duplex, nearly globular, $D = 1$–$5\,\mu m$
4	Ti-46.5Al-5.5Nb	nearly globular, $D = 1$–$5\,\mu m$
5	Ti-47Al-4.5Nb-0.2C-0.2B	duplex, nearly globular $D = 1$–$2\,\mu m$
6	Ti-45Al-8Nb-0.2C (TNB-V2)	duplex, $D = 5$–$10\,\mu m$, $D_C = 10$–$30\,\mu m$
7	Ti-45Al-4.5Nb-0.2C-0.2B (TNB-V5)	nearly globular $D = 1$–$8\,\mu m$
8	Ti-54Al	near-gamma, $D = 10$–$20\,\mu m$

D, grain size; D_C, colony size.

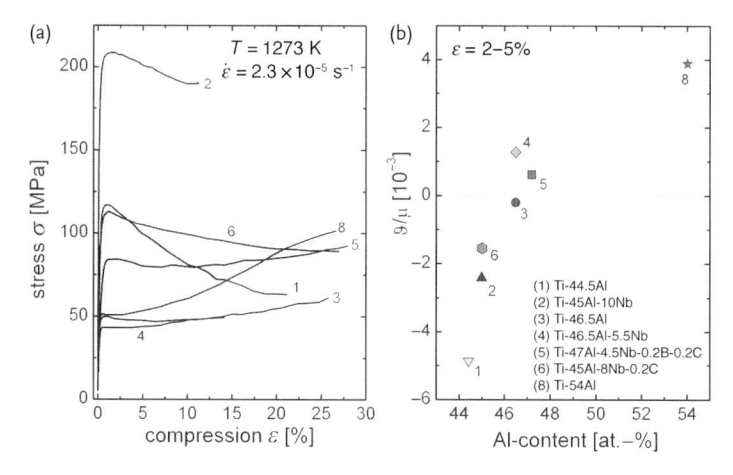

Figure 7.15 Compression behavior at $T = 1273$ K of the alloys listed in Table 7.5. (a) Flow curves of cylindrical compression samples and (b) work-hardening coefficients ϑ determined beyond yielding and normalized by the shear modulus $\mu = 56.29$ GPa at this temperature [51].

(1) Ti-44.5Al
(2) Ti-45Al-10Nb
(3) Ti-46.5Al
(4) Ti-46.5Al-5.5Nb
(5) Ti-47Al-4.5Nb-0.2B-0.2C
(6) Ti-45Al-8Nb-0.2C
(8) Ti-54Al

diffusivity (see Section 7.3). The extruded alloys 1 and 3 to 5 exhibit a wide variation in the grain size and volume content of lamellar colonies, which may explain the strength differences. In terms of this Hall–Petch argument on the stress dependence on grain size, the data for the single-phase material 8 compares well with those of alloys 3 and 4. Figure 7.15b shows the work-hardening coefficients determined in the steady-state region of work hardening beyond yielding (in most cases, at $\varepsilon = 2\%$ to 5%), normalized by the shear modulus $\mu = 56.29$ GPa at this temperature [30]. A well-expressed work softening is more or less common to all the two-phase materials that contain a significant volume of lamellar colonies. On the contrary, work hardening is observed when the microstructure is nearly globular (alloy 4) or duplex (alloy 5). This difference between the two-phase alloys is probably associated with a deformation instability, which occurs under compression [52]. In terms of a laminate model [53], the lamellar morphology may be considered as an ensemble of TiAl and Ti$_3$Al platelets. When perfectly aligned with the compression axis, these columns are highly stable under compression, as long as the axial load is below a critical value. Above this critical load, the equilibrium becomes unstable and the slightest disturbance will cause the structure to bend or buckle. Such a disturbance could develop by inhomogeneous dislocation glide and climb across the lamellae or by impinging mechanical twins. Furthermore, the elastic response of the α_2 and γ phases upon loading is significantly different [30]. Thus, the tendency toward unstable buckling will increase if there is an inhomogeneous distribution of α_2 and γ lamellae. Research by Imayev et al. [52] has shown that, in the region of highest local bending, spheroidization of the α_2 phase and recrystallization of the γ phase occur. These processes often proceed in the development of fine-grained shear bands, which, at higher degrees of deformation, traverse the whole work piece. It might be expected that deformation within shear bands can easily occur by grain-boundary sliding without significant hardening. In two-phase alloys that contain only a small volume of lamellar colonies, this deformation instability is almost absent, so that work hardening by conventional dislocation mechanisms and mechanical twinning probably becomes predominant. The single-phase γ alloy 8 exhibits the highest work-hardening rate, which at first glance could be attributed to the absence of the mechanical instability because no lamellar colonies are present. However, there are also other factors that could favor work hardening of alloy 8. As has been described in Chapter 5, superdislocations largely carry the deformation in Al-rich γ phase. The superdislocations are prone to various nonplanar dissociation reactions, resulting in extremely high friction stresses and the formation of sessile junctions. Such dislocation configurations are expected to be strong glide barriers for subsequently gliding dislocations.

Furthermore, cross-slip of extended dislocations is difficult because the partials first have to be brought together to form an unextended dislocation, before the dislocation can spread into the cross-slip plane. The effect of a dissociated core on dislocation climb is still a matter of debate. However, there are considerable speculations that jog nucleation and vacancy creation or destruction could involve a large energy barrier that makes climb slow [33]. Because cross-slip and climb are required for dynamic recovery, deformation-induced dislocation structures could be thermally more stable in Al-rich TiAl alloys than in Ti-rich alloys. This reasoning is in accordance with the observation reported by Imayev et al. [52] that the volume fraction of recrystallized grains in the microstructure of Ti-54Al deformed at 1273 K to strain $\varepsilon = 75\%$ is significantly smaller than that of $(\alpha_2 + \gamma)$ alloys.

The average values of the activation volume and of the experimental activation energy determined at strain $\varepsilon = 10\%$ are listed in Table 7.6, for the experimental details of the measurements see [51]. The activation volume was referred to the Burgers vector $\mathbf{b} = 1/2<110]$ of ordinary dislocations, which in most of the alloys

Table 7.6 Flow stresses σ, activation volumes V and experimental activation energies Q_e determined at $T = 1220$ to 1233 K on the alloys specified in Table 7.5.

Alloy	T (K)	σ (MPa)	V/b^3	Q_e (eV)
1	1226	200	18.3 ± 0.4	3.6 ± 0.4
2	1220	330	13.4 ± 0.2	3.8 ± 0.2
3	1225	143	25.6 ± 1.1	4.7 ± 0.3
4	1232	108	25.0 ± 1.5	5.1 ± 0.2
5	1224	162	17.4 ± 0.1	5.0 ± 0.3
8	1233	92	42.7 ± 1.1	5.1 ± 0.1

Data determined for strains $\varepsilon = 10\%$. V was referred to the Burgers vector of ordinary dislocations $\mathbf{b} = 1/2<110]$ [51].

investigated here determine the deformation behavior. The foregoing discussion has shown that hot-deformation of TiAl is very complex, not only involving dislocation propagation, but also dynamic recovery, recrystallization and phase transformation. These processes certainly occur sequentially with the slowest being rate controlling. Whereas a good consensus has been established in the TiAl literature that dynamic recovery is essentially controlled by dislocation glide and climb, the atomic mechanism of the dynamic recrystallization and phase transformation are only rudimentary understood. TEM examination of two-phase alloys deformed at high temperatures has shown that the lateral migration of internal boundaries is provided by the propagation of interfacial dislocations and growth ledges [54, 55]. The prominent feature is the formation of multiheight ledges perpendicular to the interfacial plane, which often grow into zones of about ten nanometers in width and apparently serve as a precursor for the nucleation of new grains [55]. In most cases these processes must be accompanied by diffusion in order to re-establish ordering, or in the case of phase transformation, to achieve the appropriate phase composition. Thus, as an approximation, all these processes will be considered as a single stress-driven and diffusion-assisted mechanism, which is supported by thermal activation. In view of this simplification, a direct correlation with existing climb models and diffusion data is tentative at best. As has been discussed in Section 6.4.2, the relatively large activation volume of $V = 13b^3$ to $43b^3$ is indicative of the superposition of dislocation climb with glide. The strength and distribution of the glide obstacles certainly depend on the constitution and microstructure, which could rationalize the observed variation of V with alloy composition.

The activation energy Q_p determined on alloys 1 and 2 is close to the self-diffusion energy of Al, which is the slower-diffusing species in TiAl. At moderately high temperatures, as in the present case, diffusion can probably be supported by antistructural disorder [46], which in TiAl alloys accommodates off-stoichiometric deviations [56], see Section 3.2. In this case the diffusivity depends on the nature of the antisite defect, that is, whether the alloy is Ti- or Al-rich. There is supporting evidence that diffusion by antistructutral bridges is much easier in Ti-rich than in Al-rich alloys [46]. This reasoning may qualitatively explain why Q_p depends on the alloy composition and why it is particularly high for the Al-rich alloy 8. As mentioned above, this high activation energy is certainly another factor that retards dynamic recovery of dislocation structures and thus stabilizes work hardening in Al-rich alloys.

Figure 7.16 combines all the flow stresses and related reciprocal activation volumes in a plot of σ versus $1/V$, which is suggested by Equation (7.3) [51]. Collectively the data for the duplex and equiaxed gamma alloys 1 and 3 to 8 fulfils relatively well the expected linear relationship. The intercept defines a small athermal stress part of $\sigma_\mu = 50$ MPa. The data for the fully lamellar alloy 2 seems not to fulfill the linear relationship defined by the duplex and equiaxed alloys. Although the data is scarce, it might be expected that the dependence of σ on $1/V$ for alloy 2 is essentially determined by its strong flow softening (Figure 7.15). More details are provided in [51].

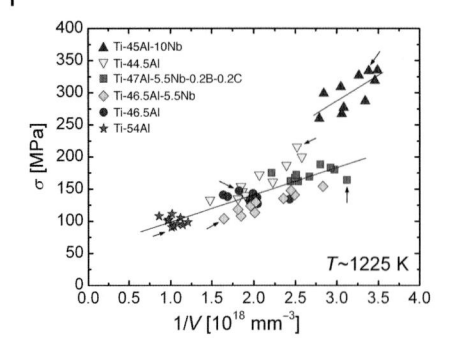

Figure 7.16 Dependence of the flow stress σ on the reciprocal activation volume 1/V determined at $T = 1225$ K for the alloys investigated. σ is the stress measured at the beginning of the relaxation for which V was also determined. The different data points belonging to an individual alloy were obtained by evaluating relaxations performed at different strains. The arrows mark the data determined at the beginning of deformation [51].

Summary: The flow behavior of TiAl alloys at elevated temperatures reflects the competition between work hardening and softening due to dynamic recovery and recrystallization. Work softening is common to all two-phase alloys that contain a significant volume of lamellar colonies. Single-phase γ alloys show a greater tendency to strain harden than two-phase alloys, presumably because dynamic recovery and recrystallization in single-phase alloys are retarded.

7.2.6
High Strain-Rate Deformation

Practical issues like damage tolerance, high-rate forming, machining, and foreign-object damage require an understanding of the material response to high strain-rate deformation. For TiAl alloys research in this field is rare. Malon and Gray [57] investigated the response of a Ti-48Al-2Nb-2Cr alloy with duplex structure upon shock loading with strain rates between $10^{-3}\,s^{-1}$ and $2000\,s^{-1}$ over the temperature range −196 °C to 1100 °C. It was demonstrated that the mechanical properties significantly changed compared to their usual quasistatic values; the material experienced a substantial increase in ultimate strength when subject to high loading rates. At room temperature, the strain-rate sensitivity of the yield stress determined between $\dot{\varepsilon}_1 = 10^{-1}\,s^{-1}$ and $\dot{\varepsilon}_2 = 2000\,s^{-1}$ was

$$(\partial \ln \sigma / \partial \ln \dot{\varepsilon})_T = 0.02 \tag{7.6}$$

This value corresponds relatively well to the strain-rate sensitivity that has been determined at lower strain rates of $\dot{\varepsilon} = 10^{-5}\,s^{-1}$ to $10^{-4}\,s^{-1}$, but otherwise comparable conditions; see Section 6.4.2. It should be noted that the strain-rate sensitivity in Section 6.4.1 was defined as $(\Delta\sigma/\Delta \ln \dot{\varepsilon})_T$. An increase of the strain rate from $10^{-3}\,s^{-1}$

to $2000 \, s^{-1}$ substantially increased the work hardening from $\vartheta/\mu = 0.037$ to 0.06. The thermally activated jog dragging and debris hardening discussed in Section 7.2.3 could rationalize this effect.

Interestingly, the stress–temperature profile determined with the highest strain rate $\dot{\varepsilon} = 2000 \, s^{-1}$ between $-196\,°C$ and $1100\,°C$ exhibited a minimum at $500\,°C$. TEM observations showed that deformation with shock-loading strain rates further increased the propensity for mechanical twinning in TiAl alloys. Twinning occurs on multiple $1/6 <11\bar{2}]\{111\}$ systems, which leads to numerous intersections of twin bands [58]. The authors pointed out that the extensive twinning could be a consequence of the particular stress state occurring in shock-loaded samples. Superimposed to the shear stress is a hydrostatic stress state, which suppresses crack nucleation. The damage evolution and cracking upon impact loading was investigated by Gray et al. [59]; the reader is referred to this paper for more details and references.

Summary: Under dynamic deformation conditions, material properties may be significantly different from their quasistatic values. Nevertheless, the yield-stress response to high strain rates superimposed at room temperature can be roughly estimated by utilizing the strain-rate sensitivity determined at low strain rates around $\dot{\varepsilon} = 10^{-5} \, s^{-1}$.

7.3
Solution Hardening

From the early 1990s until the present, a burst of research has been periodically directed on TiAl alloys with much of this work focused primarily on the optimization of alloy composition. A major aim of this research was to implement a higher degree of solid-solution strengthening. However, the addition of third elements affects the extension of the individual phase fields of the Ti–Al system (Chapter 2). Thus, microalloying alters the path by which the microstructure evolves with associated effects on the mechanical properties. This may easily mask strength variations caused by any extrinsic hardening mechanism. Thus, for an assessment of solid-solution strengthening, the effects of off-stoichiometric deviations and structural changes on the yield strength have to be considered. The advances that have been achieved by solid-solution hardening will be reviewed in this section. In view of the extensive body of literature, the frame of reference established here is restricted to a few examples, which may typify the general behavior.

7.3.1
Elemental Size Misfit of Solute Atoms with the TiAl Matrix

There are various interactions between solute atoms and dislocations. In metals, basically two mechanisms are considered. First, solute atoms that alter the lattice parameters will produce hardening that increases with the degree of size misfit.

Secondly, solute atoms that alter the shear modulus produce a hardening that increases with the modulus difference. Fleischer [60, 61] has classified the solute–dislocation interactions into strong and weak interactions according to the related increase of hardening with the solute concentration. Strong interactions arise from solutes that produce asymmetric distortions of the matrix. Such atoms interact with both edge and screw dislocations. When such asymmetric distortions are not present, the solute–dislocation interactions are weak. Substitutional solution hardening, which is mainly considered here, is generally of the soft hardening type, most often ascribed to a size misfit or modulus difference. In dilute alloys the hardening increment $\Delta\tau_s$ resulting from weak obstacles of concentration c_s is typically [61, 62]

$$\Delta\tau_s = \mu\, \varepsilon_s^{3/2}\, c_s^{1/2}/\alpha_s \tag{7.7}$$

ε_s is an appropriate sum of the size and modulus misfit and $\alpha_s \approx 700$. The value of α_s varies inversely with the unit cell volume and with the length of the Burgers vector; α_s is therefore likely to be smaller for an ordered compound than for a disordered counterpart of the structure.

Theoretical investigations have shown that the lattice parameters of ternary alloys depend on the bonding mechanism of the respective solute atom with the surrounding TiAl matrix and thus on its site occupation [63–70]. Substitution of Ti by a solute atom tends to increase the c/a ratio; whereas the opposite trend is expected if the solute occupies an Al site [66]. Song et al. [70] have briefly reviewed these calculations. Several authors [71–73] have measured the lattice parameters of several binary and ternary alloys with the base-line composition Ti-(44–50)Al+X. X designates the ternary elements Mn, Zr, Nb, Cu, and V, which were added with a total amount of 1 to 6%. The investigations showed the following trends:

i) off-stoichiometric deviation of binary alloys to higher Al contents left the lattice parameter a constant but increased the parameter c; hence, the tetragonality c/a increased [71, 73];

ii) Cr additions decreased a, c and c/a [72];

iii) Hf additions increased a and c in such a way that c/a decreased [72];

iv) Mn and Cu additions did not affect the lattice parameters [73];

v) Nb additions increase a and left c constant [72, 73];

vi) V additions reduced c and a in such a way that c/a decreased [73];

vii) Zr additions significantly reduced c and a, but c/a increased [73].

The authors associated these changes of the lattice parameters with the mechanical properties of their alloys. The relatively high yield stresses of the Zr-containing alloys were attributed to the strong change of the lattice parameters, resulting in substantial solution hardening. The authors have argued that a large c/a ratio is generally harmful for the materials ductility.

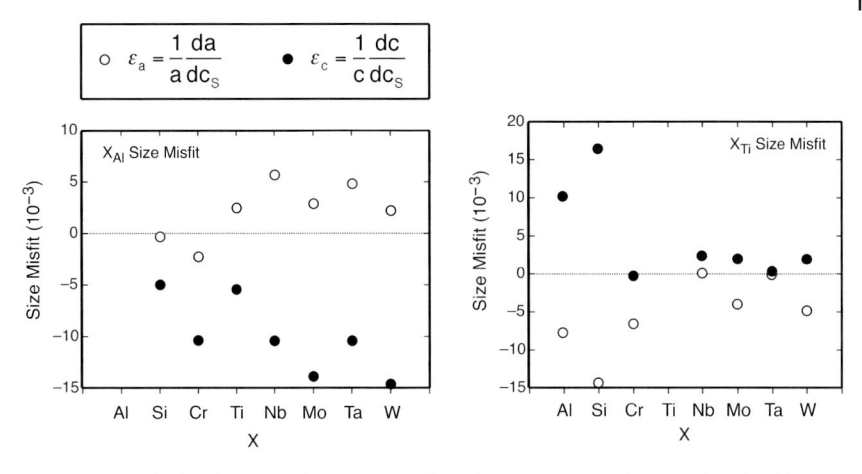

Figure 7.17 Calculated size misfit parameters for solute atoms in TiAl on (a) the Al sublattice and (b) on the Ti sublattice. Results of Woodward et al. [74], Figure redrawn.

Woodward et al. [74], utilizing first-principles calculations, have determined the size misfits of intrinsic lattice defects and several alloying elements. Figure 7.17 shows the predicted misfits along the a and c directions of the TiAl unit cell

$$\varepsilon_a = \frac{1}{a}\frac{da}{dc_S} \tag{7.8}$$

$$\varepsilon_c = \frac{1}{c}\frac{dc}{dc_S} \tag{7.9}$$

The difference between ε_a and ε_c expresses the asymmetry of the lattice distortion that is produced by a certain element. Accordingly, a strong asymmetric distortion is expected if Al is substituted by Nb, Mo, Ta, and W, or if Ti is substituted by Al or by Si. However, Woodward et al. [74] have pointed out that a large misfit is always associated with high formation energy, making such defect configurations unlikely. Thus, the potential of a certain substitutional element to harden the matrix must be seen together with its site preference (Section 3.2). Research performed by Rossouw et al. [75] have revealed that Nb, Hf, Zr, and Ta partition exclusively or preferentially at Ti sites of γ(TiAl). Elements partitioning preferentially at the Al sites include Ga, Mn, W, Mo, and Cr. It should be noted that the situation for some elements is more complex because their site preference depends on the stoichiometry of the host alloy. It is worth noting that the Al_{Ti} antisite center produces a strong asymmetric distortion. In Al-rich single-phase alloys a high density of these defects occurs, depending on the deviation from stoichiometry. Thus, Al_{Ti} antisite defects may provide a substantial hardening of Al-rich alloys. On the contrary, Ti_{Al} antisite defects, which occur in Ti-rich alloys, are relatively weak obstacles.

Summary: The potential of a solute element to harden TiAl alloys is determined by its size misfit, the asymmetry of its stress field and its site preference in the TiAl matrix.

7.3.2
Survey of Observations

In view of the problems mentioned in the introduction to this section, the most reliable information on solute effects may be extracted from investigations performed on PST crystals because in these materials the spurious effects of the microstructure can be minimized. PST crystals always have a fully lamellar microstructure, and by choosing an appropriate orientation ($\Phi = 45°$) of the loading axis with respect to the lamellae, shear deformation parallel to the lamellar interfaces can be achieved that is not affected by the lamellar interfaces but only by the relatively weak domain boundaries. For Ti-(48-49)Al PST crystals, solution hardening has been recognized when 0.6% to 1% of either V, Cr, Mn, Mo, Ta, Nb, or Zr was added [76, 77]. Among these elements, Zr, Mo, Ta, and Nb seem to be most effective for strengthening γ alloys.

Several groups have studied the effects of ternary additions on the yield strength of polycrystalline materials [72, 78–87]. In a two-phase alloy with the base-line composition Ti-48Al, Cr was found to be an effective strengthener, while Mn was not [72, 77, 82, 84]. However, research performed by Zheng *et al.* [88] on Ti-45 Al and Ti-45Al-3Cr has shown that the addition of Cr causes significant structural changes, involving formation of dispersed B2 particles, an increase in the volume fraction of γ phase and degeneration of $α_2$ lamellae. Thus, the effect of Cr may depend on the Al content. Cheng *et al.* [89] have investigated a series of TiAl alloys with the general composition Ti-44Al-8(Nb, Ta, Zr, Hf)-(0–0.2)Si-(0–1)B, which had been subjected to standard processing. The main intention of this investigation was to evaluate the potential of heavy alloying for improving the structural and mechanical performance of γ-based alloys. The authors recognized that Ta is a strong solution strengthener in this type of alloys. Atom probe field ion microscopy performed by Larson *et al.* [90] on Ti-47Al-2Cr-1Nb-0.8Ta-0.2W-0.15B has shown that the majority of the tantalum additions in the $α_2$ and γ phases remains in solution. However, Ta was found to affect the equilibrium composition of the majority phases. Furthermore, Ta does appear to change precipitation of the boride phase and seems to be responsible for the formation of a Ti(Cr, Al)$_2$ phase that contains significant amounts of Ta and W. These findings clearly underline the problems in assessing solute strengthening in two-phase alloys.

Finally, it should be mentioned that heavy alloying with elements that stabilize the high-temperature β phase is currently being explored. The main intention of these studies is to achieve additional freedom in hot-working operations and in controlling the microstructure. This aspect will be discussed in Chapter 16.

Summary: Alloying with modest levels (<3 at.% to 5 at.% in total) of metallic elements like Cr, V, Cr, Mn, Mo, Ta, Nb, and Zr provides some potential for the solution hardening of two-phase alloys. However, the hardening effects are often masked by structural changes that are simultaneously initiated by the alloying.

7.3.3
Effect of Solute Niobium

Early investigations by Huang [85] and Chen et al. [91] have shown that a significant strengthening of alloys based on Ti-45Al can be achieved when Nb is added to a level of 10 at.%. Despite the extensive body of investigations [92–94] broadly confirming this finding, there is ongoing controversy about the nature of the strengthening effect of Nb additions, that is, whether it arises from solid-solution hardening or from a change of the microstructure. In a series of publications the group around Chen et al. [95–97] has taken the view that the hardening effect is caused by solid-solution strengthening. However, there are several difficulties with this scenario. Nb has a solubility of about 20 at.% in TiAl alloys, depending on temperature [98]. Atom location by channeling-enhanced microanalysis (ALCHEMI) studies have revealed that Nb solely occupies the Ti sublattice of TiAl [75], a result that was confirmed by first-principles calculations [70, 74]. The size misfit between Ti and Nb atoms is at maximum 0.2% (Figure 7.17); thus, this seems unlikely to completely explain the observed strengthening effect. As has been discussed in Section 7.3.1, strong solute–dislocation interactions are only expected for large misfits or defects that introduce noncentrosymmetric distortions. In a recent paper, Zhang et al. [96] have proposed that, in spite of the reported strong site-occupancy preference, some Nb may go to the Al sublattice. In this environment the Nb atoms is thought to provide a maximum misfit of about 1% (Figure 7.17) and could act as an effective hardening agent. However, the effect appears to be too small to be decisive.

Systematic research performed by Paul et al. [92] and Appel et al. [93, 94] has shown that the structural changes occurring in this alloy system are of perhaps greater significance. In these studies, the origin of the hardening mechanism was investigated in that different Nb-containing alloys were compared with their binary counterparts. The hardening mechanisms occurring in these alloys were analyzed in terms of thermodynamic glide parameters. Figure 7.18 shows the temperature profiles of the yield stress and of the related reciprocal activation volumes for these alloys. The yield stress is almost independent of the Nb content, but is closely related to the Al content. In particular, no strengthening effect occurred when Nb was added to Ti-48Al or to the Al-rich single-phase alloy Ti-54Al [94]. Against the hypothesis of Nb solution hardening is also the outcome of the thermodynamic analysis. Solution hardening is a typical example of a thermally activated process, which should be manifested in the activation volume. As can be seen in Figure 7.18, this is clearly not the case; thus, the Nb_{Ti} defect center does not provide a significant glide resistance. Taken together, these factors suggest that the strengthening effect of Nb additions in Al-lean alloys is mainly caused by a change of the microstructure. In the Ti–Al system, alloying with Nb generally decreases the $\beta(B2)$- and α-transus temperatures, respectively, and contracts the α phase field

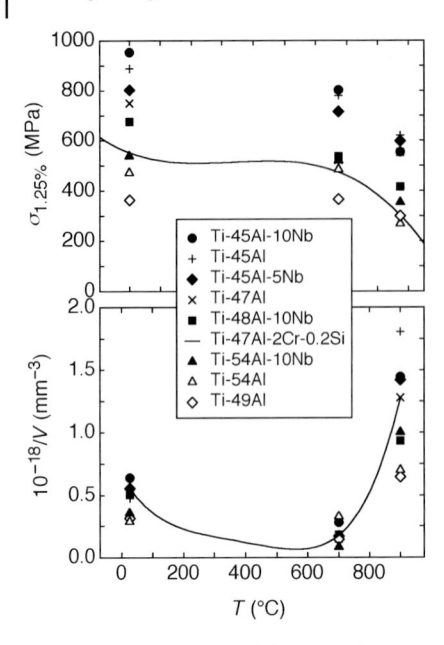

Figure 7.18 Assessment of the strengthening effect of Nb additions to TiAl alloys. Dependence of the flow stress σ and of the reciprocal activation volume $1/V$ of binary and Nb-containing alloys on the deformation temperature. The drawn lines refer to the values of a Ti-47Al-2Cr-0.2Si alloy with a near-gamma microstructure. Values measured in compression at strain $\varepsilon = 1.25\%$ and $\dot{\varepsilon} = 4.16 \times 10^{-4}$ s^{-1} [92–94].

[99]. This modification of the phase stabilities probably leads to a significant structural refinement [92–94]. Figure 7.19 demonstrates the fine scale of grains that has been observed in an extruded Ti-45Al-10Nb alloy with duplex microstructure. Electron microscope observations [92–94] have shown that the lamellar colonies in such alloys have lamellar spacings at the nanometer scale. Furthermore, as might be expected from the phase diagram, the density of α_2 lamellae is relatively high. The high-resolution micrograph shown in Figure 6.3 illustrates these features. These observations were later confirmed by Liu et al. [97] and Fischer et al. [100].

The dense arrangement of interfaces impedes dislocation glide and twinning and is undoubtedly the most important source of the high yield strength. In a later paper, Liu et al. [97] acknowledged the importance of this structural refinement for the strengthening of Nb-bearing and Al-lean alloys. Thus, the high yield stress of Nb-containing alloys can be rationalized in terms of a Hall–Petch mechanism [92–94].

However, in spite of the similarity in the mechanisms determining the yield strength and the activation volume of binary and Nb-containing alloys, several other aspects of the deformation behavior are notably different. At room temperature the Nb-bearing alloys exhibit an appreciable ductility, whereas the binary

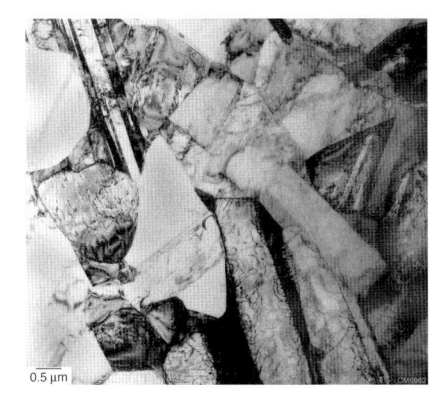

Figure 7.19 Duplex microstructure of a Ti-45Al-10Nb alloy, which had been extruded below the α-transus temperature.

alloys with the same Al content do not. This finding indicates that the activation of glide processes is supported by Nb additions. In Nb-containing alloys an abundant activation of twinning has been recognized at all deformation temperatures, as demonstrated in Figure 7.20. Thus, mechanical twinning is expected to be relatively easy in Nb-bearing alloys. As has been discussed in Section 6.3.3, there is good evidence that the intrinsic stacking fault (SISF) energy of γ(TiAl) is lowered by the Nb additions. TEM evidence supporting this view was provided by Yuan et al. [101], who recognized a correspondingly large dissociation width of the 1/2<112] dislocations in Ti-48-5Nb. However, investigations performed on Ti-(45-49)Al-10Nb alloys and the binary counterparts have show that the Al content could also affect the stacking-fault energy [102, 103]. Chen et al. [104, 105] observed a significant variation of the local composition in Ti-47Al-2Mn-2Nb + 0.8 vol.% TiB_2 with a duplex microstructure. In the γ lamellae the concentration of Al was lower and the Nb concentration was higher than those of the primary γ grains. An abundant occurrence of stacking faults was observed, in the Nb-rich lamellae, whereas stacking faults were scarce in the primary γ grains. This led the authors to believe that Nb additions reduce the stacking-fault energy of γ(TiAl).

Deformation in Nb-containing alloys is also carried by dislocation glide [92–94, 106]. This can be seen in Figure 7.21, which shows ordinary dislocation activity within a small γ domain. Similar to the process of twin propagation, the slip path of the dislocations is limited by the fine scale of the microstructure. The important point to note is that significant glide of superdislocations was recognized in Nb-containing alloys, which in binary two-phase alloys is seldom observed (Figure 7.22). Dissociated superdislocations and mechanical twins often coexist in the same grain or lamella, again suggesting that the SISF energy of γ(TiAl) is lowered by the Nb additions, and that twin nucleation originates from the superposition of extended stacking faults on alternate {111} planes. In terms of the von Mises criterion, it might be expected that the relatively homogeneous activation of

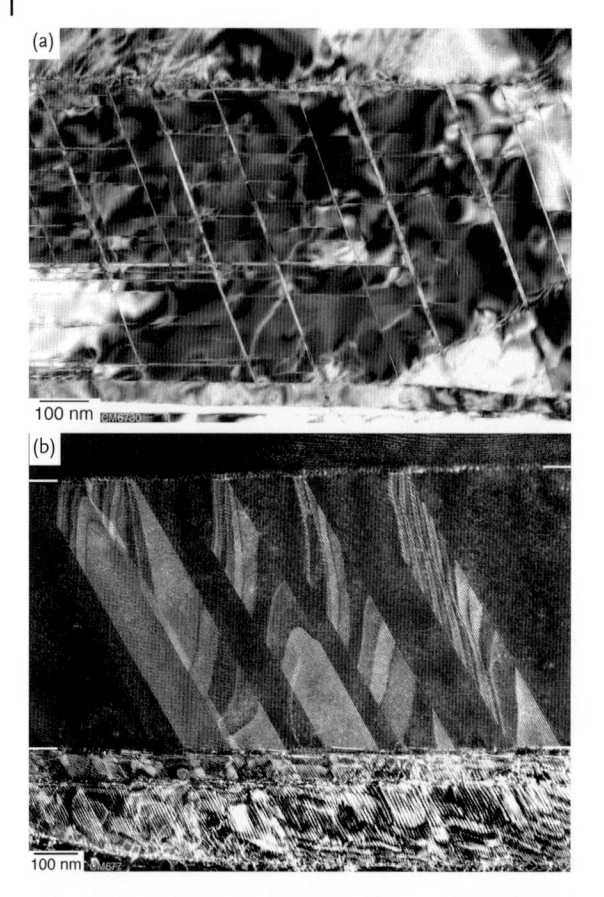

Figure 7.20 Mechanical twinning in lamellar Ti-45Al-5Nb. TEM observations made after compression to strain $\varepsilon = 3\%$. (a) Intersection of twin bands in a γ lamella observed after room-temperature compression. (b) Deformation twins generated under compression at 700 °C; twinning partial dislocations are piled up against a lamellar boundary [92].

ordinary dislocations, superdislocations and mechanical twinning is beneficial for the low-temperature ductility.

Concerning high-temperature deformation, the important point to note is that the activation enthalpy, ΔH, of the alloys containing a large amount of Nb is significantly higher than that of other alloys. The results of these mechanical measurements are listed in Table 7.7. The data agree reasonably with those of diffusion studies performed on Nb-containing alloys [46, 107], which revealed that Nb is a slow diffuser in γ(TiAl). The results imply that diffusion-assisted transport processes might be impeded in these materials, which, in terms of the climb theory is a good precondition for high-temperature strength. Another significant consequence of the reduced diffusivity is that the tendency of the material to coarsen is apparently reduced so that the very fine microstructure and the resulting glide resistance are maintained up to relatively high temperatures.

7.3 Solution Hardening

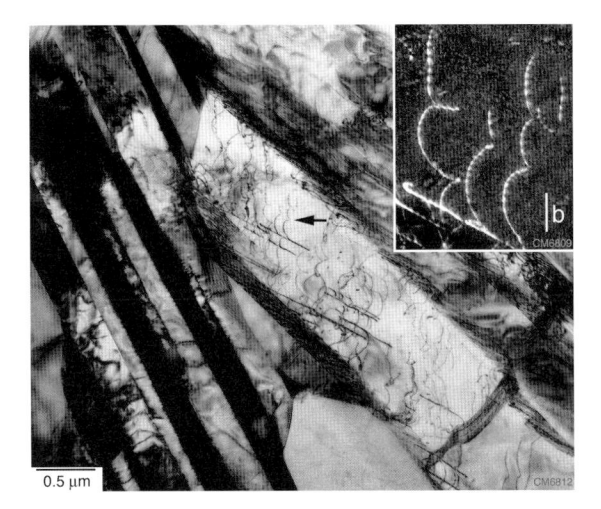

Figure 7.21 TEM micrograph of Ti-45Al-5Nb deformed at 973 K to strain $\varepsilon = 3\%$, which shows activity of ordinary dislocations within a small γ domain. Insert shows the arrowed region with extensive pinning of screw-type dislocations (pseudoweak-beam image using $g = \{111\}$ reflection in the $g/3.1g$ condition).

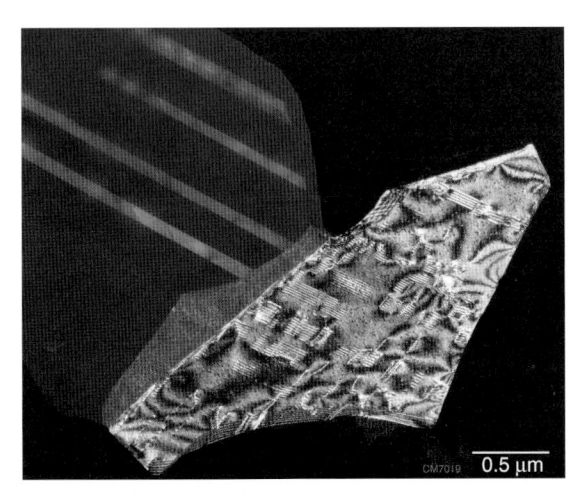

Figure 7.22 Deformation structure observed in an extruded Ti-45Al-10Nb alloy observed after tensile deformation at room temperature to a plastic strain $\varepsilon = 1.5\%$ and flow stress $\sigma = 1050$ MPa. Note the widely dissociated superdislocations, stacking faults and twins coexisting in a deformed γ grain [92].

The important point to note from Figure 7.18 is that a Nb content of 5 at.% is sufficient in order to achieve a good balance of mechanical properties. This might be important for weight saving and for tailoring microstructures. Nb is also a commonly added element because of its ability to improve oxidation resistance. Thus, TiAl alloys containing Nb additions of 5 to 10 at.% exhibit several desired

Table 7.7 Flow stresses and activation enthalpies of different TiAl alloys measured at 1100 K demonstrating the effect of Nb additions.

Alloy composition (at.%), microstructure	$\sigma_{1.25}$ (MPa)	ΔH (eV)
Ti-45Al, fully lamellar	634	3.71
Ti-45Al-10Nb, nearly lamellar	672	4.19
Ti-45Al-5Nb, fully lamellar	656	4.46
Ti-48Al-10Nb, equiaxed gamma	459	4.04
Ti-47Al-1.5Nb-1Mn-1Cr-0.2Si-0.5B, duplex	458	2.9
Ti-48Al-2Cr, nearly lamellar	360	2.04
Ti-49Al-1V-0.3C, near-gamma	476	3.07
Ti-47Al-2Cr-0.2Si, equiaxed	368	3.16
Ti-47Al-2Cr-0.2Si, lamellar	435	3.13
Ti-54Al, gamma	393	3.56

Values estimated at the beginning of deformation at $\varepsilon = 1.25\%$. $\sigma_{1.25}$, flow stress at strain $\varepsilon = 1.25\%$; ΔH, activation enthalpy. Data from [92].

attributes that have the potential for expanding the service range of titanium aluminides towards higher temperatures.

Summary: Nb-bearing alloys based on the general composition Ti-45Al-(5-10)Nb have a good balance of mechanical properties. The beneficial effect of the Nb additions can be attributed to the following factors: (i) structural refinement, (ii) reduction of the stacking-fault energy and (iii) reduction of diffusivity.

7.4
Precipitation Hardening

For TiAl alloys dispersion strengthening has been explored as a way to improve the high-temperature strength and creep resistance. The elements B, C, N, O, and Si have been investigated as additives to form hard precipitates in the TiAl phase. The driving force for the precipitation is the limited solubility of these elements in the γ matrix, which is typically at a few hundred at. ppm. As generally observed on other alloys, the shape, size and distribution of the second-phase particles are the principle factors determining the mechanical performance. In this respect, hardening by carbides has received most attention, as the optimum dispersion can be achieved by homogenization and aging procedures. Thus, hardening by carbides will be considered first.

7.4.1
Carbide Precipitation in TiAl Alloys

The group around Nemoto [108–112] has systematically investigated the precipitation reactions occurring in Ti-51Al containing 0.5 at.% C. For comparison, the maximum solubility of carbon determined in binary Ti-(46–48)Al alloys is in the range 200 to 300 at. ppm [87]. The alloys were subject to solution annealing at 1423 K and then quenched. Subsequent aging resulted in two types of precipitates. Aging at about 1023 K produces perovskite carbide Ti$_3$AlC, which seems to be homogeneously nucleated in the γ matrix. The perovskite precipitates exhibit the orientation relationships with the TiAl matrix [109–113]

$$(001)_P \parallel (001)_M, <100]_P \parallel <100]_M, [001]_P \parallel [001]_M \tag{7.10}$$

P and M designate the precipitate and the γ matrix, respectively. There is a significant lattice mismatch between the precipitates and the γ matrix, which may be expressed as

$$\varepsilon_M = \frac{2(a_M - a_P)}{a_M + a_P} \tag{7.11}$$

a_M and a_P are the lattice parameters of the matrix and the precipitate. The values estimated in the **a** and **c** directions are $\varepsilon_M^a = -0.057$ and $\varepsilon_M^c = -0.021$, respectively [110]. The lattice parameters determined by Schuster *et al.* [114] have been used for this estimation. This misfit anisotropy is probably the reason why the perovskite precipitates predominantly have a road-like morphology, with the long axis parallel to $[001]_P$. This can be understood by the following reasoning. As long as the precipitates are relatively small, the elastic strain built up between the precipitates and the matrix is nearly isotropic. However, with increasing aging time, both the matrix and the precipitate will generate more anisotropic strain fields at the interface, which will in turn influence the growth of the precipitates. Hence, in order to reduce strain energy, the perovskite precipitates preferentially grow along $[001]_P$. In an early investigation, the interface was observed to be coherent [113]. However, as has been discussed in Section 6.3.3, the nature of the interface could depend on the particle size. Semicoherent interfaces have been observed by high-resolution TEM examination of perovskite precipitates in Ti-48.5Al-0.37C [115].

Aging of the homogenized material at 1073 to 1173 K leads to the formation of the hexagonal H phase Ti$_2$AlC [112, 113]. The H phase has a plate-like morphology; the orientation relationships with the γ phase are

$$(0001)_H \parallel \{111\}_M \text{ and } <11\bar{2}0>_H \parallel <\bar{1}01]_M \tag{7.12}$$

The H phase exhibits a strong mismatch with the γ matrix. The largest misfit occurs along $[0001]_H$, which probably restrains the plate growth in this direction. In two-phase alloys, the H phase was found to preferentially precipitate at lamellar interfaces or in regions of former α$_2$ phase [116, 117]. First-principles calculations performed by Benedek *et al.* [118] have shown that the interface energy of the H phase is significantly higher than that of the perovskite phase. The authors argued that this could be why the H phase, unlike the perovskite phase, is heterogeneously nucleated. In the literature, another hexagonal phase, Ti$_3$AlC$_2$, is mentioned [119], but has not been observed so far as a precipitate in TiAl.

Summary: In carbon-containing TiAl alloys aging procedures can form ternary precipitates. The perovskite Ti$_3$AlC phase homogeneously nucleates at relatively low aging temperatures of about 1023 K. The hexagonal H phase Ti_2AlC occurs at somewhat higher temperatures of 1073 to 1173 K. The characteristic morphology of the carbides can be rationalized in terms of their lattice mismatches with the TiAl matrix.

7.4.2
Hardening by Carbides

There are various interactions between dislocations and precipitates; a gliding dislocation must either cut through the particles or penetrate the array by bowing between the obstacles. In order to account for the different mechanisms, the total flow stress may be subdivided into thermal and athermal contributions, analogous to Equations 6.24 and 6.33. Christoph et al. [120], utilizing this approach, have analyzed the strengthening effect of carbon additions after different thermal treatments. TiAl alloys with the baseline composition Ti-48.5Al were prepared by arc melting and systematically doped with carbon so that the carbon content c_C varied between 0.02 and 0.4 at.%. The following three thermal treatments were performed on these alloys:

i) hot isostatic pressing (HIP) at 1458 K and 1.4 kbar followed by slow cooling; this resulted in the formation of coarse H-phase platelets and a small amount of Ti_3AlC perovskite phase;

ii) annealing at 1523 K followed by water quenching to produce a carbon solution;

iii) following step (iii), aging at 1023 K for 4 h to form Ti_3AlC perovskite precipitates.

Figure 7.23 shows the yield strength and the reciprocal activation volume of the alloys for these three thermal treatments as a function of the carbon concentration. For the quenched materials, in which carbon is thought to be present as solute atoms (or tiny agglomerates), the flow stress was found to be nearly independent of carbon concentration c_C (Figure 7.23a). At room temperature the reciprocal activation volume $1/V$ of these materials slightly increased with c_C, which indicates that the density of short-range obstacles increases with c_C. It is therefore concluded that carbon atoms in solid solution act as weak glide obstacles that can apparently be easily overcome with the aid of thermal activation. Thus, although the mechanism is manifested in the activation volume, it is rather ineffective for hardening the material. In contrast, the flow-stress increased with c_C after the material had been aged and carbon was present as Ti_3AlC precipitates. At room temperature, the reciprocal activation volume of these materials slightly decreases with c_C. A natural explanation of this behavior is that the perovskite precipitates are glide obstacles with long-range stress fields, which cannot be overcome with the aid of thermal activation. Thus, the flow-stress increase is not associated with an increase of $1/V$. This is also suggested by the fact that the activation volumes of the aged materials are quite similar to those of conventional (undoped) two-phase titanium aluminides (Section 6.4.2). The Gibbs free energy of the aged materials at room

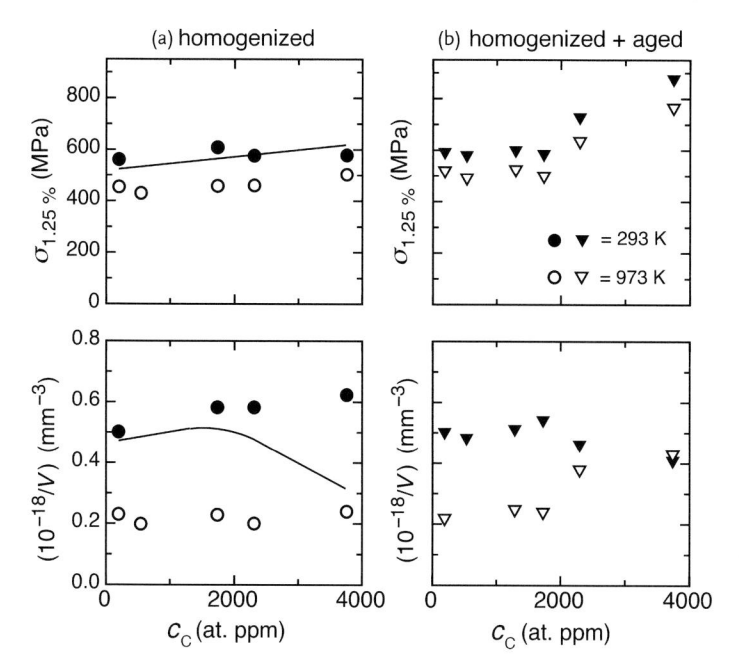

Figure 7.23 Precipitation hardening of two-phase titanium aluminide alloys with the baseline composition Ti-48.5Al-(0.02–0.4)C. Dependence of the flow stress $\sigma_{1.25}$ and of the reciprocal activation volume $1/V$ on the carbon concentration c_C. Values estimated at the deformation temperatures 293 K and 973 K and $\varepsilon = 1.25\%$. (a) Homogenized and quenched alloys with carbon in solid solution. The drawn lines refer to the room-temperature values of the materials in the as-HIPed condition, where the carbon is present as a coarse dispersion of H phase Ti$_2$AlC and a small amount of Ti$_3$AlC phase. (b) Quenched and aged materials that contain a fine dispersion of Ti$_3$AlC perovskite precipitates. Data from Christoph et al. [120].

temperature was found to be independent of carbon concentration; the average value being

$$\Delta G = 0.7 \text{ eV} \tag{7.13}$$

This value is very similar to those estimated on carbon-free two-phase alloys. The glide resistance determining the dislocation velocity in these materials has been attributed to lattice friction, jog dragging and antisite defects (Section 6.4.2). Because of the close agreement of the activation parameters, these mechanisms are also thought to control the dislocation velocity in the aged carbon-doped material. This implies that the strengthening effect of the perovskite precipitates is athermal in nature. The fact that the high flow stress of the heavily doped materials is maintained up to 973 K supports this view. Figure 7.23a also includes the values of carbon-doped materials that were tested after HIPing followed by slow cooling. Due to this treatment most of the perovskite precipitates are probably transformed into coarse particles of H phase Ti$_2$AlC [111, 113, 116, 117, 120]. In this condition the flow stress of the materials is relatively low and almost independent of carbon concentration. The related reciprocal activation volumes decrease with c_C, which clearly indicates that the Ti_2AlC particles are also acting as athermal glide obstacles. However, the coarse dispersion of this phase seems to be less effective in strengthening the material. This observation agrees with the widely accepted view that both the size and dispersion of the particles are important for effective precipitation strengthening.

The high glide resistance provided by the perovskite precipitates became evident in TEM observations performed on Ti-48.5Al-0.37C. The precipitates had an average length of $l_P = 22$ nm and a width (along <100] of $d_P = 3.3$ nm. The investigations revealed that all types and characters of perfect and of twinning partial dislocations, including superdislocations, contributed to deformation at room temperature (Figure 7.24a). This is different from deformation of undoped two-phase alloys, where deformation at room temperature is mainly provided by ordinary dislocations (Figure 7.24b), and glide activity of superdislocations is scarce. The dislocations are strongly bowed out and apparently penetrate the obstacle array along paths of easy movement. The glide resistance provided by the perovskite precipitates has been specified in terms of dislocation–point obstacle interactions [120, 121]. Figure 7.25 demonstrates the anchoring of a superdislocation with Burgers vector $\mathbf{b} = [011]$ at the precipitates. The effective shear stress τ_C acting on the pinned segments was estimated by analyzing their curvature. By comparison with the observed loop shapes, effective shear stresses of $\tau_C = 300$ MPa were typically obtained. This value can be converted into a normal stress $\sigma_C = M_T \tau_C = 900$ MPa,

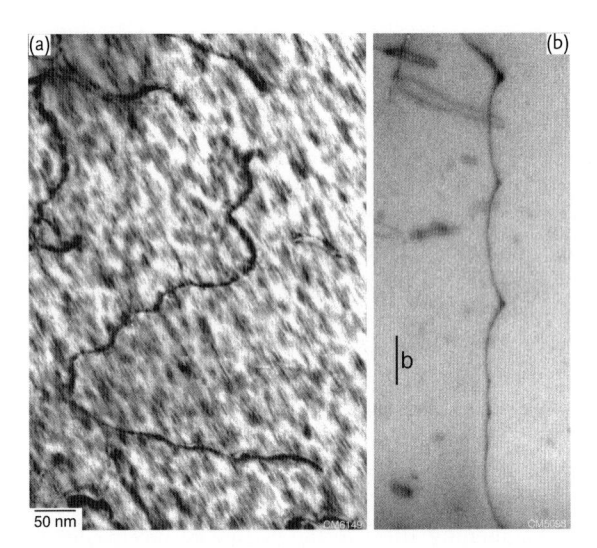

Figure 7.24 Effects of hardening by perovskite precipitates on the dislocation structure. (a) Pinning of superdislocations with Burgers vector $\mathbf{b} = [011]$ in Ti-48.5Al-0.37C. The dislocations penetrate the obstacle array deeply along paths of easy movement and apparently become immobilized at groups of unfavorably arranged particles. Note that all dislocation characters occur. (b) An ordinary screw dislocation in Ti-48Al-2Cr. The two micrographs were taken after compression at room temperature to strain $\varepsilon = 3\%$. The magnification marker and the direction of the Burgers vector hold for both micrographs.

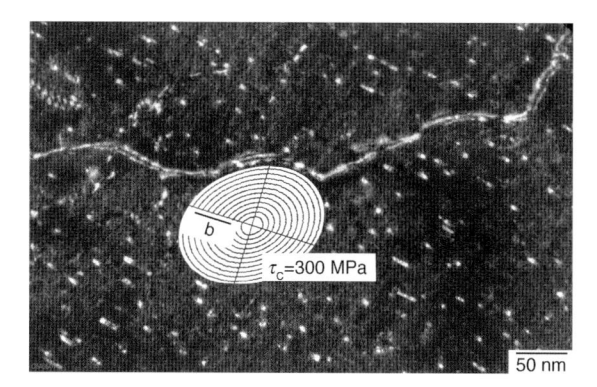

Figure 7.25 Estimation of the effective stress acting on the dislocation segments of a [011] superdislocation pinned by perovskite precipitates. Pseudoweak-beam image recorded using $g = (002)_{TiAl}$ reflection from near the [020] pole in the g/3.1g condition. Note the high density of Ti_3AlC-precipitates that are manifested by strain contrast. The insert shows line-tension configurations calculated for different stresses and projected into the foil plane. For the segment analyzed, the length $l_c = 110$ nm, the half-axis $q = 80$ nm and effective shear stress $\tau_c = 300$ MPa were determined. Data from Christoph et al. [120].

which is consistent with the flow stress $\sigma = 1000$ MPa measured after macroscopic deformation to a strain of $\varepsilon = 3\%$. $M_T = 3.06$ is the Taylor factor. The obstacle distance of the perovskite precipitates along the dislocations was $l_C = 50$ to 100 nm. The <011] superdislocations are typically bowed out at the particles through angles of $\psi = 110°$. This corresponds to interaction forces

$$f^* = 1.5 \times 10^{-8} \text{ N or } f^*/\mu b^2 = 0.57 \tag{7.14}$$

and leads to the conclusion that the obstacles are still shearable and can be overcome by the dislocations without Orowan looping.

The high glide resistance of the perovskite precipitates has also been recognized in twin structures. The twinning partial dislocations exhibit strong interactions with the precipitates and are often immobilized. The interaction of the twinning partial dislocations with the precipitates leads to a distinct twin morphology that is demonstrated in Figure 7.26a. The twinning dislocations are strongly bowed out and become immobilized at groups of unfavorably arranged particles. Such processes apparently locally impede twin growth and lead to a significant variation of the twin thickness. Thin twins are fragmented, that is, islands of untwined regions occur. Figure 7.26b is a high-resolution image of an immobilized twin. Before being finally terminated, the thickness of the twin is reduced by three (111) planes. A detailed analysis [115] has shown that the strain accommodation at the triple step occurs by the recombination of twinning partial dislocations into an emissary lattice dislocation with the Burgers vector $\mathbf{b} = 3 \times 1/6[11\bar{2}]$. Indeed, an isolated dislocation (arrowed) can be recognized close to the twin tip. The compressed image, Figure 7.26c, shows this more clearly. This dislocation can slip away, leaving behind a stress-free ledge with almost no shear discontinuity across the interface. Figure 7.27 shows the structure of a twin in the local region of a precipitate

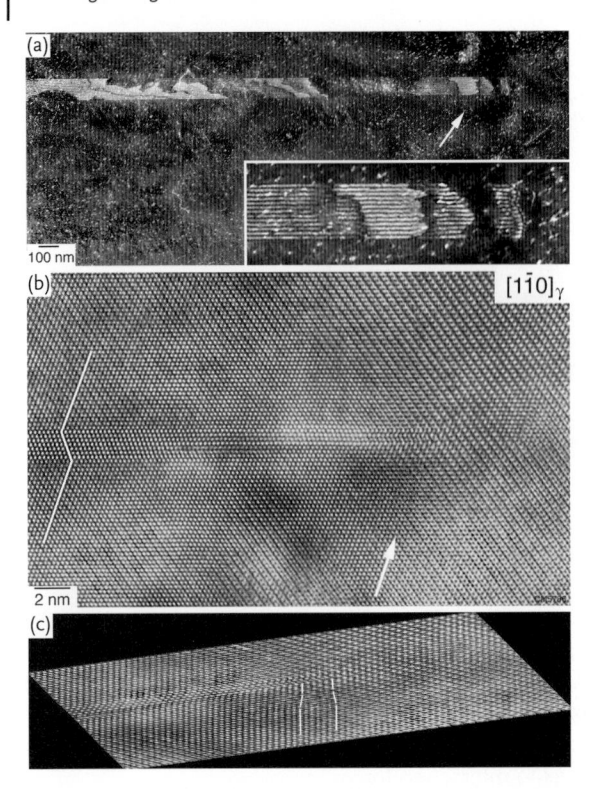

Figure 7.26 Interaction of mechanical twins with Ti$_3$AlC perovskite precipitates in an aged Ti-48.5Al-0.37C alloy observed after compression at $T = 300$ K to strain $\varepsilon = 3\%$. (a) Pinning of twinning partial dislocations by the precipitates, which are manifested by strain contrast. Note the fragmented morphology of the twin, the anchoring of the partial dislocations, and the immobilization of the twin. The insert shows the pinning of the Shockley partial dislocations by perovskite precipitates. Pseudoweak-beam image recorded using $g = (002)_{TiAl}$ reflection from near the [020] pole in the $g/3.1g$ condition. (b) Immobilization of a mechanical twin. Note the three-plane step in one of the (111) twin matrix interfaces and the isolated dislocation adjacent to the twin tip. The dislocation can more easily be recognized in figure (c), which shows the arrowed area, compressed along the $(\bar{1}\bar{1}1)$ matrix planes.

in a high-resolution micrograph. The twin appears deflected and exhibits highly disturbed twin/matrix interfaces, a morphology that might be explained by geometrical reasoning. The rod-like precipitates obliquely intersect the {111} twin habit plane (Figure 7.27a). Thus, the twinning dislocations situated on the incoherent twin/matrix interface become immobilized so that a tapered twin/matrix interface is formed (Figure 7.27b). The process leaves a highly stressed L1$_0$ structure, with faults on every third {111} plane. For more details, see [115]. From the observed pinning effects one is left with the impression that hardening by perovskites may completely embrittle an already brittle material. However, as has been described in Section 6.3.3, twin nucleation in the precipitation-hardened material seems to be relatively easy. In this way, a fine dispersion of twins can be

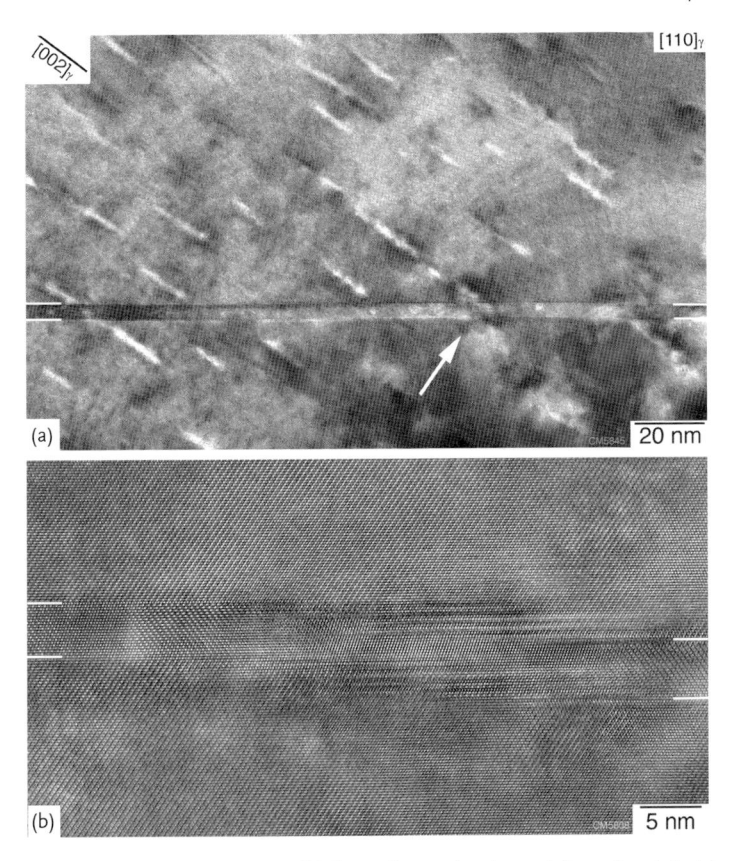

Figure 7.27 Atomic structure of deformation twins interacting with Ti$_3$AlC perovskite precipitates in Ti-48.5Al-0.37C observed after compression at $T = 300$ K to strain $\varepsilon = 3\%$. (a) Low-magnification high-resolution image showing a deformation twin intersecting rod-like perovskite precipitates aligned along [002]. (b) Higher magnification of the region arrowed in (a) showing the twin structure in the local region of the precipitate [115].

generated, which may partially compensate for the high glide resistance produced by the pinning processes.

Summary: Fine dispersions of Ti$_3$AlC perovskite precipitates form arrays of strong glide obstacles. The particles effectively pin perfect and twinning partial dislocations, resulting in a high athermal contribution to the flow stress. These characteristics provide a significant potential for improving the mechanical performance at elevated temperatures. In comparison, carbon in solid solution or in the form of coarse H phase particles is less efficient for hardening the material.

7.4.3
Hardening by Borides, Nitrides, Oxides, and Silicides

Research performed by Larson et al. [122] on a Ti-47Al-2Cr-2Nb-0.2W alloy doped with 0.15 at.% B has shown that the solubility of boron in TiAl is very low. The value measured for the γ phase was 0.011 at.% B and that for the α_2 phase <0.003 at.% B. Thus, boron was mainly concentrated in boride precipitates. Several borides have been identified; these include TiB_2, TiB and a Cr-enriched M_2B phase [123–126]. However, TiB_2 seems to be the predominant boride phase for boron levels of less than 1 at.%. The borides occur in various morphologies, as needles, ribbons, flake-like dispersoids, and blocky particles. Fine borides may be present together with coarse particles. The morphology of the borides apparently varies with production method. Boride particles have been found to refine the grain size of TiAl castings, and most research has been devoted to this task. However, little is known to what extent dispersion hardening is produced by borides. This may partially be due to the structural changes that are associated with boron doping and that could mask the effect of precipitation strengthening. However, supporting evidence of boride strengthening is suggested by the enhanced work hardening that has been observed on boron-doped alloys [127]. Figure 7.28a shows the increase of the work-hardening coefficient with boron concentration in alloys

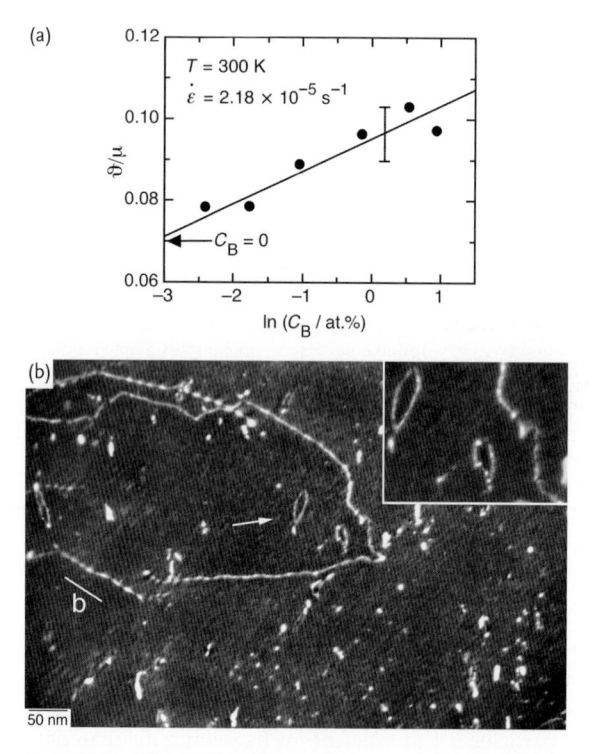

Figure 7.28 Effects of borides on the deformation behavior of two-phase alloys based on Ti-48Al-2Cr+B. (a) dependence of the normalized work-hardening coefficient $\vartheta/\mu = (1/\mu)d\sigma/d\varepsilon$ on the boron concentration, μ is the shear modulus. (b) Dislocation loops left around boride particles due to dislocation interactions according to the Orowan mechanism. Ti-48Al-2Cr-0.87B; compression at room temperature to strain $\varepsilon = 3\%$ [127].

based on Ti-48Al-2Cr. The strengthening probably arises from fine boride particles, which are impenetrable for dislocations. As suggested by Figure 7.28b, the particles may be overcome via the classical Orowan mechanism [128], which then leaves dislocation loops around the particles [129]. The loops accumulated at the particles give rise to a backstress, which, in turn, leads to an increment of strain hardening [130, 131].

Tian et al. [110, 111, 132] have characterized the types of nitride precipitates that form in Ti-(49-51)Al and categorized them as P-type (Ti$_3$AlN-perovskites) and H-type (Ti$_2$AlN) nitrides. As with the carbides, the perovskite nitrides precipitate in the γ matrix after aging the material at relatively low temperatures, around 1073 K. The perovskite phase is metastable and tends to be replaced by the H-type precipitates at higher temperatures after prolonged aging times. Also, the morphology of the nitrides was similar to that of the carbides. The fine dispersion of perovskites increased the hardness of the alloys, however, overage softening was observed after formation of the H phase.

Kawabata et al. [133] have systematically investigated the effect of oxygen addition on the mechanical properties of TiAl alloys. Three binary alloys with the compositions Ti-(50, 53, 56)Al had been doped with up to 0.69 at.% oxygen, and the deformation behavior of these alloys was investigated in the temperature range 293 to 1273 K. The effect of the oxygen additions was found to depend on the Al concentration of the alloys. For the stoichiometric alloys Ti-50Al+O, a significant increase of the yield stress by a factor of 1.5 was recognized for all test temperatures, accompanied by increased work-hardening rates. The authors ascribed these effects to solution hardening by oxygen. However, the relatively low solubility limit of oxygen in the γ phase of about 250 at. ppm [87, 134] cast doubts on this early interpretation. The hardening effect of oxygen is much weaker in Ti-56Al+O and practically absent in Ti-53Al+O. In these alloys α-Al$_2$O$_3$ particles were observed, which led the authors to believe that dispersion hardening in the Al-rich alloys occurred. As recognized in this study, the addition of oxygen seems also to refine the cast microstructure, which again makes the assessment of the strengthening effect of oxygen difficult.

Silicon additions in the range of 0.2 to 1 at.% lead to the precipitation of the ζ phase Ti$_5$(Si, Al)$_3$ [135–137]. The ζ phase has hexagonal structure and is found to be incoherent with both the α$_2$ and γ matrix [136]. Noda et al. [138] have studied the quaternary system Ti-48Al-1.5Cr-(0.2-0.65)Si. ζ phase Ti$_5$(Si, Al)$_3$ silicides were observed in the cast structure and thought to be formed according to the reaction

$$L \rightarrow \beta + Ti_5(Al, Si)_3 \tag{7.15}$$

Subsequent aging at 900 °C for 5 h led to the heterogeneous nucleation of ζ phase at the α$_2$/γ lamellar boundaries, according to the eutectoid-type reaction

$$\alpha_2 \rightarrow \gamma + Ti_5(Al, Si)_3 \tag{7.16}$$

These precipitates grew at the expense of the α$_2$ lamella. Based on systematic studies performed on ternary and quaternary alloys, Cheng et al. [89] concluded that the solubility of Si could depend on ternary alloying elements. In particular, strong silicide formers, such as Hf and Zr, may significantly reduce the Si solubility in these alloys. There is no specific information available about the strengthening mechanism of the ζ phase precipitates. However, there is good evidence that the presence of this phase could improve the creep resistance [139, 140].

Summary: Borides, nitrides, oxides, and silicides may provide precipitation hardening. An assessment of the relative importance of these precipitates is difficult because the addition of the relevant light elements often leads to structural changes that may overshadow dispersion hardening of the matrix. Coarse particles of these precipitates seem to be harmful for ductility.

7.5
Optimized Nb-Bearing Alloys

According to the discussion in Sections 7.1 and 7.2, TiAl alloys based on the composition

$$\text{Ti-45Al-(5-10)Nb-(0.2-05)C} \tag{7.17}$$

and containing fine dispersions of carbides could have a good potential to expand the service range of titanium aluminide alloys towards higher temperatures. From conventional alloys it is quite clear that hardening of the matrix alone is not necessarily a source of great strength, but is only one of many factors that have to be considered as contributing to the deformation resistance. For high Nb-containing alloys, stabilizing the very fine microstructure is also an important factor. However, fine dispersions of perovskite precipitates are susceptible to coarsening by Ostwald ripening, thereby reducing their effectiveness in impeding dislocation motion. Implementation of H phase particles could be a possible alternative because this phase is thermodynamically more stable [110, 111, 132] and apparently provides good creep resistance [140]. However, under unfavorable conditions, coarse H phase platelets are formed and located at lamellar boundaries, which tends to embrittle the material. Adapting alloy chemistry and processing can largely overcome these problems. It should be underlined that the carbon concentration must be optimized with regard to the Al content because the α_2 phase may scavenge a significant amount of carbon. Alloys based on composition (7.17) were subject to hot-working operations to convert the microstructure [93, 94, 141]. During this thermomechanical treatment, precipitation reactions occurred so that the precipitates were heterogeneously nucleated at the mismatch structures of the newly formed internal boundaries. Figure 7.29 demonstrates the effectiveness of the method by comparing the high-temperature flow behavior of carbon-free and carbon-containing alloys in the as-cast and as-extruded conditions. The flow curves exhibit a peak at low strains followed by flow softening to an ostensibly constant stress level at $\Phi = 60\%$ to 80% ($\Phi = \ln(\varepsilon + 1)$ is the true strain); these features are characteristic of dynamic recrystallization. The microstructure of the two materials was nearly lamellar in the as-cast condition and duplex in the extruded condition.

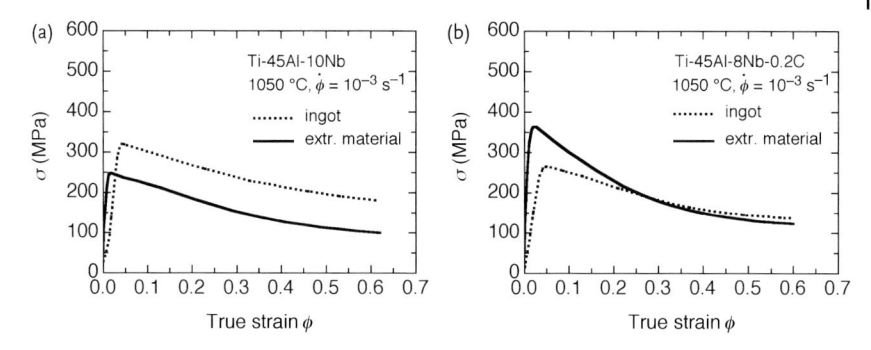

Figure 7.29 Flow curves of carbon-free and carbon-containing TiAl–Nb alloys measured on cast and extruded material under the conditions indicated. σ is the compression stress and $\Phi = \ln(\varepsilon + 1)$ is the true strain. (a) Ti-45Al-10Nb, (b) Ti-45Al-8Nb-0.2C.

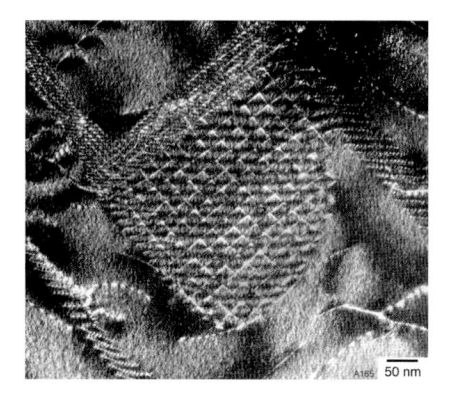

Figure 7.30 A sub-boundary observed in extruded Ti-45Al-8Nb-0.2C. Note the presence of fine carbides that decorate the dislocations.

It is generally expected (Chapter 16) that cast lamellar TiAl ingot material exhibits the higher flow stress response to hot deformation, when compared with duplex material, and this has indeed been observed for the carbon-free material. However, the reverse behavior occurs at the carbon-containing alloy: the yield stress of the extruded material is higher than that of the as-cast material. This is probably a consequence of the carbide precipitation reactions occurring during hot-working. Preheating of the cast billet at 1250 °C certainly brought carbon into solution. During subsequent extrusion and cooling, the precipitates were probably heterogeneously nucleated at the newly formed grain boundaries and interfaces. The mismatch structures present at these boundaries provide a high density of nucleation sites. This results in a fine dispersion of carbides, which apparently stabilize the microstructure. In support of this hypothesis, Figure 7.30 shows a TEM micrograph of a sub-boundary in the extruded material, in which extremely fine precipitates decorate the dislocations. This structural feature is probably the reason for

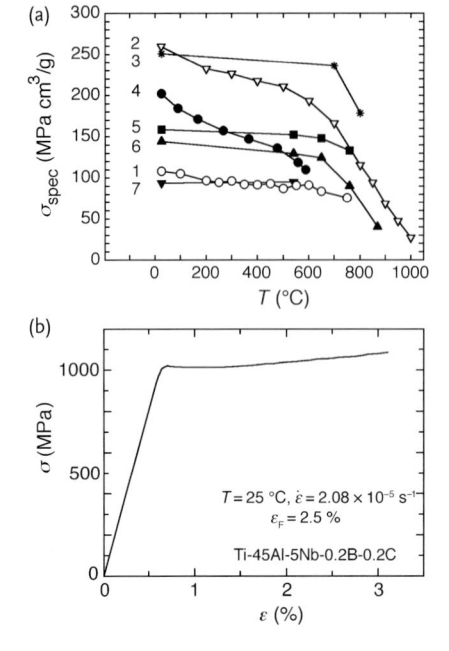

Figure 7.31 Mechanical properties of γ-based titanium aluminide alloys and conventional high-temperature alloys. (a) Temperature dependence of density-adjusted yield stress for: (1) Ti-47Al-2Cr-0.2Si (near-gamma microstructure), (2) Ti-45.6Al-7.7Nb-0.2C (nearly lamellar microstructure), (3) γ-Md (modulated microstructure). For comparison, the values of nickel-base superalloys and conventional titanium alloys are given, with (4) IMI 834, (5) René 95, (6) Inconel 718, (7) IN 713 LC. (b) Load–elongation trace of a tensile test performed at room temperature on an extruded Ti-45Al-5Nb-0.2B-0.2C (TNB-V5) alloy. Fracture occurred after a plastic strain of $\varepsilon_f = 2.53\%$.

the good high-temperature strength and creep resistance of the Ti-45Al-8Nb-0.2C alloy (TNB-V2) (Section 9.9). When compared to a simple particle dispersion, where the particles act as isolated obstacles to dislocation motion, a dispersion that stabilizes a fine dislocation substructure will be much more effective because it blocks slip in a larger crystal volume. Figure 7.31a shows the density-adjusted yield stresses of different TiAl alloys, nickel-based superalloys, and conventional titanium alloys. The diagram demonstrates the advantage that has been achieved in the design of TiAl alloys by Nb additions and implementing carbide-dispersion strengthening. The data also shows that these alloys could be an attractive alternative to nickel and titanium alloys, when the specific strength in component design is the figure merit. However, the implementation of hardening mechanisms for improving the high-temperature performance is often counter to the alloy attributes that are desirable for low-temperature ductility and damage tolerance. For example, while the addition of carbon tends to increased tensile proof stress values and creep resistance, there is a general concomitant decrease in tensile ductility. In order to illustrate the balance of mechanical properties that has been achieved with the Nb-bearing TNB alloys, Figure 7.31b shows a load elongation trace of a room-temperature tensile test that was performed on an extruded Ti-45Al-5Nb-0.2B-0.2C alloy (TNB-V5). The reliability of the room-temperature properties of the Ti-45Al-8Nb-0.2C alloy will be discussed in Section 10.7.2. The Nb-bearing and precipitation-hardened materials generally exhibit yield stresses in excess of 1 GPa, which are combined with an appreciable plastic elongation of 1% to 2%. An absolute precondition for this balance of properties is a good consolidation of the microstructure. Figure 7.6a includes the data for the γ-Md alloy with a modulated microstructure; more about this material is provided in Chapter 8.

Summary: TiAl alloys containing 5% to 10 at.% Nb and subject to hardening by carbides exhibit several desired attributes for high-temperature applications. Wrought alloys of this type could be an attractive alternative to the heavier nickel-based superalloys in intermediate ranges of stress and temperature. From the technical point of view, the advantage of these high-strength TiAl alloys is that they can be used at a fraction of their ultimate strength, thereby minimizing risk of failure.

References

1. Hall, E.O. (1951) *Proc. Phys. Soc.*, **64B**, 747.
2. Petch, N.J. (1953) *J. Iron Steel Inst.*, **174**, 25.
3. Lipsitt, H.A., Shechtman, D., and Schafrik, R.E. (1975) *Metall. Trans. A*, **6A**, 1991.
4. Huang, S.C. (1988) *Scr. Metall.*, **22**, 1885.
5. Vasudevan, V.K., Court, S.A., Kurath, P., and Fraser, H.L. (1989) *Scr. Metall.*, **23**, 467.
6. Huang, S.C. and Shi, D.S. (1991) *Microstructure/Property Relationships in Titanium Aluminides and Alloys* (eds Y.-W. Kim and R.R. Boyer), TMS, Pittsburgh, PA, p. 105.
7. Umakoshi, Y., Nakano, T., and Yamane, T. (1992) *Mater. Sci. Eng. A*, **152**, 81.
8. Nakano, T., Yokoyama, A., and Umakoshi, Y. (1992) *Scr. Metall. Mater.*, **27**, 1253.
9. Umakoshi, Y. and Nakano, T. (1993) *Acta Metall. Mater.*, **41**, 1155.
10. Koeppe, C., Bartels, A., Seeger, J., and Mecking, H. (1993) *Metall. Trans. A*, **24A**, 1795.
11. Gerling, R., Oehring, M., Schimansky, F.-P., and Wagner, R. (1995) *Advances in Powder Metallurgy and Particulate Materials*, vol. 3 (12) (eds M. Phillips and J. Porter), Metal Powder Industries Federation, APMI International, Princeton, NJ, p. 91.
12. Jung, J.Y., Park, J.K., Chun, C.H., and Her, S.M. (1996) *Mater. Sci. Eng. A*, **220**, 185.
13. Mercer, C. and Soboyejo, W.O. (1996) *Scr. Mater.*, **35**, 17.
14. Maziasz, P.J. and Liu, C.T. (1998) *Metall. Mater. Trans. A*, **29A**, 105.
15. Liu, C.T. and Maziasz, P.J. (1998) *Intermetallics*, **6**, 653.
16. Dimiduk, D.M., Hazzledine, P.M., Parthasarathy, T.A., Seshagiri, S., and Mendiratta, M.G. (1998) *Metall. Mater. Trans. A*, **29A**, 39.
17. Bohn, R., Klassen, T., and Bormann, R. (2001) *Acta Mater.*, **49**, 299.
18. Maruyama, K., Yamaguchi, M., Suzuki, G., Zhu, H., Kim, H.Y., and Yoo, M.H. (2004) *Acta Mater.*, **52**, 5185.
19. Kim, Y.-W. (1998) *Intermetallics*, **6**, 623.
20. Umakoshi, Y. and Nakano, T. (1992) *ISIJ Int.*, **32**, 1339.
21. Sun, Y.Q. (1997) *Mater. Sci. Eng. A*, **240**, 131.
22. Sun, Y.Q. (1997) *High-Temperature Ordered Intermetallic Alloys VII*, Materials Research Society Symposium

Proceedings, vol. 460 (eds C.C. Koch, C.T. Liu, N.S. Stoloff, and A. Wanner), MRS, Warrendale, PA, p. 109.
23 Sun, Y.Q. (1998) *Philos. Mag. A*, **77**, 1107.
24 Appel, F. and Wagner, R. (1993) *Physica Scripta*, **49T**, 387.
25 Appel, F., Sparka, U., and Wagner, R. (1999) *Intermetallics*, **7**, 325.
26 Appel, F. and Wagner, R. (1995) *Gamma Titanium Aluminides* (eds Y.-W. Kim, R. Wagner, and M. Yamaguchi), TMS, Warrendale, PA, p. 231.
27 Bartels, A., Koeppe, C., Zhang, T., and Mecking, H. (1995) *Gamma Titanium Aluminides* (eds Y.-W. Kim, R. Wagner, and M. Yamaguchi), TMS, Warrendale, PA, p. 655.
28 Paul, J.D.H. and Appel, F. (2003) *Mater. Trans. A*, **34A**, 2103.
29 Herrmann, D. *Diffusionsschweißen von γ(TiAl)-Legierungen: Einfluss von Zusammensetzung, Mikrostruktur und mechanischen Eigenschaften*. PhD Thesis, Technical University Hamburg-Harburg, Germany, 2009.
30 Schafrik, R.E. (1977) *Metall. Trans. A*, **8A**, 1003.
31 Kocks, U.F. and Mecking, H. (2003) *Prog. Mater. Sci.*, **48**, 171.
32 Gray, G.T., III and Pollock, T.M. (2002) *Intermetallic Compounds. Vol. 3, Principles and Practice* (eds J.H. Westbrook and R.L. Fleischer), John Wiley & Sons, Ltd, Chichester, p. 361.
33 Hirth, J.P. and Lothe, J. (1992) *Theory of Dislocations*, Krieger, Melbourne.
34 Appel, F. (1989) *Phys. Status Solidi*, **116**, 153.
35 Nabarro, F.R.N., Bassinski, Z.S., and Holt, D. (1964) *Adv. Phys.*, **13**, 193.
36 Appel, F., Lorenz, U., Oehring, M., Sparka, U., and Wagner, R. (1977) *Mater. Sci. Eng. A*, **233**, 1.
37 Wang, B.-Y., Wang, Y.-X., Gu, Q., and Wang, T.-M. (1997) *Comput. Mater. Sci.*, **8**, 267.
38 Gilman, J.J. (1962) *J. Appl. Phys.*, **33**, 2703.
39 Chen, H.S., Head, A.K., and Gilman, J.J. (1964) *J. Appl. Phys.*, **35**, 2502.
40 Kroupa, F. (1962) *Philos. Mag.*, **7**, 783.
41 Kroupa, F. (1966) *Acta Metall.*, **14**, 60.
42 Kroupa, F. (1966) *J. Phys.*, **27**, 154.
43 Bullough, R. and Newman, R.C. (1970) *Rep. Prog. Phys.*, **33**, 101.
44 Bacon, D.J., Bullough, R., and Willis, J.R. (1970) *Philos. Mag.*, **22**, 31.
45 Damask, A.C. and Dienes, G.J. (1963) *Point Defects in Metals*, Gordon and Breach, New York, NY, p. 145.
46 Mishin, Y. and Herzig, C. (2000) *Acta Mater.*, **48**, 589.
47 Kiritani, M., Sato, A., Sawai, K., and Yoshida, S. (1968) *J. Phys. Soc. Jpn*, **24**, 461.
48 Viguier, B. and Hemker, K.J. (1996) *Philos. Mag. A*, **73**, 575.
49 Sabinash, C.M., Sastry, S.M.L., and Jerina, K.L. (1995) *Mater. Sci. Eng. A*, **192–193**, 837.
50 Kim, H.E., Soon, H., and Hong, S.H. (1998) *Mater. Sci. Eng. A*, **251**, 216.
51 Herrmann, D. and Appel, F. (2009) *Metall. Mater. Trans. A*, **40A**, 1881.
52 Imayev, R.M., Imayev, V.M., Oehring, M., and Appel, F. (2005) *Metall. Mater. Trans. A*, **36A**, 859.
53 Fischer, F.D., Clemens, H., Schaden, T., and Appel, F. (2007) *Int. J. Mater. Res.*, **98**, 1041.
54 Shang, P., Cheng, T.T., and Aindow, M. (1999) *Philos. Mag. A*, **79**, 2553.
55 Appel, F. (2001) *Mater. Sci. Eng. A*, **317**, 115.
56 Yoo, M.H. and Fu, C.L. (1998) *Metall. Mater. Trans. A*, **29A**, 49.
57 Maloy, S.A. and Gray, G.T., III (1996) *Acta Mater.*, **44**, 1741.
58 Gray, G.T. (1994) *Twinning in Advanced Materials* (eds M.H. Yoo and M. Wuttig), TMS, Warrendale, PA, p. 337.
59 Gray, G.T., III, Steif, P.S., and Pollock, T.M. (2001) *Structural Intermetallics 2001* (eds K.J. Hemker, D.M. Dimiduk, H. Clemens, R. Darolia, H. Inui, J.M. Larsen, V.K. Sikka, M. Thomas, and J.D. Whittenberger), TMS, Warrendale, PA, p. 269.
60 Fleischer, R.L. (1962) *The Strength of Metals* (ed. D. Peckner), Reinhold Press, p. 93.
61 Fleischer, R.L. (1962) *Acta Metall.*, **10**, 835.
62 Fleischer, R.L. (1987) *Scr. Metall.*, **21**, 1083.
63 Vujic, D., Li, Z.X., and Whang, S.H. (1988) *Metall. Trans. A*, **19A**, 2445.

64 Erschbaumer, H., Podloucky, R., Rogel, P., Tonmittschka, G., and Wagner, R. (1993) *Intermetallics*, **1**, 99.
65 Zou, J. and Fu, L.C. (1995) *Phys. Rev.*, **B51**, 2115.
66 Wolf, W., Podloucky, R., Rogel, P., and Erschbaumer, H. (1996) *Intermetallics*, **4**, 201.
67 Pananikolaou, N., Zeller, R., and Dederichs, P.H. (1997) *Phys. Rev.*, **B55**, 4157.
68 Song, Y., Xu, D.S., Yang, R., Li, D., and Hu, Z.Q. (1998) *Intermetallics*, **6**, 157.
69 Song, Y., Yang, R., Li, D., Wu, W.T., and Guo, Z.X. (1999) *J. Mater. Res.*, **14**, 2824.
70 Song, Y., Guo, Z.X., and Yang, R. (2002) *J. Light Met.*, **2**, 115.
71 Pfullmann, Th. and Beaven, P.A. (1993) *Scr. Metall.*, **28**, 275.
72 Kawabata, T., Tamura, T., and Izumi, O. (1993) *Metall. Mater. Trans. A*, **24A**, 141.
73 Kawabata, T., Fukai, H., and Izumi, O. (1998) *Acta Mater.*, **46**, 2185.
74 Woodward, C., Kajihara, S.A., Rao, S.I., and Dimiduk, D.M. (1999) *Gamma Titanium Aluminides 1999* (eds Y.-W. Kim, D.M. Dimiduk, and M.H. Loretto), TMS, Warrendale, PA, p. 49.
75 Rossouw, C.J., Forwood, C.T., Gibbson, M.A., and Miller, P.R. (1996) *Philos. Mag. A*, **74**, 77.
76 Yamaguchi, M. and Inui, H. (1993) *Structural Intermetallics* (eds R. Darolia, J.J. Lewandowski, C.T. Liu, P.L. Martin, D.B. Miracle, and M.V. Nathal), TMS, Warrendale, PA, p. 127.
77 Yao, K.-F., Inui, H., Kishida, K., and Yamaguchi, M. (1995) *Acta Metall. Mater.*, **43**, 1075.
78 Kawabata, T., Tamura, T., and Izumi, O. (1989) *High-Temperature Ordered Intermetallic Alloys III, Materials Research Society Symposia Proceedings*, vol. 133 (eds C.T. Liu, A.I. Taub, N.S. Stoloff, and C.C. Koch), MRS, Pittsburgh, PA, p. 330.
79 Huang, S.C., and Hall, E.L. (1991) *Alloy Phase Stab. Des.*, **186**, 381.
80 Hahn, K.D. and Whang, S.H. (1989) *High-Temperature Ordered Intermetallic Alloys III, Materials Research Society Symposia Proceedings*, vol. 133 (eds C.T. Liu, A.I. Taub, N.S. Stoloff, and C.C. Koch), MRS, Pittsburgh, PA, p. 385.
81 Tsujimoto, T. and Hashimoto, K. (1989) *High-Temperature Ordered Intermetallic Alloys III, Materials Research Society Symposia Proceedings*, vol. 133 (eds C.T. Liu, A.I. Taub, N.S. Stoloff, and C.C. Koch), MRS, Pittsburgh, PA, p. 391.
82 Huang, S.C. and Hall, E.L. (1991) *Metall. Trans. A*, **22A**, 2619.
83 Huang, S.C. and Hall, E.L. (1991) *Acta Metall. Mater.*, **6**, 1053.
84 Zheng, Y., Zhao, L., and Tangri, K. (1992) *Scr. Metall. Mater.*, **26**, 219.
85 Huang, S.-C. (1993) *Structural Intermetallics* (eds R. Darolia, J.J. Lewandowski, C.T. Liu, P.L. Martin, D.B. Miracle, and M.V. Nathal), TMS, Warrendale, PA, p. 299.
86 Sabinash, C.M., Sastry, S.M.L., and Jerina, K.L. (1995) *Scr. Metall. Mater.*, **32**, 1381.
87 Menand, A., Huguet, A., and Nérac-Partaix, A. (1996) *Mater.*, **44**, 4729.
88 Zheng, Y., Zhao, L., and Tangri, K. (1996) *Mater. Sci. Eng. A*, **208**, 80.
89 Cheng, T.T., Willis, M.R., and Jones, I.P. (1999) *Intermetallics*, **7**, 89.
90 Larson, D.J., Liu, C.T., and Miller, M.K. (1999) *Mater. Sci. Eng. A*, **270**, 1.
91 Chen, G.L., Zhang, W.J., Yang, Y., Wang, J., and Sun, Z. (1993) *Structural Intermetallics* (eds R. Darolia, J.J. Lewandowski, C.T. Liu, P.L. Martin, D.B. Miracle, and M.V. Nathal), TMS, Warrendale, PA, p. 319.
92 Paul, J.D.H., Appel, F., and Wagner, R. (1998) *Acta Mater.*, **46**, 1075.
93 Appel, F., Oehring, M., and Wagner, R. (2000) *Intermetallics*, **8**, 1283.
94 Appel, F., Oehring, M., and Paul, J.D.H. (2001) *Structural Intermetallics 2001* (eds K.J. Hemker, D.M. Dimiduk, H. Clemens, R. Darolia, H. Inui, J.M. Larsen, V.K. Sikka, M. Thomas, and J.D. Whittenberger), TMS, Warrendale, PA, p. 63.
95 Zhang, W.J., Liu, Z.C., Chen, G.L., and Kim, Y.-W. (1999) *Philos. Mag. A*, **79**, 1073.
96 Zhang, W.J., Deevi, S.C., and Chen, G.L. (2002) *Intermetallics*, **10**, 403.
97 Liu, Z.C., Lin, J.P., Li, S.J., and Chen, G.L. (2002) *Intermetallics*, **10**, 653.

98 Helwig, A., Palm, M., and Inden, G. (1998) *Intermetallics*, **6**, 79.
99 Chen, G., Zhang, W.J., Liu, C.Z., and Kim, Y.-W. (1999) *Gamma Titanium Aluminides 1999* (eds Y.-W. Kim, D.M. Dimiduk, and M.H. Loretto), TMS, Warrendale, PA, p. 371.
100 Fischer, F.D., Waitz, T., Scheu, Ch., Cha, L., Dehm, G., Antretter, T., and Clemens, H. (2010) *Intermetallics*, **18**, 509.
101 Yuan, Y., Yin, K.B., Zhao, X.N., and Meng, X.K. (2006) *J. Mater. Sci.*, **41**, 469.
102 Zhang, W.J. and Appel, F. (2002) *Mater. Sci. Eng. A*, **329–331**, 649.
103 Zhang, W.J. and Appel, F. (2002) *Mater. Sci. Eng. A*, **334–331**, 59.
104 Chen, S.H., Schumacher, G., Mukherji, D., Frohberg, G., and Wahi, R.P. (2001) *Philos. Mag. A*, **81**, 2653.
105 Chen, S.H., Mukherji, D., Schumacher, G., Frohberg, G., and Wahi, R.P. (2000) *Philos. Mag. Lett.*, **80**, 19.
106 Zhang, W.J., Liu, Z.C., Chen, G.L., and Kim, Y.-W. (1999) *Mater. Sci. Eng.*, **A271**, 416.
107 Herzig, C., Przeorski, T., Friesel, M., Hisker, F., and Divinski, S. (2001) *Intermetallics*, **9**, 461.
108 Nemoto, M., Tian, W.H., and Sano, T. (1991) *J. Phys.*, **1**, 1099.
109 Nemoto, M., Tian, W.H., Harada, K., Han, C.S., and Sano, T. (1992) *Mater. Sci. Eng. A*, **152**, 247.
110 Tian, W.H. and Nemoto, M. (1993) *Philos. Mag. A*, **68**, 965.
111 Tian, W.H. and Nemoto, N. (1995) *Gamma Titanium Aluminides* (eds Y.-W. Kim, R. Wagner, and M. Yamaguchi), TMS, Warrendale, PA, p. 689.
112 Tian, W.H. and Nemoto, M. (1997) *Intermetallics*, **5**, 237.
113 Chen, S., Beaven, P.A., and Wagner, R. (1992) *Scr. Metall. Mater.*, **26**, 1205.
114 Schuster, J.C., Nowotny, H., and Vaccaro, C. (1980) *J. Solid State Chem.*, **32**, 213.
115 Appel, F. (2005) *Philos. Mag.*, **85**, 205.
116 Gouma, P.I., Mills, M.J., and Kim, Y.-W. (1998) *Philos. Mag. Lett.*, **78**, 59.
117 Gouma, P.I., Subramanian, K., Kim, Y.-W., and Mills, M.J. (1998) *Intermetallics*, **6**, 689.
118 Benedek, R., Seidman, D.N., and Woodward, C. (2003) *Defect Properties and Related Phenomena in Intermetallic Alloys, Materials Research Society Symposium Proceedings*, vol. 753 (eds E.P. George, H. Inui, M.J. Mills, and G. Eggeler), MRS, Warrendale, PA, p. 129.
119 Lopacinski, M., Puszynski, J., and Lis, J. (2001) *J. Am. Ceram. Soc.*, **84**, 3051.
120 Christoph, U., Appel, F., and Wagner, R. (1997) *Mater. Sci. Eng. A*, **239–240**, 39.
121 Appel, F., Christoph, U., and Wagner, R. (1997) *High-Temperature Ordered Intermetallic Alloys VII, Materials Research Society Symposium Proceedings*, vol. 460 (eds C.C. Koch, C.T. Liu, N.S. Stoloff, and A. Wanner), MRS, Warrendale, PA, p. 77.
122 Larson, D.L., Liu, C.T., and Miller, M.K. (1997) *Intermetallics*, **5**, 411.
123 Graef, M.D., Löfvander, J.P., McCullough, C., and Levi, C.G. (1992) *Acta Metall. Mater.*, **40**, 3395.
124 Inkson, B.I., Boothroyd, C.B., and Humphreys, C.J. (1995) *Acta Metall. Mater.*, **43**, 1429.
125 Godfrey, B. and Loretto, M.H. (1996) *Intermetallics*, **4**, 47.
126 Chen, C.L., Wu, W., Lion, J.P., He, L.L., Chen, G.L., and Ye, H.Q. (2007) *Scr. Mater.*, **56**, 441.
127 Müllauer, J. and Appel, F. (2003) *Defect Properties and Related Phenomena in Intermetallic Alloys, Materials Research Society Symposium Proceedings*, vol. 753 (eds E.P. George, H. Inui, M.J. Mills, and G. Eggeler), MRS, Warrendale, PA, p. 231.
128 Orowan, E. (1948) *Symposium on Internal Stresses in Metals and Alloys, Session III, Discussion*, Institute of Metals, London, p. 451.
129 Ashby, M.F. (1969) *Physics of Strength and Plasticity* (ed. A.S. Argon), MIT, Cambridge, MA, p. 113.
130 Brown, L.M. and Stobbs, W.M. (1971) *Philos. Mag.*, **23**, 1185.
131 Brown, L.M. and Stobbs, W.M. (1971) *Philos. Mag.*, **23**, 1201.
132 Tian, W.H. and Nemoto, M. (2005) *Intermetallics*, **13**, 1030.
133 Kawabata, T., Abumiya, T., and Izumi, O. (1992) *Acta Metall. Mater.*, **38**, 2557.

134 Nérac-Partaix, A. and Menand, A. (1996) *Scr. Mater.*, **35**, 199.
135 Tsuyama, S., Mitao, S., and Minakawa, K. (1992) *Mater. Sci. Eng. A*, **153**, 451.
136 Wang, G.-X., Dogan, B., Hsu, F.-Y., Klaar, H.-J., and Dahms, M. (1993) *Metall. Trans. A*, **26A**, 691.
137 Hsu, F.-Y., Wang, G.-X., and Klaar, H.-J.A. (1995) *Scr. Metall. Mater.*, **33**, 597.
138 Noda, T., Okabe, M., Isobe, S., and Sayashi, M. (1995) *Mater. Sci. Eng. A*, **192/193**, 774.
139 Viswanathan, G.B., Kim, Y.-W., and Mills, M.J. (1999) *Gamma Titanium Aluminides 1999* (eds Y.-W. Kim, D.M. Dimiduk, and M.H. Loretto), TMS, Warrendale, PA, p. 653.
140 Karadge, M., Kim, Y.-W., and Gouma, P.I. (2003) *Metall. Mater. Trans. A*, **34A**, 2119.
141 Appel, F., Oehring, M., Paul, J.D.H., Klinkenberg, Ch., and Carneiro, T. (2004) *Intermetallics*, **12**, 791.

8
Deformation Behavior of Alloys with a Modulated Microstructure

In an attempt to improve the balance of mechanical properties, a novel type of TiAl alloys with a composite-like microstructure has been recently developed [1, 2]. The characteristic constituents of the alloy are laths with a modulated substructure that is comprised of stable and metastable phases. The modulation occurs at the nanometer scale and thus provides an additional structural feature that refines the material. The physical metallurgy of these modulated alloys will be briefly reviewed in this chapter.

8.1
Modulated Microstructures

In TiAl-based systems three-phase equilibria of the α/α_2, the γ and the β phases can be established by choosing suitable Al contents and alloying with other metallic elements. The disordered b.c.c. lattice of the β phase provides sufficient independent slip systems and may thus directly act as a ductilizing constituent in the final microstructure. In view of this potential, large efforts have been expended in the development of β-phase-containing alloys [3–8]. However, an understanding of the evolution of constitution and microstructure in these alloys seems to be in its infancy. Also, the documentation of mechanical data for β-phase-containing alloys is scarce. The alloy design that will be described here is based on decomposition reactions of the $\beta/B2$ phase, which can be triggered by a suitable choice of the Al content and ternary additions forming the $\beta/B2$ phase. It might be expected that the evolution of the constitution does not reach thermodynamic equilibrium, thus, the number of the transformation products may be larger than expected from the phase rule. This results in complex microstructures with a multitude of stable and metastable phases. However, several possible product phases with lower crystal symmetry, such as ω, ω' or ω'', are extremely brittle and thus harmful constituents of the microstructure. This inherent complexity of the phase evolution requires a tight harmonization of alloying elements that stabilize the $\beta/B2$

Gamma Titanium Aluminide Alloys: Science and Technology, First Edition. Fritz Appel, Jonathan David Heaton Paul, Michael Oehring.
© 2011 Wiley-VCH Verlag GmbH & Co. KGaA. Published 2011 by Wiley-VCH Verlag GmbH & Co. KGaA.

phase and influence the subsequent phase decomposition. The alloys (designated γ-Md) that are considered here have the base-line composition

$$\text{Ti-(40-44)Al-8.5 Nb} \tag{8.1}$$

According to literature data [3–9], such alloys should contain a significant amount of β phase. As indicated by X-ray analysis the constitution of the alloys involves the β/B2, $α_2$ and γ phases. Additional X-ray reflections could be attributed to the presence of two orthorhombic phases with B19 structure, (oP4, Pmma and oC16, Cmcm). However, a clear association with the various orthorhombic structures reported in the literature [10, 11] was not possible because of their structural similarity. The X-ray diffraction analysis did not reveal any superlattice reflections from the β phase corresponding to B2 phase. However, the existence of the B2 phase cannot be completely ruled out because its volume fraction might be small and its ordering incomplete.

The microstructure observed after extrusion and stress-relief annealing at 1030 °C is demonstrated in Figure 8.1a by a SEM image obtained in the backscat-

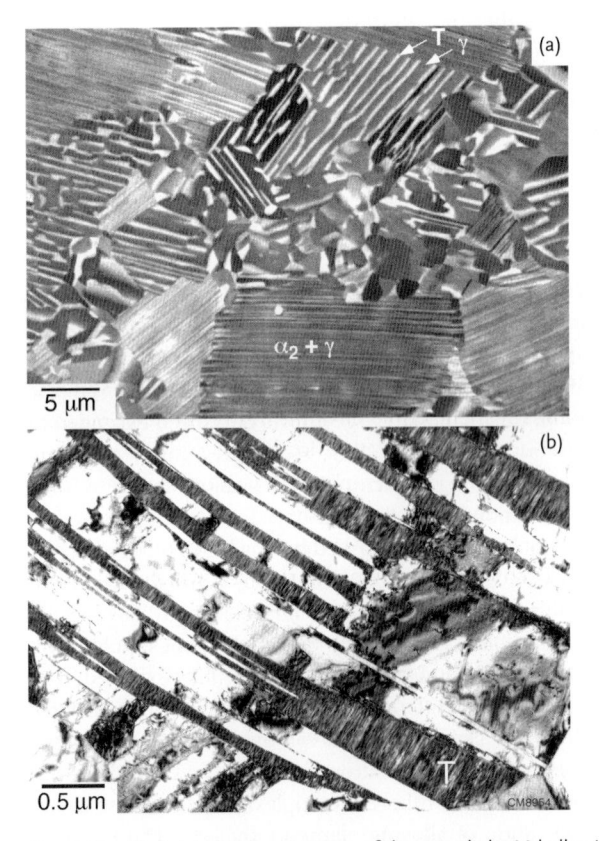

Figure 8.1 Modulated microstructure of the extruded γ-Md alloy imaged by (a) backscattered electrons in the scanning electron microscope and (b) transmission electron microscopy. T designates modulated laths.

tering electron (BSE) mode. It shows two major microstructural constituents, lamellar colonies and a constituent with a pearlite-like morphology. The contrast of the BSE image indicates that the latter constituent consists of β and γ laths, that is, it does not result from the eutectoid reaction $\alpha \to \alpha_2 + \gamma$. The microstructure is particularly fine and homogeneous with the size of the lamellar and pearlite-like colonies being about 30 μm.

Takeyama and Kobayashi [7] have reported that similar pearlite-like microstructures were formed in TiAl–V and TiAl–Mo alloys by the reaction

$$\alpha \to \beta(B2) + \gamma \tag{8.2}$$

Starting from the high-temperature β phase, such an alloy has the solid-state transformation phase sequence

$$\beta \to \beta + \alpha \to \alpha \to \beta(B2) + \gamma \tag{8.3}$$

According to current phase diagrams, at low temperatures in the Ti–Al–Nb system the β phase is ordered B2 [9, 12]. However, the presence of lamellar ($\alpha_2 + \gamma$) colonies suggests a phase-transformation sequence of

$$\beta \to \beta + \alpha \to \alpha \to \alpha + \gamma \to \alpha + \gamma + \beta(B2) \tag{8.4}$$

This sequence is in agreement with available phase diagram information [9, 12], but cannot explain the occurrence of the pearlite-like microstructure [7]. It could be that during the transformation reaction $\beta \to \beta + \alpha$, equilibrium is not achieved and that α grains of slightly different compositions are formed, which subsequently decompose along different pathways.

When examined in the transmission electron microscope, the structure shown in Figure 8.1b is observed. The characteristic features of this structure are laths with a periodic variation in the diffraction contrast, which intersperse the other constituents. As shown in Figure 8.2a, the contrast fluctuations occur at a very fine length scale with diffuse boundaries. The evidence of the high-resolution electron-microscope observations is that a single lath is subdivided into several regions with different crystalline structures with no sharp interface in between. The high-resolution micrograph in Figure 8.2b shows a lath adjacent to a γ lamella in $<101]_\gamma$ projection, which can be used as a reference. The interface between the lath and the γ phase (designated as γ/T) consists mainly of flat terraces, which are parallel to the $(111)_\gamma$ plane. Steps of different heights, often delineate the terraces.

The modulated laths are comprised of an orthorhombic constituent, which is interspersed by slabs of β/B2 and a little α_2 phase (Figure 8.3). Selected-area diffraction of the orthorhombic constituent is consistent with the B19 phase, which can be described as the orthorhombic phase (oP4) or as a hexagonal superstructure of $D0_{19}$ (hP8). In the Ti–Al system, the B19 structure has already been observed by Abe et al. [13] and Ducher et al. [14]. The B19 structure is structurally closely related to the orthorhombic phase (oC16, Cmcm) with the ideal stoichiometry Ti$_2$AlNb, which among the intermetallic compounds is remarkable for its relatively good room-temperature ductility [15].

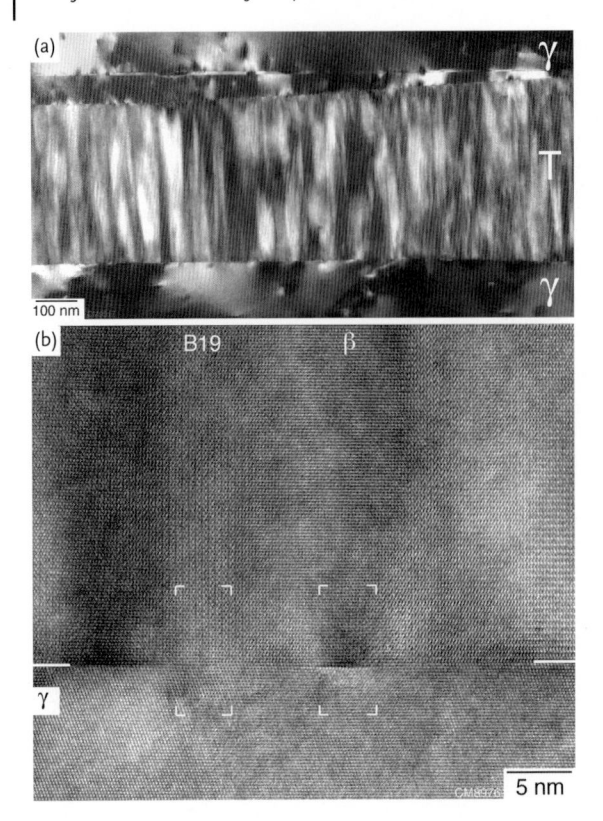

Figure 8.2 TEM analysis of the modulated microstructure. (a) A modulated lath imaged by diffraction contrast. (b) High-resolution TEM micrograph of a modulated lath adjacent to a γ lamella.

In Figures 8.2b and 8.3a the B19 structure is imaged in the $[010]_{B19}$ projection. As deduced from the high-resolution images, the orientation relationships between the constituents involved in a modulated lath with the adjacent γ are

$$(100)_{B19} \parallel \{110\}_{\beta/B2} \parallel (0001)_{\alpha2} \parallel \{111\}_{\gamma}$$

and

$$[010]_{B19} \parallel <111>_{\beta/B2} \parallel <11\bar{2}0>_{\alpha2} \parallel <1\bar{1}0>_{\gamma}; [001]_{B19} \parallel <11\bar{2}>_{\gamma} \tag{8.5}$$

In the diffraction pattern the presence of the periodic distortion is manifested by the existence of weak satellite reflections adjacent to the main reflections. The distance of the satellites from the main reflections is the reciprocal of the modulation wavelength, and the direction joining the satellites with their main reflections is parallel to the direction of the modulation vector. These observed features are reminiscent of a modulated structure, which in recent years have attracted considerable interest; for a review see [16]. A crystal structure is said to be modulated

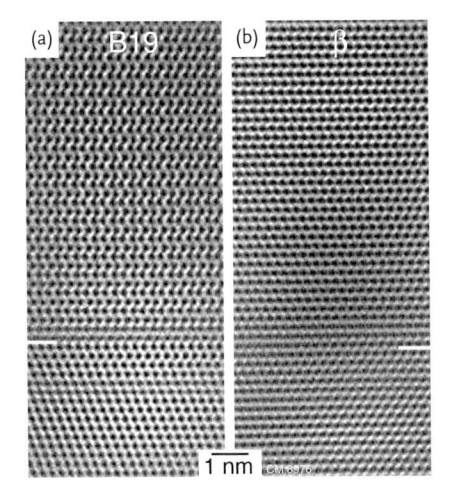

Figure 8.3 High-resolution TEM evidence of the modulated microstructure. (a) and (b) Fourier-filtered images of the areas boxed in Fig. 8.2.

if it exhibits periodicities other than the Bravais lattice periodicities. These additional periodicities arise from one or more distortions, which increase to a maximum value and then decrease to the initial value. The modulation may involve atomic coordinates, occupancy factors or displacement parameters. The strain of the discontinuities is often relieved by a continuous and periodic variation of the physical properties of the product.

Although a fundamental understanding of the structural modulation within TiAl alloys has yet to be achieved, it might be related to decomposition reactions of the high-temperature β/B2 phase. The body-centered cubic (b.c.c.) structure of the β/B2 phase is accomplished by stacking the most densely packed {011} planes in an *ABAB* sequence along <111>. The atoms of the {011} planes are in contact only along the body-diagonal direction. The interstices in these planes have a slight saddle configuration. Model building suggests that the atoms in the second layer might be able to slip a small distance to one side or the other, leading to a distortion of the cubic structure. With only small displacements, that is, closing these gaps, the $\{011\}_{bcc}$ planes become geometrically identical with the {111} and (0001) planes of the face-centered cubic (f.c.c.) or hexagonal (h.c.p.) structures, respectively. This can give rise to several b.c.c./f.c.c. and b.c.c./h.c.p. transformations, which results in a multitude of stable and metastable phases. First-principles calculations of Nguyen-Manh and Pettifor [10] and Yoo and Fu [17] have shown that the β/B2 phase existing in TiAl alloys containing supersaturations of transition metals (Zr, V, Nb) is unstable under tetragonal distortion; a shear instability that was attributed to the anomalous (negative) tetragonal shear modulus. Specifically, B2 may transform by homogeneous shear to several low-temperature orthorhombic phases, which can exist metastably. The energetically favorable transformations are [10]

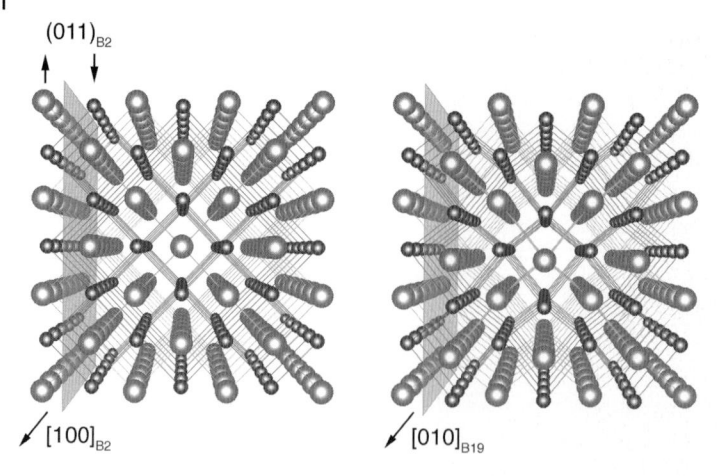

Figure 8.4 Formation of the B19 structure from the parent B2 phase illustrated by perspective views of hard-sphere models. [100] projection of the B2 phase; arrow heads mark shuffle displacements of neighboring $(011)_{B2}$ planes in opposite $[01\bar{1}]$ directions to form B19.

$$B2(Pm3m) \rightarrow B19(Pmma) \rightarrow B33(Cmcm) \tag{8.6}$$

At the atomic level, the B2 phase may transform to B19 by a shuffle displacement of neighboring $(011)_{B2}$ planes in opposite $[01\bar{1}]$ directions, as illustrated in Figure 8.4. A subsequent displacement of neighboring $(011)_{B2}$ planes in the [100] direction generates the B33 structure [10]. In the system investigated here the predominant orthorhombic phase seems to be B19. It is tempting to speculate that the modulation of the laths is triggered by a periodic variation of the composition of the parent B2 phase, as occurs during spinodal decomposition. Clearly, the mechanism requires further investigation.

Summary: In TiAl alloys with the base-line composition Ti-(40-44)Al-(5-10)Nb a novel laminate structure can be formed, consisting of multiphase laths embedded in a matrix of the γ, $\beta/B2$, and α_2 phases. The multiphase laths probably result from the decomposition of prior $\beta/B2$ phase and have a fine nanoscale substructure due to a pronounced phase discontinuity.

8.2
Misfitting Interfaces

The phase discontinuity occurring in the modulated laths results in complex mismatch structures, which, in broad terms, are similar to those described for many other misfitting solid/solid interfaces. However, there are specific features that are believed to be important for the mechanical performance of the modulated alloys; this will be considered in some detail. In the modulated laths the B19 structure is apparently formed as a domain structure, as all $<01\bar{1}>\{011\}$ shuffle displacements in the parent B2 structure are crystallographically equivalent. Specifically, the

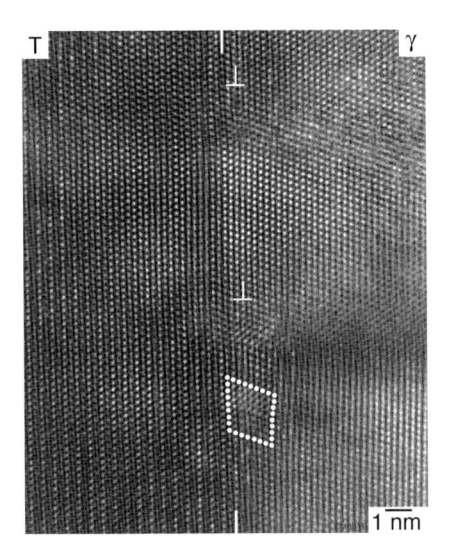

Figure 8.5 High-resolution micrograph showing misfit dislocations at an interface between a modulated lath (T) and a γ lamella. Three closely separated misfit dislocations with the projected Burgers vector **b** = 1/3[111] can be recognized by a shallow view along the γ/T interface. Note the spreading of the dislocation core.

perpendicular displacements [01$\bar{1}$](011) and [011](01$\bar{1}$) produce two B19 variants with their (100)$_{B19}$ and (001)$_{B19}$ planes parallel to each other. The misfit between these two planes is about 8%. The same degree of misfit occurs between (002)$_{B19}$ and {110}$_{\beta/B2}$, whereas (200)$_{B19}$ matches well with {110}$_{\beta/B2}$. The misfit appears to be partially relieved by interfacial dislocations that in most cases are situated in the adjacent γ phase, just in front of the lattice distortion zone in the modulated lath.

As shown in Figure 8.5, the interfacial dislocations are manifested by an additional (111)$_\gamma$ plane, parallel to the γ/T interface, and may thus be described by the projected Burgers vector $\mathbf{b}_p$ = 1/3[111]. The dislocations often have a stand-off of three to five (111)$_\gamma$ planes from the interface and exhibit a significant core spreading and thus are prone to a dissociation reaction. A possible dissociation reaction of the 1/3[111] misfit dislocations has been discussed in Section 6.3.3. The products of this reaction are 1/6 <11$\bar{2}$] twinning partials and 1/2<110] ordinary dislocations. A total reaction involving three misfit dislocations would lead to three twinning partial dislocations and one ordinary dislocation. The mechanism may explain why the misfit dislocations often serve as nucleation sites for mechanical twins (Figure 8.6), and why mechanical twinning is a prominent deformation mechanism within the γ phase of the modulated alloys (Figure 8.7).

The mismatch strains between the modulated laths and the γ phase also seem to be relieved by the formation of ledges or steps at the interface (Figure 8.8) [1, 2]. The interfaces are often facetted with atomically flat terraces parallel to {111}$_\gamma$ in between the ledges. At the terraces the orientation relationship is such that one set of close-packed planes and direction is common to the modulated lath and the

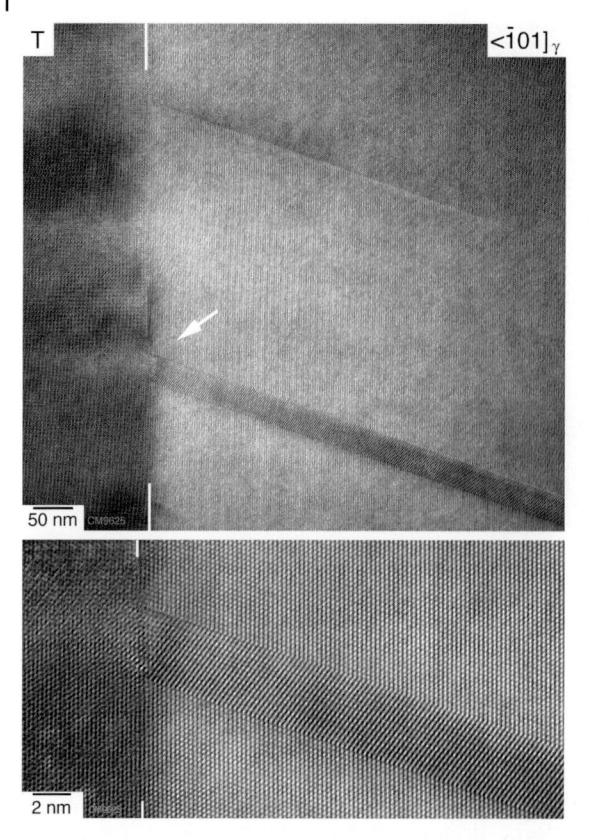

Figure 8.6 High-resolution micrograph in $[\bar{1}01]_\gamma$ projection demonstrating twin nucleation at a γ/T interface. An embryonic twin (above) and a thicker twin (below) were nucleated. The micrograph below shows the arrowed area in higher magnification. γ-Md sample, tensile deformation at room temperature in air to a plastic elongation of about $\varepsilon = 2.5\%$.

adjacent γ phase. The ledges range in scale from atomic to multiatomic dimensions. A high-index habit plane is often formed by introducing ledges spaced only a few atoms apart along the common close-packed planes. TEM examination gives the impression that the habit plane could lie parallel to any arbitrary crystal plane simply by varying the height and spacing of the ledges along the common close-packed planes. The ledge concept of misfit accommodation in phase transformations has been discussed in Section 6.1.3. However, in spite of these accommodation processes, a significant amount of strain seems to remain in the material. This is indicated by the distortion of the lattice planes and the strong strain contrast seen in most micrographs. These residual strains are taken up by elastic distortion, that is, the adjacent phases are uniformly strained to bring the atomic spacings into registry. The homogeneous strain accommodation further improves the coherency of the interfaces but generates coherency stresses. It might be expected that under

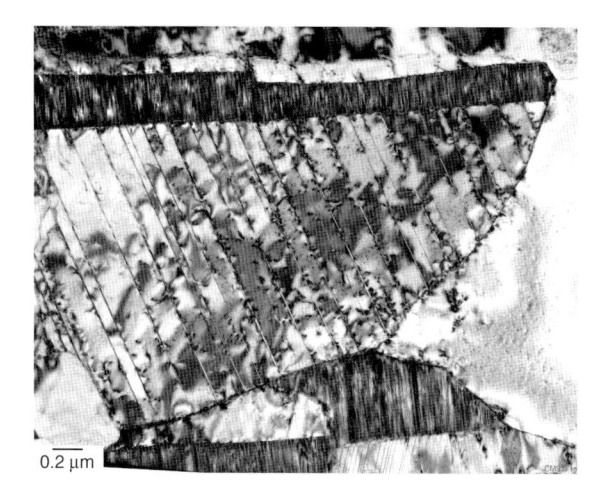

Figure 8.7 TEM micrograph showing extensive mechanical twinning in γ lamellae adjacent to modulated laths. γ-Md sample, tensile deformation at room temperature under air to a plastic elongation of about ε = 2.5%.

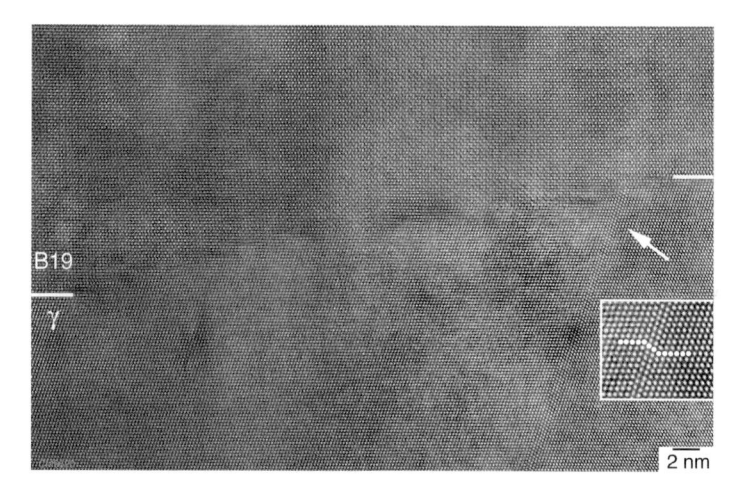

Figure 8.8 Ledged interface between the B19 phase of a modulated lamella and the γ phase; note the emission of a mechanical twin.

creep conditions these coherency stresses can relax, which could give rise to several interface-related deformation phenomena (see Sections 9.5 and 9.6).

The modulated laths can apparently further transform into the γ phase, as demonstrated in Figure 8.9. This process often starts at grain boundaries and proceeds through the formation of high ledges via distinct atomic shuffle displacements. As this transformation was frequently observed in deformed samples [1, 2], it might be speculated that the process is stress induced and provides some kind of transformation toughening.

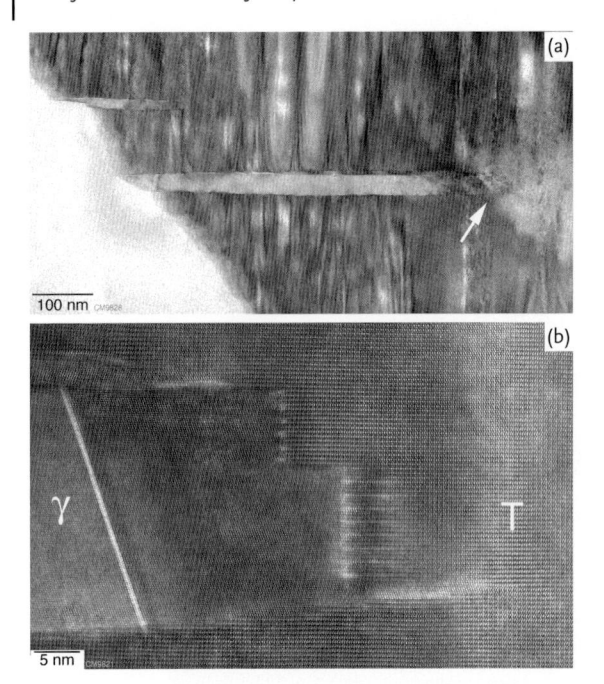

Figure 8.9 Stress-induced transformation of a modulated lamella (T) into γ phase. (a) Generation of a γ lamella at a grain boundary. (b) Higher magnification of the area marked by the arrow in (a). Note the ledges at the interfaces [2].

Summary: A single modulated lath is subdivided into several regions with different crystalline structures with no sharp interface in between the different crystallographic regions. Interfacial dislocations and ledges accommodate the lattice mismatch between the modulated laths and the adjacent γ phase. The mismatch structures apparently support twin nucleation.

8.3
Mechanical Properties

The tensile strength of the modulated γ-Md alloy in the extruded state is demonstrated in Figure 8.10 by stress–strain curves. At room temperature the yield stress is in excess of 1 GPa, with plastic elongations of about 2%. This appreciable combination of strength and ductility might be rationalized by the presence of the modulated laths, which refine the microstructure and are comprised of relatively ductile constituents like β or Ti_2AlNb. Furthermore, the misfit situation between the tweed laths and the adjacent γ phase provides a fine dispersion of dislocations and mechanical twins, which certainly supports plastic deformation. Thus, the overall microstructure of the material may be considered as a composite of relatively brittle phases like γ, $α_2$ or B2, in which ductile laths are embedded. The

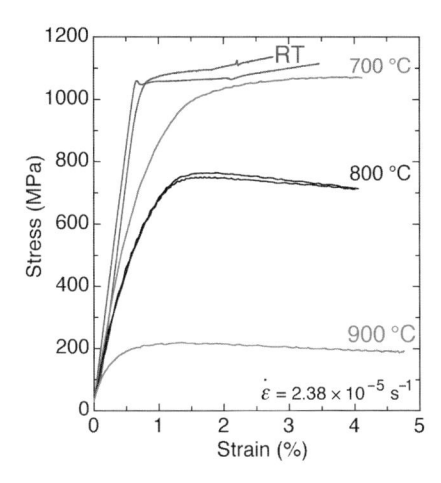

Figure 8.10 Mechanical properties of the extruded γ-Md alloy. Stress–strain curves of tensile tests performed at different temperatures [2].

phases from which the modulated laths are comprised are not separated by sharp interfaces and have crystallographic planes in continuous or similar orientation. This situation is different from that in conventional nanograined structures, where the grain orientation significantly changes and that are often relatively brittle [18]. It might be speculated that shear processes across the interfaces of the tweed laths are relatively easy and not associated with the development of high constraint stresses. There is a good strength retention up to 700 °C accompanied by weak strain hardening. At higher deformation temperatures the yield stress decreases and the stress–strain curves exhibit work softening. This observation suggests that recovery and dynamic recrystallization occur during deformation. The creep properties and fracture behavior of the γ-Md alloy are described in Sections 9.10 and 10.3.

Summary: Alloys with a modulated microstructure exhibit an outstanding balance of strength and tensile ductility combined with good temperature strength retention.

References

1 Appel, F., Oehring, M., and Paul, J.D.H. (2006) *Adv. Eng. Mater.*, **8**, 371.
2 Appel, F., Paul, J.D.H., and Oehring, M. (2008) *Mater. Sci. Eng. A*, **493**, 232.
3 Cheng, T.T., and Loretto, M.H. (1998) *Acta Mater.*, **46**, 4801.
4 Zhang, Z., Leonard, K.J., Dimiduk, D.M., and Vasudevan, V.K. (2001) *Structural Intermetallics 2001* (eds K.J. Hemker, D.M. Dimiduk, H. Clemens, R. Darolia, H. Inui, J.M. Larsen, V.K. Sikka, M. Thomas, and J.D. Whittenberger), TMS, Warrendale, PA, p. 515.
5 Kobayashi, S., Takeyama, M., Motegi, T., Hirota, N., and Matsuo, T. (2003) *Gamma Titanium Aluminides, 2003* (eds Y.-W. Kim, H. Clemens, and A.H. Rosenberger), TMS, Warrendale, PA, p. 165.

6 Jin, Y., Wang, J.N., Yang, J., and Wang, Y. (2004) *Scr. Mater.*, **51**, 113.
7 Takeyama, M., and Kobayashi, S. (2005) *Intermetallics*, **13**, 993.
8 Imayev, R.M., Imayev, V.M., Oehring, M., and Appel, F. (2007) *Intermetallics*, **15**, 451.
9 Hellwig, A., Palm, M., and Inden, G. (1998) *Intermetallics*, **6**, 79.
10 Nguyen-Manh, D., and Pettifor, D.G. (1999) *Gamma Titanium Aluminides 1999* (eds Y.-W. Kim, D.M. Dimiduk, and M.H. Loretto), TMS, Warrendale, PA, p. 175.
11 Banerjee, D. (1997) *Prog. Mater. Sci.*, **42**, 135.
12 Kainuma, R., Fujita, Y., Mitsui, H., Ohnuma, I., and Ishida, K. (2000) *Intermetallics*, **8**, 855.
13 Abe, E., Kumagai, T., and Nakamura, M. (1996) *Intermetallics*, **4**, 327.
14 Ducher, R., Viguier, B., and Lacaze, J. (2002) *Scr. Mater.*, **47**, 307.
15 Gogia, A.K., Nandy, T.K., Banerjee, D., Carisey, T., Strudel, J.L., and Franchet, J.M. (1998) *Intermetallics*, **6**, 741.
16 Haibach, T., and Steurer, W. (1996) *Acta Crystallogr. A*, **52**, 277.
17 Yoo, M.H., Zou, J., and Fu, C.L. (1995) *Mater. Sci. Eng. A*, **192–193**, 14.
18 Kumar, K.S., Van Swygenhoven, H., and Suresh, S. (2003) *Acta Mater.*, **51**, 5743.

9
Creep

Creep resistance and structural stability are important prerequisites for high-temperature application of TiAl alloys and determine the service range in competition with other structural materials. Given this importance, there has been a systematic effort to characterize creep mechanisms and to improve the creep resistance by optimizing alloy composition and microstructure. The progress that has been achieved in this field will be reviewed in this chapter. However, no attempt will be made to reiterate all that has been said before, since detailed review articles [1–6] have been published over the years, and the available experimental evidence has been documented in reasonable detail. Instead, a few examples will be presented, which typify the creep characteristics of TiAl alloys. Special attention will be given to areas where relatively recent work has in some way changed the perspective.

9.1
Design Margins and Failure Mechanisms

The service temperature currently considered for titanium aluminides is in the range 650 to 750 °C, or (0.53 to 0.59)T_m when referred to the absolute melting temperature T_m. Stresses imposed on TiAl components at these temperatures may produce a continuously increasing creep strain, even if they are below the yield point. This continuous plastic flow can result in large deformation, which exceeds the design limits of the component and may terminate in fracture by creep rupture. In Table 9.1 the approximate temperatures are listed at which creep deformation in various high-temperature materials becomes significant [7].

Designs are usually made on the basis of a maximum permissible amount of creep during the expected lifetime of a particular component. Most components are subject to stress states varying in both position and time, which makes predictions of benchmarks for critical stresses and temperatures difficult [8–12]. For example, in turbine blades the temperature and stress conditions differ with the position in the blade, and the extent of creep varies correspondingly. The highest stress usually occurs towards the midsection of the aerofoil. The maximum overall

Gamma Titanium Aluminide Alloys: Science and Technology, First Edition. Fritz Appel, Jonathan David Heaton Paul, Michael Oehring.
© 2011 Wiley-VCH Verlag GmbH & Co. KGaA. Published 2011 by Wiley-VCH Verlag GmbH & Co. KGaA.

Table 9.1 Approximate temperatures T at which creep deformation becomes significant for different metals and alloys.

Material	T (°C)	T/T_m
Aluminum alloys	205	0.54[a]
Conventional titanium alloys	315	0.30[a]
Low alloy steels	370	0.36[a]
Austenitic iron-based alloys	540	0.49[a]
Nickel- and cobalt-based superalloys	650	0.56[a]
Refractory metals	980–1450	0.40–0.45[a]
TiAl alloys based on γ(TiAl) + α$_2$(Ti$_3$Al)	650–700	0.53–0.56

a) data from [7].

creep strain allowable in turbine blades depends on the demands on dimensional stability, but is usually less than 1%.

The resulting creep rate is in the order of 10^{-8} s^{-1}, assuming a service interval of a year. Such a deformation rate is usually accomplished by thermally activated dislocation propagation. For other parts, such as roots of turbine blades, higher stresses are relevant because larger creep strains up to a few per cent can be tolerated. Stress concentrations and large local creep strains are also expected at unavoidable flaws. Under such conditions various processes become failure concerns; these involve dynamic recrystallization, grain growth, overaging of precipitate dispersions, cavitation, and creep rupture.

In high-temperature technologies Ni-based superalloys are widely used, which over the past six decades have been developed to near perfection. In view of the anticipated applications, TiAl alloys must be assessed against this standard. Dimiduk and Miracle [13] pointed out that the creep resistance of most TiAl alloys is inferior to that of nickel-based superalloys, even if the comparison is made on a density-corrected basis. This deficit in creep resistance limits the substitution of superalloys by titanium aluminides and has motivated a large body of research and development.

Summary: At temperatures above 650 °C damage of TiAl components may occur due to continuous creep. The relevant mechanisms are numerous and synergistic, depending on the operation conditions.

9.2
General Creep Behavior

The creep curves of TiAl alloys exhibit similar characteristic regions that have been observed for many other metals: primary creep in which the creep rate decreases with time, steady-state creep with a constant strain rate, and tertiary creep in which the creep rate accelerates with time [2]. Figure 9.1 demonstrates this behavior on

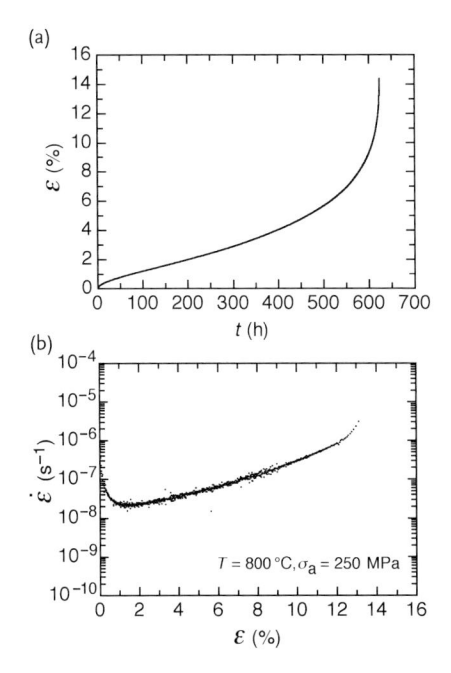

Figure 9.1 Creep behavior observed during tensile testing in air under the conditions indicated. The Ti-45Al-10Nb alloy investigated had been extruded at a temperature corresponding to the $(\alpha + \gamma)$ phase field. (a) Variation of creep strain ε with time t; (b) creep rate $\dot{\varepsilon}$ as a function of creep strain ε determined from the creep curve shown in (a).

the creep curves of an extruded Ti-45Al-10Nb alloy. When compared with pure disordered metals, the extent of steady-state creep is limited. After primary creep, the creep rate usually reaches only a minimum and then increases again with strain. In engineering test practice the minimum creep rate is often only represented by an inflection between the end of primary creep and the beginning of tertiary creep. This short duration of the minimum creep rate is inconsistent with the traditional understanding of steady-state creep. Thus, the regime with minimum creep rate is often referred to as "secondary creep". High stresses and temperatures generally reduce the extent of primary creep and practically eliminate the secondary stage, with the result that the creep rate accelerates almost from the beginning of the test.

The creep stress is usually defined as the constant load divided by the initial specimen area. Thus, for constant-load tensile creep the true stress increases as the specimen area decreases, and in compression the true stress decreases as the specimen area increases. The most important reason for the different responses in tension and compression is that diffusion-assisted dislocation climb is often the dominant deformation mechanism. Tensile stresses expand the lattice and reduce the resistance to diffusion, whereas compressive stresses reduce the lattice dimensions and increase the resistance for diffusion. This aspect could be important if during creep phase transformations occur that are associated with volume changes. The creep life at constant tensile stress is usually longer than that under constant tensile load.

9.3
The Steady-State or Minimum Creep Rate

Although the primary and tertiary stages are distinctive features of the creep curve, the creep behavior of TiAl alloys is usually discussed in terms of secondary creep. The creep rate occurring in this stage can be regarded as a competition between work hardening and dynamic recovery. Work hardening is caused by the accumulation of defects, such as dislocations, point defects or mechanical twins; this increases the stored energy of the material (Section 7.2). Dynamic recovery is driven by this energy and involves defect annihilation and rearrangement of dislocations to form low-angle grain boundaries. Both processes are accomplished by dislocation cross-slip and climb. When a balance is achieved between work hardening and recovery over a large strain interval, then the material exhibits steady-state creep. Therefore, an increase in the work-hardening rate, or a reduction in the recovery rate, can reduce the secondary creep rate. In conventional metals steady-state creep can produce extremely larges strains if instabilities, such as necking of the specimen in a tensile creep test, are avoided. Materials that crept under such conditions usually exhibit a well-developed subgrain structure.

The steady-state or minimum creep rate is usually described by the power law expression [14, 15]

$$\dot{\varepsilon} = A \left(\frac{\sigma_a}{E} \right)^n \exp\left(-\frac{Q_C}{RT} \right) \tag{9.1}$$

E is the Young's modulus at the creep temperature, n the stress exponent, Q_C is the activation energy, σ_a the applied stress, and R is the universal gas constant. A is a microstructure-related material quantity that is assumed to be constant. The stress exponent n is determined by holding the temperature constant and measuring the minimum creep rate for tests conducted at a variety of stresses, Q_C is determined by combining the minimum creep rates and the related temperatures to an Arrhenius plot. Alternatively, n and Q_C can be determined by incremental changes of stress and temperature as [16]

$$n = (\Delta \ln \dot{\varepsilon} / \Delta \ln \sigma)_T \tag{9.2}$$

$$Q_C \equiv \Delta H = kT^2 (\Delta \ln \dot{\varepsilon} / \Delta T)_\sigma \tag{9.3}$$

The advantage of this technique is the ability to determine n and Q_C from single tests. Methodical aspects of the analysis of creep transients following stress changes have been reviewed by Biberger and Gibeling [17]. Table 9.2 provides a collection of stress exponents and activation energies that have been determined on a variety of materials [1, 18–30]; for more details see [2, 31–33].

Table 9.2 Creep data of TiAl alloys. σ_a (creep stress), n (stress exponent), T (test temperature), Q_C (activation energy).

Alloy composition (at.%), microstructure	σ_a (MPa)	n	T (°C)	Q_C (kJ/mol)	Ref.
Ti-53Al-1Nb, single phase gamma	32–345	1–6	760–900	192–560	[22]
Ti-50.3Al	103–241	4.0	700–850	300	[18]
Ti-(50-53)Al, near-gamma	60–400	3.5–8	727–877	300–600	[1]
Ti-51Al-2Mn, near-gamma	280–400		500–600	440	[24]
Ti-47Al, duplex	38–138	2.3	600–900	340	[19]
Ti-48Al-2Cr, fully lamellar	150–260	7.6	800		[23]
Ti-48Al-2Nb-2Cr, duplex	103–200	3.0	705–815	300–410	[20]
Ti-48Al-2Nb-2Cr, cast	80–150	9.6	750–850	359	[28]
Ti-48Al-2V, nearly lamellar	187–420	4–4.6	760–825	320–340	[27]
Ti-48Al-5TiB$_2$, duplex	19–60	1	760–815	340	[21]
Ti-48Al-5TiB$_2$, duplex	300–625	6.0–9.0	676–760	455	[21]
Ti-46.5Al-2Cr-3Nb-0.1W, equiaxed	220–380	8	670–720	420	[26]
Ti-46.5Al-2Cr-3Nb-0.1W, lamellar	220–380	13	670–720	450	[26]
Ti-45Al, lamellar	100–300	4	827–927	360	[30]
Ti-45Al, lamellar	300–400	7	927		[30]
Ti-45Al-8Nb-0.2C, lamellar	100–300	4	827–927	390	[30]
Ti-42Al, lamellar	60–500	3.6	827		[25]
Ti-40Al-10Nb, lamellar	160–240	3.0	750	366	[29]

9.3.1
Single-Phase γ(TiAl) Alloys

Research performed by the group of Oikawa [1, 34–36] on single-phase polycrystalline Ti-(50–53)Al has shown that the dependence of the minimum creep rate on stress exhibits three distinct regimes (Figure 9.2). At the highest stresses and strain rates (regime I), the stress exponent is about $n = 4.7$, close to the value typically determined for climb-controlled creep in pure metals. However, the observation of extensive recrystallization [1] indicates that climb is not the principal creep mechanism in regime I. At lower stresses (regime II), both n and Q_C are nearly twice the values expected for climb-controlled creep. Local grain-boundary migration has been observed in this regime. Again these findings indicate that the creep rate in regime II is determined by mechanisms other than dislocation climb. Only at the lowest stresses and strain rates (regime III), does creep seem to be controlled by dislocation propagation. No evidence of recrystallization has been reported for this regime. Gorzel and Sauthoff [37] investigated compression creep of polycrystalline Ti-55Al at 1000 °C and low stresses of $\sigma_a = 10$ MPa to 23 MPa. The authors determined stress exponents of $n = 3.2$ to 4.8 and attributed this data to dislocation climb. The evaluation also showed that the contribution of Coble creep, the diffusion of matter along grain boundaries, was insignificant.

Hemker and Nix [3] have pointed out that the analysis of the minimum creep rate in terms of the Dorn equation is generally difficult because a phenomenon

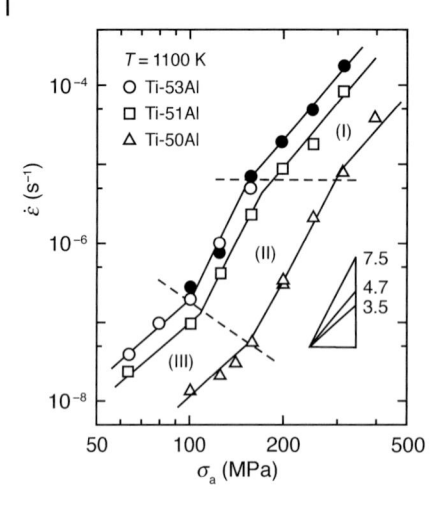

Figure 9.2 Dependence of the minimum strain rate on stress for different single-phase γ(TiAl) alloys. Data from Ishikawa and Oikawa [36], redrawn.

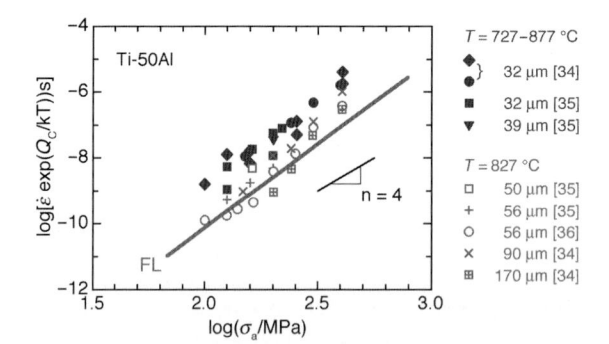

Figure 9.3 Creep data of polycrystalline Ti-50Al alloys, normalized according to Equation 9.1 utilizing an activation energy of $Q_C = 375$ kJ/mol. The grain sizes of the alloys are indicated in the legend. For comparison, line FL represents the data of several fully lamellar alloys analyzed by Parthasarathy et al. [32]. Diagram adapted from [32].

called inverse creep is observed. Primary creep is followed by a period with steadily increasing creep rate [38]. This regime is distinguished from tertiary creep because it is often followed by a period with lower creep rate; true tertiary creep is autocatalytic with continuously accelerating strain rate. Thus, for inverse creep the deformation rate depends not only on stress and temperature but also on the amount of accumulated strain. The evolution of the creep structure with strain will be described in Section 9.3.3.

It should be noted from Figure 9.2 that the minimum creep rate increases in single-phase γ alloys with increasing Al content. The effect is well documented but not well explained. It might be speculated that a higher concentration of Al_{Ti} antisite defects supports diffusion-assisted dislocation propagation. Figure 9.3 demonstrates creep data that has been determined on polycrystalline Ti-50Al alloys for different grain sizes [32, 34–36]. The data was normalized with Equation 9.1, utilizing an activation energy Q_C = 375 kJ/mol [32]. Alloys with a larger grain size tend to have lower creep rates; this is plausible because a larger grain size reduces the length of the grain boundary, where much of the creep processes may reside. However, this trend is not always observed; the alloy with a 56-µm grain size has a lower creep rate than that with a 90-µm grain size. The authors attributed this inconsistency to different levels of interstitial elements, which may have overshadowed the grain-size effect.

Summary: The minimum creep rate of single phase γ(TiAl) alloys depends in a complex manner not only on stress and temperature but also on Al content and grain size. Low creep rates seem to be accomplished by thermally activated dislocation climb. High stresses and temperatures tend to support grain-boundary sliding and dynamic recrystallization.

9.3.2
Two-Phase $α_2$(Ti$_3$Al) + γ(TiAl) Alloys

When compared with single-phase γ alloys, two-phase alloys may exhibit slower or faster creep rates, depending on either or both constitution and microstructure. Beyond the primary creep regime the creep rate in most cases was found to increase steadily. The broad variation of n and Q_C (Table 9.2) indicates that several mechanisms are involved in creep of two-phase alloys. Alloy chemistry, processing, and microstructure undoubtedly affect the minimum creep rate, which probably gives rise to the broad variation of n and Q_C. Zhang and Deevi [33], comparing the findings of several published studies, have comprehensively reviewed the effect of these factors on the creep behavior of TiAl alloys. Unfortunately, in most cases, the particular parameters that are being controlled in a given experiment represent only a fraction of those influencing creep so that comparison of various sets of experimental results is difficult. It is also difficult to conceive of an alloy deforming under creep conditions in such a manner that only one creep mechanism is operative. Furthermore, as will be shown later, there is good evidence that the substructure changes during creep, and, so too, the pre-exponential factor A may change. Thus, despite the widespread adoption of Equation 9.1, the prediction of a creep mechanism solely from the parameters n and Q_C appears tentative at best. Given this uncertainty, Equation 9.1 does not appear to be able to provide appropriate guidance for the development of creep-resistant alloys. Since creep is thermally activated and the creep mechanism can change dramatically with temperature, it is advisable to use creep data that encompasses the operating range of the component to be analyzed (especially at the high-temperature end). The creep strain rate is also a strong function of stress and it is best if the stress levels in the creep tests span the operational stress range of interest.

Under certain testing conditions, however, one or another of the various mechanisms can predominantly control the creep rate. In the domain of intermediate stresses ranging from 100 MPa to 300 MPa and temperatures of 700 °C to 900 °C, the creep rate seems to be controlled by dislocation climb. This is suggested by the estimated stress exponents of n = 3 to 4 and activation energies Q_C = 300 kJ/mol to 400 kJ/mol. The Q_C values determined under these creep conditions are somewhat higher, but reasonable close to the self-diffusion energy of Al (Section 3.3), which is the slower-diffusing species in TiAl. Analyzing literature data, Parthasarathy et al. [32] have revealed that the creep rates of fully lamellar polycrystalline alloys can collectively be represented by a normalized form of the Dorn equation, utilizing an activation energy of $Q_C = 375$ kJ/mol.

Gorzel and Sauthoff [37] have suggested that Coble creep may occur, when two-phase alloys are tested at high temperatures and low stresses. The hypothesis is based on creep studies on polycrystalline Ti-51Al with grain sizes between 40 μm and 60 μm. Tests performed at 1000 °C revealed a linear dependence of the creep rate on the grain size for $\sigma_a \leq 20$ MPa with a substantial decrease of the creep rate with increasing grain size. These findings led the authors to belief that Coble creep with negligible threshold stress is the rate-controlling mechanism under these conditions.

The effects of alloy chemistry and microstructure on the creep behavior of two-phase alloys will be considered in Sections 9.4 and 9.9.

Summary: The minimum creep rate of two-phase alloys depends, apart from temperature and stress, on the alloy chemistry, constitution and microstructure. The evaluation of the minimum creep rate in terms of the Dorn equation leads to a wide range of stress exponents and activation energies. Attempts to attribute these values to a certain creep mechanism are unsatisfactory. Engineering design should be based on multiple tests at each temperature and stress level that occur under service conditions.

9.3.3
Experimental Observation of Creep Structures

Given the uncertainties in analyzing stress exponents and activation energies, much research has been expended on defect structures in crept samples by electron microscopy in order to identify creep mechanisms. Single-phase TiAl (Ti-51Al-2Mn) examined after secondary creep at (550 to 703) °C exhibited superdislocations, faulted dipoles, pinned ordinary dislocation, and a modest number of twins [3, 38]. Several authors [3, 24, 35, 38] have recognized that the defect structure evolves throughout the creep tests. Whereas the population of superdislocations was found to be almost constant, the density of ordinary dislocations and that of the mechanical twins increased with strain. Thus, secondary creep in single-phase alloys is probably accomplished by these two latter mechanisms. The dislocations and twins are homogeneously distributed, although in rare cases weak dislocation walls have been observed [39, 40]. It is important to note that well-developed sub-boundaries have not been recognized after secondary creep of TiAl. For comparison, in the steady-state creep structure of conventional metals the dislocation density in the sub-boundaries is at least a factor 20 above that remaining inside the subgrains. The distinct absence of subgrains during secondary creep cast doubts on the validity of traditional creep models for single-phase TiAl alloys [3, 24, 38].

In two-phase alloys, secondary creep seems to be provided by ordinary dislocations, mechanical twinning or a combination of these mechanisms [26, 29, 41–45]. Morris and Lipe [41] determined an extraordinarily high stress exponent of $n = 19$ in Ti-48Al and Ti-48Al-2Mn-2Nb with duplex and lamellar microstructures, respectively, and attributed this finding to the existence of a long-range backstress. The concept of backstress, also known as threshold stress, is usually introduced in order to appropriately scale creep data so that stress exponents and activation energies conform to the expected values based on an assumed deformation mechanism. However, the physical interpretation of the backstress is often limited, and experimental methods for determining the backstress are usually indirect.

Intrinsic pinning of ordinary dislocation due to jog formation has been recognized in several studies [26, 28, 44]. The pinning mechanism involves all the details that have been described in the context of jog dragging of screw dislocations (Section 6.3.1). Based on *in-situ* TEM heating experiments, it has been has suggested that the jogs move by climb under the combined action of thermomechanical stresses and osmotic climb forces that arise from the chemical potential of excess vacancies [46, 47]. Evidence for spiral-like bypassing, pinching-off and subsequent growth of dislocation loops suggests that the jogs serve as sources for ordinary dislocations. Nix and Barnett [48] proposed that jog-dragging screw dislocations may control the creep rate at high temperatures. In this model the jog segments, which of course are edge dislocations, are dragged by climb. Karthikeyan *et al.* [49] have modified this model and adapted to the situation in TiAl. The modification consists in the incorporation of an upper bound for the height of the jogs that can be dragged by climb. Above this critical value, the jogs are not longer dragged but act as dislocation sources, as has been described in Section 6.3.1.

Malaplata *et al.* [28] performed creep studies on Ti-48Al-2Cr-2Nb at 750 °C and tensile stresses of 80 and 150 MPa. Post-mortem TEM examination of the crept samples revealed dislocation climb, arising from the nucleation and propagation of jog pairs. This view was supported by the kinematics of the dislocation propagation that the authors observed during their *in-situ* TEM experiments.

As with single-phase alloys, formation of well-developed sub-boundaries seems to be difficult during secondary creep. It should be noted that most creep studies on TiAl were carried out at relatively high stresses, resulting in high strain rates and almost short-term tests. Thus, not surprisingly, the observed mechanisms are not so different from those that have been recognized after constant strain-rate deformation at $\dot{\varepsilon} = 10^{-5}\,\mathrm{s}^{-1}$ or so.

Summary: Secondary creep of TiAl alloys produces an essential uniform distribution of ordinary dislocations and mechanical twins, the density of which increases through the life of the creep tests. Contrary to many observations made on conventional metals with high stacking-fault energy, no well-defined subgrain structure is developed. Taken together these findings give rise to the conclusion that during secondary creep a steady-state deformation structure is not attained. At intermediate temperatures and stresses there is a domain in which creep seems to be controlled by dislocation climb.

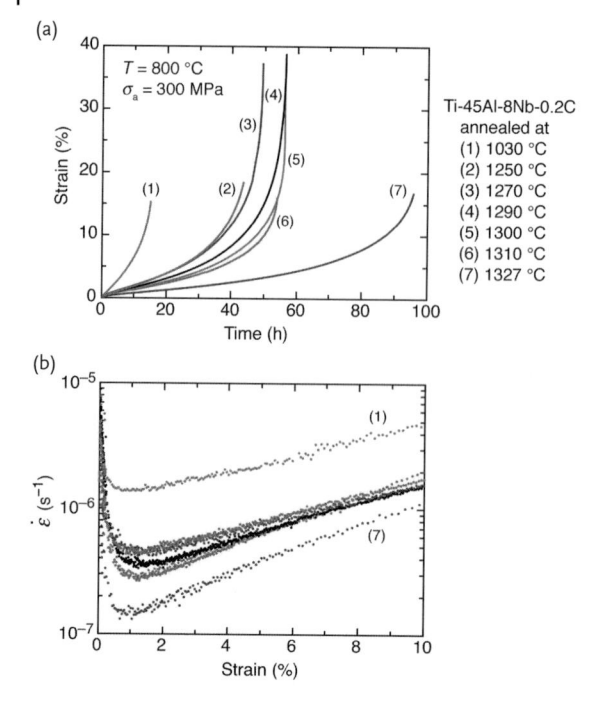

Figure 9.4 A series of creep curves demonstrating the effect of the microstructure on the creep behavior of an extruded Ti-45Al-8Nb-0.2C alloy. Different proportions of lamellar colonies were produced by the thermal treatments 1 to 7 indicated in the legend. (a) Variation of creep strain ε with time t; (b) creep rate $\dot{\varepsilon}$ as a function of creep strain ε determined from the creep curves shown in (a).

9.4
Effect of Microstructure

There is clear evidence that the microstructure exerts strong effects on the creep behavior. Fully lamellar microstructures exhibit the best creep resistance, typically reducing strain rates by at least one order of magnitude, when compared with the duplex form of the same alloy [2, 50–55]. Figure 9.4 demonstrates this behavior in more detail by a series of creep curves that were measured on different microstructural forms of a Ti-45Al-8Nb-0.2C alloy. The alloy had been extruded at 1230 °C, and subject to stress-relief annealing at 1030 °C for two hours (designated HT1). After this thermal treatment the material exhibits a banded duplex structure involving a large volume content of fine equiaxed grains and some remnant lamellae. Curve 1 represents the creep behavior of this material. The volume content of lamellar colonies in the material was then gradually increased by the subsequent transformation treatments 2 to 7, specified in the legend. The α-transus temperature of the extruded alloy was $T_\alpha = 1332$ °C. After the transformation treatment 7 the microstructure was fully lamellar. The related creep curves 2 to 7 demonstrate that the creep resistance increases as the volume content of the lamellar colonies increases. Likewise, the creep resistance of lamellar alloys is superior to that of alloys with an equiaxed microstructure. This becomes evident in Figure 9.3, in which the creep rates of equiaxed and fully lamellar alloys are compared. Morris and Leboeuf [26] have recognized a similar behavior on equiaxed and lamellar variants of a Ti-46.5Al-2Cr-3Nb-0.1W alloy.

In broad terms, the high creep resistance of the fully lamellar microstructure is a result of the fine spacing of the lamellae, which reduces the effective slip length for dislocations and twins compared to that in the duplex and equiaxed microstructures. Bartholomeusz and Wert [56] have previously suggested that the lower creep rates of lamellar alloys are the result of constraint stresses present at the interfaces. For a detailed discussion regarding the barrier strength of lamellar interfaces to dislocations and mechanical twins, see Sections 6.1.5 and 6.2.5. The observed effects of the lamellar morphology on the creep resistance may be summarized as follows.

i) The creep curves of lamellar alloys are similar to those of single-phase γ alloys that were tested at low stresses and temperatures (regime III in Figure 9.2).

ii) The minimum creep rate of lamellar alloys tends to increase if γ grains are present along the colony boundaries, an effect that was attributed to dynamic recrystallization occurring in these intercolony γ grains [57].

iii) The creep rate of fully lamellar alloys is less dependent on colony size if the grain size is larger than 100 μm [25]. Serrated colony boundaries apparently improve the creep resistance [2]. The serrations result from incursions of lamellae into neighboring colonies.

iv) Refinement of the lamellar spacing improves the creep resistance of fully lamellar alloys, as has been recognized by several groups [25, 32, 58–61]. In these studies, varying the cooling rate was used to change the lamella thickness. The effect is well established at high stresses but becomes negligible at low stresses due to the onset of dynamic recrystallization and interface sliding [25]. Crofts *et al.* [58] have pointed out in this context that it is almost impossible to vary only one microstructural parameter by heat treatment. Varying the cooling rate may not only alter the lamella thickness but also the perfection of the colony boundaries and the nature of the lamellar boundaries. Each of these factors could affect the creep rate. The phenomena are obviously complex and should not be oversimplified. It should also be noted that ultrafine lamellae are not stable at elevated temperatures; this aspect will be considered in Section 9.6.

v) In the range of 44 to 48 at.% Al, the minimum creep rate is essentially insensitive to the proportion of α_2 phase [25].

vi) Investigations performed on differently oriented PST crystals have revealed a significant anisotropy of the creep resistance [62–64], which parallels the yield-stress anisotropy that has been recognized on PST crystals in constant strain-rate tests and described in Section 6.1.5. The creep resistance was found to be highest for the 0° orientation and weakest for the 45° orientation of the lamellae against the compression axis. The beneficial effect of a suitable lamellae orientation on the creep resistance is remarkable and similar to the effects of precipitation dispersions of carbides or silicides. The anisotropy was also reflected in the activation energies [62]. The values estimated are $Q_C = 398 \text{ kJ/mol}$ for the soft 45° orientation, $Q_C = 532 \text{ kJ/mol}$ for the 0° orientation and $Q_C = 432 \text{ kJ/mol}$ for the 90° orientation. As the Q_C values in the hard orientation are not consistent with a diffusion mechanism, the authors [62] have taken the view that the creep rate for these orientations does not rely entirely on diffusion and that shear transfer through the interfaces must play a significant role. However, it is also possible that an internal backstress exists as a result of the misfit situation in the lamellar morphology (Sections 6.1.3 and 6.1.4). Kim and Maruyama [63] determined on PST crystals a remarkably low stress exponent of $n \sim 1$ when crept at 877 °C in the soft orientation and attributed this finding to mechanical twinning parallel to the lamellar interfaces. The resulting refinement of the lamellae by twinning apparently did not harden the material. The group of Yamaguchi [65–68] determined the creep properties of directionally solidified TiAl–Si alloys with a preferred orientation of the columnar lamellar colonies. The authors recognized excellent creep resistance when the lamellae in the colonies were parallel to the deformation axis. Additions of either Re, W or Mo further improved the creep performance of directionally solidified alloys [68]. However, the creep resistance significantly degraded when the lamellar planes were inclined to the load axis [67]. A similar anisotropy of the creep behavior with respect to the lamellae orientation was also recognized for Ti-47Al-2Nb-2Cr that exhibited a casting texture [69].

vii) The minimum creep rate of lamellar alloys increases progressively with stress, indicating that three distinct regimes may exist. Research performed by Wang et al. [70] at 760 °C on refined fully lamellar alloys (Ti-47Al-2Cr-2Nb and Ti-47Al-2Cr-1Nb-1Ta) have shown that the stress exponent in the low-stress regime ($\sigma_a < 300$ MPa) is $n \sim 1$ and then gradually increases to $n \sim 7$ in the high stress regime ($\sigma_a > 400$ MPa). The behavior is demonstrated in Figure 9.5. However, such stress dependence is not always observed, as indicated by the data for the nearly lamellar Ti-45Al-8Nb-0.2C alloy, which is included in the diagram. This disparity of observations might be associated with differences in processing conditions and fine details of the microstructures.

Summary: Among the various microstructures that can be established in polycrystalline ($\alpha_2 + \gamma$) alloys, fully lamellar alloys are most creep resistant. The factors that mainly determine the creep resistance are (i) the lamellae thickness, (ii) the orientation of the lamellar colonies with respect to the loading axis, (iii) the perfection of the colony boundaries, and, (iv) to some extent, the colony size.

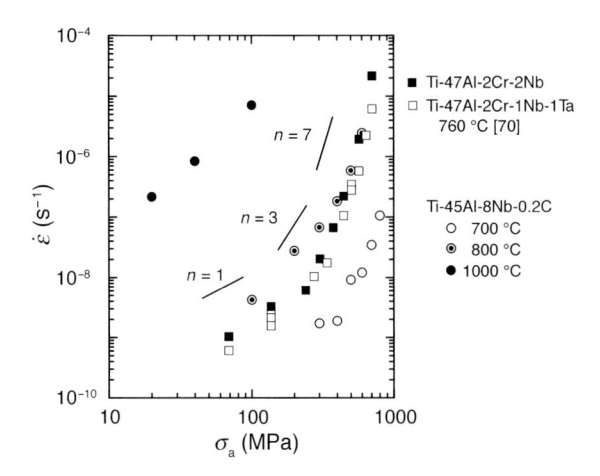

Figure 9.5 Stress dependence of the minimum creep rate of three lamellar TiAl alloys that are specified in the legend.

9.5
Primary Creep

As mentioned in Section 9.1, for certain applications, the allowable creep strain could be limited to less than 1%, thus, primary creep could be design limiting. It should be mentioned that nickel-based superalloys exhibit little, if any, primary creep. Given this importance, primary creep of TiAl alloy has attracted much attention [51, 66, 71–82]. It should be noted that some of the reports of extremely large primary creep strain are a direct result of tests at high fractions of the yield strength of the material. Primary creep consists of two components: an instantaneous strain that occurs virtually simultaneous upon loading and a primary transient where the strain rate declines with time to the minimum creep rate.

Several investigations have focused on a mechanistic understanding of primary creep. In single-phase γ(TiAl) alloys the primary transient probably reflects the exhaustion of <011] superdislocations and the intrinsic pinning of the ordinary dislocations resulting in strong work hardening [3, 24, 38]. The remaining superdislocations are probably locked by dissociation and cross-slip (Sections 5.1.3 and 5.1.4), which provides an additional element of work hardening. Loiseau and Lasalmonie [83] observed plate-like Ti_2Al precipitates in Ti-50Al and Ti-56Al after creep at 750 °C to 900 °C and stresses between 30 MPa and 200 MPa. The precipitates were heterogeneously nucleated at all types of dislocations, which led to dislocation locking. The authors suggested that this effect reduced the density of the mobile dislocations and progressively decreased the primary creep rate. Increasing the applied creep stress seemed to enhance the precipitation kinetics.

In two-phase alloys with a duplex structure extensive twinning has been recognized after primary creep [42, 84, 85]. The twins are thought to refine the

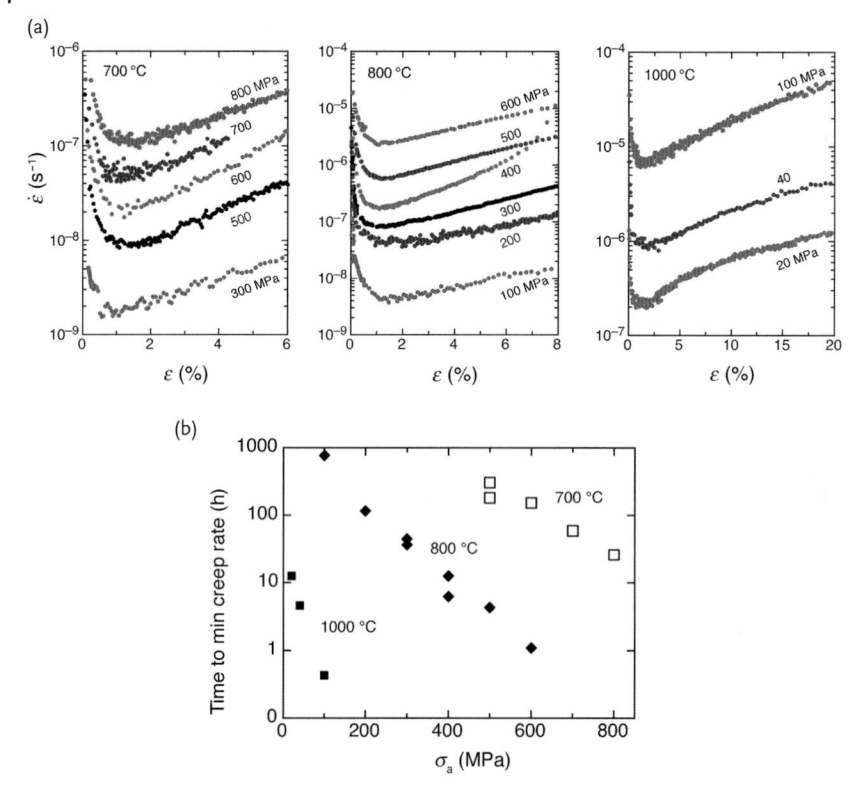

Figure 9.6 Creep behavior of the nearly lamellar Ti-45Al-8Nb-0.2C alloy observed at different temperatures and stresses. (a) Creep rates $\dot{\varepsilon}$ as function of strain ε; no steady-state creep has been observed under any of the conditions investigated. (b) Time to achieve the minimum creep rate as function of stress σ_a.

microstructure, thus providing a Hall–Petch type hardening, thereby gradually reducing the creep rate. Depending on temperature and stress conditions, the primary creep strain of fully lamellar alloys can exceed that of duplex alloys. This might be a problem for the application of an otherwise creep-resistant fully lamellar material. Figure 9.6a shows the behavior that has been observed at different temperatures and stresses on nearly lamellar Ti-45Al-8Nb-0.2C. The minimum of the creep rate occurs at about $\varepsilon = 1\%$, almost independent of temperature and stress. The creep rate rapidly decreased during primary creep to the minimum and then increased again; no steady-state creep was observed under these conditions. Figure 9.6b demonstrates that the time to achieve the minimum creep rate dramatically decreases with increasing stress and temperature.

In lamellar alloys primary creep seems to be supported by the coherency stresses and misfit structures, which are present at the lamellar interfaces (Sections 6.1.3 and 6.1.4). At low temperatures the coherency stresses acting on misfit dislocations are in equilibrium with the high glide resistance of the dislocations, thus

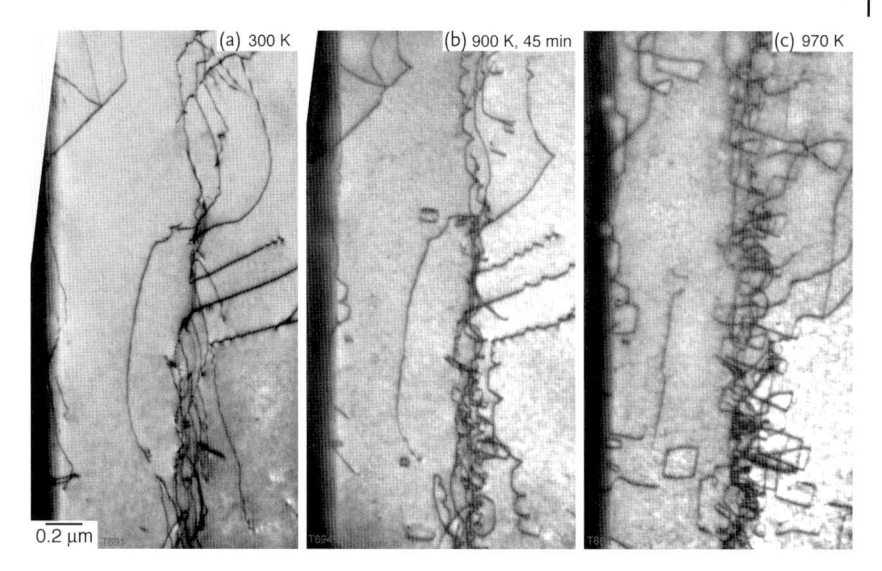

Figure 9.7 Climb of interfacial dislocations during *in-situ* heating inside the TEM. Note the emission of dislocations from the lamellar interfaces and the formation of helical dislocation configurations, which may operate as Bardeen–Herring climb sources. Nearly lamellar Ti-48Al-2Cr, predeformation at room temperature to strain around $\varepsilon = 3\%$.

impeding dislocation motion. Under creep conditions the glide resistance is reduced due to thermal activation, and the coherency stresses may relax due to dislocation emission (Section 6.3.2, Figure 6.44). In order to mimic the situation occurring during primary creep of lamellar alloys, *in-situ* TEM heating experiments have been performed; part of this study is demonstrated in Figure 9.7. During the experiment, dislocations were emitted from the interfaces and moved away by climb, as indicated by their spiral morphology. In this way, interconnected loop structures were formed, which can operate as additional glide or climb sources, depending on the applied stress. The mechanism apparently plays an important role during creep. This is indicated by the frequent occurrence of loop structures adjacent to interfaces, which has been observed in crept samples (Figure 9.8). The described dislocation emission can certainly be assisted by the creep stress. Nevertheless, the mechanism probably becomes exhausted because the misfit character of the interfaces would significantly change if the dislocation emission had proceeded for a long time. Thus, the mechanism provides a natural explanation for the decreasing creep rate of the primary transient.

Hsiung and Nieh [75] have recognized that primary creep in fully lamellar Ti-47Al-2Cr-2Nb at 760 °C and 138 MPa was provided by lamellar colonies that were favorably oriented for glide parallel to the {111} interface planes. The authors demonstrated that glide and multiplication of interfacial dislocations occurred under these conditions. Dislocation barriers produced by secondary slip and grain boundaries seemed progressively to retard the propagation of the interfacial dislocations.

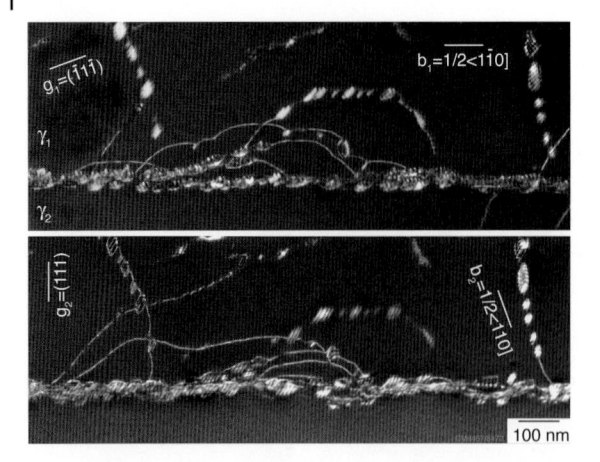

Figure 9.8 Dislocation-loop structures observed in a crept sample adjacent to a semicoherent γ_1/γ_2 interface. Two sets of ordinary dislocations were emitted with the Burgers vectors indicated. Ti-48Al-2Cr, crept at $T = 700\,°C$ under $\sigma_a = 110\,MPa$ for $t = 13\,400\,h$ to $\varepsilon = 0.46\%$.

The primary creep strain was found to increases with applied stress. Malaplate et al. [28, 86] examined duplex and nearly lamellar variants of Ti-48Al-2Cr-2Nb after primary creep at 750 °C under stresses of 80 MPa and 150 MPa. Deformation was found to be mainly controlled by jogged screw dislocations and involved jog dragging by climb and lateral glide of the jogs. Similar to the findings of Hsiung and Nieh [75], Malaplate et al. [86] recognized a remarkable heterogeneity in the deformed state, which was probably caused by structural inhomogeneities in the starting material. A higher degree of structural inhomogeneity seems to expand primary creep.

In the TiAl literature several metallurgical techniques have been reported that could increase the resistance against primary creep. These involve:

i) refinement of the lamellar spacing [51, 77, 78];
ii) aging treatments of the material prior to testing [78–80];
iii) prestraining of lamellar alloys at higher creep stresses [70, 87];
iv) prestraining in tension at constant strain rate beyond yielding [88].

The question arises, however, whether or not such a treatment can permanently reduce primary creep. The interface processes described can also be induced by thermoelastic stresses. When the constituents of a two-phase material have different or anisotropic coefficients of thermal expansion, constraint stresses are easily generated at the vicinity of the interfaces upon temperature changes. It is well documented that if these thermal stresses are high enough, dislocations are produced at the interfaces and punched into the matrix [89]. Thus, the mismatch structures of the interfaces may be regenerated if rapid temperature changes occur during service. A permanent resistance against primary creep can probably only be achieved if the interface-related dislocation mechanisms are impeded by the implementation of an additional glide resistance, as provided by precipitates for example. Several authors [90–92] have indeed recognized that precipitation of silicides and carbides at the lamellar interfaces can significantly reduce primary creep. Utilizing such a concept, Seo et al. [81] have significantly improved the creep resistance of fully lamellar Ti-48Al-2W at 760 °C. Aging of this alloy at 950 °C led to the heterogeneous nucleation of B2 particles at the lamellar interfaces, which grew with aging time. Simultaneously, decomposition of α_2 lamellae occurred and resulted in discontinuous lamellae. Based on these processes the authors described the evolution of the microstructure as

$$\alpha_2 + \gamma \rightarrow \alpha_2 + \gamma + B2 \rightarrow \gamma + B2 \tag{9.4}$$

The heterogeneous precipitation of B2 particles at the interfaces seems to retard the propagation and multiplication of interfacial dislocations and thus significantly reduces primary creep. The concept was later confirmed by a study of Zhu et al. [93] on lamellar Ti-48Al-2W. The authors pointed out that coarsening of the B2 particles upon creep could be a concern.

Summary: Primary creep is a critical concern in the design of TiAl components for high-temperature applications. In the otherwise creep-resistant fully lamellar alloys, primary creep is caused by the relaxation of constraint stresses due to the propagation and multiplication of interfacial dislocations. Structural inhomogeneities in the starting material seem to enhance primary creep.

9.6
Creep-Induced Degradation of Lamellar Structures

Numerous investigations have demonstrated that the lamellar microstructure degrades upon creep [2, 4, 5, 23, 30, 47, 94–101]. This structural instability is a serious problem for long-time service of lamellar alloys; thus, the relevant mechanisms will be discussed in this section. Driving forces for the structural changes may arise from the imparted strain energy, vacancy supersaturation, reduction of surface energy, defects in the lamellar morphology, and nonequilibrium phase composition. Since many aspects of this structural degradation are intimately linked to defect configurations at the atomic level, standard techniques of metallography are often inadequate to provide the necessary information. In the following, a few examples of TEM analysis will be presented, which demonstrate the complexity of the processes involved in the degradation of the lamellar morphology. Most of these studies have been performed on samples that had been subject to long-term tensile creep at 700 °C under relatively low stresses of 80 to 140 MPa [99, 101]. In order to avoid interference with effects caused by oxidation or corrosion, the creep tests were performed in an atmosphere of nominally 99.999% He, which was additionally purified by a liquid-nitrogen cold trap.

When compared with the microstructure of undeformed material, the lamellar interfaces in crept samples were highly imperfect. The interfaces exhibit many steps, and the lamellae are often terminated. Figures 9.9–9.12 demonstrate the initial state of structural changes that have been observed on a lamellar 120°

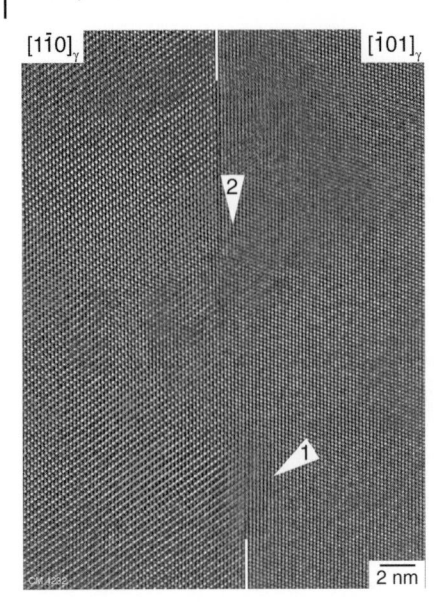

Figure 9.9 Structural changes occurring in a nearly lamellar Ti-48Al-2Cr alloy upon long-term tensile creep at $T = 700\,°C$, $\sigma_a = 140\,\text{MPa}$, $t = 5988\,\text{h}$ under a He atmosphere. The micrograph shows a 120° lamellar boundary that joins the variants $[1\bar{1}0]_\gamma$ and $[\bar{1}01]_\gamma$. Note the step of seven (111) planes present in the interface. Details 1 and 2 represent structural changes that apparently occurred during creep; these are elucidated in Figures 9.10–9.12.

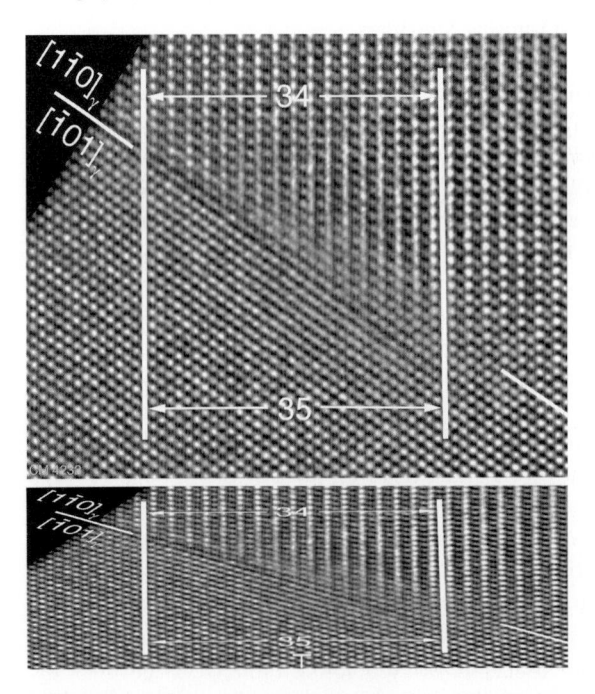

Figure 9.10 Higher magnification of detail 1 in Figure 9.9 showing a misfit dislocation with Burgers vector $\mathbf{b} = 1/2[001]$. White bars mark the interface. The figure below shows the same area compressed along the (002) planes; the dislocation symbol indicates the approximate position of the extra (002) plane.

9.6 Creep-Induced Degradation of Lamellar Structures

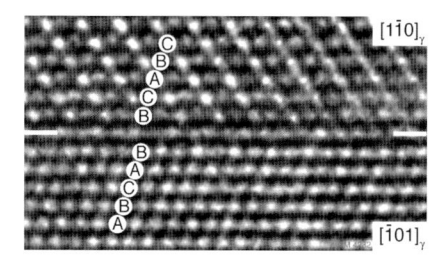

Figure 9.11 Detail 1 in Figure 9.9. Rigid-body translation of the of the γ variants joining the 120° boundary, which was probably produced by the propagation of a Shockley partial dislocation. The stacking sequence across the interface is **ABCAB X BCABC**. Most atoms in the X layer occupy an A position, that is, an intrinsic stacking fault is present at the interface.

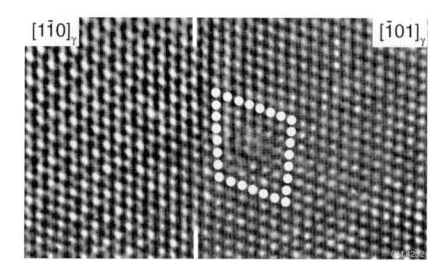

Figure 9.12 Detail 2 in Figure 9.9 showing a misfit dislocation with the resolved Burgers vector **b** = 1/3[110].

boundary joining two misfitting γ variants. Magnified images of detail 1 are provided in Figures 9.10 and 9.11. The misfit between the (002) and (020) planes of the two γ variants is accommodated by a 1/2[001] dislocations, which can be seen by the extra (002) plane (Figure 9.10). There is a rigid-body translation of the two lamellae, which was probably produced by the propagation of a Shockley partial dislocation (Figure 9.11). The stacking sequence across the interface is *ABCAB X BCABC*; however, most atoms in the *X* layer occupy an *A* position, that is, an intrinsic stacking fault is present at the interface. Figure 9.12 shows detail 2 of the interface, where an interfacial 1/3[111] dislocation is present. Taken together, the structural details observed at the interface could be linked by the reaction

$$2\times 1/2[00\bar{1}] \rightarrow 2\times 1/6[11\bar{2}] + 1/3[\bar{1}\bar{1}\bar{1}] \tag{9.5}$$

meaning that the interfacial step is the result of a decomposition reaction involving two 1/2[001] misfit dislocations. Propagation of the steps along the interface and their coalescence could explain the formation of multiple interfacial steps.

Figure 9.13 demonstrates the formation of a high interfacial step in a 60° pseudotwin boundary. The interfacial steps had often grown into broad zones, which extended over about 200 nm perpendicular to the interface; Figure 9.14 demonstrates an intermediate stage of this growth process. Multiheight ledges are commonly observed after phase transformation and growth, and several mechanisms have been proposed to explain the phenomenon [102, 103]. Analogous to these

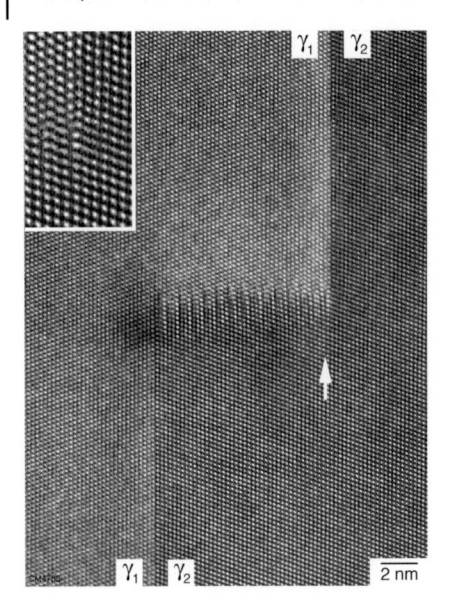

Figure 9.13 Formation of an interfacial step in a 60° pseudotwin boundary. Note the interfacial dislocation (arrowed) that is manifested by an additional {111} plane. Ti-48Al-2Cr, crept under He atmosphere at $T = 700\,°C$, $\sigma_a = 140\,MPa$, $t = 5988\,h$ to $\varepsilon = 0.69\%$.

models it is speculated that the large ledges observed in the crept TiAl alloy arise from one-plane steps, which moved under diffusional control along the interfaces and were piled up at misfit dislocations (arrowed in Figure 9.13). Once a sharp pile up is formed, the configuration may rearrange into a tilt configuration with a long-range stress field. This would cause further perfect or Shockley partial dislocations to be incorporated into the ledge. This would also explain the large height of ledges and the observation that in all cases the ledges were associated with misfit dislocations. The detailed atomic structure of the macroledges is not clear. As can be seen in Figures 9.13 and 9.14, there is a variation of the contrast in the ledges with a periodicity of three (111) planes. This is reminiscent of the 9R structure, which is a phase that probably has a slightly higher energy than the $L1_0$ ground state [104]. The formation of the 9R structure is a well-known phenomenon in many f.c.c. metals that exhibit twinning. Singh and Howe [105] have recognized the 9R structure in heavily deformed TiAl. It should be noted, however, that similar three-plane structures have been observed in a massively transformed, but undeformed Ti-48.7Al [106]. In this work, the contrast phenomena have been interpreted as arising from overlapping twin-related γ variants. Nevertheless, the macroledges are a characteristic feature in the microstructure of crept samples and represent at least a highly faulted $L1_0$ structure; the question is only whether there is a periodicity in the fault arrangement. When the macroledges grow further, it might be energetically favorable to reconstruct the $L1_0$ structure and to nucleate a new γ grain. Figure 9.15 probably demonstrates an early stage of such a process. The recrystallized grains usually have a certain orientation relationship

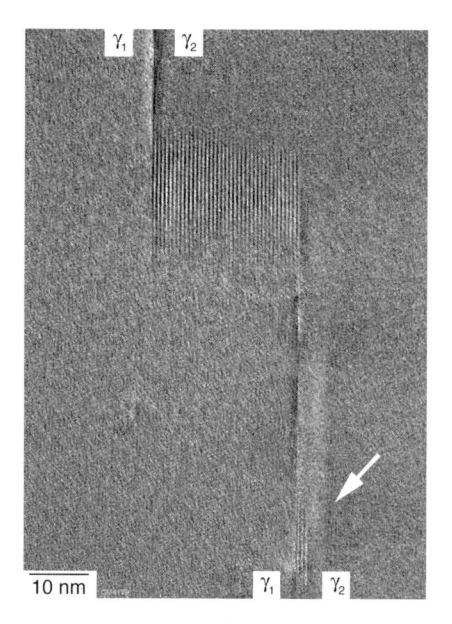

Figure 9.14 A macroledge present in a 60° pseudotwin boundary. Ti-48Al-2Cr, crept under He atmosphere at $T = 700\,°C$, $\sigma_a = 140\,MPa$, $t = 5988\,h$ to $\varepsilon = 0.69\%$.

with respect to the parent γ lamellae; Figure 9.16 indicates that the (001) planes of the recrystallized grain are parallel to the $(1\bar{1}\bar{1})$ planes in the parent lamella γ_1. There is a significant mismatch for this orientation relationship, which is manifested by a high density of ledges and dislocations at the $(001)\|(1\bar{1}\bar{1})$ interface. Recrystallization of ordered structures has been investigated in several studies; for a review see [107]. Cahn [108, 109] has given a detailed account of the relevant literature. There is a drastic reduction in grain-boundary mobility, when compared with disordered metals. Recovery of ordered alloys is also complicated by the fact that the ordered state has to be restored. In this respect it is interesting to note that the small grain shown in Figure 9.16 is completely ordered, giving the impression that the ordering is immediately established after grain nucleation or that nucleation occurred in the ordered state. This might be a consequence of the fine scale of the lamellar microstructure and the heterogeneous grain nucleation at the interfacial ledges. There are certainly crystallographic constraints exerted by the parent lamellae adjacent to the ledges, which may control nucleation and growth. Clearly, the process needs further investigation.

There is a significant body of evidence in the TiAl literature indicating that dissolution of α_2 lamellae occurs during creep [4, 5, 47, 59, 95, 100, 101, 110–113]. The phase transformation is probably driven by a nonequilibrium constitution. This is because the decomposition of α phase into lamellar $\alpha_2 + \gamma$ during cooling is sluggish and the volume fraction of γ phase is less than equilibrium. High-temperature creep is expected to promote phase transformation towards equilibrium constitution; thus, dissolution of α_2 and formation of γ occurs [99]. The

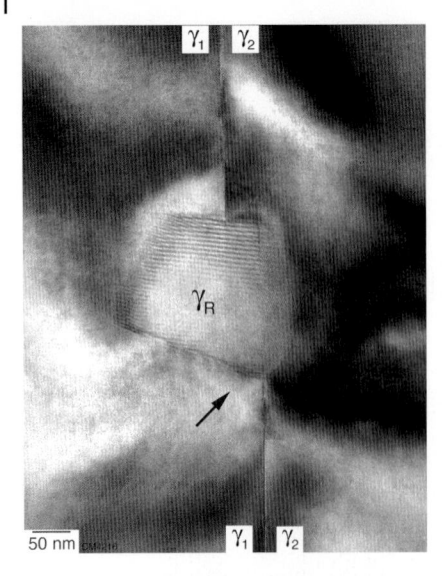

Figure 9.15 Degradation of the lamellar structure in Ti-48Al-2Cr under long-term creep at $T = 700\,°C$, $\sigma_a = 140\,MPa$, for $t = 5988\,h$ to strain $\varepsilon = 0.69\%$. Recrystallized γ grain formed at a ledge in a lamellar interface joining the gamma variants γ_1 and γ_2 with a pseudotwin orientation relationship. Note the step in the interface and the ordered state of the recrystallized grain.

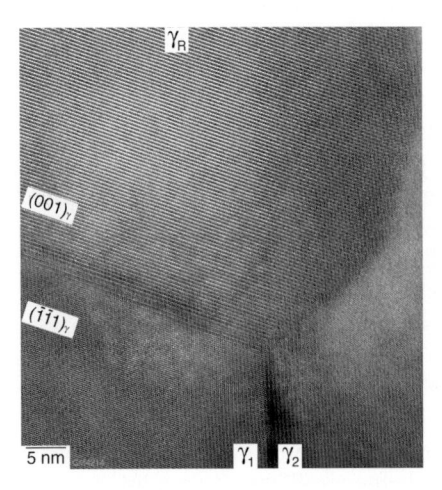

Figure 9.16 Higher magnification of the boundary triple-point marked in Figure 9.15. Note the orientation relationship $(001)\|(111)$ between the recrystallized grain and lamella γ_1.

high-resolution micrograph taken after long-term creep of a Ti-48Al-2Cr alloy shown in Figure 9.17 supports this reasoning; there is clear evidence that the density of steps at the α_2/γ interfaces is significantly higher that that at the γ/γ interfaces, meaning that the α_2 lamella dissolves, whereas the γ lamellae are relatively stable. The processes eventually end with the formation of new grains and

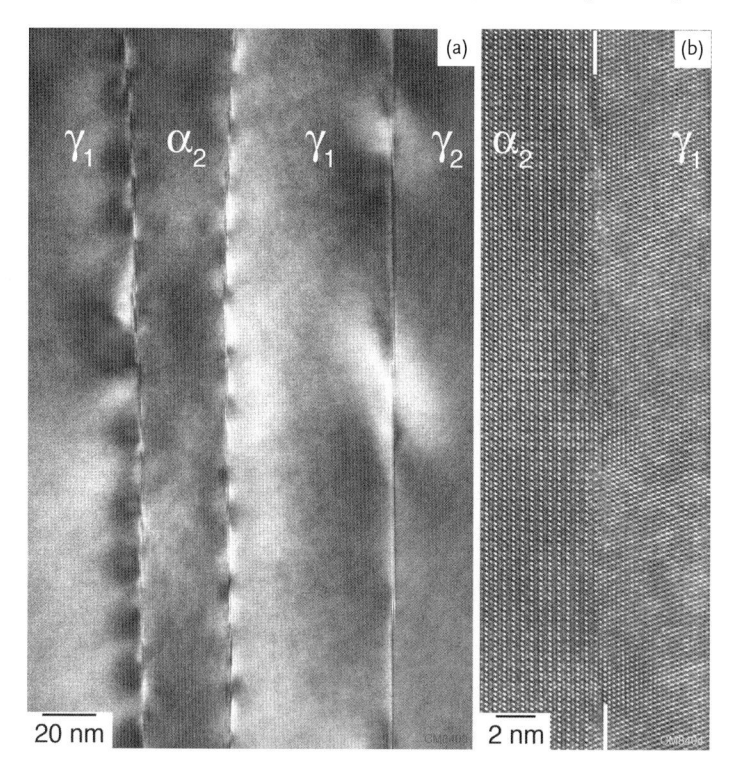

Figure 9.17 Initial stage of the $\alpha_2 \rightarrow \gamma$ phase transformation in the lamellar structure of a Ti-48Al-2Cr alloy occurring during long-term creep at $T = 700\,°C$, $\sigma_a = 110\,MPa$, $t = 13\,400\,h$ to strain $\varepsilon = 0.46\%$. (a) Low-magnification, high-resolution image of the lamellar structure. Note the significantly higher density of steps at the α_2/γ interfaces, which indicates dissolution of α_2 phase. (b) Atomic structure of one of the α_2/γ interfaces demonstrating its stepped character.

a more or less complete conversion of the lamellar morphology into a fine spheroidized microstructure. An initial stage of this phase transformation and recrystallization process is demonstrated in Figure 9.18.

The $\alpha_2 \rightarrow \gamma$ phase transformation is often associated with local deformation, as suggested by Figure 9.19. The micrograph shows two α_2 terminations that are connected by an interface, thus the α_2 lamella is partially dissolved. Two twins were emitted at one of the terminations and a dislocation with a Burgers vector out of the interface is present (Figure 9.20). The α_2 terminations have extremely small principal radii of curvature. It may be expected that elimination of such structural features reduces the surface energy and provides a driving force towards further coarsening. The interface connecting the two α_2 terminations exhibits a fault translation that corresponds to an intrinsic stacking fault. This is indicated by the stacking sequence **ABC B CAB**. The observation underlines once again the fine scale of interface processes that may occur during creep of lamellar alloys.

Figure 9.18 Phase transformation and recrystallization observed after long-term tensile creep on Ti-48Al-2Cr. Spheroidization of α_2 lamellae due to the formation of γ grains (arrowed, designated as γ_R). Creep conditions: $T = 700\,°C$, $\sigma_a = 110\,MPa$, $t = 13\,400\,h$ to strain $\varepsilon = 0.46\%$.

The $\alpha_2 \rightarrow \gamma$ transformation requires a change of both the stacking sequence and the local composition, as has been described in Section 6.1.1. However, achieving the appropriate composition requires long-range diffusion, which at a creep temperature of 700 °C is very sluggish. Low-temperature diffusion might be supported by the presence of Ti_{Al} antisite defects (Sections 3.3 and 6.4.3). In the newly formed γ phase a high density of such defects is certainly formed, in order to accommodate the excess of titanium. Thus, a substantial antistructural disorder occurs, which forms a percolating substructure. Under such conditions, the antisite defects may significantly contribute to diffusion because antistructural bridges (ASBs) are formed. As has been described in Section 6.4.3, an elementary bridge event involving one vacancy and one antisite defect consists of two nearest-neighbor jumps, which result in a nearest-neighbor displacement of two atoms of the same species. For the $\alpha_2 \rightarrow \gamma$ transformation the so-called ASB-2 mechanism [114] (Section 3.3) might be relevant, which requires only low migration energy. Diffusion may also be supported by the mismatch structures present at the interfaces. Dislocations and ledges represent regions where the deviation from the ideal crystalline structure is concentrated. These are paths of easy diffusion, which can effectively

9.6 Creep-Induced Degradation of Lamellar Structures

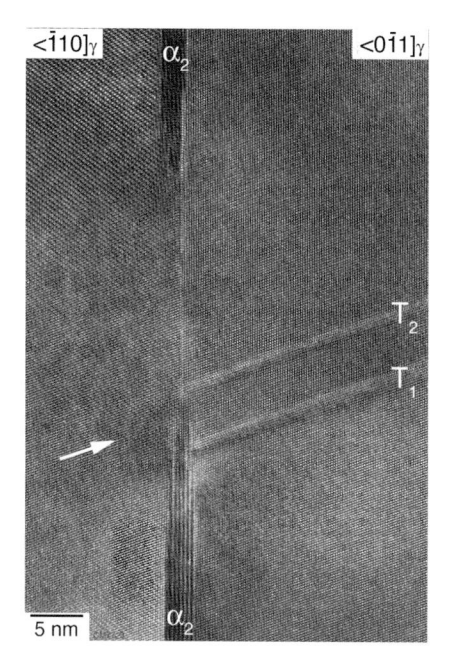

Figure 9.19 A partially dissolved α_2 lamella in Ti-46.5Al-4(Cr, Nb, Ta, B) embedded in γ phase. Creep deformation at $T = 700\,°C$, $\sigma_a = 200\,MPa$ to strain $\varepsilon = 1.35\%$. Note the two α_2 terminations that are connected by an interface and the emissions of two twins T_1 and T_2 at one of the terminations. The stacking sequence indicates a rigid-body translation of the adjacent γ lamellae. Arrowed details are shown in Figure 9.20.

support the exchange of Ti and Al atoms. One may expect all these processes to be thermally activated und supported by superimposed external stresses. In this respect, the coherency stresses present at the interfaces (see Section 6.1.4) are certainly significant because they are comparable to or even higher than the shear stresses applied during creep tests and are often associated with mismatch structures. Thus, given a nonequilibrium constitution, it is understandable that a drop in the α_2 volume content also occurs, when a lamellar alloy is subject to the same temperature–time profile without externally applied stress [115–118]. For example, Hu et al. [115] observed the complete removal of α_2 laths after long-term high-temperature exposure of Ti-48Al-2Cr-2Nb-1B at 700 °C for 3000 h.

In this respect the thermal stability of lamellar microstructures with ultrafine lamellar spacings is of interest. Such material has been designed in an attempt to achieve exceptional high strength. With this aim in mind, Boehlert et al. [113] have converted a Ti-46Al PST crystal to a fully lamellar polycrystalline form. The heat treatment applied involved an α phase solution treatment followed by aging at a temperature corresponding to the $\alpha_2 + \gamma$ phase field. This resulted in a refinement of the lamellar spacing by almost two orders of magnitude to about 20 nm. Microsamples, extracted from the refined polycrystalline material with lamellar planes

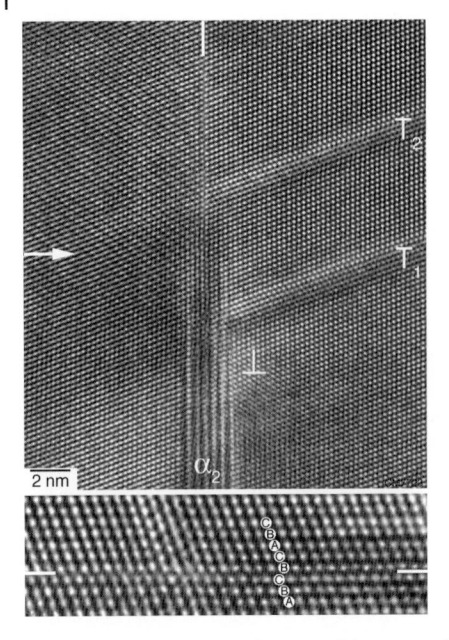

Figure 9.20 Higher magnification of the area marked in Figure 9.19 showing deformation activity in the vicinity one of the α_2 terminations. Two mechanical twins and a dislocation with a Burgers vector out of the interface plane (marked by symbol) are present. The lower image shows the stacking sequence across the interface (parallel to the arrow) that connects the two α_2 terminations. The stacking sequence indicates the presence of a fault translation.

parallel to the tensile axis exhibited a tensile yield strength of about 1100 MPa at room temperature. However, the ultrafine lamellae were not stable at elevated temperature, resulting in high minimum creep rates when compared with those of coarser PST crystals. The structural instabilities observed in the crept samples included dissolution of α_2 phase, decrease in the α_2 lamellae thickness and dynamic recrystallization. A lamellae orientation parallel to the tensile axis ($\Phi = 0$) combined with high temperatures and stresses seems to support dynamic recrystallization. Annealing experiments performed with the same temperature–time profile but without an applied load revealed dissolution of the α_2 phase but no recrystallization; this suggests that defect accumulation is required for recrystallization. Schillinger et al. [119] recognized similar behavior for polycrystalline lamellar Ti-46Al-1.5Cr-2Mo-0.25Si-0.3B with coarser lamellar spacings (35 nm to 200 nm).

It has been demonstrated that the degradation of the lamellar morphology reduces the resistance against both primary and secondary creep [2, 5, 58, 120]. Several attempts [112, 121–123] have been made to stabilize the lamellar microstructure by heat treatments performed prior to creep, which are thought to remove metastable α_2 phase, to reduce of the driving force for dynamic recrystallization or to trigger precipitation reactions. From the data present in the literature it is evident that extended aging treatments at intermediate temperatures are required in order to obtain a stable lamellar microstructure. For example, investigations performed by Huang et al. [116] using lamellar alloys based on Ti-(44–46) Al and containing other alloying elements have shown that the structural changes after 3000 h exposure at 700 °C are still at an intermediate state. Several studies [2, 70, 123] have shown that lamellar microstructures with interlocked colony boundaries are relatively stable against grain-boundary sliding and microstructural degradation. It is understood that all these procedures must be specified for the respective alloy composition.

Summary: Long-term creep of lamellar TiAl alloys leads to a degradation of the lamellar morphology into a spheroidized microstructure. The related phase transformation and recrystallization processes are driven by nonequilibrium phase compositions, stored strain energy by accumulated defects and reduction of interface area.

9.7
Precipitation Effects Associated with the $\alpha_2 \to \gamma$ Phase Transformation

The $\alpha_2 \to \gamma$ phase transformation may also lead to precipitation of interstitial elements like oxygen, nitrogen or carbon, which in technical alloys are present in significant amounts. This became evident from long-term creep experiments that have been carried out on nearly lamellar Ti-48Al-2Cr [101]. Tensile creep experiments were performed at 700 °C and stresses between 60 MPa to 140 MPa, which last for 6000 h to 13 400 h. As described in Section 9.6 particular care was taken in this study in order to avoid environmental degradation of the samples. A characteristic feature of the microstructure of crept samples is the presence of precipitates. Most, if not all precipitates are situated at isolated dislocations and at the mismatch structures of subgrain boundaries and lamellar interfaces (Figures 9.21a and b). Denuded zones were observed next to decorated defects, indicating that the precipitates were heterogeneously nucleated. The nature of the precipitates could not be identified. However, it has been established by field-ion microscopy [124] that the solubility limit of the α_2 phase for interstitial elements is at least one order of magnitude higher that that of the γ phase. It may thus be envisaged that the precipitation effects results from the $\alpha_2 \to \gamma$ transformation described in the previous section. In the newly formed γ phase, the concentration of the interstitial elements can easily exceed the solubility limit so that precipitation of oxides, nitrides or carbides may occur. The essential features of the mechanism are sketched in Figure 9.21c. As can be seen in Figure 9.21a, the precipitates effectively pin the dislocations; this could degrade the ductility the material if it is tested after creep.

Summary: Heterogeneous nucleation of precipitates containing interstitial elements may occur during creep, which is probably caused by the dissolution of metastable α_2 phase. The interstitial impurities that preferentially partition to α_2,

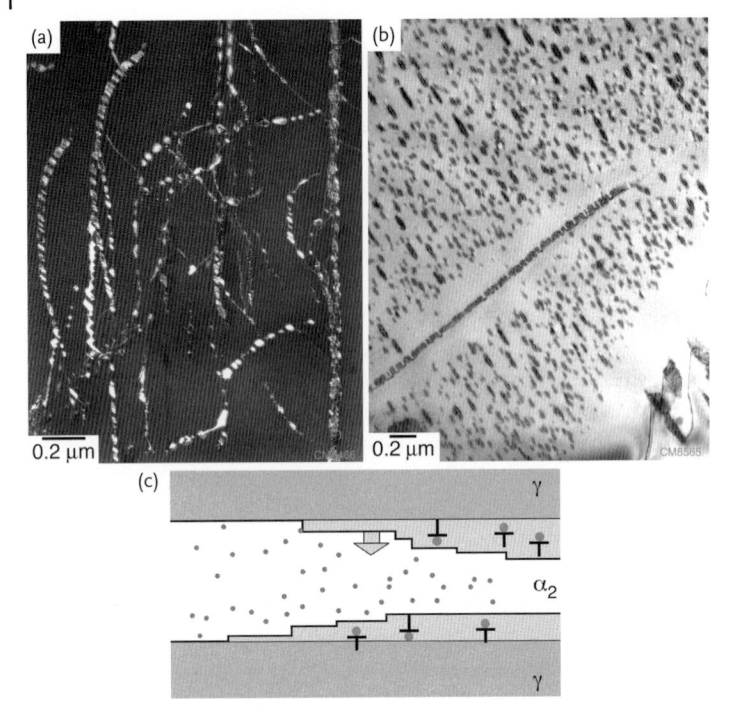

Figure 9.21 Precipitation effects observed in a Ti-48Al-2Cr alloy after long-term tensile creep at $T = 700\,°C$, $\sigma_a = 110\,MPa$, for $t = 13\,400\,h$ to strain $\varepsilon = 0.46\%$. (a) Heterogeneous nucleation of precipitates at $1/2<110]$ dislocations. (b) Denuded zones adjacent to a dislocation and a colony boundary. (c) Schematic drawing of the precipitation of interstitial elements following the $\alpha_2 \rightarrow \gamma$ transformation described in the text. The concentration of the interstitial elements dissolved in the prior α_2 phase exceeds the solubility limit of the newly formed γ phase. Thus, heterogeneous precipitation of these elements occurs at dislocations.

supersaturate the remaining α_2 and eventually get rejected to the product γ phase as precipitates.

9.8
Tertiary Creep

Extended tertiary creep is a prominent feature of many TiAl alloys typically starting at 2% to 3% strain. The rapid increase of the creep rate in this regime indicates that large strains are accumulated and that creep rupture is imminent. In polycrystalline Ti-53Al-1Nb tested at $T = 760$ to $900\,°C$, the time spent in tertiary creep was found to obey a power-law relationship with stress exponents $n = 3.5$ to 3.8 and an activation energy of $Q_C = 304\,kJ/mol$ [22]. The mechanisms that have been recognized after tertiary creep involve progressive increase of the dislocation density and dynamic recrystallization [22, 125], formation of shear bands in lamellar grains [31, 57, 126], grain-boundary sliding [127], and the formation of widely separated voids [31, 126, 127]. High stresses and temperatures seem to support void formation.

In conventional metals the onset of tertiary tensile creep is commonly associated with accumulation of damage due to void formation, necking, environmental degradation or a combination of these processes. However, tertiary creep of TiAl alloys also occurs under compression with well-controlled test atmospheres [24]. Thus, void formation and necking seem to be of minor importance for TiAl alloys; it appears more likely that the acceleration of the creep rate with the onset of tertiary creep is caused by the structural changes that have been described in Section 9.6 (a more complicated case of structural degradation occurring in modulated TiAl alloys is addressed in Section 9.10.2). The particular mechanism of strain accumulation during tertiary creep is also suggested by the large creep-failure strains of TiAl alloys, which are typically in the range of 15% to 25%. However, while this behavior suggests good resistance against creep rupture, tertiary creep is certainly a critical issue in the design of components. Given the high tertiary creep rates, the accumulated strain can easily exceed the design limits. Thus, for life predictions, the time to the onset of tertiary creep might be a more appropriate criterion.

Dlouhy et al. [128], investigating long-term creep of several two-phase alloys, have recognized that a prediction of the rupture life by the Monkman–Grant relationship [129] is not satisfactory. Beddoes et al. [2] suggested that environmental surface degradation could be important for the onset of tertiary creep. For example, the creep life of lamellar alloys in vacuum could be twenty times the rupture life in air [130]. However, this aspect has not yet been systematically addressed.

Summary: Tertiary creep accounts for a substantial portion of the total creep life of TiAl alloys. The acceleration of the creep rate in this regime is probably caused by structural degradation due to phase transformations and dynamic recrystallization. Formation and coalescence of voids seem to be delayed until the final stages of creep.

9.9
Optimized Alloys, Effect of Alloy Composition and Processing

Considerable efforts have been made to improve the creep resistance of lamellar ($\alpha_2 + \gamma$) alloys by optimizing the alloy composition. Elements such as Nb, W, Ta, Mo, and V have been added, which are thought to act as solution strengtheners and to reduce the diffusion rate, for detailed data see [2, 31–33]. In the case of W, there appears to be a minimum level required in order to enhance the creep resistance [32]; additions up to 0.2 at.% appear to be inefficient, significant improvement has been recognized for 2 at.% W [57, 125, 131]. In either case, slowing down the diffusion rate by Nb addition increases the creep resistance [132–134].

Interstitial elements such as C [135, 136], N [137] and O [2] can further decrease creep rates. Alloying with these elements often leads to precipitation of fine particles, which are heterogeneously nucleated at internal boundaries (Section 7.4). There is a broad variety of precipitates, depending on which combination of elements is present in a particular alloy. The precipitates are thought to stabilize the lamellar structure [138], to strengthen the colony boundaries [139] and to impede dislocation motion [135, 136]. The effect of Si additions is not clear. While a previous study has demonstrated a beneficial effect of Si (0.25–1 at.%) on the creep properties of fully lamellar alloys [122], an opposite effect was observed by Wang and Nieh [140].

According to the trends described in Sections 7.3 and 7.5, an alloy design based on Ti-45Al and relatively large Nb additions together with a fine dispersion of perovskite or H-phase precipitates seems to be suitable for expanding the high-temperature capability of TiAl alloys. In an attempt to optimize these alloys, the effects of Nb, B, C, and elements stabilizing the β phase were harmonized; this resulted in the definition of a new class of engineering TNB alloys [134]. The balance of mechanical properties that has been achieved in these alloys will be demonstrated in the following.

Figure 9.22 demonstrates the synergistic effects that can be achieved by the implementation of carbide precipitates in high Nb-containing alloys. The carbon addition leads to a significant improvement of the creep resistance, when compared with the undoped reference material Ti-45Al-10Nb. In view of the mechanisms described in Section 7.4.2, the improvement can be attributed to an increase of the glide and climb resistance of the dislocations by the presence of the precipitates. As has been described in Section 7.5, the carbides are preferentially hetero-

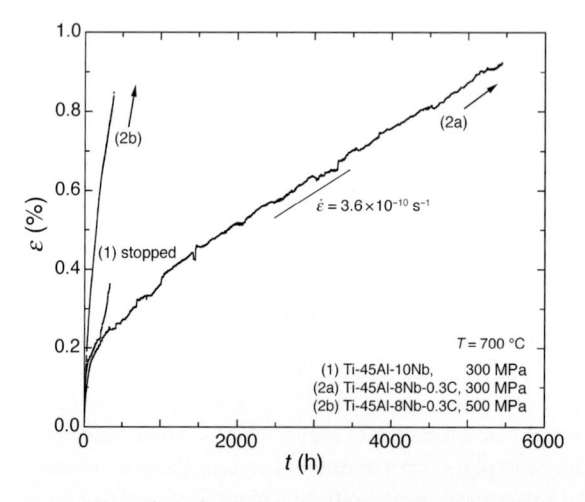

Figure 9.22 Effect of carbide precipitates on the creep behavior of high Nb-containing alloys, both with similar microstructure containing a high fraction of lamellar colonies. Extruded alloys, composition and creep conditions are given in the legend.

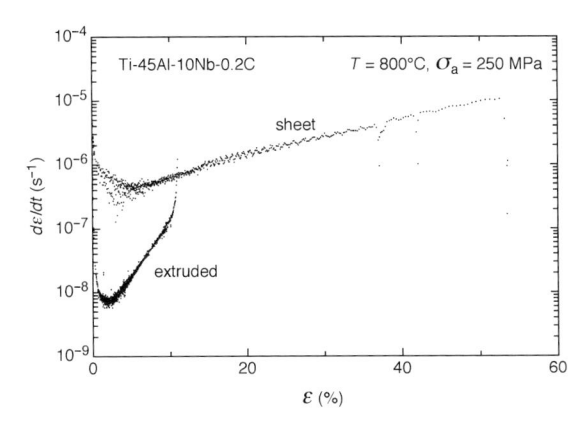

Figure 9.23 Effect of processing conditions and microstructure on the creep behavior of a Nb-bearing and carbon-doped alloy. Creep rates of extruded and sheet materials.

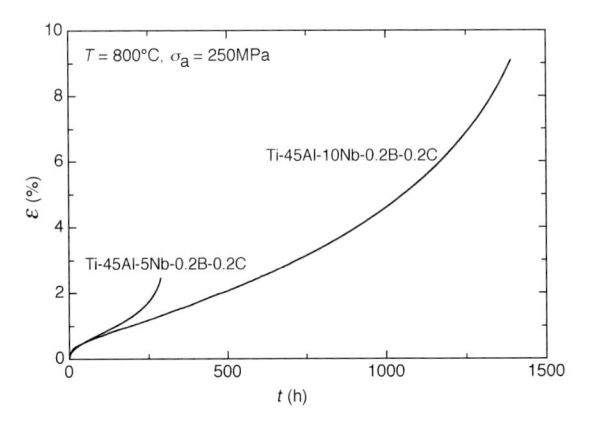

Figure 9.24 Influence of the Nb content on the creep behavior. Creep curves of extruded alloys, both with a similar duplex structure containing a high fraction of lamellar colonies.

geneously nucleated at the mismatch structures of internal boundaries so that the microstructure is stabilized. This view is supported by the fact that the creep properties strongly depend on the processing conditions and resulting microstructure. Figure 9.23 demonstrates the creep rates observed on sheet material that was produced by pack rolling an extruded billet; the data for the extruded material is given as a reference. The higher creep rate of the sheet material certainly results from its fine microstructure, which with the given C content and the sheet-rolling process conditions used, apparently could not be sufficiently stabilized. These findings clearly indicate the close correlation between alloy composition, processing, microstructure and properties.

Figure 9.24 demonstrates the beneficial effect of high Nb additions on the creep resistance by comparing alloys containing 5 and 10 at.% N. According to the

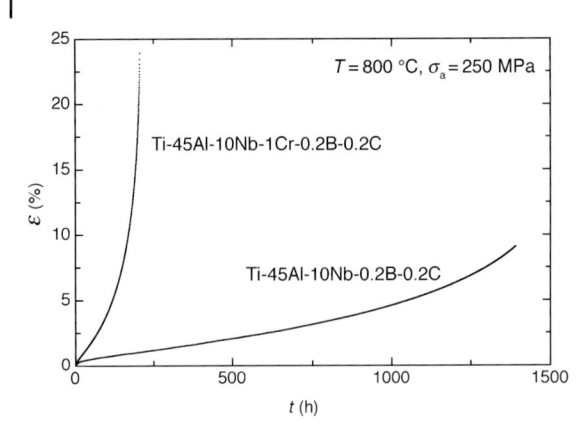

Figure 9.25 Effect of Cr addition on the creep resistance of Nb-containing TiAl alloys.

discussion is Section 7.3.3, the beneficial effect of Nb additions on the creep resistances may be attributed to two factors, the reduced diffusivity, as mentioned above, and the lower stacking-fault energy [134]. The lower diffusivity of Nb-bearing alloys makes all the processes that involve diffusion-assisted transport processes sluggish; these are dislocation climb, grain-boundary sliding, recovery, recrystallization, and coarsening of precipitation dispersions. Alloys with a lower stacking-fault energy also tend to have better creep resistance, since extended partial dislocations have difficulty both in cross-slipping and in climbing. However, further addition of elements that stabilize the β phase seems to be harmful for the creep resistance of this class of alloys. This is demonstrated in Figure 9.25 for two otherwise identical alloys, one, however, containing 1 at.% Cr [134]. It might be speculated that a significantly higher volume fraction of β phase was formed in the Cr-containing alloy, which seems to be detrimental for the structural stability and gives rise to the strong acceleration of the creep rate almost from the beginning of the test.

Taken together, alloys based on Ti-45Al-(5–10)Nb-0.2B-0.2C (designated TNB) present a high potential as creep-resistant material. Utilizing a Larson–Miller plot, one may assess the creep data from different materials against each other. To derive the Larson–Miller parameter LMP = $[(T + 273) \times (20 + \log t)]/1000$, it was assumed that the temperature and stress dependence of the minimum creep rate are given by an Arrhenius equation. For the TiAl alloys the LMP was calculated for the creep times t observed at creep strain $\varepsilon = 1\%$. The creep stresses of the TiAl alloys were normalized by an average density of $\rho = 4\,\text{g/cm}^3$. The resulting Larson–Miller plot is shown in Figure 9.26. The diagram contains the data of the well-established high-temperature alloy René 80 and that of several other TiAl alloys [134, 141–144], which are specified in the legend. TNB-V2 designates the data for an extruded high niobium-containing alloy that was strengthened by Ti$_3$Al carbides [134]. According to the author's knowledge, this material is one of the most creep-resistant polycrystalline alloys that have been presented in the litera-

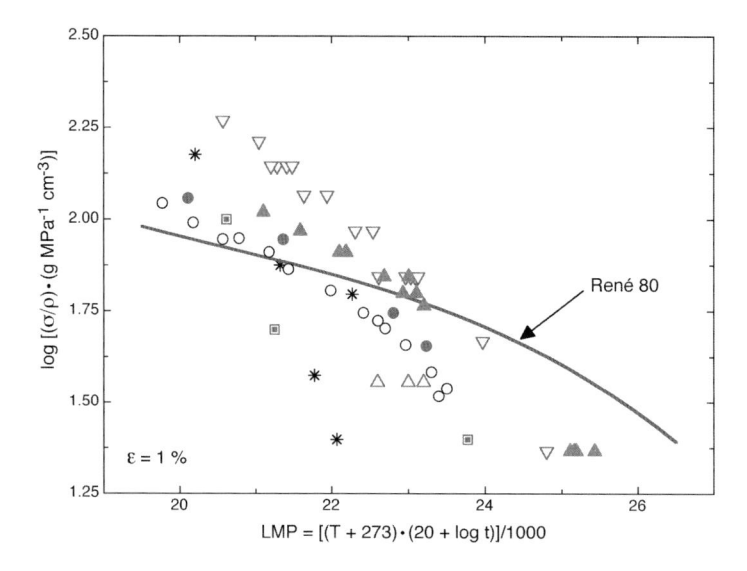

Figure 9.26 Larson–Miller plots for several TiAl alloys and the nickel-based superalloy René 80. The Larson–Miller parameter for the TiAl alloys refers to strain $\varepsilon = 1\%$.

Legend:
- ▲ Ti-45Al-5Nb-0.2B-0.2C, nearly lamellar [134]
- ▽ Ti-45Al-5Nb-0.2B-0.2C, nearly lamellar [134]
- ✻ γ-Md, duplex [144]
- △ Ti-47Al-1Cr-1Nb-1Mn-0.5B-0.2Si, lamellar [141]
- ○ Ti-46.7Al-1.3Fe-1.1V-0.35B, duplex [143]
- ● Ti-46.7Al-1.3Fe-1.1V-0.35B, lamellar [143]
- ▣ Ti-46.5Al-2Cr-2Nb-0.8Mo-0.2W-0.2Si, nearly lamellar [142]

Axes: $\log[(\sigma/\rho)] \cdot (\text{g MPa}^{-1}\,\text{cm}^{-3})$ vs. LMP $= [(T + 273) \cdot (20 + \log t)]/1000$

ture. The comparison with the data for René 80 indicates that the TNB-V2 alloy can be an attractive alternative to the superalloys in the intermediate temperature interval of 700 °C to 800 °C.

From the engineering viewpoint, the challenge is to establish a hardening mechanism without compromising desirable low-temperature properties, such as ductility and toughness. Unfortunately, in the TiAl literature information is very limited about the balance of high- and low-temperature mechanical properties that can be obtained on certain alloy systems. It is only recently that such data has been provided for the family of TNB alloys. The room-temperature tensile properties of the alloys Ti-45Al-5Nb-0.2B-0.2C (TNB-V5), Ti-45Al-8Nb-0.2C (TNB-V2) and γ-Md are demonstrated in Sections 7.5 and 8.3, respectively. Accordingly, these materials are capable of carrying tensile stresses of 1000 MPa combined with plastic elongations of $\varepsilon = 0.6\%$ to 2.5%. The fracture toughness data are presented in Section 10.4.2. Unfortunately, generally information about the balance of properties is very limited.

Summary: Based on the general composition Ti-45Al-(5–10)Nb-0.2B-0.2C, optimized alloys can be defined that exhibit good creep resistance at 700 to 800 °C combined with an appreciable room-temperature strength and ductility.

9.10
Creep Properties of Alloys with a Modulated Microstructure

In Chapter 8, a novel type of TiAl alloy (designated γ-Md) with a composite-like microstructure has been described. The characteristic constituents of the alloy are laths with a modulated substructure that is comprised of stable and metastable phases. The modulation occurs at the nanometer scale and thus provides an additional structural feature that refines the material. At room temperature, the tensile yield stress of the material is in excess of 1 GPa, with plastic elongations of about 2%. Although the material exhibits good temperature strength retention up to 700 °C, its creep resistance could be a reason for concern. Microstructural changes are perhaps more likely to occur during creep because the extremely fine modulation may degrade under the combined action of stress and temperature. The investigation of the creep resistance of this novel TiAl alloy is thus the subject of this section.

9.10.1
Effects of Stress and Temperature

Figure 9.27 shows the results of tensile creep tests performed at 700 and 800 °C and different initial stresses as strain–time and strain-rate–strain curves [144]. Under all the conditions investigated the samples deformed uniformly without necking. At intermediate temperatures and stresses the creep curve exhibited the characteristic regions that have been typically observed for TiAl alloys (Section 9.2). High stresses and temperatures accelerate the creep rate almost from the beginning of the test. However, the secondary creep regime is relatively well expressed at the test temperature 700 °C and stresses of 100 and 150 MPa, respectively. The usual procedure of comparing minimum creep rates of these tests leads to a stress exponent of $n = (\partial \ln \dot{\varepsilon} / \partial \ln \sigma)_T = 3.46$; a value that is often taken as an indication of a dislocation-climb process. No attempt has been made to determine n at higher stresses and temperatures because of the early onset of tertiary creep occurring under these conditions. For comparison with other alloys the creep date of the γ-Md alloy was included in Figure 9.26. The Larson–Miller parameter was calculated for the creep times t observed at creep strain $\varepsilon = 1\%$ and the creep stresses were normalized by the density $\rho = 4 \text{g/cm}^3$. The creep properties of the modulated γ-Md alloy are clearly inferior to those of TNB-V2. However, a relatively good creep resistance is indicated at low Larson–Miller parameters. Figure 9.26 also includes the data for two conventional β-phase-containing alloys that were characterized in [142, 143] and exhibited similar behavior. In the starting structure of these alloys the β phase was preferentially distributed along the lamellar boundaries.

9.10 Creep Properties of Alloys with a Modulated Microstructure | 347

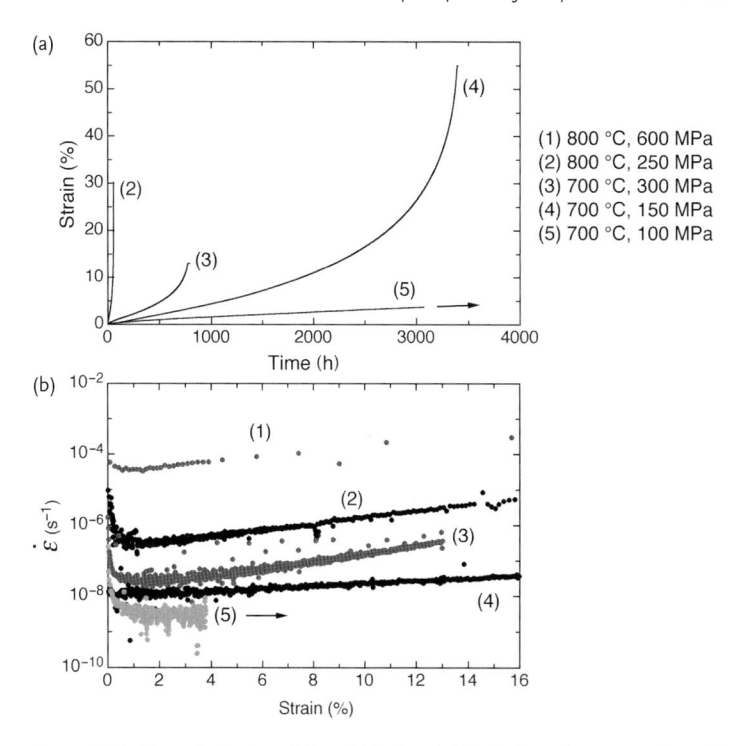

Figure 9.27 Creep behavior of the γ-Md alloy. (a) Variation of creep strain ε with time t; (b) creep rate $\dot{\varepsilon}$ as a function of creep strain ε determined from the creep curve shown in (a). The arrow indicates that the experiment is still continuing [144].

9.10.2
Damage Mechanisms

As shown in Figure 9.27, the tertiary creep period can account for a substantial portion of the total creep strain and creep life of the γ-Md alloy, as is usually the case for TiAl alloys. Thus, the early onset of this regime at higher stresses and temperatures appears to be the most serious limitation of the high-temperature capability of the γ-Md alloy. This has prompted TEM examination of the γ-Md alloy to be performed after tertiary creep [144]. The specimen investigated had been crept for 790 h at 700 °C and 300 MPa. This stress corresponds to about 30% of the yield stress measured in constant strain-rate tests at 700 °C and $\dot{\varepsilon} = 2.38 \times 10^{-5}$ s^{-1}. The sample exhibited an extended tertiary creep period and ruptured at strain of about $\varepsilon = 16.5\%$.

As shown in Figure 9.28, the most prominent feature of the crept sample is that the modulation of the laths is largely removed. There are various processes that may account for this structural change. The orthorhombic phase is seen to transform into γ phase, as demonstrated in Figure 9.29. A general observation is that two different γ variants are usually generated adjacent to a modulated lath. The

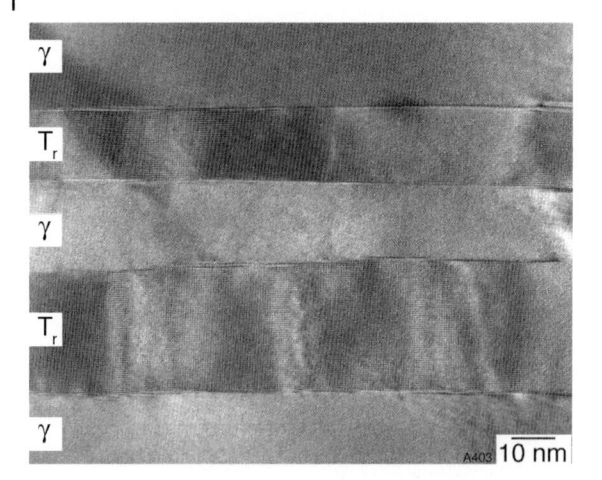

Figure 9.28 Remnant modulation of two laths T_r left in a γ-Md sample crept for 790 h at $T = 700\,°C$, $\sigma_a = 300\,MPa$ to a strain of around $\varepsilon = 16.5\%$ [144].

variants γ_1 and γ_2 shown in the micrograph sandwich a modulated lath. If the stacking sequence of the $\{111\}_\gamma$ planes of the variant γ_1 is labeled **ABC**, that of variant γ_2 is **CBA**. This inversion indicates that the atomic positions of the γ variants are twin related. Stacking of the $\{111\}_\gamma$ planes in different directions is thought to induce strain fields of opposite signs, which eventually reduce the total strain energy. It might be speculated that such a combination of shear processes makes the transformation easier. The terminated modulated laths often have small radii of curvature, as shown in Figure 9.29b. The reduction of surface energy attained by the elimination of such structural details may further support the phase transformation. It should be noted that the modulated laths also could be locally transformed into α_2 phase. As $(100)_{B19}$ planes are stacked in an **ABA** sequence, this transformation can be established by shuffles of neighboring $(100)_{B19}$ planes followed by reordering that changes the occupancies of Ti, Al and Nb atoms.

Phase transformation coupled with dynamic recrystallization appears to be another reason for the degradation of the modulated laths. As illustrated in Figure 9.30, upon creep the laths become spheroidized by the nucleation of new grains. An early stage of this phase transformation is apparently shown in Figure 9.31. The small grain at the center of the micrograph is probably the remainder of a modulated lath. The lattice is a highly distorted B19 structure imaged down the [001] direction and shows some remnant modulation.

Figure 9.32 probably shows a later stage of this process. The remnant modulated lath designated T_{r1} is embedded in between the gamma variants γ_1 and γ_2. During creep part of the lath T_{r1} was probably transformed into β phase. The orientation relationship is such that one set of close-packed planes is common to both phases, that is, the (111) plane in γ_1 runs smoothly into the $(1\bar{1}0)$ plane in β. Thus, as for many other solid/solid interfaces, there is a strong tendency for the planes and directions with highest atomic densities to align across the interface. The high-

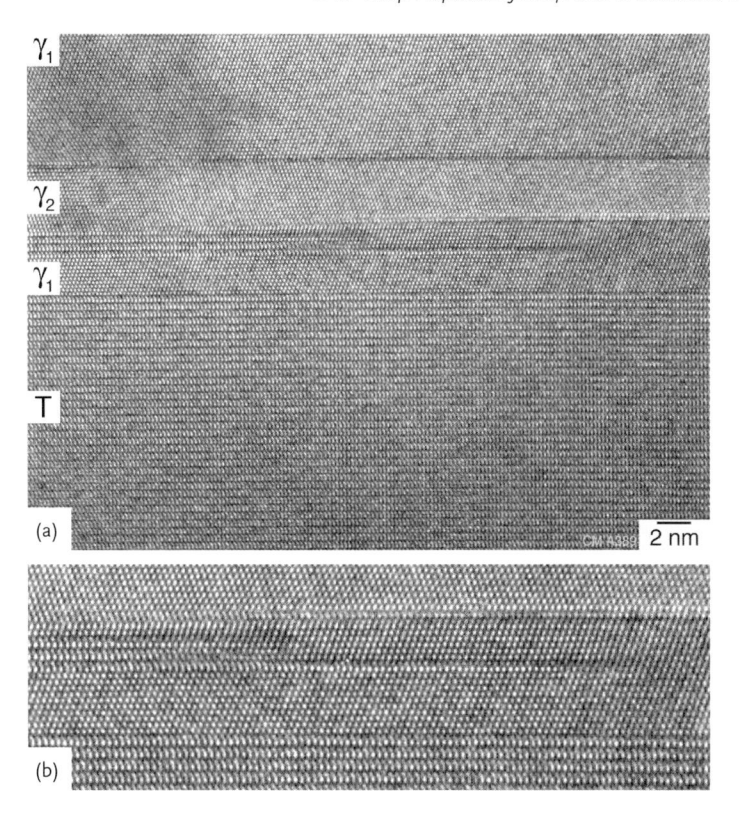

Figure 9.29 Transformation of a modulated lath into γ phase; γ-Md sample crept for 790 h at $T = 700\,°C$, $\sigma_a = 300\,MPa$ to strain of around $\varepsilon = 16.5\%$. (a) A very thin B19 lamella is sandwiched by the variants γ_1 and γ_2, which are twin related. (b) Higher magnification of the central part of the micrograph.

Figure 9.30 Multibeam TEM micrograph showing the spheroidization of a modulated lath T_r by phase transformation and dynamic recrystallization. γ-Md sample crept for 790 h at $T = 700\,°C$, $\sigma_a = 300\,MPa$ to a strain of around $\varepsilon = 16.5\%$.

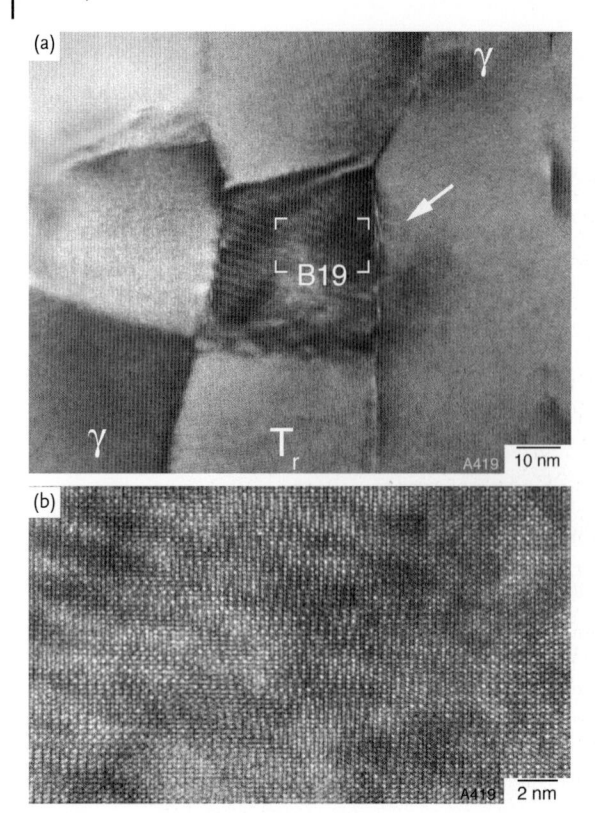

Figure 9.31 Early stage of phase transformation and dynamic recrystallization; γ-Md sample crept for 790 h at $T = 700\,°C$, $\sigma_a = 300\,MPa$ to a strain of around $\varepsilon = 16.5\%$. (a) A small grain (arrowed) probably remaining from a modulated lath T_r. (b) Higher magnification of the grain showing a highly distorted B19 structure imaged down the [001] direction. Note the almost perfect B19 structure at the right-hand side of the figure.

index interfacial plane is formed by introducing misfit dislocations only a few atomic distances apart (Figure 9.32b). The $(1\bar{1}0)_\beta$ planes include an angle of 11° with the (111) planes of the variant γ_1, which in part may accommodate the misfit between γ and β.

The prominent deformation mode observed in the crept sample is mechanical order twinning along $1/6\,<11\bar{2}]\{111\}$. A close examination of the microstructure suggests that most, if not all, γ grains are twinned, as exemplified in Figure 9.33. Twin nucleation is apparently accomplished by the same mechanism as has been described for constant strain-rate deformation in Section 8.2, that is, by dissociation reactions of misfit dislocations. It might be expected that the activation of mechanical twinning also helps to release the internal stresses present in the material. However, the observed abundance of mechanical twinning does not preclude that dislocation glide and climb could occur.

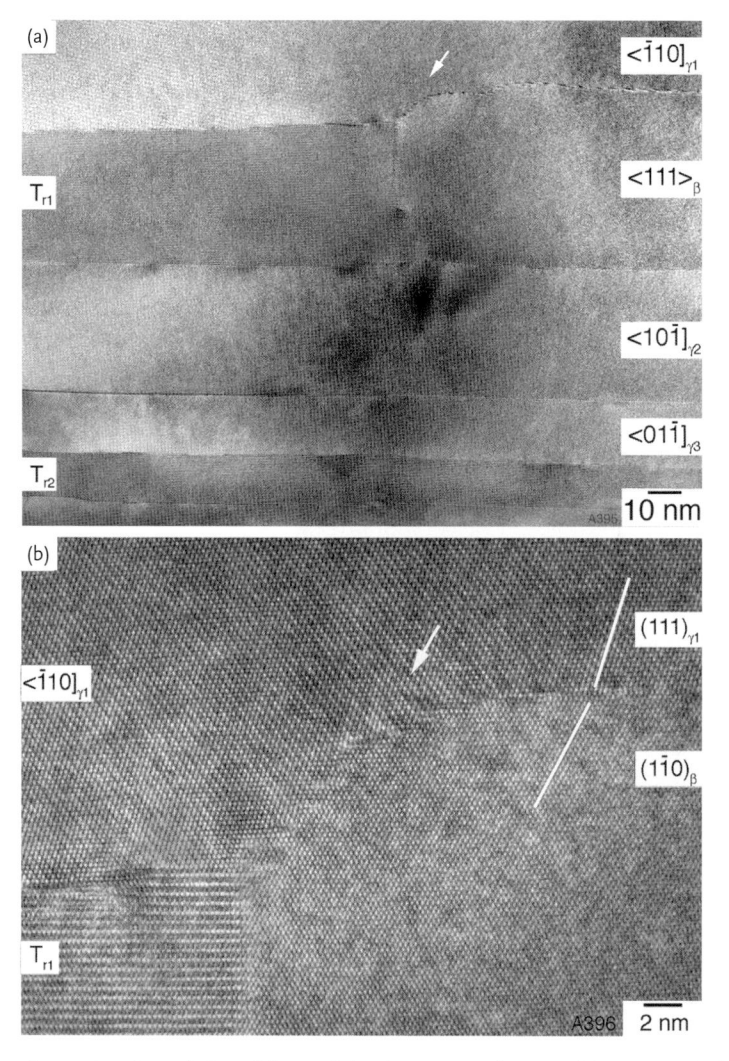

9.10 Creep Properties of Alloys with a Modulated Microstructure | 351

Figure 9.32 Degradation of the modulated structure; γ-Md sample crept for 790 h at $T = 700\,°C$, $\sigma_a = 300\,MPa$ to a strain of around $\varepsilon = 16.5\%$. (a) A new β grain is formed within a modulated lath T_{r1} that is sandwiched by the γ variants $<\bar{1}10]_{\gamma1}$ and $<10\bar{1}]_{\gamma2}$. (b) Higher magnification of the area marked in (a). Note the tendency to maintain continuity between the close-packed planes $(1\bar{1}0)_\beta$ and $(111)_\gamma$ and the misfit localization at interfacial dislocations.

Summary: The creep resistance of the modulated γ-Md alloy is mainly limited by the onset of tertiary creep. The processes associated with this behavior arise from several phase transformations towards thermodynamic equilibrium, dynamic recrystallization and relaxation of constraint stresses that exist between misfitting phases.

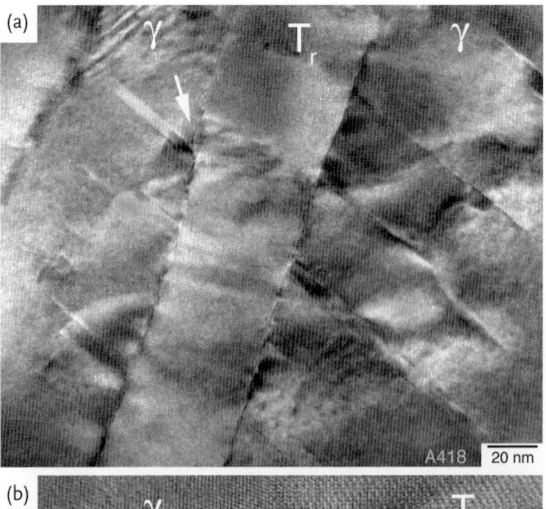

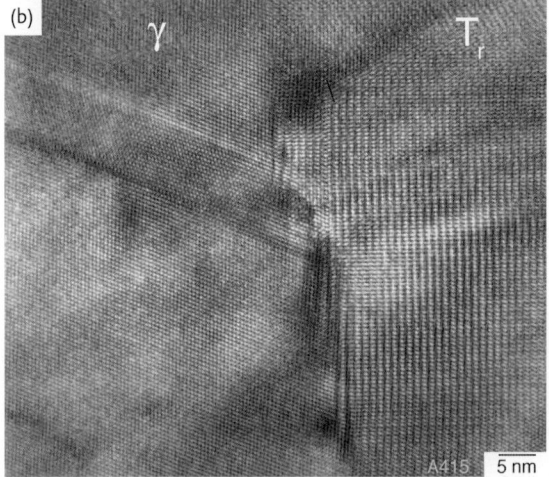

Figure 9.33 Twin structures observed in a γ-Md sample crept for 790 h at $T = 700\,°C$, $\sigma_a = 300\,MPa$ to a strain of around $\varepsilon = 16.5\%$. (a) Low-magnification, high-resolution image demonstrating mechanical twinning as the prominent deformation mechanism. (b) Higher magnification of the arrowed region in (a) showing the emission of a mechanical twin from a remnant modulated lath T_r into the adjacent γ lamella.

References

1 Ishikawa, Y., Maruyama, K., and Oikawa, H. (1993) *Structural Intermetallics* (eds R. Darolia, J.J. Lewandowski, C.T. Liu, P.L. Martin, D.B. Miracle, and M.V. Nathal), TMS, Warrendale, PA, p. 345.

2 Beddoes, J., Wallace, W., and Zhao, L. (1995) *Int. Mater. Rev.*, **40**, 197.

3 Hemker, K.J., and Nix, W.D. (1997) *Structural Intermetallics 1997* (eds M.V. Nathal, R. Darolia, C.T. Liu, P.L. Martin, D.B. Miracle, R. Wagner, and M. Yamaguchi), TMS, Warrendale, PA, p. 21.

4 Herrouin, F., Hu, D., Bowen, P., and Jones, I.P. (1998) *Acta Mater.*, **14**, 4963.

5. Bartholomeusz, M.F., and Wert, J.A. (1994) *Metall. Mater. Trans. A*, **25A**, 2371.
6. Shah, D., and Lee, E. (2002) *Intermetallic Compounds. Vol. 3, Principles and Practice* (eds J.H. Westbrook and R.L. Fleischer), John Wiley & Sons, Ltd, Chichester, p. 297.
7. Powell, G.W. (1986) *Metals Handbook*, vol. 11, 9th edn, Amer. Soc. Metals, Materials Park, p. 263.
8. Glenny, R.J., Northwood, J.E., and Burwood-Smith, A. (1975) *Int. Metall. Rev.*, **20**, 1.
9. Perrin, I. (1995) *Gamma Titanium Aluminides* (eds Y.-W. Kim, R. Wagner, and M. Yamaguchi), TMS, Warrendale, PA, p. 41.
10. Ilschner, B. (1981) *Creep and Fatigue in High Temperature Alloys* (ed. J. Bressers), Applied Science, London, p. 1.
11. Sims, C.T., and Hagel, W.C. (1972) *The Superalloys*, John Wiley & Sons, Inc., New York.
12. Wood, W.I., and Restall, J.E. Corrosion '87, Inst. Corr. Sci. Tech./NACE, 1987.
13. Dimiduk, D.M., and Miracle, D.B. (1989) *High-Temperature Ordered Intermetallic Alloys III, Materials Research Society Symposia Proceedings*, vol. 133 (eds C.T. Liu, A.I. Taub, N.S. Stoloff, and C.C. Koch), MRS, Pittsburgh, PA, p. 349.
14. Sherby, O.D., Orr, R.L., and Dorn, J.E. (1954) *Trans. AIME*, **200**, 71.
15. Barrett, C.R., Ardell, A.J., and Sherby, O.D. (1964) *Trans. AIME*, **230**, 200.
16. Evans, A.G., and Rawlings, R.D. (1969) *Phys. Status Solidi*, **34**, 9.
17. Biberger, M., and Gibeling, J.C. (1995) *Acta Metall. Mater.*, **43**, 3247.
18. Martin, P.L., Mendiratta, M.G., and Lipsitt, H.A. (1983) *Metall. Trans. A*, **14A**, 2170.
19. Kampe, S.L., Bryant, J.D., and Christodoulou, L. (1991) *Metall. Trans. A*, **22A**, 447.
20. Wheeler, D.A., London, B., and Larson, D.E. Jr. (1992) *Scr. Metall. Mater.*, **26**, 939.
21. Sadananda, K., and Feng, C.R. (1993) *Mater. Sci. Eng. A*, **170**, 199.
22. Hayes, R.W., and Martin, P.L. (1995) *Acta Metall. Mater.*, **43**, 2761.
23. Es-Souni, M., Bartels, A., and Wagner, R. (1995) *Acta Metall. Mater.*, **43**, 153.
24. Lu, M., and Hemker, K.J. (1997) *Acta Mater.*, **45**, 3573.
25. Maruyama, K., Yamamoto, R., Nakakuki, H., and Fujitsuna, N. (1997) *Mater. Sci. Eng. A*, **239–240**, 419.
26. Morris, M.A., and Leboeuf, M. (1997) *Mater. Sci. Eng. A*, **239–240**, 429.
27. Sujata, M., Sastry, D.H., and Ramachandra, C. (2004) *Intermetallics*, **12**, 691.
28. Malaplate, J., Caillard, D., and Couret, A. (2004) *Philos. Mag.*, **84**, 3671.
29. Zhan, C.-J., Yu, T.-H., and Koo, C.-H. (2006) *Mater. Sci. Eng. A*, **435–436**, 698.
30. Yamaguchi, M., Zhu, H., Suzuki, M., Maruyama, K., and Appel, F. (2008) *Mater. Sci. Eng. A*, **483–484**, 517.
31. Zhang, W.J., Spigarelli, S., Cerri, E., Evangelista, E., and Francesconi, L. (1996) *Mater. Sci. Eng. A*, **211**, 15.
32. Parthasarathy, T.A., Mendiratta, M.G., and Dimiduk, D.M. (1997) *Scr. Mater.*, **37**, 315.
33. Zhang, W.J., and Deevi, S.C. (2003) *Mater. Sci. Eng. A*, **362**, 280.
34. Takahashi, T., Nagai, H., and Oikawa, H. (1990) *Mater. Sci. Eng. A*, **128**, 195.
35. Maruyama, K., Takahashi, T., and Oikawa, H. (1992) *Mater. Sci. Eng. A*, **153**, 433.
36. Ishikawa, Y., and Oikawa, H. (1994) *Mater. Trans. JIM*, **35**, 336.
37. Gorzel, A., and Sauthoff, G. (1999) *Intermetallics*, **7**, 371.
38. Lu, M., and Hemker, K.J. (1998) *Metall. Mater. Trans. A*, **29A**, 99.
39. Lipsitt, H.A., Shechtman, D., and Schafrik, R.E. (1975) *Metall. Trans. A*, **6A**, 1991.
40. Oikawa, H. (1992) *Mater. Sci. Eng. A*, **153**, 427.
41. Morris, M.A., and Lipe, T. (1997) *Intermetallics*, **5**, 329.
42. Morris, M.A., and Leboeuf, T. (1997) *Intermetallics*, **5**, 339.
43. Wang, J.G., Hsiung, L.M., and Nieh, T.G. (1998) *Scr. Mater.*, **39**, 957.
44. Viswanathan, G.B., Vasudevan, V.K., and Mills, M.J. (1999) *Acta Mater.*, **47**, 1399.

45 Rudolf, T., Skrotzki, B., and Eggeler, G. (2001) *Mater. Sci. Eng. A*, **319–321**, 815.
46 Appel, F., and Wagner, R. (1998) *Mater. Sci. Eng.*, **R22**, 187.
47 Appel, F. (2001) *Intermetallics*, **9**, 907.
48 Nix, W.D., and Barnett, C.R. (1965) *Acta Metall.*, **13**, 1247.
49 Karthikeyan, S., Viswanathan, G.B., and Mills, M.J. (2004) *Acta Mater.*, **52**, 2577.
50 Shi, D.S., Huang, S.-C., Scarr, G.K., Jang, H., and Chestnutt, J.C. (1991) *Microstructure/Property Relationships in Titanium Aluminides and Alloys* (eds Y.-W. Kim and R. Boyer), TMS, Warrendale, PA, p. 353.
51 Huang, S.-C. (1992) *Metall Trans. A*, **23A**, 375.
52 Viswanathan, G.B., and Vasudevan, V.K. (1993) *High-Temperature Ordered Intermetallic Alloys V*, Materials Research Society Symposia Proceedings, vol. 288 (eds I. Baker, R. Darolia, J.D. Whittenberger, and M.H. Yoo), MRS, Pittsburgh, PA, p. 787.
53 Worth, B.D., Jones, J.W., and Allison, J.E. (1995) *Gamma Titanium Aluminides* (eds Y.-W. Kim, R. Wagner, and M. Yamaguchi), TMS, Warrendale, PA, p. 931.
54 Schwenker, S.W., and Kim, Y.-W. (1995) *Gamma Titanium Aluminides* (eds Y.-W. Kim, R. Wagner, and M. Yamaguchi), TMS, Warrendale, PA, p. 985.
55 Keller, M.M., Jones, P.E., Porter, W.J., and Eylon, D. (1995) *Gamma Titanium Aluminides* (eds Y.-W. Kim, R. Wagner, and M. Yamaguchi), TMS, Warrendale, PA, p. 441.
56 Bartholomeusz, M.F., and Wert, J.A. (1994) *Metall. Mater. Trans. A*, **25A**, 2161.
57 Beddoes, J., Zhao, L., Triantafillou, J., Au, P., and Wallace, W. (1995) *Gamma Titanium Aluminides* (eds Y.-W. Kim, R. Wagner, and M. Yamaguchi), TMS, Warrendale, PA, p. 959.
58 Crofts, P.D., Bowen, P., and Jones, I.P. (1996) *Scr. Mater.*, **35**, 1391.
59 Loretto, M.H., Godfrey, A.B., Hu, D., Blenkinson, P.A., Jones, I.P., and Chen, T.T. (1998) *Intermetallics*, **6**, 663.
60 Yamamoto, R., Mizoguchi, K., Wegmann, G., and Maruyama, K. (1998) *Intermetallics*, **6**, 699.
61 Chatterjee, A., Mecking, H., Arzt, E., and Clemens, H. (2002) *Mater. Sci. Eng. A*, **329–331**, 840.
62 Parthasarathy, T.A., Subramanian, P.R., Mendiratta, M.G., and Dimiduk, D.M. (2000) *Acta Mater.*, **48**, 541.
63 Kim, H.Y., and Maruyama, K. (2001) *Acta Mater.*, **49**, 2635.
64 Kim, H.Y., and Maruyama, K. (2003) *Acta Mater.*, **51**, 2191.
65 Johnson, D.R., Masuda, Y., Yamanaka, T., Inui, H., and Yamaguchi, M. (1995) *Gamma Titanium Aluminides* (eds Y.-W. Kim, R. Wagner, and M. Yamaguchi), TMS, Warrendale, PA, p. 627.
66 Johnson, D.R., Masuda, Y., Yamanaka, T., Inui, H., and Yamaguchi, M. (2000) *Metall. Mater. Trans. A*, **31A**, 2463.
67 Lee, H.N., Johnson, D.R., Inui, H., Oh, M.H., Wee, D.M., and Yamaguchi, M. (2002) *Intermetallics*, **10**, 841.
68 Muto, S., Yamanaka, T., Johnson, D.R., Inui, H., and Yamaguchi, M. (2002) *Mater. Sci. Eng. A*, **239–331**, 424.
69 Thomas, M., and Naka, S. (1999) *Gamma Titanium Aluminides 1999* (eds Y.-W. Kim, D.M. Dimiduk, and M.H. Loretto), TMS, Warrendale, PA, p. 633.
70 Wang, N.J., Schwartz, A.J., Nieh, T.G., and Clemens, D. (1996) *Mater. Sci. Eng. A*, **206**, 63.
71 Mitao, S., Tsuiyama, S., and Minakawa, K. (1991) *Microstructure/Property Relationships in Titanium Aluminides and Alloys* (eds Y.-W. Kim and R. Boyer), TMS, Pittsburgh, PA, p. 297.
72 Kim, Y.-W. (1992) *Acta Metall. Mater.*, **40**, 1121.
73 Yamaguchi, M., and Inui, H. (1993) *Structural Intermetallics* (eds R. Darolia, J.J. Lewandowski, C.T. Liu, P.L. Martin, D.B. Miracle, and M.V. Nathal), TMS, Warrendale, PA, p. 127.
74 Hayes, R.W. (1993) *Scr. Metall. Mater.*, **29**, 1229.
75 Hsiung, L.M., and Nieh, T.G. (1997) *Structural Intermetallics 1997* (eds M.V. Nathal, R. Darolia, C.T. Liu, P.L. Martin, D.B. Miracle, R. Wagner, and M. Yamaguchi), TMS, Warrendale, PA, p. 129.

76 Seo, D.Y., Bieler, T.R., and Larsen, D.E. (1997) *Structural Intermetallics 1997* (eds M.V. Nathal, R. Darolia, C.T. Liu, P.L. Martin, D.B. Miracle, R. Wagner, and M. Yamaguchi), TMS, Warrendale, PA, p. 137.

77 Parthasarathy, T.A., Keller, M., and Mendiratta, M.G. (1998) *Scr. Mater.*, **38**, 1025.

78 Beddoes, J., Seo, D.Y., Chen, W.R., and Zhao, L. (2001) *Intermetallics*, **9**, 915.

79 Seo, D.Y., Beddoes, J., Zhao, L., and Botton, G.A. (2002) *Mater. Sci. Eng. A*, **323**, 306.

80 Zhang, W.J., and Deevi, S.C. (2003) *Intermetallics*, **11**, 177.

81 Seo, D.Y., Beddoes, J., and Zhao, L. (2003) *Metall. Mater. Trans. A*, **34A**, 2177.

82 Simkins, R.J., Rourke, M.P., Bieler, T., and McQuay, P.A. (2007) *Mater. Sci. Eng. A*, **463**, 208.

83 Loiseau, A., and Lasalmonie, A. (1984) *Mater. Sci. Eng.*, **67**, 163.

84 Seo, D.Y., Bieler, T.R., An, S.U., and Larsen, D.E. (1998) *Metall. Mater. Trans. A*, **29A**, 89.

85 Seo, D.Y., and Bieler, T.R. (1999) *Gamma Titanium Aluminides 1999* (eds Y.-W. Kim, D.M. Dimiduk, and M.H. Loretto), TMS, Warrendale, PA, p. 701.

86 Malaplate, J., Thomas, M., Belaygue, P., Grange, M., and Couret, A. (2006) *Acta Mater.*, **54**, 601.

87 Nieh, T.G., and Wang, J.N. (1995) *Scr. Metall. Mater.*, **33**, 1101.

88 Augbourg, V.M., Eylon, D., Keller, M.M., Austin, C.M., and Balson, S.J. (1996) *Titanium '95: Science and Technology* (eds P.A. Blenkinson, W.J. Evans, and H.M. Flower), The Institute of Materials, London, p. 520.

89 Matthews, J.W. (1979) *Dislocations in Solids*, vol. 2 (ed. R.F.N. Nabarro), North-Holland Publishing Company, Amsterdam, p. 461.

90 Gouma, P.I., Mills, M.J., and Kim, Y.-W. (1998) *Philos. Mag. Lett.*, **78**, 59.

91 Viswanathan, G.B., Kim, Y.-W., and Mills, M.J. (1999) *Gamma Titanium Aluminides 1999* (eds Y.-W. Kim, D.M. Dimiduk, and M.H. Loretto), TMS, Warrendale, PA, p. 653.

92 Karage, M., Kim, Y.-W., and Gouma, P.I. (2003) *Metall. Mater. Trans. A*, **34A**, 2119.

93 Zhu, H., Seo, D.Y., and Maruyama, K. (2009) *Mater. Sci. Eng. A*, **510–511**, 14.

94 Es-Souni, M., Bartels, A., and Wagner, R. (1993) *Mater. Sci. Eng. A*, **171**, 127.

95 Appel, F., Christoph, U., and Wagner, R. (1994) *Interface Control of Electrical, Chemical and Mechanical Properties* (eds S.P. Murarka, K. Rose, T. Ohmi, and T. Seidel), Materials Research Society Symposia Proceedings, vol. 318, MRS, Pittsburgh, PA, p. 691.

96 Hofmann, U., and Blum, W. (1995) *Scr. Metall. Mater.*, **32**, 371.

97 Wert, J.A., and Bartholomeusz, M.F. (1996) *Metall. Mater. Trans. A*, **27A**, 127.

98 Chen, T.T. (1999) *Intermetallics*, **7**, 995.

99 Oehring, M., Appel, F., Ennis, P.J., and Wagner, R. (1999) *Intermetallics*, **7**, 335.

100 Appel, F. (2001) *Mater. Sci. Eng. A*, **317**, 115.

101 Appel, F., Christoph, U., and Oehring, M. (2002) *Mater. Sci. Eng. A*, **329–331**, 780.

102 Furuhara, T., Howe, J.M., and Anderson, H.J. (1991) *Acta Metall. Mater.*, **39**, 2873.

103 van der Merwe, J., and Shiflet, G. (1994) *Acta Metall. Mater.*, **42**, 1173.

104 Ernst, F., Finnis, M.W., Hofmann, D., Muschik, T., Schönberger, U., and Wolf, U. (1992) *Phys. Rev. Lett.*, **69**, 620.

105 Singh, S.R., and Howe, J.M. (1992) *Philos. Mag. Lett.*, **65**, 233.

106 Abe, E., Kajiwara, S., Kumagai, T., and Nakamura, N. (1997) *Philos. Mag. A*, **75**, 975.

107 Humphreys, F.J., and Hatherly, M. (1995) *Recrystallization and Related Annealing Phenomena*, Pergamon, Oxford.

108 Cahn, R.W. (1990) *High Temperature Aluminides and Intermetallics* (eds S.H. Whang, C.T. Liu, D.P. Pope, and J.O. Stiegler), TMS, Warrendale, PA, p. 245.

109 Cahn, R.W., Takeyama, M., Horton, J.A., and Liu, C.T. (1991) *J. Mater. Res.*, **6**, 57.

110 Ramanuyan, R.V., Maziasz, P.J., and Liu, C.T. (1996) *Acta Mater.*, **44**, 2611.

111 Maziasz, P.J., Ramanuyan, R.V., Liu, C.T., and Wright, J.L. (1997) *Intermetallics*, **5**, 83.
112 Chen, T.T., and Willis, M.R. (1998) *Scr. Mater.*, **39**, 1255.
113 Boehlert, C.J., Dimiduk, D.M., and Hemker, K. (2002) *Scr. Mater.*, **46**, 259.
114 Mishin, Y., and Herzig, Chr. (2000) *Acta Mater.*, **48**, 589.
115 Hu, D., Godfrey, A.B., and Loretto, M. (1998) *Intermetallics*, **6**, 413.
116 Huang, Z.W., Voice, W., and Bowen, P. (2000) *Intermetallics*, **8**, 417.
117 Karthikeyan, S., and Mills, M.J. (2005) *Intermetallics*, **13**, 985.
118 Lapin, J., and Pelachová, T. (2006) *Intermetallics*, **14**, 1175.
119 Schillinger, W., Clemens, H., Dehm, G., and Bartels, A. (2002) *Intermetallics*, **10**, 459.
120 Hayes, R.W., and London, B. (1994) *Scr. Metall. Mater.*, **30**, 259.
121 Kim, Y.-W., and Dimiduk, D.M. (1991) *JOM*, **43**, 40.
122 Noda, T., Okabe, M., Isobe, S., and Sayashi, M. (1995) *Mater. Sci. Eng. A*, **192–193**, 774.
123 Zu, H., Seo, D.Y., Maruyama, K., and Au, P. (2006) *Metall. Mater. Trans. A*, **37A**, 3149.
124 Menand, A., Huguet, A., and Nérac-Partaix, A. (1996) *Acta Mater.*, **44**, 4729.
125 Triantafillou, J., Beddoes, J., Zhao, L., and Wallace, W. (1994) *Scr. Metall. Mater.*, **31**, 1387.
126 Hayes, R.W., and McQuay, P.A. (1994) *Scr. Metall. Mater.*, **30**, 259.
127 Du, X.-W., Zhu, J., and Kim, Y.-W. (2001) *Intermetallics*, **9**, 137.
128 Dlouhy, A., Kucharova, K., and Orlova, A. (2009) *Mater. Sci. Eng. A*, **510–511**, 350.
129 Monkman, F.C., and Grant, N.J. (1956) *Proc. ASTM*, **56**, 593.
130 Huang, J.S., and Kim, Y.-W. (1991) *Scr. Metall. Mater.*, **25**, 191.
131 Martin, P.L., and Lipsitt, H.A. (1990) *Proc. 4th Int. Conf. on Creep and Fracture of Engineering Materials and Structures* (eds B. Wilshire and R.W. Evans), The Institute of Metals, London, p. 255.
132 Appel, F., Oehring, M., and Wagner, R. (2000) *Intermetallics*, **8**, 1283.
133 Zhang, W.J., Chen, G.L., Appel, F., Nieh, T.G., and Deevi, S.C. (2001) *Mater. Sci. Eng. A*, **315**, 250.
134 Appel, F., Paul, J.D.H., Oehring, M., Fröbel, U., and Lorenz, U. (2003) *Metall. Mater. Trans. A*, **34A**, 2149.
135 Blackburn, M.J., and Smith, M.P. (1982) R & D on composition and processing of titanium aluminide alloys for turbine engines. AFWAL-TR-4086.
136 Worth, B.D., Jones, J.W., and Allison, J.E. (1995) *Metall. Mater. Trans. A*, **26A**, 2961.
137 Yun, J.H., Cho, H.S., Nam, S.W., Wee, D.M., and Oh, M.H. (2000) *J. Mater. Sci.*, **35**, 4533.
138 Tsuyama, S., Mitao, S., and Minakawa, K. (1992) *Mater. Sci. Eng. A*, **153**, 451.
139 Fuchs, G.E. (1995) *Mater. Sci. Eng. A*, **192–193**, 707.
140 Wang, J.N., and Nieh, T.G. (1997) *Scr. Mater.*, **37**, 1545.
141 Appel, F., Oehring, M., and Ennis, P. (1999) *Gamma Titanium Aluminides 1999* (eds Y.-W. Kim, D.M. Dimiduk, and M.H. Loretto), TMS, Warrendale, PA, p. 603.
142 Wang, J.G., and Nieh, T.G. (2000) *Intermetallics*, **8**, 737.
143 Nishikiori, S., Takahashi, S., Satou, S., Tanaka, T., and Matsuo, T. (2002) *Mater. Sci. Eng. A*, **329–331**, 802.
144 Appel, F., Paul, J.D.H., and Oehring, M. (2009) *Mater. Sci. Eng. A*, **510–511**, 342.

10
Fracture Behavior

Poor ductility and damage tolerance at low to intermediate temperatures are the most important drawbacks for TiAl alloys and restrict their wider industrial application. Owing to this technical importance, much effort has been expended on the factors determining crack growth and fracture resistance in TiAl alloys; this work will be reviewed in the present chapter. The chapter is so organized that the most important fracture phenomena will be described first, in order to elucidate the physical background. Available toughness data will then be presented, followed by a brief discussion of their technical implications.

10.1
Length Scales in the Fracture of TiAl Alloys

As with many other materials, fracture of TiAl components is governed by several processes that occur over a wide range of length scales. Figure 10.1 is a simplified illustration of the mechanisms that probably determine the response of a crack in a lamellar $\alpha_2(Ti_3Al) + \gamma(TiAl)$ alloy at temperatures below the brittle-to-ductile transition temperature (BDT). The relevant length scales range from that of the macroscale sample dimensions to the atomic scale, encompassing the various microstructural length scales that are associated with the cracked sample geometry, lamellar colonies, individual lamella, dislocations, and twins. The imposed loading is applied at a macroscopic distance from the crack tip (Figure 10.1a), and the resulting stress is greatly magnified in the crack-tip region. Fracture mechanics has established a framework to link the macroscopic geometry and loads to microscopic fracture processes [1–4]. The stress distribution near the crack tip depends on the material properties and is often described using linear elasticity. Within these models the crack is characterized by a singularity in the stress field that decays as the inverse square root of the distance r from the crack. The strength of the singularity is described by the stress-intensity factor K

$$K = \sigma_a \sqrt{\pi a_c} \qquad (10.1)$$

σ_a is the applied stress, which acts perpendicular to the crack plane, and $2a_c$ is the crack length. K is related to the energy release rate G, which is defined as the work

Gamma Titanium Aluminide Alloys: Science and Technology, First Edition. Fritz Appel, Jonathan David Heaton Paul, Michael Oehring.
© 2011 Wiley-VCH Verlag GmbH & Co. KGaA. Published 2011 by Wiley-VCH Verlag GmbH & Co. KGaA.

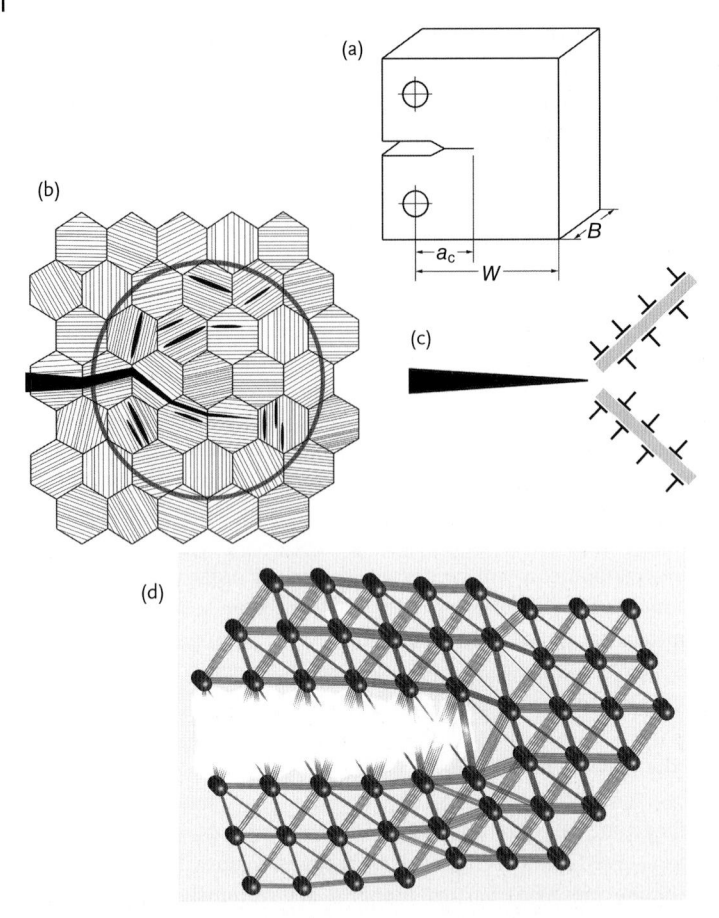

Figure 10.1 Schematic illustration of the processes determining crack propagation in a lamellar ($\alpha_2 + \gamma$) alloy. (a) A crack in a macroscopic component; (b) continuum plastic zone involving a sufficient number of grains; (c) crystal plasticity zone due to dislocation glide and mechanical twinning; (d) actual separation process at the atomic level.

done when a crack is virtually extended from a_c to $a_c + \delta a_c$ [5]. Thus, G describes the total elastic energy stored in the system, including the potential stored in the loading system, per unit area of crack advance. For plane-strain conditions, the energy-release rate is given by

$$G = \frac{K^2(1-v^2)}{E} \quad (10.2)$$

where E is Young's modulus and v Poisson's ratio. Within this model the mechanical stability of a crack, known as the Griffith criterion [6], can be formulated as a balance of a crack driving force, the energy release rate G, and the surface energy γ_s of the two freshly produced fracture surfaces. Critical values K_c and G_c refer to the condition in which a crack extends in a rapid (unstable) manner. For an infinite body in plane strain that contains a central through-thickness crack of length $2a_c$, crack extension occurs if σ_a or a_c is greater than given by the relationship

$$\sigma_a = \sqrt{\frac{2E\gamma_s}{\pi a_c (1-v^2)}} \qquad (10.3)$$

It should be noted that in the framework of this model the resistance to crack growth only arises from the produced surface energy.

At the next smaller length scale (Figure 10.1b), the effects of the microstructure become important. Factors that may affect the fracture behavior involve phase distribution and morphology of the constituents, crack interaction with phase and grain boundaries, lamellar spacing, and the elastic and plastic anisotropy of the constituents. These interactions give rise to stress and strain fluctuations with wavelengths comparable to the grain or colony size and basically reflect the dependence of the fracture resistance on the microstructure. The overwhelming part of fracture characterization in TiAl alloys has been performed at this mesoscopic scale.

At the next finer scale (Figure 10.1c), the influence of dislocation glide and mechanical twinning becomes noticeable. As the crack advances, the dislocations have to propagate and multiply. The energy that has to be supplied for these processes leads to significant energy dissipation; the related energy-release rate can be significantly higher than that expected for the creation of a new crack surface. The dislocations and twins may play an ambivalent role in the fracture process, depending on their mobility and the orientation of the slip and twinning systems with respect to the crack plane. On the one hand, local dislocation and twin plasticity may shield and blunt crack tips so that the crack driving force is reduced. In fact, fracture in crystalline metals rarely occurs by cleavage when there is large-scale plasticity. On the other hand, the local stress levels produced by dislocation pile ups or immobilized twins may exceed the cohesive strength, thus causing crack nucleation.

At the atomic scale (Figure 10.1d), the crack propagates by stretching and breaking individual bonds between the atoms. The process may exhibit a significant crystallographic anisotropy, depending on the strength and directionality of the atomic bonding. Many of the fundamental mechanisms underlying crack propagation at the atomic level are now accessible to atomistic simulation, either by means of empirical potentials or through ab-initio quantum-mechanical calculations. It is only recently that direct observations by high-resolution electron microscopy have been presented.

In closing this section, it is worth noting that crack growth and failure in TiAl alloys are determined by various processes that occur on length scales ranging from the component dimensions down to the atomic level. It is the interaction of all these processes that determines whether or not crack advance occurs and how much energy is dissipated along the crack path.

10.2
Cleavage Fracture

Low-temperature deformation of TiAl alloys usually exhibits the following characteristics:

i) unstable or catastrophic failure at applied stresses less than the general yield strength;
ii) little or no macroscopic plastic strain to failure;
iii) little fractographic evidence of local plastic strain;
iv) intragranular or intralamellar plate failure.

These attributes are often ascribed to brittle cleavage fracture [7]. In ideally brittle propagation, the crack is atomically sharp and therefore atomic potentials are important. You and Fu [8, 9], utilizing first-principles total-energy calculations, have determined the ideal cleavage strength for different intermetallic compounds. The data obtained for TiAl and Ti$_3$Al are listed in Table 10.1. The ideal cleavage energy (Griffith energy) G_c was defined as the total surface energy of the two cleaved planes, that is, $G_c = 2\gamma_s$. Table 10.1 also contains the theoretical stress-intensity factors K_{Ic}, which were calculated from the cleavage energy for a mode-I crack and the elastic compliance constants, as proposed by Shi and Liebowitz [10]. In mode-I the crack is symmetrically loaded perpendicular to the crack plane, giving rise to crack opening.

There is a remarkable anisotropy among the G_c values, which in part could be attributed to the atomic composition of the cleavage surfaces [11]. When two surfaces with different atomic composition are separated by cleavage, then the intera-

Table 10.1 Cleavage strength of TiAl and Ti$_3$Al. G_c (ideal cleavage energy), K_{Ic} (theoretical stress-intensity factors for a mode-I crack).

Alloy	Cleavage plane (hkl)	G_c (J m^{-2})	Cleavage direction [uvw]	K_{Ic} (MPa m$^{1/2}$)
TiAl	(100)	4.6	[001]	0.94
			[011]	0.89
	(001)	5.6	[010]	1.04
			[110]	1.04
	(110)	5.3	[001]	1.03
			[1$\bar{1}$0]	0.86
	(111)	4.5	[1$\bar{1}$0]	0.90
			[11$\bar{2}$]	0.94
Ti$_3$Al	(0001)	4.8	<uv.0>	0.93
Interface TiAl/Ti$_3$Al	(111) ∥ (0001)	4.65		0.94

Data from Yoo and Fu [8, 9].

tomic forces are expected to be more long-range, compared to the case when two identical surfaces are generated. This effect is probably manifested in the high values of G_c for the $(001)_\gamma$ and $(110)_\gamma$ planes, when compared with that for the $(111)_\gamma$ plane. Based on the calculated cleavage energies it may be expected that the $(111)_\gamma$ and $(100)_\gamma$ planes are the cleavage habit planes of γ(TiAl). Panova and Farkas [12], utilizing embedded atomic calculations, have essentially confirmed these findings.

Several authors have associated the low cleavage strength of TiAl with the directional bonding between Ti and Al atoms [8, 9, 13–16]. The bonding directionality is caused by the hybridization of the d-electrons and the anisotropic distribution of p-electrons. Based on first-principles calculations for the electronic structure and the binding energy, Song et al. [15, 17] have proposed that the directionality could be weakened if suitable alloying elements are substituted for Al. Addition of the 3d-transition elements V, Cr and Mn tends to enhance the cleavage strength. Addition of Si and hyperstoichiometric Al seem to be harmful. Unfortunately, no clear experimental evidence for these suggestions is available, probably, because the microstructure is simultaneously changed with composition.

Experimental evidence of cleavage-like fracture in TiAl alloys was obtained by transmission electron microscope observations on crack tips [18–21]. The investigations were performed on cracks that originated from perforations in thin TEM foils; thus, the cracked sample geometry was not defined. The method may, on the other hand, provide information that is not accessible with other techniques. Figure 10.2 shows a high-resolution image of a crack that propagated in a thin foil of a lamellar α_2(Ti$_3$Al) + γ(TiAl) alloy. The crack followed the $\{111\}_\gamma$ planes at the atomic scale and was deflected at a γ_1/γ_2 interface according to the crystallography; finally, the crack was arrested at the next α_2 lamella. The observation confirms that γ(TiAl) is prone to cleavage fracture on $\{111\}_\gamma$ planes. The crack-growth resistance for this type of cleavage fracture is probably very small. As the $\{111\}_\gamma$ cleavage plane coincides with the dislocation slip and twin-habit planes, blocked slip or twinning can easily nucleate cracks. Once nucleated, cleavage cracks may grow extremely fast to a critical length.

A similar picture probably holds for α_2(Ti$_3$Al) as indicated by the low cleavage energy of the $(0001)_{\alpha 2}$ plane (Table 10.1). Fracture toughness tests performed by Umakoshi et al. [22] on Ti$_3$Al single crystals have indeed shown that the material is prone to cleavage fracture on the $(0001)_{\alpha 2}$ basal and $\{10\bar{1}2\}_{\alpha 2}$ pyramidal planes. The authors noted that brittle fracture never occurred on the prism planes, which is reasonable because Ti$_3$Al single crystals are very ductile when pure prismatic slip is activated (Section 5.2.2). Research performed by Yakovenkova et al. [23] on Ti$_3$Al single crystals revealed cleavage on the $(0001)_{\alpha 2}$ basal plane and on $\{10\bar{1}1\}_{\alpha 2}$, $\{10\bar{1}2\}_{\alpha 2}$ and $\{10\bar{1}3\}_{\alpha 2}$ pyramidal planes, when the samples were oriented for basal slip. Cracks occurring on the basal plane were found to coalesce, so that the gross $(0001)_{\alpha 2}$ fracture plane was facetted.

Summary: A low defect tolerance is likely to exists in both single-phase γ(TiAl) and single-phase α_2(Ti$_3$Al) because cleavage fracture occurs on low-index crystallographic planes. Once nucleated, cleavage cracks may grow extremely fast to a critical length, unless no other toughening mechanism is available.

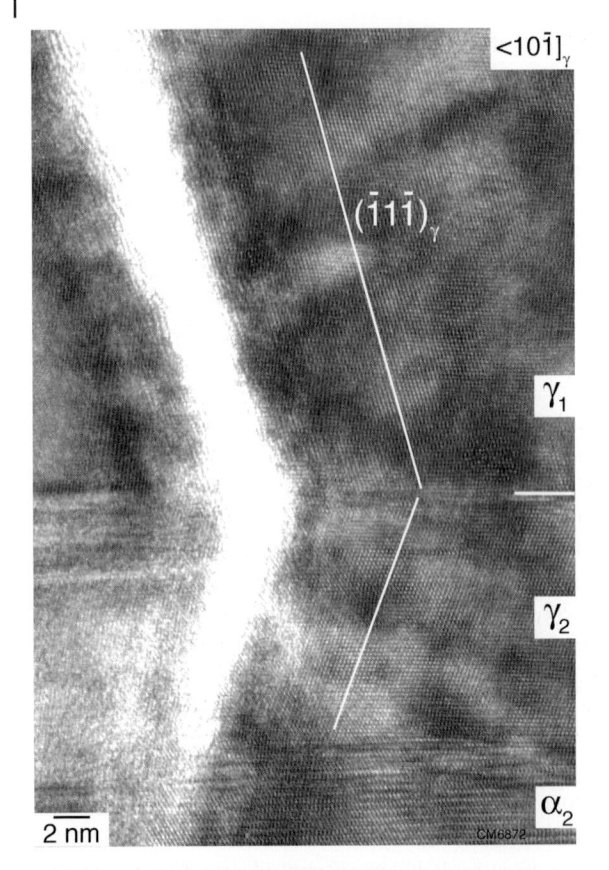

Figure 10.2 Crack propagation in the γ phase of a lamellar α_2(Ti$_3$Al) + γ(TiAl) alloy. Note the cleavage-like fracture on {111}$_\gamma$ planes, the deflection of the crack at the interface γ_1/γ_2 and the immobilization of the crack at the α_2 lamella.

10.3
Crack-Tip Plasticity

10.3.1
Plastic Zone

During crack propagation plastic deformation may occur in the crack-tip region as dislocations and twins are nucleated and emitted from the crack tip. In broad terms, brittle cleavage or ductile behavior can be predicted depending on whether the energy-release rate for brittle cleavage is less or greater than the energy-release rate for the emission of dislocations from the crack tip. If dislocation emission is energetically favorable, then spontaneous emission of dislocations occurs prior to reaching the Griffith stress, and the crack becomes blunted. The blunted crack remains trapped at its original position until the external stress is increased sufficiently to promote crack-tip damage processes that eventually lead to fracture by other mechanisms.

Irwin [24] has shown that under such conditions unstable propagation is still characterized by a critical value of the energy-release rate G_c, provided that the extent of the plastic yielding is small compared with the dimensions of the testpiece. In this case, G_c is determined partly by the production of two new surfaces and partly by the generation of plastic deformation.

Information about crack-tip plasticity in TiAl alloys is rare because most investigations have been performed by scanning electron microscopy (SEM), which is not capable of imaging the structural details at the appropriate size scale. There are a few TEM studies, which provide the necessary resolution; however, as said before, these studies suffer from thin-foil constraints. Nevertheless, the results are instructive regarding the role of dislocations, twins and interfaces in crack-tip plasticity, and are thus supplementary to SEM observations.

Figure 10.3 shows a crack in an ($\alpha_2 + \gamma$) alloy with duplex structure, as observed in the transmission electron microscope. The crack appears to be arrested at the γ/γ_T twin/matrix interface of an annealing twin. It is possible to calculate a plastic zone size at the crack tip as a function of the stress-intensity factor and the yield strength by using the crack stress-field solutions. Plastic flow may occur at loci where the shear stress exerted by the crack exceeds the yield stress. Based on this consideration Yoo et al. [9] described the plastic zone for the three fracture modes of (111)[1$\bar{1}$0] and (11$\bar{2}$)[1$\bar{1}$0] cracks in γ(TiAl) in terms of anisotropic elasticity. The crack shown in Figure 10.3 propagated accidentally in the wedge-shaped foils so that the loading conditions were unknown. Regarding these uncertainties, for simplicity the stress $\sigma_{yy} = f(r, \psi)$ of a mode-I crack in an elastically isotropic solid under plane-stress conditions was determined [25]. In the framework of this model, the stress σ_{yy} acting at a point (r, ψ) in the xy plane of a crack with a leading edge parallel to the z-axis is described by

$$\sigma_{yy} = \sigma_a \sqrt{\frac{a_c}{2r}} \left(\frac{5}{4}\cos(\psi/2) - \frac{1}{4}\cos(5\psi/2) \right) \tag{10.4}$$

σ_a is the applied tensile stress. The size of the region around the crack tip where plastic deformation occurs was determined by setting σ_{yy} equal to the yield stress σ_0. The plastic zone defined in this way was calculated for $a_c = 20\,\mu\text{m}$ in accordance with experimental observation and $\sigma_a = \sigma_0/2 = 215\,\text{MPa}$, respectively, and projected into the {1$\bar{1}$0} foil plane of Figure 10.3. In reality, the plastic zone ahead of a crack has a three-dimensional character with lobes extending above and below the plane of fracture. The shape of the calculated plastic zone is not in accordance with the experimental observation. The deformation features seen in Figure 10.3 indicate strong shear localization by mechanical twinning, which is probably a manifestation of the plastic anisotropy of the γ phase. In other words, the lack of slip systems that can be activated for a given crack orientation and stress (Section 6.2.3) hinders the formation of a more homogeneously deformed plastic zone. Atomistic studies of Panova and Farkas [12] have shown that the amount of crack-tip plasticity in γ(TiAl) may strongly depend on the crystallographic orientation of

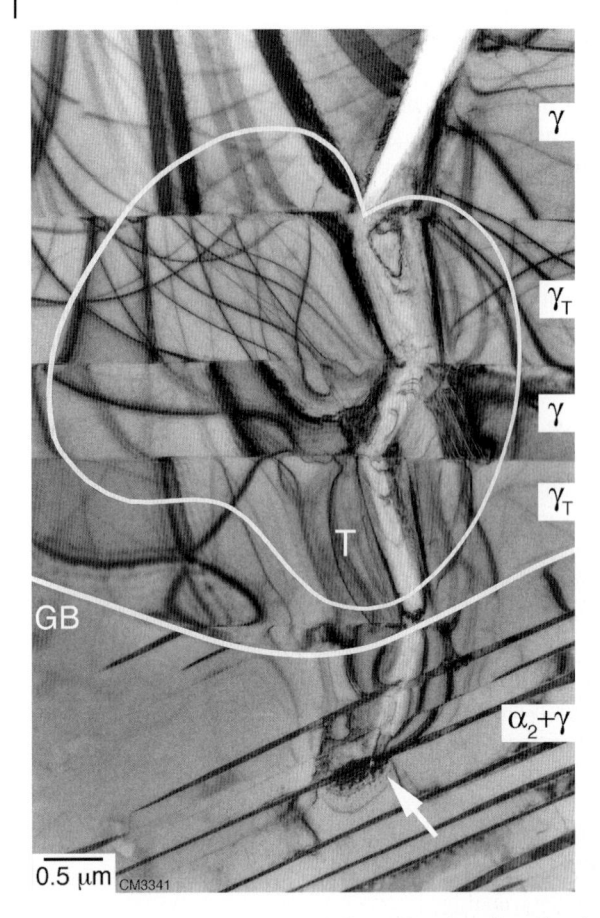

Figure 10.3 A crack in an ($\alpha_2 + \gamma$) alloy with duplex microstructure. The micrograph shows a γ grain containing a few coarse annealing twins adjacent to a colony of ($\alpha_2 + \gamma$) lamellae. The colony boundary is marked with GB. The crack propagated in the γ grain on a {111} plane and was arrested at the interface γ/γ_T formed by one of the annealing twin lamellae. Ahead of the crack tip, a plastic zone is formed that consists mainly of deformation twins (T), which propagated across the annealing twins. The main twin T is finally immobilized in the lamellar colony (arrow). For comparison, the shape of the plastic zone expected from a mode-I crack tip is drawn. Ti-48Al-2Cr, duplex microstructure.

the crack front. For crack propagation on the (111) plane, extensive emission of Shockley partial dislocations on the inclined ($11\bar{1}$) plane is expected when the crack front is parallel to <$1\bar{1}0$]. For the same crack, no significant dislocation generation was recognized when the crack front was parallel to [$11\bar{2}$]. Taken together, the phenomenon of cleavage fracture in γ(TiAl) is relatively easy to rationalize: the resistance to slip and the lack of easy slip systems restrict crack-tip plasticity, and cleavage fracture occurs as a consequence of the large tensile stress developed at the crack tip. In this context it is interesting to remember that f.c.c. metals, with their many equivalent slip systems, have not been observed to undergo cleavage.

In summary: The question of whether an atomically sharp crack can propagate in a brittle manner is decided by the competition between cleavage decohesion and crack-tip blunting and shielding by dislocation glide. In γ(TiAl), the development of crack-tip plasticity largely depends on the crack orientation with respect to the few easy slip and twinning systems available in the $L1_0$ structure.

10.3.2
Interaction of Cracks with Interfaces

In lamellar ($\alpha_2 + \gamma$) alloys, crack-tip plasticity is largely affected by various interactions of the crack with interfaces. Figure 10.4 demonstrates the typical features of crack propagation across these interfaces [18]. Trace analysis with pre-existing twins revealed that the crack followed {111} planes. Within the individual lamellae,

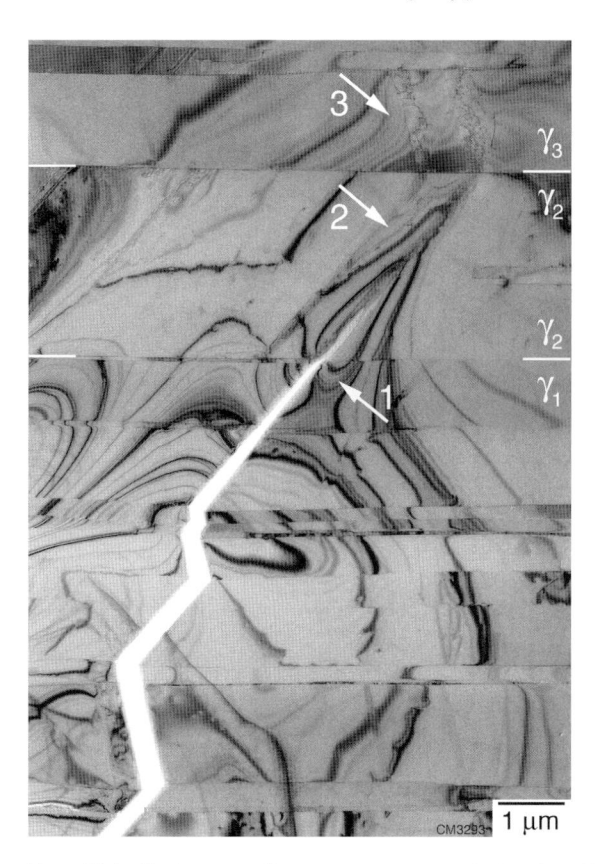

Figure 10.4 Crack propagation across lamellar interfaces in a Ti-48Al-2Cr alloy. The crack is arrested at interface γ_1/γ_2 (arrow 1). Ahead of the crack tip, two twins are generated, which are immobilized at the next semicoherent interface joining the gamma variants γ_2/γ_3 with pseudotwin orientation (arrow 2). The stress concentration in front of the twins is shielded by two slip bands of ordinary dislocations (arrow 3). Figures 10.5 and 10.6 show the marked details in higher magnification.

the crack surfaces are smooth and featureless indicating cleavage-like fracture. At the lamellar interfaces, crack deflection repeatedly occurred according to local crystallography. Thus, a more tortuous zigzag-shaped crack path was formed. Finally, the crack was arrested at a semicoherent γ_1/γ_2 interface joining gamma variants with a 120° orientation relationship (arrow 1). In front of the crack tip a remarkable deformation activity is recognizable (detail 2 in Figure 10.4). Two deformation twins are generated and immobilized at the next pseudotwin boundary γ_2/γ_3 (detail 3). The stress concentration in front of the twins is relaxed by two dislocation slip bands, which are shown in more detail in Figure 10.5. The dislocations with Burgers vector $\mathbf{b} = 1/2 < 110]$ were probably emitted from the interfacial dislocation network under the action of the incoming twins, as has been discussed in Section 6.3.2. The slip path of the dislocations is limited by the width of lamella γ_3. There is evidence of dislocation pinning, presumably by jogs. From an evaluation of the dislocation curvature, an average shear stress of $\sigma = 250\,\text{MPa}$ was estimated. For comparison the (normal) yield stress of the material is $\sigma_0 = 430\,\text{MPa}$, from which a shear stress of $\tau_0 = \sigma_0/M_T = 140\,\text{MPa}$ can be deduced; $M_T = 3.06$ is the Taylor factor. As illustrated in Figure 10.6, dislocation dipoles were trailed at ordinary dislocations as a result of cross-slip and jog formation. The high shear stress acting on the dislocations seems to initiate dislocation multiplication via multiple cross-glide. The operation of this multiplication mechanism depends on

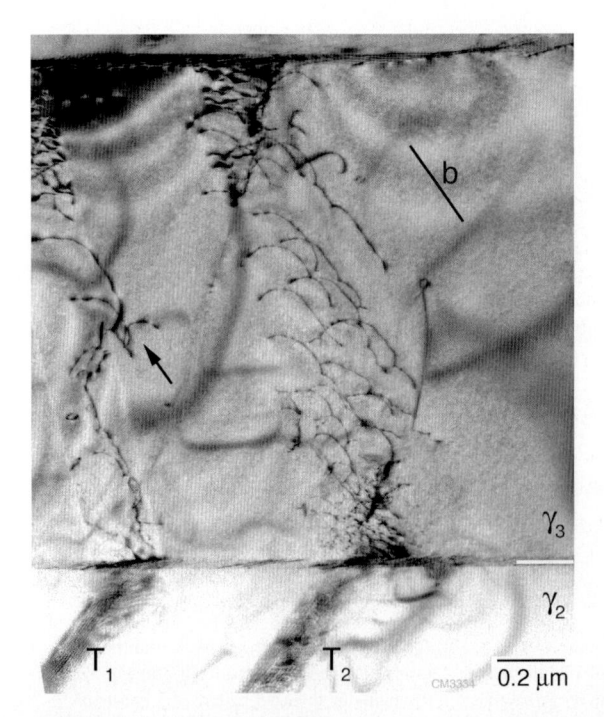

Figure 10.5 Detail 2 in Figure 10.4; two deformation twins T_1 and T_2 emitted from the crack tip are immobilized at the pseudotwin interface γ_2/γ_3. Two slip bands of ordinary dislocations were generated in lamella γ_3, which probably shield the stress concentration of the immobilized twins. Note the strong curvature of the dislocations.

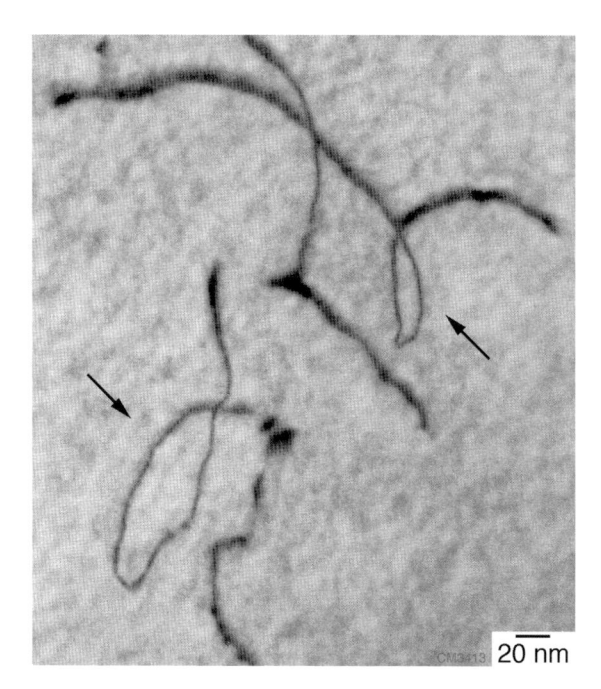

Figure 10.6 Higher magnification of the arrowed detail in Figure 10.5 showing the initial state of dislocation multiplication. Dislocation dipoles are trailed behind jogs in screw dislocations (arrowed). Because of the high stress acting ahead of the crack tip, the dipole arms could probably pass each other and may act as single-ended dislocation sources during further crack propagation.

both the jog height h and the shear stress acting on the dislocation (Section 6.3.1, Equation 6.16). The shear stress value $\sigma = 250\,\text{MPa}$ estimated from the dislocation curvature suggests that very narrow dipoles with $h > 15\mathbf{b}$ (**b** is the Burgers vector of ordinary dislocations) may operate as dislocation sources. A few dipoles (arrowed in Figure 10.6) are probably in the initial stage of multiplication; this is indicated by the fact that the edge arms have already bypassed each other.

Summary: Crack propagation across lamellar boundaries is greatly influenced by local crystallography. Within the individual γ lamellae, crack propagation occurs along {111} planes inclined to the interfaces. Crack deflection at the interfaces makes the crack path tortuous. Crack tips are shielded and blunted by interface-related emission of dislocations and twins.

10.3.3
Crack–Dislocation Interactions

Each dislocation is a source of stress and can induce an additional stress-intensity factor. Thus, any dislocation arrangement strongly modulates the crack-tip stress field. The micrographs shown in Figure 10.7 may illustrate this point [18]. Figure 10.7a shows cleavage-like propagation of a crack in the γ phase of a duplex alloy. The crack is immobilized at a colony boundary. Figures 10.7b and 10.7c are

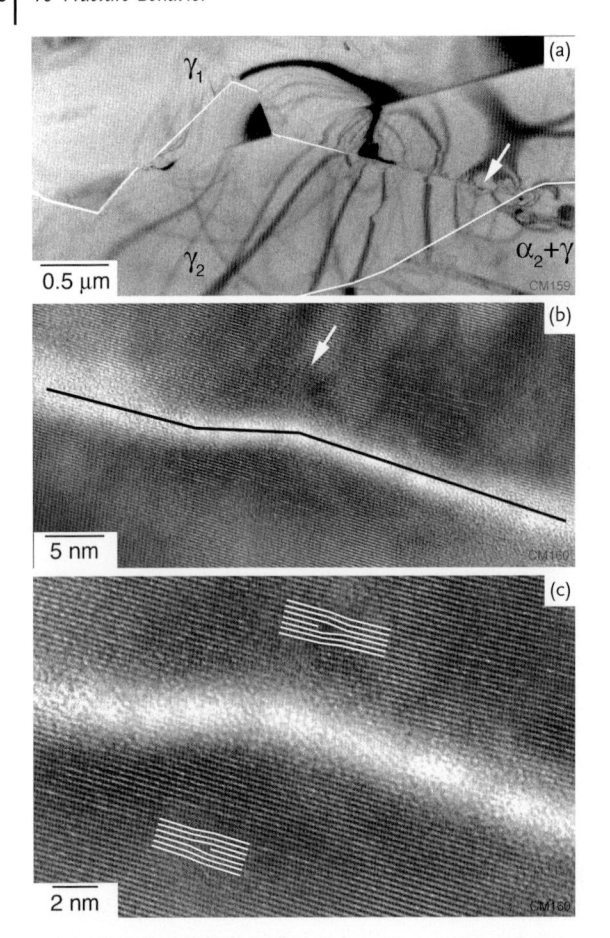

Figure 10.7 Crack–dislocation interactions. Ti-45Al-3Mn with duplex microstructure. (a) Low-magnification, high-resolution image showing cleavage-like fracture mainly on $\{111\}_\gamma$ planes and immobilization of the crack in the vicinity of a colony boundary. (b) Enlargement of the region marked with the arrow in (a), showing crack deflection from the $\{111\}$ cleavage plane near the crack tip. (c) Higher magnification of the arrowed area in (b); two dislocations in a dipole-like configuration are discernible and most likely cause deflection of the crack. Perhaps a third dislocation is present, although this could not be clearly identified. Lamellar Ti-45Al-3Mn, undeformed.

high-resolution images of the near-crack-tip region with increasing magnification. In this region the crack is still bridged, presumably by an oxide layer, which is imaged as a granular structure. Although the crack generally propagated parallel to the (111) plane, crack deflection occurred at the atomic scale. As shown in Figure 10.7c, the crack is deflected between two dislocations, which are arranged in a dipole configuration. The dislocations have a separation distance of about 25**b**. An estimation of the related shear stress acting perpendicular to the crack plane is difficult because of the thin-foil constraints. From the theory of straight dislocations [26], very high values of 500 MPa to 1000 MPa are expected for the observed configuration if bulk conditions are assumed. Thus, the observed crack deflection becomes plausible. Such a mechanism makes the crack path more tortuous and is expected to enhance the toughness. More on this subject will be presented in Section 10.4.4.

Apart from this interaction between the crack-tip and dislocation stress fields, additional energy dissipation might be expected when a crack intersects screw dislocations. At each screw dislocation the fracture surface receives a step whose height equals the Burgers vector of the dislocation. The formation of such steps requires a substantial surface energy, which could stabilize crack growth. The pattern that the steps form on the cleavage surface is known as the river pattern [27]. However, dislocation pile ups can also produce stress concentrations that tend to propagate shear in a locally unstable manner. For example, edge dislocation pile ups would produce stress concentrations that would tend to propagate a pure mode-II crack, while screw dislocations would promote a pure mode-III crack [28].

Summary: There are various interactions between propagating cracks and pre-existing dislocations, which could toughen TiAl alloys at the microscopic scale. These involve crack deflection and generation of surface steps. The potential of these mechanisms remains to be ascertained.

10.3.4
Role of Twinning

Brittle fracture is closely related to the competition between cleavage decohesion and crack-tip blunting or shielding. Stable crack growth requires the plastic zone to keep up with the cleavage crack, which is difficult when the dislocation mobility is low. As has been described in Section 6.4, the dislocation mobility in γ(TiAl) is controlled by a high Peierls stress, jog dragging and localized obstacles. This gives rise to a rate-dependent drag force and certainly impedes strain accommodation at the crack tip. Furthermore, dislocation multiplication via multiple cross-glide, which has been recognized in the crack-tip regions, needs some glide or time to take place. Plastic zones consisting of dislocation assemblies may therefore easily be outrun by fast cracks. In this respect, the activation of mechanical twinning might be important because the growth rate of twins is often a significant fraction of the elastic shear-wave velocity. The enhanced stress concentration ahead of cracks and the limited capability of plastic relaxation by dislocation glide may activate twinning, provided that a suitable twinning system is available. Figure 10.8 demonstrates an interactive case of fracture and mechanical twinning [20, 29]. The crack originated from the perforation of the thin foil and was probably subject to mixed-mode loading, which involved a significant mode-II shear component. Mode-II is the in-plane shearing or sliding mode. As in the previous cases, the overall shape of the crack indicates crack deflection at the lamellar interfaces. The crack became immobilized at position (1) within a γ grain. Figures 10.9 and 10.10 show the atomic structure at the very front of the crack-tip region (position 2 in Figure 10.8). In response to the crack stress field, a narrow twin is formed in

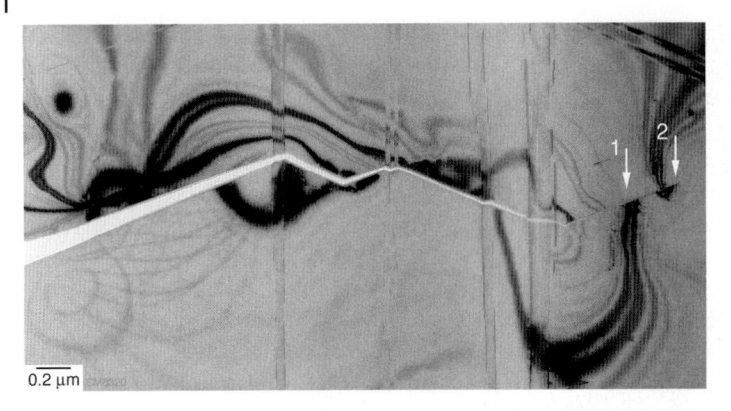

Figure 10.8 Association between fracture and mechanical twinning in a duplex Ti-46.5Al-4(Cr, Nb, Ta, B) alloy. Crack propagation along $\{111\}_\gamma$ planes and deflection at lamellar interfaces. The crack was immobilized at position (1). Details in front of the crack are shown in Figures 10.9 and 10.10, respectively.

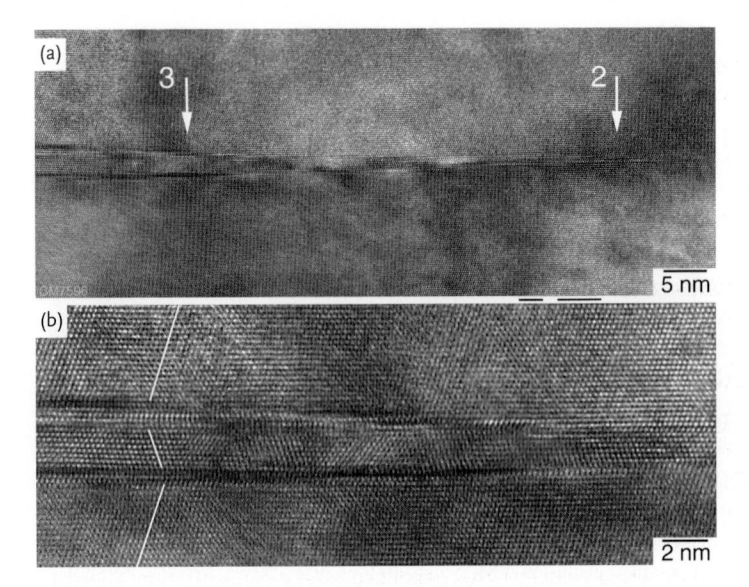

Figure 10.9 High-resolution image of detail 2 in Figure 10.8. (a) Twin formation ahead of the crack tip. (b) Higher magnification of the arrowed region 3 in (a).

front of the crack tip (Figure 10.9). The width of the twin gradually becomes narrower with increasing distance from the crack tip, as may be expected from the decay of the crack stress field. This suggests that nucleation of the twin took place sequentially as the crack advanced. Twinning seems to have preceded crack propagation as indicated by the twin configurations seen near to the crack wake (Figure 10.10). The crack wakes are facetted, which may indicate another mechanism of energy absorption that occurred during crack propagation. The amount of energy dissipation that can be obtained by these mechanisms is difficult to assess because,

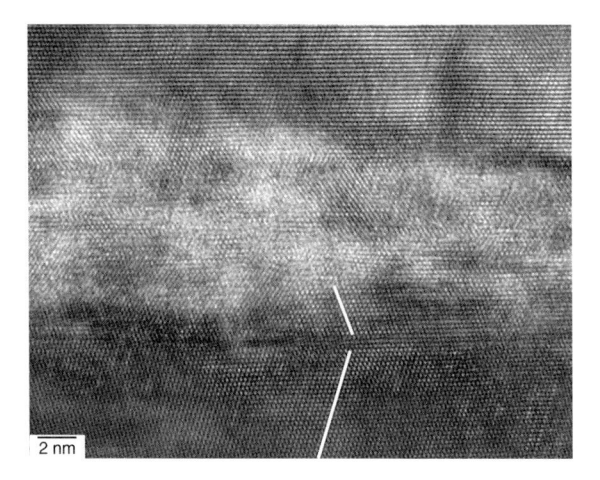

Figure 10.10 High-resolution image of the crack wake showing twin structures.

under most experimental conditions, other factors controlling crack propagation, such as microstructure, dislocation plasticity, interface decohesion, and crack deflection are certainly superimposed. Micromechanical modeling of the cleavage modes in γ(TiAl) [30] has shown that the shear stress field of a $[1\bar{1}0](111)$ mode-II crack can be shielded by $1/6<11\bar{2}](111)$ twins. However, unstable crack propagation is expected when a mode-I component is superimposed. Generally, mixed-mode crack loading tends to concentrate crack-tip plasticity onto the (111) habit plane. Yoo et al. [9] have suggested that this strain localization may support translamellar fracture along {111} planes across all types of lamellar γ/γ interfaces (Figure 10.4).

However, the TEM observations also revealed an unfortunate detriment of mechanical twinning, namely the formation of microcracks. Figure 10.11 shows the deformation structure of a duplex Ti-48Al-1.5Mn alloy, which had been deformed in tension at room temperature to failure. Deformation is mainly manifested by two twinning systems that intersect each other and intersect several lamellar boundaries. High stresses obviously occur at the twin intersections, as indicated by strain contrast and dense twin structures (see Section 6.3.4). There are several microcracks present, which seemingly originate from twin intersections. Clearly, the possibility that these cracks were formed during sample preparation by twin-jet electropolishing cannot be ruled out because selective etching at the dense defect structures may have occurred. However, against this are the facts that the micrograph was taken from a relatively thick region of the wedge-shaped sample and that cracking preferentially occurred along the twin-habit planes. Similar to this mechanism microcracks may also be produced when twins are blocked at grain or phase boundaries. *In-situ* straining experiments performed in the SEM on different two-phase alloys have revealed remarkable glide and twinning activity ahead of crack tips [31]. Microcracks are formed at grain boundaries, when the adjacent grains are unable to accommodate the shear of incoming twins. As has been described in Section 5.1.5, twinning is polarized, that is, reversal of

Figure 10.11 Deformation structure of a cast Ti-48Al-1.5Mn alloy with a nearly lamellar structure observed after tensile deformation at room temperature to failure. (a) Microcracks generated at intersecting twin bands. (b) Higher magnification of the area marked with the arrow in (a). Note the fine microcrack (arrowed) formed along a twin band.

twinning shear will not produce a twin. Furthermore, in the $L1_0$ structure only one true twinning system per {111} plane is available. Thus, for a given shear direction certain grains or lamellae will not twin. Due to these restrictions, crack-tip shielding by mechanical twinning is certainly limited to favorably oriented grains or lamellae. It may therefore be expected that mechanical twinning plays a dual role in crack-tip plasticity. Deve and Evans [32] measured an appreciable toughness in duplex Ti-50Al and associated this finding with crack-tip shielding due to mechanical twinning. Twinning occurred in both the plastic zone ahead the crack tip and within crack-bridging ligaments. However, in fine-grained materials, where a large number of grains is sampled in the process zone ahead of the crack, twinning may not be effective, as suggested by the low fracture toughness determined on such materials [33]. Thus, no clear position has emerged in the TiAl literature about the effect of mechanical twinning on the fracture toughness.

In closing, it may be noted that mechanical twinning plays a complex role in fracture of TiAl alloys. Crack-tip shielding due to mechanical twinning might be a toughening mechanism occurring at the atomic scale and within the stress-induced process zone involving several γ grains or lamellar colonies. Such processes may decelerate fast-growing cracks. However, at the same time, the $\{111\}_\gamma$ planes serve as slip planes, twin-habit planes and cleavage planes. Thus, microcracks may easily be generated at intersecting twin bands or otherwise blocked slip or twinning.

10.4
Fracture Toughness, Strength, and Ductility

Fracture toughness – the resistance to crack extension – can be very sensitive to metallurgical aspects such as constitution, microstructure, and grain orientation. A very large variability in fracture toughness has been demonstrated in a variety of studies. This section is an overview of these results, separately considering the most important factors governing toughness.

10.4.1
Methodical Aspects

As discussed in Section 10.1, the quantity K_c represents the critical value of the stress-intensity factor K for a given load, crack length and geometry required to cause fracture. Thus, K_c is called the fracture toughness. K_c may be used as a design criterion in fracture prevention. The minimum value of fracture toughness is called plane-strain fracture toughness K_{Ic}. The subscript I refers to the fact that these fractures occur almost entirely by mode-I crack opening. For K_c to be a valid failure prediction criterion, the material must be thick enough to ensure plane-strain conditions at the crack tip. Another limiting condition is that the plastic zone size at the crack tip must be small relative to the crack length as well as to the geometrical dimensions of the specimen. It should be noted that for plane-stress conditions, a much larger plastic zone exists compared to that for plane-strain conditions [5]. In view of these restrictions, standard test methods have been established for the determination of K_{Ic} values using defined specimen geometries and test procedures [34]. However, it is often difficult to strictly apply these procedures to brittle and anisotropic materials like TiAl. For example, the standards often call for the creation of a sharp crack at the notch tip of test specimens using a cyclic fatigue loading procedure. In TiAl alloys, a fatigue crack is difficult to prepare without inadvertently fracturing the specimen. Furthermore, crack propagation in TiAl is largely dictated by cleavage on low-index planes. This could be a problem in coarse-grained material because crack propagation may greatly deviate from the desired cracked sample geometry, making a proper estimation of the

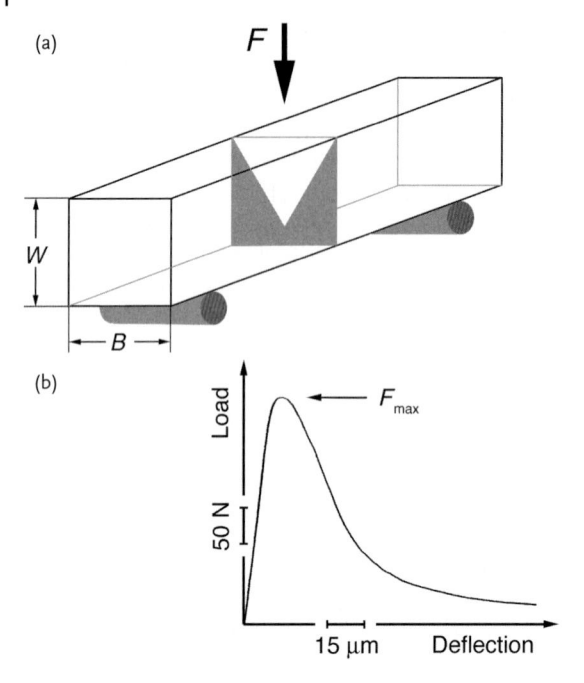

Figure 10.12 Experimental setup for the determination of the fracture toughness. (a) Three-point bending specimen with a chevron notch. (b) Load–deflection curve of a chevron-notched specimen. Cast Ti-47Al-1.5Nb-1Cr-1Mn-0.2Si-0.5B with a nearly lamellar microstructure. Testing temperature 25 °C, $v_M = 0.167\,\text{mm/min}$.

fracture toughness almost impossible. One type of specimen that has emerged to alleviate these problems is the chevron-notched specimen [34]. A chevron-notch has a triangular ligament, as illustrated for a three-point bending bar in Figure 10.12a. Upon loading, a crack should initiate at the tip of the chevron because the local stress intensity is very high. Once nucleated, the geometry of the chevron presents an increasingly larger front to the advancing crack; thus, forcing the crack to extend in a stable manner on a desired plane. Figure 10.12b demonstrates the load–deflection curve of a three-point bending test performed on a chevron-notched specimen. The maximum load F_{max} in the test is achieved when the crack grows to a critical crack length that corresponds to the minimum Y^* in the compliance function Y of the chevron sample. Beyond this point the crack becomes unstable since Y now increases with crack extension. The stress-intensity factors of the various chevron-notched specimen were determined using elasticity theory involving both numerical [35] and analytical calculations [36] along with experimental methods. The fracture toughness is calculated from the maximum load according to

$$K_{Ic}(F_{max}) = \frac{F_{max} Y^*}{B\sqrt{W}} \tag{10.5}$$

where W is the specimen height and B the specimen thickness. Y^* was determined by Munz et al. [35] for different specimen and loading geometries. In addition, the fracture toughness may also be determined using the work of fracture W_F. This is the amount of work performed in both initiating the crack at the tip of the chevron and propagating it through the chevron ligament until the specimen is unloaded. W_F is given by the area under the load–deflection curve. Heat-tinting of the partially fractured samples can be used for determining the related area A_F of crack propagation within the chevron ligament; the oxidized area of the heat-tinted specimen defines A_F. The critical energy-release rate G_{Ic} can be determined from W_F and A_F as

$$G_{Ic} = \frac{W_F}{A_F} \quad (10.6)$$

and $K(G_{Ic})$ is defined as

$$K(G_{Ic}) = \sqrt{\frac{G_{Ic}E}{1-\nu^2}} \quad (10.7)$$

A comparison of the K_{Ic} values determined using Equations 10.5 and 10.7 may provide confidence for the validity of the methods [37, 38]. Figure 10.13 demonstrates the room-temperature toughness values determined by the two methods on different TiAl alloys. The data is best fitted by the equation $K(G_{Ic}) = 0.897 K_{Ic}(F_{max}) + 0.042$, suggesting good agreement of the two data sets. This finding indicates that the various assumptions made in linear elastic fracture mechanics for determining the fracture toughness are relatively well fulfilled for the test conditions chosen.

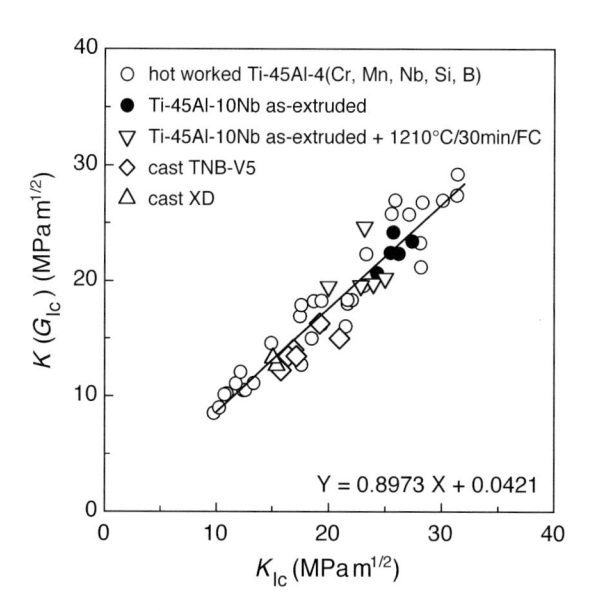

Figure 10.13 Plot of $K_{Ic}(F_{max})$ against $K(G_{Ic})$ revealing proportionality between the two values. Data from [37, 38].

Low levels of scatter were associated with the toughness determined in this way; for example, a standard deviation of 2% to 5% was found for K_{Ic}, based on eight tests.

The fracture behavior of TiAl alloys was also characterized in terms of $K_R(\Delta a)$ resistance curves, that is, the monotonic crack-growth resistance K_R as a function of crack extension Δa. The tests are usually conducted on compact-tension specimens by monotonically loading the samples under displacement control. Crack lengths were monitored using optical-telescope, compliance and electrical-potential techniques. Applied-load and crack-length measurements were used to calculate the stress intensities K_R at extension Δa. In a K_R curve, the onset of crack growth is often referred to as the initiation toughness K_i, which also corresponds to K_{Ic} when plane-strain conditions prevail. The maximum stress intensity K_{ss} in the K_R curve is often referred to as the crack-growth toughness, that is, K_{ss} represents the crack growth toughness at the onset of unstable fracture.

10.4.2
Effects of Microstructure and Texture

Within the length scale range of a few micrometers to 1 mm, the crack path during fracture of a polycrystalline sample is usually controlled by the microstructure. Two-phase ($\alpha_2 + \gamma$) alloys can be reasonably tough, although the individual constituents are quite brittle. Mitao et al. [39] have measured the room-temperature fracture toughness of binary alloys with Al contents ranging from 44 at.% to 51 at.%. The data is shown in Figure 10.14 and provides a first survey about the range

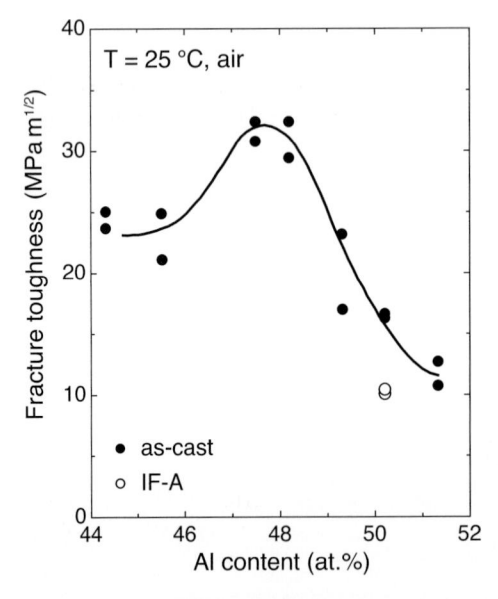

Figure 10.14 Dependence of the fracture toughness on the Al content for as-cast and isothermally forged and annealed (IF-A) samples. Diagram adapted from Mitao et al. [39].

of toughness that is available for TiAl alloys. Over the years, much effort has been expended in order to assess the effects that the microstructure has on toughness. Table 10.2 provides a collection of strength and toughness data, which were determined on different TiAl alloys at room temperature [40–45]. Appropriate thermomechanical treatments were carried out on these alloys to produce near-gamma, duplex and lamellar microstructures; a full description of the structural details can be found in the cited publications. As can be seen in Table 10.2, a wide range of K_{Ic} values can be obtained for a given alloy. The main trends evident in the toughness properties are:

i) Single-phase gamma alloys are brittle compared with $(\alpha_2 + \gamma)$ alloys and exhibit very low toughness [32, 39, 46].

ii) Fine-grained duplex materials typically have a fracture toughness of $K_{Ic} = 9$ MPa m$^{1/2}$ to 17 MPa m$^{1/2}$ at room temperature [40–45].

iii) The toughness of duplex alloys decreases with increasing volume fraction of equiaxed γ grains [41–43], as shown in Figure 10.15.

iv) Fully lamellar alloys with randomized colony orientations have a relatively high toughness of $K_{Ic} = 25$ MPa m$^{1/2}$ to 30 MPa m$^{1/2}$ [41].

v) The fracture toughness increases with increasing colony size (Figure 10.16), perhaps because of the tendency to form larger crack-bridging ligaments [41, 46–48]. However, the effect seems to become saturated for colony sizes larger than 600 μm. Rogers et al. [49] have argued that this could, in part, be due to the large colony size, since in such cases only a few colonies may be sampled in the process zone.

vi) No clear image has been established about the effect of the lamellar spacing on the fracture toughness. Chan and Kim [41] and Kim [42] observed an increase of both the initiation toughness and crack-growth resistance with decreasing lamellar spacing for several alloys (Figure 10.17). The effect was found to depend on colony size. The authors ascribed the toughening effect to more effective crack bridging by shear ligaments. Thinner lamellae hinder translamellar microcracking and linking of the microcracks with the main crack. This results in the formation of larger ligaments. However, research by Rogers et al. [49] has shown that a variation in the lamellar lath thickness between 0.43 μm and 1.68 μm (which is nearly the same range as that investigated in [41, 42]) has no statistically significant effect on the fracture toughness of fully lamellar microstructures. This disparity of observations seems to indicate that the fracture toughness sensitively depends on the fine details of constitution and microstructure, as for example the frequency distributions of lamellar spacings and colony sizes. Furthermore, it is difficult to conceive of a thermomechanical treatment that only affects the lamellar spacing while other structural parameters remain constant.

vii) Crack-growth-resistance curves (Figure 10.18) show a range of crack initiation and crack-growth-resistance behaviors [42, 43, 50–55].

Table 10.2 Mechanical properties of different TiAl alloys at room temperature.

Alloy	Composition/microstructure	σ_y (MPa)	σ_F (MPa)	K_{Ic} MPa m$^{1/2}$	Ref.
1	Ti-49.1Al, extruded below T_α, near-gamma			11$^{p)}$	[40]
2	Ti-47Al-2.6Nb-0.93Cr-0.85V, duplex	416	558	11	[41]
3	Ti-47Al-2.6Nb-0.93Cr-0.85V, lamellar	330	383	16$^{p)}$-25$^{o)}$	[41]
4	Ti-46.5Al-2.1Cr-3Nb-0.2W, duplex	465	580	10-11*	[42]
5	Ti-46.5Al-2.1Cr-3Nb-0.2W, nearly lamellar	550	685		[42]
6	Ti-46.5Al-2.1Cr-3Nb-0.2W, refined fully lamellar	475	550	21.5*	[42]
7	Ti-47.7Al-2Nb-0.8Mn + 1vol.% TiB$_2$ (XD), nearly lamellar	546	588	12-16*	[43]
8	Ti-47Al-2Nb-2Cr-0.2B, fully lamellar	426	541	18-32*	[43]
9	Ti-47Al-2Nb-2Cr, PM, lamellar	975	1010	18-22*	[43]
10	Ti-47.3Al-2.3Nb-1.5Cr-0.4V, coarse lamellar	450	525	18-29*	[43]
11	Ti-47.3Al-2.3Nb-1.5Cr-0.4V, duplex	450	590	11	[43]
12	Ti-47Al-2Cr-0.2Si, near-gamma	538		13	[44]
13	Ti-47Al-2Cr-0.2Si, nearly lamellar	517		32	[44]
14	Ti-47Al-1.5Nb-1Cr-1Mn-0.2Si-0.5B, duplex,	408	426	13$^{p)}$-17$^{o)}$	[44]
15	Ti-47Al-1.5Nb-1Cr-1Mn-0.2Si-0.5B, nearly lamellar,	408	426	23$^{p)}$-33$^{o)}$	[44]
16	Ti-49.7Al-5Nb-0.2C-0.2B, extruded below T_α, duplex			9$^{o)}$	[40]
17	Ti-46.6Al, extruded below T_α, duplex			16$^{o)}$	[40]
18	Ti-46.5-5.3Nb, extruded below T_α, duplex			11$^{o)}$	[40]
19	Ti-45Al-5Nb-0.4C-0.2B, extruded below T_α, duplex			9$^{o)}$	[40]
20	Ti-45Al-10Nb, extruded below T_α, duplex	1090	1100	18$^{p)}$-25$^{o)}$	[44]
21	Ti-45Al-10Nb, extruded above T_α, nearly lamellar	992	992	18$^{p)}$-25$^{o)}$	[44]
22	Ti-44Al, extruded below T_α, nearly lamellar			28$^{p)}$	[40]
23	Ti-45Al-8Nb-0.2C, extruded below T_α, duplex	830	980	11$^{p)}$-16$^{o)}$	[40]
24	Ti-(40-44)Al-8.5Nb, γ-Md, modulated structure	1060	1120	13$^{p)}$-18$^{o)}$	[40, 45]

* A range in values indicates *R* curve behavior, the first value corresponds to crack-initiation toughness, K_i, the second to the steady-state toughness maximum measured crack-growth resistance K_{SS}.
p) crack propagation parallel to extrusion direction,
o) crack propagation perpendicular to extrusion direction.
σ_y, yield strength; σ_F, fracture strength; K_{Ic}, fracture toughness; PM, powder metallurgy. Crosshead displacement speed 0.01 mm/min.

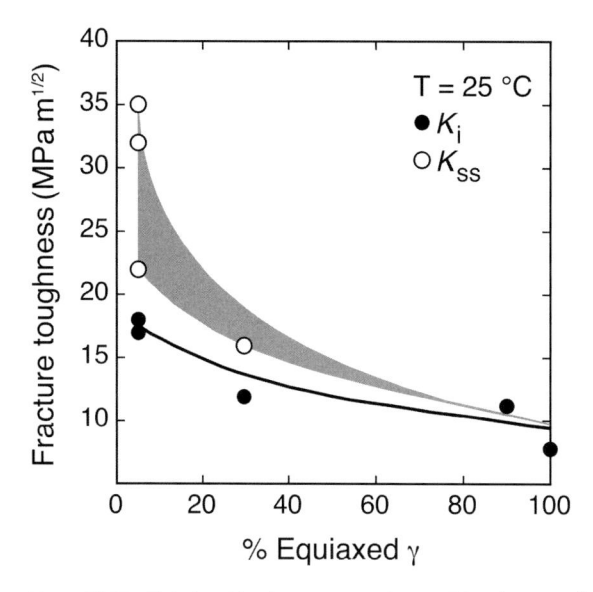

Figure 10.15 Relationships between toughness (K_i and K_{ss}) and volume fraction of equiaxed γ phase for several γ-based TiAl microstructures. Diagram based on alloys 7 to 11 in Table 10.2, redrawn from Kruzic et al. [43].

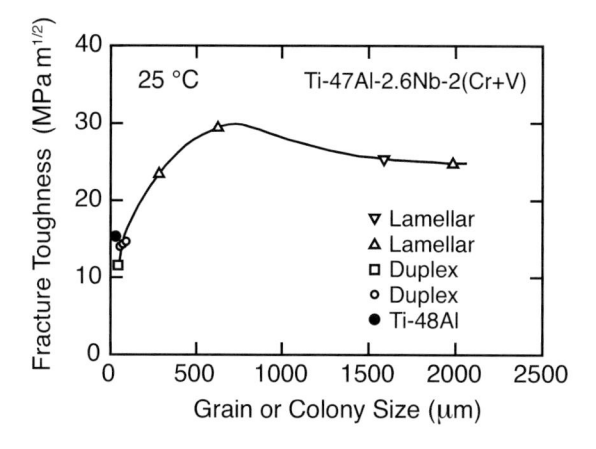

Figure 10.16 Dependence of the fracture toughness on grain and colony size for two ($\alpha_2 + \gamma$) alloys. The fracture toughness represents the asymptotic value K_{ss} of the crack-resistance curves $K_R(\Delta a)$ at the onset of unstable fracture. Data from Chan and Kim [41], redrawn.

Single-phase γ and duplex microstructures exhibit flat K_R curves, that is, the materials undergo crack advance under essentially constant K. Such behavior is usually expected for an ideally brittle material. The initiation toughness defined by the K_R curve for the near γ and duplex alloys is $K_i = 8\,\text{MPa m}^{1/2}$ to $11\,\text{MPa m}^{1/2}$. Alloys with lamellar microstructure exhibit rising K_R curves. This indicates that

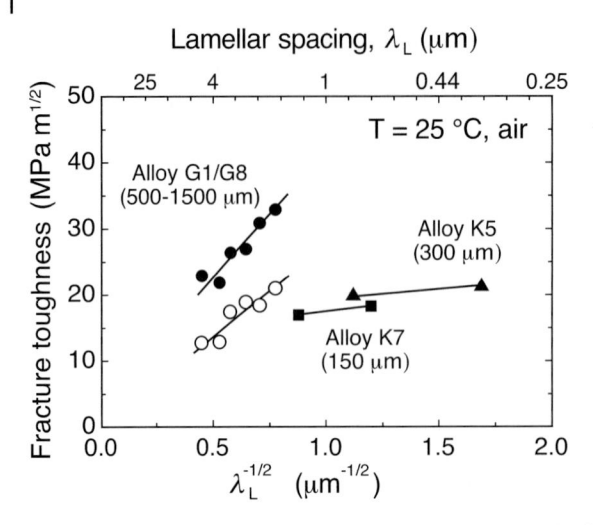

Figure 10.17 Effects of lamellar spacing and colony size on the fracture toughness of lamellar alloys. Open symbols indicate K_{Ic}, full symbols K_{SS}. Data from Kim [42], redrawn.

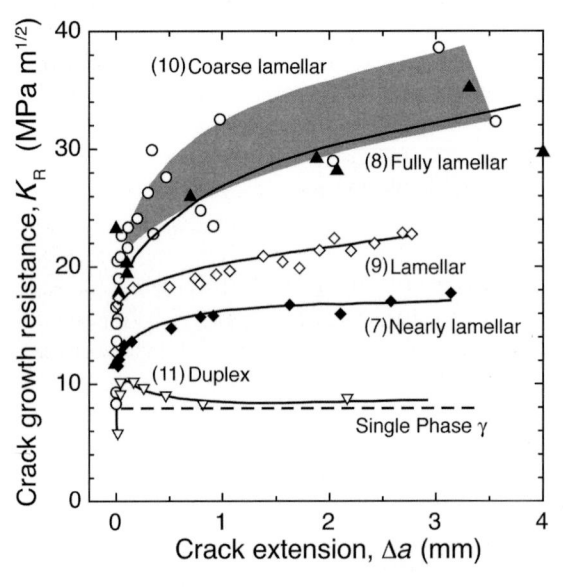

Figure 10.18 Crack-growth resistance curves for alloys 7 to 11 defined in Table 10.2 and a single-phase γ alloy. Data from Kruzig et al. [43], redrawn.

the fracture resistance increases with increasing crack extension. The effect is more pronounced for larger colony sizes; for example, an initiation toughness of $K_i \approx 18$ MPa m$^{1/2}$ and a steady-state toughness of $K_{SS} = 39$ MPa m$^{1/2}$ after 3-mm crack extension were determined on coarse-grained lamellar Ti-47.3Al-2.3Nb-1.5Cr-0.4V [43].

Metallographic inspection of broken specimens revealed different fracture features, largely depending upon the underlying microstructure. Failure of near-gamma alloys is characterized by cleavage fracture. In duplex microstructures, crack advance occurs primarily by cleavage through equiaxed gamma grains, presumably on low-index planes. Clearly the presence of γ grains degrades crack-growth resistance. A finer grain size in duplex structures produces a much less tortuous fatigue crack path, resulting in lower toughness.

The fracture behavior of the lamellar microstructure is more complex because colony orientation with respect to the crack front strongly affects local direction and resistance of crack growth. This is mainly a consequence of the plate-like morphology and the cleavage anisotropy of the constituent phases (Section 10.2). In the lamellar structure the most prominent cleavage planes, $(111)_\gamma$ and $(0001)_{\alpha_2}$, are parallel to the lamellae interfaces. Thus, crack propagation is very easy, when the crack plane is parallel to the lamellar interfaces. Research performed on PST crystals [56, 57] has shown that a very low fracture toughness of $K_{Ic} = 3.3\,\text{MPa}\,\text{m}^{1/2}$ to $4.3\,\text{MPa}\,\text{m}^{1/2}$ is characteristic for this orientation, which is close to the theoretical cleavage strength for the low-index planes of the γ and α_2 phases (Table 10.1). Examination of the fracture surface revealed cleavage of α_2 lamellae [56, 58]. Conversely, an extremely tortuous crack is produced when the crack is forced to traverse lamellar laths (translamellar cracking) [43, 46, 51]. Translamellar failure deflects mode-I crack growth, as was illustrated in Figure 10.4. Multiple cracking ahead of the main crack forms ligaments of intact material, which bridge the crack [43, 46]. The ligaments exert a closure force on the crack faces and are thought to enhance toughness. The crack-wake ligaments can be formed by interface decohesion within the same colony, interface delamination in a neighboring colony or by cracking at colony boundaries. A detailed crack-path and fractographic examination by Wang et al. [59] has shown that the operation of these mechanisms is strongly affected by the nature of the colony boundaries. The effect was found to depend on the misorientation of the adjacent lamellar packets. Colony boundaries were found to be particularly effective in immobilizing cracks if their respective lamellar planes were twisted relative to each other.

The anisotropic fracture behavior of the lamellar microstructure is also recognizable in textured material [60–62]. These investigations [60, 61] were performed on a cast Ti-48Al-2Cr alloy with a colony size of about 1 mm. Metallographic characterization revealed a high degree of preferred lamellae orientation due to the radial dendrite growth of the α phase during solidification. Using this casting texture, the direction of the crack growth relative to lamellar plate orientation could be varied, as sketched in Figure 10.19a for a three-point bending test. The measurements were performed at room temperature and revealed very low toughness when the crack propagated parallel to the interfaces. Conversely, crack growth through the lamellar interfaces resulted in much higher toughness. The related fractographs shown in Figure 10.19b underline this fact. For crack propagation along the interfaces, the load–deflection curves often exhibited "pop-in" events, meaning instantaneous load drops occurring prior to maximum load $F_{\max}$ [49]. These pop-in events are often associated with a significant reduction of the

Type I   II   III

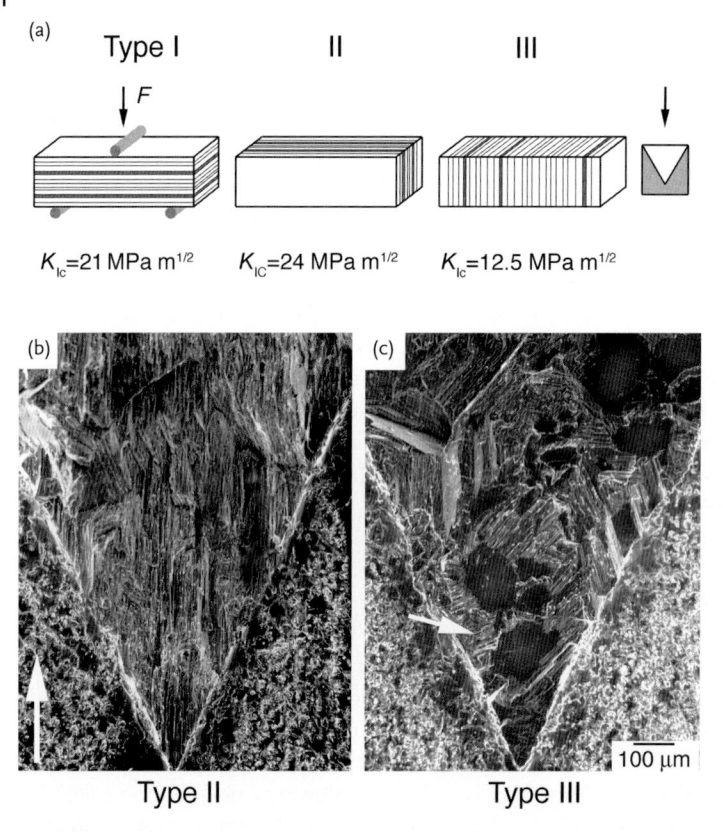

$K_{Ic}=21$ MPa m$^{1/2}$ $K_{Ic}=24$ MPa m$^{1/2}$ $K_{Ic}=12.5$ MPa m$^{1/2}$

Type II   Type III

Figure 10.19 Fracture anisotropy of the lamellar morphology. (a) Three point-bending specimens with chevron notches that are differently oriented to preferentially aligned lamellar colonies and the related room-temperature fracture toughnesses K_{Ic} [60, 61]. Ti-48Al-2Cr, nearly fully lamellar, colony size 1 mm, lamellar spacing 0.5 µm, preferred orientation of lamellae. Tests performed in air with displacement control using a crosshead displacement rate of 0.01 mm/min. (b and c) Scanning electron micrographs of fracture surfaces observed after testing chevron notched bending bars with a preferred orientation of the lamellar colonies, corresponding to Figure 10.19a for specimen types II and III [60]. Crack-growth direction is from bottom to top (arrowed in (b)). Note the plate-like failure in (c), arrowed.

specimen compliance and probably result from noncatastrophic crack growth prior to failure. A detailed fractographic examination has shown that the pop-in events result from local intralamellar plate failure occurring directly under the chevron tip. This observation is technically important because even in a randomized microstructure, areas of only slightly misaligned colonies can occur. If a crack is initiated parallel to the lamellar planes of such packets, it is very easy for the crack to extend to a critical length. Such consideration may, in part, determine if further development of large-grained fully lamellar or massively transformed microstructures will be of benefit. The anisotropic fracture behavior is also an issue for wrought processing [63]. Extruded TiAl alloys often exhibit a marked fiber texture with elongated γ grains and remnant colonies oriented parallel to the extrusion axis. Fracture surfaces parallel to the extrusion axis show many cleavage facets due to a large amount of intralamellar and transgranular fracture, whereas the crack path perpendicular to the extrusion direction is much more tortuous. This difference is reflected in the fracture toughness: the values determined for crack propagation along the extrusion direction are 20% to 50% lower than those for crack propagation perpendicular to the extrusion direction.

Summary: The room-temperature toughness of TiAl-based alloys is highly sensitive to microstructure. The toughness of equiaxed gamma and duplex microstructures is relatively low because crack propagation occurs mainly by transgranular cleavage with some intergranular failure. Lamellar alloys with a random colony orientation have appreciably higher toughness because crack growth is retarded by crack deflection, microcracking and the formation of crack-bridging ligaments. The crack-growth resistance provided by these toughening mechanisms increases with increasing crack length, which leads to a rising K_R curve. However, there is a marked fracture anisotropy: crack growth along the lamellar interfaces results in low toughness; significantly higher values are obtained when the crack is forced to traverse the lamellar laths.

10.4.3
Effect of Temperature and Loading Rate

High temperature tends to decrease the yield points of metals and to enhance plastic deformation. Thus, during crack propagation energy dissipation by plastic deformation becomes more important the higher the test temperature. In broad terms this behavior is reflected by the increase of fracture toughness with temperature (Figure 10.20). The differences between the alloy systems may be rationalized by the following arguments. The room-temperature toughness of the near-γ and duplex alloys is relatively low because cleavage fracture prevails (Section 10.2). At higher temperature, dislocations can glide and climb due to thermal activation. Deformation processes can easily spread within the plastic zone of the cracks so that the constraints due to the local slip geometry are less restrictive. Thus, the fracture mechanism changes within the individual constituents from cleavage at low temperatures to energy-absorbing ductile forms at elevated temperatures. This additional toughening probably causes the strong increase of K_{Ic} with T in near-γ and duplex alloys. Conversely; the room-temperature toughness of lamellar alloys is relatively high and largely determined by a process zone that involves several colonies (Section 10.4.1). For the temperature interval considered here, it is highly unlikely that the structure of the lamellar morphology at this mesoscopic scale is changed, that is, the major toughening mechanism is less affected by temperature. Nevertheless, at the microscopic scale, thermal activation may increase dislocation mobility so that interface-related plasticity (Section 10.3.2) in front of the crack tip is enhanced. However, the resulting toughening effect is seemingly insignificant

Figure 10.20 Dependence of the fracture toughness on the test temperature for the alloys specified in the legend. Tests performed in air using displacement control with a crosshead displacement rate of 0.01 mm/min. The numbers in the legend refer to the alloys listed in Table 10.2.

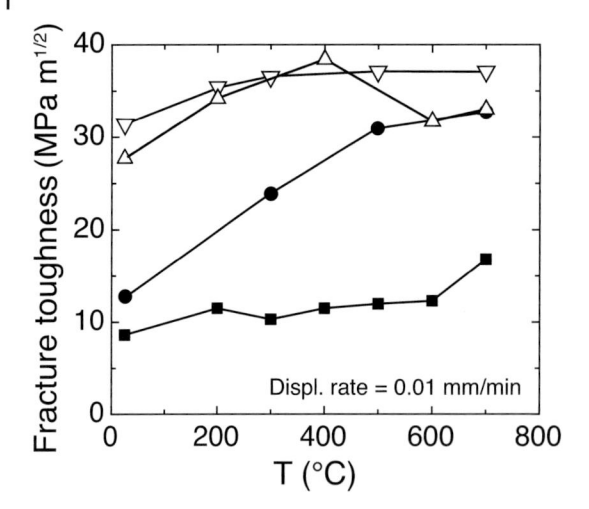

▽ Ti-47Al-2Cr-0.2Si lamellar (13)
△ Ti-44Al nearly lamellar (22)
● Ti-47Al-2Cr-0.2Si near gamma (12)
■ Ti-49.7Al-5Nb-0.2C-0.2C duplex (16)

compared to the high toughening level that is already present in lamellar alloys at room temperature. Thus, the fracture toughness of lamellar alloys is only weakly dependent on temperature.

The effect of temperature enhanced plasticity is also manifested in the fracture resistance curves $K_R(\Delta a)$ [43, 55], as demonstrated in Figure 10.21 for a nearly lamellar XD alloy with composition Ti-47.7Al-2Nb-0.8Mn-1 vol% TiB$_2$ [43]. When compared with room-temperature data, a significant increase of both the initiation and steady state toughness is seen at 600 °C. It is remarkable that a similar behavior was also observed for Ti-46.5Al-4(Cr, Nb, Ta, B) with a near γ microstructure [55]. Tests performed at 800 °C revealed ductile fracture for both lamellar and near-γ microstructures. However, this change of the fracture mode is mainly reflected in a significantly higher initiation toughness of about $K_i = 30$ MPa m$^{1/2}$ [43] to 40 MPa m$^{1/2}$ [55]. For the lamellar XD alloy no rising K_R curve was observed (Figure 10.21), which was attributed to the absence of microcracking and crack-bridging ligaments. For the two microstructures of the Ti-46.5Al-4(Cr, Nb, Ta, B) alloy, near-γ and designed lamellar, R curve behavior was recognized, but was found to be restricted to small crack extensions [55].

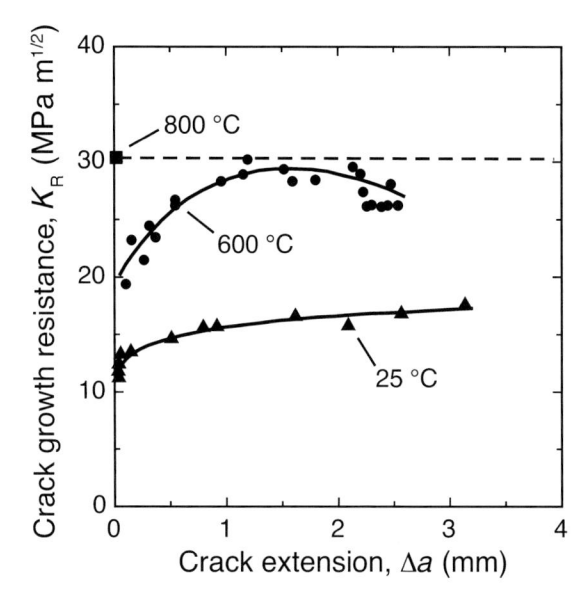

Figure 10.21 Crack-growth resistance curves determined for different temperatures on a nearly lamellar XD alloy with composition Ti-47.7Al-2Nb-0.8Mn + 1 vol% TiB$_2$ (alloy 7 in Table 10.2). Diagram from Kruzig et al. [43].

Fracture tests performed between 200 °C and 400 °C exhibited an interesting feature, which is probably associated with the dislocation dynamics within the plastic zone. The load–deflection curves of these tests show periodic pop-in events, particularly after the load maximum (Figure 10.22). The phenomenon is reminiscent of dynamic strain aging, which during monotonic deformation was observed in the same temperature interval (see Section 6.4.3). Dynamic strain aging was attributed to dislocation locking by ordered defect atmospheres comprised of Ti$_{Al}$ antisite defects and vacancies. It is tempting to speculate that discontinuous yielding also occurs within the plastic zone of crack tips. This could give rise to deceleration (if dislocations break away) and acceleration (if the dislocations are locked) of the growing crack, resulting in intermittent crack propagation.

In general, an increased strain rate tends to cause changes in K_{Ic} similar to those resulting from a decrease in temperature, that is, a rate-sensitive material is expected to have a lower fracture resistance at high crack speeds. This is because the dislocations present in the plastic zone are being outrun by a fast crack. However, the opposite behavior occurs when TiAl alloys are tested at room temperature in air, Figure 10.23 [64]. This anomalous behavior may indicate environmental embrittlement; a material susceptible to environmentally assisted crack growth is stronger at fast testing rates. In this respect, dislocation pinning is often associated with hydrogen uptake. For slowly moving cracks more time is available for hydrogen to diffuse into the material; thus, brittleness is enhanced at low loading rates. Several authors [65, 66] have proposed that the environmental

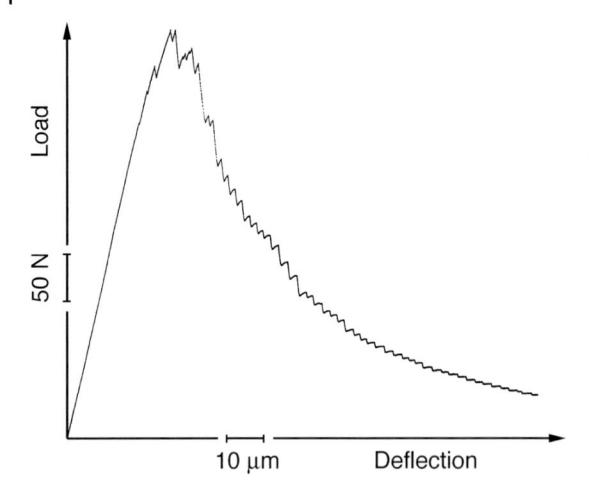

Figure 10.22 Load–deflection curve at $T = 300\,°C$ showing periodic pop-ins. Duplex Ti-47Al-1.5Nb-1Cr-1Mn-0.2Si-0.5B (alloy 14 in Table 10.2).

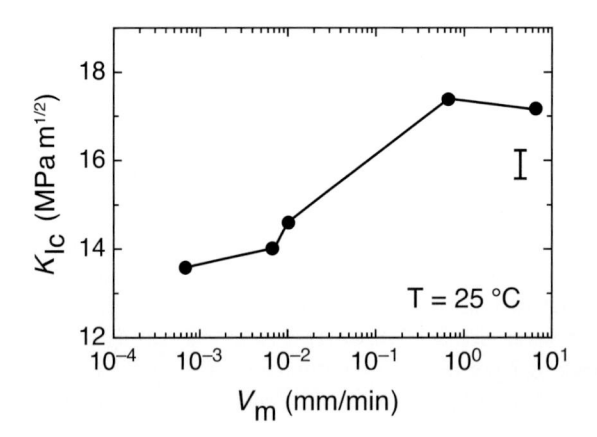

Figure 10.23 Dependence of the fracture toughness K_{Ic} on the loading rate v_m measured at $T = 25\,°C$. Ti-47Al-1.5Nb-1Cr-1Mn-0.2Si-0.5B alloy, duplex microstructure with 20% gamma grains (grain size 20 μm) and a colony size of 400 μm. Diagram from Lorenz et al. [64].

degradation of ductility of TiAl alloys is a form of hydrogen embrittlement. This interpretation appears reasonable, since Ti alloys in general tend to be susceptible to hydrogen embrittlement [67, 68], and hydrogen is an environmental constituent that could easily move into the lattice, even at room temperature. Taken together, these findings indicate that crack propagation in the intermediate temperature range of 200 °C to 400 °C is a critical concern.

Summary: With rising temperature, crack-tip shielding and blunting by plastic deformation becomes important in TiAl alloys. At a test temperature of 600 °C an appreciable increase of fracture toughness combined with R curve behavior has been recognized for all microstructures.

10.4.4
Effect of Predeformation

The microscopic observations described in Section 10.3.3 were encouraging enough to design an experiment in which the association between deformation and fracture could be assessed [20], Figure 10.24a. Large samples ($12 \times 15 \times 30\,\text{mm}^3$) of a two-phase alloy were compressed at room temperature to a strain of $\varepsilon = 4\%$ in order to produce an appreciable density of dislocations and mechanical twins.

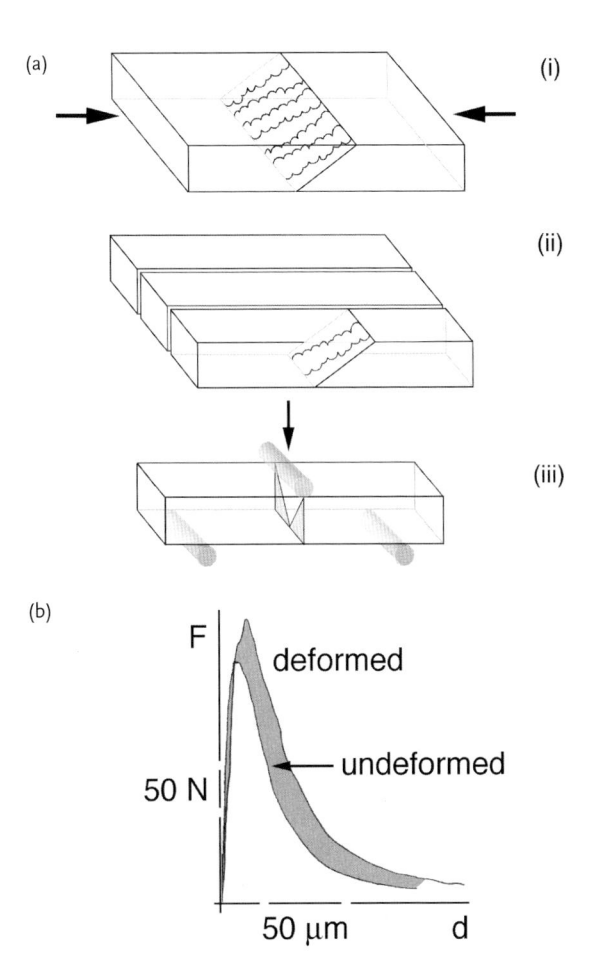

Figure 10.24 Effect of precompression on the fracture toughness. (a) Sample preparation: (i) Predeformation at room temperature, (ii) preparation of chevron-notched bending bars, (iii) measurement of the fracture toughness using three-point bending tests. (b) Load (F) against deflection (d) curves for two chevron-notched specimens demonstrating the effect of predeformation on crack resistance. Tests performed on duplex Ti-47Al-1.5Nb-1Cr-1Mn-0.2Si-0.5B at room temperature in air and a crosshead speed of $v_M = 0.01\,\text{mm/min}$. Predeformation in compression at room temperature to strain $\varepsilon = 4\%$. Data from [20].

TEM examination of these samples revealed a dislocation density of $10^7\,\text{cm}^{-2}$ to $10^8\,\text{cm}^{-2}$ and a significant amount of twinning. From the predeformed samples chevron-notched bending bars of dimensions $4.5 \times 5.5 \times 25\,\text{mm}^3$ were prepared by spark erosion and mechanical grinding. The samples were fractured in thee-point bending tests as described earlier. Figure 10.24b shows the load–deflection trace of a non-predeformed and that of a predeformed sample. Predeformation clearly increases the resistance of the material against crack propagation, which is manifested by a higher fracture toughness and a larger work of fracture. It might be speculated that this is a consequence of various possible interactions between the crack and the deformation-induced defects. While gross compression of a component is difficult to realize, local predeformation can easily be accomplished. For example, adaptation of this technique could enhance the load-bearing capacity of notched components. It should be mentioned in this context that significant improvement of the fracture toughness has been achieved by local indentations [69]. The inability of cracks, once nucleated, to open under hydrostatic stress is an obvious reason for this improvement. Thus, the method is not only an interesting laboratory technique but also a possible means of alleviating brittle fracture in highly stressed parts of a component.

Summary: Precompression appreciably enhances the fracture toughness of TiAl alloys and can probably be attributed to crack–dislocation interactions.

10.5
Fracture Behavior of Modulated Alloys

In TiAl alloys with the base-line composition Ti-(40–44)Al-(5–10)Nb (γ-Md), a novel type of *in-situ* composite can be established [45] (see Chapter 8). The characteristic constituents of this composite are laths with a modulated substructure that is comprised of several stable and metastable phases. The modulation occurs at the nanometer-scale and thus leads to a significant refinement of the material, which exhibits distinct fracture behavior.

Figure 10.25 shows a TEM micrograph of a crack, which propagated in a thin foil of the γ-Md alloy [45]. This observation, although made at relatively low magnification, shows a considerable amount of detail. The crack is deflected at a bundle of modulated lamellae. The modulated laths obviously formed crack-bridging ligaments; one of them is seen to be disrupted. Thus, the crack path becomes microscopically very tortuous with the production of highly distorted material along the fracture surface. Finally, the crack is arrested at another modulated lath. This process is apparently accompanied by crack-tip plasticity, as indicated by twin nucleation. These features contrast with the observations made on conventional alloys, which revealed pronounced cleavage fracture on $\{111\}_\gamma$ planes (Section 10.2). Thus, the modulated laths seem to toughen the material in a fashion that is reminiscent of that of ductile reinforcement in a brittle material.

This point of view is confirmed by fracture tests that were performed to determine the fracture toughness. Figure 10.26 demonstrates a load–deflection trace of

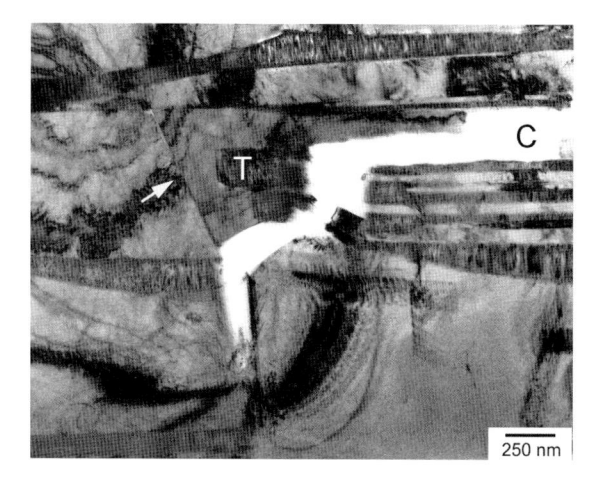

Figure 10.25 Transmission electron micrograph of a crack C in a TiAl alloy (γ-Md) containing modulated laths T. Note the crack deflection at a bundle of modulated lamellae, the blunting of the crack tip and the emission of a mechanical twin (arrowed) [45].

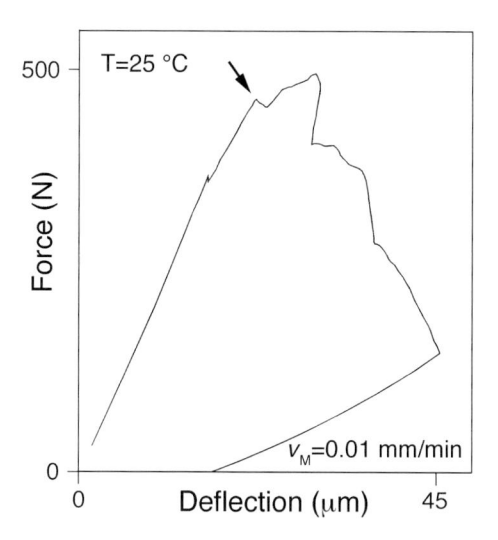

Figure 10.26 Load against deflection trace from a three-point bending test performed at room temperature on a chevron-notched specimen of the γ-Md alloy, in order to determine the fracture toughness. Extruded material, crack propagation perpendicular to the extrusion direction. Note the pop-in event occurring upon loading (arrowed).

such a test. The characteristic features are pop-in events, which apparently reflect the different crack-growth resistances of the individual constituents. These sudden drops of the bending force might be related to intralamellar plate failure when the crack propagates through regions that are almost free of modulated lamellae. The toughness values determined on extruded material of the γ-Md alloy for crack

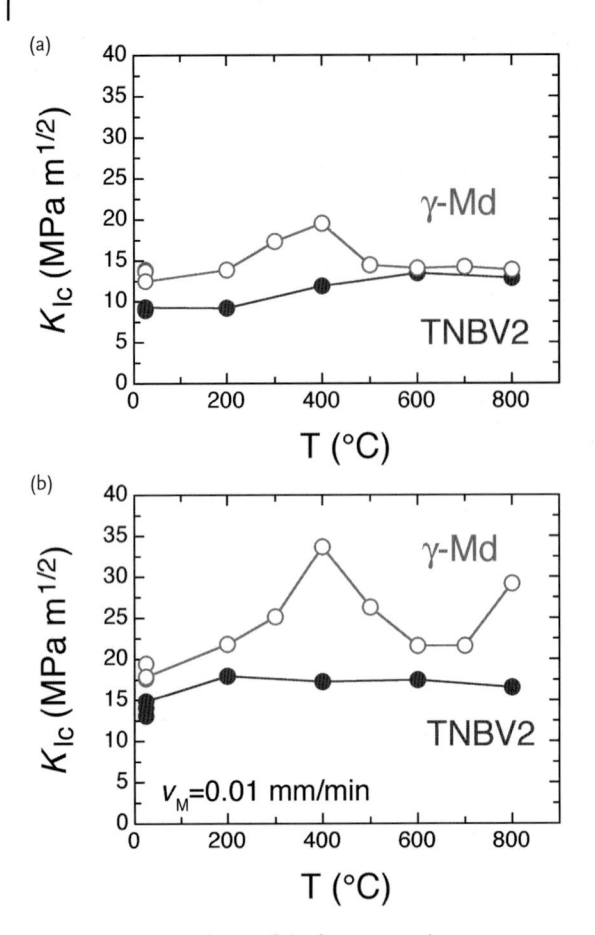

Figure 10.27 Dependence of the fracture toughness on test temperature of the extruded alloys γ-Md and Ti-45Al-8Nb-0.2C (TNB-V2) for (a) crack propagation parallel and (b) perpendicular to the extrusion direction.

propagation parallel and perpendicular to the extrusion direction are shown in Figure 10.27. K_{Ic} was calculated from the maximum load in the force–deflection curves. The toughness values of 13 MPa m$^{1/2}$ and 18 MPa m$^{1/2}$ determined for the parallel and perpendicular sample orientations are significantly higher than those determined for conventional duplex alloys (Table 10.2).

The modulated laths can apparently further transform into the γ phase, as has been discussed in Section 8.2 (Figure 8.9). This process often starts at grain boundaries and proceeds through the formation of high ledges via distinct atomic shuffle displacements. As this transformation was frequently observed in deformed samples, it might be speculated that the process is stress induced and also occurs at crack tips. Such a transformation is expected to be thermally activated; this could be the reason for the maxima in the toughness–temperature profile, which have been observed for the two testpiece orientations (Figure 10.27). Such transformations have long been known to take place in metals (austenite→martensite in ausformed steels [70]) and ceramics (partially stabilized zirconia [71]) and, likewise, may toughen the γ-Md alloy.

Summary: TiAl alloys with a modulated microstructure (γ-Md) exhibit appreciable fracture toughness, which is attributable to the crack deflection, formation of crack-bridging ligaments and perhaps phase transformations. The toughness–temperature profile exhibits a maximum at about 400 °C, which has not been observed on conventional TiAl alloys.

10.6
Requirements for Ductility and Toughness

Current materials being used in applications for which TiAl alloys are being considered have superior toughness and ductility. Figure 10.28 [72–74] shows the general trend for room-temperature fracture toughness as a function of yield strength for three major alloy classes aluminum, titanium and steel; included is the data for TiAl-based alloys. A higher yield or ultimate tensile strength generally produces a decrease in K_{Ic} data for all materials, and thus a greater susceptibility to catastrophic fracture. Although this tendency is hardly manifested for TiAl alloys, the generally low toughness sets severe limitations on engineering use. These aspects in the risk in using γ(TiAl) alloys will be briefly addressed in this section.

A minimum toughness is required so that a component can withstand processor service-induced defects. The equation

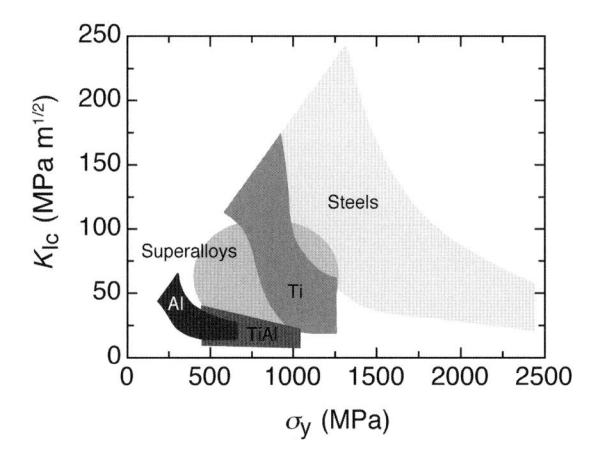

Figure 10.28 Locus of plane-strain fracture toughness versus yield strength. Data for steels, aluminum alloys and conventional Ti alloys from [72, 73]; data for superalloys from [74]; data for TiAl alloys from GKSS.

$$\sigma_a = \frac{K_{Ic}}{\sqrt{\pi a_c}} \tag{10.8}$$

shows in a simplified way how changes in fracture toughness influence the relationship between allowable nominal stress and allowable crack size. This equation is obtained from equating the stress-intensity factor K in Equation 10.1 for a through-thickness center crack in a wide plate to the fracture toughness K_{Ic}. Since the approach is based on linear elastic fracture mechanics, the allowable stress to be considered should not exceed approximately 80% of the yield strength. The allowable stress σ_a in the presence of a crack of size a_c is directly proportional to K_{Ic}, while the allowable crack size for a given stress is proportional to the square of K_{Ic}. Thus, increasing K_{Ic} has a much larger influence on allowable crack size than on allowable stress. In other words, at given K_{Ic}, a higher stress reduces the allowable crack size. Based on the desired service stress and typical defect sizes that are practical for inspection, the material must have at least this minimum toughness to obtain acceptance in the design community. While this reasoning may seem trivial; it is sometimes overseen in the assessment of mechanical data from high-strength TiAl alloys. For example, the room-temperature toughness of the extruded γ(Md) alloy is 13 MPa m$^{1/2}$ (Table 10.2). If such material would be used at a working stress of σ_a = 500 MPa, then the allowable crack size is about 200 μm. It should be noted that the yield strength of the γ-Md alloy at room temperature is 1060 MPa and that the fracture strength is 1120 MPa at a plastic tensile elongation of about 2.5%. This data, together with the fact that the working stress is less than 50% of the tensile strength, provides some confidence for the use of the material from the perspective of toughness.

One apparent paradox that is often cited in the TiAl literature is the toughness–ductility inversion that can be observed when fully lamellar and duplex microstructures are compared [41, 75]. Typically, fully lamellar microstructures can exhibit very high fracture toughness values of up to 30 MPa m$^{1/2}$, but tensile elongations are usually less than 1%. Conversely, duplex microstructures often exhibit tensile elongations of 2% to 3%, but typically toughness values are measured as 10 MPa m$^{1/2}$ to 18 MPa m$^{1/2}$. The effect was attributed to the plastic anisotropy of the lamellar morphology and the correlation between grain size and crack nucleation [42, 76]. From the practical viewpoint this means that constitution and microstructure must be carefully selected for the intended application. Another concern is the variability in toughness and ductility, which can even occur on testpieces extracted from a single cast slab or from the same extrusion or forging. Progress in reducing property scatter is likely to result from careful compositional control and optimized thermomechanical processing.

One problem that has surfaced recently is the discovery that TiAl alloys are susceptible to embrittlement when subject to exposure at moderately high temperatures [65, 66, 77–82]. Importantly, this effect appears after very short exposure times in air or in vacuum, with little or no discernible change in the bulk microstructure of the alloys investigated. The effect was attributed to the formation of a brittle surface layer [77], near-surface compositional changes [77], surface residual stresses [82, 83], oxygen [77] or hydrogen embrittlement [83, 84], or a combination of these mechanisms. However, if it is proven that hydrogen embrittlement occurs, as in conventional Ti alloys [67, 68], then it appears that considerable work remains to be done to find adequate mitigation strategies for overcoming the problem. In any event, while the phenomenon is striking and without a satisfactory explanation, the problem remains narrow in scope.

Several authors [83–87] have addressed the toughness and ductility requirements for application of low-ductility intermetallic alloys in aircraft-engine components. The authors have pointed out that even the limited ductility of TiAl alloys is likely to be acceptable in some applications. For example, a plastic strain to failure of 0.8% is sufficient to relieve contact stresses between the airfoil and the turbine disk in the attachment region, especially at radii in the dovetail. The reader is referred to these articles for more details.

Summary: Toughness requirements for TiAl-based alloys are dependent on the working stress level. Based on detectable flaw sizes and typical working stresses under service, the relatively low toughness is likely to be acceptable for several applications in high-temperature technologies.

10.7
Assessment of Property Variability

The strength, ductility and other properties of γ-TiAl are determined by many factors, the composition and processing playing a dominant role. Tensile properties are particularly sensitive to intrinsic microstructural features such as grain size, γ to α_2 volume fraction and distribution, texture and segregation [88]. External features such as surface-connected porosity and machining damage can also have a detrimental influence on such brittle materials [89]. The intrinsic low ductility of TiAl at room temperature and its variability, combined with evidence for its dependence on strained volume, suggests that a probabilistic approach to determine properties, such as strength, may be appropriate [89]. For safety-critical parts this is very sensible, as components must be designed with respect to minimum material properties rather than average values.

10.7.1
Statistical Assessment

In the field of ceramics, which also exhibit brittle behavior, the variability of fracture strength can be analyzed in a probabilistic manner using Weibull statistics as outlined in [90]. Danzer *et al.* [91] have published an overview concerning the fracture of ceramics, including a detailed description of Weibull statistics. This method ranks the strength (or other properties) on a logarithmic scale against the probability of failure on a double-logarithmic scale. In practice, a series of *n* tensile tests results in *n* different values of strength, which are then ranked from the lowest to the highest value. An approximate failure probability F_i is assigned to each test value and depends on its position in the ranking. F_i can be estimated by an equation such as [90]

$$F_i = \frac{i - 0.5}{n} \tag{10.9}$$

where i is the position in the ranking list (1 being the ranking for the lowest property value) and n is the total number of tests performed. Thus, there is a specific value of F_i for each measured strength (or other property) value. The value of F_i for each test is then plotted on the y-axis of the Weibull plot in the form of ln[ln (1/(1−F_i))] while the corresponding strength value is plotted on the x-axis as ln[strength]. The slope of the plotted test series data is referred to as the Weibull modulus m, and is a measure of the property variability. Different gradient segments within the plotted data indicate that different flaw types dominate fracture. The probability of failure $F(\sigma, V)$ is given by the equation [90]

$$F(\sigma, V) = 1 - \exp\left[-\frac{V}{V_0}\left(\frac{\sigma}{\sigma_0}\right)^m\right] \tag{10.10}$$

σ_0 is known as the characteristic strength; it is the strength at which the failure probability $F(\sigma, V) = 1 - \exp(-1)$ is approximately 63% for specimens with a volume $V = V_0$. Low values of the Weibull modulus (m) indicate a high variability in strength, meaning that large safety margins have to be made to ensure component reliability. For materials where the fracture stress σ can be at any level above zero, the graph of ln[strength] against ln[ln (1/(1−F_i))] is called a two-parameter Weibull plot. Advanced engineering ceramics show strength variations with maximum reported Weibull moduli of no higher than 35 [89].

10.7.2
Variability in Strength and Ductility of TiAl

Studies using Weibull statistics to quantify the variability in the strength and ductility of TiAl alloys are very rare, with, to the author's knowledge, only two papers being published to date. The first paper by Biery et al. [89] addressed four different investment cast alloys with variations in the casting conditions. The second study, by Paul et al. [38], considered an extruded high niobium-containing alloy that had subsequently been given a near alpha-transus heat treatment followed by precipitation hardening. With the exception of a slowly cooled coarse-grained Ti-47Al-2Cr-2Nb alloy which showed a two-parameter Weibull modulus (m) for failure strength of 19 [89]; both studies found the variability in the failure strength TiAl to be lower than that of the most advanced ceramics, with m values of between 47 and 95 reported in [89], and 53 in [38], see Figure 10.29. However, even for the cast TiAl with a Weibull modulus of 19, it should be remembered that, unlike for ceramics, some plasticity was observed.

For the cast alloys, and in particular for duplex $\alpha_2 + \gamma$ materials, it was found that fracture originated at local clusters of grains where large degrees of strain

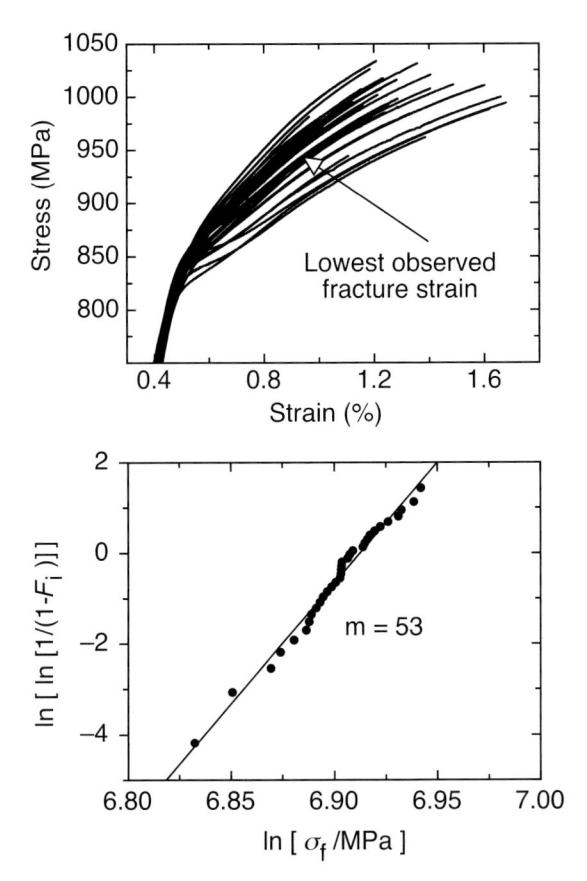

Figure 10.29 The variability in tensile behavior of 33 room-temperature tensile tests and the related two-parameter Weibull plot for fracture strength [38]. All the tensile specimens showed plasticity, the fracture strength variability had a Weibull modulus of 53. The high niobium containing TiAl alloy (Ti-45Al-8Nb-0.2C) had been extruded and subsequently heat treated near to the alpha-transus temperature and then precipitation hardened. The extent of microstructural variation within the material is shown in Figure 10.31 where the microstructures of the specimens with the highest plastic elongation to failure and the highest fracture stress are shown.

localization occurred [89]. Although the strain was distributed more uniformly in lamellar microstructures, failure was still found to be initiated in regions that experienced the highest strain. Experiments showed that the volume of material tested had an influence on the fracture strength; larger specimen volumes showing reduced strength values – similar to ceramic materials.

The variability in the mechanical behavior after yield of the cast alloys was found to be determined primarily by the microstructural scale and was influenced by cooling rate [89]. Paul et al. [38] speculated that microstructural differences between specimens had an influence on the work-hardening behavior and thus on the plastic strain to failure, Figures 10.30 and 10.31. The variability in plastic strain of

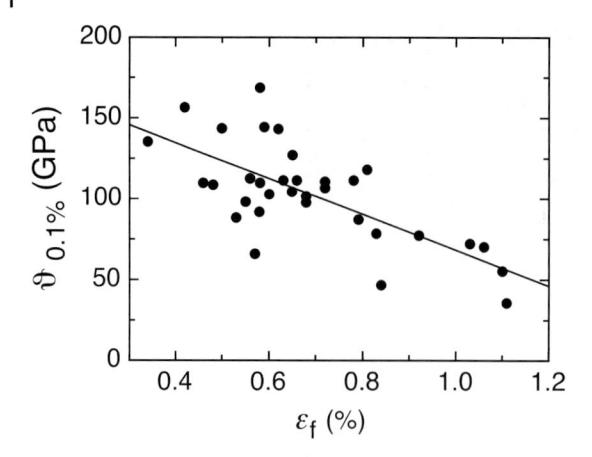

Figure 10.30 Graph showing the relationship between the plastic elongation to failure and the work-hardening rate at 0.1% plastic strain [38]. The data indicates that specimens with a high initial work-hardening rate tend to fracture before those with lower hardening rates. This was believed to result from microstructural variations from specimen to specimen, see Figure 10.31.

the extruded material was rather large, with an *m* value of 4.4 on a two-parameter Weibull plot, Figure 10.32. The data for the cast alloys was not published. Nevertheless, all of the tensile specimens made from the extruded material showed some plastic deformation. It should be mentioned that the yield strength of the extruded material was around twice that of the cast alloys. For similar levels of fracture toughness this implies the presence of much smaller (critical) defects in the extruded material.

10.7.3
Fracture Toughness Variability of TiAl

To the authors' knowledge, work concerning a statistical assessment of the fracture toughness behavior consists of only one paper [64]. This study characterized the room-temperature fracture toughness of different alloys using three-point chevron-notch bend testing. The study found that the variability in fracture toughness of fine-grained duplex materials was characterized by a Weibull modulus of 18 to 24, whereas that for coarse-grained lamellar microstructures was 7 to 10, see Figure 10.33 The graph shows that for lamellar microstructures the crack orientation with respect to the lamellae has a very significant role. The values of fracture toughness for the fine-grained duplex material were typically between 12 MPa m$^{1/2}$ and 15 MPa m$^{1/2}$, while those for the lamellar microstructure varied between 8 and 19 MPa m$^{1/2}$ for cracking parallel to lamellar interfaces and 19 to 28 MPa m$^{1/2}$ for cracking across lamellar interfaces.

Apart from tensile and fracture toughness properties probabilistic methods can also be used to describe other properties such as fatigue behavior as outlined by Weibull [90], Jha *et al.* [92] and Soboyejo *et al.* [93].

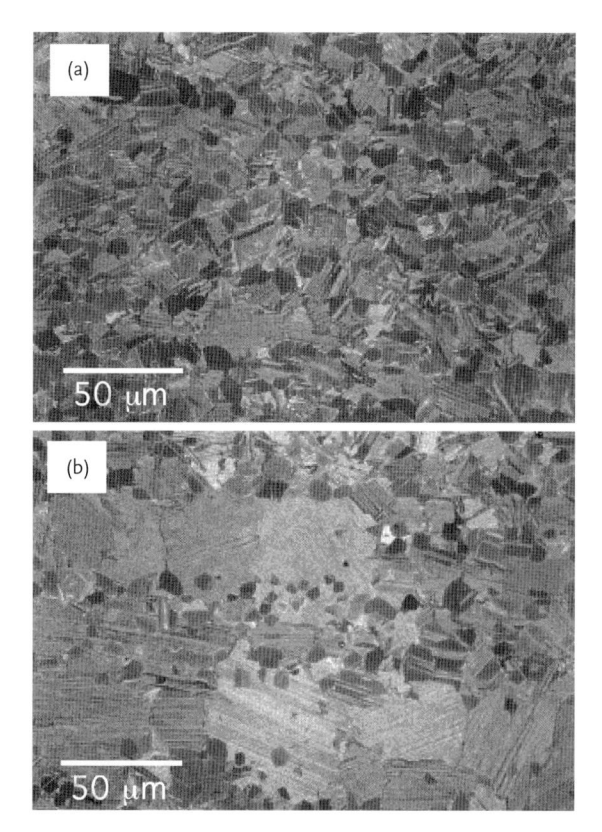

Figure 10.31 Microstructure of similarly heat-treated tensile specimens taken from different positions within an extruded high niobium-containing alloy (Ti-45Al-8Nb-0.2C) that exhibited (a) the highest plastic elongation to failure and (b) the highest strength [38].

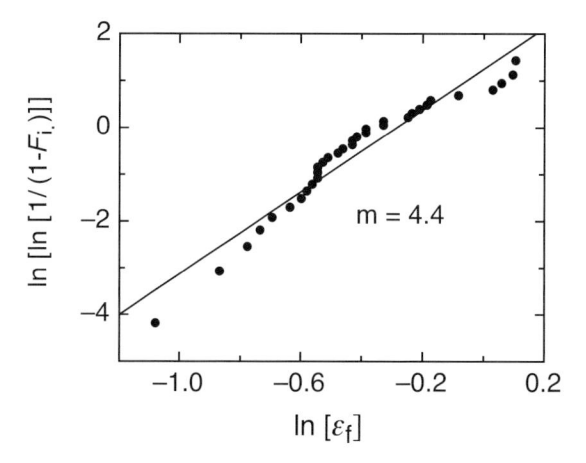

Figure 10.32 A two-parameter Weibull plot for the plastic elongation to failure of the extruded and subsequently heat-treated high niobium-containing TiAl alloy [38]. The low m value of 4.4 indicates a high variability in plastic strain to failure; nevertheless no specimens fractured during loading in the elastic regime.

10 Fracture Behavior

γ-TAB duplex
K_Q=12-15
m=18-24

K_Q=8-19
m=7

Ti-48Al-2Cr lamellar
K_Q=19-28 MPa m$^{1/2}$
m=10

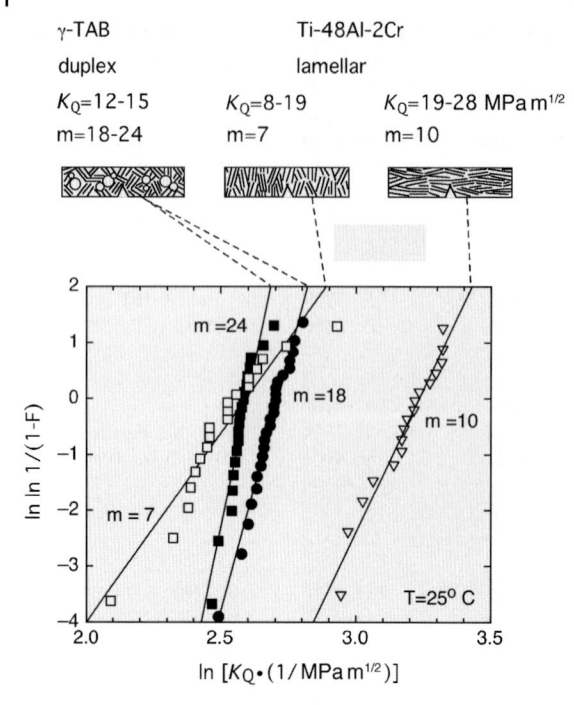

Figure 10.33 Weibull plot showing the variability of room-temperature fracture toughness (K_Q) for different duplex microstructures of the γ-TAB alloy (Ti-47Al-1.5Nb-1Cr-1Mn-0.5B-0.2Si) and for different crack orientations with regard to the lamellae orientation in Ti-48Al-2Cr [64] (diagram adapted). The toughness of the duplex microstructure varied between 12 and 15 MPa m$^{1/2}$ with a Weibull modulus that ranged between 18 and 24. The orientation of the lamellar microstructure is shown to have a large influence on the fracture toughness with cracking parallel to the lamellae being much easier than cracking across lamellae. It should be noted that the fracture toughness in this figure is given as K_Q, although this has subsequently been shown to be the same as K_{1c} [38]. v_m = 0.01 mm/min.

References

1 Ewalds, H.L., and Wanhill, R.J.H. (eds) (1984) *Fracture Mechanics*, Edward Arnold Publishers, Baltimore.
2 Anderson, T.L. (1995) *Fracture Mechanics: Fundamentals and Applications*, 2nd edn, CRC Press, Boca Raton, FL.
3 Hertzberg, R.W. (1996) *Deformation and Fracture Mechanics of Engineering Materials*, 4th edn, John Wiley & Sons, Inc., New York.
4 Schwalbe, K.-H. (1980) *Bruchmechanik Metallischer Werkstoffe*, Carl Hanser Verlag, München.
5 Irwin, G.R. (1957) *J. Appl. Mech.*, **24**, 361.
6 Griffith, A.A. (1920) *Philos. Trans. R. Soc. Ser. A*, **221**, 163.
7 Knott, J.W. (1979) *Fundamentals of Fracture Mechanics (Revised)*, Butterworth's, London.
8 Yoo, M.H., and Fu, C.L. (1992) *Mater. Sci. Eng. A*, **153**, 470.
9 Yoo, M.H., Zou, J., and Fu, C.L. (1995) *Mater. Sci. Eng. A*, **192-193**, 14.
10 Shih, G.C., and Liebowitz, H. (1968) *Fracture – An Advanced Treatise, Vol. II*

(ed. H. Liebowitz), Academic Press, New York, p. 68.
11 Yoo, M.H., and Yoshima, K. (2000) *Intermetallics*, **8**, 1215.
12 Panova, J., and Farkas, D. (1998) *Metall. Mater. Trans. A*, **29A**, 951.
13 Greenberg, B.F., Anisimov, V.I., Gornostirev, Y.N., and Taluts, G.G. (1988) *Scr. Metall.*, **22**, 859.
14 Morinaga, M., Saito, J., Yukawa, N., and Adachi, H. (1990) *Acta Metall. Mater.*, **38**, 25.
15 Song, Y., Xu, D.S., Yang, R., Li, D., and Hu, Z.Q. (1998) *Intermetallics*, **6**, 157.
16 Song, Y., Yang, R., Li, D., Hu, Z.Q., and Guo, Z.X. (2000) *Intermetallics*, **8**, 563.
17 Song, Y., Guo, Z.X., Yang, R., and Li, D. (2002) *Comput. Mater. Sci.*, **23**, 55.
18 Appel, F., Christoph, U., and Wagner, R. (1995) *Philos. Mag. A*, **72**, 341.
19 Appel, F. (2001) *High-Temperature Ordered Intermetallic Alloys IX, Mater. Res. Soc. Symp. Proc.*, vol. 646 (eds J.H. Schneibel, K.J. Hemker, R.D. Noebe, S. Hanada, and G. Sauthoff), MRS, Warrendale, PA, p. N.1.8.1.
20 Appel, F. (2005) *Philos. Mag.*, **85**, 205.
21 Appel, F., and Oehring, M. (2003) *Titanium and Titanium Alloys* (eds C. Leyens and M. Peters), Wiley-VCH Verlag GmbH, Weinheim, p. 89.
22 Umakoshi, Y., Nakano, T., and Ogawa, B. (1996) *Scr. Mater.*, **34**, 1161.
23 Yakovenkova, L., Malinov, S., Karkin, L., and Novoselova, T. (2005) *Scr. Mater.*, **52**, 1033.
24 Irwin, G. (1948) *Fracturing of Metals*, ASM, Cleveland, OH, p. 147.
25 Liebowitz, H. (1968) *Fracture – An Advanced Treatise, Vol. II*, Academic Press, New York.
26 Hirth, J.P., and Lothe, J. (1992) *Theory of Dislocations*, Krieger, Melbourne.
27 Gilman, J.J. (1958) *Trans. AIME*, **212**, 310.
28 Hirth, J.P. (1993) *Scr. Metall.*, **28**, 703.
29 Appel, F. (1999) *Advances in Twinning* (eds S. Ankem and C.S. Pande), TMS, Warrendale, PA, p. 171.
30 Yoo, M.H. (1998) *Intermetallics*, **6**, 597.
31 Simkin, B.A., Ng, B.C., Crimp, M.A., and Bieler, T.R. (2007) *Intermetallics*, **15**, 55.
32 Deve, H.E., and Evans, A.G. (1991) *Acta Metall. Mater.*, **39**, 1171.
33 Bowen, P., Rogers, N.J., and James, A.W. (1995) *Gamma Titanium Aluminides* (eds Y.-W. Kim, R. Wagner, and M. Yamaguchi), TMS, Warrendale, PA, p. 849.
34 American Society for Testing and Materials (1983) ASTM E 399-83. Standard Method for Plane Strain Fracture Toughness of Metallic Materials, Philadelphia, PA.
35 Munz, D., Bubsey, R.T., and Shannon, J.L. (1980) *J. Test. Eval.*, **8**, 103.
36 Whithey, P.A., and Bowen, P. (1990) *Int. J. Fracture*, **46**, R55.
37 Eggers, L.G. (2000) Diploma Work, University Hamburg.
38 Paul, J.D.H., Oehring, M., Hoppe, R., and Appel, F. (2003) *Gamma Titanium Aluminides* (eds Y.-W. Kim, H. Clemens, and A.H. Rosenberger), TMS, Warrendale, PA, p. 403.
39 Mitao, S., Tsuyama, S., and Minakawa, K. (1991) *Mater. Sci. Eng. A*, **143**, 51.
40 Appel, F., and Paul, J.D.H. (2011) On the fracture toughness of high Niobium containing TiAl alloys, *Intermetallics*. To be published.
41 Chan, K.S., and Kim, Y.-W. (1992) *Metall. Trans. A*, **23A**, 1663.
42 Kim, Y.-W. (1995) *Mater. Sci. Eng. A*, **192-193**, 519.
43 Kruzic, J.J., Campbell, J.P., McKelvey, A.L., Choe, H., and Ritchie, R.O. (1999) *Gamma Titanium Aluminides 1999* (eds Y.-W. Kim, D.M. Dimiduk, and M.H. Loretto), TMS, Warrendale, PA, p. 495.
44 Appel, F., Lorenz, U., Paul, J.D.H., and Oehring, M. (1999) *Gamma Titanium Aluminides 1999* (eds Y.-W. Kim, D.M. Dimiduk, and M.H. Loretto), TMS, Warrendale, PA, p. 381.
45 Appel, F., Oehring, M., and Paul, J.D.H. (2008) *Mater. Sci. Eng. A*, **493**, 232.
46 Chan, K.S., and Kim, Y.-W. (1995) *Acta Metall. Mater.*, **43**, 439.
47 Liu, C.T., Schneibel, J.H., Maziasz, P.J., Wright, J.L., and Easton, D.S. (1996) *Intermetallics*, **4**, 429.
48 Wang, J.N., and Xie, K. (2000) *Intermetallics*, **8**, 545.
49 Rogers, N.J., Crofts, P.D., Jones, I.P., and Bowen, P. (1995) *Mater. Sci. Eng. A*, **192-193**, 379.
50 Kim, Y.-W., and Dimiduk, D.M. (1991) *JOM*, **43**, 40.

51 Deve, H.E., Evans, A.G., and Shih, D.S. (1992) *Acta Metall. Mater.*, **40**, 1259.
52 Chan, K.S., and Davidson, D.L. (1993) *Structural Intermetallics* (eds R. Darolia, J.J. Lewandowski, C.T. Liu, P.L. Martin, D.B. Miracle, and M.V. Nathal), TMS, Warrendale, PA, p. 223.
53 Chan, K.S. (1995) *Gamma Titanium Aluminides* (eds Y.-W. Kim, R. Wagner, and M. Yamaguchi), TMS, Warrendale, PA, p. 835.
54 Wagner, R., Appel, F., Dogan, B., Ennis, P.J., Lorenz, U., Müllauer, J., Nicolai, H.P., Quadakkers, W., Singheiser, L., Smarsly, W., Vaidya, W., and Wurzwallner, K. (1995) *Gamma Titanium Aluminides* (eds Y.-W. Kim, R. Wagner, and M. Yamaguchi), TMS, Warrendale, PA, p. 387.
55 Pippan, R., Höck, M., Tesch, A., Motz, C., Beschliesser, M., and Kestler, H. (2003) *Gamma Titanium Aluminides* (eds Y.-W. Kim, H. Clemens, and A.H. Rosenberger), TMS, Warrendale, PA, p. 521.
56 Nakano, T., Kawanaka, T., Yasuda, H.Y., and Umakoshi, Y. (1995) *Mater. Sci. Eng. A*, **194**, 43.
57 Yokoshima, S., and Yamaguchi, M. (1996) *Acta Mater.*, **44**, 873.
58 Heatherly, L., George, E.P., Liu, C.T., and Yamaguchi, M. (1997) *Intermetallics*, **5**, 281.
59 Wang, P., Bhate, N., Chan, K.S., and Kumar, K.S. (2003) *Acta Mater.*, **51**, 1573.
60 Zhang, T. (1994) Diploma Work, Technical University Hamburg-Harburg.
61 Appel, F., Lorenz, U., Zhang, T., and Wagner, R. (1995) *High-Temperature Ordered Intermetallic Alloys VI, Mater. Res. Soc. Symp. Proc.*, vol. 364 (eds J.A. Horton, I. Baker, S. Hanada, R.D. Noebe, and D.S. Schwartz), MRS, Pittsburgh, PA, p. 493.
62 Bowen, P., Chave, R.A., and James, A.W. (1995) *Mater. Sci. Eng. A*, **192-193**, 443.
63 Oehring, M., Lorenz, U., Niefanger, R., Christoph, U., Appel, F., Wagner, R., Clemens, H., and Eberhardt, N. (1999) *Gamma Titanium Aluminides 1999* (eds Y.-W. Kim, D.M. Dimiduk, and M.H. Loretto), TMS, Warrendale, PA, p. 439.
64 Lorenz, U., Appel, F., and Wagner, R. (1997) *Mater. Sci. Eng. A*, **234-236**, 846.
65 Liu, T., and Kim, Y.-W. (1992) *Scr. Metall.*, **27**, 599.
66 Oh, M.H., Inui, H., Misaki, M., and Yamaguchi, M. (1993) *Acta Mater.*, **41**, 1939.
67 Chu, W.-Y., Thompson, A.W., and Wiliams, J.C. (1990) *Hydrogen Effects on Materials Behavior* (eds N.R. Moody and A.W. Thompson), TMS-AIME, Warrendale, PA, p. 543.
68 Briant, C.L., Wang, Z.F., and Chollocoop, N. (2002) *Corrosion Sci.*, **44**, 1875.
69 Appel, F., and Paul, J.D.H. (2011) The effect of internal stresses on the fracture toughness of TiAl alloys, *Mater. Sci. Eng. A*. To be published.
70 McEvily, A.J. Jr., and Bush, R.H. (1962) *Trans. ASM*, **55**, 654.
71 Larsen, D.C., Adams, J.W., Johnson, L.R., Teotia, A.P.S., and Hill, L.G. (1985) *Ceramic Materials for Advanced Heat Engines*, Noyes Publishing Co., Park Ridge, NJ, p. 249.
72 Pellini, W.S. (May 1972) Criteria for fracture control plans. NRL Report 7406.
73 Skinn, D.A., Gallagher, J.P., Berens, A.P., Huber, P.D., and Smith, J. (1994) *Damage Tolerant Design Handbook, A Compilation of Fracture and Crack Growth Data for High Strength Alloys*, CINDAS/Purdue University, Lafayette, IN.
74 Dimiduk, D.M. (1999) *Mater. Sci. Eng. A*, **263**, 281.
75 Kim, Y.-W. (1992) *Acta Metall. Mater.*, **40**, 1121.
76 Hazzledine, P.M., and Kad, B. (1995) *Mater. Sci. Eng. A*, **192**, 340.
77 Dowling, W.E., and Dolon, W.T. (1992) *Scr. Metall.*, **27**, 1663.
78 Kelly, T.J., Austin, C.M., Fink, P.J., and Schaeffer, J. (1994) *Scr. Metall.*, **30**, 1105.
79 Lee, D.S., Stucke, M.A., and Dimiduk, D.M. (1995) *Mater. Sci. Eng. A*, **192-193**, 824.
80 Thomas, M., Berteaux, O., Popoff, F., Bacos, M.-P., Morel, A., and Ji, V. (2006) *Intermetallics*, **14**, 1143.
81 Planck, S., and Rosenberger, A.H. (1999) *Gamma Titanium Aluminides 1999* (eds Y.-W. Kim, D.M. Dimiduk, and M.H. Loretto), TMS, Warrendale, PA, p. 791.

82 Draper, S.L., Lerch, B.A., Locci, I.E., Shazly, M., and Prakash, V. (2005) *Intermetallics*, **13**, 1014.
83 Wu, X., Huang, A., and Loretto, M.H. (2009) *Intermetallics*, **17**, 285.
84 Wright, P.K. (1993) *Structural Intermetallics* (eds R. Darolia, J.J. Lewandowski, C.T. Liu, P.L. Martin, D.B. Miracle, and M.V. Nathal), TMS, Warrendale, PA, p. 885.
85 Knaul, D.A., Beuth, J.L., and Milke, J.G. (1999) *Metall. Mater. Trans. A*, **30A**, 949.
86 Perrin, I.J. (1999) *Gamma Titanium Aluminides 1999* (eds Y.-W. Kim, D.M. Dimiduk, and M.H. Loretto), TMS, Warrendale, PA, p. 41.
87 Prihar, R.I. (2001) *Structural Intermetallics* (eds K.J. Hemker, D.M. Dimiduk, H. Clemens, R. Darolia, H. Inui, J.M. Larsen, V.K. Sikka, M. Thomas, and J.D. Whittenberger), TMS, Warrendale, PA, p. 819.
88 Biery, N.E., De Graef, M., and Pollock, T.M. (2001) *Mater. Sci. Eng. A*, **319-321**, 613.
89 Biery, N., De Graef, M., Beuth, J., Raban, R., Elliott, A., Austin, C., and Pollock, T.M. (2002) *Metall. Mater. Trans. A*, **33A**, 3127.
90 Weibull, W. (1951) *J. Appl. Mech.-Trans. ASME*, **18**, 293.
91 Danzer, R., Lube, T., Supancic, P., and Damani, R. (2008) *Adv. Eng. Mater.*, **10**, 275.
92 Jha, S.K., Larsen, J.M., and Rosenberger, A.H. (2005) *Acta Mater.*, **53**, 1293.
93 Soboyejo, W.O., Chen, W., Lou, J., Mercer, C., Sinha, V., and Soboyejo, A.B.O. (2002) *Int. J. Fatigue*, **24**, 69.

11
Fatigue

By far the most anticipated engineering applications of TiAl alloys involve components that are subjected to fluctuating or cyclic loading. The fluctuating stresses produce cumulative damage, which often results in failure with little or no warning. Fatigue cracks may even grow when the crack driving force is much smaller than that is needed for the same crack to grow under monotonic loading. Thus, the fatigue characteristics of γ(TiAl) alloys have an important implication on life prediction and design. By general definition, based on the relative sizes of the elastic and plastic components of the strain, two types of fatigue are usually distinguished. In high-cycle fatigue (HCF) the fluctuations in load and strain are relatively small, but the number of cycles during component life is measured in millions, as will happen if the component vibrates. In low-cycle fatigue (LCF) the fluctuations in load and strain are larger, but they occur infrequently, for example when an aircraft takes off or lands. Fatigue studies requiring more than 10^4 cycles fall in the high-cycle (HCF) classification, while those involving fewer than 10^4 cycles are grouped in the (LCF) category. The data that has been collected for TiAl alloys in these two fatigue domains will be reviewed in this chapter.

11.1
Definitions

For most fatigue tests performed on TiAl alloys the specimens were subjected to a sinusoidal stress–time pattern of constant amplitude and fixed frequency. Such tests are described by the following terms:

> maximum stress in the cycle σ_{max};
> minimum stress in the cycle σ_{min};
> mean stress $\sigma_m = (\sigma_{max} + \sigma_{min})/2$; (11.1)
> alternating stress amplitude $\sigma_a = (\sigma_{max} - \sigma_{min})/2$;
> stress range $\Delta\sigma_a = (\sigma_{max} - \sigma_{min})$;
> stress ratio $R = \sigma_{min}/\sigma_{max}$.

Gamma Titanium Aluminide Alloys: Science and Technology, First Edition. Fritz Appel, Jonathan David Heaton Paul, Michael Oehring.
© 2011 Wiley-VCH Verlag GmbH & Co. KGaA. Published 2011 by Wiley-VCH Verlag GmbH & Co. KGaA.

A second set of definitions relates to a specimen that contains a crack long enough so that linear elastic fracture mechanics applies, that is, the crack is long compared to structural features, such as colonies or grains. The cyclic mode-I stress intensity ΔK is the difference between the maximum and minimum value of the mode-I stress intensity

$$\Delta K = Y(\sigma_{max} - \sigma_{min})\sqrt{\pi a_c} \tag{11.2}$$

where a_c is the current crack length and Y is a dimensionless factor of value around one, the precise value of which depends upon the geometry of the sample. The threshold stress intensity range ΔK_{TH} is defined as the value of ΔK below which crack growth is not observable. This threshold typically occurs at crack-growth rates on the order of 10^{-10} m/cycle or less, as defined in ASTM Standard E647 [1]. Below ΔK_{TH} fatigue cracks are considered as nonpropagating cracks. Commonly, ΔK_{TH} is quoted for $R = 0$, which results in crack closure. However, a fatigue crack can be closed, while the remotely applied load is still tensile [2]. This closure effect was originally explained by the plastic crack-tip deformation that is produced in the tensile cycle. The material within the plastic zone is elastically constrained by the surrounding material. Upon unloading, these elastic constraint stresses preload the crack faces against each other. The crack driving force is thus correspondingly reduced on the following loading cycle. Fatigue crack growth is postulated to occur only after the crack is fully open. Thus, when data permits, an effective stress intensity is defined as

$$\Delta K_{eff} = Y(\sigma_{max} - \sigma_{OP})\sqrt{\pi a_c} \tag{11.3}$$

σ_{OP} is the stress at which the crack is open. Thus, the effective crack driving stress intensity ΔK_{eff} is lower than the nominal crack tip driving stress intensity ΔK. σ_{OP} is determined by crack-opening compliance measurements [3, 4].

The typical response of crack growth to the cyclic stress intensity is schematically illustrated in Figure 11.1. The plot of log (da/dN) versus log ΔK (or also versus ΔK_{eff}) often has a sigmoidal shape that can be divided into three major regions. Region I is the near-threshold region and indicates a threshold stress intensity ΔK_{TH}. Range II, on log scales, is often linear and well approximated by an expression of the form

$$\frac{da}{dN} = C_P(\Delta K)^n \tag{11.4}$$

which is widely known as the Paris law [5]. The empirical crack-growth coefficient C_P and n (the Paris exponent) are constants. For region II, stable crack propagation is assumed. It is worth noting that cyclic crack-resistance curves for TiAl alloys are usually determined for long cracks that are typically greater than about 3 mm in length. When propagating at low stresses, the behavior of long cracks may be accurately modeled by linear elastic fracture mechanics (LEFM). In region III the fatigue crack growth accelerates, eventually becoming catastrophic as the maximum stress intensity K_{max} approaches the fracture toughness K_{Ic} or because the cross-sectional area carrying the load becomes reduced to a point so that it can no longer support the load.

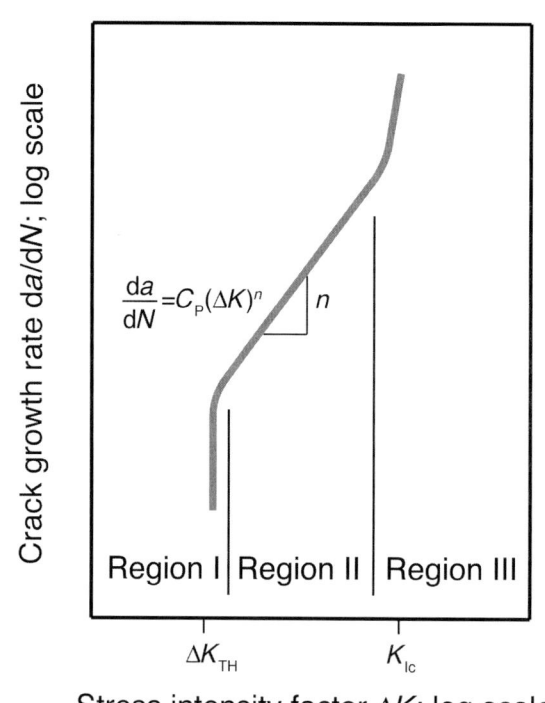

Figure 11.1 Schematic representation of fatigue-crack-growth characteristics.

The plot shows crack growth rate da/dN (log scale) versus stress intensity factor ΔK (log scale), with three regions: Region I (below ΔK_{TH}), Region II (the Paris regime where $\frac{da}{dN} = C_p(\Delta K)^n$), and Region III (approaching K_{Ic}).

An important consideration in axial fatigue is uniformity of stress and strain in the specimen gage section. A major source of the nonuniformity of gage section stress and strain is a bending moment resulting from specimen misalignment. This can arise from eccentricity in the load train, improper specimen gripping or lateral movement of the load-train components during testing because of their inadequate stiffness. For brittle materials like TiAl alloys, this misalignment can significantly shorten the fatigue life [6]. Minor differences in testing can produce significant differences in behavior. Thus, although large precautions are usually taken, scatter of the results due to testing setup are almost unavoidable.

11.2
The Stress–Life (S–N) Behavior

Basic fatigue data in the HCF range are usually displayed on a plot of cyclic stress level versus the logarithmic life, frequently called S–N curves. Conventional titanium alloys exhibit a steep reduction in the relatively short life range, leveling off to approach a stress asymptote at longer lives. This stress amplitude is called the fatigue or endurance limit; it indicates that below this stress level the material should sustain an infinite number of cycles without failure. This behavior greatly contrasts with what has been observed for TiAl-based alloys [7–14]. Although some

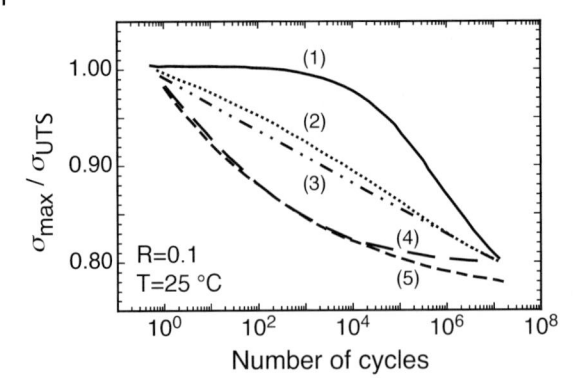

(1) Ti-47Al-2Nb-2Mn-0.8B (XD™-processed)
(2) Ti-46Al-2Cr-2.7Nb-0.2W-0.15Si-0.2C-0.1B
(3) Ti-46.5Al-2Cr-3Nb-0.2W (K5), refined fully lamellar
(4) Ti-46.5Al-2Cr-3Nb-0.2W (K5), lamellar
(5) Ti-46.5Al-2Cr-3Nb-0.2W (K5), duplex

Figure 11.2 Influence of the microstructure on the room temperature stress–life (S–N) behavior. Measurements performed in air. The fatigue strength σ_{max} was normalized by ultimate tensile strength σ_{UTS}. Data from Larsen et al. [12].

investigations revealed a decay of the fatigue strength in the short-life range of TiAl alloys, the S–N curves don't seem to approach a stress asymptote at longer lives, but seem to drop off indefinitely. For such materials there is no fatigue limit, and failure as a result of cyclic load is only a matter of applying enough cycles. Furthermore, in the long-life range, the S–N curve is relatively flat. Thus, testpiece lifetimes can vary by several orders of magnitude within a very narrow range of σ_{max}. Figure 11.2 demonstrates room-temperature data for fatigue life that was collected by Larson [12] for different alloys and microstructures. The fatigue strength σ_{max} was determined at 10^6 cycles and normalized by ultimate tensile strength σ_{UTS}. The significant effect of microstructure becomes evident in the shape of the S–N curves. From the diagram it would appear that σ_{max} is always a high fraction of σ_{UTS}. However, this could in part be a strain-rate effect because σ_{UTS} was determined at a relatively low strain rate of $\dot{\varepsilon} = 10^{-4}\,\text{s}^{-1}$ or so. Nevertheless, the high ratio of fatigue strength to ultimate strength compares favorably with that of other high-temperature materials. The S–N curve for the XD alloy is remarkable because it is nearly constant up to around 10^3 cycles and the fatigue strength is close to the ultimate strength of the material. It might be expected that the material in this region experienced general yielding and gross plastic deformation. In most cases the fracture initiation sites were observed to occur at the specimen surface. Thus, fatigue life is very sensitive to surface finish of the specimens [8]. Much of the controversy that has developed in the literature may be related to the methods used for sample preparation. More on this subject is provided in Chapter 18.

Summary: The fatigue life of TiAl alloys at room temperature is characterized by flat S–N curves. Thus, testpiece lives can vary by several orders of magnitude within a very narrow range of maximum tensile stress in the cycle.

11.3 HCF

Most fatigue data for TiAl alloys has been determined by tests performed at a stress ratio of $R = 0.1$ and a frequency of typically 20 Hz. The test temperature has been varied over the range between room temperature and 900 °C.

11.3.1 Fatigue Crack Growth

At low and intermediate temperatures, fatigue crack growth in γ(TiAl)-based alloys is generally very rapid. Figure 11.3 exemplifies the trends that have been observed by many authors [4, 9, 15–19]. The crack-growth rates are extremely sensitive to the applied stress intensity. This translates into large exponents of the Paris law

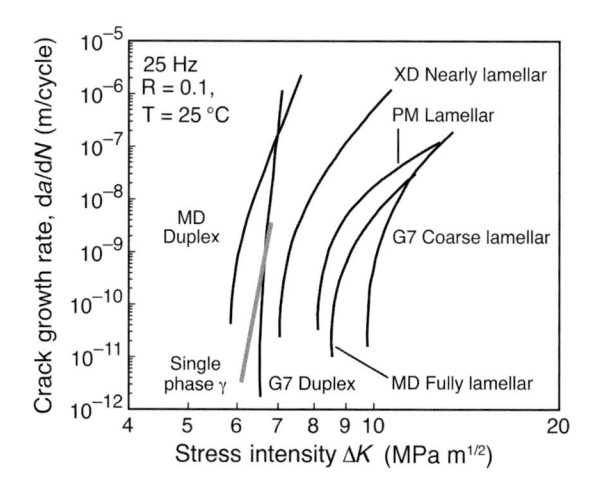

XD: Ti-47.7Al-2Nb-0.8Mn-1 Vol.% TiB_2
G7: Ti-47.3Al-2.3Nb-1.5Cr-0.4V
MD: Ti-47Al-2Nb-2Cr-0.2B
PM: Ti-47Al-2Nb-2Cr, powder metallurgy

Figure 11.3 Fatigue-crack-growth rate data determined for different TiAl alloys that are specified in the legend. In general, lamellar microstructures show superior fatigue-crack-growth resistance compared to equiaxed γ and duplex structures. Fatigue experiments performed with $R = 0.1$, 25 Hz in air at room temperature. Data from Kruzig et al. [19], redrawn.

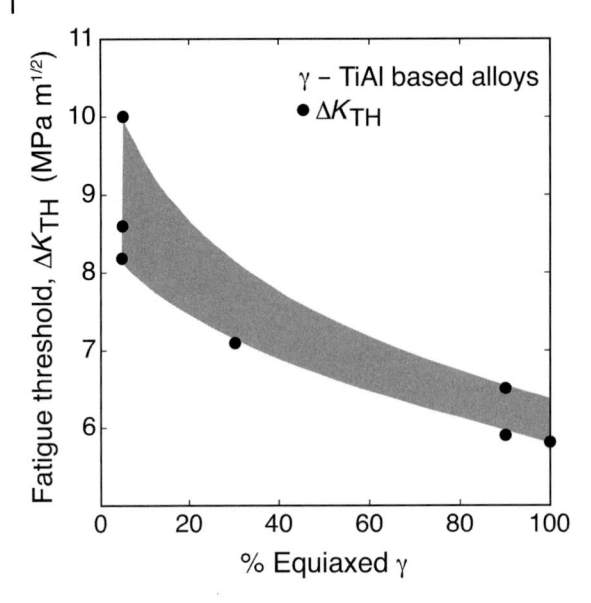

Figure 11.4 Dependence of the fatigue-crack-growth threshold ΔK_{TH} on the volume fraction of equiaxed γ grains in several alloys based on γ(TiAl). Data from Kruzic et al. [19], redrawn.

(11.4), which are 5 to 10 times higher than typical values of metallic systems. For example, the crack-growth rates for near-γ and duplex alloys span nearly six orders of magnitude for less than a 1 MPa m$^{1/2}$ change in applied stress intensity. The resulting Paris stress exponent of about $n = 50$ is similar to those measured for ceramics [17, 18]. In comparison, the Paris exponent for conventional titanium alloys is about $n = 3$.

There is a marked influence of the microstructure on the cyclic crack-growth rates, which in broad terms is reminiscent of the effects observed under monotonic loading (Section 10.4.2). Figure 11.4 shows the variation of the fatigue-crack-growth threshold ΔK_{TH} on the volume fraction of equiaxed γ grains for several alloys based on γ(TiAl) [19]. Metallographic examination of fatigue crack-surface morphologies have shown that duplex alloys predominantly fail by transgranular cleavage of γ grains, which offer low resistance to crack growth [20]. A lamellar microstructure obviously provides a substantial increase in fatigue-growth resistance, which is manifested by lower Paris exponents of $n = 10$ to 30 and significantly higher threshold intensities ΔK_{TH}. Akin to monotonic loading, in lamellar alloys the fatigue crack path depends on the colony orientation [21]. Translamellar crack advance occurs, where the crack surface is nominally perpendicular to the lamellar packets. A combination of interlamellar and intralamellar crack advance was recognized where the crack plane is parallel to the lamellar interfaces. These characteristics probably cause a significant anisotropy in the fatigue-crack-growth resistance, which was recognized by Gnanamoorthy et al. [22] on lamellar Ti-46Al with a preferred lamellae orientation. Multiple cracking ahead of the main crack was recognized; however, unlike monotonic loading, the formation of crack-bridging shear ligaments was relatively rare in fatigue. This may in part be due to the inherently small crack-opening displacement under HCF, which restricts the extent of the bridging zones to lie within 300 µm behind the crack tip, compared to a few millimeters under monotonic loading. Moreover, the bridges appear to fail relatively early at the terminus of the delaminations; thus, ligament toughening is expected to be less effective under fatigue [4, 20, 23, 24]. In any event, although reasonable fatigue-crack-growth resistance can perhaps be achieved for optimized lamellar alloys, component design, which utilizes materials with a steep crack-growth resistance curves, that is, $n \geq 10$ is difficult, unless the applied stress intensity in the component is known accurately.

Summary: At room temperature the fatigue-crack-growth rates in TiAl alloys are extremely sensitive to the cyclic stress intensity. This implies that the fraction of the total fatigue life resulting from crack propagation is very small. Lamellar alloys exhibits superior fatigue-crack-growth resistance compared to their duplex counterparts, primarily due to the beneficial shielding effects from crack deflection and microcracking ahead of the tip of the main crack tip.

11.3.2
Crack-Closure Effects

As was briefly described in Section 11.1, crack closure can significantly affect the stress intensity ΔK driving fatigue crack growth. Since Elber's early work [2], several other crack-closure mechanisms have been identified, for a review see [25]. For TiAl alloys the following mechanisms might be relevant:

i) plasticity-induced closure due to a wake of residual deformation, as already described in Section 11.1;
ii) roughness-induced closure due to fracture surface mismatch;
iii) transformation-induced closure due to volume increasing phase changes;
iv) crack-filling closure due to interfacial intrusion of fretting debris or corrosion products.

Hénaff et al. [3] have demonstrated the significance of crack closure in Ti-48Al-2Cr-2Nb with a near-γ microstructure. The authors determined the crack-growth rates for various stress ratios R at room temperature and used a compliance technique to monitor crack closure and crack opening. Figure 11.5 shows the effect of the data correction according to Equation 11.3. ΔK_{eff} correlates crack-growth behavior reasonably well, as the fatigue-crack-growth curves for the different stress ratios collapse into a single narrow scatter band. The closure correction generally shifts the crack-growth curves nearly parallel to smaller stress intensities with the result that smaller threshold intensities are determined [3, 14, 26]. Worth et al. [20] obtained similar results on duplex and lamellar forms of a Ti-46.5Al-3Nb-2Cr-0.2W alloy.

Crack-closure corrections also tend to reduce the differences in fatigue-crack-growth resistance that have been observed for the different microstructures [14, 27–30] and described in the previous section. This finding can probably be rationalized by different contributions of roughness-induced closure, categorized above under (ii). In lamellar alloys the fatigue crack path is very tortuous with highly distorted material along the fracture surface. For larger colony sizes, more serrated

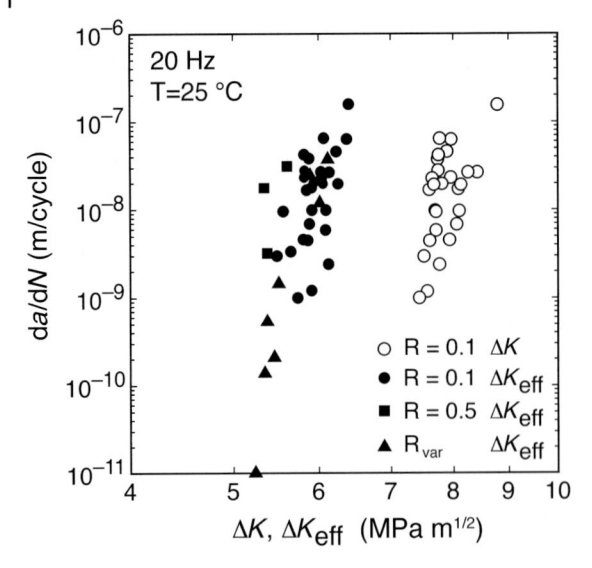

Figure 11.5 Crack-growth rates da/dN plotted as a function of the stress intensities ΔK and as function of the stress intensity ΔK_{eff} corrected for crack closure. Ti-48Al-2Cr-2Nb with an equiaxed near-γ microstructure; fatigue experiments performed in air with the R values specified in the legend; R_{var} corresponds to $R = 0.4$ to 0.6. Data from [3], redrawn.

and faceted surface morphologies are expected to occur, which lead to higher opening loads. Conversely, in equiaxed and duplex materials the crack surface is very flat, thus only limited closure effects may arise from crack-tip plasticity. It should be noted, however, that marked differences in crack-growth resistance between the microstructures remained, even if the stress intensity was corrected for crack closure. Thus, crack-closure corrections are most important for explaining the dependence of crack-growth rate on stress ratio. It might be expected that the various forms of closure can operate synergistically. For example, oxide-induced closure requires a fretting action. Therefore, closure must already exist in another form, such as plasticity-induced or roughness-induced closure. The relative contribution of the individual mechanisms may depend on microstructure, constitution and temperature. This may explain the observation that crack closure was sometimes found to be of minor importance [26]. However, there is good consensus that little or no closure effects occur in TiAl alloys when the stress ratio is $R = 0.45$ or higher [30, 31].

Summary: The stress intensity driving fatigue cracks in γ(TiAl)-based alloys can be reduced by crack-closure phenomena. For room-temperature testing, the closure effects occur at low stress ratios $R \leq 0.45$. The rough fracture surfaces of lamellar alloys also support premature closure of the crack wakes. Materials with very small grain size showed no crack closure even at low stress ratios. Crack-closure corrections shift the crack-growth curves to smaller effective stress intensities with the result that smaller threshold intensities are defined.

11.3.3
Fatigue at the Threshold Stress Intensity

Akin to brittle ceramics and other intermetallic compounds, fatigue-crack-growth rates in γ(TiAl) alloys are extremely sensitive to the stress intensity (Section 11.3.1). This implies that that the fraction of the total fatigue life resulting from crack propagation in the Paris regime is very small and not suitable for life management of components. Campbell *et al.* [32] have suggested that under such conditions, use of threshold stress intensity ΔK_{TH} (see Figure 11.1) in design could be more appropriate. Such a damage-tolerant design approach must assure that the maximum cyclic stress intensity in a component remains below ΔK_{TH}, in order to prevent rapid propagation to failure of cracks initiated at pre-existing flaws. It is worth remembering that ΔK_{TH} is the threshold for growth of long cracks (Section 11.1). However, at this low stress-intensity level, fatigue life could be significantly affected by nucleation and growth of short cracks. This is a major complication for the use of ΔK_{TH} in life prediction. Based on the general classification in [1], several types of short cracks have been distinguished. For TiAl alloys the so-called "microstructurally" short cracks are important, the length of which is comparable to the scale of the microstructure. Short cracks may propagate significantly faster than linear elastic fracture mechanics (LEFM) would predict, either because stress levels are too high and small-scale yielding conditions are exceeded, or because microstructural features seriously affect crack-growth behavior [33, 34]. Moreover, small cracks may grow at ΔK values that are below the threshold stress intensity ΔK_{TH} that was determined from the growth rate of long cracks (Section 11.1). This effectively shifts region I of the sigmoidal fatigue-crack-growth curve shown in Figure 11.1 to smaller stress intensities. Thus, life predictions based on the sigmoidal crack-growth curve, when extended to small crack behavior, may lead to nonconservative estimates. In view of this physical and technical importance, several authors have investigated the cyclic growth behavior of small cracks in TiAl alloys. In most cases, small surface cracks with a typical length of $a_S = 25\,\mu m$ to $500\,\mu m$ were introduced by electrodischarge machining and subject to different cyclic-stress intensities in order to monitor their growth rate. The results obtained in these studies on different two-phase $(\alpha_2 + \gamma)$ alloys can be summarized as follows.

i) At room temperature and the same applied ΔK levels, small surface cracks grow faster than long through-thickness cracks [23, 32], sometimes by several orders of magnitude [19]. Figure 11.6 compares the different growth rates of short and long cracks, which have been observed at room temperature in lamellar and duplex Ti-47Al-2Nb-2Cr-0.2B [32]. The anomalous behavior of short cracks might be caused by the lack of premature contact between the crack faces behind the crack tip as it advances, the opposite of that typically observed at low stress intensities for long-crack behavior. In broad terms, the growth behavior of short cracks is similar to that of long cracks at higher stress ratios.

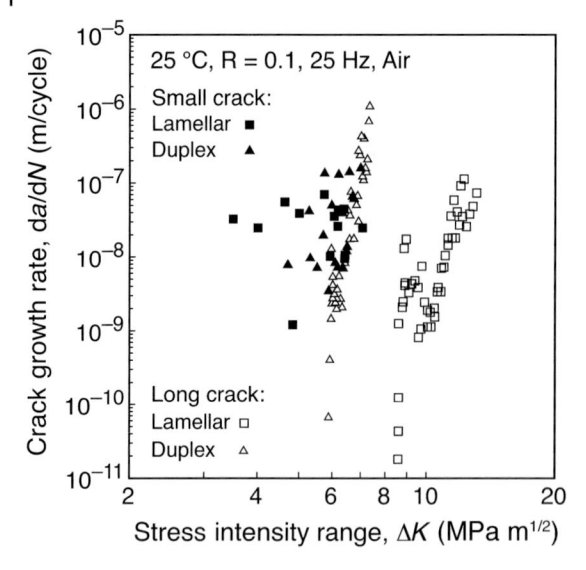

Figure 11.6 Fatigue-crack-growth rates for through-thickness long cracks (crack length $a_c > 5\,\text{mm}$) and small surface cracks (crack length $a_s = 35\,\mu\text{m}$ to $275\,\mu\text{m}$) determined in lamellar and duplex structures of Ti-47Al-2Nb-2Cr-0.2B. Data from Campbell et al. [32], redrawn.

ii) Short cracks were found to propagate at applied ΔK levels below the long-crack threshold intensity ΔK_{TH} [23, 24, 32]. This observation (together with the comments made under [i]) may be rationalized on the grounds that short cracks are cleavage cracks that propagate within single grains or colonies. This intragranular cleavage fracture occurs on low-index crystallographic planes and is characterized by a very low toughness (Section 10.2).

iii) The short-crack behavior is more pronounced in coarse-grained lamellar alloys than in fine-grained duplex alloys [23, 24].

iv) The definition of a threshold intensity below which short fatigue cracks will not propagate is more subtle than for long cracks and must be specified for the alloy microstructure. For duplex alloys, the crack-growth resistance curve determined for long cracks could be of guidance, if the data is corrected with regard to crack closure and crack-tip shielding to define an effective threshold stress intensity $(\Delta K_{\text{TH}})_{\text{eff}}$ [35, 36]. Kruzic et al. [19] have demonstrated that small surface cracks ($a_s > 25\,\mu\text{m}$) did not grow below this stress intensity level $(\Delta K_{\text{TH}})_{\text{eff}}$ in duplex Ti-47Al-2Nb-2Cr-0.2B. This behavior of duplex alloys could be a potential advantage for damage-tolerant design.

v) There seems to be no growth threshold for small cracks in lamellar alloys [4, 19, 23, 24]. Short surface cracks seem to grow whenever their length is less than the diameter of the lamellar colonies; the cleavage anisotropy of the α_2 and γ phases makes crack propagation parallel to the interfaces easy (see

Section 10.4.2). It might be expected that these cracks could be immobilized at colony boundaries for some time. This increases the likelihood that colony boundaries can serve as potential sites for crack nucleation. Further cycling provides enough damage near the crack tip for the crack to eventually grow into an adjacent colony. This causes the crack-growth rate to increase again. In this respect a smaller colony size could increase the growth resistance of small cracks. Once a short crack grows to a size that is several times larger than the colony size, colony boundaries probably do not provide sufficient constraint as they did when the crack was smaller; eventually a larger process zone involving several colonies is formed. Thus, the crack-resistance curve describing the growth of small cracks merges with that of long cracks.

vi) The absence of a definable threshold stress intensity below which small cracks will not propagate in lamellar alloys could be a concern in alloy design. While a large colony size may provide relatively good resistance against the growth of large cracks, in refined lamellar structures the effect could be outweighed by the reduced microcrack unit sizes that are relevant for the propagation of small cracks, as commented by Worth *et al.* [4].

Summary: For duplex alloys an effective threshold stress intensity can be defined below which neither short nor long cracks do grow. In lamellar TiAl alloys, short fatigue cracks grow at stress intensities well below the threshold intensity ΔK_{TH} determined for long cracks. Thus, life predictions based on the sigmoidal crack-growth curve (determined for long cracks) when extended to small crack behavior may lead to nonconservative estimates.

11.4
Effects of Temperature and Environment on the Cyclic Crack-Growth Resistance

At first glance, the fatigue crack-growth resistance of TiAl alloys appears to be relatively insensitive to test temperature in the temperature range 25 °C to 800 °C [9, 21, 29, 37–39]. Figure 11.7 shows the crack-growth rate at a constant stress intensity of $\Delta K = 10$ MPa m$^{-1/2}$ as a function of test temperature. The data was determined on a Ti-46.5Al-3Nb-2Cr-0.2W alloy with a refined lamellar microstructure [26]. The crack-growth rate gradually increases between 200 °C and 600 °C, goes through a maximum at 650 °C and then steeply declines at higher temperatures. At 800 °C, the growth rate was immeasurably small, i.e. it was lower than 10^{-10} m/cycle. Balsone *et al.* [9] have determined the Paris-law constants (see Equation 11.4) and threshold stress intensities on nominally the same alloy; this data is listed in Table 11.1. It should be noted that the crack-growth characteristic at elevated temperatures depends on frequency. This becomes evident when the data listed in Table 11.1 is compared with that determined on the same alloy utilizing a frequency of 20 Hz [40].

The fracture surfaces of samples tested in air at 800 °C or above were often too oxidized to examine details. With this reservation in mind, metallographic

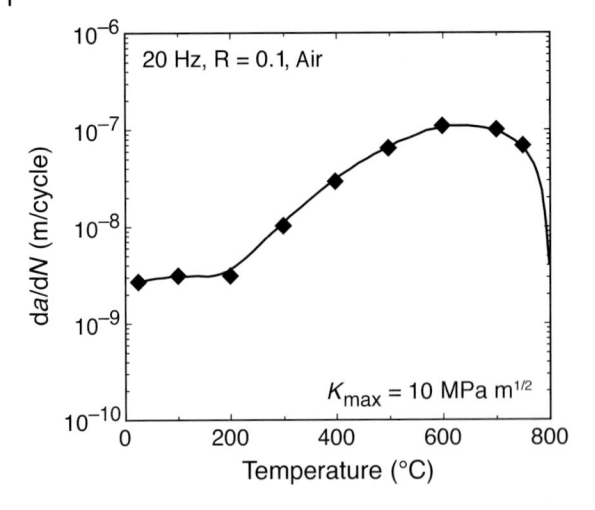

Figure 11.7 Dependence of the crack-growth rate on test temperature T at constant stress intensity $K_{max} = 10$ MPa m$^{1/2}$. Ti-46.5Al-3Nb-2Cr-0.2W (K5) with refined lamellar microstructure. Data from Rosenberger et al. [26], redrawn.

Table 11.1 Paris-law constants and threshold intensities determined for lamellar and duplex forms of a Ti-46.5Al-3Nb-2Cr-0.2W alloy.

Microstructure	T (°C)	$C_P \times 10^{19}$	n	ΔK_{TH} (MPa m$^{1/2}$)
Lamellar	23	2.89	10.5	7.8
	600	1.34	3.23	4.5
	800	1.94	3.60	9.7
Duplex	23	4.47	27.8	6.0
	600	1.62	5.96	2.5
	800	4.15	7.65	7.1

Values measured in air with a stress ratio $R = 0.1$ and a frequency of 1 Hz. C_P (crack-growth coefficient) and n (Paris exponent according to Equation 11.4); ΔK_{TH} (threshold stress intensity). Data from [9].

observations may be summarized as follows. Brittle micromechanisms of failure seem to predominate in fatigue of precracked testpieces to a temperature of 800 °C. Metallographic examination revealed mixed transgranular and intergranular features; the relative contributions of which seem to depend on the detailed experimental conditions, such as stress amplitude and frequency [7]. Similar to room-temperature fatigue, the presence of γ grains degrades the crack-growth resistance of duplex alloys, which is manifested by higher Paris exponents and lower threshold intensities, when compared with lamellar structures [21, 9].

Lamellar alloys tested at 800 °C under ultrahigh-vacuum conditions exhibited evidence of ductile intralamellar and translamellar fracture. There is good experi-

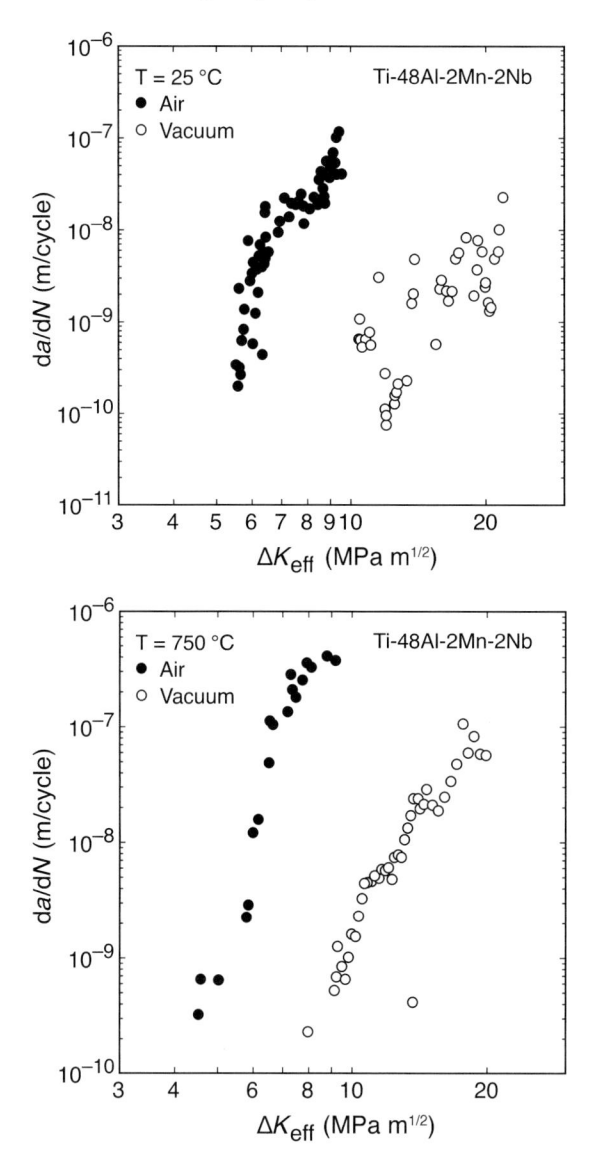

11.4 Effects of Temperature and Environment on the Cyclic Crack-Growth Resistance | 415

Figure 11.8 Influence of environment on fatigue crack growth in fully lamellar Ti-48Al-2Mn-2Nb at room temperature and 750 °C. Data from Hénaff et al. [43], redrawn.

mental evidence for a pronounced environmental influence on the growth rate of fatigue cracks [25, 41–43]. Crack-growth rates are typically faster in air by at least one order of magnitude, when compared with the growth rates determined in vacuum; Figure 11.8 shows this to be the case for lamellar Ti-48Al-2Mn-2Nb [43]. However, the embrittlement is not clearly reflected in the morphology of the fatigue fracture surface: the observed differences do not account for the large environmental influence on the crack-growth behavior [26]. Thus, the precise mechanism of environmental attack is as yet unclear. Several authors have attributed the environmental embrittlement to hydrogen released from water vapor in air, similar to the environmental embrittlement displayed by iron aluminides [41, 43]. In any case, the embrittlement is superimposed with the mechanisms that control intrinsic fatigue-crack-growth resistance. As these processes may change with temperature, analysis of high-temperature fatigue is very complex for TiAl alloys.

A striking feature of the high-temperature fatigue of TiAl alloys is the anomalous temperature dependence of the crack-growth characteristics, which has been observed between 25 °C and 800 °C in air environments [9, 29, 37, 38, 44, 45, 46]. Figure 11.9 demonstrates this effect on crack-growth resistance curves that were determined at 25 °C, 600 °C and 800 °C in air. [45]. Clearly, the threshold intensities are lowest at 600 °C and highest at 800 °C, with 25 °C data lying in-between. Crack-growth rates at 600 °C are at least two orders of magnitude higher than those at room temperature, depending on stress ratio. Balsone et al. [37] have found the highest crack-growth rates to be at 425 °C, that is, at an even lower temperature. Thus, fatigue within the intermediate temperature range of 400 °C to 650 °C seems to be the worst situation for two-phase alloys. Since these anomalous crack-growth characteristics are less prominent for fatigue tests performed in vacuum, Larsen et al. [29] have proposed that resistance to fatigue crack growth normally increases with temperature, and that the anomalous high growth rates observed at intermediate temperatures are caused by (unspecified) environmental embrittlement. On further increasing temperature, this effect is outweighed by a greater contribution of thermally assisted plasticity. It is worth noting that samples tested at 425 °C in air are not oxidized and those tested at 600 °C usually have only a light surface oxide.

McKelvey et al. [45] have proposed that the crack growth rates, rather than being accelerated at intermediate temperatures (400 °C to 650 °C), are retarded at 800 °C. This increase of the fatigue-growth resistance at 800 °C was ascribed to oxide-induced crack closure, which retards crack growth at near-threshold stress intensities ΔK_{TH}. Indeed, the oxide scale produced at 800 °C on duplex Ti-47.4Al-1.9Nb -0.9 Mn + 1 vol.% TiB_2 (XD) approached micrometer dimensions. The observed dependence of the crack-growth curves on stress ratio R (Figure 11.9) seems to support this view, since crack-closure effects are expected to be sensitive to R (Section 11.3.3). Indeed, after correction for crack closure, the crack-growth rates determined for 600 °C and 800 °C tend to overlap. However, the threshold intensity defined by closure-corrected data for 600 °C is lower than that for 25 °C. Furthermore, the argument that the crack-growth rate depends on R is not convincing because the investigation was performed on a duplex alloy containing about 70% lamellar colonies. The lamellar colonies could give rise to a marked closure effect due to crack-wake roughness, which is difficult to separate from that produced by the oxide. In view of this unclear situation it is suggested that another mechanism based on dislocation pinning may contribute to the anomalously fast crack propagation observed between 400 °C and 650 °C. In this temperature range dynamic

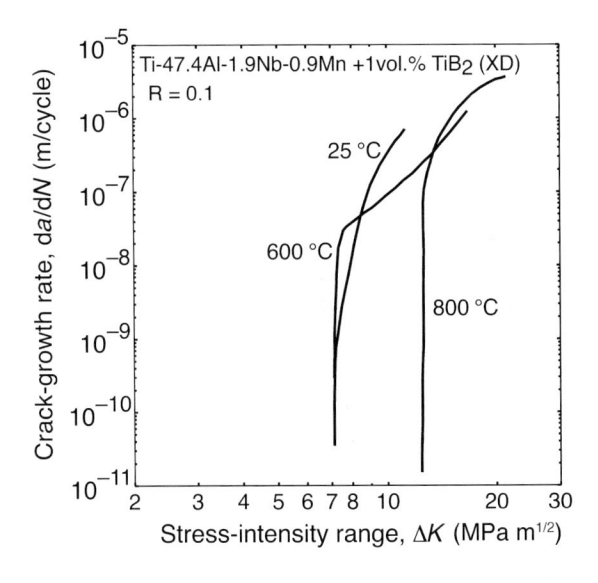

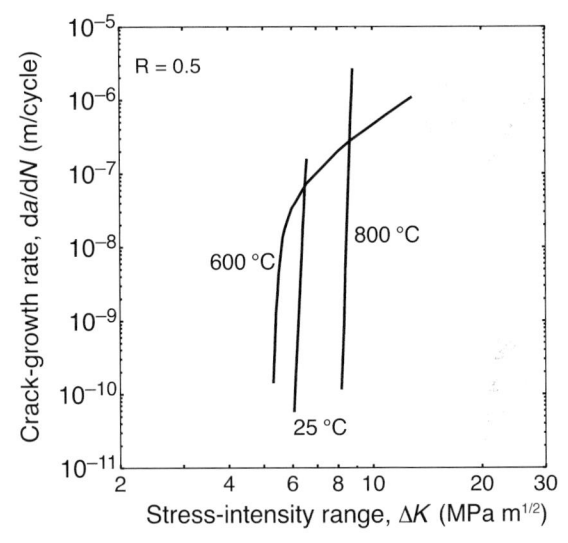

Figure 11.9 Crack-growth-resistance curves da/dN versus ΔK obtained for duplex (XD processed) Ti-47.7Al-1.9Nb-0.9Mn + 1 vol.% TiB$_2$ in air for different stress ratios $R = 0.1$ and $R = 0.5$. Note that the fatigue thresholds are lowest at 600 °C and highest at 800 °C, with 25 °C data in between. Data from [45], redrawn.

strain aging may occur due to the formation of ordered defect atmospheres around the dislocations. The species that reorient in the dislocation stress field are elastic dipoles that are composed of a Ti$_{Al}$ antisite defect and a vacancy (see Section 6.4.3). The defect atmospheres may effectively pin dislocations moving within the plastic zone of fatigue cracks. Utilizing static strain aging experiments, the optimum temperature for this pinning process was determined to be about 250 °C, which is significantly lower than the temperature where the highest fatigue-crack-growth rates occur. However, dynamic strain aging is known to depend on the strain rate because it is determined by both diffusion and dislocation velocity [47]. Under the fatigue conditions considered here with frequencies of 10 Hz to 25 Hz, the nominal strain rate is significantly higher (by a factor of 1000, or so) than that utilized in the static strain-aging experiments. This strain-rate difference may shift the optimum conditions for dislocation pinning under fatigue to higher temperatures. On further increasing the testing temperature (800 °C) the pinning process disappears because the atmospheres are dissolved. It should be noted that this pinning process is expected to occur for all two-phase alloys and is almost independent of microstructure.

Summary: The fatigue-crack-growth resistance of γ(TiAl)-based alloys appears to be relatively unaffected by test temperature in the temperature range 25 °C to 800 °C. This apparent insensitivity to temperature is probably a result of several mechanisms, which operate simultaneously. A marked environmental embrittlement that occurs over a wide temperature interval is superimposed on the intrinsic factors that govern the growth of fatigue cracks. At intermediate temperatures around 500 °C, the fatigue performance could be compromised by dislocation locking due to defect atmospheres.

11.5
LCF

11.5.1
General Considerations

Many components in high-temperature technologies are subject to occasional large mechanical or thermal transients during operation. Furthermore, local plastic deformation may occur at stress concentrations, even if the loads on a component are nominally low. These locally deformed regions are cyclically strain controlled because of the constraints imposed by the surrounding bulk of elastic material. Under these conditions, commonly referred to as LCF or strain-controlled fatigue, the total design lifetime may involve only a few hundred or a few thousand of these large strain cycles. Turbine engine blades or discs are prime examples of components subject to strain-controlled fatigue. Local areas of these components can experience high stresses and strains due to external load transfer, abrupt changes in geometry, temperature gradients, and material imperfections.

For such loading conditions, TiAl alloys should be considered with particular caution because the material exhibits little plasticity and intrinsic brittleness due to the atomic ordering and directional bonding. As described in Section 5.1, the perfect dislocations present in the majority γ(TiAl) phase have the Burgers vectors $b_{<110]} = 1/2<110]$, $b_{<011]} = <011]$ and $b_{<112]} = 1/2<11\bar{2}]$ and preferentially glide on

$\{111\}_\gamma$ planes. The superdislocations with the Burgers vectors $\mathbf{b}_{<011]}$ and $\mathbf{b}_{<112]}$ exhibit asymmetric core spreading, which makes their glide resistance sensitive to the direction of motion. Furthermore, all these dislocations are subject to high friction forces, which result in low mobility. The other significant deformation mode in γ(TiAl) is mechanical twinning along $1/6<11\bar{2}]\{111\}$. The mechanism provides auxiliary deformation modes with shear components in the c-direction of the $L1_0$ unit cell of γ(TiAl). However, unlike twinning in f.c.c. metals, there is only one twinning shear direction available per $\{111\}_\gamma$ plane. Since twinning shear is unidirectional, grains with a distinct range of orientations cannot mechanically twin. Given the anisotropy of glide and twinning, there are certain grain orientations in polycrystalline TiAl for which deformation is difficult, that is, the plastic response depends on the direction and sense of the load. For these reasons, the LCF performance of TiAl alloys is expected to be strongly associated with the micromechanisms of plasticity.

Summary: LCF is a progressive failure phenomenon brought about by cyclic strains that extend into the plastic range. The capability of TiAl alloys to sustain such loading conditions is inherently limited by the plastic anisotropy at the dislocation level and the lack of independent slip and twinning systems that can operate under reversed plastic straining.

11.5.2
Cyclic-Stress Response

There have been relatively few studies on LCF of polycrystalline TiAl alloys. Most investigations were performed on alloys with the base-line composition Ti-(46-49) Al and containing several ternary or quaternary elements. These alloys typically have a room-temperature yield stress of $\sigma_0 = 400$ MPa to 500 MPa, and the brittle–ductile transition temperature is around 650 °C to 700 °C. Because of the ductility problems described in the previous section, most investigations were performed at low strain amplitudes of a few tenths of a per cent and low strain rates of $10^{-4}\,\mathrm{s}^{-1}$, or so. The LCF performance of these alloys has been reviewed in [14]. It is only recently that LCF tests were performed on high-strength alloys [48]. Figure 11.10 illustrates the inherent difficulty of such tests for a high-strength alloy with the composition Ti-45Al-8Nb-0.2C (TNB-V2). The diagram shows the stress–strain curves of 33 room-temperature tensile tests together with the cyclic-strain amplitude used in the LCF tests. The total cyclic-strain amplitude $\Delta\varepsilon_t/2 = 0.7\%$ is well beyond yielding and only slightly smaller than the lowest tensile strain of $\varepsilon_f = 0.8\%$ measured under monotonic loading. Thus, during LCF significant plasticity and damage will be accumulated, which may easily lead to failure. At room temperature this typically occurs after several hundred to a few thousand cycles. As might be expected, the higher the cyclic strain amplitude, the shorter the fatigue life. Generally, long fatigue life is observed if the cyclic saturation stress is below the ultimate tensile strength. [49–53].

Figure 11.11a shows typical stress–strain hysteresis loops observed at $N = 300$ cycles on nearly lamellar Ti-45Al-8Nb-0.2C (TNB-V2) for three test temperatures

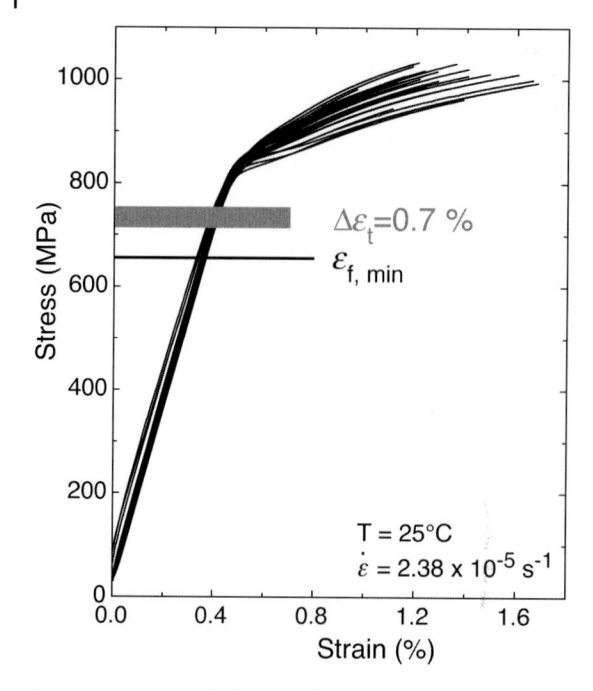

Figure 11.10 Low-cycle fatigue of TiAl alloys. Stress–strain curves for 33 room-temperature tensile tests. The total cyclic strain amplitude $\Delta\varepsilon_t/2 = 0.7\%$ indicated in the diagram corresponds to a plastic tensile strain of about $\varepsilon_{pl} = 0.25\%$. Extruded Ti-45Al-8Nb-0.2C (TNB-V2) with nearly lamellar microstructure.

[48]. The plastic strain amplitudes and the number of cycles to failure are listed in Table 11.2. The increase of plastic strain amplitude with temperature is readily visible and, at first sight, appears to be detrimental because, as mentioned above, the isothermal fatigue life is generally reduced when the strain amplitude increases. When compared with the LCF data at 25 °C and 550 °C, the higher material ductility at 850 °C probably outweighs the damage accumulation resulting from the large strain amplitude developed at this temperature. The transition from brittle to ductile deformation behavior occurs at about 750 °C for the alloy. Figure 11.11b shows the cyclic stress response curves, constructed from the tensile portion of the hysteresis loops, as a function of the number of applied cycles. Accordingly, the material exhibits cyclic hardening at room temperature, almost saturated stress response at 550 °C and cyclic softening at 850 °C. The observation essentially coincides with the data reported in the literature for other α_2(Ti$_3$Al) + γ(TiAl) alloys [14, 52–56]. Several authors [6, 54, 55, 57] have reported that the Coffin–Manson relationship [57, 58] is obeyed, according to which a linear relation exists between the logarithm of the plastic strain amplitude $\Delta\varepsilon_p$ and the logarithm of the number of cycles to failure.

Research performed by Recina and Karlsson [6, 51, 59] has shown that duplex alloys at intermediate temperatures (600 °C) have a longer fatigue life than their lamellar counterparts. This is reasonable because duplex alloys are more ductile

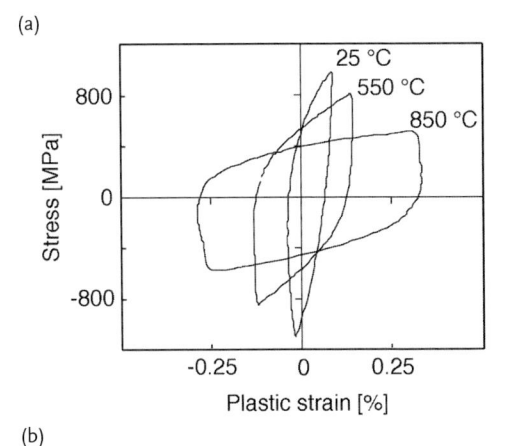

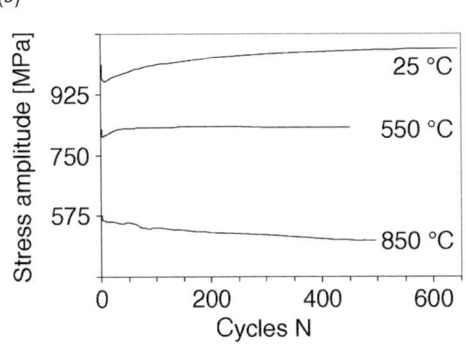

Figure 11.11 LCF characteristics observed on nearly lamellar Ti-45Al-8Nb-0.2C (TNB-V2) alloy at $R = -1$, $\Delta\varepsilon_t/2 = \pm 0.7\%$ at the temperatures indicated. (a) Stress–strain hysteresis loops observed at $N = 300$ cycles. (b) Cyclic stress response to specimen failure. Data from Appel et al. [48].

Table 11.2 Summary of LCF data measured for the Ti-45Al-8Nb-0.2C (TNB-V2) alloy with a nearly lamellar microstructure alloy at $R = -1$ and $\Delta\varepsilon_t/2 = \pm 0.7\%$.

T (°C)	$\Delta\varepsilon_p/2$ (%)	N_f
25	0.05	641
550	0.13	452
850	0.3	501

T (temperature), $\Delta\varepsilon_p/2$ (plastic strain amplitude at 300 cycles), N_f (number of cycles to failure). Data from [48].

under monotonic loading. Park et al. [56] observed an increase of the LCF life for lamellar Ti-46.6Al-1.4Mn-2Mo-0.3C with the carbon addition, as compared to the carbon-free counterpart of the alloy. The authors attributed the effect to a refinement of the lamellar spacing. Fractographs of samples subjected to LCF show all the features that have been recognized after HCF testing; thus all the features described earlier for HCF are at least qualitatively applicable to LCF as well. Likewise, there is a marked and complex influence of the test atmosphere [60].

Summary: The life of TiAl testpieces under LCF with plastic cyclic strain amplitudes of a few tenths of a per cent is limited to a several hundred cycles and is relatively insensitive to temperature. The LCF life is largely determined by the amount of inelastic strain in each cycle. Cyclic hardening occurs at room temperature, cycling softening at elevated temperature. Duplex alloys exhibit better LCF performance than their lamellar counterparts.

11.5.3
Cyclic Plasticity

Huang and Bowen [61] have used transmission electron microscopy to examine testpieces of fully lamellar Ti-48Al-2Mn-2Nb after room-temperature fatigue. A distinct feature in the fatigue structure was the presence of slip bands and fine mechanical twins. The slip bands were associated with parallel stair-rod dislocations that were thought to be the reaction products of dislocation intersections due to double slip on oblique {111} planes. The traces of the microcracks were parallel to the slip bands, which led the authors to believe that formation of slip bands preceded translamellar cracking. Gloanec et al. [62] observed a similar inhomogeneous defect structure. It should be noted, however, that other studies have revealed a relatively homogeneous distribution of dislocations and mechanical twins [63, 64].

Research of Umakoshi et al. [65–68] on polysynthetically twinned (PST) crystals has demonstrated a significant anisotropy of the fatigue properties with respect to the loading axis, which is analogous to that observed under monotonic loading (Section 6.2.5). In the so-called soft orientation (the lamellar planes are at $\Phi = 45°$ to the loading axis), cyclic hardening is relatively low. Deformation was provided by glide of ordinary dislocations, <101] superdislocations and mechanical twinning on $\{111\}_\gamma$ planes that are parallel to the interface planes. The relative contribution of these mechanisms was found to depend on the type of γ domain situated within the lamellae. Fracture often occurred along the (0001) basal plane of α_2 lamellae. In the hard orientation (the lamellar planes are parallel to the loading axis, $\Phi = 0°$), strong cyclic hardening was observed. As in the case of monotonic loading, deformation occurs on slip and twinning systems that are inclined to the lamellar planes. However, unlike monotonic loading, the cyclic properties are sensitive to how the lamellar planes are rotated around their normal with respect to the loading axis. The deformation structure was found to be very inhomogeneous within the γ lamellae, with regions of extremely high dislocation density; the fine details being dependent on the domain type. This is probably the reason for the strong cyclic hardening, which was observed for this testpiece orientation at room temperature. The fatigue life of PST crystals decreased rapidly above 600 °C due to dynamic recovery and interlamellar separation. More on this subject is reported in [65–69].

Figure 11.12 Deformation structure in samples fatigued at $T = 25\,°C$ to failure after $N = 641$ cycles with $R = -1$ and a total strain amplitude of $\Delta\varepsilon_t/2 = \pm 0.7\%$. Invisibility criteria revealed that the fast majority of the dislocations has the Burgers vector $\mathbf{b} = 1/2<110]$ [48].

It is only fairly recently that defect structures in fatigued high-strength multiphase alloys have been reported [48]; the LCF behavior of such an alloy is shown in Figures 11.10 and 11.11. Characterization of this alloy (TNB-V2) by scanning electron microscopy revealed lamellar colonies with intercolony γ, β and α_2 phase. However, examination by transmission electron microscopy (TEM) indicated that the orthorhombic B19 phase is a significant constituent of the microstructure. This multiphase constitution leads to complex fatigue processes.

TEM examination of the samples fatigued to failure at 25 and 550 °C revealed dense dislocation arrangements and debris structures that were accumulated in tangles, as demonstrated in Figure 11.12. Although the defect structure in the samples for the two test temperatures was quite similar, there were some differences. Whereas clusters of very small debris were predominant in the sample tested at room temperature, more elongated dipoles occurred in the 550 °C sample. Contrast experiments have revealed that the vast majority of the dislocations are ordinary dislocations with the Burgers vector $\mathbf{b} = 1/2<110]$, which propagated on $\{111\}_\gamma$ glide planes. Isolated dislocations are preferentially aligned along their screw orientation and frequently exhibited cross-slip, which for ordinary dislocations is easy (see Sections 5.1 and 6.3.1). There is good evidence that the cross-slip processes were initiated by intersection of the dislocations. In most, if not all, regions examined, double slip on oblique $1/2<110]\{111\}$ systems was detected. Dislocations gliding on these systems have to intersect each other. An analytical study has shown that the elastic stresses occurring between such intersecting dislocations are as high as the yield stress of the material. These stresses not only tend to immobilize the dislocations, but also induce extended cross-slip [70]. The dipole and debris structures observed after room temperature LCF can be rationalized by jog dragging, which has been described in Section 6.3.1. Dense dislocation-loop structures have also been also recognized by Nakano et al. [69] on fatigued PST crystals.

The dipoles were often found to serve as dislocation sources. The anchored dislocation segments bow out under the applied stress in a fashion similar to a classical Frank–Read mechanism. An initial state of this multiplication process is apparently shown in Figure 11.13a.

The dislocations generated during fatigue at 550 °C often exhibit a spiraled configuration (Figure 11.13b) that is reminiscent of a climb mechanism. This observation is surprising as the homologous temperature is only 0.47 T_m, a temperature at which bulk diffusion is still ineffective [71], see Section 3.3. There are two factors that may support a low-temperature climb mechanism. As suggested by the TEM observations, LCF produces a supersaturation of intrinsic point defects.

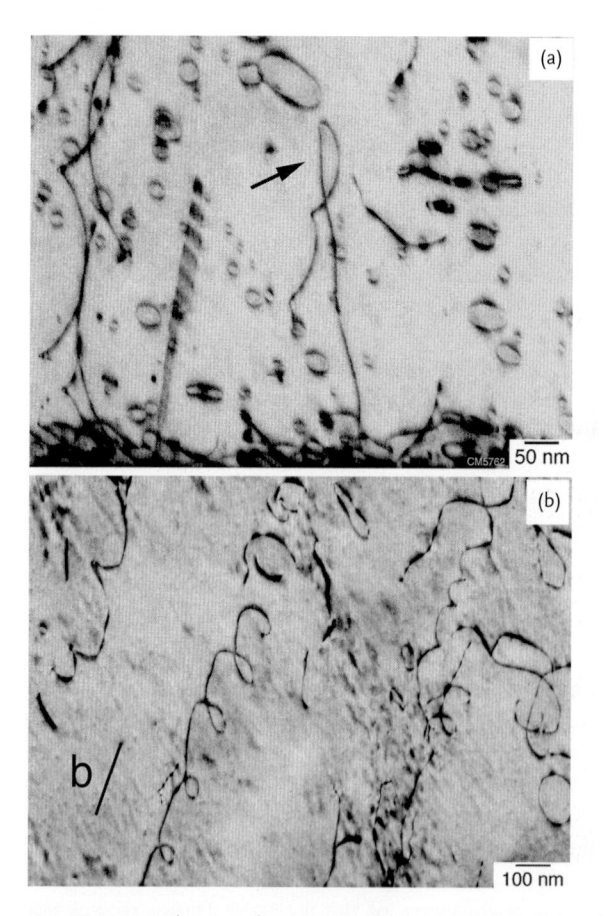

Figure 11.13 Dislocation dynamics during low-cycle fatigue at low and intermediate temperatures. (a) Initial state of a single-ended dislocation source anchored at a high jog. (b) Spiraled dislocation configuration suggesting climb. Micrographs taken after LCF at $T = 550\,°C$, $\Delta\varepsilon_t/2 = \pm 0.7\%$, $N_f = 452$ [48].

During the tests, the dislocations may climb under the combined action of the applied stress and the osmotic climb forces arising from the chemical potential of excess vacancies. It also appears possible that diffusion is supported by the presence of Ti_{Al} antisite defects, that is, Ti atoms situated on the Al sublattice (see Section 3.2). In the alloy investigated a substantial antistructural disorder is expected to occur because of the significant off-stoichiometric deviation. Different mechanisms have been proposed that involve Ti_{Al} antisite atoms associated with vacancies, which suggest that the migration energy for long-range diffusion is significantly reduced (see Section 6.4.3).

The discussion in Section 7.2.3 has shown that the dipole defects may act as additional glide obstacles. Small debris defects are expected to have short-range stress fields so that they can probably be overcome with the aid of thermal activation and probably provide a thermal-stress contribution to room-temperature work hardening under monotonic deformation. Likewise, it is expected that the dislocation dipoles and debris contribute to the cyclic hardening observed at room temperature, (Figure 11.11b).

The dipole defects are formed from small amounts of missing or additional material. Thus, these fatigue defects are expected to be unstable upon annealing (Section 7.2.3). *In situ* TEM heating experiments have indeed revealed that debris defects produced during room-temperature fatigue can easily be annealed out at moderately high temperature; part of this study is shown in Figure 11.14. Recovery seems to be accomplished by dislocation climb, as can be recognized from the details shown in Figure 11.15. The dislocations have probably climbed under the

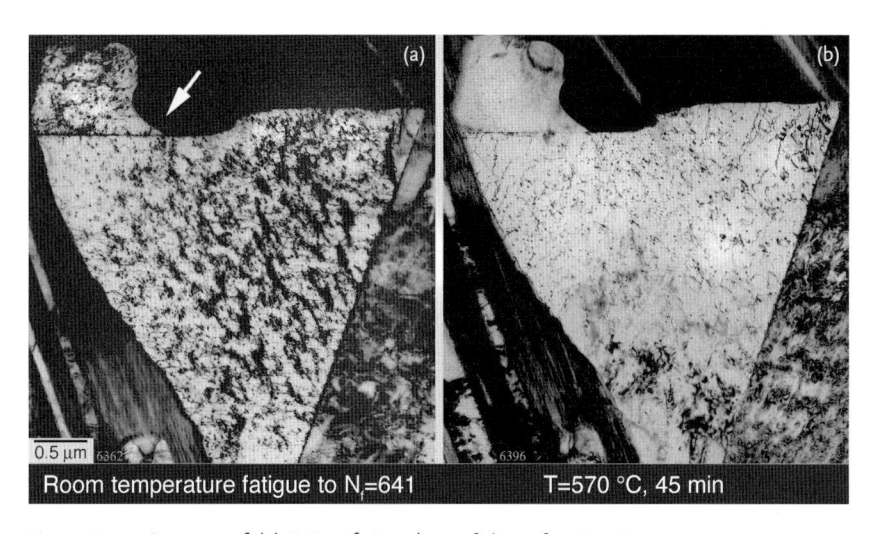

Figure 11.14 Recovery of debris in a fatigued sample observed *in situ* during heating of the sample inside the TEM; nearly lamellar Ti-45Al-8Nb-0.2C (at.%). (a) Debris structure of the sample fatigued at room temperature to failure after $N_f = 641$ cycles with a stress ratio $R = -1$ and a total strain amplitude $\Delta\varepsilon_t/2 = \pm 0.7\%$. (b) Recovered structure after annealing at 570 °C for 45 min.

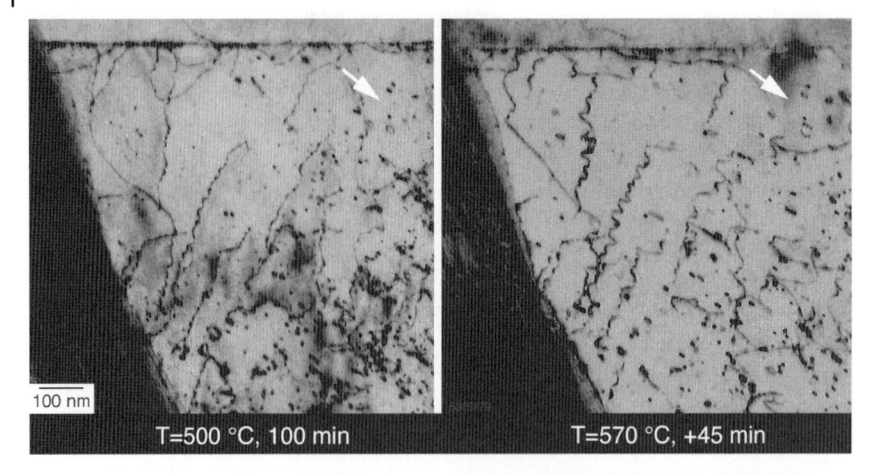

Figure 11.15 Dislocation–point-defect interactions observed during *in-situ* heating of a sample that had been fatigued at room temperature to failure at $N_f = 641$. Higher magnification of the area marked with an arrow in Figure 11.14 at two intermediate stages of recovery defined in the legend. Note the highly spiraled dislocations in the later stage of recovery.

combined action of the applied stress and the osmotic climb forces arising from the chemical potential of excess vacancies produced during room-temperature fatigue. This finding may explain why in samples fatigued at 850 °C dipoles and debris are completely absent. It is tempting to speculate that dynamic recovery has taken place, which is consistent with the cycling softening observed at this temperature (Figure 11.11b).

Summary: Strain-controlled room-temperature fatigue of TiAl alloys leads to dense structures of dislocations and debris. The debris is probably produced by jog-dragging screw dislocations and is unstable upon annealing at moderately high temperatures. The fatigue response of lamellar grains is highly anisotropic, as may be deduced from fatigue experiments performed on PST crystals.

11.5.4
Stress-Induced Phase Transformation and Dynamic Recrystallization

Another important fatigue mechanism occurring in the TNB-V2 alloy is the stress-induced transformation of the orthorhombic B19 phase, which is a significant constituent in the microstructure [48]. The B19 phase can be described as an orthorhombic structure (oP4) or as a hexagonal superstructure of α_2(Ti$_3$Al). After fatigue, very fine γ lamellae have been observed within the B19 phase, which indicates that the B19 phase is transformed into γ phase (Figure 11.16a). The transformation often starts at grain boundaries and proceeds through the formation of high ledges via distinct atomic shuffle displacements. It should be noted that the B19→γ transformation has been observed at all the fatigue test tempera-

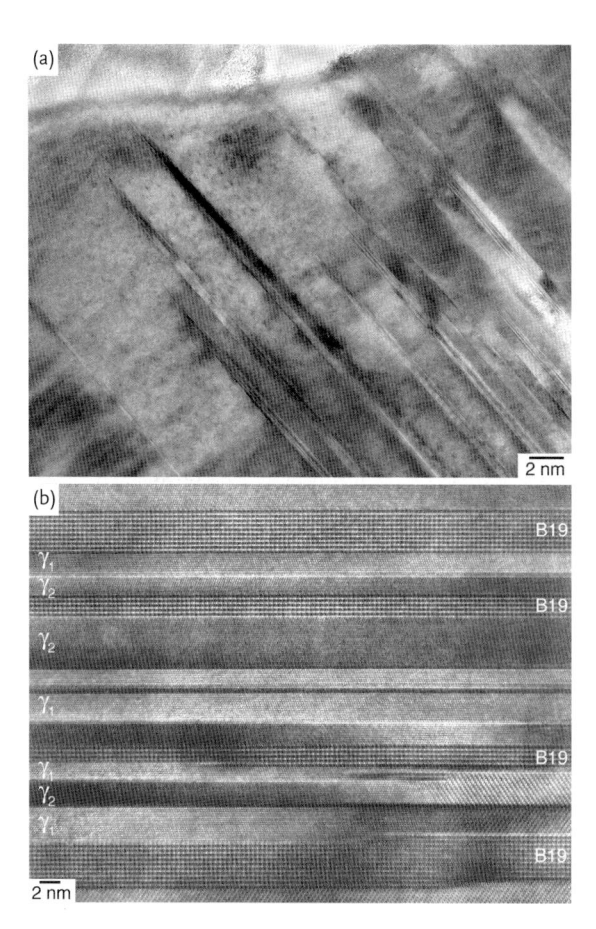

Figure 11.16 Phase transformation B19→γ during low-cycle fatigue at $T = 550\,°C$, $\Delta\varepsilon_t/2 = \pm 0.7\%$, $N_f = 452$; nearly lamellar Ti-45Al-8Nb-0.2C. (a) Low-magnification high-resolution TEM micrograph showing fine γ lamellae produced in the B19 phase. (b) Lamellar morphology consisting of fine γ lamellae adjacent to B19 phase [48].

tures investigated. The result is a lamella morphology that is comprised of extremely fine γ lamellae adjacent to B19 phase (Figure 11.16b). A general observation is that two different γ variants are usually generated adjacent to the B19 phase, as seen for almost all the γ lamellae in Figure 11.16b. If, for example, the stacking sequence of the $\{111\}_\gamma$ planes of the variant γ_1 is labeled **ABC**, that of variant γ_2 is **CBA**. This inversion of the stacking sequence is thought to induce strain fields of opposite signs, which eventually reduces the total strain energy. It might be speculated that such a combination of shear processes makes the transformation easier. Due to the difference in lattice constants between the γ and B19 phases the transformation may accommodate local strains and is thus expected to serve as a toughening mechanism.

Figure 11.17 Degradation of the lamellar morphology after fatigue at $T = 850\,°C$ $\Delta\varepsilon_t/2 = \pm 0.7\%$, $N_f = 501$; nearly lamellar Ti-45Al-8Nb-0.2C [48].

Samples fatigued at 850 °C exhibit a degradation of the lamellar morphology via the above-described phase transformation combined with dynamic recrystallization, as exemplified in Figure 11.17. High-resolution observations have shown that these processes preferentially start at deformation-induced heterogeneities by the formation of sub-boundaries and proceeds by the nucleation of nanometer-size γ(TiAl) grains (Section 8.2). It should be noted that the $\alpha_2 \rightarrow \gamma$ phase transformation was also observed in lamellar Ti-46Al-1.4Mn-2Mo after LCF at 800 °C [52].

Summary: LCF of multiphase alloys based on α_2(Ti$_3$Al) and γ(TiAl) leads to significant structural changes due to stress-induced phase transformation. At high test temperatures, this phase transformation is coupled with dynamic recrystallization and degrades the lamellar morphology.

11.6
Thermomechanical Fatigue and Creep Relaxation

Components that operate at elevated temperature often undergo cyclic temperature changes. If the thermal expansions or contractions are either wholly or partially constrained, cyclic strains and stresses can result. These temperature-induced cyclic strains produce fatigue just as if cyclic mechanical loading produced the strains. Under service conditions, temperature-induced cyclic strains are often superimposed on cyclic mechanical loading. This is referred to as thermomechanical fatigue (TMF). Concerning the time, temperature and strain paths experienced, two basic regimes can be defined. In-phase TMF implies that the maximum normal strain and the maximum temperature occur simultaneously, while out-of-phase TMF implies that the maximum normal strain coincides with the minimum temperature. It should be noted that only these two regimes result in a proportional variation of the strain and temperature. The tests performed on TiAl alloys have covered the temperature range 100 °C to 850 °C, utilizing different minimum and maximum temperatures [72–80]. The results may be summarized as follows.

i) In-phase thermomechanical fatigue leads to longer fatigue life, when compared with isothermal LCF. Not surprisingly, out-of-phase TMF is the worst condition for the fatigue life [75, 76, 78]. For example, out-of-phase TMF performed on lamellar Ti-45Al-5.1Nb-0.4B-0.25C reduced the sample life to one fifth of that observed in isothermal LCF tests [79].

ii) In-phase TMF results in negative mean stresses, whereas positive (tensile) mean stresses are developed under out-of phase testing [76, 78]. This finding was almost independent of microstructure. The generation of the mean stresses was attributed to cyclic softening at high temperature and cycling hardening at low temperature [76].

iii) The short fatigue life for out-of-phase TMF was attributed to the generation of mean tensile stresses and oxidation of the testpiece during the high-temperature periods of the test. The oxide layer was thought to provide crack nucleation sites during the low-temperature tensile cycle [76].

iv) The presence of mean stresses makes life predictions for TMF by the Manson–Coffin law [57, 58] impossible. As a step forward, other models for TMF life predictions have been discussed in [73, 76].

During high-temperature service, components such as gas-turbine blades are subjected to a combined fatigue and creep damage. For example, if the tensile phases have larger stresses or longer durations than the compressive phases, there will be additional damage by creep.

Park *et al.* [81], in addressing this problem, recognized a drastic reduction of the LCF life at 800 °C for lamellar Ti-46.6Al-1.4Mn-2Mo, when the testpieces were subject to additional creep periods of 5 min to 30 min at the peak tensile strain of the cycle. The effect was attributed to structural changes due to $\alpha_2 \rightarrow \gamma$ transformation occurring during the creep periods.

Summary: Out-of-phase thermomechanical fatigue (TMF) of alloys based on $\alpha_2(Ti_3Al) + \gamma(TiAl)$ is characterized by short lifetimes. The inphase TMF life is longer than the fatigue life occurring under isothermal LCF. A combination of LCF and creep significantly shortens the fatigue life.

References

1 American Society for Testing and Materials (2000) ASTM E647. Standard Test Method for Measurement of Fatigue Crack Growth Rates, Vol. 03.01, ASTM, West Conshohocken, PA, p. 591.

2 Elber, W. (1970) *Eng. Fract. Mech.*, **2**, 37.

3 Hénaff, G., Cohen, S.-A., Mabru, C., and Petit, J. (1996) *Scr. Mater.*, **34**, 1449.

4 Worth, B.D., Larsen, J.M., and Rosenberger, A. (1997) *Structural*

Intermetallics 1997 (eds M.V. Nathal, R. Darolia, C.T. Liu, P.L. Martin, D.B. Miracle, R. Wagner, and M. Yamaguchi), TMS, Warrendale, PA, p. 563.

5 Paris, P.C., and Erdogan, F. (1963) *J. Basic Eng. ASME Trans. Ser. D*, **85**, 528.

6 Recina, V., and Karlsson, B. (1997) *Structural Intermetallics 1997* (eds M.V. Nathal, R. Darolia, C.T. Liu, P.L. Martin, D.B. Miracle, R. Wagner, and M. Yamaguchi), TMS, Warrendale, PA, p. 479.

7 Sastry, S.M.L., and Lipsitt, H.A. (1977) *Metall. Trans. A*, **8A**, 299.

8 Trail, S.J., and Bowen, P. (1995) *Mater. Sci. Eng. A*, **192–193**, 427.

9 Balsone, S.J., Larsen, J.M., Maxwell, D.C., and Jones, J.W. (1995) *Mater. Sci. Eng. A*, **192–193**, 457.

10 Kumpfert, J., Kim, Y.-W., and Dimiduk, D.M. (1995) *Mater. Sci. Eng. A*, **192–193**, 465.

11 Vaidya, W.V., Schwalbe, K.-H., and Wagner, R. (1995) *Gamma Titanium Aluminides* (eds Y.-W. Kim, R. Wagner, and M. Yamaguchi), TMS, Warrendale, PA, p. 867.

12 Larsen, J.M. (1999) *Gamma Titanium Aluminides 1999* (eds Y.-W. Kim, D.M. Dimiduk, and M.H. Loretto), TMS, Warrendale, PA, p. 463.

13 Grange, M., Thomas, M., Raviart, J.L., Belaygue, P., and Recorbet, D. (2003) *Gamma Titanium Aluminides 2003* (eds Y.-W. Kim, H. Clemens, and A.H. Rosenberger), TMS, Warrendale, PA, p. 213.

14 Hénaff, G., and Gloanec, A.-L. (2005) *Intermetallics*, **13**, 543.

15 Venkateswara Rao, K.T., Odette, G.R., and Ritchie, R.O. (1994) *Acta Metall. Mater.*, **42**, 893.

16 Venkateswara Rao, K.T., Kim, Y.-W., Muhlstein, C.L., and Ritchie, R.O. (1995) *Mater. Sci. Eng. A*, **192–193**, 474.

17 Ritchie, R.O., and Dauskardt, R.H. (1991) *J. Ceram. Soc. Jpn.*, **99**, 1047.

18 Dauskardt, R.H. (1993) *Acta Metall. Mater.*, **41**, 2765.

19 Kruzic, J.J., Campbell, J.P., McKelvey, A.L., Choe, H., and Ritchi, R.O. (1999) *Gamma Titanium Aluminides 1999* (eds Y.-W. Kim, D.M. Dimiduk, and M.H. Loretto), TMS, Warrendale, PA, p. 495.

20 Worth, B.D., Larsen, J.M., Balsone, S.J., and Jones, J.W. (1997) *Metall. Mater. Trans. A*, **28A**, 825.

21 Bowen, P., Chave, R.A., and James, A.W. (1995) *Mater. Sci. Eng. A*, **192–193**, 443.

22 Gnanamoorthy, R., Mutoh, Y., Hayashi, K., and Mizuhara, Y. (1995) *Scr. Metall. Mater.*, **33**, 907.

23 Chan, K., and Shih, D. (1997) *Metall. Mater. Trans. A*, **28A**, 79.

24 Chan, K., and Shih, D. (1998) *Metall. Mater. Trans. A*, **29A**, 73.

25 Suresh, S. (1998) *Fatigue of Materials*, 2nd edn, Cambridge University Press, Cambridge.

26 Rosenberger, A.H., Worth, B.D., and Larsen, J.M. (1997) *Structural Intermetallics 1997* (eds M.V. Nathal, R. Darolia, C.T. Liu, P.L. Martin, D.B. Miracle, R. Wagner, and M. Yamaguchi), TMS, Warrendale, PA, p. 555.

27 Hénaff, G., Bittar, B., Mabru, C., Petit, J., and Bowen, P. (1996) *Mater. Sci. Eng. A*, **219**, 212.

28 Ueno, A., and Kishimoto, H. (1996) *Fatigue'96* (eds G. Lütjering and H. Nowak), Pergamon Press, Oxford, p. 1731.

29 Larsen, J.M., Worth, B.D., Balson, S.J., Rosenberger, A.H., and Jones, J.W. (1996) *Fatigue'96* (eds G. Lütjering and H. Nowak), Pergamon Press, Oxford, p. 1719.

30 Campbell, J.P., Venkateswara Rao, K.T., and Ritchie, R.O. (1996) *Fatigue'96* (eds G. Lütjering and H. Nowak), Pergamon Press, Oxford, p. 1779.

31 Gloanec, A.-L., Hénaff, G., Bertheau, D., Belaygue, P., and Grange, M. (2003) *Scr. Mater.*, **49**, 825.

32 Campbell, J.P., Kruzic, J.J., Lillibridge, S., Venkateswara Rao, K.T., and Ritchie, R.O. (1997) *Scr. Mater.*, **37**, 707.

33 Ritchi, R.O., and Lankford, J. (1986) *Small Fatigue Cracks*, TMS, Warrendale, PA.

34 Miller, K.J., and de los Rios, E.R. (1992) *Short Fatigue Cracks*, Mechanical Engineering Publications Limited, London, UK.

35 Campbell, J.P., Venkateswara Rao, K.T., and Ritchie, R.O. (1999) *Metall. Mater. Trans. A*, **30A**, 563.

36 Kruzic, J.J., Campbell, J.P., and Ritchie, R.O. (1999) *Acta Mater.*, **47**, 801.
37 Balsone, S.J., Jones, J.W., and Maxwell, D.C. (1994) *Fatigue and Fracture of Ordered Intermetallic Materials* (eds W.O. Soboyejo, T.S. Srivatsan, and D.L. Davidson), TMS, Warrendale, PA, p. 307.
38 Venkateswara Rao, K.T., Kim, Y.-W., and Ritchie, R.O. (1995) *Scr. Metall. Mater.*, **33**, 459.
39 Mutoh, Y., Kurai, S., Hanson, T., Moriya, T., and Zu, S.J. (1997) *Structural Intermetallics 1997* (eds M.V. Nathal, R. Darolia, C.T. Liu, P.L. Martin, D.B. Miracle, R. Wagner, and M. Yamaguchi), TMS, Warrendale, PA, p. 495.
40 Worth, B.D., Larsen, J.M., Balsone, S.A., and Worth, J.W. (1996) *Titanium'95: Science and Technology* (eds P.A. Blenkinsop, W.J. Evans, and H.M. Flower), Institute of Materials, London, p. 286.
41 Liu, C.T., and Kim, Y.-W. (1992) *Scr. Metall. Mater.*, **27**, 599.
42 Mabru, C., Háneff, G., and Petit, J. (1996) *Fatigue'96* (eds G. Lütjering and H. Nowack), Pergamon, Oxford, p. 1749.
43 Hénaff, G., Odemer, G., and Tonneau-Morell, A. (2007) *Int. J. Fatigue*, **29**, 1927.
44 Larsen, J.M., Worth, B.D., Balsone, S.J., and Jones, J.W. (1995) *Gamma Titanium Aluminides* (eds Y.-W. Kim, R. Wagner, and M. Yamaguchi), TMS, Warrendale, PA, p. 821.
45 McKelvey, A.L., Venkateswara Rao, V.K., and Ritchie, R.O. (1997) *Scr. Mater.*, **37**, 1797.
46 Planck, S., and Rosenberger, A.H. (1999) *Gamma Titanium Aluminides 1999* (eds Y.-W. Kim, D.M. Dimiduk, and M.H. Loretto), TMS, Warrendale, PA, p. 791.
47 Hirth, J.P., and Lothe, J. (1992) *Theory of Dislocations*, Krieger, Melbourne.
48 Appel, F., Heckel, Th., and Christ, H.J. (2010) *Int. J. Fatigue*, **32**, 792.
49 Dowling, W.E., Donlon, W.T., and Allison, J.E. (1991) *High-Temperature Ordered Intermetallic Alloys IV, Mater. Res. Soc. Symp. Proc.*, vol. 213 (eds L.A. Johnson, D.P. Pope, and J.O. Stiegler), MRS, Pittsburgh, PA, p. 561.
50 Hardy, M.C. (1996) *Titanium'95: Science and Technology* (eds P.A. Blenkinsop, W.J. Evans, and H.M. Flower), Institute of Materials, London, p. 256.
51 Recina, V., and Karlsson, B. (1999) *Mater. Sci. Eng. A*, **262**, 70.
52 Park, Y.S., Nam, S.W., Hwang, S.K., and Kim, N.J. (2002) *J. Alloys Compd.*, **335**, 216.
53 Gloanec, A.-L., Hénaff, G., Jouiad, M., Bertheau, D., Belaygue, P., and Grange, M. (2005) *Scr. Mater.*, **52**, 107.
54 Malakondaiah, G., and Nicholas, T. (1996) *Metall. Mater. Trans. A*, **27A**, 2239.
55 Cui, W.F., and Liu, C.M. (2009) *J. Alloys Compd.*, **477**, 596.
56 Park, Y.S., Ahn, W.S., Nam, S.W., and Hwang, S.K. (2002) *Mater. Sci. Eng. A*, **336**, 196.
57 Manson, S.S. (1954) Technical Report NACA-TR-1170 (National Advisory Committee for Aeronautics).
58 Coffin, L.F. (1954) *Trans. Am. Soc. Mech. Eng.*, **76**, 931.
59 Recina, V., and Karlsson, B. (2000) *Scr. Mater.*, **43**, 609.
60 Christ, H.-J., Fischer, F.O.R., and Maier, H.J. (2001) *Mater. Sci. Eng. A*, **319–321**, 625.
61 Huang, Z.W., and Bowen, P. (2001) *Scr. Mater.*, **45**, 931.
62 Gloanec, A.-L., Hénaff, G., Jouiad, M., and Bertheau, D. (2002) *Fatigue, Eighth International Fatigue Congress, Stockholm* (ed. A.F. Blom), EMAS, Sheffield, UK, p. 3039.
63 Srivatsan, T.S., Soboyejo, W.O., and Strangwood, M. (1995) *Eng. Fract. Mech.*, **52**, 107.
64 Chen, W.Z., Song, X.P., Quian, K.W., and Gu, H.C. (1999) *Int. J. Fatigue*, **20**, 3039.
65 Umakoshi, Y., Yashuda, H.Y., and Nakano, T. (1995) *Mater. Sci. Eng. A*, **192–193**, 511.
66 Yasuda, H.Y., Nakano, T., and Umakoshi, M. (1996) *Philos. Mag. A*, **73**, 1053.
67 Umakoshi, Y., Yashuda, H.Y., and Nakano, T. (1996) *Int. J. Fatigue*, **18**, 65.
68 Umakoshi, Y., Yashuda, H.Y., Nakano, T., and Ikeda, K. (1998) *Metall. Mater. Trans. A*, **29A**, 943.
69 Nakano, T., Yasuda, H.Y., Higashitanaka, N., and Umakoshi, Y. (1977) *Acta Mater.*, **45**, 4807.

70 Appel, F. (1989) *Phys. Status Solidi (a)*, **116**, 153.
71 Mishin, Y., and Herzig, Chr. (2000) *Acta Mater.*, **48**, 589.
72 Lee, E.U. (1994) *Metall. Mater. Trans. A*, **25A**, 2207.
73 Roth, M., and Biermann, H. (2006) *Scr. Mater.*, **54**, 137.
74 Heckel, T.K., Guerrero-Tovar, A., Christ, H.-J., and Appel, F. (2007) *Ti 2007 Science and Technology* (eds M. Niinomi, S. Akiyama, M. Hagiwara, M. Ikeda, and K. Maruyama), The Japan Institute of Metals, Sendai, p. 275.
75 Roth, M., and Biermann, H. (2008) *Int. J. Fatigue*, **30**, 352.
76 Bauer, V., and Christ, H.-J. (2009) *Intermetallics*, **17**, 370.
77 Cui, W.F., Liu, C.M., Bauer, V., and Christ, H.-J. (2009) *Intermetallics*, **17**, 370.
78 Roth, M., and Biermann, H. (2010) *Metall. Mater. Trans. A*, **41A**, 717.
79 Christ, H.-J. (2007) *Mater. Sci. Eng. A*, **468–470**, 98.
80 Brookes, P., Kühn, H.J., Skrotzki, B., Klingelhöffer, H., Sievert, R., Pfetzing, J., Peter, D., and Eggeler, G. (2010) *Mater. Sci. Eng. A*, **527**, 3829.
81 Park, Y.S., Nam, S.W., Yang, S.J., and Hwang, S.K. (2001) *Mater. Trans.*, **42**, 1380.

12
Oxidation Behavior and Related Issues

As with all materials that are used at high temperatures in oxygen-containing atmospheres, oxidation is an ever-present issue. During oxidation the base metal reacts with the surrounding atmosphere and various types of oxides/corrosion products are formed. In components, this can result in a reduction of the cross-sectional area of load-bearing material and higher stresses, and, over a longer period of time this may lead to premature component failure. Design of an oxidation-resistant alloy is usually aimed towards the establishment of a protective, slow-growing, stable oxide that is adherent to the surface and thus capable of shielding the underlying material.

Gamma titanium aluminides are more oxidation resistant than conventional titanium and alpha-two-based alloys [1]. However, within the very wide range of compositions that have been studied, ranging from low to high aluminum-containing binary alloys, ternary and multicomponent containing alloy systems; a vast range of oxidation behavior has been observed. The research performed on the oxidation of gamma and the related protection technologies has thus become so extensive that a book could probably be written on this subject alone. The aim of this chapter however is only to give a brief insight into the main features concerning the oxidation of TiAl, from a TiAl technologist's point of view. Of course such a treatment will have a different emphasis and be less detailed than how an oxidation specialist would treat the subject, but nevertheless we hope to give an overview outlining basic features with enough references so that the reader can easily find more detailed information if desired. Overview articles can be found in [2–6].

12.1
Kinetics and Thermodynamics

When a metal is held within an oxidizing atmosphere (such as air) at elevated temperature it reacts and forms a scale. While this process takes place, the weight change involved in the formation of the scale can be plotted against time. Depending upon the material and temperature, several oxidation rate laws including linear, parabolic, cubic or logarithmic may be observed, as described in various

Gamma Titanium Aluminide Alloys: Science and Technology, First Edition. Fritz Appel, Jonathan David Heaton Paul, Michael Oehring.
© 2011 Wiley-VCH Verlag GmbH & Co. KGaA. Published 2011 by Wiley-VCH Verlag GmbH & Co. KGaA.

books on oxidation. Clearly, linear oxidation kinetics provide no protection for the base material. Oxidation-resistant materials in which diffusion-controlled scale growth takes place, exhibit parabolic oxidation behavior. In this case the increase in weight (Δm in mg) for a given surface area (A in cm^2) as a function of temperature (T) and time (t) obeys a law given by:

$$(\Delta m/A)^2 = k_p t \qquad (12.1)$$

where k_p is the parabolic oxidation rate constant in mg^2 cm^{-4} s^{-1}. When k_p has been determined for a series of different temperatures, the oxidation energy Q, can be determined from the slope of a graph of $1/T$ against $\ln(k_p)$. It should be noted that the units of k_p depend on what technique has been used to measure the amount of oxidation, as explained by Birks and Meier [7]. At a given temperature, materials that exhibit low values of k_p show slower oxidation kinetics, and are thus more resistant to oxidation. Naturally, as oxidation proceeds, the rate-determining processes can also change so that "composite" oxidation behavior may be observed. For example, according to Birks and Meier [7], niobium oxidizing in air at around 1000 °C initially conforms to the parabolic rate law but exhibits linear oxidation behavior at longer exposure times. If sufficient stress is built up within the scale during oxidation then cracking and/or spalling may occur and also influence the kinetics. For further information the reader is referred to general books on oxidation including [7, 8].

For oxidation resistance in binary or multicomponent engineering alloy systems, it is necessary that selective oxidation of one alloy component provides a stable protective oxide layer. Meier [9] has examined the factors that contribute to the formation of a protective oxide scale in a binary alloy containing the elements A and B. Protective scale formation requires that the component B within the alloy is selectively oxidized and that the stability of the oxide BO is higher than that of the lowest oxide of component A. Additionally, and very importantly, the concentration of element B must be high enough to be able to supply the oxide (BO) at the surface and its flux during oxidation must be sufficient to suppress the growth of an oxide of element A after element B has been denuded from regions near to the surface due to BO oxide formation. The minimum concentration required to fulfill the second criteria is always higher than for the first criteria. For a more detailed assessment and further information on the oxidation of intermetallics the reader is referred to [9–12].

When a pure metal and its oxide are in equilibrium (activity of both the metal and oxide = 1), the oxygen partial pressure (P_{O_2}) at a particular temperature (T) can be used to calculate the standard free energy of oxide formation ΔG_T° (otherwise known as the standard enthalpy of formation) using the formula

$$\Delta G_T^\circ = RT \ln(P_{O_2}) \qquad (12.2)$$

where T is the temperature (in K) and R is the universal gas constant.

Concerning the oxidation of pure metals, Ellingham diagrams such as that shown in Figure 12.1 graphically indicate how the value of ΔG° changes as a function of temperature, and can also be used to compare the relative thermodynamic

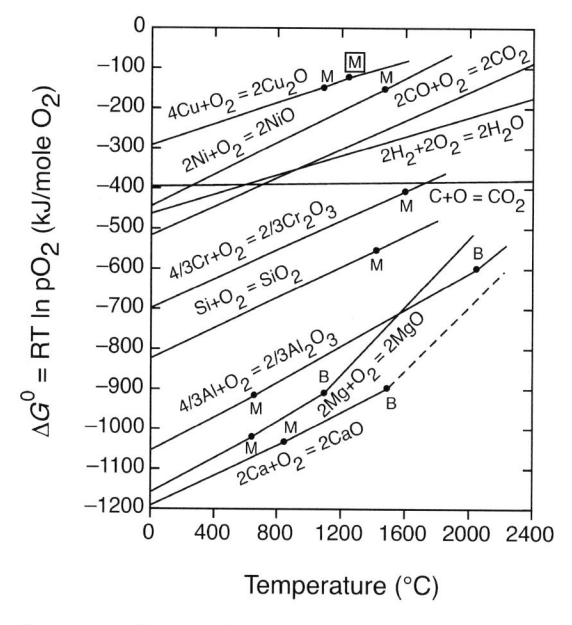

Figure 12.1 Ellingham diagram showing the standard free energy of formation of selected oxides as a function of temperature [7]. This diagram has been redrawn and some of the original data lines removed for the sake of clarity. Oxides with more-negative values of $\Delta G°$ are thermodynamically more stable than those with less-negative values. On the diagram, M and B represent melting and boiling points of metals. The M enclosed within a square frame represents the melting point of an oxide.

stability of different oxides. From a thermodynamic point of view, Figure 12.1 shows that oxides such as CaO that have high (more negative) values of $\Delta G°$ are more stable than oxides such as Al_2O_3 that have lower (less negative) values of $\Delta G°$. The oxygen equilibrium pressures of various metal–metal-oxide systems, presented by Rahmel et al. [2], are shown in Figure 12.2. The graph shows that the equilibrium partial pressure of oxygen is more or less identical over a large temperature range for both aluminum and titanium, thus indicating that the thermodynamic stability of TiO and Al_2O_3 are practically the same. As indicated in the graph, the activity of the various elements and their oxides was equal to one. In alloys, the activity of the individual alloy components (which may significantly deviate from ideal Raoultian behavior) plays an important role in determining the oxide thermodynamic stability. This is because for a specific element in equilibrium with its oxide, the oxygen partial pressure depends on its activity. Alloy design of oxidation-resistant materials needs to take into account such thermodynamic considerations. Depending on conditions and exact composition, binary Ti–Al alloys may form either titanium oxides or alumina during exposure to oxygen at high temperature. For good oxidation resistance, the formation of a continuous layer of alumina is required. The formation of titanium oxides that

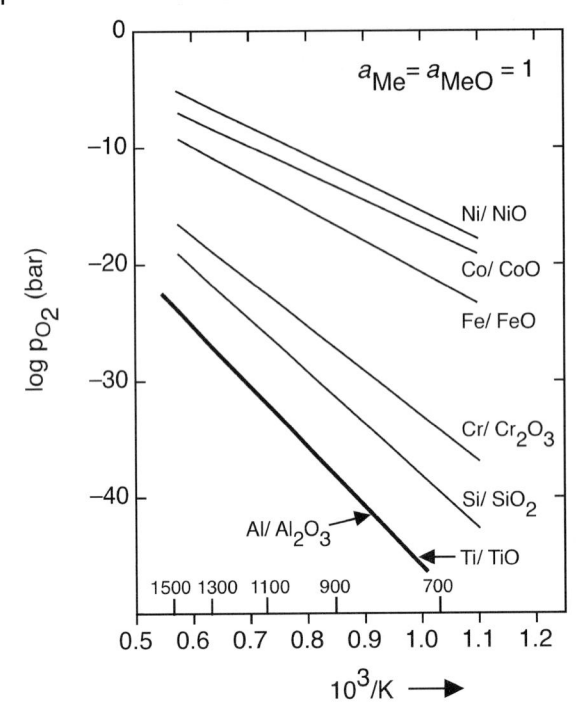

Figure 12.2 Oxygen equilibrium pressures of selected metal/oxide systems [2], showing that the oxygen equilibrium pressures of Al/ Al_2O_3 and Ti/TiO are very similar. This indicates that Al_2O_3 and TiO have very similar thermodynamic stability, thus making it difficult to easily predict the most-stable oxide. The diagram has been redrawn, based on the original.

can grow quickly and that provide only limited protection should be avoided. Phase relationships within the Ti–Al–O system have been presented in [13, 14] and are also discussed in [15–17].

With regard to the relative stability of oxides in TiAl, thermodynamic calculations have been performed on the Ti–Al–O system [17, 18]. As indicated earlier, the oxide stability of a specific alloy component requires that the activity of the alloy component be considered and refs. [17, 18] included such information. According to Eckert and Hilpert [19] Al_2O_3 is the stable oxide in binary alloys for aluminum contents as low as Ti-54Al, whereas Rahmel and Spencer [17] and Luthra [18] indicated minimum aluminum levels of Ti-61Al and Ti-55Al, respectively. Thus, for alloys in the ($\alpha_2 + \gamma$) two-phase field, TiO would seemingly be the more thermodynamically stable oxide. However, experimental phase equilibria and oxidation studies have indicated that Al_2O_3 is the more stable oxide [2, 3, 14]. With regard to these differences, the effect of dissolved oxygen seems to be important. For further information the reader is referred to [2, 20]. Jacobson et al. [20] have performed a more recent thermodynamic study that included determining the activities of Ti and Al in two binary (Ti-45Al and Ti-62Al) and three tertiary

(Ti–Al–Cr) alloys. The aim of the study was to clarify the stability of oxides for alloys in the ($\alpha_2 + \gamma$) two-phase field. It was concluded that the stabilities of TiO and Al_2O_3 are so close for alloys within the ($\alpha_2 + \gamma$) two-phase field that it was experimentally not possible to determine the more stable oxide [20]. Additionally, it was stated that secondary factors such as surface preparation, impurity content and minor microstructural variations seemingly have a significant impact on oxidation resistance.

At this point, it should be noted for clarity that according to [16], although TiO is typically used in the literature for thermodynamic considerations, in practice TiO_2 is usually found rather than TiO, as TiO rapidly oxidized to TiO_2. Additionally, when reference is made to alumina or Al_2O_3 it is usually the α-Al_2O_3 form that is being referred to, as this is almost always found to form in titanium based alloys exposed at relatively low temperatures.

12.2
General Aspects Concerning Oxidation

Apart from the effect of alloy composition and any surface treatment, other factors such as the atmosphere and surface finish can also play a role in the oxidation behavior. This section will outline the influence of such factors.

12.2.1
Effect of Composition

Similar to the mechanical properties, the oxidation behavior is highly dependent on alloy composition. According to Rahmel et al. [2] the oxidation behavior of TiAl in pure oxygen can be schematically represented as shown in Figure 12.3. Initial oxidation during period A leads to the formation of a protective layer of alumina. Oxidation during period B is marked by the growth of a mixed $Al_2O_3 + TiO_2$ scale (initial Al_2O_3 breakdown). At the start of period C, the alumina barrier layer that was present to some degree during period B has been broken down and the resulting nonprotective scale leads to breakaway oxidation. How these stages of oxidation are influenced by nitrogen within the atmosphere and niobium alloying additions is discussed in [21].

As alumina is one of the most stable oxides that exists, the aim of TiAl alloy design towards oxidation resistance is directed towards the formation and retention of a continuous Al_2O_3 layer. As explained by Rahmel et al. [2] and described below, the effect of alloying additions can be difficult to assess for a number of reasons. First, alloying additions (especially those added in significant quantities) can change the microstructure and phase constitution so that comparison with a reference composition can be difficult. For example, elements that promote or reduce the amount of α_2, could significantly affect the initial oxidation behavior during period A, without having any direct effect on the oxide. Secondly, depending on which period of oxidation is being considered, an alloying element may

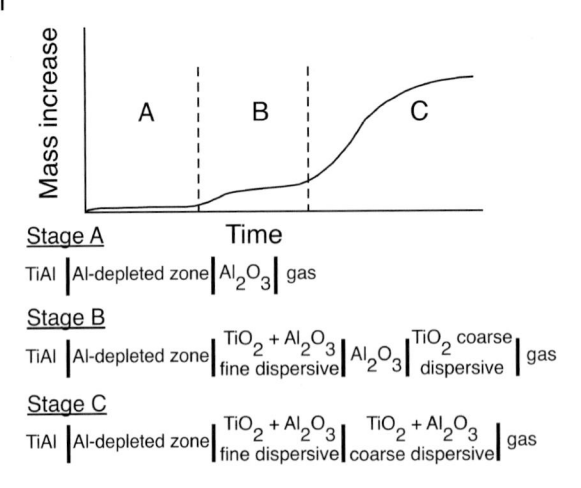

Figure 12.3 Schematic depiction of TiAl oxidation, indicating three stages of oxidation [2]. Stage A involves the growth of protective Al_2O_3. During stage B growth of a mixed $TiO_2 + Al_2O_3$ scale occurs and is associated with breakdown of the initial Al_2O_3 scale. Breakaway oxidation takes place during stage C due to complete breakdown of the Al_2O_3 scale. Redrawn diagram based on original.

have both positive and negative effects on oxidation. In this context chromium is described as being capable of extending period A oxidation, so that the time to initial breakdown of the oxide film is increased. However, the kinetics of mixed-oxide growth during period B is increased by chromium additions [2].

The oxidation of binary TiAl alloys has been discussed in [21–25]. According to Meier et al. [22], only alloys within the γ (TiAl) phase field and oxidized in pure oxygen at temperatures less than around 1000 °C form continuous alumina surface films. Quadakkers et al. [21] showed increased oxidation resistance with higher aluminum content for oxidation at 900 °C in an argon/oxygen mixture, Figure 12.4. However, it was found that the oxidation behavior did not follow a parabolic rate law, the rate constant K_p being found to increase with exposure time. This was believed to be at least partly due to the increasing localized breakdown of the thin alumina scale and subsequent growth of mixed-oxide scale protrusions after longer exposure times. It was found that reducing aluminum content led to an increased coverage of the mixed-oxide scale. Oxidation rate has also been found to depend on the size of the α_2 phase within TiAl, material containing coarse α_2 showing increased oxidation compared to material of the same composition but with more finely distributed α_2 [26].

The effect of various alloying elements on oxidation behavior has been investigated in many studies, although to what degree the important factors described above were taken into account is not always clear. For information concerning the effect of particular elements the reader is referred to the following papers: Cr, V, Si, Mo and Nb [27], Nb [28], Cu [29], Si and Nb [30], Sb [31], Nb [32] and Cu, Y, Si, Sn, Zr, Hf, V, Nb, Ta, Cr, Mo, W, Mn, Ni, and Co [33]. Further references

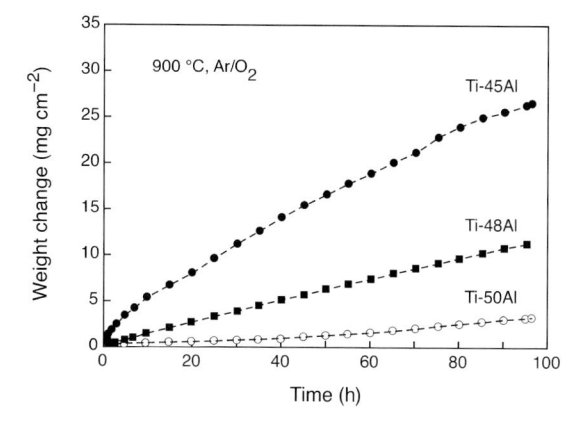

Figure 12.4 Weight-gain data for three binary alloys obtained during isothermal oxidation at 900 °C in an argon–oxygen gas mixture [21]. The graph indicates increased oxidation resistance of higher aluminum containing alloys. The data in the original graph has been replotted.

concerning the effect of elements as near-surface ion-implanted species, rather than as alloying additions, can be found in Section 12.2.7.

According to Rahmel et al. [2], the only alloying element that all workers on oxidation believe to be beneficial for oxidation resistance in air is niobium. Nickel et al. [32] believe this protection results from a decrease in the diffusion of metal and oxygen in the scale. After oxidation, it has been reported that niobium is enriched in the subsurface zone [15, 34], possibly as a result of the more rapid oxidation of Ti and Al [2]. An important aspect with regard to the addition of alloying elements is that combinations of certain elements in specific amounts can have an unexpected positive synergy that significantly improves oxidation resistance. Such effects can arise from changes in the processes required for the transport of oxygen and/or metal ions within the scale. The mechanisms involved, however, are rather complicated and outside the scope of this chapter. For further information concerning such processes the reader is referred elsewhere [2, 15, 35].

12.2.2
Mechanical Aspects of Oxide Growth

During oxidation at high temperature, oxide initially forms on the outer surface and subsequently grows and develops as a function of time, temperature and alloy composition. During and after the oxidation process, stress may generate within the oxide from either *growth stresses* and/or *thermal stresses*. The development of *growth stresses* is discussed in [7, 8, 36]. One mechanism of such stress generation depends on the relative specific volume of the oxide and metal, the nature of the oxide stress depending on the Pilling–Bedworth ratio (PBR). This is the ratio between the volume of oxide formed and the volume of metal consumed [37]. If the PBR is above unity then the stress in the oxide is compressive. The role of volume ratio, however, is only of minor importance when oxidation proceeds through an outward migration of metal ions [8]. Determination of the PBR for nickel-based aluminide alloys is presented in [38]. For TiAl oxidizing at 800 °C to 900 °C, compressive growth stresses of around 100 MPa were determined in the oxide scale [39]. However, within the same study oxide growth stresses within chlorine ion-implanted TiAl were measured to be a few GPa. With regard to the development of *thermal stress* on cooling, a compressive stress is developed in the oxide and a tensile stress in the specimen if the coefficient of thermal expansion of the metal is larger than that of the oxide, and vice versa. Cathcart [36] indicates that the surface roughness also influences stress development as rough surfaces provide better keying of the oxide to the metal and thus aid stress generation. Kofstad [8] describes work by Stringer [40] in which compressive stresses resulting from the oxidation of thin-walled tantalum tubes resulted in dilatations of up to 7% of the diameter. Thermal stresses can play an important role in scale spallation during cyclic oxidation, as discussed by Yoshihara and Kim [41] who present CTE data for some TiAl alloys. Cyclic oxidation of TiAl has been investigated by Shimizu *et al.* [42] and Haanappel *et al.* [43]. It is suggested that electrolytic chromium plating and plasma-sprayed CoNiCrAlY coatings [42] or Y additions [44] may improve the cyclic oxidation resistance.

Oxide cracking during elastic loading of a component is an important factor because if this takes place, then unprotected material will be exposed to the environment. The critical strain to achieve oxide cracking during tensile deformation has been discussed in [2, 45–47]. During tensile testing of Ti-50Al and Ti-50Al-2Nb in air at 900 °C using strain rates between $1 \times 10^{-9} \, s^{-1}$ and $3 \times 10^{-4} \, s^{-1}$, the critical strains required for scale cracking varied from 0.12% to 0.5% for Ti-50Al and 0.17% to 0.58% for Ti-50Al-2Nb, respectively [47]. However, fracture mechanics modeling indicated that these differences were likely due to different pore-size distributions within the scales. Obviously, testing at very slow strain rates requires longer times to reach a specific strain, thus allowing more time for oxide pores to form and grow. Taking this into account, the critical strain for scale cracking was believed to be independent of strain rate. Crack healing, which was dominated by growth of TiO_2, took place when straining was conducted at rates slower than a

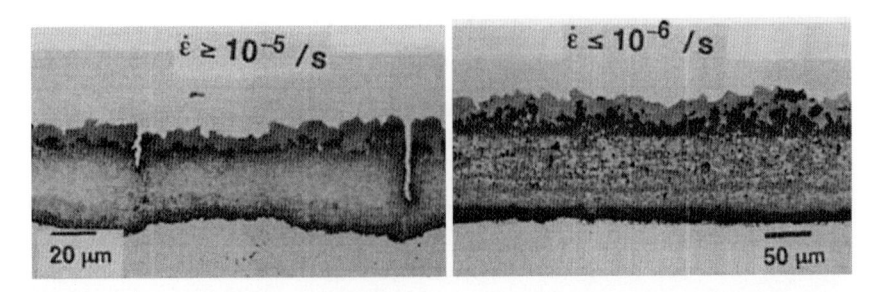

Figure 12.5 Metallographic cross sections of tensile specimens (of presumably Ti-50Al) tested at 900 °C but with different (constant) strain rates [2]. It appears that when the strain rate is below a critical alloy-dependent rate, then oxide crack healing can take place.

given critical strain rate; see Figure 12.5 [2]. This critical strain rate was thought to be $1.5 \times 10^{-6}\,s^{-1}$ for Ti-50Al and $1.9 \times 10^{-4}\,s^{-1}$ for Ti-50Al-2Nb [47]. Although not discussed here, work to determine the fracture toughness of oxide scale at temperature is presented by Bruns and Schütze [48].

12.2.3
Effect of Oxygen and Nitrogen

It has been demonstrated in a number of studies that oxidation rates are higher in air than in oxygen [22, 24, 49–51]. Becker et al. [15] have also reported this to be the case for stoichiometric TiAl exposed at 1000 °C, however, they observed the opposite behavior for the same material oxidized at 900 °C, that is faster oxidation in oxygen than air. This anomalous behavior of Ti-50Al at 900 °C was not observed by Zheng et al. [51] who observed a deleterious effect of nitrogen on oxidation resistance at 900 °C. It was also found that the oxidation behavior of Ti-48Al-5Nb at 900 °C was better in air than in an argon–oxygen gas mixture for exposure times longer than 50 h [51]. Detailed examination revealed that although the material oxidized in air had a slightly thicker oxide scale, the material oxidized under the argon–oxygen mixture showed significant internal oxidation that resulted in a larger weight gain, resulting in the observed reduction in oxidation resistance. The effect of preoxidation in high [52] and low-pressure [53] oxygen environments, has also been studied, and has been shown in some circumstances to increase subsequent resistance to oxidation.

Work has been performed to gain to understanding of the reasons behind the different oxidation rates in oxygen and air [15, 54]. Becker et al. [15] considered that in pure oxygen the partial pressure is five times higher than in air, and that reactions with nitrogen are excluded. In order to clarify the effect of these factors, oxidation tests were performed in pure oxygen, air, and oxygen–argon gas mixtures. It was found that when oxidation took place in a nitrogen-free atmosphere, oxygen partial pressure at levels between 0.01 and 1 bar had no influence on the oxidation kinetics [15]. Nwobu et al. [54] reported similar findings. Other work concerning the effect of oxygen partial pressure on oxidation behavior at 1300 K (1027 °C) is reported in [55], although here an increase of oxidation rate with decreasing oxygen partial pressure was observed. With regard to the mechanism by which nitrogen in the air enhances oxidation, work performed by Zheng et al. [51] concluded that nitrogen hampers both the formation and long-term stability of alumina in the initially formed scale through the formation of titanium- or titanium/aluminum-based nitrides. Dettenwanger et al. [56] observed the formation of an "alternating scale" containing TiN and Al_2O_3 near the metal/scale interface and an aluminum-depleted zone beneath the oxide scale. The formation of both TiN and Ti_2AlN has also been reported in [54, 57]. Rakowski et al. [50] observed the formation of an intermixed (not layered) TiN and Al_2O_3 scale and indicate that it is permeable to nitrogen. Figure 12.6 is a schematic diagram proposed by Rakowski et al. [50] for the development of oxide-scale morphology on TiAl in air at temperatures between 800 °C and 900 °C. The formation of nitrides is thermodynamically surprising as the standard free energy for the formation of

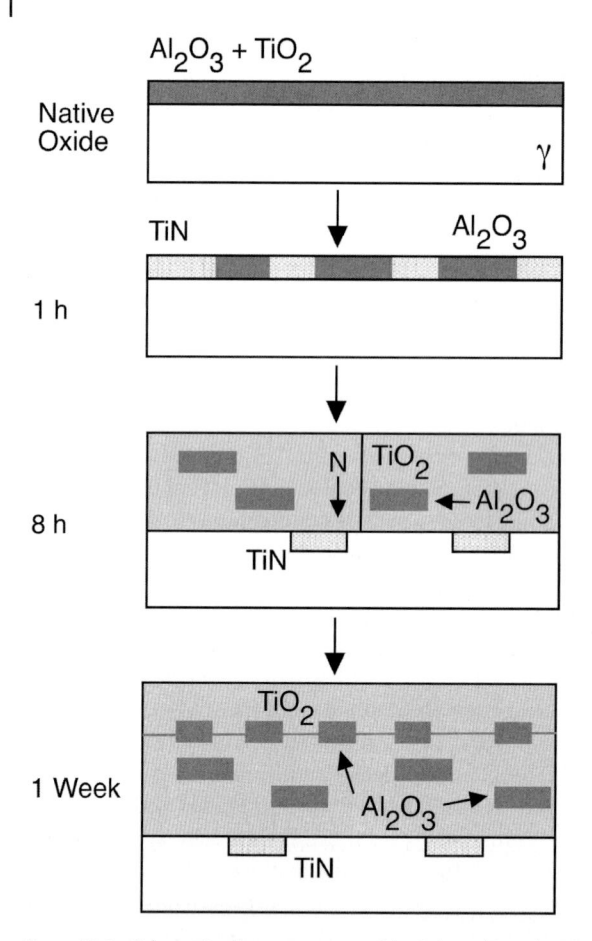

Figure 12.6 Schematic diagram proposed by Rakowski et al. [50] showing the development of TiAl oxide morphology during oxidation at 800 to 900 °C in air. Redrawn diagram based on original.

titania is more negative [57]. This is discussed in detail elsewhere [57], but indicates that the activity of nitrogen is very high compared to oxygen. According to unpublished work that is referenced in Rakowski et al. [58], TiAl does not react with nitrogen at temperatures lower than around 600 °C. Thus, it could be speculated that the oxidation behavior of TiAl in air at temperatures lower than 600 °C may be similar to that in oxygen.

Rahmel et al. [2] and Quadakkers et al. [21] take the view that nitrogen may influence the oxidation resistance both positively and negatively. In the early stages of oxidation, nitrogen reduces the oxidation resistance as the initially formed alumina-rich scale is destroyed and the titanium nitrides that are formed oxidize. This effect is particularly pronounced in alloys such as Ti-48Al and Ti-50Al that have a greater tendency to form Al-rich scales [21]. The effect of nitrogen is thought to be less significant in alloys, such as Ti-45Al, that have a lower tendency to form Al_2O_3. If, during later stages of oxidation, a near-continuous nitride layer is formed, it is thought that this could stabilize the Z-phase (see Section 12.2.5), and prevent the fast oxidation kinetics associated with internal oxidation and destruction of the outer alumina barrier layer. This beneficial effect of nitrogen was observed for a whole series of ternary niobium-containing alloys with aluminum ranging between 45 and 48 at.% and niobium between 2 and 10 at.%. For the binary alloys investigated, which contained between 45 and 50 at.% aluminum, the beneficial effect of nitrogen was thought to depend on alloy composition, exposure time and possibly temperature [21].

12.2.4
Effect of Other Environmental Factors

In industrial applications, oxidation takes place in an environment consisting not only of oxygen and nitrogen but also of combustion gases and water vapor. Thus, knowledge of oxidation behavior in such environments is of technical importance. Taniguchi et al. [59] report that the presence of water vapor leads to significantly higher oxidation rates at 827 °C (1100 K) and 927 °C (1200 K) in Ti-50Al. At 827 °C the mass gain was observed to increase with increasing water vapor content up to around 75% water vapor. Additionally, the scale became richer in TiO_2 and thicker as the fraction of water vapor increased. Although the types of oxide formed in dry and humid conditions remained the same, the scale morphologies were different. In the presence of water vapor the scale formed was characterized by a two-layer structure, with an outer TiO_2 layer that showed signs of directional growth and a porous inner layer consisting of very fine TiO_2 and Al_2O_3 grains. The thin Al_2O_3-rich layer usually formed after exposure in oxygen or air, was not observed after oxidation in a water-vapor-containing environment. Near linear oxidation kinetics were determined in the presence of water vapor, whereas parabolic behavior was observed for exposure in oxygen. Kremer and Auer [60] made similar observations for oxidation of Ti-50Al at 900 °C. Again, the outer oxide layer was mainly TiO_2 and the inner layer fine-grained Al_2O_3 and TiO_2 with some discrete Al_2O_3 particles between these two layers, however, no mention was made concerning a directional TiO_2 morphology. Internal oxidation of Al to Al_2O_3 in an Al depleted subsurface zone was observed. The oxidation rate was found to increase with increasing partial pressure of water vapor and decreasing oxygen partial [60], see Figure 12.7. Although the above studies have shown that the presence of water vapor leads to an increase in oxidation, work by Brady et al. [61] and others [62, 63] indicates that the oxidation kinetics of TiAl alloys that form a protective alumina scale are not significantly affected by water vapor. Zeller et al. [64] suggest that the influence of water vapor on oxidation kinetics may strongly depend on the relative volume contents of TiO_2 and Al_2O_3 in the scale.

While the papers by Taniguchi et al. [59] and Kremer and Auer [60] are of great interest, the oxidation temperatures employed are rather high compared to the nominal expected service temperatures for TiAl in aerospace components. The

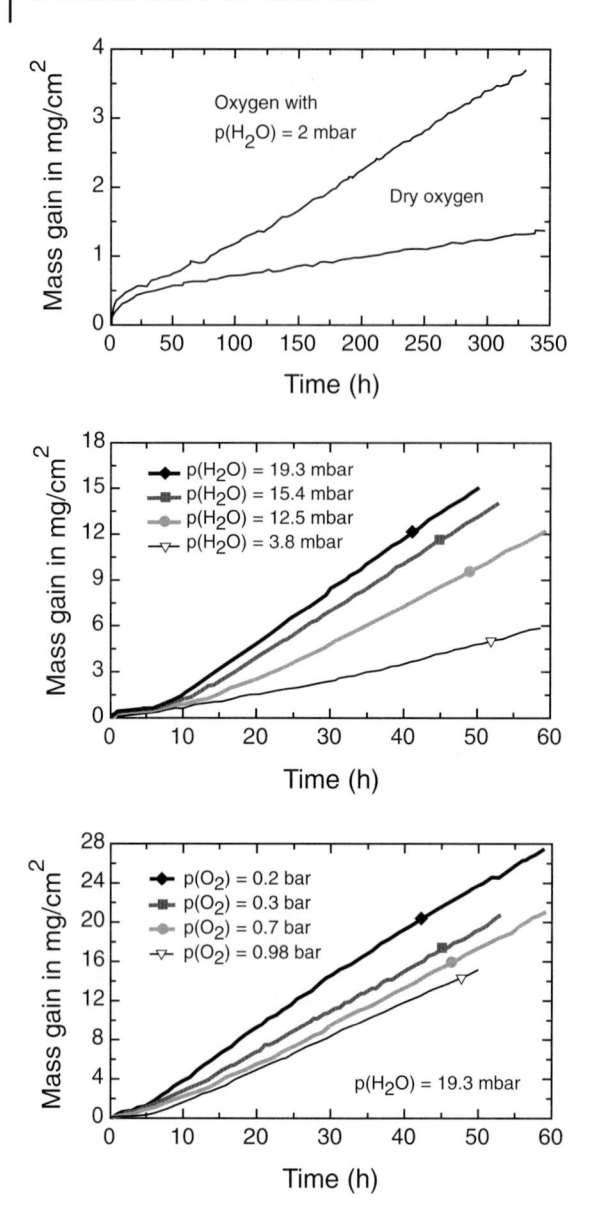

Figure 12.7 Graphs showing the effect of water vapor pressure on the oxidation kinetics of Ti-50Al during oxidation at 900 °C in oxygen and oxygen/water vapor environments [60]. Graphs have been redrawn. Lines have replaced the data points on the original graphs.

Figure 12.8 Micrographs showing different scale morphologies developed in Ti-47Al-1Cr-Si after 1000 h exposure at 700 °C in (a) dry air and (b) air with 10 vol.% water vapor [64].

work of Zeller et al. [64], which consider temperatures of 700 °C and 750 °C, may thus be more relevant with regard to component applications. In spite of the lower temperatures investigated by Zeller et al. [64], similar features to those described by Taniguchi et al. [59] and Kremer and Auer [60] were observed. Namely, increased oxidation rate (after an initial incubation period), a modified oxide morphology and the formation of an aluminum depleted subsurface zone. However, the outer TiO_2 layer had a fine needle-like morphology, see Figure 12.8, unlike that described by Taniguchi et al. [59]. In a second paper, Zeller et al. [65] showed that elevated temperature fatigue and creep behavior were also influenced by the presence of water vapor.

Most studies have considered the oxidation of TiAl alloys in air, oxygen or argon–oxygen mixtures; with the exception of Li and Taniguchi [66], almost no work has been published on oxidation within an environment similar to that encountered by real aero-engine components. As part of a turbocharger endurance test, Tetsui and Ono [67] observed that the scale formed after exposure to diesel engine exhaust gas did not develop an outer TiO_2 layer, and that a niobium-enriched layer was formed at the metal/scale interface in a high niobium-containing alloy. Additionally, the scale was thinner than expected when compared to the results of oxidation tests in air. In the study by Li and Taniguchi [66], five different TiAl alloys were cyclically exposed at 900 °C (1173 K) under a simulated combustion gas consisting of $10\%O_2$-$7\%CO_2$-$6\%H_2O$–N_2 (by volume). It was found that alloys that formed continuous Al_2O_3 layers on account of their chemistry showed excellent oxidation resistance. In fact, better resistance was observed in the simulated combustion gas than in air. This was found to be a consequence of the reduced oxygen content in the simulated combustion gas. The presence of H_2O and CO_2 did not significantly influence the oxidation behavior of these continuous Al_2O_3 layer-forming alloys, but this was not the case for Ti-50Al, which showed enhanced oxidation. The authors indicated that this is because H_2O and CO_2 play a role when TiO_2-rich scales are present, but not when Al_2O_3-rich scales are formed.

12.2.5
Subsurface Zone, the Z-Phase, and Silver Additions

After oxidation and formation of a scale on TiAl, the development of an aluminum-depleted zone underneath the scale has been reported [59, 60, 64, 68, 69]. As outlined below various names, including NCP (reported in [21, 70]), X-phase and Z-phase, have been have been given to a cubic phase found in this zone. According to Dettenwanger et al. [57], one would expect the formation of α_2 (Ti_3Al) in this subsurface layer (as has been reported by Nwobu et al. [54]), but Dettenwanger et al. observed this only after long exposure times. The two-phase aluminum-depleted zone observed after such long exposure times initially developed with the formation of a cubic phase (called X-phase) at the scale/TiAl interface, with subsequent heterogeneous nucleation of α_2 at the X-phase/TiAl interface. The thickness of the depleted layer increased after further exposure and both the X-phase and α_2 grew into the base metal. The oxygen content of the α_2 was measured to be 7 at.%, which is reported to be in good agreement with the value measured by Shida and Anada [71].

Beye and Gronsky [72] reported that the depleted layer consisted of the $Ti_{10}Al_6O$ and $Ti_{10}Al_6O_2$ phases. While the $Ti_{10}Al_6O$ phase can coexist with $Ti_{10}Al_6O_2$, they proposed that it is metastable and can transform to the more stable $Ti_{10}Al_6O_2$ phase. Becker et al. [15] report a zone of internal oxidation close to the metal/scale interface and the presence of the $(Ti, Al)_3O_2$ and $(Ti, Al)O$ phases. Dowling and Donlon [73] describe a cubic phase of composition Ti_2Al enriched with 2 to 5 wt.% oxygen after high-temperature exposure in air or vacuum. Zheng et al. [74] observed a depleted layer after oxidation that was initially composed of a single phase with a cubic lattice structure ($a = 0.69$ nm, which they called Z-phase). Later on in the oxidation process, the depleted layer also contained the α_2 phase. The approximate composition of the Z-phase was determined to be $Ti_5Al_3O_2$. Quadakkers et al. [21] report two different morphological types of the depleted layer both consisting of Z-phase-phase and oxygen-enriched α_2. Copland et al. [75] also mention the development of oxygen-enriched α_2 at the Z-phase-phase/TiAl interface, which coincided with the break up of protective Al_2O_3 and the formation of a less-protective $TiO_2 + Al_2O_3$ scale. Decomposition of the Z-phase-phase into α_2 and Al_2O_3 was also discussed.

After long exposure times Shemet et al. [76] reported the formation of internal alumina precipitates in binary alloys, that were probably formed at the Z-phase/α_2 interface. According to Shetmet et al. [77] when the subsurface layer consists of the Z-phase rather than α_2, protective alumina is formed. This seems to be in disagreement with Copland et al. [75] who concluded that exclusively Al_2O_3 scale growth couldn't be sustained on γ-based alloys that form the Z-phase in the subsurface zone. The addition of Ag to binary alloys can stabilize the Z-phase in the subscale depletion layer, and thus prevent the formation of α_2 and Ti-rich nitrides that lead to breakdown of the alumina scale in conventional TiAl alloys [77]. High chromium additions to Ag-doped alloys (compositions such as Ti-48Al-2Ag-7Cr) can show improved oxidation resistance over Ti–Al–Ag alloys. It was speculated that as a coating material Ti–Al–Ag–Cr alloys may show higher toughness than Ti–Al–Cr-based materials and higher-temperature capabilities than Ti–Al–Ag alloys [78].

12.2.6
Effect of Surface Finish

With reference to an article by Choudhury et al. [49], which indicates that the oxidation behavior of γ-TiAl may be influenced by surface preparation, Rakowski et al. [58] investigated the effect of surface finish on the oxidation behavior of Ti-50Al and Ti-48Al. It was reported that surface finish had no influence on the oxidation behavior in air due to the overwhelming influence of the "nitrogen effect". However, for exposure in oxygen, surface finish did affect the oxidation behavior, with finely polished material (to 1 µm) exhibiting much faster oxidation behavior than material prepared to a P600 grit finish. The relevant oxidation curves for Ti-50Al are shown in Figure 12.9. The surface effect is believed to be a consequence of the amount of stored energy in the surface material that results from surface preparation. Surface material with a high stored energy is believed to form a recrystallized zone that promotes the formation of an alumina film during exposure in oxygen. Two mechanisms involving short-circuit diffusion were proposed; one involving an increase of the number of grain boundaries, while the other envisaged a higher dislocation density. Increasing the diffusivity of aluminum would reduce the near-surface depletion of aluminum and thus promote alumina stability.

Copland et al. [75] confirmed the experimental findings of Rakowski et al. [58] for Ti-52Al exposed in pure oxygen, namely that a P600 surface finish exhibited improved oxidation resistance when compared to a 1-µm surface polish. The P600 surface led to growth of an Al_2O_3 scale, whereas the polished material grew a mixed $TiO_2 + Al_2O_3$ scale. It was suggested that the rougher surface promoted the nucleation of Al_2O_3, making it easier to form a continuous protective scale, and also leads

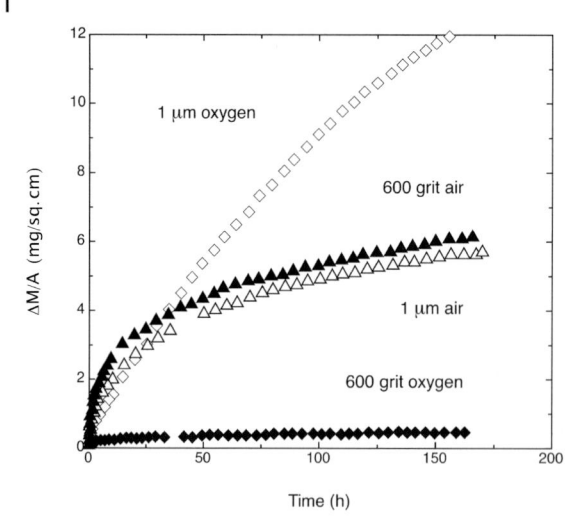

Figure 12.9 Weight gain data for Ti-50Al exposed at 900 °C in oxygen and air [58]. The specimens had either a P600 grit surface finish or a 1-μm polished surface. In air, surface preparation had little effect on oxidation behavior at 900 °C due to the nitrogen effect, however, in oxygen the polished surface showed much faster oxidation. At temperatures below 600 °C, where TiAl no longer reacts with nitrogen, it is believed that surface preparation may again have an influence on oxidation behavior [58]. The data on the original graph has been replotted.

to the development of a continuous layer of the Z-phase at the oxide/metal interface. However, after prolonged exposure, the subsurface zone consisted of Z-phase and precipitated α_2 (enriched in oxygen). This coincided with the break up of the Al_2O_3 layer and the formation of a less-protective $Al_2O_3 + TiO_2$ scale.

12.2.7
Ion Implantation

This is a surface-engineering technique that has been widely used for improving the oxidation resistance. Shalaby [79] was awarded a patent in 1997 for the use of the technique in the protection of gamma and alpha-two titanium aluminides. The technique relies on the use of an ion accelerator to implant high-energy ions into the surface of the material, under vacuum (or near-vacuum) conditions. The aim is to implant elements that locally stabilize the formation of alumina scale at the outer surface. Penetration depth profiles typically extend 0.01 μm to 1 μm into the surface with implanted-element concentrations of up to several tens of per cent being possible [80]. The implantation of ions using different energies can be used to modify/optimize their subsurface penetration distribution. For example, Figure 12.10 shows how the concentration of fluorine varies as a function of depth and implantation energy [81]. The as-implanted profile is, however, modified by oxidation and the establishment of a protective alumina layer, and after long-term oxidation, diffusion of fluorine into the base metal has been described [82].

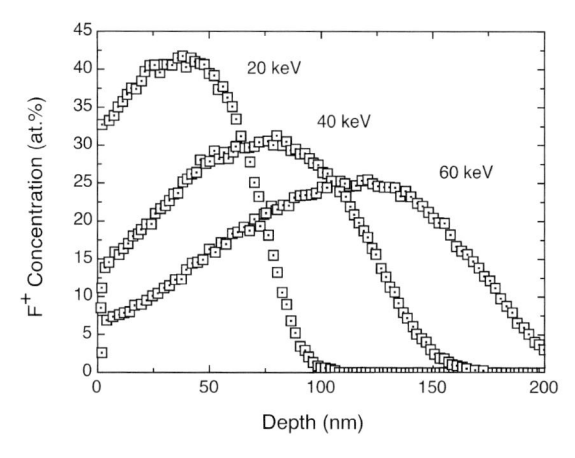

Figure 12.10 Graph showing the influence of ion implantation energy on the concentration distribution profile of fluorine ions in TiAl for a dose of 2×10^{17} ions/cm^2 [81]. The original data has been replotted in the graph.

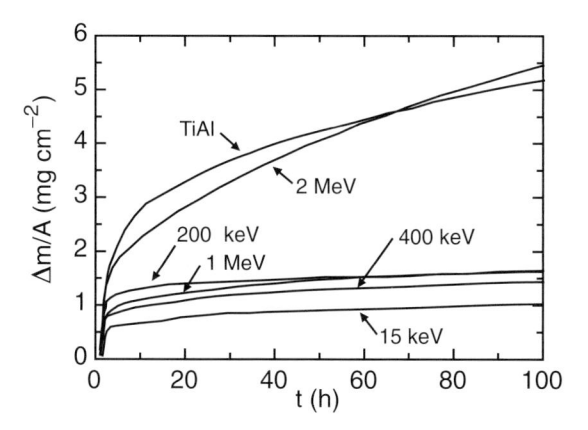

Figure 12.11 Influence of implantation energy on the oxidation kinetics of Cl-implanted TiAl at a constant implantation dose of 10^{16} Cl cm^{-2} [84]. The original figure has been replotted and the data smoothed.

Stroosnijder states that the real value of ion implantation is as a tool in the research environment, where the influence of various elements on oxidation behavior can be investigated [80]. For chlorine ions, implantation energies of 1 MeV and doses above 10^{16} cm^{-2} are reported to result in improved oxidation resistance of Ti-50Al oxidized in air at 900 °C [83]. However, as shown in Figure 12.11, the implantation energy can also influence the oxidation behavior [84]. The positive effect of chlorine and the other halogens on oxidation resistance is further described in Section 12.2.8.

One recently developed technique used for ion implantation is called "plasma-immersion ion implantation" (sometimes referred to as PIII or PI^3) which has certain advantages over so-called "beam-line ion implantation". Advantages include no line-of-sight limitations (thus making the technique suitable for industrial use on components with complex geometries), high dose rates and simple equipment that is relatively easy to use [85–87]. For further details concerning the influence of various ion-implanted species on oxidation behavior (isothermal and/or cyclic) the reader is referred to: Cl, Si [83], Br, Cl, F, I [84], Cl, F [85], Cl [87], F [88], Nb [89], Nb, Al [90], F [91], Cl [92], Si [93], Cl [94], Nb [95], Al, Si, Cr, Mo [96], Cl, P, B, C, Br [97], Nb [98], Nb, Si, Ta, W [99], Nb [100], Al, Ti, Cr, Mo, Y, Mn, Pt, Nb, Si [101] and V, Nb, Ta, W, Cr, Al, Y, Ce [102].

12.2.8
Influence of Halogens on Oxidation

In 1993 Kumagai, of Sumitomo Light Metal Industries, applied for a patent concerning the beneficial effect on oxidation resistance of halogens as alloying additions [103]. Since then, much work has been directed to obtaining an understanding of the phenomena and developing further routes of halogen incorporation, particularly by the group of Professor Schütze at DECHEMA. Indeed, several patents have been applied for that concern different ways of incorporating halogens into the surface of TiAl and TiAl components [104–108].

The mechanism by which halogen additions improve the oxidation resistance is believed to involve the formation of gaseous aluminum halide at the metal/oxide interface. This diffuses outward through the initially formed scale and oxidizes due to the increased partial pressure of oxygen nearer the surface. This process liberates aluminum atoms that are then oxidized to form a continuous protective layer of alumina. The liberated halogen atoms can diffuse back within the metal where they can again react with aluminum to form aluminum halide, and the process is repeated [84, 85]. The incredible oxidation protection offered by the use of halogens can be seen in Figure 12.12 which shows a SEM cross section of Ti-48Al-2Cr oxidized at 900 °C for 100 h. The halogen-treated side has a very thin continuous layer of alumina whereas the unprotected side has developed a thick oxide scale [84]. It has been reported that fluorine performs particularly well under cyclic oxidation conditions especially in wet environments and that it could provide protection to alloys containing more than 40 at.% aluminum [81]. Further information regarding protection through the use of halogens, including the detailed mechanism and thermodynamics can be found in [81, 109–111]. Other work concerning the influence of halogens, applied to the surface by either ion implantation, spraying, dipping or painting with halogen-containing liquids, can be found in [112–114], as well as the relevant references given in Section 12.2.7.

12.2.9
Embrittlement after High-Temperature Exposure

In most envisaged applications, components made of TiAl will be subjected to relatively high temperatures of around 650 °C to 750 °C in aero-engines and possibly up to 950 °C in turbochargers. Thus, oxidation will take place, the extent to

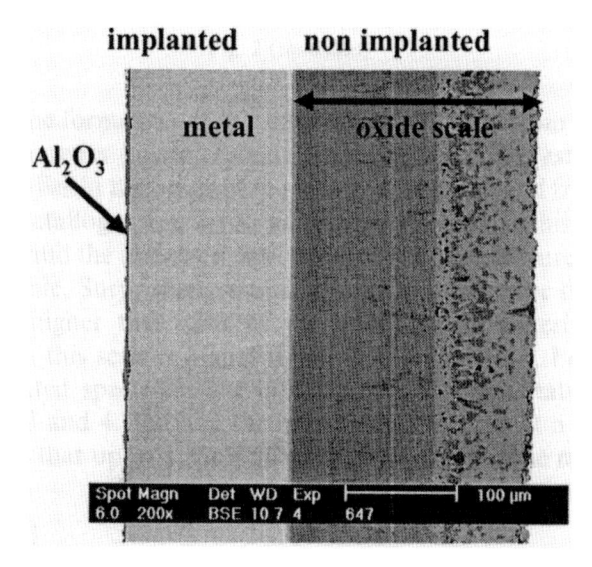

Figure 12.12 Backscattered electron image of a cross section through Ti-48Al-2Cr oxidized in air at 900 °C for 100 h [84]. The unprotected right-hand side has developed a thick oxide scale whereas the chlorine implanted (10^{16} Cl cm^{-2}, 1 MeV) left hand side is covered by a thin protective alumina scale.

which being determined by factors such as temperature, environmental atmosphere, alloy composition and the state of the component surface. It has been widely documented that the high-temperature exposure of tensile specimens can lead to a serious reduction of ductility on subsequent testing at room temperature [73, 115–122]. From the work performed it is clear that such embrittlement is related to a process localized to the oxidized outer surface. As shown in Figure 12.13a, making tensile specimens from tensile test specimen blanks that had been exposed at 800 °C for 1504 h in air, resulted in no loss of ductility. Other work has also shown this [116, 120]. Subsequent exposure of tensile specimens made from the exposed blanks, to 700 °C for 24 h, resulted in almost complete loss of ductility Figure 12.13b, although removal of a few tens of micrometers from the oxidized surface was sufficient to restore the ductility, in agreement with [73, 115, 119, 121]. Kelly et al. [117] indicate that exposure at 315 °C for 10 h is sufficient to reduce ductility. It has also been demonstrated that exposure in moderate vacuum [73, 115] or argon [115] leads to a similar degree of embrittlement as observed after exposure in air.

In spite of the work performed, the exact embrittlement mechanism still does not seem to be fully understood. To date, the diffusion of oxygen into the near-surface region [119] and the formation of a brittle surface layer [73] that initiates cracks that propagates in the presence of atmospheric moisture [115, 117, 118] have been put forward as possible explanations. Thomas et al. [116] differentiate between the mechanisms responsible for embrittlement in Ti-47Al-2Cr-2Nb after

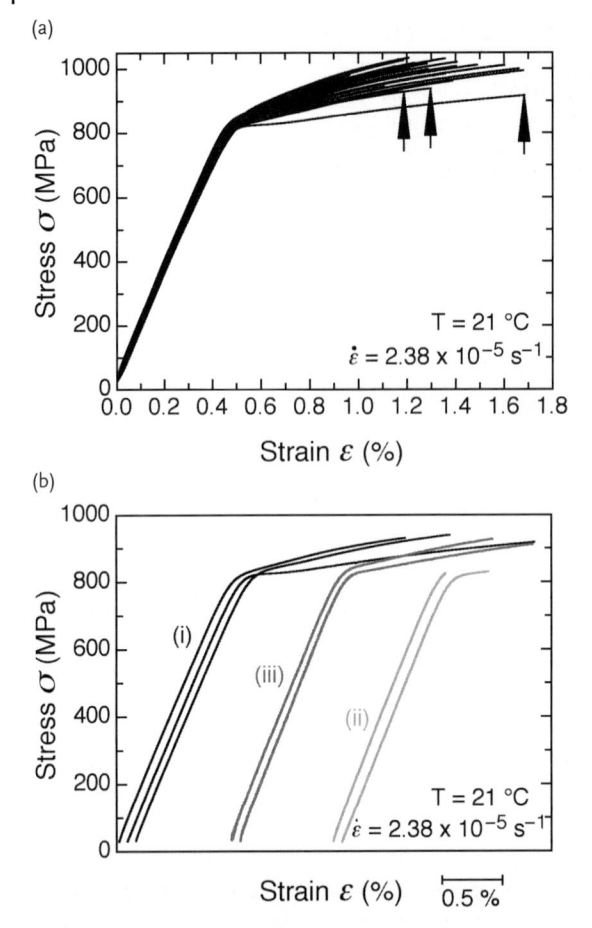

Figure 12.13 (a) Tensile-test curves showing that heat-treatment of tensile test blanks at 800 °C for 1504 h in air (the 3 arrowed curves) did not significantly change the plastic elongation compared to untreated material. The 3 tensile curves in (b) labeled (i) are the same curves as arrowed in (a). Those marked with (ii) were exposed to 700 °C for 24 h in air and showed embrittlement. Ductility was restored after removal of around 100 μm from the oxidized tensile specimens by grinding, the curves labeled (iii).

short- and long-term exposure. After long exposures, brittleness of the surface layers (oxide scale, Al-depletion zone and Nb-rich precipitates) was identified. However, after short exposure times embrittlement was thought to be related to the interaction of several factors such as an oxygen-enriched surface, reduced propagation of mechanical twinning near the surface, and a residual stress gradient due to cooling [116].

The effect of residual stress from the grinding of tensile specimens may also play a role and has been discussed by Austen and Kelly [115] and Kelly et al. [117]. The removal of 100 μm to 150 μm from a ground specimen by chemical milling

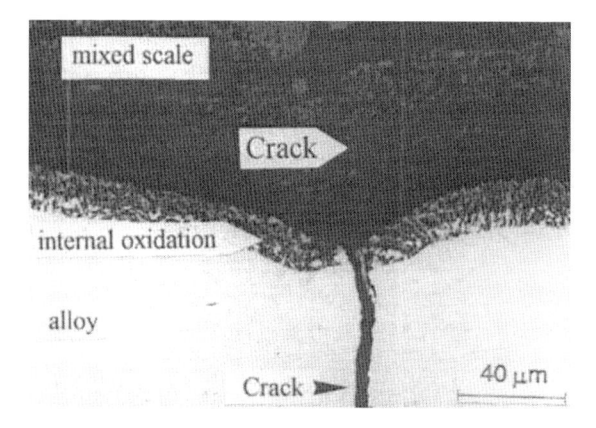

Figure 12.14 Crack in Ti-45Al that extends into the subsurface base metal after 100 h exposure to an argon–oxygen environment at 900 °C [21].

reportedly reduced or even prevented susceptibility to embrittlement after subsequent exposure. However, if less than 75 μm or more than 250 μm was removed then it was reported that the susceptibility increased. In both as-exposed [21] and also room-temperature-tested tensile specimens [117], the presence of cracks from the surface to within the bulk material after elevated temperature exposure has been observed, Figure 12.14. On subsequent room-temperature tensile testing in air, embrittlement was observed, although if testing took place under argon then near-normal ductility was obtained, despite the surface cracking [117]. Additionally, testing at high strain rates also resulted in near-normal ductility. Thus, a mechanism involving the formation of surface cracks followed by moisture-induced hydrogen embrittlement was proposed [115]. Interestingly, Yamaguchi and Inui [123] have shown a deleterious effect of hydrogen on the tensile ductility of PST TiAl, in particular when tested at slow strain rates, see Figure 12.15. Liu and Kim [124] have also shown that TiAl is prone to moisture-induced environmental embrittlement at room temperature. The effect of high-temperature exposure to hydrogen on room-temperature tensile and fatigue properties has been studied by Iino et al. [125]. For further information on the embrittlement of intermetallics through moisture in the environment, the reader is referred to Chen and Liu [126]. Recently, Wu et al. [122] have suggested that the embrittlement could arise from a significant tensile stress being developed just under the outer surface that encourages premature cracking of the subsurface material, resulting in reduced ductility. Clearly, the reasons behind embrittlement are rather complex, although as outlined above it is agreed that ductility loss results from surface-based mechanisms, with the fracture-initiation sites being at the surface [116], and that bulk material properties are more or less unaffected by high-temperature (oxidation) exposure [119–121].

Very interestingly, it has been shown that at temperatures of around 200 °C to 250 °C the ductility of exposed as-cast Ti-48Al-2Cr-2Nb is restored to values close to those determined in unexposed material tested at room temperature [115, 117],

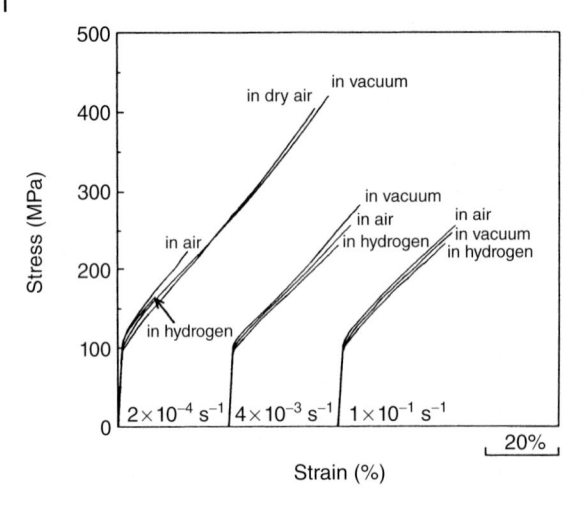

Figure 12.15 Effect of testing environment and strain rate on the room-temperature tensile ductility of PST TiAl with the lamellae oriented 31° away from the tensile axis [123]. The deleterious effect of hydrogen is evident at slow strain rates where embrittlement is particularly significant.

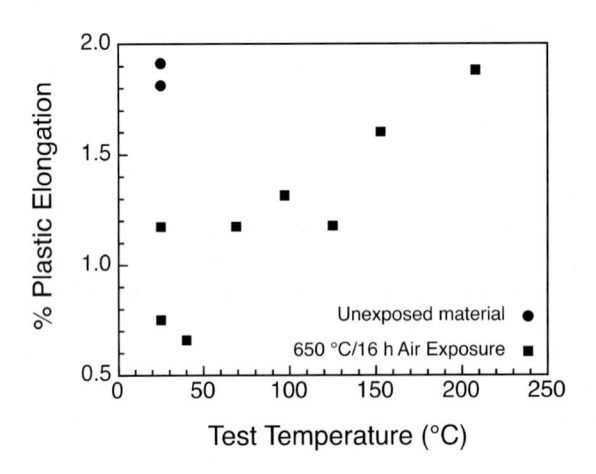

Figure 12.16 Graph showing that cast Ti-48Al-2Cr-2Nb exposed at 650 °C for 16 h regains room-temperature levels of ductility on testing at temperatures of around 200 to 250 °C [117]. The original data has been replotted.

see Figure 12.16. Thus, it is believed that the embrittlement phenomena should not be an issue during high-temperature operation but may cause difficulties during assembly and servicing of exposed components [117]. However, Li and Taniguchi [127] have presented results indicating that exposure at 900 °C of cast Ti-48Al-2Cr-2Nb and Ti-48Al-2Cr-2Fe in simulated combustion gas reduced tensile ductility during testing at 800 °C, albeit to a level remaining above 10%, and that the effect increased with exposure time. Unfortunately, unlike the case for scale morphology development [128], no work has been performed on embrittlement after very long exposure times.

With regard to the effect of alloy composition and microstructure, relatively little systematic work has been performed. Lee et al. [118] indicate that embrittlement takes place regardless of alloy composition or microstructure. Draper et al. [119] indicate that the loss of ductility after exposure is more serious for high-strength alloys containing niobium, although it was not differentiated if the high niobium content or the high strength was responsible. Additionally, for such high-strength alloys, testing in argon after exposure showed similar embrittlement compared to testing in air; a different observation from that made in [117]. Unpublished oxidation curves obtained at GKSS for various alloys at 800 °C are shown in Figure 12.17 and indicate a wide range of oxidation behavior. Data for a state-of-the-art oxidation resistant Ti-50Al-2Ag alloy presented by Niewolak et al. [129] is also shown. The surface of tensile test specimens with the same compositions, exposed for 1552 h at 800 °C, are shown in Figure 12.18. As can be seen, significant oxide spalling has taken place for some compositions, in particular for the binary Ti-45Al alloy. The γ-TAB alloy showed little oxide spalling but exhibited the second highest weight gain and developed a relatively thick scale compared to the high niobium-containing alloys. Nevertheless, upon tensile testing the γ-TAB alloy exhibited around 0.3% plastic elongation at a relatively low stress level, whereas the more oxidation-resistant alloys exhibited no plastic deformation, Figure 12.19. Thus, the degree to which strength level and oxidation resistance independently influence embrittlement seems unclear.

The room-temperature dynamic fracture toughness after exposure was measured in [119] and was found to decrease from around 22 MPa m$^{1/2}$ in as-extruded unexposed material to around 11 MPa m$^{1/2}$ after 300 h exposure at 700 °C. Draper et al. [119] report that although exposure caused embrittlement, fatigue properties both at 650 °C and room temperature were unaffected. Planck and Rosenberger [121] report that exposure of Ti-46Al-2Nb-2Cr-1Mo-0.2B at 760 °C for up to 500 h resulted in only a modest degradation of fatigue properties determined at room temperature and 760 °C, but that the fatigue properties at 540 °C were more severely affected. From other work [130], this temperature is close to a range where strain aging has been shown to take place and ductility accordingly reduced. As with tensile ductility, fatigue properties were restored after removal of the embrittled surface layer [121].

GKSS has performed some unpublished work on use of halogens to see if they are capable of reducing embrittlement. Unfortunately, the results to date have been rather disappointing. Until the exact mechanism by which embrittlement takes place is understood, it is rather difficult to implement mitigation strategies. If moisture-induced hydrogen embrittlement of surface cracks that emanate from the brittle oxide is the main mechanism; then the use of ductile coatings may be useful, as described in a patent by McKee of General Electric [131]. For further information on work performed within the United States, up to around 1996, on both the embrittlement and oxidation behavior of intermetallic compounds the reader is directed to an overview paper by Meier [132].

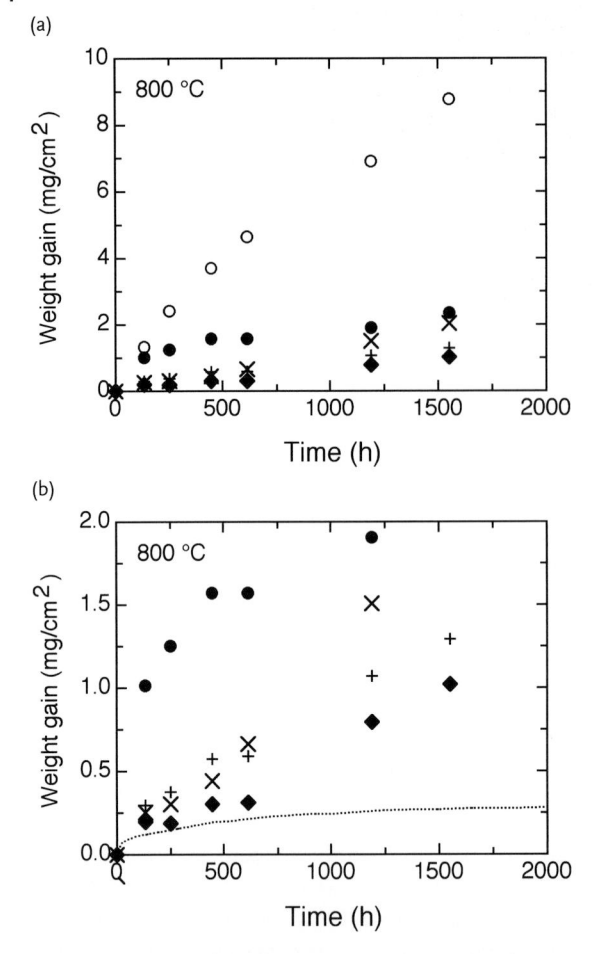

Figure 12.17 (a) Isothermal oxidation curves of various alloys (○ = Ti-45Al, ● = γ-TAB = Ti-47Al-4(Nb, Cr, Mn, Si, B), × = Ti-45Al-10Nb-0.2B-0.2C, + = Ti-45Al-8Nb-0.2C, ◆ = TNB-Co3). After being exposed at 800 °C for 1552 h in air, the γ-TAB alloy showed the second lowest resistance to oxidation but nevertheless was the only alloy that exhibited some tensile ductility on subsequent tensile testing at room temperature, Figure 12.19. Figure 12.17 (b) shows an enlarged section of (a) and includes data given in [129] for a state-of-the-art oxidation-resistant alloy of composition Ti-50Al-2Ag (dotted line).

12.2.10
Coatings/Oxidation-Resistant Alloys

In general, coating of components can be useful in two respects, namely environmental and/or thermal protection. Obviously, the coatings applied should be resistant to thermal fatigue and its coefficient of thermal expansion should be relatively close to that of the base material so that rumpling/spalling of the coating is avoided. For example, according to Chen and Lin [133] a compressive stress of

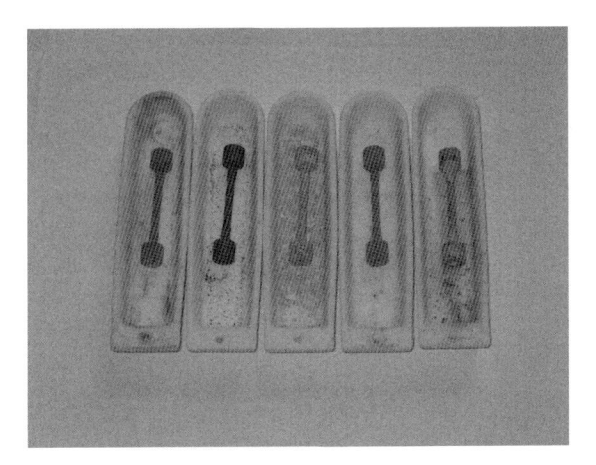

Figure 12.18 Appearance of tensile specimens after exposure in air for 1552 h at 800 °C. The composition of the specimens from left to right is: TNB-V2 (Ti-45Al-8Nb-0.2C), γ-TAB (Ti-47Al-4(Nb, Cr, Mn, Si, B)), Ti-45Al, TNB-Co3 and Ti-45Al-10Nb-0.2B-0.2C. The Ti-45Al binary composition (center) showed the worst resistance to oxidation and the most pronounced spalling behavior.

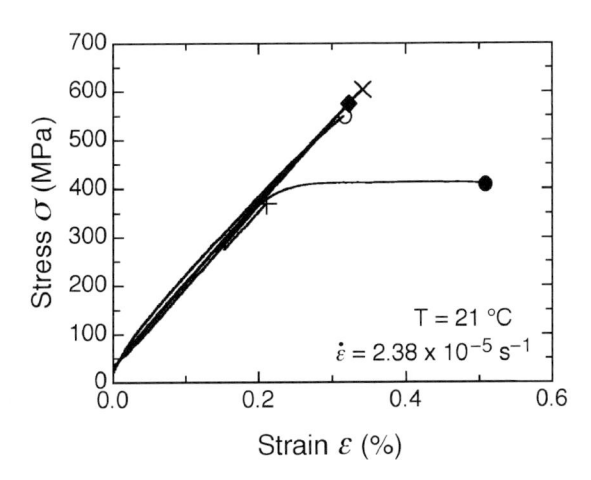

Figure 12.19 Room-temperature-tensile test curves for the specimens shown in Figure 12.18 that had been exposed in air for 1552 h at 800 °C. As can be seen, only the γ-TAB alloy exhibited any plastic deformation on loading, despite its comparatively low oxidation resistance (○ = Ti-45Al, ● = γ-TAB = Ti-47Al-4(Nb, Cr, Mn, Si, B), × = Ti-45Al-10Nb-0.2B-0.2C, + = Ti-45Al-8Nb-0.2C, ◆ = TNB-Co3).

2590 MPa can develop in a 0.002-mm thickness coating of TiC on TiAl on account of the difference in expansion coefficients, during cooling from a sputter-deposition temperature of 700 °C to room temperature.

Various methods have been employed to coat TiAl including: 1. sputtering [133–140]; 2. reactive sputtering [141]; 3. spark plasma sintering [142]; 4. plasma vapor deposition [139, 143]; 5. pack cementation [144]; 6. electroplating and pack cementation [145]; 7. enamel coating [138, 141] and 8. anodic coating [146]. Additionally, coating using aluminosilicate [147] and $CaTiO_3$ [148] in addition to a phosphoric acid treatment [149] has been investigated. An extensive list of the various techniques and coatings studied up to around 2002, including references, is presented in [16].

With regard to the development of materials that may be suitable as oxidation-resistant coatings, Brady *et al.* [150] patented a series of alloys based on Ti–Al–Cr that contain the $Ti(Cr, Al)_2$ Laves phase in a "ductile" matrix of γ-TiAl. The general composition range in which the alloys lie is Ti-(49-53)Al-(9.5-12.5)Cr. The Laves phase results in the formation of a protective alumina scale, while the TiAl matrix increases the material crack resistance when compared to the Laves phase alone. Additionally, the TiAl matrix ensures better chemical and thermal compatibility with the underlying TiAl base material [3]. According to the patent, low-pressure plasma spraying, sputtering, plasma vapor deposition, chemical vapor deposition, slurry processing and diffusion coating techniques, among others, may be used to apply the coating. It is stated that a Ti-51Al-12Cr coating was successful in protecting a Ti-48Al-2Cr-2Nb substrate [3]. At the time this paper ([3]) was published (1996), the effect of such a coating on LCF and HCF was being evaluated. Other work on the oxidation behavior of Ti–Al–Cr-based coatings is presented in [134, 135, 142]. Another family of alloys, developed at FZ Jülich, that are thought to be suitable as coating materials are based on TiAl–Ag [77, 129]. The best oxidation resistance in air at 800 °C was provided by the alloy Ti-50Al-2Ag.

Although the work performed on coatings can be considered as valuable, the question should be asked as to its current relevance for industrial applications. Halogen treatment is a very simple and effective method to inexpensively improve the oxidation resistance. In this respect, the use of expensively deposited coatings seems inappropriate. In our opinion thermal-barrier coatings may become an important theme in the future for the enhancement of operating temperatures through the use of actively cooled components. However, this requires significant obstacles to be overcome including the general acceptance and introduction of TiAl as an engine material, the production of cooling channels within components and the effect of any thermal-barrier coating on fatigue properties to be minimized. This last point is particularly important, it being known that the fatigue life of coated components is often reduced compared to the base material. Nevertheless, we believe coatings may be of real value in the area of embrittlement mitigation, although apart from [151], very few publications have been made specifically in this very important area.

12.3
Summary

During high-temperature application titanium aluminide will oxidize and a form a scale that may be able to provide protection. This, however, depends on many factors including alloy composition and microstructure, temperature, environment and surface finish. Niobium additions and the use halogen treatments can certainly facilitate the formation of a protective alumina scale. Room-temperature ductility is compromised by high-temperature exposure, although the degree to which this occurs may possibly depend on composition and/or strength level. For exposed Ti-48Al-2Cr-2Nb, room-temperature levels of ductility are obtained on retesting at around 150 °C to 250 °C, indicating that embrittlement should not be an issue at operating temperatures for this alloy. No work has been published to see if this is also the case for higher-strength alloys. Embrittlement has been shown to have little effect on fatigue properties. The use of coatings to mitigate the embrittlement problem has hardly been explored although this seems a very worthwhile area of research. Work towards oxidation-resistant and thermal-barrier coatings has been performed, although we believe this to be rather premature.

References

1 Kim, Y.-W. (1989) *JOM*, **41** (7), 24.
2 Rahmel, A., Quadakkers, W.J., and Schütze, M. (1995) *Mater. Corr.*, **46**, 271.
3 Brady, M.P., Brindley, W.J., Smialek, J.L., and Locci, I.E. (1996) *JOM*, **48** (11), 46.
4 Taniguchi, S. (1997) *Mater. Corr.*, **48**, 1.
5 Fergus, J.W. (2002) *Mater. Sci. Eng.*, **A338**, 108.
6 Yang, M.R., and Wu, S.K. (October 2003) Bulletin of the College of Engineering, National Taiwan University, No. 89, p. 3.
7 Birks, N., and Meier, G.H. (1983) *Introduction to High Temperature Oxidation of Metals*, Edward Arnold, London.
8 Kopstad, P. (1988) *High Temperature Corrosion*, Elsevier, London & New York.
9 Meier, G.H. (1989) *Mater. Sci. Eng.*, **A120**, 1.
10 Meier, G.H. (1989) *Oxidation of High-Temperature Intermetallics* (eds T. Grobstein and J. Doychak), TMS, Warrendale, PA, p. 1.
11 Meier, G.H., and Pettit, F.S. (1992) *Mater. Sci. Tech.*, **8**, 331.
12 Meier, G.H., and Pettit, F.S. (1992) *Mater. Sci. Eng.*, **A153**, 548.
13 Zhang, M.X., Hsieh, K.C., DeKock, J., and Chang, Y.A. (1992) *Scr. Metall. Mater.*, **27**, 1361.
14 Li, X.L., Hillel, R., Teyssandier, F., Choi, S.K., and Van Loo, F.J.J. (1992) *Acta Metall. Mater.*, **40**, 3149.
15 Becker, S., Rahmel, A., Schorr, M., and Schütze, M. (1992) *Oxid. Met.*, **38** (5/6), 425.
16 Leyens, C. (2003) *Titanium and Titanium Alloys* (eds C. Leyens and M. Peters), Wiley-VCH, Weinheim, p. p.187.
17 Rahmel, A., and Spencer, P.J. (1991) *Oxid. Met.*, **35** (1/2), 53.
18 Luthra, K.L. (1991) *Oxid. Met.*, **36** (5/6), 475.
19 Eckert, M., and Hilpert, K. (1997) *Mater. Corr.*, **48**, 10.
20 Jacobson, N.S., Brady, M.P., and Mehrotra, G.M. (1999) *Oxid. Met.*, **52** (5/6), 537.
21 Quadakkers, W.J., Schaaf, P., Zheng, N., Gil, A., and Wallura, E. (1997) *Mater. Corr.*, **48**, 28.
22 Meier, G.H., Appalonia, D., Perkins, R.A., and Chiang, K.T. (1989) *Oxidation of High-Temperature Intermetallics* (eds T. Grobstein and J. Doychak), TMS, Warrendale, PA, p. 185.
23 Welsch, G., and Kahvechi, A.I. (1989) *Oxidation of High-Temperature Intermetallics* (eds T. Grobstein and J. Doychak), TMS, Warrendale, PA, p. 207.
24 Perkins, R.A., Chiang, K.T., and Meier, G.H. (1987) *Scr. Metall.*, **21**, 1505.

25 Shida, Y., and Anada, H. (1993) *Mater. Trans. JIM*, **34**, 236.
26 Pérez, P., Jiménez, J.A., Frommeyer, G., and Adeva, P. (2000) *Oxid. Met.*, **53** (1/2), 99.
27 Kim, B.G., Kim, G.M., and Kim, C.J. (1995) *Scr. Metall. Mater.*, **33**, 1117.
28 Yoshihara, M., and Miura, K. (1995) *Intermetallics*, **3**, 357.
29 Dang, B., Fergus, J.W., Gale, W.F., and Zhou, T. (2001) *Oxid. Met.*, **56** (1/2), 15.
30 Maki, K., Shioda, M., Sayashi, M., Shimizu, T., and Isobe, S. (1992) *Mater. Sci. Eng.*, **A153**, 591.
31 Huang, B.Y., He, Y.H., and Wang, J.N. (1999) *Intermetallics*, **7**, 881.
32 Nickel, H., Zheng, N., Elschner, A., and Quadakkers, W.J. (1995) *Mikrochim. Acta*, **119**, 23.
33 Shida, Y., and Anada, H. (1993) *Corr. Sci.*, **35**, 945.
34 Figge, U., Elschner, A., Zheng, N., Schuster, H., and Quadakkers, W.J. (1993) *Fresenius' J. Anal. Chem*, **346**, 75.
35 Kerare, S.A., and Aswath, P.B. (1997) *J. Mater. Sci.*, **32**, 2485.
36 Cathcart, J.V. (1976) *Properties of High Temperature Alloys* (eds Z.A. Foroulis and F.S. Pettit), The Electrochemical Society, Princeton, p. 99.
37 Pilling, N.B., and Bedworth, R.E. (1923) *J. Inst. Met.*, **29**, 529.
38 Xu, C., and Gao, W. (2000) *Mater. Res. Innovat.*, **3**, 231.
39 Przybilla, W., and Schütze, M. (2002) *Oxid. Met.*, **58** (3/4), 337.
40 Stringer, J. (1968) *J. Less Common Metals*, **16**, 55.
41 Yoshihara, M., and Kim, Y.-W. (2005) *Intermetallics*, **13**, 952.
42 Shimizu, T., Iikubo, T., and Isobe, S. (1992) *Mater. Sci. Eng.*, **A153**, 602.
43 Haanappel, V.A.C., Sunderkötter, J.D., and Stroosnijder, M.F. (1999) *Intermetallics*, **7**, 529.
44 Wu, Y., Hagihara, K., and Umakoshi, Y. (2005) *Intermetallics*, **13**, 879.
45 Schütze, M. (1991) *Die Korrosionsschutzwirkung Oxidischer Deckschichten Unter Thermisch-Chemisch-Mechanischer Werkstoffbeanspruchung, Materialkundlich-Technische Reihe, Band 10* (eds G. Petzow and F. Jeglitsch), Gebr. Borntraeger Verlag, Berlin.
46 Schütze, M., and Schmitz-Niederau, M. (1995) *Gamma Titanium Aluminides* (eds Y.-W. Kim, R. Wagner, and M. Yamaguchi), TMS, Warrendale, PA, p. 83.
47 Schmitz-Niederau, M., and Schütze, M. (1999) *Oxid. Met.*, **52** (3/4), 241.
48 Bruns, C., and Schütze, M. (2001) *Oxid. Met.*, **55** (1/2), 35.
49 Choudhury, N.S., Graham, H.C., and Hinze, J.W. (1976) *Properties of High Temperature Alloys* (eds Z.A. Foroulis and F.S. Pettit), The Electrochemical Society, Princeton, p. 668.
50 Rakowski, J.M., Pettit, F.S., Meier, G.H., Dettenwanger, F., Schumann, E., and Ruhle, M. (1995) *Scr. Metall. Mater.*, **33**, 997.
51 Zheng, N., Quadakkers, W.J., Gil, A., and Nickel, H. (1995) *Oxid. Met.*, **44** (5/6), 477.
52 Yang, M.R., and Wu, S.K. (2000) *Oxid. Met.*, **54** (5/6), 473.
53 Yoshihara, M., Suzuki, T., and Tanaka, R. (1991) *ISIJ Int.*, **31** (10), 1201.
54 Nwobu, A.I.P., Flower, H.M., and West, D.R.F. (1996) *Titanium '95: Science and Technology* (eds P.A. Blenkinsop, W.J. Evans, and H.M. Flower), IOM, London, UK, p. 411.
55 Taniguchi, S., Tachikawa, Y., and Shibata, T. (1997) *Mater. Sci. Eng.*, **A232**, 47.
56 Dettenwanger, F., Schumann, E., Rühle, M., Rakowski, J., and Meier, G.H. (1995) *High-Temperature Ordered Intermetallic Alloys VI* (eds J.A. Horton, I. Baker, S. Hanada, R.D. Noebe, and D.S. Schwartz), Materials Research Society Symposium Proceedings, vol. 364, Mater. Res. Soc., Pittsburgh, PA, p. 981.
57 Dettenwanger, F., Schumann, E., Rühle, M., Rakowski, J., and Meier, G.H. (1998) *Oxid. Met.*, **50** (3/4), 269.
58 Rakowski, J.M., Meier, G.H., Pettit, F.S., Dettenwanger, F., Schumann, E., and Rühle, M. (1996) *Scr. Mater.*, **35**, 1417.
59 Taniguchi, S., Hongawara, N., and Shibata, T. (2001) *Mater. Sci. Eng.*, **A307**, 107.
60 Kremer, R., and Auer, W. (1997) *Mater. Corr.*, **48**, 35.

61 Brady, M.P., Smialek, J.L., Humphrey, D.L., and Smith, J. (1997) *Acta Mater.*, **45**, 2371.
62 Kremer, R. (1996) PhD thesis, University of Erlangen-Nürnberg.
63 Hald, M. (1998) PhD thesis, RWTH Aachen.
64 Zeller, A., Dettenwanger, F., and Schütze, M. (2002) *Intermetallics*, **10**, 59.
65 Zeller, A., Dettenwanger, F., and Schütze, M. (2002) *Intermetallics*, **10**, 33.
66 Li, X.Y., and Taniguchi, S. (2004) *Intermetallics*, **12**, 11.
67 Tetsui, T., and Ono, S. (1999) *Intermetallics*, **7**, 689.
68 Beye, R., Verwerft, M., De Hosson, J.T.M., and Gronsky, R. (1996) *Acta Mater.*, **44**, 4225.
69 Cheng, Y.F., Dettenwanger, F., Mayer, J., Schumann, E., and Rühle, M. (1996) *Scr. Mater.*, **34**, 707.
70 Lang, C., and Schütze, M. (1997) *Mater. Corr.*, **48**, 13.
71 Shida, Y., and Anada, H. (1994) *Mater. Trans. JIM.*, **35**, 623.
72 Beye, R.W., and Gronsky, R. (1994) *Acta Metall. Mater.*, **42**, 1373.
73 Dowling, W.E., and Donlon, W.T. (1992) *Scr. Metall. Mater.*, **27**, 1663.
74 Zheng, N., Fischer, W., Grübmeier, H., Shemet, V., and Quadakkers, W.J. (1995) *Scr. Metall. Mater.*, **33**, 47.
75 Copland, E.H., Gleeson, B., and Young, D.J. (1999) *Acta Mater.*, **47**, 2937.
76 Shemet, V., Hoven, H., and Quadakkers, W.J. (1997) *Intermetallics*, **5**, 311.
77 Shemet, V., Tyagi, A.K., Becker, J.S., Lersch, P., Singheiser, L., and Quadakkers, W.J. (2000) *Oxid. Met.*, **54** (3/4), 211.
78 Tang, Z., Shemet, V., Niewolak, L., Singheiser, L., and Quadakkers, W.J. (2003) *Intermetallics*, **11**, 1.
79 Shalaby, H. (Dec. 1997) *Surface protection of gamma and alpha-2 titanium aluminides by ion implantation*, U.S. Patent 5,695,827.
80 Stroosnijder, M.F. (1998) *Surf. Coat. Technol.*, **100–101**, 196.
81 Masset, P.J., Neve, S., Zschau, H.E., and Schütze, M. (2008) *Mater. Corr.*, **59**, 609.
82 Zschau, H.E., and Schütze, M. (2008) *Mater. Corr.*, **59**, 619.
83 Hornauer, U., Richter, E., Wieser, E., Möller, W., Schumacher, G., Lang, C., and Schütze, M. (1999) *Nucl. Instrum. Methods Phys. Res. B*, **148**, 858.
84 Schütze, M., Schumacher, G., Dettenwanger, F., Hornauer, U., Richter, E., Wieser, E., and Möller, W. (2002) *Corr. Sci.*, **44**, 303.
85 Yankov, R.A., Shevchenko, N., Rogozin, A., Maitz, M.F., Richter, E., Möller, W., Donchev, A., and Schütze, M. (2007) *Surf. Coat. Technol.*, **201**, 6752.
86 Conrad, J.R. (2000) *Handbook of Plasma Immersion Ion Implantation and Deposition* (ed. A. Anders), John Wiley & Sons. Inc, New York, p. 1.
87 Hornauer, U., Richter, E., Wieser, E., Möller, W., Donchev, A., and Schütze, M. (2003) *Surf. Coat. Technol.*, **173–174**, 1182.
88 Zschau, H.E., Schütze, M., Baumann, H., and Bethge, K. (2004) *Mater. Sci. Forum*, **461–464**, 505.
89 Taniguchi, S., Zhu, Y.C., Fujita, K., and Iwamoto, N. (2002) *Oxid. Met.*, **58** (3/4), 375.
90 Li, X.Y., Taniguchi, S., Zhu, Y.C., Fujita, K., Iwamoto, N., Matsunaga, Y., and Nakagawa, K. (2002) *Nucl. Instrum. Methods Phys. Res. B*, **187**, 207.
91 Zhu, Y.C., Li, Y.Y., Fujita, K., Iwamoto, N., Matsunaga, Y., Nakagawa, K., and Taniguchi, S. (2002) *Surf. Coat. Technol.*, **158–159**, 503.
92 Hornauer, U., Günzel, R., Reuther, H., Richter, E., Wieser, E., Möller, W., Schumacher, G., Dettenwanger, F., and Schütze, M. (2000) *Surf. Coat. Technol.*, **125**, 89.
93 Taniguchi, S., Kuwayama, T., Zhu, Y.C., Matsumoto, Y., and Shibata, T. (2000) *Mater. Sci. Eng.*, **A277**, 229.
94 Schumacher, G., Lang, C., Schütze, M., Hornauer, U., Richter, E., Wieser, E., and Möller, W. (1999) *Mater. Corr.*, **50**, 162.
95 Wang, W., Zhang, Y.G., Ji, V., Shi, J.Y., and Chen, C.Q. (1999) *Gamma Titanium Aluminides 1999* (eds Y.-W. Kim, D.M. Dimiduk, and M.H. Loretto), TMS, Warrendale, PA, p. 799.

96 Taniguchi, S., Uesaki, K., Zhu, Y.C., Matsumoto, Y., and Shibata, T. (1999) *Mater. Sci. Eng.*, **A266**, 267.

97 Schumacher, G., Dettenwanger, F., Schütze, M., Hornauer, U., Richter, E., Wieser, E., and Möller, W. (1999) *Intermetallics*, **7**, 1113.

98 Zhang, Y.G., Li, X.Y., Chen, C.Q., Wang, W., and Ji, V. (1998) *Surf. Coat. Technol.*, **100–101**, 214.

99 Haanappel, V.A.C., and Stroosnijder, M.F. (1998) *Surf. Coat. Technol.*, **105**, 147.

100 Taniguchi, S., Uesaki, K., Zhu, Y.C., Zhang, H.X., and Shibata, T. (1998) *Mater. Sci. Eng.*, **A249**, 223.

101 Stroosnijder, M.F., Schmutzler, H.J., Haanappel, V.A.C., and Sunderkötter, J.D. (1997) *Mater. Corr.*, **48**, 40.

102 Zhang, Y.G., Li, X.Y., Chen, C.Q., Zhang, X.J., and Zhang, H.X. (1997) *Structural Intermetallics 1997* (eds M.V. Nathal, R. Darolia, C.T. Liu, P.L. Martin, D.B. Miracle, R. Wagner, and M. Yamaguchi), TMS, Warrendale, PA, p. 353.

103 Kumagai, M., Shibue, K., Kim, M.S., and Furuyama, T. (Sept. 1995) *Product of a halogen containing Ti–Al system intermetallic compound having a superior oxidation and wear resistance*, U.S. Patent 5,451,366.

104 Schütze, M., and Donchev, A. (May 2006) *Erhöhung der Oxidationsbeständigkeit von TiAl-Legierungen durch die Behandlung mit Fluor*, German patent application DE 10 2006 024 886 A1.

105 Schütze, M., and Donchev, A. (Nov. 2003) *Verfahren zur behandlung der Oberfläche eines aus Al-Legierung, insbesondere TiAl-Legierung bestehenden Bauteiles sowie die Verwendung organischer Halogenkohlenstoffverbindungen oder in einer organischen Matrik eingebunder Halogenide*, German patent application DE 103 51 946 A1.

106 Schütze, M., and Schumacher, G. (April 2000) *Verfahren zur Erhöhung der Oxidationsbeständigkeit von Legierungen aus Aluminium und Titan*, German patent application DE 100 17 187 A1.

107 Schütze, M., and Hald, M. (July 1996) *Verfahren zur Erhöhung der Korrosionsbeständigkeit von Werkstoffen auf der Basis TiAl über die Implantation von Halogenionen in die Werkstoffoberfläche*, German patent application DE 196 27 605 C1.

108 Schütze, M., and Hald, M. (Oct. 1995) *Verfahren zur Erhöhung der Korrosionsbeständigkeit von Werkstoffen auf der Basis TiAl über die Reaktion von halogenhaltigen Verbindungen aus der Gasphase mit der Werkstoffoberfläche*, German patent application DE 195 39 305 A1.

109 Schütze, M., and Hald, M. (1997) *Mater. Sci. Eng.*, **A239–240**, 847.

110 Donchev, A., Gleeson, B., and Schütze, M. (2003) *Intermetallics*, **11**, 387.

111 Masset, P.J., and Schütze, M. (2008) *Adv. Eng. Mater.*, **10** (7), 666.

112 Donchev, A., Richter, E., Schütze, M., and Yankov, R. (2008) *J. Alloys Compd.*, **452**, 7.

113 Donchev, A., Richter, E., Schütze, M., and Yankov, R. (2006) *Intermetallics*, **14**, 1168.

114 Zschau, H.E., Schütze, M., Baumann, H., and Bethge, K. (2006) *Intermetallics*, **14**, 1136.

115 Austin, C.M., and Kelly, T.J. (1993) *Structural Intermetallics* (eds R. Darolia, J.J. Lewandowski, C.T. Liu, P.L. Martin, D.B. Miracle, and M.V. Nathal), TMS, Warrendale, PA, p. 143.

116 Thomas, M., Berteaux, O., Popoff, F., Bacos, M.P., Morel, A., Passilly, B., and Ji, V. (2006) *Intermetallics*, **14**, 1143.

117 Kelly, T.J., Austin, C.M., Fink, P.J., and Schaeffer, J. (1994) *Scr. Metall. Mater.*, **30**, 1105.

118 Lee, D.S., Stucke, M.A., and Dimiduk, D.M. (1995) *Mater. Sci. Eng.*, **A192–193**, 824.

119 Draper, S.L., Lerch, B.A., Locci, I.E., Shazly, M., and Prakash, V. (2005) *Intermetallics*, **13**, 1014.

120 Pather, R., Wisbey, A., Partridge, A., Halford, T., Horspool, D.N., Bowen, P., and Kestler, H. (2001) *Structural Intermetallics 2001* (eds K.J. Hemker, D.M. Dimiduk, H. Clemens, R. Darolia,

H. Inui, J.M. Larsen, V.K. Sikka, M. Thomas, and J.D. Whittenberger), TMS, Warrendale, PA, p. 207.
121 Planck, S.K., and Rosenberger, A.H. (1999) *Gamma Titanium Aluminides 1999* (eds Y.-W. Kim, D.M. Dimiduk, and M.H. Loretto), TMS, Warrendale, PA, p. 791.
122 Wu, X., Huang, A., Hu, D., and Loretto, M.H. (2009) *Intermetallics*, **17**, 540.
123 Yamaguchi, M., and Inui, H. (1993) *Structural Intermetallics* (eds R. Darolia, J.J. Lewandowski, C.T. Liu, P.L. Martin, D.B. Miracle, and M.V. Nathal), TMS, Warrendale, PA, p. 127.
124 Liu, C.T., and Kim, Y.-W. (1992) *Scr. Metall. Mater.*, **27**, 599.
125 Y. Iino, Gao K.W., Okamura, K., Qiao, L.J., and Chu, W.Y. (2002) *Mater. Sci. Eng.*, **A338**, 54.
126 Chen, G.L., and Liu, C.T. (2001) *Int. Mater. Rev.*, **46** (6), 253.
127 Li, X.Y., and Taniguchi, S. (2005) *Intermetallics*, **13**, 683.
128 Locci, I.E., Brady, M.P., MacKay, R.A., and Smith, J.W. (1997) *Scr. Mater.*, **37**, 761.
129 Niewolak, L., Shemet, V., Thomas, C., Lersch, P., Singheiser, L., and Quadakkers, W.J. (2004) *Intermetallics*, **12**, 1387.
130 Fröbel, U., and Appel, F. (2006) *Intermetallics*, **14**, 1187.
131 McKee, D.W. (March 1992) *Method for depositing chromium coatings for titanium oxidation protection*, U.S. Patent 5,098,540.
132 Meier, G.H. (1996) *Mater. Corr.*, **47**, 595.
133 Chen, C.C., and Lin, R.Y. (1994) *Scr. Metall. Mater.*, **30**, 523.
134 Leyens, C., Schmidt, M., Peters, M., and Kaysser, W.A. (1997) *Mater. Sci. Eng.*, **A239–240**, 680.
135 Lee, J.K., Lee, H.N., Lee, H.K., Oh, M.H., and Wee, D.M. (2002) *Surf. Coat. Technol.*, **155**, 59.
136 Chu, M.S., and Wu, S.K. (2003) *Acta Mater.*, **51**, 3109.
137 Wendler, B.G., and Kaczmarek, L. (2005) *J. Mater. Proc. Technol.*, **164–165**, 947.
138 Xiong, Y., Zhu, S., and Wang, F. (2005) *Surf. Coat. Technol.*, **197**, 322.
139 Fröhlich, M., Braun, R., and Leyens, C. (2006) *Surf. Coat. Technol.*, **201**, 3911.
140 Ebach-Stahl, A., Fröhlich, M., Braun, R., and Leyens, C. (2008) *Adv. Eng. Mater.*, **10** (7), 675.
141 Tang, Z., Wang, F., and Wu, W. (2000) *Mater. Sci. Eng.*, **A276**, 70.
142 Lee, J.K., Oh, M.H., and Wee, D.M. (2002) *Intermetallics*, **10**, 347.
143 Braun, R., Fröhlich, M., Braue, W., and Leyens, C. (2007) *Surf. Coat. Technol.*, **202**, 676.
144 Mabuchi, H., Tsuda, H., Kawakami, T., Nakamatsu, S., Matsui, T., and Morii, K. (1999) *Scr. Mater.*, **41**, 511.
145 Izumi, T., Nishimoto, T., and Narita, T. (2005) *Intermetallics*, **13**, 727.
146 Yang, M.R., and Wu, S.K. (2002) *Acta Mater.*, **50**, 691.
147 Shalaby, H. (June 1992) *Protection of gamma titanium aluminides with aluminosilicate coatings*, U.S. Patent 5,118,581.
148 Yoshimura, M., Urushihara, W., Yashima, M., and Kakihana, M. (1995) *Intermetallics*, **3**, 125.
149 Retallick, W.B., Brady, M.P., and Humphreys, D.L. (1998) *Intermetallics*, **6**, 335.
150 Brady, M.P., Smialek, J.L., and Brindley, W.J. (Nov. 1998) *Two-phase (TiAl + TiCrAl) coating alloys for titanium aluminides*, U.S. Patent 5,837,387.
151 Prasad, B.D., Sankaran, S.N., Wiedemann, K.E., and Glass, D.E. (1997) *Structural Intermetallics 1997* (eds M.V. Nathal, R. Darolia, C.T. Liu, P.L. Martin, D.B. Miracle, R. Wagner, and M. Yamaguchi), TMS, Warrendale, PA, p. 295.

13
Alloy Design

In the TiAl literature, alloy and suitable processing technology development has played a significant role for at least the last 20 years. Advances have been made, so that the strength levels achieved in technical alloys over the last few years are around double those of 20 years ago, while maintaining the same levels of room-temperature ductility. Improvement of other properties such as creep and oxidation resistance has been achieved through alloying and the use of halogens, although improvement of other properties, such as toughness, through composition rather than microstructure has met with less success. It now seems well acknowledged that there is no universal alloy composition that is suitable for all applications; rather alloy composition must be tailored together with processing to achieve the required properties for specific components.

Accurate information concerning the phase relationships in both binary Ti–Al and in multicomponent systems is very important in designing alloys and suitable heat treatments. This is because for a given alloy, properties are very dependent on microstructure, and microstructure is highly sensitive to composition. The high microstructure/composition sensitivity is perhaps one of the most challenging aspects in producing components from industrially produced ingot stock; as compositional variation within ingots can lead to finished components with a range of microstructures and thus properties. This is obviously unacceptable, especially for safety-relevant aerospace parts. A good overview concerning the influence of different alloying elements was published in 1993 [1]. This chapter will briefly outline how alloys have developed over the last 20 years and discuss the different alloy-design philosophies.

13.1
Effect of Aluminum Content

In binary alloys, the aluminum level determines the initial phase to precipitate and the subsequent phase transformations that occur on solidification. According to Huang and Hall [2] who published work in 1991 on binary alloys within the range Ti-(46-60)Al made via rapid-solidification processing, ductility is highest for duplex microstructures where the heat treatment is roughly in the middle of the

Gamma Titanium Aluminide Alloys: Science and Technology, First Edition. Fritz Appel, Jonathan David Heaton Paul, Michael Oehring.
© 2011 Wiley-VCH Verlag GmbH & Co. KGaA. Published 2011 by Wiley-VCH Verlag GmbH & Co. KGaA.

$\alpha + \gamma$ phase field, which corresponds to an α_2 volume fraction of around 10%. This is why the heat-treatment temperature at which the peak in ductility is observed increases from around 1300 °C for Ti-48 to 1400 °C for Ti-51Al, respectively, see Figure 13.1 [2]. Room-temperature ductility was believed to show a maximum at around Ti-48Al, Figure 13.2 [1]; which Kim and Dimiduk [3] report was postulated to be governed by an optimum volume ratio of γ to α_2. Contrary to today's think-

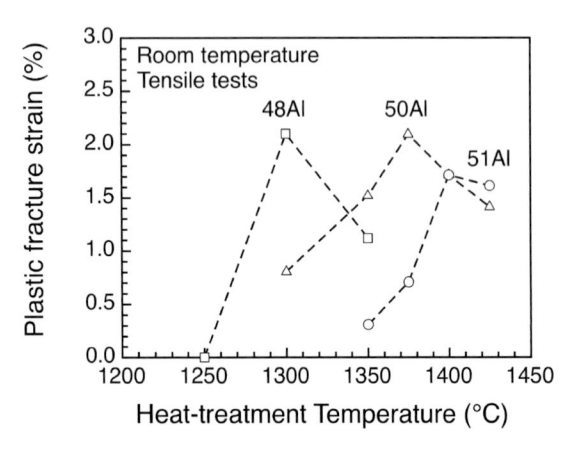

Figure 13.1 The effect of heat-treatment temperature on room-temperature tensile ductility for three binary alloys [2]. Duplex microstructures that exhibit the highest ductility result from heat treatment at temperatures near to the center of the $\alpha + \gamma$ phase field. The original data has been replotted.

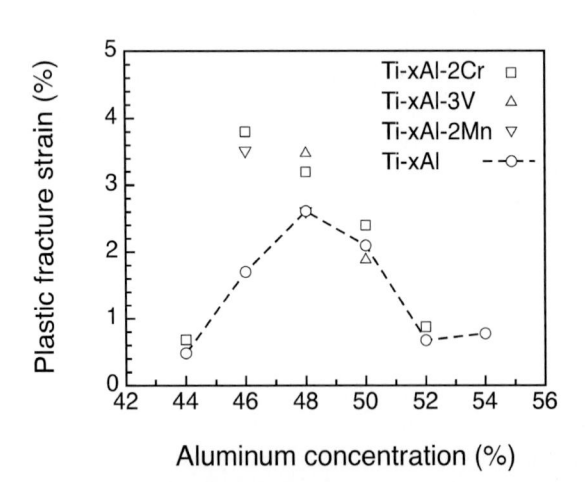

Figure 13.2 The effect of Al content on the room-temperature ductility of wrought materials [1]. According to Huang [1] only binary alloys containing between 46 and 50 at.% Al show appreciable levels of ductility. The "ductilizing" effect in two-phase alloys of 2Cr, 3V, and 2Mn additions can be observed. The original data have been replotted.

ing, Huang and Hall [2] suggested that single-phase γ had higher ductility than the "fully transformed" lamellar microstructure (see their fig. 4). Present day alloys of technical interest contain between 43 and 48 at.% Al, together with other alloying additions. It is now accepted that technical alloys should contain both the γ and α_2 phases so that dislocations at phase boundaries can contribute to deformation and oxygen within the γ phase can be scavenged by the α/α_2 phase. As explained later, some newer alloys contain around 42 to 43 at.% Al in order to increase the β/B2 phase fraction which aids hot processing.

From today's point of view its seems difficult to make any hard and fast rules concerning the effect of aluminum content on ductility as changes in processing, particularly from casting to hot extrusion have led to plastic ductility being measured in alloy compositions that would be brittle in the cast state. For example, the effect of aluminum on the properties of cast Ti-xAl-2Cr-2Nb presented by Austen and Kelly [4], see Figure 14.11, indicates peak ductility at around 48 at.% Al and very little (or no ductility) for aluminum levels of 45 at.% or less. However, at GKSS, plastic elongations of around 1.4% have been obtained for an extruded binary alloy of nominal composition Ti-44.5% Al. Of course, this ductility improvement mainly results from the refinement of the coarse-grained cast lamellar microstructure, but nevertheless illustrates the difficulties associated with the transferability of mechanical property trends.

With regard to the influence of aluminum content on strength, the situation is slightly clearer than for ductility with low aluminum-containing alloys generally being stronger, as clearly demonstrated for cast Ti-xAl-2Cr-2Nb alloys in Figure 14.11. Paul *et al.* [5] investigated the compression flow stress of a series of cast binary and ternary niobium-containing alloys and identified the increased α_2 content of low Al-containing as the probably cause of the increased strength. The effect of Al content on other mechanical properties is difficult to assess as no detailed systematic studies have been made. As described in Chapter 12, alloys containing higher aluminum contents are desirable for oxidation resistance.

13.2
Important Alloying Elements – General Remarks

Numerous elements have been added to TiAl in order to improve the properties and/or the processability. One of the first alloys of technical interest was that developed/patented by Huang of General Electric with the composition Ti-48Al-2Cr-2Nb [6]. This alloy contains 48 at.% Al, which is a level that was considered to be optimal for room-temperature ductility. The chromium addition of 2 at.% was shown to be optimum for ductility enhancement, while the 2 at.% Nb addition was sufficient to convey sufficient high-temperature oxidation resistance.

Present-day alloy compositions of technical importance can contain up to several elements and can be described by the general formula:

$$\text{Ti-}(42-49)\text{Al-}(0.1-10)\text{X} \tag{13.1}$$

with X designating the elements Cr, Nb, V, Mn, Ta, Mo, Zr, W, Si, C, Y and B. Of course, alloys containing other elements have been produced and studied in the literature, although the above compositional range covers the majority of alloys investigated. A real difficulty with multicomponent compositions is shifting of the phase boundaries, which influences the heat-treatment temperatures required for specific microstructures to be established. Thus, knowledge of the relevant composition-dependent phase relationships is of great importance. The influence of alloying elements is discussed below.

13.2.1
Cr, Mn, and V

With regard to phase-boundary shifting, Huang [1] has indicated, Figure 13.3, how additions of 2Cr, 3V and 2Mn reduce the alpha-transus temperature compared to binary alloys. The effect of such elemental additions on the ductility of wrought alloys is shown in Figure 13.2, where it can be seen that they have a "ductilizing" effect, although this effect is not observed in single-phase γ alloys [1]. The exact reasons behind the influence of these elements on ductility in two-phase alloys remains elusive and is not easy to determine as each addition influences many

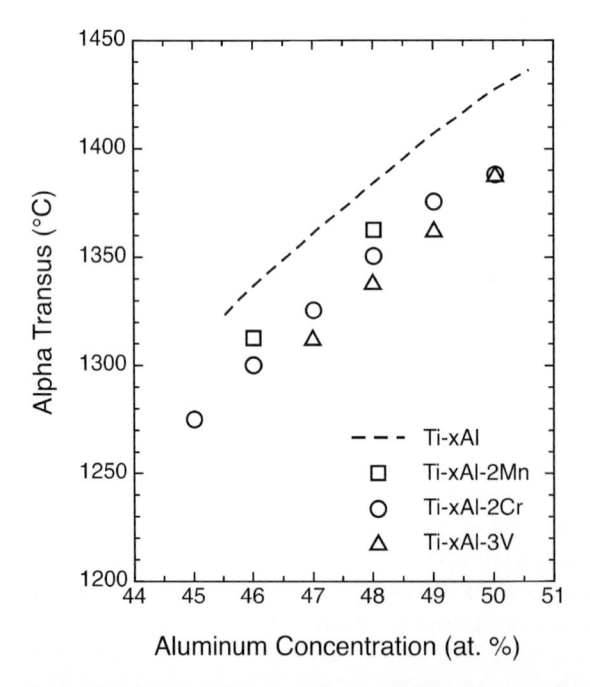

Figure 13.3 Effect of 2Cr, 3V and 2Mn additions on the alpha transus temperature [1]. As can be seen, such additions reduce the alpha-transus temperature compared to binary alloys. The original data has been replotted.

factors, including planar fault energies [7]. Kim and Demiduk [3] summarize some of the proposed mechanisms by which it has been speculated that such additions act. Further discussion can be found in a paper by Yamaguchi and Inui [8].

13.2.2
Nb, W, Mo, and Ta

The refractory elements W, Nb, Mo and Ta have been used in TiAl alloys for a number of years with 2 at.% Nb being present in the "GE alloy". Although these elements heavily segregate within cast material and lead to an increased overall density, the potential advantages of such additions on properties, particularly at high temperatures, has led to their increased use. In summarizing other work, Huang [1] indicates that additions of W, Nb Mo and Ta all have the effect of increasing oxidation resistance. More importantly increased high-temperature strength and creep resistance has also been reported for such additions. When such additions are particularly large and/or the aluminum level is relatively low, then stabilization of the β/B2 at the expense of other phases can take place due to the associated displacement of the phase boundaries. At high temperatures the β phase is soft and ductile, making hot-working operations easier to perform (Section 13.3.3), although according to Sun *et al.* [9] β/B2 reduced high-temperature strength, creep resistance and room-temperature ductility. On the contrary, other work has shown that β/B2 precipitation may increase high-temperature strength and creep resistance [10]. Thus, the situation is rather complicated and probably depends on many factors including processing, aluminum content and the level and combination of alloying elements. Interestingly Hodge *et al.* [11] note that small W additions improve creep properties in near-lamellar microstructures, whereas large additions lead to the formation of β, which results in a deterioration of the creep properties. Recently, as explained in Section 13.3.4, the use of large Ta additions has been employed in a novel technique capable of producing fine microstructures from massively transformed gamma.

13.2.3
B, C, and Si

Boron additions ranging from 0.1 to around 1 at.% can be used for grain refinement of cast alloys. Bryant *et al.* [12], working on *in-situ* composites at Martin Marietta discovered the refining effect of B when added as TiB_2 to cast TiAl alloys. The method of addition became known as the XD process and is able to produce cast microstructures with significantly refined colony sizes. The addition of elemental boron is also an effective grain refiner, forming TiB_2 in the casting. Boron refinement is particularly advantageous for cast near-net-shaped components where, until the recently developed/patented massively transformed gamma technique, microstructural refinement after casting was limited. The grain-refinement mechanism through which B additions act is rather unclear as discussed by Cheng [13], but seems to be aluminum dependent, with Al-lean alloys requiring less B to achieve effective grain refinement [14]. In low aluminum-containing alloys that solidify through the β phase field, B levels as low as 0.1 to 0.2 at.% can be very effective with cast colony sizes of about 30 μm being reported [15]. In this respect the addition of β stabilizing elements such as Nb, W and Ta can be helpful, although they also can form monoborides and effectively reduce the boron available for grain refinement. If cooling rates during solidification are slow, as is the case for thick cross sections, then long refractory metal borides can form and lead to premature failure during loading [16].

In order to improve high-temperature strength and creep resistance, the effect of carbon additions for precipitation hardening of TiAl has been studied. Recently, Appel et al. [17] have shown that finely distributed carbide precipitates can additionally act as heterogeneous sites for the nucleation of "nanotwins" during deformation, and are thus capable of releasing stress concentrations and therefore support plastic deformation. Chen et al. [18] and Tian et al. [19] were probably the first to investigate carbide-precipitation behavior in TiAl. Tian et al. [19] additionally investigated nitrogen additions. Both the carbon and nitrogen precipitates, Ti_3AlC and Ti_3AlN, were found to have a perovskite-type structure with a needle-like morphology (containing two domain types) oriented parallel to the [001] direction of the parent γ matrix. Based on this work, GKSS systematically investigated how the level of carbon additions affected the deformation behavior of quenched and aged material and developed an effective precipitation-hardening heat treatment [20, 21]. The precipitates in [20, 21] were reported to have an average length of 22 nm and a width of 3.3 nm; these values are much smaller than those reported by Tian et al. [19] of 150 to 450 nm in length and 10 to 30 nm in width. This may be due to differences in alloy composition, carbon levels and heat treatments. Tian and Nemoto [22] report that when quenched carbon-containing material is aged at 800 °C for extended periods or at higher temperatures, then plate-like precipitates of the H-phase (Ti_2AlC) form on {111} planes of the γ TiAl phase. Chen et al. [18] report dissolution of the perovskite and growth of H-phase at temperatures above 750 °C. Similar findings are reported for nitrogen-containing alloys [23]. Gouma and Karadge [24] report the precipitation of H-phase from prior $α_2$ laths that decompose to form γ. At GKSS we speculate that the precipitation of H-phase is detrimental for ductility, in particularly when coarse particles are formed.

Carbon additions are not normally used in cast alloys due to loss of ductility. The underlying reasons behind this are not known, it could be that large brittle carbides form and compromise the material, or alternatively, the material could be sufficiently strengthened by carbides but the coarse cast microstructure is unable to support the load. In this respect, hot extrusion may provide both a fine microstructure and sufficient cooling so that the carbon remains in solution and can be precipitated on subsequent aging.

Silicon is added to TiAl to improve the creep strength through precipitation-hardening/microstructural stabilization [25, 26] and for oxidation resistance [1]. The improvement in creep properties in a cast alloy of nominal composition Ti-

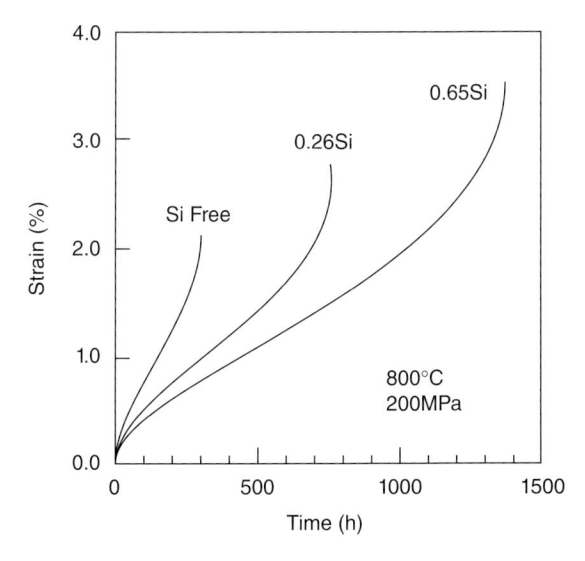

Figure 13.4 Effect of Si additions on the creep behavior of cast + aged Ti-47.5Al-1.5Cr-based alloy [26].

47.5Al-1.5Cr with additions of 0.26 and 0.65 at.% Si is demonstrated in Figure 13.4 [26]. The strengthening mechanism is dependent on the formation of Ti_5Si_3 precipitates that have been observed to form in both aged [24, 26, 27] and creep-deformed [24, 25, 27] material.

After aging at 900 °C for 5 h, Noda et al. [26] report fine precipitates less than 200 nm in size. It seems that precipitation takes place at γ/α_2 lamellar boundaries [24, 26]. Precipitates are reported at γ grain boundaries and within γ lamellae in [27]. Although not discussed, it could be that these precipitates were formed at γ/α_2 boundaries and that the α_2 phase subsequently transformed into γ as described by Viswanathan et al. [28] and Appel et al. [29]. Si additions of up to 0.65 at.% are reported not to be detrimental to the tensile properties [26], although additions higher than 1 at.% have been found to reduce creep resistance due to increased microstructural instability and enhanced dynamic recrystallization [25].

13.3
Specific Alloy Systems

Over the last 20 years very many different alloys have been studied, and thus it is difficult to discuss each composition. In general, it is possible to classify the different alloys as falling into 4 different classes: (i) "conventional alloys", (ii) high niobium-containing alloys, (iii) β-solidifying alloys, and (iv) massively transformed alloys.

13.3.1
Conventional Alloys

These alloys are typified by compositions such as Ti-48Al-2Cr-2Nb that was developed/patented by General Electric [6], Ti-48Al-2Cr that was used in many studies on rolling/deformability [30–32] and Ti-46Al-1Cr-0.2Si that was developed by Frommeyer et al. [33]. Other alloys in this group include the GKSS developed γ-TAB (Ti-47Al-4(Nb, Cr, Mn, Si, B)) and the Plansee γ-MET (Ti-46.5Al-4(Nb, Cr, Ta, B)). All the alloys in this group contain a relatively high aluminum content compared to more modern alloys, which was considered to be essential for good room-temperature ductility. At that time, ductility was considered to be the most important property and alloy design was aimed towards its optimization. The majority phases in such alloys are α_2 and γ, although B2/γ is mentioned to be present in a low interstitial (0.06 compared to 0.18 at.% O) Ti-48Al-2Cr-2Nb alloy [34]. Most work aimed towards ductility at the beginning of the 1990s was performed on cast material, and the maximum in ductility was believed to be centered at Al concentrations around 48 at.%. Remarkably, elongations of around 2.5 to 3.1% are reported for material based on cast Ti-48Al-2Cr-2Nb [34]. The strength levels determined were rather low with yield stresses of 300 to 330 MPa and fracture stresses of 445 to 477 MPa being reported [34]. In order to increase strength levels (and creep resistance), later-generation "conventional alloys" contained slightly less aluminum and additions such as B, C and Si. Nevertheless, on a density-corrected basis these alloys do not match the properties of the nickel-based superalloys and are not sufficiently oxidation resistant to extend the application temperature significantly above around 700 °C. With this aspect in mind the next class of γ-TiAl alloys contained high additions of niobium.

13.3.2
High Niobium-Containing Alloys

Chen et al. [35] published one of the first papers that demonstrated the high strength levels and good oxidation resistance of highly niobium-alloyed γ-TiAl. Accordingly Paul et al. [5] performed work on cast materials in order to clarify the strengthening mechanisms. It was found that the increased strength levels were due to athermal dislocation obstacles that were related to the reduced Al content of the alloys, not to the high Nb content. Lower Al levels lead to an increased α_2 content and structural refinement, both of which lead to a higher athermal stress contribution to deformation. The activation parameters for high-temperature deformation were determined for both binary and the equivalent Al-containing high Nb-containing ternary alloys. It was found that the activation enthalpy for deformation was somewhat higher for the high Nb-containing alloys. This is consistent with studies on the diffusion of Nb in TiAl [36], and implies that diffusion-assisted deformation processes are more difficult in such alloys. This may benefit high-temperature strength and creep resistance [37] and also impede dynamic recrystallization. Extensive twinning activity, which may have resulted from a decrease in the stacking-fault energy, was observed in deformed high Nb-containing alloys and may have led to improved ductility.

The high Nb-containing alloys developed as a result of research at GKSS are based on the composition Ti-45Al-(5–10)Nb-(0–0.5)B, C and are known as TNB. Extruded TNB of composition Ti-45Al-5Nb-0.2B-0.2C shows room-temperature strength levels of up to 1 GPa combined with around 2 to 2.5% plastic strain. Unlike the conventional TiAl alloys, such strength properties make TNB very competitive with superalloys and also extend the high-temperature capability of the material. Plansee purchased a license for TNB and took part in research projects with NASA and Pratt & Whitney using the TNB alloy. Naturally, GKSS also used TNB alloys in joint projects with partners such as Rolls Royce and DaimlerChrysler.

13.3.3
β-Solidifying Alloys

The development and advantages of β-solidifying alloys for casting was probably first recognized by Naka et al. [38]. This class of alloys has compositions that stabilize the β phase so that this is the first phase to precipitate on solidification and is the only phase present immediately after solidification is complete. This is different from peritectic alloys where after solidification the high-temperature α phase is also present to some extent within the material. This has the disadvantage that on further cooling or subsequent heat treatment where the β phase transforms to α, then the orientation of this transformed α is predetermined by the pre-existing α phase. As the γ phase has only one orientation relationship with the high temperature hexagonal α, namely $(0001)_{\alpha 2}$ // $\{111\}_\gamma$ and $<11\bar{2}0>_{\alpha 2}$ // $<1\bar{1}0>_\gamma$ then all lamellae that form from a single α grain will have the same orientation [39].

The important difference with alloys that solidify completely through β and that subsequently precipitate α on further cooling, is that the α that forms from the β can do so with up to 12 different orientation variants [39]. Thus, a large β grain can be partitioned by lamellar colonies with up to 12 different orientations. This effect, termed "crystal partitioning" [38, 39], can thus lead to significant grain refinement of the casting and significantly reduced texture. Naka et al. [38] identified elements such as W, Re and Fe as being the most effective additions to promote β solidification, followed by Mo, Cr, Nb and Ta in decreasing strength. Sun et al. [9] have also analyzed the effectiveness of different alloying elements in TiAl.

A possible concern for the β-solidifying alloys is the reported reduction in mechanical properties such as room-temperature ductility, high-temperature strength and creep resistance associated with the presence of the β/B2 phases [9]. This arises from the hard and brittle nature of the Al-rich B2 phase at room temperature, which at elevated temperatures transforms into a soft and ductile β phase. Additionally, it has been reported that β/B2 can decompose into omega (ω) phases [38, 40–42]. In a Ti-44Al-8Nb-1B alloy the amount of B2(ω) phase is almost doubled from 3 vol.% in as-cast material to 5.8 vol.% after exposure at 700 °C for

5000 h [42] and reached 9.1 vol.% after 10000 h [43]. Different crystallographic forms of ω phase are reported in a Ti-44Al-4Zr-4Nb-0.2Si-1B alloy to precipitate in network or cell structures [41]. It is believed that such phases may play a significant role in embrittlement of the material after long-term exposure [41], although it is difficult to separate the influence of only these precipitates from other microstructural changes that take place in the inherently brittle material [43].

The fine as-cast microstructure and excellent hot workability of these types of alloys have led to significant interest. Much Japanese work has been performed on phase relationships [44, 45] as well as on the processing of such alloys into parts [46–48]. Phase-diagram information is very important so that processing temperatures and heat-treatment temperatures can be adapted to composition.

The soft nature of the high-temperature β phase enables the use of conventional forging operations to make parts from alloys such as Ti-42Al-10V and Ti-42Al-5Mn. In these alloys β(B2) and γ are the thermodynamically stable phases at room temperature [46–48], unlike conventional alloys where α_2 and γ are the stable phases. Hot-worked Ti-42Al-5Mn is easy to machine compared to other TiAl alloys. Both Ti-42Al-10V and Ti-42Al-5Mn are reported to have acceptable specific creep strength. Although the tensile strength levels of the alloys are very high, the room-temperature plastic elongation to failure of 0.35% for Ti-42Al-10V [47] seems rather low for a hot-worked material.

In other work, Young-Won Kim is developing multicomponent alloys that solidify through β, and that contain enough β phase to allow conventional hot-working techniques to be employed [49]. Specifically designed heat treatments that are capable of removing the β/B2 phase will be required after processing so that acceptable high-temperature properties are attained. These types of alloy are known as "beta-gamma" alloys and exist within the composition range Ti-(42–45)Al-(2–8)Nb-(1–9)(Cr, Mn, V, Mo)-(0–0.5)(B, C) [49]. Work by Clemens et al. [50] and Wallgram et al. [51] has been performed to establish phase relationships in Ti-43Al-4Nb-1Mo-0.1B and parts have been made using near conventional hot-working procedures. Creep properties at 800 °C and 300 MPa are presented and naturally depend on the final heat treatment. Room-temperature tensile strengths of between 800 and 950 MPa were measured with total elongations to failure of up to around 2% depending on heat treatment. Unfortunately, no tensile-test curves were presented.

13.3.4
Massively Transformed Alloys

In steels it is well known that, depending on the cooling rate, different microstructures can be obtained on cooling from the austenite phase field. In this respect the Jominy test has proved very useful in determining the influence of composition and cooling rate on microstructure. Innovative work at the IRC in Birmingham has used Jominy testing to investigate the formation of microstructure in TiAl alloys, particularly concentrating on the formation of "massive gamma" (γ_m).

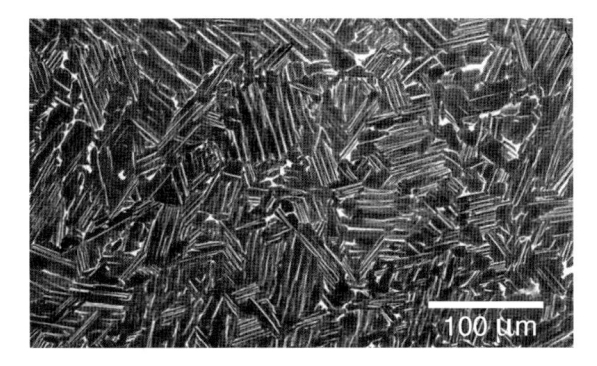

Figure 13.5 "Convoluted microstructure" developed in a 20-mm diameter bar of Ti-46Al-8Nb that was oil quenched from 1360 °C and then aged for 2 h at 1320 °C [52].

As noted by Wu and Hu [52], the formation of "massive gamma" in alloys containing 46 to 48 Al has been recognized for some years. Massive gamma (γ_m) forms in a displacive transformation without long-range diffusion and consists of heavily faulted γ phase. It develops from the high-temperature alpha phase when the cooling rate is relatively fast but not so fast as to result in metastable α (α_2). When massive gamma is subsequently heat treated within the ($\alpha + \gamma$) phase field, α precipitates on all four {111} planes of the massive gamma [53, 54] and a very fine microstructure consisting of fine alpha α_2 plates within a γ matrix is formed, see Figure 13.5 [52]. Until detailed investigations were performed by the IRC, massive gamma microstructures were usually obtained through quenching treatments, which induced high internal stresses that caused quench cracking. Thus, in spite of the fine "convoluted microstructure" that can be developed after heat treatment in the $\alpha + \gamma$ phase field, the accompanying cracking was thought likely by many investigators to hinder any useful technical application.

Nevertheless the IRC investigated the necessary range of cooling rates that lead to the formation of massive gamma and then developed/optimized compositions that form such a microstructure at slower cooling rates, closer to those encountered by real components. The research involved quantifying the degree to which different microstructures were present along the length of a Jominy specimen as a function of composition. In work based on Ti-46Al-8Nb it was shown that both the alpha grain size and the oxygen level within the alloy play a role in the formation of massive gamma. For fine-grained material the range of cooling rates over which a fully massive gamma microstructure is obtained is reduced compared to coarse-grained material [55]. Thus, boron additions have a negative effect on massive gamma formation. Microstructural refinement using this technique is therefore ideally suited to cast alloys that typically have a large grain size. High oxygen levels within the alloy were also found to suppress the massive transformation [56].

According to Hu *et al.* [57], the design of TiAl alloys that form massive gamma requires the use of elements that (i) are weak β stabilizers, (ii) partition unequally

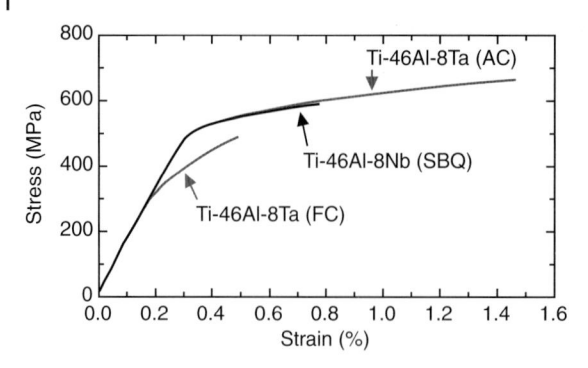

Figure 13.6 Room-temperature tensile test curves for Ti-46Al-8Nb and Ti-46Al-8Ta [58]. The sample were HIPed after furnace cooling (FC), air cooling (AC) or salt-bath quenching (SBQ). The fine microstructure developed in the air-cooled Ti-46Al-8Ta alloy gave rise to appreciable ductility. The original curves have been replotted.

between the α and γ phases, and (iii) are slow diffusers. The use of strong β stabilizers should be avoided because they can reduce or even eliminate the single-phase α field from which the material is cooled to produce the massive transformation. Elements that show preferential partitioning for α or γ are required so that significant diffusion (of the elements) is required on passing through the $\alpha + \gamma$ phase field to form the γ phase by diffusion. This is made more difficult by requiring more diffusion to take place and thus supports the transformation to massive gamma that has the same composition as the parent α phase. Massive gamma formation is promoted through the use of slow-diffusing species. Taking these considerations into account, Hu *et al.* [57] identify the element Ta as possibly being even more effective than Nb in promoting massive gamma. It was found that the susceptibility to form massive gamma increased with Ta content, so that a fully massive microstructure could be obtained throughout a 25-mm diameter rod of Ti-46Al-8Ta after air cooling. After subsequent heat treatment to produce the convoluted microstructure, tensile plastic strains to failure of up to 1.1% were obtained for 20-mm diameter air-cooled bars, see Figure 13.6 [58]. In order to protect the knowledge and processing methods three patent applications were filed [59–61].

13.4 Summary

Although the effect of the various alloying elements was established at the beginning of the 1990s, it should be apparent that alloy design has evolved over the years to meet changing property requirements and to make TiAl more competitive compared to the superalloys on a density-corrected basis. Advances in processing technologies and alloying philosophies have led to significant progress being achieved, although challenges with regard to microstructural sensitivity, property reliability, component cost and engineering trust in TiAl need to be addressed.

References

1 Huang, S.C. (1993) *Structural Intermetallics* (eds R. Darolia, J.J. Lewandowski, C.T. Liu, P.L. Martin, D.B. Miracle, and M.V. Nathal), TMS, Warrendale, PA, p. 299.
2 Huang, S.C., and Hall, E.L. (1991) *Metall. Trans.*, 22A,**427**.
3 Kim, Y.-W., and Demiduk, D.M. (1991) *JOM*, **43**, (Aug.),40.
4 Austin, C.M., and Kelly, T.J. (1993) *Structural Intermetallics* (eds R. Darolia, J.J. Lewandowski, C.T. Liu, P.L. Martin, D.B. Miracle, and M.V. Nathal), TMS, Warrendale, PA, p. 143.
5 Paul, J.D.H., Appel, F., and Wagner, R. (1998) *Acta Mater.*, **46**, 1075.
6 Huang, S.C. (Nov. 1989) *Titanium aluminium alloys modified by chromium and niobium and method of preparation*, U.S. Patent 4,879,092.
7 Woodward, C., MacLaren, J.M., and Dimiduk, D.M. (1993) *High Temperature Ordered Intermetallic Alloys V, Materials Resarch Society Symposium Proceedings*, vol. 288 (eds I. Baker, R. Darolia, J.D. Whittenberger, and M.H. Yoo), MRS, Warrendale, PA, p. 171.
8 Yamaguchi, M., and Inui, H. (1993) *Structural Intermetallics* (eds R. Darolia, J.J. Lewandowski, C.T. Liu, P.L. Martin, D.B. Miracle, and M.V. Nathal), TMS, Warrendale, PA, p. 127.
9 Sun, F.S., Cao, C.X., Kim, S.E., Lee, Y.T., and Yan, M.G. (2001) *Metall. Mater. Trans.*, **32A**, 1573.
10 Beddoes, J., Seo, D.Y., Chen, W.R., and Zhao, L. (2001) *Intermetallics*, **9**, 915.
11 Hodge, A.M., Hsiung, L.M., and Nieh, T.G. (2004) *Scr. Mater.*, **51**, 411.
12 Bryant, J.D., Christodoulou, L., and Maisano, J.R. (1990) *Scr. Metall. Mater.*, **24**, 33.
13 Cheng, T.T. (2000) *Intermetallics*, **8**, 29.
14 Hu, D. (2001) *Intermetallics*, **9**, 1037.
15 Imayev, R.M., Imayev, V.M., Oehring, M., and Appel, F. (2007) *Intermetallics*, **15**, 451.
16 Hu, D., Wu, X., and Loretto, M.H. (2005) *Intermetallics*, **13**, 914.
17 Appel, F., Fischer, F.D., and Clemens, H. (2007) *Acta Mater.*, **55**, 4915.
18 Chen, S., Beaven, P.A., and Wagner, R. (1992) *Scr. Metall. Mater.*, **26**, 1205.
19 Tian, W.H., Sano, T., and Nemoto, M. (1993) *Philos. Mag. A*, **68**, 965.
20 Christoph, U., Appel, F., and Wagner, R. (1997) *Mater. Sci. Eng.*, **A239–240**, 39.
21 Appel, F., Christoph, U., and Wagner, R. (1997) *High Temperature Ordered Intermetallic Alloys VII, Materials Research Society Symposium Proceedings*, vol. 460 (eds C.C. Koch, C.T. Liu, N.S. Stoloff, and A. Wanner), MRS, Warrendale, PA, p. 77.
22 Tian, W.H., and Nemoto, M. (1997) *Intermetallics*, **5**, 237.
23 Tian, W.H., and Nemoto, M. (2005) *Intermetallics*, **13**, 1030.
24 Gouma, P.I., and Karadge, M. (2003) *Mater. Lett.*, **57**, 3581.
25 Tsuyama, S., Mitao, S., and Minakawa, K.N. (1992) *Mater. Sci. Eng.*, **A153**, 451.
26 Noda, T., Okabe, M., Isobe, S., and Sayashi, M. (1995) *Mater. Sci. Eng.*, **A192–193**, 774.
27 Es-Souni, M., Bartels, A., and Wagner, R. (1995) *Mater. Sci. Eng.*, **A192–193**, 698.
28 Viswanathan, G.B., Kim, Y.-W., and Mills, M.J. (1999) *Gamma Titanium Aluminides 1999* (eds Y.-W. Kim, D.M. Dimiduk, and M.H. Loretto), TMS, Warrendale, PA, p. 653.
29 Appel, F., Christoph, U., and Oehring, M. (2002) *Mater. Sci. Eng.*, **A329–331**, 780.
30 Clemens, H., Rumberg, I., Schretter, P., and Schwantes, S. (1994) *Intermetallics*, **2**, 179.
31 Clemens, H., Glatz, W., Schretter, P., Koppe, C., Bartels, A., Behr, R., and Wanner, A. (1995) *Gamma Titanium Aluminides* (eds Y.-W. Kim, R. Wagner, and M. Yamaguchi), TMS, Warrendale, PA, p. 717.

32 Koeppe, C., Bartels, A., Seeger, J., and Mecking, H. (1993) *Metall. Trans.*, **24A**, 1795.
33 Frommeyer, G., Wunderlich, W., Kremser, Th., and Liu, Z.G. (1992) *Mater. Sci. Eng.*, **A152**, 166.
34 Kelly, T.J., Juhas, M.C., and Huang, S.C. (1993) *Scr. Metall. Mater.*, **29**, 1409.
35 Chen, G., Zhang, W., Wang, Y., Wang, J., Sun, Z., Wu, Y., and Zhou, L. (1993) *Structural Intermetallics* (eds R. Darolia, J.J. Lewandowski, C.T. Liu, P.L. Martin, D.B. Miracle, and M.V. Nathal), TMS, Warrendale, PA, p. 319.
36 Herzig, C., Przeorski, T., Friesel, M., Hisker, F., and Divinski, S. (2001) *Intermetallics*, **9**, 461.
37 Appel, F., Lorenz, U., Paul, J.D.H., and Oehring, M. (1999) *Gamma Titanium Aluminides 1999* (eds Y.-W. Kim, D.M. Dimiduk, and M.H. Loretto), TMS, Warrendale, PA, p. 381.
38 Naka, S., Thomas, M., Sanchez, C., and Khan, T. (1997) *Structural Intermetallics 1997* (eds M.V. Nathal, R. Darolia, C.T. Liu, P.L. Martin, D.B. Miracle, R. Wagner, and M. Yamaguchi), TMS, Warrendale, PA, p. 313.
39 Jin, Y., Wang, J.N., Yang, J., and Wang, Y. (2004) *Scr. Mater.*, **51**, 113.
40 Huang, Z.W., Voice, W., and Bowen, P. (2003) *Scr. Mater.*, **48**, 79.
41 Huang, Z.W. (2008) *Acta Mater.*, **56**, 1689.
42 Huang, Z.W., and Zhu, D.G. (2008) *Intermetallics*, **16**, 156.
43 Huang, Z.W., and Cong, T. (2010) *Intermetallics*, **18**, 161.
44 Kobayashi, S., Takeyama, M., Motegi, T., Hirota, N., and Matsuo, T. (2003) *Gamma Titanium Aluminides 2003* (eds Y.-W. Kim, H. Clemens, and A.H. Rosenberger), TMS, Warrendale, PA, p. 165.
45 Takeyama, M., and Kobayashi, S. (2005) *Intermetallics*, **13**, 993.
46 Tetsui, T., Shindo, K., Kobayashi, S., and Takeyama, M. (2002) *Scr. Mater.*, **47**, 399.
47 Tetsui, T., Shindo, K., Kobayashi, S., Takeyama M. (2003) *Intermetallics*, **11**, 299.
48 Tetsui, T., Shindo, K., Kaji, S., Kobayashi, S., and Takeyama, M. (2005) *Intermetallics*, **13**, 971.
49 Kim, Y.-W., Kim, S.-L., Dimiduk, D., and Woodward, C. (2008) *Gamma Titanium Aluminides 2008* (eds Y.-W. Kim, D. Morris, R. Yang, and C. Leyens), TMS, Warrendale, PA, p. 215.
50 Clemens, H., Chladil, H.F., Wallgram, W., Zickler, G.A., Gerling, R., Liss, K.-D., Kremmer, S., Güther, V., and Smarsly, W. (2008) *Intermetallics*, **16**, 827.
51 Wallgram, W., Schmölzer, T., Cha, L., Das, G., Güther, V., and Clemens, H. (2009) *Int. J. Mater. Res.*, **100**, (formerly Z. Metallkd.), 1021.
52 Wu, X., and Hu, D. (2005) *Scr. Mater.*, **52**, 731.
53 Abe, E., Kumagai, T., and Nakamura, M. (1997) *Structural Intermetallics 1997* (eds M.V. Nathal, R. Darolia, C.T. Liu, P.L. Martin, D.B. Miracle, R. Wagner, and M. Yamaguchi), TMS, Warrendale, PA, p. 167.
54 Kumagai, T., Abe, E., Takeyama, M., and Nakamura, M. (1997) *Scr. Mater.*, **36**, 523.
55 Hu, D., Huang, A.J., Novovic, D., and Wu, X. (2006) *Intermetallics*, **14**, 818.
56 Huang, A., Loretto, M.H., Hu, D., Liu, K., and Wu, X. (2006) *Intermetallics*, **14**, 838.
57 Hu, D., Huang, A., Loretto, M.H., and Wu, X. (2007) *Ti-2007 Science and Technology*, Proc. of the 11th World Conference on Titanium (eds M. Nimomi, S. Akiyama, M. Ikeda, M. Hagiwara, and K. Maruyama), The Japan Institute of Metals, Tokyo, Japan, p. 1317.
58 Saage, H., Huang, A.J., Hu, D., Loretto, M.H., and Wu, X. (2009) *Intermetallics*, **17**, 32.
59 Hu, D., Wu, X., and Loretto, M. (July 2004) *A method of heat treating titanium aluminide*, E.U. Patent Application EP 1 507 017 A1.
60 Voice, W., Hu, D., Wu, X., and Loretto, M. (Dec. 2006) *A method of heat treating titanium aluminide*, E.U. Patent Application EP 1 813 691 A1.
61 Huang, A., Loretto, M., Wu, X., and Hu, D. (Aug. 2007) *An alloy and method of treating titanium aluminide*, E.U. Patent Application EP 1 889 939 A2.

14
Ingot Production and Component Casting

14.1
Ingot Production

The success of TiAl to reach the market in a wide range of component forms depends critically on the understanding and tailoring of alloy composition and on subsequent processing to achieve the required mechanical properties. Perhaps the most important aspect that cannot be understated, especially for safety-critical parts such as aero and land-based gas-turbine components, is property reliability and predictability. In this respect alloy composition and processing are important factors, however, the most important prerequisite is the requirement for ingots that can be manufactured to a reproducible composition and quality on an economical basis. High-quality components can only be produced if routes that generate defect-free material, without significant microstructural and chemical inhomogeneities, are developed and understood. Ingot production is the first step in this processing chain.

It is extremely important that macrosegregation of the alloying elements (in particular aluminum) within the ingot, both end to end and center to surface, is minimal. For aluminum, a variation of less than ±0.3 at.% is probably acceptable, although this band may be larger for more "forgiving" alloys whose microstructure and properties are less sensitive to aluminum variation. Additionally, it is important that the mean aluminum content of an ingot is as close as possible to the nominal composition. If deviation of the aluminum level is larger than acceptable, due to either segregation or to the mean ingot composition being out of specification, then this can lead to the development of quite different microstructures that show different mechanical properties. This in turn gives rise to increased property variability in material obtained from a single ingot or between material obtained from different ingot batches of nominally the same composition. For these reasons ingot quality with regard to compositional control and chemical homogeneity is an extremely important issue.

The three main melting techniques that have been used to produce TiAl ingots on an industrial basis are vacuum arc melting/remelting (VAR), plasma-arc melting (PAM) and induction skull melting (ISM) [1]. Each of these production methods, which have their own specific advantages and disadvantages, will be discussed below. The electron-beam melting techniques (EBM) that are sometimes used to melt titanium alloys are not discussed here because control of the aluminum content is very difficult due to its vaporization during the melting process. This makes the EBM method unsuitable for the production of TiAl ingots. It should be mentioned that after final melting, all ingots used for hot working are usually HIPed at temperatures around 1000 °C to 1200 °C for a few hours (e.g., 4 h) under an inert gas (usually argon) at pressures of 100 MPa to 200 MPa. This is done mainly to close internal porosity. For ingots that will be used as feedstock for castings, HIPing is not necessary. One aspect that needs particular attention, especially for large ingots, is the requirement for slow cooling after solidification (or HIPing) to minimize the build up of internal stresses that can lead to cracking.

14.1.1
Vacuum Arc Melting (VAR)

This method of ingot production involves the melting (and subsequent remelting) of a consumable electrode in a high vacuum environment with an arc that is formed between a consumable electrode and a water-cooled copper mold. The processing route to produce TiAl ingots is essentially the same as that used to produce conventional titanium alloys.

Initially the alloying components (in elemental or prealloyed masteralloy form) are mixed together in the required amounts to produce the required ingot composition. The production of typical masteralloys used in VAR has been presented in [2]. The preferential loss during the melting process of elements (such as aluminum) with a high vapor pressure due to evaporation, must be taken into account at this stage by compensating for the expected losses with excess of the particular elements concerned. The homogeneously mixed components are then cold pressed to form compacts. The final electrode for the melting process may be made up from a number of such compacts that have been welded together under an inert atmosphere. This assembly is then placed within a water-cooled copper mold under vacuum and a DC arc is struck between the electrode and the copper mold. As the process proceeds, the electrode is consumed and an ingot is formed within the mold. Such an ingot (called a primary ingot) is usually referred to as $1 \times$ VAR, although this is misleading as the ingot has only been melted and not remelted. The process is shown schematically in Figure 14.1 [3].

To increase the mass and length of the primary ingot, a second cold-pressed electrode could be melted on top of the first. The chemical homogeneity of the primary ingot is not normally of satisfactory quality. For this reason, the primary ingot is then used as a consumable electrode in a second vacuum arc melting process. For conventional titanium alloys the primary ingot is inverted when used as an electrode in the second melting operation [4]. It has not been published if this is also the case for TiAl ingots. Obviously, the diameter of the water-cooled copper mold used in the second melting process must be larger than that used in the first melting process in order that the primary ingot can pass within the mold.

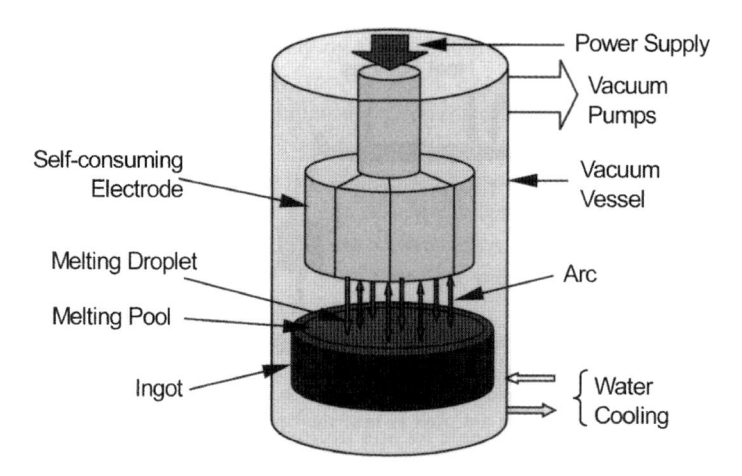

Figure 14.1 Schematic diagram of a VAR unit [3].

Once the primary ingot has been remelted, the secondary ingot is then referred to as being 2 × VAR. Depending on the ingot quality requirements, a third melting process may be performed to further improve the chemical homogeneity. For conventional titanium alloys the secondary ingot (2 × VAR) is again inverted before being remelting for a second time (3 × VAR). Such multiple melting steps have been reported not to significantly change the alloy composition of gamma titanium aluminides [5] but do improve ingot homogeneity [6]. In the literature it has been reported that the VAR technique has been used to manufacture TiAl ingots as large as 300 mm diameter (900 mm long) weighing 255 kg [5] and 355 mm diameter, weighing over 270 kg [7].

For the production of conventional titanium alloy ingots, a number of parameters are monitored or adjusted during the melting process, as discussed by Lütjering and Williams [4]. The vacuum level is continuously monitored to ensure that no contamination occurs through gas or water leaks. The melting rate is continuously adjusted to control the size (depth) of the melt pool. The optimum depth of the melt pool is alloy specific and results in reduced segregation. Most VAR units for the production of titanium alloys are equipped with electric coils at the top of the mold for electromagnetic stirring of the melt to increase ingot homogeneity, although the extent to which they are employed depends on the alloy and the ingot producer. However, there is no general agreement that this technique is beneficial or necessary. During melting of the final 25 to 35% length of a titanium ingot, the power is gradually reduced to reduce the melting rate. This is done to reduce the extent of pipe shrinkage and the formation of alpha-stabilized regions (known as a Type-II defects) at the upper end of the ingot. This results in an increase in the yield of sound material [4]. Regarding the extent to which the procedures typically used in the VAR production of conventional titanium alloy ingots are also applied in the manufacture of TiAl ingots is difficult to judge. This is because very little information has been published in the literature on the exact details, and the melting of TiAl ingots is not yet fully industrialized on a mass scale.

Since around 1999 considerable progress has been made in the production of TiAl ingots via VAR processing. Indeed, it has been stated that increases in ingot size have not led to a deterioration in macroscopic chemical homogeneity, and that the chemical homogeneity of 2 × VAR ingots in 2003 is comparable to that of 3 × VAR ingots produced in 1999 [5]. Furthermore, it was shown that the macrovariation of aluminum within a 300mm diameter 2 × VAR ingot of Ti-45Al-(5–10)Nb-C was 45.1 ±0.3 at.%. The homogeneity of niobium within the ingot was said to be comparable to aluminum. These results are very encouraging. It was also claimed that the deviation of aluminum within an ingot can generally be controlled to within ±0.5 at.% Al over an entire ingot, and to within ±0.7 at.% Al within a complete production lot [5]. However, no supporting data from a production lot was presented to justify this ±0.7 at.% Al limit. Such a concentration variation seems rather large, and implies that regions within some ingots may contain up to 1.4 at.% more (or less) aluminum than regions within other ingots of the same nominal composition. Work performed at GKSS has indicated an aluminum concentration gradient of up to around 1 at.% from the center to the surface of a large 2 × VAR ingot. For some alloy compositions such a large variation in aluminum is not acceptable, as it would give rise to problems of processing within a given (composition dependent) processing window and thereafter to an unacceptable variation in microstructure and mechanical properties. Ram and Barrett [8] have indicated that a technique developed by PCC and Oremet for the production of ingots, which ensures that larger amounts of material are molten at a given time, can result in considerably less segregation compared to conventional VAR.

When considering the defects and elemental segregation present in VAR ingots it is important to take into account how many times the material has been melted/remelted. During the initial melting of the pressed electrode many processes take place simultaneously. These processes have been referenced in [6] as: (i) melting of the constituent elements and masteralloys (endothermic reaction), (ii) formation of the main phases, that is, titanium aluminide (exothermic reaction), (iii) formation of other phases, such as grain-refining borides (exothermic reaction), (iv) alloying of the main phases, such as TiAl, with elements such as Cr, Nb, Ta, etc. (endothermic reaction) and (v) solidification of the different phases. With so many processes taking place at the same time, it is not surprising that unstable melting and solidification conditions arise and result in an inhomogeneous primary ingot that exhibits different macrostructural zones, as presented in [6]. Large inclusions of pure titanium were observed in a 120-mm diameter primary ingot of Ti-46.5Al-4(Cr, Nb, Ta, B) [6]. Such inclusions could arise from pieces of titanium being surrounded by Al-rich regions in the initial cold-compacted electrode. These can subsequently fall into the melt (as lumps) when the surrounding Al-rich regions, which have a significantly lower melting point than titanium, are melted.

The vertical arrangement of the VAR technique in this sense is a disadvantage because any high melting point inclusions present in the elemental or masteralloy components, or that form during the melting process, can remain in the ingot. Depending on their exact size, melting point, time within the melt pool and the melt pool superheat, such inclusions can dissolve into the melt if given enough time. Through optimization of the initial elemental and masteralloy components, the electrode fabrication technique and the melting parameters, pure titanium inclusions are unlikely in the final ingot. However, a large (around 1 mm diameter) titanium-enriched inclusion containing 5 wt.% more titanium than the surrounding matrix has been observed in a 3 × VAR ingot [6]. It is thought that such inclusions remain because of their incomplete melting, which arises from the very high local cooling rates that develop in the melt when the arc is temporarily short circuited due to excessive flow of molten material from the electrode into the melt.

14.1.2
Plasma-Arc Melting (PAM)

Plasma-arc melting was developed for the production of conventional titanium alloys as an alternative to the VAR process. The term PAM refers to a technique where the feedstock is melted using a plasma torch and the molten metal accumulates on a horizontal water-cooled copper hearth from where it then flows into a retractable crucible, as shown schematically in Figure 14.2 [9].

The feedstock usually consists of "hockey puck" compacts [10] made up of a blended mixture of the required amounts of elements and masteralloys, although premelted ingot material is used in further melting steps. The loss of elements during the PAM process (e.g., aluminum), which can result in the composition being out of specification in the final ingot, is less than for VAR as the whole process must take place under a (protective) atmosphere so that a plasma can be

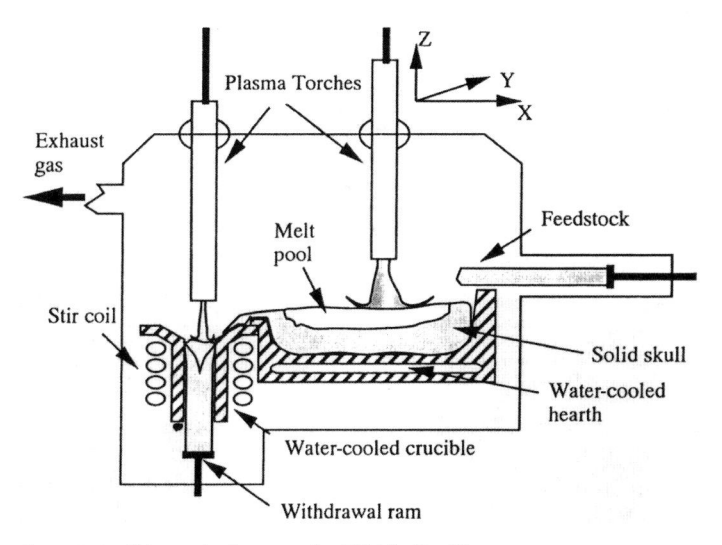

Figure 14.2 Schematic diagram of a PAM facility [9].

generated. Helium has been used as a carrier gas for the torch and within the chamber at Allvac [1]. At the IRC in Birmingham, helium has been employed for the torch with the chamber being backfilled with argon [9] that is maintained to a constant pressure of 1.1 bar [11, 12]. Helium is the preferred gas for the plasma as its higher conductivity and higher ionization potential result in a higher thermal input [13]. Before the melting process is started and the plasma arc is struck, the whole system is pumped to vacuum and then backfilled with inert gas. When a new alloy is melted, the hearth is lined with compacted "hockey pucks" of the appropriate composition that are melted before additional pucks are melted to supply the material for the ingot. During the melting process the system atmosphere is determined by a balance between the rate of plasma-torch gas input and the rate of gas removal due to pumping or through a pressure-relief valve. The system integrity can be monitored by measuring the system leak rate although continuous monitoring using a mass spectrometer is preferred [13].

As can be seen in Figure 14.2, the feedstock is melted using a plasma torch (which can be programmed to move over the hearth in a predetermined manner) and a solid layer of the alloy (referred to as the "skull") is formed between the molten metal and the copper-cooled hearth. The hearth may be up to several inches in depth. After a certain volume of feedstock has been melted, the molten metal flows over the wall of the hearth into the water-cooled copper crucible where it is heated from above with a second plasma torch. As the metal flows into the crucible, the crucible base is retracted so that continuous-casting conditions are established. The top end of the water-cooled copper casting crucible may be surrounded by electromagnetic coils (as at the IRC) so that additional stirring of the melt may take place before solidification, to ensure increased chemical homogeneity within the ingot. However, according to Blackburn and Malley [13], this constrains melting to be conducted at lower gas pressures so that the interaction between the plasma arc and the magnetic field is reduced.

Although in Figure 14.2 only one copper hearth is shown, more complex systems may employ up to 3 hearths, where the molten metal flows from one hearth to another before entering the continuous casting crucible [1]. Such a multiple hearth system has the advantage that the residence time is increased and the melt is more homogenized. Gamma TiAl ingots of up to 660 mm in diameter and 1690 kg in weight have been produced using the PAM technique [1, 14].

Plasma-arc melting is a more modern melting technique than VAR and seems to provide several advantages that have been discussed by Lütjering and Williams [4]. During melting the "skull" that is formed on the copper hearth ensures that the molten alloy is not contaminated from the hearth. Another advantage is that the time that the alloy is molten is independent of the ingot size and controllable. This enables the possibility of dissolving any oxygen- or nitrogen-rich defects without the presence of a deep melt pool, which can increase segregation. As a consequence of the molten metal passing over the hearth, any high-density inclusions can fall to the bottom of the hearth melt pool, which results in them not being present in the final ingot. The VAR process does not allow such a possibility; everything from the electrode or from reactions between different components within the electrode can be transferred to the final ingot. Another advantage that is technically less significant, but that has important economic consequences, is the ability of PAM to produce ingots with noncircular cross sections and the possibility of incorporating recycled material. Any high-density inclusions that could be present in recycled material can be removed by PAM. Additionally, it is possible to monitor and control PAM. This is not easy, however, as many of the processes are highly coupled with each other [15]. Control strategies for PAM melting are discussed in [15].

Due to the relatively low superheat generated in PAM compared to VAR, the surface condition is generally rougher for PAM ingots, which may therefore require more machining and thus increase material loss [4]. However, according to work performed at the IRC in Birmingham, the surface quality of PAM ingots is dependent on the operating parameters and in particular the amplitude of an oscillating motion that is superimposed on the ram withdrawal from the crucible, a parameter known as "dither". A smoother surface finish, comparable to VAR, was observed when the amplitude of the dither was low [9]. Other parameters that have been shown to be important for the surface quality of titanium alloys are the torch gas and the torch–ingot standoff distance (arc length), both of which relate to the ingot power density [13]. When such parameters are optimized, the surface finish of PAM ingots can be as good as VAR [13].

As mentioned in Section 14.1.1, during VAR of titanium ingots the power is gradually reduced during melting of the final 25 to 35% length of the ingot in order to reduce pipe shrinkage and alpha-stabilized regions. These types of defects are not generally observed in PAM ingots as a result of the shallow melt-pool depth [13]. Thus, in this respect PAM may offer increased ingot yield over VAR.

A number of studies have been performed to investigate PAM ingot quality. It is important to distinguish between studies that have been performed on single-, double- or triple-melted ingots. As with VAR, both the ingot quality and costs increase for multiple-melted ingots. With regard to single-melted ingots, one of the main concerns is that high melting temperature alloying elements or masteralloy particles do not fully melt and remain in the ingot. Work by Clemens [10] on single-melted ingots of 47XD (Ti-47Al-2Nb-2Mn with 0.8 vol.% TiB_2) and an ABB2 alloy containing tungsten (Ti-47Al-2W-0.5Si) found both Ti-rich particles in ingots of each composition (probably arising from unmelted sponge), and also tungsten-aluminum masteralloy particles in the ingots of the ABB2 composition, see Figure 14.3.

Interestingly, the majority of the particles were found at the bottom end of the ingots indicating that the "hockey puck" compacts, used to line the hearth, were not molten for a long enough time before pouring into the crucible. If the residence time were longer then this would certainly reduce or even eliminate the Ti-rich particles. With regard to the W-rich high-density inclusions, it was thought likely that the molten-pool depth was not deep enough for them to settle into the mushy zone at the bottom of the melt pool. If this was indeed the case, then when a suitably deep pool is established this problem should be eliminated. Additionally, the particle size of the W–Al masteralloy used to make the ABB2 composition

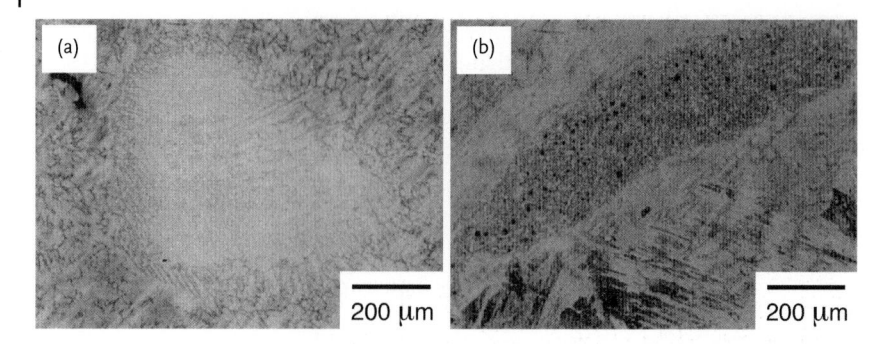

Figure 14.3 (a) An unmelted titanium sponge particle and (b) an unmelted tungsten-aluminum masteralloy particle that were found in single-melted PAM ingots [10].

ingots was found to play an important role. The use of a small particle size combined with low melting rates, rather than large particles and high melting rates, was found to result in fewer unmelted W-rich particles within the ingot.

The macrosegregation of aluminum and tungsten within the ingot from end to end was found to depend on the particle size of the W–Al masteralloy. It was suggested that if the masteralloy particle size was too fine, then the masteralloy particles could fall out of the compacts, giving rise to uncontrollable additions of aluminum and tungsten. This hypothesis was supported by the observation that the variation of these elements within the ingot tracked each other [10]. The aluminum content was found to vary by about 2.5 wt.% (equivalent to around 3.1 at.%) over the entire length of one ingot of the ABB2 composition. A second ingot of the same nominal composition, but made using a masteralloy with a larger particle size, showed very little variation in aluminum (around ±0.3 at.%) along a 100-inch long central part of the ingot [10].

Clearly the preparation of the compacts, choice of masteralloy additions and their physical characteristics are extremely important for the quality of single-melted ingots. This theme has been addressed by Godfrey and Loretto [16], who have investigated the effect on ingot quality of compacts made with fine and coarse elemental materials for a Ti-48Al-2Mn-2Nb alloy. It was found that when coarse materials were used, the elemental distribution within the compacts was very inhomogeneous with aluminum being concentrated near to the sides of the compacts and titanium towards the center. Compacts that were produced using fine elemental components showed a much more homogeneous distribution of the elements within the compacts, even though both types of compact had the same overall composition, see Figure 14.4. The extent of elemental inhomogeneity within the compacts was reflected in the homogeneity of single-melted ingots although it was much less pronounced in double-melted ingots [16].

The chemical inhomogeneities and defects such as pores that are present after single PAM cannot be removed by subsequent hot extrusion and isothermal forging, as has been shown for large-diameter ingots (43 cm and 66 cm diameter) by Porter et al. [14]. Indeed Al-rich bands that resulted from a melt stream of

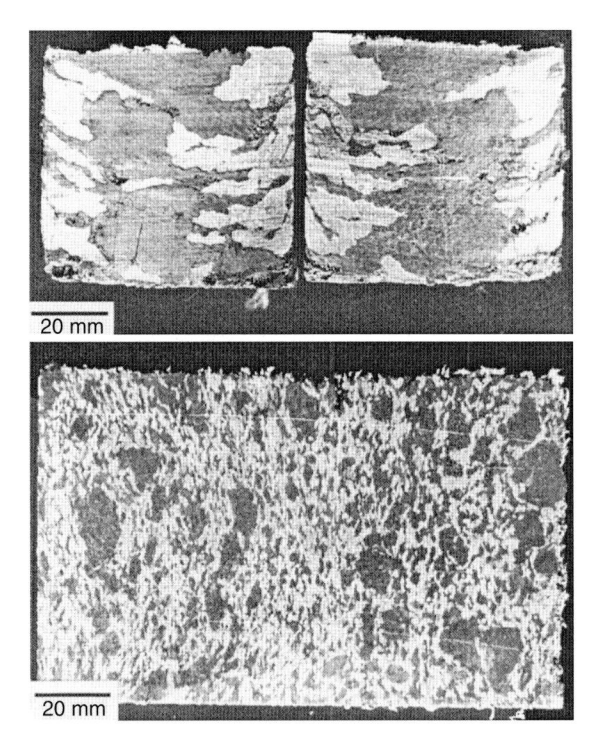

Figure 14.4 The upper diagram shows cross sections through elemental compacts made from coarse raw materials and indicates the inhomogeneous distribution of elements, with aluminum (lighter colored) having a higher concentration at the sides. The lower diagram shows a cross section through a compact made from much finer raw materials that result in more homogeneous compacts and better ingot homogeneity after primary ingot melting [16].

inhomogeneous composition, and pores of up to around 200 μm in length, which resulted from ingot center line porosity, were observed even after extrusion and isothermal forging. Ti-rich inclusions and also low-density inclusions (possibly TiN) were observed in the processed material. An inhomogeneous distribution of boron was thought to have resulted in the grain size ranging from an average value of 350 μm to up 800 μm.

The influence of multiple plasma-arc melting has been investigated in excellent work by Dowson et al. [12], Godfrey and Loretto [16], and Godfrey et al. [17]. In agreement with other studies, a very high level of compositional inhomogeneity was observed in single-melted ingots. The typical macrostructure of a PAM ingot both parallel and perpendicular to the longitudinal axis for an alloy of composition Ti-48Al-2Mn-2Nb is shown in Figure 14.5 [12]. Apart from a small chill zone at the surface of the ingot, the macrostructure consisted of large columnar grains oriented inwards and upwards, parallel to the direction of heat flow. Detailed analysis of the orientation of the lamellae, which are perpendicular to the direction of heat flow, enabled the depth of the melt pool within the ingot to be determined

(a) (b)

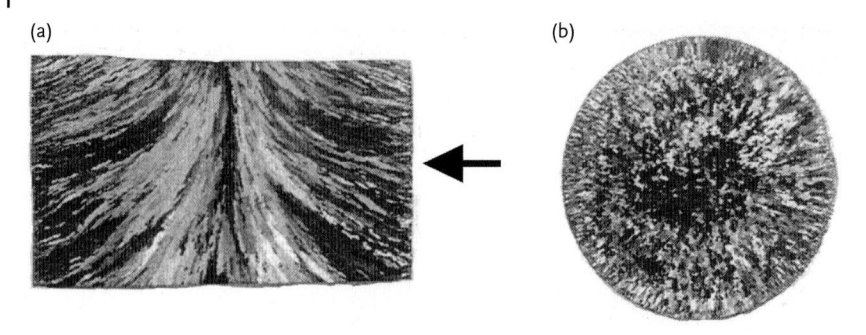

Figure 14.5 Macrographs taken (a) parallel and (b) perpendicular to the direction of PAM ingot withdrawal that show typical macroscopic structural features [12].

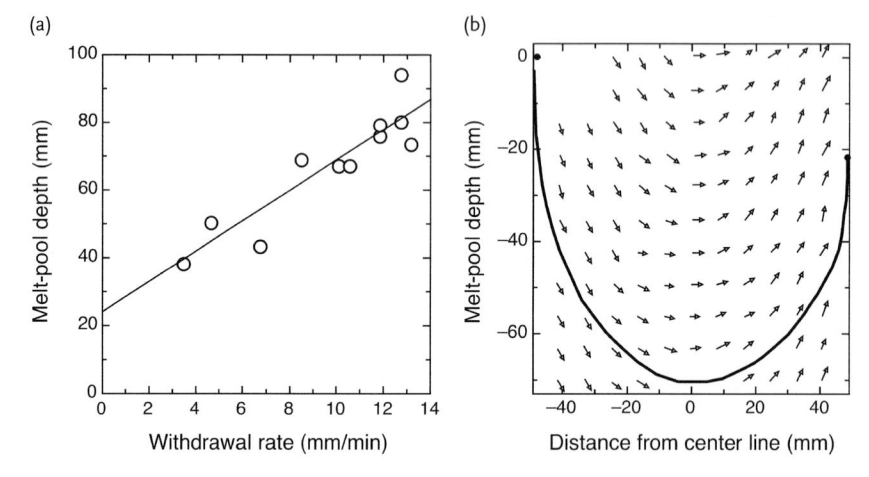

Figure 14.6 Graphs indicating how during PAM (a) the ingot-withdrawal rate influences the depth of the melt pool and (b) the depth of the melt pool varies as a function of distance from the center of the ingot [12]. The left-hand graph indicates that slower withdrawal rates lead to a shallower and flatter melt pool. The right-hand graph shows the development of an unsymmetrical melt-pool geometry possible resulting from the flow of hot molten metal into the crucible from its right-hand side. The arrows in the diagram are drawn perpendicular to the direction of columnar grain growth. Graph replotted and diagram redrawn.

for various withdrawal rates, see Figure 14.6 [12]. The graph in Figure 14.6a indicates that a slow withdrawal rate promotes a flatter shallower melt pool that should be conducive to an improved level of ingot homogeneity. Interestingly, it was found that the shape of the melt pool was not symmetrical around the central axis of the ingot, as indicated in Figure 14.6b. The authors rationalized this observation with reference to the semicontinuous nature with which the molten metal is swept by the plasma torch from the hearth into the crucible. The periodic nature of this process combined with the fact that the molten metal (which has a high temperature) flows into the crucible from only one side, is thought to establish large thermal gradients across the melt pool that influence its shape.

In single-melted PAM ingots, the shape of the melt pool has been shown to be reflected by aluminum isoconcentration lines that indicate segregation effects [12, 16]. Triple or even just double melting has been shown to significantly increase the degree of ingot homogeneity and results in no characteristic segregation patterns being detectable. Indeed the variation of aluminum within a double-melted 80-mm diameter ingot was no more than ±0.5 at.% [16].

Apart from the seemingly important aspect of optimizing the production of compacted feedstock material, Godfrey and Loretto have indicated at least three ways to increase PAM ingot homogeneity [16]:

1) double or triple melting;
2) increase the residence time within the melt pool;
3) increase the melt pool volume.

Work by GKSS on a double-melted PAM ingot of composition Ti-45Al-5Nb-0.2B-0.2C has shown remarkably little variation of aluminum and niobium between the ingot surface and ingot center at the top of the ingot, see Figure 14.7. However, as can be seen, this was not the case at the bottom of the ingot where significant chemical inhomogeneity was observed. This could arise as a result of the initial "hockey puck" compacts not being molten for a long enough time within the hearth. Whereas the mean aluminum level is around 0.6 at.% above the nominal level, the level of niobium is very close to the expected value.

14.1.3
Induction Skull Melting (ISM)

The only company to supply induction skull-melted gamma-TiAl ingots is Flowserve (formerly the Duriron company), who patented the method in 1988 [18]. As its name suggests this is also a "skull-melting technique" and avoids ceramic particle pick-up into the ingot. Figure 14.8 shows a schematic diagram of the induction skull-melting crucible arrangement [19]. As can be seen, copper segments are arranged vertically and form a cylindrical crucible with narrow slots separating neighboring segments. Each individual copper segment is cooled with circulating water, which is provided through the water-cooled copper base. An induction coil surrounds the segmented copper crucible and is used to melt the material that is placed within the crucible. Melting can be performed either under a protective gas atmosphere such as argon or under vacuum [19]. Vacuum operation could, however, lead to excessive evaporation of volatile elements such as aluminum. The loss of aluminum from TiAl during induction skull melting in relation to the partial pressure has been addressed by Yanging *et al.* [20].

Ingots are generally manufactured by a single melting operation using unalloyed charge material. After melting has taken place, the melt is vigorously mixed by the induction process. The ingots (or alternatively cast components) are cast by tilting the copper crucible and pouring the melt into ingot (or component) molds

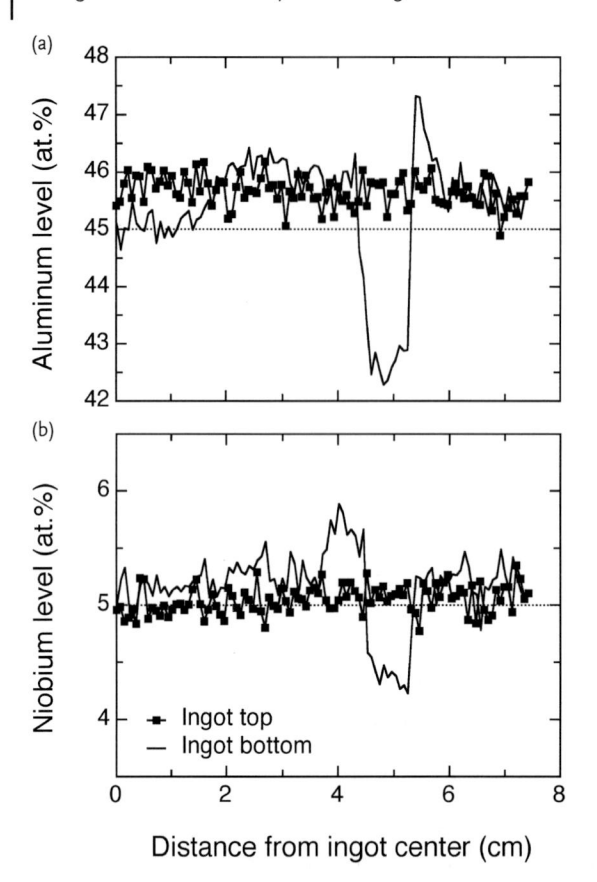

Figure 14.7 Graphs showing the distribution of (a) aluminum and (b) niobium from the center to the surface of a double-melted PAM ingot. Measurements were made using a scanning electron microscope equipped with EDX. The dotted lines show the nominal levels of aluminum and niobium for the alloy composition (Ti-45Al-5Nb-0.2B-0.2C), GKSS unpublished work.

(e.g., made of graphite), see Figure 14.9 [19]. The melt capacity of the ISM equipment at Flowserve is around 40 kg. Larger ingots can be manufactured by combining several ISM ingots to make an electrode for VAR processing, as described in a U.S. Patent [21].

The main advantage of the ISM method is that it is a clean, relatively cheap technique that is very flexible with regard to alloy composition. Additionally, the intense stirring of the melt that is achieved by the induction field should ensure good chemical homogeneity within the ingot. Flowserve offers ingot compositional guarantees of aluminum ±0.75 wt.% and ±0.5 wt.% for other elements with typical oxygen contents of around 500 to 600 ppm [19].

A main disadvantage of ISM is the limited superheat that is developed within the melt. Mi *et al.* [22] cite that the superheat generated in ISM is rarely in excess

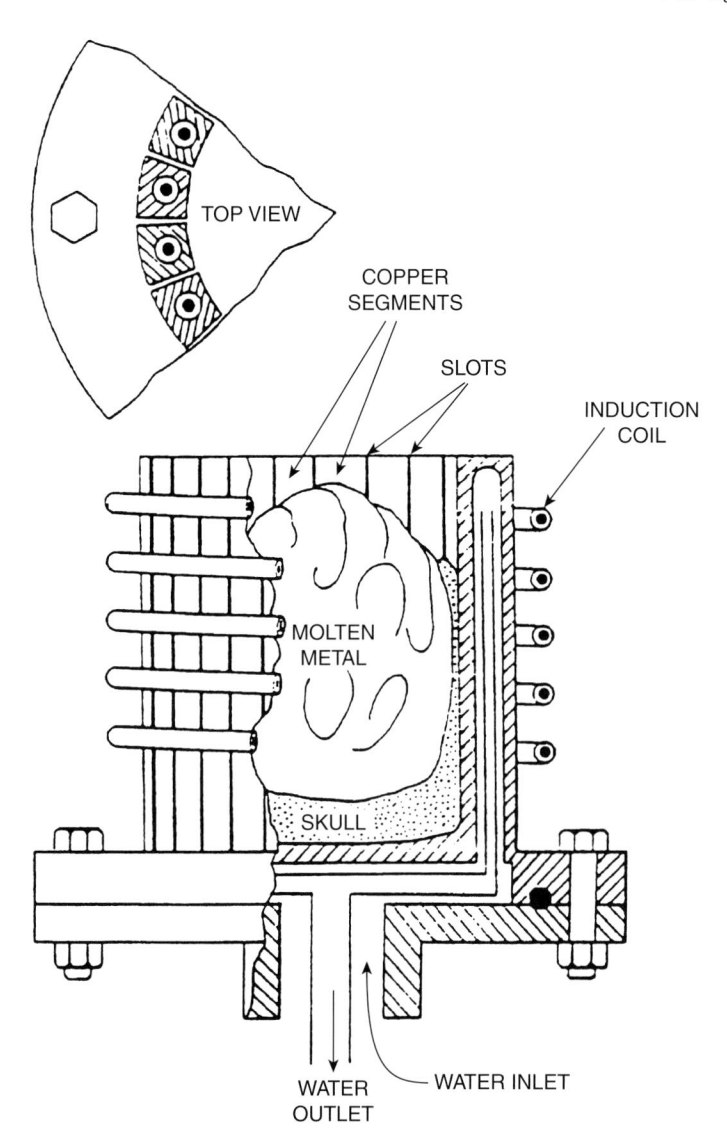

Figure 14.8 Schematic diagram of an ISM crucible [19].

of around 20 °C, which makes it necessary to "dump pour" the liquid metal from the ISM crucible into the ingot (or component) mold. This results in an uncontrolled chaotic flow of the liquid metal into the mold that generates surface break-up and can entrap bubbles in the flowing metal, with the risk of leaving thin oxide films and bubbles inside the final ingot (or component). Such oxide films are thought to be pushed by the dendrite arms towards the center of the casting during solidification where they may act as heterogeneous nucleation sites for the

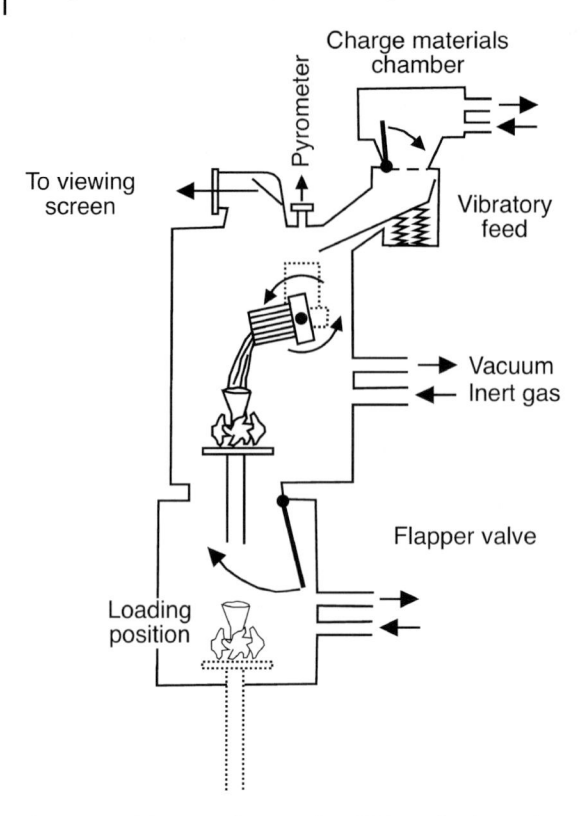

Figure 14.9 Schematic diagram of an ISM unit for the production of ingots (or cast components) [19]. The original diagram has been redrawn.

formation of centerline porosity [22]. The use of high-power ISM (also known as "levitation melting") allows the melt be levitated (pushed) away from the crucible walls, which results in reduced crucible–melt contact. This can lead to an increase of the superheat to around 60–70 °C [22] and also results in the formation of less skull material. This technique, which has a limited melt capacity of a few kilograms (5 kg is reported in [22]), has been used in connection with component casting and is further described in Section 14.2.4. Unfortunately, there is no specific data in the literature regarding the extent of elemental segregation or typical defects found in ingots produced by ISM.

14.1.4
General Comments

In order to evaluate the quality of ingots produced on an industrial scale, it is necessary to study a large number of ingots so that the variability from ingot to ingot can be assessed as well as the type of defects that are present. Ingots of some alloy compositions, especially those containing refractory elements for example, are more difficult to produce. Thus, the assessment of ingot quality in simpler alloy systems may not translate to more complex alloys. Such extensive and costly investigations have not been presented in the open literature. Published data often concerns ingots that have been investigated for defects and macrohomogeneity. However, the variability in quality from ingot to ingot for a specific alloy composition, melting process and ingot size has not been elucidated. Chemical inhomogeneity on the microscale will always be present in any ingot, large or small, and may be of the order of several at.%, but this is not considered to be a quality issue. The size of an ingot probably plays an important role in quality. The cooling rate is an important parameter, which is determined to a certain extent by ingot size, and can influence the extent of segregation and the development of internal stresses that can lead to ingot cracking. The development of thermal stresses within ingots during cooling after both casting and high-temperature heat treatment has been investigated by Alam and Semiatin [23]. The homogenization of ingots by high-temperature heat treatment, with the aim of reducing elemental inhomogeneity, has also been studied [24].

As has been discussed in Sections 14.1.1 to 14.1.3, all three melting techniques that have been used to produce TiAl ingots have their own specific advantages and disadvantages. VAR, for example, is a well-established technique in the titanium industry and must be used in the final melting of conventional titanium alloys for qualification as aerospace-grade material for the manufacture of safety-critical components such as discs [4]. This is because PAM has the disadvantage that small argon-filled pores may remain within the ingot; these can only be removed if the final melting procedure takes place under vacuum, as is the case with VAR. The potential loss of aluminum during VAR is overcome in PAM through the use of a positive pressure of argon that has been reported to result in a very low loss of volatile elements [9].

PAM has the ability to reduce or even eliminate high- and low-density inclusions such as W-rich particles or TiN particles through the possibility of controlling the hearth melt-pool depth and the residence time. This ability of PAM processing is a major advantage over VAR. However, it is believed that high-density inclusions can also be avoided in the final VAR ingot through the correct use of optimized masteralloys [2, 6], while low-density inclusions can be reduced to uncritical levels through careful preparation of the electrode. In conventional titanium alloys low-density inclusions (the so-called "hard alpha" particles) occur as a result of nitrogen and/or oxygen contamination, which can arise from a number of sources including contamination during sponge production, masteralloy production and electrode fabrication or through small air or water leaks during melting [4]. Triple VAR melting was introduced for titanium alloys to eliminate such inclusions but due to the limited superheat is not able to remove the more refractory particles [11]. Such inclusions have a serious effect on the fatigue properties of titanium alloys and there is no reason why this will not be the case if they are present in TiAl components. For further information concerning the ability of hearth techniques to produce high-quality conventional titanium ingot, the reader is referred to references [25, 26].

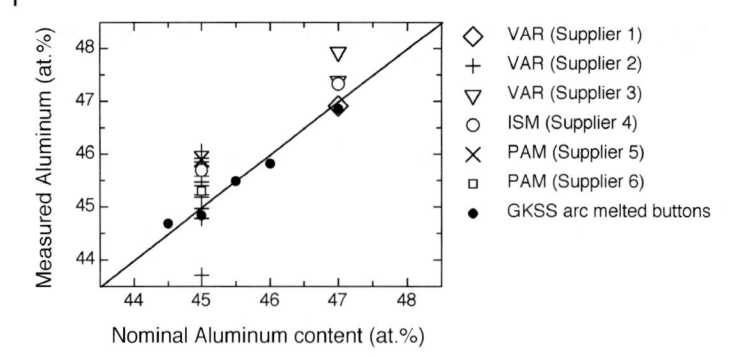

Figure 14.10 Graph indicating the difference between the measured and nominal aluminum content of TiAl ingots of different sizes and compositions obtained from different suppliers and produced using different techniques. The accuracy of the analytical measurement technique is demonstrated by the GKSS arc-melted buttons, which lie very close to their nominal levels of aluminum.

It is very difficult to make any direct quality comparisons between the different melting techniques using literature data. This is because different alloy compositions, ingot sizes and the numbers of melting steps vary from study to study. To date, no systematic investigation on ingot quality has been made using the different melting techniques for the same alloy composition and ingot size using the same prematerials. Over the last 10 to 15 years of TiAl research at GKSS, differently manufactured industrially produced ingots (of different sizes and compositions), from a variety of sources within both the United States and Europe, have been purchased for various research projects. This material has been processed and studied using a variety of techniques. Material from some of these ingots has been analyzed using scanning electron microscopy equipped with energy-dispersive X-ray analysis calibrated with standards. Figure 14.10 shows a graph of nominal aluminum content against that determined by GKSS for some of the ingots from the various suppliers. As can be seen, the aluminum levels within the ingots can be up to around 1.3 at.% different from the nominal composition. As the position within the ingot where the aluminum contents have been determined is relatively random, the graph only indicates what deviations were measured and may not indicate the maximum deviation in aluminum from the nominal level. The points in Figure 14.10 for carefully prepared GKSS arc-melted buttons (40 g ingots) all lie very close to their nominal aluminum level, thus demonstrating the accuracy of the analytical measurements. For scientific work, laboratory-prepared material may therefore be preferable as deviation from the nominal aluminum level is low. It must be mentioned that not all of the ingots purchased from each supplier are represented in Figure 14.10 and the ingots were bought over a long time period during which improvements in ingot production may have taken place. Additionally, the ingots were of different sizes and compositions, thus it is unfair to make any comparisons between the different suppliers.

Better chemical homogeneity in ingots may be achieved through the combined use of different processing methods, such as the combined use of ISM with VAR to make large ingots [21]. A new technique that may be promising combines the hearth and continuous ingot withdrawal of PAM with the ISM technique [27]. The molten metal is conditioned in a hearth and then poured into a bottomless ISM unit where intense mixing occurs, while the ingot is slowly withdrawn. It is claimed that the aluminum variation is less than ±0.5 at.%, while other metallic elements vary by less than ±0.2 at.%.

In conclusion, it must be strongly emphasized that all TiAl technologies rely upon good-quality homogeneous material which enables components with homogeneous microstructures and thus reproducible mechanical properties to be produced. If this is not possible then prealloyed powder processing may be the solution, in spite of the perceived cost disadvantages. Powder technologies are discussed in Chapter 15.

14.2 Casting

Casting is a well-developed production technology for conventional titanium alloy components as well as other materials. Gamma TiAl castings can be produced using conventional casting equipment that is currently being used for aerospace-grade titanium and nickel alloys [28]. Casting probably represents the most cost-effective method for producing near-net-shape parts from TiAl on an industrial scale. A cost breakdown for an investment cast TiAl shroud seal is given by McQuay and Larsen [28], who indicate that around a third of the cost relates to the scrap material that results from the casting process (data from 1997) but that this could be reduced to around 10% as the technology matures. McQuay and Sikka have recently (2002) written an overview concerning current casting technology [29].

The choice of casting alloy depends to a very large extent of the component to be manufactured. Generally, alloys that show good room-temperature ductility, such as the GE alloy Ti-48Al-2Cr-2Nb, do not show particularly high yield strength (around 300 MPa for the GE alloy) [30]. Thus, highly stressed components, such as turbocharger wheels, must be made from higher-strength cast alloys and the room-temperature ductility sacrificed to some extent. A list of potential applications for cast gamma in aero-engines is given in [30].

From a casting aspect, one point that needs to be considered is that alloys that have been designed for the manufacture of components using hot-working procedures such as by extrusion and forging may not be suitable as casting alloys. For example, a cast alloy of Ti-44Al-8Nb-1B with grain size of around 50 μm had a ductility of 0.3% at room temperature but of over 1% in the wrought form that had a grain size of around 70 μm [31]. Thus, grain size is not the only important factor to influence ductility. Alloys suitable for casting components must have good castability and preferably develop fine microstructures that are as free from segregation as possible. In this regard, the alloy composition plays an important role as it determines the solidification path and thus to some extent the degree of microstructural refinement, segregation and casting texture. Significant changes in the macrostructure and casting texture have been observed between Ti-47Al-2Cr-2Nb and Ti-48Al-2Cr-2Nb [32].

It has been shown that solidification through the beta phase (i.e. no peritectic reaction occurs) can lead to more refined cast microstructures that are believed to result in improved properties compared to alloys that solidify peritectically [33–36]. Naka *et al.* [33] were perhaps the first to specifically relate the solidification path with microstructural refinement and also found reduced texture in such alloys. Other work has found that beta-solidifying alloys can be both texture free and relatively segregation free [37]. Although such alloys may develop fine microstructures [35, 36] and high strengths [36], only a couple of studies have indicated room-temperature ductility above 1% [33, 34]. Naka *et al.* [33] showed that the room-temperature ductility (and creep properties at 800 °C) of the (heat-treated) beta-solidifying alloy Ti-46.6Al-2Re-0.8Si was improved compared to Ti-48Al-2Cr-2Nb (the alloys showing plastic elongations of 1.4% and 0.9%, respectively). In spite of this, no work has been presented that convincingly demonstrates the anticipated improved room-temperature ductility of (cast) beta-solidifying alloys over those that solidify peritectically. From the viewpoint of in-service mechanical properties it is necessary to remove the beta (or its decomposition product B2) phase from the final component by heat treatment [33]. It may be speculated that when this is not achieved, high room-temperature strength but low ductility (combined with reduced creep resistance) may result. Alloys that solidify through β generally contain less than a certain Al concentration, this is determined by the nature of the phase diagram and is thus composition dependent. Niobium additions for example, result in alloys with higher aluminum contents solidifying through the beta phase field. Chen *et al.* [38], indicate that a 10 at.% addition of niobium results in ternary alloys containing up to 49 at.% aluminum solidifying through beta, although in our experience this seems rather high. A list of some casting alloys and their particular attributes has been given by McQuay and Sikka [29] and is shown in Table 14.1.

As discussed in Chapter 13, one alloying element that is highly effective in refining the microstructure of cast TiAl alloys is boron. The effect was initially found using particulate titanium diboride (TiB_2) additions [46, 47] although elemental boron additions are also effective. The mechanism of grain refinement, however, remains disputed [48] and seems to be alloy dependent, with refinement being more effective for lower aluminum-containing alloys [49]. A disadvantage of this refinement method is that long boride ribbons can form (especially in thick sections) in alloys containing elements such as Nb, W and Ta, which can compromise ductility [31]. Thus, other methods capable of refining the microstructure of cast components have been developed. These include the use of rapid cooling (quenching) and/or the implementation of solid-state reactions [31, 50, 51], as discussed in Section 13.3.4.

After the choice of alloy composition has been made, the appropriate type of melting and casting process needs to be addressed. In the literature four main types of casting process have been described, namely: investment casting, gravity metal mold casting, centrifugal casting and countergravity low-pressure casting.

Table 14.1 Cast gamma engineering alloys and their attributes according to [29].

Alloy name	Composition (at.%)	Attributes	Ref.
GE 48-2-2	Ti-48Al-2Nb-2Cr	Ductility, fracture toughness	[39]
Lockheed-Martin 45XD™	Ti-45Al-2Nb-2Mn-0.8 vol.% TiB$_2$	Tensile & fatigue strength, castability	[40]
Lockheed-Martin 47XD™	Ti-47Al-2Nb-2Mn-0.8 vol.% TiB$_2$	Elevated-temperature strength, castability	[40]
Honeywell WMS	Ti-47Al-2Nb-1Mn-0.5W-0.5Mo-0.2Si	Creep resistance	[41]
ABB-Alstom, ABB-2	Ti-47Al-2W-0.5Si	Creep and oxidation resistance	[42]
GKSS TAB	Ti-47Al-1.5Nb-1Mn-1Cr-0.2Si-0.5B	Castability, property balance	[43]
Daido Steel	Ti-48Al-2Nb-0.7Cr-0.3Si	Ductility	[44]
IHI	Ti-45Al-1.3Fe-1.1V-0.35B	Castability	[45]

Each process will be described below with particular reference to relevant work presented in the literature.

14.2.1
Investment Casting

In this process, also known as the lost-wax process or precision casting, a molten metal is cast into a mold that is made of a ceramic shell. The shell is manufactured by repeatedly coating a wax pattern (copy) of the component with ceramic-based slurries that are allowed to dry before being recoated. The shell is thus made of a series of different layers all of which are necessary for successful component manufacture. The suitability of various refractory materials for the melting and casting of TiAl alloys has been investigated by Kuang et al. [52]. It was found that pure calcia (CaO), yttria and yttria-coated magnesia showed promise for melting and casting processes. Ceramic mold shells coated with an yttria face coat [31] and molds made from alumina [53, 54] have also been described in the literature.

As with all titanium aluminide alloys, the mechanical properties of cast components depend critically on the aluminum level. In this respect, work performed at GKSS has shown that the chemical homogeneity of ingot feedstock, or the extent of aluminum loss during the melting and casting process are of great importance. The graphs shown in Figure 14.11 demonstrate how aluminum variations within material with a nominal composition of Ti-48Al-2Cr-2Nb (Austin & Kelly) affect room-temperature tensile properties [30]. The tensile properties of various

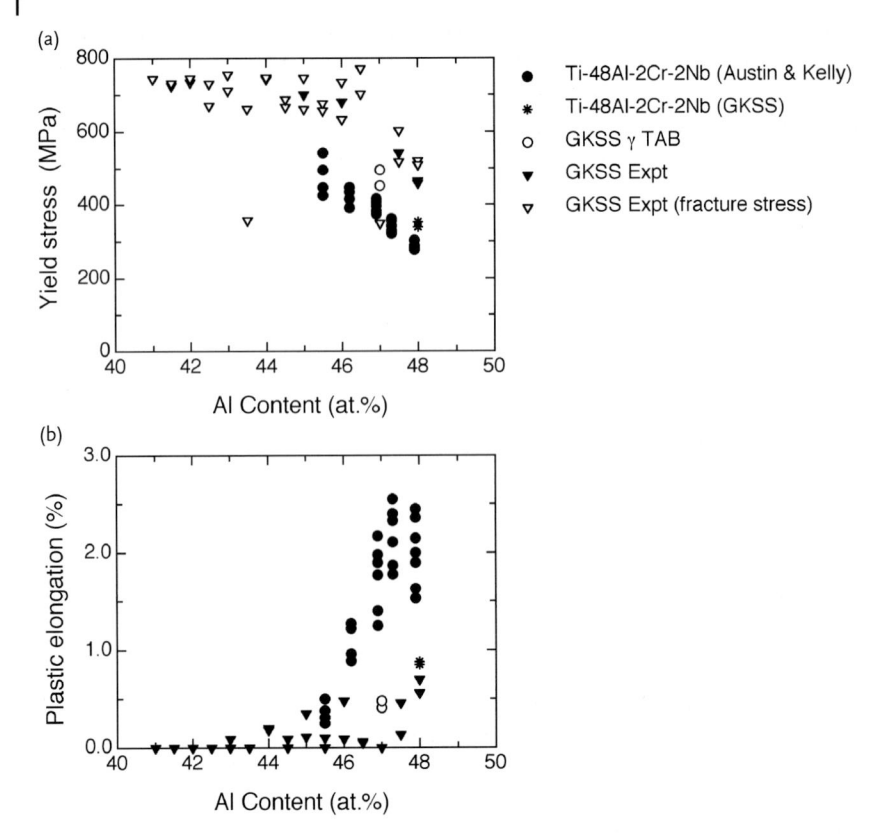

Figure 14.11 Dependence of the mechanical properties (a) the yield stress (0.2% flow stress for GKSS tested alloys) and (b) the plastic elongation to failure on the aluminum content of cast alloys. The data shown for the Ti-48Al-2Cr-2Nb alloy presented by Austin and Kelly has been replotted from data in [30], and indicates that variations in the nominal aluminum content within the Ti-48Al-2Cr-2Nb alloy can give rise to dramatic property variations. The other data is for alloys or alloy series that have been cast at GKSS using argon arc melting of carefully weighed elemental components that were melted on a water-cooled copper hearth into cylindrical bars. This material was subsequently HIPed at 1180 °C for 4 h under a pressure of 200 MPa. The casting texture may be significantly different to investment-cast bars as solidification initiates at the lower side of the bar in contact with the hearth and moves progressively towards the upper side. This may explain the difference in properties between the Ti-48Al-2Cr-2Nb alloy tested by GKSS and that tested by Austin and Kelly [30]. The yield stress for the GKSS-tested material was taken as the 0.2% flow stress and testing was performed using a strain rate of $2.38 \times 10^{-5} \, s^{-1}$. As some of the GKSS experimental alloy series (Expt) failed before reaching 0.2% plastic strain, the fracture strength of this alloy series is plotted in Figure 14.11a.

microstructures are shown for each specific aluminum level. The data from [30] indicates that the yield stress is relatively independent of microstructure for a given aluminum content, but that the aluminum level has a significant effect on both strength and ductility. These results highlight the importance of compositional control through the whole production chain, from the ingot to the cast components. Data for other alloys that were cast and tested at GKSS, are also presented in Figure 14.11.

Investment casting of TiAl using a melting technique called vacuum induction melting (VIM) has been used in work aimed at producing turbocharger wheels and turbine blades [55]. This technique is similar to ISM, the charge material being melted by induction but within a ceramic crucible rather than a copper crucible. This has the advantage that the melt does not come into contact with the cold crucible wall and thus a higher superheat can be generated within the molten metal. Disadvantages include possible ceramic particle pick-up and increased interstitial contamination. Melting was performed within a CaO-coated alumina crucible under a partial pressure of argon to minimize the loss of aluminum [55]. The crucible containing the molten TiAl was tilted to one side and the melt poured into a ceramic-based mold.

Initial casting of a turbocharger wheel indicated considerable damage in the form of cracking at positions where the blade roots attached to the inner wheel [55]. Although it was not stated whether the mold was preheated or not, it was thought that large stresses were built up due to the cast TiAl and the ceramic mold having different coefficients of thermal expansion. Heat flow and the build up of thermal stresses and strains was thus studied both experimentally and modeled with the aim of improving the quality of the castings. Successful casting of a turbine blade was achieved through the use of better-optimized casting parameters with a mold-preheating temperature of 1027 °C and furnace cooling (although some microcracking within the gamma phase of the microstructure was observed). Puzzlingly, results relating to the casting of a turbocharger wheel using better-optimized parameters were not discussed. The build up of thermal stresses during cooling of TiAl investment castings has also been investigated elsewhere [56]. The lack of reliable thermomechanical and thermophysical data and the need to incorporate the effect of component cracking during cooling on the state of stress were identified in [56]. However, since this paper was written (1995), some thermophysical data for a Ti-44Al-8Nb-1B alloy has been published [57].

The level of superheat, that is, the temperature of the molten metal in comparison to its melting point and the mold preheating temperature, are important parameters that can significantly influence the quality of a cast component. The fluidity of TiAl has been shown to closely depend on the mold preheating temperature, see Figure 14.12 [53]. Low superheats can lead to inadequate feeding that results in incomplete filling of component molds and porosity.

Increasing the superheat from around 140 °C to 180 °C has been shown to result in better mold filling and an increased yield of cast valves from around 50% to 90% [54]. This study also employed the VIM technique described earlier and used an Al_2O_3-based mold material that was preheated to 900 °C. The use of

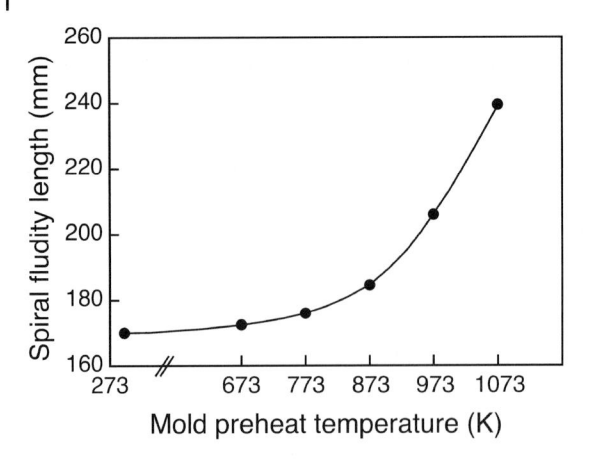

Figure 14.12 Graph showing how the castability as measured by the spiral fluidity length increases with mold preheating temperature [53]. The graph indicates that higher mold temperatures lead to better mold filling. Graph redrawn using original data.

ceramic-based crucible and mold materials does, however, pose problems relating to ceramic particle pick-up and oxygen contamination of the melt. Indeed oxygen levels of between 820 and 2900 wt. ppm (mean value of 1384 wt. ppm from 12 melts employing superheats between 140 °C and 180 °C) have been reported using such a system, with the level of contamination being proportional to the degree of superheat [54]. The effect of superheat on melt–refractory reactions and melt contamination has also been studied by Barbosa et al. [58]. Oxygen values of between 700 and 1200 ppm have been measured [59] using ceramic-free melting and casting processes (mean value of 1004 ppm from 12 melts with an unknown level of superheat), although it is not clear if the units used are wt. ppm or at. ppm. It has been shown that oxygen levels above 500 ppm (not specified if wt. ppm or at. ppm) lead to a reduction in the ductility of cast material [60]. Harding et al. [61] have investigated the effect of superheat generated though three melting techniques, namely VIM (referred to as VIR in their paper), high-power ISM (also called "levitation melting", see Section 14.2.4) and conventional ISM (Section 14.1.3), on the quality of cast components. Radiographic examination of cast bars produced via the VIM melting technique (with higher superheat) showed no large pore defects, whereas bars made using conventional ISM had numerous bubble-like defects that had a distinctly different appearance to shrinkage porosity. However, as the VIM cast material had a higher mold preheating temperature, the influence of the superheat alone could not be clearly demonstrated.

The section thickness and mold temperature both play important roles in determining the microstructure and mechanical properties of cast components as they directly influence the cooling rate. This can dramatically influence the texture, macrostructure and microstructure of as-cast and heat-treated material and possibly the solidification path [62, 63]. Slow cooling rates have been shown to favor a $\{110\}_\gamma$ casting texture, whereas faster cooling tends to favor a $\{111\}_\gamma$ texture. This means that for slower cooling rates during solidification, the {110} planes of the γ unit cell are perpendicular to the direction of heat flow (i.e. the <110> directions are parallel to the heat flow). Similarly, for faster cooling rates the $\{111\}_\gamma$ planes are perpendicular to the heat flow (i.e. the <111> directions are parallel to the heat flow). There appears to be a composition-dependent critical cooling rate, which separates the formation of the $\{110\}_\gamma$ and the $\{111\}_\gamma$ casting textures [63].

When Ti-48Al-2Cr-2Nb was cast into molds with a low (350 °C) preheat temperature (fast cooling rate) it exhibited a lamellar microstructure that decomposed during HIPing at temperatures between 1200 °C and 1260 °C. However, when the mold was preheated to 1204 °C (slow cooling rate) the lamellar microstructure was stable on HIPing [62]. Other work has also reported an influence of cooling rate on cast microstructure [40, 64] and mechanical properties [65]. The room-temperature tensile properties of fast and slowly cooled Ti-47Al-2Cr-2Nb have been presented by Biery et al. [66] who found that the cooling rate played an important role in determining the post-yield property variability, primarily through grain size. The fast-cooled material showed ductility that ranged from 1.3 to 1.9%, while the slowly cooled material exhibited values between 0.6 and 1.7%. Thus, large castings with varying section thicknesses can be expected to exhibit differing microstructures and mechanical properties as a result of the differing cooling rates within the component.

The role of component size and therefore cooling rate on the mechanical properties in cast TiAl alloys has been evaluated [31, 67]. It was shown that the tensile ductility decreased from around 0.7–0.8% to about 0.2% when the diameter of investment cast + HIPed bars, of boron-containing alloys, was increased from 15 mm to 30 mm [31], see Figure 14.13. This was thought to be due to long boride

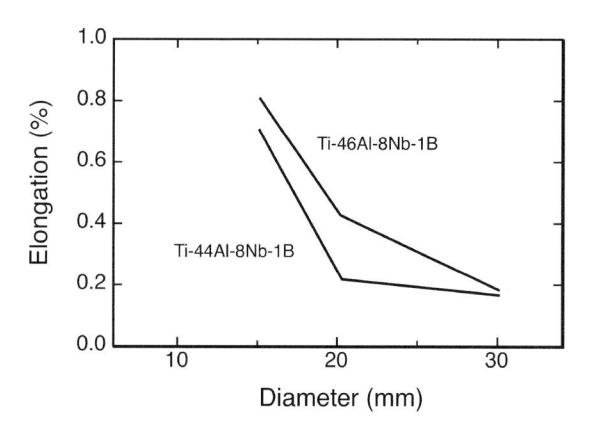

Figure 14.13 Variation of the room-temperature tensile ductility with the diameter of Ti-44Al-8Nb-1B and Ti-46Al-8Nb-1B investment cast bars after HIPing at 1320 °C/150 MPa/2 h and 1370 °C/150 MPa/4 h, respectively [31]. The graph indicates reduced ductility with increasing cast bar diameter. This is thought to be due to coarse borides forming under slower cooling conditions and leading to premature failure. Graph redrawn using original data.

ribbons being more easily formed under slower cooling rates and leading to premature failure. Long ribbons are believed to form more easily if high levels of strong boride formers such as Nb, W and Ta are present in the alloy. Boride refinement can be achieved through faster cooling rates using lower mold temperatures but this can lead to problems of mold filling during casting [31]. Kuang et al. [67] have shown that the grain size increases for larger-diameter bars and that the strength of castings with a duplex microstructure obeys a Hall–Petch relationship. The tensile ductility of cast Ti-48Al-2Nb-2Mn bars decreased from 0.9% to 0.6% in the HIPed state when the bar diameter was increased from 12 mm to 25 mm. However, after a subsequent heat treatment of 1300 °C for 24 h, ductility remained at around 0.5 to 0.6% and seemed independent of bar diameter.

The occurrence of porosity within cast components clearly has significant implications on the room-temperature properties of a relatively brittle material such as TiAl. The cooling rate has been found to influence the porosity distribution [65]. High cooling rates were found to lead to interdendritic porosity that formed as a result of insufficient feeding of the solidification shrinkage. Slower cooling rates that are promoted by higher mold temperatures led to the porosity being more localized to the surface of the casting, possibly resulting from a reaction with the (unspecified) mold material. The geometry of a casting can also play an important role in the formation of casting porosity. "Inverted carrot"- and "carrot"-shaped specimens were cast by Simpkins et al. [68], who investigated the effect of HIP pore closure and age hardening on the tensile properties of XD alloys. The inverted-carrot geometry led to significant casting porosity (as high as 30% of the mold volume) at the bottom end of the castings due to inadequate feeding, see Figure 14.14 [68]. The casting porosity did not have a significant effect on the tensile properties after HIPing; specimens taken from the upper and lower ends of the castings showed similar strengths and elongations. However, a significant effect on the primary creep resistance was observed. This was attributed to the HIP process leading to the recrystallization of fine gamma grains during the deformation of material around the pores for their closure. This gives rise to a reduction in the volume fraction of the lamellar microstructure and thus a degradation of the creep properties. Clearly, although pores may be removed by a HIP treatment, their formation needs to be avoided, not only from this microstructural aspect but also because their collapse can lead to dimples on component surfaces, see Figure 14.15 [69]. Such surface imperfections may lead to the surface quality being outside the tolerances required for aerodynamic structures. One possibility to overcome this surface-dimpling problem is to cast to slightly oversize dimensions and then chemically mill or grind to the final component dimensions. However, when such dimples are not formed, investment-cast parts can show both excellent surface quality and dimensional accuracy.

General Electric, building on previous TiAl casting experience [30] and the successful testing (over 1000 simulated flights) of a full set of 98 stage-five cast low-pressure turbine blades [70], has announced that it will use cast LPT blades based on Ti-(47–48)Al-2Cr-2Nb (made by PCC) in their next generation of aero engines (GEnx-1B) for the Boeing 787 Dreamliner that is due to enter into service in 2011

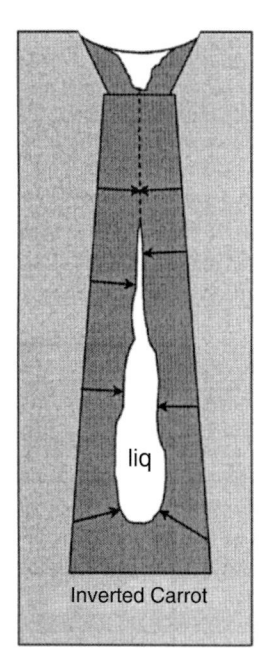

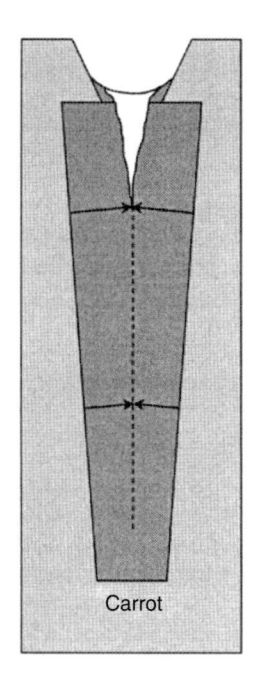

Figure 14.14 Diagram showing how porosity may develop in cast components. The inverted carrot shape has solidification fronts that meet in the upper part of the casting, entrapping the remaining liquid in the lower part, leading to large regions of porosity. The carrot-shape solidification fronts first meet at the bottom of the casting and then move smoothly up the specimen, the last liquid that solidifies does so near to the top [68].

[71, 72]. The EU has funded a program called IMPRESS that investigated the investment casting of a 40-cm long TiAl blade using casting [73].

14.2.2
Gravity Metal Mold Casting (GMM)

This method has been developed by the Howmet Corporation for the production of relatively simple-shaped components and utilizes a permanent metallic mold rather than a ceramic mold [74]. Use of a metallic (ferrous based) mold can significantly reduce the costs as the expensive and complicated manufacture of ceramic mold shells is eliminated and removal of the parts after casting from the ceramic shells is no longer necessary. Additionally, the inclusion of ceramic particles within the component and oxygen pick-up from the shell material are avoided. The gravity metal mold technique for the production of cast automotive exhaust valves is shown schematically in Figure 14.16 [74, 75]. As can be seen, ingot feedstock is melted using cold-wall crucible melting technologies and then poured into the metallic mold under vacuum. Once solidified the components are removed from the mold.

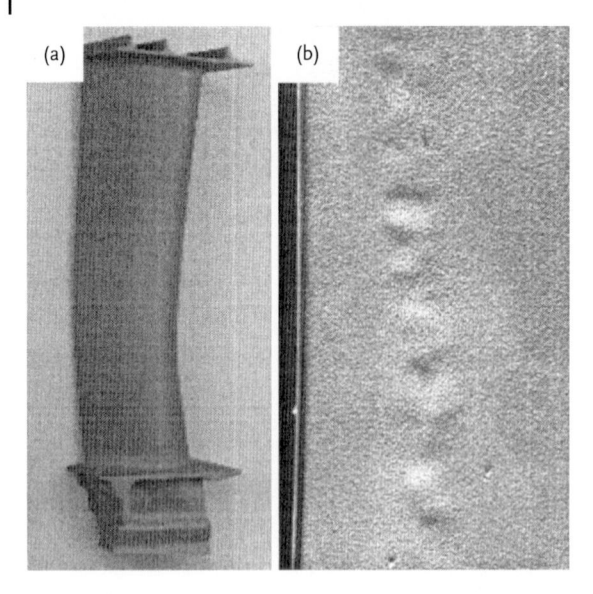

Figure 14.15 (a) A cast low-pressure (LP) turbine blade and (b) dimples on the surface of a HIPed LP turbine blade resulting from the collapse of porosity during HIPing [69].

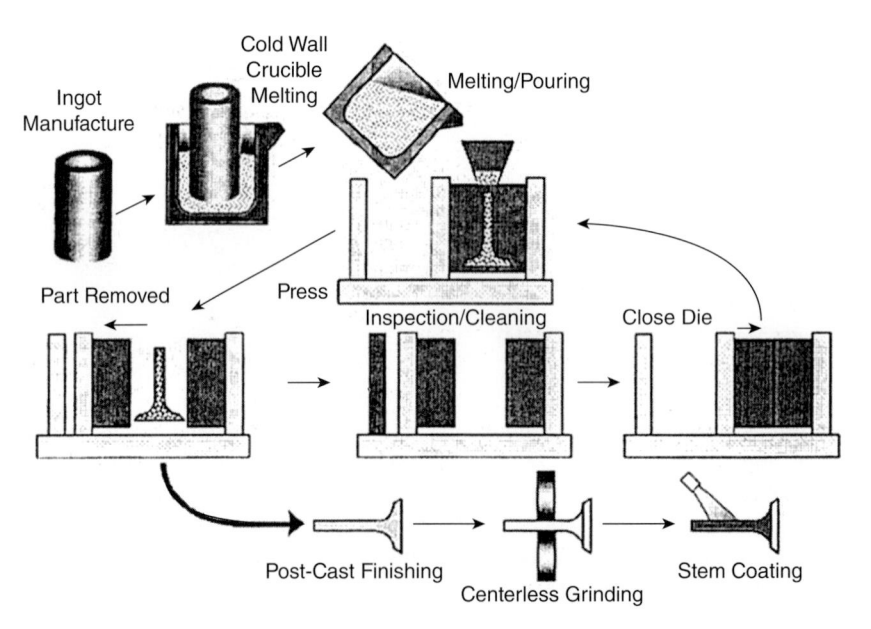

Figure 14.16 Schematic diagram showing the gravity metal mold casting process for the production of automotive exhaust valves [74, 75].

Jones et al. [75] and Eylon et al. [76] have described the production of cast automotive exhaust valves using permanent metallic molds. Melting of the feedstock was performed by consumable-arc cold hearth VAR. In the study, three different mold orientations were used, namely top gated, side gated and bottom gated. Additionally, the effect of a mold preheating temperature of 260 °C was compared to a mold that was at room temperature. Furthermore, the effect of the application of pressure during solidification was examined. This was done by:

i) a "gas boost" that involved pressurizing the casting chamber with inert gas after pouring, as described in a U.S. Patent [77];

ii) centrifugal casting (a technique that is described in Section 14.2.3);

iii) squeeze pins – these are hydraulically operated plungers which apply pressure to the in-gate and thus force liquid metal into the die; and

iv) injection – this involves pouring the metal into a shot sleeve from where it is forced under pressure into the die cavity, as described in a U.S. Patent [78].

The squeeze-pin technique was difficult to evaluate, as the time window with which the pressure needed to be applied was less than 0.5 s. At the time the paper by Jones et al. [75] was published (1995), the effect of the injection technique had not been determined. The findings presented within these papers [75, 76] established that:

i) Preheating the molds to 260 °C had no effect on mold filling or the surface quality when compared to molds filled at room temperature.

ii) The parts should be vertically oriented to ensure that porosity occurs near to the centerline.

iii) When porosity was extensive or off the centerline, then poor dimensional accuracy and increased machining costs arose after subsequent HIP processing.

iv) None of the permanent mold-casting methods produced as-cast valves that were pore-free or contained very low porosity.

v) The microstructures that developed using metallic molds were finer than those observed in investment castings of similar section thickness. This is a result of the faster solidification rate arising from the much higher thermal conductivity of metallic molds compared to the ceramic molds that are used in investment casting. The finer microstructure gave rise to an increase of room temperature flow stress.

vi) The injection technique can lead to a further refinement of the cast microstructure.

vii) The only evidence of any reaction with the mold was a 25-μm thick Al-depleted layer at the surface of the castings.

viii) Top gating was shown to result in significantly better mold filling, reduced porosity and increased dimensional accuracy of parts as compared to side-gated components. This was thought to arise because of the extra hydrostatic head pressure during solidification. Additional benefits were realized though centrifugal casting, a topic that is discussed in Section 14.2.3.

From the above discussion it is clear that gravity metal mold casting has certain advantages over the conventional investment-casting process from both costs and technical (lack of ceramic particle pick-up, oxygen contamination and refined microstructure) points of view. However, since these publications very little new information seems to have been published on the GMM technique in regard to casting TiAl alloys.

14.2.3
Centrifugal Casting

There are basically two types of centrifugal-casting techniques, horizontal and vertical. Horizontal-centrifugal casting involves rotating about a horizontal axis and vertical around a vertical axis [79]. The horizontal technique is used to cast components that have an axis of revolution such as steel pipes. The vertical-centrifugal casting technique is used to make noncylindrical or even nonsymmetrical components. Centrifugal casting feeds molten metal into rotating molds with centrifugal force, which is applied until after solidification has been completed, forcing the molten metal under a constant pressure to fill the mold. The centrifugal force increases the pressure head of molten metal and results in castings of high quality and integrity [79].

In principle, centrifugal casting could be combined with various different types of melting technique. ISM has usually been used for the studies concerning centrifugal casting of titanium aluminide that have been published in the literature, although two studies used cold-hearth VAR [75, 76]. This could be because VAR requires premelted ingot feedstock to ensure homogeneity, whereas ISM can use elemental charge material and is thus more flexible and less expensive.

The manufacture of components such as automotive exhaust valves and turbocharger wheels has been investigated [54, 59, 80–82]. The charge material used for the production of automotive valves consisted of titanium sponge or scrap, aluminum bars, elements such as niobium and chromium and low-cost masteralloys [80]. In [59, 80, 81] melting was performed by ISM and the molten metal poured into a rotating cylinder that consisted of niobium-based automotive valve molds imbedded in a steel-based surround. It should be mentioned that the valve molds were preheated to around 1000 °C prior to casting to avoid the formation of shrinkage porosity in the valve plates and upper parts of the valve stems [59]. The whole casting process took place under vacuum. Centrifugally cast TiAl automotive valves just after removal from the mold are shown in Figure 14.17 [59]. No macrosegregation was observed within the valves that were produced and the extent of interdendritic microsegregation was low so that no subsequent homogenization

Figure 14.17 Centrifugally cast automotive exhaust valves after removal from the mold [59].

heat treatment was thought to be necessary. Mechanical testing of the cast Ti-47Al-2Nb-2Cr material showed that the flow stress was around 600 MPa and plastic elongations were between 0.5 and 1.8% [59]. For more information on the mechanical properties the reader is directed elsewhere [59].

The influence of component gating has been investigated during the centrifugal casting of turbocharger wheels, with both direct and indirect gating systems being tested [82]. Indirect gating requires longer feeders and results in the molten metal entering the mold cavity in the opposite direction to the centrifugal force, which should lead to less-turbulent flow of the melt as it fills the mold. However, cold runs and incomplete filling of the turbocharger blades occurred as a result of the longer feeder length. In further casting trials it was found that in order to keep macro- and microshrinkage porosity out of the part, a very large riser and support structure were required, see Figure 14.18 [82]. The requirement for such a large amount of molten TiAl naturally increases the amount of scrap and thus cost of the cast components. In spite of the higher melt turbulence during direct gating, it was demonstrated that sound turbocharger wheels could be produced, see Figure 14.19 [82].

Typical defects that were observed in cast turbochargers are shown in Figure 14.20 [82]. Incomplete filling of the mold and cold runs arose from insufficient feeding as a result of low fluidity, Figure 14.20a. Internal porosity within the casting can result in surface distortion after HIPing, see Figure 14.20b (also see Figure 14.15). The incompletely filled or "frayed" blade tips shown in Figure 14.20c can occur when the front coat of the ceramic-based mold cannot withstand the pressure exerted by the molten metal and partly collapses. The reason for the surface bubbles or pinholes that were occasionally observed (see Figure 14.20d) was not identified. After the casting process had been optimized the quality of cast

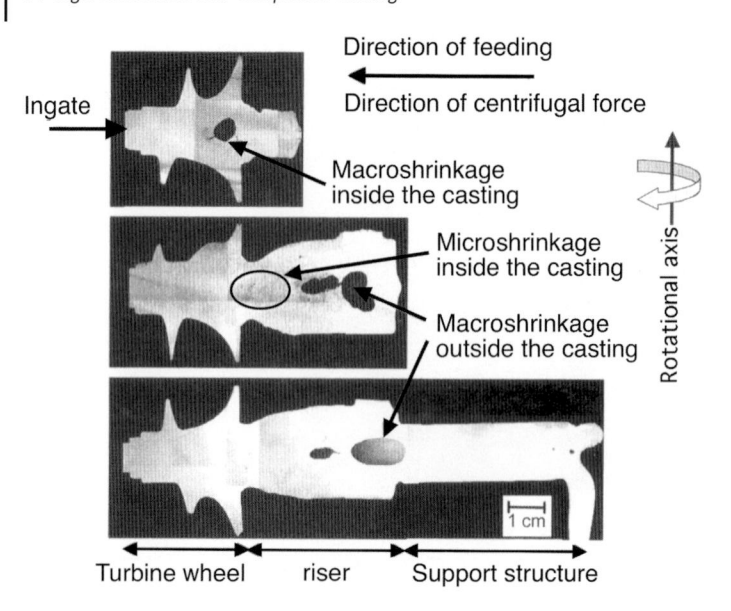

Figure 14.18 TiAl turbine wheels with riser and support structure that were centrifugally cast using indirect gating, that is the melt flows into the mold against the direction of the centrifugal force [82]. It can be seen that a large riser and support structure are required to avoid macro- and microshrinkage within the turbocharger wheel.

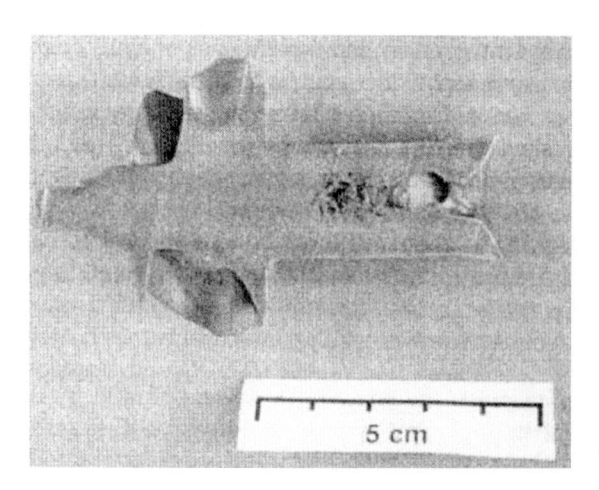

Figure 14.19 TiAl turbine wheel that was centrifugally cast using direct gating, that is the melt flows into the mold in the same direction as the centrifugal force [82]. The shrinkage is outside of the component and much less cast material is required.

components was excellent, see Figure 14.21. The blade-tip thickness of the turbochargers shown is around 0.7 mm or even less.

GKSS has taken part in two Government-funded (BMBF) programmes concerned with the production of centrifugally cast TiAl components. The first of these programmes, coordinated by GKSS, ran from August 1990 to July 1994

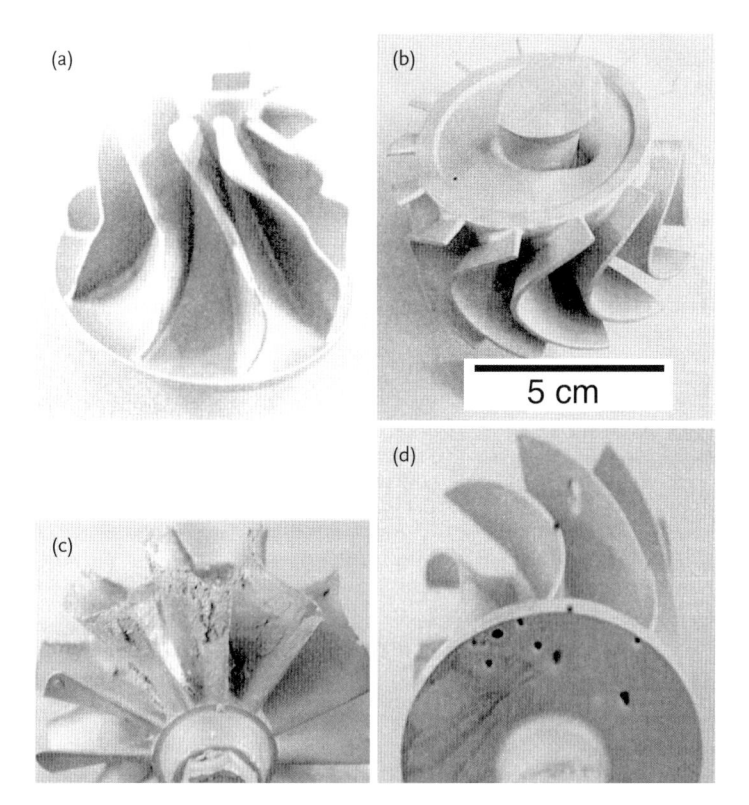

Figure 14.20 Defects observed in centrifugally cast turbocharger wheels [82]. (a) Cold runs or incomplete mold filling, (b) distortion after hot isostatic pressing due to closure of excessive macroshrinkage within the wheel, (c) incompletely filled or "frayed" blade tips due to mold front coat damage and (d) pinholes.

Figure 14.21 Centrifugally cast turbocharger wheels after process optimization. The thickness of the blade tips is around 0.7 mm and the wheel diameter around 5 cm (photo courtesy of ACCESS, Aachen, Germany).

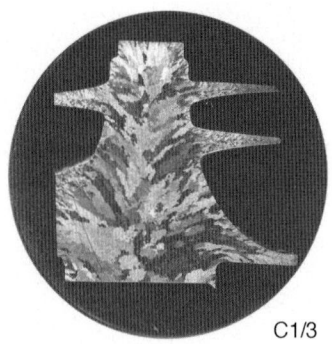

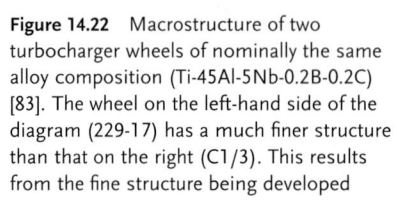

Figure 14.22 Macrostructure of two turbocharger wheels of nominally the same alloy composition (Ti-45Al-5Nb-0.2B-0.2C) [83]. The wheel on the left-hand side of the diagram (229-17) has a much finer structure than that on the right (C1/3). This results from the fine structure being developed through a beta solidification path whereas the wheel on the right, with a coarse structure, solidified peritectically through alpha. The difference in aluminum content between the two turbochargers was only around 0.3 at.%, which is less than the aluminum variation within the ingot feedstock.

and was aimed at the production of turbine components and included a short-duration engine test performed at MTU [43]. A cast alloy (GKSS γ-TAB) that exhibits a good balance of properties and a property database was developed within the project.

The second casting program that involved GKSS (February 2002 to December 2005) was coordinated by DaimlerChrysler and was aimed at the development of a TiAl casting technology (based on centrifugal casting) for complex lightweight components, specifically turbocharger wheels. One aspect that became apparent during the project was the need for high-quality melt feedstock with little variation in composition [83]. This is well demonstrated in Figure 14.22, which shows a vast difference in the macrostructure of two similarly cast turbocharger wheels of nominally the same alloy composition (TNB-V5, Ti-45Al-0.2B-0.2C). The only difference between the two turbocharger wheels was the level of aluminum. The coarse macrostructure resulted from the aluminum content being around 45.2 at.%, while the finer macrostructure developed when the aluminum content was below this level. It thus seems that when the aluminum level in the alloy was 45.2 at.%, solidification proceeded peritectically, whereas at lower levels the alloy solidified through beta, which resulted in a more refined microstructure. Adding a small amount of titanium to the aluminum-rich turbocharger and a similarly small amount of aluminum to the aluminum-poor turbocharger resulted in the macrostructural features being reversed on remelting (the effect was also reversible). This technique was used to determine the critical aluminum level that separates the different solidification paths and also indicated that it could be shifted to higher aluminum levels by increasing the niobium content within the alloy. Thus, the need for optimized alloy compositions that exhibit both macro and microstructures which are less sensitive to aluminum variation was identified and addressed.

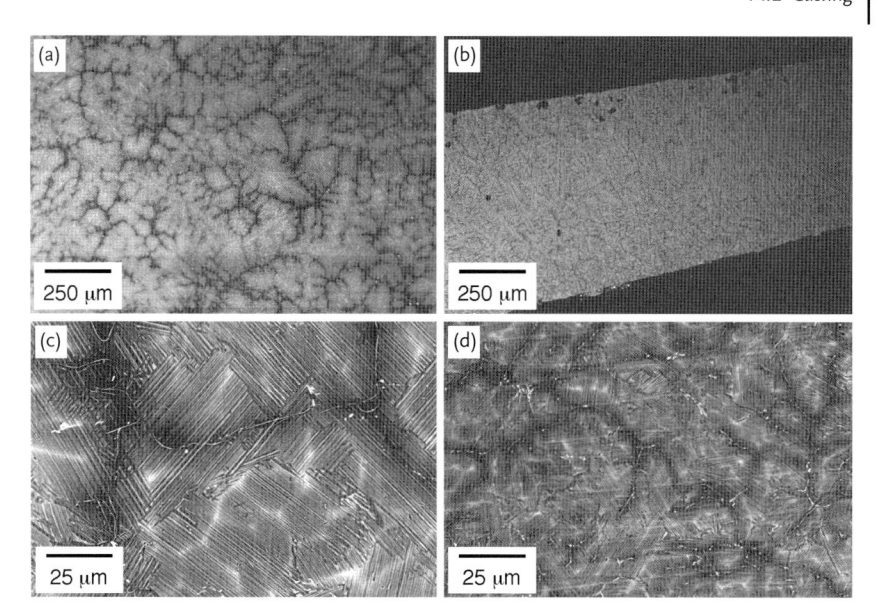

Figure 14.23 Macrostructure of (a) the central axial region and (b) the blade of a cast turbocharger wheel. One can see that the structure is much finer in the faster-cooled blade. Microstructures of (c) the central axial region and (d) the blade. As can be seen, the microstructure is finer and the borides are more broken up in the blade area [83].

As has been discussed earlier, the section thickness (and thus cooling rate) has been found to have a marked effect on the microstructure. Both the segregation and the microstructure within the thin blades were on a much finer scale than those developed in the central axial region of the turbocharger wheel, compare Figures 14.23a and c with Figures 14.23b and d, respectively. On comparing Figures 14.23c and d it can be seen that the faster cooling rate of the blades compared to the central region of the turbocharger, leads to a breaking up of the long boride ribbons in agreement with [31]. Additionally, the lamellar are much finer in the faster-cooled material.

Such large differences of microstructure within the same component will undoubtedly lead to a variation of the mechanical properties. The properties of the relatively coarse central axial region of the turbocharger wheel can be determined from conventional tensile and creep tests on cast bars with a similar diameter. However, the properties of the blade region are more difficult to determine. With this aim, 1-mm thick centrifugally cast sheets were produced by ACCESS and tested by GKSS. Microtensile specimens with a gage of around 0.5–0.7 mm in thickness, 2 mm width and 7 mm length were tested at room temperature, 700 °C and 800 °C, see Figure 14.24. Both the HIPed and un-HIPed sheets showed limited room-temperature ductility but the HIPed sheet showed enhanced ductility compared to the un-HIPed (as-cast) sheet at 700 °C. The reason for this was not determined but did not result from differences in the level of aluminum. With regard

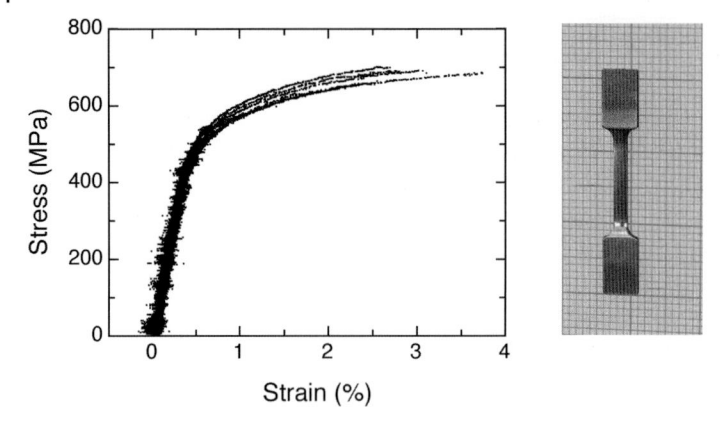

Figure 14.24 Tensile curves for 6 HIPed microtensile specimens that were tested at 700 °C. As can be seen, some ductility was measured in spite of the difficulties associated with testing, namely the small specimen size and alignment problems. A photograph of a microtensile specimen that has been placed on mm graph paper is shown to give an indication of the size of the specimens [83].

Table 14.2 Mean oxygen and nitrogen levels measured within ingot melt feedstock (2 different ingots), 6-cast turbocharger wheels and 5-cast sheets with a thickness of 1 mm.

	Oxygen content (wt. ppm)	Nitrogen (wt. ppm)
Ingot material	577–806 (21)	72–91 (21)
Cast turbocharger	1184–1445 (18)	86–120 (12)*
Cast sheet (1 mm thickness)	1143–1357 (15)	85–108 (15)

* For unknown reasons, two of the turbocharger wheels contained nitrogen levels of around 610 wt. ppm, these results are not included in the table.
The total number of measurements is given in brackets.

to the creep behavior, unlike Simpkins *et al.* [68], HIP did not seem to degrade the creep properties, possibly because there was no porosity in the cast sheets and thus no recrystallization of the microstructure. However, the minimum creep rate of the HIPed sheet material was around 8 times faster than that measured in HIPed cast cylinders of nominally the same composition, probably due to the finer microstructure.

The level of interstitial contamination (oxygen and nitrogen) during the melting and casting of the turbocharger wheels and sheet material was determined at GKSS. The level of interstitials within the melt feedstock ingot was very low. The level of oxygen within the cast turbochargers wheels and sheets doubled, to levels around 1200 wt. ppm, when compared to the level within the ingot feedstock, see Table 14.2. The amount of superheat developed during the melting process was unknown but this has been shown to influence the amount of interstitial oxygen pick-up from the ceramic melting and mold materials [54] as described in Section 14.2.1.

14.2.4
Countergravity Low-Pressure Casting

In this casting technique, a vacuum is used to manufacture components in which permeable ceramic based molds are filled with melt that is drawn upwards (countergravity) by the applied vacuum. Different variants of the technique exist. These include countergravity low-pressure casting of vacuum-melted alloys (CLV) that is suitable for alloys containing reactive metals [84]. The well-known Hitchiner process is included in this category of techniques [85]. The CLV process is shown schematically in Figure 14.25 [84]. Melting of the charge takes place within a crucible that is initially under vacuum and a hot mold is introduced into the upper chamber, which is then evacuated. After melting has been accomplished, both chambers are filled with argon gas, Figure 14.25a. The fill pipe of the hot mold is then dipped into the melt and a vacuum is applied to the upper chamber to draw the melt upwards into the component molds, Figure 14.25b. The vacuum is released when the parts have solidified; any molten metal that remains in the molds flows back into the crucible, Figure 14.25c. The advantages of this technique are that it is able to fill thin sections and make castings that are free of ceramic particles as a result of the nonturbulent flow. Additionally, the amount of scrap material is reduced as the molten feeder material is returned to the crucible and multiple components can be formed simultaneously [86].

Daido Steel have a patent application for a technique they have developed, called the Levi-Cast process, for the production of precision cast components [87]. This method combines a cold-crucible "levitation melting" process (also called high-power ISM) with countergravity casting. The "levitation melting" crucible is a high-power development of the ISM crucible, but, as mentioned in Section 14.1.3, has the advantage that the melt is pushed away from the walls of the water-cooled copper crucible (by electromagnetic forces induced in the melt) which results in

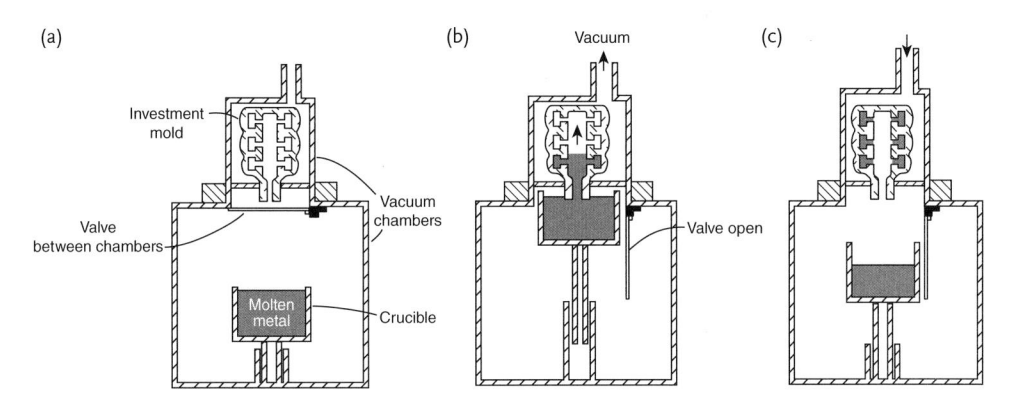

Figure 14.25 Schematic diagram showing the CLV casting process [84]. (a) The crucible charge is melted under vacuum. (b) The entire system is filled with argon and the mold fill pipe is dipped into the pool of molten metal before the upper chamber is evacuated to draw metal upwards into the mold. (c) After the components have solidified, the vacuum is released and any remaining molten metal returns to the crucible.

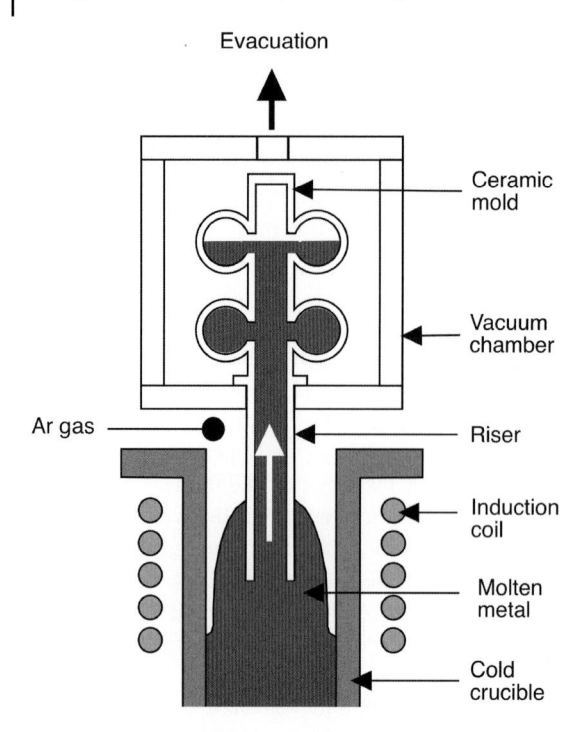

Figure 14.26 Schematic illustration of the Levi-Cast process [44]. The original diagram has been redrawn.

an increase of the melt superheat. Values of superheat up to around 300 °C are reported in [88] although only values between 60 °C and 70 °C are reported in [22]. A schematic illustration of the Levi-Cast process is shown in Figure 14.26 [44]. Melting is performed under an argon atmosphere so that loss of volatile elements is minimized. The oxygen content of cast material is reported to be around 500 ppm and blade thicknesses as thin as 0.35 mm are achievable [44]. Turbocharger wheels have been successfully made using this technique and introduced into series production by Mitsubishi Heavy Industries (MHI) [89, 90].

14.2.5
Directional Casting

This technique has not be used to cast components, but rather to produce material with an aligned lamellar microstructure. To date, it has only been used on a laboratory scale to study the influence of solidification conditions on microstructure, and to make specimens suitable for the investigation of lamellar orientation on properties. As the processing technique still has to be developed into an industrial process, it will not be discussed further in this section, rather the reader is directed to a number of publications in the literature [91–93].

14.3
Summary

A number of industrial processes are available that are capable of producing TiAl ingot material, each process having both technical and economic advantages and disadvantages. This initial stage of processing is of prime importance to the success of TiAl, with a clear requirement for ingots with reproducible composition and minimum chemical segregation. Clearly such material must be crack free, void of any inclusions and available at a competitive price. With regard to relatively small components, high-quality castings with fine tolerances are possible. For component castings made from high-strength alloys, ductility may have to be sacrificed to some extent. Large, thick-section components are more difficult to produce, generally having coarser microstructures that detrimentally influence properties. Here, alloy design is of key significance with a number of strategies being followed, as detailed in Chapter 13.

References

1 Wood, J.R. (2003) *Gamma Titanium Aluminides 2003* (eds Y.-W. Kim, H. Clemens, and A.H. Rosenberger), TMS, Warrendale, PA, p. 227.
2 Güther, V., Otto, A., Kestler, H., and Clemens, H. (1999) *Gamma Titanium Aluminides 1999* (Y.-W. Kim, D.M. Dimiduk, and M.H. Loretto), TMS, Warrendale, PA, p. 225.
3 Sibum, H. (2003) *Titanium and Titanium Alloys* (eds C. Leyens and M. Peters), Wiley-VCH, Weinheim, p. 231.
4 Lütjering, G., and Williams, J.C. (2003) *Titanium*, Springer Verlag, Berlin.
5 Güther, V., Chatterjee, A., and Kettner, H. (2003) *Gamma Titanium Aluminides 2003* (eds Y.-W. Kim, H. Clemens, and A.H. Rosenberger), TMS, Warrendale, PA, p. 241.
6 Güther, V., Joos, R., and Clemens, H. (2001) *Structural Intermetallics 2001* (eds K.J. Hemker, D.M. Dimiduk, H. Clemens, R. Darolia, H. Inui, J.M. Larsen, V.K. Sikka, M. Thomas, and J.D. Whittenberger), TMS, Warrendale, PA, p. 167.
7 Martin, P.L., Hardwick, D.A., Clemens, D.R., Konkel, W.A., and Stucke, M.A. (1997) *Structural Intermetallics 1997* (eds M.V. Nathal, R. Darolia, C.T. Liu, P.L. Martin, D.B. Miracle, R. Wagner, and M. Yamaguchi), TMS, Warrendale, PA, p. 387.
8 Ram, S.V., and Barrett, J.R. (1996) *Titanium '95: Science and Technology* (eds P.A. Blenkinsop, W.J. Evans, and H.M. Flower), IOM, London, UK, p. 88.
9 Johnson, T.P., Young, J.M., Ward, R.M., and Jacobs, M.H. (1993) *Structural Intermetallics* (eds R. Darolia, J.J. Lewandowski, C.T. Liu, P.L. Martin, D.B. Miracle, and M.V. Nathal), TMS, Warrendale, PA, p. 159.
10 Clemens, D.R. (2001) *Structural Intermetallics 2001* (eds K.J. Hemker, D.M. Dimiduk, H. Clemens, R. Darolia, H. Inui, J.M. Larsen, V.K. Sikka, M. Thomas, and J.D. Whittenberger), TMS, Warrendale, PA, p. 217.
11 Sears, J.W. (1992) *The Processing, Properties and Applications of Metallic and Ceramic Materials*, Proceedings of an International Conference held at the International Convention Centre (ICC) Birmingham, UK, 7–10th September 1992, vol. 1 (eds M.H. Loretto and C.J. Beevers), MCE Publications Ltd, p. 119.
12 Dowson, A.L., Johnson, T.P., Young, J.M., and Jacobs, M.H. (1995) *Gamma Titanium Aluminides* (eds Y.-W. Kim, R. Wagner, and M. Yamaguchi), TMS, Warrendale, PA, p. 467.

13 Blackburn, M.J., and Malley, D.R. (1992) *The Processing, Properties and Applications of Metallic and Ceramic Materials*, Proceedings of an International Conference held at the International Convention Centre (ICC) Birmingham, UK, 7–10th September 1992, vol. 1 (eds M.H. Loretto and C.J. Beevers), MCE Publications Ltd, p. 99.
14 Porter III, W.J., Kim, Y.-W., Li, K., Rosenberger, A.H., and Dimiduk, D.M. (2001) *Structural Intermetallics 2001* (eds K.J. Hemker, D.M. Dimiduk, H. Clemens, R. Darolia, H. Inui, J.M. Larsen, V.K. Sikka, M. Thomas, and J.D. Whittenberger), TMS, Warrendale, PA, p. 201.
15 Ward, R.M., Fellows, A.E., Johnson, T.P., Young, J.M., and Jacobs, M.H. (1993) *J. de Physique IV, Coll. C7, suppl. J. de Physique III*, **3**, 823.
16 Godfrey, B., and Loretto, M.H. (1999) *Mater. Sci. Eng.*, **A266**, 115.
17 Godfrey, B., Dowson, A.L., and Loretto, M.H. (1996) *Titanium '95: Science and Technology* (eds P.A. Blenkinsop, W.J. Evans, and H.M. Flower), IOM, London, UK, p. 489.
18 Stickle, D.R., Scott, S.W., and Chronister, D.J. (April 1988) *Method for induction melting reactive metals and alloys*, U.S. Patent 4,738,713.
19 Reed, S. (1995) *Gamma Titanium Aluminides* (eds Y.-W. Kim, R. Wagner, and M. Yamaguchi), TMS, Warrendale, PA, p. 475.
20 Yanqing, S., Jingjie, G., Jun, J., Guizhong, L., and Yuan, L. (2002) *J. Alloys Compounds*, **334**, 261.
21 Reed, D.S. (May 2002) *Homogeneous electrode of a reactive metal alloy for vacuum arc remelting and a method for making the same from a plurality of induction melted charges*, U.S. Patent 6,385,230 B1.
22 Mi, J., Harding, R.A., Wickins, W., and Campbell, J. (2003) *Intermetallics*, **11**, 377.
23 Alam, M.K., and Semiatin, S.L. (1993) *Processing and Fabrication of Advanced Materials for High Temperature Applications II* (eds V.A. Ravi and T.S. Srivatssan), TMS, Warrendale, PA, p. 593.
24 Semiatin, S.L., Nekkanti, R., Alam, M.K., and McQuay, P.A. (1993) *Metall. Trans.*, **24A**, 1295.
25 Shamblen, C.E. (1996) *Titanium '95: Science and Technology* (eds P.A. Blenkinsop, W.J. Evans, and H.M. Flower), IOM, London, UK, p. 1438.
26 Chinnis, W.R. (1996) *Titanium '95: Science and Technology* (eds P.A. Blenkinsop, W.J. Evans, and H.M. Flower), IOM, London, UK, p. 1494.
27 Blum, M., Jarczyk, G., Chatterjee, A., Furwitt, W., Güther, V., Clemens, H., Danker, H., Gerling, R., Sasse, F., and Schimansky, F.P. (Oct. 2006) *Method for producing alloy ingots*, U.S. Patent Application US 2006/0230876 A1.
28 McQuay, P., and Larsen, D. (1997) *Structural Intermetallics 1997* (eds M.V. Nathal, R. Darolia, C.T. Liu, P.L. Martin, D.B. Miracle, R. Wagner, and M. Yamaguchi), TMS, Warrendale, PA, p. 523.
29 McQuay, P.A., and Sikka, V.K. (2002) *Intermetallic Compounds*, vol. 3, Progress (eds J.H. Westbrook and R.L. Fleischer) John Wiley & Sons Ltd., Chichester, UK, p. 591.
30 Austin, C.M., and Kelly, T.J. (1993) *Structural Intermetallics* (eds R. Darolia, J.J. Lewandowski, C.T. Liu, P.L. Martin, D.B. Miracle, and M.V. Nathal), TMS, Warrendale, PA, p. 143.
31 Hu, D., Wu, X., and Loretto, M.H. (2005) *Intermetallics*, **13**, 914.
32 Thomas, M., Raviart, J.L., and Popoff, F. (2005) *Intermetallics*, **13**, 944.
33 Naka, S., Thomas, M., Sanchez, C., and Khan, T. (1997) *Structural Intermetallics 1997* (eds M.V. Nathal, R. Darolia, C.T. Liu, P.L. Martin, D.B. Miracle, R. Wagner, and M. Yamaguchi), TMS, Warrendale, PA, p. 313.
34 Grange, M., Thomas, M., Raviart, J.L., Belaygue, P., and Recorbet, D. (2003) *Gamma Titanium Aluminides 2003* (eds Y.-W. Kim, H. Clemens, and A.H. Rosenberger), TMS, Warrendale, PA, p. 213.
35 Jin, Y., Wang, J.N., Yang, J., and Wang, Y. (2004) *Scr. Mater.*, **51**, 113.
36 Wang, Y., Wang J.N., Yang, J., and Zhang, B. (2005) *Mater. Sci. Eng.*, **A392**, 235.

37 Küstner, V., Oehring, M., Chatterjee, A., Güther, V., Brokmeier, H.G., Clemens, H., and Appel, F. (2003) *Gamma Titanium Aluminides 2003* (eds Y.-W. Kim, H. Clemens, and A.H. Rosenberger), TMS, Warrendale, PA, p. 89.

38 Chen, G.L., Zhang, W.J., Liu, Z.C., and Li, S.J. (1999) *Gamma Titanium Aluminides 1999* (eds Y.-W. Kim, D.M. Dimiduk, and M.H. Loretto), TMS, Warrendale, PA, p. 371.

39 Austin, C.M., Kelly, T.J., McAllister, K.G., and Chesnutt, J.C. (1997) *Structural Intermetallics 1997* (eds M.V. Nathal, R. Darolia, C.T. Liu, P.L. Martin, D.B. Miracle, R. Wagner, and M. Yamaguchi), TMS, Warrendale, PA, p. 413.

40 Larsen, D., and Govern, C. (1995) *Gamma Titanium Aluminides* (eds Y.-W. Kim, R. Wagner, and M. Yamaguchi), TMS, Warrendale, PA, p. 405.

41 Seo, D.Y., An, S.U., Bieler, T.R., Larsen, D.E., Bhowal, P., and Merrick, H. (1995) *Gamma Titanium Aluminides* (eds Y.-W. Kim, R. Wagner, and M. Yamaguchi), TMS, Warrendale, PA, p. 745.

42 Lupinc, V., Marchionni, M., Onofrio, G., Nazmy, M., and Staubli, M. (1999) *Gamma Titanium Aluminides 1999* (eds Y.-W. Kim, D.M. Dimiduk, and M.H. Loretto), TMS, Warrendale, PA, p. 349.

43 Wagner, R., Appel, F., Dogan, B., Ennis, P.J., Lorenz, U., Mullauer, J., Nicolai, H.P., Quadakkers, W., Singheiser, L., Smarsly, W., Vaidya, W., and Wurzwallner, K. (1995) *Gamma Titanium Aluminides* (eds Y.-W. Kim, R. Wagner, and M. Yamaguchi), TMS, Warrendale, PA, p. 387.

44 Noda, T. (1998) *Intermetallics*, **6**, 709.

45 Nishikiori, S., Takahashi, S., and Tanaka, T. (1999) *Gamma Titanium Aluminides 1999* (eds Y.-W. Kim, D.M. Dimiduk, and M.H. Loretto), TMS, Warrendale, PA, p. 357.

46 Larsen, D.E., Kampe, S., and Christodoulou, L. (1990) *Intermetallic Metal Matrix Composites* (eds D.L. Anton, R. McMeeking, D. Miracle, and P. Martin), *Mater. Res. Soc. Symp. Proc.*, vol. 194, Materials Research Society, Pittsburgh, PA, p. 285.

47 Larsen, D.E., Christodoulou, L., Kampe, S.L., and Sadler, P. (1991) *Mater. Sci. Eng.*, **A144**, 45.

48 Cheng, T.T. (2000) *Intermetallics*, **8**, 29.

49 Hu, D. (2001) *Intermetallics*, **9**, 1037.

50 Wu, X., and Hu, D. (2005) *Scr. Mater.*, **52**, 731.

51 Wang, J.N., and Xie, K. (2000) *Scr. Mater.*, **43**, 441.

52 Kuang, J.P., Harding, R.A., and Campbell, J. (2000) *Mater. Sci. Technol.*, **16**, 1007.

53 Sung, S.-Y., and Kim, Y.-J. (2007) *Intermetallics*, **15**, 468.

54 Liu, K., Ma, Y.C., Gao, M., Rao, G.B., Li, Y.Y., Wei, K., Wu, X., and Loretto, M.H. (2005) *Intermetallics*, **13**, 925.

55 Dlouhy, A., Zemcík, L., and Válek, R. (2003) *Gamma Titanium Aluminides 2003* (eds Y.-W. Kim, H. Clemens, and A.H. Rosenberger), TMS, Warrendale, PA, p. 291.

56 Würker, L., Fackeldey, M., and Sahm, P.R. (1997) *Structural Intermetallics 1997* (eds M.V. Nathal, R. Darolia, C.T. Liu, P.L. Martin, D.B. Miracle, R. Wagner, and M. Yamaguchi), TMS, Warrendale, PA, p. 347.

57 Harding, R.A., Brooks, R.F., Pottlacher, G., and Brillo, J. (2003) *Gamma Titanium Aluminides 2003* (eds Y.-W. Kim, H. Clemens, and A.H. Rosenberger), TMS, Warrendale, PA, p. 75.

58 Barbosa, J., Ribeiro, C.S., and Monteiro, A.C. (2007) *Intermetallics*, **15**, 945.

59 Blum, M., Choudhury, A., Scholz, H., Jarczyk, G., Pleier, S., Busse, P., Frommeyer, G., and Knippscheer, S. (1999) *Gamma Titanium Aluminides 1999* (eds Y.-W. Kim, D.M. Dimiduk, and M.H. Loretto), TMS, Warrendale, PA, p. 35.

60 Nakagawa, Y.G., Matsuda, K., Masaki, S., Imamura, R., and Arai, M. (1995) *Gamma Titanium Aluminides* (eds Y.-W. Kim, R. Wagner, and M. Yamaguchi), TMS, Warrendale, PA, p. 415.

61 Harding, R.A., Wickins, M., and Li, Y.G. (2001) *Structural Intermetallics 2001* (eds K.J. Hemker, D.M. Dimiduk, H. Clemens, R. Darolia, H. Inui, J.M. Larsen, V.K. Sikka, M. Thomas, and J.D.

62 Muraleedharan, K., Rishel, L.L., Graef, M.D., Cramb, A.W., Pollock, T.M., and Gray III, G.T. (1997) *Structural Intermetallics 1997* (eds M.V. Nathal, R. Darolia, C.T. Liu, P.L. Martin, D.B. Miracle, R. Wagner, and M. Yamaguchi), TMS, Warrendale, PA, p. 215.

63 De Graef, M., Biery, N., Rishel, L., Pollock, T.M., and Cramb, A. (1999) *Gamma Titanium Aluminides 1999* (eds Y.-W. Kim, D.M. Dimiduk, and M.H. Loretto), TMS, Warrendale, PA, p. 247.

64 Larsen, D.E. (1996) *Mater. Sci. Eng.*, **A213**, 128.

65 Rishel, L.L., Biery, N.E., Raban, R., Gandelsman, V.Z., Pollock, T.M., and Cramb, A.W. (1998) *Intermetallics*, **6**, 629.

66 Biery, N., De Graef, M., Beuth, J., Raban, R., Elliott, A., Austin, C., and Pollock, T.M. (2002) *Metall. Mater. Trans.*, **33A**, 3127.

67 Kuang, J.P., Harding, R.A., and Campbell, J. (2002) *Mater. Sci. Eng.*, **A329–331**, 31.

68 Simpkins II, R.J., Rourke, M.P., Bieler, T.R., and McQuay, P.A. (2007) *Mater. Sci. Eng.*, **A463**, 208.

69 Wu, X. (2006) *Intermetallics*, **14**, 1114.

70 Austin, C.M., and Kelly, T.J. (1995) *Gamma Titanium Aluminides* (eds Y.-W. Kim, R. Wagner, and M. Yamaguchi), TMS, Warrendale, PA, p. 21.

71 Norris, G. (June 2006) Flight International Magazine.

72 Weimer, M., and Kelly, T.J. (2006) Presented at the *3rd International Workshop on γ-TiAl Technologies*, 29–31 May 2006, Bamberg, Germany.

73 Harding, R.A., Wickins, M., Wang, H., Djambazov, G., and Pericleous, K.A. (2011) *Intermetallics*, **19**, 805.

74 McQuay, P.A., Simpkins, R., Seo, D.Y., and Bieler, T.R. (1999) *Gamma Titanium Aluminides 1999* (eds Y.-W. Kim, D.M. Dimiduk, and M.H. Loretto), TMS, Warrendale, PA, p. 197.

75 Jones, P.E., Porter III, W.J., Eylon, D., and Colvin, G. (1995) *Gamma Titanium Aluminides* (eds Y.-W. Kim, R. Wagner, and M. Yamaguchi), TMS, Warrendale, PA, p. 53.

76 Eylon, D., Keller, M.M., and Jones, P.E. (1998) *Intermetallics*, **6**, 703.

77 Colvin, G.N., Ervin, L.L., and Johnson, R.F. (Feb. 1994) *Permanent mold casting of reactive melt*, U.S. Patent 5,287,910.

78 Colvin, G. (June 2000) *High vacuum die casting*, U.S. Patent 6,070,643.

79 Royer, A., and Vasseur, S. (1988) *Casting, Metals Handbook*, vol. 15, 9th edn (ed. [chairman] D.M. Stefanescu), ASM International, Metals Park, OH, p. 296.

80 Blum, M., Fellmann, H.G., Franz, H., Jarczyk, G., Ruppel, T., Busse, P., Segtrop, K., and Laudenberg, H.-J. (2001) *Structural Intermetallics 2001* (eds K.J. Hemker, D.M. Dimiduk, H. Clemens, R. Darolia, H. Inui, J.M. Larsen, V.K. Sikka, M. Thomas, and J.D. Whittenberger), TMS, Warrendale, PA, p. 131.

81 Blum, M., Jarczyk, G., Scholz, H., Pleier, S., Busse, P., Laudenberg, H.-J., Segtrop, K., and Simon, R. (2002) *Mater. Sci. Eng.*, **A329–331**, 616.

82 Baur, H., Wortberg, D.B., and Clemens, H. (2003) *Gamma Titanium Aluminides 2003* (eds Y.-W. Kim, H. Clemens, and A.H. Rosenberger), TMS, Warrendale, PA, p. 23.

83 Appel, F. (July 2006) Kontinuierliche TiAl–Gießtechnologie für komplexe Leichtbaukomponenten, BMBF project number 03N3108D, Final Reports.

84 Chandley, D. (1988) *Casting, Metals Handbook*, vol. 15, 9th edn (ed. [chairman] D.M. Stefanescu), ASM International, Metals Park, OH, p. 317.

85 Chandley, G.D. (Aug. 1991) *Apparatus and process for countergravity casting of metal with air exclusion*, U.S. Patent 5,042,561.

86 Chandley, G.D. (1999) *Mater. Res. Innovat.*, **3**, 14.

87 Noboru, D. (Jan. 1995) *Production of clean precision cast product*, Japanese Patent Application JP7016725, (in Japanese).

88 Yamada, J., and Demukai, N. (Nov. 1998) *Levitation melting method and melting and casting method*, U.S. Patent 5,837,055.

89 Tetsui, T. (2002) *Mater. Sci. Eng.*, **A329–331**, 582.

90 Abe, T., Hashimoto, H., Ishikawa, H., Kawaura, H., Murakami, K., Noda, T., Sumi, S., Tetsui, T., and Yamaguchi, M.

(2001) *Structural Intermetallics 2001* (eds K.J. Hemker, D.M. Dimiduk, H. Clemens, R. Darolia, H. Inui, J.M. Larsen, V.K. Sikka, M. Thomas, and J.D. Whittenberger), TMS, Warrendale, PA, p. 35.

91 Saari, H., Beddoes, J., Seo, D.Y., and Zhao, L. (2005) *Intermetallics*, **13**, 937.

92 Cheng, T., Mitchell, A., Beddoes, J., Zhao, L., Saari, H., and Durham, S. (2001) In *Proceedings of the 9th International Symposium on Processing and Fabrication of Advanced materials, St. Louis, USA, 9–12 Oct. 2000*, p. 159.

93 Cheng, T., Mitchell, A., Beddoes, J., Zhao, L., Saari, S., and Durham, S. (2000) *High Temp. Mater. Process.*, **19** (2), 79.

15
Powder Metallurgy

Powder metallurgy is a widely used technique within materials technology for manufacturing both metallic and ceramic components, particularly to produce parts for the automotive industry. Usually, a metal powder is loaded into a die where it is compacted, at temperatures around room temperature, into a near-net component shape. Such "green" compacts are then placed inside a furnace under a protective atmosphere at a temperature (below the melting point of the material) where sintering of the powder particles takes place so that a component with adequate mechanical properties is formed. In liquid-phase sintering, a minor constituent of the green compact becomes liquid at the sintering temperature, although if this is excessive, then the compact can lose its required shape. Sometimes, powder compaction and sintering are performed simultaneously in a process known as hot pressing or pressure sintering.

The advantage of such powder-based techniques is that high-quality net, or near-net, shape texture-free parts can be made with refined, macrosegregation free, microstructures on an economical basis. For some materials powder-based methods are far superior to casting for various reasons. These include excessive melt-mold reactions, insufficient casting superheat, hot cracking, excessive cast macrosegregation, strong casting texture or the build up of high internal stresses upon cooling. Other advantages of powder metallurgy include the possibility of making either composite materials or porous materials with controlled amounts of porosity. Additionally, the amount of waste material can be less, when compared to parts with similar properties but made using conventional hot-working processes.

With regard to γ-TiAl, basically two powder-metallurgy approaches have been reported in the literature, namely prealloyed and elemental powder technologies. The prealloyed powder approach relies on relatively expensive powder, but the properties of correctly processed material are generally far superior to those of material made using the cheaper elemental-powder technique. Additionally, the microstructures are chemically more homogeneous. Mechanical alloying has also been used to produce TiAl material, although the properties developed are unsuitable for high-temperature structural applications as a result of the extremely fine grain size. However, such material might be useful for secondary processing, such as, hot rolling or superplastic forming. The following sections describe each of the main powder-metallurgy approaches in more detail, that is, prealloyed powder, elemental powder and mechanical alloying.

15.1
Prealloyed Powder Technology

The use of this technology for the production of TiAl components requires the use of prealloyed "atomized" powder. Various atomization techniques including gas atomization, the rotating-electrode process (REP), the plasma rotating-electrode process (PREP), and rotating-disc atomization have been discussed in the literature. The quality of a powder is an important issue and is determined by a number of factors including the particle size, the particle-size distribution, the particle shape, and the degree of interstitial pick-up during the atomization process. From this point of view the different powder-production techniques have certain advantages and disadvantages. Unlike conventional metallic powders, TiAl powder is generally too hard to be cold compacted. It is therefore usually processed by hot isostatic pressing (HIP) followed by some sort of hot-working procedure. However, other techniques such as spray forming and metal injection molding (MIM) have also been investigated in the literature, as described in later sections of this chapter.

A brief description of the main methods employed in the production of prealloyed γ-TiAl powders is given below. Generally, all of the methods are capable of delivering high quantities of spherical-shaped powder. Due to the very complicated nature of all the atomization processes only basic principles will be described. For more detailed information the reader is directed to the relevant technical literature.

15.1.1
Gas Atomization

This term covers a number of processes in which an inert gas is used to break up a melt stream into metallic powder particles. After production, the prealloyed TiAl powder is highly reactive and therefore, to reduce contamination to a minimum, it must be protected from exposure to the atmosphere during subsequent processing. Within the literature, various gas atomization processes have been employed including plasma inert-gas atomization (PIGA) and electrode induction melting gas atomization (EIGA). A variation of the PIGA technique involves induction skull melting (either under vacuum or an inert gas) prior to melt atomization. This technique is employed by Crucible Research in Pittsburgh and is referred to as the titanium gas-atomizer process (TGA).

Each of the above gas-atomization techniques relies on molten metal passing through a nozzle in which the melt is disintegrated by a jet of high-velocity inert gas such as argon or helium. The nozzle design is a technologically very important feature with regard to the atomization process. Basically, two types of gas-atomization nozzle exist, namely free-fall (or open-stream) and closed-loop (or confined-feed or close-coupled) [1, 2]. As its name suggests, the free-fall type of nozzle allows the melt stream to free-fall some distance before entering the center of the atomization nozzle where melt stream disintegration takes place. By contrast, in the closed-loop type of nozzle, the melt is transferred into the high-pressure atomizing gas via a melt-delivery system. The closed-loop nozzle generally produces finer particles than the free-fall nozzle, although it is more difficult to operate [2]. For further general information regarding the effect of nozzle type on particle shape and size distribution, the reader is referred to refs. [3–7] for free-fall and [6, 8–14] for closed-loop nozzle systems. Apart from differences in atomization-nozzle design, below the level of the gas atomization nozzle the PIGA, EIGA and TGA atomization processes are essentially very similar. The main distinguishing feature between the PIGA, EIGA and TGA processes is the type of melting technique that is employed. Each of these gas-atomization processes will be described later in this subsection.

The particle size, shape and size distribution are of great importance with regard to the amount of useful powder that is produced in a single atomization run. For example in biomedical component manufacture, metal-injection molding of conventional titanium alloys requires the use of fine spherical powders with a size of less than 45 µm [15]. Additionally, such powder should show a high degree of flowability, a property that depends on many factors, including particle size, particle shape and particle-size distribution. Thus, an understanding of how the whole atomization process including the effect of nozzle design, gas pressure, melting technique, melt temperature, melt-stream diameter and other variables affect the powder quality and properties is very important. Only with such knowledge can atomization be optimized to produce the maximum yield of suitable powder. A detailed appraisal of these topics is, however, outside the intended scope of this publication and the reader is referred to [1–14, 16, 17] for further detailed information.

During atomization the particle size decreases with decreasing surface tension of the liquid metal (equivalent to a higher melt temperature) and with increasing velocity of the atomization gas. After a melt droplet falls away from the nozzle zone where the inert gas jet has caused melt-stream disintegration and droplet formation, surface-tension forces come into play. These can lead to the spheroidization of initially irregular shaped droplets, provided that they have not already solidified or that their viscosity is not too high. At this point, the melt superheat, cooling medium and droplet size are among the important parameters that determine the powder characteristics. If the time for droplet spheroidization (τ_{sph}) is longer than that for solidification (τ_{sol}) then the powder particles will tend to have an irregular shape, that is, the particles will tend to be spherical in shape if $\tau_{sph} < \tau_{sol}$. Nichiporenko and Naida [18] have considered this aspect and proposed equations for the spheroidization and solidification times, τ_{sph} and τ_{sol}. Such relationships can be useful for estimating the effect of certain variable parameters on particle shape. When optimized, gas atomization, particularly high-pressure gas atomization (HPGA), can lead to a large fraction of fine spherical-shaped particles [17]. Figure 15.1 shows TiAl powder particles that have been made by PIGA at GKSS. Small satellites, which are formed as a result of particle–particle

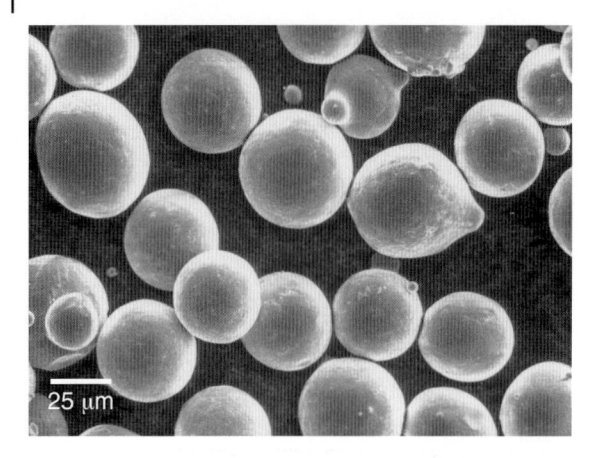

Figure 15.1 TiAl powder particles made by plasma inert-gas atomization (PIGA) at GKSS. Particle-surface morphology as well as smaller "satellite" particles that arise from particle–particle interactions during the atomization process can be seen.

interactions during the atomization process and particle surface morphology, can be seen. It has been reported that very few or no such satellites were observed in PREP powders [19], a technique discussed in Section 15.1.2.

During the gas-atomization process, and also during atomization using other techniques, the cooling and solidification rates are faster for the smaller-sized droplets. This, as discussed later, results in particle-surface morphology and microstructure being size dependent. Smaller sized particles generally exhibit reduced microsegregation and possibly even metastable phases. The type of gas and its velocity relative to the droplets also influences cooling and solidification rates. As these aspects are common to all atomization techniques that take place under a protective gas, further details will be discussed later in Section 15.1.4.

15.1.1.1 Plasma Inert-Gas Atomization (PIGA) at GKSS

The PIGA process uses a plasma torch to melt prealloyed material under a protective atmosphere (helium or argon) in a water-cooled copper crucible [20, 21]. This has the advantage that the melt is clean and does not allow any possibility for the incorporation of ceramic particles. At the bottom of the copper crucible there is a channel through which the molten metal can be transferred towards the atomization nozzle. During the melting process this melt-transfer system, which consists of a water-cooled copper funnel, is closed. When atomization is about to begin, a 150-kW induction coil around the funnel is activated and the melt flows from the crucible through the funnel, falling as a thin homogeneous melt stream through the center of the atomization nozzle, where atomization takes place [22].

The induction coil around the transfer funnel induces electromagnetic forces on the melt that constrict it into forming a thin melt stream. This has the advantage that contact with the water-cooled funnel is minimized, thus maintaining

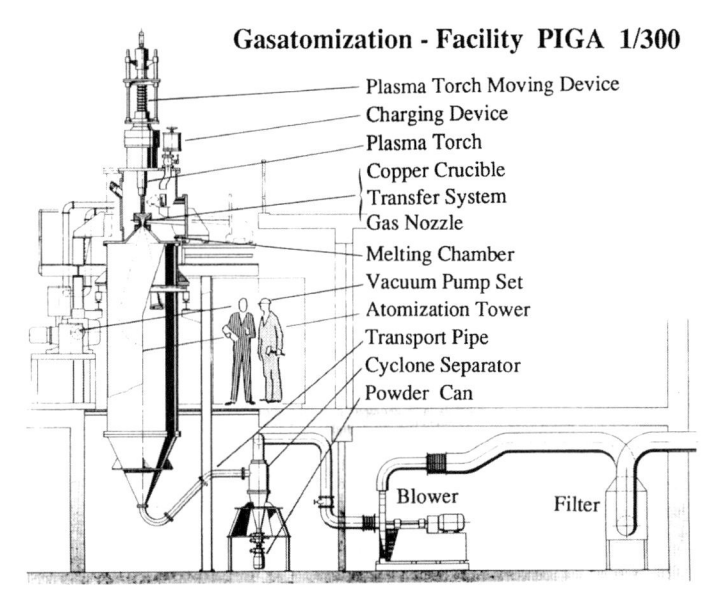

Figure 15.2 Schematic diagram of the plasma inert-gas atomization (PIGA) facility at GKSS [21].

melt superheat that influences the surface tension of the melt and thus the atomization process. Additionally, as the melt stream is constricted to a thinner diameter the mass flow of melt is altered, resulting in a higher volume fraction of smaller powder particles being produced [20].

Figure 15.2 shows a schematic drawing of the PIGA equipment at GKSS [21]. Before melting takes place, the whole system is repeatedly evacuated and flushed with dry inert gas. Additionally, the atomization tower, transport piping and cyclone separator can be heated to around 150 °C. This is done to ensure that oxygen and moisture are completely removed from the system before melting takes place. The facility at GKSS uses a 300-kW [22] helium plasma torch and is operated using an argon gas-atomization nozzle pressure of around 1.8 to 2.2 MPa. The volumetric capacity of the melting crucible is 1 liter, which is capable of containing about 4 kg of TiAl, depending on exact alloy composition. A 4 kg melt results in around 3.8 kg of powder with a diameter of less than 1 mm and a remaining skull of about 0.2 kg. The flow rate of molten metal through the nozzle is approximately 70 g/s (equivalent to 250 kg/h) [23].

Although PIGA has been optimized to produce a large fraction of fine spherical-shaped particles, it has been suggested by Prof. Mike Loretto that large particles can be produced as a result of material from the mushy zone (the layer between the skull and the melt) being dislodged and passing through the atomization nozzle [24]. Such material will have a different composition, temperature, viscosity and surface tension compared to the melt stream and thus, the atomization process and the subsequent powder produced will be influenced. After gas atomization and melt-droplet formation, the particles cool down within the atomization tower. At GKSS this tower is around 4.4 m in height and about 1.2 m in diameter. A tower of this size ensures that the droplets have sufficient time to cool down so that they remain spherical and are not deformed when they impact against the inner wall. The powder particles are separated from the atomization gas by a cyclone separator and collected in a powder can. This is subsequently sealed, ensuring that the powder has absolutely no contact with the atmosphere. Further processing such as sieving is then performed within a glove box under a protective atmosphere.

15.1.1.2 Titanium Gas-Atomizer Process (TGA)

The TGA system that has been developed by the Crucible Materials Corporation is shown schematically in Figure 15.3 [25]. Below the atomization nozzle the

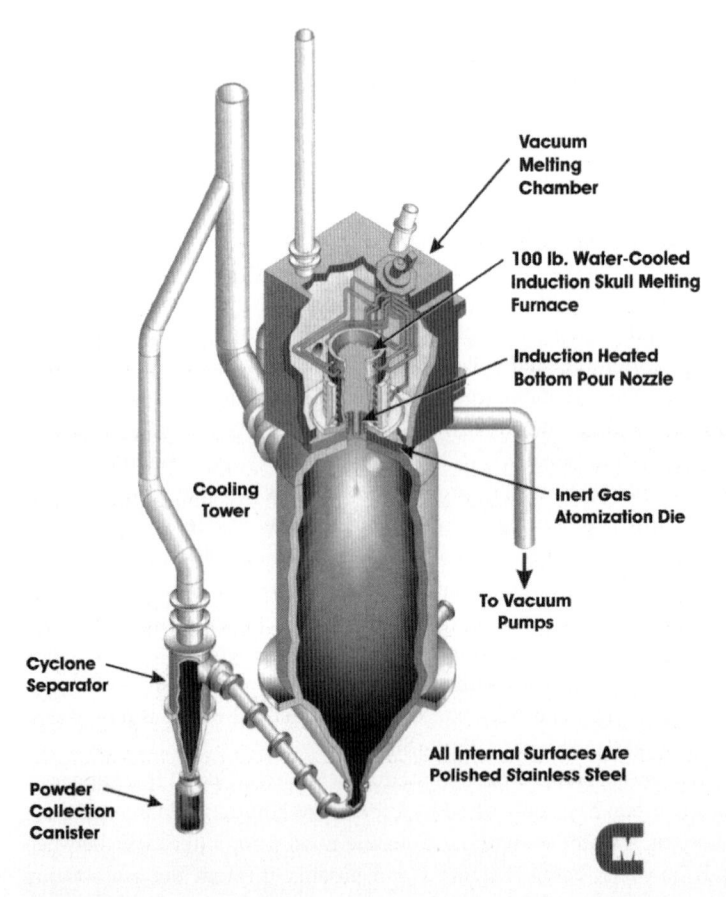

Figure 15.3 Diagram showing the "titanium gas atomizer" (TGA) developed by Crucible Materials [25].

system is essentially the same as for the PIGA (and EIGA) technique. As with PIGA (and EIGA), to ensure minimal contamination, the system is under an inert atmosphere and all internal surfaces of the atomization tower are made from polished stainless steel and extensive cleaning is performed between individual runs. In the TGA process the charge may be either prealloyed ingot material or an ensemble of the correct amount of the individual elemental components and/or master alloys. An induction skull-melting (ISM) unit (discussed in Section 14.1.3) is used to melt the charge material. The intensive stirring generated by the induction field ensures good homogenization of the melt before it is poured into a heated graphite susceptor [25]. The power applied to the ISM crucible and to the inductively heated graphite susceptor control the melt pour. Nominally, a 4-mm diameter melt stream falls from the bottom of the susceptor into high-pressure jets of argon gas that cause melt-stream disintegration and powder atomization. As with the other gas-atomization processes, the powder solidifies within the cooling (atomization) tower and is carried to a cyclone where it is collected in a gas-tight canister. Although no reference is made to melt contamination through graphite in ref. [25], the tolerance for carbon within a typical TiAl powder is given as being no higher than 0.05 wt.%. As with the PIGA technique highly spherical powder particles can be produced. However, Yolton [26] and Suryanarayana et al. [27] have reported that gas-atomized powders typically exhibit a wider particle-size distribution than both centrifugally atomized powder and PREP powder.

15.1.1.3 Electrode Induction Melting Gas Atomization (EIGA)

This technique, which was patented in 1991 [28] by Leybold and is used by companies such as TLS Technik, is shown schematically in Figure 15.4. As can be seen, the melting stock consists of a rotating consumable ingot, the end of which is melted with a cone-shaped induction coil. The rotation speed is relatively slow at around 5 revolutions per minute, which is much slower than that of the "rotating-electrode process" powder production method described later. This rotation ensures symmetrical melting of the ingot tip. Once molten, the melt accumulates at the tip of the rotating ingot and drops into the gas-atomization nozzle where it is disintegrated into droplets similarly to the PIGA and TGA techniques. As material is consumed in the process, the whole rotating ingot is continuously lowered. The melt flow rate is controlled by a combination of the power of the induction coil and the rate with which the ingot is lowered towards the coil. If the feed rate is too high then the rotating ingot will touch the induction coil, but if it is too slow then the melt rate will be reduced. Melt rates of around 50 kg/h (equivalent to 13.9 g/s) are reported to be possible (as of 2004) [23]. As only small amounts of melt are produced at any time, this method consumes relatively small amounts of power compared to the PIGA and TGA processes. It also has the advantage that the melt only comes into contact with the inert-gas atmosphere, thus completely avoiding the possibility of any contamination. However, as with the rotating-electrode process described in Section 15.1.2, the method relies on the ingot feedstock being homogeneous, otherwise powder particles with significantly different compositions may be produced.

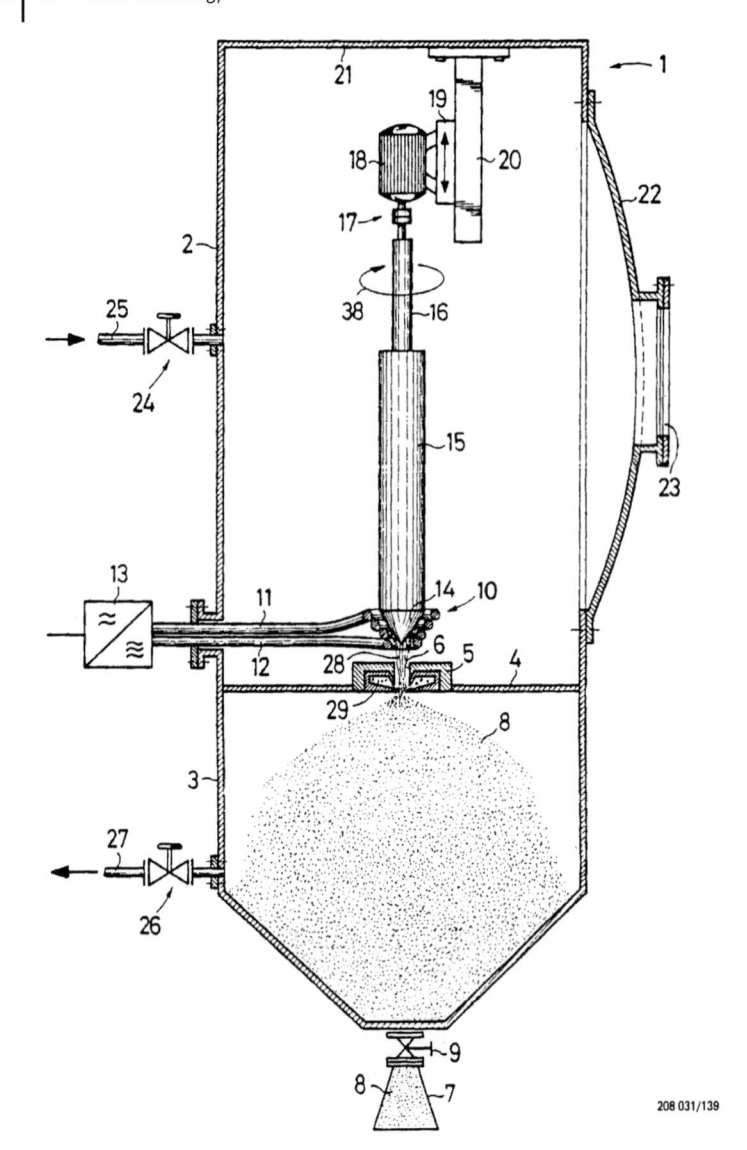

Figure 15.4 Diagram showing the operating principle of electrode induction melting gas atomization (EIGA) that was patented in 1991 [28]. The pointed end (14) of a cast ingot (15) is melted by a specially shaped induction coil (10). The stream of molten metal (28) flows directly into a gas-atomization nozzle (5) where it is atomized into powder (8). While this process takes place, the whole ingot continuously rotates slowly (38) around its longitudinal axis and is lowered into the induction coil as the ingot is consumed. It is important that the rate at which the ingot is lowered is in equilibrium with its consumption so that the position of the pointed tip with respect to the coil remains constant.

15.1.2
Rotating-Electrode Processes

It is reported that rotating-electrode processes for producing prealloyed metal powders were developed by Nuclear Metals Inc. [29–31] who gained significant expertise [32–34]. The earlier rotating-electrode process (REP) which used an electric arc, and its later development into the plasma rotating-electrode process (PREP), are usually conducted under an inert gas atmosphere and involve melting the end face of a fast-rotating ingot. The molten metal is centrifugally ejected due to centrifugal force and forms droplets that solidify into spherical powder particles within the chamber where the process takes place. Helium is the preferred medium within the chamber on account of its arc characteristics and good thermal properties, which result in fast cooling rates. The difference between the REP and PREP processes is in the design of the cathode. In the REP process the cathode consists of a cooled nonconsumable tungsten-tipped device. The PREP uses a more complex type of cathode referred to as a transferred arc plasma gun and has the advantage over the REP in that the powder produced is not contaminated with tungsten particles [34].

A diagram indicating the equipment used for the REP is shown in Figure 15.5 [32]. Nuclear Metals have operated two REP/PREP units, namely short bar and long-bar units [34]. The short-bar unit employs a prealloyed consumable bar (the

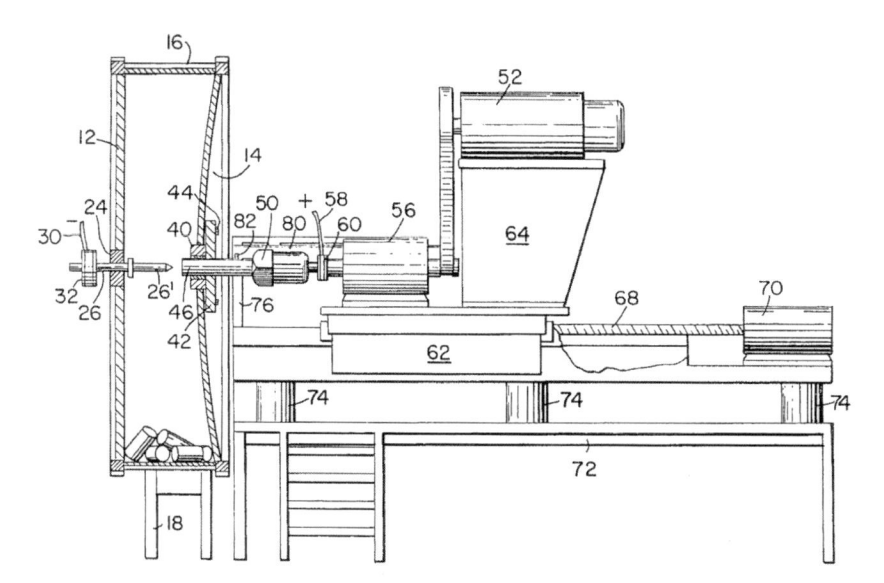

Figure 15.5 Diagram of the equipment used for the production of powder by the rotating-electrode process (REP) [32]. The alloy bar (46) is rotated at high speed (nominally 15 000 rev/min) while an arc is struck between it and the electrode (26). The arc melts the end of the bar and molten powder particle are centrifugally expelled into the enclosed chamber. A translation mechanism ensures that the end of the bar remains at a constant distance from the electrode as the bar is consumed.

anode) that can be up to 89 mm in diameter and 250 mm in length. Up to 80% of the length of a short bar is converted into powder. The long-bar unit was designed to accept ingots with lengths up to 1.83 m long and diameters of 63.5 mm. In both units the ingot axis is horizontal. The long-bar unit consists of a chamber of 2.44 m diameter in which the atomization process takes place. As the long bar is consumed in the atomization process, it is continually fed into the chamber through a seal mechanism. There is the possibility of attaching the rest stubs of consumed ingots to fresh ingots. Thus, a large amount of material can be processed using effectively almost 100% of the ingot material. In the short-bar unit, once around 80% of the ingot material has been consumed it must be replaced with a fresh ingot through a gloved access point. The long-bar unit thus enables powder to be produced on a more or less industrial scale, rather than in small batches. Both the long- and short-bar techniques have been patented [32, 33].

During the atomization process using the long-bar geometry, the nominal rotational speed of the electrode is 15 000 revolutions per minute [34]. This high speed makes it necessary to ensure that the electrode is straight and has the correct dimensions so that minimal oscillating forces are exerted on the equipment. The long-bar technique is thus suited for materials that have a high stiffness whereas the short-bar method is more suitable for low-stiffness materials as the ratio of length to diameter is reduced [34]. Small-diameter short bars can have rotational speeds of up to 25 000 revolutions per minute.

The rotational speed is a very important parameter in the atomization process and is one of the parameters that determine the particle size. The median particle diameter (d_{50}) can be qualitatively approximated using the equation [31, 34]:

$$d_{50} = \frac{K}{\omega\sqrt{D}} \tag{15.1}$$

where ω is the speed of rotation (in rad/s), D is the diameter of the electrode and K is a constant that depends on the alloy and the power of the arc. Thus, the conditions required for a given powder size can be suitably adjusted. Champagne and Angers have shown that a bimodal particle-size distribution can be obtained [35, 36]. This happens because the melt droplets at the edge of the rim form extended ligaments that break up when they are released from the disc. Tokizane et al. [37] have also observed such a bimodal distribution in TiAl powder. As mentioned earlier, the particle-size distribution in centrifugally atomized powders is generally within a smaller range than that which is typically obtained using gas atomization [26, 27, 34]. As with gas atomization, the REP and PREP are both capable of producing highly spherical powders.

Apart from contact between the molten-metal droplets and the plasma/gas, the droplets solidify before contact with other materials, thus minimizing contamination of the powder. The cooling rates experienced during PREP atomization (under helium gas) of Ti alloy (IMI-829) particles ranging from 420 μm to 37 μm in diameter have been determined to lie between 10^4 and 10^6 K/s, respectively [38]. Cooling rates as high as 10^6 K/s have also been reported for PREP atomized (under helium) TiAl alloy [39]. As with gas atomization, the cooling rate is determined by a number of factors including the particle size, the type of inert gas used within the chamber, and the speed of the molten droplets relative to the gaseous environment. As with the EIGA process, ensuring that powder particles with similar chemical composition are produced requires that the rotating electrode is of homogeneous composition, especially end to end. This, as discussed in Chapter 14, is not always easy to ensure for TiAl. Indeed, Habel et al. [40] state that gas-atomized powder that is produced from a fully molten charge shows greater homogeneity than PREP powder.

15.1.3
Rotating-Disc Atomization

This technique is based on a stream of molten metal being mechanically atomized by falling onto the center of a fast-rotating "container" (or disc) and being thrown off at the edges. The whole process takes place in a chamber under an inert atmosphere, helium being preferred over argon due to the higher resulting solidification rates. During flight the atomized melt droplets assume a spherical shape and subsequently solidify within the processing chamber.

The method has several variants depending on the method of melting and the shape of the rotating "container", which may be disc-shaped, cup-shaped, crucible-shaped, or simply a flat or concave-shaped plate. The plasma arc-melting/centrifugal atomization (PAMCA) technique described by Peng et al. [41], is one example of this method, see Figure 15.6. Typical cooling rates estimated for titanium alloy powder particles with sizes of 600 µm and 30 µm were 1.7×10^3 K/s and 1.6×10^6 K/s, respectively.

The rapid solidification rate centrifugal atomization technique (RSR) that employs a water-cooled disc that is rotated at a speed of up to 35 000 revolutions per minute is another example of rotating-disk atomization [27]. This method was patented by United Technologies [42] and has been used by Pratt and Whitney to make TiAl powder [43]. Solidification rates for the RSR method are reported to be in the range 10^4 K/s to 10^6 K/s for particles sizes of 80 µm to 25 µm [27].

As can be seen in Figure 15.6, the PAMCA technique can additionally employ helium-gas jets around the spinning disc to increase the solidification rate (as is also possible with RSR processing). After melt atomization and droplet solidification has taken place within the processing chamber, the larger powder particles fall into a collector and the smaller ones are carried with gas to cyclones where they are separated.

The disintegration of melt at the edge of a spinning disc has been analyzed by Li and Tsakiropoulos [44] who present a model for the mean particle size obtained during rotating-disc atomization. The mean particle size determined using the model shows a relatively good agreement with experimentally determined data. Increasing the speed of rotation or the disc diameter result in a reduction of the particle size. Additionally, it is shown that melts with a high ratio of density to surface tension also promote a smaller powder particle size.

One problem associated with disc atomization is the build up of a skull of solidified material on the surface of the cooled rotating disc. This can lead to problems

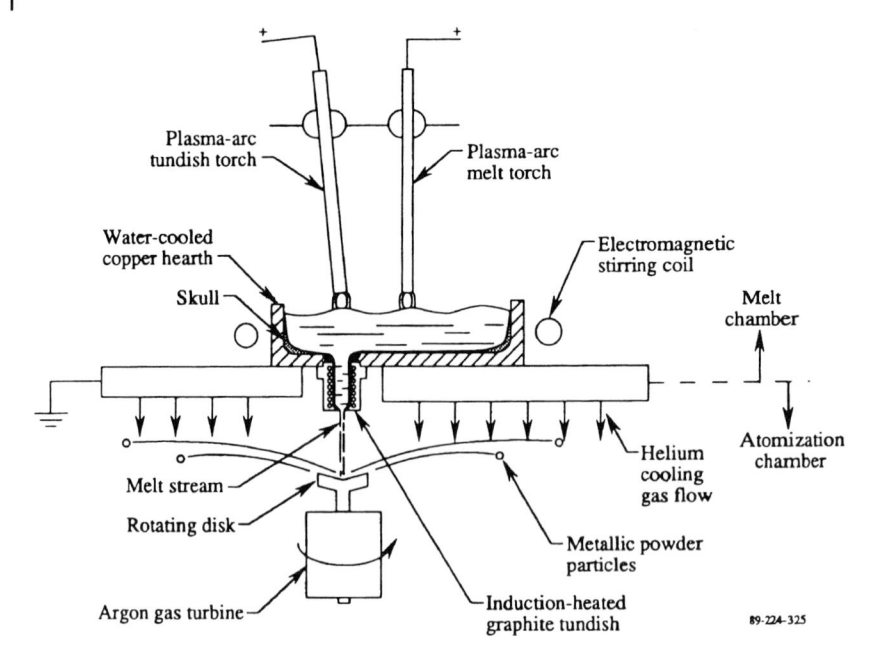

Figure 15.6 The plasma-arc melting/centrifugal atomization (PAMCA) process [41]. Plasma torches are used to melt the alloy in a water-cooled copper crucible. The molten metal is allowed to fall as a thin stream onto a very fast-rotating disk. On impact, the melt is atomized and powder is formed.

related to disc balance during rotation, which can affect the amount and quality of powder produced. Ho and Zhao [45] have modeled the heat-transfer situation at the disc in relation to skull formation. Their findings indicate that the skull can be reduced in volume by increasing the rate of melt flow, the disc temperature and the amount of melt superheat. Additionally, for the production of fine powders it is suggested that the diameter of the rotating disc should not be too large so that the development of a phenomenon known as "hydraulic jump" is prevented. This phenomenon gives rise to a build up of material at a characteristic distance from the center of the disk and leads to the formation of a donut-shaped skull on the disk. For more details the reader is referred to Ho and Zhao [45].

15.1.4
General Aspects of Atomization

The particle-size distribution is an important powder characteristic, and with this aspect in mind, the various atomizing techniques each have their advantages/disadvantages. It must be remembered, however, that for a good yield of high-quality powder (using any technique) the atomizing process must be optimized. Various methods exist for the determination of particle-size distribution and the results may depend on particle shape. Thus, it is not easy to directly compare size distributions from different research groups, who use different techniques [46]. In the TiAl literature, the size distributions that are presented have usually been determined via sieving using successively finer mesh sieves. However, it is not always clearly indicated that the distributions presented are for all particle sizes that were produced during atomization or only for particles up to a certain size, namely those below a certain mesh size. Thus, no attempt will be made here to compare size-distribution data from different authors for powders made using different techniques. With this in mind, Wegmann et al. [47] are the only authors to compare the size distribution of powders made using three different techniques, namely PIGA, EIGA, and rotating-disc atomization, for the same alloy. The results, shown in Figure 15.7, are in the form of particle-size distribution plots determined for powders sieved to below 355 μm. It can be seen that using helium instead of argon during rotating-disc atomization had little effect on particle size (argon d_{50} = 150 μm, helium d_{50} = 160 μm), although as discussed later helium increased the amount of thermally induced porosity developed after high-temperature exposure. Both EIGA and PIGA resulted in significantly finer powder compared to rotating-disc atomization, the EIGA technique producing slightly finer powder (EIGA d_{50} = 55 μm) compared to PIGA (PIGA d_{50} = 77 μm). This latter observation is believed to have resulted from EIGA providing a finer melt stream to enter the atomization nozzle. It was also noticed for powder made by rotating-disc atomization that the fraction of nonspherical particles increased with increasing particle

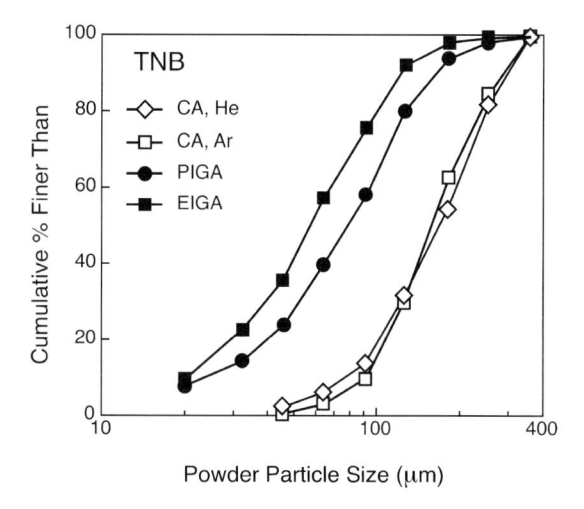

Figure 15.7 Particle-size distributions determined by Wegmann et al. [47] for powders made at GKSS by rotating-disk centrifugal atomization (CA), plasma inert gas atomization (PIGA) and electrode induction melting gas atomization (EIGA). Argon gas was employed in PIGA and EIGA. The difference between using argon and helium gas was only studied for CA, but was found to have little effect on size distribution. Replotted data.

size for particle sizes above around 90 μm. Although not very significant, Figure 15.7 seems to support the observations commented on earlier with regard to gas-atomized powders typically exhibiting a wider particle-size distribution than centrifugally atomized powders.

When compared with conventional metallurgy, the main advantage in using prealloyed powder techniques is the possibility of producing near-net-shaped components or semifinished material with minimal segregation and thus more homogeneous microstructures. On account of the fast cooling rates that can be achieved during atomization, significant undercooling of the atomized droplets can result in dendritic solidification being suppressed and cellular, segregation-free single-phase or even amorphous microstructures being developed within the individual powder particles. This leads to reduced microsegregation within the individual particles, and the possibility of obtaining metastable phases. In this context, both the particle size and the gaseous environment play crucial roles. With regard to estimating the cooling rate for the different production techniques there are two possibilities, namely calculation or experimental determination. Determination of cooling rate by calculation, is, however, not an easy task as complex heat-transfer conditions have to be considered. Peng et al. [41] have determined the solidification/cooling rate due to heat loss by radiation for titanium-alloy particles and found a reasonable agreement with experimentally determined values. Experimental determination requires measurement of the secondary dendrite-arm spacing, which is discussed elsewhere [27, 38]. Cai and Eylon [39] have determined cooling and solidification rates from the microstructural features of a TiAl PREP powder.

Based on calculations, Gerling et al. [20, 21] have presented diagrams indicating how the cooling rate depends on both the particle size and the surrounding environment see Figure 15.8. This graph shows the cooling rate of a binary Ti-49Al alloy with an initial temperature of 1477 °C (1750 K) as a function of particle size for particle diameters between 10 and 250 μm. As can be seen, the cooling rate depends on both the particle size and the relative velocity between the powder particles and the atomizing gas (v_R). Two particle–gas relative velocities have been considered for the case of helium (namely $v_R = 100$ m/s and 0 m/s). A value of $v_R = 100$ m/s corresponds to an upper limit in relative velocities during gas atomization, while $v_R = 0$ m/s corresponds to an absolute lower limit of cooling rates within a helium environment. Thus, the actual cooling rate during helium gas atomization can be expected to lie somewhere between the curves for $v_R = 100$ m/s and 0 m/s. From these curves for helium, it is seen that smaller particles experience much faster cooling than larger ones. On comparing the cooling curves for helium and argon, both with $v_R = 0$ m/s, it can be seen that the cooling rate provided by helium is significantly higher than that for argon at any given particle size. This is due to a higher heat transfer between the particles and helium, compared to that for argon; this results from differences in the thermal properties of the two gases.

The effect of a vacuum environment on the cooling rate is also presented in the graph. In this case, heat loss can only take place through radiation and thus the cooling rates are significantly reduced. Such cooling rates could be experienced

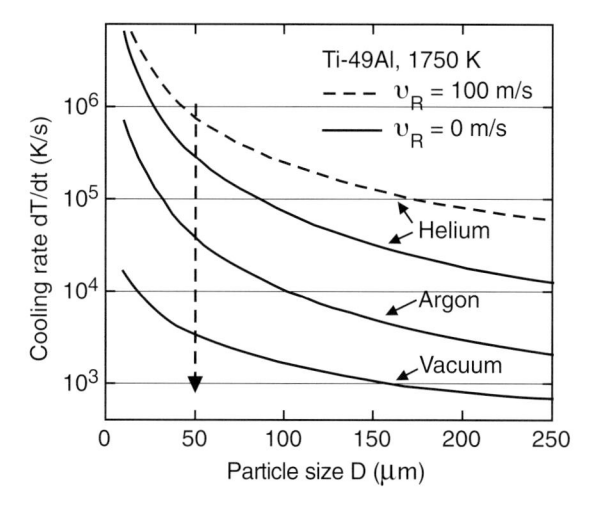

Figure 15.8 Calculated relationships between cooling rate and particle size for different environments and particle speeds within their environment [20, 21]. The graph indicates that irrespective of particle size, helium gas results in the fastest cooling rates, whereas vacuum conditions give rise to the slowest cooling rate. A high relative speed between the particles and the helium gas further increases the cooling rate. The original data have been redrawn.

using rotating-disc atomization in a chamber under vacuum. Zhao et al. [48] have suggested that such operation could provide both economical and technical advantages over gas-assisted processes in that the problems associated with gas entrapment could be virtually eliminated and that the costs associated with large-scale use of expensive gases would be reduced. However, these benefits could be offset by the more sluggish solidification, which may result in loss of aluminum from particle surfaces due to evaporation. Indeed, it has been reported that powder particles of a Ti-49Al-0.3C alloy made by using the REP with electron beam melting (i.e. under vacuum conditions) exhibited outer surfaces that were enriched in α_2, that is, denuded in aluminum [23].

From Figure 15.8 it is apparent that a significant increase in the cooling rate is afforded through the use of argon and especially helium gas (even with $v_R = 0$ m/s) compared to a vacuum environment. Indeed, Gerling et al. [21] calculated that a 40-μm diameter particle of Ti-49Al takes around 12 ms to cool from 1477 °C (1750 K) to 977 °C (1250 K) under an argon atmosphere ($v_R = 0$ m/s) and only 1 ms under helium, Figure 15.9. For larger-diameter particles these cooling times will of course be longer. Thus, diffusion-controlled phase changes and the resulting microstructural changes will be suppressed to a greater extent, and metastable phases are more likely to form, in smaller particles and when helium rather than argon is used as the atomizing gas. This is confirmed in Figure 15.10, which shows a graph of the ratio between the X-ray diffraction intensity of $\{20\bar{2}1\}_{\alpha2}$ planes over that of $\{20\bar{2}1\}_{\alpha2} + \{111\}_\gamma$ for various gas atomized powders [23]. The graph indicates that the volume fraction of (metastable) α_2 within the particles increases with

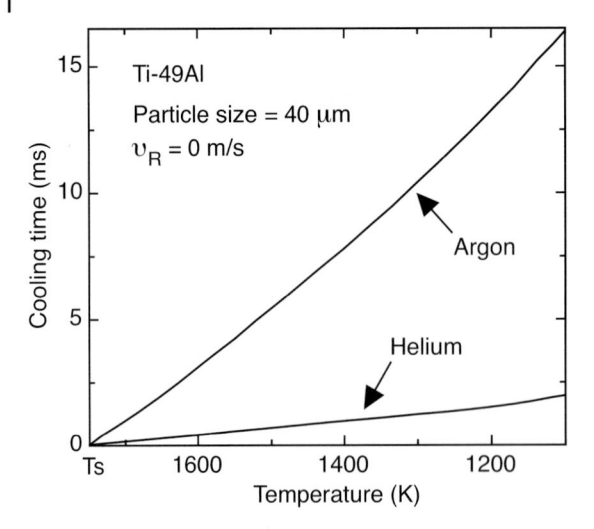

Figure 15.9 Graph showing the cooling time under a stationary argon or helium atmosphere for a 40-μm diameter particle at an initial temperature of 1750 K (1477 °C) [21]. The higher rate of cooling afforded through the use of helium rather than argon is very evident. The original data has been redrawn.

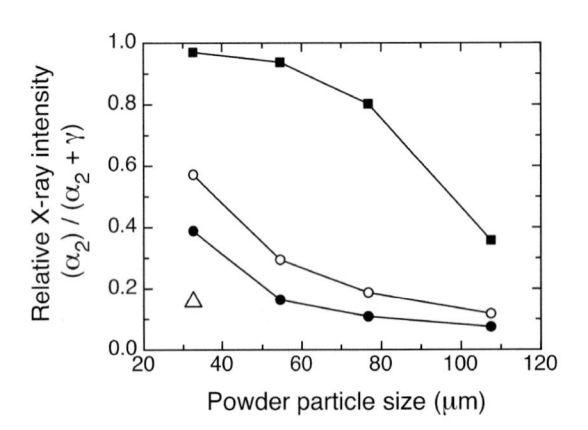

Figure 15.10 Ratio of the X-ray diffraction intensity of $\{20\bar{2}1\}_{\alpha2}$ planes compared to that of $\{20\bar{2}1\}_{\alpha2} + \{111\}_{\gamma}$ planes in various gas atomized powders as a function of particle size (replotted data) [23]. ■: TAB (Ar), ○: Ti-50Al-2Nb (He), ●: Ti-50Al-2Nb (Ar) and △: Ti-50Al-2Nb (under vacuum). High values indicate the presence of a larger amount of metastable α_2 phase within the particles. It can be seen that as the cooling rate decreases, that is, as the particle size increases, less metastable α_2 phase is formed. The higher cooling rate afforded by helium rather than argon and vacuum can be indirectly observed. The data for the TAB alloy lies well above that of the Ti-50Al-2Nb alloy, probably on account of its approximately 3 at.% lower aluminum content.

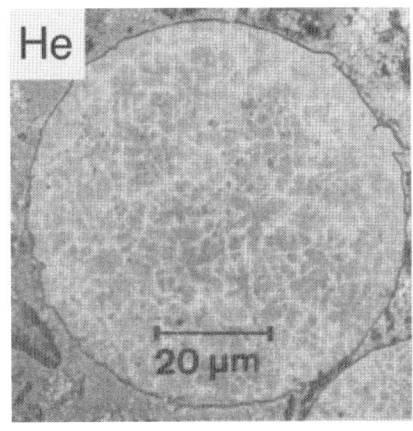

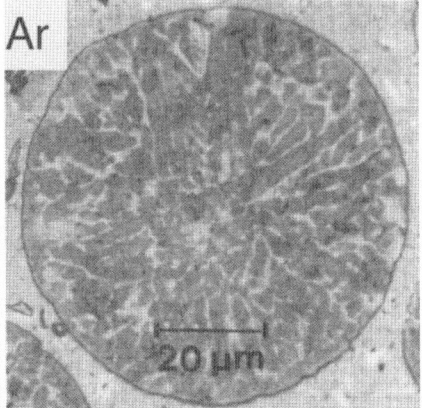

Figure 15.11 Images of Ti-50Al-2Nb powder particles within the size range 45 to 63 μm taken by light microscopy [20]. The particles on the left and right sides were made by gas atomization (PIGA) using helium and argon gases, respectively. The use of helium gas results in microsegregation on a much finer scale.

decreasing particle size and that the effect is more significant for helium than for argon gas atomization [20]. Fuchs and Hayden [19] also observed that finer particles of Ti-48Al and Ti-48Al-2Nb-2Cr, made using PREP, contained significantly more hexagonal phase than coarser particles. Similar observation have also been made by Graves et al. [49] who indicated that annealing for one hour at 600 °C was sufficient to transform metastable α_2 to the equilibrium state. Further details concerning the phase equilibria and solidification of TiAl alloys can be found in Chapters 2 and 4 of this publication and elsewhere [50, 51].

The finer microsegregation that is attained using helium rather than argon as the atomization gas can be seen in Figure 15.11 [20]. This figure shows the microstructure of two similarly sized gas-atomized particles of Ti-50Al-2Nb, one atomized using helium and the other using argon. For powder made during a single helium gas-atomization run, Figure 15.12 indicates the scale of microsegregation increases with particle size [52]. Habel et al. [53] state that the cell size does not change significantly (ranging from 2 μm to 10 μm) in argon gas-atomized powder particles less than 150 μm in size. The coarser microsegregation resulting from a much slower cooling rate afforded through use of the rotating-electrode process (REP) under vacuum is shown in Figure 15.13 for Ti-50Al-2Nb [23]. This micrograph shows powder particles of comparable size made by argon gas atomization and REP under vacuum.

In a study by Choi et al. [54] it was observed that rapidly solidified powders of Ti-50Al-1.8Nb and Ti-48Al-2.4Nb-0.3Ta (made by the rotating-electrode process (REP) and rapid solidification rate centrifugal atomization (RSR) respectively) generally exhibited a dendritic structure characteristic of α being the primary phase to form from the liquid. However, around 5% of the Ti-48Al-2.4Nb-0.3Ta powder and 1% of the Ti-50Al-1.8Nb powder showed surface relief, indicating that

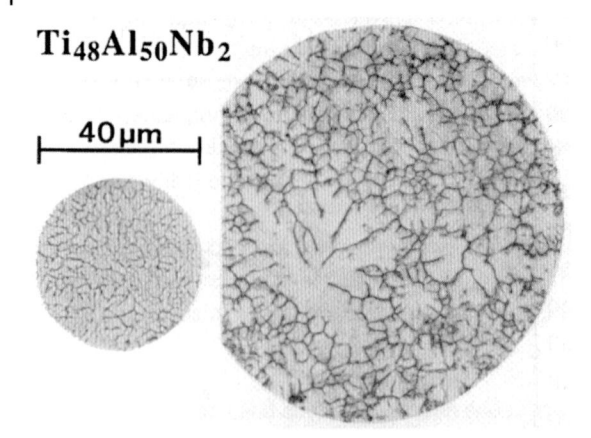

Figure 15.12 Optical microscope images of Ti-50Al-2Nb powder particles made using helium-gas atomization (PIGA) within the size range 20 to 45 μm (left) and 90 to 125 μm (right) [52]. The micrograph shows that the scale of microsegregation is much finer in the smaller particle, which results from a much faster cooling rate.

bcc-β was the primary phase to form. It was suggested that this subsequently transformed martensitically to a martensitic form of the hexagonal alpha phase, called α'. However, the authors indicated that this only took place for the very smallest particles (of less than 30 μm) and was rationalized in terms of other work [55, 56] which showed that in sufficiently undercooled binary alloys the β-phase was the primary phase to solidify.

In a study comparing PREP and gas-atomized powders of Ti-48Al and Ti-48Al-2Nb-2Cr, Fuchs and Hayden [19] observed three different types of particle surface morphology, namely "rosette", dendritic and featureless. The microstructure of "rosette"-type particles was reported to be a mixture of α and γ grains, the γ phase being within the intercellular regions. The dendritic particles exhibited hexagonal dendrites, suggesting that the disordered hexagonal α solidified as the primary phase. The particles with a featureless surface relief were only observed in powder produced using PREP (under a helium atmosphere). Depending on composition, between 20 and 40% of particles less than 75 μm in size exhibited the featureless surface. Examination of the microstructure of these particles indicated that no segregation or second phases were present, indicating that the supercooling may have been such that diffusion was suppressed during solidification. It was speculated that these particles were in fact disordered α, and not martensitic α', as no characteristic β dendritic structure was observed. In this work, it was concluded that higher cooling rates and undercooling were obtainable using PREP. Further discussion concerning the effect of rapid cooling conditions on the microstructure of TiAl can be found elsewhere [51, 55].

One of the main disadvantages of prealloyed powder, apart from cost, is the possibility of atmospheric contamination and the incorporation of atomizing gas within the powder. As explained earlier, after powder production all subsequent

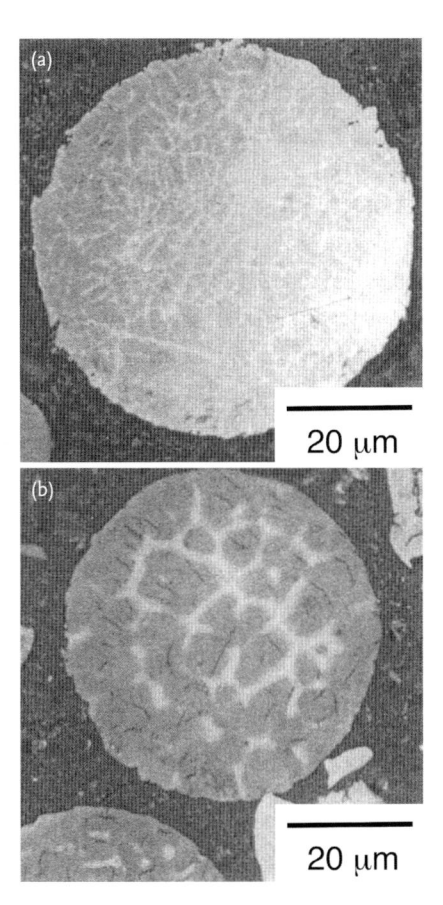

Figure 15.13 Images of Ti-50Al-2Nb powder particles within the 45 to 63 μm size fraction made using (a) argon-gas atomization and (b) the rotating-electrode process (REP) under vacuum [23]. The slower cooling particle made by the REP under vacuum exhibits significantly coarser microsegregation.

processing and handling must be performed under a protective atmosphere to minimize contamination. Work by Tönnes et al. [57] has shown that oxygen levels between 700 and 1600 wt. ppm has little effect on the tensile strength properties of HIPed Ti-48Al-2Cr-2Nb made using prealloyed powder. However, the room-temperature tensile ductility of a duplex microstructure decreased from 2.1% to 0.5% and that of fully lamellar material from 0.5% to 0.2% on increasing the oxygen level from 1050 to 1600 wt. ppm. It was thus concluded that oxygen should be limited to levels no higher than 1050 wt. ppm in order to guarantee acceptable ductility in Ti-48Al-2Cr-2Nb.

Gerling et al. [23] have investigated the variation between oxygen and nitrogen level with particle size for a Ti-45Al-7.5Nb alloy made by plasma inert-gas atomization. The results are shown in Figure 15.14 and indicate that there is very little contamination with nitrogen. Oxygen contamination is seemingly a problem that

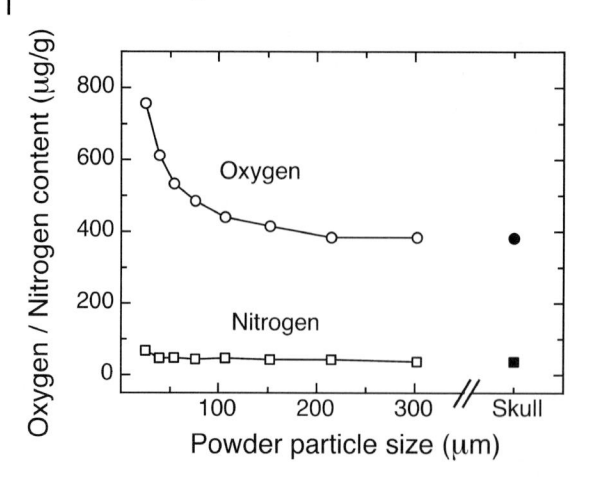

Figure 15.14 Graph showing the variation of oxygen and nitrogen content of PIGA skull material and of gas-atomized powder as a function of particle size. The nitrogen content of the powder remains at more or less the same level as that in the skull material (i.e. no contamination). The level of oxygen seems to increase as the powder particle size is reduced below around 200 μm diameter. However, Gerling et al. [23] believe that this apparent contamination resulted from the handling time in air while the powder was transferred from a protective atmosphere into the oxygen-analysis system. Data from [23], replotted.

becomes serious with decreasing particle size, in particular for particles less than around 90 μm in diameter. Yolton [26] has made similar observations. This effect with oxygen was further investigated by Gerling et al. [23], who observed that when fine powder was processed under high-purity argon prior to HIPing, then the oxygen levels in the HIPed compacts were around the same as that of the starting ingot material. The "apparent" oxygen contamination of the smaller particles was thus attributed to the time spent in air during transfer of the powder to the oxygen-analysis system. As part of the study, deliberate atmospheric exposure of the 20–32 μm and 45–63 μm power fractions for times of between 3 min and 9 h was investigated. After 9 h exposure, the oxygen level increased by around 200 μg/g and 100 μg/g for the respective powder fractions.

During the atomization process either "open" or "closed" (i.e. entrapped) gas pores can form within some of the atomized powder particles. Unlike for "closed" pores, gas from "open" pores can be removed through an adequate degassing treatment. With regard to the influence of atomization technique, Wegmann et al. [47] suggest that rotating-disc atomization produces powders with significantly more porosity than either PIGA or EIGA, and that the number of nonspherical particles increases after particle size exceeds around 90 μm. Choi et al. [54] have also noted significant internal porosity in powder made by the rotating-disc atomization (RSR technique). In contrast, powder made by the rotating-electrode process (REP) was stated as containing no internal porosity [54]. If this is a general feature, then it may certainly be one advantage of the REP over the rotating-disc and gas-atomization techniques.

Work by Gerling et al. [58] has studied the dependence of porosity size and volume fraction on particle size, for gas-atomized TiAl powder particles within the size range of 32 to 355 μm. Additionally, the argon content and the pressure of argon within pores was determined with respect to particle size. It was found that particles within the size range 32 to 45 μm had a porosity fraction of around 0.05%, whereas those in the size range 250 to 355 μm had a porosity fraction around 0.9%. Around 80% of all the pores within particles up to 355 μm had diameters less than 15 μm. The corresponding argon content with these two powder size fractions was ≈0.5 μg/g and 1.2 μg/g, respectively, thus suggesting that using finer particle-size fractions could be employed as a way of limiting argon content. A HIPed powder compact, made using particle sizes up to 250 μm contained around 0.45 μg/g argon. The argon gas pressure within the pores increased with decreasing particle size. At room temperature, the gas pressure within pores up to 15 μm in diameter was in the region of 0.17 MPa, but this reduced to levels around 0.03 MPa for pores in the size range 90 to 105 μm [58].

Depending on the temperature and pressure during a HIP treatment, the "closed" gas pores within powder particles may be compressed to such an extent that they are no longer observable within the resulting fully dense compacted material. As this process takes place under HIP conditions, the pressure within the pores may be expected to increase, assuming the absence of diffusion out of the pores. However, this pressure will remain and may lead to pore reopening on subsequent high-temperature exposure under atmospheric or reduced pressure conditions [46, 54]. As mentioned in Section 16.5.2.5, a problem associated with sheet material made from gas-atomized prealloyed powder is reduced high-temperature superplastic elongation [59]. Indeed, Appel et al. [60] observed what looked like a pore bursting from the surface of a rolled sheet after high-temperature deformation, see Figure 15.15. Eylon et al. [61] have studied the development of pores after heat treatment of HIPed conventional Ti-6Al-4V prealloyed powder (made by the REP). After exposure at 1290 °C for 2 h, it was found that pores within the HIPed material grew from sizes of around 0.1–0.5 μm to 10–50 μm. The reason for the pores in the HIPed material was not identified, although it was speculated (possibly mistakenly) that they arose due to a leak in the can or improper degassing. Such porosity that develops during elevated-temperature exposure is known as "thermally induced porosity" or TIP.

TIP is now believed to result from gas that is present within the HIPed compacts. As outlined above, this gas originates from closed pores in particles that are formed during the atomization process, but that are fully closed after a suitable HIP treatment. The effect of atomization technique on the phenomenon of TIP in TiAl has been studied in detail by Wegmann et al. [47]. The study investigated the effect of exposure at 1390 °C for three atomization techniques (PIGA, EIGA and rotating-disc atomization under argon) and two alloy compositions (Ti-48.9Al and Ti-46Al-9Nb). It was considered that after a HIP treatment of 1280 °C/4 h/200 MPa, the porosity level within all the compacts was below the detection limit of 0.01 vol%, although some pores were occasionally observed. After exposure for up to four hours, the largest thermally induced pore observed had a size of 65 μm. The use of helium gas did not reduce the TIP problem, indeed pore formation has

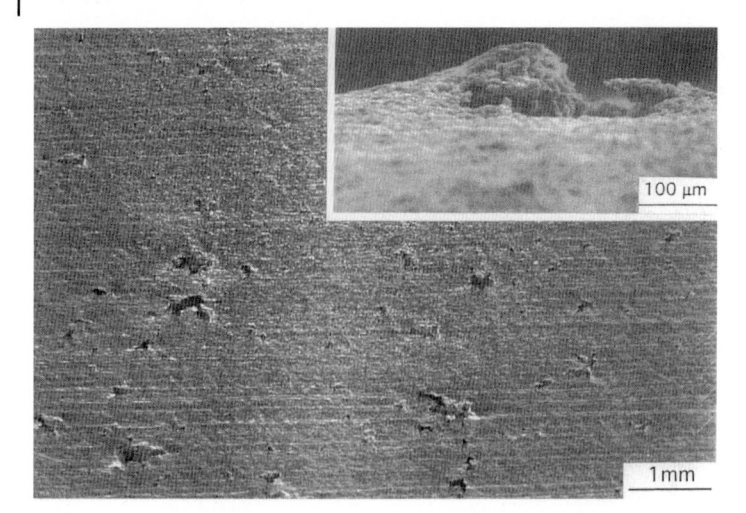

Figure 15.15 Appearance of cavities in a rolled sheet of Ti-47Al-1.5Nb-1Cr-1Mn-0.5B-0.2Si that was made from HIPed gas atomized prealloyed powder. The sheet was tensile tested at 1000 °C using a strain rate of $1.4 \times 10^{-4}\,\text{s}^{-2}$ to failure at 150% strain [60]. Such powder-based sheets show reduced high-temperature superplastic elongation due to the formation of cavities. It is believed that these cavities arise from entrapped gas that expands at high temperatures. The insert shows the raised periphery of a surface cavity, as if a bubble had burst out from just under the surface.

been reported to be even more pronounced than when using argon gas atomization [23, 47, 62]. In order to reduce the problem of TIP in technical applications, it has been suggested that powders of less than 180 μm in diameter made using PIGA with argon gas be employed [23, 47].

TIP in TiAl after high-temperature exposure has also been observed by Yolton et al. [25], Choi et al. [54] and Shong et al. [63]. It was observed that larger thermally induced pores resulted from both low HIPing temperatures and low pressures [63]. To the current author's knowledge TIP has not been investigated after long-term heat treatment at nominal temperatures to which TiAl components such as blades could be exposed during application. This is technically very important, as such porosity could lead to a degradation of component properties after long-term use. According to Yolton et al. [25], TIP generally occurs when the as-HIPed material is heated to temperatures above the original HIP temperature, which for TiAl alloys is significantly higher than nominal component operating temperatures.

15.1.5
Postatomization Processing

After atomization and powder collection, further handling should only be performed under a protective atmosphere within a glove box. The first operation is to separate the powder into the various size fractions. This is done by sieving using successively finer meshes, as described elsewhere [46]. What happens thereafter depends upon how the powder is to be processed. It could be HIPed to form a near-net-shaped component, or more usually to form a compact for subsequent hot-working procedures. Alternatively the powder could be used in processes such as metal injection molding or in rapid prototyping methods such as laser melt deposition. However, for TiAl these techniques have only been investigated by a limited number of groups and are seemingly in their infancy.

15.1.5.1 Hot Isostatic Pressing (HIPing), Hot Working, and Properties

The HIPing process involves the encapsulation of powder within a suitably shaped container that is then evacuated, sealed and isostatically compressed at elevated temperature using inert high-pressure gas. The container is usually made from thin commercially pure titanium sheet. Cheaper iron-based sheet is not normally used because a eutectic reaction occurs between iron and titanium at 1085 °C [64], which can lead to local melting. As the packing density of any powder within a container is less than unity, the volume reduction after HIPing to full density must be taken into account when near-net-shape components are being manufactured. Container evacuation is normally performed during a heating/cooling cycle that involves gradual heating to around 400 °C and remaining at this temperature for around 4 h. The vacuum level reached within the can at the end of the degassing cycle is usually around $3 - 4 \times 10^{-5}$ mbar. After subsequent cooling, the evacuated container is sealed. Hot isostatic pressing is usually performed [23] within the temperature range of 1000 °C to 1300 °C under gas pressures of between 150 MPa and 200 MPa for 2- to 4-h periods. Obviously, the alloy composition and the HIP temperature employed will influence the microstructure of the HIPed material.

In an attempt to explain a lower than expected volume fraction of α_2 in HIPed material, Zhang et al. [65] speculated that under high pressure, T_α may be increased or that the kinetics of the γ to α transformation are significantly slower compared to those under atmospheric pressure. Huang et al. [66] have shown that the higher than expected volume fraction of γ in HIPed material results from its lower "atomic volume" compared to the α_2 phase. Habel and McTiernan [67] indicated that HIPing Ti-46Al-2Cr-2Nb at 1200 °C for 2 h (100 MPa), resulted in a fine near-gamma microstructure with a grain size of around 5 µm. Slightly higher temperatures resulted in slightly coarser duplex microstructures containing a small amount of β, while HIPing at 1300 °C gave rise to a nearly lamellar microstructure with a 35-µm colony size and a coarse lamellar spacing. In other work, it has been shown that additions of boron are able to reduce the extent of grain growth during HIPing [23]. The effect that HIP temperature has on microstructure can have a significant effect on mechanical properties [65]. Additionally, the powder fraction used can also influence the microstructure, it being observed that HIPed compacts made from fine powder contained more α_2 than those made from coarser powder [65]. During a study of phase transformations, Zhang et al. [68] developed a wide variety of microstructures in Ti-48Al-2Mn-2Nb through different post-HIP heat treatments. The work also indicated that excessive impurity levels (3000 ppm oxygen and 0.11 wt.% carbon compared to 800 ppm oxygen and 0.03 wt.% carbon) could lead to a 20 °C to 30 °C increase in T_α.

The whole aim of a HIP treatment is to consolidate the powder into a seemingly fully dense product. However, when the powder contains gas within entrapped porosity it is theoretically not possible to achieve full density. Using a formula proposed by Ashby [69], it has been shown that for an initial pore pressure of 0.1 MPa and a HIP pressure of 182 MPa, an ultimate relative density of 0.99997 can be achieved in material with a relative density of greater than 0.95 before consolidation [54]. Shong et al. [63] have used TEM to show the presence of submicrometer pores at prior particle boundaries after HIPing at temperatures up to 1000 °C. They indicate that porosity at a given HIPing pressure reduces with increasing HIP temperature. Work by Choi et al. [54, 70] included studying the densification of TiAl powders during HIPing and determining consolidation curves such as that shown in Figure 15.16. HIP maps considering the various densification mechanisms were also generated using computer programs that were developed by Ashby [69]. The calculated maps were generally in good agreement with experimental results. For further information the reader is referred to refs. [54, 69, 70]. Information relating to powder densification is an important aspect with regard to the production of near-net-shape products. With this in mind, Abondance et al. [71] have performed numerical modeling for the HIPing

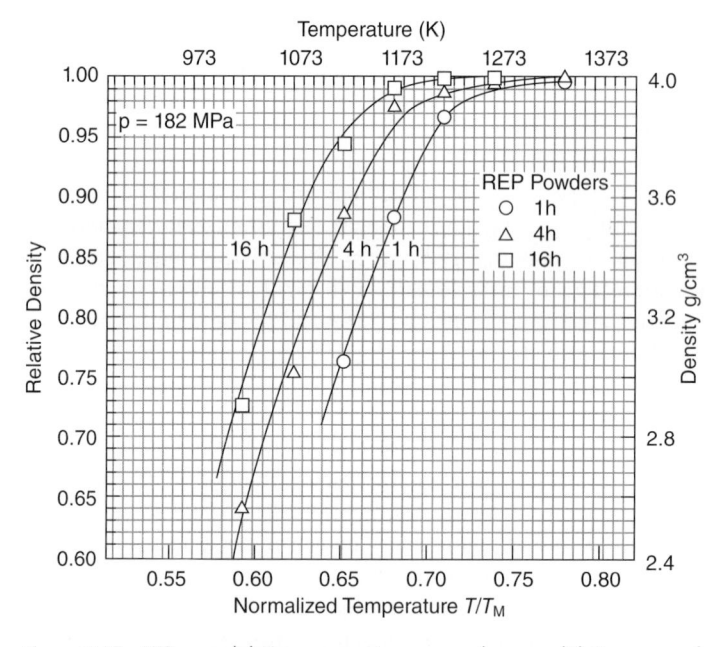

Figure 15.16 HIP consolidation curves at a pressure of 182 MPa for Ti-48Al-2.4Nb-0.3Ta powders made by rapid solidification rate (RSR) processing [54]. The graph shows how powder consolidation proceeds as a function of HIP temperature for various HIP times. It can be seen that higher temperatures lead to faster consolidation.

of conventional Ti-6Al-4V alloy powder and have shown that deviations in geometry were generally less than ±0.1 mm when compared to the corresponding real HIPed parts. Unfortunately the size of the parts was not mentioned. Similar work, also for Ti-6Al-4V powder, involved a FEM being incorporated into a CAD module that was successfully used to predict final component geometry to within a 2% discrepancy [72]. Plansee has demonstrated the use of near-net-shape HIPing technology for the production of automobile valve preforms [73].

As with ingot-based technologies, the mechanical properties of powder-based material depend on many factors including microstructure, which naturally depends on the alloy composition, the HIP temperature and any subsequent processing/heat-treatment procedures. To make any general comparisons between the mechanical properties of ingot- and powder-based materials is difficult as, strictly speaking, this can only be done by comparing alloys of the same composition that have been treated/hot worked in the same way and have similar microstructures. Additionally, the ingot should be of reasonably good chemical homogeneity, which at present is often not the case.

HIP-compacted TiAl can be hot worked using the same techniques as outlined for ingot material. Semiatin et al. [74] have presented work comparing the hot compression behavior of cast ingot and HIPed powder made from a Ti-48Al-2Cr-2Nb alloy. Both materials exhibited similar dependencies of flow stress with temperature and strain rate. Flow softening that resulted mainly from dynamic recrystallization was also observed. However, the peak flow stress was much lower in the powder material compared to the ingot material due to its fine grain size. This suggests that the powder metallurgy route would facilitate a reduced load during any isothermal forging procedures. Thus, similar to the case of wrought TiAl from the ingot route, HIPed powder seems to be very suitable for use in secondary hot-working processes, such as closed-die forging [74]. Fuchs [75] has made a similar study on ingot- and powder-based Ti-48Al-2Nb-2Cr and made similar conclusions. Beddoes et al. [76–78] have also made studies characterizing the high-temperature compression behavior of powder-based materials and have presented forgeability maps [76]. With regard to minimum forging temperature, successful forging at 850°C to an 80% reduction has been reported to lead to crack-free material with superplastic properties at temperatures as low as 900°C [23, 79, 80].

At the TMS annual meeting in 2003, Dimiduk et al. [81] reported that due to the long timescale for the production of wrought products and the need for defect control, the US Air Force was at that time focusing work on powder-based processing technologies. However, with reference to the experience gained during powder processing of nickel-based alloys and with reference to other work on gamma alloys [82–85], Dimiduk et al. [81] raised a note of caution, despite early encouraging signs. Nevertheless, Yolton et al. [25] and Das et al. [86] have presented work on large HIPed powder compacts of 263 kg and 209 kg, respectively. Whereas Yolton et al. [25] detected no signs of pores after HIPing using radiography (minimum detection size 30 μm), Das et al. [86] found defects using ultrasound (that were subsequently characterized) within a forged pancake. Habel et al. [53]

showed that material forged at 1260 °C required further heat treatment in order to improve the ductility from 0.1% to 1.1%. In other work by Habel and McTiernan on Ti-46Al-2Cr-2Nb, it was shown that HIP temperature (and thus microstructure) plays an important role, with up to 1.7% elongation being measured after HIPing at 1300 °C but only 0.2% elongation after HIPing at 1200 °C [67]. An inverse relationship between the tensile elongation and hardening rate (similar to that plotted by Paul et al. [87] for extruded ingot), which indicates the influence of microstructure on ductility, was also established. Tönnes et al. [57] obtained yield stresses of around 370 MPa and tensile strengths of 460 MPa combined with plastic elongation of around 2% for Ti-48Al-2Cr-2Nb with a microstructure containing between 30 to 40% lamellar colonies with a size of 100 μm. These values of strength and ductility compare favorably with those given by Austin and Kelly [88] for cast Ti-48Al-2Cr-2Nb.

Mechanical properties of powder-based TiAl processed near T_α have been reported within the literature [89, 90]. Possibly some of the best tensile properties were measured by Liu et al. [90] in powder that had been extruded at temperatures above T_α. Yield stress of values of 971 MPa combined with a tensile elongation of 1.4% in Ti-47Al-2Cr-2Nb were reported. However, in the paper it is not stated whether the powder was HIPed or not before extrusion, possibly implying that un-HIPed powder was extruded within the can, as described by Roberts and Ferguson [91]. Extrusion at temperatures above T_α gave rise to a fully lamellar microstructure with 65 μm colony size [90]. Extrusion at temperatures within the $\alpha + \gamma$ phase field gave rise to extremely fine duplex microstructures, but the ductility of the duplex microstructures was inferior to that of the fine fully lamellar material. This is reported by Liu et al. [90], with reference to other work [92–95], to be contrary to the usual behavior for cast materials and was thought to be a possible consequence of higher porosity in the duplex microstructure of materials extruded at lower temperatures. The remarkably good properties exhibited by the material extruded at a temperature within the α phase field were suggested to arise from the extremely fine colony size and the fine interlamellar spacing (0.1 μm) combined with the unique ultrafine morphology of the α_2 lamellae.

With reference to other work [96–98], Semiatin et al. [29] state that HIPed compacts and HIPed + hot-worked powder products can exhibit mechanical properties that are equal or even superior to those of wrought ingot material. There are, however, not very many studies that simultaneously examine the behavior of ingot and powder metallurgy materials with the same composition. Das et al. [86], Thomas et al. [99] and Malaplate et al. [100] have investigated the properties of alloys with similar compositions employing each route. The powder approach led to more homogeneous microstructures and higher strength and ductility [86]. The property variability was narrower for powder-based material due to increased homogeneity [99]. Additionally, the property scatter of HIPed powder compacts could be reduced by using particle-size fractions with a narrower size range. With regard to creep, Malaplate et al. [100] indicated that both the extent and duration of primary creep were larger in cast material due to the more heterogeneous nature of the microstructure.

15.1.5.2 Laser-Based Rapid-Prototyping Techniques

Rapid-prototyping techniques employing the use of lasers can produce near-net-shape components from prealloyed powder without the need for any hard tooling and as such can result in considerable time and cost savings. Due to the difficulties associated machining, such fabrication methods have their advantages, especially for geometrically complicated shapes.

Kruth *et al.* [101] discuss the basis powder metallurgical aspects of "selective metal powder sintering" (SMS). In this technique the powder is spread as a thin layer, around 0.4 mm in thickness, on top of a container by a deposition system. A laser beam is then scanned over the powder in a predetermined manner to achieve sintering. This is repeated one layer after another until a three-dimensional part (called green product) is made. The part may have significant porosity so a post-treatment is necessary.

Another process, known as "laser forming" (LF) or "direct metal deposition" (DMD), involves introducing powder into a laser beam where it is melted, deposited and rapidly solidified [102, 103]. Figure 15.17 shows a typical chamber setup, in this case for the equipment used at Penn State University [103]. The laser melt deposition process takes place within a chamber that is purged with high-purity argon that flows upward from a diffuser plate at the bottom of the chamber [103]. The Penn State chamber employs a focused 14-kW CO_2 laser the focal point of which can be translated in the X, Y and Z directions according to the desired shape of the component to be made. The powder is supplied to the focal point via a feeder and a carrier gas. This is an example of a fixed target and chamber system.

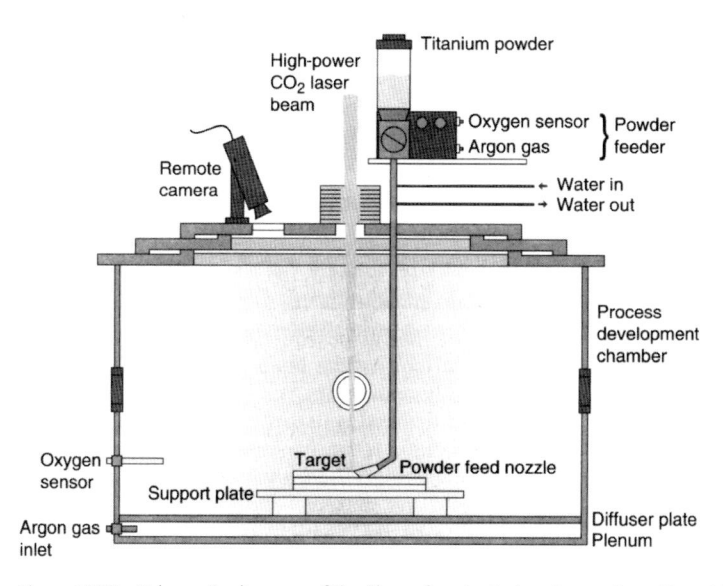

Figure 15.17 Schematic diagram of the "laser forming" chamber at Penn State University that is used to make 3D parts from prealloyed powder [103].

"Direct laser fabrication" (DLF) [104, 105], "laser melt deposition" (LMD) [106] and "laser-engineered net shaping" (LENS) [107] are processes very similar to the laser forming process described above, but employ a laser with a fixed focal point. In the DLF and LENS processes the powder nozzle and laser beam delivery column move together within an integrated unit in the Z direction with the substrate moving in the X and Y directions. In the LMD process the substrate moves in the X, Y and Z directions and the integrated powder/laser unit is fixed.

Relatively large parts can be produced using such types of processing. Moll et al. [103] have manufactured two plates of Ti-47Al-2Cr-2Nb, the larger one was 20 cm long by 3.2 cm wide and 15 cm in height. During processing, each individually deposited layer was about 1 mm thick. The larger plate (which was made in three steps, with intermediate cooling to room temperature) contained several cracks that were between layers where deposition had been restarted on substrate that was at room temperature. Such cracking was thought to arise from the build-up of thermal stresses and it was proposed that it might be necessary to maintain substrate temperature during processing. The second but smaller plate (only 7 cm in height) was made using continuous deposition and contained no cracks. The layered nature of the parts was apparent, with each individually deposited layer containing columnar and equiaxed grains of varying size as a result of variations in the solidification rate [103]. The processed material had a very fine fully lamellar microstructure. Occasionally pores up to 10 μm in size caused by entrapped argon and/or shrinkage were observed within the material. Very importantly, the composition of the plates was nearly identical to the powder, indicating no loss of aluminum or oxygen contamination during processing. With regard to properties, the plate material was reported to compare favorably with ingot and powder materials processed in more conventional ways and exhibited a strength of around 500 MPa at room temperature combined with 1 to 2% elongation [103].

Srivastava et al. [104, 105, 108] have also observed extremely fine microstructures with an interlamellar spacing of 50 to 100 nm in Ti-48Al-2Mn-2Nb processed by DLF. It is believed that the small grain size and lamellar spacing result from an increased number of lamellae nuclei [108]. The complex structure in such laser-manufactured materials is thought to arise as a consequence of rapid cooling combined with reheating/remelting and the associated compositional heterogeneities [108]. Parameters such as powder-size fraction, laser scan rate, powder feed rate and the efficiency of thermal extraction from freshly deposited material by the substrate will all have an influence on the microstructure [104, 109]. Indeed, a variety of microstructures could be obtained by varying the processing parameters, although postprocessing heat treatment was necessary to increase the microstructural homogeneity [104]. Laser-based techniques may thus be well placed for processing TiAl alloys and repair of components, although at the moment most published work on powder concerns conventional processing techniques. According to Loretto et al. [109] this type of processing may find use in niche applications, such as blade tipping. In later private discussions, it was envisaged that whole blades may be made through such techniques and could be flying within the next few years. For further information regarding these types of rapid processing methods the reader is referred to [110–112].

15.1.5.3 Metal Injection Molding (MIM)

This is a near-net-shape technology that combines the flexibility of powder metallurgy with the design flexibility of plastic molding. For a full description of the MIM process the reader is directed elsewhere [113, 114]. The method can be used to make parts with complex geometries, which exhibits isotropic mechanical properties. It is particularly suited for materials that are difficult to machine or parts that have small wall thicknesses (<1 mm). Holes with bore diameters down to around 0.4 mm can also be made within parts [85].

The MIM process involves a number of processing steps including: production of metal powder, mixing, molding, debinding and final sintering [114]. With regard to the powder used, much finer powders are used than for conventional powder metallurgy. This is because sintering requires diffusion to take place, the rate of which is inversely proportional to the square of the powder diameter. Shrinkage and densification thus take place much faster when the particle size is small and any pores that remain within the sintered part are reduced in size [114]. Erickson [114] suggests the use of particles with diameters ranging between 0.5 µm and 20 µm, although Gerling *et al.* [85] and Whitton [15] suggest particles with a diameter of less than 45 µm may be acceptable.

Compared to other areas of processing-related research within the TiAl community, it has been reported by Gerling *et al.* [23] that there has only been relatively little work performed on MIM of TiAl [115–117]. Therefore this section will mainly concentrate on the published work of GKSS that employed gas-atomized prealloyed powders. Initially, the powder is combined with a binder during a mixing stage. In work by Gerling *et al.* [118] it was found that a binder consisting of a mixture of polyethylene, paraffin and stearic acid was the most suitable. For the powder used, the volume fraction of binder in the feedstock required to achieve the desired rheological properties was 32 vol.%. Mixing was performed in a Z-blade mixer at 120 °C. The feedstock was injection molded into cube-shaped parts with dimensions of 40 mm × 20 mm × 7 mm. An injection pressure of 420 bar was used and the temperature of the feedstock and molding die were 90 °C and 45 °C, respectively. A two-step debinding process was employed involving the use of hexane at 40 °C for removal of the paraffin. This was followed by a thermal treatment, under vacuum, at temperatures between 250 °C and 400 °C for cracking and vaporization of the polyethylene. Subsequent sintering was performed at 1360 °C for 3.5 h under pressures of 10^{-5} mbar, 300 mbar of argon and 900 mbar of argon. Some parts were subsequently additionally HIPed at 1300 °C for 2 h at 200 MPa [118].

Compared to earlier work by the same authors that was presented elsewhere [119], it is reported that optimization of the MIM process resulted in a reduction of the oxygen and nitrogen pick-up by around 500 µg/g and 100 µg/g, respectively [118]. This is encouraging although the levels of around 1350 to 1600 µg/g for oxygen contamination measured in the as-sintered (under argon) materials are only slightly higher than those determined in as-cast components, see Table 14.2. As can be seen, the nitrogen levels of around 100 µg/g are comparable to those in cast components. However, the carbon content of the sintered MIM parts was up to around 10 times higher than that which was measured in powder that had been HIPed in the conventional manner. This results from the use of organic-based components within the binder. HIPing the sintered compacts resulted in no significant increase in the levels of oxygen or carbon, although an 8 to 27 times increase in nitrogen was identified. The reason for this increase was not clearly explained.

Sintering at 1360 °C for 3.5 h under vacuum resulted in aluminum evaporation from the outer surface that led to an 8 at.% reduction in aluminum [118]. This gave rise to a surface casing with a fully lamellar microstructure, whereas the center of the parts had a near-gamma microstructure. A transition layer between the aluminum depleted fully lamellar outer surface and the near-gamma central region was observed. It should be mentioned that the observation of a near-gamma microstructure in the γ-TAB alloy that was used in the study is rather surprising. This is because T_α for the correct composition of this alloy is around 1360 °C and one would therefore expect a fully/near-fully lamellar microstructure to develop.

Sintering under a 300-mbar partial pressure of argon resulted in aluminum evaporation being much less pronounced. Using a 900-mbar partial pressure resulted in the outer surface having a similar microstructure to the central core. Naturally, performing sintering under argon gas has consequences and these were manifested as increased levels of argon within the sintered parts. The argon levels increased from 0.3 µg/g for material that was sintered under vacuum to 1.6 µg/g for sintering at 300 mbar and to 5.9 µg/g for sintering at 900 mbar. Despite the possible effects that increased argon may have, it was deemed that processing under such conditions is justified on account of the reduced aluminum evaporation [118]. A possible solution to the problem of aluminum evaporation could involve performing sintering under a partial pressure of aluminum vapor. However, to our knowledge no such investigations have ever been performed.

Neglecting surface effects, the microstructure of the parts sintered under argon were comparable [118]. The porosities of the argon-sintered specimens were also very similar, although detailed examination indicated that the extent of porosity was slightly greater in the material sintered under a 900-mbar argon partial pressure. After subsequent HIPing at 1300 °C for 2 h under a pressure of 200 MPa, the porosity of the material sintered under 900 mbar of argon was close to zero. In the material sintered under 300 mbar of argon this was not the case. The extent of porosity was reduced to 2.6% (around 65% of its initial value) while the pore-size distribution remained relatively unchanged. This was explained by the closure under HIP conditions of "closed pores", while "open pores" (and their size distribution) remained open and unchanged. However, the reason behind the existence of such open porosity in the 300-mbar sintered condition and not in the 900-mbar sintered condition was unexplained [118]. The microstructure was not significantly changed by the additional HIPing treatment.

15.1 Prealloyed Powder Technology

Table 15.1 Room-temperature tensile properties of sintered metal injection molded (MIM) TiAl. The properties of alloys processed in other ways are also presented for comparison purposes.

Alloy	Processing	$\sigma_{0.2}$ (MPa)	σ_f (MPa)	ε_{total} (%)	$\varepsilon_{plastic}$ (%)	Ref.
Ti-48Al-2Cr-2Nb	MIM	–	265	0.3	–	[117]
Ti-48Al-2Mo	MIM	–	471	0.42	–	[116]
Ti-47.4Al-2.6Cr	MIM	–	317–329	1.8–2.0	–	[115]
γ-TAB[a]	MIM	–	120–260	–	0	[119]
γ-TAB[b]	MIM + HIP[d]	410	430	–	0.6	[119]
γ-TAB	MIM, 300[c]	324	347	–	0.59	[118]
γ-TAB 300[c]	MIM, 300[c] + HIP[d]	344	359	–	0.42	[118]
γ-TAB 900[c]	MIM, 900[c]	328	351	–	0.59	[118]
γ-TAB 900[c]	MIM, 900[c] + HIP[d]	398	412	–	0.45	[118]
Extruded γ-TAB	Extruded + HT1	404	426	–	1.0	[120]
Extruded γ-TAB	Extruded + HT2	572	738	–	1.7	[120]
Cast γ-TAB	Cast + HIP[d]	500	550	–	1.25	[121]
Ti-48Al-2Cr-2Nb	Cast	278–303	–	–	1.5–2.4	[88]

a) Data for sintering at 2 to 4 h at 1410 °C.
b) Data for 4 h sintering at 1410 °C.
c) Sintered under an argon partial pressure in mbar.
d) Subsequently HIPed at (1300 °C/2 h/200 MPa).
HT1 = 1030 °C/2 h/slow cool; HT2 = 1360 °C/18 min/oil quench + 800 °C/6 h/slow cool; γ-TAB: Ti-47Al-1.5Nb-1Cr-1Mn-0.5B-0.2Si.

The tensile properties of alloys made using MIM and published in the literature are given in Table 15.1. Data for alloys processed using different techniques is also given for comparison. It is difficult to draw any firm conclusions between different alloys, as the mechanical properties of γ-TiAl are both alloy and microstructure dependent. However, on comparing the γ-TAB alloy made using MIM it seems that a postsinter HIP is capable of increasing both the flow and fracture stresses. The best combinations of strength and ductility for MIM material were presented in [119]. The strength values presented compare well with those of an extruded γ-TAB with a fine-grained microstructure [120] (HT1), and those of cast Ti-48Al-2Cr-2Nb [88]. The strength properties are, however, inferior to those of a cast γ-TAB alloy HIPed at 1300 °C/2 h/200 MPa to establish balanced properties [121]. The values of ductility in MIM materials are generally lower than those of alloys processed by casting or extrusion, and may be difficult to increase significantly as long as contamination remains high [118]. While the technique could have some potential for manufacturing very complex geometries, further work is required to investigate if the mechanical properties can be further improved.

15.1.5.4 Spray Forming

This is a process that can be performed immediately after atomization has taken place, thus reducing the possibility of interstitial contamination in the material produced. Dowson et al. [122] have described two variations of the process. The first deposition method can take place under argon at atmospheric pressure or more preferably under near-vacuum conditions (<200 mbar – referred to as low-pressure centrifugal spray deposition, LPCSD), and uses centrifugal atomization to deposit powder into a ring-shaped-type structure. The technique (termed gas-assisted spray deposition: GSD) uses gas atomization to deposit material onto a substrate that may be preheated. The two spray-forming variations are shown schematically in Figure 15.18 [122]. For GSD the substrate could be a rotating steel drum or a flat sheet [122]. Other work has been reported on using EIGA for powder atomization with immediate deposition of the powder onto a rotating plate that

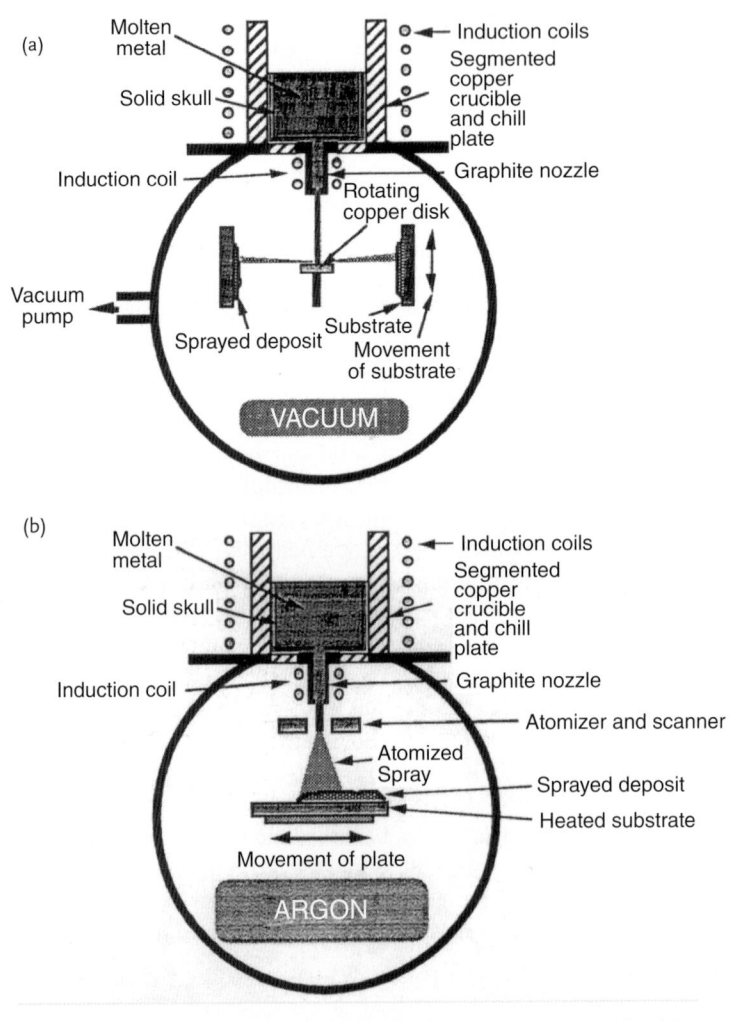

Figure 15.18 Schematic diagrams showing the (a) low-pressure centrifugal deposition (LPCSD) and (b) the gas-assisted spray deposition (GSD) processes [122].

could be moved both horizontally and vertically [123–125]. Using this EIGA-based technique, deposition rates of up to 0.48 kg/min (8 g/s) were investigated in 1999 [125], although the technique is now (reported 2004) capable of melting rates up to 0.83 kg/min [23]. In comparison, the deposition rates of 6 to 13.5 kg/min (dependent on nozzle diameter) reported by Young *et al.* [126], who used an induction skull-melting technique (as shown in Figure 15.18), are much higher.

Apart from porosity, probably two of the most important factors to be considered in spray-formed material are the degree of contamination and microstructure. According to work performed at IRC in Birmingham on Ti-48Al-2Mn-2Nb, there is no nitrogen pick-up and oxygen contamination is minimal, ranging from 600 to 700 wt. ppm in the charge material to around 750 wt. ppm in deposited material [122, 126]. Gerling *et al.* [23] state that the oxygen and nitrogen content of spray-processed material can be kept to levels of around 390 µg/g and 30 µg/g, respectively, that is, similar to those in ingot material. The Birmingham work also indicated that there was little change in composition with respect to Ti, Al, Mn and Nb after the spray forming of Ti-48Al-2Mn-2Nb [126].

As pointed out by Gerling *et al.* [23] the microstructure of deposited material depends on the thermal regime during and after spraying. In this respect, the deposition rate (i.e. the rate of heat supplied to the substrate) and the type of spray-forming technique play a role. For example, Dowson *et al.* [122] observed that the microstructure of material made by LPCSD depended on the distance between the rotating disc and the substrate. A short distance of 200 mm gave rise to a cored columnar/dendritic microstructure with interdendritic shrinkage porosity; all indicating that the particles were molten during deposition. However, increasing the distance to 400 mm (effectively reducing the heat supplied) gave rise to an equiaxed microstructure, indicating that heat removal by the substrate was sufficient to provide enough cooling for equiaxed grain growth. One may speculate that segregation problems could be reduced by using gas-assisted deposition rather than LPCSD, because the particles are cooled to a greater extent during flight. Li and Lavernia [127] describe fine fully lamellar microstructures in GSD spray formed γ-TiAl alloys. However, Gerling *et al.* [23] report that a near-gamma microstructure often develops in a thin layer close to an initially cold substrate, but that a lamellar structure then develops after a certain distance from the substrate. Such variations in microstructure result from differences in cooling rate, as discussed above and elsewhere [123, 125]

One aspect that is inevitable using GSD is the incorporation of the atomizing gas within the spray-deposited material. Gerling *et al.* have measured argon contents of around 4 ±0.5 µg/g in deposits of sprayed TiAl [85] and believe that such levels are indicative of optimized conditions [23]. It must be stated, however, that this level is up to ten times more than that measured in gas-atomized powders [85]. Of course, using a spraying technique such as LPCSD should result in much-reduced argon contamination, although actual values for comparison are not given in either ref. [122] or ref. [126]. The greatly reduced levels of porosity and gas entrapment are reported by Young *et al.* [126] to be the major advantages of LPCSD over GSD. The distribution of porosity in spray-formed rings made using LPCSD under either near-vacuum conditions (<200 mbar argon) or 1.1 bar of argon was not uniform; porosity tending to be concentrated in bands close to the inner surface of the ring [126]. Despite the presence of shrinkage porosity, which as mentioned earlier results from limited cooling of the particles during flight, the overall level of porosity was reduced when spraying was performed under near-vacuum conditions. Young et al. [126] concluded that LPCSD is superior to GSD for the manufacture of spray-formed material.

According to Lavernia et al. [128] the porosity within spray-formed material can arise from three mechanisms: solidification shrinkage, interstitial regions that remain between deposited particles, and that arising from gas entrapment within particles during atomization. The final porosity level is determined by the sum of each contribution, which of course depends on exact processing conditions. Gerling et al. [123], Schimansky et al. [124] and Liu et al. [125] have all observed porosity variations within GSD compacts with increased porosity observed within the initial 10 to 11 mm thickness of material next to the substrate (which they refer to as "Region 1"). Depending on exact processing conditions and alloy composition, porosity levels of up to around 8% were determined for material within "Region 1", while maximum levels were less than 4.5% outside "Region 1" [124]. With respect to the pore-size distribution outside "Region 1", Schimansky et al. [124] state that 90% of the pores were smaller than 20 μm. For Ti-48.9Al, the extent of porosity outside "Region 1" has also been found to depend upon the deposition rate (melt flow rate), with a minimum porosity of around 1% resulting from a melt flow rate of 7 g/s and an atomization pressure of 0.5 MPa [125]. Melt flow rates of 8 g/s and 4.6 g/s gave rise to porosities of around 2.4% and 5.2%, respectively. Increasing the atomization gas pressure from 0.5 MPa to 1.0 MPa at a constant melt flow rate of 6 g/s increased the porosity from around 3% to 6% [125]. It is believed that the higher porosity resulted from a reduced substrate temperature that arose from the reduced mean particle size that is obtained when a higher atomization pressure is used and the associated increase of in-flight heat loss [123]. Pores as large as 65 μm were reported in as-sprayed Ti-48.9Al [123, 129].

HIPing material that was spray-formed using deposition rates of 6 g/s or higher reduced the porosity to levels around 0.06% [129]. However, in cases where the melt flow rate was too low, HIPing could not significantly reduce porosity because of its "open" nature [129]. Heat treatment after HIPing was used to modify the microstructure, the use of very high temperatures (up to 1390 °C for 0.3 h) gave rise to only a slight increase in porosity (to 0.18%), see Section 15.1.4 for further information on porosity in TiAl powder products.

Isothermal forging of small billets made from spray-formed compacts has been discussed in a number of papers [23, 85, 129, 130]. Forging of binary Ti-48.9Al [129, 130] and multicomponent [130] alloys to a 77% reduction has been performed at 1100 °C using initial strain rates of around $2 \times 10^{-3} s^{-1}$. It was found that forging transformed the initial lamellar microstructure of the Ti-48.9Al spray-formed material into a fine-grained (2.9 μm) near-gamma microstructure. The maximum grain size observed in as-forged material was 12 μm [130]. In comparison, forging of the multicomponent γ-TAB alloy led to the initially duplex microstructure transforming to a near-gamma microstructure with gamma grains of 1.9 μm and a maximum grain size of 5.5 μm [130].

The porosity in the binary alloy was reduced from 1.0% in the as-sprayed condition to 0.04% in the as-forged condition. Similarly, the porosity of the γ-TAB alloy reduced from 2.0% to 0.03% after forging. After a subsequent stress-relief heat treatment for 2 h at 1030 °C, the grain sizes were 4.9 μm for the binary alloy and 2.2 μm for the γ-TAB. Within the binary alloy some grains grew to around 44 μm in size, while the heat-treated γ-TAB showed grains no larger than 7.5 μm. In both alloys no increase in porosity was observed after the stress-relief heat treatment [130]. It was found that spray-formed Ti-48.9Al was essentially texture free, although isothermal forging to a 77% reduction resulted in texture development [131]. Upon subsequent heat treatment at 1030 °C it was observed that the as-forged texture became "sharper" with the maximum intensity increasing from 3.3 to 3.8.

As is the case for all γ-TiAl materials, microstructure and alloy composition play a very important role in determining the properties of spray-formed material. Fracture toughness has been investigated by Johnson *et al.* [132] who determined a fracture toughness of around 18 MPa m$^{1/2}$ for Ti-48Al-2Mn-2Nb. Compressive creep has been studied by Li *et al.* [133] and Li and Lavernia [127]. Selected room-temperature tensile values for various alloys and conditions are given in Table 15.2. The reader is referred to the relevant papers for extended processing details. At 800 °C a tensile elongation of 120% has been determined for γ-TAB in a spray-formed + forged + relieved condition [130]. Such material may be suitable for superplastic forming at temperatures around 1000 °C [23].

Table 15.2 Room-temperature tensile properties of some spray-formed TiAl alloys.

Alloy	Condition	Microstructure	$\sigma_{0.2}$ (MPa)	σ_f (MPa)	$\varepsilon_{plastic}$ (%)	Ref.
Ti-48.9Al	Spray formed	Near lamellar	378	466	1.3	[124, 130]
Ti-48.9Al	Spray formed	Near lamellar	441	484	0.6	[123]
Ti-48.9Al	Spray formed + forge + SR	Near-gamma	488	517	2.0	[130]
γ-TAB	Spray formed	Duplex	609	658	0.7	[124]
γ-TAB	Spray formed	Duplex	563	586	0.5	[123]
γ-TAB	Spray formed + forge + SR	Near-gamma	–	691	0.1	[130]
Ti-48Al-2Cr-2Nb	Spray formed	Near lamellar	477	527	0.6	[123]

SR = stress relieved at 1030 °C/2 h; γ-TAB: Ti-47Al-1.5Nb-1Cr-1Mn-0.5B-0.2Si.

15.1.5.5 Sheet/Foil Production through (i) HIP of Cast Tapes and (ii) Liquid-Phase Sintering

The manufacture of TiAl sheets through isothermal hot rolling of ingot or prealloyed powder-based materials is covered in Chapter 16. However, other so-called less-expensive ways to produce sheet from prealloyed powder, that do not include hot rolling, have also been investigated. In particular HIPing of cast tapes [134, 135] and liquid-phase sintering of powder using high-power infrared heating have been studied [136]. Both methods involve making a slurry containing prealloyed powder, as described in refs. [134–136]. The slurry is subsequently cast into thin slabs that are allowed to dry to form a "green" tape.

(i) HIPing of Cast Tapes The dimensions of the cast tapes that were put inside the HIP cans were around 5×2 cm with a thickness of around 600 μm [134]. This relatively small size was only limited by the size of the equipment used in the work. Through the use of a heating schedule, the binder was removed from the "green" tape within the HIP can (under flowing argon). After binder removal, the HIP can was allowed to cool slowly to room temperature. Prior to being sealed, the HIP cans were heated to 300 °C under vacuum conditions [135]. The earlier work presented by Rahaman et al. [134] employed PREP powder, whereas the later work by Adams et al. [135] used gas-atomized powder. In both papers HIPing was performed at a temperature of 1100 °C for 15 min at 130 MPa pressure. Heating and cooling rates of around 20 K/min were employed in both cases. The thin sheets (foils between 250 and 300 μm in thickness) that were produced (grain size ≈ 3.2 μm) were removed from the mild-steel HIP cans using acid dissolution followed by oxidation at 700 °C to remove a protective tantalum foil interlayer. The carbon content of the foils had increased by around 0.04 wt.% as compared to the starting powder (0.09 wt.%), but oxygen levels were significantly higher at 0.44 wt.% compared to 0.08 wt.% in the powder. This contamination was believed to have taken place during removal of the tantalum foil as oxygen levels in the sintered material were only 0.02 wt.% higher those in the powder (0.08 wt.%). For completeness, it should be mentioned that Rahaman et al. [134] worked on an alloy of composition Ti-49.2Al-2.6Nb-0.3Ta while Adams et al. [135] worked on Ti-46.6Al-2.2Nb-1.3Cr-0.3Mo-0.2B-0.3C.

(ii) Liquid-Phase Sintering The work of Rivard et al. [136] involved the use of cast tapes of composition Ti-48Al-2Cr-2Nb that were originally 122 cm × 10.2 cm and 0.75 mm in thickness, which were subsequently cut into smaller 10.2 × 10.2 cm squares. These were then stacked on top of one another, heated (1 h at 85 °C) and then pressed together (using 3.94 MPa pressure) for 15 min so that a 4-ply laminate was formed. The binder was removed and sintering of both the as-cast tape and also the laminated material performed using heat-treatment schedules that included holding at 1000 °C for 1 h under vacuum conditions (6.7×10^{-2} mbar). High-intensity infrared processing of both the processed tape and laminate was conducted within a water-cooled, atmosphere-controlled, box; equipped with a 750-kW tungsten plasma arc lamp. The processing conditions used were based on modeling work that was described within the same paper [136]. The resulting through-thickness microstructure was found to depend on the processing time and radiation intensity.

For given processing conditions the modeling predicted three distinct layers for the processed cast tape, with the top layer being heated to a temperature above the liquidus temperature, while the bottom layer remained in the sintered state. Such layers were observed within real material processed under similar conditions. A fine-grained layer was observed between the top and bottom layers. This is believed to result from material in this region reaching temperatures within the mushy zone, that is, between the solidus and liquidus temperatures. Similar layered microstructures were also found in the processed laminate material. Modeling indicated that it could be possible to adjust the processing conditions so that only two-layer microstructures are formed, although in reality problems with bubble formation were encountered [136].

A direct comparison of sheet/foils made using these novel techniques with isothermally rolled material has not been made. Certainly, for thin sheets/foils where heat losses are large, the difficulties associated with rolling are significant. Thus, these powder-based methods may in this respect be useful. However, the extent of contamination, porosity and microstructural variation may limit the mechanical properties attainable.

15.1.5.6 Spark Sintering

Spark sintering is a generic name that covers a number of processes such as plasma-assisted sintering, pulsed electric-current sintering, electroconsolidation (electric-pulse-assisted consolidation) and spark plasma sintering [137]. According to Matsugi *et al.* [138] the sintering techniques can be classified into three groups according to the period of current flow. Resistance heating employs time periods of $10–10^4$ s, pulse sintering $10^{-3}–10$ s and electric-discharge sintering $10^{-6}–10^{-3}$ s. Such methods have the potential to achieve rapid densification with the advantage of minimal grain growth. The basic principle of the technique involves the simultaneous application of current through compacted powder under vacuum conditions. During spark plasma sintering heating rates as high as 1000 K/min are reported to be achievable [137]. For further information concerning this type of consolidation, the reader is referred to a comprehensive review article written by Munir *et al.* [137].

With regard to γ-TiAl, work has been presented on both the sintering of prealloyed and elemental powders. "Self-propagating high-temperature synthesis" of elemental powders to produce "prealloyed" material that was subsequently ground for use in spark sintering has been described [138, 139]. The use of mechanical alloying to produce material subsequently consolidated to TiAl using spark sintering is described in [140–142]. With regard to the sintering of "conventional" prealloyed powder (i.e. made from the atomization of molten alloy), only a limited number of papers have been published [143, 144]. In both studies different types of microstructure, including fine duplex and fully lamellar could be developed according to the processing temperature and alloy composition. While the mechanical properties reported in ref. [143] were determined from bending tests, Couret *et al.* [144] performed real tensile and creep testing. Naturally, the properties were determined primarily by microstructure. For a duplex microstructure, room-temperature tensile tests showed up to 2.5% plastic elongation combined with yield and fracture stresses of around 570 MPa and 690 MPa, respectively. The lamellar microstructure showed 0.6% plastic elongation with yield and fracture stresses of 450 MPa and 560 MPa. Under creep conditions of 700 °C/300 MPa the lamellar microstructure showed a minimum creep rate of $10^{-8} s^{-1}$, which was a factor of around 20 slower than that measured for a duplex microstructure [144]. From the results obtained, Couret *et al.* [144] believe that such spark sintering techniques (they used spark plasma sintering) may be promising for the production of near-net-shape TiAl components for structural applications. This technique may offer the possibility of upscaling and shows the best properties of the "exotic" powder processing methods. It should however be stressed that further work needs to be performed in this area to verify the attractive properties and to investigate their variability.

15.1.6
Summary

Section 15.1 covers the various methods employed to make prealloyed γ-TiAl powder and discuss each technique's advantages and disadvantages. At present, gas-atomized powder seems to be the most readily available. However, if in the future it is shown that the incorporation of argon within powder leads to a degradation of properties after exposure to normal component operating temperatures, then this situation may change. It has already been shown that thermally induced porosity can develop at extremely high temperatures. Additionally, isothermally rolled sheet made from gas-atomized prealloyed powder has inferior superplastic properties to that made from ingot, see Chapter 16.

The room-temperature mechanical properties of conventionally hot-worked powder-based material can be comparable to those of ingot-based materials when processing is optimized. An evaluation of prealloyed powder materials is given in ref. [145]. The great advantage of powder over ingot is the much higher chemical and macrostructural homogeneity, also in hot-worked material. Although it has not been demonstrated to date through the use of methods such as Weibull statistics, this should lead to better property reproducibility and thus improved component predictability and reliability; all of which are very important aspects for critical components. Das [86] has shown better properties and fewer defects in a large forged disc made from powder as compared to a disc made from ingot of a similar composition. Through near-net processing, powder also offers the possibility of increasing the yield of material in final products, which obviously has economic implications. However, any decisions by component manufactures and end-users made regarding powder against ingot will depend on a multitude of technical, economical and supply-chain factors. Self-propagating high-temperature synthesis of prealloyed powder seems an interesting area worthy of further development.

15.2
Elemental-Powder Technology

The production of TiAl from elemental Ti, Al and other minor alloying elements (in the form of powders) can be achieved using processing methods such as reactive sintering or mechanical alloying. It is generally believed that these methods led to material with inferior mechanical properties compared to material made using prealloyed powders or ingot. Thus, these techniques will only be discussed briefly, the reader being directed to the relevant papers for further detailed information.

15.2.1
Reactive Sintering

The basic reactive sintering process uses titanium and aluminum powders that are mixed (preferably under vacuum) in a specific ratio to give the desired alloy composition [146]. Ternary elements such Cr, Si, Mn, V, Nb etc. or nonmetallics such as TiB_2 may also be added at this stage for the production of multicomponent alloys [147]. After subsequent compaction into the desired shape, the compacted material is heated to a temperature where a reaction between the elemental components takes place and intermetallic phases are formed. According to Oddone and German [146] during the heating of such material to 1000 °C, the aluminum melts at 660 °C and then reacts highly exothermically with titanium. Accordingly, this can result in a self-sustaining reaction with the temperature exceeding the melting point of TiAl. Additionally, the enhanced diffusion and capillary reaction between the melt and unmelted particles is likely to result in good densification. Apart from such liquid-phase sintering, heat treatment may alternatively be performed at temperatures below 660 °C in what is called solid-state sintering [147, 148]. However, as diffusion is required for phase formation and is comparatively very slow in solid-state sintering, liquid-phase sintering is the preferred technique [147]. Reactive sintering has the advantage that low-cost materials are used and that the unsintered compacts can be easily deformed on conventional equipment at low temperatures into near-net shapes, thus overcoming problems associated with the limited workability of γ-TiAl.

Oddone and German [146] report that early Russian work performed on reactive sintering of TiAl showed swelling of the compacts after sintering that was possibly associated with both the cracking of an intermetallic layer around the titanium particles and Kirkendall porosity. Later work by Wang et al. [147, 149, 150] has investigated phase formation and porosity development during sintering. This work was generally performed on cold-compacted material that was subsequently worked by cold extrusion. However, cold-compaction by swaging of mixed powders within a tube followed by sintering under an external pressure has also been reported [151]. Heavily cold extruded material results in better quality material after sintering, as the starting material is more homogeneous [147, 152].

Among others [153–155], Wang/Dahms et al. [147, 149, 156, 157] have discussed the reactions that can take place during sintering. The formation of a series of nonequilibrium intermetallic phases during the sintering process was identified. The exact nature of the heating cycle also has an influence on the phases present after sintering. Yang et al. [148] have found that the amount of aluminum remaining after solid-state sintering during hot pressing was inversely proportional to the applied pressure. A 45 MPa pressure combined with sintering for 22 h at 630 °C was necessary to completely eliminate aluminum from green compacts made from powder with particles sizes between 44 and 149 µm. "Dense" ($\gamma + \alpha_2$) material was produced after sintering at 630 °C for 22 h (45 MPa pressure applied for the first 5 h) followed by 2 h at 1250 °C under a pressure of 45 MPa [148]. During liquid-phase sintering, $TiAl_3$ is the first phase to form at titanium-aluminum interfaces, followed by $TiAl_2$ between the titanium/$TiAl_3$ interfaces. Further reactions leading to the formation of TiAl and Ti_3Al also take place. However, at thermodynamic equilibrium only the TiAl (γ) and Ti_3Al phases should be present. For further details on phase formation the reader is referred to the above-mentioned literature.

As mentioned above, Kirkendall porosity can form in sintered material. This arises due to the large difference in the diffusion coefficients between titanium and aluminum; more aluminum diffusing into titanium particles within the compact as titanium into aluminum. The extent of porosity thus depends on many factors, including diffusion distances, sintering temperature and type of sintering. The sintering process can be carried out either under vacuum conditions (so-called "pressureless" sintering) or under the application of external pressure. Wang and Dahms have investigated the dependence of porosity on extrusion ratio and sinter temperature [149, 150]. As can be seen in Figure 15.19, for "pressureless" sintering the level of porosity decreases significantly as the sinter temperature is increased; the effect is particularly marked for low-extrusion ratios [150]. For solid-state "pressureless" sintering at 600 °C, Kirkendall porosity was seen to develop mostly at the aluminum/titanium interfaces. It is believed that the pore size is roughly the size of the aluminum fibers that developed during extrusion. Thus, increasing the extrusion ratio should result in a decrease of pore size, which as seen in Figure 15.20 was indeed the case. Further details are given in [150].

The effect of sintering extruded material under HIP conditions of 1350 °C/4 h/ 200 MPa (without encapsulation for cost reduction) followed by "pressureless" sintering was also investigated [150]. The application of isostatic pressure during the sintering process suppresses the Kirkendall porosity that develops and can thus reduce the overall porosity level [149]. Additionally, "pressureless" sintering followed by HIP was also studied [150]. The results are shown in Figure 15.21 and indicate that HIP sintering results in reduced porosity compared to "pressureless" sintering, particularly for material extruded to lower reduction ratios. For an

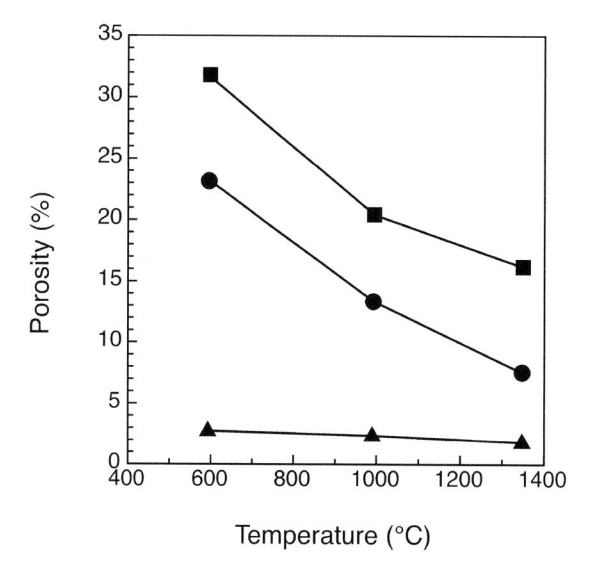

Figure 15.19 Variation of the porosity after "pressureless" sintering for 6 h at 600 °C, 1000 °C and 1350 °C for material extruded to reduction ratios of 350 (▲), 25 (●) and 17 (■) [150]. It can be seen that high extrusion ratios and increasing the sintering temperature significantly reduces the porosity, in particular for material deformed to lower reductions. Replotted data.

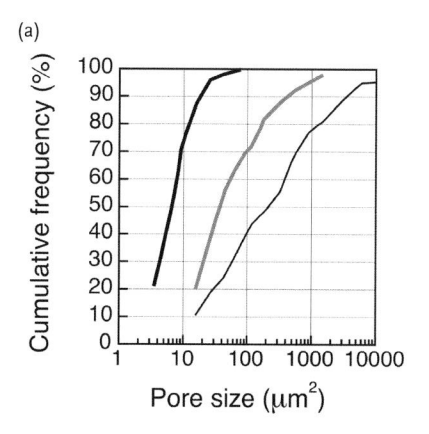

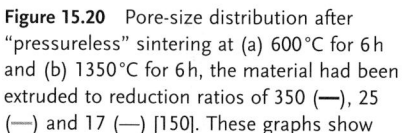

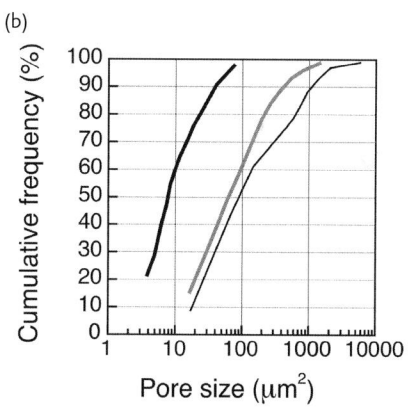

Figure 15.20 Pore-size distribution after "pressureless" sintering at (a) 600 °C for 6 h and (b) 1350 °C for 6 h, the material had been extruded to reduction ratios of 350 (—), 25 (—) and 17 (—) [150]. These graphs show that the size of porosity decreases as the extrusion ratio is increased. This occurs as a result of the microstructural refinement of more heavily deformed material. Redrawn data.

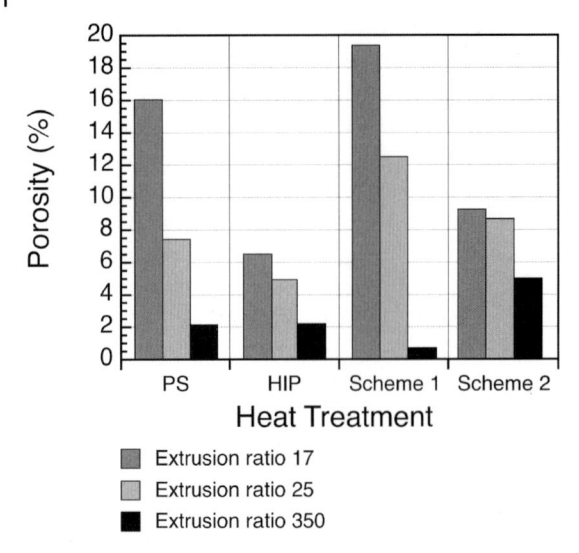

Figure 15.21 Porosity of specimens given different sintering treatments [150]. PS = "pressureless" sintering = 1350 °C/6 h. HIP = HIP sintering = 1350 °C/4 h under 200 MPa pressure. Scheme 1 = PS + HIP. Scheme 2 = HIP + PS. The diagram shows that HIP sintering results in reduced porosity compared to "pressureless" sintering, especially for lightly extruded material. For heavily extruded material scheme 1 results in lower porosity than scheme 2. For less heavily extruded material the case is reversed. Redrawn data.

extrusion ratio of 350, the final porosity was the same for both HIP and "pressureless" sintered material. For this extrusion ratio, a sintering scheme involving "pressureless" sintering followed by HIPing was found to result in nearly pore-free material (scheme 1). However, this was not the case for extrusion ratios of 17 and 25, here HIPing followed by "pressureless" sintering (scheme 2) resulted in lower porosity levels. In this work, the effect of heating rate during sintering was not studied, however, other work on an extruded TiAl-Mn alloy has shown that higher heating rates resulted in reduced porosity levels [158].

As with conventionally melted TiAl alloys, contamination, especially with oxygen is an important issue for mechanical properties. The oxygen content of reactive sintered material depends on both the quality of the starting material and the atmosphere under which processing is performed. Wang et al. [159] report oxygen levels of 3300 ppm (given as 3300 μg/g in [160]) when mixing, compaction and extrusion are performed in air, although through the use of a degassing procedure levels of around 800 μg/g are attainable [160]. Yang et al. [148] report extremely low postsinter (hot-pressing) oxygen levels of between 100 to 450 ppm.

With regard to microstructure, Wang et al. [159] investigated the effect of both composition and heat treatment on microstructure and properties in binary alloys. Not surprisingly, the microstructure depends on both the aluminum content and the maximum temperature to which the reactive sintered material is exposed.

Depending on subsequent heat-treatment temperature, either "dual phase" or duplex microstructures were developed. Due to the inhomogeneous distribution of elemental titanium and aluminum in the cold-extruded material before sintering, equilibrium microstructures were not obtained. "Dual-phase" microstructures consisting of seemingly fully lamellar islands within a near-gamma (or possibly duplex) matrix were presented by Morgenthal et al. [161] for Ti-48Al-2Cr-2Nb HIP sintered at 1250 °C/2 h/200 MPa. Even after a heat treatment of 1320 °C/1 h + 1000 °C/10 h, gamma (or duplex) areas were observed in the microstructure between lamellar colonies [161].

As explained earlier, the advantage of reactive sintering is that after mixing of the powders, near-net-shape components can easily be formed as the material is very workable prior to sintering. In this way, the difficult-to-work nature of TiAl can be overcome. Since the mid-1990s a number of papers have been published that have used mixed elemental powders and canned hot extrusion for consolidation [162–164]. In such processing the reactive sintering step takes place either before hot extrusion or before/during hot extrusion. This is not the same as the earlier method where cold extrusion of the mixed powder compacts is more or less dissociated from the sintering process. In this more recently developed simultaneous reactive sintering/extrusion processing, it has been shown that indirect extrusion of the mixed powders leads to much smaller grain sizes than when direct extrusion is employed [165]. After hot extrusion, further heat treatment can be applied to further promote phase stabilization and establish the required microstructure that depends on heat treatment and composition. One problem that has been observed in reactive sintered material made from elemental powder mixtures, is the presence of only partially reacted refractory particles such as Mo [166] and Nb. Such particles can lead to the local stabilization of nonequilibrium phases such as β and can also act as sites for the development of Kirkendall porosity [166]. In this context, the use of refractory-alloyed powder is advantageous as the refractory is not in its elemental form. The use of such alloyed aluminum powder has also been shown to reduce the extent of porosity in reactively sintered cold extruded material [167]. This is believed to result from hardening of the aluminum alloy and the resulting finer microstructure of the extruded material compared to when unalloyed elements are used [150, 167].

15.2.1.1 Mechanical Properties of Reactive Sintered Material

The tensile properties of reactive sintered material are generally believed to be inferior to those developed in material made in other ways. To some extent this is certainly a consequence of the porosity that is almost unavoidable in sintered material and the much higher contamination with oxygen. Data presented by Wang et al. [159] indicate tensile fracture stresses of around 200 MPa in Ti-43Al and 160 MPa in Ti-47Al. These values of fracture stress are much lower (around 50% less) than the yield stresses presented by Austin and Kelly [88] for cast Ti-(44–48)Al-2Cr-2Nb alloys. Later work presented by Wang and Dahms [147] indicates improved room-temperature elongations of up to 0.6% combined with yield stresses of up to 320 MPa in reaction-sintered Ti-48Al.

The mechanical properties of extruded presintered/extrusion-sintered material have been presented in several publications [162–164, 168–170]. The tensile properties presented are generally extraordinarily good, especially tensile ductility, with room-temperature elongations of up to 3.8% combined with a yield strength of 481 MPa being reported [162]. Such levels of ductility are significantly higher than those measured in material sintered after cold extrusion, and are comparable with some values determined for conventionally extruded ingot or prealloyed powder material. Such processing is therefore extremely interesting for the production of low-cost components with acceptable mechanical properties. Unfortunately, no other major groups working on TiAl have employed such processing within their research and therefore worldwide confirmation of such good properties remains lacking.

15.2.1.2 Manufacture of Reactively Sintered Components/Parts

The main advantage of conventional reactive sintering is the possibility of forming near-net-shape parts prior to the formation of intermetallic phases. For example, sheet/foil has been made using two variations of reactive sintering. Wang and Dahms [149] proposed extrusion of mixed elemental powders followed by cold rolling to sheet/foil prior to reactive sintering. Morgenthal *et al.* [171] describe a similar technique. An alternative way to produce foils, which includes cold roll bonding of thin (around 0.070 mm to 0.075 mm thick) alternating foils of aluminum and titanium followed by reactive sintering, has been investigated by a group from the University of Alabama [172–175]. Unfortunately apart from hardness [172, 175], mechanical properties such as tensile strength of the sintered material were not determined in these studies. The processing of a niobium-containing ternary alloy sheet, through the incorporation of niobium foil is described in [176].

Schneider *et al.* [177] describe the manufacture of automotive engine valves. Their technique involved double cold extrusion of cold-compacted mixed elemental powders followed by forging and machining to near-net-shape geometry, prior to reactive sintering and final machining. Although some heat-treated conditions gave rise to material with totally brittle properties, a heat-treatment of 1400 °C for 0.3 h was reported to lead to material with some ductility ($\approx$0.6% elongation) and strength (around 350 MPa).

15.2.2
Summary

Reactive sintering of elemental components is an easy way to produce intermetallic alloy from relatively cheap starting materials and does not require the use of highly specialized equipment such as an isothermal forge. There is a great benefit from being able to work the compacted mixed powders into near-net-shape components before the formation of difficult to work intermetallic phases. The near-net shape of the components reduces the amount of machining required, and thus end cost. Although it is possible to reduce the formation of porosity through optimized processing, the mechanical properties always seem to be compromised by remnant porosity and high interstitial contamination. The use of reactive sintering either prior to, or during hot extrusion, has been shown to result in remarkable mechanical properties but has only been investigated/demonstrated by a limited number of groups.

15.3
Mechanical Alloying

Mechanical alloying was originally developed to produce oxide dispersion strengthened superalloys, based on nickel- and iron-alloy systems. Further development has extended its range of usage, so that synthesis of a variety of equilibrium and nonequilibrium material from prealloyed powders or blended elements is possible. For a comprehensive overview on mechanical alloying and milling the reader is referred to Suryanarayana [178]. With regard to TiAl, little work has been performed in the recent past, primarily because of the absence of any room-temperature ductility in the processed material. Thus, this section will only briefly discuss the topic.

TiAl-based material has been made via mechanical alloying of elemental materials and also prealloyed alloyed powders. Hashimoto *et al.* [179] found that the production of mechanical-alloyed TiAl was drastically improved when elemental powders were milled within a nitrogen-containing atmosphere. After vacuum sintering, the milled powder showed a nanometer-grain-sized microstructure. In order to reduce the final costs, the same group later performed mechanical alloying of relatively large (mm-sized) Ti sponge and recycled Al chips combined with pulse discharge sintering [180].

Mechanical alloying of prealloyed powders has been performed at GKSS [181–183]. The work performed was based on mechanical alloying of prealloyed Ti-48.9Al and Ti-37.5Si powders as well as pure silicon powder. Consolidation was achieved by a HIP treatment at temperatures ranging from 750 °C to 1000 °C, and resulted in fine nanometer-grain-sized material. In room-temperature compression, a Ti-45Al-2.4Si alloy with a grain size of 170 nm showed brittle behavior with a fracture strength of around 2680 MPa [181]. For larger grain sizes, ranging from 194 to 390 nm, ductile behavior was observed, with compressive ductility ranging from about 3.5 to 7.2%, respectively. The highest reported fracture stress was 2930 MPa for a Ti-46Al-5Si alloy with an average grain size of 160 nm. The elevated-temperature properties of such submicrometer-grained TiAl materials are presented in [182]. It was shown that the materials became very soft at temperatures above 500 °C, and that tensile strains of over 175% were obtained for a Ti-45Al-2.4Si alloy at 800 °C. Indeed, strain-rate sensitivities of up to 0.48 were reported at 800 °C. The goal was to be able to produce consolidated nanometer grain-sized material that could easily be formed to component shapes using conventional hot-working procedures. Such material was successfully produced [183]. Although not stated in the publications, the ultimate aim was to further process this material through heat treatment in order to establish conventional TiAl microstructures and the associated mechanical properties. This important last step was unfortunately never performed.

References

1. Ting, J., Peretti, M.W., and Eisen, W.B. (2002) *Mater. Sci. Eng.*, **A326**, 110.
2. Ting, J., and Anderson, I.E. (2004) *Mater. Sci. Eng.*, **A379**, 264.
3. See, J.B., and Johnston, G.H. (1978) *Powder Technol.*, **21**, 119.
4. See, J.B., Runkle, J.C., and King, T.B. (1973) *Metall. Trans.*, **4**, 2669.
5. Dube, R.K., Koria, S.C., and Subramanian, R. (1988) *Powder Metall. Int.*, **20** (6), 14.
6. Singer, A.R.E., Coombs, J.S., and Leatham, A.G. (1974) *Modern Developments in Powder Metallurgy*, vol. 8 (eds H.H. Hausner and W.E. Smith), MPIF, Princeton, NJ, p. 263.
7. Singh, D., Koria, S.C., and Dube, R.K. (2001) *Powder Metall.*, **44** (2), 177.
8. Lawley, A. (1992) *Atomisation – The Production of Metal Powders*, MPIF, Princeton, NJ.
9. Lavernia, E.J., and Wu, Y. (1996) *Spray Atomisation and Deposition*, John Wiley & Sons, Ltd, Chichester, UK.
10. Leiblich, M., Caruana, G., Torralba, M., and Jones, H. (1996) *Mater. Sci. Technol.*, **12**, 25.
11. Subramanian, C., Mishra, P., and Suri, A.K. (1995) *Int. J. Powder Metall.*, **31** (2), 137.
12. Uslan, T., Saritas, S., and Davies, T.J. (1999) *Powder Metall.*, **42** (2), 157.
13. Unal, A. (1987) *Mater. Sci. Technol.*, **3**, 1029.
14. Lubanska, H. (1970) *JOM*, **22** (Feb), 45.
15. Whitton, T. (2005) *BONEZone*, **Winter**, 97.
16. Pilch, M., and Erdman, C.A. (1987) *Int. J. Multiphase Flow*, **13**, 741.
17. Anderson, I.E., and Terpstra, R.L. (2002) *Mater. Sci. Eng.*, **A326**, 101.
18. Nichiporenko, O.S., and Naida, Y.I. (1968) *Soviet Powder Metall. Met. Ceram.*, **67**, 509.
19. Fuchs, G.E., and Hayden, S.Z. (1992) *Mater. Sci. Eng.*, **A152**, 277.
20. Gerling, R., Schimansky, F.P., and Wagner, R. (1993) *Titanium '92 Science and Technology* (eds F.H. Froes and I.L. Caplan), TMS, Warrendale, PA, p. 1025.
21. Gerling, R., Schimansky, F.P., and Wagner, R. (1992) *Advances in Powder Metallurgy & Particulate Materials – 1992, Vol. 1 (Powder Production and Spray Forming)* (eds J.M. Capus and R.M. German), MPIF, Princeton, NJ, p. 215.
22. Gerling, R., Schimansky, F.P., and Wagner, R. (1994) *Progress in atomising high melting intermetallic titanium based alloys by means of a novel plasma melting induction guiding gas atomisation facility (PIGA)* (ed. D. François), Les editions de physique, Les Ulis, France, p. 387.
23. Gerling, R., Clemens, H., and Schimansky, F.P. (2004) *Adv. Eng. Mater.*, **6**, 23.
24. Loretto, M. (2007) Private e-mail communication.
25. Yolton, C.F., Kim, Y.-W., and Habel, U. (2003) *Gamma Titanium Aluminides 2003* (eds Y.-W. Kim, H. Clemens, and A.H. Rosenberger), TMS, Warrendale, PA, p. 233.
26. Yolton, C.F. (1990) *PM in Aerospace and Defence Technologies*, vol. 1 (ed. F.H. Froes), MPIF, Princeton, NJ, p. 123.
27. Suryanarayana, C., Froes, F.H., and Rowe, R.G. (1991) *Int. Mater. Rev.*, **36** (3), 85.
28. Hohmann, M., and Ludwig, N. (1991) *Einrichtung zum Herstellen von Pulvern aus Metallen*, German Patent DE 4102 101 A1, Jan. 1991.
29. Semiatin, S.L., Chesnutt, J.C., Austin, C., and Seetharaman, V. (1997) *Structural Intermetallics 1997* (eds M.V. Nathal, R. Darolia, C.T. Liu, P.L. Martin, D.B. Miracle, R. Wagner, and M. Yamaguchi), TMS, Warrendale, PA, p. 263.
30. Seetharaman, V., and Semiatin, S.L. (2002) *Intermetallic Compounds, Vol. 3,*

31 Roberts, P.R. (1989) *Advances in Powder Metallurgy*, vol. 3 (eds T.G. Gasbarre and W.F. Jandeska), MPIF, Princeton, NJ, p. 427.
32 Kaufmann, A.R. (1974) *Production of pure, spherical powders*, U.S. Patent 3,802,816, April 1974.
33 Kaufmann, A.R. (1963) *Method and apparatus for making powder*, U.S. Patent 3,099,041, July 1963.
34 Klar, E., Roberts, P.R., Fox, C.W., Patterson, R.J., and Ray, R. (1993) *Atomization, ASM Handbook (Formerly 9th Edition, Metals Handbook, Vol. 7, Powder Metallurgy)* (ed. E. Klar) (Coordinator), ASM International, Metals Park, OH, p. 25.
35 Champagne, B., and Angers, R. (1980) *Int. J. Powder Metall. Powder Technol.*, **16**, 359.
36 Champagne, B., and Angers, R. (1981) Modern developments in powder metallurgy. *Proceedings of the 1980 International Powder Metallurgy Conference, Washington, DC*, vol. 12 (eds H. Hausner, H. Antes, and G. Smith), MPIF, Princeton, NJ, p. 83.
37 Tokizane, M., Fukami, T., and Inaba, T. (1991) *ISIJ Int.*, **31**, 1088.
38 Osborne, N.R., Eylon, D., and Froes, F.H. (1989) *Advances in Powder Metallurgy*, vol. 3 (eds T.G. Gasbarre and W.F. Jandeska), MPIF, Princeton, NJ, p. 213.
39 Cai, X.Z., and Eylon, D. (1996) *Titanium '95: Science and Technology* (eds P.A. Blenkinsop, W.J. Evans, and H.M. Flower), IOM, London, UK, p. 455.
40 Habel, U., Yolton, C.F., and Moll, J.H. (1999) *Gamma Titanium Aluminides 1999* (eds Y.-W. Kim, D.M. Dimiduk, and M.H. Loretto), TMS, Warrendale, PA, p. 301.
41 Peng, T.C., London, B., and Sastry, S.M.L. (1989) *Advances in Powder Metallurgy*, vol. 3 (eds T.G. Gasbarre and W.F. Jandeska), MPIF, Princeton, NJ, p. 387.
42 Holiday, P.R., and Patterson, R.J. (1978) *Apparatus for producing metal powder*, U.S. Patent 4,078,873, Mar. 1978.
43 Larson, D.J., Liu, C.T., and Miller, M.K. (1999) *Mater. Sci. Eng.*, **A270**, 1.
44 Li, H., and Tsakiropoulos, P. (1998) *Int. J. Non-Equilibrium Process*, **11**, 55.
45 Ho, K.H., and Zhao, Y.Y. (2004) *Mater. Sci. Eng.*, **A365**, 336.
46 Ullrich, W.J., Frock, H.N., Berg, R.H., Pao, M.A., Hubbard, J.L., and Parsons, D.S. (1993) *Particle Size and Size Distribution, ASM Handbook (Formerly 9th Edition, Metals Handbook, Vol. 7, Powder Metallurgy)* (ed. E. Klar) (Coordinator), ASM International, Metals Park, OH, p. 214.
47 Wegmann, G., Gerling, R., and Schimansky, F.P. (2003) *Acta Mater.*, **51**, 741.
48 Zhao, Y.Y., Jacobs, M.H., and Dowson, A.L. (1998) *Metall. Mater. Trans.*, **29B**, 1357.
49 Graves, J.A., Perepezko, J.H., Ward, C.H., and Froes, F.H. (1987) *Scr. Metall.*, **21**, 567.
50 McCullough, C., Valencia, J.J., Levi, C.G., and Mehrabian, R. (1989) *Acta Metall.*, **37** (5), 1321.
51 Hall, E.L., and Huang, S.C. (1990) *Acta Metall. Mater.*, **38**, 539.
52 Gerling, R., Schimansky, F.P., and Wagner, R. (1993) *Materials by Powder Technology. Ptm'93, Vol. 5 (Superalloys and Intermetallics)* (ed. F. Aldinger), DGM, Oberursel, Germany, p. 379.
53 Habel, U., Das, G., Yolton, C.F., and Kim, Y.-W. (2003) *Gamma Titanium Aluminides 2003* (eds Y.-W. Kim, H. Clemens, and A.H. Rosenberger), TMS, Warrendale, PA, p. 297.
54 Choi, B.W., Deng, Y.G., McCullough, C., Paden, B., and Mehrabian, R. (1990) *Acta Metall. Mater.*, **38**, 2225.
55 McCullough, C., Valencia, J.J., Levi, C.G., and Mehrabian, R. (1990) *Mater. Sci. Eng.*, **A124**, 83.
56 Valencia, J.J., McCullough, C., Levi, C.G., and Mehrabian, R. (1989) *Acta Metall.*, **37**, 2517.
57 Tönnes, C., Rösler, J., Baumann, R., and Thumann, M. (1993) *Structural Intermetallics* (eds R. Darolia, J.J. Lewandowski, C.T. Liu, P.L. Martin, D.B. Miracle, and M.V. Nathal), TMS, Warrendale, PA, p. 241.

58 Gerling, R., Leitgeb, R., and Schimansky, F.P. (1998) *Mater. Sci. Eng.*, **A252**, 239.

59 Clemens, H., Kestler, H., Eberhardt, N., and Knabl, W. (1999) *Gamma Titanium Aluminides 1999* (eds Y.-W. Kim, D.M. Dimiduk, and M.H. Loretto), TMS, Warrendale, PA, p. 209.

60 Appel, F., Clemens, H., Glatz, W., and Wagner, R. (1997) *High Temperature Ordered Intermetallic Alloys VII, MRS Symposium Proceedings*, vol. 460 (eds C.C. Koch, C.T. Liu, N.S. Stoloff, and A. Wanner), MRS, Warrendale, PA, p. 195.

61 Eylon, D., Schwenker, S.W., and Froes, F.H. (1985) *Metall. Trans.*, **16A**, 1526.

62 Rabin, B.H., Smolik, G.R., and Korth, G.E. (1990) *Mater. Sci. Eng.*, **A124**, 1.

63 Shong, D.S., Kim, Y.-W., Yolton, C.F., and Froes, F.H. (1989) *Advances in Powder Metallurgy*, vol. 3 (eds T.G. Gasbarre and W.F. Jandeska), MPIF, Princeton, NJ, p. 359.

64 Murray, J.L. (1992) *Binary Alloy Phase Diagrams*, vol. 2, 2nd edn (eds T.B. Massalski, H. Okamoto, P.R. Subramanian, and L. Kacprzak), ASM International, Materials Park, OH, p. 1785.

65 Zhang, G., Blenkinsop, P.A., and Wise, M.L.H. (1996) *Titanium '95: Science and Technology* (eds P.A. Blenkinsop, W.J. Evans, and H.M. Flower), IOM, London, UK, p. 542.

66 Huang, A., Hu, D., Loretto, M.H., Mei, J., and Wu, X. (2007) *Scr. Mater.*, **56**, 253.

67 Habel, U., and McTiernan, B.J. (2004) *Intermetallics*, **12**, 63.

68 Zhang, G., Blenkinsop, P.A., and Wise, M.L.H. (1996) *Intermetallics*, **4**, 447.

69 Ashby, M.F. (1987) The modelling of hot isostatic pressing. *Proc. Lulea Conf. on HIPing, University of Lulea, Sweden 1987*, p. 29.

70 Choi, B.W., Marschall, J., Deng, Y.G., McCullough, C., Paden, B., and Mehrabian, R. (1990) *Acta Metall. Mater.*, **38**, 2245.

71 Abondance, D., Dellis, C., Baccino, R., Bernier, F., Moret, F., De Monicault, J.M., Guichard, D., and Stutz, P. (1996) *Titanium '95: Science and Technology* (eds P.A. Blenkinsop, W.J. Evans, and H.M. Flower), IOM, London, UK, p. 2634.

72 Yuan, W.X., Mei, J., Samarov, V., Seliverstov, D., and Wu, X. (2007) *J. Mater. Process. Technol.*, **182**, 39.

73 Clemens, H., Eberhardt, N., Glatz, W., Martinz, H.P., Knabl, W., and Reheis, N. (1997) *Structural Intermetallics 1997* (eds M.V. Nathal, R. Darolia, C.T. Liu, P.L. Martin, D.B. Miracle.R. Wagner, and M. Yamaguchi), TMS, Warrendale, PA, p. 277.

74 Semiatin, S.L., Cornish, G.R., and Eylon, D. (1994) *Mater. Sci. Eng.*, **A185**, 45.

75 Fuchs, G.E. (1997) *Metall. Mater. Trans.*, **28A**, 2543.

76 Beddoes, J., Zhao, L., Immarigeon, J.P., and Wallace, W. (1994) *Mater. Sci. Eng.*, **A183**, 211.

77 Beddoes, J., Zhao, L., and Wallace, W. (1994) *Mater. Sci. Eng.*, **A184**, L11.

78 Beddoes, J., Zhao, L., Au, P., and Wallace, W. (1995) *Mater. Sci. Eng.*, **A192–193**, 324.

79 Wegmann, G., Gerling, R., Schimansky, F.P., Clemens, H., and Bartels, A. (2002) *Intermetallics*, **10**, 511.

80 Gerling, R., Schimansky, F.P., and Clemens, H. (2003) *Gamma Titanium Aluminides 2003* (eds Y.-W. Kim, H. Clemens, and A.H. Rosenberger), TMS, Warrendale, PA, p. 249.

81 Dimiduk, D.M., Martin, P.L., and Dutton, R. (2003) *Gamma Titanium Aluminides 2003* (eds Y.-W. Kim, H. Clemens, and A.H. Rosenberger), TMS, Warrendale, PA, p. 15.

82 Gouma, P.I., and Loretto, M.H. (1996) *Titanium '95: Science and Technology* (eds P.A. Blenkinsop, W.J. Evans, and H.M. Flower), IOM, London, UK, p. 550.

83 Loretto, M.H., Hu, D., and Godfrey, A. (1997) *High Temperature Ordered Intermetallics VII, MRS Symposium Proceedings*, vol. 460 (eds C.C. Kock, C.T. Liu, N.S. Stoloff, and A. Wanner), MRS, Warrendale, PA, p. 127.

84 Godfrey, A., Hu, D., and Loretto, M.H. (1997) *Mater. Sci. Eng.*, **A239**, 559.

85 Gerling, R., Clemens, H., Schimansky, F.P., and Wegmann, G. (2001)

85 *Structural Intermetallics 2001* (eds K.J. Hemker, D.M. Dimiduk, H. Clemens, R. Darolia, H. Inui, J.M. Larsen, V.K. Sikka, M. Thomas, and J.D. Whittenberger), TMS, Warrendale, PA, p. 139.
86 Das, G. (2006) Presented at the 3rd International Workshop on γ-TiAl Technologies, 29th to 31st of May 2006, Bamberg, Germany.
87 Paul, J.D.H., Oehring, M., Hoppe, R., and Appel, F. (2003) *Gamma Titanium Aluminides 2003* (eds Y.-W. Kim, H. Clemens, and A.H. Rosenberger), TMS, Warrendale, PA, p. 403.
88 Austin, C.M., and Kelly, T.J. (1993) *Structural Intermetallics* (eds R. Darolia, J.J. Lewandowski, C.T. Liu, P.L. Martin, D.B. Miracle, and M.V. Nathal), TMS, Warrendale, PA, p. 143.
89 Fuchs, G.E. (1995) *High Temperature Ordered Intermetallic Alloys VI, MRS Symposium Proceedings*, vol. 364 (eds J. Horton, I. Baker, S. Hanada, R.D. Noebe, and D.S. Schwartz), MRS, Warrendale, PA, p. 799.
90 Liu, C.T., Maziasz, P.J., Clemens, D.R., Schneibel, J.H., Sikka, V.K., Nieh, T.G., Wright, J., and Walker, L.R. (1995) *Gamma Titanium Aluminides* (eds Y.-W. Kim, R. Wagner, and M. Yamaguchi), TMS, Warrendale, PA, p. 679.
91 Roberts, P.R., and Ferguson, B.L. (1991) *Int. Mater. Rev.*, **36** (2), 62.
92 Kim, Y.-W. (1992) *Acta Metall. Mater.*, **40**, 1121.
93 Kim, Y.-W. (1989) *JOM*, **41** (7), 24.
94 Kim, Y.-W. (1994) *JOM*, **46** (7), 30.
95 Huang, S.C. (1993) *Structural Intermetallics* (eds R. Darolia, J.J. Lewandowski, C.T. Liu, P.L. Martin, D.B. Miracle, and M.V. Nathal), TMS, Warrendale, PA, p. 299.
96 Eylon, D., Cooke, C.M., Yolton, C.F., Nachtrab, W.T., and Furrer, D.U. (1993) *Plansee Seminar '93* (eds H. Bildstein and R. Eck), Plansee, Reutte, Austria, p. 552.
97 Clemens, H., Glatz, W., Schretter, P., Yolton, C.F., Jones, P.E., and Eylon, D. (1995) *Gamma Titanium Aluminides* (eds Y.-W. Kim, R. Wagner, and M. Yamaguchi), TMS, Warrendale, PA, p. 555.
98 Fuchs, G.E. (1993) *High Temperature Ordered Intermetallic Alloys V, MRS Symposium Proceedings*, vol. 288 (eds I. Baker, R. Darolia, J.D. Whittenberger, and M.H. Yoo), MRS, Pittsburgh, PA, p. 847.
99 Thomas, M., Raviart, J.L., and Popoff, F. (2005) *Intermetallics*, **13**, 944.
100 Malaplate, J., Thomas, M., Belaygue, P., Grange, M., and Couret, A. (2006) *Acta Mater.*, **54**, 601.
101 Kruth, J.P., Van De Schueren, B., Bonse, J.E., and Morren, B. (1996) *Ann. CIRP*, **45** (1), 183.
102 Moll, J.H., and McTiernan, B.J. (2000) *Metal Powder Rep.*, **55** (1), 18.
103 Moll, J.H., Whitney, E., Yolton, C.F., and Habel, U. (1999) *Gamma Titanium Aluminides 1999* (eds Y.-W. Kim, D.M. Dimiduk, and M.H. Loretto), TMS, Warrendale, PA, p. 255.
104 Srivastava, D., Chang, I.T.H., and Loretto, M.H. (1999) *Gamma Titanium Aluminides 1999* (eds Y.-W. Kim, D.M. Dimiduk, and M.H. Loretto), TMS, Warrendale, PA, p. 265.
105 Srivastava, D., Chang, I.T.H., and Loretto, M.H. (2001) *Intermetallics*, **9**, 1003.
106 Qu, H.P., and Wang, H.M. (2007) *Mater. Sci. Eng.*, **A466**, 187.
107 Zhang, X.D., Brice, C., Mahaffey, D.W., Zhang, H., Schwendner, K., Evans, D.J., and Fraser, H.L. (2001) *Scr. Mater.*, **44**, 2419.
108 Srivastava, D., Hu, D., Chang, I.T.H., and Loretto, M.H. (1999) *Intermetallics*, **7**, 1107.
109 Loretto, M.H., Horspool, D., Botten, R., Hu, D., Li, Y.G., Srivastava, D., Sharman, R., and Wu, X. (2002) *Mater. Sci. Eng.*, **A329–331**, 1.
110 Keicher, D.W., and Miller, W.D. (1998) *Metal Powder Rep.*, **53** (12), 26.
111 Abbott, D.H., and Arcella, F.G. (1998) *Metal Powder Rep.*, **53** (2), 24.
112 Lewis, G.K., and Schlinger, E. (2000) *Mater. Design*, **21**, 417.
113 German, R.M., and Bose, A. (1997) *Injection Moulding of Metals and Ceramics*, MPIF, Princeton, NJ.
114 Erickson, A.R. (1993) *Injection Moulding, Metals Handbook*, vol. 7, 5th

edn (ed. E. Klar) (Coordinator), ASM International, Metals Park, OH, p. 495.
115 Terauchi, S., Teraoka, T., Shinkuma, T., Sugimoto, T., and Ahida, Y. (2001) Development of production technology by metallic powder injection molding for TiAl-type intermetallic compound with high efficiency. *Proceedings of the 15th Int. Plansee Seminar 2001* (eds G. Kneringer, P. Rödhammer, and H. Wildner), (Plansee, Reutte, Austria, 2001), p. 610.
116 Katoh, K., and Masumoto, A. (1995) Powder metallurgy of Ti-Al intermetallic compounds by injection molding. *Proceedings of the 6th Symposium on High Performance Materials for Severe Environments, Tokyo*, 49.
117 Shimizu, T., Kitajemin, A., Kato, K., and Sao, T. (2000) *Proceedings of 2000 Powder Metallurgy World Congress* (eds K. Kosuge and H. Nagai), Japan Society of Powder and Powder Metallurgy, Kyoto, Japan, p. 292.
118 Gerling, R., Aust, E., Limberg, W., Pfuff, M., and Schimansky, F.P. (2006) *Mater. Sci. Eng.*, **A423**, 262.
119 Gerling, R., and Schimansky, F.P. (2002) *Mater. Sci. Eng.*, **A329–331**, 45.
120 Oehring, M., Lorenz, U., Appel, F., and Roth-Fagaraseanu, D. (2001) *Structural Intermetallics 2001* (eds K.J. Hemker, D.M. Dimiduk, H. Clemens, R. Darolia, H. Inui, J.M. Larsen, V.K. Sikka, M. Thomas, and J.D. Whittenberger), TMS, Warrendale, PA, p. 157.
121 Müllauer, J., Appel, F., and Wagner, R. (1997) *Proc. of the 5th European Conf. on Adv. Materials and Processes and Applications (Euromat '97), Vol. 1 (Metals and Composites)* (eds L.A.J.L. Sarton and H.B. Zeedijk), Netherlands Society for Materials Science, Zwijndrecht, NL, p. 1/167.
122 Dowson, A., Jacobs, M.H., Young, J.M., and Chen, W. (1995) *Gamma Titanium Aluminides* (eds Y.-W. Kim, R. Wagner, and M. Yamaguchi), TMS, Warrendale, PA, p. 483.
123 Gerling, R., Liu, K., and Schimansky, F.P. (1999) *Gamma Titanium Aluminides 1999* (eds Y.-W. Kim, D.M. Dimiduk, and M.H. Loretto), TMS, Warrendale, PA, p. 273.
124 Schimansky, F.P., Liu, K.W., and Gerling, R. (1999) *Intermetallics*, **7**, 1275.
125 Liu, K.W., Gerling, R., and Schimansky, F.P. (1999) *Scr. Mater.*, **40**, 601.
126 Young, J.M., Jacobs, M.H., Duggan, M., and Dowson, A.L. (1996) *Titanium '95: Science and Technology* (eds P.A. Blenkinsop, W.J. Evans, and H.M. Flower), IOM, London, UK, p. 2641.
127 Li, B., and Lavernia, E.J. (1997) *Structural Intermetallics 1997* (eds M.V. Nathal, R. Darolia, C.T. Liu, P.L. Martin, D.B. Miracle, R. Wagner, and M. Yamaguchi), TMS, Warrendale, PA, p. 331.
128 Lavernia, E.J., Ayers, J.D., and Srivasta, T.S. (1992) *Int. Mater. Rev.*, **37**, 1.
129 Gerling, R., Schimansky, F.P., Wegmann, G., and Zhang, J.X. (2002) *Mater. Sci. Eng.*, **A326**, 73.
130 Wegmann, G., Gerling, R., Schimansky, F.P., and Zhang, J.X. (2002) *Mater. Sci. Eng.*, **A329–331**, 99.
131 Staron, P., Bartels, A., Brokmeier, H.G., Gerling, R., Schimansky, F.P., and Clemens, H. (2006) *Mater. Sci. Eng.*, **A416**, 11.
132 Johnson, T.P., Jacobs, M.H., Ward, R.M., and Young, J.M. (1993) *Proceedings of the Second International Conference on Spray Forming* (ed. J.V. Wood), Woodhead Publishing Ltd., Cambridge, p. 183.
133 Li, B., Wolfenstine, J., Earthman, J.C., and Lavernia, E.J. (1997) *Metall. Mater. Trans.*, **28A**, 1849.
134 Rahaman, M.N., Dutton, R.E., and Semiatin, S.L. (2003) *Mater. Sci. Eng.*, **A360**, 169.
135 Adams, A.G., Rahaman, M.N., and Dutton, R.E. (2008) *Mater. Sci. Eng.*, **A477**, 137.
136 Rivard, J.D.K., Sabau, A.S., Blue, C.A., Harper, D.C., and Kiggans, J.O. (2006) *Metall. Mater. Trans.*, **37A**, 1289.
137 Munir, Z.A., Anselmi-Tamburini, U., and Ohyanagi, M. (2006) *J. Mater. Sci.*, **41**, 763.
138 Matsugi, K., Ishibashi, N., Hatayama, T., and Yanagisawa, O. (1996) *Intermetallics*, **4**, 457.
139 Matsugi, K., Hatayama, T., and Yanagisawa, O. (1999) *Intermetallics*, **7**, 1049.

140 Sun, Z.M., Wang, Q., Hashimoto, H., Tada, S., and Abe, T. (2003) *Intermetallics*, **11**, 63.
141 Sun, Z.M., Hashimoto, H., Tada, S., and Abe, T. (2003) *Gamma Titanium Aluminides 2003* (eds Y.-W. Kim, H. Clemens, and A.H. Rosenberger), TMS, Warrendale, PA, p. 349.
142 Calderon, H.A., Garibay-Febles, V., Umemoto, M., and Yamaguchi, M. (2002) *Mater. Sci. Eng.*, **A329–331**, 196.
143 Kothari, K., Radhakrishnan, R., Wereley, N.M., and Sudarshan, T.S. (2007) *Powder Metallurgy*, **50** (1), 21.
144 Couret, A., Molénat, G., Galy, J., and Thomas, M. (2007) *Advanced Intermetallic-Based Alloys, MRS Symposium Proceedings*, vol. 980 (eds J. Wiezorek, C.L. Fu, M. Takeyama, D. Morris, and H. Clemens), MRS, Warrendale, PA, p. 389.
145 Zhao, L., Beddoes, J., Au, P., and Wallace, W. (1997) *Adv. Perform. Mater.*, **4**, 421.
146 Oddone, R.R., and German, R.M. (1989) *Advances in Powder Metallurgy*, vol. 3 (eds T.G. Gasbarre and W.F. Jandeska), MPIF, Princeton, NJ, p. 475.
147 Wang, G.X., and Dahms, M. (1993) *Structural Intermetallics* (eds R. Darolia, J.J. Lewandowski, C.T. Liu, P.L. Martin, D.B. Miracle, and M.V. Nathal), TMS, Warrendale, PA, p. 215.
148 Yang, J.B., Teoh, K.W., and Hwang, W.S. (1997) *Mater. Sci. Technol.*, **13**, 695.
149 Wang, G.X., and Dahms, M. (1993) *JOM*, **45** (5), 52.
150 Wang, G.X., and Dahms, M. (1993) *Metall. Trans.*, **24A**, 1517.
151 Taguchi, K., and Ayada, M. (1995) *Gamma Titanium Aluminides* (eds Y.-W. Kim, R. Wagner, and M. Yamaguchi), TMS, Warrendale, PA, p. 619.
152 Leitner, G., Dahms, M., Jaenicke-Rössler, K., Schultrich, S., and Wang, G.X. (1994) Reaction sintering of titanium aluminides at different contact areas Ti/Al. *Proc. 1994 Powder Metallurgy World Congress*, vol. 2 (ed. D. François) Les editions de physique, Les Ulis, France, p. 1229.
153 van Loo, F.J.J., and Rieck, G.D. (1973) *Acta Metall.*, **21**, 61.
154 van Loo, F.J.J., and Rieck, G.D. (1973) *Acta Metall.*, **21**, 73.
155 Shibue, K., Kim, M.S., and Kumagai, M. (1991) *Proc. Int. Symp. on Intermetallic Compounds* (eds O. Izumi, M. Kikuchi, and N. Honjo), JIM, Sendai, Japan, p. 833.
156 Bohnenkamp, U., Wang, G.A., Jewett, T.J., and Dahms, M. (1994) *Intermetallics*, **2**, 275.
157 Dahms, M., Jewett, T.J., and Michaelsen, C. (1997) *Z Metallkd.*, **88**, 125.
158 Yang, S.H., Kim, W.Y., and Kim, M.S. (2003) *Intermetallics*, **11**, 849.
159 Wang, G.X., Dahms, M., and Dogan, B. (1992) *Scr. Metall. Mater.*, **27**, 1651.
160 Dahms, M. (1994) *Adv. Perform. Mater.*, **1**, 157.
161 Morgenthal, I., Neubert, X., and Kieback, B. (1994) Reactive HIP-sintering of extruded Ti48Al2Cr2Nb elemental powder mixture. *Proc. 1994 Powder Metallurgy World Congress*, vol. 2, (ed. D. François) Les editions de physique, Les Ulis, France, p. 1259.
162 Lee, I.S., Hwang, S.K., Park, W.K., Lee, J.H., Park, D.H., Kim, H.M., and Lee, Y.T. (1994) *Scr. Metall. Mater.*, **31**, 57.
163 Lee, T.K., Mosunov, E.I., and Hwang, S.K. (1997) *Mater. Sci. Eng.*, **A239–240**, 540.
164 Cho, H.S., Nam, S.W., Hwang, S.K., and Kim, N.J. (1997) *Scr. Mater.*, **36**, 1295.
165 Wu, Y., Park, Y.W., Park, H.S., and Hwang, S.K. (2003) *Mater. Sci. Eng.*, **A347**, 171.
166 Kim, J.K., and Hwang, S.K. (1998) *Scr. Mater.*, **39**, 1205.
167 Shibue, K. (1991) *Sumitomo Light Met. Tech. Rep.*, **32** (2), 95.
168 Wu, Y., Hwang, S.K., Nam, S.W., and Kim, N.J. (2003) *Gamma Titanium Aluminides 2003* (eds Y.-W. Kim, H. Clemens, and A.H. Rosenberger), TMS, Warrendale, PA, p. 177.
169 Kim, J.K., Kim, J.H., Lee, T.K., Hwang, S.K., Nam, S.W., and Kim, N.J. (1999) *Gamma Titanium Aluminides 1999* (eds Y.-W. Kim, D.M. Dimiduk, and M.H. Loretto), TMS, Warrendale, PA, p. 231.
170 Park, H.S., Park, K.L., and Hwang, S.K. (2002) *Mater. Sci. Eng.*, **A329–331**, 50.

171 Morgenthal, I., Kieback, B., Hübner, G., and Nerger, D. (1994) Preparation of Ti-Al-foils by roll compaction of elemental powders. *Proc. 1994 Powder Metallurgy World Congress, vol. 2 (Les editions de physique, Les Ulis, France)*, p. 1247.

172 Luo, J.G., and Acoff, V.L. (1999) *Gamma Titanium Aluminides 1999* (eds Y.-W. Kim, D.M. Dimiduk, and M.H. Loretto), TMS, Warrendale, PA, p. 331.

173 Chaudhari, G.P., and Acoff, V.L. (2003) *Gamma Titanium Aluminides 2003* (eds Y.-W. Kim, H. Clemens, and A.H. Rosenberger), TMS, Warrendale, PA, p. 287.

174 Luo, J.G., and Acoff, V.L. (2004) *Mater. Sci. Eng.*, **A379**, 164.

175 Luo, J.G., and Acoff, V.L. (2006) *Mater. Sci. Eng.*, **A433**, 334.

176 Zhang, R., and Acoff, V.L. (2007) *Mater. Sci. Eng.*, **A463**, 67.

177 Schneider, D., Jewett, T., Gente, C., Segtrop, K., and Dahms, M. (1997) *Structural Intermetallics 1997* (eds M.V. Nathal, R. Darolia, C.T. Liu, P.L. Martin, D.B. Miracle, R. Wagner, and M. Yamaguchi), TMS, Warrendale, PA, p. 453.

178 Suryanarayana, C. (2001) *Progr. Mater. Sci.*, **46**, 1.

179 Hashimoto, H., Abe, T., and Sun, Z.M. (2000) *Intermetallics*, **8**, 721.

180 Sun, Z.M., Wang, Q., Hashimoto, H., Tada, S., and Abe, T. (2003) *Intermetallics*, **11**, 63.

181 Bohn, R., Klassen, T., and Bormann, R. (2001) *Acta. Mater.*, **49**, 299.

182 Bohn, R., Klassen, T., and Bormann, R. (2001) *Intermetallics*, **9**, 559.

183 Fanta, G., Bohn, R., Dahms, M., Klassen, T., and Bormann, R. (2001) *Intermetallics*, **9**, 45.

16
Wrought Processing

Wrought processing is a standard processing practice for many metals that is used to shape the material and to control its microstructure and properties. For a rather brittle material like γ titanium aluminide alloys, wrought processing appears to be particularly useful to achieve the full potential of these alloys due to the necessity to improve the mechanical properties through microstructural manipulation. Reliability of the mechanical properties with respect to minimum values has to be guaranteed for any structural application, and this is certainly supported by increasing the resulting microstructural homogeneity and elimination of casting defects. Furthermore, casting seems to be restricted to a maximum component size because quality problems increase with the size of the cast parts. The production of large γ titanium aluminide components will therefore probably require processing routes involving powder or wrought processing.

Despite the obvious benefits of wrought processing, the development of hot working practices has only slowly emerged since the first work in the late 1980s [1–9]. On the one hand this has been caused by the specific working behavior of ordered intermetallic alloys, which include limited ductility and susceptibility to cleavage fracture up to high temperatures, significant plastic anisotropy making the material prone to flow localization, and slow recrystallization kinetics. In addition, γ titanium aluminide ingot material usually suffers from pronounced segregation of alloying elements, shrinkage porosity, coarse microstructures, and texture; all of which further reduce the workability. Additionally, conventional thermomechanical processing equipment cannot be used for primary processing of most γ titanium aluminide alloys and suitable industrial equipment for isothermal or canned processing is not readily available. This is noteworthy because γ titanium aluminide alloys are often rather complex multiphase alloys and much empirical work has to be expended to explore the relationships between composition, processing conditions, microstructure and resulting properties.

The difficulties associated in the development of wrought-processing routes has stimulated extensive research over the last 15 years covering a vast range of fundamental and applied aspects, as reviewed in several publications [10–20]. To date, the research efforts have resulted in the development of complete hot-working routes, which involve a variety of wrought operations including for example

Gamma Titanium Aluminide Alloys: Science and Technology, First Edition. Fritz Appel, Jonathan David Heaton Paul, Michael Oehring.
© 2011 Wiley-VCH Verlag GmbH & Co. KGaA. Published 2011 by Wiley-VCH Verlag GmbH & Co. KGaA.

rolling, die forging and superplastic forming that have matured to a state close to that of routine industrial processes.

16.1
Flow Behavior under Hot-Working Conditions

16.1.1
Flow Curves

As already mentioned, the workability of γ titanium aluminide alloys is principally low and, thus, cold or warm working operations cannot be applied even if the microstructures are fine and uniform. For conventional materials the hot-working range is usually defined by a homologous transition temperature of 0.6 T_m, with T_m being the absolute melting temperature [21]. For γ TiAl alloys this is around 800 °C and is roughly equivalent to the brittle–ductile transition temperature T_{bd}. However, most γ TiAl alloys can only be successfully worked at significantly higher temperatures. The range of temperatures and strain rates suitable for successful hot-working operations is usually determined through compression testing, which is considered to be a standard type of bulk workability test [22]. Although in compression testing the average stress state is similar to that in many bulk working processes [23], it has to be noted that the stress state during testing sensitively depends on the geometry of the specimen and the friction between the specimen and the die [23]. Therefore, some care should be taken, if the results of such tests are to be compared, or even applied to industrial forging operations.

An appreciable number of investigations on hot-compression behavior can be found in the literature where cylindrical specimens with a volume in the cubic centimeter range have been deformed. In some of these studies the use of different lubricants was reported [1, 24–29]. Through such testing, flow curves can be determined from which valuable information with respect to the required industrial press capacity, possible die materials and the deformation mechanisms that occur during hot working can be obtained. Such investigations can be utilized to establish workability maps, which define a range of conditions where sound, crack-free deformation can be achieved. Equally important, the dependence of strain, strain rate and temperature on the evolution of the microstructure has been characterized [7, 30–34].

Nobuki and Tsujimoto [7] have presented one of the most extensive studies in this area. This research work significantly contributed to a better understanding of the hot-working behavior of γ titanium aluminide alloys, primarily because the authors investigated the influence of Al content on the hot-working behavior in binary alloys, which is of great importance regarding the development of wrought alloys. The authors observed that all flow curves exhibited a peak in the flow stress followed by flow softening to an ostensibly constant stress level at a true strain above approximately 0.6. This behavior is characteristic of dynamic recrystallization and has also been found in numerous other investigations [3, 4, 6, 11, 25, 30,

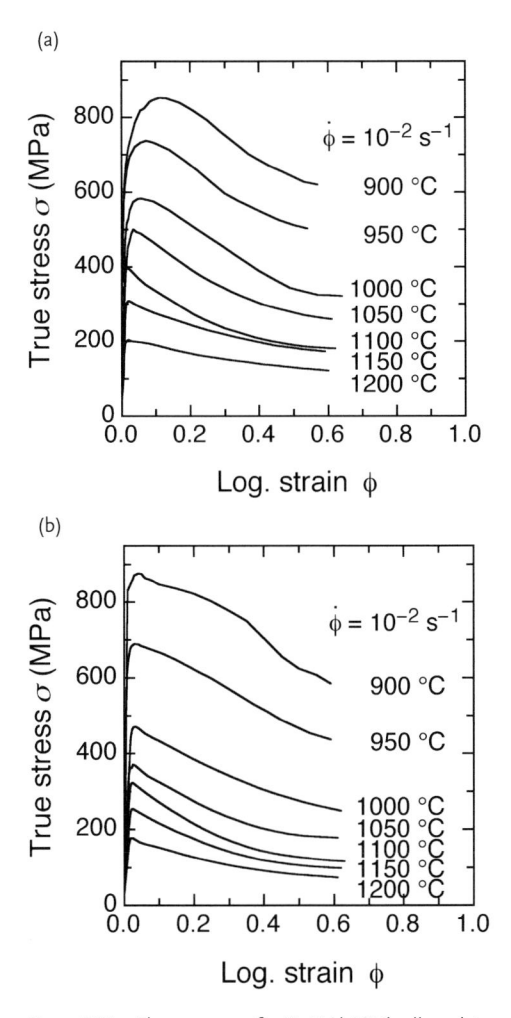

Figure 16.1 Flow curves of a Ti-45Al-10Nb alloy, determined using cylindrical specimens of 18 mm diameter and 30 mm height tested under the conditions indicated. (a) Ingot material, (b) extruded material.

31, 33–35]. Figure 16.1 shows flow curves obtained for the alloy Ti-45Al-10Nb in two conditions. As can be seen, the true strain at which the peak in the flow stress occurs depends on the applied deformation conditions, and can be as high as 0.4 [4]. However, flow softening may not only arise due to dynamic recrystallization but might also be caused by adiabatic heating of the specimen, flow localization or kinking and reorientation of lamellae during deformation. Indeed, Semiatin et al. [3, 31] have shown that adiabatic heating contributes to flow softening. The authors performed isothermal hot-compression tests on Ti-48Al-2.5Nb-0.3Ta at a strain rate of $10 \, s^{-1}$ and a temperature of 1235 °C. After deforming the specimens

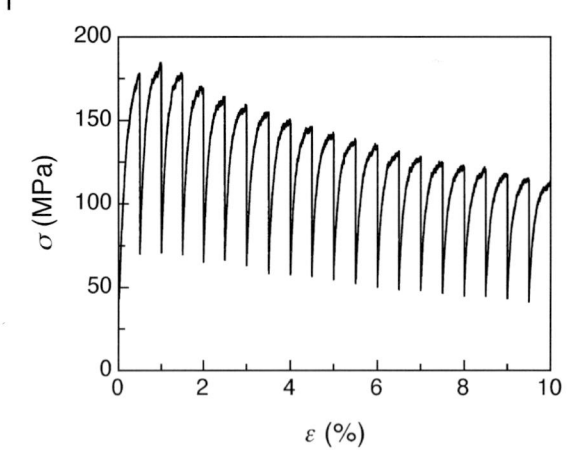

Figure 16.2 Tensile test stress–strain curve for extruded Ti-45Al-8Nb-0.2C material with a duplex microstructure [36]. The testing was conducted using a strain rate of $2.35 \times 10^{-5}\,\mathrm{s}^{-1}$ at a temperature of 1000 °C. The test was periodically interrupted for at least 5 min to measure the relaxation kinetics.

to a strain beyond the peak strain the tests were stopped for a certain period to allow dissipation of deformation-generated heat. Subsequently, straining was continued, which resulted in a steep rise of the stress to a value similar to the peak stress. The resulting stress increase approximated to the value calculated from the temperature difference between deformation heating and the nominal test temperature [3]. Accordingly, at the high strain rate applied, which roughly corresponds to that achieved in industrial rolling or extrusion processes, adiabatic heating played a major role with respect to flow softening. However, flow softening was also observed for much lower strain rates where one would not expect significant adiabatic heating. Figure 16.2 shows a tensile curve obtained for Ti-45Al-8Nb-0.2C using a strain rate of $2.35 \times 10^{-5}\,\mathrm{s}^{-1}$ [36]. Deformation was repeatedly interrupted for at least 5 min to conduct strain relaxation tests. Subsequently, constant strain-rate deformation was continued. The resulting stress–strain curve clearly exhibits flow softening. In this case adiabatic heating could not have caused flow softening due to the low strain rate and the long dwell periods. Since it has been shown in several investigations that dynamic recrystallization is initiated around the peak strain [12, 37, 38] flow softening in this case seems to have occurred due to dynamic recrystallization. In this context, it is certainly interesting to note that Fröbel and Appel [39] observed repeated recrystallization/work-hardening cycles during hot-torsion testing of a Ti-46.5Al-5Nb alloy as shown in Figure 16.3. Such oscillations have been found on many metals [40–42] and unambiguously prove that dynamic recrystallization results in flow softening. However, according to Imayev et al. [38] the relationship between flow softening and dynamic recrystallization depends on several factors. The authors conducted hot-compression tests on different binary and multicomponent alloys using identi-

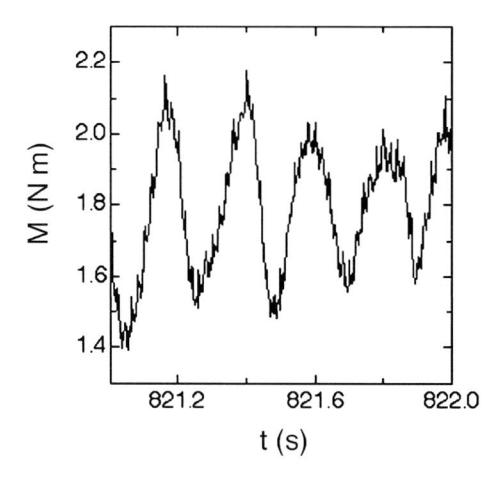

Figure 16.3 Dynamic recrystallization of Ti-46.5Al-5Nb during torsional deformation at $T = 1320\,°C$ with a torsional shear strain rate of $\dot{\phi} = 2.1 \times 10^{-2}\ \text{s}^{-1}$ [39]. The torsional momentum M displayed against time clearly exhibits oscillations that can be accounted for by multiple recrystallization cycles.

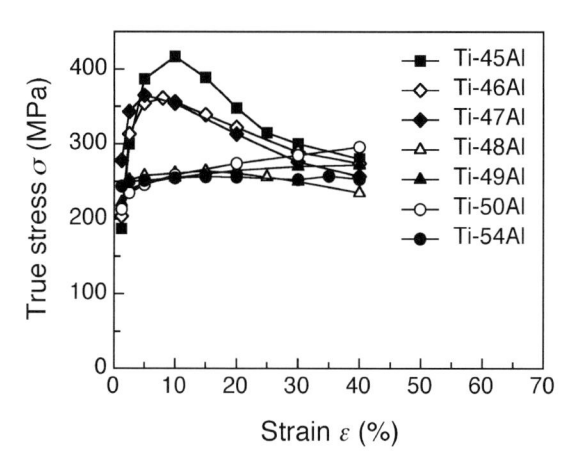

Figure 16.4 Compression testing flow curves obtained for different binary TiAl alloys at $T = 1000\,°C$ and a strain rate $\dot{\varepsilon} = 5 \times 10^{-4}\ \text{s}^{-1}$ [38]. The alloys were prepared by nonconsumable argon-arc melting and subjected to hot-isostatic pressing at $1240\,°C$ prior to testing.

cal specimen geometries and deformation conditions. A rather slow initial strain rate of $5 \times 10^{-4}\,\text{s}^{-1}$ was applied in the tests. Alloys with Al concentrations up to 48 at.% exhibited strong flow softening, whereas for Al concentrations from 50 to 54 at.% no flow softening was observed for strains up to 40% (Figure 16.4). However, other studies have shown that slight flow softening can be expected at higher strains in similar alloys under similar deformation conditions [4, 43]. This behavior can be partly explained by the different initial microstructure of the alloys, as

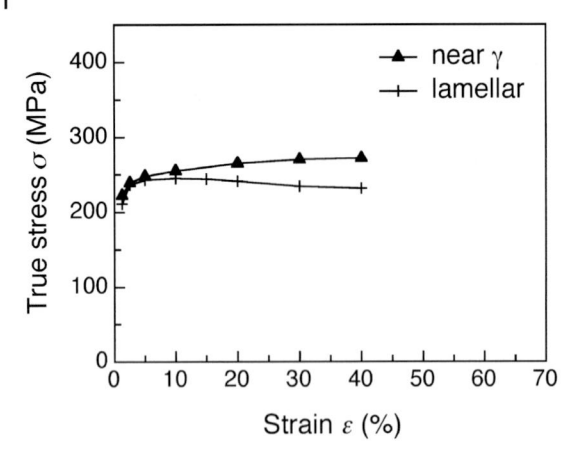

Figure 16.5 Compression testing flow curves obtained for the alloy Ti-49Al at $T = 1000\,°C$ and a strain rate $\dot{\varepsilon} = 5\times 10^{-4}\,\text{s}^{-1}$ [38]. The alloy was prepared by nonconsumable argon-arc melting and subjected to hot-isostatic pressing at $1240\,°C$. The material was tested with a near-γ microstructure that resulted from hot-isostatic pressing and also after an additional heat treatment at $1340\,°C$ that resulted in a lamellar microstructure.

becomes evident by comparing flow curves for Ti-49Al in two different microstructural conditions (Figure 16.5). In the material with a near-γ microstructure weak work hardening was observed, however, the lamellar material showed flow softening. The origin of these results will be discussed in detail in the next section, here it should only be mentioned that lamellar materials are prone to bending and kinking of lamellae, which leads to flow localization and reorientation of the lamellae. Seetharaman and Semiatin [25] have concluded, in agreement with the later finite-element calculations of Schaden et al. [44], that these processes can give rise to strong flow softening, as is observed in Figure 16.4 for the alloys with an Al content below 49 at.% and a lamellar microstructure. Such a behavior is not unique to γ titanium aluminide alloys [22]. For example, Ponge and Gottstein [45] have shown that in Ni_3Al significant flow softening results from dynamic recrystallization due to the localization of deformation in recrystallized necklace regions if these regions form a continuous three-dimensional network through the specimen. In summary, dynamic recrystallization certainly contributes to flow softening in γ titanium aluminide alloys but cannot solely account for it, rather it results from a combination of several factors, involving most notably flow localization that is partly initiated by dynamic recrystallization, and adiabatic heating during deformation.

16.1.2
Constitutive Analysis of the Flow Behavior

In the work of Nobuki and Tsujimoto [7] mentioned above, as well as in many other investigations, a pronounced variation of the peak flow stress with strain rate and temperature was found. This is not surprising considering the deformation behavior at temperatures around the brittle/ductile transition. However, under hot-working conditions, which involve significantly higher temperatures, different deformation mechanisms could become active. For a constitutive analysis of the flow stress variation with strain rate and temperature the empirical equation

$$A(\sinh[\alpha\sigma])^n = \dot{\varepsilon}\exp(Q_{HW}/kT) \tag{16.1}$$

has been found to adequately describe the hot-working behavior of many metals [40, 42, 46–49]. In this equation A, α, the stress exponent n and the apparent activation energy Q_{HW} are material constants. k is the Boltzmann constant. The right-hand side of Equation 16.1 is the Zener–Hollomon parameter

$$Z \equiv \dot{\varepsilon}\exp(Q_{HW}/kT) \tag{16.2}$$

that combines the strain rate and the temperature to a temperature-compensated strain rate [40, 42, 46–49]. For low stresses, $\alpha\sigma < 0.8$, Equation 16.1 reduces to the power law, that is, the Dorn equation

$$Z = A(\sinh[\alpha\sigma])^n \approx A'\sigma^{n'} \tag{16.3}$$

whereas for high stresses, $\alpha\sigma > 1.5$, the exponential law

$$Z = A(\sinh[\alpha\sigma])^n \approx A''\exp(\beta\sigma) \tag{16.4}$$

is approximated.

In several studies in the literature, beginning with the work of Nobuki and Tsujimoto [7], it was found that a constitutive analysis based on Equation 16.1 satisfactorily describes the flow behavior of γ titanium aluminide alloys. Examples are given in Figure 16.6, which shows the variation of the peak stress with the Zener–Hollomon parameter for different ingot and extruded materials, together with curves fitted according to Equation 16.1. In Figure 16.7 the data for one alloy is shown as a function of temperature and strain rate. From the constitutive analysis a set of parameters A, α, n and Q_{HW} was obtained in a number of studies for different alloys. These results are listed in Table 16.1, which is an extended version of the data collected by Kim et al. [34] in a comprehensive study on the high-temperature deformation of γ titanium aluminide alloys. As can be seen in this table, except for the alloy Ti-43.6Al-5V, the stress exponent n and the stress multiplier α do not vary considerably among the investigated cast alloys with average values of $n = 3.4 \pm 0.4$ and $\alpha = (4.5 \pm 0.9) \cdot 10^{-3}\,\mathrm{MPa^{-1}}$. However, the pre-exponential factor A and the apparent activation energy Q_{HW} systematically depend on the Al content. In the study by Kim et al. [34] it was found that Q_{HW} increased linearly with the ratio of the Ti to Al concentration. This dependence results in an activation energy of 3.5 eV for an alloy with equal Ti and Al concentrations. Further, the authors observed that the pre-exponential factor depended exponentially on the apparent activation energy, as has been reported for microalloyed steels of different compositions [34, 55]. Due to this dependence the peak stress for all alloys considered lie on a common curve if they are plotted against the normalized

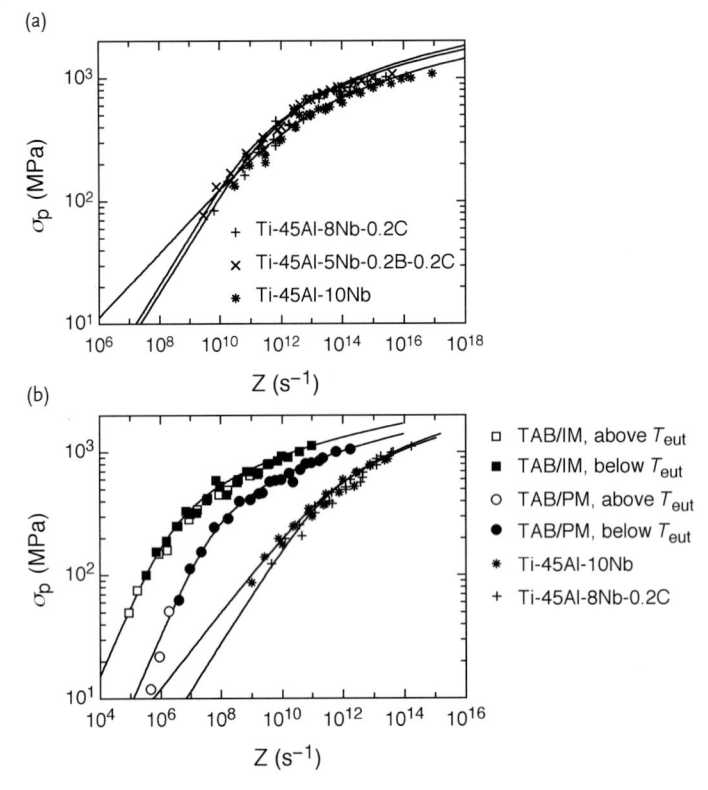

Figure 16.6 (a) Dependence of the peak stress of the compression flow curves (see Figure 16.1) on the Zener–Hollomon parameter $Z = \dot{\phi}\exp(Q_{HW}/kT)$ for ingot material of three TiAl alloys. The apparent activation energy Q_{HW} as well as the fitted curves were determined according to the constitutive Equation 16.1; the resulting activation parameters are given in Table 16.1. (b) Dependence of the peak stress of the compression flow curves (see Figure 16.1) on the Zener–Hollomon parameter $Z = \dot{\phi}\exp(Q_{HW}/kT)$ for three extruded TiAl alloys. The apparent activation energy Q_{HW} as well as the fitted curves were determined according to the constitutive Equation 16.1; the resulting activation parameters are given in Table 16.1. T_{eut}: eutectoid temperature, TAB: Ti-47Al-1.5Nb-1Cr-1Mn-0.2Si-0.5B; IM, PM: the initial materials prior to extrusion had been obtained by hot-isostatic pressing of ingot/powder material. Prior to testing, the compression specimens of the TAB composition and the alloys based on Ti-45Al exhibited near-γ and nearly lamellar microstructures, respectively.

Zener–Hollomon factor Z/A rather than against the Zener–Hollomon factor Z, as shown in Figure 16.6 [34]. In summary, the hot-compression behavior of γ titanium aluminide alloys in either cast or cast and heat-treated conditions can be described by the constitutive Equation 16.1.

It seems interesting to compare the deformation parameters derived by fitting experimental data to Equation 16.1 with those determined directly from creep or constant strain-rate experiments. For the alloy Ti-45Al-8Nb-0.2C, Hoppe and

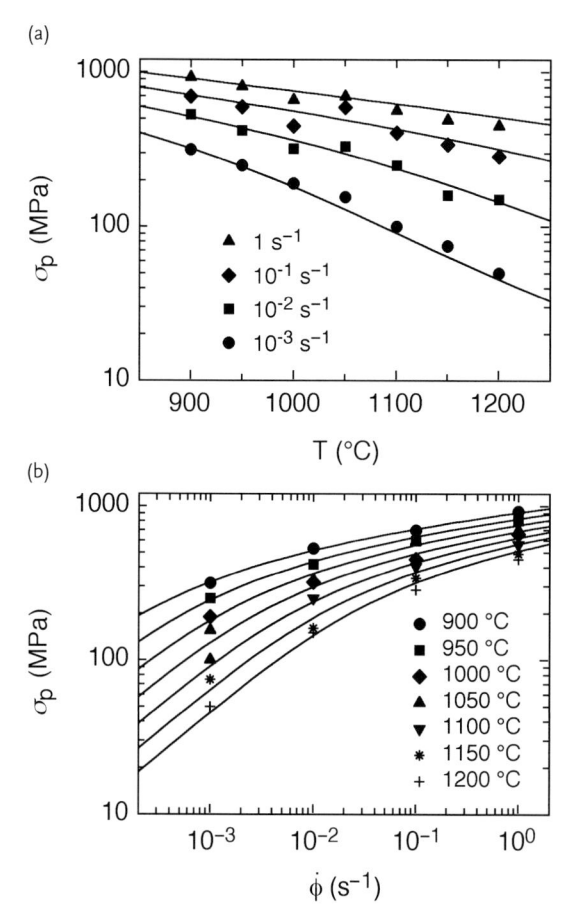

Figure 16.7 (a) Dependence of the peak stress of the compression flow curves (see Figure 16.1) on temperature for extruded TAB alloy (ingot material) with a fine near-γ microstructure. The fitted curves were determined according to the constitutive Equation 16.1; the resulting activation parameters are given in Table 16.1. The same peak flow stress data is shown in Figure 16.6b as function of the Zener–Hollomon parameter and in Figure 16.7b as function of the strain rate. TAB: Ti-47Al-1.5Nb-1Cr-1Mn-0.2Si-0.5B, IM. (b) Dependence of the peak stress of the compression flow curves (see Figure 16.1) on logarithmic strain rate for extruded TAB alloy (ingot material) with a fine near-γ microstructure. The fitted curves were determined according to the constitutive Equation 16.1; the resulting activation parameters are given in Table 16.1. The same peak flow-stress data is shown in Figure 16.6b as function of the Zener–Hollomon parameter and in Figure 16.7a as a function of the temperature.

Appel [36] determined an apparent activation energy of $Q = 3.36\,\text{eV}$ in load-relaxation experiments at 800 °C and the same value in combined strain-rate and temperature-cycling tests. As shown in Table 16.1, the constitutive analysis resulted in a significantly higher apparent activation energy of $Q_{HW} = 3.9\,\text{eV}$ for this alloy. For binary single-phase alloys of compositions Ti-50Al [56], Ti-53.4 Al [57] and

Table 16.1 Activation parameters for hot-deformation behavior as determined from compression tests according to the constitutive Equation 16.1. A: pre-exponential factor, α: stress multiplier, n: stress exponent, Q_{HW}: apparent activation energy.

Composition	Reference	Processing	Initial microstructure	T (°C)	$\dot{\varepsilon}(s^{-1})$	A (s^{-1})	α (10^{-3} MPa^{-1})	n	Q_{HW} (eV)
Ti-43Al	[50]	cast	lamellar	927–1207	7.5×10^{-4}–7.5×10^{-1}	1.67×10^{17}	4.95	2.94	5.5
Ti-47Al-2V	[50]	cast	lamellar	927–1207	7.5×10^{-4}–7.5×10^{-1}	3.60×10^{15}	3.61	3.60	4.8
Ti-51Al	[50]	cast	γ	927–1207	7.5×10^{-4}–7.5×10^{-1}	1.53×10^{13}	4.90	3.63	4.3
Ti-52Al	[50]	cast	γ	927–1207	7.5×10^{-4}–7.5×10^{-1}	6.33×10^{12}	4.56	3.74	4.1
Ti-43.8Al	[7]	cast	lamellar	927–1327	1×10^{-3}–1×10^{-1}	1.41×10^{22}	6.32	3.13	7.0
Ti-44.9Al	[7]	cast	lamellar	927–1327	1×10^{-3}–1×10^{-1}	1.30×10^{16}	4.52	3.74	5.1
Ti-48.2Al	[7]	cast	duplex	927–1327	1×10^{-3}–1×10^{-1}	5.52×10^{10}	4.01	3.70	3.6
Ti-49.5Al	[7]	cast	γ	927–1327	1×10^{-3}–1×10^{-1}	3.68×10^{9}	5.57	3.03	3.4
Ti-50.2Al	[7]	cast	γ	927–1327	1×10^{-3}–1×10^{-1}	1.09×10^{11}	4.79	3.57	3.7
Ti-47Al-1V	[30]	cast	nearly lamellar	1000–1200	1×10^{-3}–1×10^{0}	2.93×10^{13}	3.69	3.8	4.2
Ti-43.6Al-5V	[51]	cast	nearly lamellar	800–1150	3×10^{-4}–1×10^{-1}	2.24×10^{11}	9.67	1.57	3.8
Ti-47Al-2Cr-4Nb	[52]	cast	near-γ	1000–1200	1×10^{-3}–1×10^{-1}	4.26×10^{8}	4.24	3.1	3.1
Ti-48Al-2Cr-2Nb	[9]	cast	duplex	975–1200	3×10^{-3}–1×10^{-1}	2.7×10^{9}	5.38	2.7	3.4
Ti-48Al-2Cr-2Nb	[24]	cast	lamellar	950–1220	1×10^{-4}–5×10^{1}	3.36×10^{13}	6.02	3.0	4.2
Ti-49.5Al-2,5Nb-1.1Mn	[53]	cast	near-γ	1000–1250	1×10^{-3}–1×10^{-1}	8×10^{10}	3.2	3.9	3.4
Ti-46Al-2W	[34]	cast	nearly lamellar	1000–1200	1×10^{-3}–1×10^{-1}	1.31×10^{14}	4.37	3.6	4.7
Ti-48Al-2W	[34]	cast	near-γ	1000–1200	1×10^{-3}–1×10^{-1}	2.35×10^{12}	3.6	3.7	4.1
Ti-45Al-10Nb	[54]	cast	nearly lamellar	900–1200	1×10^{-3}–1×10^{0}	2.98×10^{11}	3.24	3.8	3.9
Ti-45Al-5Nb-0.2B-0.2C	[54]	cast	nearly lamellar	900–1200	1×10^{-3}–1×10^{0}	6.18×10^{10}	4.10	2.6	3.7
Ti-45Al-8Nb-0.2C	[54]	cast	nearly lamellar	900–1200	1×10^{-3}–1×10^{0}	6.67×10^{11}	3.66	2.7	3.9
Ti-45Al-10Nb	[54]	IM, wrought	nearly lamellar	950–1200	1×10^{-3}–1×10^{0}	5.87×10^{10}	2.84	3.3	3.5
Ti-45Al-8Nb-0.2C	[54]	IM, wrought	nearly lamellar	950–1200	1×10^{-3}–1×10^{0}	4.91×10^{10}	3.32	2.6	3.7
Ti-47Al-4.2(Nb,Cr,Mn,Si,B)	[54]	IM, wrought	γ	900–1200	1×10^{-3}–1×10^{0}	6.40×10^{5}	6.51	1.8	2.3
Ti-47Al-4.2(Nb,Cr,Mn,Si,B)	[54]	PM, wrought	γ	900–1250	1×10^{-3}–1×10^{1}	1.58×10^{7}	6.08	1.8	2.6

Ti-49Al [58] Beddoes *et al.* [59] analyzed the creep behavior in the temperature range 707–927 °C on the basis of the Dorn equation and determined an activation energy of 3.2 eV. Again, higher apparent activation energies of 3.4, 3.7, 4.1 and 4.3 eV were obtained for similar alloys by the constitutive analysis of hot-compression tests (Table 16.1). In the vast body of literature regarding the hot-working behavior of metals, apparent activation energies often exceed those determined for creep or diffusion by about 20% [48]. Such an increase was observed if not just dynamic recovery but also dynamic recrystallization was involved in the hot-deformation behavior [48]. As the structural changes resulting from dynamic recrystallization affect the flow stress, an activation energy only is determinable for very rapid temperature changes over incremental strains. In the compression tests, no temperature cycling was applied, but the temperature was varied from specimen to specimen. Since the microstructures may initially differ to some degree for different preheats and evolve at different rates, the determination of activation energy is influenced by dynamical recrystallization and thus depends on the test conditions. This partially explains the considerable scatter in material parameters listed in Table 16.1. Furthermore, it is hard to imagine that deformation behavior of complex multiphase materials like most of the γ titanium aluminide alloys is governed by a single atomic process, thus making the determined apparent activation energies difficult to interpret.

Despite these difficulties, it is tempting to consider the parameters obtained in relation to high-temperature deformation mechanisms. Except for the alloy Ti-43.6Al-5V, the stress exponents for cast materials lie between 2.6 and 3.8 (Table 16.1), which indicates that deformation might be governed by dislocation climb [60–63] in the low-Z regime, where power-law behavior is approximated. For dislocation climb, an activation energy identical to that of self-diffusion is expected [60–63]. Herzig *et al.* [64] measured activation energies of 2.60 eV and 2.58 eV for Ti self-diffusion in Ti-53Al and Ti-56Al, respectively. With respect to the activation energy of Al self-diffusion, the reader is referred to Chapter 3.3, here it need only be mentioned that Mishin and Herzig [65] and Herzig *et al.* [66] determined a value of 3.04 eV in Ti-54Al. Thus, the apparent activation of 3.2 eV mentioned above for creep of binary single-phase γ alloys is in reasonable agreement with the activation energy of diffusion, if it is assumed that diffusion of the slower-moving species, that is, Al, controls the climb rate. As discussed in the preceding paragraph, the higher apparent activation energies determined in rather fast hot-deformation tests could be attributed to dynamic recrystallization. It should be mentioned here that the apparent activation energy obtained through constitutive analysis is only reduced by about 0.2 eV if the temperature-dependent Young's modulus E is considered, that is, if σ is replaced by $\sigma/E(T)$ in Equation 16.1. The temperature dependence of the elastic modulus thus only plays a minor role with respect to the activation energy. In conclusion, the stress exponents and activation energies obtained are consistent with the assumption that dislocation climb and dynamic recrystallization govern the flow behavior of binary single-phase alloys.

It has been mentioned already that the apparent activation energy increases for decreasing Al content (see Table 16.1). This results from the strong flow softening that occurs in binary alloys with Al contents of less than 50 at.% [38]. As reported in the context of flow-softening behavior, in such alloys bending and kinking of lamellae is observed at the beginning of hot deformation. This results in flow localization and reorientation of lamellae. As the size and fraction of lamellar colonies increases with decreasing Al content these processes become more severe and certainly affect the apparent activation energy. With respect to the influence of phase constitution on the hot-working behavior, the question arises whether the disordering of the α_2 phase above the eutectoid temperature significantly affects the flow stress. Figure 16.6b shows the peak stress against the Zener–Hollomon parameter determined for an alloy containing 47 at.% Al (TAB) in hot compression tests. As can be seen, the peak stresses measured below and above the eutectoid temperature clearly fall on a common curve in each case. Thus, a significant influence of disordering of the α_2 phase on the flow stress can be excluded for alloys with moderate fractions of α_2 phase. This finding coincides with Nobuki and Tsujimoto [7] who observed no change in the deformability when the eutectoid temperature was exceeded. Interestingly, it has been observed that even relatively small fractions of the β phase have a strong effect on hot-working behavior and particularly result in small stress exponents (see alloy Ti-43.6Al-5V in Table 16.1) and excellent high-temperature workability [51]. The hot-working behavior of β-phase-containing alloys will be discussed in more detail in Section 16.3.2.

Up to this point, only the flow behavior of cast or cast and heat-treated alloys has been discussed. However, there is a clear influence of processing and the resulting microstructure on the deformation behavior during subsequent hot working. This becomes evident from Table 16.1, which also contains activation parameters for wrought-processed ingot and powder materials with a fine-grained γ microstructure. For these alloys, stress exponents of 1.8 and apparent activation energies of 2.3 eV and 2.6 eV were determined. The difference compared to the cast alloys might be explained by the contribution of grain-boundary sliding to hot deformation, which would result in both lower stress exponents and lower activation energies compared to dislocation climb. This assumption is supported by an investigation of Imayev et al. [67] on the superplastic deformation of Ti-48Al-2Cr-2Nb with a submicrocrystalline microstructure. On the basis of the Dorn equation, the authors determined strain-rate sensitivities ($m = 1/n$) greater than 0.5 and an apparent activation energy of 2.0 eV, which is close to that expected for grain-boundary diffusion [67, 68]. In this respect it is notable, that Kim et al. [69] have shown that grain-boundary sliding significantly contributes to hot deformation if the microstructure is refined below a certain grain size. The authors performed load-relaxation tests and analyzed the results in the framework of inelastic deformation theory. When the material was only deformed to small strains, no indications of grain-boundary sliding were found. After deformation to large strains, grain refinement due to dynamic recrystallization was observed and the contribution of grain-boundary sliding to deformation was found to be significant.

In summary, the flow behavior for a relatively wide range of γ TiAl alloys can be well described by the empirical Equation 16.1. This constitutive equation therefore might serve as a basis for modeling different aspects of the hot-working behavior, for example, calculating hot-rolling and forging forces. The activation parameters obtained are consistent with the view that dislocation climb and dynamic recrystallization are involved in the hot-deformation behavior. In addition, the activation parameters indicate that grain-boundary sliding contributes to deformation at high temperatures in fine-grained materials. However, an unambiguous identification of deformation mechanisms in the hot-working temperature range is not possible using the applied analysis and requires dedicated experiments.

16.2
Conversion of Microstructure

During hot working of many metals dynamic recrystallization occurs and a completely new grain structure, microstructural morphology and texture can be obtained. For this reason, an understanding of dynamic recrystallization and its kinetics is of significant importance for tailoring the final microstructure. Moreover, dynamic recrystallization generally enhances the workability of materials, for example by separating incipient grain-boundary cracks from the grain boundaries [14, 70] and, thus, its kinetics has to be considered in the selection of wrought-alloy compositions and processing conditions. During hot working of TiAl alloys, conversion of the microstructure is usually not only caused by dynamic recrystallization but also determined by phase transformations, spheroidization and refinement/coarsening of phase constituents, and other processes like the interaction of dislocations with phase boundaries [71–74]. All these processes are determined by reaction kinetics, thus giving rise to a very complex correlation between alloy composition, phase constitution and hot-working parameters. Many details of the microstructural evolution during hot working are not yet fully understood, which makes processing often unpredictable and impedes process development.

Recrystallization of ordered phases is expected to be difficult for basically two reasons [75–77]. First, the ordered state has to be restored and, secondly, there is a drastic reduction in grain-boundary mobility. For the ordered phase Cu_3Al, it has been observed that recrystallization was severely slowed down if the temperature was reduced from above to below the ordering temperature [75, 78]. This difference in recrystallization kinetics was attributed to a retardation of the grain-boundary mobility by a factor of 100. Similar observations regarding the grain-boundary mobility were also made for other ordered phases, for example, Ni_4Mo [75, 77] and $(Co_{78}Fe_{22})_3Al$ [76]. It is further notable that recrystallization in the ordered state was faster, if deformation was carried out in the ordered and not in the disordered state [77]. This difference in recrystallization behavior was attributed to the higher stored energy in the ordered state.

16.2.1
Recrystallization of Single-Phase Alloys

Investigations on the recrystallization behavior of γ TiAl alloys are rare in the literature, and this is particularly true for single-phase alloys. Recently an interesting study by Hasegawa *et al.* [43] investigated texture formation during dynamic recrystallization of Ti-52Al and the low stacking-fault energy f.c.c. metals Ni and Cu. Polycrystalline specimens of these materials were deformed in compression at different temperatures and strain rates and the texture formation was analyzed in relation to the Zener–Hollomon parameter Z. For low values of Z, the development of a strong texture was observed for the TiAl alloy with a pole density up to 10, whereas the f.c.c. metals showed only weak textures. In contrast, after deformation at high Z values similar pole densities between 3 and 4 were found for all materials. For the TiAl alloy the maximum pole density in inverse pole figures of the compression plane occurred at an orientation of $\{032\}$, as was found in earlier investigations [79–81]. This orientation is rotated about $10°$ from $\{011\}$ towards $\{010\}$. In Ni and Cu, the maximum pole density tended to accumulate around an orientation $10°$ away from $\{011\}$ towards $\{001\}$, for which no maximum was found in the TiAl alloy in agreement with the earlier investigations. In all materials, bulging of grain boundaries was observed in the low-Z regime, while in the high-Z regime it was found that new grains were nucleated through dynamic recrystallization. Grain-boundary bulging, in which new grains are formed by the local strain-induced migration of grain boundaries [41, 45], is a common process that occurs during dynamic recrystallization of many metals. It was also observed in earlier studies of TiAl alloys, for example, Ti-50Al [82]. Figure 16.8 shows an example of this process. The difference in the maximum pole density between the materials after deformation in the low-Z regime could be explained by the frequency of twinning during grain-boundary migration. In the TiAl alloy twinning was only rarely observed but it was frequently found in the f.c.c. metals. When twinning occurs repeatedly during grain-boundary migration, new orientations are generated and a strong deformation texture cannot be formed. From the results of

Figure 16.8 Scanning electron micrograph taken in the backscattering electron mode showing bulging at the boundary between two γ grains [83]. Ti-47Al-1.5Nb-1Cr-1Mn-0.2Si-0.5B extruded in the $(\alpha + \gamma)$-phase field at $T = 1250\,°C$. The extrusion direction is horizontal in the image.

Hasegawa et al. [43] it can be concluded that dynamic recrystallization in single-phase γ TiAl alloys shows some similarities with low stacking-fault energy f.c.c. metals, although the deformation mechanisms are quite different. The main difference between the materials was the low propensity for twinning in the TiAl alloys, which is a specific feature of TiAl alloys with Al contents above 50 at.% (see Chapter 5).

16.2.2
Multiphase Alloys and Alloying Effects

In two- or multiphase alloys with Al contents less than 50 at.%, the processes of microstructure conversion during hot working are much more complex. Information on the early stages of dynamic recrystallization and the phase transformations that occur during deformation have been obtained from a high-resolution electron microscopy study devoted to microstructural changes on the atomic scale after creep deformation [71–73]. In this work, samples of a Ti-48Al-2Cr alloy with a significant fraction of lamellar colonies had been deformed almost eight orders of magnitude slower than usual in hot working. Nevertheless, some qualitative information on the conversion of the lamellae to a spheroidized microstructure under hot-working conditions could be deduced. The recrystallization and phase-transformation processes were found to be closely related to the mismatch structures at the lamellar interfaces. A prominent feature was the formation of multiple-height ledges perpendicular to the γ/γ interfaces, which often had grown in zones of over 10 nm in width. The atomic arrangement in these zones is reminiscent of the 9R structure; a phase that probably has a slightly higher free energy than the $L1_0$ phase [71–73]. As these zones grow further, it might become energetically favorable to nucleate a new γ grain at the interface. The new grains are completely ordered, giving the impression that the ordered state is immediately established after grain nucleation or that nucleation occurred in the ordered state. During creep, dissolution of the $α_2$ phase has been observed, the mechanism of which is quite complex: not only must the stacking sequence be changed, but also the local chemical composition has to be adjusted by long-range diffusion. There is ample evidence that these processes are associated with the propagation of ledges and enhanced self-diffusion along the cores of misfit dislocations [71–73]. Such ledges may then be preferable sites where spheroidization of the $α_2$ phase begins at larger strains.

In systematic studies, Imayev et al. [37, 38] investigated the effects of constitution on the hot-compression behavior and on the evolution of the microstructure. In this work, binary alloys with Al contents between 45 and 54 at.% and also some engineering alloys were deformed in the ($α_2$ + γ)-region. The microstructure of the alloys in the initial state contained two constituents; lamellar colonies and equiaxed γ grains. In the binary alloys the fraction of each microstructural constituent systematically varied with increasing Al concentration from a fully lamellar to a near-γ microstructure. As already discussed, the flow curves for Ti-rich alloys exhibit a broad peak at low strains ($ε = 5–10\%$), followed by flow softening. In contrast, alloys with higher Al concentrations showed work hardening for strains up to 40% under the same deformation conditions (Figure 16.4). The dependence of microstructural evolution on strain during hot compression was studied in detail for the binary alloys Ti-47Al and Ti-49Al, and engineering alloys containing similar Al levels. In general, the recrystallization/globularization processes that took place during deformation resulted in fine-grained equiaxed microstructural constituents of recrystallized γ grains and globularized α_2 particles with grain sizes below 10 μm (Figure 16.9). Similar results have been found in numerous other investigations [3, 4, 6, 7, 11, 12, 14, 15, 18, 30, 32, 33]. In the studies of Imayev et al. [37, 38] the fraction of recrystallized/globularized grains increased as deformation proceeded (Figure 16.10). However, no substantial recrystallization

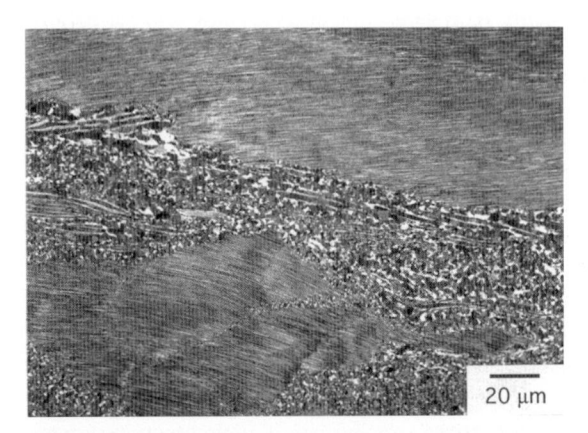

Figure 16.9 Scanning electron micrograph taken in the backscattering electron mode from a Ti-46Al alloy deformed in compression at $T = 1000\,°C$ and a strain rate of $\dot{\varepsilon} = 5 \times 10^{-4}\,s^{-1}$ to $\varepsilon = 50\%$ [38]. The deformation axis lies vertically in the micrograph. Note the heterogeneous nucleation of recrystallized grains at prior colony boundaries.

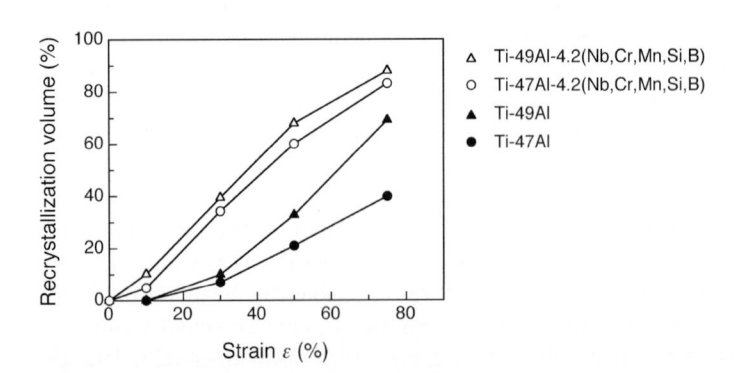

Figure 16.10 Dependence of the volume fraction of recrystallized/globularized grains on strain ε for binary and engineering alloys [38]. Deformation in compression at $T = 1000\,°C$ and $\dot{\varepsilon} = 5 \times 10^{-4}\,s^{-1}$.

occurred below strains of 10%. Similar recystallization/globularization kinetics were observed by Semiatin et al. [12] in a two-phase alloy and by Fukutomi et al. [80] in a single-phase alloy. The kinetics approximately followed an Avrami (sigmoidal) behavior [12] as first described by Kolmogorov [84], Johnson and Mehl [85] and Avrami [86] for static recrystallization. This type of recrystallization kinetics is usually observed for metals not only during static [87] but also during dynamic recrystallization [40]. According to the model of Luton and Sellars [88] the evolution of the volume fraction X of dynamically recrystallized grains can be described by the equations

$$X(\varepsilon \leq \varepsilon_c) = 0 \tag{16.5}$$

and

$$X(\varepsilon > \varepsilon_c) = 1 - \exp(-k_x ([\varepsilon - \varepsilon_c]/\dot{\varepsilon})^q) \tag{16.6}$$

with ε_c being the critical strain for initiation of dynamic recrystallization and k_x and q being constants. In agreement with this model, Semiatin et al. [12] found that after deformation to the same strain, the fraction of recrystallized/globularized grains was higher when the strain rate was decreased (Figure 16.11). Further, an increase in the fraction of recrystallized grains was also observed for increasing deformation temperatures as shown by Fukutomi et al. [80] and Salishchev et al. [89]. Thus, with respect to recrystallization/globularization kinetics, lower strain rates or higher temperatures are advantageous; an aspect that is particularly important for primary processing of ingot material with low workability. Interestingly, in the studies of Imayev et al. [37, 38] (Figure 16.12) there is a marked effect of the Al concentration on the recrystallization/globularization kinetics after hot-compression testing at 1000 °C. The largest volume fraction of recrystallized grains occurred in alloys with nearly stoichiometric composition, while for higher and lower Al concentrations, lower volume fractions of the material were recrystallized. Taken together, the results suggest that during hot working the conversion of microstructure is significantly affected by the phase constitution.

In the studies of Imayev et al. [37, 38], dynamic recrystallization was mostly initiated at prior grain/colony boundaries where the constraint exerted by neighboring grains is apparently greatest (Figure 16.9). The recrystallized/globularized zones proceed to form a necklace structure during further deformation, an example of which is shown in Figure 16.13. This has also been observed in many other studies [7, 14, 30, 35, 90–93]. Whether or not this heterogeneous formation of recrystallized grains leads to an overall homogeneous recrystallization, depends on the initial grain/colony size, as demonstrated by the recrystallization kinetics determined by Semiatin et al. [12] on TiAl alloys with two lamellar colony sizes (Figure 16.11). The alloys in the studies by Imayev et al. [37, 38] with Al contents of 48–50 and 54 at.% had a duplex or near-gamma microstructure, respectively, with relatively small grain sizes of 50–100 μm, which probably provided a considerable number of initiation sites for recrystallization.

In contrast, the Ti-rich alloys with Al concentrations of 45–47 at.% had a coarse-grained lamellar or nearly lamellar microstructure with colony sizes up to 2000 μm

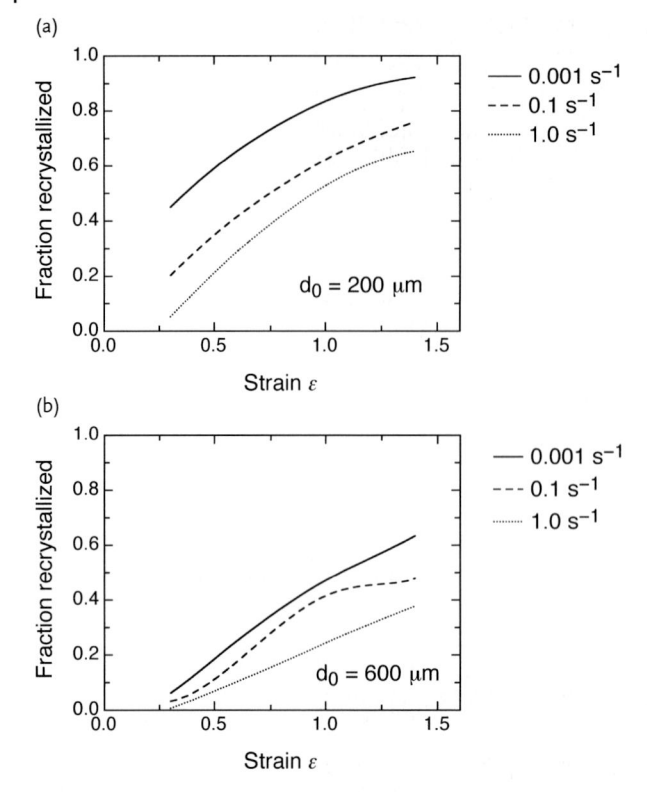

Figure 16.11 Dependence of the volume fraction of recrystallized/globularized grains on strain ε for the alloy Ti-45.5Al-2Cr-2Nb [12]. Deformation in compression at $T = 1093\,°C$ and various strain rates $\dot{\varepsilon}$. Prior to testing the samples had been processed to yield lamellar microstructures with colony sizes of either (a) 200 μm or (b) 600 μm.

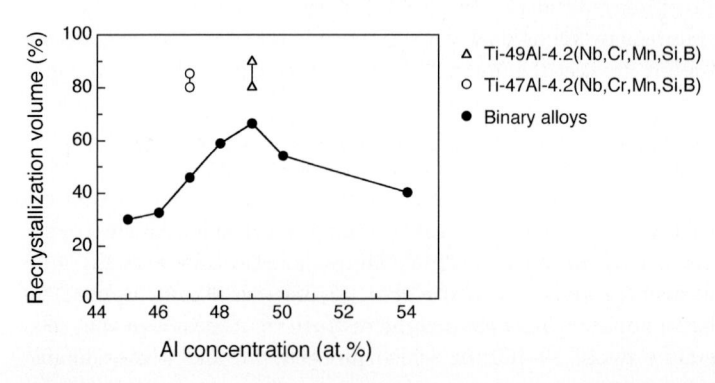

Figure 16.12 Dependence of the volume fraction of recrystallized/globularized grains on the aluminum content for binary and engineering alloys [38]. Deformation in compression at $T = 1000\,°C$ and $\dot{\varepsilon} = 5 \times 10^{-4}\,s^{-1}$ to a strain of $\varepsilon = 75\%$.

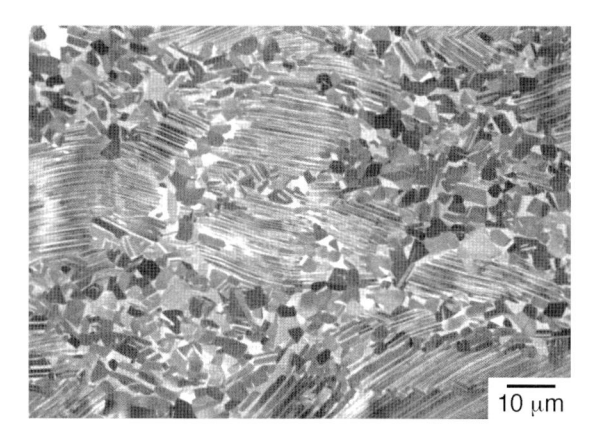

Figure 16.13 Scanning electron micrograph taken in the backscattering electron mode from a Ti-45Al-8Nb-0.2C alloy deformed in compression at T = 1200 °C using a logarithmic strain rate of $\dot{\phi} = 1 \times 10^{-3}$ s^{-1} to $\phi = 0.6$. The deformation axis lies vertically in the micrograph. Note the nucleation of recrystallized grains at prior colony boundaries forming a necklace structure. Prior to testing the material had been extruded close to the α-transus temperature to produce a near-lamellar microstructure.

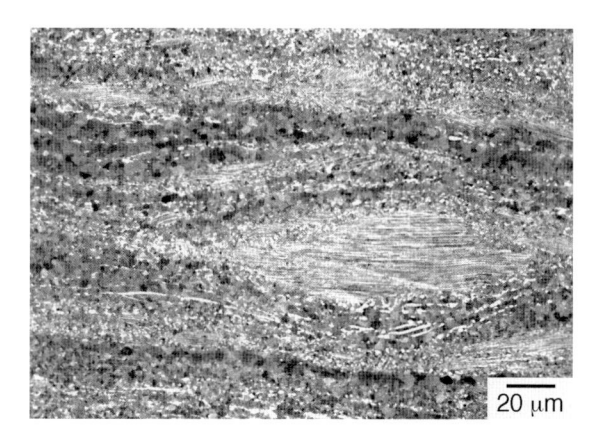

Figure 16.14 Scanning electron micrograph (backscattered electron mode) of the banded microstructure observed in a Ti-46Al alloy specimen after deformation at $T = 1000$ °C using $\dot{\varepsilon} = 5 \times 10^{-4}$ s^{-1} to a strain $\varepsilon = 75\%$ [38]. The deformation axis lies vertically in the image.

in the initial condition. In these materials highly localized shear bands were formed during deformation along prior colony boundaries (Figure 16.14). The bands consisted of an array of very fine equiaxed γ grains of 2–3 μm in size, or a mixture of equiaxed α$_2$ and γ grains with a typical grain size of 0.5–2 μm. These bands often traversed the entire workpiece. After formation of these bands, subsequent deformation is likely to preferentially occur by grain-boundary sliding. Thus, outside the shear bands the amount of imparted strain energy is relatively low, which makes recrystallization sluggish. Within the bands the local shear strain often exceeded 100%, as can be recognized from the width of the bands and the relative displacement of structural features. This mechanism not only leads to the typical inhomogeneous banded microstructure shown in Figure 16.14, but could also be responsible for cracking that has been observed at the bulged surface of the hot-worked specimens. Besides recrystallization the occurrence of banded microstructures is often caused by chemical inhomogeneities within TiAl alloys, as will be described later in more detail. The development of shear bands may explain why Ti-rich alloys show strong flow softening, despite their sluggish recrystallization kinetics (see Figure 16.12). Strain localization and shear-band formation were observed in several studies on hot working of TiAl and were recognized as critical issues for the microstructural evolution as well as the workability [7, 12, 14, 25, 35, 92, 93]. This aspect will be further addressed in a later section of this chapter. The importance of the initial microstructure is also demonstrated by the data obtained for boron-containing alloys shown in Figure 16.12. Boron is known to significantly refine the as-cast microstructure [13, 94, 95], which probably gives rise to the enhanced recrystallization kinetics, when compared with the equivalent binary alloys (Figure 16.12). Besides microstructural refinement, the addition of B in concentrations above 0.03 at.% results in the formation of boride particles [96] and might aid the recrystallization kinetics due to the accumulation of dislocations at boride particles [37], as was recognized by Nobuki et al. [50].

In a study on the creep deformation of Ti-48Al-2Nb-2Cr in the temperature range 727–927 °C, Hofmann and Blum [92] obtained similar findings as Imayev et al. [37, 38] for hot working. Using TEM the authors found that a necklace structure formed at the beginning of creep deformation along the boundaries of lamellar colonies. It consisted predominantly of subgrain boundaries, although it remained open whether nucleation of new grains was involved in dynamic recrystallization. The frequent formation of subgrains during deformation at 900 °C was also observed in another TEM study [97]. As described earlier for single-phase alloys, the necklace structure that formed during creep in the work of Hofmann and Blum [92] might be typical for dynamic recrystallization in the low-Z regime. The subgrains were obviously formed in regions where deformation was localized and were observed even after very low total strains of the order of 2%. During further deformation, coalescence of the fine-grained equiaxed bands into a contiguous network was observed. The bands grew continuously in thickness with increasing strain by (sub)grain-boundary migration and by the formation of new (sub)grains in adjacent lamellar colonies. The authors could show that the equiaxed recrystallized microstructural constituent was a steady-state structure with a crystallite size that inversely depended on the stress, as is generally the case for steady-state subgrain structures [92].

A commonly held concept of hot working is that dynamic recovery and recrystallization are triggered by heterogeneities in the deformed state. These two processes involve climb of the dislocations and will only occur when there is sufficient thermal activation to allow long-range diffusion. In this context, the effect of the Al content on the deformation and diffusion mechanisms will be discussed in the following. It is now well established that the γ(TiAl) phase of two-phase alloys deforms by glide of ordinary dislocations and superdislocations. In addition, mechanical twinning along 1/6<11$\bar{2}$] occurs. The relative contribution of the individual mechanisms to deformation depends on the aluminum concentration, the content of ternary elements and the deformation temperature. For the γ phase in two-phase alloys, it is now well established that glide of superdislocations is difficult so that deformation is mainly provided by ordinary dislocations and order twinning. In contrast, the deformation of the γ phase in Al-rich alloys occurs by glide of superdislocations. For two-phase alloys the tendency to deform by twinning is enhanced if the deformation temperature is increased. There is good evidence that the flow stress of these materials is essentially determined by the interaction of perfect and twinning partial dislocations with lamellar interfaces. The lamellar spacing generally decreases with decreasing Al content [98, 99], thus, the high peak stress of the Ti-rich material becomes plausible. Work hardening of these alloys is due to the increase in dislocation and twin density and the mutual elastic interaction of perfect and twinning partial dislocations propagating on parallel and oblique {111} planes.

There are various processes that might lead to heterogeneities during deformation. In the deformed state, multipole structures of ordinary dislocations have frequently been observed. The dislocations are probably trapped because of their elastic interaction and local recombination by cross-glide. Thus, the multipoles are immobile and strong glide obstacles [100–102]. It is well documented in the literature that these processes give rise to high local stresses [100–102]. As soon as multiple twinning systems with nonparallel shear vectors are activated, twin intersections become particularly important since they represent a deformation-induced locking mechanism. High-resolution electron microscope observations have shown that lattice rotations of 8–10° occur in the intersection zone [103]. The resulting misfit is accommodated by dense dislocation arrangements, which are similar to tilt boundaries. Such structural features are, together with the high internal stresses, probably very conducive to the initiation of recovery and recrystallization.

In two-phase alloys, the release of strain energy is probably relatively easy. The ordinary dislocations have a compact core [100], thus, climb of these dislocations is kinematically not impeded by dissociation into partial dislocations. Tangles of ordinary dislocations are expected to recover rapidly because dislocations with orthogonal Burgers vectors do not interact to form stable junctions. It was also shown by electron microscopy that the deformation-twin structures in two-phase alloys could easily be annealed out [102, 103]. With respect to the composition dependence of dynamic recrystallization the specific diffusion behavior of ordered alloys also needs to be considered. In two-phase TiAl alloys, diffusion is probably supported by a significant chemical disorder of the γ phase. There is good evidence that off-stoichiometric deviations in Ti-rich alloys are accommodated by the formation of Ti_{Al} antisite defects, that is, Ti atoms situated on the Al sublattice, as explained in more detail in Chapter 6. The essence of the arguments is that the phase separation upon cooling described by the phase diagram often cannot be established within the constraints of processing routes because the kinetics of the α/γ transformation is sluggish [104, 105]. This can result in a significant fraction of Ti_{Al} antisite defects. At sufficiently high concentration, the antisite defects form a percolating substructure of antistructural bridges along which long-range diffusion can occur [64, 105]. The point to note is that the migration of Ti_{Al} antisite defects is expected to require an activation energy of only about 0.7 eV [64]. This is significantly lower than the self-diffusion energy of Ti in TiAl, Q_{sd} = 2.6 eV, determined at high homologous temperatures [64]. Thus, via antistructural bridges, high diffusion rates can probably be established at the low deformation temperature used in the studies of Imayev et al. [37, 38]. This would enhance diffusion-assisted recovery and recrystallization of Ti-rich alloys compared to alloys with more than 50 at.% Al, but cannot explain the slow recrystallization kinetics of alloys with less than 48 at.% Al (Figure 16.12). However, it is well established that material flow-softening characteristics are prone to plastic instabilities and the formation of shear bands, which strongly concentrate the deformation. This issue will be discussed in detail in the next section. Taken together, it becomes understandable why the volume fraction of recrystallized grains in two-phase materials is relatively small.

The deformation and recrystallization behavior of Al-rich single-phase alloys may be rationalized by the following arguments. As mentioned earlier, deformation of these alloys is mainly provided by superdislocations, which are often subject to extended decomposition and dissociation reactions. The interaction and reaction of these dislocations give rise to the formation of stable junctions [100], which certainly provide strong glide barriers. Also, dislocation dissociations hinder climb and cross-slip of dislocations, which are the key elements for the formation of sub-boundaries. More on this subject is found in Chapter 5. The relatively large grain size of Al-rich ingot material is certainly detrimental for grain nucleation. Furthermore, diffusion in Al-rich alloys is probably more sluggish, when compared with Ti-rich alloys. The migration of the relevant Al_{Ti} antisite defects requires an activation energy of about 1.7 eV [64], which is significantly higher than that for Ti_{Al} defects and makes diffusion via antistructural bridges ineffective. These processes result in work hardening accompanied with reduced recrystallization. In summary, the best recrystallization/globularization behavior occurs when the rates of work hardening, recovery and recrystallization are in equilibrium, as manifested by a steady-state flow stress. According to Figures 16.4, 16.5 and 16.12, this condition is met by alloys with Al contents of 49 at.%.

One of the major goals of hot working is to achieve grain, or more generally, microstructural refinement. For dynamic recrystallization of metals, it has been shown that equilibrium between the nucleation and growth of grains establishes a steady-state grain size, which essentially depends on the plateau stress achieved at large strains [40]. The steady-state grain size d_{ss} is related to the plateau stress σ_{ss} according to the empirical relation [40, 42]

$$\sigma_{ss} = M d_{ss}^{-p} \tag{16.7}$$

with $0.5 \leq p \leq 1$ and an empirical constant M. This relation has been explained by the dependence of the flow stress on the subgrain size ($\sigma \propto d^{-1}$) and on the grain size ($\sigma \propto d^{-0.5}$) [42]. Following Mecking and Gottstein [40], if sufficiently high strain rates and low deformation temperatures are applied, dynamic recrystallization can be used in principle to obtain fine grain sizes that are hardly attainable by static recrystallization. In several studies, relationships satisfying Equation 16.7 with grain-size exponents between 0.5 and 1 have been observed for TiAl alloys after hot working or creep deformation [4, 89, 92]. Under appropriate conditions even submicrometer-sized grains were achieved via dynamic recrystallization [67, 106]. Such microstructures are particularly useful for subsequent superplastic deformation. Because both the recrystallization kinetics and the workability are reduced for increasing strain rates and decreasing working temperatures, multi-step working usually has to be applied to obtain very fine grain sizes.

16.2.3
Influence of Lamellar Interfaces

The lamellar morphology exhibits a significant plastic anisotropy, when the orientation of the lamellae with respect to the deformation axis is considered [98, 107, 108]. Such behavior may significantly affect the conversion of lamellar microstructures during hot working. The influence of texture was investigated in the studies performed by Imayev et al. [37, 38] on samples of Ti-48Al-2Cr with preferred lamellae orientations of $\alpha = 0°$, 45° and 90° to the deformation axis. These samples were deformed in hot compression at 1000°C and a strain rate of $10^{-3}\,\text{s}^{-1}$. Figure 16.15 shows the flow curves obtained for the three lamellar orientations investigated. The highest flow stress was found for the $\alpha = 0°$ lamellae orientation (lamellae parallel to deformation axis), whereas specimens with orientations of $\alpha = 90°$ and $\alpha = 45°$ exhibited intermediate and the lowest flow stresses, respectively. There was also a marked influence on the softening rate, which was highest for

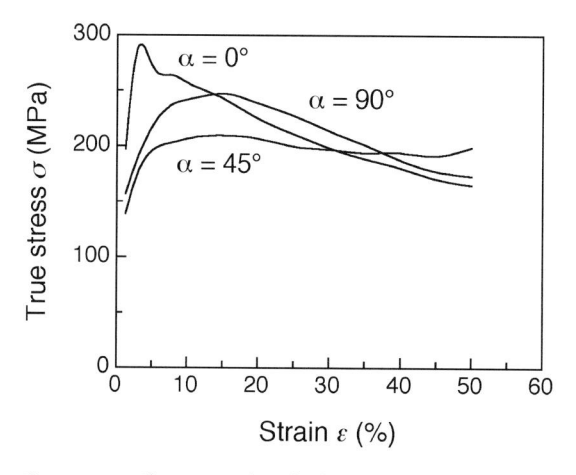

Figure 16.15 Flow curves for cylindrical compression samples with preferred orientations α of the lamellae against the deformation axis [38]. Ti-48Al-2Cr tested at $T = 1000°C$ and $\dot{\varepsilon} = 10^{-3}\,\text{s}^{-1}$.

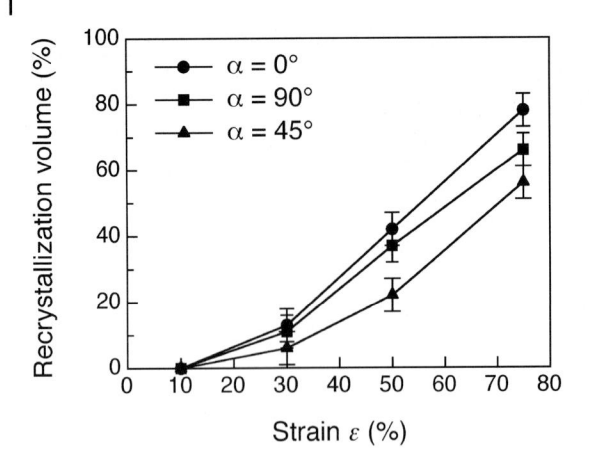

Figure 16.16 Dependence of the volume fraction of recrystallized/globularized grains on strain ε, for specimens with different preferred orientations of the lamellae (angle α) with respect to the deformation axis [38]. Ti-48Al-2Cr, deformation in compression at $T = 1000\,°C$ and $\dot{\varepsilon} = 10^{-3}\,s^{-1}$.

$\alpha = 0°$ and lowest for $\alpha = 45°$. For all orientations, the strain-rate sensitivity $m\,(= d\ln\sigma/d\ln\dot{\varepsilon})$ increased during deformation from $m = 0.15$–0.2 to about $m = 0.3$, which apparently reflects the progressive grain refinement. Metallographic investigations revealed that for all orientations, the initial cast microstructure was partly converted by hot deformation into a fine-grained equiaxed microstructure with γ grain sizes in the range of 2 to 5 µm and an α_2 particle size below 1 µm. The volume fraction of recrystallized/globularized grains increased most rapidly for the $\alpha = 0°$ orientation, while specimens with $\alpha = 45°$ exhibited the slowest development of recrystallized/globularized grains (Figure 16.16). For the $\alpha = 0°$ lamellae orientation, extensive cross-twinning was found after deformation to a strain of 10%. In this state the lamellae were already slightly bent. Dense dislocation structures and subgrain boundaries were observed in such bent lamellae areas using TEM (Figure 16.17a). These areas are, together with colony boundaries, the preferred sites for the nucleation of recrystallized grains. Indeed, the nucleation of recrystallized grains in front of blocked twinning at lamellar interfaces was directly observed after hot deformation of a lamellar alloy (Figure 16.18) in a study devoted to slip/twin interaction of TiAl alloys [109]. A significant contribution of twinning to the nucleation of recrystallized grains was also deducted from the observation that the fraction of twins during dynamic recrystallization correlated with the fraction of recrystallized grains [11]. On further deformation, strong bending of the $\alpha = 0°$ lamellae and a significant change of their orientation with respect to the deformation axis occurred (Figure 16.17b), a finding that has often been reported in the literature [12, 25, 30, 32, 33, 93, 110]. As can be seen in this image, the γ lamellae have already recrystallized to a significant extent after deformation to 50% strain. Spheroidization of the α_2 lamellae was also observed

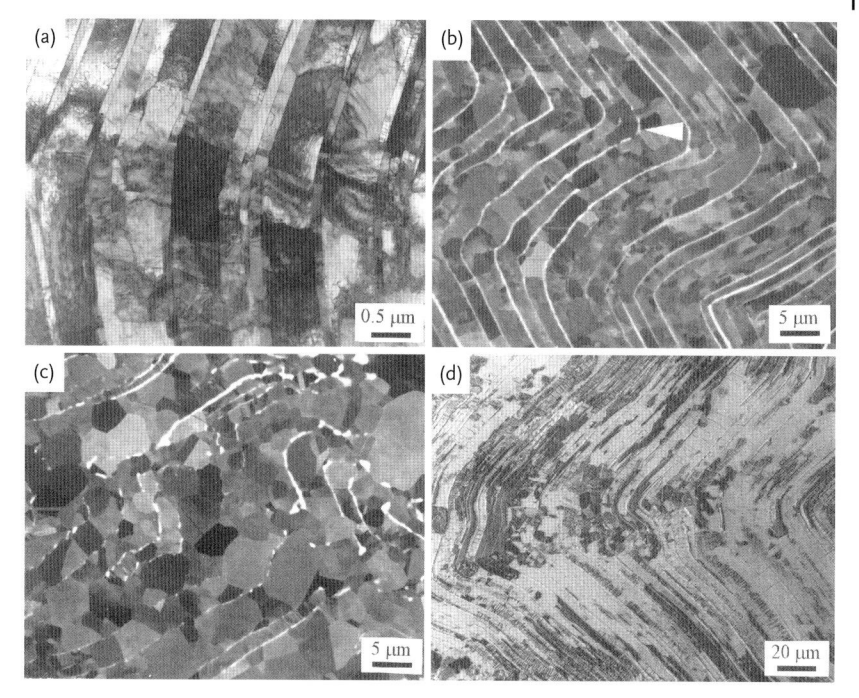

Figure 16.17 Conversion of the lamellar microstructure by buckling and kinking of lamellae. Ti-48Al-2Cr with the $\alpha = 0°$ preferred lamellae orientation with respect to the compression axis [38]. Deformation at $T = 1000\,°C$ and $\dot{\varepsilon} = 10^{-3}\,s^{-1}$ to different strains ε. (a) TEM micrograph showing the initial state of lamellae bending after deformation to strain $\varepsilon = 10\%$. Note the high internal stresses, which are manifested by strain contrast. (b) Extensive kinking of the lamellae observed after deformation to strain $\varepsilon = 50\%$. Note the recrystallization in the γ lamellae and the appearance of isolated α_2 particles in a strongly bent area (arrowed). Scanning electron micrograph taken in the backscattering electron mode. (c) Later stage (higher local strain) of the processes described in (b), $\varepsilon = 50\%$. Note the spheroidization of the α_2 lamellae. (d) Initial stages of shear-band formation at kinked lamellae, deformation to a strain $\varepsilon = 75\%$. Optical micrograph.

at this stage. Here, it is worth mentioning that the first spheroidized α_2 particles were found in the vicinity of areas where lamellae bending had occurred (Figure 16.17b). In regions of higher local strain, most of the α_2 lamellae were globularized and an almost complete conversion of the initial lamellar microstructure to an equiaxed one had taken place (Figure 16.17c). The bending of lamellae often resulted in the development of shear bands (Figure 16.17d), as already noted. Strain localization in lamellar alloys can be extremely pronounced and generates very inhomogeneous microstructures with highly localized bands of recrystallized grains (Figure 16.19). Figure 16.20 shows another example of the formation of shear bands. In this case a shear strain of 1.3 was estimated from the width of the band and the displacement of interrupted lamellae. In summary, bending and kinking of lamellae have been shown to play an important role in the hot-working

Figure 16.18 TEM micrograph showing deformation twinning in a Ti-48Al-2Cr alloy after elevated-temperature deformation [109]. The specimen was deformed by tensile testing at $T = 700\,°C$ to a strain $\varepsilon = 8.9\%$. The tensile axis lies vertically in the micrograph. Note the immobilization of twins at lamellar interfaces. The inset shows the region designated by the arrow at a higher magnification, and demonstrates the early stages of recrystallization. The process starts in front of immobilized twins at the interfaces.

and recrystallization behavior for the $\alpha = 0°$ lamellae orientation, and more generally, for γ(TiAl) alloys.

Twinning and cross-twinning was also observed on samples with the $\alpha = 90°$ lamellae orientation, but to a lesser extent than for the $\alpha = 0°$ orientation in agreement with observations by Nobuki and Tsujimoto [7]. This behavior probably reflects the fact that the 90° lamellae orientation is less favorable for twinning and may explain why the recrystallization kinetics were slower for this orientation. As expected, no significant bending occurred for this lamellae orientation.

For the soft orientation ($\alpha = 45°$), deformation twins with habit planes inclined to lamellar boundaries were only rarely detected after small strains, and there were

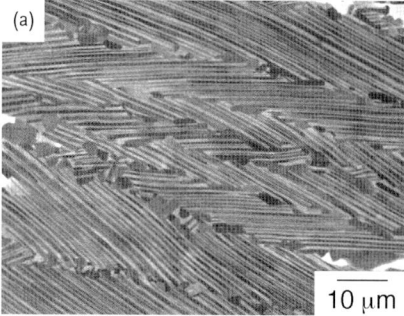

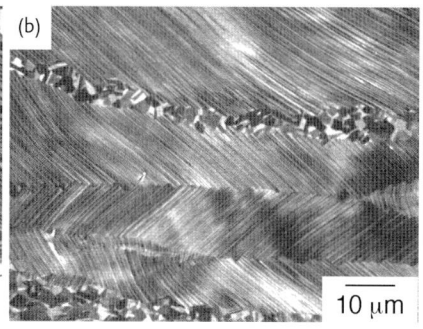

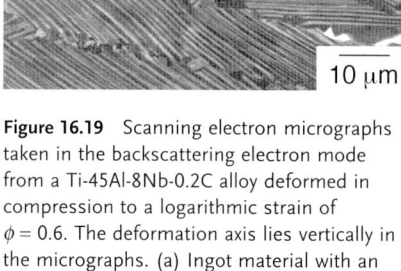

Figure 16.19 Scanning electron micrographs taken in the backscattering electron mode from a Ti-45Al-8Nb-0.2C alloy deformed in compression to a logarithmic strain of $\phi = 0.6$. The deformation axis lies vertically in the micrographs. (a) Ingot material with an initial lamellar microstructure deformed at $T = 1200\,°C$ and a logarithmic strain rate $\dot{\phi} = 1\times 10^{-3}\ s^{-1}$. (b) Extruded material with a nearly lamellar microstructure tested at $T = 1150\,°C$ and a logarithmic strain rate $\dot{\phi} = 1\times 10^{-3}\ s^{-1}$. Note the occurrence of pronounced shear localization in both micrographs and different states of recrystallization in the kink bands.

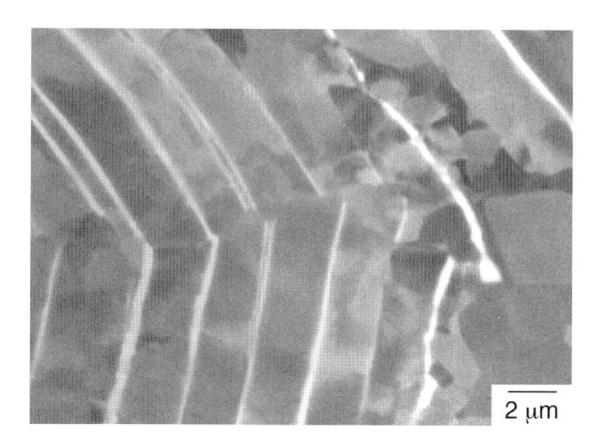

Figure 16.20 Scanning electron micrograph taken in the backscattering electron mode from a Ti-48Al-2Cr alloy deformed in compression at $T = 1000\,°C$ and a strain rate of $\dot{\varepsilon} = 1\times 10^{-3}\ s^{-1}$ to $\varepsilon = 50\%$. The material had an aligned lamellae orientation with $\alpha = 0°$, that is, the lamellae were parallel to the deformation axis, which lies vertically in the micrograph. From the displacement of interrupted lamellae and the width of the shear band in this image, a local shear strain of 1.3 was estimated.

no indications for cross-twinning. Deformation twinning might have occurred parallel to the lamellar interfaces, but this could not be unambiguously verified by metallographical investigations. For higher strains a slight bending of the lamellar interfaces was observed and the orientation of lamellae slowly changed with strain. After deformation to a strain of 75%, a considerable fraction of the specimen had retained the lamellar microstructure but with the lamellae orientation nearly perpendicular to the deformation axis. Recrystallization and globularization processes were mainly found at colony boundaries, in contrast to the other orientations.

The observed variation of the peak stress with the lamellar interface orientation (Figure 16.15) is probably a consequence of the plastic anisotropy of lamellar structures. For the $\alpha = 0°$ and $90°$ orientations the principal shear direction is inclined to the lamellar boundaries so that shear processes across the lamellae are initiated. Lamellar interfaces are known to be very effective barriers for all characters of perfect and partial dislocations [100, 107, 109, 111, 112]. Thus, the slip path of dislocations and twins is essentially limited by the width of the lamellae. For this reason the formation of dislocation pile-ups and immobilized twins at the interfaces is a common feature in strained material with these lamellae orientations [111]. High constraint stresses occur ahead of the immobilized shear bands at the interfaces, thus more slip systems are activated, which often provides nucleation sites for recrystallization. On the other hand, in alloys with lamellar interfaces at 45° to the deformation axis, dislocation glide occurs parallel to the lamellar interfaces and dislocation pile-ups are only expected at domain or colony boundaries. This may lead to a reduced density of nucleation sites for recrystallization when compared with the two "hard" lamellae orientations, and thus may explain the slower recrystallization kinetics for this orientation. Summarizing, it is concluded that the interaction of dislocations and deformation twins with lamellar interfaces provides nucleation sites for recrystallization, and thus plays a significant role in the recrystallization kinetics of lamellar TiAl alloys.

It is interesting to note that the conversion of the lamellar microstructure apparently involves an element of deformation instability, which is manifested by lamellae kinking and is reminiscent of the buckling of load-carrying structures. In terms of a laminate model, the lamellar morphology may be considered as an ensemble of TiAl and Ti$_3$Al plates. When perfectly aligned along the compression axis, these "columns" are highly stable under compression, as long as the axial load is below a critical value. Above this critical load, the equilibrium becomes unstable and the slightest disturbance will cause the structure to buckle. In the lamellar structure such a disturbance might arise from dislocation pile-ups or deformation twins laterally impinging on the lamellae interfaces. Furthermore, there is a significant difference between the elastic response of the α_2 and γ phases upon loading. Thus, the tendency for unstable buckling is expected to increase if there is an inhomogeneous distribution of α_2 and γ lamellae. Figure 16.17 gives supporting evidence for this mechanism and illustrates how the kinking of the lamellae may be accomplished. The process probably starts with local bending of the lamellae. From the curvature and thickness of the lamellae local strains can be deduced, which are often larger than 10%. Subsequent recrystallization of these subgrains apparently occurs in such a way that lamellae kinking is the result. In the regions of highest local bending, spheroidization and dissolution of the α_2 phase occur, which suggests that the phase transformation is supported by stress. Buckling and kinking of the lamellae often results in the formation of shear bands, the detrimental effects of which on the conversion of the microstructure have already been discussed. From a mechanical point of view, buckling failure does not depend on the yield strength of the material, but only on the dimensions of the structure and the elastic properties. Thus, in lamellar alloys of a given lamellar spacing, the tendency for buckling and shear-band formation should mainly be determined by the axial length of plates, that is, by the colony size. In agreement with this view, Semiatin and coworkers [12, 113] have shown that through heat treatment to increase the colony size, the recrystallization kinetics are retarded and the degree of flow softening enhanced. The relationship between the colony size and the tendency for buckling also explains why the recrystallization is slow for Ti-rich alloys (see Section 16.2.2), which often exhibit coarse lamellar microstructures.

With respect to buckling and kinking of lamellae during compressive deformation, a micromechanical modeling study by Dao et al. [114] is very helpful in understanding the deformation processes that occur within lamellar microstructures. For a detailed description of the calculations and the underlying constitutive theory, the reader is also referred to related publications [115, 116]. These authors performed a two-dimensional finite-element polycrystal analysis of the stress–strain behavior of lamellar and nearly lamellar model microstructures. The essential anisotropies of the slip and/or twinning deformation modes of the two constituent phases, α_2 and γ, were included in the model calculations. The reference configuration of the microstructure containing 27 lamellar colonies with randomly oriented lamellar interfaces is shown in Figure 16.21 together with the initial finite-element mesh. The simulations were carried out for room-temperature deformation at a strain rate of $10^{-3} \, \text{s}^{-1}$. Figure 16.22 shows one result of this study,

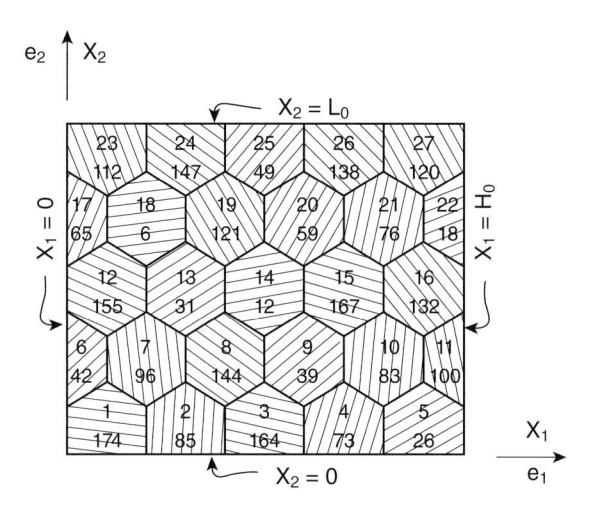

Figure 16.21 Micromechanical modeling of the stress-strain behavior of lamellar and nearly lamellar model microstructures performed by Dao et al. [114] using a two-dimensional finite element polycrystal analysis. The diagram shows the reference configuration of the 27 lamellar colonies representing the randomly generated polycrystalline aggregate. The orientations ψ^n shown are selected in the $0° \leq \psi^n \leq 180°$ range.

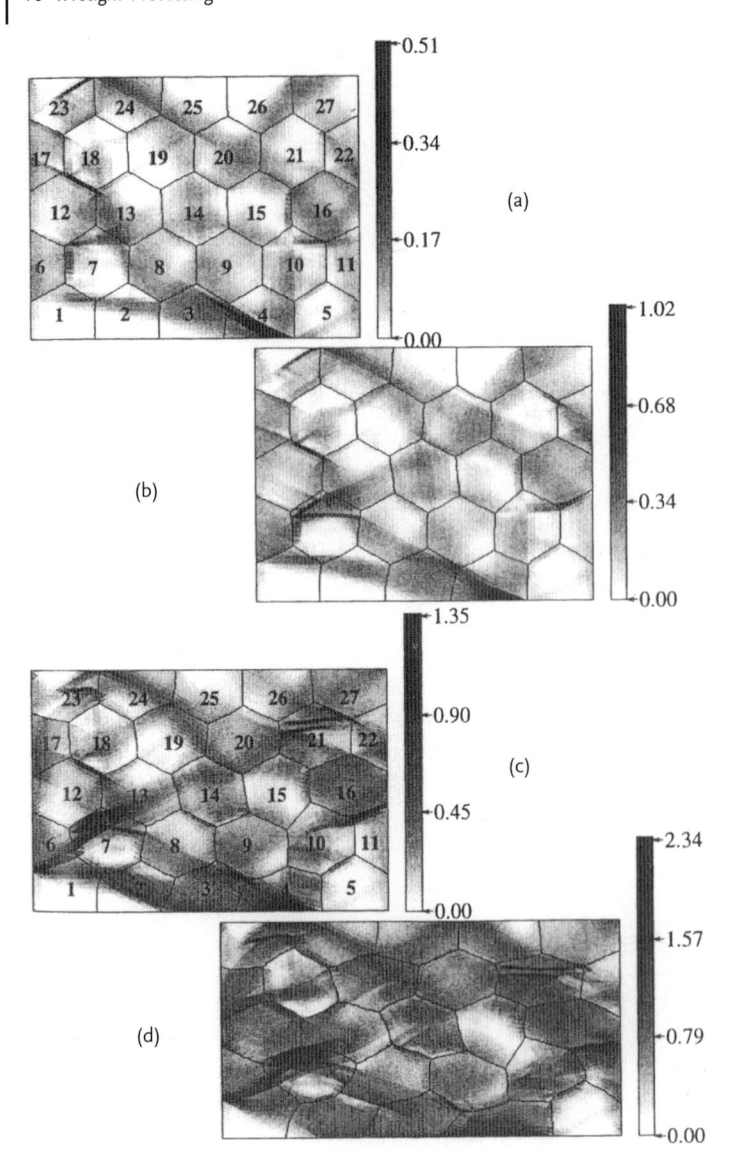

Figure 16.22 Micromechanical modeling of the stress–strain behavior of lamellar and nearly lamellar model microstructures performed by Dao et al. [114] using a two-dimensional finite-element polycrystal analysis. Accumulated sum of slip deformation developed in the fully lamellar microstructure at (a) 2.5%, (b) 5%, (c) 10% and (d) 20% macroscopic strain. For the reference configuration and the finite-element mesh see Figure 16.21.

in which the accumulated local sum of slip events is displayed for a fully lamellar microstructure after compressive deformation to different macroscopic strains. It is obvious from this figure that the deformation is highly nonuniform in fully lamellar microstructures and that localized shearing is present even for modest macroscopic strains. However, the development of macroscopic shear bands across the entire microstructure was not observed in this work. Figure 16.22 also shows that localized bands were developed at boundaries between colonies that were closely aligned (parallel) with lamellae colonies in the soft orientation, whereas in colonies with lamellae nearly parallel to the compression axis, kink bands occurred as described above for high-temperature deformation. The study convincingly supports the conclusion that highly localized deformation near colony boundaries, as well as the development of kink bands within the colony interiors are specific features associated with the deformation of two-phase lamellar TiAl alloys. In the kink bands the shear is localized in a band that is almost perpendicular to the lamellae. At sites of concentrated deformation fracture might be initiated during room-temperature deformation and recrystallization be triggered at higher temperatures. Large colony sizes certainly favor these processes. Similar results have been obtained in more recent finite-element modeling studies that employed a deformation temperature of 1000 °C and a lamellar microstructure [44]. The calculations showed the formation of initial buckling waves as well as shear concentration that resulted from kinking at imperfections inside lamellar colonies and at colony boundaries. The results are strikingly reminiscent of those observed experimentally (see Figures 16.17, 16.19 and 16.20). In conclusion, coarse-grained lamellar microstructures are prone to deformation instabilities and shear-band formation; these processes can result in major problems during the wrought processing of TiAl alloys.

16.2.4
Microstructural Evolution during Hot Working above the Eutectoid Temperature

Up to now, the conversion of the microstructure during hot working has almost exclusively been considered for the ($\alpha_2 + \gamma$)-phase region, that is, below the eutectoid temperature, which depending on the alloying additions lies in a range of 1115–1200 °C. Below the eutectoid temperature the equilibrium volume fractions of the α_2 and γ phase do not strongly vary with temperature, and in most TiAl alloys the volume fraction of the α_2 phase is low ($\leq$15%). Above the eutectoid temperature, the α_2 phase disorders to the α phase, and the volume fraction of the α phase strongly increases with temperature. Thus, recrystallization and globularization processes in the α phase become important with respect to microstructural evolution. Additionally, the width of α lamellae can be expected to increase with temperature, and thus the propensity for kinking and bending of lamellae is probably reduced. In the literature, the recrystallization behavior of the α phase has rarely been examined. This might be due to the fact that α grains transform into lamellar colonies on cooling after working at temperatures high in the ($\alpha + \gamma$)-field, so that a direct observation of recrystallization in the α phase is difficult. However, from a number of investigations that involved hot extrusion or forging, it can unambiguously be concluded that the α phase can be completely recrystallized at temperatures near or in the α single-phase field [10, 13, 117–121]. After hot working at such temperatures, the materials revealed newly formed lamellar or duplex microstructures. Compared to cast microstructures, significantly refined colony sizes in the range of 30 to 200 µm were observed, even if fully lamellar materials were obtained [13, 18, 118–121]. TEM studies of alloys extruded above the α-transus temperature exhibited lamellar microstructures with straight and regularly spaced alternating α_2 and γ lamellae and some (<5 vol.%) small γ grains at colony boundaries [118, 119]. The lamellae were free of imperfections such as kinks or sub-boundaries [118, 119] suggesting that the lamellar microstructure was formed after deformation and that recrystallization took place in the single α-phase field. In the alloy Ti-45Al-10Nb that was extruded midway between the eutectoid and the α-transus temperature, Zhang et al. [122] observed a completely recrystallized equiaxed microstructural constituent as well as deformed remnant lamellae. Such a partially recrystallized microstructure does not significantly differ from those obtained after hot working in the ($\alpha_2 + \gamma$)-region [7, 32, 123]. This also holds for the orientation of the remnant lamellae, which was reported by Zhang et al. [122] to be aligned perpendicular to the main deformation direction. TEM investigations showed that the remnant lamellae were converted into bands of fine grains, for which misorientations of up to 16° were determined with respect to the initial orientation. The authors also observed newly formed γ grains that had grown into adjacent remnant α lamellae. This process of phase-boundary bulging obviously occurred during cooling after extrusion and supported the breakdown of the lamellar structure [122]. The authors concluded that raising the working temperature and processing at relatively high strain rates, as is usually the case during extrusion, could reduce the fraction of remnant lamellar colonies. This conclusion was partially confirmed in an empirical study on the recrystallization/globularization kinetics of a Ti-45Al-8Nb-0.2C alloy that was deformed in compression within the temperature range 1150–1330 °C [54]. The evolution of the microstructure with temperature is shown in Figure 16.23. In specimens deformed to a strain of $\varepsilon = 40\%$, the recrystallization/globularization volume fraction increased with increasing temperature, whereas for $\varepsilon = 80\%$ a significant increase was only observed at temperatures between 1300 and 1330 °C (Figure 16.24). The acceleration of the recrystallization kinetics with temperature observed for the lower strain is not necessarily related to the disordering of the α_2 phase, because such a temperature dependence also was found below the eutectoid temperature [80, 89]. However, the diagram also shows that both recrystallization was significantly enhanced and a complete conversion of the microstructure could be achieved when working was performed in the single α-phase field (Figure 16.23). This might be understood by the absence of lamellae in the microstructure, the detrimental effect of which has been discussed earlier. The benefits and disadvantages of wrought processing in the α field will be described in more detail in Section 16.3.2.

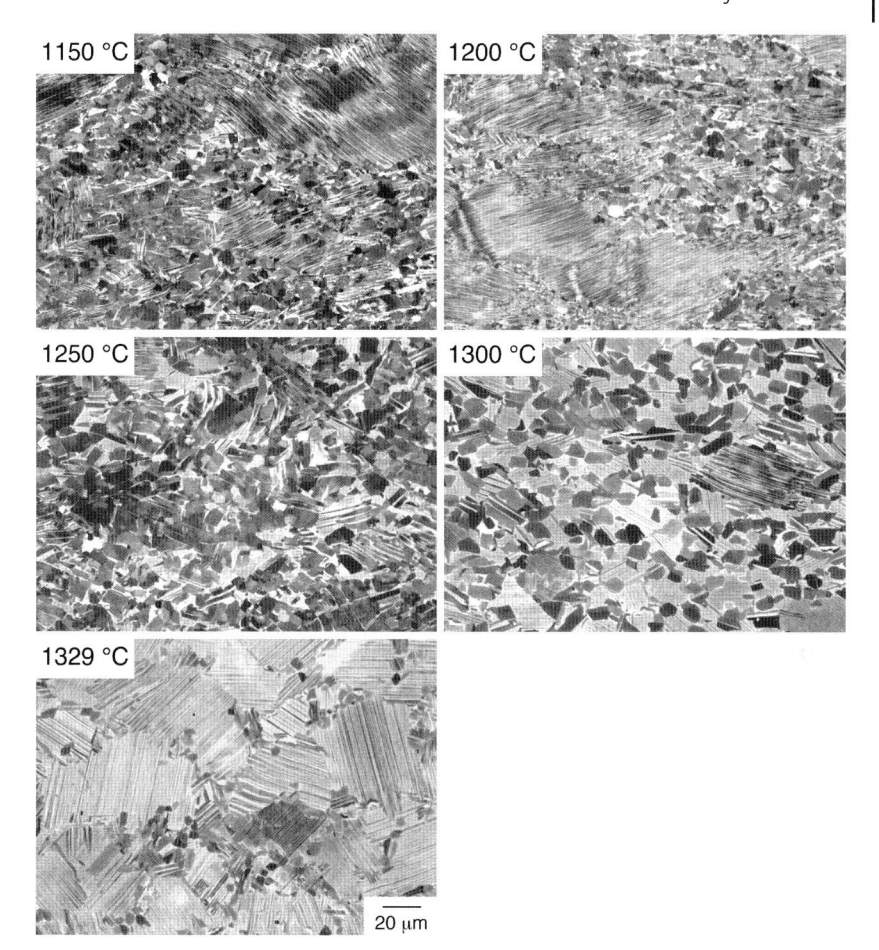

Figure 16.23 Scanning electron micrographs taken in the backscattering electron mode from Ti-45Al-8Nb-0.2C ingot material deformed in compression at the temperatures indicated and a strain rate of $\dot{\varepsilon} = 1.4 \times 10^{-3}$ s^{-1} to $\varepsilon = 40\%$. The deformation axis lies vertically in the micrographs.

16.2.5
Technological Aspects

From a technological point of view, it has to be emphasized that the ingot microstructure is a key factor with respect to recrystallization kinetics. Coarse microstructural constituents as well as inhomogeneities in both the microstructure and the composition are difficult to remove by wrought processing and can result in flow localization and incomplete recrystallization. Apparently, good recrystallization behavior is obtainable for alloys with Al contents in the range of 48–50 at.% Al that may contain grain-refining coagulants and/or dispersoids. Unfortunately

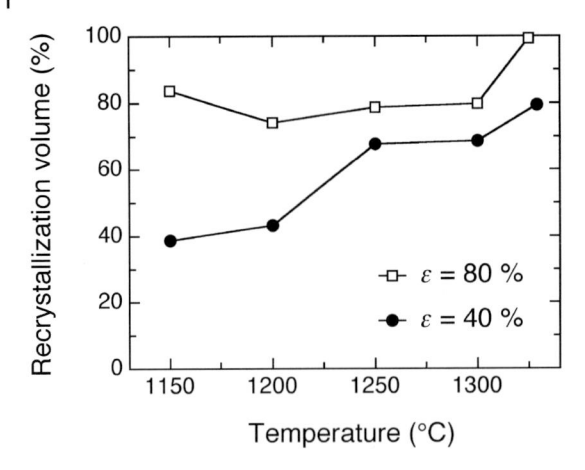

Figure 16.24 Dependence of the volume fraction of recrystallized grains or colonies on deformation temperature for Ti-45Al-8Nb-0.2C ingot material. Deformation in compression at a strain rate of $\dot{\varepsilon} = 1.4 \times 10^{-3}$ s^{-1} to the strains indicated.

such alloys have a relatively low strength. Thus, some attention should be paid in further alloy development to the identification of higher-strength alloys, which exhibit fine microstructures even in large ingots. Corresponding alloy-design concepts have been proposed by Naka et al. [124], Küstner et al. [125], Jin et al. [126], Wang et al. [127] and Imayev et al. [128]. Further attempts have been aimed at different alloy constitutions, for example, alloys with significant fractions of the β phase in the hot-working temperature range. Kobayashi et al. [129] and Takeyama and Kobayashi [130] have developed an alloy concept that is based on the transformation path $\beta + \alpha \Rightarrow \alpha \Rightarrow \beta + \gamma$. These alloys can be worked in the β + α phase field and subsequently transformed to fine lamellar microstructures.

Regarding the anisotropic hot-working behavior it might be concluded that an orientation of lamellar interfaces at 90° to the main deformation direction is best suited in order to obtain high volume fractions of recrystallized grains, since in this case the recrystallization kinetics are rather fast and the development of shear bands can be suppressed to a certain degree. Due to the radial symmetry of the casting texture of cylindrical ingots, this preferable lamellae orientation can easily be realized in primary processing through hot-extrusion of whole ingots. If possible, it also seems beneficial to change the working direction during wrought processing because new recrystallized grains are formed through kinking of remaining lamellae and localized deformation in existing shear bands might be reduced. Koeppe et al. [32] found that increased microstructural homogeneity could be obtained when multistep forging was performed without changing the working direction. However, in their experiments a banded equiaxed microstructure had already been formed after the first forging step and an intermediate annealing; thus, remnant lamellae played no role with respect to microstructural evolution during subsequent hot working steps.

16.3
Workability and Primary Processing

16.3.1
Workability

The general low workability of γ titanium aluminide alloys makes wrought processing expensive and restricts the availability of industrial equipment, for example, due to the required temperatures, (low) strain rates, press capacities and die materials. It is therefore important to examine the working window in terms of temperatures, strain rates and strains and to understand its dependence on alloy composition and microstructure. As described in Section 16.1, this has been done in a number of studies by testing cylindrical specimens in compression followed by inspection of the deformed specimens. In these studies, workability maps have been determined that define regions that allow crack-free working in terms of strain rate and temperature. Figure 16.25 shows examples of such workability maps that have been established for different alloys, microstructures and applied strains. As can be seen, there are remarkable differences with respect to the conditions for successful hot working. The workability certainly depends on many factors such as the geometry of the test specimen, the lubrication conditions and the strain; however, the microstructure apparently has a major influence on the workability. This becomes obvious from the results of Singh et al. [131] and Srinivasan et al. [132], included in Figure 16.25, who compared fine-grained powder material with coarse-grained cast material of nearly the same alloy composition under identical test conditions. The microstructural differences resulted in a shift of the line separating sound from cracked material by temperature of up to 200 K and by more than two orders magnitude of the strain rate. Other studies have also shown that the workability of γ alloys can be significantly improved by microstructural refinement [35, 134–136].

In the study mentioned earlier (Section 16.1), Nobuki and Tsujimoto [7] investigated the hot-working behavior of a series of binary cast alloys by compression testing over a wide range of strain rates and temperatures. The workability maps obtained in this study (Figure 16.26) clearly showed that the workability systematically depends on the Al content of the alloys. The alloy with the highest Al concentration exhibited the widest range of conditions for crack-free working, while for decreasing Al content cracking occurred at higher temperatures and lower strain rates. As described in Section 16.1.2, Nobuki and Tsujimoto [7] used the constitutive Equation 16.1 to fit the peak stress of the flow curves to the strain rate and temperature and then used these results to plot isostress contours onto the workability maps. Two examples of such workability maps are shown in Figure 16.27 from which it can be concluded that cracking occurs at a critical stress for each alloy. The critical stress amounted to 250–300 MPa for Al contents of 48.2–50.2 at.% and to approximately 150 MPa for Al concentrations of 43.8–44.9 at.% [7]. The authors explained the dependence of workability on the Al concentration by the variation of the cast microstructure with Al content. In the range of Al

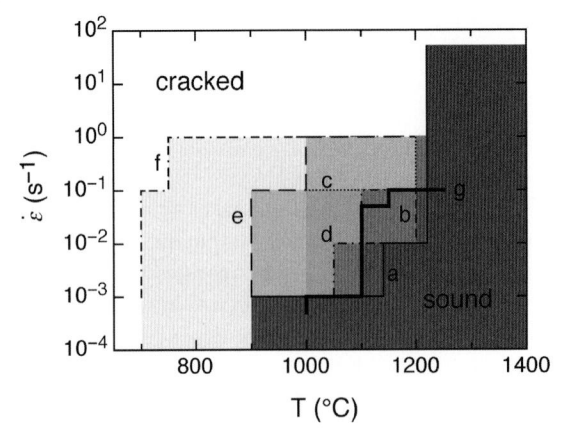

Figure 16.25 Hot-workability maps for different γ TiAl alloys. The lines show conditions up to which sound deformation was possible during isothermal hot compression. (a–f) similar alloys with Al concentrations of 47–49 at.%, (g) Ti-45Al-(8–9)Nb-W-B-Y. (b) and (c): same alloy, height reduction of 80% and 40%, respectively. (e) and (f): similar alloys, (e) cast material with a coarse microstructure and (f) a fine-grained powder material. (a) Ti-49Al-2Mn-2Nb, as-cast material, lamellar microstructure. Specimen diameter 6 mm, specimen height 8 mm, lubricated with a lead-based glass below 1140 °C and a silica-based glass above 1140 °C, 50% height reduction [24]. (b) and (c) Ti-47Al-1V, cast material, homogenized 24 h at 1200 °C, nearly lamellar microstructure. Specimen diameter 6 mm, specimen height 9 mm, height reduction (b) 80% and (c) 40% [30]. (d) Ti-47Al-2Cr-1Nb, cast material, homogenized 8 h at 1100 °C, HIP 4 h/1250 °C/170 MPa. Nearly lamellar microstructure, 400 μm colony size, specimen diameter 8 mm, specimen height 12 mm, mica sheet lubricant, 40% height reduction [29]. (e) Ti-48Al-2Nb-2Cr, as-cast material, nearly lamellar microstructure, colony size 50–500 μm. Specimen diameters 6.35–19 mm, height to diameter ratio 1.5, graphite-based lubricant, 50% height reduction [131, 132]. (f) Ti-47Al-2Nb-2Cr, powder material, equiaxed microstructure, grain size 3–20 μm. Specimen diameters 6.35–19 mm, height to diameter ratio 1.5, graphite-based lubricant, 50% height reduction [131, 132]. (g) Ti-45Al-(8–9)Nb-W-B-Y, cast material, HIP, nearly lamellar microstructure, colony size 70 μm. Specimen diameter 10 mm, specimen height 16 mm, 50% height reduction [133].

contents investigated, the size of the lamellar colonies in the cast material increased with decreasing Al concentration that leads to strong flow localization and slow recrystallization kinetics as discussed in Sections 16.2.2 and 16.2.3. Both processes certainly have a detrimental effect on the workability and could explain the observed composition dependence of the workability. Here, it is of interest to note that according to Equation 16.1, the strain-rate sensitivity m can be written as a function of the flow stress,

$$m = \frac{1}{n\alpha\sigma(\dot{\varepsilon},T)} \tanh(\alpha\sigma(\dot{\varepsilon},T)) \tag{16.8}$$

In this equation, the stress exponent n and the stress mulitiplier α are constants. Thus, for a critical constant flow stress $\sigma(\dot{\varepsilon},T)$, above which cracking occurs, the

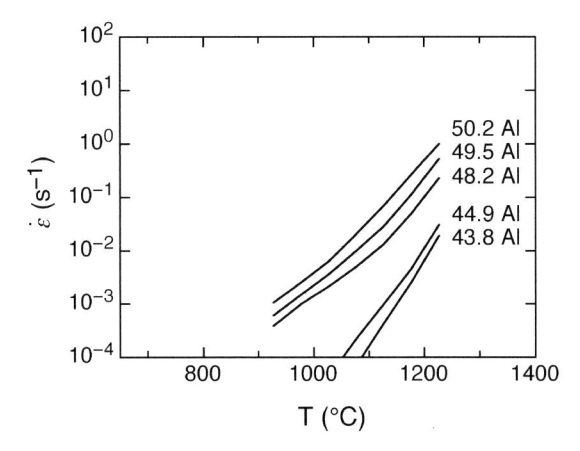

[Graph: ε̇ (s⁻¹) vs T (°C), showing curves for 50.2 Al, 49.5 Al, 48.2 Al, 44.9 Al, 43.8 Al]

Figure 16.26 Hot-workability maps for different binary alloys as determined by Nobuki and Tsujimoto [7] using isothermal hot-compression tests. The lines show conditions below which sound deformation was possible up to a true strain of 1.0. The cast materials had been homogenized for 24 h at 1200 °C. The microstructures varied from fully lamellar for low Al concentrations over duplex to near-γ microstructures for high Al contents. The mean grain/colony size ranged between 130 and 560 μm, with the smallest grain/colony size observed for the Ti-48.2Al alloy. The cylindrical compression specimens had a diameter of 18 mm and a height of 24 mm prior to testing.

strain-rate sensitivity $m(\dot{\varepsilon},T)$ is also constant and could be the critical parameter. In other words the isostress contours shown in the workability maps of Figure 16.27 correspond to contours of constant m. From the results derived using a constitutive analysis on the basis of Equation 16.1, the strain rate m was calculated for several materials including those investigated by Nobuki and Tsujimoto [7]. The parameters used for the calculations are listed in Table 16.1. As shown in Figure 16.28 a strain-rate sensitivity $m = 0.2$–0.25 corresponds to the critical flow stresses reported by Nobuki and Tsujimoto [7]. In the following discussion the underlying mechanisms of failure during hot working will be discussed and fracture criteria identified. This will help give a better understanding of the workability maps discussed above and elucidate the role of factors such as stress and strain rate.

From a general point of view workability problems can be grouped into fracture-controlled failures such as hot-shortness and triple-point cracking or grain boundary cavitation and flow-localization-controlled failures that may give rise to gross shear fracture and microstructural inhomogeneities [22]. Here, it is worthwhile to recall the low cleavage stress and cohesive strength of both the γ and the α_2 phases and the resulting susceptibility to crack nucleation in γ alloys [15, 137–139]. In the γ phase, the {111} planes serve as both dislocation glide planes and twin-habit planes. As the {111} planes exhibit a low cleavage energy, blocked slip or twinning may lead to crack nucleation [18]. Thus, the deformation behavior as well as the plastic anisotropy play a role with respect to workability together with the various

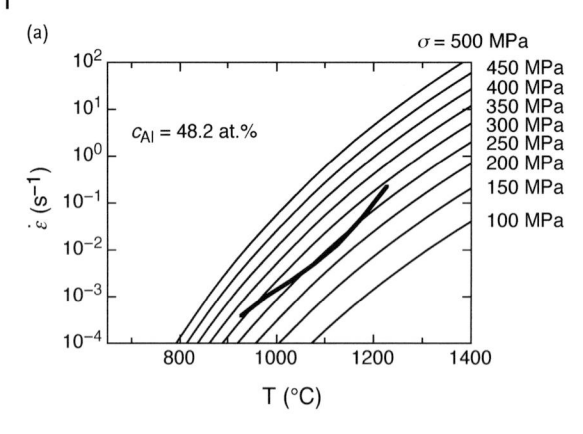

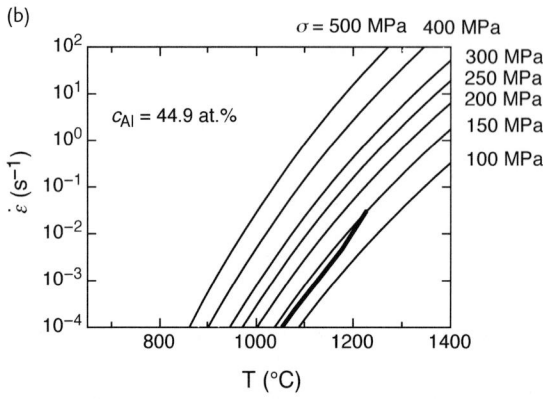

Figure 16.27 Hot-workability maps for two binary alloys as determined by Nobuki and Tsujimoto [7]. The thick lines show conditions up to which sound deformation was possible during isothermal hot compression to a true strain of 1.0. For more details regarding the specimens see Figure 16.26. In addition to the fracture loci lines of constant flow stress are shown in the diagrams that were obtained by fitting measured peak stresses in the flow curves to the constitutive Equation 16.1 and calculating the dependence of flow stress on temperature and strain rate [7]. (a) Ti-48.2 Al, relatively fine microstructure with 30 vol.% lamellar colonies, (b) Ti-44.9 Al, coarse fully lamellar microstructure.

recovery and recrystallization processes during hot deformation. Although it is directly evident that an understanding of the different failure processes and their dependence on alloy composition, microstructure and processing conditions is particularly important for the design of hot-working routes, with the exception of the superplastic deformation of fine-grained material with equiaxed microstructures, workability criteria have only been developed by Semiatin and coworkers [12, 14, 35, 140–142].

In an attempt to understand the mechanisms of both brittle and ductile failure, Seetharaman and Semiatin [35, 142] investigated the hot-tensile behavior of a Ti-49.5Al-2.5Nb-1.1Mn alloy in both the cast and wrought conditions. The alloy

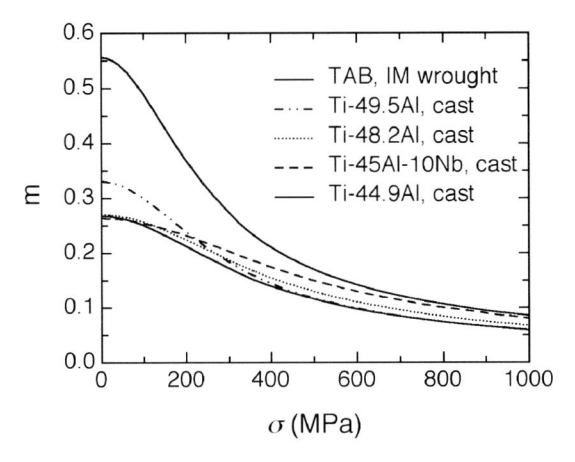

Figure 16.28 Dependence of the strain-rate sensitivity $m = d\ln\sigma/d\ln\dot{\varepsilon}$ on flow stress, according to the constitutive Equation 16.1 (see Equation 16.8). For this calculation the fitting parameters given in Table 16.1 were used. Please note that some of the alloys have significantly lower flow stresses than the maximum stresses shown here. TAB: Ti-47Al-1.5Nb-1Cr-1Mn-0.2Si-0.5B, ingot material (IM), extruded at 1250 °C, annealed 2 h/1030 °C/furnace-cooled, near-γ microstructure with a mean grain size of 3.3 μm. Ti-44.9Al, Ti-48.2Al and Ti-49.5: cast materials with fully lamellar or duplex microstructures investigated by Nobuki and Tsujimoto [7] (see Figure 16.26 and 16.27).

showed a rather sharp transition from brittle failure that was characterized by growing wedge-shaped cracks, to ductile behavior in which microvoid initiation and growth occurred. For each microstructural condition, the brittle-to-ductile transition temperature increased with increasing strain rate. From an Arrhenius analysis of the brittle–ductile transition temperature, the authors concluded that dynamic recrystallization caused the change in failure mode. This conclusion was verified by microstructural examinations. For the cast material (grain size 125 μm) an activation energy of 2.95 eV was determined, while for the wrought material (grain size 35 μm) a value of 1.96 eV was obtained. The latter value indicates a significant contribution to plastic flow by grain-boundary sliding accommodated by grain-boundary diffusion or intragranular slip [142].

In this work [35, 142] fractographical examinations showed that for high strain rates and low to intermediate temperatures, brittle intergranular fracture occurred. For lower strain rates a significant amount of wedge cracking and cavitation was observed. The lamellar colony boundaries were found to be the preferred sites for wedge-crack nucleation. Wedge cracks are commonly initiated at grain boundaries and triple points by grain-boundary sliding if stress concentrations are not effectively relieved by processes as matrix slip, grain-boundary migration or grain-boundary diffusion [142]. Alternatively, at higher strain rates wedge cracks can be formed by the intersection of slip bands/deformation twins with grain boundaries accompanied by grain-boundary sliding. In another study, Semiatin and Seetharaman [143] observed wedge cracking after isothermal hot-compression testing not only at bulged free surfaces where secondary tensile stresses had developed, but also in the center of the specimens where the stress state was essentially compressive. Here, it is notable that the modeling study of Dao et al. [114], already described in Section 16.2.3, has shown that for lamellar materials very high hydrostatic tensile stresses can occur at certain colony boundary sites when the material is deformed in compression. Further, the authors found strong localization of deformation at some colony boundaries. Thus, it can be explained why wedge cracks are initiated at such boundaries during hot deformation under conditions where grain-boundary sliding is active. The occurrence of wedge cracking is detrimental, even if it does not lead to gross fracture of the workpiece during hot working because such cracks result in serious degradation of the mechanical properties [142]. Deformation conditions under which wedge cracks can evolve should therefore be avoided during hot-working operations.

Based on their experimental work and on models for intergranular fracture in literature Semiatin and Semiatin [143], Semiatin et al. [14] and Seetharaman and Semiatin [142] proposed that wedge cracking occurs if a critical value of $\sigma_p \sqrt{d}$ is exceeded with σ_p being the peak stress and d the grain/colony size. For cast material of the alloy Ti-49.5Al-2.5Nb-1.1Mn, Semiatin and Seetharaman [141] found the critical value of $\sigma_p \sqrt{d}$ to be in the range 1.5–3 MPa m$^{1/2}$, whereas for the same alloy in the wrought condition as well as for the alloy Ti-46.6Al-2.7Nb-0.3Ta $\sigma_p \sqrt{d}$ had a critical value of approximately 1 MPa m$^{1/2}$ [35, 143]. The authors also observed a slight temperature dependence of the critical value of $\sigma_p \sqrt{d}$. By applying this criterion to the results of Nobuki and Tsujimoto [7] for different alloys and colony/grain sizes, values of 3.1 to 4.9 MPa m$^{1/2}$ are obtained that are in rough agreement with the results of Semiatin and coworkers [14, 35]. However, since Nobuki and Tsujimoto [7] determined the workability limits by the occurrence of free-surface cracking and did not consider wedge crack initiation, it remains an open question whether wedge cracking actually contributed to failure during their hot-working experiments.

As noted above, Seetharaman and Semiatin [35, 142] observed cavitation in specimens deformed under ductile failure conditions at higher temperatures and lower strain rates. The void fraction strongly decreased with distance from the fracture surface, but a certain fraction of voids was present throughout the entire deformed region of the specimens. Obviously, cavity formation, growth and coalescence control the free-surface fracture under these conditions. Seetharaman et al. [144] found that fracture at the bulged surface of pancakes during forging could be well predicted by the maximum tensile work criterion

$$\int_0^{\bar{\varepsilon}_f} \sigma_{max} d\bar{\varepsilon} = \text{constant} \tag{16.9}$$

proposed by Cockroft and Latham [145]. In this equation σ_{max} is the highest local tensile stress, $\bar{\varepsilon}$ the true strain and $\bar{\varepsilon}_f$ the true strain at fracture. The authors validated this criterion by measuring the critical tensile work in tension testing and comparing it with the calculated work that was expended during forging due to secondary tensile stresses at the equator of the bulge [14, 144].

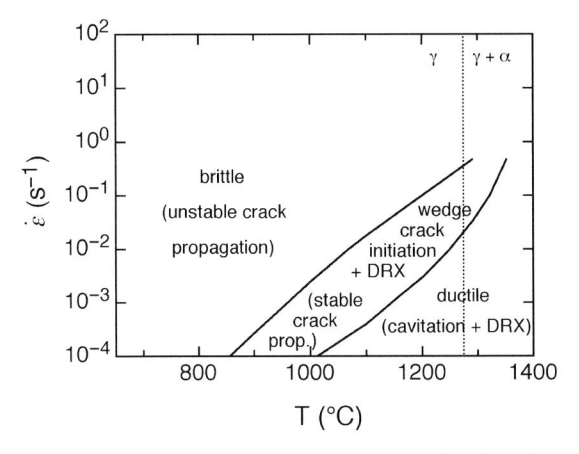

Figure 16.29 Processing map showing different failure mechanism regimes that occurred during hot working of Ti-49.5Al-2.5Nb-1.1Mn in the cast + HIP condition [142]. The microstructure of the material consisted of equiaxed γ grains with a mean grain size of about 125 μm and 10–15 vol.% lamellar colonies. According to its Al content, the alloy is believed to be in the γ single-phase field up to the temperature indicated in the diagram. α_2 phase originating from casting that does not correspond to the thermodynamic equilibrium was nevertheless, still present after hot working.

On the basis of their analysis of the workability and the workability criteria they developed, Seetharaman and Semiatin [14] constructed a processing map for a cast γ alloy in which regimes of different failure mechanisms are shown (Figure 16.29). This map displays regions of wedge crack-induced brittle failure, wedge-crack/cavitation-induced ductile failure as well as rupture; it illuminates the experimentally determined workability maps presented in Figures 16.25 and 16.26. The authors recommended that hot-working operations be performed in the region marked as ductile in order to avoid not only gross fracture but also the formation of wedge cracks that are detrimental for mechanical properties. It should be mentioned that flow localization has not been considered in their processing map. This will be discussed below.

In Section 16.2 extensive evidence has been presented that γ alloys are prone to flow localization, even in the absence of frictional or chilling effects. Flow localization in particular arises during hot working of alloys with lamellar microstructures and results in nonuniform microstructures. This seriously affects the mechanical properties of wrought material. In extreme cases, shear bands might traverse the whole workpiece and become macroscopically visible. Figure 16.30 shows an extruded bar of the alloy Ti-45Al-8Nb-0.2C after the canning material was removed. The steel can had been separated from the TiAl workpiece by a Mo foil, and thus the extruded assembly could be easily dismantled. On the surface of the TiAl bar, grooves can be seen that run parallel to the extrusion direction and clearly indicate that deformation was nonuniform during extrusion. The grooves can be

Figure 16.30 Extruded bar of the alloy Ti-45Al-8Nb-0.2C with surface grooves indicating nonuniform deformation. Prior to extrusion the ingot had been canned in a steel capsule with a Mo foil separating the capsule from the TiAl workpiece. Extrusion was performed at 1230 °C to an area reduction ratio of 14:1. After extrusion the canning material could easily be removed.

understood as resulting from localized shear processes that manifest themselves at the workpiece/can interface. Similar observations also were made after the compression testing of small cylindrical specimens and forging of larger workpieces [140, 143, 146]. The nonuniform deformation can give rise to surface roughness and markings, shearing of the test pieces (Figure 16.31), and in severe cases leads to cracking (Figures 16.32 and 16.33). Imayev et al. [146] observed that bands of fine grains, which were attributed to localized recrystallization, had been formed near the bulged free surface of a large Ti-45Al-10Nb forging. In these bands cavities and cracks were found. Obviously, the deformation had been concentrated in the bands, which resulted in the formation of voids and cracks and even induced premature failure (Figure 16.33). In alloys with an Al content of or below 45 at.% combined with high Nb additions, bands of the β(B2) phase often occur and form a percolating substructure throughout the material. It might be expected that this phase has a relatively low yield stress and carries most of the deformation. Deformation may therefore preferentially occur along bands of the β(B2) phase. This significantly increases the workability of alloys with sufficiently high amounts of the β(B2) phase [147]. Nevertheless, in materials containing lower levels of β(B2), shear localization and cracking can occur during deformation [28]. During extrusion gross fracture due to flow localization usually is prevented by the superimposed hydrostatic compression; however, cracking along shear bands has been observed in extruded material and might deteriorate the mechanical properties (Figure 16.34).

According to Semiatin and Jonas [21] and Semiatin et al. [14], the occurrence of flow localization can be estimated by the so-called flow localization or α parameter

$$\alpha = (\gamma - 1)/m \tag{16.10}$$

Figure 16.31 Compression specimens after isothermal hot deformation to a logarithmic strain of 0.6 ($\varepsilon = 0.45$). The specimens had been lubricated with a glass lubricant. (a) and (b): Ti-45Al-10Nb ingot material which after hot-isostatic pressing at 1260 °C had a lamellar microstructure with a colony size of 500 µm containing bands of the β/B2 phase. Initial specimen diameter 18 mm, initial specimen height 30 mm. (a) $T = 1150$ °C, $\dot{\varepsilon} = 10^{-1}$ s^{-1}, (b) $T = 1200$ °C, $\dot{\varepsilon} = 10^{-2}$ s^{-1}. (c) and (d): TAB alloy Ti-47Al-1.5Nb-1Cr-1Mn-0.2Si-0.5B, extruded at 1250 °C, annealed 2 h/1030 °C/furnace-cooled, near-γ microstructure with a mean grain size of 3.3 µm. Initial specimen diameter 12 mm, initial specimen height 20 mm. (c) $T = 1000$ °C, $\dot{\varepsilon} = 10^{-1}$ s^{-1}, (d) $T = 1050$ °C, $\dot{\varepsilon} = 10^{-2}$ s^{-1}.

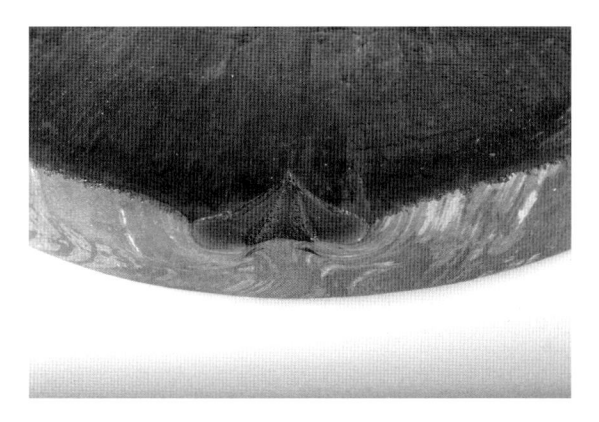

Figure 16.32 Pancake of Ti-45Al-5Nb-0.2B-0.2C (diameter 165 mm, height 22.2 mm) that was isothermally forged at 1150 °C and a strain rate of 5×10^{-3} s^{-1} to a strain of 80%. The material had been hot extruded prior to forging. The image shows cracking that might have occurred due to strain localization during forging.

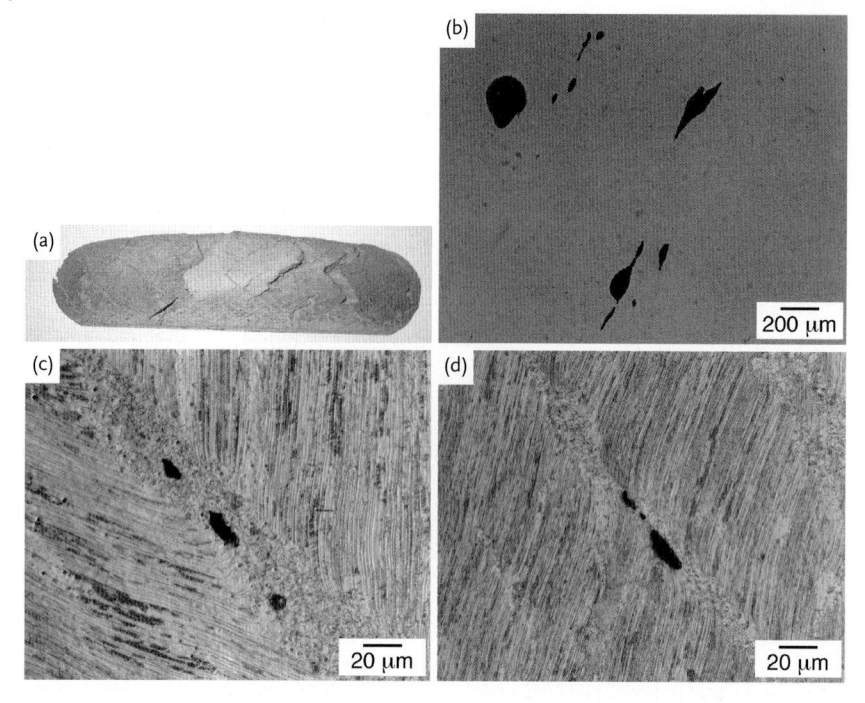

Figure 16.33 Failure of Ti-45Al-10Nb during isothermal forging at 1050 °C and a strain rate of $10^{-3}\,s^{-1}$. (a) Severe cracking on the bulged surface, (b–d) optical micrographs of shear bands, wedge cracks and voids, (b) after mechanical polishing, (c) and (d) after electropolishing and etching. The forging direction is vertical [146].

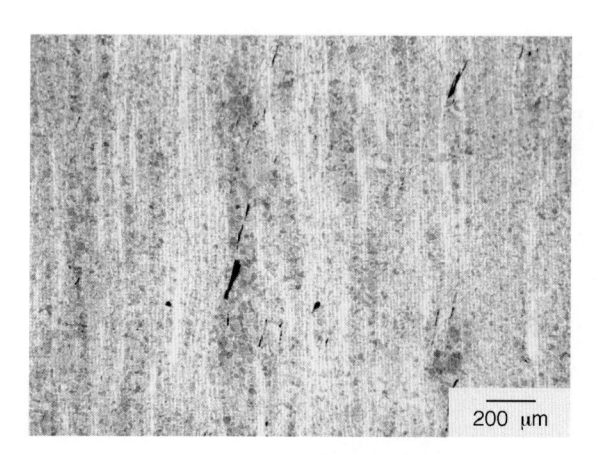

Figure 16.34 Cracking along shear bands in extruded material of Ti-45Al-8Nb-0.2C (extrusion temperature 1230 °C).

with γ being the rate of flow softening $\gamma = (1/\sigma)(d\sigma/d\varepsilon)|_{\dot{\varepsilon},T}$ and m the strain-rate sensitivity. A necessary condition for the onset of flow localization is $\alpha = 0$, that is, $\gamma = 1$. However, noticeable flow localization usually takes place when the value of α is larger than 5 [14, 21]. From Equation 16.10 it is directly evident that strong flow softening and low strain-rate sensitivities promote flow localization. Considering the general dependence of γ and m on temperature and strain rate, working at high temperatures and low velocities has an advantage with regard to reducing flow localization. Semiatin et al. [141] determined values of the flow-localization parameter that exceeded 5 for Ti-45.5Al-2Cr-2Nb at working temperatures below 1150 °C. At this temperature, shear bands and shear cracks developed during extrusion but they were not observed for higher extrusion temperatures. The authors found lower values of the α parameter if the material was worked in the wrought rather than the cast condition. Thus, the occurrence of flow localization can be well predicted on the basis of the $\alpha \geq 5$ criterion, as has been found for Ti alloys [14]. The study also confirms the influence of the microstructure on flow localization; however, other factors such as nonuniform temperature and phase distributions have also to be considered with respect to inhomogeneous flow of γ alloys (see Section 16.1.1).

Summarizing, failure during hot working of γ alloys is governed by brittle fracture, wedge-crack initiation and flow localization up to relatively high temperatures. This, in particular, severely restricts the working window of cast materials with coarse and inhomogeneous microstructures to low strain rates and very high temperatures. On the other hand, the workability of γ alloys sensitively depends on the composition and the microstructure; hot-working operations such as sheet rolling or superplastic forming are possible if the processing steps and conditions are suitably adapted to the material. Unfortunately, there is no easy rule for the design of hot-working routes and much work has to be expended to find optimized alloys and working conditions.

16.3.2
Ingot Breakdown

With respect to the wrought processing of γ TiAl alloys, primary hot working of the usually coarse-grained, segregated and textured ingot material certainly is the most critical step. This, for instance, becomes apparent from Figure 16.25, which has been discussed in previous sections. As indicated by the various workability maps shown in Section 16.3.1, such materials have to be worked at temperatures higher than 1000 °C. Thus, conventional hot working with cold or mildly heated dies cannot be applied without canning of the billet. This is because either die chilling or the required high strain rates would result in cracking. Recently, however, alloys with high amounts of β phase have been developed and can withstand conventional forging, but their good workability is potentially obtained at the expense of high-temperature strength and creep resistance [147–151]. These alloys will be later discussed in more detail. Generally for TiAl alloys, isothermal processing has to be applied or the workpieces must be canned (quasi-isothermal working). From the aforementioned workability maps it can be seen that for isothermal forging, only relatively low strain rates can be used as the maximum temperature is limited by the die material. For the commonly employed TZM alloy (Mo-0.5 wt.%Ti-0.08 wt.% Zr-0.02 at.% C) this is around 1150 °C. In the literature concerning wrought processing of γ alloys, nothing has been reported on the use of die materials capable of higher working temperatures. The higher working temperatures that are commonly used for hot extrusion can therefore only be applied if the workpieces are canned. The workability maps shown in Figures 16.25–16.27 and 16.29, indicate that the strain rate and especially the temperature should be precisely controlled in order to avoid fracture during hot working. One should also bear in mind that the flow stress of γ alloys is strongly dependent on the strain rate and in particular on temperature, as can be seen in Figure 16.27 and has been discussed in Section 16.1. Control of the working temperature seems to be critical if conventional canned hot working is applied; for this reason attention should be paid to the can design and its heat insulation. Even small variations of about 20 K in the temperature distribution across the workpiece during hot working might result in microstructural inhomogeneities and induce cracking at the surface. For an industrial wrought-processing route, neither the resulting property variations or in the worst case the scrap material are acceptable.

Since casting porosity is not closed by forging [33], the ingot material is usually compacted by hot-isostatic pressing for some hours at a temperature in the (α + γ) region under a pressure of 150–250 MPa [12, 152, 153] prior to hot working. Often the ingot material is also subjected to a homogenization heat treatment [12, 15, 152–154]. This appears to be extremely important due to the strong dependence of mechanical properties, as well as workability (see Figure 16.26), on the alloy composition and homogeneity. Most γ alloys solidify through one or two peritectics, which results in pronounced microsegregation since among other reasons no single-phase field is traversed directly after solidification. Variations in Al concentration of up to 10 at.% on a scale of 50 µm to several mm have been observed in small ingots [155, 156]. Such long-range segregation was found for a low cooling rate of about 30 K min^{-1} [156]. For industrially fabricated ingots with diameters of at least 120 mm microsegregation of around 3 to 5 at.% Al was determined on a scale of several hundred µm (Figure 16.35) [156–159]. With respect to homogenization it is notable that in a high Nb-containing alloy, Ti-45Al-10Nb, the variation of the slowly diffusing element Nb was around 3 at.% over the same length scale.

Such chemical segregation results in microstructural inhomogeneity that is detrimental for the workability and persists after hot working [31, 154, 157, 158]. Furthermore, the constitution of the alloy varies locally and coarse-grained regions can occur after heat treatments or hot-working steps. Thus, there is no doubt that at least in peritectically solidifying alloys a homogenization treatment would be extremely useful. Semiatin and McQuay [160] studied the homogenization kinetics of Ti-47Al-2.5Nb-0.3Ta and found that heat treatment at temperatures above the α-transus temperature (as high as 1400 °C) was needed to remove microsegregation. As found by Koeppe et al. [32], such high-temperature treatments indeed

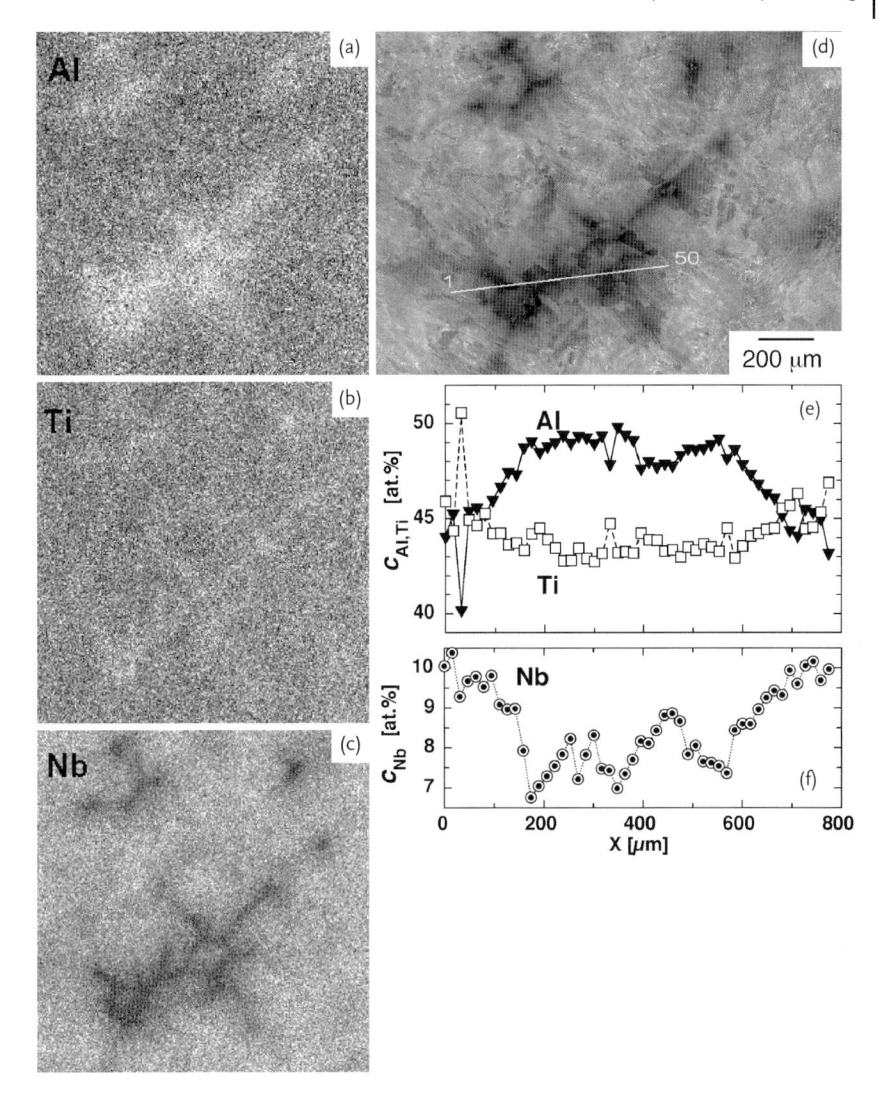

Figure 16.35 Dendritic segregation in a 150-kg ingot (diameter 220 mm, produced by vacuum arc remelting) of the alloy Ti-45Al-10Nb after hot-isostatic pressing (4h/1220 °C/200 MPa) [157]. (a–c) X-ray maps for the elements Al, Ti and Nb taken using an EDX analysis system in a scanning electron microscope. (d) Backscattered electron image of the same region as shown in (a–c). (e) and (f) Quantitative EDX line scans recorded along the line indicated in (d).

result in microstructures that are significantly more homogeneous after hot working. Homogenization treatments of this nature, however, caused significant coarsening of the lamellar colonies [160] and thus their benefit is questionable with respect to both, workability and the recrystallization kinetics during forging (see Figure 16.11). Dimiduk et al. [15] concluded that subtransus homogenization treatments have a modest beneficial effect on workability, while those in the single α-phase field are actually detrimental. The authors also noted that homogenization might be successful at even higher temperatures within the (α + β) region so that the β phase can be used to control the α-grain size. In agreement with Semiatin and McQuay [160], Brossmann et al. [158, 161] observed in Ti-45Al-10Nb that annealing times longer than 1 h at temperatures above the α-transus were required to remove segregation on a scale larger of 10 μm. As the homogenization temperature of 1420 °C that was used is located in the (α + β) phase field, no significant microstructural coarsening occurred [158, 161]. In summary, it seems that for peritectically solidifying alloys the segregation can be reduced by heat treatment, but it can only be removed at very high temperatures. In particular, the addition of refractory alloying elements makes homogenization more difficult to achieve [160]. A promising approach is to perform heat treatments in the (α + β) region, but this would require optimized alloy chemistry in order to avoid too much β phase [15]. The situation is different for alloys that solidify solely through the β phase, since only modest microsegregation occurs [155]. Wrought processing of such alloys will be discussed later in this section.

Typical conditions for large-scale industrial isothermal and quasi-isothermal (canned) forging are $T = 1000$ to $1200\,°C$ and logarithmic strain rates in the range 10^{-3} to $5 \times 10^{-2}\,s^{-1}$. Under these conditions billets of 50 kg can be forged to a strain reduction of 5:1 without failure, an example is shown in Figure 16.36 [157, 162]. Kim [163] even reported the successful forging of a 210-kg billet produced by powder metallurgy. More technological details on crack-free industrial or near-industrial forging, together with literature references are found in Table 16.2. After forging, Semiatin et al. [123] reported that macrostructures for both isothermally and quasi-isothermally forged pancakes were relatively uniform and exhibited only small "dead-metal" zones. The microstructure that developed under these conditions typically appeared as a more or less homogeneous partially recrystallized microstructure with all the inhomogeneities described in previous sections. These inhomogeneities comprised of remnant lamellar colonies containing kink bands, shear bands, necklace structures as well as bands containing both fine and coarse grains that can be attributed to chemical inhomogeneity. Examples of microstructures after industrial forging are shown in Figures 16.37 and 16.38. The fraction of globularized microstructure in forged material has been found to be fairly high [12, 29, 30, 32, 157, 169], although fractions as low as 30% have also been reported [169]. In the globularized microstructural constituent, grain sizes between 2–4 μm in boride-containing alloys have been observed, while values of 12–18 μm [32], 2.7–30 μm [172] and 20 μm [29] have been reported for B-free alloys. Please note that the finer microstructure obtained in the boron-containing alloy can be attributed to the finer cast microstructure of such alloys and their recrystallization behavior, as explained in Section 16.2.2. Compared to isothermal forging, conventional forging using canned workpieces typically resulted in finer, more uniform microstructures [12]. This technique expands the processing window by decreasing the minimum temperature, increasing the highest strain rate, and increasing the maximum strain under which deformation occurs without observable macro-

Figure 16.36 Isothermal forging (preheat temperature 1150 °C, die temperature 1050 °C) of a 270-mm diameter billet machined from a HIPed ingot of Ti-47Al-1.5Nb-1Cr-1Mn-0.2Si-0.5B. The resulting forged pancake with a diameter of 581 mm (80% height reduction) is shown in the lower micrograph [157].

scopic failure. Thus, by canned forging, a larger amount of strain energy can be imparted, which is certainly beneficial for homogeneous dynamic recrystallization. However, due to the flow-stress mismatch between the TiAl and the can materials employed (stainless steel [12, 29, 133], low carbon steel [169]) and also heat-transfer effects, nonuniform flow of the can and workpiece can take place and thus attention has to placed on the can design. Jain *et al.* [173] have applied finite-element modeling in order to find an optimized can design and to identify suitable processing conditions. The modeling results were validated by forging trials. The work

Table 16.2 Processing conditions for successful industrial or near-industrial forging of TiAl alloys (H: height).

Alloy	Workpiece dimensions	Forging conditions	Reference
Ti47Al-1V		Isoth., $5 \cdot 10^{-4}\,s^{-1}$, 1150 °C, 80%	[30]
Ti-48Al-2.5Nb-0.3Ta		Isoth., $1 \cdot 10^{-3}\,s^{-1}$, 1105 °C, 75%	[31]
Ti-48Al2Nb-2Cr		Isoth., $3 \cdot 10^{-3}\,s^{-1}$, 1175 °C, 80%	
Ti-47Al-1Cr-1V-2.5Nb	Ø 70 mm × H 100 mm	Isoth., 1180 °C, 88%	[164]
Ti-48Al-2Cr	up to Ø 190 mm	Conv., can material not given, $1\,s^{-1}$, 1125–1360 °C, 85%	[33, 152, 165]
Ti-48Al, Ti-48Al-2Cr	Ø 35 mm × H 80 mm	Isoth., $10^{-3}\,s^{-1}$, 920–1350 °C, 70–78%	[32]
Ti-47Al-2Cr-2Nb	Ø 82 mm × H 40, 103	Isoth., $10^{-3}\,s^{-1}$, 1050 °C, 62, 85%, Isoth., $10^{-2}\,s^{-1}$, $10^{-3}\,s^{-1}$ 1150 °C, 62, 85%,	[166]
Ti-47Al-2Cr-2Nb-1Ta			
Ti-46Al-2Cr-2Nb-1Ta			
Ti-47Al-2Cr-4Ta			
Ti-45Al-5Nb-1W			
Ti-45.5Al-2Cr-2Nb	Ø 66 mm × H 76 mm	Isoth., $1.5 \times 10^{-3}\,s^{-1}$, 1150 °C, 83%, corner breaking, TZM dies, boron nitride lubricant Conv., stainless steel can, heat isolation, $2.5 \times 10^{-1}\,s^{-1}$, 1150, 1200, 1250 °C, 83% controlled dwell technique, flat tool steel dies, Deltaglaze 69 glass lubricant, uncanned preform: severe cracking	[123]
	Ø 60 mm × H 86 mm		
Ti-47Al-2Cr-2Nb	Ø 70 mm × H 50 mm, Ø 70 mm × H 100 mm	Isoth., 1st step $1.7 \times 10^{-3}\,s^{-1}$, 1175 °C, 50%	[167, 168]
		Intermediate heat treatment 48 h/1200 °C, 2nd step $1.7 \times 10^{-3}\,s^{-1}$, 1175 °C, 75%	
Ti-47Al-1.5Nb-1Cr-1Mn-0.2Si-0.5B (nominal composition), actual Al content 45 at.%	Ø 270 mm × H 250 mm	Isoth., $10^{-3}\,s^{-1}$, preheat 1150 °C, die 1050 °C, 80%	[162]

Table 16.2 (Continued)

Alloy	Workpiece dimensions	Forging conditions	Reference
Ti-42Al-5Mn	Ø 150 mm × H 150 mm	Conventional, uncanned billet, no heat isolation, preheat temperature 1300 °C, die temperature 100–200 °C, 30 s transport time between furnace and press, 10–20 mm/s ram speed (6.7×10^{-2}–$1.3 \times 10^{-1} s^{-1}$), 67%	[147, 148]
Ti-45.2Al-3.5(Nb Cr, B) Ti-44.2Al-3(Nb Cr, B)	Ø 70 mm × H 120 mm	1st step conv., low carbon steel can, $5 \times 10^{-2} s^{-1}$, α + γ phase field, 2nd step isoth., 10^{-3}–$10^{-2} s^{-1}$, α_2 + γ phase field, glass lubricant Total strain 87.5%	[169]
Ti-47Al-2Cr-1Nb	Ø 88 mm × H 122 mm	Isoth., stainless steel can, 1.2–$5 \times 10^{-3} s^{-1}$, 1150–1200 °C, 1st step 40%, 2nd step 50%, intermediate heat treatment, 30 min/1150 °C	[29]
Ti-41Al-4Nb-6.6V-2Cr-0.2B		Conventional, preheat temperature 1300 °C, 16 mm/s ram speed, 70%	[170]
Ti-45Al-(8-9)Nb-(W,B,Y)	Ø 115 mm × H 127 mm Ø 115 mm × H 175 mm Ø 115 mm × H 245 mm	Conv., stainless steel can, α + γ phase field, 75%	[133]
Ti-Al-Mo-V (high Al)	Ø 22 mm × H 235 mm	Conventional, uncanned billet, no heat isolation, preheat temperature 1250 °C, die temperature 800 °C, 3 mm/s ram speed ($8.6 \times 10^{-2} s^{-1}$), 75%	[171]

contains useful thermophysical data, as also found in a previous publication by Semiatin et al. [123]. The authors used TiAl billets of 60 mm in diameter and 86 mm in height that were canned into pipes with a wall thickness of 4.8 mm [173]. Prior to canning, the billets were insulated in order to reduce heat transfer to the can, to prevent metallurgical reactions between the TiAl and the stainless steel can and also to aid workpiece removal after forging. End caps with thicknesses of 6.4 mm or 12.7 mm were used to seal the cans. After heating to temperatures

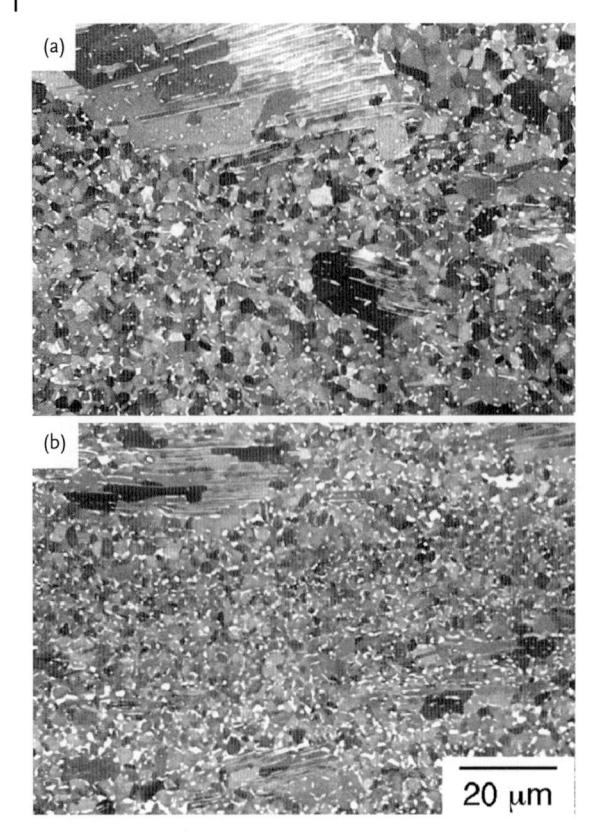

Figure 16.37 Backscattered electron micrographs showing the microstructure of Ti-45Al-10Nb ingot after isothermal forging. The forging direction is vertical in the micrographs. (a) Isothermally forged at 1100 °C to strain $\varepsilon = 65\%$. (b) Canned TiAl (stainless steel cans) isothermally forged at 1000 °C to strain $\varepsilon = 75\%$.

of 1150 °C or 1250 °C the canned billets were allowed to cool for a dwell period of 40 s. The dwell period served the purpose of reducing the flow-stress mismatch between can and billet by allowing the can to preferentially cool. For most simulations and forging trials a strain rate of $0.3\,\text{s}^{-1}$ was used. The simulations as well as the laboratory forgings showed that the uniformity of deformation depended in a subtle way on the forging parameters and the can design. The optimal can geometry had end caps that were chamfered and flush with the outer wall of the can. The thicker end caps did not improve the quality of the forging. Because of the complex deformation and heat-transfer/generation processes involved in canned forging, an optimum strain rate exists for a given can design. The optimized conditions that were identified allowed quite uniform conventional (nonisothermal) forging of the alloy Ti-45.5Al-2Cr-2Nb. However, even under these conditions recrystallization of the lamellar structure was incomplete, although the microstructure was more homogeneous compared to that obtained

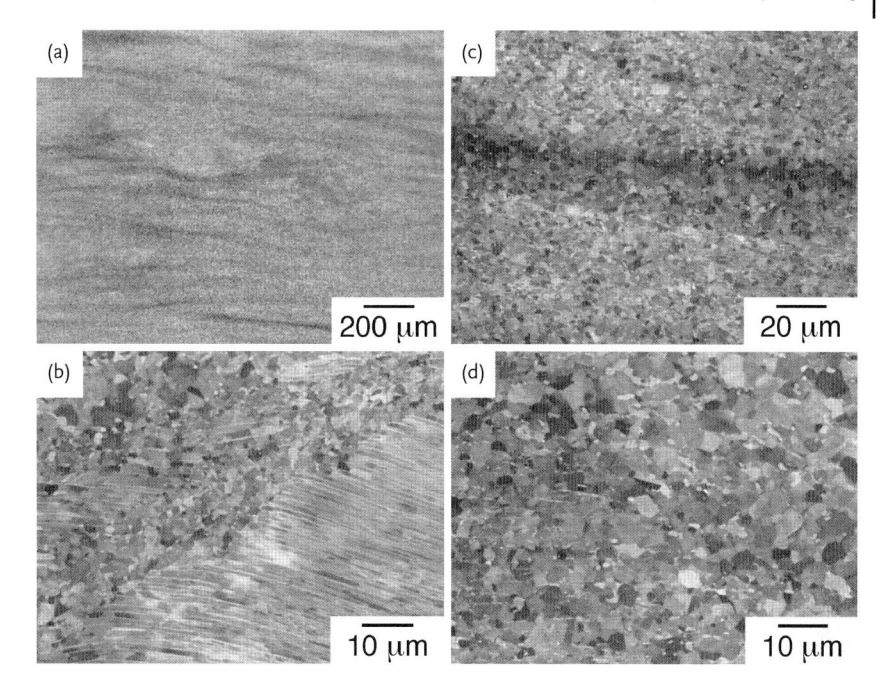

Figure 16.38 Backscattered electron micrographs showing the microstructure of Ti-45Al-10Nb ingot after two-step isothermal forging at 1150 °C to a total strain $\varepsilon = 88\%$. The forging direction is vertical in the micrographs. (a) Image taken at low magnification showing relatively good microstructural homogeneity, (b–d) higher magnification images of selected areas of the microstructure shown in (a).

after isothermal forging and contained significantly fewer remnant lamellar colonies [12, 123].

Forging practices have evolved with the aim of increasing recrystallization/ globularization and obtaining more homogeneous microstructures. Semiatin et al. [12, 123] introduced a short dwell period during forging as well as the use of a two-step forging process with an intermediate heat treatment. Both practices were intended to induce static recrystallization and resulted in improved microstructural homogeneity compared to standard single-step forging. In a TEM study of a Ti-45Al-10Nb alloy that had been extruded in the ($\alpha + \gamma$) region, Zhang et al. [122] could directly show that the fraction of recrystallized γ grains increased from 36% to 62% through a 1150 °C annealing treatment performed after extrusion. Similarly, Carneiro and Kim [174] observed static recrystallization during aging at 1000 °C of extruded alloys with the compositions Ti-45Al-6Nb-0.4C-0.2Si, Ti-45Al-6Nb-0.2B-0.4C-0.2Si and Ti-45Al-1Cr-6Nb-0.3W-0.3Hf-0.2B-0.4C-0.2Si. Figure 16.38 shows the microstructure after two-step isothermal forging of the alloy Ti-45Al-10Nb. Although some remnant lamellae and banded structures perpendicular to the forging direction can be observed, the microstructure was quite homogeneous on a macroscopic scale. If more than two forging steps were utilized then the microstructural homogeneity could be even further improved, as demonstrated by Koeppe et al. [32], Fujitsuna et al. [175] and Imayev et al. [68]. Attention has to be paid to avoid extensive grain growth at temperatures in single-phase regions and thus the processing conditions have to be adapted for each alloy [32, 68, 175]. As an example of the development of a complete forging route for ingot breakdown, Imayev et al. [169] developed a relatively cheap two-step forging process for the β-solidifying alloys (Ti-45.2Al-3.5(Nb,Cr,B) and Ti-44.2Al-3(Nb,Cr,B)). The process included quasi-isothermal (canned) forging in the (α + γ) field, an intermediate annealing treatment just below the eutectoid temperature, and final isothermal forging in the (α_2 + γ) field. The intermediate heat treatment almost completely eliminated the remnant lamellar colonies, and also induced numerous annealing twins. The authors speculated that recrystallization during annealing was stimulated by phase-transformation processes driven by the nonequilibrium condition obtained after forging. The two-step forging process resulted in material that exhibited very good workability during subsequent sheet rolling [169, 176].

Fully recrystallized microstructures can be obtained by the so-called α-forging process [12, 123]. As has already been reported in Section 16.2.4, complete recrstallization is achieved by forging in the α field to a strain of 80%. The practice described by Semiatin et al. [12, 123] comprised of preheating the billet to a temperature high in the (α + γ) phase field, rapid cooling to a temperature low in the (α + γ) field and then forging. Due to the high cooling rate, the α phase is retained in a metastable condition. The technique is, however, restricted to relatively small billets so as to maintain sufficiently high cooling rates. Alternatively, α-forging can be carried out by heating the billet to a temperature above the α transus and then forging on colder dies. Kim and Dimiduk [13] reported that fully lamellar microstructures with colony sizes between 120 μm and 600 μm could be obtained using this route, depending on composition and processing conditions. The lamellae are not inherited from the initial cast microstructures but newly formed during the cooling process after forging, as described in Section 16.2.4. In a later publication a colony size as small as 25 μm was reported by Kim [163] for an α-forged K5 alloy (Ti-46Al-3Nb-2Cr-0.3W-0.2B-0.4(C, Si). This material had a fully recrystallized, fine lamellar microstructure, and an exceptional tensile ductility of 3.3% combined with a yield strength of around 680 MPa at room temperature. These results indicate that very good microstructural homogeneity can be obtained through α forging. The process must be carefully designed, so that the canned billet has a temperature well above the α-transus temperature at the end of forging, but on the other hand does not remain too long in the α field after forging in order to avoid grain coarsening [163]. Thus, α-forging seems to be particularly suited for the production of TiAl discs, while for other component geometries α extrusion might be more appropriate. However, it has to be noted that wrought processing in the α field is only beneficial if no subsequent hot-working steps are to be applied, since the lamellar microstructure obtained has a lower workability in comparison to duplex or equiaxed microstructures that occur after working in the (α + γ) or (α_2 + γ) phase fields.

Significant progress in improving the hot-workability of γ(TiAl) alloys was obtained by Tetsui and coworkers, who developed an alloy that contains the β phase over a wide temperature range [147, 148]. The alloy with the composition Ti-42Al-5Mn, exhibits the phase reactions $\beta \rightarrow \beta + \alpha \rightarrow \beta + \alpha + \gamma \rightarrow \beta + \alpha_2 + \gamma \rightarrow \beta + \gamma$ when cooling from the high-temperature β phase field to room temperature. The β phase in this material is disordered at high temperatures, but adopts the chemically ordered B2 structure at low temperatures. The authors found that the presence of the β phase enormously improved the workability compared to other γ alloys, and even made it possible to conventionally forge uncanned ingot material on rather cold dies (100–200 °C) [147, 148]. This opens up new possibilities with respect to TiAl manufacturing routes, since both the expensive isothermal forging or canning of billets is no longer necessary. Also, the limited size of semifinished material resulting from primary ingot breakdown being limited by available ingot dimensions and press capacity can be overcome because components can be forged directly from ingot [147, 148]. The authors were able to conventionally forge billets of a diameter and height of 150 mm to 50-mm thick crack-free pancakes, and also produced rectangular bars and plates with varying thicknesses using open-die forging. Prior to forging, the billets were preheated to 1300 °C and then forged with a strain rate around $10^{-1}\,\text{s}^{-1}$. Further details of the forging conditions can be found in Table 16.2. However, due to the presence of the β phase, the high-temperature strength of the alloy was inferior to other γ alloys, but significantly exceeded that of a high-temperature Ti alloy (Ti-6242) [148]. As noted by the authors, it might be expected that the high-temperature strength of this alloy can be improved by further alloy development. This could be directed towards minimizing the volume fraction of β phase at the intended service temperature. Similar alloy design concepts were also pursued by Nobuki *et al.* [51], Kim *et al.* [170], Kremmer *et al.* [149], Clemens *et al.* [150], Zhang *et al.* [171] and Wallgram *et al.* [151], who also demonstrated that relatively large amounts of the β phase could significantly improve the workability, which made conventional forging possible [149–151, 170, 171]. In the work of Kremmer *et al.* [149], Clemens *et al.* [150] and Wallgram *et al.* [151], conventional forging was not used for primary ingot breakdown but rather for closed-die forging of extruded material. Nevertheless, the results confirmed that for specific alloy compositions, hot-working routes that involve conventional forging could be used, thus making wrought processing easier and cheaper.

In addition to forging, canned hot extrusion has often been used for primary ingot breakdown, and has turned out to be an effective primary processing procedure. Extrusion is generally performed at high strain rates in order to reduce temperature losses, while the billet is pressed in the rather cold tooling. Usually, ram speeds of 15–50 mm/s are used [12, 18] from which strain rates of $1-3.5\,\text{s}^{-1}$ are estimated if customary billet diameters and extrusion ratios are used. The estimated effective mean strain rate during extrusion can be obtained by the formula [177]

$$\dot{\bar{\varepsilon}} = 6\, v_{\text{ram}} (\ln R)/D_c, \qquad (16.11)$$

with v_{ram} being the ram speed, R the extrusion ratio (cross-sectional area of the container divided by the total cross-sectional area of the extrudate), and D_c the container diameter. Due to the high strain rates employed, preheat temperatures in the (α + γ) or the α phase field have to be chosen considering the workability as well as the available extrusion-press capability. As both the extrusion die and container are usually only mildly heated (e.g., to 500 °C), canning of the billet is indispensable. Conventional Ti alloys, or commercial purity Ti as well as steels may be used as canning materials, with a preference for Ti alloys at temperatures beyond 1250 °C [12]. However, Ti alloys are prone to weld with the container wall, and it has been reported that stainless steel can successfully be used at temperatures well above 1300 °C [18]. Further working conditions include extrusion ratios between 4 and 20, container diameters of 85–330 mm and the use of streamline or conical die geometries [10, 12]. The cross section of the extruded product can be varied by employing dies with round or rectangular cross sections (with aspect ratios up to 6), or using dies with special profiles [10, 12, 178].

At the extrusion temperature, the can materials have significantly lower flow stresses than the γ(TiAl) billet. This flow-stress mismatch is often as high as 300 MPa and leads to inhomogeneous deformation or even cracking [179]. As a result, the diameter of the core of the extrudate is not constant over its entire length [120]. Due to the strong casting texture that is usually present in ingot material, the core can anisotropically deform and become elliptical [180]. In addition to the nonuniform deformation, a nonuniform temperature distribution across the cross section of the billet can develop during canned extrusion [10]. As described in Section 16.1.1, significant deformation heating can arise in TiAl alloys when high strain rates are used for hot extrusion. Semiatin [10] performed FEM simulations of the temperature distribution and evolution during extrusion of a 60-mm diameter TiAl billet that was encapsulated in a Ti-6Al-4V can with a wall thickness of 5 mm. In the calculations, the billet was assumed to have been preheated to 1300 °C and then been extruded to a ratio of 6 : 1 at a ram speed of 25 mm/s. Under these conditions, a temperature increase of 30–40 K occurred at the center of the billet while at the outer diameter the temperature fell by 60 K [10]. Such temperature differences can be one source of microstructural variations through the cross section of the extrudate. In addition, a nonuniform temperature distribution results in inhomogeneous flow over the cross section (so-called flow pattern C) [177], which is another source of microstructural inhomogeneities. In extreme cases cracking might occur along shear bands, as shown in Figure 16.34. Considering the FEM calculations, it is concluded that the selection of can geometry and extrusion conditions (temperature, strain rate, extrusion ratio and heat transfer conditions) are critical with respect to the production of a sound product and its microstructural homogeneity.

The problems that arise due to flow-stress mismatch between can and billet, resulting from temperature losses, can be largely overcome by using thermal insulation to reduce the heat transfer from the workpiece to the can. This enables controlled dwell periods between preheating and extrusion to be employed [10, 179, 181]. Effective heat insulation has been achieved using woven silica fabric or metallic foils or sheets, for example from Mo [141]. These can also serve as diffusion barriers between the can and the billet, which is necessary when steel cans are used, so that melting through reaction between the differing materials is avoided. For extrusion temperatures above 1000 °C, as is usually the case for γ(TiAl) alloys, heat losses are mainly caused by radiation; however, the incorporation of radiation shields within the can design can prevent heat loss from the billet due to both radiation as well as conduction [182]. Taking advantage of these methods, extrusion processes have been widely used for γ(TiAl) ingot breakdown [10, 12, 17, 120, 121]. The high hydrostatic pressures involved should allow forming of virtually any composition. Unfortunately, in the literature little detailed information on extrusion conditions and can geometries can be found. As a rule of thumb, for strain rates around $1 s^{-1}$ and extrusion ratios of 10:1 the preheat temperature should not be below 1230 °C in order to avoid too strong flow localization. Under these conditions a wide variety of TiAl alloys canned within steel capsules of up to 220 mm in diameter have been successfully extruded [2, 10, 12, 16–18, 120, 121, 152, 157, 166, 174, 183–188]. As an example, Figure 16.39 shows extruded sections

Figure 16.39 Sections of extruded Ti-45Al-(5-10)Nb-X bars that were obtained by extruding 80 kg (canned) ingots through differently shaped dies to cross-sectional reductions of 6:1 to 11:1. The extruded bars had a length of 6–8 m.

that were obtained from 80-kg ingots of Ti-45Al-(5-10)Nb-X that had been extruded to cross-sectional reductions of 6:1 to 11:1 using differently shaped dies, including those with a rectangular geometry [17, 18, 157]. The resulting bars had lengths of 6 to 8 m.

After extrusion, almost fully recrystallized microstructures are observed for extrusion ratios greater than 6:1 [17, 18, 120, 121, 157, 174, 187–189] although minor fractions of unglobularized remnant lamellar colonies were also found [186, 189]. The microstructure can be varied between equiaxed and fully lamellar by careful selection of the extrusion temperature (Figure 16.40), although a typical "banded" morphology parallel to extrusion direction often remains. Such bands typically consist of fine and coarse grains or of lamellar colonies and equiaxed grains. The fine-grained bands contain both α_2 und γ grains, whereas the coarser bands consist only of the γ phase, disregarding borides or other impurity phases like oxides. In the fine-grained bands, Oehring et al. [120] observed particularly fine grain sizes of 0.7 μm in a boron-containing alloy. Grain sizes of less than 3 μm [187] and 1 to 3 μm [188] have also been reported. Even in the coarser bands, relatively small grain sizes of 3.9 μm [120], 10 to 40 μm [187] and 5 to 20 μm [188] were observed. In extruded material with nearly or fully lamellar microstructures remarkably refined lamellar colony sizes of 22 to 65 μm [118, 184, 185], 100 to 200 μm [13], 58 μm [120] and 35 μm [174] combined with fine lamellar spacings were found. The formation of bands within the microstructure is associated with significant variations in the local chemical composition (Figure 16.41). These arise due to segregation during solidification and are extremely difficult to remove through homogenization heat treatments, as described above. This observation has been made in several studies in the literature [156–158, 161, 174] and provides supporting evidence that dynamic recrystallization during hot working is strongly affected by local composition. The coarse-grained bands probably originate from prior Al-rich interdendritic regions, where no α_2 phase was present. Thus, grain growth following recrystallization is not impeded by particles of the α or α_2 phase. On the contrary, fine-grained bands or bands of lamellar colonies are formed from the Al-depleted regions of former dendrite cores.

It is obvious that temperature differences over the billet cross section, as reported above, will lead to corresponding inhomogeneities of the lamellar fraction in the

Figure 16.40 Backscattered electron micrographs showing the microstructure of Ti-45Al-8Nb-0.2C ingot after hot extrusion in steel cans, the extrusion axis is horizontal. (a) and (b) Extrusion ratio 11:1, rectangular cross section of the extrudate with an aspect ratio of 1:3.3, preheat temperature 1230 °C, ram speed 12 mm/s, diameter of the canned billet 220 mm. The micrographs in (a) and (b) show the microstructure after extrusion of two ingots with the same nominal composition. The difference in the microstructures results from differences between the actual composition of the ingots. The higher lamellar fraction in (b) indicates a lower Al content than for (a). (c) Extrusion ratio 11:1, rectangular cross section of the extrudate with an aspect ratio of 1:3.3, preheat temperature above the α-transus temperature of 1320 °C, ram speed 12 mm/s, diameter of the canned billet 220 mm.

16.3 Workability and Primary Processing | 631

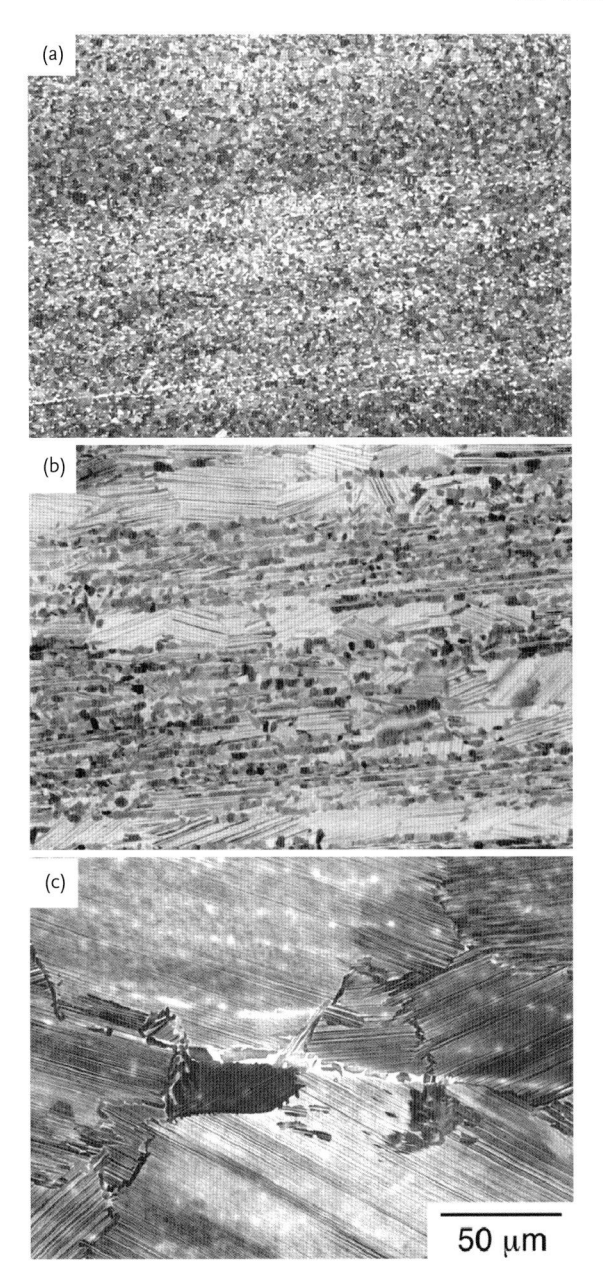

(a)

(b)

(c)

50 μm

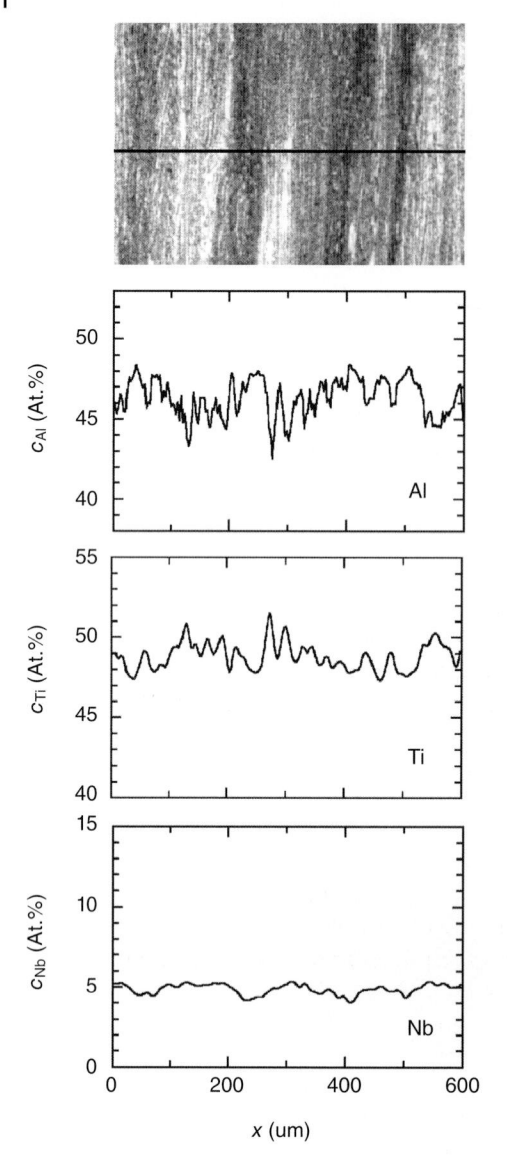

Figure 16.41 Backscattered electron micrograph showing the microstructure of Ti-45Al-5Nb-0.2C-0.2B after hot extrusion and quantitative EDX line scans recorded along the line indicated in the micrograph. The material had been extruded to a reduction ratio of 15:1 using a preheat temperature of 1230 °C and a ram speed of 12 mm/s. The initial diameter of the canned billet was 175 mm.

microstructure, but the problem can be minimized by using larger billets (>100 mm diameter) and heat insulation between the can and work-piece. In the literature, variations of grain size ranging from 6 to 14 µm [183] as well as of the texture [83] over the cross sections of extrusions have been observed, and can also be ascribed to temperature variations. Fully lamellar materials obtained by extrusion (so-called α-extruded or thermomechanically processed lamellar materials [13]) exhibit colony sizes ≤ 100 µm and small lamellae spacings < 200 nm, resulting in an excellent combination of room-temperature strength (630 to 1080 MPa), ductility (1 to ~4% plastic strain at fracture) and fracture toughness (22 to 34 MPa m$^{1/2}$ [13, 17, 28, 118, 120, 121, 157, 184–186, 190, 191]. The small colony size of these materials results from dynamic recrystallization of the α phase during working, whereas the small lamellar spacing arises due to the rather high cooling rates after extrusion. It should be noted that appreciable mechanical properties have been found in quite a number of alloy compositions with Al contents from 45 to 48 at.% and the addition of several alloying elements. The addition of boron has been found to result in larger lamellar spacings compared to B-free alloys [118, 120, 184]. This can be explained by the small undercooling, at which γ lamellae are precipitated from the α phase in B-containing alloys, as demonstrated by Kim and Dimiduk [96]. In conclusion, α-extrusion certainly represents significant progress with respect to ingot breakdown since semifinished material of a desired cross section with outstanding room temperature mechanical properties can be easily and routinely manufactured on a large scale. However, as with α-forging this processing procedure is only useful if no further hot-working steps will be performed to manufacture the final product.

The structural and chemical homogeneity of extruded products can be improved by higher reduction ratios or multistep processing [16], the latter, however, imposes severe constraints on the geometry of components. Through a double-step extrusion process, a total extrusion ratio of 100 to 200 was obtained and resulted in a further refinement of the microstructure along with an increase in microstructural homogeneity [16]. Nevertheless, inhomogeneities manifested as banding in the microstructure could not completely be eliminated. Draper *et al.* [192] have reported that two-step extrusion gave rise to excellent mechanical properties, however, they did not exceed the properties obtained using single-step extruded materials as given above. Attempts have been made to overcome the problem of restricted product geometry after multistep extrusion by equal-channel angular extrusion [10]. In this method the workpiece is extruded through an angular channel, with the channel angle determining the imparted strain energy. Other advantages of the technique are moderate working pressures, and the ability to control crystallographic texture and mechanical anisotropy during multipass processing by judicious rotation of the workpiece between working passes. Large-scale processing utilizing the ECAE technique is, however, still in its infancy.

Although little documented in the literature, forgings or extrusions of γ(TiAl) alloys often exhibit significant mechanical anisotropy. The pancake shown in Figure 16.36 was assessed by comparing the strength properties in the axial and the radial directions. Within the pancake, the banded microstructure is radially oriented in correspondence to the material flow during forging. The tensile strength determined for deformation in the axial direction was almost 10% higher than that in the radial direction. However, the samples with a radial orientation tended to exhibit larger tensile elongation. According to Mecking and Hartig [180] and Morris and Morris-Muñoz [193] this mechanical anisotropy could result from the crystallographic texture that is found in γ(TiAl) alloys forged by upsetting [79–81] or be due to an anisotropy of the microstructure [194, 195]. Mecking and Hartig [180] concluded from their study that in globular γ(TiAl) alloys, the anisotropy of plastic deformation occurs solely due to the texture of the material. However, the above-mentioned pancake exhibited a strongly anisotropic microstructure with a relatively high fraction of remnant lamellar colonies having a preferential orientation with lamellar interfaces perpendicular to the forging direction. This anisotropy is reflected in the mechanical properties and is particularly apparent on comparing the anisotropy of the fracture toughness. Relatively low values of K_{Ic} = 10 to 12 MPa m$^{1/2}$ were determined for crack propagation in the radial direction, whereas values of K_{Ic} = 16 to 20 MPa m$^{1/2}$ were measured for crack propagation in the axial direction [162]. In view of the susceptibility of γ(TiAl) alloys to cleavage, the preferred orientation of lamellar colonies may provide easy crack paths for propagation in the radial direction. Similarly to forged material, mechanical anisotropy has also been observed for extruded material. Oehring *et al.* [120, 121] observed no difference in the room-temperature compression strength for specimens taken either parallel or perpendicular to the extrusion direction in material with an equiaxed microstructure. This result indicates that the rather weak texture in the material did not lead to an anisotropic yielding behavior. However, the fracture toughness was 17 MPa m$^{1/2}$ for crack propagation perpendicular to the extrusion direction, while a value of only 13 MPa m$^{1/2}$ was determined when the notch plane was parallel to the extrusion direction. This difference was attributed to the anisotropic microstructure of the material, which besides banding parallel to the extrusion direction also exhibited elongated grains along the extrusion direction [120, 121]. In extruded material with the same composition and a lamellar microstructure, anisotropy of the mechanical properties was generally more pronounced; the flow stress in compression being significantly higher if the specimen orientation was parallel to the extrusion direction compared to the perpendicular orientation. From texture measurements, it was concluded that the lamellae were preferentially oriented parallel to the extrusion direction, although the texture was rather weak [83, 120, 121]. Extensive investigations on PST crystals in the literature have shown that loading parallel to lamellar interfaces is a "hard" deformation mode because the lamellar interfaces act as athermal glide obstacles [98, 107, 111, 196]. Thus, the higher flow stress for deformation parallel to the extrusion direction can be easily understood by the anisotropy of the lamellar microstructure. For deformation perpendicular to the extrusion direction, there is no preferential orientation of lamellar interfaces, as can be concluded from the fiber texture in the extrusion direction [83] and this results in a lower flow stress. Analogously, the orientation dependence of the fracture toughness can be ascribed to the anisotropic nature of the lamellar morphology. For cracks propagating perpendicular to the extrusion direction, the fracture toughness was significantly higher compared to cracks moving along the extrusion direction [120, 121]. It is known that the lamellar microstructure provides easy crack paths parallel to lamellar interfaces [197], which explains the experimental finding. In summary, the significant anisotropy in mechanical properties of wrought-processed material results predominantly from microstructural anisotropy. This not only comprises of the anisotropic behavior of the lamellar morphology but also anisotropic grain shapes, remnant lamellae and banding in the microstructure. As shown in Figures 16.42 and 16.43, the tensile properties of extruded material critically depend on the orientation of the loading axis with respect to the extrusion direction or, more generally, to the main direction of material flow during hot working. This dependence has to be accounted for in the design of wrought-processing routes for components.

Summarizing, hot extrusion is the easiest way to convert γ(TiAl) ingots of appreciable size to almost completely recrystallized material with acceptable or even good microstructural homogeneity. Both extrusion above the α-transus temperature as well as below followed by a heat treatment close to the α field have been found to result in excellent mechanical properties [13, 17, 118, 157, 190, 192, 199], including fatigue and creep properties [74, 192]. When subtransus-processed material is heat treated after hot working, favorable lamellar microstructures with fine colony sizes are obtained if excessive α-grain growth during the heat treatment is inhibited. Following Kim and Dimiduk [13] this can be achieved by alloying with B or by choosing a heat-treatment temperature in the ($\alpha + \beta$) phase field, however, moderate colony sizes can also be obtained by specific supertransus heat treatments. Considering the broad variety of processing schemes that have been utilized for hot extrusion and more generally for ingot breakdown, an assessment within the framework of tensile data might be useful in order to identify optimized processing routes. In particular, valuable information might be gained if differently processed materials of the same composition are compared. Table 16.3 contains a collection of such data. As is seen in this table, the best combination of strength and ductility is obtained for α processing, in particular α extrusion. However, the table also demonstrates that very attractive properties can be achieved in material processed below the α transus temperature, including forged material. The table further defines a range of properties that are obviously hard to surpass, and also shows that a certain range of tensile properties can be reproducibly attained. It is also interesting to note that the scatter of the mechanical properties in one batch of extruded material was found to be relatively low, as demonstrated by the fact that for 33 tensile specimens a minimum plastic elongation to fracture of 0.34% was obtained [199] (see Chapter 10 and Figure 16.43). Thus, sufficient microstructural and chemical homogeneity can be achieved in extruded material to ensure an acceptable reproducibility of the mechanical properties, at least if the material is processed on a pilot scale and the starting material is of good quality. Industrial processing of larger batches will be more sensitive with respect to reproducibility, but no major studies have emerged in the literature on this issue. This also holds for the property variability of forged ingot material, and thus, a

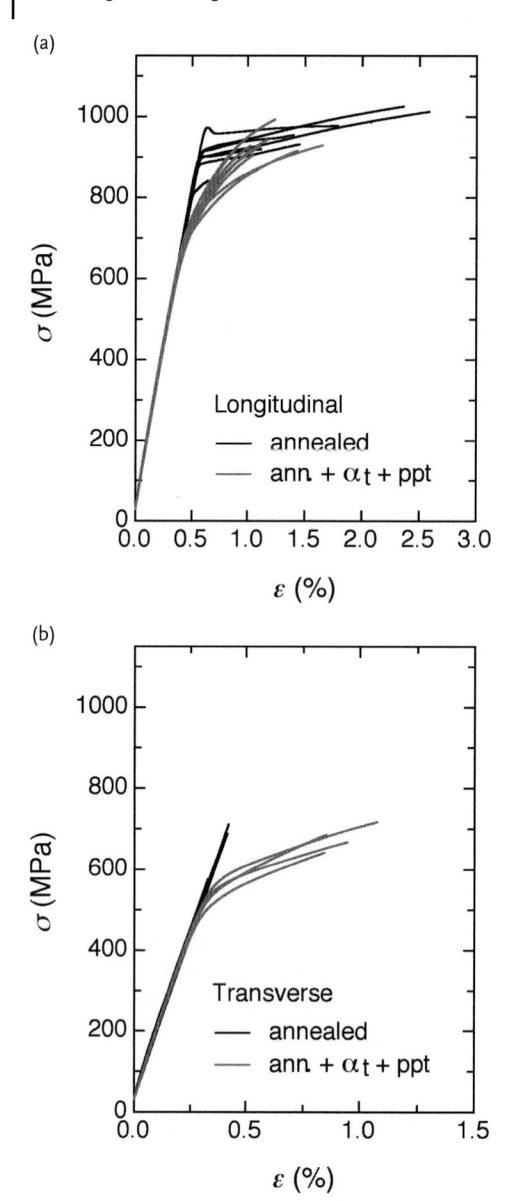

Figure 16.42 (a) Room-temperature tensile stress–strain curves for the alloy Ti-45Al-8.5Nb-0.2W-0.2B-0.1C-0.05Y extruded at 1222 °C to a reduction ratio of 4.6:1 [198]. The specimens were taken parallel to the extrusion direction (longitudinal orientation). After extrusion, the material was given a stress-relieve heat treatment for 2 h at 1030 °C followed by furnace cooling (annealed condition). Some material was further heat treated at 30 min/1329 °C/air cooled (α_t) and subsequently precipitation hardened at 6 h/800 °C/furnace cooled (ppt). (b) Room-temperature tensile stress–strain curves determined for the same material as in (a) but with the specimen orientation being perpendicular to the extrusion direction (transverse orientation) [198].

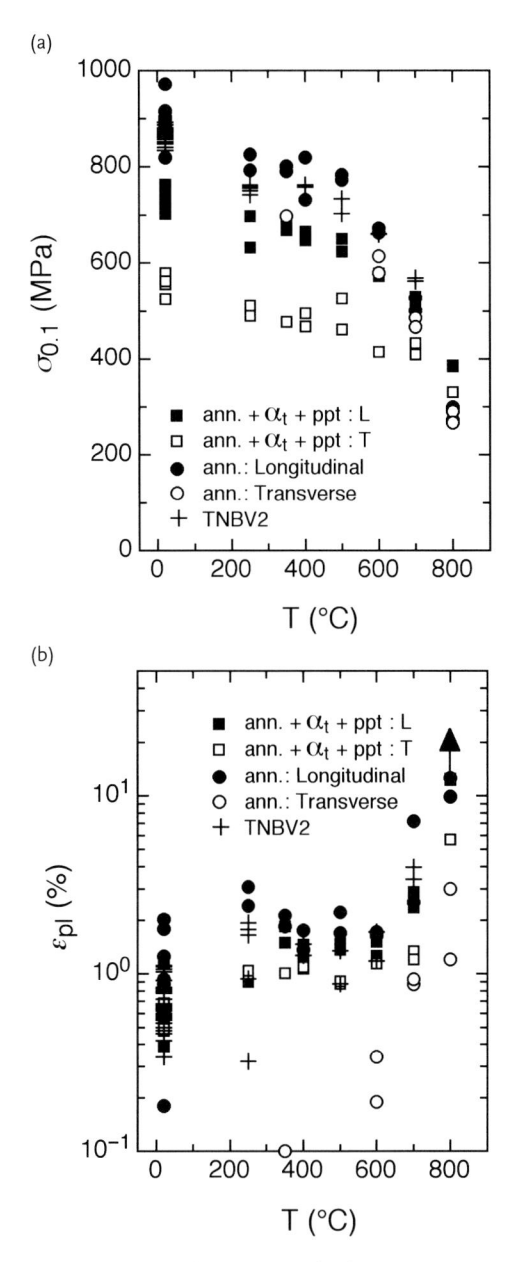

Figure 16.43 Tensile-testing results for Ti-45Al-8.5Nb-0.2W-0.2B-0.1C-0.05Y extruded at 1222 °C to a reduction ratio of 4.6:1 [198]. The various heat-treated conditions of the material are denoted in the following way: ann.: stress-relieved heat treatment 2 h/1030 °C/furnace cooled, α_t: 30 min/1329 °C/air cooled, ppt: 6 h/800 °C/furnace cooled. L and T designate the specimen orientation, longitudinal (tensile axis parallel to the extrusion direction) and the transverse (perpendicular to the extrusion direction). TNBV2: alloy Ti-45Al-8Nb-0.2C, extruded at 1230 °C to a reduction ratio of 11:1 and subjected to the following heat treatments: 2 h/1030 °C/furnace cooled + 30 min/1310 °C/air cooled + 6 h/800 °C/furnace cooled, longitudinal orientation. (a) Yield strength at 0.1% plastic strain, (b) plastic elongation at fracture.

Table 16.3 Tensile and fracture toughness testing of forged and extruded TiAl alloys.

Alloy	Processing	Microstructure	Orientation	$\sigma_{0.2}$ (MPa)	σ_{max} (MPa)	$\varepsilon_{fr, pl}$ (%)	K_{Ic} (MPa m$^{1/2}$)	Ref.
Ti-47Al-1V	IM, forged, 1150 °C, 80%	Equiaxed microstructure	Perp.	426	500	2.0		[30]
Ti-47Al-1Cr-1V-2.5Nb	IM, forged, 1180 °C, 88%, HT 1350 °C	FL, cs 250–500 μm	Perp.	508	588	1.1	22.8 (par.)	[164]
	IM, forged, 1180 °C, 88%, HT 1330 °C	NL cs 70–140 μm γ gs 10–20 μm	Perp.	511	702	2.8		[164]
	IM, forged, 1180 °C, 88%, HT 1280 °C	Duplex gs 15–40 μm	Perp.	421	557	3.8	12.9 (par.)	[164]
	IM, forged, 1180 °C, 88%, HT 1000 °C	NG γ gs 5–100 μm α_2 gs 1–5 μm	Perp.	485	562	2.9		[164]
Ti-46Al-2Nb-2Cr	IM, forged, α + γ, HT T_α − 25 K	Duplex cs 100 μm gs 15–20 μm	Perp.	460		1.8		[200]
	IM, forged, α + γ, HT T_α − 50 K	Duplex cs 30 μm	Perp.	470		1.9		[200]
Ti-45Al-2Cr-2Nb-1Ta	IM, forged, α + γ, HT T_α − 25 K	Duplex cs 100 μm gs 15–20 μm	Perp.	660		1.1		[200]

Alloy	Processing	Microstructure	Orient.	YS	UTS	El. (%)	Ref.
Ti-47Al-2Cr-2Nb	IM, forged, α + γ, HT $T_\alpha - 50$ K	Duplex cs 30 μm	Perp.	650		1.2	[200]
	IM, forged, 1175 °C, 75%, HT 1330 °C	NL	Perp.	429	474	1.6	[167]
	IM, 2 step forged, 1175 °C, 50%, 75%, HT 1330 °C	NL	Perp.	462	486	1.4	[167]
	PM, forged, 1175 °C, 75%, HT 1350 °C	FL	Perp.	380	433	1.6	[167]
	IM, extr., 1300 °C, 16:1, HT 1350 °C	NL	Par.	439	458	1.2	[186]
	PM, extr., 1300 °C, 16:1, HT 1350 °C	NL	Par.	486	512	1.2	[186]
	IM, extr., 1350 °C, 16:1	NL, cs 40 μm	Par.	592	766	3.0	[186]
	PM, extr., 1350 °C 16:1	NL, cs 40 μm	Par.	582	793	3.8	[186]
	PM, extr., α field, 16:1 HT 900 °C	FL, cs 65 μm ls 100 nm	Par.	971	1005	1.4	[118]
Ti-47Al-2Cr-2Nb-0.15B	IM, extr., α field HT 900 °C	FL, ls 140–325 nm	Par.	666	844	4.5 29.9 perp.	[185]
Ti-46Al-2Cr-2Nb-0.15B	IM, extr., α field, HT 900 °C	FL, cs 22 μm	Par.	811	1010	4.7	[185]

(Continued)

Table 16.3 (Continued)

Alloy	Processing	Microstructure	Orientation	$\sigma_{0.2}$ (MPa)	σ_{max} (MPa)	$\varepsilon_{fr,\,pl}$ (%)	K_{Ic} (MPa m$^{1/2}$)	Ref.
Ti-47Al-1.5Nb-1Cr-1Mn-0.2Si-0.5B (nominal composition), actual Al content 45 at.%	IM, forged, 1150°C, 80%	Equiaxed microstructure, Remnant lamellar colonies	Par.	715	736	0.9	11.4 perp. 17.0 par.	[201]
	IM, forged, 1150°C, 80%, HT 1330°C, FC	FL	Par.	424	550	1.1	18.1 perp. 24.6 par.	[201]
	IM, forged, 1150°C, 80%, HT 1350°C, OQ	FL	Perp. Par.	815 711	936 791	0.4 0.35	23.5 perp. 27.5 par.	[201]
Ti-47Al-1.5Nb-1Cr-1Mn-0.2Si-0.5B	IM, extr., 1250°C, 7:1, HT 1030°C	Equiaxed, gs 3.3 µm	Par.	404	426	1.0	13.3 par. 17.0 perp.	[120, 121]
	IM, extr., 1250°C, 7:1, HT 1380°C	NL, cs 150 µm	Par.	405	501	1.8		[120, 121]
	IM, extr., 1250°C, 7:1, HT 1360°C/OQ	FL, cs 130 µm	Par.	572	738	1.7		[120, 121]
	IM, extr., 1380°C, 7:1,	NL, cs 58 µm, ls 340 nm	Par.	502	632	2.1	22.7 par. 33.6 perp.	[120, 121]

Alloy	Processing	Microstructure	Orientation			Ref.
	IM, extr., 1250°C, 7:1, HT 1030°C	FL, cs 150 µm	Par. to extr. direction, perp. to forging direction	381	494	0.8	[120, 121]
	Forged, HT 1360°C/FC, blade root					
	IM, extr., 1250°C, 7:1, HT 1030°C	FL, cs 41 µm, ls 50 nm	Par. to extr. direction, perp. to forging direction	614	679	1.3	[120, 121]
	Forged, HT 1360°C/OQ, blade foil					
Ti-45Al-5Nb-0.2C-0.2B	IM, extr., 1250°C, 7:1, HT 1030°C	Duplex	Par.	1018	1085	2.5	[157]
Ti-45Al-5Nb-0.2C-0.2B	IM, 2 step extr., 1250°C, 100:1	Duplex	Par.	1040	1130	1.3	[192]
Ti-46Al-3Nb-2Cr-0.3W-0.2B-0.4(C, Si)	IM, forged, α field	FL, cs 25 µm	Perp.	680	820	3.3	[163]
Ti-42Al-10V	IM, extr., 1260°C, 5.3:1	72% lam. col., cs 9 µm, ls 40 nm, B2 grains	Par.	1265	1334	0.35	[202]

IM: Ingot metallurgy; PM: powder metallurgy; HT: heat treatment; FC: furnace cooled; OQ: oil quenched; FL: fully lamellar; NL: nearly lamellar; NG: near-γ; B2: β phase with ordered B2 structure; gs: grain size; cs: colony size; ls: lamellar spacing.

For tensile testing the specimen orientation is given with respect to the forging or extrusion direction, for fracture toughness testing the orientation of the notch plane or plane of gross fracture is indicated with respect to the forging or extrusion direction.

comparison between the reproducibility of forged and extruded material is currently not possible.

16.4
Texture Evolution

In the preceding chapters the anisotropic mechanical properties of γ(TiAl) alloys have been extensively described. This anisotropy arises from both the intrinsic properties of the γ-TiAl phase with its tetragonal $L1_0$ structure, which for example is reflected in a strong anisotropy of Young's moduli, and also from microstructural anisotropy, the lamellar microstructure being the most prominent example. The lamellar microstructure consists of stacks of parallel platelets with crystallographically oriented interfaces resulting in a strong anisotropy of lamellar materials with a preferred lamellae orientation. For these reasons, the texture of γ alloys and its evolution during processing is of particular significance with respect to the mechanical properties.

The texture development in γ-TiAl alloys during hot working is characterized by some peculiarities in comparison to the behavior of solid-solution phases. These include the occurrence of perfect and superdislocations, which not only result in significantly different critical resolved shear stresses but also have different core structures influencing the deformation, recovery and recrystallization behavior. The complexity of processes during high-temperature deformation is increased by twinning, which is a prominent deformation mode in γ(TiAl) alloys. Moreover, recrystallization of intermetallic phases is generally impeded by the low mobility of grain boundaries and the requirement that the ordered state has to be restored on primary recrystallization. As γ alloys are usually worked in temperature ranges where either the γ, the α and γ or only the disordered hexagonal solid-solution α phase are present, not only does the texture evolution of the different phases at temperature have to be considered, but also the phase transformations that occur on cooling have to be taken into consideration.

The texture determination of γ alloys faces some experimental problems that mainly originate from the fact that the c/a ratio of the tetragonal structure is close to 1. This results in overlapping of a number of peaks, for example the {002} and {200} reflections [203, 204]. In addition, the superlattice reflections exhibit weak intensities and some occur at very low Bragg angles, which makes texture measurements using X-rays even more difficult. The related experimental difficulties have been discussed in detail by Schillinger et al. [204]. However, these problems can be overcome through the use of parallel-beam optics for conventional X-ray tubes, thus enabling pole figures for the closely adjacent {002} and {200} reflections of the γ phase as well the weak reflections of the α_2 phase to be determined [204–206]. It is also noteworthy that in TiAl alloys, neutron diffraction exhibits a particular behavior that is associated with the scattering characteristics of the Ti and Al atoms. The absolute values of the scattering amplitudes of these elements are nearly equal, however, the incident and scattered waves are in phase for Al atoms but out of phase for Ti atoms, that is, the coherent scattering length, b, has nearly the same magnitude for Al and Ti but is positive for Al and negative for Ti (b_{Al} = + 0.345 × 10^{-12} cm, b_{Ti} = −0.337 × 10^{-12} cm). Considering the nuclear structure factor,

$$F_{nuc}(\mathbf{h}) = \sum_{j=1}^{N} b_j \cdot \exp(2\pi i \cdot \mathbf{h} \cdot \mathbf{x}_j) \qquad (16.12)$$

with **h** the reciprocal lattice vector (h, k, l), $\mathbf{x}_j$ the position of atom j in the unit cell, and N the number of atoms in the unit cell; for the equiatomic ordered γ(TiAl) phase this leads to an enhancement of some (h, k, l) diffraction peaks and to the cancellation of other peaks compared to X-ray diffraction. The reduced number of reflections may be beneficial when multiphase assemblies of different phases that exhibit lattice planes with very similar spacing have to be analyzed, as is the case for engineering alloys containing the $α_2$(Ti$_3$Al) and γ(TiAl) phases as major constituents. Texture analysis using neutron diffraction has further advantages over X-rays, since most elements have an extremely low absorption for thermal neutrons. Thus, the whole body of a sample can be examined and not merely its surface [207, 208]. This is particularly important for alloys with coarse microstructures as present in TiAl castings for example.

In the following discussion, textures will be described using pole figures or inverse pole figures. With respect to the representation of textures in general, and description of the complete texture using the orientation distribution function (ODF), the reader is referred to the work of Verlinden *et al.* [209]. The representation of texture in γ alloys is explained in detail by Schillinger *et al.* [204].

Before discussing texture evolution during hot working, texture formation in cast material will be briefly described as the cast condition is often the initial state of material to be wrought processed and its texture might have an influence on further texture development. In the literature, such studies are relatively rare, despite the importance of texture with respect to mechanical properties. This certainly is due, in part, to the fact that preferred orientations of lamellae can already be seen in optical micrographs. In cast alloys the γ and $α_2$ lamellae are often primarily aligned nearly parallel to the axis of a cast ingot, that is, perpendicular to the direction of heat extraction during solidification. Such a preferred orientation of lamellae is commonly observed for γ alloys solidifying through the α phase and can be easily understood by the growth of α phase dendrites in the [0001] direction on solidification [210]. On further cooling γ lamellae are formed within the α phase and subsequently the remaining α platelets transform into ordered $α_2$ lamellae. The crystallographic alignment of γ lamellae is described by the Blackburn orientation relationship (Equation 4.1) which results in the observed preferred orientation of lamellae parallel to the chill surface.

Indeed, texture investigations have shown that in cast ingot material of alloys with an Al content of 47 to 48 at.% {111} fiber textures occur with the fiber axis perpendicular or nearly perpendicular to the ingot axis [83, 211]. Figure 16.44 shows pole figures determined by neutron diffraction from an ingot of the alloy

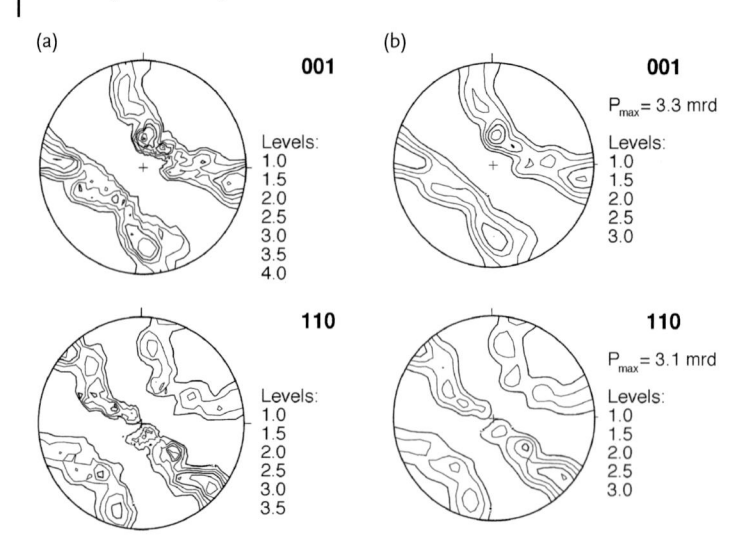

Figure 16.44 Pole figures from a Ti-47Al-1.5Nb-1Cr-1Mn-0.2Si-0.5B (TAB) ingot specimen taken about 10 mm of the ingot rim, determined using neutron diffraction [83]. The projection plane is perpendicular to the ingot axis. (a) Background-corrected raw data pole figures. (b) The same pole figures after smoothing using a Gaussian algorithm.

Ti-47Al-1.5Nb-1Cr-1Mn-0.2Si-0.5B (TAB) [83]. Despite the rather large volume of material that contributed to the measurement, the pole figures in Figure 16.44a have clearly been determined from an insufficient number of grains due to the coarse-grained nature of the material. Thus, the substructure of the girdles is not statistically significant and the maximum pole densities P_{max} probably represent the contribution from only a few large grains inside the orientation girdles. After applying a smoothing procedure to the measured pole figures that reduced the effect of coarse grains, it could be concluded from the resulting filtered pole figures (Figure 16.44b) that the observed orientation distribution could be considered as a fiber texture. For quantitative texture analysis, the orientation distribution function (ODF) was calculated in this work [83] using the iterative series expansion method proposed by Dahms and Bunge [212] and by Dahms [213]. The investigations are further detailed in Figure 16.45, which illustrates the (001) pole figure recalculated from the experimental data [83]. For all intensity levels a relatively good agreement with the measured pole figure (compare Figure 16.45 with 16.44) can be recognized. Because it could not be measured for the reasons associated with neutron diffraction discussed earlier, the (111) pole figure (Figure 16.45b) was calculated from the experimental data. According to this calculation the fiber axis has the <111> orientation, that is, the {111} planes of the γ phase are roughly parallel to the ingot axis, in agreement with the solidification texture mentioned above. Depending on the position of the sample inside the ingot, the orientation of the fiber axis can change, for example due to variations in the direction of heat flow. In the VAR ingot manufacturing process, radial heat flow is only expected near the

(a)   **001**

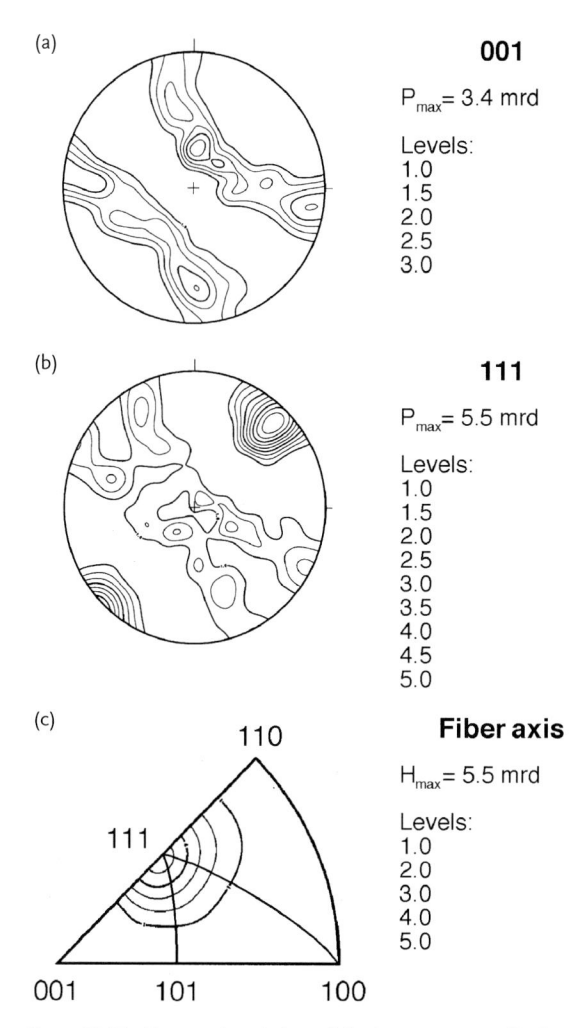

P$_{max}$ = 3.4 mrd

Levels:
1.0
1.5
2.0
2.5
3.0

(b)   **111**

P$_{max}$ = 5.5 mrd

Levels:
1.0
1.5
2.0
2.5
3.0
3.5
4.0
4.5
5.0

(c)   110   **Fiber axis**

H$_{max}$ = 5.5 mrd

111

Levels:
1.0
2.0
3.0
4.0
5.0

001   101   100

Figure 16.45 Texture description of the ingot specimen shown in Figure 16.44 after calculation of the ODF. Recalculated (a) (001) and (b) (111) pole figures (projection plane perpendicular to the ingot axis). (c) Inverse pole figure for the fiber axis, which has coordinates of 85° tilting and 316° rotation [83].

chill surface. The melt from the electrode is continuously falling parallel to the ingot axis onto the upper end of the ingot. Thus, a temperature gradient will be generated along the length of the ingot and the heat-flow direction will thus have an axial component near the ingot axis. The inverse pole figure (Figure 16.45c) was plotted exactly in the direction of the fiber axis, which has the coordinates 85° tilting and 316° rotation. This result can be explained by the fact that the sample was not taken directly from the rim of the ingot (90° tilting position) but from close to the rim. In summary, the solidification texture of α-phase-solidifying alloys discussed above has been confirmed by experimental measurements. However, it should be mentioned that during solidification De Graef et al. [214] observed that the growth of α phase with a [0001] orientation parallel to the heat-flow direction only occurred for relatively fast cooling rates; whereas for slow cooling rates the α phase grew with a <11$\bar{2}$0> orientation, resulting in γ lamellae being parallel to the direction of heat flow. It is also interesting to note that for binary alloys containing up to 49 at.% Al, β is the first phase to solidify and not α [215]. Thus, a texture originating from the β phase would be expected. However, this is clearly not the case, even for the alloy Ti-46.5Al [214]. Johnson et al. [216] and Küstner et al. [125] suggested that peritectic α phase is nucleated from the melt, and not from the primary β phase, thus explaining the lack of texture related to the β phase. On further cooling, the α phase will grow into the primary β phase and suppress further nucleation of α from β. In this way, the secondary α phase determines the final texture of the material. Indeed, Küstner et al. [125] observed an entirely different texture in cast Ti-45Al compared to Ti-48Al that was attributed to complete solidification via the β phase.

As already noted in the introductory remarks, the evolution of texture during hot working of TiAl alloys is undoubtedly determined by several factors. These involve the nature of the deformed state, crystallite rotation during deformation, recrystallization and phase transformations. It is usually difficult to select hot-working conditions in such a manner that only one of these processes is operative. However, under certain testing conditions, one or another of the various processes can be predominant. Thus, the main characteristics of the above-mentioned processes are considered separately in order to see whether or not they are reflected in the overall microstructure and texture.

A commonly held concept regarding the large-scale deformation of polycrystalline material is that the grains are rotated in a direction that allows multiple slip through activation of the most favorable slip or twinning systems [217]. This results in orientations of high symmetry with respect to the main deformation axis. High-temperature deformation of two-phase alloys is very complex due to the broad variety of constitution and microstructure, which are governed by the high-temperature phase fields. Thus, for an assessment of deformation-induced microstructures the different alloy constitutions have to be considered.

Information about the deformation characteristics of aluminum-rich α(Ti) is rare, particularly when this phase exists in equilibrium with γ(TiAl). For the following discussion the deformation modes of pure titanium were assumed, which are <11$\bar{2}$0>{10$\bar{1}$0} prismatic glide, <10$\bar{1}$1>{10$\bar{1}$2} twinning and to a much lesser extent <11$\bar{2}$0>{10$\bar{1}$1} pyramidal glide [218]. The critical resolved shear stress for <11$\bar{2}$0>{0001} basal glide in pure α-Ti was found to be significantly higher than for prismatic and pyramidal glide [219].

The deformation modes of the γ(TiAl) phase with its tetragonal $L1_0$ structure are discussed in detail in Chapter 5. The possible Burgers vectors are b_1 = <010] and b_2 = 1/2 <$\bar{1}$10] for ordinary dislocations, and b_3 = <0$\bar{1}$1] and b_4 = 1/2<11$\bar{2}$] for superdislocations. Shear processes along b_2, b_3 and b_4 occur exclusively on the octahedral {111} planes. In addition, order twinning along 1/6<11$\bar{2}$]{111} occurs.

In contrast to fcc metals, there is only one distinct true twinning direction per {111} plane that does not alter the ordered $L1_0$ structure. Nevertheless, in principle the von Mises criterion for a general plastic shape change is satisfied in TiAl alloys, because the $L1_0$ structure provides more than five independent slip systems. However, the relative ease and propensity of these deformation modes depends on the aluminum concentration, the content of ternary elements and the deformation temperature. For two-phase alloys it is now well established that the activation of superdislocations is difficult, when compared with 1/2 < 110] ordinary dislocations. This is probably a consequence of the high lattice friction of the superdislocations and their complex decomposition and nonplanar dissociation, which make glide and climb difficult. The described deformation behavior is undoubtedly a major restriction for the deformability of polycrystalline material. In grains that are unfavorably oriented for glide of 1/2 < 110] dislocations, high constraint stresses can be developed due to the shape change of adjacent grains. The significance of these stresses was illustrated by modeling the yield behavior of polycrystalline aggregates [220, 221] and directly demonstrated by electron-microscope observations [100]. The constraint stresses are believed to assist the superdislocations to overcome the high lattice friction so that glide processes of these dislocations can be locally initiated. An inability to meet this condition may result in premature failure. On the other hand, mechanical twinning also provides shear components in the c direction of the tetragonal unit cell and the propensity for twinning increases with temperature [100, 137]. At high temperatures, strain accommodation can also occur by climb of ordinary dislocations out of their {111} glide planes, which is certainly beneficial for a general shape change of polycrystalline material and reduces the Taylor factor. Furthermore, at temperatures higher than 1000 °C, <010] dislocations may contribute to deformation [100]. Thus, under hot-working conditions the plastic anisotropy of γ(TiAl) is probably reduced and relatively large strains might be realized by the propagation of ordinary dislocations and order twins.

The above discussion suggests the following states of deformation for hot-working temperatures within the (α + γ) field. For extrusion, a preference of the γ grains to orient with the <111> direction parallel to the extrusion direction might be expected. This would ensure that the imposed radial reduction of the workpiece can be accomplished by two 1/2 < 110]{111} slip systems that provide shear by ordinary dislocations perpendicular and inclined to the sample axis. A substantial population of γ grains would also be expected with an <001] orientation parallel to the extrusion axis, because in this orientation radial shear could also be provided by two 1/2 < 110]{111} slip systems. It is interesting to note that these two grain orientations are the major components of the double-fiber texture shown in Figure 16.46, which was observed for the alloy Ti-47Al-1.5Nb-1Cr-1Mn-0.2Si-0.5B (γ-TAB) extruded below the α-transus temperature [83]. It is also worth mentioning that the texture after extrusion is relatively weak, which might be attributed to recrystallization. The initial casting texture has been completely destroyed after extrusion as becomes obvious by considering the texture of extruded powder material of the same alloy (Figure 16.46).

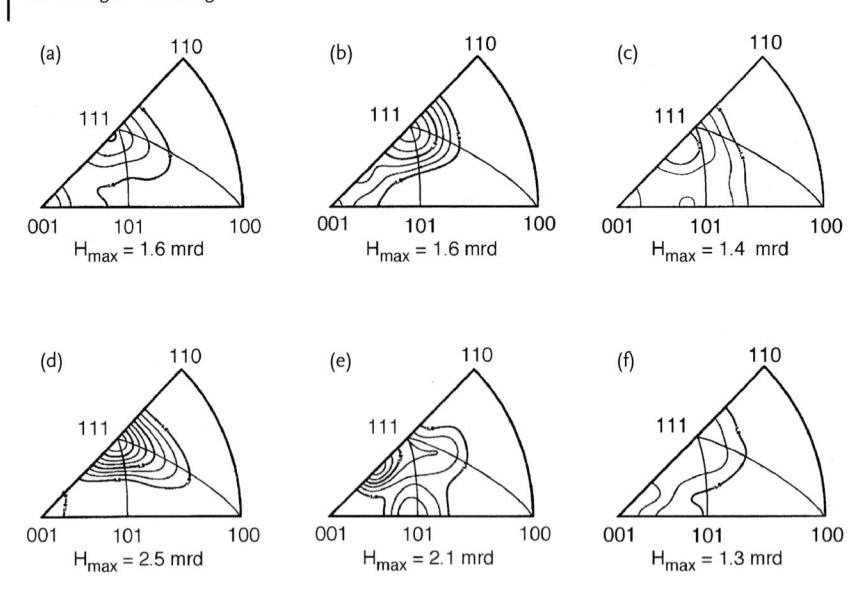

Figure 16.46 Texture of extruded Ti-47Al-1.5Nb-1Cr-1Mn-0.2Si-0.5B (TAB) material determined using neutron diffraction. Inverse pole figures for the extrusion direction. (a) Ingot material extruded at 1250 °C, (b) ingot material extruded at 1250 °C and annealed at 1030 °C, (c) powder material extruded at 1250 °C, (d) ingot material extruded at 1300 °C, (e) ingot material extruded at 1380 °C, (f) ingot material extruded at 1380 °C and annealed at 1030 °C [83].

Clearly, a more elaborate analysis is required in order to better assess the contribution of different deformation mechanisms to texture development during wrought processing and to separate the effect of deformation from other factors. In order to investigate the influence of deformation on texture evolution and to avoid dynamic recrystallization, Hartig et al. [203] performed a study in which the alloy Ti-50Al was deformed in compression at 400 °C. During testing, a large hydrostatic pressure was superimposed to avoid brittle failure at the relatively low testing temperature. The alloy consisted mainly of the γ(TiAl) phase with a 5 to 10% volume fraction of the α_2 phase. Thus, the deformation characteristics of the γ phase in this alloy might be considered as representative for two-phase binary TiAl alloys. After compression testing to a strain of 46%, the texture could be roughly described as a <110] + <101] double-fiber texture. In the inverse pole figure, texture components for the compression axis near to <101] (at approximately <302]) and <110] were visible, as well as a scattered component midway between <100] and <110]. Similar textures, containing some or all of these components, were observed in two-phase TiAl alloys after compression at temperatures up to 1200 °C in a number of studies [79–81, 211, 222–224].

Hartig et al. [203] simulated the texture development during compression using a Taylor model and compared the measured texture with modeling. They concluded that both ordinary and superdislocations contributed to deformation and

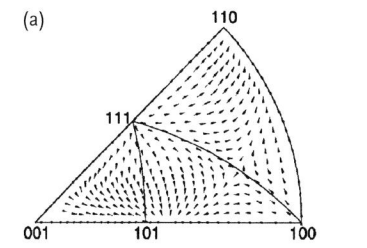

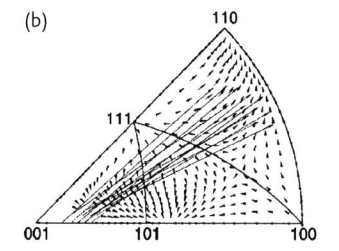

Figure 16.47 Orientation flow map of the compression direction calculated using the Taylor model and assuming CRSS ratios between ordinary and superdislocations of $\tau_o/\tau_s = 0.8$ and that between twinning and superdislocations of $\tau_t/\tau_s = 0.2$ [223]. (a) $\varepsilon = 0.05$, (b) $\varepsilon = 0.15$.

texture development. However, deformation twinning was not incorporated in the modeling. More recently, the deformation behavior of polycrystalline TiAl alloys was quantitatively assessed in a series of studies by analyzing the single-crystal yield surface in the framework of the Taylor model [81, 204, 211, 220]. The authors considered a wide range of yield stresses for the individual deformation modes and finally determined the flow-stress ratios $\tau_o/\tau_s = 0.7$–0.9 and $\tau_t/\tau_s = 0.2$–0.9 as optimum values [81, 204, 211, 220, 223]. The subscripts o, s and t refer to ordinary dislocations, superdislocations and twinning, respectively. Figure 16.47 shows orientation flow maps for compression calculated by Stark et al. [223] assuming CRSS ratios $\tau_o/\tau_s = 0.8$ and $\tau_t/\tau_s = 0.2$. The map exhibits orientation flow towards <110] and <302], in agreement with the experiments and also reflects the observed orientation band between <100] and <110]. However, despite this agreement, the question remains as to why these CRSS ratios are significantly higher than those concluded from electron-microscopy investigations [100, 225]. This discrepancy has been explained by the strong work hardening of TiAl alloys that even after small strains would lead to similar shear stresses for different deformation modes; and thus the CRSS ratios may not markedly be reflected in the texture evolution [220]. Summarizing, the textures observed after compression and extrusion, as described above, can be regarded as deformation textures and understood by considering the deformation modes of the γ(TiAl) phase.

From the foregoing discussion it may be concluded that the textures observed after hot working in the (α + γ) field are predictable from the deformation modes of the γ phase. However, after hot working the material usually is recrystallized, which indicates that recrystallization and phase transformation can be equally important for texture evolution, a point that warrants further consideration. These topics have primarily been studied on sheet-rolled material. Attention was also paid to sheet rolling as it often results in a characteristic anisotropy of mechanical properties [81, 97, 180, 195, 225, 226]. In Figure 16.48 the texture of the alloy Ti-46.5Al-4(Cr, Nb, Ta, B) after sheet rolling and primary annealing at 1000 °C is shown [83]. The texture of this sheet was much stronger (maximum orientation density for (001) planes was 7.3 times higher than random) when compared to extruded material of a similar alloy (shown in Figure 16.46), and can be described

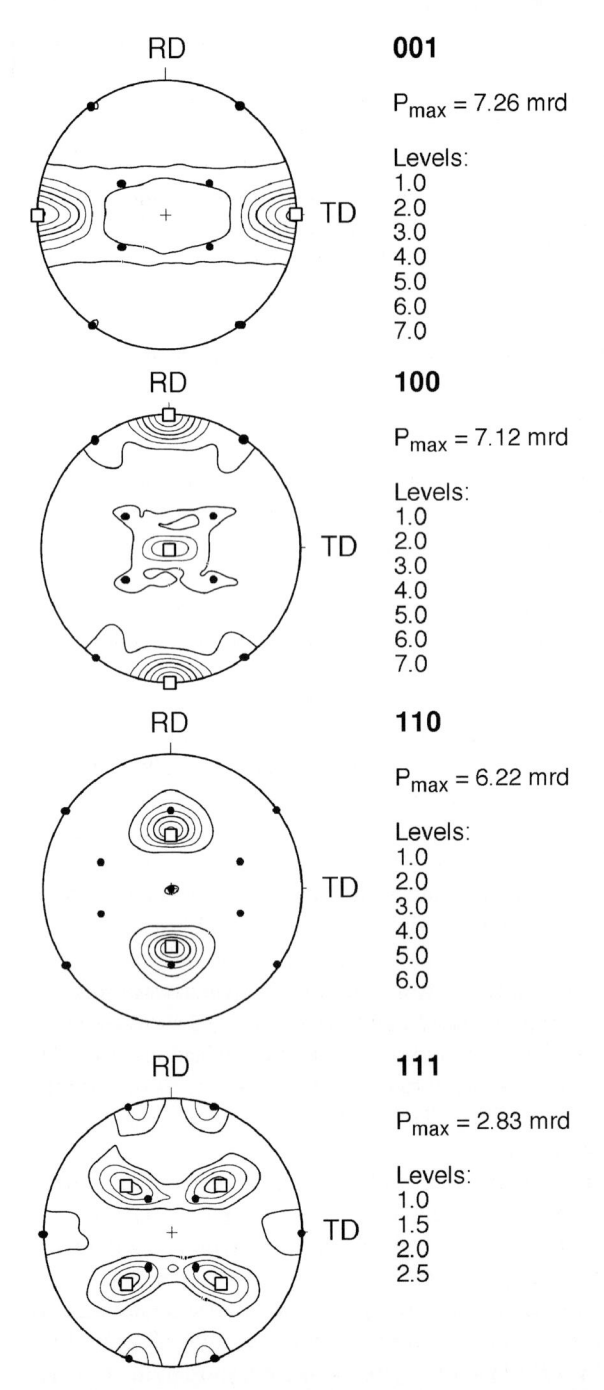

Figure 16.48 Texture of rolled sheet-material determined using neutron diffraction. Recalculated pole figures for Ti-46.5Al-4(Cr, Nb, Ta, B) that was processed by hot-isostatic pressing prealloyed powder, sheet rolling and primary annealing at 1000 °C [83]. In addition, the ideal modified cube texture (010)[100] (square symbols) and the ideal brass texture ($1\bar{1}0$)[111] (filled circles) are indicated.

001

P_{max} = 7.26 mrd

Levels:
1.0
2.0
3.0
4.0
5.0
6.0
7.0

100

P_{max} = 7.12 mrd

Levels:
1.0
2.0
3.0
4.0
5.0
6.0
7.0

110

P_{max} = 6.22 mrd

Levels:
1.0
2.0
3.0
4.0
5.0
6.0

111

P_{max} = 2.83 mrd

Levels:
1.0
1.5
2.0
2.5

as a modified cube texture. The modified cube texture has previously been reported in rolled γ-TiAl sheets [81, 195, 203] and is characterized by the <001] c-axes lying in the transverse direction and the <100] a-axes lying parallel to the rolling direction. In addition to the modified cube texture, a further texture component was present in this material, as can be seen from the (111) pole figure in Figure 16.48, which shows that the (111) planes are aligned nearly perpendicular to the transverse direction of the sheet. Here, it should be noted that the material was rolled in the two-phase (α + γ) phase field. The γ lamellae, which are formed on subsequent cooling, have a clear orientation relationship with the hexagonal α phase and are not expected to show a modified cube texture. Thus, texture components originating from the α phase should be present and may explain the above-mentioned additional texture component. This will be considered more closely in the following. During primary annealing of the sheet material the γ lamellae are mostly dissolved at the expense of growing γ grains [83], therefore this texture component will be decreased in pole density figure and the modified cube texture of the γ grains become more dominant.

In contrast to the Ti-46.5Al-4(Cr, Nb, Ta, B) rolled material, sheet of the high Nb-containing alloy Ti-45Al-8Nb-0.2C revealed a weaker texture in the rolled state [83]. The pole figures for this sheet material (Figure 16.49) do not show a cube texture and, in particular, no differences between the preferential alignment of the a- and c-axes. The texture of the high Nb-containing TiAl was nearly identical to that of a Cu-6 wt.% Zn alloy that had been cold rolled to 92.6% [227–229]. In Cu-Zn alloys with increasing Zn content, a continuous texture variation is observed in cold-rolled sheets that changes from the copper-type texture in pure Cu to the ideal brass-type texture for Cu-30 wt.% Zn [229]. Thus, the texture of the high Nb-containing TiAl sheet material can formally be described as a combination of the deformation textures of pure Cu and the ideal brass type. A similar type of rolling texture has been reported by Koeppe *et al.* [230], who found a mixture of the brass, the copper, the {123}<63$\bar{4}$>-S, and the modified cube texture components in primary annealed Ti-48Al-2Cr sheet. Some or all of these texture components have been observed in rolled TiAl sheets by a number of other investigations [204, 211, 223, 231]. Since the brass and the Cu textures are deformation textures, it might be suspected that sheet of this alloy (Ti-45Al-8Nb-0.2C) has not been sufficiently recrystallized during hot rolling. However, the texture of this material remained stable during primary annealing for 2 h at 1000 °C (Figure 16.49), which is about 0.7 of the absolute melting temperature. In this respect, it is important to note that this alloy contained C, which precipitates in γ TiAl alloys as carbides of the Ti_2AlC or Ti_3AlC phases (see Section 7.4.1).

The above-mentioned textures, that are closely related to cold-rolled fcc metals, can be described by fibers connecting the various components [204, 205, 223]. The α fiber runs from the brass to the Goss orientation and the β fiber connects the brass and the copper through the S component. Due to the tetragonality of the γ phase, these fibers split into two nonsymmetrically equivalent components [204, 205, 223]. A more detailed description is given by Schillinger *et al.* [204], Stark *et al.* [223] and Verlinden *et al.* [209]. Bartels *et al.* [205] and Schillinger *et al.* [204]

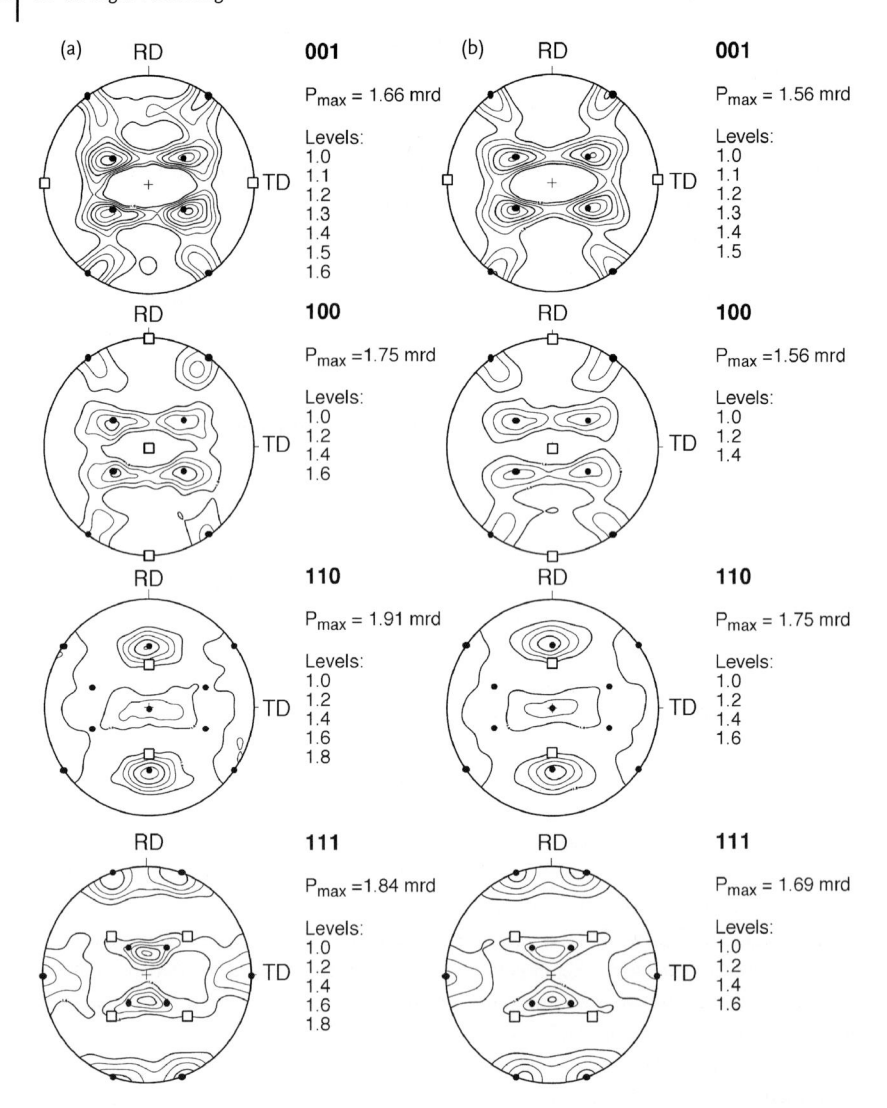

Figure 16.49 Texture of rolled sheet-material determined using neutron diffraction. Recalculated pole figures for Ti-45Al-8Nb-0.2C that was processed by hot extruding an ingot through a rectangular die followed by sheet rolling [83]. In addition, the ideal modified cube texture (010)[100] (square symbols) and the ideal brass texture (1$\bar{1}$0)[111] (filled circles) are indicated. (a) As-rolled condition, (b) after primary annealing for 2 h at 1000 °C.

employed the single-crystal yield surface model, which was mentioned earlier, to understand the occurrence of the different components and were able to simulate similar textures to those that were experimentally observed. In these simulations the earlier-cited ratios between the CRSS of the different deformation modes were used. On the basis of the simulations the rolling textures that are reminiscent of those of cold-rolled fcc metals, can indeed be understood as deformation textures of the γ phase.

The sheet material, which developed the texture shown in Figure 16.49, consisted predominantly of lamellar colonies (volume fraction of 85%) [83]. This implies that during rolling a corresponding volume fraction of the hexagonal α phase was present, the texture of which should determine the texture of the γ lamellae according to the Blackburn relationship (Equation 4.1). However, at first glance no texture components originating from the α phase were present in these sheets, since the observed modified-cube and copper/brass-type textures indicate texture formation in the γ phase. It might be speculated that texture components originating from the α phase might be weak and thus are not visible. Another explanation could be that texture components originating from the α phase are very similar to the copper/brass textures. The brass texture contains an ideal component $(1\bar{1}0)$ [111] [229], that is, the $(\bar{1}\bar{1}1)$ planes of the γ phase lie in the plane defined by the rolling and the normal directions. This ideal component is seen in the inverse-pole figures (Figure 16.50) for this sheet material, which demonstrate that the rolling direction is aligned preferentially along <112>. The occurrence of this rolling texture can be understood by considering the deformation of the α phase. During rolling, the α grains should deform by prismatic glide so that lengthening and thinning of the rolling preform can be realized. This would lead to two substantial populations of α grains that are characterized by the $<10\bar{1}0>$ and $<2\bar{1}\bar{1}0>$ directions being oriented parallel to the rolling direction and the basal planes lying perpendicular to the sheet plane. Although this type of texture has not been reported for rolled α-Ti sheets, it has been observed after the rolling of Ti alloys, for example Ti-5 at.% Al-2.5 at.% Sn, in the α field [232]. Upon cooling of the rolled TiAl sheet, these α grains are transformed into two populations of γ lamellae with the <112> and <110> directions oriented parallel to the rolling direction and the {111} planes lying perpendicular to the plane of the sheet. As demonstrated in Figures 16.48 and 16.49, one of these two components can indeed be recognized in the final texture, in addition to the cube texture that originates from the γ grains that were present in the samples during hot working. Thus, the occurrence of the brass texture can also result from deformation of the α phase. Schillinger et al. [204] and Stark et al. [223] determined the texture of the $α_2$ phase after sheet rolling and observed two main texture components, the so-called basal and transversal components. The basal component is characterized by the basal planes of the $α/α_2$ phase lying perpendicular to the normal direction of the sheet; whereas for the transversal component, the basal planes are perpendicular to the transverse sheet direction and the $<1\bar{1}00>$ directions lie parallel to the rolling direction [223]. The transversal component is exactly the texture component that will lead to the brass component in the γ phase. Stark et al. [223] have argued that the transversal component in the $α/α_2$ phase occurs due to codeformation of the α and the γ phase during sheet rolling. However, the sheet material discussed here contained only 15% γ phase during rolling and, thus, it is questionable whether the texture development can be accounted for only by codeformation of the two phases.

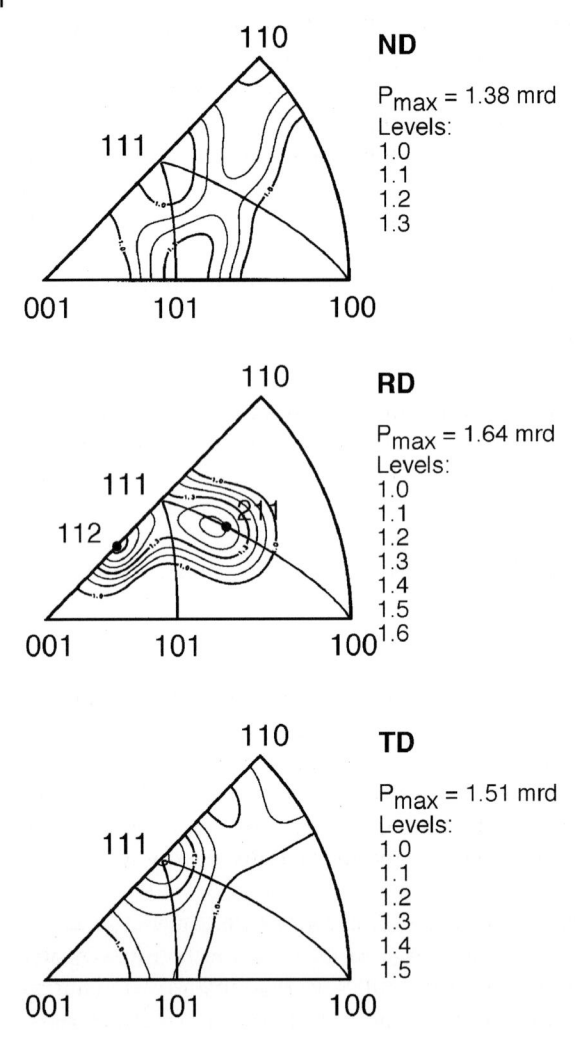

Figure 16.50 Inverse pole figures of Ti-45Al-8Nb-0.2C sheet material that was processed by hot extruding an ingot through a rectangular die followed by sheet rolling [83] (see Figure 16.49).

With respect to texture evolution during extrusion above the α-transus temperature, a deformation texture would be expected by which the radial reduction of the workpiece (imposed by the funnel-shaped die) can be realized by symmetrical prismatic glide with a c-axis orientation perpendicular to the extrusion direction. This condition is fulfilled in two cases. One of the prism planes can be either perpendicular to the extrusion direction or inclined at an angle of 60°. The first case would give rise to a <10$\bar{1}$0> deformation texture that would have the extrusion direction alignment. During subsequent cooling of the workpiece from the α phase field, γ lamellae are formed according to the Blackburn relationship (Equation 4.1), that is, the {111} planes of the new γ lamellae are parallel to the basal planes of the prior α grains. In the final texture these new γ lamellae should be manifested by a significant population of γ lamellae with the <112> direction oriented parallel to the extrusion direction. In the second case, that is, prism planes inclined at 60° to the extrusion direction, one would expect a <$2\bar{1}\bar{1}0$> texture of the α grains to be developed, and the subsequent phase transformation should lead to a <110> texture of the γ grains aligned along the extrusion direction. As demonstrated in Figure 16.46, these two components are indeed characteristic features that distinguish the textures obtained after hot working in the single-phase α field from those found after processing in the (α + γ) phase field. In this case, codeformation of the α and γ phases can be excluded as no γ phase is present during extrusion within the α phase field. Therefore, it may be concluded that prismatic glide plays a significant role with respect to texture development of the α phase in TiAl alloys, although other factors such as codeformation and recrystallization may have an influence. In order to summarize the influence of the constitution on texture formation, the texture evolution due to extrusion at different temperatures and the subsequent phase transformations is schematically illustrated in Figure 16.51.

From the foregoing discussion it may be concluded that the textures observed after hot working in the (α + γ) field are primarily related to the deformation modes of the γ phase. However, in most of the examples discussed the material was fully recrystallized after hot working, indicating that recrystallization can be equally important for texture evolution. A point that seems to be significant for texture development is that new grains grow by bulging, as demonstrated in Figure 16.8. A characteristic feature of grain-boundary bulging is that the new grains have a similar orientation to the old grains from which they have grown. Thus, the operation of this mechanism is expected to result in a recrystallization texture that is closely related to the deformation texture. As noted in Section 16.2.1, grain-boundary bulging where new grains are formed by the local strain-induced migration of grain boundaries [45] is a common process that occurs during dynamic recrystallization of many metals. It was also observed in earlier studies in TiAl alloys, for example, Ti-50Al [82] and Ti-52Al [43]. In view of these considerations, it appears plausible that the textures observed after hot working in the (α + γ) field are closely related to the operating deformation modes. However, it seems difficult to understand the occurrence of the modified-cube texture in the same way. There was an earlier, general belief that the cube texture in fcc metals mainly resulted from recrystallization either due to oriented nucleation or oriented growth of cube-oriented material [233–237]. Work by Hjelen *et al.* [235] and Doherty *et al.* [236] has demonstrated that cube-oriented material mainly nucleates from transition bands. According to Schillinger *et al.* [204] the modified cube texture is the ideal orientation in TiAl alloys that allows plastic deformation during sheet rolling exclusively by glide of ordinary dislocations. The problem is still a matter of debate

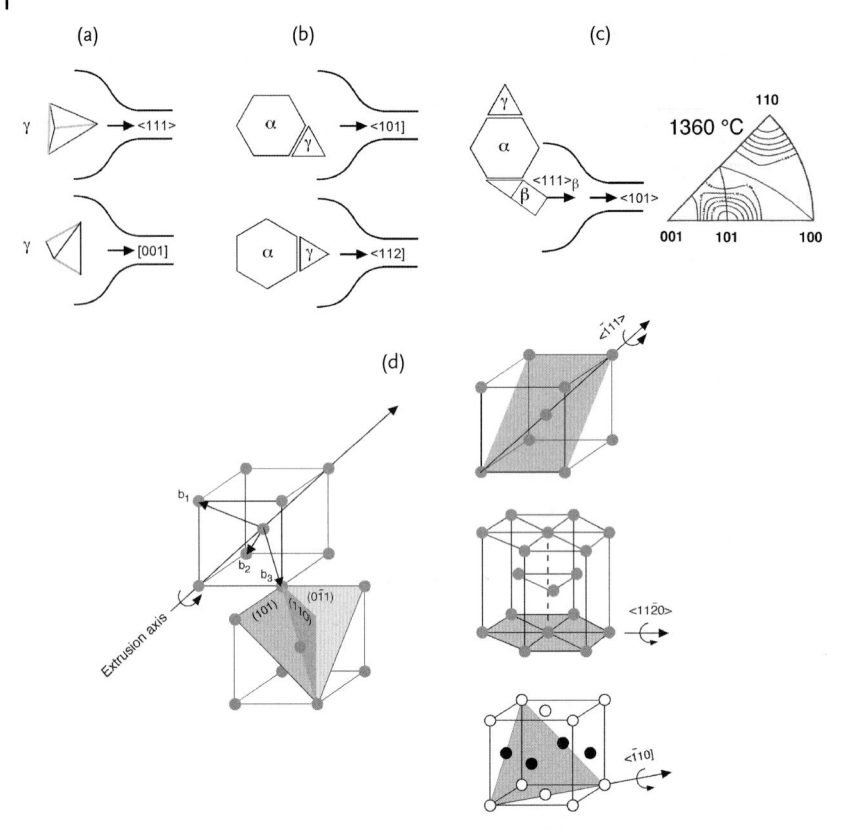

Figure 16.51 Evolution of textures during extrusion at different temperatures. Schematic drawing of the operating slip systems to realize the shape change imposed by the funnel-shaped die and subsequent phase transformations. (a) Extrusion in the (α + γ) phase field; the texture components are mainly determined by γ grain orientations that accomplish the radial reduction of the workpiece by symmetrical double glide along ½<1$\bar{1}$0]{111} (see Figures 16.46a, c and d). (b) Extrusion in the single-phase α field; the texture components are determined by prismatic glide of α(Ti) and the α → γ phase transformation occurring upon cooling. (c,d) Extrusion of the alloy Ti-45Al-10Nb in the β phase field. The texture is determined by symmetrical multislip of the β phase along ½<111>{110} and subsequent phase transformations.

and controversy [204, 205]. However, contrary to the oriented-growth hypothesis is the fact that the texture of as-rolled TiAl sheets is not sharpened by subsequent annealing (Figures 16.49 and 16.52).

Summarizing, the evolution of microstructure and texture during hot working of two-phase titanium aluminide alloys is governed by the operating deformation modes, recrystallization and phase transformations. The relative contribution of these factors depends on alloy composition and processing temperature. The structural features observed after hot working in the (α + γ) phase field essentially

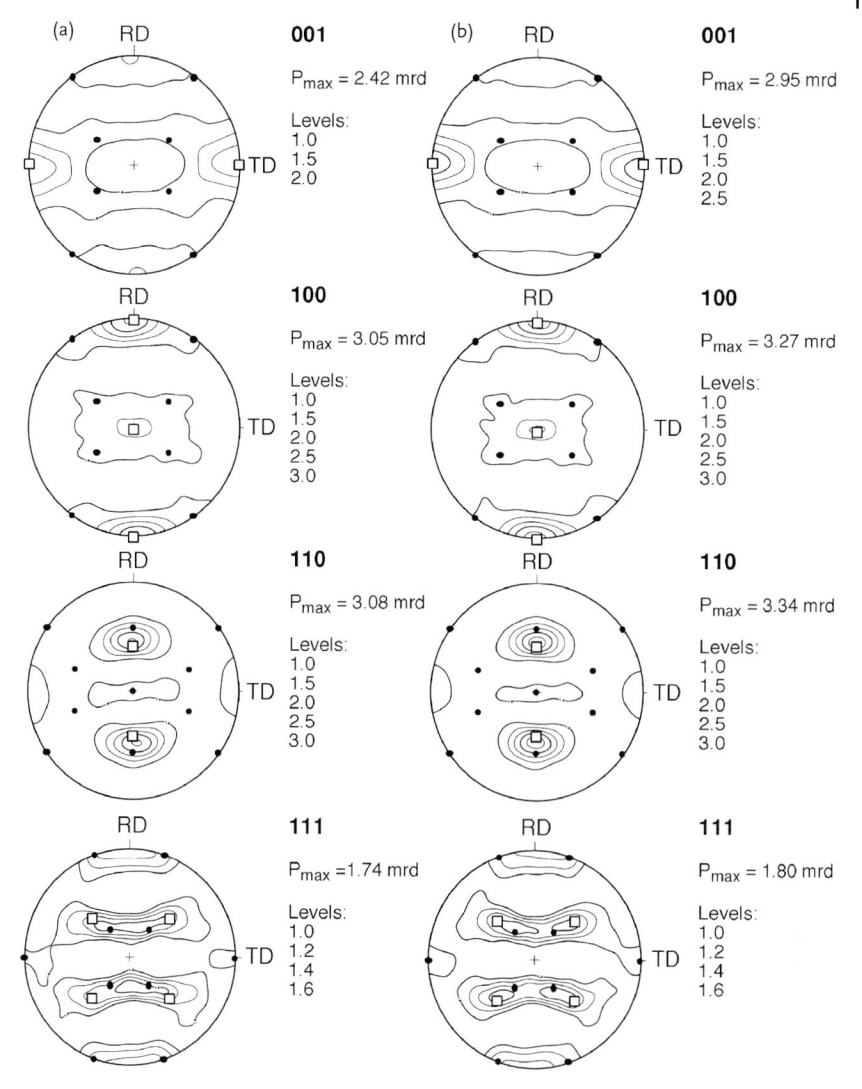

Figure 16.52 Texture of rolled sheet-material determined using neutron diffraction. Recalculated pole figures for Ti-45Al-10Nb that was processed by hot extruding an ingot through a rectangular die followed by sheet rolling [83]. In addition, the ideal modified cube texture (010)[100] (square symbols) and the ideal brass texture $(1\bar{1}0)[111]$ (filled circles) are indicated. (a) As-rolled condition, (b) after primary annealing for 2 h at 1000 °C.

reflect the kinematics of the propagation of ordinary dislocations and mechanical order twinning that are required to accomplish the imposed shape change. Hot working in the single α phase field leads to textures that are largely governed by the deformation modes of the hexagonal α lattice and the subsequent α → γ phase transformation that occurs upon cooling.

16.5
Secondary Processing

16.5.1
Component Manufacture through Wrought Processing

In general, the microstructural refinement obtained in primary wrought-processed material reduces the susceptibility to cracking during any secondary hot-working steps and thus allows the material to be worked into useful product forms by, for example, closed-die forging or rolling. In addition, the conversion of the cast lamellar microstructure to a fine-grained equiaxed microstructure with improved uniformity, decreases the tendency for flow localization and is certainly helpful with respect to die filling and the dimensional accuracy of the desired component. In particular, for safety-critical applications, multistep hot-working routes appear to be attractive since microstructural and chemical defects might be removed to some extent, and a specific microstructure and texture be obtained. In this respect, γ alloys do not differ very much from Ni-based superalloys (see Couts and Howson [238]). Furthermore, for many applications where cost is a primary issue, near-net-shape processing could reduce the manufacturing costs compared to machining parts from primary-processed material. Despite the engineering relevance of developing complete manufacturing routes, most of the published work on wrought processing was devoted to ingot breakdown and relatively little information can be found in the open literature regarding practical forging of γ alloys. Nevertheless, the literature on primary processing is often useful with respect to practical details and specified processing conditions. Kim [239] was probably the first to report on the successful closed-die forging of γ-TiAl into a near-net-shape part. Later, Furrer et al. (1995) developed a wrought-processing route that involved isothermal closed-die forging to produce demonstration airfoils of around 8 cm in length. The manufacture route included isothermal upset forging of a cylindrical ingot to a disc, cutting and contouring preforms from the disc, closed-die forging (using temperatures of 1165 to 1204 °C and strain rates between $8.3 \times 10^{-3} s^{-1}$ and $8.3 \times 10^{-2} s^{-1}$), heat treatment and final abrasive trimming of the blade. The authors applied the processing route to several alloys on the basis of Ti-47Al-(2-4)Nb-(1-2)Cr, some of which contained B and Si. They observed very uniform microstructures across the blades, both in the as-forged condition and after heat treatment. Within the framework of primary processing, it has already been mentioned that the final heat treatment is critical if fully lamellar microstructures are to be formed. In this respect, it is interesting to note that Furrer et al. [240] found that both Si and B were beneficial in the development of controlled colony-size lamellar structures, as apparently silicides and borides retard the growth of α grains during the final heat treatment. Alternative ways to control the colony size were discussed in Section 16.3.2. Summarizing, the work of Furrer et al. [240] demonstrated that near-net-shape parts with fine and uniform microstructures could be fabricated by closed-die forging.

The hot-working route described by Furrer et al. [240] has been modified in respect of the component geometry and the primary processing procedure. Knippscheer et al. [241] reported on a fabrication route for automotive valves involving multistep extrusion followed by hot bulging to form a forging preform that was finally quasi-isothermally forged. The production was carried out entirely on industrial equipment. The valves exhibited a homogeneous, fine microstructure and were successfully tested in car engines [241]. Sommer and Keijzers [242] have developed variants of this processing route in which backextrusion followed by friction or inertia welding of a larger diameter puck to the stem were used as a second processing step prior to isothermal forging. Kestler et al. [243] used hot extrusion followed by forging of the stem to produce car-engine valves. In this way it was possible to retain the lamellar microstructure in the head of the valve, which has to withstand the highest temperature, whereas a fine-grained equiaxed microstructure in the stem was formed during forging. The authors determined tensile data for 150 independent valve production lots from the alloy Ti-46.5Al-4(Cr, Nb, Ta, B) and found that the plastic elongation to fracture was consistently above 1%, demonstrating the reproducibility of the manufacturing route. The valves were routinely produced after 2000 by Plansee AG (Austria) and Sinterstahl GmbH (Germany) and used in high-performance car and motorcycle engines [243]. Within the framework of two joint projects involving Leistritz Turbinenkomponenten (Remscheid, Germany, formerly Thyssen Umformtechnik), Rolls-Royce Deutschland (Dahlewitz, Germany), GfE (Nürnberg, Germany) together with GKSS Research Center, the complete processing technology for the production of high-pressure aero-engine compressor blades from different γ alloys has been explored [17, 28, 120, 121, 157, 244]. The processing route involved hot-isostatic pressing of ingot material, hot extrusion, stress-relief annealing, decanning and machining of forging preforms, multistep closed-die forging, and machining as well as electrochemical milling for the final shaping operation, as schematically shown in Figure 16.53. In the first development cycle the alloy Ti-47Al-1.5Nb-1Cr-1Mn-0.2Si-0.5 B was used and more than 200 blades were successfully produced (Figure 16.54). The forgings showed that the material possessed an excellent die-filling capability, as also observed for other γ alloys [12]. After forging, a substantial improvement of the microstructural homogeneity was observed, when compared to the as-extruded material, in both the blade foil (Figure 16.55) and the root. Prior to electrochemical milling the forged performs were subject to a final heat treatment at the α-transus temperature that resulted in a very homogeneous lamellar microstructure with a mean colony size of 130 µm in the blade foil (Figure 16.55). Mechanical properties of this alloy in different stages of the processing route can be found in Table 16.3. Similar processing technology was used for much larger ingots (also including a route without forging) and resulted in the successful manufacture of several hundred blades from the high Nb-containing alloy Ti-45Al-8Nb-0.2C. The blades fulfilled the component requirements in terms of mechanical properties, but the question remains whether sufficient reproducibility over large batches can be achieved at acceptable costs. In conclusion, the production of turbine components by wrought processing still poses some

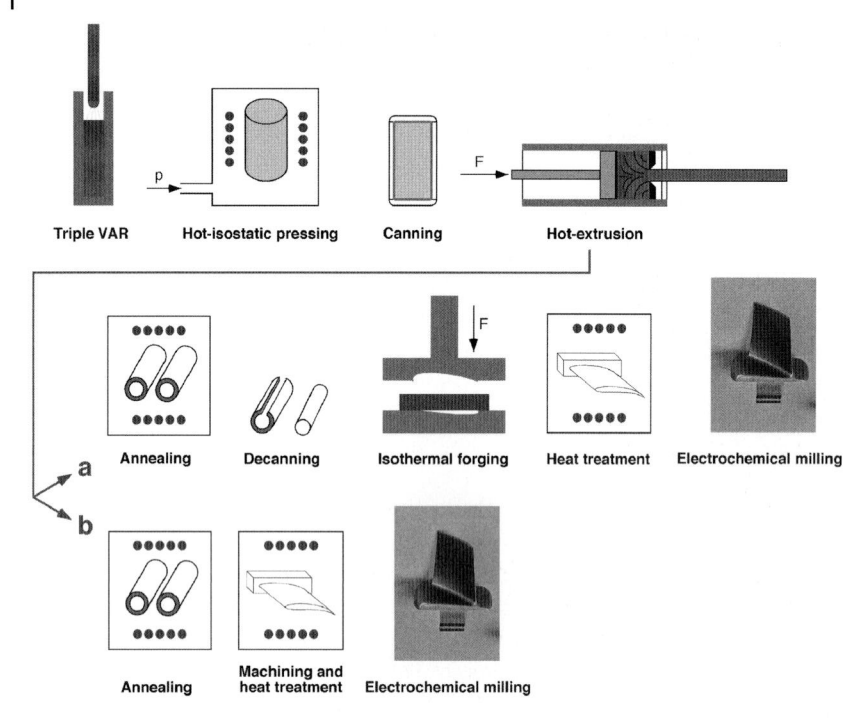

Figure 16.53 Schematic drawing illustrating two processing routes utilized for the production of high-pressure aero-engine compressor blades [157].

Figure 16.54 High-pressure compressor blades for an aero-engine manufactured via the hot-working route illustrated in Figure 16.53a [157].

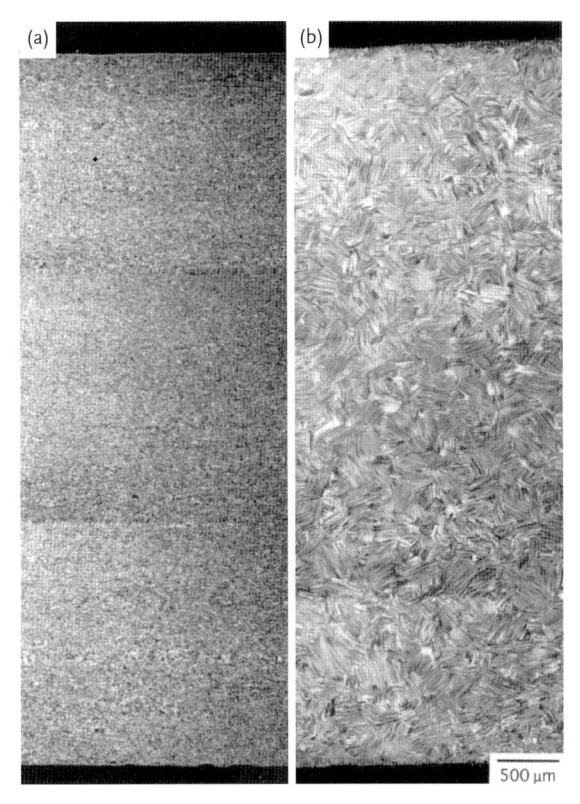

Figure 16.55 Microstructure in the blade section of a high-pressure compressor airfoil manufactured from Ti-47Al-1.5Nb-1Cr-1Mn-0.2Si-0.5B (TAB) via the hot-working route illustrated in Figure 16.53a. (a) Microstructure after stress-relief annealing, (b) microstructure after a heat treatment at the α-transus temperature.

problems in particular regarding reproducibility, but appears promising from a technical point of view.

Usually, forging processes are simulated by finite-element modeling in order to select proper forging conditions and die geometry design, however, for TiAl such information has rarely been published. Brooks et al. [245] experimentally studied the isothermal closed-die forging of 180-mm long aerofoils from Ti-48Al-2Nb-2Cr-1B and compared the forging experiments with modeling results. The forging was performed using TZM tooling at 1125 °C and a constant strain rate of $10^{-3}\,s^{-1}$. The final strains ranged from around 0.5 in the root block to 2.5 in regions adjacent to the flash. The preforms were coated with Acheson DAG3707 lubricant while Acheson Boron Nitride (DAG5710) was used on the dies and work-pieces as a parting agent. The authors developed a material model that not only described the flow softening behavior during hot deformation, but also the microstructural evolution. The latter was represented via an internal state variable in a constitutive equation. The model could reasonably predict the actual loading press forces and the component shape as well as the microstructure, in terms of the volume fraction of recrystallized grains, within different sections of the airfoil.

In most cases, a hot-worked component has to be machined during the last stage of the manufacturing process, for example by milling, drilling or grinding. In general, for machining conventional equipment can be used, however, care has to be taken with respect to the low ductility of γ alloys and the occurrence of surface cracks. The problems that are encountered when γ alloys are machined involve breakout (drilling), splintering at edges (milling), overheating (grinding) and local deformation due to overclamping [153]. To a large extent the problems can be overcome by using suitable machining tools and parameters as found in the literature [246, 247], enabling complex parts be machined [153]. With respect to finishing a component, electrochemical milling offers an alternative to machining. The process is able to produce complex shapes, with high metal removal rates, and provides a high surface finish with the absence of residual stresses, surface hardening or microcracks [248]. On the other hand, it has been found that compression stresses and surface hardening generated by shot-peening, significantly increase the fatigue strength of the material [249]. A similar effect can also be obtained by roller burnishing, which is generally superior to shot peening, because it imparts a deeper subsurface zone of plastic deformation [250]. However, surface hardening has been found to recover at temperatures of 650 to 750 °C, and is thus less effective for components that will be used in high-temperature service applications [249].

As has been described in Section 16.3.2, the development of alloys that contain the β phase over a large temperature range show significantly improved workability compared to other γ alloys [147, 148]. The authors were able to conventionally forge uncanned ingot material and produced rectangular bars, pancakes and plates with varying thicknesses [147, 148]. Details of the forging conditions can be found in Table 16.2. Such alloys allow the manufacture of complex components in a single forging step followed by machining. This is very attractive with respect to costs and to size limitations that result from primary processing. The authors further found that the machinability of such alloys was substantially better than for conventional alloys and corresponded to that of Ni-based superalloys [147, 148]. Kremmer et al. [149], Clemens et al. [150] and Wallgram et al. [151] used a similar alloy design concept for the manufacture of turbine blades using hot extrusion followed by conventional forging, and demonstrated that a robust industrial forging process can be established for such alloys. These results are certainly encouraging in view of the potential for further alloy development that may allow the combination of robust wrought processing with low costs and good high-temperature strength. However, as mentioned in Chapter 13, high levels of β/B2 phase in such alloys can lead to a deterioration of the mechanical properties.

16.5.2
Rolling – Sheet Production and Selected Mechanical Properties

Sheet TiAl could be used to construct large hollow blades or aerospace structures such as thermal protection systems for spacecraft or aero-engine exhaust nozzles.

For such applications, properties such as strength, ductility, creep resistance and of course formability need to be optimized.

To date, the most successful and widely used method to produce TiAl sheet is based on hot rolling, which generally uses material that has already been processed in some way. Cast material is usually not used because of its coarse inhomogeneous microstructure, but in principle it could be used if it had the correct form and sufficient workability. The hot rolling of encapsulated cast material has been proposed as a relatively low-cost alternative route [251]. Direct strip casting is an alternative technique that has been used to make sheets of 100 mm width and 1 to 2 mm in thickness, which were subsequently used to study texture and microstructural evaluation [252].

Fujitsuna *et al.* [175] have published work illustrating a route that involves casting TiAl into a sheet geometry and then further processing the cast material by isothermal rolling. An isothermal rolling mill was constructed in 1991 to investigate this method and its suitability for the production of sheet for subsequent superplastic forming. In this work it was demonstrated that a 300-mm long sheet of Ti-46Al (rolling temperature 1100 °C, strain rate $1 \times 10^{-3} \, s^{-1}$) could be produced. The material was deemed unsuitable for superplastic applications due to a coarse mainly lamellar microstructure that contained microstructural inhomogenities.

At present it is generally accepted that hot-worked ingot material or HIPed prealloyed powder compacts are the best starting materials for sheet production as a result of their much better chemical homogeneity, fine microstructure and better deformation behavior. To date, the only company that has been able to produce sheet on an industrial scale is Plansee AG in Austria who use an "advanced sheet-rolling process". Unfortunately, due to the commercial sensitivity of sheet rolling, very little information has ever been published in the open literature on the detailed technological aspects. These details are of utmost importance if hot rolling of sheet is to be successful. Semiatin and coworkers [253] have nevertheless published high-quality work that provides an insight into the necessary conditions. Due to the mechanical properties of TiAl, that is, the high strength, limited tensile ductility and limited workability, hot rolling to sheet is only possible within a relatively narrow parameter window and is thus not an easy process.

16.5.2.1 Pack Rolling

Pack rolling has been investigated by Semiatin *et al.* [253] who analyzed the temperature transients that arise across the pack during pack transfer from the furnace to the rolling mill and during the rolling process. During pack transfer from the furnace to the mill the temperature transients were numerically modeled taking into account radiation from the pack, heat conduction within the pack layers and heat transfer across the pack cover/workpiece interfaces. Modeling was performed for a number of hypothetical packs consisting of Ti-6Al-4V pack covers and a near-gamma titanium aluminide workpiece.

The pack geometries considered had total thicknesses of 11.2, 6.8, 4.8 and 3.5 mm and the workpiece thickness was considered to be 40% of the total pack thickness. In the modeling, the furnace temperature was set to 1316 °C, the

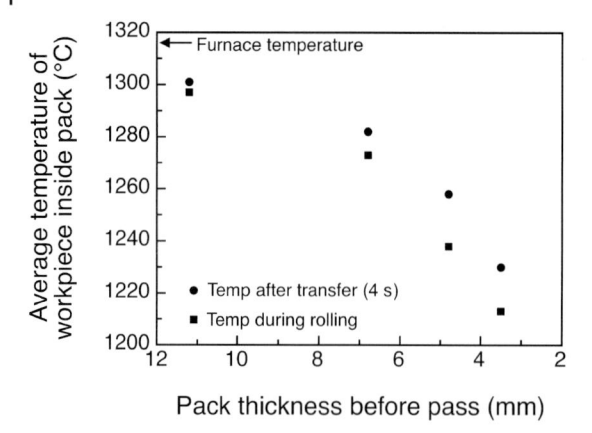

Figure 16.56 Average temperature of the TiAl workpiece within the pack (i) ● after a 4-s transfer time from the furnace (before rolling) and (ii) ■ during rolling in the middle of the deformation zone. The data plotted has been determined by Semiatin et al. [253]. The graph indicates that heat loss is much more significant for thin section sheet. For any thickness sheet, the heat loss during the time within the rolls is less than that during transfer from the furnace. Data in original paper given in a tabular form.

temperature of the rolls was 150 °C, the transfer time between the furnace and rolling mill was 4 s, the speed of the rolls was about 110 mm/s (≈ 6.6 m/min) and the reduction per pass was between 8 and 12%. Using a roll diameter of 406 mm and the assumed rolling speed and reduction, deformation within the rolling gap was determined to take place within approximately 0.1 s. Due to the low rolling reductions employed the effect of deformation heating was ignored. The average temperature of the workpiece through the pack both after transfer and during rolling was determined by Semiatin et al. [253] using numerical analysis and is shown graphically in Figure 16.56. This figure shows that the temperature of the TiAl workpiece falls during both transfer and during rolling and that the effect becomes more significant as the pack thickness is reduced. Additionally, whatever the sheet thickness, the temperature loss during the time within the roll gap is less than that during the 4-s transfer from the furnace. The temperature-distribution curves determined through a 3.5-mm thick pack prior to roll-gap entry, at the middle of the deformation zone (between the rolls) and on exiting the rolling mill are shown in Figure 16.57 [253]. This figure clearly illustrates that a temperature gradient can develop between the TiAl workpiece and the pack material after transfer and that this becomes more significant as rolling progresses. This temperature gradient is extremely important with respect to the ability to successfully roll TiAl to sheet.

A technique to produce TiAl sheet material was patented by TIMET [254] in the early 1990s, and relies on the TiAl being enclosed within a canning material with a layer of insulation between the can and TiAl. Indeed, the work by Semiatin and coworkers [253], discussed above, explains the scientific background to the patent.

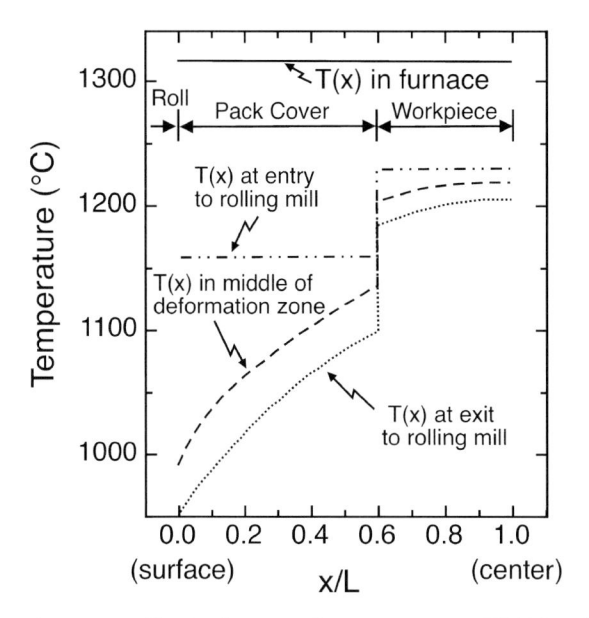

Figure 16.57 The development of temperature gradients within a 3.5-mm thick pack consisting of Ti-6Al-4V covers and a gamma TiAl workpiece as predicted using numerical analysis. The initial furnace temperature $T(x)$ was 1316 °C, and a 4-s transfer time to the rolling mill was assumed. Exact details regarding the figure and the calculations involved in its determination are given in [253]. Diagram redrawn based on original.

Although no information is given in the patent on the type of insulating material, the present authors have been able to successfully roll small specimens using a layer of ceramic-wool-like insulation such as fibrofrax. The canning material is required to shield the TiAl from oxidation and to aid in the TiAl deformation process. The difference of flow stress between the can and the TiAl is of importance. If the flow stress of the canning material is significantly lower than that of the TiAl at the rolling temperature, then appreciable secondary stresses can develop that result in tensile stresses being imposed on the TiAl within the roll gap. If conditions are unfavorable, that is, the flow-stress mismatch is high, the speed of the rolls is too fast, the deformation reduction is high or the TiAl has significantly cooled (important for thin sheet), then the secondary stresses can lead to both the initiation and growth of wedge cracking or even complete failure across the TiAl within the can. The insulation within the can has the effect of allowing the can to cool while minimizing heat loss from the TiAl, this results in a reduction in the flow-stress mismatch, thus widening the processability window.

The pack-rolling method has the advantage that conventional rolling equipment can be employed, which is of great importance for industrial companies who do not want to invest in extremely costly isothermal rolling equipment that needs to work under a protective atmosphere.

16.5.2.2 Rolling Defects

Rolling of an isothermally forged Ti-48Al-2.5Nb-0.3Ta alloy ($T_\alpha \approx 1360\,°C$) under near conventional (roll temperature between 65 and 250 °C) and near-isothermal (roll temperature 870 °C) rolling conditions was investigated by Semiatin et al. [140]. The preheat and reheat temperatures used were between 1260 and 1370 °C, and the imposed strain rates employed, which depend upon the rolling speed and reduction per pass, were $1\,s^{-1}$ or $5\,s^{-1}$. Rolling was found to be successful using intermediate temperatures, that is, 1315 °C, and unsuccessful at the higher and lower temperatures used, irrespective of the strain rate. Failure was generally of a brittle nature at lower rolling temperatures with intergranular cracking. At high temperatures (1370 °C, i.e. above T_α) wedge-cracking defects were observed that probably resulted from grain-boundary sliding and separation of large α grains. The lack of global deformation, even at such high temperatures, may at first seem surprising, however, it should be remembered that the α phase present at such temperatures is hexagonal, and that deformation may only be achieved on a restricted number of crystallographic planes (see Section 16.4). For the alloy studied by Semiatin et al. [140], rolling was successful at a temperature that was within the upper half of the α + γ phase field, that is, during rolling the majority phase was α but there was also a significant amount of γ. At this temperature, the material is sufficiently deformable to be successfully rolled and the excessive growth of the alpha grains, that leads to grain-boundary sliding and separation, is retarded by the gamma phase. Thus, to a certain extent, rolling temperature can be considered as an alloy-composition-determined parameter.

In addition to the correct rolling temperature, other parameters also need to be optimized that is, the rolling speed should not be too fast as this would lead to overcritical deformation rates and cracking, as described above. Wurzwallner et al. [165] and Clemens et al. [255, 256] suggest rolling speeds of less than 10 m/min. Additionally, relatively small rolling reductions of no more than 10–15% should be employed. Thus, to achieve large reductions and thin sheet, multipass rolling must be performed with intermediate reheating of the pack. A technical problem that occurs as a consequence of the slow rolling speeds, particularly during the rolling of thin sheet, is heat loss from the pack, principally through radiation. When this is severe, the sheet temperature may fall too low and cracking will occur. The effect of heat loss from the pack on the rolling pressure has been investigated by Semiatin and Seetharaman [257] and can be seen in Figure 16.58. This figure demonstrates that the average rolling pressure increases with the pass number, that is, as the sheet thickness is reduced. Of course, the heat loss from the pack material that is last to enter the roll gap is higher than that from the front end. Hence, cracking due to excessive heat loss can be expected to take place, initially at the back end of the pack and then move forward towards the front end of the pack as the sheet thickness reduces. The heat-loss effect can be minimized by reducing the distance between the furnace and roll gap to a minimum. The use of an optimized sheet-rolling process is also important from a microstructural viewpoint. Clemens et al. [256] have shown that the amount of α_2 and γ in the as-rolled microstructure can vary significantly along the

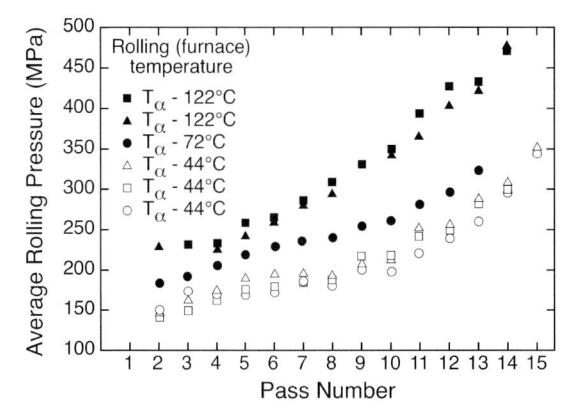

Figure 16.58 Rolling pressure data for Ti-45.5Al-2Cr-2Nb during multipass rolling with intermediate reheating to the rolling (furnace) temperature. It can be seen that as the sheet becomes thinner (increasing number of rolling passes) the rolling pressure increases due to heat loss through both radiation and contact with the rolls. Also, the rolling pressure is reduced by rolling at temperatures near to the alpha-transus temperature, and increases significantly for relatively small temperature reductions. For more details regarding this figure the reader is referred to [257]. Replotted data.

length of a long sheet if the rolling process with regard to temperature loss is not optimized.

16.5.2.3 Industrial Production of Sheet

Conventional Pack, High-Temperature, Near-Isothermal Rolling In spite of the difficulties discussed above that are associated with rolling TiAl, Plansee AG have been able to produce sheets with uniform microstructure with dimensions of up to 1800 × 500 × 1.0 mm. They report that further upscaling in sheet length and width is technically possible from the rolling standpoint, but that hot-worked ingot prematerial of sufficient size and homogeneity may be difficult to obtain. Indeed, in this respect sheet produced from HIPed prealloyed powder compacts may be advantageous. This route to produce TiAl sheet was first investigated around 1994 by Eylon et al. [258] and Clemens et al. [259]. Sheet made using texture-free prealloyed powder has often been used in basic research to investigate the development of texture during rolling.

"Low-Temperature" Near-Isothermal Sheet Rolling In an attempt to reduce the processing costs and thus the price of sheet, Russian work on sheet rolling has concentrated on a "low-temperature" rolling process [176, 260]. In this work, rolling is reported to be possible using pack soak temperatures of between 750 °C and 1000 °C. The rolling speeds and reductions per pass employed, result in effective deformation rates around $10^{-1} s^{-1}$. The authors claim that rolling of Ti-50Al can be successful at temperatures between 750 °C and 850 °C for material with a grain size of 0.4 µm and between 750 °C and 1000 °C for a 5-µm grain size. The approach has also been demonstrated on technically relevant alloys such as Ti-45.2Al-3.5(Nb, Cr, B) and Ti-44.2Al-3.5(Nb, Cr, B) [176]. This technique also has the advantage, as a consequence of the lower temperatures employed, that 18/10 stainless steel could be used as a canning material that is reported to be less expensive than the canning material used in the conventional high-temperature isothermal pack rolling process [260]. Sheet made in this way is believed to show very good superplastic properties as compared to the high-temperature, near-isothermally pack-rolled-sheet material [261]; however, the size of the sheets produced to date using this technique is considerably smaller (a sheet size of $200 \times 120 \times 1.7$ mm is reported in [261]). At present, this "low-temperature" technique only seems to be on a laboratory scale.

The secret behind the success of this "low-temperature" rolling technique is the use of extremely fine-grained material for rolling. This material is produced from ingot that has undergone multistep forging at ever decreasing temperatures. Initial forging is performed at temperatures within the $\alpha + \gamma$ phase field using strain rates of around $5 \times 10^{-2} s^{-1}$ with final forging taking place at temperatures within the $\alpha_2 + \gamma$ phase field using strain rates between $10^{-3} s^{-1}$ and $10^{-2} s^{-1}$ [176]. The fine-grained material that is produced exhibits superplastic properties and is key to being able to successfully roll at "low temperatures".

16.5.2.4 Mechanical Properties of Sheet

As the microstructure of the sheet after rolling and decanning is not in thermodynamic equilibrium a primary annealing treatment of around 1000 °C for 2 h is performed. This brings the microstructure into closer thermodynamic equilibrium and simultaneously reduces residual stresses and flattens the sheet before final grinding. Material in this state is referred to as being in the "primary annealed" condition. As with other forms of TiAl, the mechanical properties of sheet depend to a great extent on the alloy composition and the microstructure developed during the final heat treatment. Sheet produced from HIPed powder is certainly of interest from both the technical and basic research standpoints. However, one major disadvantage between sheet manufactured from prealloyed powder (produced by gas atomization) compared to sheet from ingot is the presence of pores. These can form, as discussed in Section 15.1.4, as a result of entrapped gas within the powder particles. Indeed, pores elongated along the rolling direction have been observed in primary annealed Ti-46Al-9Nb PM sheet and they can have lengths of several micrometers [262]. Such defects can give rise to premature failure, particularly during superplastic forming processes.

In the literature, there are a number of papers that describe the properties of sheet [165, 255, 256, 259, 262–265] made from both prealloyed powder and from hot-worked ingot material. The properties of sheet from alloys such as Ti-48Al-2Cr and Ti-47Al-2Cr-0.2Si manufactured from ingot has been studied by Clemens et al. [255]. For Ti-48Al-2Cr, three different types of microstructure, namely primary annealed, near-gamma and duplex were established through different heat-treatments and then tested ($\dot{\varepsilon} = 1 \times 10^{-4}$ s^{-1}). Room-temperature "fracture elongations" in the range 2.5% to 5.1% were determined for 1.5-mm thick primary-annealed sheet. Flow stresses of 427 MPa, 361 MPa and 338 MPa were measured for the primary-annealed, near-gamma and duplex microstructures, respectively. The room-temperature properties were mainly determined by the microstructural homogeneity and grain size; which was around 15 µm for the primary annealed microstructure, 27 µm for the near-gamma and 34 µm for the duplex microstructure (which had around a 50% volume fraction of lamellar colonies). Ultimate tensile strength levels of the sheets were in the range of 465 to 558 MPa. At 700 °C, elongations to fracture of over 60% were determined combined with fracture stresses between 496 and 600 MPa. The authors state that there was no effect of testing orientation on the mechanical properties of primary-annealed Ti-48Al-2Cr sheet [263]. However, a considerable strength anisotropy was measured between 600 °C and 900 °C in [266]. The ultimate tensile strength in the rolling direction was about 200 MPa higher at 700 °C and around 250 MPa higher at 800 °C than that measured in the transverse direction ($\dot{\varepsilon} = 8 \times 10^{-5}$ s^{-1}).

Concerning the properties of sheet made from the latest generation of high-strength alloys, such as TNB, only a few papers have been published to date addressing the structural characterization and tensile properties [262] and creep properties [264, 265]. Unpublished work has indicated tensile strength of around 1000 MPa and plastic elongation to failure of over 1% for sheet in the primary annealed condition made from extruded ingot.

16.5.2.5 Superplastic Behavior

One potential use of sheet material is for hot- or superplastic-forming processes for the production of sheet structures. For such processes to be successful the sheet material must show superplastic properties. Such properties are shown by a number of materials, usually at elevated temperatures under certain strain-rate conditions. The microstructure must be fine (usually multiphase) and dynamic recovery must be strong so that work hardening is small or nonexistent. Additionally, the thermal component of the flow stress must be strongly strain-rate sensitive. The flow stress (σ) can be approximated from the power law: $\sigma = K(\dot{\varepsilon})^m$, where K is a constant and $\dot{\varepsilon}$ is the strain-rate. m is the so-called strain-rate sensitivity exponent ($m = d\ln\sigma/d\ln\dot{\varepsilon}$), and is usually determined by strain-rate ramping experiments performed at different temperatures. For a material to show superplastic properties, m should have a value of greater than 0.4 to 0.5 although values above 0.3 have also been quoted in the literature [263, 267].

A number of papers have been published that include information on the superplastic behavior of TiAl [152, 176, 230, 255, 256, 261, 263, 266, 267]. Most of the work has been performed on "conventional TiAl compositions" such as Ti-48Al-2Cr [230, 255, 263] and Ti-47Al-2Cr-0.2Si [9, 21, 22] (i.e. alloys not containing high Nb additions). Additionally, data on gas-atomized powder-based sheet [266], and on material produced by "low-temperature" pack rolling of forged ingot is available [176, 261]. The main result of all these studies is that TiAl sheet can show superplastic properties under certain conditions. Values of m ranging from around 0.4 at a strain rate of $1\times 10^{-3}\,s^{-1}$ to 0.63 at $4\times 10^{-5}\,s^{-1}$ have been presented for Ti-47Al-2Cr-0.2Si in the primary-annealed condition at 1100 °C [152, 256, 267]. At 1000 °C the respective values of m were around 0.36 and 0.43. Such values at this temperature are important as they indicate that equipment used to form conventional titanium alloys may also be used for TiAl-based sheet.

Elongation values of 165% and 210% are reported for primary-annealed Ti-48Al-2Cr at 1000 °C and 1100 °C, respectively ($\dot{\varepsilon}=5\times 10^{-4}\,s^{-1}$) [263]; the corresponding strain-rate sensitivity exponents at these temperatures and strain rate were 0.25 and 0.35. Under the same strain-rate conditions, at 1200 °C the elongation was over 220% and the strain-rate sensitivity exponent was 0.55 [263]. Even at 700 °C, elongations of between 40% and 90% were obtained in Ti-48Al-2Cr ($\dot{\varepsilon}=1\times 10^{-4}\,s^{-1}$) [230] depending on microstructure. Clemens et al. [255] state that superplastic behavior was observed in Ti-48Al-2Cr at temperatures above 950 °C and also demonstrated superplastic forming processing on both laboratory and industrial scales. Primary-annealed Ti-47Al-2Cr-0.2Si sheet exhibited elongation values of up to 190% at 1000 °C ($\dot{\varepsilon}=2\times 10^{-4}\,s^{-1}$) [267]. Gas-pressure dome-bulging experiments (performed by McDonnel Douglas) at 1000 °C on this material resulted in strains of around 600% at the top of the dome (reported by Clemens et al. [256]), although it is not reported if a backpressure was applied to suppress porosity formation. More information regarding the forming of sheet into structures can be found in [152, 230, 251, 255, 256, 263, 267].

Superplastic behavior of Ti-45.2Al-3.5(Nb, Cr, B) sheet produced by the "low-temperature" pack-rolling method, has been presented in [176, 261]. At 900 °C, the elongation of specimens tested parallel to the rolling direction was around 240%, compared to roughly 160% for the transverse direction (initial strain rate of $10^{-3}\,s^{-1}$). The respective 50% flow stress levels were roughly 280 and 305 MPa. At 1000 °C the respective elongation values were around 350% and 330%, while at 1100 °C the elongation in both directions increased to around 550%. The authors believe that such "low-temperature" rolled sheet shows better superplastic behavior over the temperature range 950 to 1100 °C than "conventionally" rolled sheet, and is also more cost effective [176].

The formation of porosity during superplastic forming of sheet is an important failure mechanism. Pore formation is generally less extensive in ingot-based sheet material than sheet produced from gas-atomized powder and results in significantly better high-temperature superplastic elongation [152, 153, 256]. Pore formation and grain-boundary decohesion are easier in sheet produced from gas-atomized prealloyed powder, thus reducing the superplastic elongation, despite such sheet exhibiting strain-rate sensitivity exponents that are comparable to those of ingot-based material. Such pores could have been formed as a result of argon gas being trapped within the HIPed powder particles, a subject that is addressed in Section 15.1.4 (also see Figure 15.15). In this respect, sheet manufactured through the ingot route may be preferred if the required superplastic deformability is large.

However, even within ingot-based material, pores can form, particularly at high strains. Such cavities are thought to originate at second-phase precipitates; Ti_5Si_3 particles in Ti-47Al-2Cr-0.2Si sheet for example [267].

16.5.3
Novel Techniques

Wrought processing of TiAl ingot material is possible by a number of techniques and is necessary to achieve improved mechanical properties through a refinement of microstructure and a reduction of chemical inhomogeneity. The manufacture of large components from ingot material presents significant challenges because conventional hot-working procedures, particularly those that impart substantial amounts of work, lead to the dimensions of the material in at least one direction being reduced, so that the resulting material is not large enough for component manufacture. Additionally, the current status of large-ingot manufacture is not able to guarantee sufficiently homogeneous material that is completely defect free. Thus, even if conventional hot working can be used to make a large component, it cannot be assured that no significant defects will be present or that the microstructure will be sufficiently refined in the final component.

These problems have been addressed by GKSS, and work has been undertaken to develop a manufacturing technique that can guarantee the production of large, relatively homogeneous, "defect-free" components. The novel technique developed, for which patents have been approved, is described below.

16.5.3.1 Manufacture of Large "Defect-Free" Components

As described above, the manufacture of large components from ingot material presents problems that to date have not been satisfactorily resolved. With this in mind, GKSS had the goal to develop a technique that could be employed to make large parts with refined microstructures, optimized chemical homogeneity and guaranteed to result in "defect-free" material. The manufacture of a "large disc" was chosen as a technology-demonstrator platform.

Although not discussed in detail here, the method involved a combination of hot extrusion, followed by isothermal forging, joining of the forged discs followed by a second isothermal forging step. The beauty of the technique is that after the initial forging, the plates can be nondestructively examined and only material of sufficiently high quality is joined and subsequently forged to make the final product. Details of the technique will be submitted for publication in the near future. Three discs of varying sizes were made during the program, Figure 16.59. The microstructure of the second largest disc in the as-forged state + near-alpha-transus heat-treated state is shown in Figure 16.60; room-temperature tensile tests curves performed on material from this disc are shown in Figure 16.61. It is interesting to see that, despite the extensive hot deformation, the strength levels and plastic tensile elongations are not significantly better than good-quality extruded material, tested parallel to the extrusion direction. It should be mentioned that one

Figure 16.59 The largest TiAl disc (≈ 32 cm in diameter and 4 cm thick) made using the novel manufacturing procedure that GKSS has developed.

Figure 16.60 Microstructure of the second largest disc in the as-forged + near-alpha-transus heat-treated state. As can be seen, the microstructure was rather fine.

test, not shown in Figure 16.61a, showed no plastic elongation, breaking at a stress of 500 MPa. On further investigation, this premature failure was due to a TiC inclusion that probably had formed during ingot manufacture. This was rather disappointing, but it should be appreciated that in this demonstrator program, rather limited effort was spent for NDE of the primary forged material. Nevertheless, the project achieved its aim in demonstrating the feasibility of the idea. Very encouragingly, after the near-alpha-transus heat treatment, the strength and plasticity were nearly isotropic. For further information concerning the technique, the reader is directed to [268].

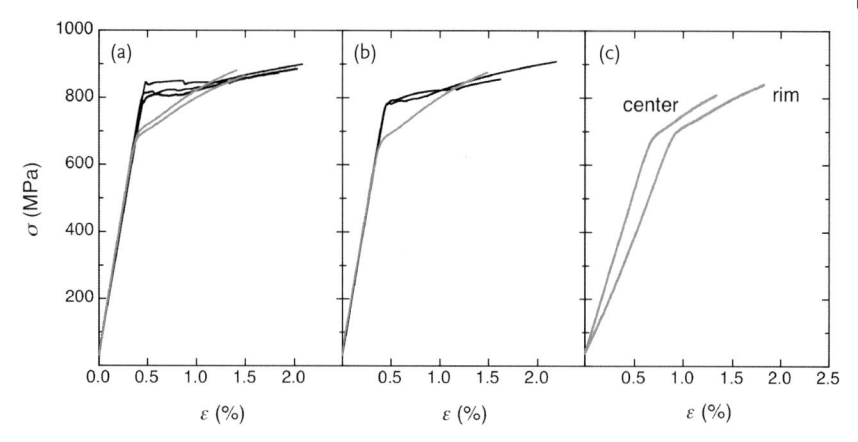

Figure 16.61 Room-temperature tensile tests taken from the second largest disc that was made in the GKSS technology demonstrator program. Black lines indicate that the forged material was annealed prior to testing, whereas the gray lines are for material that had been heat treated near to the alpha-transus temperature. (a) Curves for specimens that were taken from close to the disc rim and oriented perpendicular to the radial direction. (b) Curves for specimens that were parallel to the radius, that is, parallel to the material flow during forging.
(c) Displacement-controlled tensile test curves (no extensometer) of specimens that were taken from near to the rim or central region of the disc, parallel to the forging axis, that is, in the through-thickness direction. Such specimens in the as-forged + stress relieved condition are not shown in the graph because they broke on elastic loading at stresses of around 650 MPa; similar to annealed extruded material tested in the transverse direction (Figure 16.42b).

References

1 Nobuki, M., Furubayashi, E., and Tsujimoto, T. (1989) *Proc. 1st Japan International SAMPE Symposium* (eds N. Igata, I. Kimpata, T. Kishi, E. Nakata, A. Okura, and T. Uryu), Society for the Advancement of Material and Process Engineering, Chiba, Japan, p. 163.

2 Feng, C.R., Michel, D.J., and Crowe, C.R. (1989) *High-Temperature Ordered Intermetallic Alloys III*, vol. 133 (eds C.T. Liu, A.J. Taub, N.S. Stoloff, C.C. Koch) *Mater. Res. Soc. Symp. Proc.*, Mater. Res. Soc., Pittsburgh, PA, p. 669.

3 Semiatin, S.L., Frey, N., Thompson, C.R., Bryant, J.D., El-Soudani, S., and Tisler, R. (1990) *Scr. Metall. Mater.*, **24**, 1403.

4 Fukutomi, H., Hartig, C., and Mecking, H. (1990) *Z. Metallkunde*, **81**, 272.

5 Kim, Y.-W., and Kleek, J.J. (1990) *PM 90 – World Conference on Powder Metallurgy*, vol. 1, The Institute of Metals, London, p. 272.

6 Imayev, R.M., and Imayev, V.M. (1991) *Scr. Metall. Mater.*, **25**, 2041.

7 Nobuki, M., and Tsujimoto, T. (1991) *ISIJ (Iron and Steel Institute of Japan) Int.*, **31**, 931.

8 Beaven, P.A., Pfullmann, T., Rogalla, J., and Wagner, R. (1991) *High-Temperature Ordered Intermetallic Alloys IV*, vol. 213 (eds L.A. Johnson, D.P. Pope, J.O. Stiegler, and *Mater. Res. Soc. Symp. Proc.*), Mater. Res. Soc., Pittsburgh, PA, p. 151.

9 Shih, D.S., and Scarr, G.K. (1991) *High-Temperature Ordered Intermetallic Alloys IV*, vol. 213 (eds L.A. Johnson, D.P. Pope, J.O. Stiegler, and Mater.

10. Semiatin, S.L. (1995) *Gamma Titanium Aluminides* (eds Y.-W. Kim, R. Wagner, and M. Yamaguchi), TMS, Warrendale, PA, p. 509.
11. Salishchev, G.A., Imayev, R.M., Imayev, V.M., and Shagiev, M.R. (1995) *Gamma Titanium Aluminides* (eds Y.-W. Kim, R. Wagner, and M. Yamaguchi), TMS, Warrendale, PA, p. 595.
12. Semiatin, S.L., Chesnutt, J.C., Austin, C., and Seetharaman, V. (1997) *Structural Intermetallics 1997* (eds M.V. Nathal, R. Darolia, C.T. Liu, P.L. Martin, D.B. Miracle, R. Wagner, and M. Yamaguchi), TMS, Warrendale, PA, p. 263.
13. Kim, Y.-W., and Dimiduk, D.M. (1997) *Structural Intermetallics 1997* (eds M.V. Nathal, R. Darolia, C.T. Liu, P.L. Martin, D.B. Miracle, R. Wagner, and M. Yamaguchi), TMS, Warrendale, PA, p. 531.
14. Semiatin, S.L., Seetharaman, V., and Weiss, I. (1998) *Mater. Sci. Eng.*, **A243**, 1.
15. Dimiduk, D.M., Martin, P.L., and Kim, Y.-W. (1998) *Mater. Sci. Eng.*, **A243**, 66.
16. Clemens, H., and Kestler, H. (2000) *Adv. Eng. Mater.*, **2**, 551.
17. Appel, F., Wagner, R., and Oehring, M. (2000) *Intermetallics*, **8**, 1283.
18. Appel, F., Clemens, H., and Kestler, H. (2002) Forming, in *Intermetallics Compounds*, vol. 3, Progress (eds J.H. Westbrook and L. Fleischer), Chapter 29, John Wiley & Sons, Ltd, Chichester, UK, p. 617.
19. Appel, F., and Oehring, M. (2003) *Titanium and Titanium Alloys* (eds C. Leyens and M. Peters), Wiley-VCH, Weinheim, p. p.89.
20. Kestler, H., and Clemens, H. (2003) *Titanium and Titanium Alloys* (eds C. Leyens and M. Peters), Wiley-VCH, Weinheim, p. p.351.
21. Semiatin, S.L., and Jonas, J.J. (1984) *Formability and Workability of Metals*, American Society for Metals, Metals Park, OH.
22. Dieter, G.E. (1988) Introduction, in *Forming and Forging, Metals Handbook*, vol. 14, 9th edn (ed. Chairman S.L. Semiatin), ASM International, Metals Park, OH, p. 363.
23. Dieter, G.E. (1988) Workability tests, in *Forming and Forging, Metals Handbook*, vol. 14, 9th edn (ed. Chairman S.L. Semiatin), ASM International, Metals Park, OH, p. 373.
24. Davey, S., Loretto, M.H., Evans, R.W., Dean, T.A., Huang, Z.W., Blenkinsop, P., and Jones, A. (1995) *Gamma Titanium Aluminides* (eds Y.-W. Kim, R. Wagner, and M. Yamaguchi), TMS, Warrendale, PA, p. 539.
25. Seetharaman, V., and Semiatin, S.L. (1996) *Metall. Mater. Trans.*, **27A**, 1987.
26. Mohan, B., Srinivasan, R., and Weiss, I. (1995) *Gamma Titanium Aluminides* (eds Y.-W. Kim, R. Wagner, and M. Yamaguchi), TMS, Warrendale, PA, p. 587.
27. Hu, Z.M., and Dean, T.A. (2001) *J. Mater. Proc. Technol.*, **111**, 10.
28. Appel, F., Oehring, M., Paul, J.D.H., Klinkenberg, C., and Carneiro, T. (2004) *Intermetallics*, **12**, 791.
29. Huang, Z.H. (2005) *Intermetallics*, **13**, 245.
30. Fujitsuna, N., Ohyama, H., Miyamoto, Y., and Ashida, Y. (1991) *ISIJ (Iron and Steel Institute of Japan) Int.*, **31**, 1147.
31. Semiatin, S.L., Frey, N., El-Soudani, S.M., and Bryant, J.D. (1992) *Metall. Trans.*, **23A**, 1719.
32. Koeppe, C., Bartels, A., Seeger, J., and Mecking, H. (1993) *Metall. Trans.*, **24A**, 1795.
33. Clemens, H., Schretter, P., Wurzwallner, K., Bartels, A., and Koeppe, C. (1993) *Structural Intermetallics* (eds R. Darolia, J.J. Lewandowski, C.T. Liu, P.L. Martin, D.B. Miracle, and M.V. Nathal), TMS, Warrendale, PA, p. 205.
34. Kim, H.Y., Sohn, W.H., and Hong, S.H. (1998) *Mater. Sci. Eng.*, **A251**, 216.
35. Seetharaman, V., and Semiatin, S.L. (1997) *Metall. Mater. Trans.*, **28A**, 2309.
36. Hoppe, R., and Appel, F. (2006) Internal report, GKSS Research Centre, Geesthacht, Germany.
37. Imayev, R.M., Salishchev, G.A., Imayev, V.M., Shagiev, M.R., Kuznetsov, M.R., Appel, F., Oehring, M., Senkov, O.N., and Froes, F.H. (1999) *Gamma*

Titanium Aluminides 1999 (eds Y.-W. Kim, D.M. Dimiduk, and M.H. Loretto), TMS, Warrendale, PA, p. 565.
38 Imayev, R.M., Imayev, V.M., Oehring, M., and Appel, F. (2005) *Metall. Mater. Trans.*, **36A**, 859.
39 Fröbel, U., and Appel, F. (2007) *Metall. Mater. Trans.*, **38A**, 1817.
40 Mecking, H., and Gottstein, G. (1978) *Recrystallization of Metallic Materials* (ed. Dr. F. Haessner), Riederer Verlag, Stuttgart, p. 195.
41 Sakai, T., and Jonas, J.J. (1984) *Acta Metall.*, **32**, 189.
42 McQueen, H.J., and Jonas, J.J. (1975) *Treatise of Materials Science and Technology*, vol. 6 (ed. R. Arsenault), Academic Press, New York, San Francisco, London, p. 393.
43 Hasegawa, M., Yamamoto, M., and Fukutomi, H. (2003) *Acta Mater.*, **51**, 3939.
44 Schaden, T., Fischer, F.D., Clemens, H., Appel, F., and Bartels, A. (2006) *Adv. Eng. Mater.*, **8**, 1109.
45 Ponge, D., and Gottstein, G. (1998) *Acta Mater.*, **46**, 69.
46 Sellars, C.M., and Tegart, W.J.M.G. (1966) *Mém. Sci. Rev. Métall.*, **63**, 731.
47 Jonas, J.J. (1969) *Acta Metall.*, **17**, 397.
48 McQueen, H.J., and Ryan, N.D. (2002) *Mater. Sci. Eng.*, **A322**, 43.
49 Zener, C., and Hollomon, J.H. (1944) *J. Appl. Phys.*, **15**, 22.
50 Nobuki, M., Hashimoto, K., Takahashi, J., and Tsujimoto, T. (1990) *Mater. Trans. Japan Inst. Met.*, **31**, 814.
51 Nobuki, M., Vanderschueren, D., and Nakamura, M. (1994) *Acta Metall. Mater.*, **42**, 2623.
52 Kim, H.Y., Sohn, W.H., and Hong, S.H. (1997) *Light-Weight Alloys for Aerospace Applications IV* (eds E.W. Lee, W.E. Frazier, N.J. Kima, and K. Jata), TMS, Warrendale, PA, p. 195.
53 Seetharaman, V., and Lombard, C.M. (1991) *Microstructure/Property Relationships in Titanium Aluminides and Alloys* (eds Y.-W. Kim and R.R. Boyer), TMS, Warrendale, PA, p. 237.
54 Oehring, M., Lorenz, U., and Appel, F. (2005) unpublished work, GKSS Research Centre, Geesthacht, Germany.
55 Medina, S.F., and Hernandez, C.A. (1996) *Acta Mater.*, **44**, 137.
56 Takahashi, T., Nakai, H., and Oikawa, H. (1989) *Mater. Trans. JIM*, **30**, 1044.
57 Takahashi, T., Nakai, H., and Oikawa, H. (1989) *Mater. Sci. Eng.*, **A114**, 13.
58 Bartolomeusz, M.F., Yang, Q., and Wert, J.A. (1993) *Scr. Metall. Mater.*, **29**, 389.
59 Beddoes, J., Wallace, W., and Zhao, L. (1995) *Int. Mater. Rev.*, **40**, 197.
60 Weertman, J. (1955) *J. Appl. Phys.*, **26**, 1213.
61 Weertman, J. (1957) *J. Appl. Phys.*, **28**, 362.
62 Weertman, J. (1957) *J. Appl. Phys.*, **28**, 1185.
63 Weertman, J., and Weertman, J.R. (1983) *Physical Metallurgy, Part II*, 3rd edn (eds R.W. Cahn and P. Haasen), North Holland Physics Publishing, Amsterdam, The Netherlands, p. 1309.
64 Herzig, C., Przeorski, T., and Mishin, Y. (1999) *Intermetallics*, **7**, 389.
65 Mishin, Y., and Herzig, C. (2000) *Acta Mater.*, **48**, 589.
66 Herzig, C., Friesel, M., Derdau, D., and Divinski, S.V. (1999) *Intermetallics*, **7**, 1141.
67 Imayev, V.M., Salishchev, G.A., Shagiev, M.R., Kuznetsov, A.V., Imayev, R.M., Senkov, O.N., and Froes, F.H. (1999) *Scr. Mater.*, **40**, 183.
68 Imayev, R.M., Salishchev, G.A., Senkov, O.N., Imayev, V.M., Shagiev, M.R., Gabdullin, N.K., Kuznetsov, A.V., and Froes, F.H. (2001) *Mater. Sci. Eng.*, **A300**, 263.
69 Kim, J.H., Shin, D.H., Semiatin, S.L., and Lee, C.S. (2003) *Mater. Sci. Eng.*, **A344**, 146.
70 Cahn, R.W. (1983) *Physical Metallurgy, Part II*, 3rd edn (eds R.W. Cahn and P. Haasen), North-Holland Physics Publishing, Amsterdam, p. 1595.
71 Appel, F., Oehring, M., and Ennis, P.J. (1999) *Gamma Titanium Aluminides 1999* (eds Y.-W. Kim, D.M. Dimiduk, and M.H. Loretto), TMS, Warrendale, PA, p. 603.
72 Appel, F. (2001) *Mater. Sci. Eng.*, **A317**, 115.
73 Appel, F. (2001) *Intermetallics*, **9**, 907.

74 Appel, F., Paul, J.D.H., Oehring, M., Fröbel, U., and Lorenz, U. (2003) *Metall. Mater. Trans.*, **34A**, 2149.
75 Cahn, R.W. (1990) *High Temperature Aluminides and Intermetallics* (eds S.H. Whang, C.T. Liu, D.P. Pope, and J.O. Stiegler), TMS, Warrendale, PA, p. 244.
76 Cahn, R.W., Takeyama, M., Horton, J.A., and Liu, C.T. (1991) *J. Mater. Res.*, **6**, 57.
77 Baker, I. (2000) *Intermetallics*, **8**, 1183.
78 Hutchinson, W.B., Besag, F.M.C., and Honess, C.V. (1973) *Acta Metall.*, **21**, 1685.
79 Hartig, C., Fukutomi, H., Mecking, H., and Aoki, K. (1993) *ISSJ (Iron and Steel Institute of Japan) Int.*, **33**, 313.
80 Fukutomi, H., Nomoto, A., Osuga, Y., Ikeda, S., and Mecking, H. (1996) *Intermetallics*, **4**, S49–S55.
81 Bartels, A., Hartig, C., Willems, S., and Uhlenhut, H. (1997) *Mater. Sci. Eng.*, **A239–240**, 14.
82 Imayev, R.M., Kaibyshev, O.A., and Salishchev, G.A. (1992) *Acta Metall. Mater.*, **40**, 581.
83 Brokmeier, H.-G., Oehring, M., Lorenz, U., Clemens, H., and Appel, F. (2004) *Metall. Mater. Trans.*, **35A**, 3563.
84 Kolmogorov, A.N. (1939) *Izv. Akad. Nauk USSR – Ser. Matemat.*, **1**, 355.
85 Johnson, W.A., and Mehl, R.F. (1939) *Trans. Metall. Soc. AIME*, **135**, 416.
86 Avrami, M. (1939) *J. Chem. Phys.*, **7**, 1103.
87 Doherty, R.D., Hughes, D.A., Humphreys, F.J., Jonas, J.J., Juul Jensen, D., Kassner, M.E., King, W.E., McNelley, T.R., McQueen, H.J., and Rollett, A.D. (1997) *Mater. Sci. Eng.*, **A238**, 219.
88 Luton, M.J., and Sellars, C.M. (1969) *Acta Metall.*, **17**, 1033.
89 Salishchev, G.A., Imayev, R.M., Senkov, O.N., Imayev, V.M., Gabdullin, N.K., Shagiev, M.R., Kuznetsov, A.V., and Froes, F.H. (2000) *Mater. Sci. Eng.*, **A286**, 236.
90 Imayev, R.M., Kaibyshev, O.A., and Salishchev, G.A. (1992) *Acta Metall. Mater.*, **40**, 589.
91 Seetharaman, V., Semiatin, S.L., Lombard, C.M., and Frey, N.D. (1993) High-temperature ordered intermetallic alloys V, *Materials Research Society Symposium Proceedings*, vol. 288 (eds I. Baker, R. Darolia, J.D. Whittenberger, and M.H. Yoo), Mater. Res. Soc., Pittsburgh, PA, p. 513.
92 Hofmann, U., and Blum, W. (1999) *Intermetallics*, **7**, 351.
93 Kim, H.Y., and Hong, S.H. (1999) *Mater. Sci. Eng.*, **A271**, 382.
94 Larsen, D.E., Kampe, S., and Christodoulou, L. (1990) Intermetallic matrix composites, *Materials Research Society Symposium Proceedings*, vol. 194 (eds D.L. Anton, R. McMeeking, D. Miracle, and P. Martin), Mater. Res. Soc., Pittsburgh, PA, p. 285.
95 Graef, M.D., Löfvander, J.P.A., McCullough, C., and Levi, C.G. (1992) *Acta Metall. Mater.*, **40**, 3395.
96 Kim, Y.-W., and Dimiduk, D.M. (2001) *Structural Intermetallics 2001* (eds K.J. Hemker, D.M. Dimiduk, H. Clemens, R. Darolia, H. Inui, J.M. Larsen, V.K. Sikka, M. Thomas, and J.D. Whittenberger), TMS, Warrendale, PA, p. 625.
97 Morris-Muñoz, M.A., and Morris, D.G. (1999) *Intermetallics*, **7**, 1069.
98 Umakoshi, Y., Nakano, T., and Yamane, T. (1992) *Mater. Sci. Eng.*, **A152**, 81.
99 Zghal, S., Naka, S., and Couret, A. (1997) *Acta Mater.*, **45**, 3005.
100 Appel, F., and Wagner, R. (1998) *Mater. Sci. Eng.*, **R22**, 187.
101 Appel, F., Sparka, U., and Wagner, R. (1999) *Intermetallics*, **7**, 3325.
102 Paul, J.D.H., and Appel, F. (2003) *Metall. Mater. Trans.*, **34A**, 2103.
103 Appel, F. (1999) *Advances in Twinning* (eds S. Ankem and C.S. Pande), TMS, Warrendale, PA, p. 171.
104 Appel, F., Christoph, U., and Oehring, M. (2002) *Mater. Sci. Eng.*, **A329–331**, 780.
105 Fröbel, U., and Appel, F. (2002) *Acta Mater.*, **50**, 3693.
106 Imayev, R., Shagiev, M., Salishchev, G., Imayev, V., and Valinov, V. (1996) *Scr. Mater.*, **34**, 985.
107 Fujiwara, T., Nakamura, A., Hosomi, M., Nishitani, S.R., Shirai, Y., and Yamaguchi, M. (1990) *Philos. Mag. A*, **61**, 591.

108 Inui, H., Oh, M.H., Nakamura, A., and Yamaguchi, M. (1992) *Acta Metall. Mater.*, **40**, 3095.
109 Appel, F., and Wagner, R. (1994) *Twinning in Advanced Systems* (eds M.H. Yoo and M. Wuttig), TMS, Warrendale, PA, p. 317.
110 Chan, K.S., and Kim, Y.-W. (1993) *Metall. Trans.*, **24A**, 113.
111 Yamaguchi, M., and Umakoshi, Y. (1990) *Prog. Mater. Sci.*, **34**, 1.
112 Umakoshi, Y., and Nakano, T. (1993) *Acta Metall. Mater.*, **41**, 1155–1161.
113 Seetharaman, V., and Semiatin, S.L. (2002) *Metall. Mater. Trans.*, **33A**, 3817.
114 Dao, M., Kad, B.K., and Asaro, R.J. (1996) *Philos. Mag. A*, **74**, 569.
115 Kad, B.K., Dao, M., and Asaro, R.J. (1995) *Mater. Sci. Eng.*, **A192/193**, 97.
116 Kad, B.K., Dao, M., and Asaro, R.J. (1995) *Philos. Mag. A*, **71**, 567.
117 Kim, Y.-W. (1994) *JOM (J. Metals)*, **46**, 30.
118 Liu, C.T., Maziasz, P.J., Clemens, D.R., Schneibel, J.H., Sikka, V.K., Nieh, T.G., Wright, J., and Walker, L.R. (1995) *Gamma Titanium Aluminides* (eds Y.-W. Kim, R. Wagner, and M. Yamaguchi), TMS, Warrendale, PA, p. 679.
119 Maziasz, P.J., and Liu, C.T. (1998) *Metall. Mater. Trans.*, **29A**, 105.
120 Oehring, M., Lorenz, U., Niefanger, R., Christoph, U., Appel, F., Wagner, R., Clemens, H., and Eberhardt, N. (1999) *Gamma Titanium Aluminides 1999* (eds Y.-W. Kim, D.M. Dimiduk, and M.H. Loretto), TMS, Warrendale, PA, p. 439.
121 Oehring, M., Lorenz, U., Appel, F., and Roth-Fagaraseanu, D. (2001) *Structural Intermetallics 2001* (eds K.J. Hemker, D.M. Dimiduk, H. Clemens, R. Darolia, H. Inui, J.M. Larsen, V.K. Sikka, M. Thomas, and J.D. Whittenberger), TMS, Warrendale, PA, p. 157.
122 Zhang, W.-J., Lorenz, U., and Appel, F. (2000) *Acta Mater.*, **48**, 2803.
123 Semiatin, S.L., Seetharaman, V., and Jain, V.K. (1994) *Metall. Mater. Trans.*, **25A**, 2753.
124 Naka, S., Thomas, M., Sanchez, C., and Khan, T. (1997) *Structural Intermetallics 1997* (eds M.V. Nathal, R. Darolia, C.T. Liu, P.L. Martin, D.B. Miracle, R. Wagner, and M. Yamaguchi), TMS, Warrendale, PA, p. 313.
125 Küstner, V., Oehring, M., Chatterjee, A., Güther, V., Brokmeier, H.-G., Clemens, H., and Appel, F. (2003) *Gamma Titanium Aluminides 2003* (eds Y.-W. Kim, H. Clemens, and A.H. Rosenberger), TMS, Warrendale, PA, p. 89.
126 Jin, Y., Wang, J.N., Yang, J., and Wang, Y. (2004) *Scr. Mater.*, **51**, 113.
127 Wang, J.N., Yang, J., and Wang, Y. (2005) *Scr. Mater.*, **51**, 329.
128 Imayev, R.M., Imayev, V.M., Oehring, M., and Appel, F. (2007) *Intermetallics*, **15**, 451.
129 Kobayashi, S., Takeyama, M., Motegi, T., Hirota, N., and Matsuo, T. (2003) *Gamma Titanium Aluminides* (eds Y.-W. Kim, H. Clemens, and A.H. Rosenberger), TMS, Warrendale, PA, p. 165.
130 Takeyama, M., and Kobayashi, S. (2005) *Intermetallics*, **13**, 993.
131 Singh, J.P., Tuval, E., Weiss, I., and Srinivasan, R. (1995) *Gamma Titanium Aluminides* (eds Y.-W. Kim, R. Wagner, and M. Yamaguchi), TMS, Warrendale, PA, p. 547.
132 Srinivasan, R., Singh, J.P., Tuval, E., and Weiss, I. (1996) *Sripta Mater.*, **34**, 1295.
133 Xu, X.J., Lin, J.P., Wang, Y.L., Lin, Z., and Chen, G.L. (2006) *Mater. Sci. Eng.*, **A416**, 98.
134 Imayev, R.M., Imayev, V.M., and Salishchev, G.A. (1993) *Scr. Metall. Mater.*, **29**, 713.
135 Imayev, V.M., Imayev, R.M., and Salishchev, G.A. (2000) *Intermetallics*, **8**, 1.
136 Imayev, V.M., Imayev, R., Kuznetsov, A.V., Senkov, O.N., and Froes, F.H. (2001) *Mater. Sci. Technol.*, **17**, 566.
137 Yoo, M.H., Fu, C.L., and Lee, J.K. (1994) *Twinning in Advanced Materials* (eds M.H. Yoo and M. Wuttig), TMS, Warrendale, PA, p. 97.
138 Appel, F., Christoph, U., and Wagner, R. (1995) *Philos. Mag. A*, **72**, 341.
139 Yoo, M.H., and Fu, C.L. (1998) *Metall Trans.*, **29A**, 49.

140 Semiatin, S.L., Vollmer, D.C., El-Soudani, S., and Su, C. (1990) *Sripta Metall. Mater.*, **24**, 1409.

141 Semiatin, S.L., Segal, V.M., Goetz, R.L., Goforth, R.E., and Hartwig, T. (1995) *Scr. Metall. Mater.*, **33**, 535.

142 Seetharaman, V., and Semiatin, S.L. (1998) *Metall. Mater. Trans.*, **29A**, 1991.

143 Semiatin, S.L., and Seetharaman, V. (1997) *Scr. Mater.*, **36**, 291.

144 Seetharaman, V., Goetz, R.L., and Semiatin, S.L. (1991) *High-Temperature Ordered Intermetallic Alloys IV*, vol. 213 (eds L.A. Johnson, D.P. Pope, J.O. Stiegler) *Mater. Res. Soc. Symp. Proc.*, MRS, Pittsburgh, PA, p. 895.

145 Cockroft, M., and Latham, D. (1968) *J. Inst. Metals*, **96**, 33.

146 Imayev, R.M., Imayev, V.M., Kuznetsov, A.V., Oehring, M., Lorenz, U., and Appel, F. (2004) *Ti-2003, Science and Technology, Proc. of the 10th World Conference on Titanium* (eds G. Lüthjering and J. Albrecht), Wiley-VCH, Weinheim, Germany, p. 2317.

147 Tetsui, T., Shindo, K., Kobayashi, S., and Takeyama, M. (2002) *Scr. Mater.*, **47**, 399.

148 Tetsui, T., Shindo, K., Kaji, S., Kobayashi, S., and Takeyama, M. (2005) *Intermetallics*, **13**, 971.

149 Kremmer, S., Chladil, H.F., Clemens, H., Otto, A., and Güther, V. (2007) *Ti-2007, Science and Technology, Proc. of the 11th World Conference on Titanium* (eds M. Ninomi, S. Akiyama, M. Ikeda, M. Hagiwara, and K. Maruyama), The Japan Institute of Metals, Tokyo, Japan, p. 989.

150 Clemens, H., Chladil, H.F., Wallgram, W., Zickler, G.A., Gerling, R., Liss, K.-D., Kremmer, S., Güther, V., and Smarsly, W. (2008) *Intermetallics*, **16**, 827.

151 Wallgram, W., Schmölzer, T., Cha, L., Das, G., Güther, V., and Clemens, H. (2009) *Intern. J. Mater. Res. (formerly Z. Metallkd.)*, **100**, 1021.

152 Clemens, H., Eberhardt, N., Glatz, W., Martinz, H.-P., Knabl, W., and Reheis, N. (1997) *Structural Intermetallics 1997* (eds M.V. Nathal, R. Darolia, C.T. Liu, P.L. Martin, D.B. Miracle, R. Wagner, and M. Yamaguchi), TMS, Warrendale, PA, p. 277.

153 Clemens, H., Kestler, H., Eberhardt, N., and Knabl, W. (1999) *Gamma Titanium Aluminides 1999* (eds Y.-W. Kim, D.M. Dimiduk, and M.H. Loretto), TMS, Warrendale, PA, p. 209.

154 Semiatin, S.L., McQuay, P., Stucke, M., Kerr, W.R., Kim, Y.-W., and El-Soudani, S. (1991) *High-Temperature Ordered Intermetallic Alloys IV*, vol. 213 (eds L.A. Johnson, D.P. Pope, J.O. Stiegler) *Mater. Res. Soc. Symp. Proc.*, MRS, Pittsburgh, PA, p. 883.

155 Küstner, V., Oehring, M., Chatterjee, A., Clemens, H., and Appel, F. (2004) *Solidification and Crystallization* (ed. D. Herlach) Wiley-VCH, Weinheim, Germany, p. 250.

156 Sternitzke, M., Appel, F., and Wagner, R. (1999) *J. Microsc.*, **196**, 155.

157 Appel, F., Brossmann, U., Christoph, U., Eggert, S., Janschek, P., Lorenz, U., Müllauer, J., Oehring, M., and Paul, J. (2000) *Adv. Eng. Mater.*, **2**, 699.

158 Brossmann, U., Oehring, M., and Appel, F. (2001) *Structural Intermetallics 2001* (eds K.J. Hemker, D.M. Dimiduk, H. Clemens, R. Darolia, H. Inui, J.M. Larsen, V.K. Sikka, M. Thomas, and J.D. Whittenberger), TMS, Warrendale, PA, p. 191.

159 Chen, G.L., Xu, X.J., Teng, Z.K., Wang, Y.L., and Lin, J.P. (2007) *Intermetallics*, **15**, 625.

160 Semiatin, S.L., and McQuay, P.A. (1992) *Metall. Trans.*, **23A**, 149.

161 Brossmann, U., Oehring, M., Lorenz, U., Appel, F., and Clemens, H. (2001) *Z. Metallkd.*, **92**, 1009.

162 Müllauer, J., Appel, F., Eggert, S., Eggers, L., Janschek, P., Lorenz, U., and Oehring, M. (2000) *Intermetallics and Superalloys, EUROMAT 99*, vol. 10 (eds D.G. Morris, S. Naka, and P. Caron), Wiley-VCH, Weinheim, p. 265.

163 Kim, Y.-W. (2004) *Niobium High Temperature Applications* (eds Y.-W. Kim and T. Carneiro), TMS, Warrendale, PA, p. 125.

164 Kim, Y.-W. (1992) *Acta Metall. Mater.*, **40**, 1121.

165 Wurzwallner, K., Clemens, H., Schretter, P., Bartels, A., and Koeppe,

C. (1993) *High-Temperature Ordered Intermetallic Alloys V*, vol. 288 (eds I. Baker, R. Darolia, J.D. Whittenberger, M.H. Yoo) *Mater. Res. Soc. Symp. Proc.*, Mater. Res. Soc., Pittsburgh, PA, p. 867.

166 Martin, P.L., Rhodes, C.G., and McQuay, P.A. (1993) *Structural Intermetallics* (eds R. Darolia, J.J. Lewandowski, C.T. Liu, P.L. Martin, D.B. Miracle, and M.V. Nathal), TMS, Warrendale, PA, p. 177.

167 Fuchs, G.E. (1995) *Gamma Titanium Aluminides* (eds Y.-W. Kim, R. Wagner, and M. Yamaguchi), TMS, Warrendale, PA, p. 563.

168 Fuchs, G.E. (1997) *Mater. Sci. Eng.*, A239–A240, **584**.

169 Imayev, V.M., Imayev, R.M., and Kuznetsov, A. (2003) *Gamma Titanium Aluminides* (eds Y.-W. Kim, H. Clemens, and A.H. Rosenberger), TMS, Warrendale, PA, p. 311.

170 Kim, Y.-W., Kim, S.-L., Dimiduk, D.M., and Woodward, C. (2006) Presented at The 3rd Intern. Workshop on γ-TiAl Technologies, Bamberg, Germany, May 29–31, 2006.

171 Zhang, J., Becker, M., Appel, F., Leyens, C., and Viehweger, B. (2008) *Structural Intermetallics for Elevated Temperatures* (eds Y.-W. Kim, D. Morris, R. Yang, and C. Leyens), TMS, Warrendale, PA, p. 265.

172 Bartels, A., Koeppe, C., and Mecking, H. (1995) *Mater. Sci. Eng.*, **A192**, 226.

173 Jain, V.K., Goetz, R.L., and Semiatin, S.L. (1996) *J. Eng. Ind.*, **118**, 155.

174 Carneiro, T., and Kim, Y.-W. (2005) *Intermetallics*, **13**, 1000.

175 Fujitsuna, N., Miyamoto, Y., and Ashida, Y. (1993) *Structural Intermetallics* (eds R. Darolia, J.J. Lewandowski, C.T. Liu, P.L. Martin, D.B. Miracle, and M.V. Nathal), TMS, Warrendale, PA, p. 187.

176 Imayev, R.M., Imayev, V.M., and Kuznetsov, A. (2003) *Gamma Titanium Aluminides 2003* (eds Y.-W. Kim, H. Clemens, and A.H. Rosenberger), TMS, Warrendale, PA, p. 265.

177 Kraft, F., and Gunasekera, S. (2005) *Metalworking: Bulk Forming, Metals Handbook*, vol. 14A (ed. S.L. Semiatin), ASM International, Materials Park, OH, p. 421.

178 Güther, V., Janschek, P., Kerzendorf, G., Lindemann, J., Schillo, E., Viehweger, B., and Weinert, K. (2005) (in German) *Spanende Fertigung–Prozesse, Innovationen, Werkstoffe* (ed. K. Weinert), Vulkan-Verlag, Essen, Germany, p. 363.

179 Semiatin, S.L., and Seetharaman, V. (1994) *Scr. Metall. Mater.*, **31**, 1203.

180 Mecking, H., and Hartig, C. (1995) *Gamma Titanium Aluminides* (eds Y.-W. Kim, R. Wagner, and M. Yamaguchi), TMS, Warrendale, PA, p. 525.

181 Semiatin, S.L., Seetharaman, V., Goetz, R.L., and Jain, V.K. (1994) Controlled dwell extrusion of difficult to work alloys, U.S. Patent 5,361,477, Nov. 1994.

182 Appel, F., Lorenz, U., Oehring, M., and Wagner, R. (1997) Vorrichtung zur Kapselung von Rohlingen aus metallischen Hochtemperatur-Legierungen, German Patent DE 197 47 257 C2, Oct. 1997.

183 Seetharaman, V., Malas, J.C., and Lombard, C.M. (1991) *High-Temperature Ordered Intermetallic Alloys IV*, vol. 213 (eds L.A. Johnson, D.P. Pope, J.O. Stiegler) *Mater. Res. Soc. Symp. Proc.*, Mater. Res. Soc., Pittsburgh, PA, p. 1889.

184 Liu, C.T., Schneibel, J.H., Maziasz, P.J., Wright, J.L., and Easton, D.S. (1996) *Intermetallics*, **4**, 429.

185 Liu, C.T., and Maziasz, P.J. (1998) *Intermetallics*, **6**, 653.

186 Fuchs, G.E. (1998) *Metall. Mater. Trans.*, **29A**, 27.

187 Wesemann, J., Frommeyer, G., and Kruse, J. (2001) *Intermetallics*, **9**, 273.

188 Jiménez, J.A., Ruano, O.A., Frommeyer, G., and Knippscher, S. (2005) *Intermetallics*, **13**, 749.

189 Martin, P.L., Hardwick, D.A., Clemens, D.R., Konkel, W.A., and Stucke, M.A. (1997) *Structural Intermetallics 1997* (eds M.V. Nathal, R. Darolia, C.T. Liu, P.L. Martin, D.B. Miracle, R. Wagner, and M. Yamaguchi), TMS, Warrendale, PA, p. 387.

190 Liu, C.T., Wright, J.L., and Deevi, S.C. (2002) *Mater. Sci. Eng.*, **A329–331**, 416.

191 Sikka, V.K., Carneiro, T., and Loria, E.A. (2003) *Gamma Titanium Aluminides 2003* (eds Y.-W. Kim, H. Clemens, and A.H. Rosenberger), TMS, Warrendale, PA, p. 219.

192 Draper, S.L., Das, G., Locci, I., Whittenberger, J.D., Lerch, B.A., and Kestler, H. (2003) *Gamma Titanium Aluminides 2003* (eds Y.-W. Kim, H. Clemens, and A.H. Rosenberger), TMS, Warrendale, PA, p. 207.

193 Morris, D.G., and Morris-Muñoz, M.A. (2000) *Intermetallics*, **8**, 997.

194 Bartels, A., and Uhlenhut, H. (1998) *Intermetallics*, **6**, 685.

195 Bartels, A., Kestler, H., and Clemens, H. (2002) *Mater. Sci. Eng.*, **A329–331**, 153.

196 Yamaguchi, M., Inui, H., and Ito, K. (2000) *Acta Mater.*, **48**, 307.

197 Yokoshima, S., and Yamaguchi, M. (1996) *Acta Mater.*, **44**, 873.

198 Paul, J.D.H. (2007) unpublished work, GKSS Research Centre, Geesthacht, Germany.

199 Paul, J.D.H., Oehring, M., Hoppe, R., and Appel, F. (2003) *Gamma Titanium Aluminides 2003* (eds Y.-W. Kim, H. Clemens, and A.H. Rosenberger), TMS, Warrendale, PA, p. 403.

200 Martin, P.L., Jain, S.K., and Stucke, M.A. (1995) *Gamma Titanium Aluminides* (eds Y.-W. Kim, R. Wagner, and M. Yamaguchi), TMS, Warrendale, PA, p. 727.

201 Müllauer, J., Eggers, L., and Appel, F. (2000) *Advances in Mechanical Behaviour, Plasticity and Damage, EUROMAT 2000*, vol. 2 (eds D. Miannay, P. Costa, D. François, and A. Pineau), Elsevier, Amsterdam, p. 1339.

202 Tetsui, T., Shindo, K., Kobayashi, S., and Takeyama, M. (2003) *Intermetallics*, **11**, 299.

203 Hartig, C., Fang, X.F., Mecking, H., and Dahms, M. (1992) *Acta Metall. Mater.*, **40**, 1883.

204 Schillinger, W., Bartels, A., Gerling, R., Schimansky, F.-P., and Clemens, H. (2006) *Intermetallics*, **14**, 336.

205 Bartels, A., Schillinger, W., Graßl, G., and Clemens, H. (2003) *Gamma Titanium Aluminides 2003* (eds Y.-W. Kim, H. Clemens, and A.H. Rosenberger), TMS, Warrendale, PA, p. 275.

206 Stark, A., Bartels, A., Gerling, R., Schimansky, F.-P., and Clemens, H. (2006) *Adv. Eng. Mater.*, **8**, 1101.

207 Bunge, H.J. (1989) *Text. Microstr.*, **10**, 265.

208 Brokmeier, H.-G. (1999) *Text. Microstr.*, **33**, 13.

209 Verlinden, B., Driver, J., Samaidar, I., and Doherty, R.D. (2007) *Thermo-Mechanical Processing of Metallic Materials, Pergamon Materials Series*, vol. 11 (ed. R.W. Cahn), Elsevier, Amsterdam.

210 Johnson, D.R., Inui, H., and Yamaguchi, M. (1996) *Acta Mater.*, **44**, 2523.

211 Bartels, A., and Schillinger, W. (2001) *Intermetallics*, **9**, 883.

212 Dahms, M., and Bunge, H.J. (1989) *J. Appl. Cryst.*, **22**, 439.

213 Dahms, M. (1992) *J. Appl. Cryst.*, **25**, 258.

214 De Gaef, M., Biery, N., Rishel, L., Pollock, T.M., and Cramb, A. (1999) *Gamma Titanium Aluminides 1999* (eds Y.-W. Kim, D.M. Dimiduk, and M.H. Loretto), TMS, Warrendale, PA, p. 247.

215 McCullough, C., Valencia, J.J., Levi, C.G., and Mehrabian, R. (1989) *Acta Metall.*, **37**, 1321.

216 Johnson, D.R., Inui, H., and Yamaguchi, M. (1998) *Intermetallics*, **6**, 647.

217 Calnan, E.A., and Clews, C.J.B. (1950) *Philos. Mag.*, **41**, 1085.

218 Grewen, J. (1973) *3ème Colloque Européen sur les Textures de Déformation et de la Recristallisation des Métaux et leur Application Industrielle, Proc. of the Conf. Held at Pont-À-Mousson* (ed. R. Penelle), Société Française de Métallurgie, Paris, France, p. 195.

219 Churchman, A.T. (1954) *Proc. Royal Soc. A*, **226**, 216.

220 Mecking, H., Hartig, C., and Kocks, U.F. (1996) *Acta Mater.*, **44**, 1309.

221 Kad, B.K., and Asaro, R.J. (1997) *Philos. Mag. A*, **75**, 87.

222 Skrotzki, W., Tamm, R., Brokmeier, H.-G., Oehring, M., Appel, F., and Clemens, H. (2002) *Mater. Sci. Forum*, **408–412**, 1777.

223 Stark, A., Bartels, A., Schimansky, F.-P., and Clemens, H. (2007) *Advanced Intermetallic-Based Alloys*, vol. 980 (eds J. Wiezorek, C.L. Fu, M. Takeyama, D. Morris, H. Clemens) *Mater. Res. Soc. Symp. Proc.*, Mater. Res. Soc., Pittsburgh, PA, p. 359.

224 Stark, A., Bartels, A., Schimansky, F.-P., Gerling, R., and Clemens, H. (2008) *Structural Aluminides for Elevated Temperatures* (eds Y.-W. Kim, D. Morris, R. Yang, and C. Leyens), TMS, Warrendale, PA, p. 145.

225 Morris, M.A., Clemens, H., and Schlög, S.M. (1998) *Intermetallics*, **6**, 511.

226 Schillinger, W., Lorenzen, B., and Bartels, A. (2002) *Mater. Sci. Eng.*, **A329–331**, 644.

227 Hu, H., Sperry, P.R., and Beck, P.A. (1952) *Trans. AIME*, **194**, 76.

228 Merlini, A., and Beck, P.A. (1955) *Trans. AIME*, **203**, 385.

229 Wassermann, G., and Grewen, J. (1962) *Texturen Metallischer Werkstoffe*, Springer-Verlag, Berlin.

230 Koeppe, C., Bartels, A., Clemens, H., Schretter, P., and Glatz, W. (1995) *Mater. Sci. Eng.*, **A201**, 182.

231 Mecking, H., Seeger, J., Hartig, C., and Frommeyer, G. (1994) *Mater. Sci. Forum*, **157–162**, 813.

232 Dillamore, I.L., and Roberts, W.T. (1965) *Metall. Rev.*, **10**, 271.

233 Beck, P.A., and Hu, H. (1966) *Recrystallization, Grain Growth and Textures* (ed. H. Margolin), ASM, Cleveland, OH, p. 393.

234 Dillamore, I.L., and Katoh, H. (1974) *Met. Sci.*, **8**, 73.

235 Hjelen, J., Ørsund, R., and Nes, E. (1991) *Acta Metall. Mater.*, **39**, 1377.

236 Doherty, R.D., Kashyab, K., and Panchanadeeswaran, S. (1993) *Acta Metall. Mater.*, **41**, 3029.

237 Barrett, C.S., and Massalski, T. (1980) *Structure of Metals*, 3rd edn. Pergamon, Oxford, UK.

238 Couts, W.H., and Howson, T.E. (1987) *Superalloys II* (eds C.T. Sims, N.S. Stoloff, and W.C. Hagel), J. Wiley, New York, p. 441.

239 Kim, Y.-W. (1991) *JOM (J. Metals)*, **43 (August)**, 40.

240 Furrer, D.U., Hoffman, R.R., and Kim, Y.-W. (1995) *Gamma Titanium Aluminides* (eds Y.-W. Kim, R. Wagner, and M. Yamaguchi), TMS, Warrendale, PA, p. 611.

241 Knippscheer, S., Frommeyer, G., Baur, H., Joos, R., Lohmann, M., Berg, O., Kestler, H., Eberhardt, N., Güther, V., and Otto, A. (2000) *EUROMAT 99, Symp. B1, Materials for Transportation Technology* (ed. P.J. Winkler), Wiley-VCH, Weinheim, p. 110.

242 Sommer, A.W., and Keijzers, G.C. (2003) *Gamma Titanium Aluminides 2003* (eds Y.-W. Kim, H. Clemens, and A.H. Rosenberger), TMS, Warrendale, PA, p. 3.

243 Kestler, H., Eberhardt, N., and Knippscheer, S. (2003) *Niobium High Temperature Applications* (eds Y.-W. Kim and T. Carneiro), TMS, Warrendale, PA, p. 167.

244 Roth-Fagaraseanu, D. (2003) *Niobium High Temperature Applications* (eds Y.-W. Kim and T. Carneiro), TMS, Warrendale, PA, p. 199.

245 Brooks, J.W., Dean, T.A., Hu, Z.M., and Wey, E. (1998) *J. Mater. Proc. Technol.*, **80–81**, 149.

246 Aust, E., and Niemann, H.-R. (1999) *Adv. Eng Mater.*, **1**, 53.

247 Aspinwall, D.K., Dewes, R.C., and Mantle, A.L. (2005) *CIRP (Collège International pour la Recherche Productique) Ann.*, **54**, 99.

248 Clifton, D., Mount, A.R., Jardine, D.J., and Roth, R. (2001) *J. Mater. Proc. Technol.*, **108**, 338.

249 Lindemann, J., Buque, C., and Appel, F. (2006) *Acta Mater.*, **54**, 155.

250 Glavatskikh, M., Lindemann, J., Leyens, C., Oehring, M., and Appel, F. (2008) *Structural Aluminides for Elevated Temperatures* (eds Y.-W. Kim, D. Morris, R. Yang, and C. Leyens), TMS, Warrendale, PA, p. 111.

251 Das, G., Bartolotta, P.A., Kestler, H., and Clemens, H. (2003) *Gamma Titanium Aluminides 2003* (eds Y.-W. Kim, H. Clemens, and A.H. Rosenberger), TMS, Warrendale, PA, p. 33.

252 Matsuo, M., Hanamura, T., Kimura, M., Masahashi, N., and Mizoguchi, T.

(1991) *Microstructure/Property Relationships in Titanium Aluminides and Alloys* (eds Y.-W. Kim and R. Boyer), TMS, Warrendale, PA, p. 323.

253 Semiatin, S.L., Ohls, M., and Kerr, W.R. (1991) *Scr. Metall. Mater.*, **25**, 1851.

254 Wardlaw, T., and Bania, P. (1990) Pack assembly for hot rolling, U.S. Patent 4,966,816, Oct. 1990.

255 Clemens, H., Glatz, W., Schretter, P., Koppe, C., Bartels, A., Behr, R., and Wanner, A. (1995) *Gamma Titanium Aluminides* (eds Y.-W. Kim, R. Wagner, and M. Yamaguchi), TMS, Warrendale, PA, p. 717.

256 Clemens, H., Glatz, W., Eberhardt, N., Martinz, H.-P., and Knabl, W. (1997) *High Temperature Ordered Intermetallic Alloys VII*, vol. 460 (eds C.C. Koch, C.T. Liu, N.S. Stoloff, and A. Wanner), *Mater. Res. Soc. Symp. Proc.*, Mater. Res. Soc., Pittsburgh, PA, p. 29.

257 Semiatin, S.L., and Seetharaman, V. (1994) *Metall. Mater. Trans. A.*, **25A**, 2539.

258 Eylon, D., Yolton, C.F., Clemens, H., Schretter, P., and Jones, P.E. (1994) *Proc. PM'94-Powder Metallurgy World Congress*, Les Editions de Physique, Paris, France, p. 1271.

259 Clemens, H., Glatz, W., Schretter, P., Yolton, C.F., Jones, P.E., and Eylon, D. (1995) *Gamma Titanium Aluminides* (eds Y.-W. Kim, R. Wagner, and M. Yamaguchi), TMS, Warrendale, PA, p. 555.

260 Sagiev, M.R., and Salishchev, G.A. (2003) *Gamma Titanium Aluminides 2003* (eds Y.-W. Kim, H. Clemens, and A.H. Rosenberger), TMS, Warrendale, PA, p. 339.

261 Imayev, V.M., Imayev, R.M., Kuznetsov, A.V., Sagiev, M.R., and Salishchev, G.A. (2003) *Mater. Sci. Eng.*, **A348**, 15.

262 Gerling, R., Bartels, A., Clemens, H., Kestler, H., and Schimansky, F.P. (2004) *Intermetallics*, **12**, 275.

263 Clemens, H., Rumberg, I., Schretter, P., and Schwantes, S. (1994) *Intermetallics*, **2**, 179.

264 Bystrzanowski, S., Bartels, A., Clemens, H., Gerling, R., Schimansky, F.P., Dehm, G., and Kestler, H. (2005) *Intermetallics*, **13**, 515.

265 Bystrzanowski, S., Bartels, A., Clemens, H., Gerling, R., Schimansky, F.P., Kestler, H., Dehm, G., Haneczok, G., and Weller, M. (2003) *Gamma Titanium Aluminides 2003* (eds Y.-W. Kim, H. Clemens, and A.H. Rosenberger), TMS, Warrendale, PA, p. 431.

266 Kestler, H., Clemens, H., Baur, H., Joos, R., Gerling, R., Cam, G., Bartels, A., Schleinzer, C., and Smarsly, W. (1999) *Gamma Titanium Aluminides 1999* (eds Y.-W. Kim, D.M. Dimiduk, and M.H. Loretto), TMS, Warrendale, PA, p. 423.

267 Das, G., and Clemens, H. (1999) *Gamma Titanium Aluminides 1999* (eds Y.-W. Kim, D.M. Dimiduk, and M.H. Loretto), TMS, Warrendale, PA, p. 281.

268 Paul, J.D.H., Oehring, M., and Appel, F. (2004) Verfahren zur Herstellung von Bauteilen oder Halbzeugen, die intermetallische Titanaluminid-Legierungen enhalten, sowie mittels des Verfahrens herstellbare Bauteile, European Patent Application EP 1568486 A1, Feb. 2004.

17
Joining

17.1
Diffusion Bonding

Solid-state diffusion bonding provides a means of joining TiAl alloys without melting of the base materials. The major advantages of the technique are [1]:

i) low degree of microstructural disruption;

ii) minimum distortion and deformation, and hence, accurate dimensional control;

iii) small (or even negligible) temperature gradients result in minimum distortion and hence, accurate dimensional control;

iv) thin and thick sections can be joined to each other;

v) large surfaces can be more effectively joined compared to conventional welding;

vi) cast, wrought and sintered powder products and dissimilar materials can be joined;

vii) diffusion bonding can easily be combined with superplastic forming;

viii) a bonding aid can be used, such as an interface foil or coating, to either facilitate bonding or prevent the creation of brittle phases between dissimilar materials.

The disadvantages of diffusion bonding are the relatively long joining time and high cost of equipment due to the combined requirement of high temperature and joining pressure in vacuum. These debits limit component dimensions and affordable manufacturing. Nevertheless, in view of the above-described technical potential, there has been a systematic effort towards diffusion bonding TiAl alloys to each other [2–5] or with other materials [6]. The bulk of this work was focused on technological aspects and feasibility. The process parameters sensitively depend on composition, microstructure, mechanical properties, and diffusivity of the

Gamma Titanium Aluminide Alloys: Science and Technology, First Edition. Fritz Appel, Jonathan David Heaton Paul, Michael Oehring.
© 2011 Wiley-VCH Verlag GmbH & Co. KGaA. Published 2011 by Wiley-VCH Verlag GmbH & Co. KGaA.

Table 17.1 Composition and microstructure of the alloys involved in diffusion bonding studies. D (grain size), D_C (colony size) [7].

Alloy	Composition [at.%]	Microstructure
1	Ti-44.5Al	Duplex $D = 1–5\,\mu m$ $D_C = 20–30\,\mu m$
2	Ti-45Al-10Nb (TNB)	Fully lamellar $D > 100\,\mu m$
3	Ti-46.5Al	Duplex, nearly globular $D = 1–5\,\mu m$
4	Ti-46.5Al-5.5Nb	Nearly globular $D = 1–5\,\mu m$
5	Ti-47Al-4.5Nb-0.2C-0.2B	Duplex, nearly globular $D = 1–2\,\mu m$
6	Ti-45Al-8Nb-0.2C (TNB-V2)	Duplex, nearly globular $D = 5–10\,\mu m$ $D_C = 10–30\,\mu m$
7	Ti-45Al-5Nb-0.2C-0.2B (TNB-V5)	Nearly globular $D = 1–8\,\mu m$
8	Ti-54Al	Globular γ $D = 10–20\,\mu m$

alloys to be joined. These effects will be illustrated in the following sections utilizing data from a recently published study involving a large variety of TiAl alloys [7].

17.1.1
Alloy Compositions and Microstructures

The composition, thermomechanical treatment and microstructure of the alloys investigated are listed in Table 17.1. These alloys are marked in the $T = 1273\,K$ isothermal section of the Ti-Al-Nb phase diagram [8, 9] shown in Figure 17.1, which also illustrates the equilibrium constitution present under the bonding conditions usually applied in this study. Chemical analysis by hot-gas extraction revealed impurity contents of 100 ppm nitrogen and 600 ppm oxygen by weight. Alloys 2 and 4–7 contain a significant amount of Nb, which is an important alloying element in the design of modern TiAl alloys because it improves the oxidation resistance (see Section 12.3.1). Furthermore, Nb bearing alloys with a relatively low Al content exhibit high yield stresses combined with good temperature strength retention (Section 7.5). These attributes certainly affect diffusion bonding. Alloys 6 and 7 represent a third alloy generation that is currently being considered for technical applications. The single-phase γ alloy 8 was involved as a reference

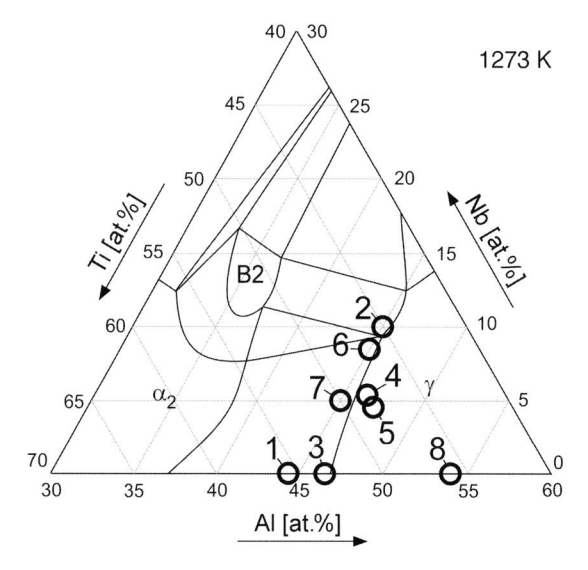

Figure 17.1 Isothermal section of the Ti-Al-Nb-phase diagram [8, 9] designating the alloys investigated.

17.1.2
Microasperity Deformation

The surfaces of a diffusion-bonding couple are never perfectly clean or perfectly smooth; thus, the area of metal-to-metal contact between the faying surfaces is limited to a relatively few microasperities. Under an applied stress these asperities deform as long as the surface area of contact is such that that the yield strength of the material is locally exceeded. Apart from temperature and stress, the extent of this deformation depends on the work-hardening behavior of the metal. If the process temperature is above the recrystallization temperature of the alloy, then work hardening is no longer a concern. It is obvious that asperity deformation depends on the surface roughness: a rougher surface will enhance shear deformation.

Deformation occurring during diffusion bonding is, almost by definition, a stress-driven, diffusion-assisted process. The interaction of load, temperature and time along with material selection, component geometry and void coalescence creates a wide range of synergistic complexity that determines bonding parameters, such as stress, temperature and time. Loads may be monotonic as in a constant strain rate test, or steady as in creep tests. Stress relaxation can occur, which in broad terms is a redistribution of plastic and elastic deformation, while the total strain is held constant. Due to its thermally activated nature, the deformation

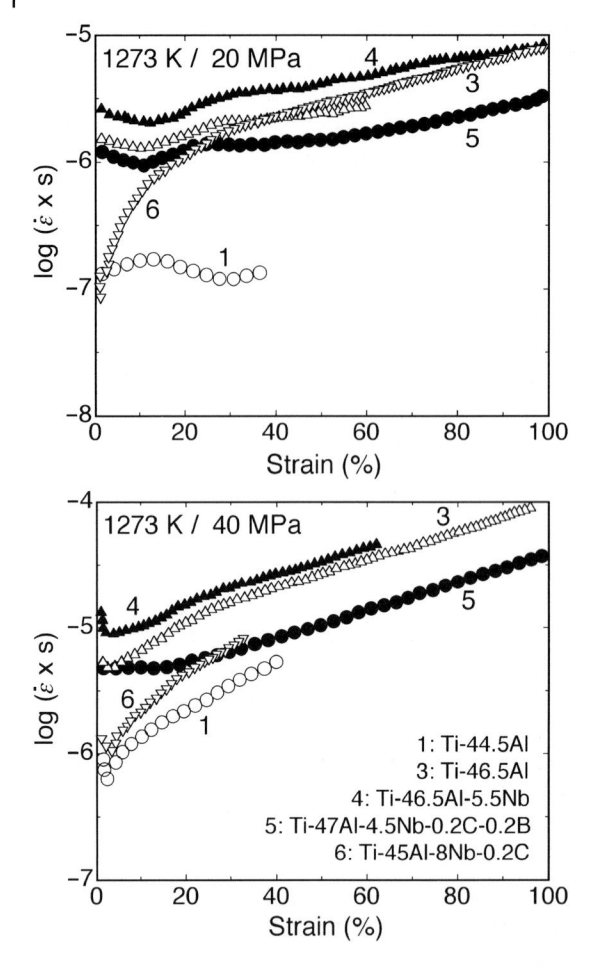

Figure 17.2 Creep behavior of the alloys 1 and 3 to 6 (Table 17.1) shown as strain rate $\dot{\varepsilon}$ versus strain ε plots. $T = 1273\,K$, $\sigma = 20$ and $40\,MPa$.

mechanism can change dramatically with temperature and stress. In any event, coalescence of the contacting surfaces must be produced with loads below those that would cause macroscopic deformation of the part. This requires a tight optimization of bonding stress and temperature.

The alloys involved in the diffusion bonding tests (Table 17.1) exhibit a broad variation of mechanical properties with composition and microstructure. The high-temperature flow behavior of these alloys was described in Section 7.2.5 (Figure 7.15). Figure 17.2 demonstrates the creep behavior for a few of these alloys at relatively low stresses and 1000 °C. The creep rate accelerates almost from the beginning of the test, that is, tertiary creep is the predominant regime. Not surprisingly, a higher stress exacerbates this effect. The mechanisms associated with tertiary creep (see Section 9.8) are a progressive increase of dislocation density, dynamic recrystallization and the formation of shear bands in lamellar grains. In the late stages of tertiary creep, cavity growth is thought to occur, which eventually leads to creep rupture. The resulting structural changes and damage processes are certainly a major concern for a successful bonding of TiAl components. Thus, in order to avoid large-scale deformation of the components, the bonding parameter should be chosen in such a way that tertiary creep is largely avoided, that is, the bonding time should be smaller than the time for the onset of tertiary creep, which is typically 20 h at 1000 °C and applied stress of 20 MPa. As diffusion bonding is often coupled with superplastic forming, the maximum elongation obtained under tensile creep is of interest. Figure 17.2 clearly indicates that the fine-grained materials 4 to 6 probably exhibit the best preconditions for such forming procedures as, in spite of the likely embrittlement in air, creep strains in excess of 140% to 170% can be achieved. Thus, there is no hard and fast rule for predicting the mechanical parameters for diffusion bonding TiAl alloys, as the flow and diffusion behavior at moderately high temperatures strongly depend on the constitution and microstructure. Because of the early onset of tertiary creep, associated with dramatic structural changes, the bonding stress should, in any case, only be a small portion of the yield stress. The bonding conditions will be specified for the examples to be demonstrated.

In summary: The principal diffusion-bonding parameters, temperature and stress, depend on yield strength, work-hardening behavior and creep resistance of the mating alloy coupons. The bonding conditions should be chosen so that coalescence of contacting surfaces is produced by asperity deformation, but without gross deformation of the component. The effects of bonding temperature and bonding stress are synergistic; at higher temperature less stress is required and vice versa.

17.1.3
Diffusion Bonding; Experimental Setup

For the bonding tests, cylindrical specimens of 4 mm height and 4 mm diameter were prepared by wire-electrodischarge machining. Immediately before bonding, the end faces were ground to make them parallel and to remove surface layers. The final surface finish was performed with #1200 grid emery paper followed by ultrasonically cleaning in acetone. The bonding tests were carried out under a vacuum of 10^{-5} mbar using a load-controlled deformation machine, as described in [7]. The samples were heated to the bonding temperature of 1273 K at a rate of 20 K/min. At this temperature the bonding stress and time were varied between $\sigma = 20$ MPa to 100 MPa and $t = 0.25$ h to 2 h. The bonded couples were furnace cooled to room temperature within 1.5 h.

17.1.4
Metallographic Characterization of the Bonding Zone

The bonding experiments described here were performed at $T = 1273$ K and with a compressive stress of $\sigma = 20$ MPa, which is well below the yield stress of all the

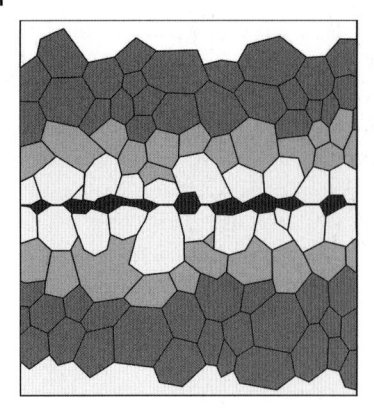

Figure 17.3 Schematic illustration of the process zone observed after diffusion bonding at relatively low stresses. BL – fine-grained bond layer at the former contact plane of the diffusion couple, DRX – region consisting of relatively large recrystallized grains, RD – initial bulk material but with remnant plastic deformation.

alloys. Thus, asperity plastic deformation occurred near to the contact plane and resulted in removal of the surface roughness and intimate matching of the opposing interfaces. This gives rise to a significant deformation gradient perpendicular to the contact plane, that is, the amount of deformation decreases with increasing distance from the contact plane. This gradient, together with the unavoidable contamination of the contact surfaces with oxygen or nitrogen, leads to a process zone, which, in broad terms, is characteristic of all the two-phase alloys investigated. As illustrated in Figure 17.3, three distinct regions are developed:

i) a fine-grained bond layer at the former contact plane of the diffusion couple, designated BL;
ii) a region made up of relatively large recrystallized grains; designated DRX;
iii) a region that is identical with the starting bulk material, but exhibits remnant plastic deformation; designated RD;
iv) occasionally, pores are present at the bond layer, the size and distribution of which depend on alloy composition process conditions.

Figure 17.4 demonstrates the bonding zone in Ti-46.5Al (alloy 3), which apart from some details, is characteristic of all the two-phase alloys investigated. The process zone is almost flat with a uniform thickness of about 20 μm, in spite of the significant structural inhomogeneity present in the starting material (Figure 17.4a). The bond layer consists of fine stress-free grains (Figures 17.4b and c), which were identified by EDX and EBSD analysis as α_2 phase (Figures 17.5 and 17.6). The formation of α_2 phase probably results from the unavoidable contamination of the contact surfaces with oxygen. It is well documented in the literature [12] that a very small amount of oxygen can stabilize the α_2 phase. This finding is consistent with an early investigation of Godfrey *et al.* [13] performed on diffusion-bonded Ti-48Al-2Mn-2Nb. The orientation analysis illustrated in Figure 17.6 gives the

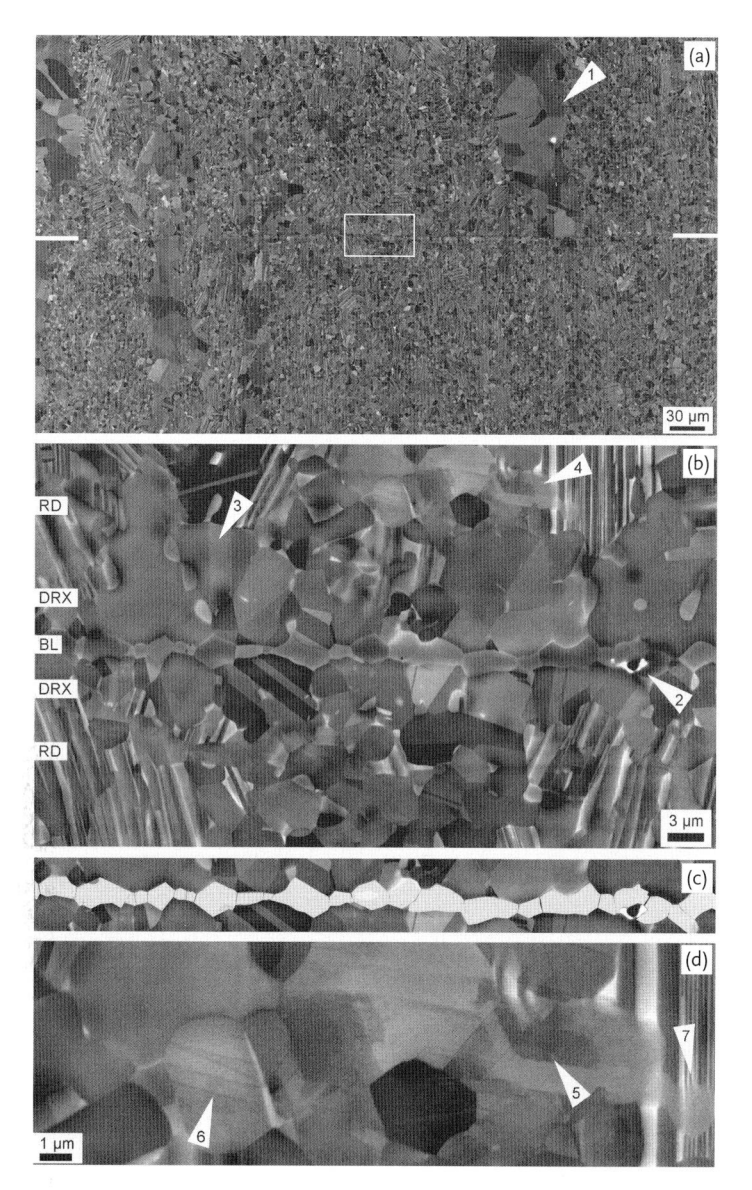

Figure 17.4 Cross section of a bond in Ti-46.5 Al (alloy 3) formed at $T = 1273$ K, $\sigma = 20$ MPa and $t = 2$ h. Scanning electron micrographs taken in the backscattered mode. (a) Low-magnification image showing the gross structure of the bond; the horizontally oriented bonding layer is marked by white bars. Note the inhomogeneity of the starting material that is manifested by a banded structure parallel to the extrusion direction (vertical in the micrograph) involving remnant lamellae, fine-grained regions and large γ grains (arrow 1). (b) Higher magnification of the area boxed in (a) showing the bonding layer BL, the recrystallized region DRX and the region with remnant deformation RD in more detail. Note the pore in the bonding layer (arrow 2) and the annealing twins in the γ grains of the DRX region (one of them is marked by arrow 3). (c) Bonding layer consisting of fine-grained α_2 phase as identified by EBSD and EDX analysis. (d) Higher magnification of the area marked by arrow 4 in (b) showing recovery (arrow 5) and mechanical twinning of γ grains situated in the RD zone (arrow 6). Note the γ grain growing into the pre-existing lamellar colony (arrow 7) [7].

(a)

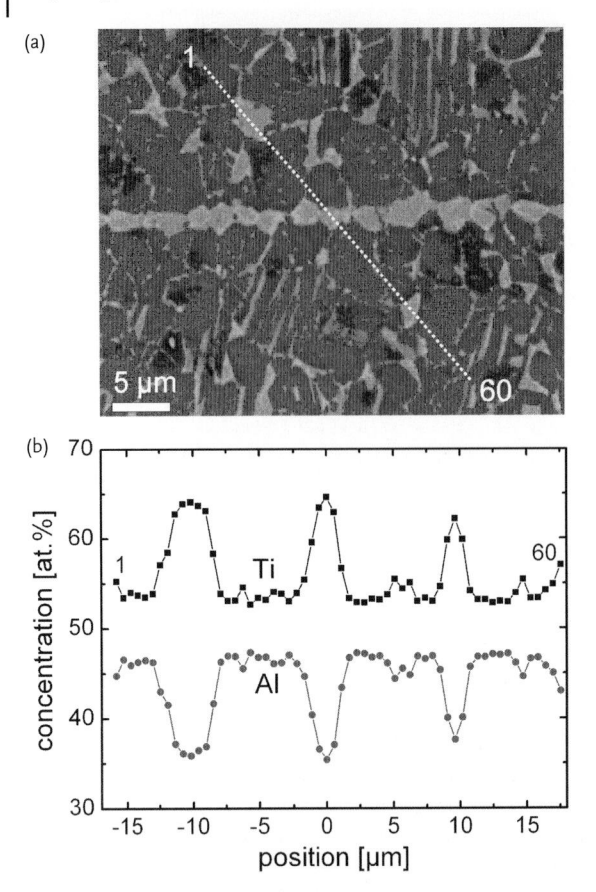

(b)

Figure 17.5 Variation of the Ti and Al concentration across the bonding interface in alloy 3, Ti-46.5Al, indicating α_2 phase at the bond interface. Bonding conditions $T = 1273$ K, $\sigma = 20$ MPa, $t = 2$ h [7].

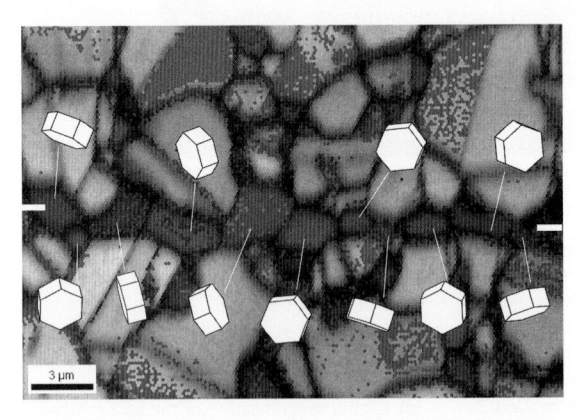

Figure 17.6 EBSD analysis of the bonding layer showing fine-grained α_2 phase. Note the preferred α_2 grain orientation that enables glide on prism planes [7].

impression that most of the newly formed α_2 grains have an orientation that is suitable for prismatic glide. This data could reflect the well-known plastic anisotropy of the α_2 phase, according to which the activation of prismatic slip along $1/3<11\bar{2}0>\{10\bar{1}0\}$ is by far the easiest, followed by basal and pyramidal slip (Section 5.2.2). It might be speculated that nucleation and growth of new α_2 grains are controlled by the deformation constraints operating during bonding, in that the preferred grain orientation ensures strain accommodation on the most favorable slip system.

In an earlier investigation [14] it was shown that the α_2 phase at the bond layer is formed at the expense of the Ti content of the adjacent regions, which needs long-range diffusion. The Ti transport could be supported by the antistructural disorder, as already mentioned in Section 3.3. The Ti_{Al} antisite defects present in Ti-rich alloys are thought to be associated with Ti vacancies. In this way, antistructural bridges are formed, which permit vacancies to migrate without disordering the crystal structure [15]. As the relevant migration energy is low, this mechanism could operate at the relatively low bonding temperature used in the experiments, conventional bulk diffusion by Ti vacancies being ineffective. Furthermore, Ti transport could be accomplished by diffusion along the various internal boundaries present in the fine-grained materials. In this respect the lamellar interfaces are probably important because deviation from the ideal crystal structure occurs and dense arrangements of misfit dislocations are present (Section 6.1.3). There is ample evidence of enhanced diffusion along such defects with an activation energy about half that for conventional bulk diffusion [16]. Due to the transport processes described, the Ti content of the pre-existing α_2 and β phases situated next to the bonding layer gradually decreases and eventually falls below the critical composition required for their existence. Thus, these phases transform into γ(TiAl). The effect is most pronounced in lamellar colonies that are in contact with the bonding line, presumably because pipe diffusion along the lamellar interfaces is significant. A common observation supporting this mechanism is that the newly formed α_2 grains are connected with pre-existing α_2 lamellae, which in a sense feed the chemically driven generation of α_2 phase at the bonding layer.

The DRX region consists of relatively large γ grains, which are often interspersed by annealing twins (Figure 17.7). The annealing twins take the form of parallel-sided lamellae that are bounded by $\{111\}_\gamma$ twinning planes and are often observed at the corners of grains. The contact plane of the annealing twins facet onto another $\{111\}$ plane belonging to the same zone as the contact plane; faceting onto random planes was not seen. These observations suggest that the annealing twins initiate from grain boundaries or grain-boundary nodes, and that the twins grow by the accumulation of the facets (Figure 17.4, arrow 3 and Figure 17.7, arrows 1). The annealing twins produce new orientations in the recrystallized grains; this may place new slip systems in a favorable position relative to the stress axis so that plastic deformation occurs more readily inside the twin than in the parent crystal outside the twin. Thus, the annealing twins could accommodate constraint stresses that are set up by surrounding grains.

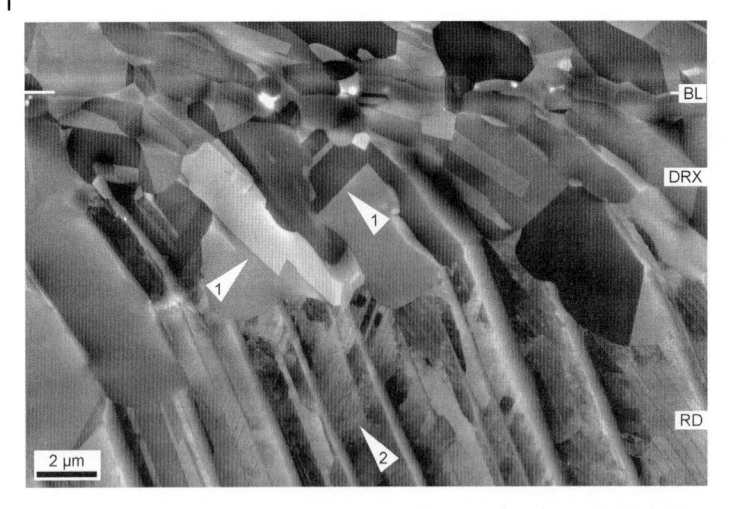

Figure 17.7 Structural details of the DRX and RD zones. Note the orientation of annealing twins (arrows 1) in γ grains of the DRX zone, indicating grain orientations that are favorable for glide on {111} planes. White bars mark the bonding layer. Note the sharp transition between the DRX and RD zone and the mechanical twins (arrow 2) within γ lamellae of the RD zone demonstrating remnant deformation. Ti-45Al-10Nb (alloy 2); bonding conditions $T = 1273\,\text{K}$, $\sigma = 20\,\text{MPa}$, $t = 2\,\text{h}$ [7].

In the RD zone marked in Figure 17.7 mechanical twins can also be seen. Strictly speaking, a mechanical twin is indistinguishable from an annealing twin because it has the same twinning elements, that is, the K_1 twinning plane and the twinning shear η_1 are identical (Section 5.1.5). However, mechanical twins are much finer than annealing twins, often lenticular and occur in clusters of variable thickness (Figure 17.7 arrow 2). It might be expected that recrystallization is triggered by the localized deformation occurring close to the contact surfaces. Since a critical deformation is necessary in order to initiate dynamic recrystallization, it appears plausible that the process is limited to a relatively narrow region adjacent to the contact zone. Since the recrystallized grains exhibit neither strain contrast nor deformation features, such as mechanical twins and dislocations, the grains are considered to be stress free.

The orientation of the recrystallized γ grains could not be determined because the c/a ratio of the tetragonal unit cell of the $L1_0$ structure is too close to unity (typically $c/a = 1.02$). Thus, the tetragonal distortion of the Kikuchi pattern is too small to distinguish the c and a directions. The signal-to-noise ratio of the pattern is also too low to allow detection of superlattice bands associated with the $L1_0$ ordering. However, as shown in Figure 17.7, the {111} habit plane of the annealing twins seen in these grains is often oriented at an angle of 30° to 50° to the bonding plane, which indicates that one of the {111} glide planes is favorably oriented for strain accommodation. Thus, as with the α_2 grains in the bonding layer, the orientation of the γ grains in the DRX zone seems to be determined by the deformation constraint set up during bonding. Conversely, this finding indicates that the

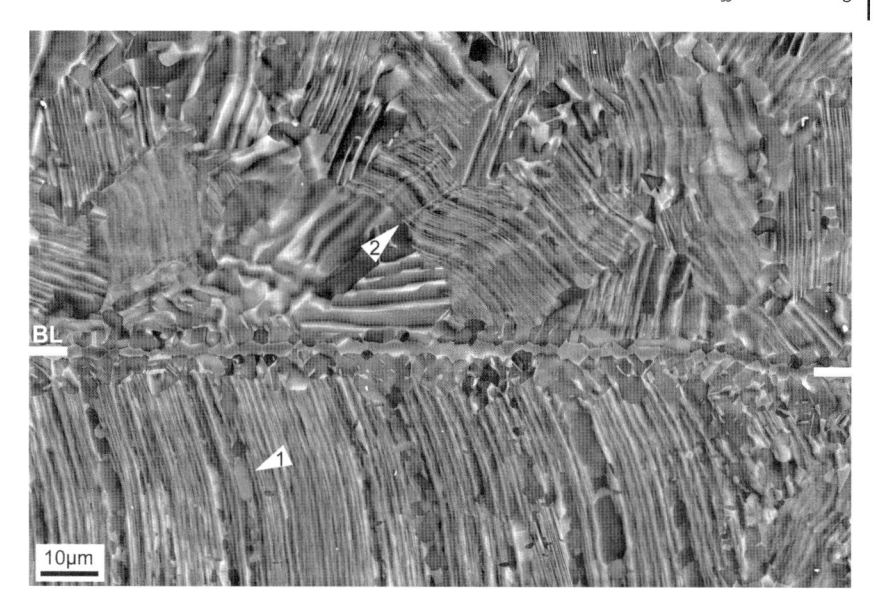

Figure 17.8 Buckling of lamellae in the process zone of diffusion-bonded Ti-44.5Al (alloy 1). Bonding conditions $T = 1273$ K, $\sigma = 60$ MPa, $t = 2$ h. Bending of lamellae promotes the recrystallization of new γ grains (arrow 1), kinking leads to the formation of fine-grained shear bands (arrow 2) [7].

formation of annealing twins makes a particularly significant contribution to the final orientation distribution of the grains in the DRX zone.

Unfortunately, no information could be deduced about the mechanism that governs the nucleation of the recrystallized γ grains. Dynamic recrystallization in lamellar morphologies is certainly triggered by buckling and kinking of lamellae, which can be observed in alloy 1 after diffusion bonding at a slightly higher bonding stress of 60 MPa. The high internal stresses that exist in zones of bent lamellae seem to promote the formation of new γ grains (Figure 17.8, arrow 1). Also, an early stage in the evolution of fine-grained shear bands, which have been described by Imayev *et al.* [17], can be seen in a highly bent lamellar colony (arrow 2). However, these mechanisms do not explain the nucleation of γ grains in the DRX zone of globular alloys. The recrystallization kinetics in such alloys seems to be sluggish, when compared with the formation of the α_2 grains at the bonding layer. This is suggested by the microstructure observed after a short bonding time but otherwise identical conditions (Figure 17.9). The characteristic feature of this bond is a fully developed bonding layer of α_2 grains, which is in direct contact with a well-expressed deformation zone, that is, practically no dynamic recrystallization occurred outside the bonding layer. This observation can be understood in terms of the driving forces for the formation of the BL and DRX zone present in lamellar and globular two-phase alloys. It is well established in the literature that the energy change associated with phase transformation is very much higher than that arising from the energy stored during deformation [18]. Hence, formation of α_2 phase at

Figure 17.9 Cross section of a bond in Ti-46.5 Al (alloy 3) formed at $T = 1273$ K, $\sigma = 20$ MPa and $t = 15$ min. Note the frequent occurrence of pores and the fully developed α_2 bonding layer BL, which is in direct contact with the deformation zone RD [7].

the bonding line due to oxygen contamination is common to both microstructures. However, as has been discussed above, dynamic recrystallization in lamellar alloys is strongly supported by the deformation instability of the lamellar colonies, resulting in bending and kinking of the lamellae followed by dynamic recrystallization. These processes are almost absent in the globular alloys; thus dynamic recrystallization can only be triggered by heterogeneities in the deformed state, resulting in sluggish kinetics. Hence, at short bonding times, the microstructure is only converted at the bonding layer, where a high chemical driving force exists. Outside the bonding layer the deformation structure is almost preserved.

The characteristic features of the RD zone are mechanical twins (Figures 17.4 and 17.7) and sub-boundaries (Figure 17.10), the latter of which indicates that the deformation structure is partially recovered. There is also evidence for dislocations in the interior of the subgrains, which were occasionally imaged by electron-channeling contrast [19]. Figure 17.10b shows a regular array of dislocations lying roughly end-on with one end of each dislocation exiting the sample and probably forming a low-angle boundary. The nature of these dislocations could be very complex because in γ(TiAl) dislocations of two or more Burgers vectors may react to form grain-boundary networks [20, 21]. A misorientation angle of $\theta = 0.6°$ is estimated from the dislocation separation distance of about 25 nm, if for the sake of simplicity the Burgers vector of ordinary dislocations $\mathbf{b} = 1/2 < 110] = 0.283$ nm is assumed. This value is typical for all the sub-boundaries investigated in this manner and is consistent with the EBSD analysis, which samples a larger area. Misorientations between regions within a single grain are of interest because they can give information about the way in which dislocations are stored during plastic deformation and recovery. The reason for this grain fragmentation is that the number and selection of the operating slip systems can locally vary, and so too, the

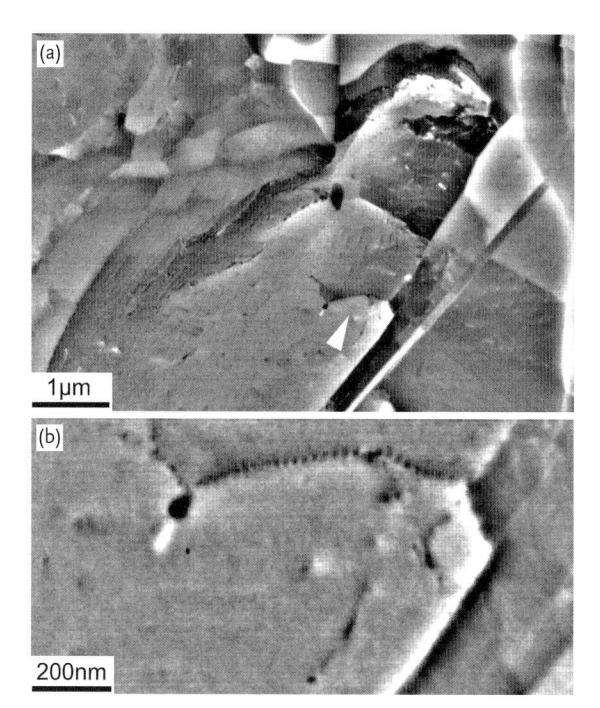

Figure 17.10 Deformation characteristics in the RD zone of diffusion-bonded Ti-46.5Al-5.5Nb (alloy 4). Bonding conditions $T = 1273$ K, $\sigma = 20$ MPa, $t = 2$ h. (a) Sub-boundaries in the RD zone situated 12 μm away from the bonding layer. (b) Higher magnification of the detail arrowed in (a) showing a regular array of dislocations lying roughly end-on with one end of each dislocation exiting the sample and forming a low-angle boundary [7].

driving forces for recovery. Most, if not all, γ grains comprising the DRX zone exhibit a very small orientation variation of less than 0.3° [7]. Although such small values are at the limit of the angular resolution of the utilized spectrometer, repeated orientation measurements have revealed the reliability of the data. This is also supported by the fact that the misorientations estimated from the dislocation content of individual small-angle grain boundaries fall into the range of misorientations determined by EBSD. Thus, it is concluded that the fragmentation of the grains is produced by very small-angle grain boundaries. These findings suggest that the deformation substructure under the bonding conditions used has been largely consumed by dislocation annihilation and the formation of sub-boundaries; thus further grain nucleation will be difficult. However, there is certainly a driving force to form fewer, more highly misoriented boundaries as recovery proceeds [22].

With increasing distance of the γ grains from the bonding layer, the orientation spread increases up to 1°, which suggests that the misorientation of the small-angle boundaries increases with increasing distance of the γ grains from the bond layer [7]. It might be speculated in this context that the dislocation content of a low-angle boundary is reduced, if it is in contact with a high-angle grain boundary, a situation that frequently occurs at the transition between the DRX and RD zones. As has been suggested by Jones et al. [23], this dislocation absorption could be the early stage of subgrain rotation. The mobility of sub-boundaries is expected to be controlled by the glide and climb resistance of the dislocations constituting the boundary [16]. The ordinary dislocations of γ(TiAl) have a compact core because of the high energy of the complex stacking fault (CSF) that would be involved in the dissociation (see Section 5.1.3). Thus, cross-slip and climb are relatively easy, which tend to promote coarsening of the subgrain structure. Taken together these observations reflect the competition between recovery and recrystallization, which is characteristic of high-temperature deformation. The factors that promote dynamic recrystallization are:

i) a lamellar morphology in the starting material;
ii) a high amount of stored energy produced by asperity deformation;
iii) a low dislocation mobility, which retards dynamic recovery.

In summary: During diffusion bonding of $(\alpha_2 + \gamma)$ alloys a three-layer process zone is typically developed that involves a fine-grained layer of α_2 phase at the former contact plane of the diffusion couple, a region made up of relatively large recrystallized grains, followed by a region of deformed bulk material. This process zone structure reflects the combined influences of oxygen contamination at the bonding surfaces and dynamic recrystallization induced by asperity surface deformation.

17.1.5
Effect of Alloy Composition

In the Ti-Al system the location of the phase boundaries is significantly affected by the addition of ternary or higher alloying elements [8, 9]. Thus, the phase evolution during any thermomechanical treatment depends on the alloy composition. Not surprisingly, these effects are also reflected in the observed bonding structures. As a first example, Figure 17.11 demonstrates the process zone in the single-phase γ alloy 8 with the composition Ti-54Al, which had been bonded under the same conditions as the two-phase alloys described earlier. The characteristic feature of the bond structure is that no α_2 layer formed at the bonding plane. This could be rationalized by the relatively high Al content of the alloy and the very low oxygen and nitrogen solubility of the γ phase [24]. Contamination of the contact surfaces by these elements apparently leads to the precipitation of oxides and nitrides, which are often situated at the bonding layer or at grain boundaries next to the contact plane (Figure 17.11b). There are distinct DRX and RD zones, which exhibit the same features as has been described for the two-phase alloys. However, the DRX zone is significantly broader than that observed in two-phase alloys. A possible explanation for this difference is that no α_2 phase is present in alloy 8, which in two-phase alloys certainly impedes grain growth. Under the conditions used, the bond is quite incomplete, as indicated by the high density of voids and pores at the initial joint interface. There are two likely explanations for this. First, the significant work hardening of the alloy may hinder the diffusion couple coming

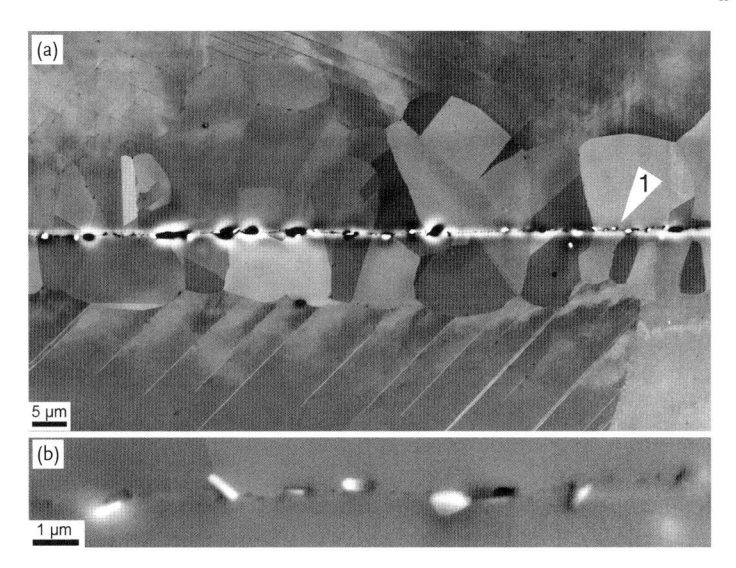

Figure 17.11 Diffusion bonding of the single-phase γ(TiAl) alloy 8 (Ti-54Al) at $T = 1273\,\text{K}$, $\sigma = 20\,\text{MPa}$ and $t = 2\,\text{h}$. (a) Cross section of the bond showing pores and voids at the joint interface, regions of recrystallized γ grains DRX, and regions with remnant deformation RD. (b) Higher magnification of detail 1 marked in (a) showing fine precipitates at the joint interface [7].

into full contact at the joint interface. Secondly, diffusion in Al-rich alloys is sluggish, when compared with Ti-rich alloys. Clearly, sound bonding of this type of alloy requires higher temperatures.

As a second example, the influence of Nb will be discussed, this is often added at levels of 5 to 10 at.% in Al-lean alloys. In the Ti-Al system alloying with Nb generally decreases the β(B2)- and α-transition temperatures, respectively, and contracts the α-phase field [25]. Nb is known to solely occupy the Ti sublattices of the γ and α_2 phases with a very small size misfit. There are two factors that could be detrimental for the diffusion bonding of Nb-bearing alloys: Nb is a slow diffuser in γ(TiAl) [15], and the alloys often exhibit good high-temperature strength and creep resistance (Figures 7.15 and 17.2). Nevertheless, sound bonds were obtained under the usual bonding conditions ($T = 1273\,\text{K}$, $\sigma = 20\,\text{MPa}$, $t = 2\,\text{h}$) as demonstrated in Figure 17.12, for a bond in Ti-46.5Al-5.5Nb (alloy 4). When compared to alloy 3 (Figures 17.4 and 17.9) that is the binary counterpart to alloy 4, the most important difference is the lower amount of α_2 phase at the bonding layer [26]. A reasonable explanation of this observation is not at hand, but it could be speculated that Nb additions slightly enhance the solubility of oxygen in the γ and α_2 phases so that the driving force for the nucleation of new α_2 phase at the bonding layer is reduced, when compared with binary alloys. From the engineering point of view, partial elimination of the brittle α_2 phase from the bonding layer is certainly beneficial for bond quality.

In some cases, growth of γ grains across the joint interface occurs (arrowed in Figure 17.12). Another difference to binary two-phase alloys is the high density of

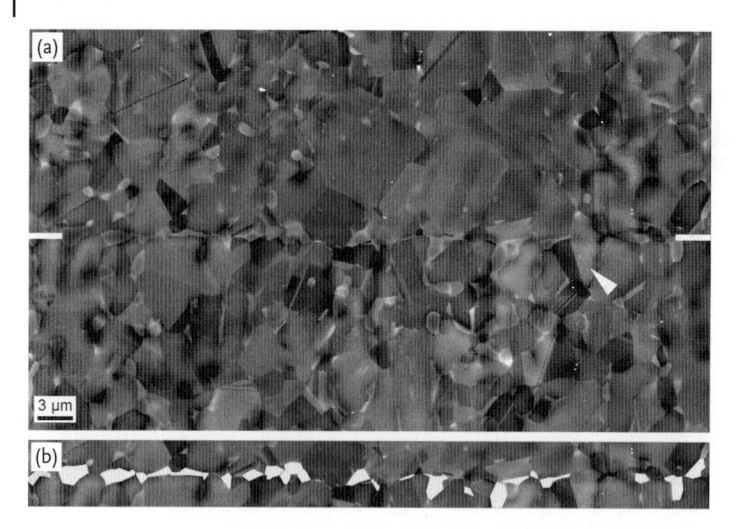

Figure 17.12 Diffusion bonding of alloy 4 (Ti-46.5Al-5.5Nb) at $T = 1273$ K, $\sigma = 20$ MPa and $t = 2$ h. (a) Cross section of the bond. (b) Phase distribution at the bonding layer showing disperse α_2 phase [7].

annealing twins in the γ grains comprising the DRX zone of Nb-bearing alloys. For f.c.c. metals it is well established that the propensity for twinning strongly depends on the stacking-fault energy. This is because the surface energy of a twin boundary is closely related to that of the stacking fault and a large part of the work to form a twin goes into creating its boundaries. In f.c.c. metals it is often the case that the stacking-fault energy decreases with increasing concentration of a substitutional solute so that twinning becomes more and more important as the solute concentration increases. Following these classical arguments, it might be expected that the stacking-fault energy is reduced by Nb additions. As described earlier, the annealing twins can accommodate the constraints that are associated with the phase transformations and structural changes occurring in the process zone during diffusion bonding. More details about alloying effects on diffusion bonding of TiAl alloys is provided in [7].

In summary: The bonding behavior of two-phase alloys is relatively independent of minor ternary metallic additions. However, Nb additions of 5 at.% to 10 at.% significantly reduce both the width of the process zone and the amount of α_2 phase formed at the bonding layer. Single-phase γ(TiAl) alloys are difficult to bond because of their reduced diffusivity and inability to dissolve the oxygen present at the contact surfaces prior to bonding.

17.1.6
Influence of Bonding Time and Stress

The evolution of the process zone with bonding time has been investigated on alloy 3 for bonding at 1273 K, 20 MPa and bonding times between 15 and 120 min.

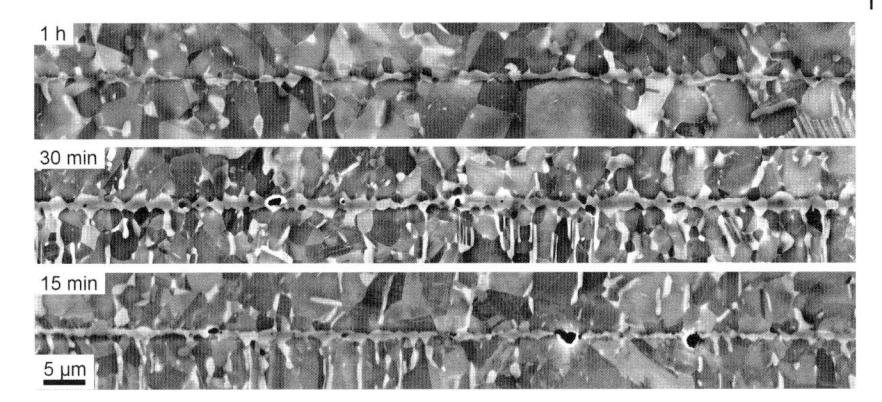

Figure 17.13 Effect of bonding time on the structure of the process zone. Ti-46.5Al (alloy 3), bonded at $T = 1273$ K, $\sigma = 20$ MPa and the times indicated. Note the fully developed α_2 layer and the presence of many pores after the shortest bonding time of $t = 15$ min. Also note the gradual development of the DRX zone with time [7].

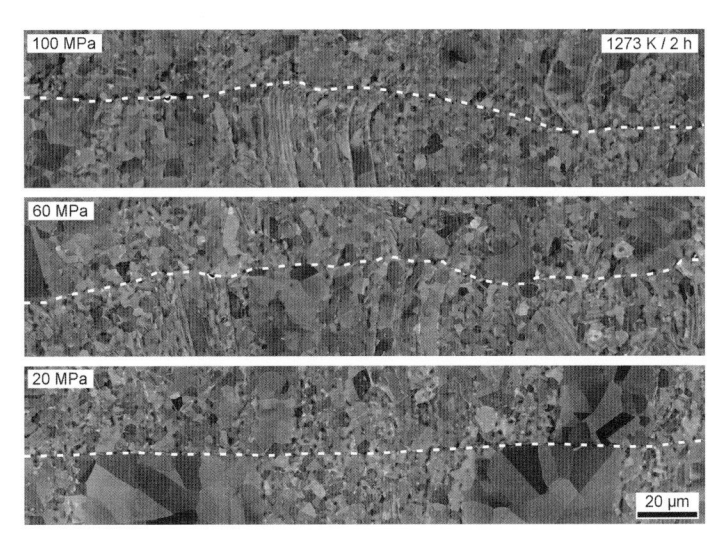

Figure 17.14 Effect of bonding stress on the structure of the process zone. Ti-46.5Al-5.5Nb (alloy 4), bonded at $T = 1273$ K, $t = 2$ h and the stresses indicated. Note the waviness of the bonding plane at higher stresses due to the impression of lamellar packets into globular microstructural constituents [7].

Part of this study is demonstrated in Figure 17.13 by a series of micrographs. Bonds developed after short bonding times generally suffer from many pores and voids. Clearly, for successful bonding of two-phase alloys at 1273 K and 20 MPa, bonding times of at least 60 min are needed.

Figure 17.14 demonstrates the effect of the bonding stress on the structure of the process zone that was observed in two-phase alloys. The characteristic feature is the waviness of the bonding plane, which gradually increases with stress. The micrographs imply that the lamellar packets were impressed into globular constituents; an observation that could be rationalized by the different glide resistance of these two morphologies. It is well established in the TiAl literature that lamellar packets have an extremely high yield strength when they are loaded parallel to the lamellae (see Section 6.1.5), as seen in Figure 17.14. Thus, up to relatively high stresses, such lamellar packets are not plastically deformed and are impressed into weaker surrounding material. The waviness of the bonding plane could be beneficial for the bond quality because it provides some pegging between the bonded materials. To take advantage of this technically, a tight optimization of the bonding parameters is required in order to avoid large-scale deformation of the component parts.

17.1.7
Mechanical Characterization of the Bonds

Figure 17.15 shows microhardness profiles that were measured across the bond line at two different positions. Comparison of these profiles with the related indentation positions indicates that the bonding process does not significantly affect the

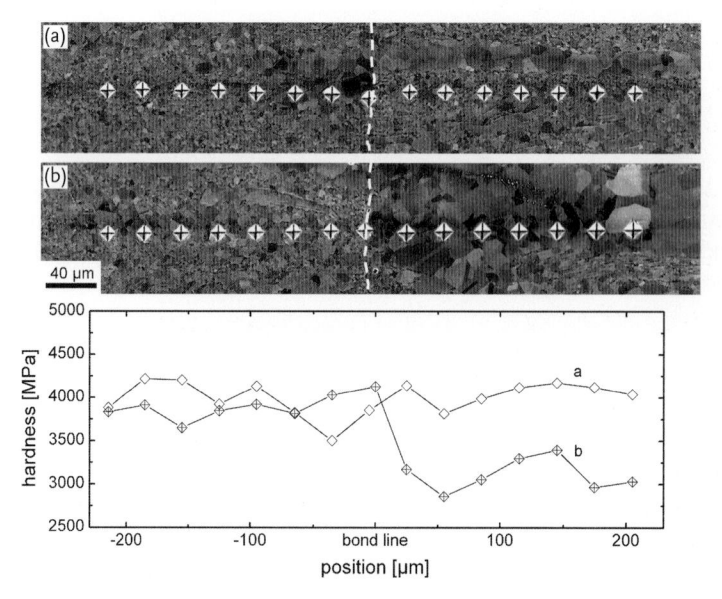

Figure 17.15 Microhardness profiles measured at the positions (a) and (b) across the bonding plane (indicated by the dotted white line). Note the change of the hardness profile (b) with the variation of the microstructure [7].

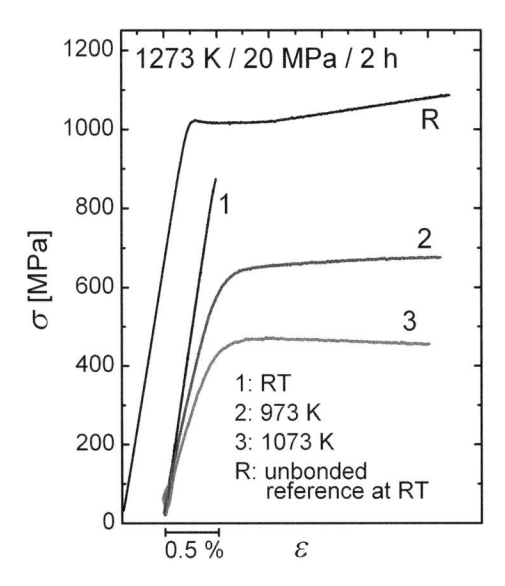

Figure 17.16 Example stress–strain curves of tensile tests performed on Ti-45Al-5Nb-0.2C-0.2B (alloy 7, TNB-V5) bonded at $T = 1273$ K, $\sigma = 20$ MPa and $t = 2$ h [7].

hardness. The variation of the microhardness observed on profile (b) is probably caused by the change of microstructure. The indentations on the right-hand side of this profile have sampled relatively coarse-grained material. According to Hall–Petch arguments (see Section 7.1), such material should exhibit a significantly lower hardness than that of the fine-grained material situated at the left-hand side of the profile.

The results of the tensile tests performed on bonded samples of alloy 7 are shown in Figure 17.16. At room temperature, the sample fractured at the bonding plane at stresses, which are typically 85% of the yield stress of the parent reference material [10, 11]. At the deformation temperatures of 973 K and 1073 K, the bonded samples exhibited appreciable ductility. The related yield stresses are practically identical with those of the parent material at these temperatures, that is, bonding did not significantly impair the tensile strength or ductility of the material at the intended service temperature. However, because of the very fine microstructure developed at the bonding plane, creep resistance of the bonds could be a concern, which is an obvious subject for future investigation.

In summary: TiAl samples joined by diffusion bonding exhibit a room-temperature tensile strength of about 85% of the room-temperature yield strength, when loaded perpendicular to the bonding plane. There is no marked change of the microhardness across the bonding interface.

17.2
Brazing and Other Joining Technologies

Although most research on joining TiAl has concentrated on diffusion bonding, some work has been performed using other techniques such as brazing, friction welding, laser welding and fusion welding. It should be remembered that the joining temperature and cooling rate after joining can have a significant effect on the mechanical properties via their influence on the microstructure and the development of residual stress and cracking at the joint interface.

17.2.1
Brazing and Transient Liquid-Phase Joining

Two types of brazing are described in the literature, vacuum brazing and infrared brazing. According to Shiue et al. [27] infrared brazing has the advantages of faster production rates and lower costs compared to conventional vacuum brazing. Additionally, as the infrared rays can be focused so that only local heating takes place, the microstructure of the rest of the workpiece remaining unchanged. Often, TiCuNi-based alloys (such as Ti-15Cu-15Ni and Ti-15Cu-25Ni [wt.%]) have been used for vacuum brazing [28, 29] and infrared brazing [30, 31]. According to Tetsui [32] brazing filler must show good wettability, high-temperature strength, sufficient toughness and not lead to embrittlement of the TiAl. The melting temperature of the brazing alloy should also not be too high so that the microstructure of the base material is not significantly altered. Das et al. [28] report the use of three-layered foils consisting a CuNi middle layer with pure titanium layers on either side, yielding a composition of Ti-15Cu-15Ni (wt.%). The vacuum-brazing conditions were 0.021 MPa stress 1010 °C/30 min + 918 °C/4 h followed by rapid helium cooling. Tetsui [32] investigated brazing fillers based on (all wt.%): Ag-28Cu, Au-18Ni, Au-50Cu, Au-12.5Ag-12.5Cu, Ag-10Pd-31.5Cu, Ti-33Ni and Ni-15Cr-3.5B using brazing temperatures between 779 °C and 1100 °C. Similarly, Sirén et al. [33] investigated Ti-15Cu-15Ni, Cu-3Si-2.25Ti-2Al, Au-3Ni-0.6Ti, Ni-20Cr-10Si and Pd-40Ni. According to Shieu et al. [27], who studied pure silver brazing filler, Ag-based brazing alloys have been successfully used to join conventional titanium alloys and have the advantage of reduced brazing temperature. Brazing temperatures of 779 °C to 899 °C were used for Ag-28Cu (wt.%) braze in [32]. Additionally, the Ti-Ag-based intermetallic phases that form are ductile compared to the brittle intermetallic phases that form when using Ti- and Ni-based brazes. For infrared brazing, brazing temperatures varied between 1100 °C and 1200 °C but the brazing times used were considerably shorter, ranging from 30 to 60 s [30]. To join Ti-50Al to Ti-6Al-4V infrared brazing times between 180 and 1200 seconds were used but at lower temperatures of 930 °C to 970 °C [31].

Apart from brazing, so-called transient liquid-phase (TLP) joining of TiAl has also been investigated. As explained by Butts and Gale [34] this method is similar to fluxless vacuum brazing; also depending on the formation of a liquid interlayer with a melting point (initially) well below that of the substrate. However, the melting temperature of the layer on reheating is similar to that of the substrate, thus making the technique suitable for joining components to be used at elevated temperatures. Seemingly, the term brazing has sometimes been used in the literature for what was actually TLP joining. TLP alloys have been made from different ratios of TiAl and Cu powder as described in [34]. The mixed powders were placed on the substrate surfaces to be joined with thicknesses of between 350 μm to 500 μm, bonding was conducted in vacuum at 1150 °C for times of up to 10 min.

When investigated, the mechanical properties of joints have been assessed using overlap shear tests or microhardness testing. Infrared brazing using pure silver filler has been reported to give maybe the best value of bond strength (385 MPa) [32]. Xu et al. [29] report a bending strength (610 MPa) essentially the same as that of the base material after brazing using Ti-15Cu-15Ni filler and a subsequent postbrazing heat treatment. Nevertheless, as with other investigations, an increase in microhardness of the joint area was observed. The formation of hard (brittle) intermetallic phases at joint interfaces is a general feature of brazing technologies. With regard to the TLP joining of TiAl using TiAl-Cu powder mixtures, initial results seem rather promising as bonds with microstructures and mechanical properties somewhat similar to the parent metal are reported for a 100:1 TiAl:Cu powder ratio. Additionally, the formation of hard Cu-rich phases could be avoided.

From an industrial point of view the joining of TiAl to other common engineering materials is also of interest. For example, turbocharger wheels have to be joined to the steel shaft. Noda et al. [35] describes a cost-efficient method, called induction brazing. The study used Ag-35.2Cu-1.8Ti and Ti-15Cu-15Ni (wt.%) filler between the turbocharger and the steel shaft and the assembly was heated locally, via an induction coil, under a slight pressure in an argon atmosphere. The tensile strength of the Ag-based filler joints is reported to be around 320 MPa at room temperature and 310 MPa at 500 °C. These values are around 50% of the base TiAl material. In another study, vacuum induction bonding of TiAl to steel was performed using a multilayered foil that consisted of a Ti foil in contact with the TiAl, then a V foil with a Cu foil in contact with the steel [36]. The room-temperature tensile strength is reported to be up to 420 MPa, close to that of the TiAl. The joint was also reported to be free of intermetallic compounds and other brittle phases. Direct joining of the TiAl to the steel, that is, diffusion bonding, resulted in strengths of 170–185 MPa.

In summary: Some success has been achieved in the joining of TiAl to TiAl and TiAl to other materials such as steels. The choice of filler alloys is extremely important, as the formation of brittle intermetallic phases should be avoided. Unlike diffusion bonding, increased hardness across the interface is a common feature after brazing. Embrittlement either after processing or long-term usage must be avoided or at least minimized. The reproducibility and degradation of joint strength at both ambient and elevated temperatures are areas that still have to be fully addressed.

17.2.2
Other Techniques

Apart from brazing and transient liquid-phase joining, a number of other methods have been studied, but on a rather limited scale. Rotational friction welding of TiAl to AISI 4140 steel is described by Lee *et al.* [37] and room-temperature strength levels of up to 375 MPa were measured when a Cu insert layer was used in-between the TiAl and the steel. Linear friction welding was investigated by Baeslack *et al.* [38] who managed to obtain crack-free welds when the weld cooling rate was reduced by using a combination of a high input energy and reduced applied force. The weld zone had an extremely fine microstructure that resulted in an increase of hardness compared to the base material. Postweld heat treatment resulted in microstructural recrystallization and a normalization of hardness values across the weld. Electron-beam welding can give crack-free material if the constraint in the workpieces is low and the cooling rate properly selected so that the α phase can fully decompose [39]. If either the cooling rate is too fast or the constraint between the workpieces high, then the development of thermal stresses leads to cracking. Bartolotta and Krause [40] reported the successful production of crack-free electron-beam welds up to 25 cm long and 15 mm in thickness using cast Ti-48Al-2Cr-2Nb. The method involves heating the workpieces in a controlled atmosphere to elevated temperatures, presumably above the brittle to ductile transition temperature. The material is then slowly removed from the heating arrangement and electron-beam welded. After the welding bead has been laid down the material is reheated to its original temperature. The production of crack-free welds on a regular basis without difficulty is reported. Welding of sheet resulting in tensile joint strengths almost as high as the base material (470 MPa) is also reported in [41]. Laser spot welding is also described [41]. Inert-gas tungsten arc-welding deposition has been reported as a repair technology that has been successfully implemented for Ti-48Al-2Cr-2Nb [40]. We speculate that the use of such techniques that require the TiAl to be melted may be restricted to alloys where the properties are relatively insensitive to somewhat coarser microstructures, that is, low strength alloys, or to alloys that show significant grain refinement on solidification. For further information on joining technologies and the type of demonstration parts that have been made the reader is referred to the review-type articles [40, 41].

References

1 Kazakov, N.F. (1981) *Diffusion Bonding of Materials*, Pergamon Press, New York.
2 Nakao, Y., Shinozaki, K., and Hamada, M. (1991) *ISIJ Int.*, **10**, 1260.
3 Yan, P., and Wallach, E.R. (1993) *Intermetallics*, **1**, 83.
4 Glatz, W., and Clemens, H. (1997) *Intermetallics*, **5**, 415.
5 Cam, G., Clemens, H., Gerling, R., and Kocak, M. (1999) *Intermetallics*, **7**, 1025.
6 Holmquist, M., Recina, V., Ockborn, J., Pettersson, B., and Zumalde, E. (1998) *Scr. Mater.*, **39**, 1101.
7 Herrmann, D., and Appel, F. (2009) *Metall. Mater. Trans. A*, **40A**, 1881.

8 Hellwig, A., Palm, M., and Inden, G. (1998) *Intermetallics*, **6**, 79.
9 Kainuma, R., Fujita, Y., Mitsui, M., Ohnuma, I., and Ishida, K. (2000) *Intermetallics*, **8**, 855.
10 Appel, F., Lorenz, U., Oehring, M., Sparka, U., and Wagner, R. (1997) *Mater. Sci. Eng. A*, **233**, 1.
11 Appel, F., Oehring, M., and Wagner, R. (2000) *Intermetallics*, **8**, 1283.
12 Kattner, U.R., Liu, J.C., and Chang, Y.A. (1992) *Metall. Trans. A*, **23A**, 2081.
13 Godfrey, S.P., Threadgill, P.L., and Strangwood, M. (1995) *High-Temperature Ordered Intermetallic Alloys VI, Materials Research Society Symposia Proceedings*, vol. 364 (eds J.A. Horton, I. Baker, S. Hanada, R.D. Noebe, and D.S. Schwartz), MRS, Pittsburgh, PA, p. 793.
14 Buque, C., and Appel, F. (2002) *Int. J. Mater. Res.*, **8**, 784.
15 Mishin, Y., and Herzig, C. (2000) *Acta Mater.*, **48**, 589.
16 Hirth, J.P., and Loth, J. (1992) *Theory of Dislocations Theory of Dislocations*, Krieger, Melbourne.
17 Imayev, R.M., Imayev, V.M., Oehring, M., and Appel, F. (2005) *Metall. Mater. Trans. A*, **36A**, 859.
18 Humphreys, F.J., and Hatherly, M. (1995) *Recrystallization and Related Annealing Phenomena*, Pergamon, Oxford.
19 Simkin, B.A., and Crimp, M.A. (1997) *High-Temperature Ordered Intermetallic Alloys VII, Materials Research Society Symposium Proceedings*, vol. 460 (eds C.C. Koch, C.T. Liu, N.S. Stoloff, and A. Wanner), MRS, Pittsburgh, PA, p. 387.
20 Hazzledine, P.M. (1998) *Intermetallics*, **6**, 673.
21 Paidar, V. (2002) *Interface Sci.*, **10**, 43.
22 Ørsund, R., and Nes, E. (1989) *Scr. Metall.*, **23**, 1187.
23 Jones, A.R., Ralph, B., and Hansen, N. (1979) *Proc. Roy. Soc. London A*, **368**, 345.
24 Menand, A., Huguet, A., and Nérac-Partaix, A. (1996) *Acta Mater.*, **44**, 4729.
25 Chen, G.L., Zhang, W.J., Liu, Z.C., Li, S.J., and Kim, Y.-W. (1999) *Gamma Titanium Aluminides 1999* (eds Y.-W. Kim, M.H. Loretto, and D.M. Dimiduk), TMS, Warrendale, PA, p. 371.
26 Appel, F., Paul, J.D.H., Oehring, M., and Buque, C. (2003) *Gamma Titanium Aluminides 2003* (eds Y.-W. Kim, H. Clemens, and A.H. Rosenberger), TMS, Warrendale, PA, p. 139.
27 Shiue, R.K., Wu, S.K., and Chen, S.Y. (2004) *Intermetallics*, **12**, 929.
28 Das, G., Bartolotta, P.A., Kestler, H., and Clemens, H. (2003) *Gamma Titanium Aluminides 2003* (eds Y.-W. Kim, H. Clemens, and A.H. Rosenberger), TMS, Warrendale, PA, p. 33.
29 Xu, Q., Chaturvedi, M.C., Richards, N.L., and Goel, N. (1997) *Structural Intermetallics 1997* (eds M.V. Nathal, R. Darolia, C.T. Liu, P.L. Martin, D.B. Miracle, R. Wagner, and M. Yamaguchi), TMS, Warrendale, PA, p. 323.
30 Lee, S.J., and Wu, S.K. (1999) *Intermetallics*, **7**, 11.
31 Shiue, R.K., Wu, S.K., Chen, Y.T., and Shiue, C.Y. (2008) *Intermetallics*, **16**, 1083.
32 Tetsui, T. (2001) *Intermetallics*, **9**, 253.
33 Sirén, M., Bohm, K.H., Appel, F., and Koçak, M. (1999) Manufacture and Characterisation of Brazed Joints in γ-TiA. *Proc. Welding Conf. LUT Join '99: Int. Conf. on Efficient Welding in Industrial Applications (ICEWIA, Lappeenranta, August 1999)*, p. 12.
34 Butts, D.A., and Gale, W.F. (2003) *Gamma Titanium Aluminides 2003* (eds Y.-W. Kim, H. Clemens, and A.H. Rosenberger), TMS, Warrendale, PA, p. 605.
35 Noda, T., Shimizu, T., Okabe, M., and Iikubo, T. (1997) *Mater. Sci. Eng.*, **A239–240**, 613.
36 He, P., Feng, J.C., Zhang, B.G., and Qian, Y.Y. (2003) *Mater. Char.*, **50**, 87.
37 Lee, W.B., Kim, Y.J., and Jung, S.B. (2004) *Intermetallics*, **12**, 671.
38 Baeslack, W.A. III, Broderick, T.F., Threadgill, P.L., and Nicholas, E.D. (1996) *Titanium '95: Science and Technology* (eds P.A. Blenkinsop, W.J.

Evans, and H.M. Flower), IOM, London, p. 424.
39 Chaturvedi, M.C., Xu, Q., and Richards, N.L. (2001) *J. Mater. Proc. Technol.*, **118**, 74.
40 Bartolotta, P.A., and Krause, D.L. (1999) *Gamma Titanium Aluminides 1999* (eds Y.-W. Kim, D.M. Dimiduk, and M.H. Loretto), TMS, Warrendale, PA, p. 3.
41 Tabernig, B., and Kestler, H. (2003) *Gamma Titanium Aluminides 2003* (eds Y.-W. Kim, H. Clemens, and A.H. Rosenberger), TMS, Warrendale, PA, p. 619.

18
Surface Hardening

18.1
Shot Peening and Roller Burnishing

Shot peening is a cold-working process in which the surface of a component is blasted with small spherical media called shot. Each piece of shot striking the metal acts as a tiny hammer, imparting a small indentation or dimple on the surface. A compressive layer is formed by a combination of subsurface compression developed at the Hertzian impression combined with lateral displacement of the surface material around each of the dimples formed [1], as illustrated in Figure 18.1. When the dimples overlap, the entire surface is effectively elongated. The material below the surface resists this expansion; this produces a compensating stress field that drives the layer of deformed material into compression. The peening intensity is related to the amount of kinetic energy transferred from the shot stream to the workpiece. The peening intensity is usually quantified by the so-called Almen test [2]. The method consists of peening a standardized test strip that is clamped to a mounting fixture. Once the fixture is removed, the Almen strips will curve towards the direction from which the shot came. The Almen intensity is by definition the arc height of the Almen strip followed by the Almen strip designation, N, A or C, which indicates the strip thickness. Complete procedures and specifications of intensity measuring equipment can be found in the SAE standards SAE-J442 and SAE-J433 [3, 4]. Shot peening coverage is defined as the ratio of the area covered by peening indentations to the total surface area treated. From conventional metals it is known that coverage should never be less than 100% as fatigue and stress corrosion cracks can develop in the non-peened area that is not encased in residual compressive stress.

Surface hardening can also be produced by roller burnishing (another term widely used in Europe is deep rolling). Processing is typically performed on axially symmetrical components. Either roller or ball tools are used to deform the surface layer with repeated cycles and a fine feed per pass; in this way high levels of cold work are produced. It is worth remembering that TiAl alloys generally exhibit strong work hardening at room temperature (Section 7.2), which provides a significant potential for surface hardening.

Gamma Titanium Aluminide Alloys: Science and Technology, First Edition. Fritz Appel, Jonathan David Heaton Paul, Michael Oehring.
© 2011 Wiley-VCH Verlag GmbH & Co. KGaA. Published 2011 by Wiley-VCH Verlag GmbH & Co. KGaA.

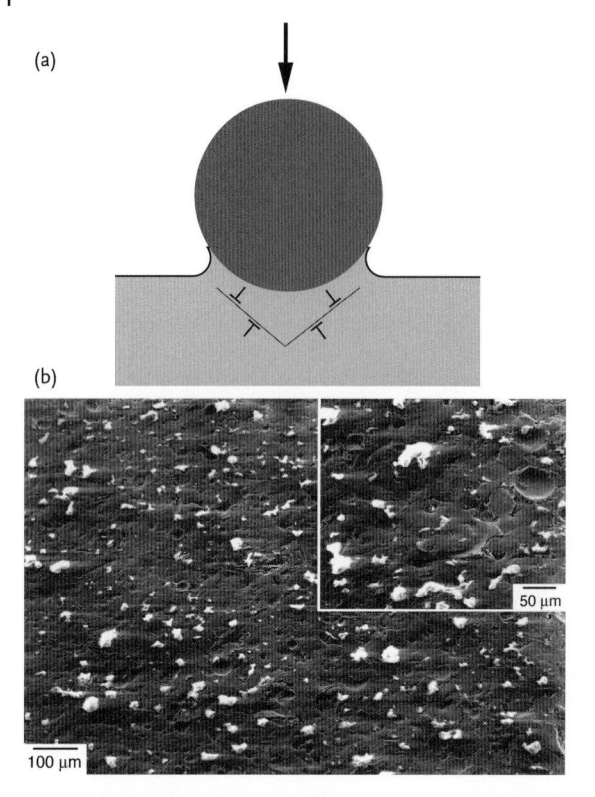

Figure 18.1 (a) Schematic illustration of mechanical yielding at the point of impact of a piece of shot on a metal surface. Below the surface, the elastically constrained material tries to restore the surface to its original shape producing a hemisphere of cold-worked material that is highly stressed in compression. (b) Scanning electron micrograph of a shot-peened surface of nearly lamellar Ti-45Al-8Nb-0.2C. Note the deep dimples and the incorporation of zirconia particles into the TiAl substrate. Peening performed at room temperature with an Almen intensity of 0.40 mm N to full coverage. Micrograph D. Herrmann [1].

18.2
Residual Stresses, Microhardness, and Surface Roughness

The compressive residual stresses on the surface can prevent crack initiation and propagation as well as closure of pre-existing cracks when they are within the depth of the compressive zone. However, surface hardening inevitably increases the microhardness and surface roughness, both of which are equally important for the component performance. For example, during fatigue, small variations in the surface morphology can provide stress concentrations sufficient to promote crack initiation even at modest plastic-strain amplitudes. Hence, the process of crack initiation is quite sensitive to the conditions at the free surface. In the

Table 18.1 Structural and mechanical characteristics of the alloys investigated before shot peening. $\sigma_{0.2}$ (tensile yield stress), σ_{UTS} (ultimate tensile stress), ε_f (fracture strain).

Alloy/Symbol	Microstructure	$\sigma_{0.2}$ (MPa)	σ_{UTS} (MPa)	ε_f (%)
Ti-47Al-1.5 Nb-1Cr-1Mn-0.2Si-0.5B				
1 ●	Fully lamellar, colony size 41 μm, lamellar spacing 50 nm	834	967	0.9
2 △	Fully lamellar, colony size 130 μm, lamellar spacing 1.5 μm	447	597	1.6
Ti-45Al-10Nb				
3	Fully lamellar	1000	1050	1

present section the general characteristics of surface hardening documented in the TiAl literature [5–10] will be illustrated, utilizing data that has been obtained on the alloys listed in Table 18.1. Shot peening (SP) was performed by means of an injector-type system using spherical zirconia-based ceramic shot with an average diameter of 0.5 mm. The Almen intensity was varied from 0.08 mm N to 0.61 mm N. All peening procedures were done to full coverage, that is, the original surface area was completely obliterated by shot-peening dimples [9]. The structural and mechanical characteristics of the alloys are described in Table 18.1. To assess the effectiveness of this surface treatment the fatigue performance peened and unpeened testpieces will be compared.

Initial verification of the shot-peening process requires the establishment of saturation conditions with respect to the Almen intensity. Figure 18.2 demonstrates the change of fatigue life with increasing Almen intensity for the two structural variants 1 and 2 of Ti-47Al-1.5 Nb-1Cr-1Mn-0.2Si-0.5B (Table 18.1). Shot peening clearly enhances the fatigue life by two to three orders of magnitude [9, 10]. With regard to the Almen intensity the effect becomes saturated sooner for material (2), which has the lower yield stress.

The change of the surface parameters associated with shot peening is demonstrated in Figure 18.3. Not surprisingly, shot peening increases the surface roughness, when compared with the electrolytically polished reference samples (Figure 18.3a), however, the roughness parameters of the two microstructures are similar. There is apparently no overpeening effect for higher Almen intensities up to 0.40 mm N, meaning, no surface erosion occurred. High Almen intensities may produce excessive surface roughness and excessive residual tensile stresses in the core of the workpieces. In view of these findings, all the further shot-peening procedures were performed at an Almen intensity of 0.40 mm N. Due to the process-induced plastic deformation, the microhardness in the near-surface regions significantly increases with a maximum value directly at the surface (Figure 18.3b). When compared with the bulk values, the increase was 45% for

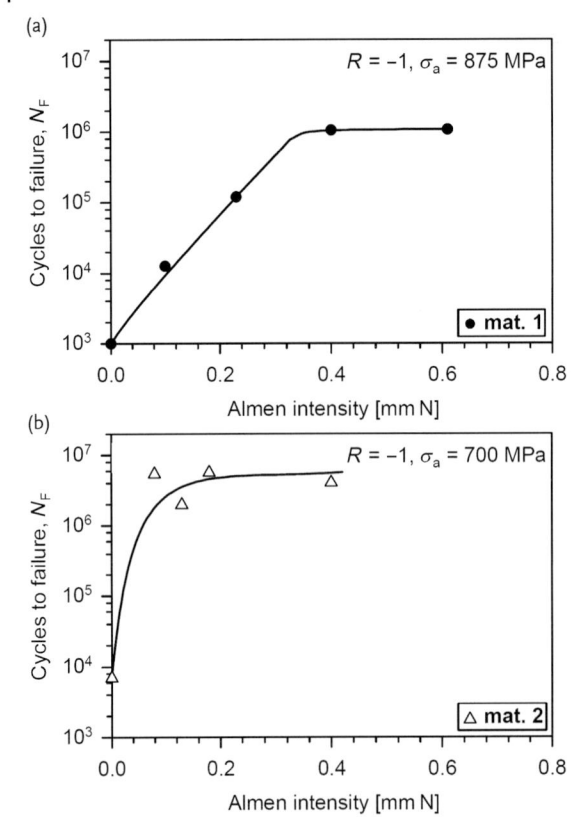

Figure 18.2 Dependence of the fatigue life on the shot-peening intensity for the alloy Ti-47Al-1.5Nb-1Cr-1Mn-0.2Si-0.5B with the structural variants 1 and 2 described in Table 18.1. Fatigue tests performed at room temperature in air; σ_a stress amplitude. Data from Lindemann et al. [9].

the fine lamellar alloy 1 and 100% for the coarse lamellar material 2. According to this data, the depth of the process zone induced by the surface deformation was roughly 200 μm for the fine and 300 μm for the coarse lamellar microstructure. The observation probably reflects the different resistance of the materials to plastic flow as documented in Table 18.1. In the weaker alloy 2, plastic deformation is certainly more intense and penetrates deeper into the material. Figure 18.3c shows the profiles of the residual compressive stresses beneath the surface and clearly indicates that the depth of the compressive zones is similar in the two materials. The higher compressive stress values measured on alloy 2 are probably a consequence of the thicker and more intensively deformed surface skin, which has to be elastically restrained by the subsurface material. It should be noted that essentially the same characteristics of surface hardening were recognized on a high Nb-containing alloy with the composition Ti-45Al-9Nb-0.2C (TNB-V2) [10].

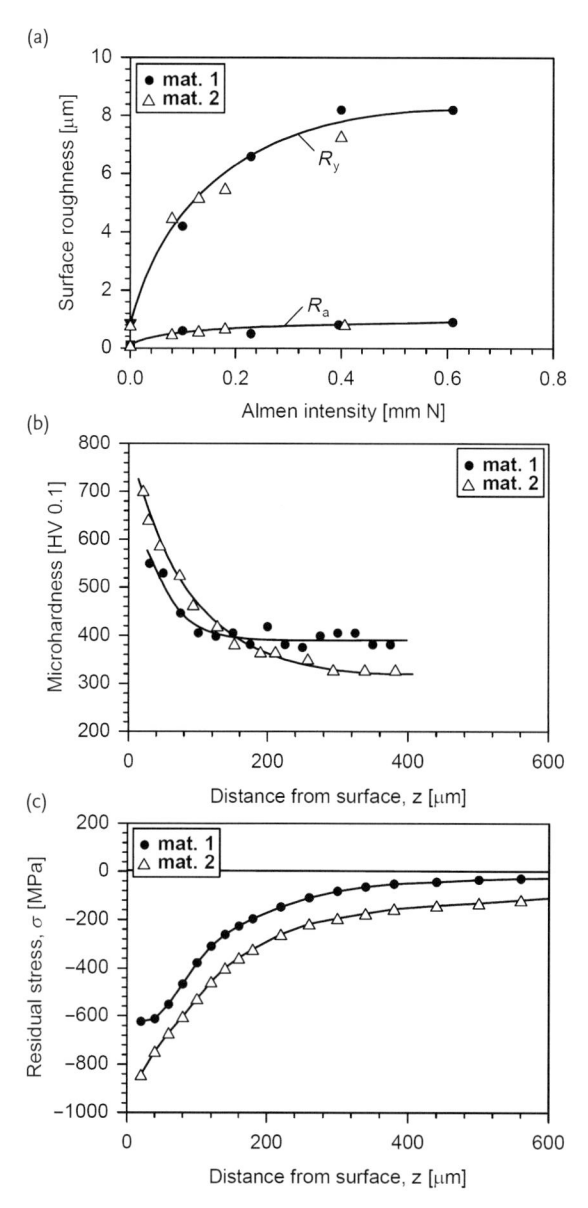

Figure 18.3 Change of the surface parameters due to shot peening at room temperature observed on materials 1 and 2. (a) Variation of the surface roughness with the Almen intensity, R_a arithmetic mean, R_y maximum value. (b) Microhardness/depth profiles measured after shot peening with an Almen intensity of 0.40 mm N. (c) Variation of the residual compressive stresses with increasing distance from the surface observed after shot penning with an Almen intensity of 0.40 mm N. Data from Lindemann et al. [9].

18.3
Surface Deformation Due to Shot Peening

In summary: Shot peening of TiAl alloys increases both the surface roughness and near-surface hardness. Compressive stresses of a few hundred MPa are also introduced and extend to a depth of about 200 μm beneath the surface.

The structural changes induced by shot peening are demonstrated for the two materials in Figures 18.4 and 18.5 [9]. There is a clearly marked deformation zone underneath the peened surface, where structural changes have taken place. This zone extends over a length scale of about 10 μm in alloy 1 and 80 μm in alloy 2, again reflecting the different yield stresses of the two materials. In any case, the deformed layer is significantly smaller than the compressive zone indicated by the microhardness and the stress profiles shown in Figure 18.3. To maintain an equilibrium of internal forces inside the specimen, the compensating compressive stresses should extend beyond the depth of the deformed layer. At the microscopic

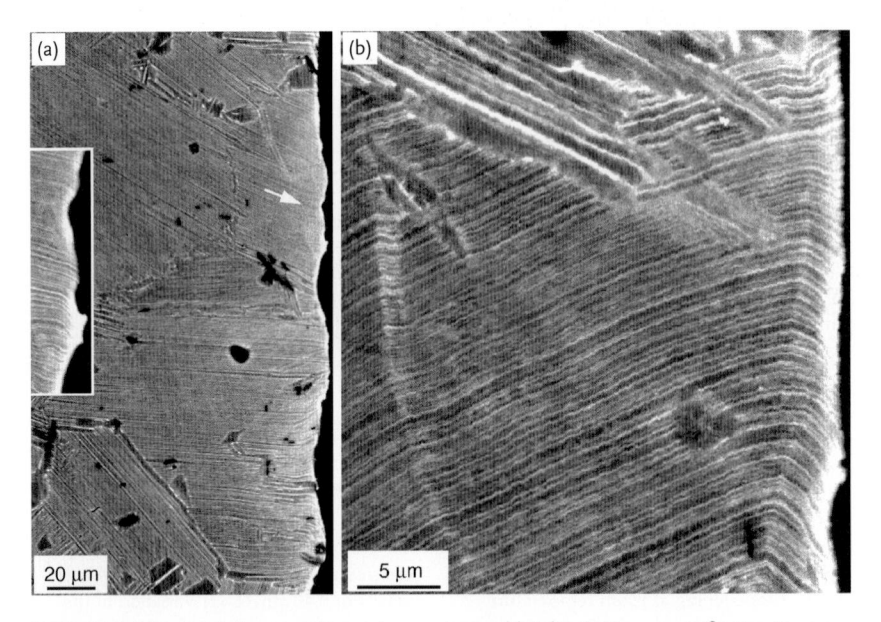

Figure 18.4 Structural changes observed on material 1 subject to shot peening at room temperature with an Almen intensity of 0.4 mm N. Cross-sectional view of a round sample imaged by backscattered electrons. The peened surface is at the right-hand side of the micrographs; the α_2 lamellae are imaged bright. (a) Low-magnification image showing the relatively rough shot-peened surface. The insert shows the region marked by the arrow in higher magnification. (b) Kinking and bending of lamellae in the subsurface region. Micrographs from Lindemann et al. [9].

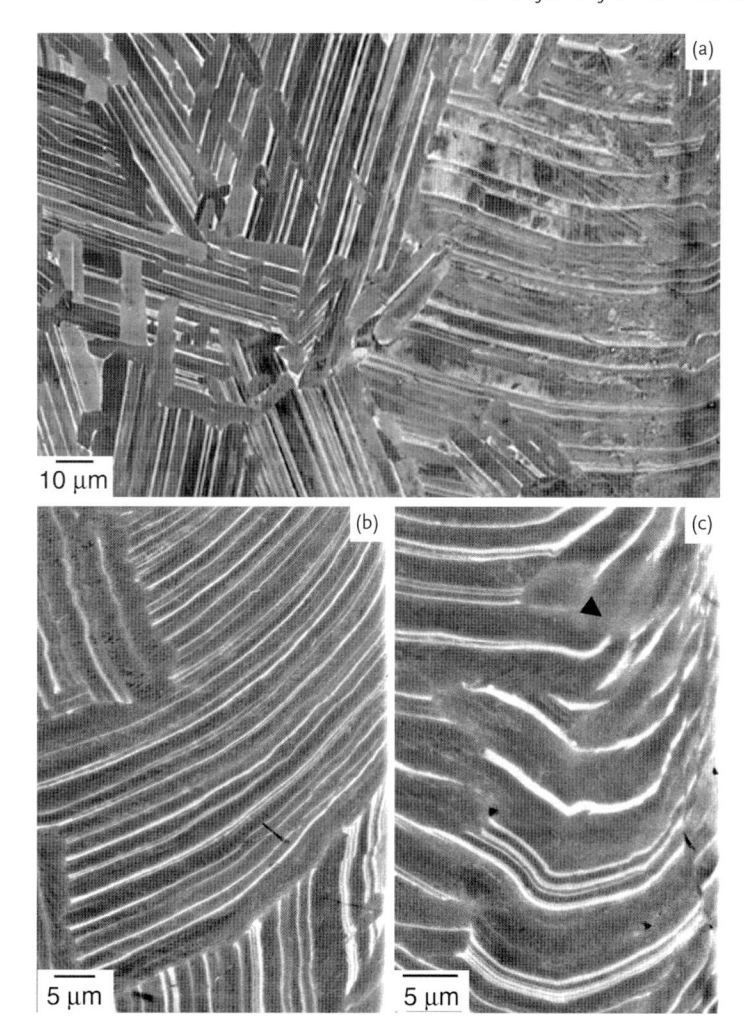

Figure 18.5 Structural changes observed on material 2 (Table 18.1) subject to shot peening at room temperature with an Almen intensity of 0.4 mm N. Cross-sectional view of a round sample imaged by backscattered electrons. The peened surface is always at the right-hand side of the micrographs; the α_2 lamellae are imaged bright. (a) Near-surface region showing a heavily deformed material layer separated from undeformed bulk material. (b) Bending of lamellae underneath the surface. (c) Speroidization and separation of α_2 lamellae at positions of strong bending and kinking of the lamellae. Micrographs from Lindemann et al. [9].

scale there is a remarkable difference in the surface roughness of the two materials; although this could not be detected by the roughness measurements (Figure 18.3a), but is evident from Figure 18.4a. The peened surface of material 1 exhibits many fine dimples, which are not seen on material 2. This observation suggests that materials with lower yield stresses are more capable to absorb the impact of the shot particles by plastic deformation. This again demonstrates that the peening parameters for brittle materials have to be carefully chosen.

Directly beneath the surface bending and kinking of the lamellae occurs, as demonstrated for the two materials in Figures 18.4b and 18.5. It is interesting to note that the overall curvature of the lamellae is unique to the layer at the sample periphery and related to the rotation direction of the sample, an observation that can be rationalized by a simple kinematical consideration. Although the shot stream is perpendicularly oriented to the surface, a tangential component is superimposed by the rotation of the sample. Thus, the net impulse of shot particles is inclined to the surface. Microscopically, bending of the γ lamellae can be realized through the accumulation of an excess number of mechanical twins and dislocations of the same sign, as has been observed in various related bending phenomena. In alloy 2, the curvature (20 μm to 100 μm) and thickness (up to 5 μm) of the γ lamellae indicate a minimum local strain of 5% to 25%. There are good arguments that the α_2 lamellae were only elastically bent. This is because the α_2 lamellae are very thin (about 100 nm) and as the α_2 phase in two-phase alloys scavenges interstitial impurities like oxygen, nitrogen or carbon. Thus, significant solution hardening of the α_2 phase can occur, which makes plastic deformation difficult (Section 6.2.2). From the above-quoted data and the Young's modulus $E = 146$ GPa of α_2(Ti$_3$Al) [11], a local stress of about 150 MPa to 700 MPa can be deduced, which, given the uncertainties of the assumptions, is consistent with the data shown in Figure 18.3c. α_2 lamellae that are nearly parallel and close to the surface often exhibit a wavy appearance (Figure 18.5b), which indicates the strong mechanical impact of the bombarding particles.

Using backscattered electron (BSE) contrast, the deformation in alloy 2 also becomes apparent through the intensive mechanical twinning (Figure 18.5). Twinning apparently occurs on all three $\{111\}_\gamma$ planes that are inclined to the $(111)_\gamma \| (0001)_{\alpha 2}$ interfacial planes. The strong strain contrast, which is observed in untwinned regions, also suggests that dislocation glide contributes to deformation.

It is only recently that TEM examination of the peened surface region has been undertaken [12]. These investigations were performed on alloy 3, which represents a high-strength TNB alloy. Electropolished coupons of these alloy were shot peened as described in Section 18.1. Target preparation was performed mainly from the backside of the peened surface utilizing a combination of jet polishing and ion milling. This allowed positioning of the electron-transparent region to regions within 3 μm to 10 μm beneath the peened surface. This position corresponds to the distance where the residual stress and the hardness are highest. TEM examination of these samples revealed a largely destroyed microstructure. Figure 18.6 shows a remnant γ lamella with only a few small islands of yet perfect crystal structure. The α_2 phase was found to be heavily deformed (Figure 18.7), which can be recognized in the compressed version of the micrograph shown below. Deformation is manifested by an offset of the $(0001)_{\alpha 2}$ basal planes. This indicates that dislocation glide with a Burgers vector component perpendicular to the basal plane occurred, which under conventional deformation is extremely difficult (Section 5.2.2). A characteristic deformation feature of the γ phase is a dense structure of dislocation loops, which in diffraction contrast appear as white con-

18.3 Surface Deformation Due to Shot Peening | 715

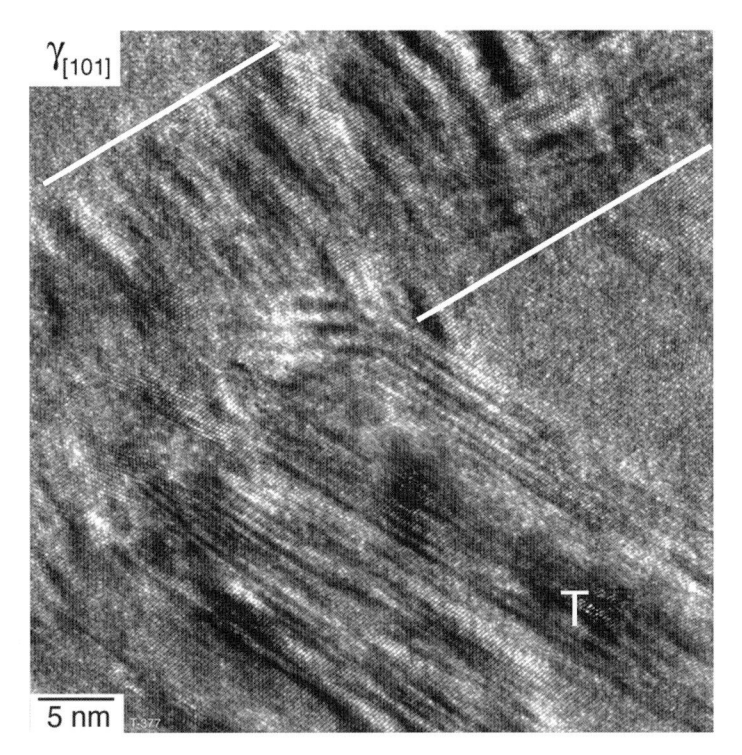

$\gamma_{[101]}$

5 nm

Figure 18.6 High-resolution electron micrograph showing remnant lamellar structure in the shot-peened surface layer. Note the twin band (T), which was apparently emitted from the lamellar interface. Lamellar Ti-45Al-10Nb (alloy 3, Table 18.1) shot peened at room temperature with an Almen intensity of 0.4 mm N.

trast dots (Figure 18.8a). Examination by high-resolution microscopy revealed that these loops originate from dislocations that are arranged in a dipole configuration (Figure 18.8b).

Mechanical twinning of the γ lamellae is another prominent deformation mechanism, as shown in Figure 18.9a. The twinning partial dislocations seem to interact with the dislocation loops and point-defect clusters; an estimation based on Frank's rule [13] indicates a 50% energy reduction, which provides a strong binding of the defects. Thus, the twins appear decorated by the point-defect clusters (Figure 18.9b). It is important to mention that foil preparation utilizing focused ion beam cutting (FIB) was unsuitable because fine structural details were lost due to excessive ion damage and recovery.

In summary: Shot peening of TiAl alloys produces a heavily deformed surface layer with a thickness of 10 μm to 80 μm depending on the microstructure and yield stress of the substrate alloy. Deformation is characterized by intensive glide and mechanical twinning, involving all potential slip systems available in the major phases α_2(Ti$_3$Al) and γ(TiAl). On the mesoscopic scale, buckling and kinking of the lamellae manifest deformation.

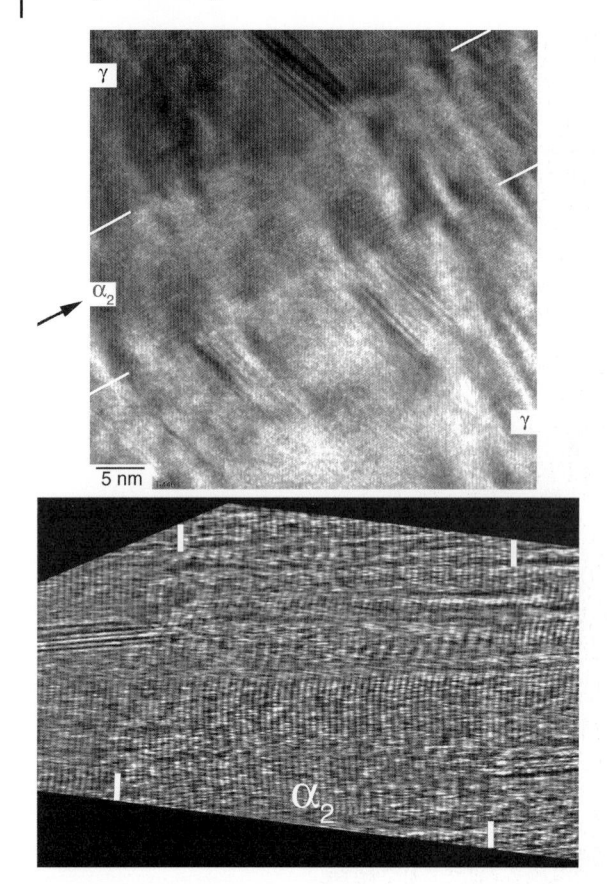

Figure 18.7 A heavily deformed α_2 lamella present in the shot-peened surface layer. Lamellar Ti-45Al-10Nb (alloy 3, Table 18.1) shot peened at room temperature with an Almen intensity of 0.4 mm N. The lower figure is the same micrograph compressed along the $(0001)_{\alpha2}$ basal planes, as arrowed. Note the offset of the $(0002)_{\alpha2}$ planes indicating that glide with a **c**-component Burgers vector occurred.

18.4
Phase Transformation, Recrystallization, and Amorphization

The deformation of the surface layer is accompanied by a significant conversion of the microstructure involving dynamic recrystallization and phase transformation. This is evident in the SEM images by the very fine grains that are nucleated in the γ lamellae and by the spheroidization and dissolution of α_2 lamellae (Figure 18.5c) [9]. The phase transformation starts with a splitting of α_2 lamellae at positions of strong bending or kinking, where the elastic stresses are highest. More details of these processes were obtained by TEM examination. Accordingly, a prominent damage mechanism is amorphization. As shown in Figure 18.10a,

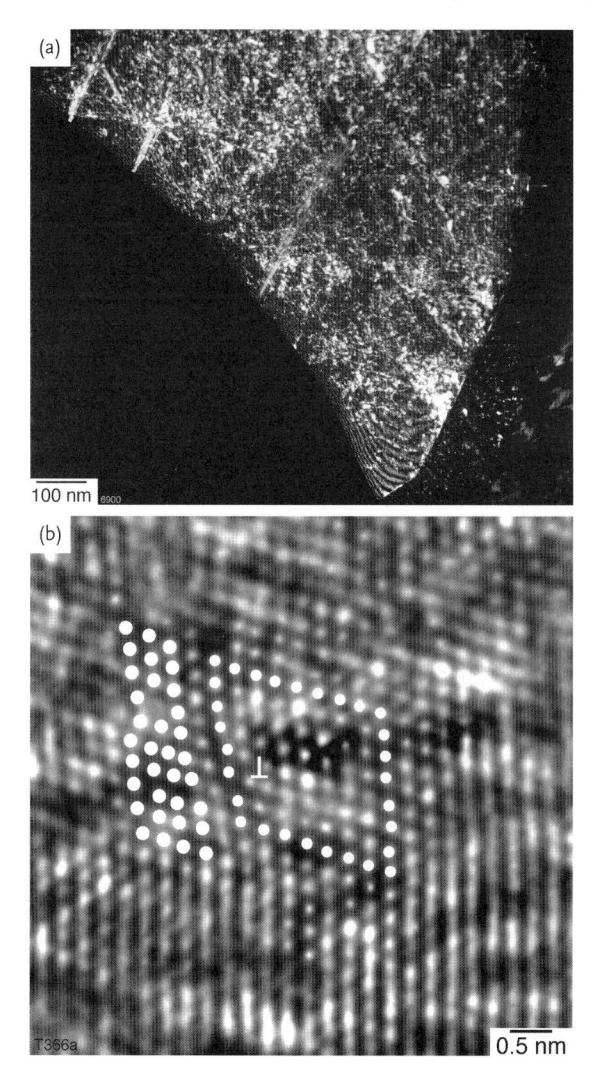

Figure 18.8 Point-defect structures observed in shot peened lamellar Ti-45Al-10Nb. (a) TEM weak-beam micrograph showing dislocation loops and point-defect clusters in heavily deformed γ phase. (b) A dislocation dipole situated in a heavily deformed γ lamella image by high-resolution electron microscopy.

nanocrystalline grains are embedded in an almost featureless amorphous phase; Figure 18.10b demonstrates the gradual loss of crystallinity towards to the adjacent amorphous phase. There seems to be a significant mismatch between the crystalline and amorphous phases, which is indicated by a systematic array of like dislocations (Figure 18.11). From the dislocation separation distance it may be concluded that the mismatch is at least 2% to 5%.

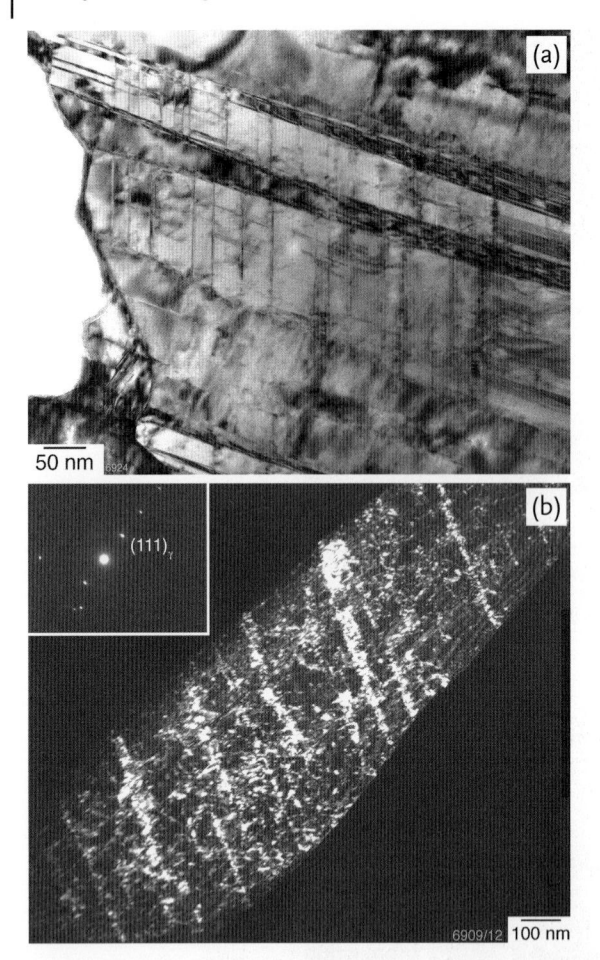

Figure 18.9 Deformation structure within the shot-peened surface layer. Lamellar Ti-45Al-10Nb (Alloy 3 in Table 18.1), shot peened at room temperature with an Almen intensity of 0.4 mm N. (a) Deformation twins in a γ lamella. (b) Mechanical twins decorated with dislocation debris and point-defect clusters.

The observation of an amorphous phase is surprising. However, there are a few arguments that make its existence plausible. First, in the surface layer the material undergoes severe plastic deformation. This introduces various defects, raises the free energy, and creates fresh surfaces due to the formation of slip steps and localized cracking. Secondly, there is certainly a substantial pick up of nitrogen, oxygen, and perhaps hydrogen because the shot peening was performed in air. Thus, several metastable nitride, oxide and hydride phases can be formed. The presence of these interstitial elements in the α_2 and γ phases may favor their amorphization. Unfortunately, the nature of the crystalline grains embedded into the amorphous phase could not be determined, but it might be speculated that they are oxide,

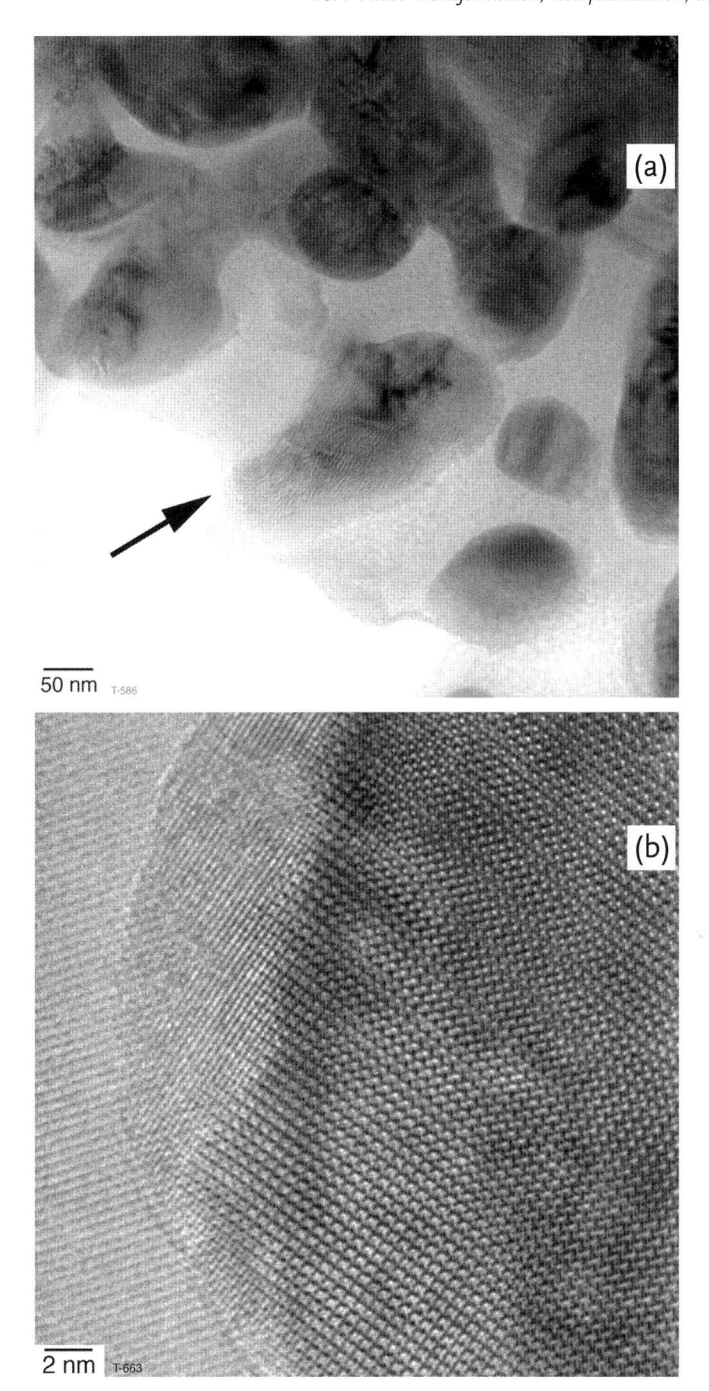

Figure 18.10 Partial amophization in the shot-peened surface layer. Lamellar Ti-45Al-10Nb (Alloy 3 in Table 18.1), shot peened at room temperature with an Almen intensity of 0.4 mm N. (a) Crystalline grains embedded in amorphous phase. (b) Gradual loss of crystallinity of a grain adjacent to the amorphous phase.

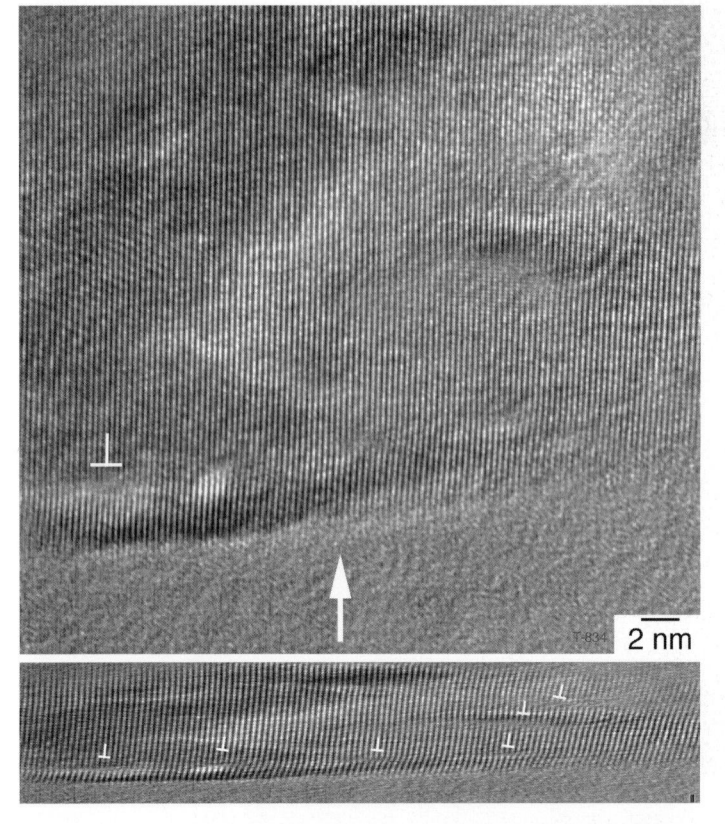

Figure 18.11 Misfit between the amorphous and crystalline surface phases produced by shot peening. Note the arrangement of like dislocations at the interface. The compressed image below shows these details more clearly. Lamellar Ti-45Al-10Nb (Alloy 3 in Table 18.1), shot peened at room temperature with an Almen intensity of 0.4 mm N.

nitride or hydride phases. The formation of one of these crystalline surface phases could be an intermediate state before amophization eventually starts. All these phenomena are well known from ball milling of Ti-Al alloys [14], a process that is similar to shot peening. A detailed analysis of the phase evolution has shown that ball milling is associated with a significant chemical disorder and the formation of metastable phases [15]. However, the mechanistic details are largely unknown, thus a few details will be presented, which are believed to be of general importance for the amorphization of crystalline phases. The initial process is probably the accumulation of dislocations, which are often arranged in dipole or multipole configurations, as demonstrated in Figure 18.12. The compressed image reveals significant bending of the lattice planes, which suggests high internal stresses. Taken together, the multipole configuration and the high internal stresses make such defect configuration prone to further structural changes. A remarkable feature of the crystalline surface phase is antiphase boundaries (APB), as seen in

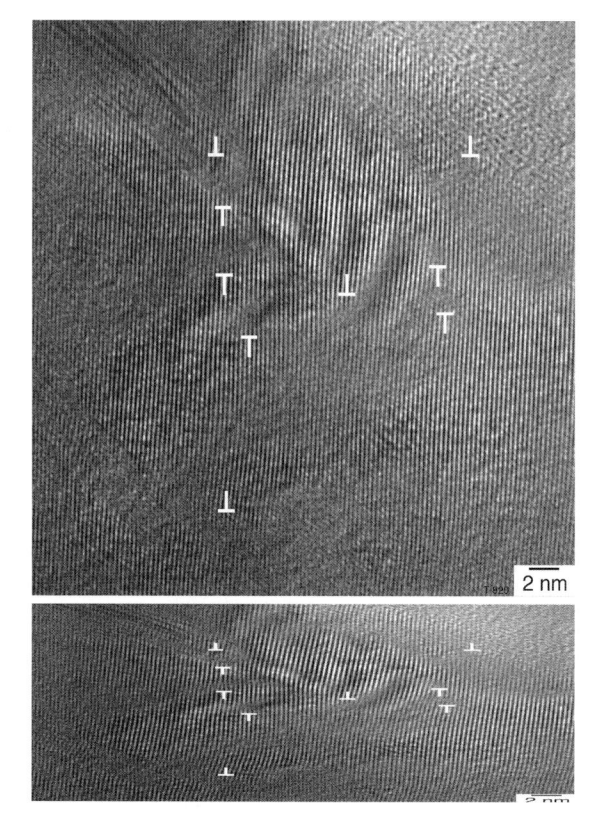

Figure 18.12 Dislocation accumulation in a crystalline grain embedded into the amorphous phase. The dislocations are arranged in dipole and multipole arrays. Note the bending of the lattice planes, which indicates high internal stresses. The compressed image below shows these details more clearly.

Figure 18.13. The APBs provide a local loss of order and are often associated with adjacent amorphous phase. Thus, it is speculated that the formation of APBs represent the initial stage of amorphization.

In summary: Shot peening of TiAl alloys leads to significant structural changes in the cold-worked surface layer. These involve $\alpha_2 \rightarrow \gamma$ phase transformation, dynamic recrystallization, amorhization, and formation of a yet unknown nanocrystalline phase, perhaps, due to the pick up of oxygen, nitrogen or hydrogen.

18.5
Effect of Shot Peening on Fatigue Strength

The fatigue performance that is obtained after shot peening is demonstrated by fatigue life (S-N) curves in Figure 18.14 [9]. The increase of the fatigue strength

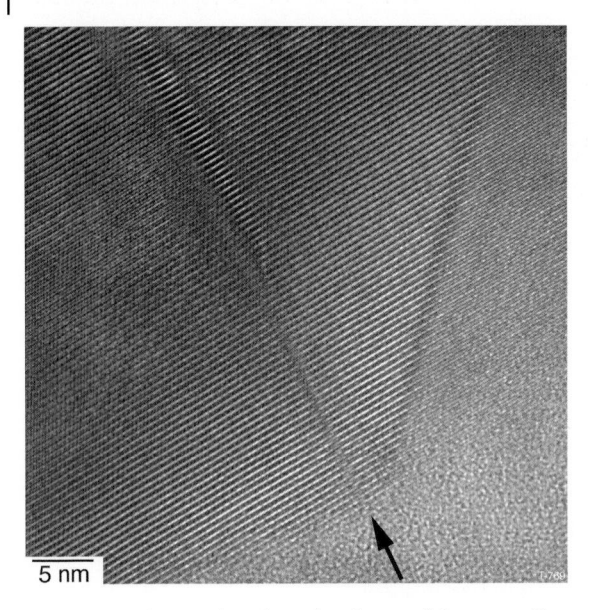

Figure 18.13 An antiphase boundary (arrowed) in a crystalline grain embedded into the amorphous surface layer produced by shot peening.

for 10^7 cycles is 300 MPa for alloy 1 with fine lamellae and 125 MPa for alloy 2 with coarse lamellae, when compared with their electropolished reference state. This difference is of particular interest because at first sight, it does not seem to be in keeping with the commonly held concept that the material with the higher compressive stresses should exhibit the better fatigue performance. In the two materials, the compressive layers are certainly deep and strong enough so that the damaging effects of notches and flaws are shifted to subsurface regions in correspondence to the depth of the compressive zone. This has been indeed observed [9]. It is tempting to speculate that under these conditions the fatigue behavior is largely governed by the microstructure and the relevant deformation mechanisms occurring in the bulk material. Lamellar TiAl alloys generally exhibit superior fatigue strength, when compared with their duplex or equiaxed counterparts (see Section 11.3.1). There is good evidence that initiation and propagation of fatigue cracks in lamellar materials are determined by the plastic anisotropy of the colonies and the cleavage fracture on {111} planes [16]. Thus, in unfavorably oriented colonies, high constraint stresses can be developed, at which internal cracks may nucleate. This, together with the easy crack propagation parallel to the interfacial {111} planes, certainly limits the fatigue resistance of lamellar material. The fracture surfaces of the fatigued samples often exhibited flat and easily identifiable initiation sites at internal lamellar colonies confirming this view. As described in Section 11.3.1, there is no unique view in the TiAl literature about the effects of colony size and lamellar spacings on the fatigue behavior because it is often difficult to vary these two structural parameters independently of each other. However,

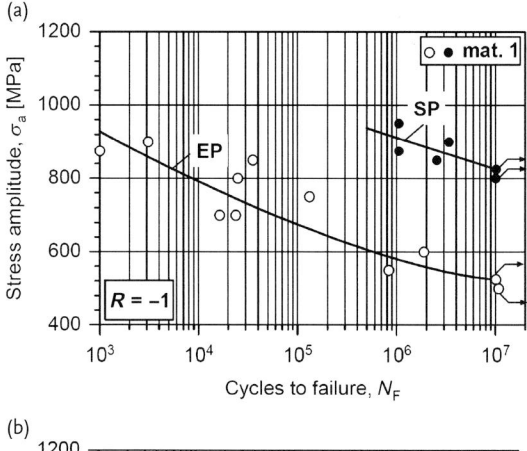

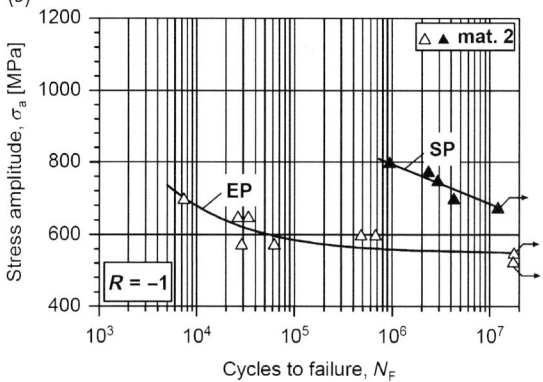

Figure 18.14 Influence of shot peening on the fatigue behavior; EP-electropolished reference samples, SP-samples that were shot peened at room temperature with an Almen intensity of 0.4 mm N; (a) alloy 1, (b) alloy 2. Data from Lindemann et al. [9].

in cases were this variation was possible, it has been recognized that the fatigue performance is increased by small colony sizes and fine lamellar spacings. Thus, the better response of alloy 1 to shot peening may be ascribed to its refined lamellar structure. There is good supporting evidence that the fatigue crack-growth resistance of TiAl alloys degrades by environmental attack (Section 11.4). In this respect, a shift in crack initiation sites to the interior of the specimens might be beneficial because the fatigue cracks then propagate in a more benign environment. It should be noted that essentially the same effects of surface hardening on the fatigue performance were recognized on a high Nb-containing alloy with the composition Ti-45Al-9Nb-0.2C (TNB-V2) [10].

Finally, it should be noted that roller burnishing [10] is equally effective at introducing deep residual compressive stresses that appreciably improve the fatigue performance. Several authors [5, 8, 17] have also noted that conventional turning and grinding generally harden the material to depths ranging up to 200 μm, depending on machining conditions. The compressive stresses developed in machined samples have almost the same value as found after shot peening [8].

In summary: Shot peening of TiAl samples produces a crack initiation resistant layer in the outer 10 μm to 80 μm. The presence of the compressive stresses shifts crack initiation to the sample interior. Taken together, these factors greatly improve the fatigue life at room temperature.

18.6
Thermal Stability of the Surface Hardening

Unfortunately, the surface hardening achieved by shot peening is prone to thermal recovery, which is detrimental in view of the anticipated high-temperature application of TiAl alloys. A 50-h anneal at 650 °C leads to a significant reduction of the microhardness, compressive residual stress and fatigue strength, as demonstrated in Figure 18.15 [9]. In the fine lamellar alloy 1, the recovery even leads to the formation of tensile stresses, which are particularly detrimental for the fatigue life. It should be mentioned that this thermal instability was also recognized when the surface hardening was performed by roller burnishing [16]. Thermal relaxation of surface hardening is a widely observed phenomenon in conventional metals, but

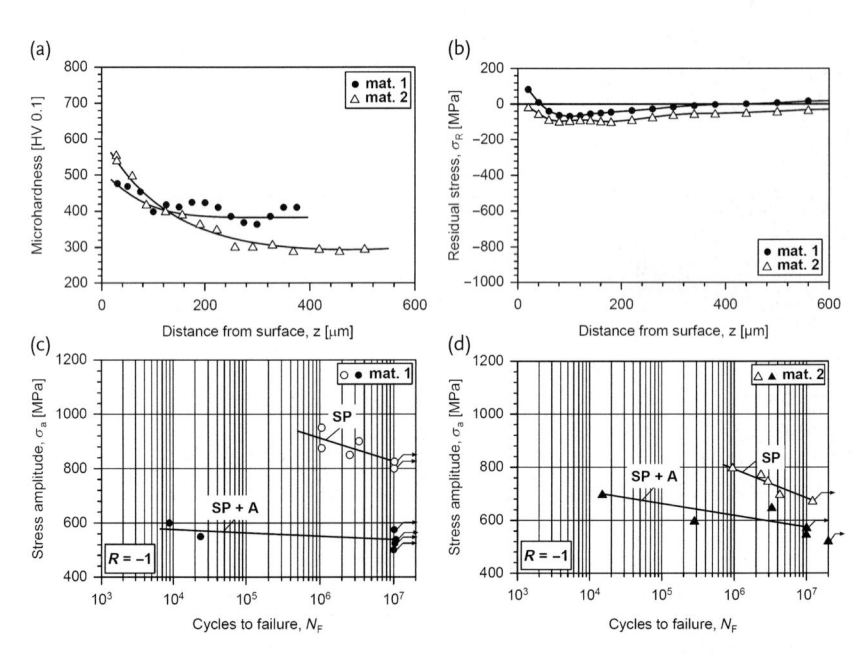

Figure 18.15 Recovery behavior of the surface hardening produced by shot peening of materials 1 and 2 after a 50-h anneal at 650 °C; (a) microhardness, (b) residual compressive stress, (c, d) fatigue strength. SP-shot peened samples, SP+A-shot peened plus annealed samples. Data from Lindemann et al. [9].

at present there is insufficient knowledge about the mechanisms involved. Concerning TiAl alloys, a particular problem is that relaxation of the surface hardening occurs at relatively low temperature, where conventional diffusion mechanisms are still ineffective. The thermal relief of the compressive surface stresses parallels the recovery of the work hardening obtained by conventional compression tests (see Section 7.2.4). Furthermore, a common feature in the defect structure of the two deformation modes is the presence of dislocation debris and point-defect clusters. These defects represent a small crystal volume and are expected to be easily annealed out. Recovery of surface hardening may additionally be supported by the high internal stresses and the presence of a significant point-defect supersaturation. Stress relief may also result from structural changes occurring in the mixed amorphous and crystalline phases. As already mentioned the crystalline grains have a dense dislocation structure, which appears to be highly unstable. Recombination of these dislocations was observed *in situ* during TEM observation, as demonstrated in Figure 18.16 by a sequence of high-resolution micrographs. It

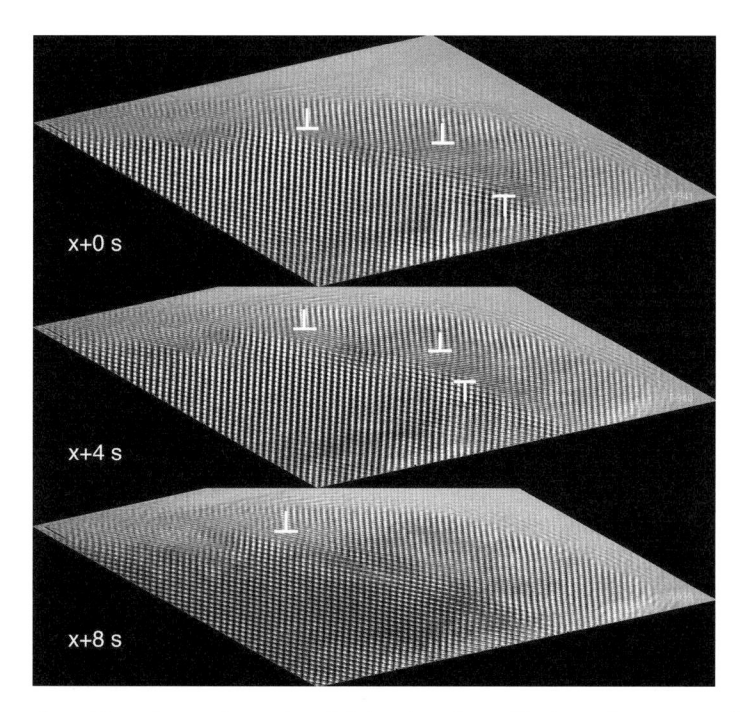

Figure 18.16 *In-situ* observation of dislocation recovery within a crystalline grain embedded into the amorphous phase of the shot-peened surface layer. Three dislocations are marked by symbols, stage (i). The isolated dislocation on the left-hand side is relatively immobile. The two other dislocations are in a dipole configuration and propagate towards each other and are about to annihilate in stage (ii). Stage (iii) shows the situation after annihilation of the dipole dislocations; this is indicated by the continuous trace of the lattice planes. Observations made at room temperature with an acceleration voltage of 300 kV. It should be noted that the micrographs were slightly compressed along the vertical direction in order to make the dislocations readily visible.

seems rather unlikely that these dislocations propagated solely under the influence of image forces because they have a large Burgers vector component within the foil plane. The dislocation motion could be driven by the internal stresses that are present in the nanosized grains or by the constraint stresses that obviously exist between the crystalline and amorphous phases (Figure 18.11). Similarly, lattice transformation has been observed *in situ*, giving rise to the assumption that the crystalline phase embedded within the amorphous phase is thermally highly unstable [12].

Clearly, the thermal instability of the surface hardening restricts a broader application of shot peening for TiAl components. However, there are several applications for components that operate at intermediate temperature and require good fatigue strength; connection rods in automotive engines are a prime example. Shot peening may also be used to restore the fatigue debits produced by electro-discharge machining (EDM), an economical method that is widely applied in the manufacture of TiAl components. The heat generated to discharge molten metal results in a solidified recast layer on the base material. This layer can be brittle and may exhibit high tensile stresses, which are harmful for the fatigue performance. The question arises as to whether the problems associated with the thermal instability could be overcome when the surface hardening is produced with minimal surface damage. For example, laser shock peening introduces a residual stress field by propagating shock waves. The shock waves result from a plasma burst that originates from a concentrated laser pulse. The primary benefit of the process is a very deep compressive layer with minimal cold working. Thus, thermal relaxation of laser-induced stresses could be less than that from mechanical shot peening.

In summary: Surface hardening induced by shot peening or roller burnishing significantly recovers upon annealing at temperatures of 650 °C to 750 °C. Thus, shot peening is an effective means for improving the fatigue performance for applications at intermediate temperature, but is less effective for components that will be used in high-temperature service.

References

1 Herrmann, D. (2009) *Diffusionsschweißen von γ(TiAl)-Legierungen: Einfluss von Zusammensetzung, Mikrostruktur und mechanischen Eigenschaften*. PhD thesis. Technical University Hamburg-Harburg, Germany.

2 Almen, J.O., and Black, P.H. (1963) *Residual Stresses and Fatigue in Metals*, McGraw-Hill, Toronto, p. 64.

3 SAE Surface Enhancement Division (2008) SAE Standard J442. *Test Strip, Holder, and Gage for Shot Peening*.

4 SAE Surface Enhancement Division (2003) SAE Standard J443. *Procedures for Using Standard Shot Peening Test Strip*.

5 Jones, P.E., and Eylon, D. (1999) *Mater. Sci. Eng. A*, **263**, 296.

6 Lindemann, J., Roth-Fagaraseanu, D., and Wagner, L. (2001) *Structural Intermetallics 2001* (eds K.J. Hemker, D.M. Dimiduk, H. Clemens, R. Darolia, H. Inui, J.M. Larsen, V.K. Sikka, M. Thomas, and J.D. Whittenberger), TMS, Warrendale, PA, p. 323.

7 Lindemann, J., Roth-Fagaraseanu, D., and Wagner, L. (2003) *Gamma Titanium Aluminides, 2003* (eds Y.-W. Kim, H. Clemens, and A.H. Rosenberger), TMS, Warrendale, PA, p. 509.
8 Wu, X., Hu, D., Preuss, M., Withers, P.J., and Loretto, M.H. (2004) *Intermetallics*, **12**, 281.
9 Lindemann, J., Buque, C., and Appel, F. (2006) *Acta Mater.*, **54**, 1155.
10 Lindemann, J., Glavatskikh, M., Leyens, C., Oehring, M., and Appel, F. (2007) *Titanium-2007 Science and Technology* (eds M. Niinomi, S. Akiyama, M. Hagiwara, M. Ikeda, and K. Maruyama), The Japan Institute of Metals, Sendai, p. 1703.
11 Schafrik, R.E. (1977) *Metall. Trans. A*, **8A**, 1003.
12 Appel, F. (2001) A High-Resolution Electron Microscope Study of Defect Structures in TiAl Alloys Produced by Shot-Peening, *Acta Mater*. To be published.
13 Hirth, J.P., and Lothe, J. (1992) *Theory of Dislocations*, Krieger, Melbourne.
14 Suryanarayana, C. (2001) *Prog. Mater. Sci.*, **46**, 1.
15 Klassen, T., Oehring, M., and Bormann, R. (1997) *Acta Mater.*, **45**, 3935.
16 Huang, Z.W., and Bowen, P. (1999) *Gamma Titanium Aluminides 1999* (eds Y.-W. Kim, D.M. Dimiduk, and M.H. Loretto), TMS, Warrendale, PA, p. 473.
17 Zhang, H., Mantle, A., and Wise, M.H.L. (1996) *Titanium'95: Science and Technology* (eds P.A. Blenkinson, W.J. Evans, and H.M. Flower), Institute of Materials, London, p. 497.

19
Applications, Component Assessment, and Outlook

19.1
Aerospace

19.1.1
Aircraft-Engine Applications

Gilchrist and Pollock [1] have presented a design case study for LPT blades made from cast TiAl, where the implementation of gamma could potentially lead to significant engine weight savings. Not only are the gamma blades lighter than their superalloy counterparts; but also a follow-on weight saving in the disc can be made as the centrifugal forces exerted on the disc by the lighter blades are reduced. The conservative nature of engine design would of course not simultaneously combine such major blade developments and lean discs designs, as this would have very significant consequences if the gamma blades had to be removed from the engine platform and replaced with conventional heavier nickel-based blades. Nevertheless, successful implementation in blade applications would be the first step towards such a goal.

Most published work on engine applications has concerned low-pressure (LPT) and high-pressure turbine (HPT) blade applications. GE has achieved a major success with TiAl implementation and certification in the new GEnx-1B engine that has been developed to power the Boeing 787 Dreamliner that is due to enter service in 2011. At the TMS annual meeting in New Orleans (2008), Dr. Tom Kelly of GE announced in his presentation that the last 2 stages of the GEnx-1B low-pressure turbine would be made from cast blades of "48-2-2" (casting by PCC). It was also mentioned that the alloy "48-2-2" is "only a name" possibly indicating that the composition used to make the LPT blades could be somewhat different. It is believed that these blades will be cast oversize and then machined to their final geometry, which of course will result in additional costs. Nevertheless, with the successful introduction of gamma into the GEnx-1B engine and its certification, a large "confidence barrier" with respect to both technical and economic issues has been broken, although it remains to be seen how gamma performs under long-term normal day-to-day service conditions. GE has invested much time, effort and resources into their gamma technology and is the first engine

Gamma Titanium Aluminide Alloys: Science and Technology, First Edition. Fritz Appel, Jonathan David Heaton Paul, Michael Oehring.
© 2011 Wiley-VCH Verlag GmbH & Co. KGaA. Published 2011 by Wiley-VCH Verlag GmbH & Co. KGaA.

producer to have placed enough trust in the material for its implementation. Anyone interested or involved in gamma TiAl can only wish GE success, as failure (especially for technical reasons) would be a major setback from which gamma may take a long time to recover.

This great achievement by GE has been built on previous TiAl casting experience [2] and the successful testing (over 1000 simulated flights) of a full set of 98 stage-five cast low-pressure turbine blades in a CF6-80C2 engine [3]. After this endurance test there were no major signs of blade damage, although it has been reported that several blades showed cracking during assembly [1]. It is also reported that the full set of 98 blades was reinstalled on a later engine test and performed without incident for another 500 cycles. The program demonstrated satisfactory component properties for tensile strength, high-cycle fatigue, dovetail low-cycle fatigue, impact resistance, wear, and hot corrosion [4].

Work on high-pressure turbine blades has mainly been performed in Germany within nationally funded programmes involving Rolls Royce (Germany), GKSS and other companies and institutes. Specifically, two programmes have been undertaken, both of which involved extrusion of ingots followed by isothermal forging to near-net shape and then electrochemical milling. The first program successfully produced over 200 HPT blades mostly from a conventional strength TiAl alloy, Figure 16.54 [5]. After production, however, the planned engine test was not performed. The second program built upon the knowledge gained in the first and employed large ingots of a high-strength, high niobium-containing alloy. Currently, work to produce engine parts by casting alone and casting followed by forging is being performed within Germany.

Apart from blades, other components within the engine have been investigated or are envisaged. These include: stator vanes [6], exhaust components [6], combustor casings [6], radial diffusers that control the deceleration of compressor charge gas into the combustor [4, 7], transition duct beams [3, 7], and turbine-blade dampers [8].

According to Rugg [9] significant potential exists in replacing Fe- and Ni-based stator vanes with gamma TiAl. This would result in large weight savings and could involve both stator vanes that operate at relatively low temperatures (300 to 400 °C) as well as those that operate at higher temperatures. Figure 19.1 shows the different stages of manufacture of a cast HP4 stator vane presented by Walker

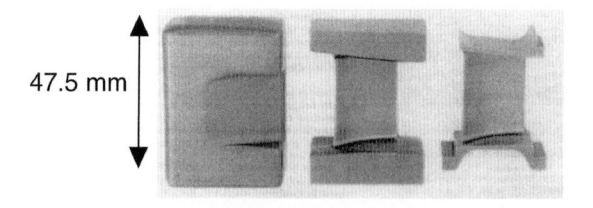

Figure 19.1 Diagram showing the various stages of manufacture of cast HP4 stator vanes [6]. From left to right: cast preform, post electrochemical milling, and the final machined stator vane.

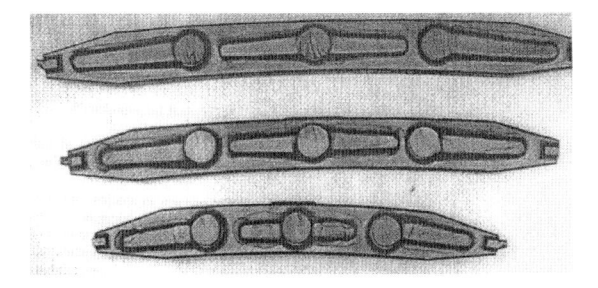

Figure 19.2 Cast transition duct beams for the GE90 engine [3]. The castings are about 15 cm long; such parts have been cast from TiAl and performed well in factory engine tests.

and Glover [6]. It is reported that such compressor vanes and blades were tested in the "CAESAR" program and that over 1000 engine cycles were successfully completed [6].

GE has also identified and tested transition duct beams made of TiAl, Figure 19.2 [3]. These beams stiffen a flow-path panel within the engine against loads induced by engine stalls [3]. Such components are up to 20 cm in length and have a width of up to 6 cm [7]. Although this is a high strain-rate loading application, where gamma may not seem an obvious choice of material, factory tests of beams installed on two GE90 engine are reported to have performed well [3].

Volvo has investigated the use of TiAl as a high-pressure turbine damper material [8]. The idea was to replace the conventional nickel-based material with a cast Ti-47Al-2Nb-2Mn + 0.8 vol.% TiB_2 XD alloy. As part of the program, an engine test was performed using TiAl for 76 dampers and also 19 of the 76 conventional high-pressure turbine blades. The blades were instrumented with strain gauges so that dynamic blade stresses could be determined. A 214-h endurance test was conducted. The results indicated that at low excitation frequencies a 50% reduction of dynamic stress could be achieved through the use of gamma dampers, although no improvement was observed at higher frequencies. No serious problems with the use of gamma were reported and potential for improved high-frequency damping capability through redesign of the damper geometry was thought possible.

19.1.2
Exotic Aerospace Applications

Apart from conventional aircraft engine applications a number of exotic aerospace applications have been proposed. Bartolotta and Krause [10] have reviewed the use of TiAl in future supersonic high-speed civil transport applications. Due to the need for reduced exhaust and noise pollution, it is envisaged that several components within a high-speed civil-transport propulsion system could be made from TiAl, including the divergent flap, nozzle sidewall and other structures at the back of the engine. In addition to castings, panels and structures made from sheet could play an important role. Within technology development programmes, representative parts have been manufactured and tested, with satisfactorily results being obtained [10–12]. Draper *et al.* [12] address the use of TiAl sheet structures in a hypersonic scramjet engine. In such an application gamma could potentially offer a 25 to 35% weight reduction over nickel-based superalloys in stiffener structures within the inlet, combustor and nozzle sections. Plansee, using the Gamma Met PX alloy that was developed by GKSS under the name TNB, fabricated three scramjet engine flap subelements, one of which was subsequently loaded to destruction. The actual failure load was 13% higher than the predicted value.

19.2
Automotive

Increased legislation with regard to improved fuel consumption and reduced emissions from motor vehicles requires reductions in weight, noise and pollution. In order to achieve such goals the motor industry is trying to downsize the conventional combustion engine and also increase engine performance and efficiency [13]. Thus, increased combustion gas temperatures of up to 1050 °C for petrol and 850 °C for diesel engines are required, combined with increased gas pressures and engine rotation speeds [13]. To help achieve these goals lightweight high-temperature-resistant components such as engine valves, turbocharger wheels and connection rods are required within the internal combustion engine. The manufacture and implementation of such components represents the second most important application area for TiAl alloys.

Manufacture of engine valves has been discussed in a number of papers with both casting [14–20] (Figure 14.17) and wrought material [19–22] being described. Del West, which is one of the largest producers of engine valves, has investigated manufacture via powder but found inferior levels of fatigue strength [20], while other papers note that the scatter in fatigue properties of powder-based material can be rather large [21, 23]. The benefits of lightweight valves in combustion engines are higher fuel economy, better performance and reduced noise and vibration [24, 25]. The reader is directed to the literature for an explanation of how TiAl brings about such benefits [24, 25] and information concerning how a valve operates within the engine [26].

According to Wünsch *et al.* [25], TiAl could be considered for both intake and exhaust valve applications. Intake valves can reach temperatures of up to around 600 °C while exhaust valves can reach temperatures in excess of 800 °C [25]. Thus, while intake valves could be made from titanium alloy, TiAl has a real opportunity to replace the conventional exhaust valve steels (21-4N, 21-2N), and the wrought nickel-based superalloys (Inconel 751) that are used in high-performance/heavy-duty engines [25]. The speed of the engine valve-train is limited by the mass of the intake valves, which are larger than the exhaust valves. Thus, TiAl can lead to improved engine performance when used as an intake valve material, even though titanium alloys are also suitable from a property-envelope standpoint [25]. Compared to steel valves, gamma offers a mass reduction of around 49%. Although ceramic valves made of Si_3N_4 offer a 57% reduction in mass, there are concerns about high manufacturing costs and its intrinsic brittle behavior [25].

During operation, engine exhaust valves are subject to harsh conditions including high temperatures, corrosive environments, cycling and creep loading and wear [24]. Such conditions can lead to a number of potential failure modes at different positions of the valve. During normal operation the highest stresses are developed in the valve underhead radius and the seat area [24], and are in the range of 35 to 70 MPa. However, this stress level depends not only on the valve geometry but also on the combustion gas pressure and the valve-seat alignment. In the transitional region between the stem and the valve head stresses can reach 20 MPa, but when there is significant misalignment bending stress levels of up to 250 MPa may develop [24]. In this respect, Sommer and Keijzers [20] of Del West explain that in environments where the valve is subjected to some degree of deflection rather than just axial loading, titanium alloys on account of their lower Young's modulus are better suited because fatigue stresses are reduced.

For TiAl to be successfully implemented as a valve material it must fulfill a series of requirements and characteristics as indicated in Figure 19.3 and described by Gebauer [26]. Wear is a problem that can take place at the tip due to frictional contact with the rocker arm and along the valve stem due to contact with the valve guide or cylinder head [25]. A number of papers including [14, 18–20, 24–27] mention the use of wear-resistant coatings. Compared to cast and near-net-shaped powder valves, wrought material offers higher strength (improved fatigue) combined with

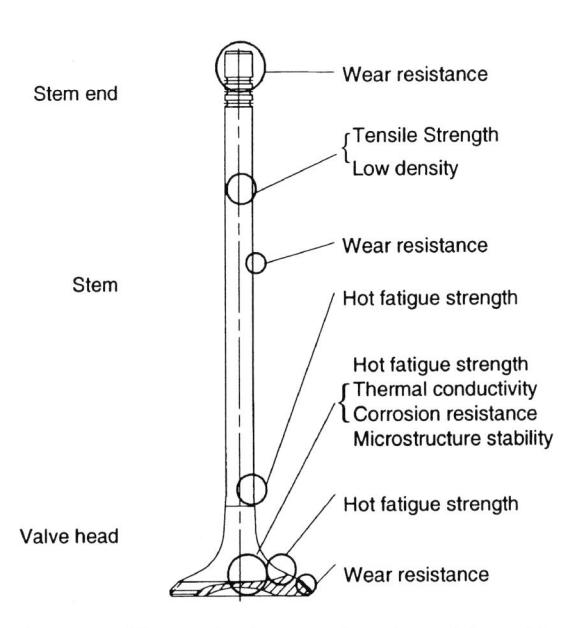

Figure 19.3 Diagram showing an engine-valve and the position specific material property requirements [26].

improved ductility. Plansee developed wrought processing combining extrusion/ double extrusion and hot upsetting [19, 21]. This technique resulted in excellent properties and was commercialized for the production of engine valves for use in Formula 1 racing [22]. Despite its technical success, FIA rule changes banned its use from 2006. Thus, while wrought valves are suitable for high-performance applications, they are comparatively expensive [13]. Although cast valves perform well in less-demanding applications, the possibility of porosity, which has been seen in cast valves [15, 18], can reduce fatigue performance [15].

A number of papers describe results of engine tests using TiAl valves [13, 14, 18, 19, 24, 25, 27–29]. Encouragingly, some of these papers have been coauthored by representatives of end-user companies including Ford [24, 25], General Motors [25, 29], and Daimler Benz/Chrysler [13, 19, 27]. Engine tests of 80 000 miles [25] and 140 000 km [13, 27] are reported. Increases in fuel efficiency of 2% [29] and in power of 8% [30] have been reported through the introduction of TiAl valves. Additionally, the number of engine revolutions before "valve jumping" occurs has been reported to increase by over 1000 rpm to levels above 14 000 rpm [17, 28]. Wear, which has sometimes been found after testing, seems the most serious problem. Various authors have classified the tip wear found as ranging from noncritical [13, 27] to extensive [24] although coating the stem and protecting the valve tip with a lash cap has been said to remedy this wear problem [24]. Minor valve-seat damage due to the indentation of hard carbon particles has also been observed [27]. While no valve failures have been reported during operation, damage during assembly and disassembly has been noted [25]. Nevertheless, according to Gebauer [26] TiAl valves are almost ready for introduction, and a cost-effective process for commercial mass production exists [14]. However, at the current time it is believed that no TiAl valves are in series production.

The second main area of interest for the application of TiAl within the automotive sector has been as a turbocharger wheel material in diesel engines, Figure 14.21. McQuay [31] reports that nearly every major turbocharger and diesel engine manufacture has successfully tested TiAl turbochargers including ABB, Honeywell Garrett and Toyota. DaimlerChrysler [13] and Mitsubishi [32–34] have also been heavily involved in development programmes. At present, the nickel-based superalloy Inconel 713C is most commonly used to manufacture diesel engine turbocharger wheels [28]. The implementation of TiAl as an alternative material would result in a reduction of particulates within the exhaust gas and reduced turbo-lag resulting from reduced inertia. A shorter response time between pressing down on the car accelerator and the car starting to accelerate is thus possible [17, 28]. The better mileage and reduced emissions would lead to a more environmentally friendly car [33]. Another benefit includes reduced noise and vibration resulting from the resonant frequencies of the rotor being shifted to higher levels [13]. An analysis of material-property requirements for a turbocharger is given in ref. [13]. Maximum service temperatures of 750 °C for diesel engines and 950 °C for petrol engines, combined with high stresses result in the need for an alloy with excellent creep resistance to maintain aerodynamic shape and good oxidation resistance. A room-temperature plastic strain to fracture of above 1% is reported to be required [13], however, in our experience such a cast alloy is unlikely to meet the other property criteria and thus a compromise towards less ductile alloys is probably necessary. After manufacture, the turbocharger wheels must be joined to the rotor shaft; for conventional turbocharger materials this is usually done by friction welding or electron-beam welding [17]. For information regarding this aspect in relation to TiAl turbochargers the reader is referred to refs. [17, 34, 35] and Section 17.2.

Results of testing cast turbocharger wheels are described in [13, 32, 33, 35]. Baur [13] reports improved performance including reduced turbo-lag when using TiAl in a Mercedes Benz C-Class C220 cdi car, particularly at low engine speeds, although the exact details are not disclosed. Compared to Inconel 713C, Noda [17] reports that using TiAl resulted in a 16% faster response time for turbocharger acceleration from 34 000 to 100 000 rpm and a 26% improvement to speeds of 170 000 rpm. The maximum speed that the TiAl turbocharger reached was 10 000 rpm faster than that of the Inconel 713C. In burst tests, a 47-mm diameter turbocharger wheel exceeded speeds of 210 000 rpm; which is 124% of the rated speed [17]. In other tests, the tip speed at bursting of an Inconel 713C was determined to be around 500 m/s while for TiAl, tip speeds in excess of 620 m/s were attained without bursting [35].

Apart from having to meet the necessary strength, fatigue, creep and oxidation resistance requirements; resistance to erosion and foreign-object damage are also important as during normal operation the blade tips can reach speeds of 420 m/s (1512 km/h) [32]. It has been shown that a high niobium-containing alloy has much better resistance to erosion compared to low niobium-containing TiAl. Figure 19.4 [33] shows that the tip of a blade made from a low niobium-containing

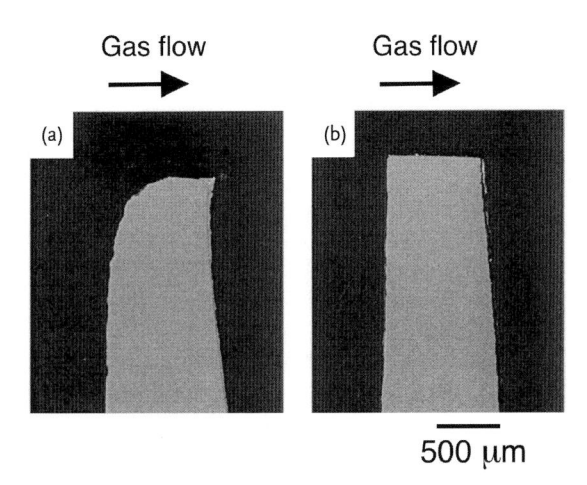

Figure 19.4 Backscattered electron micrographs of TiAl turbocharger wheel blades after testing under similar conditions [33]. The low niobium-containing alloy blade (a) shows significant loss of material due to erosion, whereas the blade made from a high niobium-containing alloy (b) shows little if any material loss.

alloy was significantly eroded, whereas the high niobium-containing alloy showed very little damage, possibly as a result of the very fine fully lamellar microstructure. With regard to oxidation, the engine test environment has been found to be less aggressive than normal atmospheric conditions, possibly as a result of material deposited during operation stabilizing a protective alumina layer [32], different gas partial pressures [33] or differences in microstructure/surface preparation between oxidation specimens exposed in air and the actual cast turbochargers used in the engine test [33].

The various engine tests have confirmed the suitability and advantages offered by TiAl as a turbocharger wheel material; and Mitsubishi Heavy Industries (MHI) has introduced this technology in the Lancer Evolution series of cars made by the Mitsubishi Motors Corp [34, 35]. Abe *et al.* [34] report the sale of over 8000 turbochargers between 1998 and 2000 while Tetsui [35] reports over 5000 between 1999 and around 2002 and Wu [36] reports over 20 000 cars being equipped with TiAl turbochargers in 2003. Although the costs remain higher than for conventional materials and other ways of improving the turbocharger performance exist [13], this introduction of TiAl is very satisfying from a technical viewpoint. Abe *et al.* [34] explain that the introduction of TiAl turbocharger wheels in newly developed petrol engines where the petrol is directly injected into the cylinder would significantly increase the number turbocharger wheels required, which should reduce costs. The use of turbocharger technology in other application areas, ship engines for example, could also increase the requirement for TiAl. McQuay [31] reported the successful application of a large cast TiAl turbocharger in the diesel engine of a ferry. Other components within the automotive engine that could be made from TiAl include piston heads, con-rods and rocker arms [13]. However, apart from unpublished work performed in Germany to make con-rods, it is not believed that any programmes on these components have been undertaken.

19.3
Outlook

At the time of writing this section, early 2011, TiAl alloys are about to be used in the real world within GE's commercial jet engine, the GEnx-1B. Qualification and certification procedures have been successfully completed and the late delivery of the Boeing Dreamliner aircraft is the only reason why TiAl is not already flying commercially. With this achievement and assuming that no major problems are encountered in service, then the future of TiAl within aero-engines seems assured and its utilization can only increase. This is of course good for all companies and establishments that are involved in TiAl as it translates into increased business and the need for suitably qualified and knowledgeable technical advisors and researchers. A successful aero-engine blade application would certainly further encourage work towards blade applications in power-generation turbines. Here, the introduction of TiAl can lead to significant efficiency improvements and thus make a real contribution to more environmentally friendly, resource-lean electricity production. In both of these fields TiAl will always be of interest, but will only be used when the supply chain can provide components on a cost-competitive basis with consistent microstructures and properties that are adequate for the application over the long term.

With regard to automotive combustion engines TiAl has already been used, albeit on a relatively small scale, in Formula 1 racing and by Mitsubishi as a turbocharger wheel material. However, in our view, the long-term use of TiAl in such applications is not assured. This is because of the need for, and the development of, greener alternative modes of power for ground transport, such as hydrogen/alcohol powered fuel cells and battery-driven engines. When these alternative technologies reach the market on a large scale then the days of TiAl within the automotive sector may be numbered.

References

1 Gilchrist, A., and Pollock, T.M. (2001) *Structural Intermetallics 2001* (eds K.J. Hemker, D.M. Dimiduk, H. Clemens, R. Darolia, H. Inui, J.M. Larsen, V.K. Sikka, M. Thomas, and J.D. Whittenberger), TMS, Warrendale, PA, p. 3.

2 Austin, C.M., and Kelly, T.J. (1993) *Structural Intermetallics* (eds R. Darolia, J.J. Lewandowski, C.T. Liu, P.L. Martin, D.B. Miracle, and M.V. Nathal), TMS, Warrendale, PA, p. 143.

3 Austin, C.M., and Kelly, T.J. (1995) *Gamma Titanium Aluminides* (eds Y.-W. Kim, R. Wagner, and M. Yamaguchi), TMS, Warrendale, PA, p. 21.

4 Schafrik, R.E. (2001) *Structural Intermetallics 2001* (eds K.J. Hemker, D.M. Dimiduk, H. Clemens, R. Darolia, H. Inui, J.M. Larsen, V.K. Sikka, M. Thomas, and J.D. Whittenberger), TMS, Warrendale, PA, p. 13.

5 Appel, F., Brossmann, U., Christoph, U., Eggert, S., Janschek, P., Lorenz, U., Müllauer, J., Oehring, M., and Paul, J.D.H. (2000) *Adv. Eng. Mater.*, **2**, 699.

6 Walker, N.A., and Glover, N.E. (2001) *Structural Intermetallics 2001* (eds K.J. Hemker, D.M. Dimiduk, H. Clemens, R. Darolia, H. Inui, J.M. Larsen, V.K. Sikka, M. Thomas, and J.D. Whittenberger), TMS, Warrendale, PA, p. 19.

7 Loria, E.A. (2000) *Intermetallics*, **8**, 1339.

8 Pettersson, B., Axelsson, P., Andersson, M., and Holmquist, M. (1995) *Gamma Titanium Aluminides* (eds Y.-W. Kim, R. Wagner, and M. Yamaguchi), TMS, Warrendale, PA, p. 33.

9 Rugg, D. (1999) *Gamma Titanium Aluminides 1999* (eds Y.-W. Kim, D.M. Dimiduk, and M.H. Loretto), TMS, Warrendale, PA, p. 11.

10 Bartolotta, P.A., and Krause, D.L. (1999) *Gamma Titanium Aluminides 1999* (eds Y.-W. Kim, D.M. Dimiduk, and M.H. Loretto), TMS, Warrendale, PA, p. 3.

11 Das, G., Bartolotta, P.A., Kestler, H., and Clemens, H. (2003) *Gamma Titanium Aluminides 2003* (eds Y.-W. Kim, H. Clemens, and A.H. Rosenberger), TMS, Warrendale, PA, p. 33.

12 Draper, S.L., Krause, D., Lerch, B., Locci, I.E., Doehnert, B., Nigam, R., Das, G., Sickles, P., Tabernig, B., Reger, N., and Rissbacher, K. (2007) *Mater. Sci. Eng.*, **A464**, 330.

13 Baur, H., Wortberg, D.B., and Clemens, H. (2003) *Gamma Titanium Aluminides 2003* (eds Y.-W. Kim, H. Clemens, and A.H. Rosenberger), TMS, Warrendale, PA, p. 23.

14 Blum, M., Choudhury, A., Scholz, H., Jarczyk, G., Pleier, S., Busse, P., Frommeyer, G., and Knippscheer, S. (1999) *Gamma Titanium Aluminides 1999* (eds Y.-W. Kim, D.M. Dimiduk, and M.H. Loretto), TMS, Warrendale, PA, p. 35.

15 Marino, F., Guerra, M., Rebuffo, A., Rossetto, M., and Vicario, V. (2003) *Gamma Titanium Aluminides 2003* (eds

Y.-W. Kim, H. Clemens, and A.H. Rosenberger), TMS, Warrendale, PA, p. 531.
16 Blum, M., Busse, P., Jarczyk, G., Franz, H., Laudenberg, H.J., Segtrop, K., and Seserko, P. (2003) *Gamma Titanium Aluminides 2003* (eds Y.-W. Kim, H. Clemens, and A.H. Rosenberger), TMS, Warrendale, PA, p. 9.
17 Noda, T. (1998) *Intermetallics*, **6**, 709.
18 Badami, M., and Marino, F. (2006) *Int. J. Fatigue*, **28**, 722.
19 Hurta, S., Clemens, H., Frommeyer, G., Nicolai, H.P., and Sibum, H. (1996) *Titanium '95: Science and Technology* (eds P.A. Blenkinsop, W.J. Evans, and H.M. Flower), IOM, London, UK, p. 97.
20 Sommer, A.W., and Keijzers, G.C. (2003) *Gamma Titanium Aluminides 2003* (eds Y.-W. Kim, H. Clemens, and A.H. Rosenberger), TMS, Warrendale, PA, p. 3.
21 Clemens, H., Kestler, H., Eberhardt, N., and Knabl, W. (1999) *Gamma Titanium Aluminides 1999* (eds Y.-W. Kim, D.M. Dimiduk, and M.H. Loretto), TMS, Warrendale, PA, p. 209.
22 Kimberley, W. (2006) Automotive Design and Production, June 2006.
23 Eberhardt, N., Lorich, A., Jörg, R., Kestler, H., Knabl, W., Köck, W., Baur, H., Joos, R., and Clemens, H. (1998) *Z. Metallkd.*, **89**, 772. In German.
24 Dowling, W.E., Donlon, W.T., and Allison, J.E. (1995) *High Temperature Ordered Intermetallic Alloys VI, MRS Symposium Proceedings*, vol. 364 (eds J. Horton, I. Baker, S. Hanada, R.D. Noebe, and D.S. Schwartz), MRS, Warrendale, PA, p. 757.
25 Hartfield-Wünsch, S.E., Sperling, A.A., Morrison, R.S., Dowling, W.E., and Allison, J.E. (1995) *Gamma Titanium Aluminides* (eds Y.-W. Kim, R. Wagner, and M. Yamaguchi), TMS, Warrendale, PA, p. 41.
26 Gebauer, K. (2006) *Intermetallics*, **14**, 355.
27 Baur, H., and Joos, R. (2000) *Intermetallics and Superalloys, Euromat 99*, vol. 10 (eds D.G. Morris, S. Naka, and P. Caron), Wiley-VCH, Weinheim, Germany, p. 384.
28 Tetsui, T. (1999) *Curr. Opin. Solid State Mater. Sci.*, **4**, 243.
29 Eylon, D., Keller, M.M., and Jones, P.E. (1998) *Intermetallics*, **6**, 703.
30 Loria, E.A. (2001) *Intermetallics*, **9**, 997.
31 McQuay, P.A. (2001) *Structural Intermetallics 2001* (eds K.J. Hemker, D.M. Dimiduk, H. Clemens, R. Darolia, H. Inui, J.M. Larsen, V.K. Sikka, M. Thomas, and J.D. Whittenberger), TMS, Warrendale, PA, p. 83.
32 Tetsui, T. (1999) *Gamma Titanium Aluminides 1999* (eds Y.-W. Kim, D.M. Dimiduk, and M.H. Loretto), TMS, Warrendale, PA, p. 15.
33 Tetsui, T., and Ono, S. (1999) *Intermetallics*, **7**, 689.
34 Abe, T., Hashimoto, H., Ishikawa, H., Kawaura, H., Murakami, K., Noda, T., Sumi, S., Tetsui, T., and Yamaguchi, M. (2001) *Structural Intermetallics 2001* (eds K.J. Hemker, D.M. Dimiduk, H. Clemens, R. Darolia, H. Inui, J.M. Larsen, V.K. Sikka, M. Thomas, and J.D. Whittenberger), TMS, Warrendale, PA, p. 35.
35 Tetsui, T. (2002) *Mater. Sci. Eng.*, **A329–331**, 582.
36 Wu, X. (2006) *Intermetallics*, **14**, 1114.

Subject Index

a

Acoustic emissions   204–206
Activation energy (apparent)
– creep   316–317
– constant strain rate deformation   218, 219, 282, 285
– hot working   581–584
– recovery of work hardening   264–268
– strain ageing   222–226
Adiabatic heating   575, 576
Alloy design concept   465–477, 606, 627, 662
Alloying elements   12
– B   18
– Nb   16, 471, 472, 614, 618, 651, 659
– O   19
Amorphization   716
Antiphase boundary   8, 55, 72
Antisite defects   27, 196
Applications
– Formula 1   2, 734, 737
– GEnx-1B   2, 3, 729
– high pressure turbine (HPT)   659, 660, 729, 730
– low pressure turbine (LTP)   3, 502, 729
– valves   503–507, 732, 733, 734
Argon contamination of TiAl powder   541
Athermal stress   234–240
Atomisation
– EIGA   522, 527
– PIGA   522, 524
– PREP   522, 529, 538
– REP   522, 529, 539
– rotating disc   531
– TGA   522, 526
Atom location by channeling-enhanced microanalysis (ALCHEMI)   28
Atom probe analysis   19
Avrami kinetics   589

b

Backdiffusion   41, 42
Banding (during peritectic solidification)   42
Basket-weave microstructure   36, 48, 50, 51
Bending of lamellae   597, 600
Breakaway oxidation   437
Break-down of Schmid law   99, 160
β-phase   5, 8, 11, 301–303, 473, 606, 617, 662
β/α transformation   35, 36, 44, 47, 48, 50–51
β solidifying alloys   36–38, 47, 48, 50, 51, 473
B2 ordering   8
B2 (phase, variant)   5, 8
Blackburn orientation relationship   34
Brazing   702–703
Buckling of lamellae   597, 600, 601
Burgers orientation relationship   35
Burgers vectors
– α_2(Ti$_3$Al)   106–112
– β/B2   115–118
– γ(TiAl)   72–73

c

Can design
– (for forging)   621–625
– (for extrusion)   628–630
Cavitation   609, 611–613
Castability   500, 595
Casting
– centrifugal   506–511
– countergravity   513
– directional   514
– gravity metal mold   503
– Hitchiner   513
– investment   497–499
– Levi-cast   513
– turbocharger wheel   499, 506–511, 514, 732, 734–736

Gamma Titanium Aluminide Alloys: Science and Technology, First Edition. Fritz Appel, Jonathan David Heaton Paul, Michael Oehring.
© 2011 Wiley-VCH Verlag GmbH & Co. KGaA. Published 2011 by Wiley-VCH Verlag GmbH & Co. KGaA.

Cellular morphology 42
Cellular transformation 50
Cleavage strength 360–361
Columnar grains, structures 34, 36, 45
Constitutional undercooling 42, 43, 46
Constitutive analysis 578–585
Constitutive equation 579, 607, 608
Convoluted microstrucure 64
Constraint stresses 149–156, 162–166
Cooling rate of particles 529, 530, 531, 534, 535
Crack tip plasticity 362–367
Creep
– activation energy 316–317
– Coble creep 317, 320
– creep properties of modulated alloys 346–352
– creep resistant alloys 341–346
– creep structures 320–321
– damage mechanisms 329–339, 347–352
– design margins 313–314
– effect of microstructure 322–325
– general creep behavior 314–315
– inverse creep 318
– minimum creep rate 316–317
– phase transformations 335–336, 347–352
– power law expression 316–317
– precipitation effects 339
– primary creep 325–329
– PST crystals 323–324
– dynamic recrystallization 323, 331–336, 348–351
– secondary creep 316–317
– single-phase alloys 317–319
– steady-state creep 316–317
– stress exponent 316–317
– tertiary creep 340–341
– two-phase alloys 319–322
Cross slip 73–74
Crystallographic data 7, 9, 12, 13
Crystal structures 6, 9, 13

d

Deformation behavior of $\alpha_2(Ti_3Al)$
– anomalous yield behavior 113–114
– basal slip 110–111
– prismatic slip 106–110, 168
– pyramidal slip 106–110
Deformation behavior of α phase 646
Deformation behavior of $\beta/B2$ phase 114–118
Deformation behavior of $\gamma(TiAl)$
– anomalous yield behavior 95–106
– effects of orientation and temperature 95–106
– reversibility tests 105–106
– Schmid factors 95–97
– tension/compression asymmetry 99
Deformation behavior of two-phase ($\alpha_2 + \gamma$) alloys
– effect of temperature 165–169, 171–173
– independent slip systems 169–171
– plastic anisotropy 156–161
– channeling of plastic deformation 157–161
– micromechanical modeling 161–164
– plastic incompatibility between α_2 and γ phases 168, 169
– role of α_2 phase 168–169, 172
Deformation heating (during extrusion) 628
Deformation twinning in $\gamma(TiAl)$ 90–92
Dendrite cores 39
Dendrite size 44, 45
Dendritic morphology 42
Density 26
Diffusion
– diffusivity data 26, 29–30
– antistructural bridge (ASB) mechanism 29–30
– Darken-Manning equation 29
Diffusion bonding
– bonding parameters 685–687
– dynamic recovery 694–696
– dynamic recrystallization 688–693
– effect of alloy composition 696–698
– formation of α_2 phase 688–691
– influence of bonding time and stress 698–700
– microasperity deformation 685, 688–692
– phase transformation 689–691
– porosity 696–697
– process zone 687–696
– mechanical characterization of bonds 700–701
– microhardness profile 700–701
Dilatometry 57
Discontinuous coarsening 49
Dislocations
– in $\alpha_2(Ti_3Al)$ 106–112
– in $\beta/B2$ 115–118
– in $\gamma(TiAl)$ 72–73
– ordinary dislocations 72, 76, 81
– superdislocations 73, 86–89
– superpartial dislocations 73, 86–89
Dislocation climb 232–234, 321, 583–585
Dislocation dipoles 178–181

Dislocation dissociation
– non-planar dissociations in γ(TiAl) 85–89
– planar dissociations in γ(TiAl) 81–85
– Shockley partial dislocations 73, 81, 84–85, 87–89, 91–94, 99, 106
Dislocation-interface interactions 157–158, 173–178
Dislocation locking
– defect atmospheres 222–232
– Kear-Wilsdorf locks 86–87
– roof barriers 86
– stair-rod dislocations 88
Dislocation mobility 207–232
Dislocation multiplication 178–186
Dislocation sources 178–184
Displacive transformation 56
Domain boundary 54
Duplex microstrucure 53, 60
Dynamic recrystallization 323, 331–336, 348–351, 427, 428, 585-606, 686–692
Dynamic strain ageing 222

e

Elastic constants 25–26
Elastic incompatibility between α_2(Ti$_3$Al) and γ(TiAl) phases 26
Electrical resistivity 26
Electrochemical milling 659, 662
Ellingham diagram 434, 435
Environmental embrittlement 392–393, 450–455
Equiaxed microstructure 45, 60
Eutectoid temperature 52, 60
Eutectoid transformation 11
Extrusion 627–642
– α extrusion 633
– conditions 628
– double-step 633
– equal-channel angular extrusion (ECAE) 633

f

Fatigue
– anisotropy of fatigue properties 422
– behavior of short cracks 411–413
– crack closure effects 404, 409–410
– cyclic crack resistance curve 415–418
– cyclic plasticity 422–426
– cyclic stress intensity 404
– cyclic-stress response 419–421
– effect of microstructure 408–409
– effects of stress ratio 409–410, 416
– effects of temperature and environment 413–418

– fatigue at threshold stress intensity 404, 411–413
– fatigue crack growth 407–409
– high-cycle fatigue (HCF) 407–418
– low-cycle fatigue (LCF) 403, 418–428
– Paris exponent 404, 407–408, 414
– Paris law 404, 407–408, 414
– phase transformations 426–428
– recovery of fatigue structures 425–426
– S-N curve 405–406
– stress-life behavior 405–406
Feathery structures 49, 60–62
Flow curves 574–578, 595
Flow localization 613–617
– parameter 614
Flow map 649
Flow softening 269, 574–578, 594
Forging 617–627
– α forging 626
– conditions 622, 623
– closed-die forging 658, 659
– practices 625, 626
Fracture behavior
– cleavage fracture 360–362
– crack-bridging shear ligaments 377, 381
– crack-dislocation interactions 367–369
– crack-growth resistance 376–381
– effects of dynamic strain ageing 385–386
– effects of loading rate and temperature 383–386
– effects of microstructure and texture 376–383
– effects of predeformation 387–388
– energy release rate 357, 358
– fracture anisotropy of lamellar alloys 381–383
– fracture toughness 373–376, 378, 396
– Griffith criterion 358–359
– length scales in fracture 357–359
– microcracks 371–372
– modulated alloys 388–391
– plastic zone 362–367
– PST crystals 381
– role of twinning 369–373
– statistical assessment 396–398
Free energy of activation 210

g

γ surface 76, 77, 136
Generation of dislocations
– Bardeen-Herring climb sources 183, 184
– interface related dislocation sources 184–186
– multiple cross glide 181–184
Gibbs free energy of activation 210

Glide resistance
– bulging and collision of kinks 218–221
– dislocation climb 232–234
– effect of temperature 207–217
– impurity-related defects 221
– jog dragging 220–221
– Peierls stress 221
Grain boundary allotriomorph 50, 54, 55
Grain boundary bulging 586, 655
Grain boundary mobility 585
Grain boundary sliding 584
Grain refinement 12, 44–49, 249–254, 469, 473, 475
Grain refining
– agent 44, 45
– effect 18, 44, 45, 249–251
– mechanism 44–49
Growth direction of dendrites 34, 35
Growth rate (of lamellae) 59, 60
Growth restriction factor 46
Gulliver-Scheil equation 33, 40, 41, 42

h
H-phase 470
Heterogeneous nucleation 45, 47, 57
Hall-Petch relation 249–251
High strain-rate deformation 272–273
High-temperature deformation
– activation energy 270–271
– work-hardening coefficient 269
Homogenization heat treatment 618–620
Hydrogen embrittlement 386, 453–455

i
Incoherent interface 63
Inoculant 46
Ingot breakdown 617–642
Ion implantation 448–450
Interfacial disconnection 56, 143–146
Internal stress
– incremental unloading 235–239
– effect of temperature 239
Interstitial contamination 500, 512

j
Jog dragging 178–181
Joining 683–704

k
Kink bulging 218, 221
Kinking of lamellae 597, 600, 601, 693, 712–714
Kirkendall porosity 559, 560
K_R-curve 377, 379, 380

l
Lack of independent slip systems 94, 112–114, 118
Lamellar boundaries
– α_2/γ boundary 126–128
– domain boundary 129–136
– 120° order fault 129–133
– pseudo-twin boundary 129–133
– true twin boundary 129–133
Lamellar interfaces
– atomic structure 126–133
– coherency stresses 56, 149–156
– diffraction analysis 132–135
– disconnections 143–145
– interfacial energies 136–139
– influence on recrystallization 595–603
– interfacial steps 142, 145
– misfit dislocations 144
– orientation relationships 52–56, 127, 129, 131–134
– rigid-body translation 126, 137–138
– rotational misfit 147–149
– semicoherent interfaces 139–149
Lamellae width 57, 58, 59
Lamellae spacing 57, 59, 60
Lamellar reaction, transformation 52–60
Lamellar microstructures 52–60
Lattice occupancy 12, 26, 27, 122
Lattice parameters 9, 10, 13
Lattice friction 75, 221
Ledges in phase transformation 50, 587
– migration 56
– mechanism 56, 60
Line tension of dislocations 73, 153
Liquid-phase joining 702–703
Long-period superstructures 11
Long-range stress 208, 234–239
Loss of coherency 141
Lubricant for hot working 574, 622, 623, 661

m
Machining 662
Manufacture
– of airfoils 658–662
– of automotive valves 659
– of components 658–662, 671
– of large parts 671
Martensitic transformation 50
Massive transformation 49, 50, 57, 58, 60, 63–64, 474
Mechanical alloying 565
Mechanical twinning in γ(TiAl)
– complementary twinning (antitwinning) 90–92

– effect of alloy composition 187–188
– effect of temperature 196–197
– order twinning 90–95
– reciprocal (conjugate) twinning 90
– thermal stability of twin structures 206
– twinning elements 90
– twin intersections 197–204
– twin nucleation 186–197
Mechanical (deformation) twinning in b.c.c. metals 116
Mechanical properties of hot worked material
– anisotropy 633–642
– of forged and extruded material 633–642
– of sheet material 669
Melting
– high-power ISM 492, 500
– ISM 479, 489–492, 527
– levitation melting 492, 500, 513
– PAM 479, 483–489
– VAR 479–483
– VIM 499
Microsegregation 39, 42
Microsegregation within particles 524, 534, 537
Microstructure formation map 43, 44
Mixed boundaries 54
Models for phase transformation
– nucleation and constitutional undercooling 42, 43
– Zener 60
Modeling of wrought processing
– of stress-strain behavior 601–603
– of forging 621–625, 661
Modulated alloys
– mechanical properties 310–311, 346–347
– modulated microstructure 301–306
– structure of interfaces 306–310

n

Near γ microstructure 60
Necklace structure 589, 592
Nearly lamellar microstructure 53, 60
Neutron diffraction 35, 642–643
Nucleation in phase transformation
– rate (of lamellae) 58, 59
– sympathetic nucleation 56, 62
– site 55
– theory 58, 59

o

Octahedral sites 19
One-dimesional antiphase superstructures (1d APS) 8, 11

Orientation distribution function (ODF) 643, 644
Orientation relationship
– Blackburn 34, 53
– Burgers 35
Orientation variants 35, 53
Ordering reaction 50, 52
Oxidation, effect of:
– Ag additions 446, 447
– atmosphere 441, 443–446
– elements 438, 450
– halogens 450
– ion implantation 448–450
– surface finish 447, 448
Oxidation embrittlement 450–455
Oxygen equilibrium pressure 434
Oxidation rate laws 433
Oxidation resistant coatings 456–458
Oxide stress 439, 440

p

Pack rolling 663
Particle size distribution 523, 527, 530, 532, 533, 534
Partitioning coefficient 12
Partitioning of alloying elements 12, 28
Peak stress (in flow curves) 574–579
Pearson symbol 6, 9, 13, 17, 18
Peierls stress 75, 101, 114, 116, 118, 221
Peritectic alloys 36, 38, 42, 44
Peritectic reaction 7, 8, 33, 36, 44
Peritectic solidification 36, 42
Peritectic transformation 36, 39
Peritectoid reaction 8
Perovskite precipitates 283–284
Phase diagram
– binary Ti-Al 5–8, 34
– calculated binary Ti-Al 8
– thermodynamic evaluation 6, 8, 17, 18
– ternary 14–18
– Ti-Al-B 14, 19, 46, 47
– Ti-Al-C 14
– Ti-Al-Cr 14, 15
– Ti-Al-Fe 14, 15
– Ti-Al-Mn 14, 15
– Ti-Al-Mo 14, 16
– Ti-Al-N 14
– Ti-Al-Nb 14, 16, 17
– Ti-Al-O 14, 19
– Ti-Al-Si 14
– Ti-Al-Ta 14, 16
– Ti-Al-V 14, 15
– Ti-Al-W 14, 16

Phase-field simulation
– of solidification   44
– of lamellae formation   56–57
Phase transformations   33–70
– during creep   333–340
– during diffusion bonding   688–691
– during shot peening   716–721
– during fatigue   426–428
– in modulated alloys   303–306, 309–310
Pilling-Bedworth ratio   439
Planar faults in $\alpha_2(Ti_3Al)$
– basal planes   110–111
– prismatic planes   108–109
– pyramidal planes   110
Planar faults in $\gamma(TiAl)$
– antiphase boundary (APB)   72, 73, 77–78
– complex stacking fault (CSF)   77
– superlattice extrinsic stacking fault (SESF)   78
– superlattice intrinsic stacking fault (SISF)   77–78
Planar fault energies
– in $\alpha_2(Ti_3Al)$ 108–110
– in $\gamma(TiAl)$ 79–80
Planar growth (solidification)   42, 43
Plastic anisotropy   156–161
Point defects   27–28
– vacancies   29–30
– interstitial atoms   29
Portevin-LeChatelier effect   222
Polysynthetically twinned (PST) crystals   132–133, 156–161, 276
Pores in powder   540
Powder-based prototyping
– direct laser fabrication   548
– laser-engineered net shaping   548
– laser forming   547
– laser melt deposition   547, 548
– MIM   549
Powder HIPing   543, 556
Powder oxygen/nitrogen contamination   539, 540
Property variability   393, 501, 546
Precipitation hardening   282–292
Prenucleation stage   52, 54, 57
Primary solidifying phase   33, 39, 45
PST crystal   276

r

Reactive sintering   559
Recovery of work hardening   262–268
Recrystallization
– of single-phase alloys   585–587
– of multiphase alloys   587–595

– kinetics   585, 589
– nucleation of recrystallized grains   596
Requirements for ductility and toughness   391–393
Resistivity measurements   57
Roller burnishing   662, 707, 723
Rolling   662–669
– isothermal   663
– "low-temperature" rolling   667, 668
– near-isothermal   667
– pack rolling   663–665
– speed   664–666
Rotational order fault   53, 54

s

Segregation   618–620
Schmid factor   95–97
Semicoherent interfaces   139–149
Scheil analysis   40
Scheil equation   33, 40, 41, 42
Shear bands   591, 592, 600, 601, 603, 613–617, 687, 693
Shot peening
– Almen intensity   707
– effect on fatigue strength   662, 721–724
– amorphization   716–721
– microhardness   708–712
– phase transformation   716–721
– residual stresses   708–712
– surface deformation   712–716
– surface roughness   708–712
– thermal stability of surface hardening   724–726
Single-crystal yield surface   649
Site occupancy   12
Slip systems
– in $\alpha_2(Ti_3Al)$ 106–112
– in $\beta/B2$ 115–118
– in $\gamma(TiAl)$ 72–73
Slip transfer through lamellae   173–178
Solidification   33–49
– direct solidification   36
– directional solidification   43
Solid solution hardening
– effect of alloying elements   274, 277
– effect of solute Nb   277–282
– size misfits   275
Solubility of interstial elements   12, 19, 28, 168
Space group   6, 9, 10, 13, 17, 18
Spark sintering   557
Specific heat   26
Specific modulus   1
Specific strength   1
Spray forming   551

Stacking fault   52, 54, 73–78
Static strain ageing
– activation energy   226–227
– antisite defects   228–229
– effect of alloy composition   225–228
– kinetics   225–226, 230–231
– yield-point return technique   222, 223
Strain rate cycling tests   212–213
Strain localization   592
Strain partitioning between α_2(Ti$_3$Al) and γ(TiAl) phases   168–169
Strain rate sensitivity   211–213, 584, 608, 609, 669
Stress exponent
– creep   316–320
– hot working   583, 584
Stress intensity factors   360–361
Stress relaxation tests   214–215
Strengthening mechanisms
– debris hardening   259–262
– grain refinement   249–251
– interface strengthening   252–254
– precipitates   282–295
– solution hardening   273–282
– work hardening   254–259
Strukturbericht designation   6, 9, 10, 13, 17, 18
Successive twinning   63
Superdislocations
– in α_2(Ti$_3$Al)   106–112
– in B2   116–118
– in γ(TiAl)   73
– superpartial dislocations   73, 87–89
Superplastic forming   669–670
Supersoft glide mode   139
Surface hardening   707
Surface relief   56
Surface stresses produced by shot peening   708–712

t

Taylor factor   170
Taylor model   648, 649
Temperature cycling tests   213
Temperature sensitivity   211
Texture
– brass type   651–653
– copper type   651
– evolution during hot working   646–657
– evolution during extrusion   647, 648, 654, 655
– modified cube texture   651, 655, 657
– of cast material   34–36, 643–646
– of sheet material   649–654
Thermal expansion coefficients   26, 440

Thermal stress   209
Thermally activated deformation
– activation energies   209–210, 218–219
– activation parameters   211, 218–219
– activation volume   209–210, 218–219
– athermal stress   208, 234, 239
– dislocation velocity   207–208
– effective stress   208
Thermally induced porosity   541
Thermal conductivity   26
Thermodynamic data   7
Thermomechanical fatigue   428–429
Thermophysical constants   25–30
Threshold stress intensity   404, 411–413
Ti$_2$Al$_5$ 8
Ti$_{2+x}$Al$_{5-x}$ 6, 10
Ti$_3$Al$_5$ 6, 10, 17
Ti$_5$Al$_{11}$ 8
TiB   46
TiB$_2$ 45, 46
Ti$_3$B$_4$ 46
T$_0$ temperature   50, 52
Torsional deformation   577
TTT curves   57, 58, 60

u

Undercooling
– constitutional undercooling   42, 43, 46
– nucleation undercooling   42–44, 46
– solutal undercooling   46

v

von Mises criterion   94, 112–114, 118, 169–171

w

Wedge-shaped cracks   611–613, 616, 666
Weibull statistics   393–394, 396–398
Widmanstätten colonies   47–49, 60–62
Widmanstätten plates   50, 53, 64
Workability   607–617
– criteria   610–617
– map   574, 607–613
– tests   574

y

Yield point phenomena   211–212, 222–223
Yield stress anomaly   95–96, 98–106, 112–113

z

Z phase   446
Zener-Hollomon parameter   579, 580, 584, 586
Zener model   60